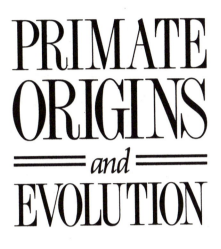

PRIMATE ORIGINS and EVOLUTION

PRIMATE ORIGINS

and

EVOLUTION

A phylogenetic reconstruction

R.D. Martin

Animal illustrations by Anne-Elise Martin

PRINCETON UNIVERSITY PRESS
Princeton, New Jersey

Published by Princeton University Press,
41 William Street, Princeton, New Jersey 08540

Library of Congress Cataloguing-in-Publication Data

Martin, R.D. (Robert D.), 1942–
    Primate origins and evolution: a phylogenetic reconstruction/
R. D. Martin: illustrated by Anne-Elise Martin.
        p.   cm.
    Includes bibliographical references.
    1. Primates—Evolution.   2. Primates. Fossil.   3. Phylogeny.
    I. Martin, Anne-Elise.   II. Title.
    QL737.P9M32   1990
    599'.03—dc20   89–36955
    ISBN 0–691–08565–X   CIP

Printed in Great Britain at the University Press, Cambridge

Contents

List of tables

Preface

The basic strategy of the book is to provide an overall survey of the evidence relating to primate evolution. Following an introduction to living primates in Chapter 1 and to their likely fossil relatives in Chapter 2, the key methodological issues relating to phylogenetic reconstruction and classification are tackled in Chapter 3. Having established that primate evolution can be properly understood only against the general background of mammalian evolution, this topic is reviewed in Chapter 4, along with a discussion of the importance of continental drift. An appendix to that chapter deals with the important issue of the scaling effects of body size. Chapter 5 then introduces the question of the relationships of tree-shrews, as a test-case for the reconstruction of the earliest origins of the primates during the adaptive radiation of the mammals. Subsequent chapters review various kinds of evidence, including dentitions and dietary adaptations (Chapter 6), the skull and major sense organs (Chapter 7), the central nervous system (Chapter 8), reproductive biology (Chapter 9), locomotor adaptations (Chapter 10) and chromosomal and molecular evidence (Chapter 11). Finally, an attempt is made in

Chapter 12 to draw all of the findings together into a provisional synthesis.

The idea for this book originally grew out of my doctoral thesis on the evolutionary relationships of the tree-shrews, an undertaking that was completed in 1967. In the course of the wide-ranging research that proved to be necessary for that thesis, it became apparent that several issues fundamental to the reconstruction of phylogenetic relationships among primates had remained largely neglected. In particular, there seemed to be no well-established methodological framework for phylogenetic reconstruction, and there was an obvious need for synthesis of findings from many different fields of primate biology. The literature on the evolutionary history of primates is packed with bulky treatises apparently based on the belief that an overwhelming tide of detailed information from one subject area (often couched in obscure terminology) is in itself sufficient to sweep us towards a reliable phylogenetic reconstruction. My conviction that this approach is seriously mistaken led me in 1969, as a freshly appointed and overly optimistic lecturer in biological anthropology at University College London, to embark on what turned out to be a mammoth undertaking.

In the event, the writing of this book has taken some 20 years. (Indeed, it is perhaps no coincidence that one of my main interests in recent years has turned out to be the evolution of the particularly long gestation periods of primates!) Quite apart from the increasing need to juggle time between numerous individual research projects, a heavy university teaching load and work on the synthesis for this book, a suite of new developments required continual revision of the original plan.

In the first place, a veritable explosion of research on primates during the 1960s and 1970s vastly increased the range of material that had to be covered. Secondly, as part of this general explosion, the pressing need to incorporate information on chromosomes, proteins and DNA became ever more obvious. Quite apart from extending the range of comparative information available for the reconstruction of phylogenetic relationships among primates, studies of molecular evolution gave an entirely new impetus to the development of tree-building methods. Finally, both at the molecular level and at the morphological level, the great importance of developing suitable quantitative methods of analysis also became clearly apparent. Against this background of exploding information and proliferating methodological developments, the task of achieving a worthwhile synthesis at times seemed simply insurmountable.

My greatest inhibition in striving for the synthesis presented in this book was the certain knowledge that treatment of individual topics would necessarily have to be relatively superficial in order to achieve the necessary breadth of coverage. Nevertheless, in order to achieve the overall aim of the book, a compromise had to be reached somewhere between the depth of specialist knowledge within individual subjects needed for accuracy and the sheer mass of knowledge that had to be absorbed across subjects. There is, of course, the danger that an all-embracing synthesis will entail the cost of major factual inaccuracy ('Never mind the quality – feel the width'). On the other hand,

there is no hope of developing a convincing picture of primate evolution unless a broad synthesis is achieved. At present, there is still far too little communication between subject areas and there is a particularly wide gulf between morphological and molecular approaches. Curiously, there is even relatively little communication between those who study chromosomes and those who study genes and gene products.

It is also worth noting that there is a widespread tendency to regard molecular approaches to phylogenetic reconstruction as in some way inherently superior to 'classical' approaches (essentially based on morphology). This tendency is encouraged by the pejorative labelling of comparative morphology as a 'soft science', in contrast to 'hard sciences' such as molecular biology. Nothing could be farther from the truth. One of the merits of an overall synthesis including both morphological and molecular approaches is that it permits a realistic assessment of their relative contributions. Although it is true that molecular features, because of their relative simplicity, are more amenable to quantitative analysis, there are at least as many problems involved in the assessment of molecular evidence as in the analysis of morphological evidence. Indeed, the review of molecular evidence relating to primate evolution provided in this book shows that most of the resulting phylogenetic trees have at some stage been selected on the basis of pre-existing conclusions based on morphological evidence. Chromosomal and molecular evidence should be warmly welcomed as an additional dimension to the data base used to conduct phylogenetic reconstructions, but we must be aware of its limitations as well as recognizing its strengths.

This book, then, presents a hard-won synthesis aimed at the identification of lasting general principles. Hopefully, these principles will remain valid even if my lack of detailed specialist knowledge in individual areas may have led to oversimplification or even factual errors. In defence of the broad approach taken, I would cite Dudley Morton (1924, p. 4), who provided an early review of the primate foot that still stands as one of the classic papers in the field of primatology:

> If one figuratively makes use of a high-power glass to examine the subject he is immediately lost in a maze of minor differences which become magnified in importance; he loses all recognition of the few great and important differences or, if not, is unable to examine them comprehensively, or to obtain a proper sense of their proportion because of the very narrowness of his field of vision. The use of a low-power lens is necessary for the establishment of the correct relationship and proportion of the various parts; afterward the smaller details can be studied more effectively.

In fact, the most rewarding contribution that could be made by this book, serving as such a 'low-power lens', would be if it actually does enable specialists to focus more effectively on crucial aspects of their own research. In particular, I hope that the broad reconstruction of primate evolution that I have carried out will provide a more reliable basis for those who are concerned specifically with the reconstruction of human evolution. There is still much to be learned about our own evolutionary history and a proper comparative background in the general evolutionary biology of primates is one of the main prerequisites for successful advances in the future.

There are, of course, many people who deserve my heartfelt thanks for contributions of one kind or another along the tortuous path that has led to the publication of this book. First and foremost, I would like to thank my wife, Anne-Elise Martin, who not only produced most of the illustrations for the book but also made it possible for me to cope with the vast amount of work involved in preparing the text. I must also acknowledge a great debt to various outstanding teachers at school and university who helped me to develop my fascination for and understanding of biology, especially Alistair Hardy, Konrad Lorenz, Harold Pusey, N. ('Dai') Rees and Niko Tinbergen. A special note of thanks is due to the late Nigel Barnicot, who played an important guiding role at an important stage in my university career and in the early development of this book.

The text of the book has drawn heavily on the writings of numerous authorities who have contributed to the field of primatology, most notably John Buettner-Janusch, W.K. Gregory, Johannes Hürzeler, Wilfrid Le Gros Clark (aptly described by Matt Cartmill as 'the first primatologist'), W.D. Matthew, John and Pru Napier, W.C. Osman Hill, O.J. Lewis, David Pilbeam, Alfred Sherwood Romer, Elwyn Simons, George Gaylord Simpson and Frederick Wood Jones. Although I do not agree with these giants in all respects, it must be said that it is far easier to survey the field when standing on their shoulders.

The preparation of the text benefited directly from the participation of many friends and colleagues. I would most particularly like to acknowledge the help of Sarah Bunney, who devoted care and attention above and beyond the call of duty in subediting the manuscript. In the final stages, her expertise not only helped to remove many flaws but also turned the exercise into a valuable learning experience. I would also like to record my heartfelt thanks to my sister, Valerie Leslie, who did much of the typing of the manuscript and greatly eased the way towards the final text with impeccable typescript and repeated injections of much-needed encouragement. I am also very grateful to my secretary Elsbeth Rüegg, who helped in many ways in the final stages of completion of the book, and to Brigitte Tanner and Brigitte Zimmerli for their skilled bibliographic assistance over the past three years. Last, but not least, I would like to thank my editors at Chapman and Hall. Alan Crowden provided advice and support in the earlier stages; Bob Carling, Sharon Duckworth and Alison Jesnick guided the manuscript through to press with great skill and dedication.

Several people read all or part of the manuscript and they deserve my warm thanks for their help: Leslie Aiello, John Allman, Peter Andrews, Fred Brett, Matt Cartmill, Christopher Dean, Adrian Friday, Peter Jewell, Christopher and Elizabeth Pryce, Jeffrey

Schwartz, Ian Tattersall and Peter Toye. Matt Cartmill and Peter Jewell must be singled out for special praise, as they provided detailed comments on the entire manuscript leading to many improvements. All of these people have helped to expunge many inadvertent errors along the way. I suppose that, following tradition, I must now accept sole responsibility for any flaws that remain.

Numerous other friends and colleagues have contributed in various ways, both directly and indirectly, over the years: Simon Bearder, Pierre Charles-Dominique, David Chivers, Glenn Conroy, Christopher Cook, Michael Day, Alan Dixson, Frances D'Souza, Gérard Dubost, John Eisenberg, John Fleagle, Philip Gingerich, Terry Harrison, Paul Harvey, David Hewett-Emmett, Marcel Hladik, Harry Jerison, Alison Jolly, Patrick Luckett, Ann MacLarnon, Gilbert Manley, Lawrence Martin, Theya Molleson, Jean-Jacques Petter, Jon Pollock, Len Radinsky, Alison Richard, June Rollinson, Caroline Ross, Jeffrey Schwartz, Alain Schilling, Bob Sussman, Dietrich von Holst, Alan Walker, Bernard Wood. I owe them all my sincere thanks for advice and companionship.

Another source of help that must be acknowledged is the input from generations of students at University College London, Yale University and the University of Zürich. Through penetrating questions and a liveliness born of youthful enthusiasm, they helped to hone my ideas and to raise my flagging spirits at times when the end of the tunnel seemed to be out of sight. I would also like to express my thanks to various grant-giving bodies that have helped me to develop my research in the general field of primate biology: the Boise Fund; the British Council; the German Academic Exchange Service; the Medical Research Council, UK; the Royal Society, London; the Science Research Council, UK (now the Science and Engineering Research Council); the Smithsonian Institution, Panama; the Wellcome Trust, UK. I hope that they, along with other readers of this book, will find that their investment was worthwhile.

R.D. Martin
Zurich
July 1989

Chapter one

A survey of living primates

With any group of organisms containing numerous living representatives, such as the primates, the first step to be taken in reconstructing phylogenetic relationships is to scrutinize and compare the living (extant) species. In historical terms, this step preceded any other because recognition of 'natural groups' of living organisms in classifications, which culminated in the influential publications of Linnaeus, antedated by over a century Darwin's publication of the theory of evolution by natural selection. Only within the context of such a theory could the fossilized, fragmentary remains of past organisms be incorporated into a combined scheme based on a network of inferred relationships. Analysis of living forms is also the logical first step for modern phylogenetic reconstruction. The most important reason for this, as will be argued in detail in Chapter 3, is that reconstruction of relationships should be based on inference of hypothetical ancestral stages. Fossils (see Chapter 2) are best incorporated into the overall scheme as a second step, and their inclusion at this later stage can be used as a valuable test of hypothetical ancestral conditions, according to the degree of compatibility observed.

There are, it must be admitted, alternative viewpoints on this fundamental issue. For example, the cladistic school (see Chapter 3) generally favours the tenet that living and fossil forms should be treated equally in phylogenetic reconstructions, taking both together at the outset. For strict adherents of this school, the major tests of the reliability of a particular reconstruction reside in the internal consistency of the phylogenetic relationships proposed and in the selection of relationships requiring a minimum of evolutionary change (namely, by applying the principle of parsimony). A quite different view is that fossil evidence should take precedence over evidence derived from living forms in the influence of phylogenetic relationships. A persuasive recent exponent of this approach is Gingerich (1979; see also Gingerich and Schoeninger, 1977). His 'stratophenetic' procedure is based on the concept that evolutionary sequences are best reconstructed with respect to the changes observed in series of fossils viewed against a background of strict geological succession. For the analysis of evolutionary change within a particular group of fossils contained in a well-defined geological context, this approach has considerable merit. But for the reconstruction of primate evolution overall, it is subject to severe limitations. The viewpoint adopted here is that inference of phylogenetic relationships should take into account both the special nature of fossil evidence (as recognized by Gingerich) and the considerable limitations of the fossil record (as recognized, implicitly or explicitly, by adherents of the cladistic school).

Given a complete fossil record, the task of phylogenetic reconstruction should be considerably facilitated, even though, with very few exceptions, only the most resistant 'hard parts' of an organism are preserved in fossil form. But, given the actual, very patchy state of the fossil record (see Chapter 2), phylogenetic reconstruction depends heavily on a process of inference utilizing all available information in an appropriate framework. It is worth listing in detail the major deficiencies of the information currently available for fossil primates:

1. There are enormous gaps in the fossil record even for the best-documented groups. At present, these gaps can be bridged only by postulating hypothetical evolutionary sequences.
2. Even when there is a fairly rich fossil assemblage covering a given geological period, the identification of evolutionary sequences purely on the basis of temporal succession implies that the assemblage can be regarded as fully representative. Yet the catchment area of a typical fossil site is so minute in comparison with the geographical range of a typical modern species of mammal that preservation of a complete sequence *in situ* is most unlikely.
3. Because of limitations (1) and (2), one must be extremely wary of assigning a direct ancestral position to any individual fossil species. The probability of finding an actual ancestral species, rather than a representative from the many extinct side branches of the primate tree, is very small indeed. For this reason, assignment of fossils to places in a phylogenetic tree is little different, in principle, to the assignment of living species;

both depend on inference of hypothetical ancestral stages.

4. The number of characters that can be preserved in identifiable form for any given fossil species is severely circumscribed. The fossilizable 'hard parts' of primates are teeth, jaws, skulls and postcranial bones (roughly following that order for frequency of occurrence in the fossil record), and, even then, only superficial characters of those parts may be accessible to study. Occasionally, characters of 'soft parts' may be reflected by fossilized remains – as with casts of the inside of the braincase (endocranial casts or endocasts) that reveal the surface structure of the brain (see Chapter 8) – but these are exceptional. It is important to remember that with living species it is possible to relate the numerous observable anatomical characters to physiological, behavioural and ecological features that vastly increase our understanding of evolutionary adaptation. In particular, it is only with living species that functional aspects can be studied directly. Finally, for many characters of living species, such as biochemical and chromosomal features (see Chapter 11), virtually no relevant information can be obtained from the fossil record.
5. It is often forgotten that the initial identification of any fossil is, in practice, essentially a matter of **classification** rather than of **phylogenetic reconstruction** (see Chapter 3). In the absence of clearly formulated hypotheses defining ancestral stages, the allocation of a new fossil find must depend primarily on overall morphological resemblance. New fossils are usually allocated to taxonomic groups on the basis of broad similarity either to extant species or (more commonly) to previously allocated fossil forms. The latter approach can, in fact, give rise to a 'domino effect', with any similarity to living representatives of a given taxonomic group becoming increasingly tenuous as more and more fossils are added by a process of classificatory accretion. The more fragmentary the fossil, the greater the dangers of misinterpretation.

Indeed, typical overconcentration on dental features in the allocation of mammalian fossils (though understandable because of the prevalence of isolated teeth in the fossil record) has made things worse. Many fossils have been treated as being even more fragmentary than they are. Hence, the initial classification of a fossil within a particular group, such as the primates, must be followed by detailed analysis of its likely phylogenetic links.

In short, primate palaeontology is to a large extent concentrated on the practicalities of the discovery, dating, description and provisional classification of fossils, all of which must be accomplished before any realistic attempt at phylogenetic reconstruction can be made. This is not to deny that fossil evidence must play an integral part in phylogenetic reconstruction. In particular, as has been observed by McKenna (1966): 'The fossil record has its defects, but it alone separates what could have happened from what actually did.' That said, it must be emphasized that a much broader approach is necessary for the effective reconstruction of primate evolution. The fossil evidence represents but one of many vital parts in the overall scheme. Because of the limitations listed above, it cannot stand alone. The ultimate aim of all those concerned with the reconstruction of primate evolution must surely be to incorporate both living and fossil forms into a coherent framework of hypothetical ancestral stages that takes account of as much available information as possible.

From the standpoint that phylogenetic reconstruction logically begins with comparisons of extant species, it is necessary first of all to survey the living primates that provide the starting point of this exercise. This survey will be followed (Chapter 2) by a review of fossil primates, as a summary of the material on which one may draw to test hypotheses of phylogenetic relationships among the living forms. In later chapters that deal with individual functional systems of primates, a standard procedure will be followed in which living species are compared

to establish phylogenetic hypotheses, then the fossil evidence is used to test and refine those hypotheses. At this point, in order to avoid complexities of nomenclature (which are covered in Chapter 3) formal taxonomic names will be restricted to species and genera. Larger groupings will be provisionally labelled with common names in order to circumvent initially the difficult issue of the relationship between classificatory terminology and phylogenetic hypotheses. It is to be understood, therefore, that such groupings do not necessarily represent phylogenetic units; additional detailed analysis is required in order to infer the existence of the latter. It should also be noted straightaway that tree-shrews are not regarded a priori as being specifically related to primates and they will not therefore be included in this survey of living and fossil primates. The question of this enigmatic group will be considered in detail in Chapter 5, and in subsequent chapters much emphasis will be placed on tests of the hypothesis that tree-shrews are connected in some way with the evolutionary radiation of primates. As with the fossil evidence, tree-shrews provide a useful test case for assessing the validity and/or implications of phylogenetic hypotheses derived from comparison of primates *sensu stricto*.

THE LIVING PRIMATES

It is now widely accepted that the living primates (excluding tree-shrews) fall into six 'natural groups' that are defined by specific patterns of geographical distribution as well as by shared similarities (Fig. 1.1). Because speciation through geographical separation probably provides the main basis for evolutionary divergence and because the process of continental drift progressively modified the pattern of contacts between the major continental landmasses throughout at least the main period of adaptive radiation of the mammals (see Chapter 4), it is to be expected that the geographical subdivisions among living primates are relevant to broad evolutionary divisions. The following six 'natural groups' have, indeed, been reflected in one way or

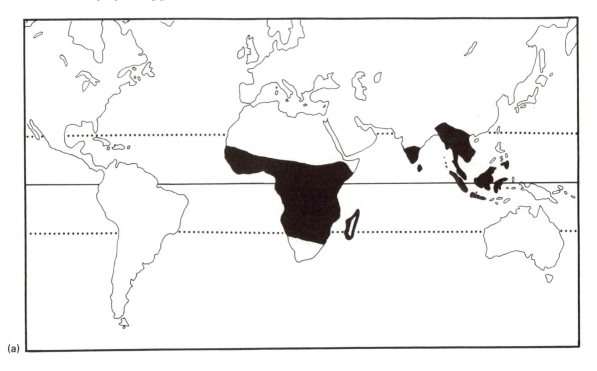

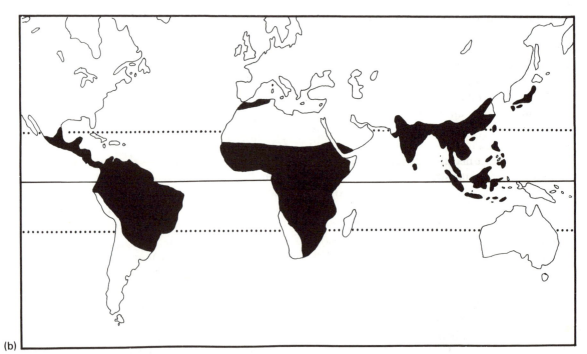

Figure 1.1. Geographical distribution of the six major natural groups of living non-human primates: (a) prosimians (lemurs, loris group and tarsiers); (b) simians (monkeys and apes). Note the more restricted distribution of the prosimians.

another in all major classifications of the primates:

1. Lemurs (confined to Madagascar).
2. The loris group (Africa, South Asia and South-East Asia).
3. Tarsiers (restricted to certain islands in South-East Asia).
4. New World monkeys (confined to South and Central America).
5. Old World monkeys (Africa, South Asia and South-East Asia).
6. Apes and humans (Africa, South Asia and South-East Asia, at least until the latter stages of human evolution).

The present geographical ranges of these six groups, shown in Fig. 1.1, also demonstrate the important fact that living primates – other than the human species – are almost entirely restricted to tropical and subtropical zones of the world (see also Richard, 1985). This suggests that primates have always been adapted to warm climatic conditions and have not been subjected to the extreme seasonal fluctuations found at the higher latitudes. It should also be noted that the first three groups are commonly referred to collectively as '**prosimians**' on the grounds that they appear to be generally more primitive in morpho-

logical terms than the remaining three groups, which are collectively called '**simians**' (monkeys, apes and humans). Throughout the following text, the terms 'prosimians' and 'simians' will be used to refer to these two higher-level groupings, once again on the understanding that they do not necessarily represent cohesive phylogenetic units. Fig. 1.1 also shows that the geographical distribution of prosimians is even more strictly limited to the warmer regions of the world than that of the simians.

The six 'natural groups' of primates will now be considered in turn, taking the genus as the main level for description and using common names wherever possible, for the reasons outlined above. In each case, basic information will be provided on general features allied to the more circumscribed characteristics covered in detail in Chapters 5–10. The features concerned are:

1. Body size, which is of central importance for primate adaptations, as shown in Chapter 4. (Note: body weights for many primate species are listed in Table 8.1.)
2. Characteristic activity pattern (nocturnal, crepuscular or diurnal).
3. Dietary characteristics.
4. Broad locomotor category.

Table 1.1 Alphabetical list of genera of living primates (from Napier and Napier, 1967, with the omission of tree-shrews and with other minor modifications)

Genus	No. of species	Common name	Natural group
1. *Allenopithecus*	1	Allen's swamp monkey	OW (1)
2. *Allocebus*	1	Hairy-eared dwarf lemur	LE (1)
3. *Alouatta*	5	Howler monkeys	NW (2)
4. *Aotus*	1	Owl monkey	NW (2)
5. *Arctocebus*	1	Angwantibo	LO (2)
6. *Ateles*	4	Spider monkeys	NW (2)
7. *Avahi*	1	Avahi	LE (4)
8. *Brachyteles*	1	Woolly spider monkey	NW (2)
9. *Cacajao*	3	Uakaris	NW (2)
10. *Callicebus*	3	Titi monkeys	NW (2)
11. *Callimico*	1	Goeldi's monkey	NW (2)
12. *Callithrix*	8	Marmosets	NW (1)
13. *Cebuella*	1	Pygmy marmoset	NW (1)
14. *Cebus*	4	Capuchin monkeys	NW (2)

Table 1.1 *Continued*

Genus	No. of species	Common name	Natural group
15. *Cercocebus*	3	Terrestrial mangabeys	OW (1)
16. *Cercopithecus*	21	Guenons	OW (1)
17. *Cheirogaleus*	2	Dwarf lemurs	LE (1)
18. *Chiropotes*	2	Bearded sakis	NW (2)
19. *Colobus*	5	Guerezas	OW (2)
20. *Daubentonia*	1	Aye-aye	LE (5)
21. *Erythrocebus*	1	Patas monkey	OW (1)
22. *Galago*	6	Bushbabies	LO (1)
23. *Gorilla*	1	Gorilla	AM (2)
24. *Hapalemur*	2	Gentle lemurs	LE (3)
25. *Homo*	1	Human	AM (3)
26. *Hylobates*	7	Gibbons	AM (1)
27. *Indri*	1	Indri	LE (4)
28. *Lagothrix*	2	Woolly monkeys	NW (2)
29. *Lemur*	5	True lemurs	LE (3)
30. *Leontopithecus*	3	Golden lion tamarins	NW (1)
31. *Lepilemur*	1*	Sportive lemur	LE (2)
32. *Lophocebus*	2	Arboreal mangabeys	OW (1)
33. *Loris*	1	Slender loris	LO (2)
34. *Macaca*	12	Macaques	OW (1)
35. *Mandrillus*	2	Drill/mandrill	OW (1)
36. *Microcebus*	3	Mouse lemurs	LE (1)
37. *Miopithecus*	1	Talapoin	OW (1)
38. *Nasalis*	1	Proboscis monkey	OW (2)
39. *Nycticebus*	2	Slow lorises	LO (2)
40. *Pan*	2	Chimpanzees	AM (2)
41. *Papio*	5	Baboons	OW (1)
42. *Perodicticus*	1	Potto	LO (2)
43. *Phaner*	1	Fork-crowned lemur	LE (1)
44. *Pithecia*	2	Sakis	NW (2)
45. *Pongo*	1	Orang-utan	AM (2)
46. *Presbytis*	14	Langurs	OW (2)
47. *Propithecus*	2	Sifakas	LE (4)
48. *Pygathrix*	1	Douc langur	OW (2)
49. *Rhinopithecus*	2	Snub-nosed langurs	OW (2)
50. *Saguinus*	22	Tamarins	NW (1)
51. *Saimiri*	2	Squirrel monkeys	NW (2)
52. *Simias*	1	Pagai Island langur	OW (2)
53. *Tarsius*	4	Tarsiers	TA
54. *Theropithecus*	1	Gelada baboon	OW (1)
55. *Varecia*	1	Variegated lemur	LE (3)

Key to 'natural groups':
LE = Madagascar lemurs: (1), mouse/dwarf lemurs; (2), sportive lemurs; (3), true lemurs/ gentle lemurs; (4), indri subgroup; (5), aye-aye.
LO = loris group: (1), bushbabies; (2), loris subgroup.
TA = tarsiers.
NW = New World monkeys: (1), marmosets/tamarins; (2), true monkeys.
OW = Old World monkeys: (1), guenon subgroup; (2), leaf-monkeys.
AM = apes and humans: (1), lesser apes; (2), great apes; (3), human.

* Recent chromosomal evidence indicates that there are several different species of *Lepilemur*.

Table 1.2 Numbers of living genera and species of primates (from Napier and Napier, 1967 with the omission of tree-shrews and with other minor modifications)

Group	No. of genera	%	No. of species	%
Malagasy lemurs	12 + 6 = 18	30	21 + *14* = 35	18
Loris group	5	8	11	6
Tarsiers	1	2	4	2
New World monkeys	16	26	64	32
Old World monkeys	16	26	73	37
Apes and humans	5	8	12	6
Prosimian subtotal	18 + 6 = 24	39	36 + *14* = 50	25
Simian subtotal	37	61	149	75
Totals overall	55 + 6 = 61	100	185 + *14* = 199	100

Numbers in italics refer to subfossil lemurs (see Tattersall, 1982). Percentage figures include subfossil lemurs where appropriate.

5. Outline of breeding patterns.
6. Nature of social groupings.
7. Special distinguishing characteristics.

The genera covered are listed in alphabetical order in Table 1.1, along with the corresponding common names, and further details on each genus can be obtained from the valuable handbook compiled by Napier and Napier (1967; see also Napier and Napier, 1985). A numerical breakdown of numbers of genera and species in each 'natural group', based on the same source, is given in Table 1.2.

The Malagasy lemurs

Despite their restriction to one of the smallest geographical areas covered by the six 'natural groups' of primates, the lemurs of Madagascar are remarkably diverse; they account for 30% of recent primate genera and 18% of the species (Table 1.2) and provide a classic example of **adaptive radiation** (Martin, 1972a; Tattersall and Schwartz, 1974; Eisenberg, 1981). For example, in terms of dental characters lemurs are more diverse than any other living group of primates (see Chapter 6). To a large extent, this diversity can be attributed to the relatively low level of competition from other mammalian groups. Apart from the primates, only four other mammalian orders are properly endemic to

Madagascar: insectivores (tenrecs), rodents, carnivores and bats. Arguably, lemurs have radiated to fill ecological niches occupied by other groups of mammals on mainland Africa. It is certainly true that all other extant primate groups occupy geographical areas containing potential competitors from a far greater variety of mammalian groups.

Despite their considerable diversity, lemurs share one consistent feature – they all breed according to a strictly seasonal pattern. In other respects there are major differences between the lemur genera, but one generalization can be made in terms of broad biological features. As body size increases, there are general trends such as a shift from nocturnal to diurnal habits, a transfer from heavy reliance on animal food to concentration on plant food and a transition from so-called 'solitary' existence to formation of well-defined social groups. Five fairly distinct subgroups of lemurs can be recognized (see also Petter, Albignac and Rumpler, 1977): (1) mouse and dwarf lemurs; (2) sportive lemurs; (3) true and gentle lemurs; (4) indri subgroup; and (5) aye-aye.

Mouse and dwarf lemurs

The mouse and dwarf lemur subgroup includes the smallest of the lemurs. All of the species in this subgroup are nocturnal, omnivorous (in the sense that both animal and plant food items of varied character are consumed), nest-living and typically 'solitary' when active (see later). The smallest species of all are the lesser mouse lemurs, *Microcebus murinus* (Fig. 1.2) and *M. rufus*, with a body weight of about 60 g. Lesser mouse lemurs are widespread in Madagascar and typically occupy areas of secondary forest, moving along fine branches or lianes and occasionally descending to the ground to hop along with an unusual frog-like motion. These lemurs are very generalized in terms of locomotion and diet, feeding on a great variety of animals and plants and showing a wide range of locomotor patterns. The lesser mouse lemur is one of the few primate species that constructs a nest. Although it commonly occupies available tree hollows, it also builds spherical nests with leaves if required. Its reproduction is also somewhat unusual among primates, in that a female can give birth to as many as three infants at a time and they are transported one at a time in the mother's mouth (e.g. when the nest is disturbed) rather than clinging to her fur. Although lesser mouse lemurs are typically seen alone at night, they have a fairly elaborate social system based on nest-sharing groups of females with overlapping home ranges. There is a distinction between isolated peripheral males and central males whose ranges overlap with those of females (Martin, 1972b).

The other members of the mouse and dwarf lemur subgroup have the same general characteristics, except that body weight extends up to 440 g for the fork-crowned lemur (*Phaner furcifer*). However, recent field studies (Charles-Dominique, Cooper *et al.*, 1980) have shown that each species has some special features that

Figure 1.2. Grey lesser mouse lemur, *Microcebus murinus*. Head and body length: 12.5 mm. (After photographs taken by the author.)

Figure 1.3. Fat-tailed dwarf lemur, *Cheirogaleus medius*. Note the substantial girth of the tail, which is used as a depot for stored fat. Head and body length: 195 mm. The body weight of this species is around 180 g. (After a photograph taken by the author.)

are particularly relevant to survival when only limited food is available. For example, *Phaner* has specialized gum-feeding adaptations; and Coquerel's mouse lemur (*Microcebus coquereli*) has an even stranger dietary speciality in that it licks exudates from the rear ends of certain insect species. Lesser mouse lemurs are able to store fat in their tails to help tide them over the dry season and this ability has been developed to an extreme in the dwarf lemurs (*Cheirogaleus* spp.), most notably in the appropriately named fat-tailed dwarf lemur (*Cheirogaleus medius*; Fig. 1.3). Virtually nothing is known about the small hairy-eared dwarf lemur (*Allocebus trichotis*) other than that it has some adaptations for gum-feeding (e.g. development of sharp tips on the nails – see Chapter 10).

Sportive lemurs

Sportive lemurs (*Lepilemur* spp.; Fig. 1.4) are also nocturnal and relatively small in body size,

weighing from 600 to 800 g according to species. Nevertheless, in contrast to the mouse and dwarf lemurs, they have some extreme specializations that are quite remarkable in such small-bodied primates. They are essentially folivores (leaf-eaters) and the energy obtained from their food may be boosted by caecotrophy: symbiotic bacteria in the enlarged caecum (a blind pouch at the junction of the small and large intestines – see Chapter 6) break down the resistant leaf material, and it has been reported that sportive lemurs periodically ingest faeces containing the output from the caecum (Charles-Dominique and Hladik, 1971; but see Russell, 1977). Sportive lemurs also exhibit a specialized pattern of locomotion typically involving leaping between vertical supports with the body held upright against each support between leaps ('vertical-clinging-and-leaping': Napier and

Figure 1.4. Sportive lemur, *Lepilemur mustelinus*. Note the powerful hindlimbs used in vertical-clinging-and-leaping. Head and body length: 280 mm. (After a photograph taken by the author.)

Walker, 1967; see Chapter 10). On the rare occasions when they descend to the ground, they hop (bipedally) on their hindlimbs with the forelimbs held high off the ground. Perhaps because of their ability to subsist on a diet of leaves, sportive lemurs are comparatively widespread in Madagascar and occur in several different types of forest. Their reproduction follows the predominant primate pattern in that a female has one infant at a time and usually carries it clinging to her fur. On the other hand, social organization is similar to that of nocturnal mouse and dwarf lemurs. Sportive lemurs use tree hollows or dense tangles of vegetation as nests, they usually move around singly at night and exhibit a very limited degree of social contact

ensured through range overlap between females (mothers and daughters) and individual males (Charles-Dominique and Hladik, 1971).

True and gentle lemurs

The third subgroup of Malagasy lemurs contains the medium-sized true lemurs (*Lemur* spp.; *Varecia*) and gentle lemurs (*Hapalemur* spp.). They range in body size from the common gentle lemur (*Hapalemur griseus*, 800 g) through the members of the genus *Lemur* (e.g. *Lemur catta*, 2.2 kg) to the variegated lemur (*Varecia variegata*, 3.3 kg). The best-known representative of this subgroup is the ringtail lemur, *Lemur catta* (Fig. 1.5), which is diurnal like

Figure 1.5. Ringtail lemur, *Lemur catta*. This is the only prosimian primate to show significant terrestrial behaviour. Head and body length: 425 mm. (After photographs taken by the author.)

most of the other true and gentle lemurs. (The common gentle lemur and possibly one or two of the true lemur species tend to be more crepuscular, while the mongoose lemur, *Lemur mongoz*, has even been described as fully nocturnal in part of its range; Sussman and Tattersall, 1976.)

The ringtail lemur has been the subject of several detailed field studies (e.g. A. Jolly, 1966, 1972a; Sussman, 1974, 1977). It is essentially a fruit-eating (frugivorous) species, although its diet includes a fair proportion of leaves as well; as far as is known no animal food is taken in the wild. Ringtails are unusual among prosimians generally in that they spend a significant proportion (about 25%) of their activity time feeding or moving about on all fours (quad-rupedally) at ground level, much like certain forest-living Old World monkey species. Although the locomotion of ringtails is typically quadrupedal, occasionally they are vertical-clingers-and-leapers among vertical supports. There is typically one infant at birth, although twins are seen more frequently than with other diurnal lemur species; the infant is carried on the mother's fur from birth onwards. Probably because of a combination of diurnal habits and a propensity for terrestrial locomotion, *Lemur catta* forms the largest social groups among the prosimians, numbering up to two dozen individuals under favourable conditions (A. Jolly, 1966, 1972a; Sussman, 1974, 1977). Such social groups usually include several adults of both sexes, but there is a tendency for most males to be peripheral (Budnitz and Dainis, 1975). This is reminiscent of the 'dispersed' system found in several nocturnal prosimian species.

Although the members of the true lemur subgroup generally have a rather similar morphology they show some diversity in behaviour – as with the patterns of activity noted above, which range from nocturnal to diurnal. All of the species other than the ringtail are essentially arboreal, but there is still some diversity in terms of diet and locomotion. The gentle lemurs (*Hapalemur* spp.) are specialized for feeding on bamboo (Petter, Albignac and Rumpler, 1977), whereas the mongoose lemur (*Lemur mongoz*) is unusually a nectar-feeder (Sussman and Tattersall, 1976). In contrast to the generalized quadrupedalism of *L. catta*, there is a well-developed tendency for vertical-clinging-and-leaping in *Hapalemur griseus* and a more restricted type of cursorial quadrupedal locomotion in *Varecia variegata*. Patterns of reproduction and social organization are also fairly diverse among the true and gentle lemurs. The variegated lemur (*V. variegata*) gives birth to multiple litters (usually twins) in a nest of some kind and the mother carries the infants in her mouth, rather than clinging to her fur. *Hapalemur griseus* shows an intermediate condition that involves initial use of a nest and mouth-carriage for its single infant, followed by carriage of the infant on the fur, as is typical for *Lemur* species (which do not use nests) from birth onwards. Variation in social organization ranges from formation of monogamous family groups, as reported for *Varecia* by Petter, Albignac and Rumpler (1977) and for *Lemur mongoz* in northwestern Madagascar by Sussman and Tattersall (1976), to the large, multi-male social groups described above for *L. catta*.

Indri subgroup

The next subgroup, containing the indri (*Indri indri*) and its relatives, is somewhat more uniform but still shows appreciable divergence in behaviour between species. Members of this subgroup range in size from the nocturnal avahi (*Avahi laniger,* body weight about 1 kg), through the diurnal sifaka species (*Propithecus* spp.; approximate weight of *P. verreauxi*, 3.4 kg) to the diurnal indri (weight about 6.3 kg), which is the largest of all the extant lemur species. Extensive field studies have been conducted on the common sifaka (*Propithecus verreauxi*; A. Jolly, 1966; Richard, 1974, 1977) and on the indri (Pollock, 1975, 1977), but relatively little is known of the natural habits of the nocturnal avahi (Ganzhorn, Abraham and Razanahoera-Rakotomalala, 1985). All members of the indri subgroup are essentially plant-feeders and they include a large proportion of leaves in their diets;

Figure 1.6. The indri, *Indri indri*. This species has particularly powerful hindlimbs, required for vertical-clinging-and-leaping, and is unusual in lacking a tail. Head and body length: 605 mm. (After a photograph taken by J.I. Pollock.)

they are all typically arboreal vertical-clingers-and-leapers, moving by bipedal hopping during travel along broad horizontal branches and during their (infrequent) excursions across the ground. The indri (Fig. 1.6) is unique among the extant lemurs in that it lacks a tail and therefore lacks the fine balancing actions that this organ would otherwise provide during locomotion through the trees. All species in this subgroup give birth to a single infant, which is carried clinging to the parent's fur from birth onwards, and none of them uses a nest. There is some distinction among species in social organization, in that sifakas form social groups of quite varied composition, commonly containing more than one adult male, whereas the indri consistently forms monogamous family groups. The indri is notable for its melodious calls, which seem to

play a part in territorial demarcation and in group cohesion (Pollock, 1975). It is suspected that the avahi also forms family groups (Petter, Albignac and Rumpler, 1977; Pollock, 1979; personal observation), but more detailed study is required to confirm this point.

Aye-aye

The aye-aye (*Daubentonia madagascariensis*; Fig. 1.7), is so distinctive that it must be granted a subgroup of its own. This is by far the largest nocturnal prosimian species (body weight about 2.8 kg) and it has one of the smallest areas of geographical distribution, being confined to parts of the coastal rainforest of northeastern Madagascar. Its peculiar appearance and habits have long attracted attention. Unlike all other

Figure 1.7. The aye-aye, *Daubentonia madagascariensis*. Note the thin middle finger used as a probe in feeding. Head and body length: 395 mm. (Adapted from photographs in Petter, Albignac and Rumpler, 1977.)

living prosimian species, the aye-aye has claws rather than flat nails on all digits other than its big toes, and the middle digit of each hand is attenuated to form a thin probe. (Contrary to numerous misleading statements in the literature, the middle finger of the hand is not longer than the other fingers; as in prosimians generally, the longest digit of the hand is the fourth, not the third – see Chapter 10). The aye-aye also has continuously growing, rodent-like incisors (a feature unique among living primates). These peculiarities seem to be related to the aye-aye's dietary specialization: it feeds extensively on the larvae of wood-boring beetles and other insects, breaking into their galleries in tree trunks with its incisors and using the thin middle finger to pulp and extract the larvae. As might be expected from the presence of claws on

the hands and feet, the aye-aye is exclusively quadrupedal, and the claws undoubtedly help the animal cling to broad trunk surfaces while it feeds. Despite its large body size, the aye-aye constructs a large, spherical leaf-nest (Petter and Peyrieras, 1970). Unfortunately, no detailed field study has yet been conducted on this highly unusual lemur species, but it seems that the female has a single infant, which is carried on her fur from birth onwards, and that there is an essentially 'solitary' lifestyle.

Cartmill (1974a) has presented a well-argued proposal that the aye-aye has, in effect, filled in Madagascar a feeding niche occupied by woodpeckers and comparable bird species in most other areas of the world. The case is greatly strengthened by the fact that in the only other land areas of significant size where woodpeckers and similar birds are absent – part of the Australasian region containing New Guinea and the Cape York Peninsula of Australia – there is a smaller-bodied marsupial counterpart of the aye-aye. This marsupial (*Dactylopsila* spp.) resembles the aye-aye in being nocturnal and in having an attenuated digit on each hand and conspicuously well-developed incisor teeth. Quite apart from providing support for Cartmill's hypothesis, this also provides one of the most clear-cut illustrations of the way in which similar ecological opportunities (in this case, the availability of wood-boring larvae untapped by other predators) can produce **convergent evolution** of remarkably similar adaptations in widely separate animal groups (see Chapter 3). There can be no doubt that *Daubentonia* (a primate) and *Dactylopsila* (a marsupial) must have developed their shared peculiarities independently long after the divergence between marsupials and placentals occurred in the evolution of mammals. Indeed, it has now been shown that *Heterohyus* – a member of the early Tertiary fossil family Apatemyidae – also had rodent-like incisors and thin, long fingers, providing another mammalian parallel to the aye-aye (von Koenigswald and Schierning, 1987).

Loris group

Members of the loris group are, without exception, nocturnal in habits and for this reason very little was known about their behaviour until relatively recently. This gap in our knowledge has now been filled quite adequately, to a large extent thanks to the detailed field studies conducted by Charles-Dominique (1977a) on five different species living in close proximity (sympatrically) in the equatorial rainforest of Gabon. Apart from revealing the distinguishing characteristics of the five species, these studies have demonstrated subtle differences in their ecological niches that permit them to coexist.

Even superficial examination reveals a marked division of the loris group into two subgroups: (1) the bushbabies, which are restricted to mainland Africa, and (2) the lorises of South and South-East Asia and the potto and angwantibo

of Africa. All the bushbabies have powerful hindlimbs and typically have an active lifestyle involving very little further morphological specialization. By contrast, lorises, the potto and the angwantibo (constituting the 'loris subgroup') have several marked specializations associated with a discreet, slow-moving pattern of locomotion (Walker, 1969a). Members of this subgroup never leap and they depend on direct passage from one branch to another to such an extent that Charles-Dominique (personal communication) was able to keep a potto for many months on a network of branches suspended from the ceiling without the need for a proper cage. In all species in this subgroup, the tail has been reduced to a stump, the second digits of the hand and foot have been truncated in association with a powerful pincer-like development of these grasping extremities, and there are numerous special adaptations of the limbs and musculature (e.g. in the blood supply; Suckling, Suckling and Walker, 1969) for slow, deliberate movement and powerful grasping, as opposed to fleet locomotion. In fact, it has now been found that all members of this subgroup have much lower basal metabolic rates (see Chapter 6) than would be expected for typical primates of their body size, indicating that the sluggish locomotion of lorises, the potto and the angwantibo probably represents just part of an overall energy-saving lifestyle.

Bushbabies

The bushbabies can be characterized to a large extent by description of a single species studied in considerable detail by Charles-Dominique (1972, 1977) – Demidoff's bushbaby (*Galago demidovii*; Fig. 1.8). This is the smallest of the extant bushbaby species, with a body weight of about 65 g. It is broadly comparable in its general behaviour and ecology to the lesser mouse lemurs (Charles-Dominique, 1972; Martin, 1972a). The diet is very broad ('omnivorous') and includes a range of animal prey in addition to various fruits and even gums. A wide range of locomotor patterns is exhibited, such as clambering among fine branches, quadrupedal running, occasional

Figure 1.8. Demidoff's bushbaby, *Galago demidovii*. This is the smallest and least specialized of bushbaby species. Head and body length: 125 mm. (After a photograph by A.R. Devez in Charles-Dominique, 1977.)

vertical-clinging-and-leaping and bipedal hopping along the ground. As with the lesser mouse lemur, arboreal locomotion among fine branches and lianes is predominant and Demidoff's bushbaby frequently constructs spherical leaf-nests, although it can also be found sleeping during the daytime in dense tangles of foliage. Reproduction is generally similar to that of the lesser mouse lemurs, in that a nest is used and the infant is carried in the mother's mouth, but, unlike the lesser mouse lemurs, Demidoff's bushbaby has only one infant at a time. Finally, the pattern of social organization shown by *G. demidovii* is very similar to that of *Microcebus murinus*, with females sharing nests and having overlapping home ranges, and there is a comparable division between peripheral and central males, with only the latter showing substantial overlap between their home ranges

and those of females. As with the lesser mouse lemurs, Demidoff's bushbabies are typically seen alone when active at night, but there is a complex system of social interactions based on overlap of home ranges and occasional nocturnal encounters between individuals.

In morphological terms, the various bushbaby species are closely similar, although they range in size from the diminutive *G. demidovii* (Fig. 1.8) through the medium-sized species (e.g. *G. senegalensis* and *G. alleni*, body weights about 230 g and 250 g, respectively) to the relatively large thick-tailed bushbaby (*G. crassicaudatus*, body weight about 1.3 kg). Moreover, there are numerous differences between the bushbaby species in details of behaviour, notably with respect to diet and locomotion. For example, Allen's bushbaby (*G. alleni*, Fig. 1.9) feeds predominantly on a diet of fallen fruit with a small supplement of animal prey, whereas *G. elegantulus* is a specialized gum-feeder (Charles-Dominique, 1977a). In fact, the common name for *G. elegantulus* – 'needle-clawed bushbaby' – is derived from the small, sharp projections from the tips of the nails that permit this species to cling head downwards to broad trunk surfaces while feeding on gum exudates. (Incidentally, a similar adaptation of the nails is found in the gum-feeding fork-crowned lemur, *Phaner furcifer*, of Madagascar.) *Galago senegalensis moholi* in South Africa also feeds extensively on gum, but supplements its diet with arthropod prey to a greater extent than does the needle-clawed bushbaby (Bearder and Martin, 1980a), while *G. crassicaudatus* in the same region feeds on a mixture of gums, fruits and arthropods (Bearder and Doyle, 1974).

There is a comparable degree of variation in typical locomotor patterns. *Galago alleni* and, to a lesser extent, *G. senegalensis*, rely predominantly on vertical-clinging-and-leaping, in contrast to the versatile but predominantly quadrupedal locomotion of *G. demidovii*. At the other extreme, *G. crassicaudatus* surprisingly leaps very little, although it has the same development of powerful hindlimbs as other bushbaby species (Bearder and Doyle, 1974). This is perhaps one of the best examples among

Figure 1.9. Allen's bushbaby, *Galago alleni*. This species is the most consistent vertical-clinger-and-leaper among bushbabies, and its powerful hindlimbs bear witness to this locomotor speciality, as in a number of lemur species. Head and body length: 260 mm. (After a photograph by A.R. Devez kindly supplied by P. Charles-Dominique.)

the primates of a behavioural limitation that confines the use in practice of a particular morphological framework. Taking the bushbabies together, it is highly unlikely that even very detailed examination of skeletal evidence would have led to prediction of the diversity of typical locomotor repertoires found in field studies. In particular, if the skeleton of *G. crassicaudatus* were known only from a fossil specimen, it is almost certain that it would have been classified, erroneously, as a typical vertical-clinger-and-leaper.

There is also some degree of variation among the bushbaby species in reproductive characteristics. Although most bushbaby species use

nests and give birth to a single infant, *G. senegalensis moholi* in South Africa can give birth to twins twice a year (Bearder and Doyle, 1974). Further, the otherwise typical pattern of infant carriage in the mother's mouth is modified in *G. crassicaudatus*, in that older infants can be carried on the fur (Bearder and Doyle, 1974). Patterns of social organization also differ between bushbaby species. Whereas all patterns can be regarded as variations on the theme of 'solitary' ranging, with social contacts ensured through overlap of home ranges (as in *G. demidovii*), there are differences of detail. For instance, adult males in *G. alleni* are extremely intolerant of other males (Charles-Dominique, 1977b), such that the pattern of social organization can be termed a 'dispersed harem' system, whereas males exhibit far more mutual tolerance in *G. senegalensis moholi* (Bearder and Martin, 1980b), producing a 'dispersed multi-male system'. Further, for *G. crassicaudatus* at least there is a tendency towards formation of cohesive social groups, with individuals maintaining quite close contact during nocturnal activity.

Loris subgroup

It is essentially in relation to their sluggish locomotor activity that members of the loris subgroup are distinguished *en bloc* with respect to the bushbaby subgroup. Their characteristic slow quadrupedal progression through trees, reminiscent more of the deliberate gait of chamaeleons than of the active leaping of bushbabies, provides a focal point for many of their special adaptations. Like bushbabies, members of the loris subgroup are omnivorous, although each species has its own dietary peculiarities. For example, in Gabon the potto (*Perodicticus potto*) subsists primarily on fruit with supplements of gum and arthropod prey, whereas the angwantibo (*Arctocebus calabarensis*; Fig. 1.10) feeds mainly on arthropods with the addition of some fruit but no gum (Charles-Dominique, 1977a). No species in the loris subgroup makes use of a nest, perhaps because slow locomotion rules out return to a general morphological terms. Only two species – the slender loris (*Loris tardigradus*) and the slow loris (*Nycticebus coucang*) – occur outside Africa, inhabiting limited areas of the Indian subcontinent and of South-East Asia, respectively. The cohesiveness of the loris group as a whole and the relatively small number of species contained therein are doubtless attributable to the intensity of competition from other mammalian groups in Africa and Asia. In comparison to the lemurs of Madagascar (a far smaller geographical area), the extant members of the loris group represent a quite minor adaptive radiation. In particular, no surviving species of this group are diurnal, perhaps because bushbabies, lorises, the potto and the angwantibo occur in the same forest areas as diurnal monkeys and apes.

Tarsiers

The living tarsiers, recently effectively reviewed in Niemitz (1984a), are all confined to island areas in South-East Asia (Borneo, Sumatra, Sulawesi and the Philippines). Until now, it has been customary to recognize three species in the single genus *Tarsius*, but a fourth species (*Tarsius pumilus*) has been recognized recently by Musser and Dagosto (1987). These four species differ only in minor details, except that *T. pumilus* is a pygmy form, and the Bornean tarsier (*Tarsius bancanus*; Fig. 1.11) can be taken as representative. All tarsiers are nocturnal, and they are all quite small in size (weight range: 60–200 g). They are specialized vertical-clingers-and-leapers, showing the same development of powerful hindlimbs as occurs in sportive lemurs, in members of the indri subgroup and in bushbabies. In tarsiers, this pattern of locomotion is associated with their preference for movement between the thin, vertical trunks of saplings in the lowest stratum (0–3 m) of the forest areas they inhabit (Fogden, 1974; MacKinnon and MacKinnon, 1980). In association with their predominant use of vertical supports, tarsiers show a special adaptation of pressing the distal part of the tail against the tree trunk as an extra prop (Sprankel, 1965; Fig. 1.11). Surprisingly,

Figure 1.10. Angwantibo, *Arctocebus calabarensis*. The infant is carried on the mother's fur from birth onwards, except that she 'baby-parks' when foraging at night, as is the case with other members of the loris subgroup and with several lemur species. Head and body length: 210 mm. (After a photograph in Manley, 1974.)

fixed nest site at the end of the night, and in association with this the infant is always carried clinging to the mother's fur (except when she 'baby-parks' – see Chapter 9). As far as is known, all species have a 'solitary' pattern of movement at night and there is very little overlap of home ranges between adults of the same sex. The pattern established for *Perodicticus potto* would seem to be typical, with the home range of each adult male overlapping the ranges of only one or two adult females (Charles-Dominique, 1977a). Once again, slow locomotion seems to have placed limits on behaviour, in this case restricting possibilities for social contact.

Despite their divergent locomotor specializations, bushbabies, lorises, the potto and the angwantibo form a very close-knit group in

Figure 1.11. Bornean tarsier, *Tarsius bancanus*. This picture shows the well-developed hindlimbs used in vertical-clinging-and-leaping and the application of the tail (naked except for a tuft of hair at the tip), against the trunk as an additional support. Head and body length: 130 mm. (After a photograph in Fogden, 1974.)

(Niemitz, 1979). Females typically have a single infant at birth; they do not make or use a proper nest and the infant is reportedly carried in the mother's mouth during nocturnal activity. 'Baby-parking' may occur during nocturnal foraging, but by day the infant probably clings to the mother's fur as she rests in a suitable retreat, such as a shadowy cavity between buttress roots (Le Gros Clark, 1924a; MacKinnon and MacKinnon, 1980).

With respect to social organization, a brief field study by Fogden (1974) suggested that *Tarsius bancanus* exhibits a pattern of social contact through overlapping ranges resembling that found with other nocturnal prosimians, such as lesser mouse lemurs and Demidoff's bushbabies. However, Niemitz (1984b) subsequently reported pair-living tendencies in *T. bancanus* and a longer-term field study of *Tarsius spectrum* by MacKinnon and MacKinnon (1980) also indicated a rudimentary family-group structure. Most groups observed by the MacKinnons seemed to consist of monogamous pairs with their offspring and the adult male and female of a pair participate in 'duet' calling. Nevertheless, tarsiers typically forage alone at night, regrouping at dawn with the aid of a gathering call before returning to the sleeping site. Further, Crompton and Andau (1986, 1987) have used radiotracking to study range use by *T. bancanus* and they found no evidence of pair-living.

As will emerge later, tarsiers occupy a special position among the primates, in that they are generally intermediate between lemur and loris groups on the one hand, and monkeys, apes and humans on the other. However, exact resolution of the phylogenetic position of tarsiers is hindered by the survival of so few living representatives. It is worth noting that those researchers who have concentrated on the characters of living tarsiers have tended to favour a specific phylogenetic relationship with the simian primates (e.g. Hubrecht, 1908; Wood Jones, 1929; Luckett, 1975; Martin, 1975, 1978; Noback, 1975; Beard and Goodman, 1976), whereas those who have been concerned primarily with the palaeontological evidence

the diet of tarsiers seems to be composed exclusively of animal prey; no observer has yet reported the ingestion of plant food of any kind (see Niemitz, 1979, 1984a) and this confirms results from the examination of stomach contents (Davis, 1962). Arthropods constitute the major part of the diet, but tarsiers will also prey on small vertebrates, including venomous snakes (e.g. Simpson, 1945; Simons, 1972) have often questioned any such relationship.

New World monkeys

New World monkeys, the first of the three 'natural groups' of simian primates, bear a relationship to other simians somewhat comparable to that existing between lemurs and other prosimians. In geographical terms, the monkeys of South and Central America are more isolated than the other simians of the Old World and their phylogenetic separation from them must date back well into the past. Nevertheless, competition between primates and other mammalian groups is more intense in South America than in Madagascar and this has generally restricted the adaptive radiation of New World monkeys. For example, it would seem that in South and Central America the various nocturnal marsupial species occupy ecological niches comparable to those occupied by certain nocturnal prosimians in parts of the Old World. Nevertheless, New World monkeys account for about 30% of the modern genera and species of primates (Table 1.2). The simians as a group are almost exclusively diurnal in habits, and it is in the New World that the only nocturnal monkey lives – the owl monkey (*Aotus*; see Fig. 1.14). Overall, the New World monkeys typically differ from their Old World counterparts in having flat noses with well-separated nostrils, and for this reason they are commonly referred to as 'platyrrhines'. As with the loris group, platyrrhine monkeys are sharply divided into two subgroups: (1) marmosets and tamarins; (2) true New World monkeys.

Marmosets and tamarins

The diurnally active marmosets and tamarins are, as a group, the smallest of the New World monkeys (weight range: 70–550 g) and they are peculiar in having claws on all digits except the big toes. They and the aye-aye are the only living primates to possess claws on most digits. It seems likely that, as with the aye-aye, their sharp claws permit them to move quadrupedally along broad trunk surfaces, where small grasping extremities are relatively inefficient. Marmosets and tamarins are quite widespread in forested areas of South and Central America and many different species have been identified. They are all omnivorous, consuming both animal prey (mainly arthropods) and small plant items such as fruits. Yet, despite this generalized diet, they are unique among living primates in having a reduced number of molar teeth (see Chapter 6). Recently, the distinction between marmosets (*Callithrix* spp.; *Cebuella pygmaea*) and tamarins (*Saguinus* spp.; *Leontopithecus* spp.) has been linked to an important dietary difference. Marmosets typically feed extensively on sap and gums (Kinzey, Rosenberger and Ramirez, 1975; Coimbra-Filho and Mittermeier, 1977; Stevenson, 1978) and make holes in the bark of trees to tap this food source (in contrast to gum-feeding prosimians, which rely on gum exudates generated by insect invasion of tree trunks). Tamarins, on the other hand, do not rely so heavily on sap and gums (if at all in the case of some species) and they certainly do not make holes in tree surfaces. This dietary difference corresponds to a distinction between marmosets and tamarins in terms of the development of the anterior teeth in the lower jaw (see Chapter 6). This, however, seems to be the only major morphological and behavioural difference separating marmosets and tamarins. Basic features of locomotion, reproduction and social organization appear to be generally consistent throughout the marmoset/tamarin subgroup.

Marmosets and tamarins are distinctive in that they regularly produce twins and live in social groups that have commonly been interpreted as monogamous. The reproduction of all species in the subgroup is peculiar in two main ways. First, the twins share the same placental circulation system during embryonic development (Wislocki, 1939), a fact which is all the more remarkable because the twins originate from separate fertilized eggs and may therefore be of different sexes. Second, infants from birth to weaning are carried not primarily by the mother but predominantly by the father or by some other member of the social group (Fig. 1.12). Despite

Figure 1.12. Golden lion tamarin, *Leontopithecus rosalia*. As with all other tamarins and marmosets, twin births are the rule and the male of a pair carries the infants most of the time. The golden lion tamarin is one of the largest species in the marmoset/tamarin subgroup with a body weight of about 550 g. Head and body length: 300 mm. (After a photograph by P. Coffey of the Jersey Wildlife Preservation Trust.)

their relatively small body size, marmosets and tamarins do not build nests, although they may make use of available tree hollows. For instance, the pygmy marmoset (*Cebuella pygmaea*) has a body weight of only 70 g or so, and therefore counts among the smallest living primates, yet it does not use a nest and the twin infants are typically carried from birth onwards by an adult other than the mother.

Studies conducted in captivity on a variety of marmoset and tamarin species have tended to indicate a basic pair-living habit for these primates. Breeding colonies have generally only been successful when the animals have been housed as adult pairs, with or without their offspring, and various forms of 'pair-bonding behaviour' have been described. It has even been found from endocrinological investigations conducted with laboratory colonies of the common marmoset (*Callithrix jacchus*) that when two adult females are housed together in the same cage only one of them will exhibit normal hormonal cycles indicating fertility (Abbott, 1984). Further, it has been argued (e.g. see Kleiman, 1977a) that close involvement of the male in parental care is intimately connected with a pair-living habit in marmosets and tamarins. Accordingly, it has become common practice to regard marmosets and tamarins as prime examples of monogamy in primates. As twins are typically produced at each birth and as males and females may take about 2 years to reach full breeding condition, it is possible for a breeding pair to produce up to three successive litters before the first set of infants have attained sexual maturity. Hence, one might expect natural family groups of marmosets and tamarins to include between two and eight individuals, depending on the stage reached by the breeding pair concerned, and one would expect to find only one fully adult male and female in each group.

Until relatively recently, no detailed field studies had been conducted on marmosets and tamarins, so most of the major features of reproduction and social organization had been inferred from observations of captive animals. This situation has now changed (e.g. see Kleiman, 1977b; Rylands, 1981; Terborgh, 1983) and it would appear that some revision of the simple family group concept is required. For instance, a detailed study by Dawson (1977) of Geoffroy's tamarin (*Saguinus oedipus geoffroyi*; body weight about 480 g), involving marked animals in some cases followed by radiotracking, has shown that subadults may transfer between groups. Thus, although a particular group may have the size expected for a family group, the individuals therein may not be related to one another in the manner often assumed. Further, there have been reports of stable groups of other species exceeding the expected maximum size of eight and it has even been reported that some groups contain more than one breeding adult male (e.g. Garber, Moya and Malaga, 1984), possibly with more than one breeding female (Goldizen and Terborgh, 1986).

Although marmosets and tamarins can generally be sharply demarcated from the other New World monkeys, there is one intermediate species – Goeldi's monkey (*Callimico goeldii*; Fig. 1.13) – that combines diagnostic features of both groups. This species, which is relatively rare and has a restricted geographical distribution around the upper tributaries of the River Amazon, has a body weight equal to that of the largest of the tamarin species (about 500 g). It similarly has claws on all digits except the big

Figure 1.13. Goeldi's monkey, *Callimico goeldii*. Unlike the true marmosets and tamarins, this species produces only one infant at a time. Head and body length: 195 mm. (After a photography by P. Coffey of the Jersey Wildlife Preservation Trust.)

toes. However, it typically produces only one infant at birth, rather than twins, and the mother plays a more important part in the initial carriage of the infant than is typical for marmosets and tamarins. Because of its intermediate features, Goeldi's monkey has been alternatively classified either as a somewhat aberrant member of the marmoset/tamarin subgroup (e.g. Napier and Napier, 1967) or as an equally unusual member of the true New World monkey subgroup (e.g. Simpson, 1945; Simons, 1972; see also the classification in Table 3.1). Hershkovitz (1977) has even taken the step of classifying Goeldi's monkey in its own family (see also Heltne, Wojcik and Pook, 1981). The intermediate position of *Callimico* between the two main subgroups of New World monkeys is neatly expressed by the fact that it has kept the full complement of molar teeth, but the hindmost molars are considerably reduced, approximating the complete loss of these teeth seen in marmosets and tamarins. Goeldi's monkey has yet to be studied extensively in the field, but preliminary observations (Heltne *et al.*, 1981) indicate that it is similar in diet and general locomotor habits to marmosets and tamarins, taking a mixture of plant and animal (mainly arthropod) food and typically having a cursorial quadrupedal method of locomotion. Once again, it is customary to keep Goeldi's monkey in pairs in captivity and it is generally assumed that this species is monogamous, but evidence from both captive colonies and wild populations indicates that more than one female may breed in a group, and groups observed in the wild may exceed the size expected for true family units (Heltne *et al.*, 1981; B. Carroll, personal communication).

True New World monkeys

The second major subgroup of New World monkeys – the true monkeys – are markedly larger, on average, than marmosets and tamarins (weight range: 750 g–15 kg). Unlike marmosets and tamarins, they do not have claws on any of their digits and single births are the rule. Further, there is no sign of reduction of the hindmost molar teeth. Locomotion is typically quadrupedal in the smaller-bodied species, but suspension beneath branches during locomotion and feeding becomes more common with increasing body size. In association with this trend, the largest of the true New World monkey species have developed prehensile tails that serve essentially as a fifth limb in arboreal progression and suspension. In other respects, however, they are very much like Old World monkeys, and no species other than the owl monkey uses a nest of any kind.

As noted above, the owl monkey (*Aotus trivirgatus*; Fig. 1.14) is distinctive in being

Figure 1.14. Owl monkey, *Aotus trivirgatus*. As with marmosets and tamarins, the male carries the infant extensively, but only begins to do so regularly when the infant is about a week old (Dixson and Fleming, 1981). The unusual nocturnal habits of this species of New World monkey are reflected by the relatively large size of the eyes. Head and body length: 360 mm. (After a photograph taken by M. Lyster of the Zoological Society of London.)

typically nocturnal in habits, although in the southern part of the large geographical range of this species, which extends from Colombia to northern Argentina, diurnal activity has been reported (Rathbun and Gache, 1980). The owl monkey is a relatively small form (body weight range 750 g–1 kg) and it has a mixed diet of fruit and animal food (particularly arthropods). It has a typical cursorial, quadrupedal pattern of locomotion, but its nocturnal ranging behaviour is somewhat limited (see Wright, 1981) in association with the fact that the owl monkey, like the potto and slow loris, has a much lower basal metabolic rate than expected for a mammal of that body size (see Chapter 6). Both field observations (Wright, 1981) and experience of managing colonies in captivity (Dixson, 1982) indicate that owl monkeys typically live in monogamous family groups. As with marmosets, tamarins and Goeldi's monkey, the male is heavily involved with infant carriage, but only begins to carry the infant after it has spent the first week of life exclusively on the mother.

Titi monkeys (*Callicebus* spp.) are generally regarded as being quite closely related to the owl monkey, although they are exclusively diurnal in habits. Like the owl monkey, they have a wide geographical distribution in South America and they have a similar mixed diet of fruit and animal food. Titis are smaller than owl monkeys (weight range: 1.1–1.5 kg), but they are more active, as they combine quadrupedal running with agile leaping between trees. A pair-living habit seems to be typical under natural conditions (Mason, 1968) and, as with *Aotus*, the male is intensively involved in infant carriage after the first week.

Squirrel monkeys (*Saimiri* spp.) share with titis the distinction of being the smallest of the true New World monkeys (weight range: 750–950 g). Like titis, they have a mixed plant and animal diet, but they are probably more insectivorous. They are essentially quadrupedal in habits and leap less than titis. Squirrel monkeys differ from all the New World monkey species mentioned so far in that the female carries her single infant without any assistance from the male. This is probably associated with the fact that *Saimiri* typically lives in quite large social groups containing several adult males ('multi-male groups'). Most field workers have reported social groups containing 20–40 individuals (e.g. Thorington, 1968a; Baldwin and Baldwin, 1981; Terborgh, 1983), but there have been several reliable accounts of much larger groups containing hundreds of squirrel monkeys travelling together (in extremely noisy fashion) through primary rainforest in Amazonia. Squirrel monkeys are unusual among the true New World monkeys in that their pattern of breeding is strictly seasonal and males increase markedly in body weight before the onset of the mating season (the 'fatted male syndrome'). No convincing explanation has been provided for such strict seasonal breeding in a species that typically inhabits primary rainforest with only limited seasonal fluctuations.

Capuchin monkeys (*Cebus* spp.; Fig. 1.15) are medium-sized New World monkeys (weight range: 2.5–3.0 kg), with a moderately developed grasping (prehensile) capacity of the tail. They are quite widely distributed in the forests of South and Central America, where they move around in multi-male troops of up to 35 individuals (see Freese and Oppenheimer, 1981). As with the other true New World monkeys covered thus far, their diet consists of a mixture of fruits and animal food, but capuchins are remarkable for the ingenuity they show in food collection. Some capuchins have been reported to open oil-palm nuts by striking them against

Figure 1.15. Black-capped capuchin, *Cebus apella*. Note the use of the tail as a fifth limb in maintaining posture. The prehensile condition of the tail is only weakly developed in *Cebus*, in comparison with *Alouatta, Ateles, Brachyteles* and *Lagothrix*. Head and body length: 420 mm. (After photographs taken by the author.)

suitable 'anvil' sites in trees and they are adept at seeking out concealed arthropods of various kinds (see Antinucci and Visalberghi, 1986). The weak prehensile capacity of the tail is frequently used for clambering in trees and for maintaining posture when at rest, but during quadrupedal cursorial progression the tail is carried with little contact with the branches. As with squirrel monkeys, the single infant is carried by the mother from birth onwards without any assistance from the male.

Howler monkeys (*Alouatta* spp.) are considerably heavier than the other New World monkeys described above, with an average body weight of 5.5–6.6 kg, according to species. In fact, there is a significant difference in body weight between males and females in all species. For example, in the mantled howler monkey (*Alouatta palliata*), which is the species best known from field studies, the adult male typically weighs about 7.3 kg, whereas the adult female usually weighs in the region of 5.8 kg. Such 'sexual dimorphism' in body weight (the male being about 25% heavier than the female in this case) is uncommon among New World monkeys, but it is fairly frequent and often more extreme in Old World monkey and ape species (see later). The howler monkeys, which are widespread in South and Central America, are also outstanding among New World monkeys in that they feed extensively on leaves, taking roughly equal proportions of leaves and fruits in their diet. Although howlers and woolly spider monkeys (see later) are the most folivorous of the New World monkey species, they are not as specialized in that direction as the leaf-monkeys of the Old World (see later). As Milton (1980) has shown in an exemplary field study, a clear preference is shown by *Alouatta palliata* for fruits when they are abundantly available, and young leaves from a variety of tree species are commonly selected rather than mature leaves. Accordingly, howlers lack the special adaptation of the stomach found in Old World leaf-monkeys. The prehensile capacity of the tail is well developed in howler monkeys, being used intermittently during locomotion and frequently during feeding and resting. Locomotion above

branches is typically quadrupedal, but howlers spend a great deal of their time suspended beneath branches, probably because of their relatively large body weights. The prehensile tail is particularly important during feeding on fruit and young leaves at the ends of branches ('terminal branch feeding'; Grand, 1972), where the suspended weight of the body can be provided with better support in a relatively precarious position. Howlers live in medium-sized multi-male troops usually containing between 10 and 20 individuals and females carry their single infants from birth onwards without any cooperation from males (Carpenter, 1934; Milton, 1980).

Spider monkeys (*Ateles* spp.), the woolly spider monkey (*Brachyteles arachnoides*) and woolly monkeys (*Lagothrix* spp.) form a fairly close-knit set of species that, like the howler monkeys, are characterized by quite heavy body weights and the possession of fully prehensile tails. *Ateles* has a typical body weight of about 7.6–9.0 kg according to species, *Lagothrix* weighs about 6.2 kg and *Brachyteles arachnoides* – the heaviest of the New World monkeys – weighs about 13.5 kg (females, ~12 kg; males, ~15 kg). In *Ateles*, there is no sexual dimorphism in body weight and in fact it has occasionally been reported that *Ateles* females may weigh marginally more than males. In *Lagothrix* and *Brachyteles*, however, there is mild sexual dimorphism in body weight, with males weighing some 20% more than females. In spider monkeys and woolly monkeys, as in howler monkeys, the underside of the last third of the tail bears a naked area of skin that, at least in some cases, has been shown to be adapted for tactile sensitivity. Thus, the fully prehensile tail can be said to act as a fifth limb in several respects (see Chapter 10). Like howlers, spider monkeys and woolly monkeys run quadrupedally above branches, but they spend much of their time suspended beneath branches, notably during 'terminal branch feeding', assisted by the prehensile tail. All of these species, except *Brachyteles arachnoides* (Milton, 1984; Streier, 1987), are more frugivorous in habits than the howlers, and this is reflected in more active movement and general agility. To date, only *Ateles* and *Brachyteles* have been studied intensively in the wild, but reports indicate that all spider monkey and woolly monkey species live in multi-male troops containing 10–20 individuals as a rule. A detailed field study of one spider monkey species, *Ateles belzebuth* (Klein and Klein, 1977), has indicated that group size may vary according to the availability of suitable fruit, so this flexibility could account for the wide range of group sizes reported for *Ateles* in the literature.

The remaining true New World monkeys are the sakis (*Pithecia* spp.; *Chiropotes* spp.) and the uakaris (*Cacajao* spp.). These are medium-sized monkeys, with body weights ranging from about 1.7 kg for *Pithecia* to 2.6–2.8 kg for *Chiropotes* and 2.9–3.9 kg for *Cacajao*. Sakis and uakaris are basically quadrupedal and have bushy tails that lack any sign of a prehensile capacity. Indeed, *Cacajao* has a noticeably short tail compared with other New World monkeys. The sakis are predominantly frugivorous, whereas uakaris eat some leaves in addition to a basic diet of fruit. All saki and uakari species live in relatively small groups. *Pithecia* has been reported to live in quite small groups that may represent monogamous families (Buchanan, Mittermeier and van Roosmalen, 1981), whereas *Carajao* and *Chiropotes* live in quite large groups that may contain over 30 animals and typically include more than one adult male (Fontaine, 1981; van Roosmalen, Mittermeier and Milton, 1981). However, there is no evidence for any of the saki or uakari species that the male plays any part in infant carriage.

Before turning to the Old World monkeys, which can to some extent be regarded as Afro-Asian counterparts of the true New World monkeys, it is worth commenting on some special features of the New World monkeys as a group. These primates are restricted to forest areas in equatorial and subequatorial South and Central America and all species are essentially arboreal. There are no species adapted for terrestrial life in more open-country areas, which are prevalent in the southern part of South America, and this provides a sharp contrast with the Old World monkeys, many of which are conspicuously terrestrial. Perhaps in association with this, there are no cases of really extreme sexual dimorphism in body size (i.e. with males much larger than females), although *Alouatta*, *Brachyteles* and *Lagothrix* are mildly dimorphic. Further, there are no predominantly folivorous species among the New World monkeys. Most species are primarily frugivorous and only in howler monkeys and woolly spider monkeys does the proportion of leaves in the diet approximately match the proportion of fruit. Hence, the extreme folivorous adaptations of the leaf-monkey of the Old World have no real counterpart among the New World monkeys. Perhaps the consistent restriction of New World monkeys to an arboreal existence accounts for the development of prehensile tails in the heaviest species – a notable specialization that is not found in equally heavy Old World monkey species, even in species for which arboreal life is the rule.

Old World monkeys

The monkeys of Africa, southern Asia and southeastern Asia form a remarkably cohesive group in basic morphological terms (e.g. see Schultz, 1970), and this is reflected in their inclusion in a single family (Cercopithecidae). All members of this family have relatively close-set, downward-pointing nostrils. This feature gave rise to the term 'catarrhine monkeys', distinguishing them from the 'platyrrhine monkeys' of the New World. (In fact, both the monkeys and the apes of the Old World share this feature and they are commonly referred to collectively as 'catarrhines'; see Table 3.1). Further, all of the Old World monkeys have **ischial callosities**, which are hardened cutaneous sitting-pads underlain by broadened, roughened **ischial tuberosities** on the ischial bones of the pelvic girdle (see Miller, 1945; Washburn, 1957; Vilensky, 1978). Gibbons and siamangs (see later) are the only other living primates to exhibit such callosities. In terms of such features as the structure of the postcranial skeleton and the form of the brain, the Old World monkeys are far less diverse than any other major primate group

(Schultz, 1970), and they have a remarkably uniform pattern of dentition (see Chapter 6). Nevertheless, as with some other 'natural groups' of primates (loris group; New World monkeys), there is a fairly clear division identifiable between two subgroups of Old World monkeys: (1) the guenon subgroup, containing the macaques, mangabeys, baboons, drills, geladas, guenons and patas monkeys; and (2) the leaf-monkey subgroup, containing the langurs, proboscis monkeys and colobus monkeys. Members of the guenon subgroup are characterized by the possession of cheek pouches (Murray, 1975), and leaf-monkeys have a specialized sacculated stomach containing symbiotic bacteria that break down the relatively indigestible cell walls of leaves (Bauchop and Martucci, 1968; Kuhn, 1964, see Chapter 6). Although this latter distinction accounts for the fact that the colobine monkeys generally consume a greater proportion of leaves in their diets, predominant leaf-eating is not a feature common to all colobine species and there are several cercopithecine monkey species that eat a large proportion of leaves but have an unspecialized stomach. Representatives of both Old World monkey subgroups are found in all three regions of their overall range (Africa, S. Asia and S.E. Asia), but the extant cercopithecines are most prevalent in Africa, and the extant colobines predominate in Asia. The Old World monkeys as a group are highly adaptable, occupying both forest and open country throughout their range. Overall, the Old World monkeys constitute one of the most successful 'natural groups' of modern primates in numerical terms, accounting for 26% of the genera and 37% of the species. Further, certain macaque species inhabit regions with some of the most extreme conditions tolerated by primates other than humans. The barbary macaque (*Macaca sylvanus*; Drucker, 1984) lives in highland areas of North Africa where snow covers the ground for part of the year, and the same is true of the rhesus macaque (*Macaca mulatta*) in northern areas of the Indian subcontinent (Richard, 1985). The Japanese macaque (*Macaca fuscata*; Suzuki, 1965) is now renowned for bathing in thermal springs in an environment with sub-zero ambient temperatures in winter.

Guenon subgroup

Among the living representatives of the guenon subgroup, the guenons themselves (*Cercopithecus* spp., *Allenopithecus* and *Miopithecus*), the patas monkey (*Erythrocebus patas*) and the mangabeys, baboons, drills and geladas (*Cercocebus* spp., *Papio* spp., *Mandrillus* spp. and *Theropithecus gelada*) are all restricted to sub-Saharan Africa. The only members of the guenon subgroup to occur outside this geographical area are the macaques (*Macaca* spp.) and the so-called 'black ape' of Celebes (*Macaca niger*). A single macaque species, the Barbary macaque, occurs in North Africa, and the remaining macaques range through India, Sri Lanka and South-East Asia. The most outstanding feature of the Old World monkeys, in comparison to all other primate groups considered so far, is that many species have become adapted for a predominantly terrestrial life (although most return to the trees to sleep at night). Terrestrial feeding and locomotion are particularly common among species inhabiting wooded savanna in Africa. The guenons are generally forest-living, arboreal species; but the vervet monkey (*Cercopithecus aethiops*) is a widespread, semi-terrestrial occupant of sparsely wooded areas in sub-Saharan Africa, and the patas monkey (*Erythrocebus patas*) is an almost exclusively terrestrial East African species living in relatively open savanna. Among the other species in the guenon subgroup, particularly in macaques and baboons, terrestrial life is relatively common. The most extreme case is doubtless that of the gelada baboon of Ethiopia (*Theropithecus gelada*), which feeds and moves exclusively on the ground, commonly shuffling along on its hindquarters, and sleeps on cliff faces during the night (Dunbar, 1977). This species rarely, if ever, ventures into the trees.

Members of the guenon subgroup are generally quite large-bodied, but there is a wide range of body sizes extending from the diminutive talapoin (*Miopithecus talapoin*) with a body weight of about 1.2 kg to the bulky chacma baboon (*Papio ursinus*) with a body weight for adult males of about 29 kg and for adult females of around 15 kg. But most species have body weights in the range 3–10 kg and a fairly typical representative is the pig-tailed macaque (*Macaca nemestrina*; Fig. 1.16), with adult body weights of 10 kg and 5.6 kg for males and females, respectively. Species in the guenon subgroup commonly show **sexual dimorphism** in body weight and in extreme cases (particularly among predominantly terrestrial species) adult males weigh about twice as much as adult females. This is the case for the chacma baboon cited above and for *Papio anubis* (Fig. 1.17), *P. cynocephalus*, *P. hamadryas* and *Erythrocebus patas*. The larger body size of males is typically associated with the possession of significantly larger canine teeth.

Although species in the guenon subgroup have the same general morphology, they have different dietary preferences, which are broadly correlated with differences in body size. Smaller species, such as the talapoin and the moustached guenon (*Cercopithecus cephus*), consume arthropod prey in appreciable quantities in addition to such readily digestible plant items as fruits. By contrast, the larger-bodied species tend to eat a greater proportion of leaves or other plant matter requiring considerable mastication and/or digestion. For instance, the gelada baboon moves slowly about in a squatting position while feeding on grass blades, supplemented by rhizomes and a variety of seeds and arthropods. Dunbar (1983) has stated that gelada baboons are the only real grazers among primates. Some of the less-specialized large-bodied species, such as the olive baboon, have been observed preying on other mammalian species (e.g. bush-pig and duiker). Overall, the feeding behaviour of monkeys in the guenon

Figure 1.16. Pig-tailed macaque, *Macaca nemestrina*. Here, a subadult male is grooming a young female that initiated the interaction by 'presenting' her sexual swelling. Note the reduction of the tail that gives this species its name. Head and body length: 530 mm. (After a photograph taken by the author.)

Figure 1.17. Olive baboon, *Papia anubis*. This species shows extreme sexual dimorphism, with the adult male (shown here) weighing almost twice as much as the adult female (21.9 kg versus 12.4 kg). The two infants shown playing alongside the adult male have dark coats, distinguishing them from older members of the species. Such differential coloration of infants is common among species of Old World monkeys, particularly leaf-monkeys. Head and body length of the male: 560 mm. (After a photograph taken by the author.)

subgroup can be described as 'opportunistic' in that diets are in most cases quite varied and often require extensive foraging.

Locomotion among species in the guenon subgroup can be broadly described as quadrupedal, but there is again considerable versatility of movement and some degree of differentiation between species. Those species that spend a large proportion of their time on the ground typically exhibit a modification of the hand that permits locomotion with the fingers used as struts (**digitigrady**). This condition, seen in *Erythrocebus, Macaca, Mandrillus, Papio, Theropithecus* and the more terrestrial of the mangabeys (*Cercocebus* spp.), represents a small development in the direction taken by ungulates (hoofed mammals), where typically only one or

two upright digits of the hand and foot persist to maintain contact with the ground in locomotion. All exclusively arboreal species of the guenon subgroup, along with some of the terrestrial species, have retained a full-length tail, though none has developed a prehensile function (as noted above). In the more terrestrial species, however, the tail has commonly been reduced in length, as, for example, in the pig-tailed macaque (Fig. 1.16) and the gelada baboon.

Reproductive patterns are fairly uniform within the guenon subgroup. All species typically have a single infant at birth and carriage of the infant on the mother's fur for a considerable time after birth is characteristic. No species constructs or uses a nest of any kind. By contrast, there is considerable variety in patterns of social organization, although all species form foraging and sleeping units of some type. Guenons (*Cercopithecus* spp.) are generally noted for the formation of so-called 'harem groups', containing only one adult male in addition to several adult females plus subadults, juveniles and infants of both sexes (Gautier-Hion, 1978; Struhsaker, 1969). Surplus adult males sometimes form 'bachelor groups' occupying largely separate ranges. This type of social organization is broadly typical of *Cercopithecus* species, excluding the vervet monkey (*C. aethiops*) found in wooded savanna regions, and it is also found in *Erythrocebus patas* (Hall, 1965). It has been reported, however, that the forest-living De Brazza monkey (*Cercopithecus neglectus*) lives in monogamous groups (Gautier-Hion and Gautier, 1978). Some of the baboons, such as *Papio hamadryas* and the gelada baboon, also form harem and bachelor male groups, although this tendency is obscured by the formation of larger herds incorporating both kinds of group (Kummer, 1968; Dunbar, 1984). A similar pattern of herd-formation from basic subgroups may also be present in drills and mandrills (Gartlan, 1970). However, many monkey species in the guenon subgroup form proper 'multi-male groups' containing several adults of both sexes in addition to subadults, juveniles and infants. This is true of the talapoin, the vervet, macaques, most baboons (such as *Papio anubis*; Fig. 1.17) and the mangabeys. The multi-male troops of these species may number from a few dozen, as is usual in macaques and mangabeys, to more than a hundred individuals, as in the talapoin, ranging over an extensive, shared home range (see Richard, 1985 for a review).

Leaf-monkeys

In comparison with the guenon subgroup, leaf-monkeys are even more uniform in most of their characteristics. With the exception of the African colobus monkeys (*Colobus* spp.), they are confined to the Indian subcontinent and South-East Asia and they are predominantly arboreal, with the exception of some populations of the hanuman langur (*Presbytis entellus*). The range of body size is relatively limited, extending from the olive colobus (*Colobus verus*; body weight about 4.3 kg) to the proboscis monkey (*Nasalis larvatus*; body weight of an adult male about 21 kg). Sexual dimorphism is generally present, but is usually less pronounced than in terrestrial monkeys of the guenon subgroup. The only exception to this is the proboscis monkey, in which the adult male weighs more than twice as much as the female. The black-and-white colobus (*Colobus guereza*; Fig. 1.18), which is typically arboreal like other leaf-monkey species, shows only moderate dimorphism; the adult male weighs roughly 25% more than the adult female (10.2 kg versus 8.1 kg). Other leaf-monkey species are generally even less sexually dimorphic. Although all species of leaf-monkey

Figure 1.18. Two female black-and-white colobus monkeys (*Colobus guereza*), with one holding an infant in a typical resting posture. Head and body length of the male: 590 mm. (After photographs taken by the author.)

include some proportion of leaves in their diets, only a few species (e.g. *Nasalis larvatus, Colobus guereza* and *Presbytis senex*) can be regarded as specialized leaf-eaters. Most species include some fruit in the diet and other dietary items such as seeds and flowers have been recorded for various species. As with the other Old World monkeys, leaf-monkeys have an essentially quadrupedal locomotion. The only locomotor specialization of note is that in *Colobus* species the thumb is considerably reduced in size, in connection with their frequent use of suspensory locomotion. The other leaf-monkeys in Asia and South-East Asia also show frequent arboreal suspension, although they lack any marked reduction of the thumb. The term 'Old World semi-brachiation' was hence coined for the leaf-monkeys as a group (Napier and Napier, 1967; see Chapter 10), but it has been criticized by Mittermeier and Fleagle (1976). This tendency to exhibit suspensory locomotion provides a further broad distinction between leaf-monkeys and the guenon subgroup.

As with the guenon subgroup, all leaf-monkeys have a broadly similar pattern of reproduction. There is usually a single infant at birth, and it is carried on the fur of an older animal (typically the mother) from birth onwards. No species of leaf-monkey constructs or uses a nest, even in connection with reproduction. Detailed field studies of social behaviour and ecology have been largely confined to the African colobus monkeys and to Asiatic langurs of the genus *Presbytis* (see Clutton-Brock and Harvey, 1977; Richard, 1985). As with the cercopithecine monkeys, some species form large multi-male troops, such as the red colobus (average group size, 40–50 individuals; Clutton-Brock, 1975; Struhsaker, 1975), and others form relatively small harem groups and bachelor male groups, such as the black-and-white colobus (*Colobus guereza*; Struhsaker and Oates, 1975), the purple-faced langur (*Presbytis senex*; Rudran, 1973) and the Nilgiri langur (*Presbytis johnii*; Poirier, 1969). The contrast in group size and other aspects of social organization between the black-and-white colobus and the red colobus in fact provides one

of the most striking instances of behavioural difference between closely related primate species (Clutton-Brock, 1974). The situation is even more unusual in the hanuman langur (*Presbytis entellus*), because in some areas this species forms harem groups and bachelor male groups, whereas in others multi-male groups are formed (see Jay, 1965). For the remaining leaf-monkey genera detailed field studies have yet to be conducted. Preliminary observations by Kern (1964) indicate that proboscis monkeys live in multi-male groups containing about 20 individuals, but very little is known about the natural habits of the douc langur (*Pygathrix nemaeus*), snub-nosed langurs (*Rhinopithecus* spp.) or the Pagai Island langur (*Simias concolor*; see Tilson, 1977). There is, however, good evidence that the Mentawai langur (*Presbytis potenziani*) forms groups consisting of a monogamous pair and their offspring (Tilson and Tenaza, 1976).

Apes and humans

The sixth major 'natural group' of the living primates is a relatively small category containing humans and their closest zoological relatives – the apes. Apes and humans together account for only 8% of the genera and 6% of the species of modern primates and in this respect are scarcely more successful than the loris group (Table 1.2). Humans and apes are united by numerous morphological characteristics, the most obvious external features being the barrel-shaped chest and the absence of a tail. Like the monkeys (with the exclusion of *Aotus*), all apes are diurnal, but they are relatively large in body size compared with monkeys.

The living representatives can be neatly divided into three subgroups: (1) the lesser apes; (2) the great apes; and (3) humans. The great apes are undoubtedly more closely related to humans than are the lesser apes, and for this reason they have received more attention in the literature dealing with the most recent phases of human evolution.

Lesser apes

The lesser ape subgroup, which has been effectively reviewed by Preuschoft, Chivers *et al.* (1984), comprises the gibbons (e.g. *Hylobates moloch*; Fig. 1.19) and the siamang (*Hylobates syndactylus*). As the name implies, they are the smallest of the apes, with body weights ranging from 5.4–6.7 kg for the gibbon species to 10.7 kg for the siamang. Sexual dimorphism is virtually non-existent with respect to body weight, although males tend to have somewhat larger canine teeth than females. Lesser apes are confined to rainforest areas of the South-East Asian region, with the siamang occurring only on the Malayan peninsula and Sumatra. Both gibbons and the siamang are essentially arboreal and mainly dependent on plant food, with

Figure 1.19. Silvery gibbon, *Hylobates moloch*. Note the long arms and the gracile hands, typical of all lesser apes. As with some other species of lesser apes and numerous species of Old World monkeys, the colour of the infants' coat is different from that of the adults. Head and body length: 520 mm. (After photographs taken by the author.)

gibbons concentrating on fruit and supple-
menting the diet with leaves and occasional insect
prey and the siamang concentrating somewhat
more on leaves. The most notable feature of
lesser apes resides in their extremely long arms,
relative to body size, associated with their
frequent use of arm-swinging locomotion.
Indeed, gibbons and the siamangs are the only
living primates to exhibit **true brachiation**, in
which the body may be propelled by the arms
alone, with a phase of free flight through the air
(see Chapter 10). In conjunction with this
specialized use of the arms, lesser apes move
bipedally along broad semi-horizontal trunks and
across the ground, with their long arms held clear
of the ground or (occasionally) used as crutches.
They differ from great apes in possessing ischial
callosities, a feature that they share only with the
Old World monkeys. Reproduction follows the
typical simian pattern, with a single infant born as
a rule and carriage from birth onwards on the fur
of the mother. The male siamang carries the
infant once it has reached 1 year of age. Social
groupings are monogamous family groups in both
gibbons (Ellefson, 1968) and the siamang
(Chivers, 1974), with group size usually reaching
a maximum of five before the oldest maturing
offspring of the breeding pair leaves the group
(see also Preuschoft *et al.*, 1984). All lesser ape
species have characteristic loud (usually
melodious) vocalizations that seem to serve a
spacing function between family groups. The
combination of monogamy with melodious, long-
carrying vocalizations also occurs in the indri of
Madagascar, which is similarly an inhabitant of
rainforest, and this provides another remarkable
case of convergent evolution in primates.

Great apes

The great apes include the chimpanzees and
gorillas of Africa, together with one species now
confined to island areas in South-East Asia – the
orang-utan of Borneo and Sumatra (*Pongo
pygmaeus*; Fig. 1.20). The gorilla is regarded as a
single species (*Gorilla gorilla*; Fig. 1.21), but two
chimpanzee species are recognized – the common
chimpanzee (*Pan troglodytes*; Fig. 1.22) and

Figure 1.20. Orang-utan, *Pongo pygmaeus*. These
great apes are the largest extant mammals with an
essentially arboreal lifestyle. The female shown here
is carrying her infant in a typical primate fashion,
clinging to her fur. Note the reduction in the size of the
thumb. Head and body length of the female: 785 mm.
(After a photograph taken by the author.)

the pygmy chimpanzee (*P. paniscus*). All great
apes have relatively long arms and may use arm-
swinging locomotion to varying degrees, but none
performs true brachiation. This is undoubtedly a
reflection of their heavier weight, which ranges
from about 39 kg in *Pan paniscus* to about 150 kg
in an adult male gorilla. The two chimpanzees
and the gorilla show a great deal of terrestrial
activity, particularly when travelling, whereas
the orang-utan is predominantly arboreal even
when moving. Indeed, the orang-utan is unusual
in that it is by far the heaviest primarily arboreal
primate (or arboreal mammal, for that matter),
with fully adult males weighing in excess of 70 kg,

Figure 1.21. Gorilla, *Gorilla gorilla*. Typical social groups have a single silverback male (top right) and several adult females. The gorilla is essentially terrestrial in habits, although some activities are performed in trees (notably by smaller individuals). There is extreme sexual dimorphism: adult males weigh 150 kg and are about 170 cm high, whereas adult females weigh 80 kg and are about 145 cm high. (After photographs taken by P. Coffey of the Jersey Wildlife Preservation Trust.)

although adult females weigh only 37 kg. All great apes typically support themselves with their arms when moving on the ground, and they may occasionally move bipedally over relatively short distances. When orang-utans move across the ground using their arms for support, they use the outer margins of their hands as the point of contact with the ground ('fist-walking'), whereas chimpanzees and gorillas support themselves on their knuckles ('knuckle-walking').

There is considerable dietary diversity among great apes. Common chimpanzees are predominantly frugivorous (Goodall, 1968; Wrangham, 1977), but they eat a wide range of items

Figure 1.22. Common chimpanzee, *Pan troglodytes*. As with the other great apes, infants are suckled for more than 2 years and may be carried by the mother for at least that period of time. There is only moderate sexual dimorphism: adult males weigh about 52 kg and adult females weigh about 40 kg. Head and body length is 850 mm in males and 780 mm in females. (After a photograph in Albrecht and Dunnett, 1971.)

including termites (Goodall, 1968; Hladik, 1973; McGrew, 1974) and even quite large mammals including other primates such as baboons and red colobus monkeys (Teleki, 1973). Orang-utans are similarly predominantly frugivorous though they also eat leaves and bark (MacKinnon, 1974; Rijksen, 1978). Gorillas, on the other hand, are exclusively vegetarian, with mountain gorillas feeding primarily on ground-level plants (e.g. wild celery), bamboo shoots and various roots (Schaller, 1963; Fossey and Harcourt, 1977; Goodall, 1977). Recent reports indicate that gorillas in West African rainforest may be more frugivorous than the mountain gorillas of East Africa (Tutin and Fernandez, 1985).

As with lesser apes, great apes typically have

a single infant at birth. The infant is carried on the mother's fur from birth onwards and may eventually ride on other adults before becoming independent. By contrast, social organization shows considerable diversity among the great apes. The orang-utan is virtually solitary in habits. Fully adult males, which have conspicuous beards, throat-sacs and cheek-flanges, usually range alone and the only groups observed other than mother–infant pairs are typically temporary associations between adult females and/or young animals (Rodman, 1973; MacKinnon, 1974; Rijksen, 1978). Social contacts between adult males and females depend largely on overlap in ranging areas, and this is reminiscent of the situation in various nocturnal prosimians. Gorillas are intermediate with respect to size of social groups, with an average of a dozen individuals to a group. Fully adult males have a large patch of grey hair on their backs ('silverback' males); there is frequently only one such male in each social group, whereas others may be solitary. The gorilla pattern of social organization is thus quite close to a harem system, with surplus 'bachelor males'. Younger, 'blackback' males are tolerated by the silverback male(s) in the social group (Schaller, 1963; Harcourt, 1979). The most complex system of social organization is that of the common chimpanzee. It has now been recognized (Reynolds and Reynolds, 1965; Nishida, 1979; Wrangham, 1979) that *Pan troglodytes* lives in large social communities with as many as 80 members, which fragment into feeding and travelling bands varying in size from just a few to three dozen individuals. Within the bands, there is a tendency for mothers with off-spring to associate with one another and for individual males to form temporary associations. This flexibility in band composition permits individual chimpanzees to experience social encounters with many other members of the community in the course of time. The communities themselves tend to be quite discrete units with distinct territorial boundaries, and there is some evidence (e.g. Bygott, 1972; Goodall, Bandora *et al.* 1979) that members of a band from one community may kill and even eat an infant from another, if an unprotected mother is encountered in a boundary zone.

Great apes, then, exhibit a surprising variety of patterns of social organization with chimpanzees, in particular, showing considerable complexity and flexibility. This variety is rendered even more striking by the fact that two of the great ape species, the orang-utan and the gorilla, are markedly sexually dimorphic, with adult males roughly twice as heavy as adult females, whereas the two chimpanzees are only moderately dimorphic. Such diversity among great apes alone clearly indicates that we must be very cautious indeed about using any single great ape species as a model for the evolution of human behaviour, despite the relatively close zoological relationship between great apes and the human species.

Humans

This brings us to the last species to be covered in this survey of living primates, namely our own species *Homo sapiens*. Clearly, because of our unique ability to modify our environmental circumstances, recognition of the fundamental biological attributes of the human species is both more difficult and less immediately informative than for other primates. Nevertheless, *Homo sapiens* originated within the evolutionary radiation of the primates and it is essential to examine the consistent biological features of human beings in order to reconstruct our evolutionary origins. Although it is true that our species has acquired certain unique features as it has evolved, these features have emerged during a continuous process of evolutionary divergence from great apes and we must reconstruct the approximate sequence of events if we are to achieve a full understanding of our own nature. There are, of course, many writers who would draw a sharp boundary between 'man' and 'animals', as if there were more difference between a human being and a chimpanzee than between a chimpanzee and an amoeba. But one of the main lessons to be drawn from evolutionary theory is that there is a fundamental continuity between all living organisms and that it

is impossible to define a realistic boundary between human beings and great apes when evolutionary change over time is considered (see also Chapter 3). However special we may feel ourselves to be, the human species arose as an integral part of the animal kingdom and there is much of value to be gained from a detailed appreciation of our own biological origins.

The human species is both geographically widespread and very variable in biological terms. Indeed, the modern geographical distribution of *Homo sapiens* is so extensive, far outstripping that of any other primate species, that for the sake of clarity it was necessary to exclude the human species from the distribution maps given in Fig. 1.1. Whereas primates are typically tropical and subtropical in their distribution, the human species has now expanded to occupy all of the climatic zones of the world. Thus, it is scarcely surprising that humans show quite a wide range of biological variation. *Homo sapiens* has been described as a 'polytypic species' and much attention has been paid to different forms of racial variation. There are, of course, some immediately obvious features (e.g. skin colour and body shape) that differ between human populations, but the biological importance of such racial distinctions has generally been exaggerated. Such relatively mild variation within any non-human primate species would attract scant scientific attention and it is primarily for cultural reasons that racial variation in the human species has achieved prominence. Recent genetical studies have in fact shown that there is little basis for a simple typological approach to racial variation in *Homo sapiens* (Jones, 1986). Although it is possible to identify several localized effects with respect to certain genetic features, the overall geographical distribution of genetic markers shows no clear pattern compatible with the concept of relatively discrete racial groups. Analyses conducted by Latter (1980) and Nei (1982) show, to the contrary, that about 90% of the total genetic diversity of the human species exists as differences between individuals in a localized population (tribe or nation) and that about 6% exists as differences between populations (e.g. between the English and the French). Divergence between the classically recognized 'racial groups' accounts for only 10% or so of the total genetic diversity of *Homo sapiens*. The major lesson to be learnt from this genetic evidence is that, contrary to popular belief, the human species is remarkably uniform. Current morphological and genetic evidence indicates that the modern members of the species *Homo sapiens* are all derived from an ancestral stock that existed about 200 000 years ago and that the present worldwide distribution of the species has been achieved in a comparatively short space of time (Stringer and Andrews, 1988).

Data on worldwide variation in human body weight and stature (Roberts, 1953) provide confirmation for the essential uniformity of the human species. For example, the overall mean body weight for human males is 56.8 kg and the range for population averages is only 39.7–77.3 kg. This range of variation is somewhat greater than that found for the most widely distributed non-human primate species, but this is scarcely surprising in view of the far greater geographical range of the human species. The stature of human males shows an even more limited degree of variation than body weight, with an overall mean of 163.4 cm and a range of population averages of 142.2–175.9 cm. Roberts (1953) showed that much of the variation in human body weight and stature (for females as well as for males) can be linked to geographical variation in mean annual temperature. The two parameters are negatively correlated with temperature ($r = -0.60$ for body weight; $r = -0.35$ for stature), although further analysis showed that the primary association exists between body weight and temperature. Hence, a good part of the variation in body size between human populations can be attributed directly to the influence of local climatic conditions, with average body weight generally declining with increasing environmental temperature.

It should also be noted that the human species is mildly but significantly sexual dimorphic in body weight, with males typically weighing about 18% more than females. This is comparable to the degree of sexual dimorphism in body weight

found in chimpanzees but far less than the level in gorillas or orang-utans.

For many other biological features of the human species, it is somewhat difficult to define a 'typical' condition, because cultural variation has generated great diversity. It is, for instance, almost impossible from a study of modern human beings to define a characteristic human diet, although the human dentition differs quite markedly from that of the great apes and most other primates in several important respects (see Chapter 6). The 'natural diet' of human beings is far from obvious even if the focus is restricted to present-day gathering-and-hunting populations, and, overall, only the vague term 'omnivorous' would seem to be applicable. Similarly, there is considerable variation in patterns of social organization and mating arrangements among modern human populations. Indeed, it has been aptly observed that human beings may show virtually the entire range of patterns found in other primate species. In the case of human diets, social systems and mating patterns, it is really necessary to draw on the comparative information available for other primate species in order to make inferences about the likely 'natural' state for the human species.

Nevertheless, there are some constant biological features of human natural history that may be recognized and cited in comparison with other primates. For instance, as with other primate species, both the gestation period and the natural lactation period are relatively constant (at 9 months and 2½ years, respectively) and, in common with most other primates, the human female typically bears one infant at a time. However, the characteristic primate pattern of carriage of the infant clinging to the adult has been rendered impossible by two specific developments that have taken place during human evolution. First, human beings have lost the covering of bodily hair to which the infant could cling. Second, from birth onwards human beings are no longer able to grasp effectively with their feet (see Chapter 10), so the human infant no longer has the natural ability to cling to the parent's body with all four extremities.

The loss of the grasping adaptation of the foot is related to the unique pattern of locomotion that has emerged in human evolution and that constitutes a major defining feature of the human species – striding bipedalism (see Chapter 10). The highly unusual character of this pattern of locomotion alone suggests that the natural context of human evolution was extraordinary and one of the greatest challenges facing those who would reconstruct human evolutionary history is identification of the special environmental factors that favoured the emergence of unique features such as bipedal locomotion.

In addition to the unusual morphological changes that have taken place in the evolution of the human species, such as extensive remodelling of the dentition, transformation of the entire skeleton to meet the special demands of bipedal locomotion and considerable increase in the size of the brain (see Chapter 8), there are also some unique behavioural features that may be regarded as biological universals. The most outstanding of these is undoubtedly **language**, which sets *Homo sapiens* apart from all other primate species and contributes to many of the cultural aspects of human life that combine to obscure our biological origins. There is now considerable evidence that there is a species-specific, biological basis for language acquisition in human beings and the emergence of this undoubtedly represents one of the most important developments in human evolution. However, despite all of the striking features that now set human beings apart from other primates, it must be remembered that there was no single point in time when the last ape-man gave rise to the first 'real' humans. The evolutionary emergence of the human species has been a continuous process, occurring as an integral part of organic evolution on Earth, and it is one of the aims of this book to provide a basic framework for exploration of the details of that emergence.

Chapter two

The primate fossil record

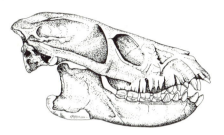

As with the preceding survey of living primates, the main aim of this introductory review of fossil forms currently regarded as allied to the primates is to present a broad outline. The various morphological character systems are covered in greater detail in Chapters 6–10. The review here is not exhaustive and for more detailed information the reader is referred to Piveteau (1957) and to the detailed recent surveys provided by Simons (1972) and by Szalay and Delson (1979). Indeed, a major distinction will be drawn here between two categories of primate fossil species; only the second category will be important for the purposes of this book.

More than 65% of fossil primate species recognized at present by palaeontologists are based on extremely fragmentary remains, consisting at most of isolated partial jaws and quite often only of a few teeth. In such cases, the characters discernible on fossil specimens suffice only for the formulation of the hypothesis that the species concerned are related to a particular group of primates. Beyond that, the specimens cannot yield much further reliable information regarding primate evolution (see Chapter 3). Because of the resulting danger that the few characters discernible on such **fragmentary fossils**

exhibit only convergent similarities to particular primate groups, they should be viewed with special caution. The problem of 'classificatory accretion' noted in Chapter 1 is particularly acute when fragmentary fossils are added one by one in a chain, on the basis of limited dental resemblances. At the end of this process, one can be faced with a situation in which an early fossil form is treated as a primate although its teeth bear no close resemblance to any single living primate species. In short, inclusion of fragmentary fossils as primates is merely a provisional act of classification; any conclusions with respect to phylogenetic relationships reached on the basis of certain dental features only should really await the discovery of more substantial fossil remains. By contrast, a small minority (less than 35%) of known fossil specimens are sufficiently well represented that one can go beyond provisional inclusion within the primates. The characters used in establishing a reasonable case for relationship to primates do not exhaust all the information yielded by such **substantial fossils** (which ideally include both cranial and postcranial remains) and the total set of characters discernible can be utilized for the refinement of phylogenetic hypotheses without circularity of reasoning. The following review of fossil primates is therefore concentrated almost entirely on substantial fossil species that should provide the backbone of any attempt to generate reliable outline hypotheses of primate evolution. It should be noted, however, that even fragmentary fossil evidence may be of special value in specific contexts, notably in examining patterns at individual fossil sites where large numbers of fragmentary specimens can be collected in well-defined sequences (e.g. see Gingerich, 1979).

CHANCE AND BIAS IN FOSSILIZATION

The fossil record of primate evolution is also fragmentary in another sense in that, even taking the most optimistic view of the number of known species of fossil primate, only an extremely patchy representation of the adaptive radiation of the primates is available for scrutiny. This fragmentary picture of primate evolution is not merely disappointing but positively misleading because of the bias in fossilization and because of a common tendency among primate palaeontologists and interested outsiders to treat the record as less fragmentary than it is in fact. Study of the processes by which animal remains become fossilized (**taphonomy** or *Biostratinomie*), a discipline of relatively recent origin, has much to offer in improving our approach to interpreting the fossil record (Müller, 1951; Behrensmeyer and Hill, 1980; Shipman, 1981), and it is worth commenting here on some of the most elementary points.

Fossilization of animal remains can take place in a variety of ways, but it most commonly involves mineral consolidation of 'hard parts' such as bones and teeth. Among mammals, there are a few notable exceptions, such as the preservation of virtually entire mammoths in permafrost, the survival of fossil skin and hair from recently extinct South American ground sloths, and the occurrence of mammalian hair in amber from the early Cenozoic. It is also sometimes possible to find special fossil evidence, such as mineralized faecal pellets (coprolites) and impressions of footprints (Abel,

1935). Indeed, fossilized footprints in East Africa provide the earliest evidence of bipedal walking in human evolution, alongside preserved tracks of contemporary Old World monkeys (Leakey and Hay, 1979). But the vast bulk of the fossil evidence of mammalian evolution comes from preserved teeth, jaws, skulls and limb bones. Fossilization of such 'hard parts' takes place through the replacement of organic matter with inorganic deposits derived from alkaline salts (e.g. calcium carbonate, phosphate or silicate). In order for this process to take place, the remains of a dead mammal must be rapidly buried in a suitable (usually alkaline) substrate – for example, on the shores of shallow seas or lakes, or in debris accumulating in caves or crevices. Otherwise, the animal corpse will be quickly broken down by the action of scavengers, bacterial decay, acid conditions and weathering.

This requirement for rapid burial in an appropriate substrate introduces a considerable selective element into fossilization and in general there is likely to be a bias against tree-living or flying animals and towards terrestrial species. Even if favourable conditions for fossilization exist, mammalian remains usually become dismembered and dispersed, particularly by flowing water. Because large bones are broken down more slowly and because dismemberment increases the likelihood of further breakdown, it is easy to see that large-bodied mammals are more likely to be fossilized effectively than are small-bodied mammals. Only in exceptional circumstances – as in accumulations of small bones derived from raptor pellets – is the fossilization of small-mammal remains favoured. The general trend against fossilization of small mammals doubtless provides one of the major reasons why the early fossil history of mammals (see Chapter 4) is still so poorly documented, compared with both the earlier evolutionary radiation of mammal-like reptiles and the later radiation of placental mammals.

During the fossilization of mammalian remains, any cavities present in the 'hard parts' may be filled with sediment after the breakdown of the original soft tissue, and this can lead to the formation of natural casts. Consolidated casts of the braincase cavity (endocranial casts or **endocasts**) are relatively common in mammalian fossils; in some cases the skull itself may have subsequently broken away to leave the cast in virtual isolation. Such internal casts of cavities can yield additional valuable information, particularly with respect to the morphology of the mammalian brain (see Edinger, 1929; Jerison, 1973; Chapter 8).

Problems may also arise after initial deposition and mineralization, when the remains may be crushed and distorted by the forces exerted by subsequent layers of sediment, or dispersal to a secondary site may occur after erosion. Secondary transfer of a fossil to a deposit of different age can occasionally result in misleading attributions, particularly when a relatively small timescale is involved, as in the later stages of human evolution.

Obviously, there are some chance factors involved in fossilization in addition to systematic sources of bias such as selection against small-bodied arboreal mammals. Chance and bias play a part not only in the fossilization process but also in the likelihood of fossil discovery. Geological deposits containing fossils must be exposed by a combination of erosion and tectonic processes for them to be accessible to fossil-hunting. In the past, the excavation of mammalian fossils has depended heavily on quarrying activities involving manual labour and on many man-hours of amateur fossil-collecting. It is only relatively recently that mammalian palaeontologists have deliberately set out on novel fossil-hunting projects involving detailed advance planning and reference to established geological guidelines, and even so it is relatively rare to learn of completely new sites being opened up without any previous history of quarrying and/or chance finds. In any event, the element of chance involved is so great that large areas of uncertainty are likely to remain with us despite systematic attempts to find fossils in carefully selected areas. For example, to take the entire continent of Africa, the only well-documented fossil site revealing the representation of terrestrial vertebrates during the first 60% of the Tertiary period (Palaeocene–Oligocene epochs inclusive)

is that of the Fayum badlands in Egypt. It would obviously be rash to take the fossils from this site as satisfactorily indicative of evolutionary developments in the entire African continent over the period 24–66 million years (Ma) ago. Similarly, very little is known of the fossil history of early mammals in South-East Asia (e.g. see McKenna, 1963), although in this case the relative lack of palaeontological investigation in that area, rather than the paucity of suitable exposures of sedimentary deposits, is probably the predominant factor. As has been pointed out by Thenius and Hofer (1960), Europe and North America are the only major continental areas of the world that have been investigated for fossils systematically (see also: Savage and Russell, 1983).

Overall, then, the extent of chance and bias in fossilization and exposure to discovery is of sobering dimensions and great care must be exercised in drawing conclusions from the imperfect fossil record available to us. Thenius and Hofer (1960) estimated that around 10 000 mammalian species were known from the fossil record, which compares with about 4000 now recognized in modern faunas (Corbet and Hill, 1980). On the one hand, this large number of fossil species reflects the prodigious efforts of palaeontologists in collecting and cataloguing the fossil specimens involved (although it may also reflect occasional over-enthusiasm in the naming of new species on the basis of limited, fragmentary evidence). On the other hand, it is important to see this number in terms of the total number of mammalian species that have existed since mammals first made their appearance. Even on the conservative assumption that the adaptive radiation of mammals began 200 Ma ago and that the number of species has increased regularly until the present day, a modal survival time of 1 Ma per species (see Bush, Case *et al.*, 1977; Stanley, 1976, 1978; Raup and Stanley, 1978) would suggest a total of about 400 000 species. Thus, the total number of known species of fossil mammal reported by Thenius and Hofer amounts to less than 3% of the minimum number of species that have probably existed at one time or another. For primates of modern aspect, a

Table 2.1 Numbers of genera and species of fossil primates of modern aspect for different Cenozoic epochs* (data from Szalay and Delson, 1979). Note: Subfossil lemurs of Madagascar are not included in this analysis because they are regarded as part of the modern fauna (see Chapter 1)

Epoch	Duration (Ma)	No. of genera	No. of species	Genera/Ma	Species/Ma
Plio-Pleistocene	3 + 2 = 5	18 (14)	43 (26)	2.6 (2.0)	6.1 (3.7)
Miocene	19	27 (15)	46 (16)	1.4 (0.8)	2.4 (0.8)
Oligocene	12	6 (3)	11 (3)	0.5 (0.3)	1.8 (0.3)
Eocene	19	46 (11)	81 (12)	2.9 (0.7)	5.1 (0.8)
Palaeocene	11	0 (0)	0 (0)	0 (0)	0 (0)
Total	66	97 (43)	181 (57)	1.5 (0.7)	2.8 (0.9)

* Main numbers = total numbers, including fragmentary fossils; numbers in parentheses = substantial fossils only.

more precise calculation is possible, based on the total of 185 species calculated for the modern fauna (see Table 1.2) and totals for fossil species derived from the comprehensive survey published by Szalay and Delson (1979), as listed in Table 2.1.

Assuming the most recent likely date for primate origins as 65 Ma ago and a uniform increase in numbers of primate species throughout the Cenozoic era, one can estimate (on the same basis as above) that a minimum of about 6000 primate species have existed at some time or another (Fig. 2.1). The total number of fossil primate species recognized on the basis of fragmentary fossil evidence (Table 2.1) is 181, so at best we have a 3% sample of the minimum total of fossil primate species. If we restrict the calculation to primate species of modern aspect represented by substantial fossil evidence, however, the total falls to 57 and a more realistic assessment is that our sample effectively only amounts to 1%. Further, as is suggested in Chapter 4, it is quite possible that primates originated a great deal earlier than the date of 65 Ma ago taken above; it follows that the sample of fossil primate species known to us represents an even smaller fraction of the total. Primates are typically small- to medium-sized mammals of predominantly arboreal habits and it is therefore hardly surprising that, despite heightened palaeontological interest in primates because of their relevance to human evolution, we still have such a pitifully small sample of the primate fossil record. Thus, although fossils are rightly

regarded as valuable and indispensable assets for attempts to reconstruct primate phylogeny, we certainly cannot regard palaeontological data as adequate in themselves for this task.

Table 2.1 is also instructive in several other respects. First, it should be noted that for the Eocene epoch only 15% of fossil primate species are known from substantial remains, whereas over later epochs the figure progressively increases from 30 to 60%. Second, the Oligocene epoch (justifying the derivation of its name: *oligo* = few) is particularly poorly represented, compared with the Eocene, and the more recent Miocene epoch is still not as well known as the Eocene in terms of genera identified per million years.

THE TIMESCALE AND PHYLOGENETIC RELATIONSHIPS

The fragmentary nature of the primate fossil record is particularly relevant to the question of ancestor–descendant relationships. Clearly, the outstanding advantage of fossil evidence resides in the fact that only palaeontological information can indicate a **timescale** for primate evolution and it is for this reason, above all, that we must make the most of primate fossils, despite the severe limitations with respect to sampling of both characters and species. Reliable dating of fossil deposits is, of course, an important first step; but it is beyond the scope of this text to deal with the complex issues involved. For general accounts of fossil dating techniques, the reader is referred to

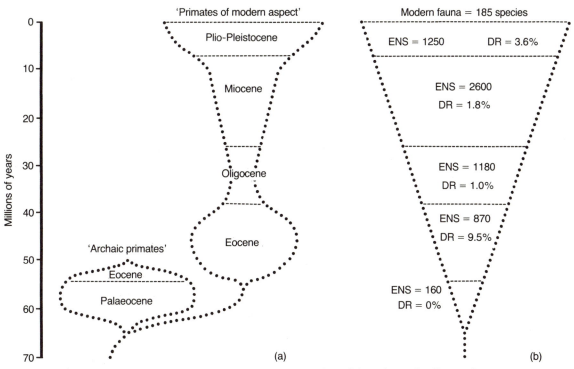

Millions of years

'Primates of modern aspect'

Plio-Pleistocene

Miocene

Oligocene

Eocene

'Archaic primates'

Eocene

Palaeocene

Modern fauna = 185 species

ENS = 1250   DR = 3.6%

ENS = 2600
DR = 1.8%

ENS = 1180
DR = 1.0%

ENS = 870
DR = 9.5%

ENS = 160
DR = 0%

(a)   (b)

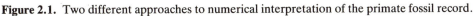

Figure 2.1. Two different approaches to numerical interpretation of the primate fossil record.

(a) Representation of primate abundance through the Cenozoic era as a direct reflection of the numbers of fossil species discovered for each epoch. This suggests that primates of modern aspect first appeared in the Eocene and then underwent a major adaptive radiation, followed by a drastic decline in the Oligocene and then by a gradual recovery to the present day.

(b) Simplistic model assuming divergence from a single ancestral species 65 million years (Ma) ago with regular expansion in numbers of species (ENS) up to the present total of 185 and with each species surviving for an average of 1 Ma. According to this view, there have been marked fluctuations in the discovery rate (DR) for fossil species. In particular, the recovery rate for the Eocene (9.5%) is very high–higher than for the most recent Plio-Pleistocene epoch. It is noteworthy that world climates were marked by a general, significant increase in temperature during the Eocene (Buchardt, 1978; Wolfe, 1978), which may have permitted primates to expand outside the areas to which they were both previously and subsequently confined by a requirement for tropical/subtropical conditions.

It is striking that only 43 primate species have been identified for the Plio-Pleistocene epoch (duration: 5 Ma), whereas 185 modern primate species are known. It is also surprising that no fossil primates of modern aspect are known for the Palaeocene epoch. Both features confirm the inference that the fossil record represents a relatively poor sample and that the simple model shown in (b) may be closer to the real situation. (Figure reprinted from Martin, 1986a, with kind permission of Cambridge University Press.)

basic reference works on the subject (Eicher, 1968; Hamilton, 1965; Libby, 1965; Zeuner, 1958). Savage (1975) has discussed the geological timescale specifically in relation to primates and a very useful survey of primate fossil sites with currently accepted dates is provided by Szalay and Delson (1979).

For present purposes, it will be taken for granted that the major primate fossils have been adequately dated. This, however, is only the first step; it must be emphasized that it is the fossil deposits that are dated; dates of **divergence points** in the evolutionary tree of primates can only be inferred. In this respect it is particularly misleading to overlook the fragmentary nature of the primate fossil record, especially when attempting to fit a large proportion of known fossil species into directly ancestral positions. Given an effective sample of less than 1% of pre-existing primate species (namely, the sample composed of substantial fossils), the chances of finding an actual ancestral form, rather than a side-branch species, are vanishingly small. How, then, can we infer dates of divergence in the evolutionary tree of the primates? The only really reliable procedure we can adopt is to set **minimum dates** for divergence points. For instance, if we have substantial fossil remains of a species closely resembling modern tarsiers (and distinct from any other major primate group), and if those remains come from a fossil site reliably dated at 55 Ma old, then we can justifiably conclude that the tarsier lineage was established at least 55 Ma ago. By contrast, it is extremely difficult to establish a **maximum date** for the origin of the tarsier lineage. It might be argued that there are no fossil remains of even putative primate species known from deposits older than 67 Ma, but, given a low percentage representation of species in the fossil record and the relative paucity of primate species early in the adaptive radiation of the group, this is not a particularly convincing argument. At present, the common practice would be to date the origin of primates of 67 Ma ago. In other words, minimum dates are often equated with actual dates for divergence points in the primate evolutionary tree. In fact, this applies not only to primate phylogeny but to vertebrate evolutionary trees in general and the likelihood is that divergence dates given are generally far too recent. The problem is that there would seem to be no really objective procedure available for estimating actual dates of divergence from the fossil evidence. On probabilistic grounds, we should expect some margin to be added to the date established for the first known fossil indicating that a particular divergence had already taken place.

Present information indicates that living organisms appeared on Earth at least 3300 Ma ago and recognizable multicellular organisms can be traced back to at least 700 Ma ago. These organisms are all known from the Precambrian eras, which were originally believed to predate the emergence of living organisms. (This provides just one reminder of the way in which accepted divergence dates are pushed back into the past as new fossil evidence accumulates.) However, the main fossil record of life on Earth remains confined to the three most recent eras: the Palaeozoic (590–250 Ma ago), the Mesozoic (250–66 Ma ago) and the Cenozoic (66 Ma ago to the present time). Each of these eras is sub-divided into periods. The evolution of mammals apparently began only in the Mesozoic, so the Palaeozoic periods need not concern us further. The first recognizable mammals documented from the fossil record emerged during the Mesozoic era, on the boundary between the first period (the Triassic) and the second (the Jurassic) around 210 Ma ago. The great adaptive radiation of mammals therefore began at least 210 Ma ago (see Chapter 4) and must have been well under way during the third and final period of the Mesozoic (the Cretaceous). The first placental mammals probably appeared by the beginning of the Cretaceous, 140 Ma ago. The adaptive radiation of the placental mammals, in so far as it is reflected in the fossil record, is largely confined to the Cenozoic era, which is accordingly called the Age of Mammals. (Note: The main part of the Cenozoic, extending up to 1.7 Ma ago, is also known as the Tertiary period.) The known substantial fossils of primates are restricted to the Cenozoic era, and it is therefore this period of time (lasting 66 Ma) that is of immediate interest in a survey of the primate fossil record. The Cenozoic is subdivided into six epochs: the Palaeocene (66–55 Ma ago), the Eocene (55–36 Ma ago), the Oligocene (36–24 Ma ago), the Miocene (24–5 Ma ago), the Pliocene (5–1.7 Ma ago) and the Pleistocene (1.7 Ma ago until Recent times, which are also referred to as the Holocene).

It is somewhat difficult to review the substantial primate fossils before a discussion of the major functional systems (e.g. dentition, locomotor apparatus) and before a consideration of procedures for phylogenetic reconstruction. First, it is impossible to provide much morphological detail and such information must be given later (in Chapters 6–10). Second, provision of such a review exposes one of the many elements of circularity involved in phylogenetic reconstruction: Assessment of primate evolution on the basis of available fossil evidence is obviously influenced by prior decisions as to which forms should be regarded as primates (e.g. see Simons, 1972; Szalay and Delson, 1979) and what their likely relationships are to one another. Because the aim here is to avoid at the outset complexities of formal nomenclature and of phylogenetic reconstruction (which are covered in depth in Chapter 3), the procedure adopted will be to refer to the major primate fossils in provisional groups, as listed in Table 2.2. This is done on the understanding that such groups may need reassessment in the light of detailed examination of possible phylogenetic relationships. (In line with this, the common names indicated in Table 2.2, for purposes of reference, are purely descriptive.) It is also helpful at this stage to give some provisional indication of the likely phylogenetic affinities of these fossil groups, as shown in Fig. 2.2, on the similar understanding that this is done essentially for descriptive purposes.

Before proceeding with the review of the substantial fossils evidence of primates, one final point remains to be established. As will emerge in later chapters, inclusion of the plesiadapiform group of fossils in the primates is a somewhat controversial issue. They are discussed here because many primate palaeontologists regard them as primates and because, as the earliest putative relatives of living primates, they provide an interesting fossil test case for assessing the implications of hypotheses concerning primate origins. However, it is now widely accepted that – whether or not plesiadapiform fossil species are included as primates – there is a considerable gap between them and the remaining fossil and living primates. Remane (1956a) referred to the plesiadapiforms as 'Subprimates' and Simons (1972) has emphasized the gap between them and typical primates by referring to the plesiadapiforms as 'archaic primates' and to the remaining primates as 'primates of modern aspect'. Similarly, Szalay and Delson (1979) refer to the 'primates of modern aspect' as 'Euprimates' to distinguish them from the

Table 2.2 Major groups of substantial fossil primates, including 'archaic primates'

Common name used in text	Formal family nomenclature*
'Archaic primates' (plesiadapiforms)	Plesiadapidae; Paromomyidae; Picrodontidae; Carpolestidae; Microsyopidae (?); Saxonellidae (?)
Early Tertiary 'lemuroids'	Adapidae
Miocene 'lorisoids'	Included in Lorisidae
Subfossil lemurs of Madagascar	Megaladapidae; others included in Lemuridae; Indriidae; Daubentoniidae
Early Tertiary 'tarsioids'	Omomyidae
Fayum Oligocene simians	Included in Cercopithecidae; Hylobatidae; Pongidae
Oligocene/Miocene New World monkeys	Included in Cebidae
Miocene/Pliocene Old World monkeys	Included in Cercopithecidae
Miocene small apes	Included in Hylobatidae
Miocene/Pliocene large apes	Included in Pongidae
Pliocene/Pleistocene hominids	Included in Hominidae

* See Table 3.1.

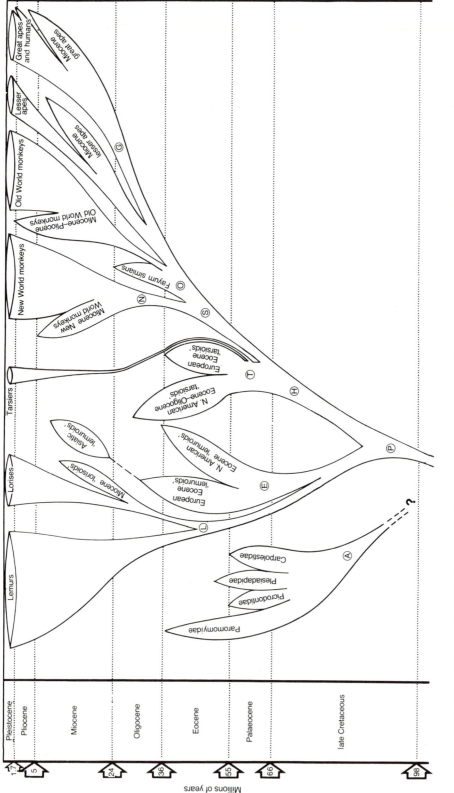

Figure 2.2. Provisional outline phylogenetic tree for primates, including the basic fossil groups listed in Table 2.2. Key: A, 'archaic primate' stock; P, 'primate of modern aspect' stock; E, Eocene 'lemuroid' stock; L, lemur/loris stock; H, 'tarsioid'/simian stock; T, 'tarsioid' stock; S, simian stock; N, New World simian stock; O, Old World simian stock; G, great ape stock. For reasons explained in the text, the main points of divergence in the primate tree are set markedly earlier than is customary in the literature. The date of the hypothetical ancestral primate stock (excluding 'archaic primates') is estimated at 90–100 Ma ago, rather than at 67 Ma ago (the date of the first fossil currently recognized as a 'primate'. Note also the early divergence indicated between lemurs and lorises, on the one hand, and tarsiers and simians, on the other. Dates for the beginning of each epoch are taken from Haq and van Eysinga (1987).

plesiadapiform group (see also: Schmid, 1978; MacPhee, Cartmill and Gingerich, 1983; Gingerich, 1986a).

THE PLESIADAPIFORM GROUP ('ARCHAIC PRIMATES')

The known plesiadapiform fossils are essentially confined to the Palaeocene, although they probably emerged during the late Cretaceous and one family (Paromomyidae) survived well into the Eocene epoch. Many of the fossil species, and two of the commonly recognized four families (Plesiadapidae, Paromomyidae, Carpolestidae and Picrodontidae), are known only from teeth and jaw fragments. Indeed, only two genera and three species are known from substantial fossil remains. Some authors include among the archaic primates the family Microsyopidae, for which some limited cranial evidence is available (Mac Phee *et al.*, 1983), and others include the family Saxonellidae, known only from fragmentary jaws and teeth (Rose, 1975; Szalay and Delson, 1979).

In fact, the suggestion of an affinity between plesiadapiforms and primates was made at a time when the former were known solely from teeth and jaws, so the entire group consisted of fragmentary fossils when the crucial proposal concerning their possible phylogenetic relationships was made (Teilhard de Chardin, 1921). This remained true when the proposed phylogenetic affinity to primates was consolidated by Simpson's (1935a) detailed discussion of *Plesiadapis*. Although the plesiadapiforms have since been reconsidered in some detail, notably by Gingerich (1971, 1975a, 1976b, 1986a), the relatively recent availability of substantial fossil material (see below) calls for a completely fresh reappraisal of their likely phylogenetic relationships. Plesiadapiforms were only 'dental primates' initially and this legacy has remained with us (see Chapter 6).

The plesiadapiforms provide a particularly good example of the dangers of relying on fragmentary evidence in drawing conclusions about the early evolution of primates – namely,

with the earliest recognized genus *Purgatorius* (allocated to the family Paromomyidae). The earliest species, *Purgatorius ceratops*, is known only from a right lower molar tooth found in late Cretaceous deposits in North America. In the paper announcing the discovery (Van Valen and Sloan, 1965), this single tooth was hailed as providing evidence for 'the earliest primates'. It is, of course, conceivable that this tooth came from a primate species; but considerably more evidence is required to generate a convincing case. The name of this fossil is indeed appropriate: from the point of view of phylogenetic reconstruction (rather than provisional classification) a zoological equivalent of 'purgatory' is the ideal place for this tooth until we have a skull and, hopefully, some postcranial bones to permit a realistic assessment of its affinities. Yet this single tooth has occasionally been regarded as providing evidence not only that the primates originated in the late Cretaceous, but that their place of origin was North America. Neither of these conclusions is warranted by available evidence (see also Savage and Russell, 1983). A later (Palaeocene) species of *Purgatorius* (*P. unio*) is recognized from some teeth and even a few partial jaws (Clemens, 1974), but the evidence is still too fragmentary to be considered as anything more than tentative. Simons (1972) notes that these teeth of *Purgatorius unio* are 'surprisingly similar' to those of the contemporary condylarth *Protungulatum*, and this gives some food for thought about the reliability of allocations based on dental evidence alone. One can only concur with the conclusion of Savage and Russell (1983): 'We are not ready to accept the one-tooth record of *Purgatorious* . . . as valid documentation of latest Mesozoic primates, although the evolutionary trend toward the great Cenozoic radiation of primates must have been under way by late Cretaceous time.'

The best known of the plesiadapiform fossils is *Plesiadapis* itself. This genus was present both in North America and in Europe during the Palaeocene epoch and several species are recognized on the basis of dental evidence (Gingerich, 1976b). Two species have been

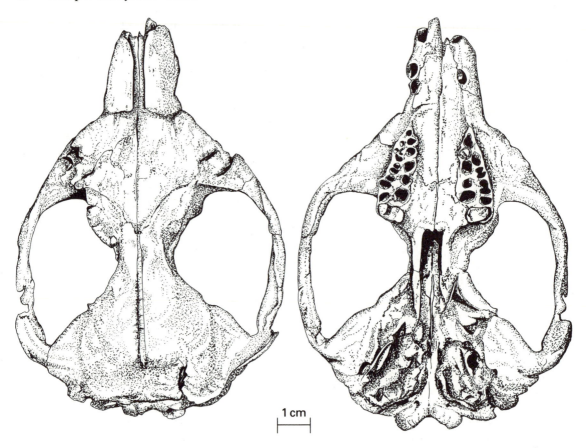

1 cm

Figure 2.3. Dorsal and ventral views of a crushed skull of the 'archaic primate' *Plesiadapis tricuspidens*. Note the large sockets for the upper anterior incisors and the pronounced gap (diastema) separating the anterior teeth from the cheek teeth. Crushing of the skull has greatly distorted the braincase and face of this specimen. (After Russell, 1964.)

documented in detail on the basis of substantial fossil remains: *Plesiadapis tricuspidens* in Europe (Russell, 1964; Gingerich, 1976b, 1986a) and *P. gidleyi* in North America (Simpson, 1935a,b; Szalay, Tattersall and Decker, 1975). Despite their present wide geographical separation, the European and North American species are very similar and can reasonably be traced to a common ancestor existing in the early Palaeocene or late Cretaceous. It is clear from the crushed skull of *P. tricuspidens* (Fig. 2.3) and from the reconstruction of the skeleton produced by Tattersall (1970; Fig. 2.4) that sufficient characters have been preserved to permit quite detailed consideration of the possible affinities of

this Palaeocene genus (see also Gingerich, 1986a). An associated skull and partial skeleton of *Plesiadapis cookei* reported by Gunnell and Gingerich (1987) have further enriched our knowledge of this genus.

The remaining species of 'archaic primate' are largely known from dental remains, although partial skulls (essentially confined to upper jaws) are known in a few cases. No associated postcranial elements seem to be known for any genera other than *Plesiadapis*. The only skull worthy of note from a genus other than *Plesiadapis* is that of the mid-Palaeocene *Palaechthon nacimienti*, a member of the family Paromomyidae. This skull has been carefully

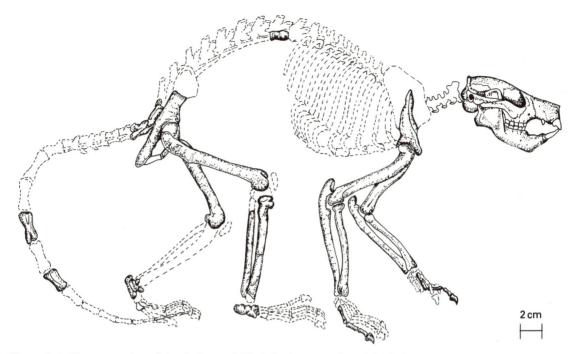

2 cm

Figure 2.4. Reconstruction of the skeleton of *Plesiadapis tricuspidens*. Stippled elements are known from actual fossil evidence; white elements with dashed outlines have been reconstructed. Conspicuous features are the clawed digits and rodent-like dentition. (After Tattersall, 1970.)

reconstructed (Kay and Cartmill, 1974, 1977; Fig. 2.5) and detailed study has yielded enough information for this to be regarded as a substantial fossil. For accounts of the more fragmentary remains of other plesiadapiforms, the reader is referred in particular to Simpson (1955), Szalay (1968a, 1972a), Rose (1975) and Bown and Rose (1976). The family Picrodontidae (Szalay, 1968a) is of particular interest, as this is a case where dental resemblance to other plesiadapiforms is tenuous and resemblance to any 'primates of modern aspect' is virtually non-existent (a prime example, perhaps, of 'classificatory accretion').

EARLY TERTIARY 'LEMUROIDS'

The species referred to here as early Tertiary 'lemuroids' are all commonly allocated to the single family Adapidae. They are largely confined to the Eocene epoch, but some of the European species persist into the early Oligocene and (largely dental) remains of *Indraloris* and *Sivaladapis* from the Miocene Siwalik deposits of India and Pakistan have been allocated on quite convincing grounds to the family Adapidae (Gingerich and Sahni, 1979, 1984). More recently, teeth and jaw fragments from Miocene deposits in Yunnan (China) have been allocated to the adapid genus *Sinoadapis* (Wu and Pan, 1985a; Pan and Wu, 1986). The affinities of this form remain unclear although there is a general resemblance to *Sivaladapis*. Most of the known species in the family are, however, limited to the Eocene. In spite of their antiquity, these Eocene species provide one of the best-documented components of the entire primate fossil record, and include one of the finest collections of substantial fossil remains. Yet, as has been aptly pointed out by Szalay and Delson (1979, p. 105), the known wealth and diversity of Eocene 'lemuroid' material 'represent the tip of an

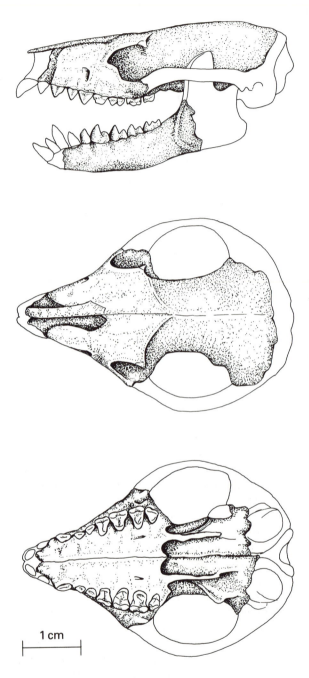

Figure 2.5. Lateral, dorsal and ventral views of the reconstructed skull of the 'archaic primate' *Palaechthon nacimienti*. Conventions as in Fig. 2.4. (After Kay and Cartmill, 1977.)

iceberg, a mere glimpse of what must have been an extremely varied and long-lasting adaptive radiation' (see Fig. 2.1). The survival of members of the 'lemuroid' group at least into the Miocene in Asia bears witness to this.

Eocene 'lemuroids' are known from both Europe and North America and substantial fossil remains are available from both regions, although the only reliably associated skeletal remains come from North American fossil sites. Originally, all of the European species were included in the subfamily Adapinae, and the North American species were allocated to the subfamily Notharctinae. Recently, however, some authors (e.g. Szalay and Delson, 1979) have modified this approach by, for example, including European material allocated to *Pelycodus eppsi* within the subfamily Notharctinae and by including the North American *Mahgarita stevensi* within the Adapinae. As *Pelycodus eppsi* is known only from fragmentary jaws and teeth and *Mahgarita* is known only from two crushed skulls and lower jaws, this modification may be somewhat premature and will not be followed here.

Until recently, only three European Eocene 'lemuroid' species were represented by substantial fossil remains: *Adapis parisiensis*, *Adapis magnus* and *Pronycticebus gaudryi*. (*Adapis magnus* is now often referred to a separate genus, *Leptadapis*; but this change in nomenclature is not necessary on present evidence). All three species are known reliably only from skulls (in the case of *Pronycticebus*, only a single skull), although numerous isolated postcranial elements have been allocated to both *Adapis* species in the literature (see Chapter 10). The European Eocene 'lemuroids' were extensively reviewed some time ago by Stehlin (1912, 1916) and have recently been subjected to detailed re-examination by Gingerich (1977a,b,c, 1981a – see also Gingerich and Martin, 1981; Gingerich and Simons, 1977). The genus *Adapis* (Fig. 2.6) is particularly interesting for two reasons. First, there is evidence of a progressive trend towards decreased body size throughout the Eocene, particularly when the genus *Protoadapis* (known only from teeth and jaw fragments) is included in

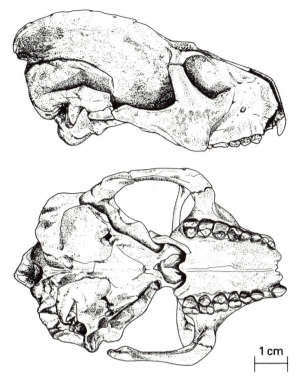

1 cm

Figure 2.6. Lateral and ventral views of the skull of the Eocene 'lemuroid' *Adapis parisiensis*. According to the interpretation made by Gingerich (1981a), there is pronounced sexual dimorphism in *Adapis* species; this is a male specimen.

the comparison, although this may not actually amount to an evolutionary sequence through the known fossil species. Second, Gingerich (1981a) has presented quite convincing evidence that *Adapis* species were sexually dimorphic. This is the earliest evidence as yet for the occurrence of sexual dimorphism in primates and it is particularly noteworthy because no modern prosimian species shows any marked degree of dimorphism. *Pronycticebus gaudryi* (Fig. 2.7) remains an enigma, in that this species is still known only from a skull and partial lower jaw recovered from late Eocene deposits and first described by Grandidier (1904). This skull has been re-examined in detail by Le Gros Clark (1934a), by Simons (1962a) and by Szalay (1971) and, despite the somewhat surprising lack of other fossil material attributed to *Pronycticebus*, it yields

sufficient information to be considered a substantial fossil.

Although Dagosto (1983) has conducted a valuable study of scattered postcranial bones attributed to *Adapis* species, the general lack of associated postcranial remains for *Adapis* and *Pronycticebus* remains a problem. However, this gap in our knowledge has been to some extent bridged by the recent discovery of three specimens from the early Eocene Messel deposits of Germany (von Koenigswald, 1979, 1985; Franzen, 1983, 1987). Unfortunately, two specimens consist only of rear portions of the skeleton. With the skulls missing and because no isolated primate teeth are known from Messel (where one tends to find relatively complete specimens or nothing), these two partial skeletons cannot be allocated to any of the known Eocene 'lemuroid' species with any confidence. Nevertheless, the skeletons have several features indicating that they both come from primates of some kind, and species of the family Adapidae are (on size grounds) at present the only likely candidates. The third specimen, by contrast, consists of the anterior portion of a skeleton, including a moderately well-preserved skull, from an individual probably weighing about 500 g. It has been allocated to the genus *Europolemur* (Franzen, 1983, 1987). Because the three specimens differ in size, it is not yet clear whether all can be allocated to this genus; but it is reasonable to group them together provisionally. Thus, the three specimens will be included as substantial fossils in later discussions (Chapter 10) as the 'Messel adapids'.

The North American Eocene 'lemuroids' (subfamily Notharctinae) are better known than their European relatives, because the fossils come from deposits with a carefully established stratigraphic background and because associated, virtually complete postcranial skeletons are known for two of the three main genera. Gregory (1920a) produced a quite outstanding monograph on *Notharctus*, which remains to the present day a centrepiece of the literature on North American Eocene 'lemuroids', and useful general reviews of the entire group have been produced by Gazin

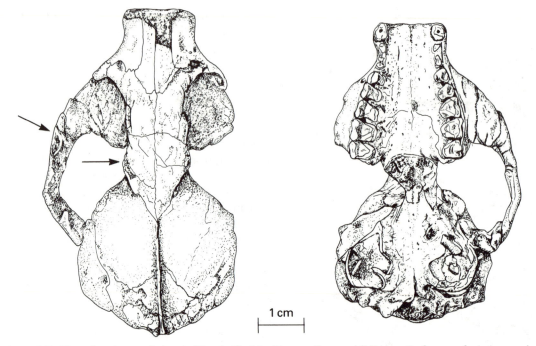

1 cm

Figure 2.7. Dorsal and ventral views of the skull of the Eocene 'lemuroid' *Pronycticebus gaudryi*. Arrows indicate points where the bony bar behind the eye (see Chapter 7) has been broken away during fossilization.

(1958). Simons (1963), Gingerich and Simons (1977) and Covert (1985). *Notharctus* is the best-known genus among the North American 'lemuroids'. The skull of this middle Eocene form (Fig. 2.8) is remarkably similar in its general outlines to that of a modern true lemur, with the striking exception that the braincase is considerably smaller. Gregory (1920a) provided reconstructions of the postcranial skeleton of *Notharctus*, based on a fairly complete collection of associated skeletal elements (Fig. 2.9), that further underline the similarity to modern lemurs. Two species of *Notharctus* are known from substantial fossil remains: *N. tenebrosus*, comparable in overall body size to the mongoose lemur, *Lemur mongoz* (Fig. 2.8), and *N. robustior*, comparable in body size to the modern varie-gated lemur, *Varecia variegata*, of Madagascar. *Smilodectes gracilis*, a contemporary middle Eocene species, is also known from virtually complete skulls and skeletons found in deposits

separate from those that have yielded *Notharctus* remains. In its dentition and postcranial morphology, *Smilodectes* is superficially quite similar to *Notharctus*, but its skull form is distinctive and resembles that of *Propithecus* rather than that of *Lemur*. Comparison of the *Smilodectes* skeleton with that of modern *Propithecus verreauxi* reveals the interesting fact that, although the relative proportions of fore-limbs and hindlimbs are similar (see Chapter 10), the Eocene fossil species had markedly shorter limbs in comparison to overall body size (Fig. 2.10).

Notharctus is particularly interesting in that it comes from fossil deposits in which it is possible to trace a sequence through an accurately determined stratigraphic record covering a period of some 7 Ma (Gregory, 1920a; Gingerich and Simons, 1977). It is accordingly possible to link *Notharctus* species to earlier Eocene 'lemuroid' species allocated to the genus

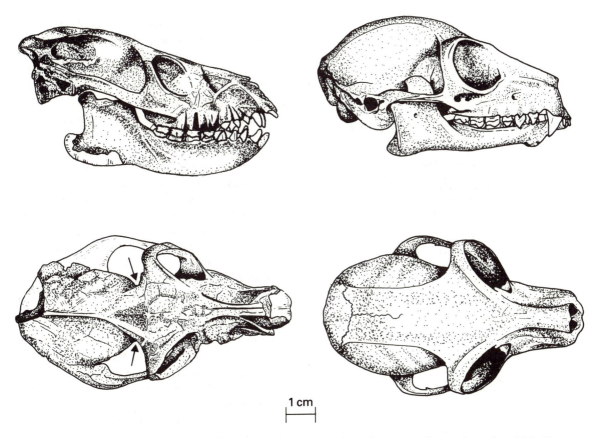

Figure 2.8. Lateral and dorsal views of the skulls of the 'lemuroid' *Notharctus tenebrosus* from the middle Eocene of North America (left) and of the modern *Lemur mongoz* from Madagascar (right). Note the markedly smaller braincase of the Eocene 'lemuroid' reflected (among other things) by the more pronounced postorbital constriction (arrowed). (After Gregory, 1920a.)

Pelycodus (including *Cantius*, see below). Unfortunately, *Pelycodus* is known only from relatively fragmentary remains. Although some scattered postcranial elements, a partial skeleton (Rose and Walker, 1985) and fragmentary skulls exist, discussion in the literature of the affinities of *Pelycodus* has been essentially confined to dental features (Simons, 1972; Gingerich and Simons, 1977; Szalay and Delson, 1979). Thus, for practical purposes, *Pelycodus* remains a fragmentary fossil until more information about its cranial and postcranial characters becomes assimilated.

This limitation is particularly relevant to the suggestion that fragmentary jaws found in early deposits of Europe may be allocated to the species *Pelycodus eppsi* (see Szalay and Delson, 1979). Simons (1962a) originally allocated the fragmentary specimens concerned to the separate genus *Cantius*, and this distinction from the North American forms was maintained by Russell, Louis and Savage (1967). More recently, Gingerich and Haskin (1981; see also Gingerich, 1986b) concluded that the European form and most North American species should all be allocated to the genus *Cantius*. Given the

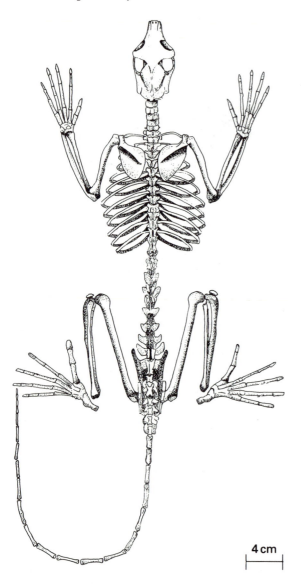

4 cm

Figure 2.9. Reconstructed skeleton of the Eocene 'lemuroid' *Notharctus tenebrosus*. Note that the hindlimbs are markedly longer than the forelimbs (After Gregory, 1920a.)

Smilodectes do not differ greatly from one another in terms of dental morphology so the genus *Pelycodus*, as defined here, may well prove to contain some quite distinct forms. By the same token, it is also unwarranted to regard *Pelycodus* as the common ancestor of all other North American and European Eocene 'lemuroids'. Nevertheless, despite the relatively fragmentary nature of the North American *Pelycodus* material (at least as far as discussion in the literature is concerned), it is probably reasonable to include this genus in discussions of the relationships of *Notharctus* species and *Smilodectes*, as a fairly continuous fossil record is available. For instance, it seems clear that – contrary to the situation with *Protoadapis* and *Adapis* in Europe – the sequence *Pelycodus–Notharctus/Smilodectes* was associated with a trend towards increase in body size. This is evident even within the individual genera *Pelycodus* and *Notharctus*. Despite the present paucity of information on *Pelycodus*, the fossil record of the North American Eocene 'lemuroids' provides us with an extremely valuable insight into the early adaptive radiation of the primates and the stratigraphic sequence is by far the best available until comparatively recent times.

The North American assemblage of substantial fossil remains from Eocene 'lemuroids' was recently expanded by the discovery and description of a distinctive new species, *Mahgarita stevensi* (Wilson and Szalay, 1976), based on a badly crushed skull and a lower jaw. On dental grounds, Wilson and Szalay have suggested that this species resembles European Eocene 'lemuroids' rather than the *Pelycodus/Notharctus/Smilodectes* group of North America and they have accordingly included *Mahgarita* in the subfamily Adapinae. However, given the relatively unspecialized dental characteristics of *Pelycodus* from the early Eocene (see Gingerich, 1986b), there remains a distinct possibility that the late Eocene *Mahgarita* has come to resemble European species through convergent evolution. This possibility can be tested adequately only if more complete fossil evidence becomes available. A second skull and a lower jaw of

fragmentary nature of the evidence, it is probably best to refer all specimens provisionally to the genus *Pelycodus*, which was accordingly represented in both North America and Europe during the early Eocene. However, it should be noted that the separate genera *Notharctus* and

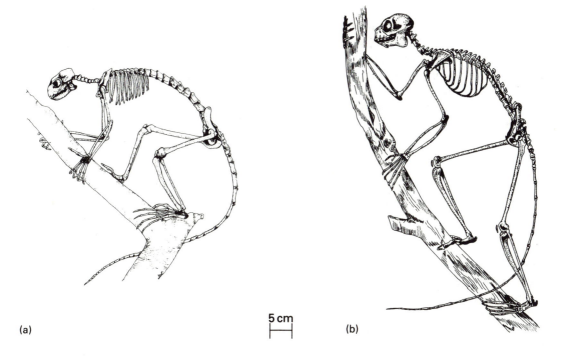

5 cm

(a)

(b)

Figure 2.10. Comparison of the reconstructed skeleton of the Eocene 'lemuroid' *Smilodectes gracilis* (a) with that of a modern lemur *Propithecus verreauxi* (b). Note the markedly shorter limbs in the Eocene fossil species, even though the two species are similar in overall body size and in the characteristic of possessing longer hindlimbs than forelimbs.

Mahgarita stevensi have been reported by Wilson and Stevens (1986), but this has as yet yielded no additional clarification. For present purposes, *Mahgarita* will be referred to simply as a North American Eocene 'lemuroid' and its dental affinities will be discussed in Chapter 6.

Before concluding this section on Eocene 'lemuroids', mention should be made of three very fragmentary specimens that raise the very tentative possibility that this group was also present in Asia and Africa in the Eocene. An upper jaw fragment bearing the last four cheek teeth has been discovered in late Eocene deposits of China and allocated to the species *Lushius quinlinensis* (Chow, 1961). Chow regarded this fragmentary fossil as a 'tarsioid', and was followed in this interpretation by Simons (1972); but Szalay and Delson (1979) have indicated that *Lushius* may instead be an Eocene

'lemuroid'. Szalay (1970) has also suggested that *Amphipithecus mogaungensis* from late Eocene deposits of Burma is a member of this group. This species is known only from a lower jaw fragment bearing three cheek teeth and virtually all other authors have regarded it as the oldest known Old World simian (see Simons, 1972). Its affinities therefore remain very dubious. Finally, Sudre (1975) has allocated *Azibius trerki* of North African Eocene deposits to the 'lemuroid' group. This species is known only from a lower jaw fragment bearing cheek teeth with somewhat unusual features. Gingerich (1977b) has compared this specimen with another apparent Eocene 'lemuroid' species from Europe, *Anchomomys gaillardi* (also known only from dental features and therefore predisposing to 'classificatory accretion'). He regards *Azibius* as part of the Eocene 'lemuroid' group. Szalay and

Delson (1979), however, have not accepted this interpretation. Until we have substantial fossil remains of *Lushius, Amphipithecus* and *Azibius* (not to mention *Anchomomys*), we cannot draw any reliable conclusions regarding the presence of Eocene 'lemuroids' in Asia or Africa. Africa, in particular, is extremely poorly known for Palaeocene and Eocene fossil mammals in general and future developments must be awaited with interest.

The apparent survival of early Tertiary 'lemuroids' into the Miocene epoch in India (Gingerich and Sahni, 1979, 1984; Wu and Pan, 1985a; Pan and Wu, 1986), is striking, as this indicates that at least some lineages of adapid primates continued virtually undocumented throughout the Oligocene. The relevant Miocene deposits of India are about 9–14 Ma old and are therefore considerably more recent than the Eocene deposits from which most fossil 'lemuroids' have been recovered. However, as yet only teeth and jaws are known for *Sivaladapis* and *Indraloris*, so little more can be said at this stage. It should be noted, however, that Gingerich and Sahni estimated the body weight of *Sivaladapis*, on dental grounds, as 5–6 kg. This fits well into the likely size range of Eocene 'lemuroids'. In China, the adapid *Sinoadapis* is found in the late Miocene, indicating survival until only 8 Ma ago.

MIOCENE 'LORISOIDS'

It is important to note that we as yet have no substantial fossil remains for prosimian primates living in Africa before mid-Miocene times (18–20 Ma ago). However, it is equally important to note that there are no known fossil sites in Africa for the Palaeocene and Eocene epochs (covering the period 36–66 Ma ago) that have yet yielded definite primate remains, and, for the Oligocene epoch (24–36 Ma ago), there is only the single site of Fayum in Egypt. A single upper molar tooth tentatively identified as a lorisid has now been reported from the Fayum (Simons, Bown and Rasmussen, 1986). It is, perhaps, somewhat surprising that until recently the Fayum site yielded only simian primate fossils (see later); but

prosimians are generally quite small relative to simians, so sampling bias obviously works against them. It would clearly have been rash to assume that no prosimians existed in Africa before 20 Ma ago, merely because we originally had no fossil evidence for them, and it should not be forgotten that we also have no trace of African prosimian fossils over the past 18 million years either, despite the relatively intense fossil-hunting that has taken place in East Africa. Yet, today, prosimians are almost as widespread in Africa as simians, although the number of species involved is much smaller.

Remains of 'lorisoid' primates have been collected from isolated localities in East Africa, notably from Songhor and Rusinga in Kenya. The major cranial and dental evidence was originally reviewed by Le Gros Clark and Thomas (1952) and by Le Gros Clark (1956). Simpson (1967) later provided an overall review of this material and Walker (1970, 1974a) reported additional postcranial evidence leading to subsequent overviews (Walker, 1974b, 1978). The postcranial material is, unfortunately, fragmentary and only associated by implication with the cranial and dental remains, but Walker's allocations on size grounds would seem to be reasonable, particularly since the foot bones (see Chapter 10) so convincingly fit the primate pattern. Only two skulls of East African Miocene 'lorisoids' are known. That of *Mioeuoticus* sp. (Fig. 2.11) is the most complete and counts as a substantial fossil; that of '*Komba robustus*' consists largely of a natural endocast with a few of the more posterior bony elements still attached (notably in the ear region) and, because no teeth are present, its affinities must remain extremely uncertain in the absence of more complete material. A third genus, *Progalago*, is represented by partial jaws and (by inference) some postcranial material, as are *Mioeuoticus* and *Komba*. This material, considered as a whole, may be regarded as provisionally representative of East African Miocene 'lorisoids', which were all relatively small-bodied forms.

Miocene lorisid material has also been recently reported from the Miocene Siwalik deposits of Pakistan and India (L.L. Jacobs, 1981; MacPhee

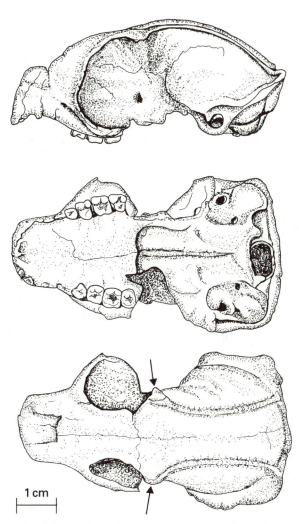

Figure 2.11. Lateral, ventral and dorsal views of the skull of the Miocene 'lorisoid' *Mioeuoticus* sp. Note the broken surfaces resulting from loss of the post-orbital bar (arrowed) during fossilization.

1 cm

SUBFOSSIL LEMURS OF MADAGASCAR

In addition to the modern array of lemur species present in Madagascar, there is a considerable number of subfossil species (at least 15), known from sites dating back no further than about 3000 years (Walker, 1967a, b; Tattersall, 1982) – hence the term 'subfossil'. Indeed, these species are so recent that, as Walker has pointed out, they should be considered to be an integral part of the modern lemur fauna. Recent extinction of these subfossil lemur species (most of which were larger in body size than surviving lemurs) probably resulted from a combination of climatic change and the first colonization of the island by human inhabitants. Early accounts of these subfossil species were provided by Forsyth Major (1893, 1896, 1900). Lorenz von Liburnau (1899, 1900), Standing (1908) and Lamberton (1934, 1936, 1938, 1939). In addition to the recent surveys provided by Simons (1972) and Szalay and Delson (1979), Tattersall (1973, 1982) has provided very useful reviews of the subfossil lemurs and Walker (1967a, 1974a) has produced comprehensive accounts of postcranial evidence and locomotor adaptations. Taking the modern and subfossil lemur species together, it seems likely that 2000 years ago Madagascar was occupied by more than 40 lemur species exhibiting an adaptive diversity even more striking than we can see today. As noted by Fleagle (1978): 'The size range of primate species from Madagascar (including subfossil species whose extinction is probably due to the appearance of humans on the island) equals the combined size range of both prosimians and higher primates elsewhere in the world.'

Some of the subfossil lemur species can be allocated to genera represented among the modern fauna (e.g. *Varecia, Hapalemur* and *Daubentonia*), but most belong to large-bodied genera no longer extant (*Mesopropithecus, Archaeolemur, Hadropithecus, Palaeopropithecus, Archaeoindris* and *Megaladapis*). *Archaeolemur* is notable because its skull is superficially monkey-like and, as a result, it was initially regarded as providing evidence of a simian invasion of Madagascar (Forsyth-Major,

and Jacobs, 1986). The only species described to date is *Nycticeboides simpsoni*, which is known from skull fragments and some skeletal elements as well as from teeth and jaws. However, the available evidence is still too limited for *Nycticeboides* to be regarded as a substantial fossil. The fragments that have been reported indicate that *Nycticeboides* was relatively small-bodied (maximum 500 g) and had relatively large eyes.

1896; Lorenz von Liburnau, 1899), although it is undoubtedly an integral member of the adaptive radiation of Malagasy lemurs. *Palaeopropithecus, Archaeoindris* and *Megaladapis* are all very large-bodied forms whose size range overlapped with that of the modern great apes of Africa and South-East Asia.

Apart from the extremely valuable infor-mation that they provide in morphological terms, greatly expanding our understanding of the adaptive radiation of the Malagasy lemurs, the subfossil species are also instructive in another sense. As suggested above, the subfossil lemurs must undoubtedly be seen as part of the recent adaptive array of lemurs, yet it is a striking fact that very few of the modern genera, and

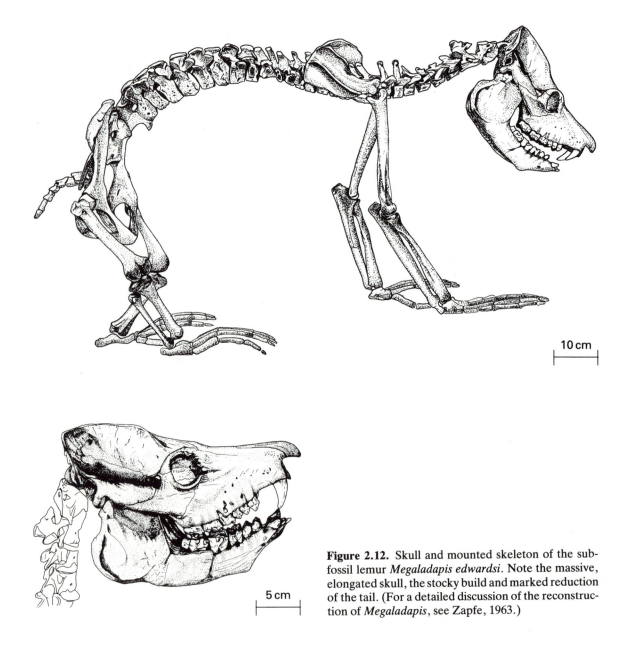

Figure 2.12. Skull and mounted skeleton of the sub-fossil lemur *Megaladapis edwardsi*. Note the massive, elongated skull, the stocky build and marked reduction of the tail. (For a detailed discussion of the reconstruction of *Megaladapis*, see Zapfe, 1963.)

particularly very few of the modern species, have been reported from the subfossil sites. To some extent, this is probably because such finds would have evoked little interest; but it would seem nevertheless that even in these relatively recent sites there was considerable selection during the fossilization process against smaller-bodied species. For instance, species of the modern mouse and dwarf lemur group (and most notably the lesser mouse lemurs) are extremely wide-spread in Madagascar today, yet Lamberton (1939) cites only one skull attributable to *Microcebus* in all of his very thorough accounts of his excavations in Madagascar. One only needs to imagine these deposits subjected to an additional 50 Ma of geological transformation to see the distorted picture of the modern lemur fauna of Madagascar that would have been left for us to discover and interpret from a comparable Eocene context.

Most of the subfossil lemur species are known largely from skulls and jaws, although fairly complete associated skeletons have been found for *Megaladapis* (Fig. 2.12) and *Archaeolemur*. Walker (1967a, 1974a) has allocated numerous postcranial bones to the various subfossil species, often on grounds of size and in relation to what is known of the relatively few postcranial elements reliably associated with skulls, and overall the postcranial anatomy of the subfossil lemurs can be regarded as reasonably well established. A postcranial skeleton of a larger-bodied fossil relative of the modern aye-aye, allocated to *Daubentonia robusta*, was described by Lamberton (1934), but no skull is known of this subfossil aye-aye.

Archaeolemur and *Hadropithecus* have attracted particular interest not only because of the monkey-like appearance of their skulls (Fig. 2.13), but because C.J. Jolly (1970a) has proposed that *Hadropithecus* has adaptations for 'small-object feeding' (e.g. seed-eating). Jolly suggested that dental contrasts between *Hadropithecus* (Fig. 2.14) and *Archaeolemur* paralleled contrasts between gelada baboons and other baboons (Jolly, 1970b) that he regarded as significant with respect to early human origins (see also Chapter 6). This is but one illustration of

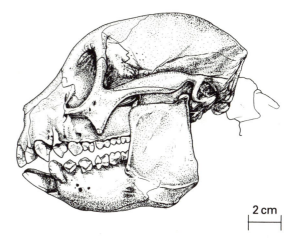

Figure 2.13. Lateral view of the skull of the subfossil lemur *Archaeolemur*. The general monkey-like appearance of this skull led to suggestions in the earlier literature that simian primates, as well as lemurs, were once represented in Madagascar (Forsyth Major, 1896).

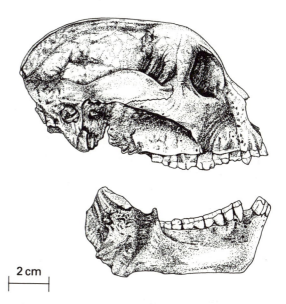

Figure 2.14. Lateral view of the skull of the subfossil lemur *Hadropithecus*. C.J. Jolly (1970a) has proposed that the dental apparatus has adaptations for 'small-object feeding', paralleling similar adaptations discernible in gelada baboons and early hominids (see Chapter 6). (Drawing after Lamberton, 1938).

the particular value of including subfossil lemur species in consideration of the adaptive radiation of the Malagasy lemurs (Martin, 1972b; Tattersall and Schwartz, 1974).

In conclusion, it should be emphasized that, before the subfossil lemurs, we have no fossil evidence whatsoever to document the evolution of lemurs within Madagascar. As it seems likely that lemurs have inhabited Madagascar since early Tertiary times, this gap in our knowledge provides yet another convincing illustration of the extremely fragmentary picture of primate evolution provided by the known fossil record.

EARLY TERTIARY 'TARSIOIDS'

As with the approximately contemporary fossil 'lemuroids', early Tertiary 'tarsioids' are known from both North America and Europe, on the basis of fairly substantial fossil evidence. Recently, a single jaw fragment from the Oligocene deposits of the Fayum (Egypt) has been identified as a 'tarsioid', *Afrotarsius*, and this indicates that early Tertiary 'tarsioids' may have occurred in Africa (Simons and Bown, 1985). Two additional teeth, from the Fayum have also been identified as 'tarsioid' (Simons, Bown and Rasmussen, 1986). Many species currently recognized as belonging to the 'tarsioid' group are similarly known only from dental fragments, particularly in North America, and it is perhaps for this reason that formal taxonomic nomenclature has in the past been particularly labile and confusing. Simons (1972) allocates virtually all of the European species to the family Microchoerinae, which he includes within the family Tarsiidae along with the living tarsiers. The North American species are allocated to the family Anaptomorphidae (containing the two subfamilies Anaptomorphinae and Omomyinae) along with the European *Teilhardina*. Szalay and Delson (1979), while basically agreeing on allocation of fossil 'tarsioids' to the three subfamilies Microchoerinae, Anaptomorphinae and Omomyinae, prefer to incorporate them all in the single fossil family Omomyidae, as distinct from the modern Tarsiidae (this is the solution adopted here; see Table 3.1). Szalay and Delson

concur in regarding *Teilhardina* as related to North American 'tarsioids', but allocate this genus to the Anaptomorphinae rather than to the Omomyinae. Hürzeler (1946, 1948) regarded *Teilhardina* as related to the other European 'tarsioids', which would seem to be rather more likely on geographical grounds, especially as Szalay and Delson regard this genus as essentially primitive in terms of dental morphology.

The confusion in nomenclature is not helped by the fact that all three subfamily names listed above are based on fragmentary fossils rather than on the relatively few genera known from substantial fossil remains. *Anaptomorphus* and *Omomys* are known only from jaw fragments and the same would be true of *Microchoerus* were it

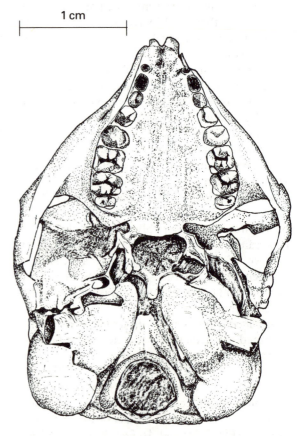

1 cm

Figure 2.15. Ventral view of the skull of the Eocene 'tarsioid' *Necrolemur antiquus*. This skull resembles that of *Tarsius* in its bell-shaped dental arcade, in the general structure of the ear region and in several other morphological details. (After Gregory, 1920a.)

not for the existence of an isolated frontal portion of the skull allocated to this genus. For the reasons given in Chapter 3, the classification provided by Simons (1972) has been followed as far as possible, but the continuing confusion over relationships within this group renders use of the single family name Omomyidae and of the common name 'fossil tarsioids' preferable as an interim measure. Although it has been common practice in the literature to regard the Omomyidae as definitely allied to modern tarsiers (e.g. Simons, 1972; Szalay and Delson, 1979; Gingerich, 1981b), this interpretation has been increasingly questioned in recent years (e.g. Cartmill, 1980; Schmid, 1982, 1983; Aiello, 1986). It would seem that the Eocene 'tarsioids', at least, are not as closely related to modern tarsiers as has been commonly suggested and the possibility must be considered that there is no direct phylogenetic relationship (Hürzeler, 1948; Schmid, 1982).

Without a doubt, the best-known genus among the fossil 'tarsioids' is the European genus *Necrolemur*, which is known from several fairly well-preserved skulls from the same French Eocene deposits (Phosphorites du Quercy) that have yielded skulls of *Adapis parisiensis* and

A. magnus (see Schlosser, 1907 and Simons, 1961a). Szalay (1976) has also referred some fragmentary postcranial material to this genus. The skull of *Necrolemur* (Fig. 2.15) is generally remarkably similar to that of modern tarsiers, although it differs significantly in several morphological details. Following a detailed morphological account given by Hürzeler (1948), the skull has been reconstructed and discussed at length by Simons and Russell (1960), who provide a valuable comparison with the skull of *Tarsius* (Fig. 2.16). All of the European 'tarsioids' are confined to the Eocene epoch. *Nannopithex* comes from somewhat earlier deposits than *Necrolemur* (middle Eocene versus late Eocene) and is known from a crushed skull and parts of a hindlimb (described by Simons, 1972, as the 'oldest known partial skeleton of an Old World primate, other than that of *Plesiadapis*') as well as several complete lower jaws (see also Weigelt, 1933). It may therefore be regarded as a substantial fossil, even though less is known about the skull than for *Necrolemur*. *Pseudoloris*, which is an approximate contemporary of *Necrolemur* in the late Eocene, is known only from upper and lower jaws, also from the Phosphorites du Quercy, but it is of

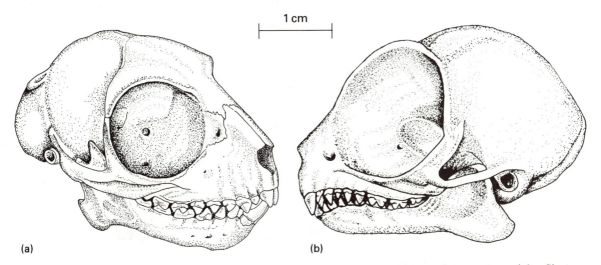

Figure 2.16. Comparison of (a) the reconstructed skull of the Eocene 'tarsioid' *Necrolemur antiquus* (after Simons and Russell, 1960) with (b) the skull of a modern tarsier (*Tarsius bancanus*). Note: although the reconstruction of the *Necrolemur* skull indicates that there was no central gap between the upper incisors, recent research (Schmid, 1982; Aiello, 1986) shows that a gap was in fact present in Eocene 'tarsioids (see Chapter 6).

62   *The primate fossil record*

particular interest because the known parts bear a very close resemblance to modern *Tarsius*, closer than any other fossil tarsioid described to date (with the possible exception of *Afrotarsius*; see Simons and Bown, 1985). The only other European 'tarsioid' known from reasonably significant fossil remains is the early Eocene *Teilhardina belgica*, which has been linked to North American 'tarsioids' by Simons (1972) and by Szalay and Delson (1979), as explained above. *Teilhardina* is known only from upper and lower dentitions and from some very fragmentary postcranial remains (Teilhard de Chardin, 1927; Quinet, 1966; Bown, 1976; Szalay, 1976), so its affinities must remain questionable until further fossil evidence is forthcoming. Dashzeveg and McKenna (1977) have recently described an early Eocene specimen from Mongolia consisting of a lower jaw fragment bearing four cheek teeth. They identify this specimen as a 'tarsioid' (*Altanius*), but the resulting extension of the geographical range of known fossil 'tarsioids' into Asia must be regarded as quite provisional until more substantial evidence is obtained. Rose and Krause (1984) have re-examined *Altanius* and have concluded that it could be either a 'tarsioid' or a plesiadapiform. This suggests that the dentition is essentially primitive and that its affinities are quite uncertain as yet.

Unfortunately, the North American fossil 'tarsioids' are very poorly represented in terms of substantial fossil remains. The only Eocene skull known is the partial skull of *Tetonius homunculus* from very early Eocene deposits. As the contemporary 'lemuroid' *Pelycodus* is known only from much more fragmentary skull remains, the skull of *Tetonius* is the earliest yet known for any primate of modern aspect. This skull was first described by Wortman (1903/1904) and a reconstruction has been produced by Szalay (1976). *Hemiacodon gracilis*, which was probably comparable in size to the European *Necrolemur*, is known from upper and lower jaws, a skull fragment and some postcranial elements found (in some cases) in the same sediments (Simpson, 1940; Gazin, 1958; Szalay, 1976), so there is just sufficient evidence available for this species to be included as a substantial fossil. The only other

skull known for North American 'tarsioids' is that of *Rooneyia viejaensis* (Fig. 2.17), which comes from very early Oligocene deposits of West Texas (Wilson, 1966). This skull is more complete than that of *Tetonius*, so it provides the best evidence we have of cranial anatomy in North American 'tarsioids' (see also: Hofer and Wilson, 1967; Szalay and Wilson, 1976). It should also be noted that an early Miocene species, *Ekgmowechashala philotau*, known only from a few lower jaw fragments, has been allocated to the North American 'tarsioid' group (Szalay and Delson (1979) place it in its own subfamily). This fossil species has dental characteristics almost as bizarre as its name, and without more substantial evidence it is difficult to regard its inclusion within the fossil 'tarsioids' as anything more than extremely tentative. Certainly, one should hesitate before accepting the conclusion that 'tarsioids' survived as late as the Miocene in North America.

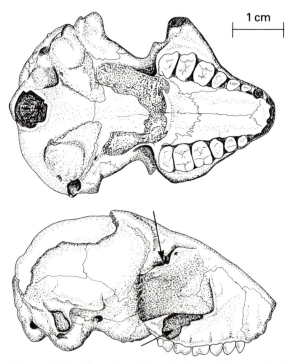

Figure 2.17. Ventral and lateral views of the skull of the early Oligocene 'tarsioid' *Rooneyia viejaensis*. Arrows indicate broken postorbital bar. (After Wilson, 1966.)

All of the known skulls of fossil 'tarsioids' (*Necrolemur, Nannopithex, Tetonius* and *Rooneyia*) can reasonably be regarded as resembling modern tarsiers, notably in peculiarities of the anterior dentition and in the morphology of the ear region. As noted above, it has therefore been widely accepted that *Tarsius* shared a common ancestry with these early Tertiary forms. However, it is important to emphasize the interesting fact that *Tarsius* appears in certain respects (e.g. morphology of cheek teeth) to be more primitive than any of the substantial Eocene/Oligocene fossil forms. This indicates, firstly, that *Tarsius* is probably not directly descended from any of the known substantial fossil species and, secondly, that divergent specialization of 'tarsioids' was well under way during the Eocene. It should also be emphasized that, despite the apparent diversity of 'tarsioid' species in both Europe and North America during the Eocene, the only possible candidate found anywhere in the Old World to bridge the gap of almost 40 Ma between the last known Eocene fossil 'tarsioids' and modern tarsiers is the fragmentary *Afrotarsius* from the Fayum. Once again, the large gaps in the record are striking.

OLIGOCENE FAYUM SIMIANS

The discovery and interpretation of Oligocene fossil simians constitute one of the major triumphs of modern primate palaeontology. This research also provides one of the starkest illustrations of the limitations of the known fossil record of the primates. The Fayum deposits in Egypt were initially explored by a professional fossil collector, Richard Markgraf, and his discoveries of jaw fragments led to early papers by Schlosser (1910, 1911) and by Remane (1921), among others (see also Kälin, 1961a, b, 1962). But the first substantial primate fossils from that site were discovered as a result of a series of expeditions organized by Simons (1962b, 1965, 1966, 1974a, b). These expeditions were then interrupted for some time because of the political situation in the Middle East, but they have recently been resumed and further important

discoveries have been made (Fleagle, Kay and Simons, 1980; Kay and Simons, 1980; Simons, Kay and Fleagle 1980; Kay, Fleagle and Simons, 1981; Fleagle and Simons, 1982a,b; Fleagle and Kay, 1983; Simons and Kay, 1983; Simons, 1985, 1986, 1987). Fleagle, Bown *et al.* (1986) currently recognize 10 different simian species from the material so far recovered from the Fayum deposits.

The Fayum badlands contain river channel deposits that are continuously eroded by wind action, thus exposing fossils for surface collection. The deposits have now been redated (Fleagle, Bown *et al.*, 1986) and it would seem that all of the fossil primates are derived from early Oligocene strata older than 31 Ma. As has been noted above, there is no primate fossil site known in Africa before the Fayum (with the exception of the North African Eocene site that yielded the questionable primate *Azibius*). We therefore have no firm evidence as to the course of primate evolution in the entire continent of Africa over the first 30 Ma or so of the Tertiary period. Indeed, the first fossil primates known from Africa south of the Sahara date back only to the early Miocene, some 24 Ma ago. For this reason, one should be wary of relying too heavily on the Fayum fossils for substantiation or refutation of hypotheses regarding primate evolution in Africa. For instance, numerous characters of the vast majority of these Oligocene primates indicate that they were definitely related to simians, rather than to prosimians. Simons (1972) initially made much of the apparent absence of prosimian fossils from the Fayum deposits and stated:

> The absence of lemurs or their forebears from these deposits cannot be attributed to the nonrecovery of fossils of small mammals of all sorts, including lemurs, owing to unfavourable depositional conditions for small vertebrates: other, smaller mammalian jaws, for instance of rodents, occur in these sites in some abundance.

This argument was used to throw doubt on the suggestion that Malagasy lemurs were derived from a stock represented in Africa during the early Tertiary. In the first place, however, apparent absence of prosimians from the relatively tiny area constituted by the Fayum deposits cannot be equated with absence of prosimians from the entire continent of Africa during the Oligocene. Second, several factors might have acted against fossilization of prosimians. It is likely that any prosimian species would have been both small-bodied and arboreal. Rodents and insectivores, by contrast, are commonly terrestrial and fossilization of jaws of these species may be favoured by their deposition in the faecal pellets of predators. (During a recent field study of *Galago senegalensis*, Simon Bearder and the author collected a large quantity of genet faecal pellets from a localized 'latrine'; see Andrews and Nesbit Evans (1983). Dozens of rodent and insectivore jaws and numerous other vertebrate remains were contained therein, but only two small jaw fragments of the bushbaby were found, even though the genet was a major predator of *Galago senegalensis* in the study area. Owls in the same area feed on rodents and insectivores, leaving skeletal remains of them in their regurgitated pellets, but do not seem to prey on bushbabies.) In any event, the recent discovery of the apparent 'tarsioid' *Afrotarsius* in the Fayum deposits (Simons and Bown, 1985) demonstrates that it may be very difficult to find remains of small-bodied prosimians. This find came only after many years of patient searching in the Fayum badlands. Similarly, evidence of the small-bodied simian *Qatrania* (described as 'marmoset-sized'; Simons and Kay, 1983) has only recently come to light, and isolated teeth referred to the Anaptomorphinae and to the Lorisidae have since been reported (Simons, Bown and Rasmussen, 1986).

Six different genera of Fayum simian primates were originally recognized by Simons (1972): *Aegyptopithecus*, *Aeolopithecus*, *Apidium*, *Oligopithecus*, *Parapithecus* and *Propliopithecus*. The genus *Aeolopithecus* is no longer recognized (see later), but the new genus *Qatrania* has been recognized on the basis of a jaw fragment and two isolated teeth (Simons and Kay, 1983). Only two of the five remaining

genera (*Aegyptopithecus* and *Apidium*) are known from substantial fossil remains: the other genera are known only from fragmentary jaws and teeth, although quite a number of these have been found for *Parapithecus*. (*Parapithecus* and *Apidium* together are the best represented of the Fayum simians in terms of jaw fragments.) *Oligopithecus*, clearly a small-bodied form, is of interest in that it comes from an earlier level of the Fayum deposits, but it is unfortunately known only from one partial lower jaw (Kay *et al.*, 1981). It has unusual, relatively primitive, dental characteristics and Gingerich (1977b) has in fact suggested that it is an adapid. The only reasonably complete skulls known are those of *Aegyptopithecus zeuxis* (Fig. 2.18), which is now known from four cranial specimens (Simons, 1985, 1987). *Apidium* is also known from three partial fragments of the frontal region of the skull. Because of the scattered nature of the fossil specimens in the Fayum, there are no associated skeletal remains, but numerous isolated post-cranial elements have been described (Fleagle,

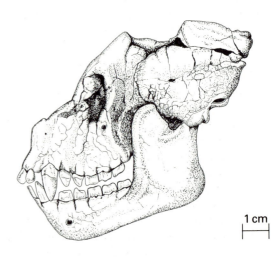

Figure 2.18. Lateral view of the skull of the early Oligocene simian *Aegyptopithecus zeuxis* (after Simons, 1967). This is the earliest known skull of a simian primate (Fayum deposits of Egypt). It is the most complete specimen of *Aegyptopithecus* known to date. Three incomplete skulls discovered subsequently indicate that there is some variation in the shape of the face (Simons, 1982).

Simons and Conroy, 1975; Conroy, 1976a, b; Fleagle and Simons, 1982a, b). On grounds of size and relative frequency in the fossil collections, Conroy (1976a) allocated most of the postcranial elements to *Aegyptopithecus* and to *Apidium*, thus confirming their status as the only substantial fossil genera from the Fayum on present evidence. More recently, some post-cranial elements have been allocated to *Parapithecus* and to *Propliopithecus*.

It must be noted, incidentally, that Szalay and Delson (1979) transferred material from *Aeolopithecus* and *Aegyptopithecus*, on dental grounds, to the genus *Propliopithecus*. Whatever justification may be provided for this proposal, by dental similarities between *Aegyptopithecus* and *Propliopithecus*, it is most unfortunate in view of widespread usage of the name *Aegypto-pithecus* in the literature in relation to the only known skulls from the Fayum, and this transfer, which was not accepted on morphological grounds by Kay, Fleagle and Simons (1981), will not be followed here. '*Aeolopithecus*', which was originally known only from one fairly complete but partially decomposed mandible, is now best regarded as included within *Propliopithecus*, as this is accepted by Kay *et al.* (1981) following the discovery of additional material; but because it still ranks as a fragmentary fossil it need not concern us further.

Despite (or, more realistically, because of) the relatively fragmentary nature of the Fayum fossils and the very limited geographical coverage of Africa during the Oligocene, Simons (1967) regarded the species identified as representing all of the major lineages of evolution of simians in the Old World. *Parapithecus* was seen as an early relative of Old World monkeys, *Apidium* was interpreted as ancestral to *Oreopithecus* (see later), '*Aeolopithecus*' was tentatively proposed as an early gibbon, and the *Aegyptopithecus/Propliopithecus/Oligopithecus* assemblage was seen as related to the origins of great apes and humans. Simons even suggested initially (1965; see also Pilbeam, 1967) that *Propliopithecus* was an early hominid, although he later (1972) decided to 'shelve' this particular idea. Given the fact that, as will be discussed in later chapters, the

proposed relationships require independent evolution of several shared characters of Old World monkeys and apes, and given the tiny probability that all of the major developments in Old World simian evolution would have become encapsulated in the only known fossil site for the first 40 Ma of the Tertiary in Africa, such an interpretation surely somewhat overstretched the available evidence. A more conservative interpretation with which some primate palaeontologists have agreed is that *Apidium* and *Parapithecus* are related to Old World monkeys, while *Aegyptopithecus*, *Propliopithecus* and (perhaps) *Oligopithecus* are related to apes (Simons, 1972, 1985; Kay, 1977). Others have rejected the first proposition while accepting the second (Delson and Andrews, 1975; Szalay and Delson, 1979); but this still requires independent evolution of several characters shared by all modern Old World simians. It is noteworthy that the Fayum primates generally exhibit numerous similarities to New World monkeys, both in cranial morphology and in postcranial anatomy (Conroy, 1976a). Although this may amount to nothing more than shared retention of primitive simian characteristics, it does weaken the proposition that the Fayum primates represent an integral part of the Old World simian radiation leading to the modern array of species. Accordingly, the view is now gaining ground that the common ancestor of the Fayum simians diverged before the evolutionary radiation of the modern Old World monkeys apes and humans (e.g. see Kay *et al.*, 1981; Andrews, 1985; Fleagle, 1986). Whatever the case may be, there is general agreement that the Fayum simians fall into two groups. The first contains *Apidium* and *Parapithecus* along with the further genus recognized by Simons and Kay (1983), *Qatrania*. The second contains *Aegyptopithecus*, *Plio-pithecus* and (possibly) *Oligopithecus*.

As a final note on the Fayum Oligocene simians, it should be mentioned that the fragmentary fossil *Pondaungia*, from late Eocene deposits of Burma, is commonly (but not universally) regarded as an early simian (Simons, 1972; Szalay and Delson, 1979; Rosenberger, 1986). If this identification could be confirmed,

Pondaungia would hence be the earliest known simian primate. As this genus is known only from a total of three jaw fragments, containing five weathered teeth altogether, the intriguing possibility of early simians inhabiting Asia during the late Eocene must, however, remain a matter for speculation.

OLIGOCENE/MIOCENE NEW WORLD MONKEYS

The known fossil record of New World monkeys for South and Central America is, frankly, rather disappointing in many respects. All of the substantial primate fossils from North America are, as indicated above, included either among the 'lemuroids' or among the 'tarsioids' of the early Tertiary. The origins of the New World monkeys therefore remain a mystery, as the only substantial fossils are known from the late Oligocene and the Miocene, and by that stage they exhibit numerous characteristics clearly allying them with the modern forms. Indeed, without exception the substantial fossils that have been described resemble true New World monkeys (family Cebidae) rather than the marmosets and tamarins, both in general body size and in detectable morphological characters. A few isolated teeth from Miocene deposits in Colombia have recently been linked to marmosets and tamarins, partly on grounds of their relatively small size (Setoguchi and Rosenberger, 1985); but this is evidence of the most fragmentary, and therefore least reliable, kind.

As has been succinctly pointed out by Ciochon and Corruccini (1975), there are two main opposing hypotheses regarding the phylogenetic relationships between the true New World monkeys and the marmoset/tamarin group. Some authors, notably Hershkovitz (1970, 1977), regard marmosets and tamarins as primitive, whereas others, such as Gregory (1922). Hoffstetter (1969), Rosenberger (1977) and Ford (1980), regard marmosets and tamarins as secondarily specialized dwarf forms. Ciochon and Corruccini favoured the first hypothesis; Szalay and Delson (1979) supported the second,

which seems to be gaining ground rapidly in the literature. Either way, as no substantial fossil relatives of marmosets and tamarins are known, palaeontology at present has little to offer in resolving this issue. It should be reiterated here that marmosets and tamarins are generally much smaller than true New World monkeys and early relatives of the former were probably far less likely to become fossilized for this reason alone.

The fossil record of New World monkeys is also disappointing in terms of geographical representation. All of the known fossils come from Patagonia, Bolivia and Colombia (see Rose and Fleagle, 1981), leaving a vast Amazonian block constituted by Brazil and several adjacent countries totally unrepresented. Yet New World monkey species are currently concentrated in Amazonia, and that is where most marmoset and tamarin species live. We therefore have much still to learn about the fossil history of New World monkeys. Most of the available evidence has been effectively reviewed by Stirton (1951), Hershkovitz (1970), Ciochon and Chiarelli (1980), Rosenberger (1977, 1979, 1980) and Rose and Fleagle (1981).

To date, only one genus has been reported from South America for 'early' Oligocene times and the date of the material has now been reduced to about 25 Ma (Fleagle, Powers, *et al.*, 1987), corresponding to the late Oligocene. Hoffstetter (1969) described the species *Branisella boliviana* on the basis of a single fragment of an upper jaw, containing only three intact cheek teeth, from the Bolivian Andes. Subsequently, a further three jaw fragments have been reported (Rosenberger, 1981a; Wolff, 1984), such that both upper and lower cheek teeth are now known. The conserved teeth certainly appear to be very primate-like, and it is probably reasonable to conclude from *Branisella* that New World monkeys were represented in South America at least 25 Ma ago. However, as a very fragmentary fossil, this Bolivian form has little more to tell us about the evolution of New World monkeys.

The remaining well-documented fossils of New World monkeys fall fairly clearly into two groups in terms of both geography and antiquity.

Dolichocebus, Homunculus and *Tremacebus* are recognized from late Oligocene/early Miocene deposits (about 18–26 Ma old) in Patagonia, whereas *Stirtonia, Cebupithecia* and *Neosaimiri* have been recorded from upper Miocene deposits (about 13–15 Ma old) in Colombia. Recently, further *Aegyptopithecus*-like molars provisionally attributed to *Dolichocebus* and to *Tremacebus* have been reported from the early Miocene deposits of Patagonia (Fleagle and Bown, 1983; Fleagle, 1985). In addition, Fleagle, Powers *et al.* (1987) have discovered jaws, teeth and postcranial remains of perhaps four other species. One has been formally named as *Soriacebus ameghinorum* on the basis of jaws and teeth. Of the Patagonian fossils, *Dolichocebus gaimanensis* (first described by Kraglievich, 1951) was originally known only from a shattered and distorted cranium with just the sheared bases of the cheek teeth present. As an isolated and fragmented specimen, it has comparatively little to tell us about the evolutionary radiation of New World monkeys (but see Rosenberger, 1979). *Homunculus patagonicus* (Ameghino, 1891; Bluntschli, 1931) and *Tremacebus harringtoni* (Rusconi, 1933; Hill, 1962; Hershkovitz, 1970) are known from somewhat more informative, but nevertheless rather limited material. *Homunculus* is known from two incomplete lower jaws, a facial fragment bearing some cheek teeth and a few postcranial elements including a complete femur and radius. *Tremacebus* is known from one almost complete skull (Fig. 2.19) bearing some cheek teeth, and it should be noted that Simons (1972) included this skull in *Homunculus*, emphasizing its fairly close similarity to the known skull fragment from that genus. Szalay and Delson (1979), following Hershkovitz (1974a), suggest that *Tremacebus* has relatively large eyes and might have been nocturnal like the modern owl monkey (*Aotus*). However, the orbits are much larger, relative to overall skull size, in the latter and in the absence of quantitative data it is difficult to assess the relative size of the eyes in *Tremacebus*.

Stirtonia, Cebupithecia and *Neosaimiri* were all described by Stirton (1951) and by Stirton and Savage (1951). *Stirtonia* and *Neosaimiri* were

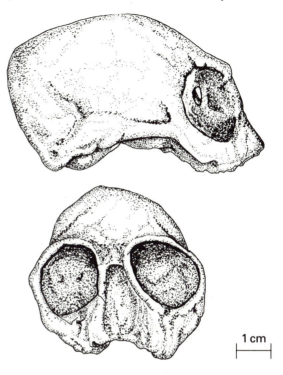

Figure 2.19. Lateral and anterior views of the only known skull of the Miocene New World simian *Tremacebus harringtoni*. This is, to date, the most complete skull of any fossil New World monkey species. (After Hershkovitz, 1974a.)

each originally known only from a single lower jaw, although some more material has now been attributed to *Stirtonia* (Kay, Madden *et al.*, 1987), and therefore count only as fragmentary fossils. *Neosaimiri* (as its name implies) bears a fairly close resemblance, both in terms of size and in terms of those characters that are preserved, to the modern squirrel monkey (*Saimiri*). *Stirtonia*, on the other hand, was probably the largest of the known fossil New World monkeys, falling (according to Szalay and Delson, 1979) within the size range of modern howler monkeys. Indeed, it seems likely that *Stirtonia* is a close relative of howler monkeys (Setoguchi, 1986; Kay, Madden *et al.*, 1987). *Cebupithecia sarmientoi*, the third Columbian form, is known from a fairly large number of postcranial elements as well as from skull fragments (upper jaws with teeth, incomplete lower jaw, part of a basicranium) and might just qualify as a substantial fossil. *Cebupithecia* may be closely related to modern saki monkeys. None of these middle Miocene forms is very complete, however, and proposed relationships to modern species must therefore be regarded as fairly tentative.

Renewed fossil-hunting in the Miocene La Venta site of Columbia has recently yielded several fragmentary primate fossils (Luchterhand, Kay and Madden, 1986; Setoguchi, Shigehara *et al.*, 1986). In addition to the diminutive *Micodon,* interpreted as a relative of the marmosets and tamarins (Setoguchi and Rosenberger, 1985), the genus *Kondous* has been recognized as a possible relative of the modern spider monkey (Setoguchi, 1985; but see Kay, Madden *et al.*, 1987) and fragmentary material from a large-eyed form has been placed in the same genus as the modern owl monkey, *Aotus* (Setoguchi and Rosenberger, 1987). Further, a single lower jaw comparable in size to that of a squirrel monkey has been allocated to the new genus *Mohanamico* (Luchterhand *et al.*, 1986). The teeth show similarities to saki monkeys and suggest a specific relationship to this group. The new fossil material has reinforced the interpretation that several modern lineages of New World monkeys can be traced back at least as far as the middle Miocene.

This survey of New World monkey fossils would not be complete without mention of the subfossil species *Xenothrix macgregori* from Jamaica. Like the subfossil lemur species of Madagascar, this New World monkey species became extinct relatively recently, perhaps as a result of human colonization of Jamaica, and should be considered as an integral part of the modern New World monkey fauna. The species is known from a single incomplete jaw (Williams and Koopman, 1952), which suggests that the body size was comparable to that of modern howler monkeys (see also Rosenberger, 1977). Further, Rimoli (1977) has reported on a single upper jaw fragment from the Dominican Republic, allocated to *Saimiri bernensis* and dated at about 4000 years old. As noted by Rose and Fleagle (1981), *Xenothrix* and *Saimiri*

bernensis (the latter being about twice the size of modern squirrel monkeys – see Chapter 1) bear witness to an intriguing Caribbean primate fauna in quite recent times. A femoral fragment from Jamaica recently attributed to the New World monkey group (Ford and Morgan, 1986) reinforces this view.

MIOCENE/PLIOCENE OLD WORLD MONKEYS

Although early fossil evidence of New World monkeys may be described as disappointing, it is rich in comparison to that for Old World monkeys. Only two well-identified primate genera possibly related to the adaptive radiation of Old World monkeys are known from deposits older than about 10 Ma (late Miocene), and both – *Victoriapithecus* and *Prohylobates* – are known only from very fragmentary remains. Dental dimensions indicate that the species concerned were quite large-bodied, weighing 7–20 kg (Fleagle, 1986). *Prohylobates* was described on the basis of three fragments of the lower jaw found in early Miocene deposits (about 18 Ma old) of Egypt by Fortau (1918) and later reviewed by Simons (1969). Delson (1979) subsequently attributed a lower jaw fragment bearing two molar teeth, discovered in deposits of similar age in Libya, to the same genus. *Victoriapithecus* is known from numerous fragmentary jaws and isolated teeth together with some postcranial elements derived from deposits in East Africa dated at about 15–20 Ma (von Koenigswald, 1969; Pickford, 1987). M.G. Leakey (1985) has reported fragmentary jaws and teeth from early Miocene deposits in Buluk, Kenya, provisionally allocated to the genus *Prohylobates*. On the basis of this material, she suggests that the generic distinction between *Prohylobates* and *Victoriapithecus* may be unwarranted. The early Miocene material, along with substantial evidence from later Old World monkey fossils, has been effectively reviewed by Maier (1970), Delson (1975a, b), Simons and Delson (1978), Szalay and Delson (1979) and Pickford (1987), in particular.

Szalay and Delson tentatively suggest that *Victoriapithecus macinnesi* might be an early relative of the leaf-monkeys, while '*Victoria-pithecus*' *leakeyi* might be an early relative of the guenon subgroup and conclude that it is reasonable to propose that these two major groups of Old World monkeys had therefore become distinct by middle Miocene times. Such a suggestion, however, stretches the interpretation of fragmentary fossil evidence to its limits. Pilbeam and Walker (1968) also reported on an isolated fragment of frontal bone and a single upper molar from the Napak deposits of East Africa (19–20 Ma old) as possible evidence of early Old World monkeys. However, Radinsky (1974a) subsequently indicated that the frontal fragment came from a hominoid species and it is now believed that this fragment came from *Micropithecus*, an Old World simian not directly related to the origin of the monkeys (P.J. Andrews, personal communication). Szalay and Delson (1979) allocate the molar to *Victoria-pithecus* (see also Pickford, 1987). Overall, the fossil record can tell us very little about the early origins of Old World monkeys and it is certainly unjustifiable to conclude from present evidence that this group originated in Africa, although Simons (1972) suggested a possible relationship between *Victoriapithecus/Prohylobates* and the Oligocene Fayum genus *Parapithecus*.

In contrast to the paucity of early Miocene fossils, quite a number of Old World monkey fossils can be reliably identified from the late Miocene onwards (10 Ma ago to the present), although they are essentially restricted to Africa and Europe. A useful review of much of this material, oriented particularly towards possible origins of baboons, was provided by C.J. Jolly (1967) and Pickford (1987) has produced a valuable summary. One feature that should be noted at once is that, as with modern Old World monkeys (particularly species in the guenon subgroup), sexual dimorphism is common among these later, well-documented fossil forms. There is also fairly general agreement that some of these fossil forms should be regarded as related to modern leaf-monkeys, whereas others are allied to the guenon subgroup. In both cases, there are fossil species that have been allocated to modern genera (e.g. *Presbytis*, *Colobus*,

Macaca, *Cercocebus*, *Papio* and *Theropithecus*), thus emphasizing the general close resemblance to modern Old World monkeys; but there are also numerous genera that are no longer represented among the modern forms.

Among the fossils thought to be allied to modern leaf-monkeys, the best represented is *Mesopithecus* (Wagner, 1839), which is known from a score of fairly complete skulls and large numbers of postcranial bones (von Beyrich, 1861). A recent study of dental enamel prisms in Old World monkeys has confirmed that *Mesopithecus* resembles leaf-monkeys, particularly langurs, rather than members of the guenon subgroup (Dostal and Zapfe, 1986). The genus is best known from the late Miocene Pikermi deposits of Greece (9–10 Ma old) and Gaudry (1862, 1878) published an early reconstruction of the skeleton based on material from that source (Fig. 2.20). Fragmentary remains from later (Pliocene) deposits of Europe extend the likely time range of *Mesopithecus* to perhaps 2 Ma ago and the likely geographical range to include Russia, Bulgaria, Yugoslavia and Iran. Beyond that, however, these fossils add little to our understanding of *Mesopithecus*. The skull closely resembles that of modern leaf-monkeys in both dental and other respects (see Verheyen, 1962; Vogel, 1968), but the form of the postcranial skeleton and the associated fauna typical of open country found in the Pikermi deposits together suggest that *Mesopithecus* was essentially terrestrial, unlike any modern leaf-monkey genus. *Dolichopithecus*, which is also known from Pliocene deposits (3–5 Ma old) of South and Central Europe (see Déperet, 1889), is represented by several fairly complete skulls and some postcranial elements. Szalay and Delson (1979) suggest that this genus is probably related to *Mesopithecus*, and it similarly exhibits marked adaptations for terrestrial life.

Fossil leaf-monkeys have also been found in Africa, the best-known genus being *Cerco-pithecoides*, which has been recovered both in South Africa (Mollet, 1947; Maier, 1971a) and in East Africa (Leakey and Leakey, 1973b). *Cerco-pithecoides*, which comes from relatively recent Plio-Pleistocene deposits (1–3 Ma old), is known

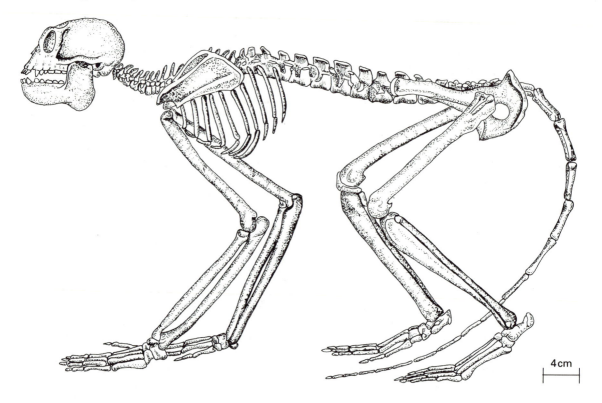

4 cm

Figure 2.20. Reconstruction of the late Miocene leaf-monkey, *Mesopithecus pentelici*. Unlike its presumed modern relatives, this monkey had a stocky skeleton and was probably terrestrial in habits. (After Gaudry, 1878.)

from several skulls from South African sites and from one skull and lower jaw from East Africa, but no postcranial bones can be reliably attributed to this genus. *Paracolobus* is known from a well-preserved skull and skeleton found fully articulated at Lake Baringo in East Africa (Leakey, 1969) and from a skull subsequently found in the Omo deposits, also in East Africa (Eck, 1977). All the known material is of late Pliocene age (2–3 Ma ago). Finally, there is a single, nearly complete skull from early Pliocene deposits of Egypt (Stromer, 1913), dated at about 5 Ma old, that has been allocated to the genus *Libypithecus*. The size of the skull suggests that *Libypithecus* was probably comparable to the modern patas monkey in overall body size. Fossils allocated to the leaf-monkey group also include some late Miocene partial jaws and isolated teeth from Pakistan that have been pro-

visionally classified as *Presbytis sivalensis*, suggesting a direct link with modern langurs (*Presbytis* spp.) of South and South-East Asia.

Known Plio-Pleistocene relatives of the Old World monkey guenon subgroup also exhibit a wide geographical distribution. Among the genera no longer represented in the modern array of Old World monkeys, the best documented is *Parapapio*. This genus is known from a variety of South African sites that have yielded a fairly large sample of reasonably complete skulls (Jones, 1937; Maier, 1971b). Unfortunately, because of the nature of the South African sites (secondarily exposed cave breccias containing only disarticulated bones), no postcranial elements can be reliably allocated to *Parapapio*. Additional, far more fragmentary material allocated to *Parapapio* has been found in East Africa, notably a lower jaw reported by

Patterson (1968), but this has not included any postcranial elements. Taking the South and East African material together, *Parapapio* is known for the period of around 2–4 Ma ago. The skulls from this genus certainly look quite baboon-like, and they indicate that sexual dimorphism was well marked as in modern baboons. Simons (1972) suggested that *Parapapio* might be ancestral to *Papio*, but a convincing case has yet to be made. There are also fossil specimens, including fairly complete skulls, from the late Pliocene and early Pliocene of East and South Africa that can be allocated to the modern baboon genus *Papio* (see Szalay and Delson, 1979), including *Papio baringensis* (Leakey, 1969). Therefore, *Parapapio* may be best seen as a parallel derivative from the stock that gave rise to *Papio*.

Two genera of late Pliocene/early Pleistocene Old World monkeys showing affinities to the modern guenon subgroup are apparently restricted to South Africa: *Dinopithecus* (Broom, 1937) and *Gorgopithecus* (Broom and Robinson, 1949). The former is known from a sample of skulls from the Swartkrans site, and the latter is known from one damaged skull and some jaws from Kromdraai. For reasons given above, no postcranial remains can be allocated confidently to these two genera and their affinities remain somewhat uncertain. However, as with *Parapapio*, it seems that they are probably quite closely allied to baboons. (Useful reviews of the South African material are given by Freedman and Stenhouse, 1972 and by Freedman, 1976). All of these possible baboon relatives are quite large in size, probably falling in the upper size range of the modern species; but the most spectacular fossil baboons of all are found among those allocated to the modern gelada genus *Theropithecus* (Fig. 2.21). Fossil gelada remains were initially reported from East Africa and allocated to the genus *Simopithecus* by Andrews (1916), but subsequent finds and discussions have extended the geographical range to include South Africa and have led to common agreement to include the fossils in the genus *Theropithecus* (C.J. Jolly, 1972; Maier, 1972; Leakey and Leakey, 1973a; Eck, 1977). Jolly (1972) has

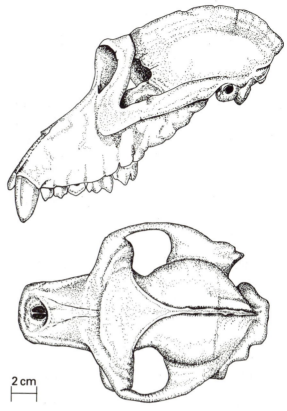

Figure 2.21. Lateral and dorsal views of the skull of a giant fossil gelada baboon, *Theropithecus oswaldi*, from the Plio-Pleistocene deposits of East Africa. (After C.J. Jolly, 1972.)

discussed the large-bodied fossil geladas in detail and provides cogent arguments for the interpretation that they were terrestrially adapted occupants of waterside habits. Numerous skulls are known, and associated postcranial remains are fairly common. The skull and almost complete skeleton of a giant fossil gelada from Olduvai Bed IV have been used to produce a reconstruction that suggests that this species, *Theropithecus oswaldi* (Fig. 2.21), had reached the body size of a modern gorilla. It is notable that the arms are markedly longer than the legs in these fossil geladas, which are known over a period of about 3 Ma leading up to quite recent times. With respect to Africa, it should also be mentioned that some fragmentary fossil specimens have been allocated to the modern mangabey genus, *Cercocebus*; but only one damaged skull is known from the South African site of Makapansgat (Eisenhart, 1975), and this allocation remains rather tentative.

Two genera from outside Africa that contain no living representatives but may be related to the modern guenon subgroup are *Procynocephalus* and *Paradolichopithecus*. *Procynocephalus* (Schlosser, 1924), is known from jaws and teeth and from a few associated postcranial bones (Teilhard de Chardin, 1938) from the late Pleistocene of China and northern India. *Paradolichopithecus* is known from late Pliocene deposits of Eurasia and fairly complete skulls have been found as well as some associated postcranial bones (Delson and Plopsor, 1975). The latter indicate that *Paradolichopithecus* was terrestrial. For geographical reasons it seems likely that both *Procynocephalus* and *Paradolichopithecus* are related to macaques, but the case has yet to be convincingly argued on morphological grounds. The genus *Macaca* itself is quite well represented in fossil deposits dating back perhaps as far as 6 Ma, but unfortunately the specimens concerned are typically fragmentary. Fragmentary evidence indicates that *Macaca* occurred over a wide geographical area in Pliocene and Pleistocene times, including North Africa, Europe (as far north as the Netherlands), northern India, China and Java. Most of the specimens are allocated to the modern Barbary macaque species *Macaca sylvanus*, but four extinct species can be recognized. *Macaca libyca* and *M. 'flandrini'* are known from very fragmentary remains; the only extinct species known from substantial fossil remains is *Macaca anderssoni* of China, but these have yet to be fully interpreted. *Macaca palaeindica* from northern India is known only from two jaws and must also count as a fragmentary fossil. Thus, despite the large geographical range and the relatively long timespan that one might infer for fossil *Macaca*, we actually have relatively little hard evidence relating to the history of this genus.

It is, of course, somewhat surprising that we have so little fossil evidence before 10 Ma ago to document the origins of the Old World monkeys, which constitute a large proportion of the modern array of primates inhabiting the Old World. One might, perhaps, conclude from this that the Old World monkeys had a very recent origin; but argument from negative evidence coming from a highly imperfect record does not carry much weight. It seems far more likely that we need to find more fossil sites covering a more representative geographical area (including West Africa and South-East Asia) in order to fill the gaps in our knowledge. We can still say little in concrete terms about the divergence between the leaf-monkeys and the guenon subgroup. This task is rendered quite difficult because the modern Old World monkey species exhibit relatively little morphological divergence (see Chapter 1), and this increases the need for substantial, rather than fragmentary, fossils. It should also be noted that no fossils yet found have been suggested as direct relatives of the primarily forest-living guenons (*Cercopithecus* spp. etc.), although these are particularly well represented in the modern fauna. The fossil Old World monkeys recorded to date are all quite large in body size, and most were probably terrestrial. One possibility is that we are confronted with the usual bias against smaller-bodied and forest-living species in the fossilization process, but there is also mounting evidence that the Old World monkeys may have originated from open-country forms.

MIOCENE SMALL APES

The term 'lesser ape', as applied to the Old World fossil record, may be interpreted in several different ways. In identifying fossil 'lesser apes', we may be influenced essentially by body size, in which case it is preferable to refer simply to 'small apes'. Referring to fossil forms as 'lesser apes' implies some direct connection with the modern lesser apes, gibbons and the siamang. Indeed, it has been proposed by various authors that certain small-bodied fossil apes, at least, do bear a specific phylogenetic relationship to the gibbon group. This was the view taken by Simons (1972) and further developed in the survey conducted by

Simons and Fleagle (1973). Alternatively, as has since been concluded by Fleagle (1984), it can be proposed that fossil 'small apes' are related only in a general way to the evolutionary radiation of the modern apes and humans and that they are not directly linked to the origin of any particular modern forms, such as gibbons. It is really in this sense that Szalay and Delson (1979) allocated some relevant fossils to the taxonomic group that includes apes and humans (superfamily Hominoidea). But it is still a moot point whether the known Miocene small apes resemble modern gibbons and the siamang merely in body size and associated features, or whether a definite phylogenetic relationship is involved in some cases. In fact, the entire picture of the evolutionary radiation of the Old World simians is currently in turmoil. The view taken here (following Harrison, 1982 and Andrews, 1985) is that fossil 'small apes' are best regarded as early representatives of this radiation whose exact phylogenetic relationships remain to be established. As it is the Old World monkeys, rather than the apes, that exhibit the most specialized dental characteristics (see Chapter 6), there is a strong possibility that many fossil 'small

apes' are not specifically related to any group of modern apes.

The earliest known, and still the best-documented, genus of small fossil apes is *Pliopithecus* (Fig. 2.22). This is a gibbon-sized form from European middle Miocene deposits (9–14 Ma ago), with a body weight estimated at 5.9–8.5 kg from molar dimensions (Fleagle and Kay, 1985). Szalay and Delson (1979), following Hürzeler (1954), recognize as many as six species in this genus, but the only substantial fossil remains are those of *Pliopithecus vindobonensis* (Zapfe, 1958, 1960). Skeletal remains from three different individuals of this species, known only from a single fossil site in Czechoslovakia, permit fairly reliable reconstruction of the skull and skeleton. Fragmentary fossil remains (mainly upper and lower jaws bearing teeth) of *Pliopithecus* have been reported from a wide geographical range in Europe, including France, Spain, Germany, Austria and possibly Hungary, in addition to Czechoslovakia (Hürzeler, 1954; Kowalski and Zapfe, 1974; Ginsburg, 1975; Kretzoi, 1975). However, more substantial remains are required to establish whether more than one species existed.

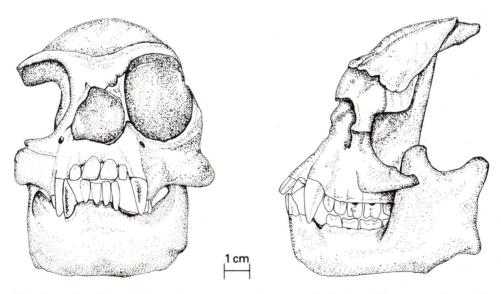

1 cm

Figure 2.22. Frontal and lateral views of a reconstruction of the skull of the Miocene small ape *Pliopithecus vindobonensis*. (After Zapfe, 1960.)

Several authors have suggested that *Pliopithecus* was rather like the spider monkey (*Ateles*) in its postcranial skeleton and it certainly lacked the most obvious specializations seen in modern lesser apes, such as pronounced elongation of the forelimbs (see Ciochon and Corruccini, 1977). Ankel (1965) in fact suggested, on the basis of the apparent size of the canal passing through the sacrum, that *Pliopithecus* had a tail and Simons (1972) subsequently referred to 'discovery' of a tail in this fossil genus. Zapfe, however, found no evidence of tail bones among the substantial fossil remains of *Pliopithecus vindobonensis* and Ankel's suggestion remains to be confirmed, especially in view of likely distortion after death of the sacrum that was measured (A. MacLarnon, 1987).

The only other fossil small apes of any antiquity come from a variety of East African middle Miocene deposits (dated about 14–23 Ma old) and from late Miocene deposits of Yunnan Province, China. The initially available East African material was allocated to the genus *Limnopithecus* by Hopwood (1933) and eventually two species were recognized: *Limnopithecus legetet* and *L. macinnesi* (Le Gros Clark and Leakey, 1950, 1951). However, Andrews and Simons (1977), following a detailed review of Miocene ape fossils of East Africa by Andrews (1973), established the genus *Dendropithecus* for the second species, which hence became *Dendropithecus macinnesi*. Andrews proposed that the first species was in fact more closely related to the genus *Dryopithecus* (see later) and '*Limnopithecus*' was accordingly reduced to a subgenus of *Dryopithecus*. Subsequently, however, it has emerged that *Limnopithecus* is unlikely to be closely related to *Dryopithecus* after all and it is best regarded as a fossil small ape of uncertain affinities, like *Dendropithecus* and *Pliopithecus*. For present purposes, the relationships of *Dendropithecus* and *Limnopithecus* are of little import, as both genera are known only from fragmentary remains. *Limnopithecus* is known almost exclusively from jaws and teeth and the only well-known, reliably associated postcranial

elements come from *Dendropithecus macinnesi* (Le Gros Clark and Thomas, 1951). No skulls are known for either genus, although an anterior skull fragment described by Fleagle (1975) may conceivably come from *Limnopithecus*.

Study of the limited postcranial material available for *Dendropithecus macinnesi* has, as with *Pliopithecus*, evoked comparisons with the New World monkey *Ateles* (e.g. Simons, 1972; Szalay and Delson, 1979). However, a biomechanical investigation conducted by Preuschoft and Weinmann (1973) yielded no definite conclusions regarding the affinities of *Dendropithecus*. Ferembach (1958) interpreted *Dendropithecus* as no more than a smaller version of *Dryopithecus* and ruled out any direct relationship with modern gibbons and siamangs. By contrast, Le Gros Clark and Thomas (1951) and Simons (1963) regarded *Dendropithecus* (about the size of a modern siamang – see Fleagle and Kay, 1985) as directly related to modern lesser apes (see also Simons and Fleagle, 1973). Andrews (1973) initially interpreted *Dendropithecus* as a possible relative of the common ancestor of the modern gibbons and the siamang, but subsequently Delson and Andrews (1975) concluded that there were no shared specializations to indicate such a link. The consensus view has now shifted against a specific phylogenetic relationship between *Dendropithecus* and modern lesser apes. Fleagle (1984) has instead suggested that fragmentary remains of the relatively new fossil forms *Micropithecus* from East Africa and *Dionysopithecus* from China may perhaps be linked to gibbons.

A fossil small ape has also been reported from late Miocene deposits of China. The material includes a partial skull, several jaw fragments and isolated teeth and has been allocated to the species *Laccopithecus robustus* (Wu and Pan, 1984, 1985b). *Laccopithecus* is similar in size and in certain morphological features to both *Pliopithecus* and to modern gibbons but its affinities have yet to be established.

The phylogenetic relationships of fossil small apes have generally remained enigmatic because of the poor availability of material, not only in terms of fossil specimens and geographical representation but also with respect to numbers of species. Only *Pliopithecus vindobonensis* can be regarded as a substantial fossil, although *Dendropithecus, Laccopithecus* and *Turkanapithecus* (see later) are sufficiently well represented to permit some useful comparisons in due course. We still do not know whether any of these primates had tails, yet loss of the tail is one of the key defining features of modern lesser apes, great apes and humans. A recent review of small fossil apes (Harrison, 1982) has confirmed suspicions that none of the fossil small apes hitherto described is directly related to gibbons and the siamang. (However, the possibility remains that the Chinese *Laccopithecus*, which was reported subsequently, may be related to modern lesser apes). Indeed, Harrison made an intriguing suggestion regarding relationships between Old World monkeys and apes. As a result of the pervasive influence of the concept of the 'phylogenetic scale' in discussions of primate evolution (see Chapter 3), it has been common to make the implicit assumption that monkeys are necessarily more primitive than apes, because modern apes are more closely related to humans. However, as has been indicated above, the Old World monkeys have distinctive specializations in their own right and Harrison (1982) makes a good case for the hypothesis that modern Old World monkeys may be derived from forms that, on grounds of body size and dental features, would be classified as 'small apes'. This new interpretation may, to some extent, fill the apparent yawning gap in the fossil record of the evolution of Old World monkeys, in that their early ancestors would have been identified as 'small apes'. Nevertheless, we are still left with the daunting problem that there are no known substantial Old World monkey fossils older than 10 Ma.

MIOCENE/PLIOCENE LARGE APES

As with the 'small ape' category, fossils grouped here as 'large apes' are included at least partly on grounds of overall body size. Most of them approximate the modern great apes in known dimensions, although a few may have been as small as the modern siamang (i.e. about 11 kg). Without exception, they probably lacked tails, but actual fossil evidence is still lacking for most species. In association with their relatively large body size, all of the species included here were probably adapted for suspensory locomotion, in so far as they were arboreal, although some species at least were doubtless adapted for varying degrees of terrestrial locomotion.

One unusual fossil primate included here as a 'large ape' represents a special case and will therefore be discussed first. The single species *Oreopithecus bambolii* was originally described by Gervais (1872) on the basis of a nearly complete lower jaw recovered from late Miocene deposits in northern Italy (dated at about 8 Ma old). Both Gervais (1872) and Schlosser (1887) suggested that *Oreopithecus* had 'monkey-like' teeth, but Forsyth Major (1880) and Schwalbe (1915) suggested a possible link with human evolution. This interpretation was subsequently developed by Hürzeler (1949, 1958, 1968), who discovered and described an almost complete, crushed specimen of *Oreopithecus bambolii*, making this one of the best-documented primate species in terms of the availability of substantial remains (Fig. 2.23). In view of the relatively rich fossil material available for *Oreopithecus*, it is somewhat surprising that there is no real consensus regarding the affinities of this Old World primate from the Miocene. It has been regarded in turn as a relative of Old World monkeys, of great apes, or of humans. Straus (1961, 1963) concluded that *Oreopithecus* was more closely allied to modern apes, but Simons (1972) and Szalay and Delson (1979) have essentially concurred in regarding this fossil as an aberrant relative of Old World monkeys (see Delson, 1986 for a recent review). As has been noted earlier, Simons originally suggested a specific phylogenetic connection between the Oligocene Fayum simian *Apidium* and *Oreopithecus*.

The basic problem is that the dentition of *Oreopithecus* is decidely unusual in certain respects, whereas the postcranial skeleton exhibits typical ape-like features, such as relatively elongated arms and complete loss of

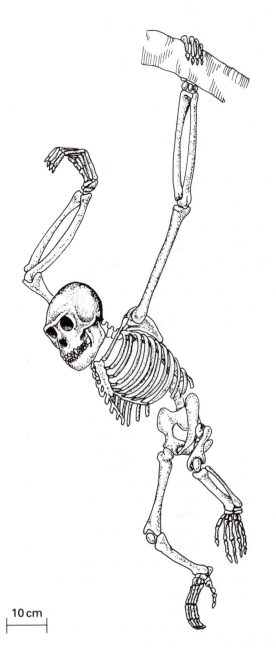

10 cm

Figure 2.23. Reconstruction of the late Miocene large ape *Oreopithecus bambolii*. Note the elongated arms and the absence of a tail. (After Simons, 1964.)

the tail. Indeed, the pelvis shows certain striking resemblances, at least superficially, to that of humans. Hence, several investigators who have concentrated on dental features of *Oreopithecus* (e.g. Szalay and Berzi, 1973; Delson and Andrews, 1975; Rosenberger and Delson, 1985) have tended to emphasize its monkey-like affinities, whereas those who have paid close attention to postcranial characters (e.g. Riesenfeld, 1975; Harrison, 1986) have been impressed by the resemblances to apes. Nonetheless, some authors who have studied dental features (e.g. Butler and Mills, 1959; Grine, Krause and Martin, 1985; Butler, 1986) have also concluded that *Oreopithecus* is more closely allied to apes than to monkeys. Unaccountably, *Oreopithecus*, one of the few available substantial fossils documenting the evolution of Old World simians, has been neglected to a remarkable degree in the recent literature. For instance, this genus is barely mentioned in a recent major review of ape and human evolution (Ciochon and Corruccini, 1983). In his concluding chapter (1983, p. 385), Ciochon even hails this regrettable lapse as a matter for celebration: '*Oreopithecus* receives hardly a mention throughout the 29 preceding chapters, and rightly so, since this "swamp ape" has little bearing on current theories of ape and human origins.'

The affinities of *Oreopithecus* have yet to be established reliably through a comprehensive analysis of all of its characteristics, but it is included here as a 'large ape' because the balance of available evidence surely indicates a link with the apes and humans rather than with the Old World monkeys. In fact, there is some disagreement in the literature over the likely body weight of *Oreopithecus* and hence about the use of the adjective 'large'. On the basis of the length of the humerus Aiello (1981b) inferred a body weight of 48.6 kg with 95% confidence limits of 30.6–77.3 kg. This agrees well with the statement by Schultz (1960), based on an overall assessment of the skeleton of *Oreopithecus*, that the body weight was at least 40 kg. Further, Stern and Jungers (1985) and Jungers (1987) have inferred a body weight of 32–33 kg from postcranial dimensions. Yet, on the basis of molar tooth dimensions, Gingerich, Smith and Rosenberg (1982) estimated the body weight of *Oreopithecus* as about 15 kg, with 95% confidence limits of about 13.5–17.5 kg (see also Fleagle and Kay, 1985). This vast discrepancy in inferred body weights is uncommon and the most likely explanation is that *Oreopithecus* had unusually small molar teeth for its body size. Indeed, there is a strong possibility that aberrant molar specialization had taken place, with reduction in size being accompanied by morphological change. If this explanation is correct, *Oreopithecus* provides an interesting parallel to the later stages of human evolution, during which there was a reduction in the size of molar teeth relative to body weight (see Chapter 6). It should also be noted that, as the overall bodily dimensions of *Oreopithecus* indicate that it was at least as big as a modern chimpanzee, it was a 'large ape', rather than a 'medium-sized primate', as suggested by Szalay and Delson (1979).

The remaining fossil large apes, which have been recovered predominantly from Miocene deposits of Eurasia and East Africa, were for some time by general consent regarded as members of a single adaptive array. Most of the species concerned were allocated by Simons (1972) to the subfamily Dryopithecinae. This taxonomic grouping has been provisionally retained here, particularly because a basic feature common to virtually all species is possession of the distinctive '*Dryopithecus* pattern' (or 'Y5 pattern') on the lower molar teeth (see Chapter 6 and Fig. 6.22). However, in this case a classificatory change may eventually prove to be unavoidable, because the phylogenetic relationships of fossil large apes originally allocated to the Dryopithecinae are turning out to be considerably more complex than was once thought (e.g. see Andrews, 1985). Successive major reviews of this group of primate fossils (Simons and Pilbeam, 1965; Pilbeam, 1969; Andrews, 1973, 1978a, b; Ciochon and Corruccini, 1983) have considerably clarified the likely phylogenetic relationships among the fossil species concerned, although the situation still

remains quite labile because of the largely fragmentary nature of the fossil evidence involved. For a long period of time, only one fairly complete skull was known for the entire group (see below and Fig. 2.24) and it is only recently that skulls of varying degrees of preservation have been recovered for other species. Similarly, for most species there is no reliably allocated postcranial evidence.

It has now emerged fairly clearly that the Miocene 'large apes' here included in the subfamily Dryopithecinae can be generally divided into three groups: (1) genera with thin enamel on their cheek teeth (e.g. *Proconsul* and *Rangwapithecus*); (2) species with somewhat thickened enamel (*Dryopithecus*); and (3) those with definitely thick enamel (*Gigantopithecus*, '*Ouranopithecus*', '*Ramapithecus*' and *Sivapithecus*). This difference has considerable implications for dental and dietary adaptation (see Chapter 6) and it is particularly interesting because the human evolutionary lineage is characterized by thick-enamelled cheek teeth.

The genus *Dryopithecus* was first established for two lower jaw fragments from a French Miocene site by Lartet (1856). Further lower jaw specimens were reported by Gaudry (1890) and by Harlé (1899) from French sites. These, together with a femur shaft originally described by Lartet – which is obviously of dubious affinities and of relatively little value for comparative purposes – constitute the bulk of the material documenting the European species *Dryopithecus fontani*. Other, even more fragmentary, material of comparable age is known from Germany, Austria and Spain, including a few isolated postcranial bones. The Spanish material (see de Villalta and Crusafont-Pairo, 1944; Crusafont-Pairo and Hürzeler, 1961) has provided the main support for recognition of a second species referred to as *Dryopithecus laietanus* by Simons (1972) and as *Dryopithecus brancoi* by Szalay and Delson (1979). However, all of the European material, which probably covers a time range of 10–14 Ma ago, is so fragmentary that one can do little more than conclude that the genus *Dryopithecus* was represented in Europe during the Miocene by species that were probably somewhat smaller than modern chimpanzees (see Gingerich, Smith and Rosenberg, 1982).

The remaining species included here as thin-enamelled members of the subfamily Dryopithecinae have all been described on the basis of material derived from East African fossil sites covering the period of about 14–22 Ma ago. As has been noted above in the discussion of fossil small apes, the genus *Limnopithecus* was at one stage recognized as a subgenus of the genus *Dryopithecus*. Similarly, a number of other forms have at some stage been included as subgenera of *Dryopithecus* (Andrews, 1973; Szalay and Delson, 1979). This applies, for example, to the genera *Proconsul* and *Rangwapithecus*. *Rangwapithecus* is known only from relatively fragmentary, predominantly dental specimens (e.g. see Andrews, 1974) and need not concern us further here. The only really substantial evidence is known for *Proconsul*, for which at least three species can be recognized: *P. africanus*, *P. major* and *P. nyanzae*.

Proconsul africanus, with an estimated body weight of about 11 kg (Aiello, 1981b; Walker, Falk *et al.*, 1983), remains the best-known species (Le Gros Clark and Leakey, 1951; Walker and Pickford, 1983; Walker *et al.*, 1983). A partial skull accompanied by a lower jaw has been used for two successive reconstructions (Davis and Napier, 1963; Walker *et al.*, 1983; see Fig. 2.24) and an associated skeleton originally described by Napier and Davis (1959) has been re-examined together with supplementary material by Walker and Pickford (1983). The skeleton, which was originally largely confined to components of the forelimb and is now almost complete, is the most complete of any Miocene 'large ape' other than *Oreopithecus*. A lively discussion has surrounded it, particularly with respect to whether it exhibits monkey-like adaptations for quadrupedal locomotion or ape-like features related to suspension (Lewis, 1971a, b, 1972a, 1972b; Preuschoft, 1973; O'Connor, 1976; Corruccini, Ciochon and McHenry, 1976; Morbeck, 1975; Walker and Pickford, 1983). As with the Miocene 'small apes', suggestions of resemblances to New World

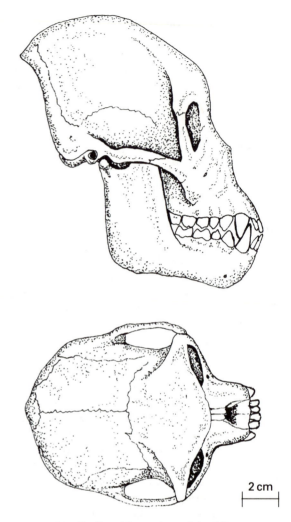

Figure 2.24. Skull and lower jaw of the Miocene ape *Proconsul africanus*. This is the most complete skull known for any fossil 'large ape' included in the subfamily Dryopithecinae. (After Walker, Falk *et al.*, 1983.)

monkeys such as spider monkeys (*Ateles*) have been common, but no clear picture has as yet emerged regarding the likely locomotor behaviour of *Proconsul africanus*. As it was apparently closer in body size to the modern siamang than to any exant great ape, it is probably not strictly accurate to regard this species as a 'large ape'. However, *Proconsul nyanzae* and *P. major* were quite large-bodied species, the latter probably being larger than the two modern chimpanzees (Andrews, 1978a; Pickford, 1986), so it is arguably appropriate to regard the genus *Proconsul* as belonging in the category of 'large apes'. Nevertheless, *Proconsul* lacked the extreme specializations for brachiation found in modern lesser apes and it is therefore distinct from all of the modern apes. In fact, there is now increasing evidence that *Proconsul africanus*, despite certain dental similarities to *Dryopithecus*, is actually quite distinct in phylogenetic terms and may not be directly linked to the adaptive radiation of the great apes and humans. Accordingly, there is no longer any justification for regarding *Proconsul* as a subgenus of *Dryopithecus*.

The remaining species of *Proconsul* are known only from relatively fragmentary remains. *Proconsul major* is known primarily from jaws and teeth, in rather limited number, although a vertebra from the lumbar region (Walker and Rose, 1968) and an ankle bone or talus (Day and Wood, 1968) have been attributed to this species (see also Le Gros Clark and Leakey, 1950, 1951). It would seem that *Proconsul major* was about the size of a modern gorilla, but further evidence is sorely needed to document this species. *Proconsul nyanzae*, which was probably about the size of a modern chimpanzee, is also known from limited material, although part of a hindlimb skeleton has recently been allocated to this species by Walker and Pickford (1983).

The remaining well-known Miocene 'large ape' species are all characterized by thick-enamelled cheek teeth (see L.B. Martin, 1985). The genus *Sivapithecus*, which is still represented largely by jaws and teeth, was originally reported from the Siwalik deposits of northern India and Pakistan (Pilgrim, 1927). Most authors have recognized two species for the Siwaliks (e.g. see Szalay and Delson, 1979): *Sivapithecus indicus*, described as approximating the modern gorilla in body size (Simons and Pilbeam, 1971), and the smaller *S. sivalensis*. However, there is some doubt about the validity of the distinction between these two species. Andrews (1983) recognized only a single, highly sexually dimorphic species, *S. indicus*, for India and Pakistan, whereas L.B. Martin and Andrews

(1984) include all of the material under the species name *S. sivalensis* (which has priority). Kay and Simons (1983a), on the other hand, have united *S. sivalensis* with *Ramapithecus* (see later), thus expanding the former species in a different manner. Kay (1982) has recently recognized a third species of *Sivapithecus*, *S. simonsi*, on the basis of three jaw fragments from the Indo-Pakistan deposits. Much of the relevant fossil evidence is very fragmentary and at present only a single species of *Sivapithecus*, *S. sivalensis*, can be recognized from the above-mentioned material with any confidence.

Additional jaws and teeth of *Sivapithecus* have recently been found in the Siwalik deposits along with some postcranial material attributed to this genus on size grounds (Pilbeam, Meyer *et al.*, 1977). The new material has been further expanded by the discovery of a partial face (Pilbeam, 1982; Ward and Pilbeam, 1983). Accordingly, *Sivapithecus* from the Indian subcontinent has now been almost upgraded from the status of a fragmentary fossil to that of a substantial fossil. Other dental material allocated to *Sivapithecus* has been found in Turkey, Greece and possibly Hungary and Czechoslovakia, such that two additional species have been proposed: *S. meteai* and *S. darwini*. The latter is very poorly documented, but a reasonable sample of *S. meteai* has been obtained – including a very informative palate – from Turkey (Andrews and Tobien, 1977; Andrews and Tekkaya, 1980). Finally, cranial material from Lufeng in China was originally allocated to the genus *Sivapithecus* by Wu (1984) but this has now been allocated to the new genus *Lufengpithecus* (Wu, 1987). Its relationship to the *Sivapithecus* material has yet to be explored in detail.

Jaws and teeth found in late Miocene deposits in Macedonia (de Bonis, Bouvrain, *et al.*, 1974; de Bonis and Melentis, 1976, 1977a, b, 1978, 1980) have been attributed to the species *Ouranopithecus macedoniensis*. It is generally agreed that specimens of *Ouranopithecus* are quite close in known morphological details to those of *Sivapithecus* (and to those of *Ramapithecus*; see later). Szalay and Delson (1979) have, in fact, included the Macedonian form within the species *S. meteai* (see also L.B. Martin and Andrews, 1984); but de Bonis (1983) presents evidence indicating that this step may not be justified. Clear resolution of this problem must await the discovery of more substantial fossil evidence. At present, the material currently attributed to *Sivapithecus* from Eurasian fossil sites (dated about 9–15 Ma ago) is very largely restricted to jaws and teeth and any conclusions about relationships among species must remain provisional, especially in view of the likely presence of marked sexual dimorphism (see Andrews, 1983).

The second genus of Miocene 'large apes' with thick-enamelled cheek teeth, *Ramapithecus*, was also first recognized among specimens from the Siwaliks of northern India and Pakistan (Lewis, 1934). This particular genus has attracted a great deal of attention because of a series of claims, initiated by Lewis, that *Ramapithecus* bears a special relationship to the evolution of humans (Simons, 1961b, 1972, 1976a, b; Simons and Pilbeam, 1965; Pilbeam, 1969). A useful summary of all the arguments eventually mustered to support the proposal of a phylogenetic link between *Ramapithecus* and hominids was provided by Tattersall (1975). At the time that these claims were made with greatest insistence, however, *Ramapithecus* was known only from very fragmentary evidence (a few partial jaws and teeth) derived first from the late Miocene deposits of the Siwaliks (*R. punjabicus*) and subsequently from earlier Miocene deposits of East Africa (*R. wickeri*). Specimens from these two areas combined yield a total age range for *Ramapithecus* of 8–15 Ma ago, and this genus was widely quoted in textbooks and in popular accounts as the 'earliest hominid'. Accordingly, the inferred date of divergence between great apes and humans was extended back to at least the middle Miocene, over 15 Ma ago. The known geographical range of *Ramapithecus* has now been expanded to include Turkey ('*Sivapithecus alpani*' of Tekkaya, 1974; see also Andrews and Tobien, 1977), and – initially on the basis of isolated teeth, but subsequently with cranial evidence – possibly China (Wu, 1957, 1958, 1984). Until recently, however, *Rama-*

pithecus was essentially known from isolated jaw fragments and teeth and was no more than a 'dental hominid'.

In fact, *Ramapithecus* now provides a very good object lesson in the dangers of heavy reliance on fragmentary evidence in the reconstruction of primate evolution. Increasing availability of fossil material has progressively led to a shift away from the hypothesis that *Ramapithecus* was the 'earliest hominid'. The high point of the hypothesis that this genus marked the beginning of the human lineage came with the first tentative jaw reconstructions conducted by Simons (1961b), which indicated, among other things, that this genus had small-crowned canine teeth and a rounded (parabolic) dental arcade as in modern humans (see Chapter 6). These reconstructions were based on the then known Siwalik material. Subsequently, a more far-reaching reconstruction carried out by Walker and Andrews (1973), based on later East African finds of *Ramapithecus wickeri*, indicated that the canines projected more than Simons had suggested and that the dental arcade was actually parallel-sided as in modern great apes. At about the same time, the first serious doubts about an early divergence date for the great apes and humans were being raised because the degree of biochemical similarity between modern humans and African great apes was greater than expected. Attempts to use molecular data to produce a 'clock' to time the major events in primate evolution (see Chapter 12) indicated very recent divergence dates for the African apes and humans and the status of *Ramapithecus* as the 'earliest hominid' became particularly suspect. Continued work on the 'molecular clock' has repeatedly thrown doubt on suggestions of a divergence as far back as 15 Ma ago between African great apes and humans. Inevitably, this led to a great deal of rethinking by those studying human evolution, but it must be said that conclusions based on the 'molecular clock' should not be accepted uncritically (see Chapter 12).

More recently, the special status originally accorded to *Ramapithecus* has been increasingly questioned by primate palaeontologists themselves, because of the growing realization that *Sivapithecus, Ramapithecus, Gigantopithecus* and *Australopithecus* all share with humans a set of dental characteristics related to the possession of thick-enamelled cheek teeth. The resulting reassessment of the initial divergence between great apes and humans has become a major overhaul for two reasons. First, it was discovered that the modern orang-utan also has relatively thick-enamelled cheek teeth and distinctive orang-like features were subsequently found both in the face of *Sivapithecus indicus* (Pilbeam, 1982; Preuss, 1982) and in the palate of *S. meteai* (Andrews and Tekkaya, 1980; Andrews and Cronin, 1982). Second, there has been increasing acceptance of the idea that *Sivapithecus* and *Ramapithecus* are very similar to one another, if not actually congeneric. Greenfield (1979) suggested inclusion of '*Ramapithecus*' in the genus *Sivapithecus* because of this close similarity and Kay and Simons (1983a), for instance, now include '*Ramapithecus*' species in *Sivapithecus sivalensis*. Whether this step is fully justifiable is neither here nor there; the essential point is that '*Ramapithecus*' is so close to *Sivapithecus* in dental features, at least, that it is no longer possible to consider '*Ramapithecus*' alone as a candidate for human ancestry. As the known facial and palatal features of *Sivapithecus* are closer to those of the orang-utan than to those of the African great apes or humans, it now seems quite unlikely that '*Ramapithecus*' is linked to hominid evolution (Andrews and Cronin, 1982). Recognition of this crucial point is in large part due to the discovery of more substantial material attributed to *Sivapithecus*. It must be said, however, that '*Ramapithecus*' – as a fragmentary fossil genus – should never have been regarded with any degree of confidence as an early hominid. To this day, the '*Ramapithecus*' material has retained the status of a fragmentary fossil. Accordingly, even the present growing consensus that '*Ramapithecus*' is congeneric with *Sivapithecus* and is related to the orang-utan, not to humans, must remain entirely provisional in the absence of substantial fossil evidence.

The remaining long-established genus of thick-enamelled Miocene large apes, *Gigantopithecus*, is unusual both in terms of its size and because it survived until quite recent times. In fact, the first evidence for this genus came from isolated teeth purchased in pharmacies in China (von Koeingswald, 1935, 1952; Weidenreich, 1945). Subsequently, three fairly complete lower jaws and large numbers of teeth were collected by Chinese investigators, who were able to confirm that *Gigantopithecus* was still present in China about 1 Ma ago. (Pei, 1957; Wu, 1962; Chang, Wu and Liu, 1973; Hsu, Han and Wang, 1974). The Chinese species, *Gigantopithecus blacki*, is still known only from lower jaws and isolated teeth and therefore counts (despite its obvious great body size, estimated at 300 kg by Simons, 1972) as a fragmentary fossil.

Even more fragmentary evidence (one jaw and some teeth) of a second species, '*Gigantopithecus bilaspurensis*' (now referred to as *Gigantopithecus giganteus*: Szalay and Delson, 1979; Kay and Simons, 1983a) was discovered more recently in the Siwalik deposits (Simons and Chopra, 1969; Simons and Ettel, 1970) Further fossil evidence, including the only known postcranial element (a distal fragment of a humerus), has been reported for this species by Pilbeam, Meyer *et al.* (1977). Remains of *Gigantopithecus giganteus* date back to about 8 Ma ago, so there is a considerable temporal gap between the Siwalik specimens and those found in China. The Siwalik species may have been somewhat smaller in body size than the Chinese species, but it was still undoubtedly larger than the modern gorilla. Opinions regarding the phylogenetic relationships of *Gigantopithecus* have varied considerably. It was suggested by de Bonis and Melentis (1976) that this genus may be related to '*Ouranopithecis macedoniensis*' from Greece (see also de Bonis, 1983), and several other authors (e.g. Simons, 1972; Szalay and Delson, 1979) have interpreted *Gigantopithecus* as a somewhat aberrant member of the Miocene 'large ape' radiation (see also Pilbeam, 1970; Corruccini, 1975a, b). However, other investigators (Frayer, 1973; Eckhardt, 1973, 1975; Robinson and Steudel, 1973) have suggested a link with human evolution. Probably, as with '*Ramapithecus*', it is reasonable – in the absence of substantial fossil evidence – to conclude that *Gigantopithecus* is just one part of the radiation of Miocene thick-enamelled great apes from which the human lineage may have diverged at some point.

Three additional fossil apes of uncertain affinities – *Turkanapithecus kalakolensis, Afropithecus turkanensis* and *Simiolus enjiessi* – have recently been reported from the new Miocene site of Kalodirr in Kenya, provisionally dated at about 16–18 Ma ago (Leakey and Leakey, 1986a,b, 1987). *Turkanapithecus*, reported to be comparable in body size to a modern colobus monkey (i.e. about 11 kg), is known from a partial skull, a mandible and some postcranial remains. The skull, in particular, has certain features indicating that *Turkanapithecus* is very distinct from the other apes of comparable size previously reported from East Africa. *Simiolus*, known from jaws, teeth and some postcranial elements, was somewhat smaller than *Turkanapithecus* and is apparently closer to previously known small apes of East Africa. *Afropithecus* is a larger-bodied ape species apparently possessing thick-enamelled teeth. It is a rather unusual, long-muzzled form known from a partial skull, some isolated partial jaws and teeth and a few postcranial bones. In some respects, *Afropithecus* resembles *Sivapithecus* and a phylogenetic connection may be possible. The deposits from which it comes would indicate a relatively early origin for thick-enamelled large apes. However, the likely affinities of *Afropithecus* can be determined only by detailed study.

PLIOCENE/PLEISTOCENE FOSSIL HOMINIDS

Evolution of the human lineage during the Pliocene and Pleistocene has, of course, been covered extensively by a series of authors and there is little need to consider more than the basic issues here. Good general surveys of the fossil evidence of human evolution are provided by Le Gros Clark (1967), Tattersall (1970), Pilbeam (1972), Wood (1978), Delson (1985) and Tobias (1985), and valuable reviews of the later stages of human evolution are provided by Howell (1967, 1969, 1978), Tobias (1973), Coppens, Howell, *et al.* (1976), C.J. Jolly (1978) and Smith and Spencer (1984). It should be noted at the outset that substantial fossil remains are known for all of the species listed below (a quite unusual situation with respect to the primate fossil record generally), but that there is virtually no fossil evidence relating to human evolution, other than a few fragments of dubious affinities, before about 3.8 Ma ago. The preceding period of human evolution therefore remains a complete mystery and an unfortunate major gap exists whatever view one takes of the time of divergence of hominids and great apes.

Fossil hominids are now commonly allocated to only two genera: *Australopithecus* (including '*Paranthropus*') and *Homo*, which very approximately succeed one another in time, although there is some overlap. *Australopithecus africanus* was the first species described on the basis of the well-known juvenile skull with natural partial endocast from Taung, South Africa (Dart, 1925), and this was later joined by specimens of the more heavily built *Australopithecus robustus* (Broom, 1938), which some authors would place in the separate genus *Paranthropus*. Subsequently, further finds attributable to both species, including fairly complete skulls and some postcranial elements, were found at a number of South African sites. All of the fossil deposits concerned consisted of cave breccias for which no really reliable dates can be obtained, because of the lack of material suitable for radioactive dating techniques. However, it is generally believed, on the grounds of indirect evidence (e.g. faunal association) that *Australopithecus africanus* existed somewhat earlier (perhaps 2–3 Ma ago) than *A. robustus* (perhaps 1–2 Ma ago). The distinction between the two *Australopithecus* species is reasonably clear for the South African sites and it has become customary, largely for this reason, to refer to 'gracile' and 'robust' australopithecines as two distinct categories. Robust australopithecines now commonly allocated to the species *A. boisei* (=‘*Paranthropus boisei*’) were later found in Olduvai Gorge, East Africa

(Leakey, 1959; Tobias, 1967) and they have now been identified in a variety of East African Plio-Pleistocene sites. A very early skull of *Australopithecus boisei*, dated to 2.5 Ma ago, has now been reported (Walker, Leakey *et al.*, 1986).

More recently, recognition of a new gracile australopithecine species '*Australopithecus afarensis*' (which may, in fact, intergrade with *Australopithecus africanus*) has been proposed on the basis of material from Hadar, Ethiopia and Laetoli, Tanzania, dated at up to 3.75 Ma ago, (Johanson, White and Coppens, 1978; Johanson and White, 1979). The somewhat younger Hadar material has attracted much attention because it includes a substantial part of the skeleton of one individual associated with cranial and dental remains. This is the only properly associated skeleton so far known for any *Australopithecus* species. The Laetoli site is also very interesting because of the discovery of fossilized animal tracks (dated at about 3.5 Ma ago) including footprints that were very probably made by a bipedal hominid (Leakey and Hay, 1979; Day and Wickens, 1980). Thus, the earliest known evidence of the evolution of human bipedal locomotion is provided not by fossilized bones but by a permanent record of the actual behaviour involved! Later postcranial elements from the *Australopithecus* sites in South Africa and East Africa, apart from the Hadar skeleton, have been inferred to derive from either *Australopithecus africanus* or *A. robustus*. Robinson (1972) believed that the two species were distinguished by quite different locomotor adaptations, but this interpretation was questioned by some later authors (e.g. McHenry, 1975; Day, 1985). More fossil material is required to settle this issue.

All of the substantial fossil evidence of *Australopithecus* discovered to date has come from South and East Africa, and it has become common in the literature to conclude that 'man evolved in Africa'. However, this conclusion has been accepted far too readily. Sites of comparable age and circumstance in other areas of the world (notably South-East Asia) are few in number and far less effort has been devoted to hominid fossil-hunting activities outside Africa.

Despite this, some very fragmentary jaws and teeth of uncertain affinities have been reported (see Szalay and Delson, 1979), and much more work and fossil discovery will be required to establish whether *Australopithecus* species were confined to Africa and represented the only hominids of their time.

It is widely accepted that the modern human genus *Homo* was derived at some point from gracile (rather than robust) australopithecine stock. A possible candidate for the intermediate stage in this process is *Homo habilis* from East Africa (Leakey, Tobias and Napier, 1964). This species is now known from several fairly well-preserved skulls, some referred postcranial elements and a partial skeleton (Johanson, Masao *et al.*, 1987). There has been considerable discussion about the validity of this species, which lived between 1.5 and 2.5 Ma ago, with some authors relegating it to *Australopithecus* and others insisting on its inclusion in *Homo* (e.g. see Pilbeam, 1972); but it is only to be expected that an intermediate form should overlap in morphology both with earlier gracile australopithecines and with later *Homo* species (this is the 'palaeospecies problem' – see Chapter 3). Much of the discussion about the exact affinities of *Homo habilis* has therefore been relatively unproductive.

The first definite fossil species of *Homo*, *Homo erectus*, was originally reported from Java (Dubois, 1894) and subsequently from China, first from very fragmentary dental evidence (Black, 1927) and then from fairly complete skulls (Weidenreich, 1943). Subsequent cranial and postcranial material has been reported from East Africa, South Africa, North Africa, Germany and Central Europe, indicating that the species *Homo erectus* had a fairly wide geographical range, extending far outside Africa, during the period 0.5–1 Ma ago (see Howells, 1973 for a review). Until recently, the postcranial skeleton of *Homo erectus* was very poorly known and evidence was confined to isolated limb bones; but the discovery of an almost complete skeleton dated at about 1.6 Ma ago in East Africa (Brown, Harris *et al.*, 1985) has now radically improved the situation. At some stage

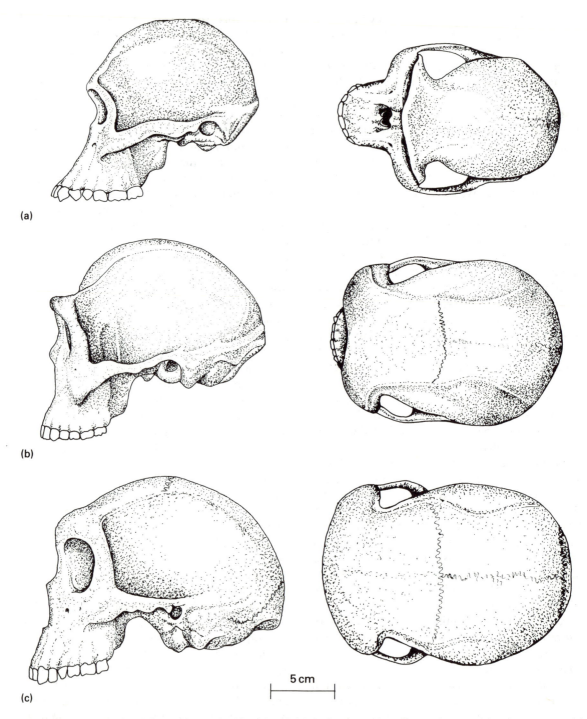

(a)

(b)

(c)

5 cm

Figure 2.25. Lateral and dorsal views of African hominid skulls: *Australopithecus africanus* (a), *Homo erectus* (b) and *Homo sapiens* (c). Note the progressive expansion in size of the braincase and the gradual reduction in the size of the jaws, relative to the overall size of the skull. (After Howell, 1978.)

near the end of the documented fossil history of *Homo erectus*, the modern species *Homo sapiens* arose and subsequently spread throughout the world, probably reaching the New World and Australasia relatively recently. Classic Neanderthalers are now widely recognized as a separate subspecies, *Homo sapiens neanderthalensis*, distinct from the modern human subspecies *Homo sapiens sapiens*, and they may in fact represent a distinct species. (For detailed discussions of Neanderthalers, see: Howells, 1974; Stringer, 1974; Trinkhaus, 1981; papers in Smith and Spencer, 1984).

Despite the particularly concentrated effort that has been devoted to investigation of our closest fossil relatives, much of human evolution remains undocumented. By the time of the first substantial fossil evidence, about 3.5 Ma ago, dental patterns had already achieved much of their present form and some type of upright walking (bipedalism) had apparently already emerged. Fossils in the known human lineage do exhibit a progressive expansion of the braincase (Fig. 2.25), so there is direct evidence of one of the most important features of human evolution. It must be noted, however, that *Australopithecus* species (particularly the gracile forms) were generally smaller in body weight than modern *Homo sapiens* and great apes, and when the effects of body size are properly taken into account (Chapter 8) it emerges that the initial stages of human brain size expansion also took place before the known fossil record of human evolution. What more eloquent expression of the limitations of the primate fossil record can one expect than the admission that the actual origins of all the crucial adaptations in human evolution are still shrouded in uncertainty?

Chapter three

Classification and phylogenetic reconstruction

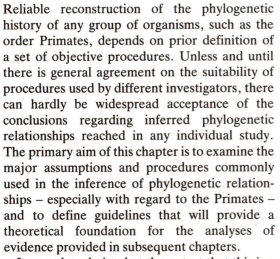

Reliable reconstruction of the phylogenetic history of any group of organisms, such as the order Primates, depends on prior definition of a set of objective procedures. Unless and until there is general agreement on the suitability of procedures used by different investigators, there can hardly be widespread acceptance of the conclusions regarding inferred phylogenetic relationships reached in any individual study. The primary aim of this chapter is to examine the major assumptions and procedures commonly used in the inference of phylogenetic relationships – especially with regard to the Primates – and to define guidelines that will provide a theoretical foundation for the analyses of evidence provided in subsequent chapters.

It must be admitted at the outset that this is a field in which the requirements of objective and realistic scientific analysis are extremely difficult to meet and have yet to be satisfied. Although increased objectivity has now been assured in the contributory areas from which evidence is drawn, involving the description, measurement, statistical evaluation and interpretation of a wide variety of biological parameters, there is a considerable element of residual subjectivity and/or uncertainty in the overall inference of

phylogenetic relationships. In fact, it is undoubtedly the case that no completely reliable conclusions concerning phylogenetic relationships can be made with the miscellany of procedures currently available. The best that can be achieved at present is optimization of the various stages involved in phylogenetic inference, with the aim of ensuring maximal objectivity throughout. A vital step in this direction is provided by clear recognition of the central theoretical issues.

In fact, before even attempting to define the major issues involved in **phylogenetic reconstruction**, it is necessary to recognize a fundamental distinction between the operations required for this exercise and those needed for **classification** (Martin, 1981a). These two areas are intimately linked; but, both historically and conceptually, there are also important differences between them. In the first place, it is perfectly possible for a biological classification to exist in the absence of any explicit evolutionary theory. The classificatory system proposed by Linnaeus (1758), which provided a valuable basis for modern biological classifications of species, was published about a century before the Darwin/Wallace hypothesis of evolution through natural selection (e.g. see Darwin, 1859).

Linnaeus, far from consciously presaging evolutionary theory, believed in the intrinsic fixity of species and actually classified the various fossils known to him as parts of a separate scheme for minerals. Secondly, although there are obvious connections between classification and phylogenetic reconstruction, there are significant theoretical aspects that divide them. It is true that Linnaeus, by ordering the biological world into an effective classificatory framework, produced a suggestive pattern that paved the way for subsequent recognition of the evolutionary principle, but he did not generate a phylogenetic scheme as such. It is only since Darwin's time that there has been a complex and continuing interplay between classification and phylogenetic reconstruction, which has resulted in a perhaps understandable confusion of the two fields. Of course, any modern biological classification should take into account prevailing interpretations of evolutionary relationships. Conversely, an outline classification of some kind is equally indispensable for phylogenetic reconstruction; but this is only one of the necessary elements, as will emerge below.

Nevertheless, it is an unfortunate fact that there is at present almost universal confusion between the two operations of classification and phylogenetic reconstruction. This is reflected by confusion over basic terminology, to the extent that there is no widely recognized term that refers unequivocally to the science of phylogenetic reconstruction. All of the well-known terms have been used to refer at least in part to the production of classifications rather than to phylogenetic trees, and there is also confusion the other way. For example, the term **taxonomy** should refer largely or exclusively to operations associated with classification. The word itself is directly linked to the term *taxon* (plural = *taxa*), which is used to refer to a grouping of organisms at any level in a hierarchical classification. Simpson (1961) has defined taxonomy as 'the theoretical study of classification, including its bases, principles, procedures and rules'. However, because most classifications are intended to reflect in some way prevailing ideas regarding phylogenetic relationships within the group of organisms concerned, an understanding of principles of phylogenetic reconstruction is ideally required for the successful pursuit of taxonomy. Perhaps for this reason, many authors have used the term 'taxonomy' in an extremely loose sense, sometimes virtually equating taxonomy with phylogenetic reconstruction. Even more confusion has been generated by the term **systematics**, which has been variously used to cover classification, phylogenetic reconstruction or a combination of both. Simpson (1961) defined systematics in a very broad sense as 'the scientific study of the kinds and diversity of organisms and of any and all relationships among them', while specifically noting that he did not intend 'relationships' to be understood in a narrow phylogenetic sense. Several other authors have used the term 'systematics' in an even more general sense to refer to the ordering of natural phenomena (Hennig, 1979; Eldredge and Cracraft, 1980). Given such broad definitions and the wide variation in practical usage, it is perhaps best to take the term 'systematics' as an inclusive category incorporating both phylogenetic reconstruction on the one hand and the theory and practice of classification (i.e. taxonomy) on the other.

It is also important to note that, somewhat surprisingly, the need for clearly defined theoretical principles has been more extensively discussed with respect to classification than in relation to phylogenetic reconstruction. Indeed, many texts bearing labels suggesting that they provide clear theoretical guidelines for inference of phylogenetic relationships ('phylogenetics'; 'phylogenetic systematics'; 'phylogenetic patterns') in the event turn out to be partially or even predominantly concerned with issues involved in classification (e.g. Hennig, 1979; Eldredge and Cracraft, 1980; Wiley, 1981). Provision of clear guidelines specifically for the reliable reconstruction of phylogenetic relationships, regardless of any concern with classification, is not usually given the attention that it deserves. Admittedly, there is still much to be done with respect to clarifying the principles of classification. As recently as 1959, Cain bemoaned the lack of theoretical guidelines in the area of taxonomy (classification):

> Much practice is intuitive instead of explicit and exact, and at least in this country most taxonomists learn their trade more by imitation than by professional training.

> Is it not extraordinary that young taxonomists are trained like performing monkeys, almost wholly by imitation, and that only in the rarest cases are they given any instruction in taxonomic theory?

These remarks could have been directed with equal or greater force at those concerned with the inference of phylogenetic relationships. Indeed, the situation is somewhat worse in the field of phylogenetic reconstruction, partly because the need for a distinct set of theoretical principles has not been so widely recognized and partly because of a widespread misconception that there is a necessary identity between taxonomic principles and principles of phylogenetic reconstruction. The important point is that, despite the existence of areas of overlap, both classification and phylogenetic reconstruction should be based on separately formulated, logically compelling principles if there is to be any real progress beyond the present confused and confusing state of affairs.

THE DISTINCTIVENESS OF CLASSIFICATION

Although classification and phylogenetic reconstruction are obviously interdependent and generally draw on similar lines of evidence, it is essential to recognize that their central concerns are distinct. Taking any phylogenetic tree illustrating an inferred pattern of relationships within a group of organisms (such as the Primates), there are two logically separate questions that may be posed:

1. How can one be sure that the phylogenetic tree accurately reflects the pattern of evolutionary relationships among the organisms concerned?
2. Given the evolutionary relationships

indicated by the tree, what form of classification would best serve the needs of those who will use it?

Question 1 refers to phylogenetic reconstruction, while question 2 refers to classification. The fact that the logical separation between these two questions was not clearly respected in much of the earlier literature is explicable largely on the grounds that the underlying theoretical principles were not yet clearly established. However, in the recent literature there is one school of thought that closely associates classification with phylogenetic reconstruction as a matter of policy and thus does not recognize the need for clear separation of the two issues. A central tenet of this **cladistic school** of classification, which has been heavily influenced by the writings of Hennig (1950, 1965, 1979), is that any classification should directly reflect the inferred pattern of phylogenetic relationships among members of the group of organisms concerned (hence the alternative label 'phylogenetic systematics' for this school of thought). That is to say, groupings at various levels in the classification are determined by inferred groupings within the phylogenetic tree (see also Eldredge and Cracraft, 1980; Wiley, 1981). The cladistic school, in other words, requires that all taxa in a classification should be **monophyletic** in the strict sense defined by Hennig: 'A monophyletic group is a group of species descended from a single ("stem") species and which includes all species descended from this stem species.'

This contrasts with the approach favoured by the more traditional school of **evolutionary systematics**, of which Simpson (1931a, 1945, 1961) and Mayr (1942, 1969, 1974) are two leading proponents. This school holds that a classification should be compatible with the consensus of opinion regarding phylogenetic relationships among the organisms involved, but should not be based exclusively on the inferred pattern of branching points within the phylogenetic tree. Such a classification may still contain some taxa that are strictly monophyletic, but not all taxa will be so. It must also be noted that there is a third major school of taxonomy, based directly on the assessment of similarities between species, which does not recognize the need to link classification to inferred phylogenetic relationships at all. This is the school of **numerical taxonomy** (Sneath and Sokal, 1973). As with evolutionary systematics, numerical taxonomy does not require that taxa should be monophyletic.

The object of any classification of organisms is, of course, the construction of a hierarchical scheme of categories of increasing rank that constitutes, in effect, a multi-level 'filing system' for the names of species. The particular structure of any classificatory scheme depends not only on the categories represented in the filing system but also on the strategy adopted for grouping organisms into those categories. In fact, virtually all modern schools of biological classification use essentially the same set of basic taxonomic categories and they differ predominantly with respect to the philosophy according to which organisms are allocated to these categories. It is therefore convenient to deal first with the typical hierarchical categories of a biological classification and thereafter with different philosophies of taxonomic allocation.

Classificatory categories

The fundamental unit of modern biological classifications is universally accepted to be the species (but see later for divisions below the species level) and the construction of taxonomic hierarchies begins with the Linnean **binomial system** of nomenclature. In this, each species is given two names with Latin or Greek roots reflecting the first two levels of the classification. For example, the human species bears the binomial title *Homo sapiens*: the first (generic) name denotes the genus and is always written with a capital letter, while the second (specific) name indicates the species and is always written with a small letter. There is also a modern convention that generic and specific names should be printed in contrasting type to distinguish them from other taxonomic category names. The combination of a particular generic name and a particular specific name is unique for any given species, although a genus may contain more than one species and a given specific name may be used for species belonging to different genera. The genus *Homo* contains only one living species, *Homo sapiens*, but many primate genera include two or more living species. For instance, there are two species of chimpanzee – the pygmy chimpanzee (*Pan paniscus*) and the common chimpanzee (*Pan troglodytes*); the guenon genus *Cercopithecus* is an extreme case among primates and contains over 20 species (see Napier and Napier, 1967). Among living primates, it is relatively unusual for the same specific name to be used for members of different genera, but there are some exceptions; for example, the black spider monkey of the New World bears the name *Ateles paniscus* and thus possesses the same specific name as the pygmy chimpanzee, *Pan paniscus*.

Thus far, the discussion has related only to living species. Linnaeus, of course, dealt essentially with living species as he did not recognize the prior existence of other species in an evolutionary sense. Following acceptance of the theory of evolution by natural selection, it has become necessary to include all fossil species in biological classifications. In principle, this can be achieved without radical modification of the procedures involved in classification. It is possible to give binomial labels to extinct species from the fossil record and to incorporate them in a **taxonomic hierarchy**. In practice, however, special problems arise with the incorporation of fossil species into a taxonomic hierarchy, as will be seen from the following discussion. It must be remembered that the Linnean system of nomenclature was designed for what was seen to be a static set of living species and that it is now used to cope with a combination of living and fossil species explicitly linked within a dynamic framework of organic evolution.

Above the species level, single names are used for individual taxonomic categories and there are a few relatively simple rules governing the construction of some of these names. Simpson (1961) lists a total of 21 taxonomic categories that may be used in a classification, but in practice fewer categories are usually specified for the sake of simplicity. In his classification of mammals, for example, Simpson (1945) used the following hierarchical sequence of taxonomic categories (passing from the largest category down to the smallest):

Class
 Subclass
  Order
   Suborder
    Infraorder
     Superfamily
      Family
       Subfamily
        Genus
         Species

Although there are no strict rules governing the composition of names for the highest taxonomic categories, the following standard endings have been adopted for mammals for the three categories immediately above the genus level:

Superfamily: – oidea
Family: – idae
Subfamily: – inae

In addition, it should be noted that an initial capital letter is used for all formal names for the higher categories from the genus upwards.

There is, of course, much more to the formal procedures involved in classification, but that is not the concern of this book. For further information, the reader is referred to the comprehensive discussions of taxonomic procedure provided by Cain (1954, 1959), Mayr (1969) and Simpson (1945, 1961). Suffice it to say here that specific rules governing the formal procedures of classification have been laid down in the *International Code of Zoological Nomenclature* (Ride, Sabrosky *et al.*, 1985). One of the primary aims of these rules is to guarantee stability of nomenclature wherever this is feasible, so great stress is laid on the retention of particular names once they have become established. Accordingly, the principle of priority is respected for specific and other taxonomic names, except where use of a name that does not have priority has become so prevalent that change would be counter-

productive. An International Commission on Zoological Nomenclature exists to adjudicate in cases where there is doubt about the correct course to be followed.

Classificatory strategies

Having dealt briefly with the formal procedures involved in classification, it is now possible to return to the question of the alternative strategies that may be followed in allocating species to higher taxonomic categories. As has been indicated above, there is a fundamental difference in approach between the cladistic and evolutionary schools of classification. This can be illustrated with a hypothetical example (Fig. 3.1), which indicates how a cladistic classification would relate to a phylogenetic tree. At first sight, there would seem to be a considerable attraction in this strict cladistic approach in that allocation of taxa (e.g. species) to a hierarchical sequence of taxonomic categories is dictated by the sequence of branching relationships in the inferred phylogenetic tree. Taking the simplified diagram

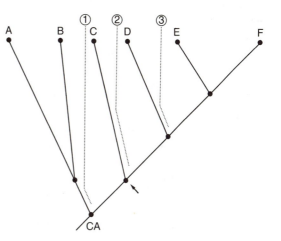

Figure 3.1. Hypothetical diagram ('phylogenetic tree') depicting the inferred phylogenetic relationships linking six modern taxa (A, B, C, D, E and F), descended from a common ancestral stock (CA). Numbered dotted lines (1, 2 and 3) indicate three successive branching points (nodes), as required by a cladistic classification. (See text and Fig. 3.2 for explanation of the arrow.)

in Fig. 3.1, the first major taxonomic subdivision of the overall group of modern taxa would be conducted at the first branching point (node), marked CA, to yield two high-level, strictly monophyletic taxonomic groups: (A + B) versus (C + D + E + F). The latter group would be further subdivided into monophyletic taxonomic categories of successively lower rank, with a second division between (C) and (D + E + F) followed by a third division between (D) and (E + F). On the face of it, then, there would seem to be a clear prescription for the form of a cladistic classification accompanying such a tree. Further, that prescription should, in principle, remain consistent regardless of whether the taxa concerned are extant forms or fossil forms.

By contrast, with the traditional approach taken in evolutionary systematics, it is potentially possible to produce a variety of different formal classifications for any inferred phylogenetic tree – particularly when fossils are included – as there is no simple rule governing the relationship between the classification and the inferred pattern of branches in the tree. That inferred branching pattern constitutes only one of the criteria used in an evolutionary classification and considerably more scope is allowed for the exercise of individual judgement by the classifier. In fact, the main criterion that would be used for an evolutionary classification based on a phylogenetic tree of the kind shown in Fig. 3.1 would be the overall **degree of divergence** between taxa. For strictly morphological characters, this becomes the concept of **morphological divergence** (Simpson, 1961; Mayr, 1974). Although the amount of divergence between any two taxa is obviously related in some way to the recency of their common ancestry (i.e. the relative location in the tree of their last shared node), it is also affected by the **rate of evolution** of the character(s) concerned. In other words, degree of divergence is a mixed criterion that takes into account not only the branching relationships within a phylogenetic tree but also rates of evolution along individual branches. It should be noted at once that many of the differences between cladistic and evolutionary classifications therefore arise because rates of

evolution differ between lineages. If this were not the case, degree of divergence would be determined essentially by time of divergence and (other things being equal) a classification based on character divergence would tend to resemble quite closely a classification based only on branching relationships within the phylogenetic tree. But marked variation in the rate of evolution of characters, at least above the molecular level (see Chapter 11), is regarded as an established fact by virtually all biologists. Hence, as has been noted by Mayr (1974), a classical classification takes into account at least one identifiable biological factor – evolutionary rate – that is known to be important but is not considered in a cladistic classification.

The branching diagram given in Fig. 3.1 does not take into account variation in rates of change, but it is possible to imagine the implications. One can, for example, consider a situation in which taxa A, B and C have evolved slowly, whereas taxa D, E and F have evolved rapidly in comparison (perhaps because of some particular change in the lineage leading from the arrowed node to the common ancestor of D, E and F). The net effect of this will be that, in terms of overall divergence, C will have remained more similar to A and B, even though it shared a common ancestry (arrowed) with (D + E + F). A classification based on degree of divergence between taxa would therefore group C with A and B, giving a primary division of the taxa into (A + B + C) versus (D + E + F). Such an arrangement has the advantage that it groups together taxa that are more similar in overall appearance, but it has the disadvantage that it makes imposition of standardized criteria difficult. For instance, morphological characters are not easy to quantify relative to one another and it is accordingly difficult to assess overall divergence. In addition, any assessment of divergence will naturally depend upon the range of characters considered by the classifier.

It is now necessary to consider in more detail the form and implications of a phylogenetic tree of the kind shown in Fig. 3.1. At the simplest level, such a diagram may just indicate the presence of inferred branching points (**nodes**), in which case many authors would refer to the diagram as a **cladogram**. In line with this, the special emphasis placed on inferred branching relationships accounts for the adjective 'cladistic' applied to the school of classification that links taxonomic divisions directly to monophyletic groups. Indeed, some proponents of the cladistic school would argue that a sharp distinction should be drawn between simple branching cladograms and **phylogenetic trees**, as the latter may incorporate additional information of various kinds (e.g. Eldredge and Cracraft, 1980). However, other adherents to the cladistic school of classification (e.g. Wiley, 1981) and many evolutionary systematists do not recognize the need for such a distinction and the view is taken here that the single term 'phylogenetic tree' suffices for all attempted reconstructions of phylogenetic relationships. In addition to indicating branching relationships, a diagram such as that shown in Fig. 3.1 might also indicate some assessment of the relative timing of nodes and the incorporation of accurately dated fossils might permit the assignment of approximate dates to some of the nodes. In some cases, an attempt may be made to go still further to indicate degrees of divergence among the modern descendants (A–F in Fig. 3.1) by horizontal spacing. But this very possibility underscores the subjective element involved in overall assessments of divergence. It is quite possible, for example, that C would prove to be closer to B in terms of a slow rate of evolution of one set of characters (e.g. skull morphology), but closer to A in terms of slow evolution of another set (e.g. reproductive morphology). Thus, a genuinely comprehensive attempt to combine branching patterns and degrees of divergence into a single phylogenetic tree would inevitably involve a multidimensional model of some kind. Two-dimensional phylogenetic diagrams that are intended to convey assessments of degree of divergence can only be relatively crude, condensed representations, so classifications based on them may perhaps be criticized for this reason.

An entirely separate issue is raised by the concept of **compatibility** between a phylogenetic

tree and a classification of the group of organisms concerned. On first examination, it might seem that an evolutionary classification of (A + B + C) versus (D + E + F) would be compatible with the phylogenetic tree shown in Fig. 3.1, whereas a classification of (A + B + D + E + F) versus (C) would be incompatible. However, it is important to note that it is perfectly possible to alter the form of the tree by rotating branches without changing the fundamental pattern of relationships (i.e. the relative positions of the nodes). Taking the tree shown in Fig. 3.1, rotation can be carried out at the node indicated by the arrow, yielding the alternative tree shown in Fig. 3.2. No real change has occurred in the branching relationships and the strategy followed by a cladistic classification would be precisely the same. With a classical classification, on the other hand, the alternative form of the tree shown in Fig. 3.2 might be used to reflect the fact that C is the most divergent of the six taxa concerned (i.e. the lineage from the arrow to C has evolved at a markedly faster rate). In this case, a classification into (A + B + D + E + F) versus (C) would be

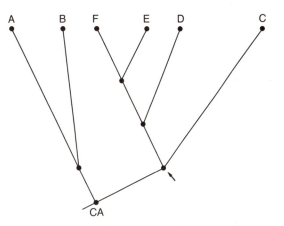

Figure 3.2. By rotating part of the phylogenetic tree shown in Fig. 3.1, at the node indicated by the arrow, it is possible to produce a different pattern that conserves the original branching sequence but may permit a different representation of divergence between taxa. In this case, taxon C might be a very divergent taxon because of more rapid evolution along the lineage leading from the arrow to C.

justifiable according to the logic of an evolutionary classification. In fact, it is possible to produce 32 different versions of the phylogenetic tree shown in Fig. 3.1 by rotating branches at the nodes. The general formula for the number of different possibilities permitted by a tree with n two-way branching points is 2^n, so a more complex tree with 20 branching points would permit more than a million alternative representations on the basis of branch rotation alone. On the one hand, this provides a great deal of scope for incorporating information about degree of divergence in a phylogenetic tree; on the other, it introduces greater complexity and potential subjectivity into the process of evolutionary classification. It is necessary to select – using criteria that are difficult to define in a rigorously objective fashion – just one of the many possible diagrams that best expresses in only two dimensions the total information available.

There is an additional complication in the choice of a particular representation of a set of inferred phylogenetic relationships in that there is an implicit convention in the literature that the 'most primitive' modern descendant from a given ancestral stock should be shown on the left of the tree (i.e. taxon A in Figs 3.1 and 3.2), with the sequence of modern taxa from left to right indicating increasing degrees of departure from the common ancestral condition. Accordingly, many phylogenetic trees depicting the evolution of primates show modern humans at the extreme right (equivalent to taxon F in Fig. 3.1) and the 'most primitive' living primates (usually lemurs, equivalent to taxon A in Fig. 3.1) at the extreme left. Even with a cladistic approach, it remains open to the classifier to select one of the many possible representations of a particular set of branching relationships to accompany a published classification, so a subjective element may intervene, at least in the eyes of a reader accustomed to seeing a primitive–advanced spectrum from left to right along the top of the tree. Although the logic of cladistic classification implies that *any* representation of the inferred pattern of branching relationships could be selected without in any way affecting the

outcome, in practice such random selection does not usually occur when a particular form of the inferred phylogenetic tree is published.

The traditional evolutionary approach to classification is therefore very flexible, in that a phylogenetic tree may be depicted in a large number of different ways to reflect assessments of degree of divergence and to permit compatibility between the classification and the inferred phylogenetic relationships. Nevertheless, even with such flexibility, there eventually comes a point when an evolutionary classification will also become incompatible with a given inferred pattern of branching points. For instance, whatever tree one might select from the 32 possibilities permitted by rotating branches in the pattern underlying Figs 3.1 and 3.2, there would always be incompatibility with a classification based on an initial division into (A + B + F) versus (C + D + E). Similarly, it is obvious from Figs 3.1 and 3.2 that D and E should not be excluded from any taxon that includes C and F. But the fact remains that there are many possible classifications of the classical type that might be compatible with a given phylogenetic tree (depending on the assessment of evidence made by the classifier), whereas in principle only one cladistic classification is permissible. Proponents of the cladistic school claim this latter feature as a major advantage, especially since it may be argued that concepts such as 'morphological divergence' typically involve some degree of subjective assessment. Little wonder, then, that there are now many adherents to the cladistic school, convinced that this approach offers greater objectivity and rigour in the quest for improved classifications. The cladistic prescription for derivation of a classification from a given inferred phylogenetic tree seems relatively straightforward, so why do many authors continue to favour the traditional evolutionary approach to classification?

In fact, there are a number of serious **practical problems** involved in cladistic classification and they greatly outweigh the apparent advantages of increased rigour. The first, and most important, problem with cladistic classifications concerns the reliability of phylogenetic reconstruction, which has frequently been taken for granted. A great deal of the continuing intense debate about the relationship between classification and phylogenetic relationships (e.g. Mayr, 1974 versus McKenna, 1975) has been conducted with the implicit assumption that techniques for phylogenetic reconstruction do not themselves constitute a major problem. Indeed, discussion of classification has tended to obscure, rather than clarify, the need for a really penetrating examination of the objectivity with which evolutionary relationships can be reconstructed in the first place. There is little point in having a seemingly objective prescription for translating an inferred phylogenetic tree into a classification if the reliability of phylogenetic inference is itself open to serious question. It is often forgotten that phylogenetic trees represent hypotheses, not established fact, and that it is in the very nature of scientific investigation that such hypotheses should be continually subjected to further testing and refinement (Hennig, 1979; Wiley, 1981). Hennig has rightly referred to the 'endless task of checking, correcting and rechecking' that faces those concerned with phylogenetic inference. It should always be remembered that any representation of evolutionary relationships within a given group of organisms is an **inferred** phylogenetic tree. At present, the element of uncertainty involved in phylogenetic reconstruction is so pronounced that it rarely (if ever) happens that two authors investigating the same group of organisms will reach the same conclusions regarding their phylogenetic relationships. If classifications are made strictly dependent on inferred branching patterns, as is specifically prescribed by the cladistic school, they must of necessity exhibit the same wide range of variation from author to author. What, then, is the net result of replacing evolutionary classifications by cladistic classifications? Variability due to subjective assessments of degree of divergence is replaced by a potentially far greater degree of variability caused by idiosyncracies in the hypothetical reconstruction of phylogenetic relationships.

In discussions of alternative strategies of classification, it is often forgotten that classi-fications must be practical as well as reflect inferred patterns of phylogenetic relationship (Warburton, 1967). Biological collections in natural history museums and elsewhere must be arranged and catalogued, for reference purposes, according to some comparatively simple and widely understandable system. Such collections must be accessible to a wide range of people whose concerns will in most cases not be confined to the narrow issue of inferred phylogenetic relationships and whose main interests will best be served by a stable, widely used classification. In short, a classification provides a key for information retrieval, so it cannot conveniently be subjected to rapidly repeated revision as is required by the cladistic prescription of maintaining direct correspondence with the hypothetical reconstruction of phylogenetic relationships. More than this, a biological classification provides a set of linguistic labels with which we can specify and discuss living organisms, both individually and collectively. Simpson (1961, 1963) specifically identified 'linguistic convenience' as a factor to be considered in formulating classifications.

All languages that are in regular use tend to undergo gradual change; but a fundamental characteristic of languages is their relative stability, without which effective communication would become impossible. So it is with classi-fications as well: some change is necessary to keep pace with changing circumstances, but too much change results in barriers to efficient communication. Eldredge and Cracraft (1980) have pertinently observed: 'Words, or perhaps more strictly, the meanings of words, are themselves classifications.' Discussion of objective reconstruction of the phylogenetic relationships within a group of organisms (e.g. the Primates) is rendered exceedingly difficult when the constitution of a subgroup labelled with a particular name varies from author to author. It can be argued, therefore, that the require-ment for relative stability in classifications is paramount and that it matters little whether a given classification fails to reflect likely phylogenetic relationships, nor whether some degree of subjectivity has intervened in the formulation of the classification. Despite his commitment to cladistic classification, Wiley (1981) has acknowledged that: 'It is usually more important to provide easy access to a collection than to attempt to keep step with changing ideas.' The important thing is that a readily understandable set of linguistic terms should be available for discussion (among other things) of the discrete issue of establishing objective rules for the reconstruction of phylogenetic relationships. To this end, far more is to be achieved by accepting a standard classical classification in a spirit of compromise than by seeking after an ideal, objective prescription for classification.

A particularly striking example of linguistic confusion arising from reclassification according to cladistic principles is provided by the case of the great apes and humans (Fig. 3.3). In classical approaches to the classification of the Primates (e.g. Simpson, 1945, 1963; Simons, 1972), it has been standard practice to allocate all of the great apes (orang-utan, chimpanzees and gorilla) to the family Pongidae and to place humans in a separate family, Hominidae. At one time, many investigators believed that the great apes constituted a strict monophyletic unit, so this classification would also have been consistent

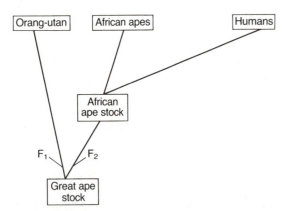

Figure 3.3. Outline tree summarizing the modern consensus view regarding the phylogenetic relationships between the great apes and humans. F_1 and F_2 represent two early fossil forms, one related to the lineage leading to the orang-utan and the other related to the lineage leading to the African apes and humans. (Figure reprinted from Martin (1981) by kind permission of the Institute of Biology, London.)

with cladistic principles. However, as can be seen from Fig. 3.3, the prevailing modern interpretation of available evidence is that the orang-utan is less closely related to humans than are the African great apes (chimpanzees and gorilla). A cladistic classification based upon the phylogenetic reconstruction indicated in Fig. 3.3 would therefore have to incorporate a taxonomic separation between the two categories: (orang-utan) and (African apes + humans). Although the principle to be followed here is in itself quite clear, there are alternative practical solutions. Hence, even with a cladistic classification there is no unique prescription for modifying a pre-existing taxonomic scheme to take account of new interpretations of phylogenetic relationships. One proposed solution has been to allocate the orang-utan to the family Pongidae and unite the African apes and humans in the family Hominidae (Goodman, 1975; Dene, Goodman and Prychodko, 1976a; Seuánez. 1979; Eldredge and Cracraft, 1980). Another solution has been to put all of the great apes and humans into the subfamily Homininae and to divide this into two 'supertribes', one containing the orang-utan and the other containing the African great apes and humans (Schwartz, Tattersall and Eldredge, 1978). Thus, although cladistic principles may dictate the divisions expressed in a classification, they do not dictate the taxonomic rank at which those divisions are made, nor do they dictate the labels used for particular taxonomic groups. In both cases cited, however, the meaning of the family name Hominidae, and hence of the derived common name 'hominid', has been radically altered.

This is not even the end of the matter, as some authors have gone on to claim that there is also a substantial body of evidence that the chimpanzees may be more closely related to humans than is the gorilla (a potential refinement conveniently ignored in Fig. 3.3). For any cladistic classification to reflect this interpretation as well, further modifications would be required. Such modifications would inevitably require either alteration of the meaning of established higher category names or additional proliferation of taxonomic categories.

With a traditional evolutionary approach to classification, by contrast, no adjustment is necessary despite such shifts in prevailing views of the evolutionary relationships between the great apes and humans. The human lineage has clearly diverged relatively rapidly, such that great apes still closely resemble one another in various morphological characters, whereas humans are distinctively different from all of them. On grounds of overall divergence, therefore, it is justifiable to place the human species in the separate family Hominidae, regardless of the actual branching relationships involved with respect to the great apes. Thus, the evolutionary classification retains its validity in the face of changing views of phylogenetic relationships, while the cladistic approach dictates the introduction of a series of new classifications and brings about repeated changes in the meaning of previously widely understood terms.

It should also be noted that, because great apes resemble one another quite closely by virtue of their relatively slow divergence from the initial common ancestral stock linking them to humans ('great ape stock' in Fig. 3.3), there is a certain value in having a formal taxonomic term to refer to them collectively. Even though it is now generally accepted that the African great apes may be more closely related to humans than is the orang-utan, many comparisons that are conducted still involve contrasts between all great apes on the one hand and humans on the other. With an evolutionary classification of the Primates, it is possible to conduct these comparisons in terms of Pongidae versus Hominidae. In other words, regardless of any traditional usage of these terms, it is a moot point whether it is more useful for classificatory divisions to reflect inferred phylogenetic relationships or to reflect major shifts in terms of overall evolutionary change.

Whatever the theoretical justification for cladistic classifications may be, the introduction of such classifications of the great apes and humans has in practice played havoc with communication. In the existing literature, there has been widespread use of the term 'hominid' to refer to modern humans and to fossils directly related to the human lineage. Until recently, the meaning of the term 'hominid' was therefore unequivocal, apart from the fact that there was naturally some disagreement about which of the earlier fossil forms could be confidently regarded as related to the human lineage. Now, however, there is widespread confusion. In some cases, the term 'hominid' is used in its former, restricted sense; in others, it is used to refer to the African apes and humans together; in yet others it is applied to even larger groupings. Rampant classificatory upheaval has destroyed the previously universal comprehensibility of statements such as: 'A parabolic dental arcade is a characteristic of all hominids.' We must now cope with such puzzling observations as that of Szalay and Delson (1979) that gibbons 'must be considered an early offshoot from the hominid stock'. Indeed, Seuánez (1979), while accepting Goodman's (1975) inclusion of the African great apes in the Hominidae, felt driven to add a footnote defining a 'hominid' as a member of the human lineage!

Given that stability of taxonomic terms is so important for effective communication, considerable care must be shown when any modifications to an existing classificatory scheme are proposed. In fact, as has been noted above, there is at least some room for the exercise of personal judgement here, whatever classificatory strategy is used. Returning to the example provided in Fig. 3.3, for instance, one may accept the cladistic strategy of basing a classification directly on the inferred branching relationships without necessarily accepting the particular choice of taxonomic labels discussed above. Use of the family name Hominidae for groups containing some or all of the apes in addition to humans (which has been directly responsible for much of the resulting confusion) represents a choice made by the classifier, rather than a step dictated by cladistic philosophy as such. In the face of the linguistic confusion that is generated by alteration of the meaning of the family name 'Hominidae', it is perfectly possible to find a solution that is compatible with cladistic principles, but which respects established usage of that particular name. For example, Boaz and

Cronin (1985) have suggested that a new taxonomic category – the hyperfamily – should be used to reflect the specific phylogenetic relationship between the African great apes and humans. Although this represents a definite step in the direction of proliferation of taxonomic categories, thus illustrating another drawback of the cladistic approach, it does show that alternative solutions are possible.

The fact that some authors have nonetheless chosen to change the meaning of the family name Hominidae reflects insufficient concern for stability of nomenclature. This is, for instance, the case with the classification proposed by Szalay and Delson (1979). Although these authors adopted a mixed strategy, allowing some flexibility in the degree to which their classification was made to fit the inferred pattern of phylogenetic branching points, they nevertheless chose to use certain taxonomic terms in a radically altered fashion. In particular, they included in the family Hominidae not only the African apes and humans but also the orang-utan and even the gibbons. Thus, according to their classification, the term 'hominid' can refer to any of the apes as well as to humans. Inadequate concern for taxonomic stability is therefore not restricted to those who favour strict cladistic classification; interference with essentially serviceable classificatory schemes was rife even before the advent of cladistic philosophy. The problem with the cladistic philosophy, however, is that it dictates such modification, whereas with the traditional evolutionary approach any modification of classifications is very largely the product of personal initiative. Clearly, classifications must change over time to maintain broad compatibility with accumulated knowledge, but it is vital to keep such changes to the absolute minimum in the interests of clear communication between biologists. This is a requirement that should be respected by all classifiers, regardless of their philosophical leanings.

Apart from the major problem of mandatory taxonomic instability, the cladistic approach to classification proves on closer examination to have several other significant flaws. One of these

is potential proliferation of taxonomic levels in cases where there is a rich branching sequence in the inferred phylogenetic tree. A strict application of cladistic logic would require that a new level in the taxonomic hierarchy should be introduced for every successive dichotomy in the tree. Clearly, a richly branching tree would require an unacceptably high number of taxonomic categories, so it is inevitably necessary to relax the criterion that classification should directly match the tree. In other words, it is virtually impossible to apply the strict cladistic approach in practice and some form of 'fudging' is required. For instance, the classifier can ignore certain branching points and base the classification on the remaining nodes of the tree. Alternatively, isolated fossils or groups of fossils may be placed in a special taxonomic category known as a 'plesion' that may occupy any rank (e.g. see Schwartz, Tattersall and Eldredge, 1978; Wiley, 1981). In fact, the concept of the 'plesion' may even be extended to include relatively primitive extant groups as well as fossils (Eldredge and Cracraft, 1980).

Another approach is to make use of features of a classification other than specified taxonomic categories convey information. Hennig (1979), for example, proposed a scheme involving long numerical prefixes and Wiley (1981) mentions a proposal to use indentation of names in a printed classification to indicate levels of subordination. Such schemes are, however, very cumbersome in practice and undetected printing errors could create havoc. Eldredge and Cracraft (1980) and Wiley (1981) have opted instead for such special devices as the **sequencing convention** discussed by Nelson (1974), which involves listing of taxa to indicate the sequence of branching relationships. All such conventions, however, have the combined effect of diluting the original cladistic prescription and of increasing the complexity of cladistic classification. Many of the proposed measures discussed above require some assessment of the relative importance of nodes in a phylogenetic tree, such that a cladistic classification is rendered less obviously objective than is often claimed. Any cladistic classification must involve a trade-off between proliferation of taxonomic levels and 'fudging' and hence any attempt to maximize the application of cladistic logic will inevitably generate a comparatively unwieldy classification.

There is also a special problem involved with the incorporation of fossils into a cladistic classification. It is one of the strengths of the cladistic approach that fossils are, in principle, treated in the same way as living species; no fossil is treated at the outset as a direct ancestor of any living representative. (But this has more advantages for phylogenetic reconstruction than for classification, as will be shown later.) A problem arises, however, with very early fossil derivatives from recently separated lineages, as is illustrated in Fig. 3.3. In this phylogenetic tree, F_1 represents an early fossil relative of the orang-utan, and F_2 represents a contemporary fossil relative of the lineage leading to the African great apes and humans. The logic of a cladistic classification of the kind indicated above prescribes that F_1 should be classified with the orang-utan (e.g. in the family Pongidae), and F_2 should be classified with the African great apes and humans (e.g. in the family Hominidae). Yet it is inevitable that these early fossil forms will be extremely similar to one another, as both have only recently diverged from the common great ape stock (Fig. 3.3). Indeed, if the fossil specimens concerned are fragmentary (as is most often the case) it is unlikely that the characters preserved would provide sufficient evidence to indicate clearly whether their affinities lie more with the orang-utan or more with the African great apes and humans. It is, for instance, quite possible that a branching pattern such as that shown in Fig. 3.3 might be based on characters not preserved in fragmentary early fossil relatives (e.g. on chromosomal and molecular evidence). Once again, the traditional evolutionary approach to classification is better placed to deal with such a problem. In this case, both early fossil forms would be placed in the family Pongidae, because this family contains all species that show relatively limited divergence from the great ape stock. Only fossils that exhibit some clear evidence of relationship to the human lineage need to be treated differently by inclusion in the family Hominidae. There is also the point that fragmentary early fossils are the most subject to changing interpretations, particularly when additional fossil evidence leads to better documentation. With a cladistic classification, changing interpretations of early fossil forms present considerable problems (unless some kind of 'fudging' operation is conducted); with an evolutionary classification, the main problem would be deciding which fossils really justify inclusion within the family Hominidae, rather than relegation to the catch-all category Pongidae.

A classification of Primates

Overall, despite the apparent attractions of the cladistic approach in terms of more consistent procedures for allocating species to higher taxonomic categories, there are numerous practical reasons for preferring the traditional evolutionary approach. It is an unavoidable, if (to some) unpalatable, fact of life that the search for a 'perfect', objective strategy of classification is a wild-goose chase. Hence, it is better to adopt a compromise solution that at least has the merit of permitting the vital property of relative stability, thus ensuring that a readily understandable set of terms will be available for discussing (among other things) the reconstruction of phylogenetic relationships. With this aim in mind, this text will follow, with only the most minor modifications, an established classification in the traditional mould (Table 3.1). The choice of this classification (following Simons, 1972) has been influenced by several considerations. In the first place, it was designed by an authority on living and fossil primates with extensive experience of the complexities of primate taxonomy. Secondly, Simons himself served the interests of taxonomic stability by departing only in certain details from a previous classificatory scheme for the order Primates that had been set out by Simpson (1945) within the framework of a general classification of the mammals. Simpson's scheme has been widely followed and the formal taxonomic names used are therefore well established in the literature.

Table 3.1 Formal classification of the order Primates (based on Simons, 1972*); common names for taxonomic groups used in Chapters 1 and 2 are given in parentheses for ease of reference

Class: **Mammalia** (mammals)
  Subclass: **Eutheria** (placental mammals)
    Order: **Primates** (primates)
      Suborder: **Prosimii** (prosimians)
        Infraorder: **Plesiadapiformes**[†] ('archaic primates')
          Family: **Plesiadapidae**[†] (plesiadapids)
                Paromomyidae[†] (paromomyids)
                Carpolestidae[†] (carpolestids)
                Picrodontidae[†] (picrodontids)
                [?**Microsyopidae**[†] (microsyopids)]
                [?**Saxonellidae**[†] (saxonellids)]
        Infraorder: **Lemuriformes**[†] ('lemuroids')
          Superfamily: **Adapoidea**[†]
            Family: **Adapidae**[†] (early Tertiary 'lemuroids')
              Subfamily: **Adapinae**[†] (European 'lemuroids')
                  Notharctinae[†] (North American 'lemuroids')
          Superfamily: **Lemuroidea** (Malagasy lemurs)
            Family: **Cheirogaleidae** (mouse and dwarf lemurs)
                  Lemuridae (true lemurs and sportive lemurs)
                  Indriidae (indri group)
                  Daubentoniidae (aye-aye)
                  Megaladapidae (subfossil giant lemurs)
        Infraorder: **Lorisiformes** ('lorisoids')
          Superfamily: **Lorisoidea**
            Family: **Lorisidae** (loris group)
               Subfamily: **Lorisinae** (loris subgroup)
                    Galaginae (bushbabies)
        Infraorder: **Tarsiiformes** ('tarsioids')
          Superfamily: **Tarsioidea**
            Family: **Omomyidae**[†] (early Tertiary 'tarsioids')
              Subfamily: **Omomyinae**[†] (omomyines)
                  Microchoerinae[†] (microchoerines)
                  Anaptomorphinae[†] (anaptomorphines)
            Family: **Tarsiidae** (tarsiers)
      Suborder: **Anthropoidea** (simians)
        Infraorder: **Platyrrhini** (New World simians)
          Superfamily: **Ceboidea** (New World monkeys)
            Family: **Cebidae** (true New World monkeys)
              Subfamily: **Cebinae** (capuchins and squirrel monkeys)
                  Aotinae (owl monkeys and titi monkeys)
                  Atelinae (spider and woolly monkeys)
                  Alouattinae (howler monkeys)
                  Pithecinae (sakis)
                  Cebupithecinae[†] (cebupithecines)
                  Callimiconinae (Goeldi's monkey)
            Family: **Callitrichidae**[‡] (marmosets and tamarins)
        Infraorder: **Catarrhini** (Old World simians)
          Superfamily: **Cercopithecoidea**
            Family: **Cercopithecidae** (Old World monkeys)
              Subfamily: **Cercopithecinae** (guenon group)
                  Colobinae (leaf-monkeys)
                  Parapithecinae[†] (Oligocene monkeys)

Superfamily: **Hominoidea** (apes and humans)
Family: **Oreopithecidae**[†] (oreopithecids)
Hylobatidae (lesser apes)
Subfamily: **Pliopithecinae**[†] (fossil small apes)
Hylobatinae (gibbons)
Family: **Pongidae** (great apes)
Subfamily: **Dryopithecinae**[†] (fossil large apes)
Ponginae (modern great apes)
Family: **Hominidae** (hominids)
Subfamily: **Australopithecinae**[†] (ape-men)
Homininae (humans)

---

* There are some minor modifications from Simons (1972). The most important are as follows. He excludes the Microsyopidae and Saxonellidae from the 'archaic primates', but several other authors do not. The mouse and dwarf lemurs are here placed in the separate family Cheirogaleidae, rather than as a subfamily of Lemuridae. The Microchoerinae are placed in the Omomyidae rather than in the Tarsiidae. The Oreopithecidae are placed with the apes (hominoids), rather than with the Old World monkeys (cercopithecoids).

† Indicates taxa consisting exclusively of fossil forms.

‡ The family name for marmosets and tamarins is commonly misspelt as Callithricidae. The International Code of Zoological nomenclature prescribes that a genus name ending in -thrix yields a family name ending with -trichidae.

Finally, several specific features of the classification proposed by Simons (1972), which represent changes from Simpson's original classification, accord well with the conclusions reached in this book regarding phylogenetic relationships and therefore facilitate discussion. The exclusion of the tree-shrews from the Primates is a case in point, and it is also useful to have a formal taxonomic distinction between New World monkeys and all Old World simians (i.e. Platyrrhini versus Catarrhini). Such relatively minor modifications between a taxonomic scheme published in 1945 and one published in 1972 admirably reflect the need for compromise between the requirement for stability of terminology and the requirement to keep pace with those changes that render the original scheme incompatible with prevailing interpretations of phylogenetic relationships.

Having thus arrived at a decision regarding the choice of a classification for reference purposes, it is now possible to turn to a discussion of the concepts and methods involved in phylogenetic reconstruction. However, because there has been a general tendency to confuse classificatory concepts with concepts necessary for phylogenetic reconstruction, this discussion must begin with a concept that has a certain utility for evolutionary classification, but which has tended to obscure rather than clarify the attempt to reconstruct primate origins.

THE PHYLOGENETIC SCALE

One of the most widespread notions in the literature on animal evolution is that living organisms can be arranged on an ascending scale of complexity, ranging from the most primitive to the most advanced. This concept of the 'Great Chain of Being' (also known as the **Scala naturae** or the *échelle des êtres*) has deep philosophical roots, as has been demonstrated by Lovejoy (1960). Indeed, it seems likely that a tendency to classify things in a way that reflects assumptions about increased significance is a fundamental attribute of the human mind (see also Knight, 1981). As with classification, the concept of the *Scala naturae* is not necessarily linked to ideas of biological evolution. This concept was invoked long before the theory of evolution by natural selection was formulated and it has played an important part in many non-biological subjects. Indeed, it seems likely that the concept of the 'Great Chain of Being' may be intimately linked to the general human endeavour of classification. It is doubtless true that the task of classification is considerably facilitated by some simple principle – such as graded complexity or significance – that permits some initial organization of data. In fact, the concept of the **grade** (Huxley, 1958), which is an integral feature of the classical approach to classification as portrayed by Simpson (1961), is directly involved in the classification of the Primates set out by Simpson (1945) and subsequently modified by Simons (1972). The first subdivision of the order Primates into two suborders (Prosimii and Anthropoidea; Table 3.1) reflects the general view that the prosimian primates (lemurs, lorises and tarsiers) represent a lower grade of evolution than the simian primates (monkeys, apes and humans). The concept of a transition from a prosimian grade to a simian grade and its relationship to primate classi-fication are illustrated in Fig. 3.4. In a similar way, the Insectivora are generally held to repre-

sent the most primitive grade among placental mammals generally (see Chapters 4 and 5).

Figure 3.4 illustrates one of the advantages of the grade concept, as applied to primate classification. As has been discussed above in relation to the evolution of great apes and humans, the classification of early fossil forms always presents problems. The same point applies to early fossil derivatives from the ancestral primate stock. In Fig. 3.4, fossil *x* is an early relative of the lineage leading to lemurs and lorises, and fossil *y* is an early relative of the lineage leading to tarsiers and simians. With a grade approach to classification, it is possible to allocate both fossil forms to the suborder Prosimii. Although this circumvents the issue of the specific phylogenetic relationships of those

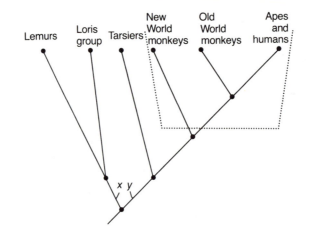

Figure 3.4. Outline phylogenetic tree for the order Primates illustrating likely phylogenetic relationships between the six major natural groups. The dotted line indicates the approximate demarcation of the suborder Anthropoidea (monkeys, apes and humans) from the suborder Prosimii (lemurs, lorises and tarsiers) in the traditional evolutionary classification followed here (see Table 3.1.). The modern simian primates are regarded as belonging to a higher 'grade' than the modern prosimians; that is, they are believed to have diverged to a greater extent from the ancestral primate stock.

   The positioning of the dotted line illustrates the point that it is conceivable, with this approach to classification, that the common ancestor of the Anthropoidea might be included in the Prosimii (see also Simpson, 1961). Numerous authors have, in fact, suggested that the New World monkeys and the Old World monkeys, apes and humans attained the 'simian grade' independently and that the common ancestor of the simians was still relatively primitive (i.e. still occupied the 'prosimian grade'). However, evidence to be presented in later chapters suggests that the common ancestor of the simians already had some distinct features that would have justified inclusion in the Anthropoidea. This is just as well, because exclusion of the common ancestor of the simians from the suborder Anthropoidea would render the latter 'polyphyletic' and hence raise additional problems with respect to the philosphy of evolutionary classification.

   The taxa *x* and *y* represent early fossil forms related to the two diverging lineages derived from the initial ancestral primate stock.

fossils, this is in fact an advantage. The grade-based classification provides for a relatively stable terminology that can be used for subsequent discussion of the possible phylogenetic relationship of the early fossil forms (always assuming that the fossils themselves provide sufficient information to make such a discussion worthwhile). This is particularly important in the case of relatively fragmentary fossils (see Chapter 2).

As a device associated with classification, the grade concept is arguably beneficial. It is generally true for any group of living animals that some species will be found to have evolved faster, overall, than others. Therefore, an initial division of living species into members of a relatively primitive grade and members of a relatively advanced grade often makes sense, in that it reflects differential degrees of divergence from the ancestral condition. In practical terms, such a grade distinction permits comparatively easy construction of a taxonomic key. Further, as has been shown, it is possible to allocate all early fossil forms to the more primitive grade, along with the most primitive surviving species. In this respect, vagueness is a distinct advantage of the grade concept as used in the classical type of classification, permitting compatibility with many hypothetical phylogenetic trees.

By contrast, when we turn to the question of **phylogenetic reconstruction**, it emerges that the imprecision of the grade concept is definitely counterproductive. In the search for a reliable account of phylogenetic history, it is necessary to construct explicit hypotheses about the actual branching relationships involved. Whereas it is not crucial how the fossil species x and y in Fig. 3.4 are classified, as long as there is a consistent terminology available for discussion of them, in phylogenetic reconstruction the attempt must at least be made to allocate them to probable positions on the primate phylogenetic tree. Failing that, if the fossils concerned are too fragmentary to provide adequate information to permit any valid inference, they should be omitted from the analysis. Phylogenetic reconstruction relies upon the precise assessment of **characters** and imprecise concepts of overall

primitiveness of entire organisms do not permit objective evaluation.

This brings us to the important point that many influential texts dealing with primate evolution (e.g. Le Gros Clark, 1959; Buettner-Janusch, 1966; Napier and Napier, 1967; Napier 1971; Hill, 1972) have implicitly or explicitly used the concept of the *Scala naturae* to discuss primate evolution, as opposed to simply using it as a device to facilitate classification. Although the 'phylogenetic scale' has a certain value as an illustrative model for introducing the living primates to readers (Fig. 3.5), if taken too literally the model leads directly to the treatment of living species as 'frozen ancestors' (an apt term coined by the late N.A. Barnicot). Regardless of the caveats that may be attached to the *Scala naturae* when it is used to illustrate primate evolution, it is all too easy to slip into the habit of equating living forms with actual ancestral stages. For example, it is commonly stated that members of the order Insectivora are the most primitive of the living placental mammals and they are therefore typically regarded as occupying the first rung on the *Scala naturae* leading to *Homo sapiens* (see Fig. 3.5). This leads on to the tendency to assume that among living placental mammals the living insectivores must in every respect be the most primitive. Similarly, among primates it is commonly assumed that the prosimians are uniformly more primitive than the simians. One effect, therefore, of the concept of the 'phylogenetic scale' has been to encourage investigators to use single representatives of inferred evolutionary 'stages' in attempts to reconstruct evolutionary history. This insidious influence of the concept of the *Scala naturae* has been described in detail by Hodos and Campbell (1969) and by Martin (1973b). It is essential to recognize that there are no uniformly primitive living species, in the sense of being 'frozen ancestors'; there are only primitive characteristics.

Loose use of the terms 'primitive' and 'advanced' in relation to entire organisms has been a feature of the literature ever since the theory of evolution by natural selection was first proposed. There is still a popular misconception

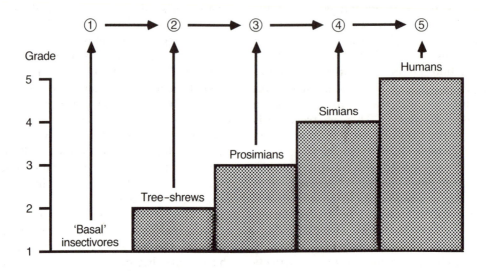

Figure 3.5. Illustration of the relationship between the grade concept, as used in classification, and the *Scala naturae*, which is often mentioned in discussions of primate evolution. The living groups indicated on the staircase-like ascending scale are commonly taken as constituting a graded series (left), approximately equivalent to the stages of evolution that culminated in the evolution of humans (= grade 5). If the grade concept is interpreted too literally, living representatives of the various grades may be misinterpreted as forming an actual evolutionary sequence (circled numbers 1–5 linked by arrows at the top of the diagram). There is an obvious anthropocentric bias in such an approach. (Figure reprinted from Martin (1973b) by kind permission of the Zoological Society of London and Academic Press, London.)

that humans are descended from the great apes in the sense that modern great apes provide a faithful representation of our early ancestors (i.e. that living great apes can be equated with the 'great ape stock' in Fig. 3.3). What Darwin actually proposed was that humans and the modern great apes are descended from a common ancestor. Successful reconstruction of the characteristics of that ancestor requires the development of objective techniques of analysis and can only be hindered by the use of simplistic concepts of successive evolutionary grades. One cannot, for example, assume, because the 'primitive' modern great apes have large canine teeth, that this was necessarily the ancestral condition for the great apes and humans. A detailed character-by-character analysis is required to establish the features present in the

common ancestral stock of the great apes and humans.

In fact, modern developments in the analysis of molecular data (see Chapter 11) have demonstrated quite convincingly that change tends to occur along all lineages both in the genetic material (DNA) and in the principal products thereof (proteins). At the molecular level, evolutionary change seems to take place almost inevitably with the passage of time and in this context, at least, it does not make sense to think in terms of 'frozen ancestors'. It may, however, be a special feature of the molecular level of organization that rates of change are comparatively steady (see Chapter 11). It is conceivable that differential rates of evolution affecting higher levels of organization (from the chromosomes upwards) may lead to the common

preservation of individual primitive morphological, physiological or behavioural characters
in living species. In any particular organism,
by contrast, there will usually be a combination
of retained primitive features and more recently
developed, advanced features (**mosaic evolution**). Hence, given the almost constant process
of change documented at the molecular level and
given the mosaic effect of evolution of individual
characters above the molecular level, it is clear
that no living species can be taken as a direct
equivalent to an actual ancestral stage.

Inadvertent treatment of living species as
'frozen ancestors' in discussions of primate
evolution is by no means uncommon. Indeed,
even such an authority as Simpson has
occasionally lapsed into referring to living forms
as equivalents of actual ancestral stages, notably
in discussing the tree-shrews (Simpson, 1950):

> Even primate re-radiation would be entirely
> possible if present primates were wiped out
> and their empty ecological niches continued
> to exist, because living tree shrews have no
> specializations that would clearly exclude
> approximate repetition of such a radiation.

This view has been stated even more directly
by Schultz (1969):

> . . . there is an excellent justification for
> treating the modern tree shrews as 'living
> fossils'.

Quite apart from the fact that considerable
reservations may be expressed about the
hypothesis that primates are derived from an
ancestor resembling a tree-shrew (see Chapter
5), it is highly improbable that any early offshoot
from the ancestral primate stock would have
remained essentially unmodified by evolutionary
processes over a period of at least 65 million
years. Nevertheless, the notion of the living
primates as constituting a graded series reflecting
actual ancestral stages has achieved a wide
currency. Indeed, it is well established in public
consciousness, as is illustrated superbly by the
following poem by Miles Gibson (reprinted from
the anthology *Permanent Damage* (1973) with

the kind permission of the author and of the
publishers, Eyre Methuen Ltd. © Miles Gibson):

Tree
on the top branch
sits a naked man
bruised with wind
and hugely alone

on the branches below
squat impossible dreams
the half-ape-man
and the not-yet-man

beneath them
laze the true ape
and the true monkey

directly below
comes the lemur
and at the bottom of the tree
on the last branch of all
hangs the skull of a shrew

from somewhere called distance
a spike of rain arrives
and punches out the hood of leaves

someone sniffs the air
someone rattles his teeth
someone yawns
someone shouts
look out behind you

the man screams
falls off
everybody
moves up one

In the context of phylogenetic reconstruction,
it is simply not acceptable to begin with a priori
assumptions that some species are uniformly
primitive while others are uniformly advanced.
For this reason, the concept of the 'phylogenetic
scale' (Fig. 3.5) must be treated with
considerable care. This is not to say, however,
that there is no biological basis for a very
rudimentary gradation among living primates.
There are, in fact, good reasons why this
gradation should exist in a very general sense. In

the first place, most prosimian primates are nocturnal in habits, whereas simian primates are, with only one exception (the owl monkey, *Aotus*), diurnally active. As it is likely that the ancestral primates were nocturnal in habits (e.g. see Chapter 7), it is only to be expected that nocturnal prosimians should have remained generally more primitive than diurnal simians. Among the prosimians, there are further reasons why one might expect most Malagasy lemurs to have remained rather more primitive than lorises or tarsiers. Lemurs have been isolated in Madagascar at least for most of the Cenozoic era and they have probably been subjected to less intense competition than prosimians in Africa or Asia because only a few other mammalian groups (insectivores, bats, rodents and carnivores) managed to colonize Madagascar. Similarly, among simians New World monkeys have been somewhat more isolated than Old World monkeys and apes throughout their evolutionary history, so a somewhat slower overall pace of phylogenetic change is to be expected. Finally, the human lineage has clearly been subjected to rather unusual selection pressures that have resulted in a striking divergence from Old World monkeys and apes. Thus, the *Scala naturae* as applied to primates does have some basis in biological reality and it is for this reason useful for the construction of a classification of the traditional evolutionary type. When it comes to the detailed reconstruction of phylogenetic history, however, we must dispense with such imprecise notions and become actively engaged in a detailed analysis of individual characters. The first step in this direction is provided by an examination of the fundamental processes of phylogenetic change.

SPECIATION AND PHYLOGENETIC CHANGE

It has been noted above that it is now generally accepted that the basic unit of any biological classification is the **species**. Originally, the species was defined by reference to some ideal type, in line with the Linnean interpretation of species as essentially unchanging entities. Following the advent of evolutionary theory,

however, a much more dynamic view of species gradually became established, although the species is still regarded as the primary unit in evolution as well as in classification (Cain 1954; Mayr, 1969). Emphasis is now generally placed on species as natural populations of individuals within which there is reproductive continuity. Mayr (1969) provided the following simple definition of the species: 'Species are groups of interbreeding populations reproductively isolated from other such groups.'

Although there are numerous practical difficulties involved in the strict application of this definition (e.g. see Cain, 1954), it has found wide acceptance. It is now widely held that modern species should be recognized, as far as possible, on the basis of fertile interbreeding within natural populations. More recently, Paterson (1973, 1982, 1985) has modified Mayr's definition in a seminal way to suggest that membership of a species can be defined on the basis of the possession of a specific mate-recognition system. This concept underlines the fact that there is selection for successful breeding within a species as well as selection against breeding between populations that differ significantly in genetic constitution. It also has the advantage that a species can be defined independently, rather than in terms of barriers separating it from other species. Reproductive continuity is hence the key feature in most current definitions of species and the modern theory of population genetics rests essentially upon the tenet that gene flow can occur relatively freely within any species population, while definite barriers to gene flow exist between species.

Given a definition of a living species as an overall population of potentially interbreeding individuals, there are clearly two rather distinct processes through which evolutionary change may conceivably occur. First, it is possible that natural selection might bring about a gradual change in the gene pool of a species population over time. This process has been termed **anagenesis** by Huxley (1957). Second, it is possible that an original species population might be subdivided in some way, such that interbreeding between the daughter populations is effectively prevented. Natural selection acting in different ways on such daughter populations (by chance and/or because of different environmental conditions) will gradually lead to significant divergence between them. Eventually, therefore, the daughter populations will become reproductively isolated and hence achieve the status of separate species. Even if such daughter populations are reunited before divergence has led to complete reproductive incompatibility, subsequent natural selection against hybrids (if these are at some disadvantage) may actually reinforce the process of reproductive isolation. It is generally accepted – at least in the case of mammals – that the first stage in the establishment of separate daughter species usually requires some geographical subdivision of an original interbreeding population (Dobzhansky, 1950; Mayr, 1963). In other words, speciation in mammals has probably typically taken place in an allopatric fashion, although other forms of speciation may perhaps be found with certain other groups of organisms (see Bush, 1975). **Speciation** – the formation of new species – is the basic process by which branching can take place in any phylogenetic tree and Huxley (1957) has accordingly referred to divergence through speciation as **cladogenesis**.

The two basic processes of gradual change within lineages and of divergence through speciation, as commonly described by modern biologists such as Mayr (1963), are illustrated in Fig. 3.6. In this diagram, an original species A gives rise to a later species B through gradual change within a lineage, and the species population B subsequently becomes subdivided and eventually yields two daughter species C and D. Several points can be made in relation to this simple illustration. For instance, it is clear that a transitional phase must be involved in the development of reproductive isolation. After the subdivision of the originally cohesive breeding population B into two subpopulations (B' and B''), there is likely to be a preliminary phase (indicated by the vertical dotted line in the tree on the left of Fig. 3.6) during which barriers to reproduction are incomplete. If geographical

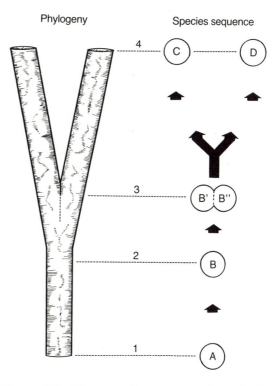

Figure 3.6. Diagrammatic representation of the process of speciation, occurring by the eventual subdivision of an original interbreeding species population (A) into two separate, non-interbreeding daughter populations (C and D). The phylogenetic sequence is illustrated on the left; the classificatory sequence is shown on the right. Key to stages:

1. Ancestral species population (palaeospecies A).
2. Intervening species population, arising from anagenetic change within the evolving lineage (palaeospecies B).
3. Stage of geographical subdivision of the lineage into two daughter populations (subspecies B' and B'). Initiation of cladogenesis.
4. Modern species populations (biospecies C and D), exhibiting full reproductive isolation. Each biospecies would be characterized by a specific mate-recognition system. Completion of cladogenesis.

separation of the two subpopulations is only short-lived, therefore, it is quite possible that speciation will not occur. For the process of speciation to be completed, some kind of threshold of change must be crossed while the two subpopulations are isolated. Because the separation of daughter species may thus require a (perhaps considerable) period of cumulative change, it is only to be expected that a cross-sectional sample of species populations at any point in time (e.g. the present day) will include some that are in the process of subdivision. Hence, not all present-day natural populations will clearly fit the biological definition of a species as a potentially interbreeding population and intermediate cases will be identifiable. For this reason, there is provision for varying levels of subdivision of species in modern taxonomic schemes and the most commonly used category is that of the **subspecies**. (The formal name of a subspecies is composed by adding a third name to the binomial species name. For example, Neanderthal man has commonly been allocated to the subspecies *Homo sapiens neanderthalensis* to reflect the interpretation that there was incomplete reproductive isolation from the population to which modern humans belong, *Homo sapiens sapiens*.)

The existence of intermediate stages in the process of speciation underlines a major distinction between classification and phylogenetic reconstruction reflected in Fig. 3.6. Any classification relies on the existence of **discontinuities** between categories, whereas a phylogenetic tree incorporates both continuity (genetic transmission from parent to offspring along all branches) and discontinuity (divergence between branches brought about by speciation). It is a fundamental feature of evolutionary theory that both continuity and discontinuity are involved. By contrast, a classification must rely above all on the recognition of discontinuities between taxa and, in cases where no natural discontinuity is present, it is frequently necessary to erect artificial boundaries. This leads to some difficulties in the matching of a classification (right of Fig. 3.6) to a phylogenetic tree (left of Fig. 3.6). At any given time, a horizontal cross-section through a phylogenetic tree will tend to emphasize **diversification** (Hanson, 1977), with a large number of separately identifiable species and relatively few intermediate cases involving populations in the process of speciation. In any consideration of evolution over time, however,

the element of **transformation** (Hanson, 1977) is predominant. Linnaeus, of course, had relatively little difficulty in defining discontinuities in his classification, since he was concerned only with the cross-section of living species. But since Darwin's time there has inevitably been a somewhat uneasy coexistence between Linnean classificatory schemes and inferred patterns of phylogenetic relationships.

One of the best illustrations of the basic incompatibility between discrete taxonomic categories and the biological reality of phylogenetic change is provided by an inherent problem of defining fossil species (**palaeospecies**). As is shown in Fig. 3.6, it is at least theoretically possible that an original ancestral species population A may persist for a considerable period without subdivision. Gradual evolutionary change in the species population over time may nevertheless bring about considerable changes in the typical form and function of the members thereof. At some arbitrary point in time, we might therefore recognize a new species (B) on the grounds of recognizable differences from the ancestral condition. Obviously, palaeospecies A and B cannot be distinguished on grounds of reproductive incompatibility, yet the observed morphological difference between A and B may be equal to (or greater than) that found between the modern species C and D. If change can occur along lineages as well as through divergence between lineages, the 'palaeospecies problem' is bound to arise (Cain, 1954). In order to classify fossil 'species' that are thought to be parts of a continuous lineage, it is necessary to erect an arbitrary division in an evolving sequence to provide an artificial equivalent to the real biological barriers existing between contemporary species populations. One outcome of this is that, wherever the taxonomist decides to draw the arbitrary boundary between two successive palaeospecies (A and B in Fig. 3.6), the first members of species B must be the offspring of the last members of species A. Although this problem is, in practice, alleviated by the marked patchiness of the fossil record (see Chapter 2), the theoretical difficulty remains. In fact, because all classificatory boundaries must be drawn somewhere in relation to the continuous branching pattern of a phylogenetic tree, the problem applies at every taxonomic level. For instance, in classificatory terms the very first mammals must have been the offspring of mammal-like reptiles (see Chapter 4). This clash between the biological reality of the dynamic, continuous process of phylogenetic change and the effect of arbitrary subdivision to meet the practical needs of classification provides just one indication of the spurious nature of arguments about the 'objectivity' of alternative classificatory procedures.

Thus far, this discussion of the processes of evolutionary change (involving both anagenesis and cladogenesis) has reflected what was, until recently, the consensus view. This derives from a well-established school of thought, which may be termed the **gradualist school**, based on the modern synthesis between Darwinian evolutionary theory and population genetics. The gradualist view is that relatively slow evolutionary change occurs both within and between species lineages as a result of the continuous action of natural selection. In other words, there is a single evolutionary process that accounts for both anagenesis and cladogenesis. This does not rule out the possibility of differential **rates of evolution** in different populations, but it does rule out the possibility of any qualitative distinction between the process concerned in speciation and the process involved in change over time within a lineage. This gradualistic model was directly challenged by Eldredge and Gould (1972), who suggested an alternative interpretation of the way in which evolutionary change progresses (see also Gould and Eldredge, 1977). They argue that most evolutionary change occurs relatively rapidly at the time of speciation and that there is little or no change over time within species lineages. Because this alternative model suggests that cohesive species populations typically tend to remain in genetic equilibrium over time and that evolutionary change occurs in short, pronounced bursts when new species become established, it has been termed **punctuated equilibrium**. The essential contrasts between the gradualist and

punctationist models of evolutionary change are illustrated in Fig. 3.7.

Although the punctuated equilibrium model of evolutionary change obviously reflects a radically different view from that traditionally advocated by supporters of the gradualistic model, it is not necessarily the case that a difference in mechanism is required. One could convert the gradualist interpretation to a punctationist one simply by proposing major disparities in rates of evolution. The model of punctuated equilibrium could, in principle, be generated by proposing that evolutionary change within lineages (anagenesis) is extremely slow, whereas evolutionary change associated with speciation (cladogenesis) is comparatively very rapid. It is commonly suggested that such radical differences in rates of evolution could arise because speciation typically takes place when a relatively small, peripheral subpopulation becomes isolated (see Bush, 1975; Hennig, 1979; Eldredge and Cracraft, 1980; Wiley, 1981). One might reasonably expect such a small subpopulation to change very rapidly, both because of particularly strong selection in an extreme, peripheral environment and because of accentuation of chance shifts in gene frequencies. Accordingly, one special feature of any punctuationist model is that, following a speciation 'event', there may be continuation of the original ancestral species, represented by the surviving core population, while it is only the isolated peripheral sub-population that becomes a distinct new species (Fig. 3.7). This contrasts with the standard gradualist interpretation, according to which both daughter populations are commonly thought to become new species. Nevertheless, there is no obligatory reason for proposing that speciation involves a special kind of evolutionary change qualitatively different from the progressive shift in gene frequencies that is thought to occur through the action of natural selection on individual species populations over time. Yet several authors (e.g. Stanley, 1979, 1981) have proposed that evolutionary change during speciation (**macroevolution**) is quite distinct from the mild shifts in gene frequencies known to occur within natural species population

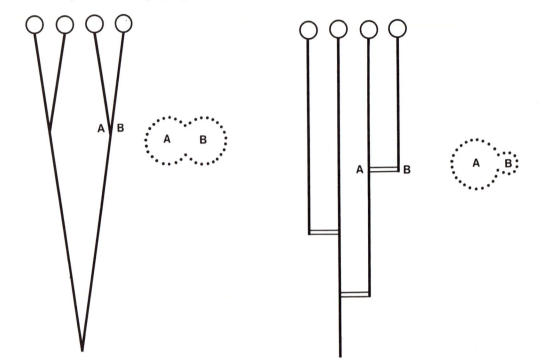

Figure 3.7. Alternative models of speciation. Left: In the **gradualist model**, evolutionary change occurs both during speciation (cladogenesis) and within lineages (anagenesis). An original species population (right branch of the tree) becomes subdivided into two daughter populations, A and B, both of which develop into new species, distinct from the ancestral species. Right: In the **punctuationist model**, a relatively small daughter population (B) becomes isolated from the original species population (A). The former develops into a new species by means of a relatively rapid speciation 'event' (indicated by the horizontal white bar), while the latter remains unchanged. Little or no evolutionary change takes place within individual species lineages ('stasis' instead of anagenesis).

(**microevolution**). It has even been suggested that one should resurrect under the heading of macro-evolution something akin to the 'hopeful monster' concept proposed by Goldschmidt (1940) to account for the origin of major new adaptations.

It is true that the process of speciation has never been directly observed. Perhaps, because of the timespan involved, it is not reasonable to expect that any significant part of the process can ever be observed. For this reason we cannot dismiss out-of-hand proposals that speciation may involve unusual macroevolutionary changes. All that can be said at present is that there is no convincing evidence for such major changes, whereas there is abundant evidence of minor shifts of gene frequencies within species

populations in response to natural selection. For example, Stanley (1981) has proposed that macroevolution may involve substantial re-arrangement at the chromosomal level, as opposed to modifications at the level of the gene, but there seems to be no compelling evidence to support this proposal.

One of the main reasons advanced for formulation of the punctuated equilibrium model has been the empirical observation that numerous fossil species seem to retain relatively constant morphology over substantial periods of geological time (morphological **stasis**). Such evidence from the fossil record figured prominently in the original paper by Eldredge and Gould (1972) and this point was argued particularly forcefully by Williamson (1981) on the basis of an unusually good sequence of fossil molluscs from Cenozoic deposits in East Africa. Whereas other investigators have tended to argue that the general absence of obvious transitional forms from the fossil record is attributable to the patchy nature of the record, supporters of the punctuationist model suggest that the record is more complete than commonly believed and that transitional forms are very rarely found because speciation takes place relatively rapidly. In fact, it seems likely that there are indeed substantial gaps in the fossil record, at least as far as mammals are concerned (see Chapter 2), so the absence of many transitional forms is hardly surprising. A few fossil sites with a relatively good sequence of mammalian fossils are known and these should, in principle, provide valuable test cases. In practice, however, interpretations of the evidence are found to differ. There are some mammalian palaeontologists (e.g. Gingerich, 1976a, 1977a, 1984a), who believe that the fossil record at such sites does provide evidence of gradual evolution both within and between lineages. But other palaeontologists (e.g. Eldredge and Cracraft, 1980; Schankler, 1981) interpret exactly the same kind of fossil sequences as evidence for punctuated equilibrium. This underlines the fact that there is considerable room for personal interpretations of the fossil record, especially where the number of available specimens is somewhat limited. This is particularly true of the relatively sparse hominid fossil record, which has been cited by some authors (e.g. Eldredge and Gould, 1972; Stanley, 1981; Rightmire, 1986) as providing a good example of punctuated equilibrium, but which may equally well be interpreted in traditional gradualist terms (Cronin, Boaz *et al.*, 1981; Wolpoff, 1984, 1986). There is also the problem that the degree of definition in the fossil record is extremely coarse (measured at least in terms of tens of thousands of years) in comparison to the timescale familiar to the geneticist. Thus, the process of speciation may seem to be an 'event' in the eyes of the palaeontologists, but in the eyes of the geneticist it may involve hundreds or even thousands of generations and hence allow plenty of time for the accumulation of numerous individual shifts in gene frequencies (Jones, 1981). As yet, there is no convincing evidence from the fossil record to indicate that speciation involves any radically different process of evolutionary change other than, perhaps, a temporary acceleration in rate.

There is, in fact, a special problem inherent in attempts to decide the choice between gradualism and punctuated equilibrium on the basis of available fossil evidence. This problem arises from the typically fragmentary nature of fossil evidence (see Chapter 2) and involves a form of reductionism that has particularly adverse implications for the supposed evidence for stasis. Because most mammalian fossils are fragmentary, the only relatively continous sequences found at the best fossil sites consist essentially of dental evidence. In other words, one can only confirm or deny the presence of gradual change in specific dental features. Commonly, a simple parameter such as the length of the first lower molar tooth is taken and the change in the mean value of this parameter over time is examined (e.g. Gingerich, 1976a, 1977a; Schankler, 1981). Even if a given dental parameter is found to remain virtually constant within a lineage over time ('stasis'), this does not indicate that the organism as a whole has remained unchanged. Hence, it is in practice exceedingly difficult to determine from the fossil record whether a species has remained intrinsically unchanged over a long period of time. Williamson (1981), for instance, was only able to examine external measurements of the shells of the molluscs involved in his study and he was therefore able to demonstrate stasis merely in the shells of these molluscs, not in the organisms as a whole. Mosaic evolution (i.e. the evolution of individual characters of an organism at markedly different rates) is a well-recognized phenomenon, following be Beer (1958). It would therefore be entirely conceivable for the internal characters of a molluscan species to change gradually over time without any detectable accompanying change in the form of the shell.

Convincing evidence for such mosaic evolution in dental features has been provided by Rose and

Bown (1984) for early Eocene fossil 'tarsioids' from sediments covering a continuous period of approximately 4 million years (Ma). These authors were able to show, on the basis of a relatively complete sequence of dental specimens of *Tetonius* and *Pseudotetonius*, that virtual stasis in one area of the dentition (the main cheek teeth) was accompanied by pronounced but gradual change in another (the anterior lower dentition). There was a further mosaic effect in the evolution of the anterior lower dentition, in that the individual changes took place at different times. By contrast, with another 'tarsioid' lineage from the same deposits, involving a transition from *Teilhardina* to *Tetonoides*, gradual change was documented for the cheek teeth. In the first case, that of *Tetonius* and *Pseudotetonius*, an author working only on the cheek teeth might have inferred that there was stasis rather than gradual change in these tarsioids. It must therefore be concluded that the fossil record, because of its inevitably fragmentary nature, cannot provide conclusive evidence of stasis in entire organisms over time. The model of evolution through the process of punctuated equilibrium relates to entire organisms, not to individual parts thereof (e.g. external or internal skeletal components) and cannot be adequately tested with incomplete evidence. Hence, while it is probably true to state that palaeontologists have discovered rather more stasis in individual characters than might have been expected on the basis of neo-Darwinian concepts of the process of evolutionary change, it is at present unnecessary to invoke a special kind of mechanism for the process of speciation (Maynard Smith, 1981).

It should be noted that the punctuationist model of evolutionary change has some practical implications with respect to some of the issues concerning classification discussed above. Indeed, some authors have linked the model of punctuated equilibrium directly to cladistic philosophy (e.g. Eldredge and Cracraft, 1980). For instance, the 'palaeospecies problem' identified in Fig. 3.6 only arises if it is assumed that evolutionary change can take place within a species population over time (anagenesis). If, instead, species populations exhibit stasis once

they have become established (see Fig. 3.7), it is in theory unnecessary to subdivide a continuous lineage into successive species. In other words, because of the relatively rapid occurrence of speciation 'events', one would not expect to find transitional forms in the fossil record and palaeospecies should be identifiable as discrete entities. It is also to be expected from the punctuationist model that a species should remain continuously identifiable in the fossil record until it becomes extinct. That is to say, it is theoretically possible that, given the appropriate conditions, a species could remain unchanged over vast periods of geological time. For instance, Eldredge and Cracraft (1980) state that a species might persist for 100 Ma or even longer. If such long-term stasis of individual species is possible, this could have considerable implications for the occurrence of 'living fossils'. Hennig (1979) in fact rejected the distinction between anagenesis and cladogenesis, but nevertheless accepted that a certain amount of change could occur within a species lineage over time, likening this to the metamorphosis that can occur within the lifespan of certain insects. Despite the possibility of such change, Hennig maintained that a single lineage should not be subdivided into separate 'palaeospecies'. Certain more recent adherents to the cladistic school have gone even further by ruling out completely the notion of anagenesis on the grounds that microevolution within species lineages generates negligible evolutionary change. However, the special features of the punctuated equilibrium model of evolutionary change apply essentially at the species level and they do not affect, for example, the classificatory dilemma that arbitrary boundaries between higher taxa must be drawn somewhere. At some point in time, even with a punctuationist model, mammal-like reptile parents must have given rise to the first mammals as their offspring.

Returning now to the specific topic of phylogenetic reconstruction, there can be no doubt that – whatever the actual mechanism involved – the basic process of divergence between species illustrated in two extreme forms in Fig. 3.7 must be regarded as the point of

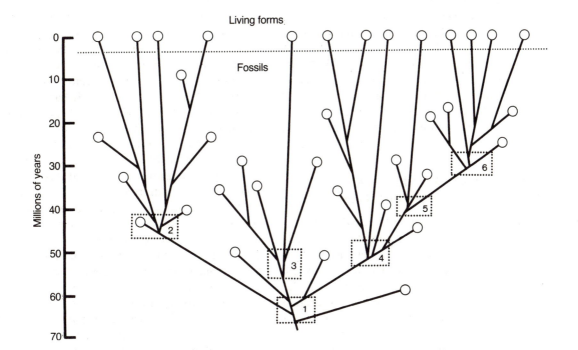

Figure 3.8. A hypothetical phylogenetic tree, illustrating the relationship between living species, palaeospecies and ancestral stocks. All known species are indicated by circles. The ancestral stocks (dotted boxes numbered 1–6) have been defined on the basis of their relevance to major evolutionary radiations. For this tree, box 1 represents the common ancestral stock and the other boxes represent later, more specific ancestral stocks. Each ancestral stock covers a period of several million years and includes a number of lineages in addition to the actual ancestral species population. Note that no known palaeospecies has been indicated as occupying a directly ancestral position, reflecting the view taken here that the very fragmentary nature of the fossil record rarely permits the discovery of a directly ancestral form. However, one of the ancestral stocks (box 2) has been indicated as including a palaeospecies practically indistinguishable from a directly ancestral species. (Figure reprinted from Martin (1973b) with the kind permission of the Zoological Society of London and Academic Press, London.)

departure. The neo-Darwinian theory of evolution by natural selection has the clear implication that all living primate species have been derived by the process of speciation from ancestral species populations. Ultimately, the origin of primates must be traceable to a single ancestral species population (Fig. 3.8). Ideally, therefore, the 'ancestral primates' should be envisaged as members of a single potentially interbreeding population and all of the common features of living primates must be derivable from genetic features of that ancestral species population (the 'stem species' of Hennig, 1979). However, there is a practical problem of

'resolving power' in the identification of such an ancestral species. In addition to the methodological problems involved in inferring ancestral characteristics (see later), there is the practical limitation that fossil forms are typically fragmentary. It can therefore be argued that it is preferable to use the concept of the **ancestral stock** (Martin, 1973b; Fig. 3.8), denoting a group of related species within a loosely defined time bracket. Although the actual ancestral species will be included, reference to an 'ancestral stock' amounts to recognition of the fact that it is practically impossible to single out that species. Given the uncertainties involved in phylogenetic construction, postulation of ancestral 'stocks' rather than 'species' avoids the suggestion of a degree of accuracy far superior to that which can actually be attained. Thus, although Eldredge and Cracraft (1980) have argued that any hypothetical ancestor above the species level is 'illogical', there are good reasons for defining an ancestral stock as a reconstructed average condition for all of the species contained in a given time bracket.

ANALYSIS OF CHARACTERS FOR PHYLOGENETIC RECONSTRUCTION

When it comes to the actual process of evaluating characters for the reconstruction of phylogenetic relationships, it is necessary to seek accuracy in two very distinct senses. Clearly, it is necessary to ensure that any characters involved in the evaluation process are described as accurately as possible (factual accuracy). However, it is equally necessary to strive for accurate interpretation of those characters as indicators of phylogenetic relationships (methodological accuracy). Factual accuracy will, of course, be a central concern in the reviews of evidence presented in Chapters 6–11. At this point, by contrast, the immediate concern is with the methodology of phylogenetic reconstruction. The fundamental step in any phylogenetic analysis is assessment of the degrees of similarity shared by individual species contained in the group under comparison. The assumption must, of course, be made that there is some genetic basis for those similarities, although in the vast majority of cases there is no direct check on the genetic identity of shared similarities. Whether or not the characters assessed are expressed in numerical terms, all phylogenetic reconstructions must imply some inference of the degree of overall genetic similarity between species, based on the overall degree of resemblance in observable features ranging from the behavioural to the biochemical. Yet, even if it is accepted that such resemblance reflects genetic similarity, there are some special problems involved in the inference of common ancestry.

It must be emphasized at the outset that

numerical taxonomy, which involves overall assessments of similarity, is explicitly concerned with classification rather than with phylogenetic reconstruction (Sneath and Sokal, 1973). For this reason, the procedures used in numerical taxonomy are not directly relevant in the present context (see also Wiley 1981). It has long been recognized that a straightforward calculation of the degree of observable similarity, at any level, between species will not normally yield an accurate picture of phylogenetic relationships. The primary reason for this is that it is relatively common for different species to acquire similarity because of independent acquisition of particular features. Such independent acquisition may occur by chance, particularly at the level of the gene (see Chapter 11); but in many cases it takes place in response to comparable selection pressures. It is, for instance, a common occurrence for aquatic vertebrates – regardless of their specific phylogenetic origins – to develop fins and streamlining of the body contours. Among mammals, a particularly good example of such independent development of similar adaptations is provided by the evolution of continuously growing incisor teeth adapted for gnawing. Only one of the living primates, the aye-aye (see Fig. 6.9), has this feature, but it is found in a variety of other living mammals, including rodents, lagomorphs (rabbits, hares and pikas) and certain marsupials. It was mentioned in Chapter 1 that the aye-aye shows a suite of similarities with a marsupial (*Dactylopsila*) because of convergent adaptation for feeding on wood-boring larvae. Because of this possibility of independent evolution of similar features, it is necessary to distinguish both theoretically and practically between similarities inherited through genetic continuity from a common ancestor (**homologous** similarities) and those acquired independently (**analogous** similarities), using the terminology that was originally proposed by Owen (1848) without any evolutionary connotation.

Independent acquisition of similarity by two or more species is commonly referred to as **convergence**, but in cases where the species concerned are quite closely related, the term

parallelism may be used. It has been argued that the special category of parallelism may be needed to fit cases where the characters of a common ancestor provided a predisposition towards the later, independent development of similar adaptations in separate lineages (Simpson, 1961). This may be refined to the proposal that parallelism should be regarded as the development of similar adaptations following divergence from an immediate common ancestor (Hennig, 1979; Hecht and Edwards, 1977). Conceivably, two separate organisms or groups of organisms derived from an immediate common ancestral stock might progressively develop a series of similarities in response to comparable selection pressures. For example, it is commonly held that many of the similarities between modern New World simians and modern Old World simians were acquired by parallel evolution subsequent to the divergence of these two primate groups from a common ancestral stock that was still at a prosimian level in terms of its characters. In this case, it could be argued that the New World and Old World simians never actually diverged in any real morphological sense, but followed essentially parallel courses in their evolution. Nevertheless, it is still possible to imagine a complete spectrum of intermediate cases linking this potential example of 'parallelism' to obvious 'convergence', involving the independent development of similarities by organisms that are otherwise very distinct from one another. For practical purposes of reconstructing phylogenetic relationships, therefore, there is no clear theoretical boundary between parallelism and convergence and it is preferable to follow the view expressed by Harrison and Weiner (1964) that 'parallelism is merely the limiting case of convergence' (see also Eldredge and Cracraft, 1980).

An overall distinction between homologous and analogous (convergent + parallel) similarities is hence essential as one component of successful phylogenetic reconstruction, because only the former provide evidence of genetic continuity. Simpson (1945) in fact indicated that this is really the only major distinction that is required for the recognition of phylogenetic affinity:

> Animals may resemble one another because they have inherited like characters, homology, or because they have independently acquired like characters, convergence. On the average, two animals with more homologous characters in common are more nearly related, their ancestral continuity is relatively more recent, than two animals with fewer.

Although this statement may be true in a very general sense, it overlooks an important distinction that has become increasingly recognized largely because of the influence of Hennig (1950, 1979). Even if it is possible for all analogous similarities to be excluded during analysis, it is by no means true that the closeness of the phylogenetic relationship between any two species will be simply proportional to the number of shared homologous similarities. The reason for this is illustrated in Fig. 3.9. If rates of evolution differ between lineages, it is predictable that two modern species from relatively slowly evolving lineages (A and B in Fig. 3.9) will tend to share a greater proportion of homologous characters, even though one of these species (B) actually shared a more recent common ancestor with a more rapidly evolving lineage leading to the modern species C. If rates of evolution are approximately the same in all lineages, this problem does not arise and Simpson's statement (above) will then accurately reflect the resulting pattern of distribution of homologous characters among descendant species. If, on the other hand, there are substantial differences in rates of evolution between lineages, the problem identified in Fig. 3.9 arises and Simpson's statement will not be valid. As was succinctly noted by Bigelow (1958): 'Unequal rates of evolution will produce cases in which overall basic similarity does not correspond with recency of common ancestry.' The potential confounding effect of differential evolutionary rates precludes phylogenetic reconstruction purely on the basis of degrees of homologous similarity and this is the primary

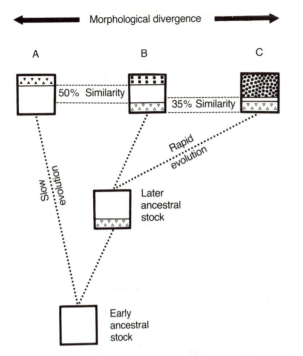

Figure 3.9. The proportions of homologous characters shared by living species (A, B and C) cannot be used as a reliable basis for inferring phylogenetic relationships if rates of evolution differ between lineages. In this hypothetical diagram, convergent similarities have been excluded so as to avoid additional complexity. Primitive features in the 'early ancestral stock' are indicated by the white area within the box. Any developments of new features in later forms are indicated by differential shading of part(s) of the original white area.

In the slowly evolving lineage leading to species A, relatively few novel features have been developed (▲). Rather more rapid evolution has taken place in the lineage leading to the 'later ancestral stock', resulting in a comparable degree of accumulation of novel features (△). Thereafter, B has diverged from the later ancestral stock at a moderate rate, with further moderate accumulation of novel features (■), while C has diverged far more rapidly and has therefore acquired many new novel features (●).

In terms of overall sharing of homologous similarities, A and B are closer together, with 50% similarity. However, all of these similarities are based on primitive features retained from the early ancestral stock. Although B and C are less similar to one another in terms of overall homologous similarity (only 35% similarity), they show 25% similarity because they have retained derived homologous features from the later ancestral stock. Accurate reconstruction of the phylogenetic relationships between A, B and C therefore depends upon the identification of the derived features developed in the later ancestral stock. Note: A cladistic classification would require a primary division between A and (B + C), whereas in a classical classification A and B would be classified together, reflecting their lesser degree of morphological divergence. (Figure reprinted from Martin (1975) by kind permission of Plenum Press, New York.)

reason why the approach adopted in numerical taxonomy is largely irrelevant to phylogenetic reconstruction (see also Hennig, 1979).

The significance of differential evolutionary rates was explicitly recognized by Harrison and Weiner (1964), who referred to the retained similarity between slowly evolving lineages as **patristic similarity** and to the special similarity between lineages attributable to relatively recent common ancestry as **cladistic similarity**. Following this terminology, the resemblance between species A and B in Fig. 3.9 is patristic, and the resemblance between B and C is predominantly (but not exclusively) cladistic. There is a limitation in the use of this terminology in that species B does share with species C some similarities retained from the initial ancestral stock. This being the case, it is preferable to refer to individual **characters**, rather than to entire organisms, in making distinctions between these two different kinds of similarity. For any group of species, there must always be an initial common ancestral stock ('early ancestral stock' of Fig. 3.9). This initial stock would have possessed a particular set of features that could subsequently be retained or modified in different lineages. Any features retained from the initial stock by later descendants will be homologous, but they will provide no information about the subsequent branching relationships within the phylogenetic tree. These initial ancestral features may be labelled **primitive** or, in the terminology of Hennig (1979), **plesiomorphic**. By contrast, new features developed in later ancestral stocks (as shown in Fig. 3.9) do reflect branching relationships within the phylogenetic tree, insofar as they are retained by their descendants. Homologous features retained from a later stock may be labelled **derived** (Mayr, 1969), or **apomorphic** (Hennig; 1979). Clearly, the terms 'primitive' and 'derived' are relative and depend on the actual tree considered. For instance, derived features that distinguish the ancestral stock of primates from the ancestral stock of all placental mammals are primitive features in the more restricted context of primate evolution. Accordingly, the stocks to which the terms 'primitive' and 'derived' are applied in any comparison must always be specified.

Recognition of this crucial distinction between **shared primitive (symplesiomorphic)** and **shared derived (synapomorphic)** homologous similarities represents the most significant, lasting contribution made by the school of cladistic analysis founded by Hennig. Whereas the proposals of the cladistic school regarding classification may be challenged on several grounds (see above), for phylogenetic reconstruction it is absolutely essential to recognize that, because of the potential confounding effect of differential evolutionary rates, specific features must be assigned to successive ancestral stages. Hypothetical reconstruction of ancestral stocks is therefore not an option, as suggested by Cain and Harrison (1960); it is a methodological necessity. With the hypothetical case illustrated in Fig. 3.9, for example, reconstruction of the phylogenetic tree for the three species A, B and C would require effective recognition both of the primitive features of the early ancestral stock and of the derived features of the later ancestral stock. Retention of primitive homologous characters from the initial stock provides no information about the subsequent branching pattern; it is only retention of derived homologous characters from the later ancestral stock that provides the necessary information. In any tree with a number of internal branching points, the relevance of any homologous characters for phylogenetic reconstruction depends on the location of the ancestral stock in which they originally emerged.

One simple illustration of the importance of the distinction between shared primitive homology (**symplesiomorphy**) and shared derived homology (**synapomorphy**) is provided by a character that has been said to indicate a phylogenetic affinity between tree-shrews and primates. Living tree-shrews and some living primates (prosimians) possess an accessory structure beneath the tongue, called the sublingua, which might reasonably be regarded as a shared homologous feature. Living insectivores (*sensu stricto*, see Chapter 5) uniformly lack the sublingua. It might therefore be concluded that the presence of a sublingua indicates that tree-shrews and primates shared a specific common ancestor. Wood Jones (1929), for instance, regarded the sublingua of tree-shrews as 'definitely lemurine in development' and Fiedler (1956) cited the sublingua as an important diagnostic character linking tree-shrews to primates. Le Gros Clark (1959) also mentioned the possession of a serrated sublingua as a feature indicating an affinity between tree-shrews and primates. However, as an essential first step in evaluating the importance of this feature, it must be established whether the sublingua is likely to have been a primitive feature of placental mammals generally, or whether it was more probably developed later, perhaps as a derived feature in a later ancestral stock of tree-shrews and primates. In fact, a detailed survey of mammalian tongues conducted by Sonntag (1925) showed that the sublingua is present in several other mammalian groups in addition to tree-shrews and prosimian primates. For example, among placental mammals it is present in rodents and it is also identifiable (though only moderately developed) in elephant-shrews. Further, a sublingua of some kind is found in marsupials. Hence, possession of the sublingua as such is not a unique shared, derived feature of tree-shrews and primates.

Given that a sublingua is found in several mammalian groups, the possibility must be considered that the sublingua is not, after all, a uniformly homologous structure and has been developed convergently in several different mammalian groups. If this is indeed the case, possession of the sublingua does not provide reliable evidence of specific phylogenetic affinity between tree-shrews and primates. An alternative possibility, however, is that the sublingua was already present in the common ancestral stock of marsupials and placental mammals (see Fig. 4.6) and that it has been retained as a primitive homology in various descendant groups. This was, indeed, the interpretation made by Sonntag (1925) and accepted by Le Gros Clark (1959). On balance, the evidence does suggest that the presence of a sublingua in tree-shrews and prosimian primates is a shared primitive mammalian feature that in itself provides no support for the hypothesis that they shared a specific common ancestry. It should be noted that there have been accompanying suggestions that tree-shrews instead share certain specific refinements of the sublingua with lemurs and lorises (e.g. Le Gros Clark 1959), thus indicating a similarity going beyond the mere retention of a primitive mammalian feature. However, Schneider (1958) has clearly shown that the sublingua of tree-shrews is, in fact, quite unlike that of most lemurs and lorises and lacks the characteristic serrations of the tip that have often been cited.

The evaluation of all such cases of shared similarity for the purposes of phylogenetic reconstruction depends on the logical distinctions indicated in Fig. 3.10. For any successful attempt at phylogenetic reconstruction, it is necessary to find a way of separating analogous similarities, primitive homologous similarities and derived homologous similarities (Fig. 3.11). In other words, *in the reconstruction of phylogenetic relationships it is not sufficient to rely on a simple distinction between apparently homologous similarities and apparently analogous similarities.*

It is one thing to recognize at a theoretical level the requirements for reliable reconstruction of phylogenetic relationships; it is quite another to fulfil them in practice. In the first place, the entire field is bedevilled with circularity (see Hull, 1967; Martin, 1973b). One cannot, for instance, set out to reconstruct hypothetical ancestral stocks using a definition of convergence based on the independent acquisition of similarities (Sneath

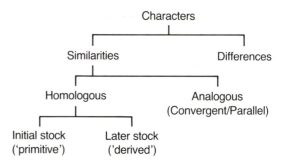

Figure 3.10. A theoretical breakdown of the distinctions between characters required for successful reconstruction of phylogenetic relationships.

and Sokal, 1973). In order to recognize similarities that have been acquired independently following the divergence of a set of species from an original ancestral stock, it is necessary to know at the outset what characters that stock possessed. In practice, the phenomenon of convergence is usually recognized either (1) when characters that exhibit superficial similarity are found on closer examination to differ in finer detail, or (2) when apparently convincing similarities between any two taxa are found to be out of step with the overall pattern of relationships based on an analysis of many characters. The first category ('overt convergence') can be excluded before the analysis of similarities among species in a comparison, while the latter ('concealed convergence') can only be recognized, if at all, as an inevitable outcome of that analysis. It is a very common experience, in the reconstruction of alternative phylogenetic trees for a given group of organisms, to find that all trees require at least some convergent development of similarities. In many cases, especially when quite complex characters are involved, it is likely that convergent similarity that has arisen in response to comparable selection pressures will be incomplete and therefore identifiable on close examination. This point has been stated succinctly by Butler (1939) in relation to the development of convergent similarities in mammalian cheek teeth: 'Convergence can result only from the action of selection on organs of similar function, and it makes the organs similar

Radiation giving rise to various living forms

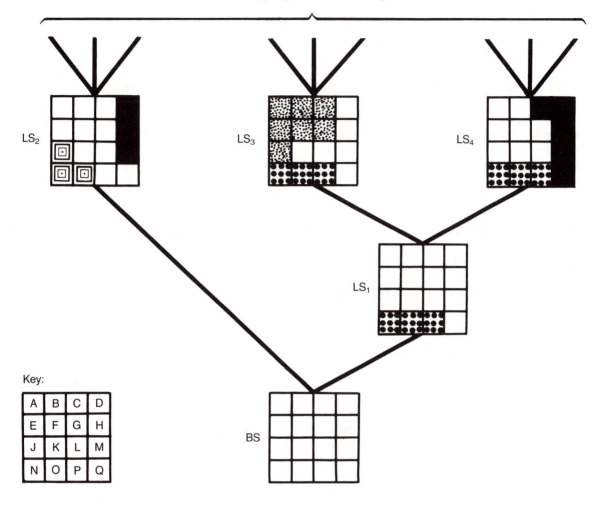

Figure 3.11. Hypothetical illustration of the distinction between primitive homology, derived homology and convergence in a phylogenetic tree. In the initial ancestral stock (BS), 16 original characters are indicated by blank boxes (see key for identification by letter). In four later stocks derived from BS (LS$_1$, LS$_2$, LS$_3$ and LS$_4$), some of these original characters have been modified in various ways. The following kinds of similarity can be identified:

1. Primitive homology: represented by any blank boxes retained in the later stocks.
2. Convergence: because of a similarity in environmental conditions, LS$_2$ and LS$_4$ have independently developed special adaptations (black boxes); three of these (D, H and M) are effectively similar in the two stocks.
3. Derived homology: changes in characters N, O and P in stock LS$_1$ persist in LS$_3$ and LS$_4$, reflecting their derivation from LS$_1$.

A superficial examination of stocks LS$_2$, LS$_3$ and LS$_4$ would suggest a link between LS$_2$ and LS$_4$, as they share 10 similarities, while LS$_2$ shares five similarities with LS$_4$ and only two with LS$_2$. Even if the effect of convergent evolution can be excluded, LS$_2$ and LS$_4$ remain the most similar (seven shared homologous similarities). Successful reconstruction of the real phylogenetic relationships depends on identification of the three derived features (N, O and P) shared by LS$_3$ and LS$_4$. (Figure reprinted from Martin (1973b) by kind permission of the Zoological Society of London and Academic Press, London.)

to the extent that their functions are similar, but no further.'

In some cases, of course, it is possible that convergence may take place by chance rather than as a result of the action of similar selection pressures. This is particularly true at the molecular level (see Chapter 11), where a strict limitation on the number of possible changes dictates a significant level of chance convergence. At higher levels of biological organization, however, Butler's dictum is likely to apply quite widely. There are two corollaries of this observation: first, any character should be examined in as much detail as possible, in order to discover any differences of detail that are masked by superficial similarity; second, inference of homology will clearly be more reliable for complex characters than for simple characters, as the latter may more easily be affected as a whole by a given set of selection pressures.

In many cases the recognition of convergent evolution as the agent responsible for superficial similarity between species amounts to a distinction between 'habitus' and 'heritage', as defined by Gregory (1920b):

> More recently acquired characters and special adaptations to the final life habits constitute what I call the 'habitus' or adaptive form of an animal, while characters which it has inherited from its ancestors in an earlier stage of evolution, and before the assumption of its present life habits, make up its 'heritage'. Both sets of characters are adaptive, but the first set are adapted to its present life habits; and the second set were adapted to the different life habits of its ancestors, and have been retained because they are still useful, or because they have not yet been eliminated by later adaptations.

This applies aptly to the first example of convergence mentioned above, where representatives of distinct groups of vertebrates such as fish and dolphins exhibit superficial similarity in terms of streamlining of the body for aquatic life (the 'habitus'), but possess fundamentally different underlying skeletal structures. Whereas a fish has retained its skeletal structure from persistently aquatic ancestors, the dolphin has undergone conversion from a terrestrial mammalian condition and therefore has a set of characters (its 'heritage') reflecting this original land-living ancestry. This example illustrates the important principle, recognized in many classifications devised in the century before the advent of Darwinian evolutionary theory, that internal characters generally provide a more reliable guide to the affinities between species than external characters (see Knight, 1981).

Having thus excluded certain similarities between species as probable results of convergence, one can proceed to analyse the remaining similarities on the provisional assumption that these are likely to be homologous. In other words, it may be taken as a working hypothesis that features that appear to be entirely similar in two or more species are homologous. (This is close to the view stated by Sneath and Sokal (1973) that similarity is equivalent to homology in operational terms.) Of course, in certain cases where two species (or groups of species) have diverged relatively recently from a common ancestor and have then proceeded to develop many similar morphological characters in parallel, it is unlikely that these will ever by recognized as independently acquired. In practical terms, this may not influence the pattern of branching established through phylogenetic analysis. Nevertheless, there is an important corollary in that actual times of phylogenetic divergence are generally likely to be earlier than those inferred from the morphological characters of fossils (see Chapter 12).

The starting point for any phylogenetic analysis must inevitably be an existing classification in which organisms are already provisionally grouped on the basis of overall similarity and in which there is usually some reflection of previous conclusions regarding likely evolutionary relationships. For present purposes, the chosen starting point is that the living members of the order Primates, as recognized in the classification of Simons (1972), are probably members of a strictly monophyletic group of mammals and are hence likely to be descended from a definable ancestral stock. Further, it will be taken as a working hypothesis that the six major 'natural groups' among the living primates – lemurs, lorises, tarsiers, New World monkeys, Old World monkeys, apes + humans – are also likely to represent lower-level strict monophyletic groups derived from later stocks within the primate phylogenetic tree. Given this initial set of working assumptions, there are three possible sources of error: (1) the overall classification may include species that do not belong in a strict monophyletic group; (2) the classification may exclude species that actually belong to a strict monophyletic group appropriately definable as Primates; and (3) similar inappropriate inclusion or exclusion may apply to the six 'natural groups' of living primates. Because of the first two potential sources of error, it is mandatory to consider a wider range of species than that which forms the primary focus of attention. For example, primate evolution must be considered against the background of the evolution of the placental mammals in general and should ideally include reference to marsupials, fossil mammal-like reptiles and the stem reptiles as well (see Chapter 4). As the essence of successful phylogenetic reconstruction resides in identification of characters present in individual ancestral stocks, one of the most important checks on the reliability of any inferred ancestral primate features is provided by compatibility with the inferred features of the earlier ancestral placental mammal stock (see later).

From this preliminary basis, all phylogenetic reconstruction – from the biochemical to the behavioural – ideally depends on the construction and testing of hypotheses concerning both the occurrence of the main branching points in the tree and the assignment of particular characters to them. It is important to emphasize here that it is both possible and desirable to construct alternative hypotheses regarding the occurrence and characterization of

branching points for any given set of species. Much depends on the reliability with which primitive features can be inferred for the initial ancestral stock (i.e. the ancestral primate stock in the case of the phylogenetic tree of primates), and some kind of testing of alternative hypotheses is required (see later).

The core of phylogenetic reconstruction resides, of course, in the selection and evaluation of **unit characters** for comparison in the group of species concerned. Every species can be envisaged as possessing a large number of definable characters, each of which can potentially exhibit one or more kinds of transformation (alternative **character states**) in other species. It is possible to take part or all of a given protein (e.g. myoglobin; see Chapter 11), an entire chromosome (see Chapter 11), a particular morphological structure (e.g. the placenta; see Chapter 9), a given physiological parameter (e.g. basal metabolic rate; see Chapter 6) or an identifiable behaviour pattern (e.g. locomotor activity; see Chapter 10) as a standard character for comparison between species. Phylogenetic inferences would then be based on the pattern of distribution of alternative states of the selected characters among the species in the sample. However, certain difficulties arise with respect to definition of the characters themselves. First of all, there are procedural problems of defining unit characters (Hecht and Edwards, 1977) and of selecting a range of characters to be used in phylogenetic analysis. Whatever approach is taken, some element of subjectivity (perhaps leading to bias) is bound to intrude, although in principle the problems should decline as the number of characters considered increases. In particular, it is a well-recognized principle that 'unit characters' that are closely associated in the same functional complex should not be treated as independent indicators of phylogenetic relationships (e.g. see Simpson, 1961; Harrison and Weiner, 1964).

The difficulties involved in defining the characters themselves are compounded by confusion over use of the adjective 'homologous'. The statement that any two species 'share a homologous character' can have at least two distinct meanings:

1. The unit character is equivalent and comparable between the two species, though it may exhibit different states (Remane, 1956b; Hanson, 1977; Hecht and Edwards, 1977).
2. The two species exhibit the same state of that character (Martin, 1973b).

It is, for instance, possible to regard a given protein molecule, such as myoglobin, as a homologous character shared by a number of species, even though the actual composition of the molecule might vary between them. On the other hand, when comparing those species one might restrict use of the term 'homologous' to the common possession of particular features (states) within that molecule (see Chapter 11). Because of this potential source of confusion, Bock (1973) has suggested that the level of homology should always be specified. One useful step in this direction is to confine use of the term 'homologous character' to a general indication of homology (e.g. possession of a particular protein or morphological organ) and to use the term 'homologous character state' to indicate more specific similarities. In line with this terminology, it can be stated that *the aim of phylogenetic analysis is to identify homologous characters and to interpret the pattern of distribution of homologous character states*.

As has been noted above, use of the adjective 'homologous' for characters shared by two or more species implies that a common genetic basis has been retained from an ancestral form. However, it is not possible to specify genetic continuity as part of an operational definition of homology (Hull, 1967; Martin, 1973b), because in most cases we cannot study the actual genetic basis of unit characters. Even at the molecular level, a major obstacle to reliable identification of genetic continuity is presented by the phenomenon of gene duplication (see Chapter 11). It is therefore necessary to make assumptions regarding the common genetic basis of unit characters (inference of character homology) and conclusions about the genetic

origins of particular transformations of those characters will emerge only after phylogenetic analysis (inference of character state homology). In practice, the decision to regard particular unit characters as homologous in a given range of species is based on a combined assessment of spatial and functional relationships to other characters, special qualities and – in certain cases – the occurrence of a series of intermediate forms suggesting transformation from a common basis (see Remane 1956b; Hanson, 1977). Among vertebrates, the brain can be regarded as a homologous unit character in all species, as it occurs in the same position in the body, protected by the skull and linked to the major sense organs according to a common pattern, as it consists of special cells (neurones) and as comparisons of brains in different vertebrate groups (fish, amphibians, reptiles, birds and mammals) reveal numerous transitional stages. Similarly, possession of the myoglobin molecule would appear to be a homologous character in vertebrates, as it always occurs in muscle in a particular structural relationship to other identifiable molecules, as it is relatively uniform in terms of the number of component parts (amino acids) and as variations in the structure of the molecule are relatively continuous throughout the vertebrates. In the case of protein molecules, however, standard criteria for recognizing homology will not exclude gene duplication.

At this point, it is necessary to take a decision about the treatment of fossil species. Fossil species represent a special case. On the one hand, there is a severe limitation in that they exhibit a relatively small and variable number of characters suitable for comparative analysis (see Chapter 2). On the other hand, fossils are uniquely valuable in that (given accurate dating of the sites from which they are derived) they can yield information on the approximate timescale of evolutionary divergence. Fossil evidence can be used in a general way to assist in the inference of the primitive condition, because the average condition for fossils from earlier time-horizons is bound to be more primitive than the average modern condition (see later). In addition, fossils may yield fairly direct evidence of convergent evolution in certain cases (Hennig, 1979). However, because of the special limitations on the information content of individual fossil species, they are best incorporated into phylogenetic reconstructions after comparison of the living species has yielded preliminary hypotheses about overall relationships. In this way, information on fossil species can to some extent be used as a **test** of the reliability of ancestral character states inferred for alternative phylogenetic trees. A suitable basic procedure for reconstruction of phylogenetic relationships through character analysis can therefore be defined as follows:

1. Identification of unit characters to be examined in a specified group of organisms.
2. Comparison of living forms, leading to the construction of alternative phylogenetic hypotheses with specification of ancestral stocks and of character states attributed to them.
3. Inclusion of known fossil forms in the competing phylogenetic reconstructions and preliminary establishment of the corresponding timescales.
4. Selection of the most probable phylogenetic tree, taking into account as much supplementary information as possible.

Having decided on the unit characters to be used (step 1), it is possible to proceed to an analysis of the distribution of alternative character states among the living species under comparison (step 2). Such analyses have been conducted in several different ways, but from the literature concerned primarily with morphological characters some practical guidelines ('rules of thumb') have gradually emerged. (For two particularly useful recent discussions of some of the issues, see Cartmill (1981) and Duncan and Stuessy (1984).) The main guidelines are as follows:

Irreversibility. A widely cited principle in the literature dealing with phylogenetic reconstruction is 'Dollo's Law' of irreversibility of evolutionary change (Dollo, 1893; Abel,

1929). This states that a structure lost during evolution will not reappear in its original form. Dollo illustrated his principle by pointing out that marsupials that have secondarily returned to ground-living habits retain features betraying the tree-living adaptations of the ancestral marsupial. This principle is generally recognized to be valid for complex characters, but it is clearly unreliable for simple characters. For example, at the molecular level it is quite possible that a simple structural feature of a protein that has changed in evolution may revert to its original form (see Chapter 11). Similarly, it is obviously possible that a given simple morphological feature that has been lost may reappear, especially if the developmental basis for that feature has been retained. Further, if the evolution of a character involves a transition from a relatively simple primitive state to a more complex derived state, reversal to the primitive state could easily occur by arrest of the derived development leading to greater complexity. Hence, the principle of irreversibility cannot be applied in a blanket fashion, but it is likely to hold good for many complex changes between species.

Universal character states. If a given character state is common to *all* living species in a given group, it is very likely that it was present in the initial ancestral stock of that group. For instance, all mammals have hair and mammary glands. It is therefore highly likely that their common ancestor possessed both features. Similarly, all living primates have a bony strut forming the lateral margin of the eye socket (i.e. the postorbital bar; see Chapter 7) and it is reasonable to infer that this feature was present in the common ancestral stock. Indeed, in practice this conclusion cannot be avoided objectively unless there is overwhelming fossil evidence indicating the general absence of that character state in undoubted early relatives of the group. In other words, universality of a given character state among living representatives might arise through multiple parallelism, but this will only be recognizable if there is very convincing fossil evidence available. In any case, the absence of the modern universal character state from early fossil species – even when these are relatively numerous – may always be interpreted in one of two ways. Instead of concluding that the character state was not present in the ancestral stock that gave rise to the modern species, it may be inferred that the fossil species diverged before the emergence of an ancestral stock that already possessed the character state found in all the living species. In practice, this possibility of alternative interpretations is unlikely to have a great influence on the overall accuracy of phylogenetic reconstruction; but it will introduce an additional element of uncertainty to the assignment of fossil species to phylogenetic trees and hence to the inference of times of divergence.

Common equals primitive. As a corollary of the interpretation that universal characters are likely to be primitive, it may be argued that, even when there are two or more alternative character states represented among the species compared, the most common character state is likely to be the primitive condition. (This is the principle of **commonality** (*sic*) recognized by Kluge and Farris, 1969.) For instance, in all living primates except *Homo sapiens* the foot has a divergent big toe adapted for grasping. As there is only one exception, it seems reasonable to propose that the common ancestor of primates possessed a grasping big toe and that this feature has been lost during human evolution (see Chapter 10). But in cases where the frequency of the commonest character state is not so overwhelming the application of this rule is by no means as convincing. It is a very common, though certainly not universal, feature of living primates that the upper molar teeth bear four main cusps apiece (Chapter 6). Some living primates, by contrast, have only three cusps on their upper molars. The principle 'common equals primitive' would indicate that four-cusped molars were present in the ancestral primates. This, however, conflicts with abundant evidence from other sources that three-cusped molars are primitive.

Although it is probably true in a very general sense that common character states are likely to

be primitive, there are several identifiable exceptions to the rule. In the first place, the relative frequencies of particular character states will depend heavily on the architecture of the phylogenetic tree. It is possible, for instance, that an initial ancestral stock might give rise to a species-poor, slowly evolving lineage and to a substantially modified, later ancestral stock yielding a rich diversity of descendant species. In such a case, character states developed in the later ancestral stock are likely to be far more abundantly represented among living representatives than the alternative character states present in the initial ancestral stock and retained by derivatives of the slowly evolving lineage. A good illustration of this effect is provided by members of the horse group (order Perissodactyla). Most living perissodactyls have a single, hoof-bearing terminal digit on both forelimbs and hindlimbs. However, a few perissodactyls (i.e. tapirs) instead have three functional digits on their forelimbs and hindlimbs. The rule 'common equals primitive' would in this case lead to the conclusion that the presence of a single digit on all four limbs was the primitive condition. This conclusion is, however, contradicted by an abundance of other evidence (see later). Obviously, the reliability of the rule will also decrease as the number of alternative character states increases and as their relative frequencies among living representatives become more balanced. In fact, with an increase in the number of alternative character states found in living species, there is an increasing likelihood that the primitive condition is no longer represented among them. In view of all of these limitations on the applicability of the rule 'common equals primitive', it must obviously be used with considerable caution.

Outgroup comparison. An alternative procedure for inferring the primitive character state for a given group of species is provided by the **outgroup** principle. Having conducted an analysis of the character states represented in the primary group of species, alternative hypotheses regarding the initial ancestral condition can potentially be 'tested' by reference to a separate, but relatively closely related, group. For example, character states inferred for the ancestral stock of living primates could be assessed by reference to another order of mammals traditionally regarded as quite separate (e.g. insectivores, rodents or carnivores). Similarly, character states inferred for the ancestral stock of the placental mammals could be tested through comparison with marsupials. As the two groups concerned in such a comparison must have shared an even earlier common ancestral stock, it is theoretically possible to obtain an additional perspective on primitive character states inferred for the group of primary interest and perhaps to reach a decisive answer in cases where other rules do not yield unequivocal results. An illustrative case is provided by the snout region in primates. In living lemurs and lorises, the nostrils are surrounded by an area of moist, naked skin (the rhinarium; see Chapter 7), whereas in living tarsiers and simians the nostrils are surrounded by hairy skin (rhinarium absent). It is on this basis that lemurs and lorises are collectively called **strepsirhines** and tarsiers and simians are labelled **haplorhines**. The rule 'common equals primitive' cannot really be applied here as both conditions are quite common. However, the living insectivores might be taken as the 'outgroup' for comparison. Presence of the rhinarium is universal among living insectivores, so it is reasonable to conclude that this feature was present in their common ancestral stock. This being the case, it seems quite likely that a rhinarium would still have been present in the ancestral primates and was lost as a derived feature of a later ancestral stock that gave rise to tarsiers and simians.

The outgroup principle has been given particular prominence in much of the literature on cladistic analysis (e.g. Hennig, 1979; Eldredge and Cracraft, 1980). In fact, a restricted form of the outgroup principle is often recommended. In this, it is prescribed that the outgroup selected for comparison should be the **sister group**; that is, the next most closely related group of organisms external to the primary group of species considered. Wiley (1981) has stated this restricted form of the outgroup principle as follows:

> Given two characters that are homologous and found within a single monophyletic group, the character that is also found in the sister group is the plesiomorphic character whereas the character found only within the monophyletic group is the apomorphic character.

It is, for instance, commonly accepted or implied that the insectivores are the sister group of the primates and there is little doubt that the marsupials constitute the sister group of the placental mammals. There are, however, two problems attached to the outgroup principle in general and to the sister group principle in particular. In the first place, recognition of another group of species as an 'outgroup' implies that the overall phylogenetic tree containing both the primary group and the outgroup has already been established, at least in outline. Recognition of the 'sister group' requires an even more precise prior knowledge of the overall phylogenetic tree. Hence, recognition of outgroups (and especially of sister groups) involves circular reasoning in that reconstruction of the phylogenetic tree for one group of species is made dependent upon prior understanding of a more extensive phylogenetic tree including at least one other group (see Cartmill, 1981). This is associated with the second problem, which is that the choice of an appropriate outgroup, and certainly the identification of the sister group, is not always clear-cut. It is, for example, not obvious which mammalian order would be the appropriate sister group for cetaceans (whales and dolphins) and similar uncertainty exists with respect to bats. Even for primates, it is not known with any certainty whether the appropriate sister group would be insectivores, tree-shrews (see Chapter 5), 'flying lemurs' or bats. Because of these uncertainties, the outgroup principle – especially in its restricted form as the sister-group principle– must also be used subject to reservations.

In fact, the most reliable test for inferred ancestral features of primates is provided by comparison with the inferred ancestral condition for placental mammals generally, rather than by a restricted comparison with one other order of mammals (the supposed 'sister group'). This is particularly necessary in cases where a feature has been lost by some of the species in a comparison. Wiley (1981) specifically noted that the outgroup rule is unreliable when loss of a feature, rather than acquisition, is concerned. An apt example is provided by the case of the sublingua discussed above. Among primates, the sublingua is present in prosimians and absent in simians, whereas it is uniformly absent from insectivores (*sensu stricto*). An exclusive comparison of primates with insectivores following the sister-group principle as stated by ·Wiley (above) would lead to the conclusion that the sublingua was absent in the common ancestral stock. Yet, as has already been mentioned, Sonntag's broad comparative study (1925) of the tongue in mammals indicated that a sublingua was present as the primitive condition.

In cases where the features involved in a sister-group comparison differ by transformation, rather than loss, the application of this principle is clearly more reliable. If character states A and B are found in the group of primary interest and only character state A is found in the sister group, then the most economical solution is to propose that character state A was primitive for both groups. However, there are a number of problems even where character transformations are concerned. In the first place, it is relatively rare to find that members of the sister group possess only a single state of the character concerned (i.e. exhibit a universal character state). Further, in cases where comparisons are confined to living forms, it is quite possible that the primitive condition may be no longer preserved in any living representative. The existence of alternative character states in the sister group as well as in the group of primary interest, together with the possibility that the primitive condition is no longer represented, will obviously render the application of the outgroup principle more complex. This added complexity

combines with the potential problem of 'parallel' evolution in relatively closely related groups discussed above. Rigid application of the sister-group principle will tend to lead to erroneous identification of primitive homologies in cases where similar character states have been developed in parallel in sister groups. A particularly clear example of these combined difficulties is provided by consideration of relative brain size in living insectivores and primates (see Chapter 8). Application of the sister-group principle would lead to the correct inference that there was a relatively small brain in the common ancestor of insectivores and primates, but the size of that brain would be overestimated because even insectivores have undergone some increase in relative brain size during their evolution. That is to say, no living insectivore or primate preserves the primitive condition with respect to relative brain size and there has been multiple parallelism in the expansion of brain size.

The outgroup principle is particularly important in cases where there is no fossil record and even more so when quantitative distances are involved, as with biochemical evidence (Chapter 11). A special case of the use of the outgroup principle is found in the interpretation of molecular data. Here, an 'unrooted' tree for a given group of animals (e.g. primates) can be 'rooted' (i.e. the common ancestral node can be fixed) by taking any more distantly related species as an outgroup. Virtually any other mammalian species or group would serve the purpose for a primate tree.

Transformation series. It is sometimes possible to infer the existence of a probable sequence through which a series of alternative character states might have developed (Hennig, 1950, 1979; Hanson, 1977). The special term **morphocline** has been coined for such a transformation series (Maslin, 1952). However, this term is doubly unsuitable because it implies that transformation series are restricted to morphological characters and because the term 'cline' already has a well-established and totally different meaning in zoogeography. (A cline is a gradual change in one or more observable characters over the geographical range of a species population.) Further, sequences of alternative character states may exhibit a branching arrangement, rather than a simple linear organization. For all of these reasons, the term **transformation series** is preferable (Hennig, 1979). Interpretation of a transformation series in phylogenetic reconstruction involves two logically distinct elements: (1) recognition of a restricted sequence for conversions from one character state to another; and (2) identification of the evolutionary point of origin in the sequence. With only four alternative character states (A,B,C and D), there are 48 possible evolutionary sequences, given that these character states may form a simple chain in any order and that the ancestral condition might be located anywhere in the sequence. The first step is to establish the order of the species in the transformation series, perhaps as follows:

$$A \rightleftharpoons C \rightleftharpoons B \rightleftharpoons D$$

A good example of such a series is provided by sequential rearrangements of individual chromosomes (Chapter 11).

The second step is to establish which of the four alternative character states is the primitive one (assuming that the primitive condition is still represented among the species under comparison and that the transformation series is therefore complete). In the terminology of Maslin (1952), this is the recognition of the **polarity** of the transformation series ('morphocline polarity'). In the absence of definitive criteria for establishing the primitive condition, it is arguably preferable to examine all possibilities in the construction of alternative phylogenetic hypotheses. For this exercise, recognition of basic transformation series at least provides some information about possible sequences of change in character states and hence narrows down the number of possibilities to be considered. If, on the other hand, the attempt is made to determine the primitive condition and hence the 'polarity' of a transformation series as a preliminary to phylogenetic reconstruction, one or more of the

other guiding principles listed here may be of value.

The logical antecedent. In certain cases, it may be justifiable to argue that a particular character state is likely to be primitive on straightforward logical grounds. For instance, it is reasonable to postulate that evolution will tend to lead to an increase in brain size rather than a decrease and that mammals with large brains are therefore derived from ancestors with smaller brains. As a general rule, the principle of the logical antecedent will depend on the assumption that simplicity is a criterion of primitiveness. This assumption is not always reliable (Hecht and Edwards, 1977). For example, parasites typically undergo secondary simplification in association with their reliance on host species for many of their basic needs. In addition, loss or reduction of individual features is a relatively common occurrence in evolution. Hence, as with the other rules discussed above, the principle of the logical antecedent should be used only as a general guideline.

The logical antecedent may also be based on functional or other considerations in cases where one condition necessarily precedes another. For example, development of a larynx requires the prior development of lungs, and bees must necessarily have evolved after the evolution of flowers (M. Cartmill, personal communication).

The ontogenetic antecedent. Developmental evidence provides one of the most powerful available aids for separating primitive from derived character states and every attempt should be made to include ontogenetic information wherever possible. Such evidence was specifically recognized by Hennig (1950, 1979) in his principle of **ontogenetic character precedence**. Unfortunately, the potential value of developmental information was for some time underestimated as a result of negative reaction to Haeckel's sweeping statement of the principle of **recapitulation** in his 'biogenetic law'.

Ontogeny is the short and rapid recapitulation of phylogeny. . . . During its own rapid development . . . an individual repeats the most important changes in form evolved by its ancestors during their long and slow palaeontological development (Haeckel, 1866, as quoted by Gould, 1977).

Numerous objections can be made to such a simplistic interpretation of ontogeny as a recapitulation of phylogeny (e.g. see de Beer, 1958). Indeed, this interpretation is closely linked to the insidious notion of the *Scala naturae* discussed earlier in this chapter. Haeckel's concept of recapitulation represents a special extension of the *Scala naturae*, going beyond the usual idea that living species form a series of 'frozen ancestors' to imply that they are 'frozen' into embryological sequences as well. However, the original formulation by von Baer (1828) of the relationship between ontogeny and phylogeny was more reasonable and his four laws of development (as quoted by Gould, 1977) provide lasting and valuable guidelines for the interpretation of developmental evidence:

1. The general features of a large group of animals appear earlier in the embryo than special features.
2. Less general characters are developed from the most general, and so forth, until finally the most specialized appear.
3. Each embryo of a given species, instead of passing through the stages of other animals, departs more and more from them.
4. Fundamentally, therefore, the embryo of a higher animal is never like (the adult of) a lower animal, but like its embryo.

It is possible to paraphrase von Baer's interpretation briefly with the statement that the embryonic development of a particular species recapitulates embryonic stages of its ancestors, not the adult stages. The entire subject of the relationship between ontogeny and phylogeny has been re-examined in detail by Gould (1977), who places particular emphasis on evolutionary changes in developmental timing that result in parallels between stages of ontogeny and stages of phylogeny.

As long as ontogenetic evidence is considered

carefully, it can serve as a powerful tool for the reconstruction of phylogenetic history. A particularly striking example is provided by interpretation of the associated evolution of the jaw hinge and middle-ear mechanism in mammals (see Chapter 4). The developmental sequence in a modern mammal quite closely reflects evolutionary transitions that have now been extensively documented from the fossil record. Ontogenetic evidence also clearly supports the inference that mammals are derived from an ancestral stock in which both the forelimbs and the hindlimbs bore extremities with five digits apiece (pentadactyl limbs; see Chapter 10). In this latter case, the principle of the ontogenetic antecedent clearly conflicts with the principle 'common equals primitive' with respect to the evolution of the limbs in perissodactyls (see discussion above).

Incidentally, it should be noted that developmental evidence provides an equally powerful tool for the recognition of homologous characters. As convergent evolution is most likely to affect the later stages of development of a character, detailed similarity in the development of a given character in two species provides more convincing evidence of homology than mere similarity between them in the completed adult expression of that character. Further, it is sometimes the case that the distinction between homology and analogy is obscured by secondary changes later in development. For instance, component bones of the skull may fuse together without obvious trace in adult mammals and only developmental evidence can reveal the true origins of certain structures. To take just one case, the utilization of developmental evidence has greatly clarified reconstructions of the evolution of the bony auditory bulla in mammals (see Chapter 7).

The palaeontological antecedent. As noted above, fossil evidence can, in a very general way, indicate the direction of transition of certain character states in evolution. One can reasonably infer that the average condition exhibited by fossils of a given group at a particular time horizon in the fossil record will be more primitive than the average condition for relatives of that group at a more recent time horizon (e.g. the present). This principle was recognized by Hennig (1950, 1979) in his concept of **geological character precedence**. For instance, because early fossil representatives of the class Mammalia tend to possess relatively large numbers of teeth, whereas more recent fossil mammals and (particularly) extant forms tend to have smaller numbers, it seems likely that any transformation series should begin with a large complement of teeth as the primitive character state. Conversely, fossil evidence suggests that the crowns of molar teeth have generally tended to become more complex during mammalian evolution. Butler (1982) cites Cope's observation that almost all of the early Palaeocene mammal species from the Puerco Beds of New Mexico had upper molars with only three principal cusps whereas many species from the Eocene and more recent deposits had quadratic upper molars bearing four main cusps. It can therefore be concluded that, as a general rule, there has been a reduction in the numbers of teeth accompanied by an increase in the complexity of cheek teeth in mammalian evolution (see Chapter 6). Another example, in which this palaeontological principle leads to the same conclusion as the logical antecedent, is provided by fossil evidence showing that all early mammals had relatively small brains (Jerison, 1973; see also Chapter 8). Similarly, the palaeontological principle agrees with the ontogenetic principle in indicating that mammals are derived from an ancestral stock characterized by pentadactyl limbs.

These last two cases are just two illustrations of the important point that particular hypotheses about primitive character states may be supported by more than one guiding principle. Indeed, such mutual reinforcement is particularly valuable, because – as with the other guiding principles – the palaeontological antecedent must be applied cautiously. Among other things, it is essential to survey a large range of species for each geological stage, as conclusions based on just a few species can easily be subject to bias. In the limiting case of a comparison between a single early fossil species

and a single later fossil (or living) species, it cannot be concluded that all character states exhibited by the former are necessarily more primitive than their counterparts in the latter. It is an inevitable consequence of differential evolutionary rates that, in any time horizon, some species with a relatively large number of primitive character states will be contemporaries of other species with many advanced character states.

The concept of the palaeontological antecedent is closely allied to the concept of evolutionary **trends**. Insofar as this concept is used in a very general, empirical sense, this is all well and good. There is undoubtedly great biological significance in the general statements that numbers of teeth tend to decrease during mammalian evolution and that brain size tends to increase. However, it must be recognized that an empirical generalization is involved and that this does not by any means imply some kind of definite directional tendency in evolution (i.e. **orthogenesis**). It is one thing to observe that a particular type of cumulative evolutionary change seems to be favoured by prevailing selection pressures; it is quite another to suggest that there is some inherent directional tendency in evolutionary change.

Empirical laws of association. In certain circumstances, it may be possible to differentiate between alternative hypotheses concerning ancestral stocks by examining the relationships between characters that have initially been considered in isolation. Typical examples are provided by the empirical observations, based on living mammals, that small-bodied species are both typically nocturnal and predominantly insectivorous (see Chapter 4), and that species with relatively large brains tend to have particular forms of placentation (see Chapter 9). As many biological characters are influenced in predictable ways by body size (see Chapter 4), it is also possible to make some inferences regarding adaptations to be expected for an ancestral form with an approximately specified body size. In this connection, it should be noted that the palaeontological antecedent indicates that early mammals were uniformly quite small in body size and that there has been a general trend for individual mammalian groups to diversify over a wide range of body sizes. A special case of association between characters is provided by correlation between transformation series (Maslin, 1952; Hennig, 1979), which may provide corroboration of phylogenetic hypotheses in certain cases. Here, too, body size is likely to play an important part in influencing correlation between character states.

If the concept of empirical laws of association is interpreted too broadly, it might be seen as providing support for the common assumption that there is a consistent association between primitive character states (Kluge and Farris, 1969). In other words, it may be assumed that a species that is primitive in certain respects is likely to be primitive in other respects as well. However, in its exteme form the notion of general association between primitive character states is nothing less than an alternative formulation of the suspect idea of 'frozen ancestors' already discussed. Among other things, a rigid supposition of association between primitive character states would conflict with the widely observed phenomenon of mosaic evolution. Hence, empirical laws of association must be defined in a very narrow and precise sense in order to preclude any suggestion that one can distinguish between primitive and advanced organisms. Even if it is found that a given species exhibits a particular complex of primitive character states, this does not mean that there will be no derived states identifiable for other characters. It is, for example, conceivable that a species occupying an ecological niche fairly close to that occupied by the initial ancestral species of the group may preserve a cluster of primitive states in certain characters. Conversely, descendant species that have invaded new **adaptive zones** (Simpson, 1961; Van Valen, 1971; Jerison, 1973), distinct from that occupied by the ancestral species, are likely to exhibit numerous associated derived character states. Nevertheless, a species occupying an ecological niche comparable to that occupied by the ancestral species of a group is likely to show at least some

derived character states for several reasons related to the dynamic nature of phylogenetic change. For instance, changes in other local species are likely to arise in the course of time and these will, in effect, alter the parameters of the ecological niche of a given species. Further, any favourable mutation that arises in a species will inevitably be subjected to positive selection even in the rare case of an ecological niche that remains relatively stable over long periods of time.

Extrinsic information. In addition to the principles discussed so far, which essentially involve **intrinsic** biological properties of the species compared, it is possible to make use of certain categories of information of an **extrinsic** kind (Hennig, 1979). Relevant information could include, for example, ecological and geographical data. Hennig refers to the assessment of such environmental factors as constituting the 'chorological method'. Several authors have, in fact, specifically stated that such extrinsic information should not be used in the reconstruction of phylogenetic relationships, but should only be considered later in discussion of the broader implications of inferred phylogenetic trees (e.g. Schaeffer, Hecht and Eldredge, 1972; Eldredge and Tattersall, 1975; Hecht, 1976; Hecht and Edwards, 1977). Hecht and Edwards (1977) concluded that 'such extrinsic data can tell us very little about the genealogical relationships of the organism' and this view is shared by Eldredge and Cracraft (1980). Yet, although it is true that extrinsic information is different in kind from intrinsic information, in that it does not directly reflect genetic continuity or discontinuity, it can nevertheless provide valuable additional insights.

A good illustration of the value of extrinsic information is provided by reconstructions of the evolutionary relationships of the Malagasy lemurs. It has been suggested above that the lemurs should, in the first instance, be assumed to represent a monophyletic 'natural group', and this is the procedure that will be adopted in later chapters of this book (see also Martin, 1972b). By contrast, several authors have concluded that the lemurs do not constitute a strict monophyletic group. In particular, it has been repeatedly suggested that the mouse and dwarf lemurs (family Cheirogaleidae) are more closely related to lorises (family Lorisidae) than to other lemurs (Cartmill, 1975; Hoffstetter, 1974; Mahé, 1972, 1976; Schwartz, Tattersall and Eldredge, 1978; Szalay and Katz, 1973; Tattersall, 1982; Tattersall and Schwartz, 1974). Indeed, there are numerous apparently homologous similarities between cheirogaleids and lorisids (Charles-Dominique and Martin, 1970), but the two alternative interpretations of the degree of phylogenetic relationship between these two taxa arise because of uncertainty about the inferred status of those similarities (similarities inferred to be primitive or convergent in the first case; inferred to be derived in the second). Assessments of the evidence have in most cases been based on intrinsic characters, notably features of skull morphology. However, the pattern of geographical distribution is also directly relevant to the question of the phylogenetic relationships between lemurs and lorises.

As explained in Chapter 1, all of the living lemur species are confined to Madagascar. The Mozambique Channel now acts as a major geographical barrier to the migration of terrestrial vertebrates between Africa and Madagascar, and it is undoubtedly the case that the lemurs have been isolated on Madagascar for some considerable period of time. Geological evidence suggests both that the separation between Madagascar and Africa has increased (probably irregularly) during primate evolution and that some degree of separation was probably present at the very beginning of the evolutionary radiation of the primates (Martin, 1972b; Tattersall, 1982; Tattersall and Schwartz, 1974). Hence, some assessment of the likely geographical relationships between Africa and Madagascar throughout the Tertiary period would provide additional evidence regarding the probability that the Malagasy lemurs represent a monophyletic group. Further, as it would appear that there has been a filtering effect in the migration of terrestrial vertebrates to

Madagascar and as the Mozambique Channel was probably present in some form throughout the period relevant to primate evolution, it can be assumed that there has always been a relatively low probability of migration across the channel, although that probability has now declined virtually to zero. Cartmill (1975) provides a very effective review of the relevant features of skull morphology in lemurs and lorises, in which he examines several alternative phylogenetic reconstructions and assesses their relative merits. On the basis of the intrinsic features of the skull considered, he concludes that the most economical hypothesis is that the mouse and dwarf lemurs are, indeed, more closely related to the lorises than to other lemurs and that one of the dwarf lemurs (*Allocebus*) is particularly close to the lorises. However, this hypothesis necessitates back-migration of the ancestor of the lorises from Madagascar to Africa and thus requires at least two crossings of the Mozambique Channel (one in each direction), whereas the hypothesis that the lemurs and lorises constitute two strictly monophyletic groups requires only one crossing. As Cartmill has observed, the choice between the two competing hypotheses depends not only on an assessment of total amounts of evolutionary change in intrinsic characters of the skull, but also on some assessment of probability of successful crossings of the Mozambique Channel. It is not justifiable to reject the latter kind of evidence merely on the grounds that it is 'extrinsic' in kind.

It is also important to note that the likelihood of successful migration will depend on a general ecological factor – namely the level of competition from other species that a migrant species encounters on arrival. For various reasons (e.g. the presence of a greater variety of vertebrate groups), such competition was probably much fiercer in Africa than in Madagascar throughout the Tertiary period. Migration of primates from Africa to Madagascar would therefore have been far more probable at any time than migration from Madagascar to Africa. Accordingly, any phylogenetic hypothesis that requires back-migration of lemurs from Madagascar to Africa (e.g. see

Szalay and Katz, 1973) is subject to a further reservation.

It is therefore clear that, in the final assessment of any phylogenetic hypothesis, both intrinsic and extrinsic factors should be incorporated. This is particularly important, as phylogenetic analysis should not be restricted to the narrow issue of reconstructing branching points between taxa. The ultimate aim must surely be to reveal as much as possible of the phylogenetic history of any group of species, including ecological and geographical aspects. It is, of course, true that the inclusion of extrinsic information may render the search for objectivity more difficult and will certainly further complicate the development of appropriate quantitative techniques for phylogenetic reconstruction. On the other hand, there is little point in oversimplifying the biological context of phylogenetic relationships to such an extent that misleading answers are generated. Given that extrinsic information is distinctly different from intrinsic data reflecting actual genetic continuity, perhaps the best solution is to construct a preliminary set of alternative phylogenetic hypotheses purely on the basis of intrinsic data and then to use extrinsic information as an additional 'test' for selecting the optimal reconstruction. This is the procedure that is recommended here.

Several authors have, in fact, taken the analysis of extrinsic factors considerably further. Hennig (1979) cites the principle of 'chorological progression', in which phylogenetic diversification is equated with the progression of species in geographical space, and the associated concept of **vicariance biogeography** has been developed as a companion to phylogenetic reconstruction (Platnick and Nelson, 1978; Rosen, 1978, 1979; Wiley, 1981). However, although it is clearly desirable to establish a defined procedure for the evaluation of geographical and ecological evidence, there are numerous complicating factors involved in the interpretation of biogeographical evidence and as yet formalized analysis has thrown little additional light on primate evolution.

As has become clear from the above enumeration

of guiding principles for separating primitive from derived character states, none can be regarded as absolutely reliable. It is quite common to find that a particular character state identified according to one principle as primitive will appear to be derived in the light of another. Conversely, if two or more principles suggest the same interpretation for a given character state, then this can be taken as valuable confirmation. It is therefore advisable to use as many of the above principles as possible for phylogenetic reconstruction and to bear firmly in mind the fact that the procedure is always inferential. Phylogenetic reconstruction, at least as presently conducted (and probably unavoidably), depends on assessment of probabilities, not on establishment of certainties. Indeed, it must be emphasized that phylogenetic analysis cannot in fact lead to the separation of derived from primitive homologies; it can only lead to the specification of **inferred primitive** and **inferred derived** character states. In explicit recognition of this limitation, the scheme indicated in Fig. 3.12 provides a simple flow diagram of the overall procedure that might be adopted in a comprehensive approach to the reconstruction of the phylogenetic history of a group such as the Primates.

PHYLOGENETIC TREE-BUILDING

Eventually, it is necessary to assimilate all of the information considered into an 'optimal' phylogenetic tree that most aptly reflects the balance of probabilities established. Clearly, the process of overall assimilation must be as objective as possible and it may be argued that, in the long run, the most satisfactory approach must be one that converts the available data into a form appropriate for computerized analysis. Ideally, any phylogenetic trees that are generated should be justifiable in quantitative terms and alternative trees should be amenable to some kind of statistical treatment that will rank them according to formal criteria of acceptance. Indeed, there have been many attempts already to arrive at computerized solutions to the problems of phylogenetic reconstruction; but,

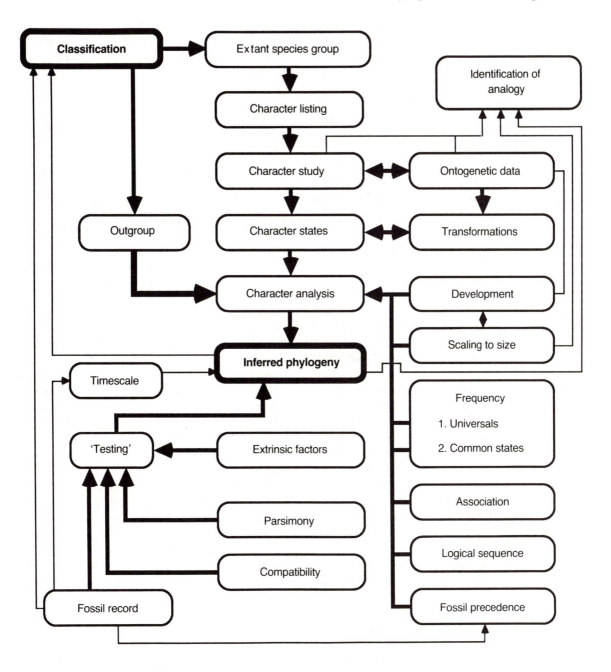

Figure 3.12. Basic flow diagram indicating the main procedures advocated for phylogenetic reconstruction using traditional biological evidence. (The use of chromosomal and biochemical evidence is a special case, discussed in Chapter 11.)

because of a widespread failure to appreciate the true complexity of the task, it cannot be claimed with confidence that any of these attempts has been crowned with real success. At present, it is difficult to see how the heterogeneous categories of information outlined above can be reduced effectively to any homogeneous mathematical form. Further, if the logical problems inherent in phylogenetic reconstruction have not been clearly formulated at the outset, no computer program – however sophisticated – can produce a reliable solution. In particular, the nature and scale of the difficulties generated by variation in rates of evolutionary change have frequently been greatly underestimated. Unfortunately, published computer-generated phylogenetic reconstructions tend to acquire a mantle of respectability regardless of the validity of the assumptions inherent in the reconstruction procedures. In particular, it must be emphasized that any technique that relies purely on the assessment of degrees of shared similarity without any recognition of the problem of differential evolutionary rates is bound to be suspect even if highly sophisticated mathematical techniques are applied to tree-building. As elsewhere in biology, there is a common tendency to believe that any form of quantification necessarily leads to superior results. It is worth recalling here the words of Le Gros Clark (1959): 'The well-known aphorism "science is measurement" is strictly true, but it by no means follows that all measurement is scientific.'

In addition to the problems involved in mathematical abstractions of the basic data and of the guiding principles of phylogenetic reconstruction, there are several specific obstacles involved in the process of tree-building itself. The first of these resides in the practical limitation that, despite the availability of modern computer techniques, it is physically impossible to conduct a direct assessment of all possible trees even for quite small numbers of species. As Felsenstein (1978) has pointed out, this difficulty is often overlooked by those concerned with phylogenetic reconstruction and taxonomy. If N species are included in a given comparison and

only two-way branching from a single ancestor is allowed (dichotomous nodes), the number of possible phylogenetic trees (T) is given by the formula:

$$T = \prod_{n=2}^{N} (2n - 3)$$

(i.e. T is the product of successive odd integers from 1 to $2n - 3$).

If the restriction to dichotomous branching is relaxed and the possibility of multiple branching is permitted (polyvalent nodes), the number of possible trees is even greater (Felsenstein, 1978). The astronomical increase in the number of possible trees with the number of taxa included in a comparison is illustrated in Fig. 3.13. The formula for trees with dichotomous nodes indicates that with a small group of only nine taxa the number of possible trees would be more than two million and that with 20 taxa the number would have risen to almost 10^{22}. With polyvalent nodes, the number of possible trees for 20 taxa is more than a hundred times greater, approaching 10^{24} (Felsenstein, 1978). As there are some 55 genera of living primates (Napier and Napier, 1967; Table 1.1), it is obvious that the direct evaluation of all possible branching patterns is out of the question for primate genera, even without attending to the need to consider other mammal groups as well. Such calculations of the numbers of possible phylogenetic trees are of little direct benefit, other than indicating the scale of the problems involved in phylogenetic reconstruction and the obvious need for some method of inference that does not require the assessment of all possible trees. Felsenstein (1978) has aptly stated: 'The principal uses of these numbers will be to double-check algorithms and notation systems, and to frighten taxonomists.'

Because it is out of the question to evaluate all possible branching patterns linking taxa in any group of appreciable size, it is essential to carry out some preliminary clustering of the species concerned. In any case, many of the possible trees that could be constructed for a given set of species would be clearly nonsensical and it would be very wasteful to calculate the amounts of

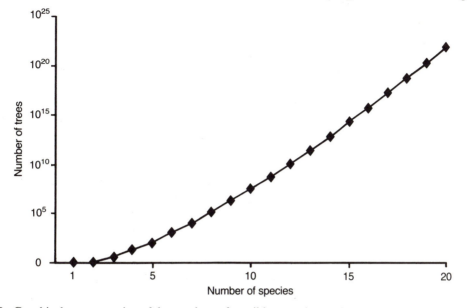

Figure 3.13. Graphical representation of the numbers of possible rooted trees that can be generated for 1–20 taxa, with a restriction to dichotomous nodes.

evolutionary change required for every possible tree. The traditional approach is to begin with the 'natural groups' indicated by existing classifications and to examine alternative arrangements both within and between them in a step-by-step manner. A more standardized approach, applicable where differences in character states can be relatively easily converted into numerical form (as with biochemical data; see Chapter 11), is the construction of a bivariate matrix of 'distances' between pairs of taxa. From this matrix, it is possible to identify preliminary clusters of relatively similar species and to proceed with tree-building in a stepwise fashion. This is the basis for many computer-based methods for inference of phylogenetic trees using numerical data and is a standard feature of numerical taxonomy (see Sneath and Sokal, 1973). However, such an approach has several pitfalls, the most important being that it does not formally allow for the existence of significantly different rates of evolution and hence does not cater for the vital distinction between primitive and derived character states. In addition, it has been proved that it is impossible to construct a workable algorithm for the identification of

the optimal tree. Accordingly, a tree inferred directly from a matrix of pair-wise character state distances between taxa must be refined in some way. One common procedure is to take this initial tree and to alter the positions of individual taxa on a trial-and-error basis in order to seek improvement. Whether this procedure actually permits identification of the 'optimal' tree is a moot point.

Conversion of character state differences between taxa into numerical form introduces another problem that has been widely recognized. This concerns the differential **weighting** of various characters and character states. Some authors (e.g. Sneath and Sokal, 1973; McKenna, 1975) have advocated equal weighting, on the grounds that any form of weighting must introduce an element of subjectivity into phylogenetic reconstruction. However, as has been recognized by Hecht and Edwards (1977), the concept of equal weighting is illusory. The initial selection of characters to be subjected to analysis amounts to an involuntary form of weighting and, as has been shown, there are several good reasons for attributing greater importance to complex chracters. It is also evident that acquisition of a new character or character state is likely to be considerably more significant in genetic terms than suppression thereof. In practice, most authors engaged in phylogenetic reconstruction use some form of weighting and it is by no means clear that arbitrary numerical representation of relative value is superior to the intuitive concentration on specific characters that has generally characterized traditional biological approaches.

Associated with the problem of weighting is the general influence of **body size**, which has already been mentioned and which is discussed in detail in the appendix to Chapter 4. In comparing character states between two species of different body sizes, it is essential to draw some distinction between character states that reflect the scaling constraints imposed by size and those that reflect some special adaptation that would still be present if the two species had the same body size. In other words, it is necessary to separate the scaling effects of body size from the special adaptations of individual species. The latter should clearly carry far more weight in phylogenetic reconstruction, because multiple parallelism has occurred in changes in body size during mammalian evolution. The necessary separation can be achieved in an empirical fashion by means of allometric analysis (see Chapter 4), which permits comparison of residual values for quantified character states after 'removal' of the effects of body weight. In the long run, any overall reconstruction of ancestral stages in a phylogenetic tree should ideally include some approximate specification of ancestral body sizes. Alternatively, any list of character states for a given ancestral stock should be set on a sliding scale indicating the predictable, progressive changes expected in relation to body size. As yet, however, the need to take into account the scaling effects of body size has been relatively neglected in attempts to reconstruct phylogenetic history (e.g. see Simpson, 1961; Hennig, 1979; Szalay and Delson, 1979; Eldredge and Cracraft, 1980; Wiley, 1981) and much remains to be done in this direction.

Mention has been made above of attempts to seek 'improvements' in preliminary phylogenetic trees, and this is just part of the wider question of choosing between alternative hypothetical phylogenetic trees. It is now necessary to deal with the question of the choice of overall criteria for deciding to accept one of several competing hypothetical phylogenetic reconstructions and to reject the alternatives. The most commonly cited principle, particularly with respect to inference of patterns of phylogenetic relationship from molecular data, is that of economy or **parsimony** (Moore, Barnabas and Goodman, 1973; Goodman, Moore, *et al.*, 1974; Cartmill, 1975, 1981; Holmquist, 1976; Panchen, 1982; Felsenstein, 1983; Kluge, 1984). Although there are various ways in which the parsimony principle may be applied in practice, it essentially requires that the attempt should be made to minimize the total amount of evolutionary change postulated for a given phylogenetic tree. In its simplest form, the parsimony principle is therefore an extension of Occam's razor: 'It is vain to do with more what can be done with fewer.' Biologists engaged in phylogenetic reconstruction have generally respected the need to maintain economy in inferring relationships, without necessarily applying the criterion of parsimony in any strict, quantitative sense. However, as an adjunct to computer tree-building techniques, increasingly explicit use is being made of the principle of parsimony for assessing the relative merit of competing phylogenetic reconstructions. In addition, parsimony is commonly cited as the overriding principle to be adopted in cladistic approaches to phylogenetic reconstruction, even in cases where differences between species are difficult to quantify. Wiley (1975) has stated that the principle of parsimony must be applied in phylogenetic analysis 'not because nature is parsimonious, but because only parsimonious hypotheses can be defended by the investigator without resorting to authoritarianism or apriorism'. This view is echoed by Eldredge and Cracraft (1980), while Kluge (1984) has rightly emphasized the need to distinguish between 'evolutionary parsimony' and 'methodological parsimony'.

It is undoubtedly true that, as a general rule,

it is reasonable to discount hypothetical phylogenetic reconstructions that are obviously uneconomical in terms of the pattern of changes suggested. On the other hand, it has become abundantly clear that the principle of parsimony should not be applied too strictly, especially when it is implied that evolutionary change has necessarily taken the most direct course possible. One reason for this has emerged from reconstructions of phylogenetic relationships based on protein sequence data (see Chapter 11). It has been found that radically different hypothetical phylogenetic trees may differ only marginally in terms of the total evolutionary change inferred. Whereas it is probably reasonable to discount many trees that require considerably more evolutionary change than is required by the most parsimonious tree, it is probably not justifiable to conclude that the latter should necessarily be selected in preference to a tree that requires only a slight increase in the total evolutionary change. There is, however, a more fundamental reason for rejecting strict application of the parsimony principle to calculations of total evolutionary change. This arises from the problem of **differential evolutionary rates** illustrated in Fig. 3.9. At least in some cases, beginning with the terminal taxa of an original tree exhibiting marked differences in evolutionary rate between lineages, it is possible to arrive at a more parsimonious, but erroneous, reconstruction (Fig. 3.14). Indeed, in the case shown in Fig. 3.14 the more parsimonious reconstruction not only entails erroneous inference of the primitive features of the initial ancestral stock; it also creates the illusion of constancy of evolutionary rate along all lineages. It is doubtless the case that such problems are reduced both by increasing the number of taxa included in a particular reconstruction and by maximizing the information used for each taxon. In a richly branching tree based on a large data set, it is to be expected that some of the confounding effects of differential evolutionary rates might be mitigated, though this will depend on the specific branching architecture of the tree. Nevertheless, although a phylogenetic hypothesis that is obviously extravagant with

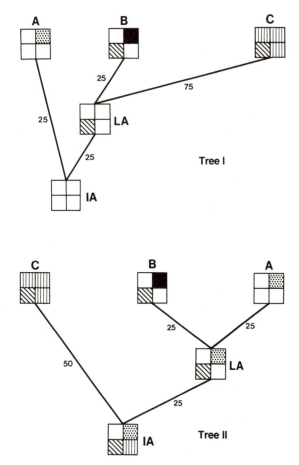

Figure 3.14. Illustration of the potential confounding effect of differential evolutionary rates in application of the principle of parsimony to the selection of an 'optimal' phylogenetic tree. Tree I shows the real phylogenetic relationships between three extant taxa and the characteristics of the relevant ancestral stocks. (For the sake of simplicity, each box represents 100 characters, which change in blocks of 25.) As in Fig. 3.9, there are marked differences in evolutionary rates between lineages and the real tree requires a total of 150 character changes. Tree II below shows the 'optimal' reconstruction based on the principle of parsimony. This tree requires a total of only 125 character changes and is therefore more parsimonious that the real tree. Incidentally, in this case the more parsimonious (but false) reconstruction not only misrepresents the initial ancestral condition, but creates the illusion of a constant rate of change along all branches. (Figure reprinted from Martin (1981a), with kind permission of the Institute of Biology, London.)

respect to the amount of evolutionary change postulated should be rejected, one must in all cases be wary of strict application of the parsimony principle as an isolated criterion.

Another criterion that may be applied in assessing the relative merits of competing phylogenetic hypotheses is the principle of **compatibility**. Wiley (1981) has stated that overall cohesiveness provides one of the most important tests of any phylogeny. Apart from being generally economical in terms of the amount of evolutionary change inferred, a phylogenetic tree should also show cohesiveness in several respects. In the first place, any phylogenetic tree must show **internal consistency**. That is to say, any individual species that appears to be relatively primitive or relatively advanced in certain character states in comparison with one set of species must emerge in the same light in other comparisons. In reconstructions involving the specification of character states for individual ancestral stocks, following the general approach suggested in Fig. 3.12, internal consistency should be assured as a matter of course. However, in cases where character states are not specified individually and overall measures of 'distance' between species are used (as, for example, with immunological data; see Chapter 11), the principle of internal consistency is quite important. When such overall measures of 'distance' are used, it is standard procedure to construct a bivariate matrix of differences between pairs of species and to seek the most parsimonious tree that best expresses the general pattern. One can, for instance, treat the numerical 'lengths' of branches within the tree as variables in a set of simultaneous equations. When this is done, it is commonly found that no single solution provides a perfect fit to the data matrix; but it is usually assumed that the optimal tree is the one that provides the closest fit. In principle, the degree of concordance between a tree and the original matrix of 'distances' can provide an additional check on the suitability of the tree (see also Sneath and Sokal, 1973); but there are various ways in which such compatibility might be defined (e.g. minimal

total deviation from a perfect fit versus the most consistent fit over all branches).

A different form of concordance is enshrined in the requirement that the character states inferred for any hypothetical ancestral stock should be mutually compatible and should constitute a functionally feasible assemblage. Because there have been comparatively few explicit attempts to reconstruct ancestral stages, this principle has been somewhat neglected; but there is considerable scope for its application in the final assessment of hypothetical phylogenetic trees. If the ultimate aim is to gain an understanding of the biological properties of ancestral forms, rather than simply obtaining an outline branching pattern, then the principle of character state compatibility must assume particular importance. It would, for example, be very revealing to explore the issue of functional compatibility and feasibility in relation to hypothetical reconstructions of ancestral molecules (see Chapter 11).

The concept of compatibility has also been invoked with respect to a form of character analysis that depends upon the identification of 'cliques' of characters in the species involved in a comparison (Estabrook, 1979; Gardner and La Duke, 1979). In such 'compatibility analysis', species are grouped on the basis of clusters of shared characters. However, it would seem that this approach ignores the distinction between primitive and derived features and that it is therefore possible that species might be grouped together on the basis of clusters of shared primitive features. This being the case, it is unlikely that clique analysis will prove to be of any real benefit to the reconstruction of phylogenetic relationships (see also Wiley, 1981).

There is yet another sense in which compatibility may be sought, and that is in the degree of compatibility between hypothetical phylogenetic trees generated from different data sets. Instead of incorporating all available data into a single tree-building approach (as is implied in Fig. 3.12), it might be preferable to analyse the data in separate sets. This is, in fact, what tends to happen in practice, as biologists commonly deal with only one specialist field at a time. A biologist working, say, on the mammalian skull may pay little attention to data from other areas of mammalogy, such as reproductive biology, and there is a particularly wide gulf between those studying traditional morphological evidence and those concerned with chromosomal and biochemical data. In addition, it is actually desirable to analyse evidence from different fields separately, as the problems of relative weighting become particularly acute when radically different kinds of characters (e.g. amino acid sequence of a protein versus structure of the placenta) are involved. Therefore, in addition to developing methods for examining the internal consistency of individual phylogenetic trees, it is also necessary to develop methods for assessing the degree of concordance between trees based on different data sets. Ruse (1979) has referred to this kind of concordance as consilience. It might, for example, be possible to draw up a short list of a given number (e.g. a dozen) of the most parsimonious trees obtained with each data set in isolation and then to identify a single tree that best fits all data sets. There has already been some work on this, but much remains to be done before a standardized procedure can be adopted.

Chapter four

Adaptive radiation of mammals

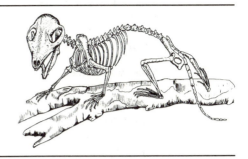

Primate evolution is one, relatively minor, part of the extensive adaptive radiation of the placental mammals since Cretaceous times (Lillegraven, 1972; Eisenberg, 1981). The first clearly recognizable placental mammal fossils are known from the late Cretaceous and, as has been mentioned in Chapter 2, the subsequent Cenozoic era covering the past 66 million years (Ma) is commonly known as the Age of Mammals. During the latter part of the Cretaceous period and throughout the Cenozoic, the known adaptive radiation of the placental mammals produced a great variety of species attributable to 35 orders, 16 of which have living representatives (Simpson, 1945; Lillegraven, 1972). Obviously, it is necessary to have some general understanding of this adaptive radiation in order to place primate evolution in its proper perspective. In fact, for the reasons outlined in Chapter 3, there are certain fundamental questions involved in the reconstruction of primate evolution that can be answered only against a background of fairly detailed information on the early origins of the placental mammals generally. For instance, early 'advances' in primate evolution can only be defined in relation to the hypothetical ancestral condition for placental mammals, and only on this basis can one confidently allocate fossil forms to the order Primates or decide on such issues as the affinities of tree-shrews (see Chapter 5). Indeed, in order to reach a clear formulation of the evolutionary origins of any single order of placental mammals it is also necessary to take into account the evolution of the monotremes and the marsupials (see Table 4.1) and hence the earliest origins of all mammals from reptilian stock. In these terms, the first roots of primate phylogenetic reconstruction must date back to more than 300 Ma ago, when the first recognizable mammal-like reptiles made their appearance.

DISTINCTIONS BETWEEN LIVING MAMMALS AND REPTILES

Reproduction

Mammals were defined initially on the basis of characteristics shared by living forms. As noted earlier, the inclusion of fossil forms in such a definition is a more recent development that followed on from formulation of the theory of evolution by natural selection. Accordingly, definitions of mammals originally concentrated

Table 4.1 Major groups of living mammals (after Simpson, 1945 and Lillegraven, 1972)

Orders	Common names
Subclass Prototheria	
Monotremata	Monotremes (spiny anteaters; platypus)
Subclass Theria	
Infraclass Metatheria	
Marsupialia*	Marsupials ('pouched mammals')
Infraclass Eutheria[†]	Placental mammals
Insectivora	Insectivores (shrews, hedgehogs, moles, etc.)
Dermoptera	'Flying lemurs' (colugos)
Chiroptera	Bats
Primates	Primates
Edentata	Edentates (sloths, anteaters and armadillos)
Pholidota	Pangolins
Lagomorpha	Rabbits, pikas and hares
Rodentia	Rodents (squirrels, rats, mice, etc.)
Cetacea	Dolphins and whales
Carnivora[‡]	Carnivores
Tubulidentata	Aardvarks
Proboscidea	Elephants
Hyracoidea	Hyraxes
Sirenia	Sea-dogs and dugongs
Perissodactyla	Odd-toed hoofed mammals (e.g. horses)
Artiodactyla	Even-toed hoofed mammals (e.g. ruminants)

* Some recent authors (e.g. Ride, 1970; Tyndale-Biscoe, 1973; Kirsch, 1977a) differ from Simpson in dividing the marsupials into separate orders.
† This name has given rise to the common name 'eutherian mammals' as a synonym for 'placental mammals'.
‡ Some modern authors (e.g. Corbet and Hill, 1980) separate the seals, sea-lions and their relatives in a distinct order, Pinnipedia.

on non-fossilizable features of reproduction, which have been particularly important in mammalian evolution. The word 'mammal' itself refers to the universal occurrence of milk-producing **mammary glands** and associated suckling behaviour among living mammals. This central characteristic was underlined by the introduction of the term Mammalia by Linnaeus (1758) and its subsequent retention in all major works dealing with the classification of the vertebrates. It is also significant that, with the exception of the few surviving egg-laying monotreme species, female mammals give birth to live young (**vivipary**) after the development of a diminutive fertilized egg within the female tract. In association with this, the egg contains very little yolk (**microlecithic** condition) and sustenance for the developing offspring comes more directly from the mother, first by diffusion into the embryonic system within the uterus and later in the form of milk provided by **suckling**. To some extent the marsupials represent a condition intermediate between monotremes and placentals in that the egg contains some yolk and a shell membrane is usually present throughout gestation, but they also possess placentation (Kirsch, 1977b). In addition to suckling, another obvious character common to all living mammals, but not directly documented by the fossil record, is the possession of at least some **hair** in the body covering, as opposed to the uniform coat of scales in living reptiles or of feathers in living birds.

Metabolism

There seems, in fact, to be an indirect link between the two universal features of modern mammals identified above, namely the possession of hair and the development of mammary glands. This link has to do with the energy relationships of the mammalian body. The evolution of hair in mammals was probably associated with the development of at least some degree of independent control of body temperature. Living reptiles typically show fluctuations of body temperature, to some degree passively following variations in environmental temperature (**poikilothermy**) except where they are able to seek out particular temperature conditions. By contrast, all mammals can control their body temperatures directly at least to some extent and typically maintain a reasonably constant level despite variation in environmental temperatures (**homeothermy**). Most living placental mammals can maintain their body temperatures at an elevated level in the region of 37 °C over a wide range of ambient temperatures. The most important aspect of this distinction between mammals and reptiles is that the former are equipped with the metabolic capacity to generate their own body heat (at a considerable cost) whereas the latter can regulate their body temperatures only to a limited degree by producing heat from general muscular activity and by moving between different temperature zones of the habitat (behavioural thermoregulation). For this reason, many authors (e.g. see McNab, 1978a) prefer to use the terms **endothermy** (heat generated internally) and **ecothermy** (heat derived from external sources) to refer to the primary metabolic distinction between mammals and reptiles. In other words, the homeothermy of modern mammals can be resolved into two main components: endothermy and efficient thermoregulation.

With respect to **thermoregulation**, mammals possess such accessory aids as shivering and sweating to buffer their body temperatures against metabolic limitations and/or marked fluctuation in environmental temperatures (Young, 1957). The typical mammalian body covering of hair acts to limit heat loss by trapping an insulating layer of air against the body surface and the rate of heat outflow from the body can be controlled by flattening or raising the hair (differential piloerection). On the other hand, if excess body heat is generated it can be dissipated by means of special mechanisms such as increase in blood flow to the body surface, sweating and (in most species) panting. The majority of mammals possess **sweat glands** at least on some parts of their skin and these are usually accompanied by numerous **sebaceous glands** that produce a greasy waterproofing secretion (sebum). Development of these skin glands doubtless took place in parallel with the development of body hair in mammals, as part of the overall adaptation for homeothermy. As it also seems likely that mammary glands evolved as modified moisture-secreting skin glands (Long, 1969, 1972), it is reasonable to conclude that the evolution of hair was a necessary precursor of the evolution of mammalian milk production (**lactation**) and hence of suckling.

Cranial and dental morphology

Many additional fundamental differences between living mammals and living reptiles are found when cranial and skeletal structures are compared in detail. These differences are generally associated with the primary functions of feeding, respiration, locomotion, the sense of smell (olfaction) and hearing, and have far-reaching implications. Perhaps the most striking differences involve the functional anatomy of the jaws and the mechanism of the middle ear. Living reptiles have several component bones in each half of the lower jaw, typically including along with the **dentary** six other postdentary bones (angular, articular, coronoid, prearticular, splenial and surangular). Of these lower jaw bones, it is the **articular** that forms the typical reptilian jaw hinge with the **quadrate** of the skull, as is also the case in fish, amphibians and birds.

By contrast, living mammals have only one bone in each side of the lower jaw, the dentary, and the **condyle** of this bone articulates directly with the **squamosal** of the skull to form the typical mammalian jaw hinge (Fig. 4.1). The mammalian dentary is also distinctive in that it has a prominent ascending **coronoid process**, anterior to the condyle, for the attachment of powerful jaw muscles. Detailed embryological studies have shown (following Reichert, 1837 and Gaupp, 1913) that the dentary on each side of the mammalian lower jaw is homologous with the well-developed anterior bone of the same name in the reptilian lower jaw. Further, these studies have shown that some of the original bony elements 'liberated' from the lower jaw during mammalian development are incorporated into the middle-ear mechanism. Reptiles, in common with amphibians and birds, have a single small bone known as the **columella** that transmits sound vibrations from the eardrum (**tympanic**

membrane) to the auditory mechanism of the inner ear. In all living mammals this function is served by a chain of three small bones (**ossicles**). The ossicle nearest the inner-ear mechanism is homologous with the columella of amphibians, reptiles and birds, and because of its stirrup-like shape it is called the **stapes**. According to Reichert's theory (1837), the other two mammalian ear ossicles are derived from the former components of the reptilian jaw hinge. The quadrate has become transformed into the **incus** (anvil), which is the middle of the three ossicles, and the articular from the lower jaw has become the **malleus** (hammer), which ensures contact with the tympanic membrane (Fig. 4.2). Together, the three ossicles in the ear of the modern mammal form a flexible chain (malleus–incus–stapes) whose sound-conducting properties can be modified subtly by contraction or relaxation of tiny muscles attached to the ossicles. In addition, the mammalian ear ossicles

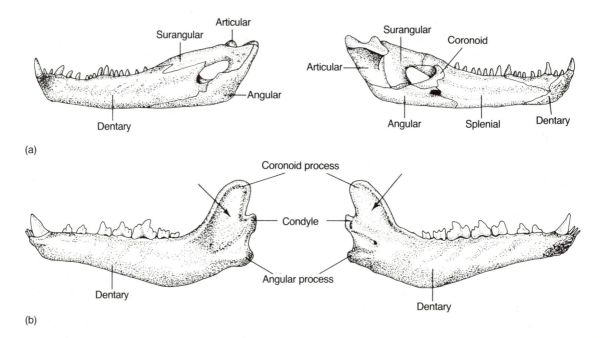

(a)

(b)

Figure 4.1. External (left) and internal (right) views of the lower jaw in a modern reptile (alligator (a)) and in a modern mammal (dog (b)). In the reptile, the lower jaw articulates with the skull though the articular bone, whereas in the mammal all of the postdentary bones have been lost and it is the articular condyle of the dentary that articulates with the skull. Note the well-developed ascending coronoid process in the mammalian jaw (arrowed). (After Romer, 1955.)

play an enhanced part in impedance matching (Kemp, 1982). The conversion of low-pressure, air-borne sound waves to high-pressure, fluid-borne waves within the ear mechanism requires that the eardrum should be larger than the membrane in the aperture (fenestra ovalis) leading to the inner ear. The lever ratios between the three ear ossicles of mammals serve to increase the pressure transmitted from the eardrum to the inner ear. This represents a considerable refinement over the condition found in other terrestrial vertebrates (amphibians, reptiles and birds).

Mammals are also distinctive in that the tympanic membrane is housed in a special bony element known as the tympanic or **ectotympanic** bone, which is most probably derived from another postdentary bone originally present at the rear end of the reptilian lower jaw – the **angular** (Allin, 1975). In amphibians, reptiles and birds the tympanic membrane is typically located at or near the surface of the skull (see Fig. 4.2) and there is no special bony supporting element comparable to the ectotympanic of mammals. Finally, most modern mammals have developed some kind of ventral bony shield beneath the middle-ear cavity containing the ossicles. There is occasionally nothing more than a ring-like, virtually horizontal ectotympanic element supporting the eardrum, as in the modern monotremes and some insectivores (see Fig. 4.3), in which a limited floor for the middle ear is provided by a membrane of fibrous connective tissue (Moore, 1981). But in most living marsupials and placentals there is an **auditory bulla** beneath the ear region of the skull. The bulla is usually ossified and is typically associated with a more vertical alignment of the ectotympanic bone and the eardrum contained therein. In many, but not all, mammals the ventral floor of this bulla includes one or more additional ossified elements referred to as **entotympanics** (van Kampen, 1905; van der Klaauw, 1931; Moore, 1981). Formation of a ventral shield beneath the middle ear seems to have occurred in parallel in different lines of evolution of marsupials and placentals, with the result that the bony pattern of the auditory bulla

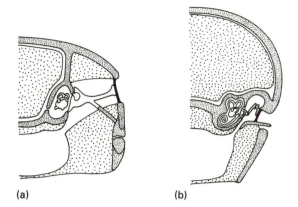

(a)   (b)

Figure 4.2. The ear mechanism in an early terrestrial vertebrate (a) and in a typical mammal (b). Coarse stippling: soft tissues; fine stippling: bones of skull and lower jaw; white: ear ossicles and inner ear apparatus; black: tympanic membrane (eardrum). (After Romer, 1966.)

tends to be distinctive, and therefore diagnostic, for different groups (see Chapter 7). For instance, living marsupials differ from living placental mammals in that the **alisphenoid** bone of the skull (Fig. 4.3) plays a major part in the formation of the bulla. Further, among living placental mammals there are major differences between groups in the relative importance of other bony elements (ectotympanic, ento-tympanics, basisphenoid and petrosal) in the development of the wall of the bulla (Novacek, 1977a; Moore, 1981). Primates, in fact, seem to be unique among placental mammals in that the **petrosal** bone contributes most to the bulla wall (see Chapter 7).

Changes in the jaw structure of mammals are associated with radical modification of the teeth in connection with the increasing importance of **mastication** of food in mammalian evolution. It seems likely that the development of the dentary/squamosal jaw hinge in mammals was related primarily to the need for increased power and control of jaw movement. Modern reptiles and birds typically swallow their food whole or pull it to pieces before ingesting it, and in many reptiles (including some fossil dinosaurs) there is a special mechanism (cranial kinetism) that permits temporary dislocation of the rear jaw bones for

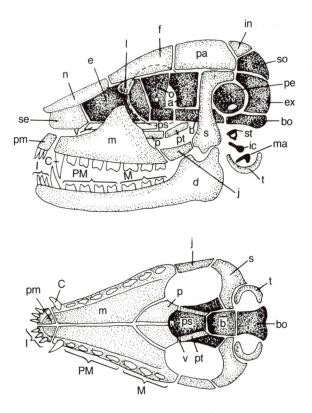

Figure 4.3. The major bony elements and dentition in the typical mammalian skull: above, lateral view; below, ventral view. Heavy stippling: bones replacing cartilage of the original chondrocranium; light stippling: dermal bones of more superficial origin. The dentition exhibits the generally accepted maximal dental formula for placental mammals: $I_3^3 C_1^1 P_4^4 M_3^3$. Abbreviations: a, alisphenoid; b, basisphenoid; bo, basioccipital; C, canines; d, dentary; e, ethmoid; ex, exoccipital; f, frontal; I, incisors; ic, incus; in, interparietal; j, jugal (malar); l, lacrimal; M, molars; m, maxilla; ma, malleus; n, nasal; o, orbitosphenoid; p, palatine; pa, parietal; pe, petrosal; pm, premaxilla; PM, premolars; ps, presphenoid; pt, pterygoid; s, squamosal; se, nasal septum; so, supraoccipital; st, stapes; t, ectotympanic; v, vomer. (Simplified from Kühn, 1957.)

the swallowing of large food items (e.g. animal prey). Mammals, on the other hand, can (with very few exceptions, such as anteaters and baleen whales) fragment their food between ingesting it into the mouth and swallowing. Increased stability of the jaw mechanism plays a necessary part in this, but the most significant feature is the typical development of a single row of teeth with different shapes and functions on each side of the upper and lower jaws (**heterodonty**). In particular, living mammals typically possess complex cheek teeth bearing several projections or **cusps**. (Dolphins, with their simple conical

cheek teeth, are one of the few exceptions to this rule.) By contrast, reptiles usually have simple conical teeth that can grasp, pierce and rip, but not masticate. This latter property permits mammals generally to break up food items into relatively small pieces, thus allowing the more rapid digestion of swallowed food. As Olson (1959) has pointed out, one of the key differences between reptiles and mammals lies in the increased activity levels of mammals; accelerated digestion is a prerequisite of increased activity. It is also necessary to have relatively rapid digestion to provide the considerably increased amounts of energy required for the maintenance of a constant, elevated core body temperature. Therefore, the evolution of heterodonty and a modified jaw mechanism in mammals was doubtless linked to the evolution of homeothermy/endothermy, at least during the latter stages of refinement. Perhaps the most extreme contrast with the mammalian condition is provided by reptiles (e.g. boa constrictors) that simply swallow large animal prey (with the aid of cranial kinetism). They can then lie up for days to digest their meal slowly without the demands for rapid energy release that would be imposed by maintenance of a constant, elevated body temperature. (Incidentally, it should be noted that birds meet their need for increased amounts of energy by fragmenting food in the gizzard.)

Mammalian heterodonty is associated with another sharp distinction from the reptilian condition. In reptiles, the simple conical teeth are replaced in waves throughout life (Osborn, 1973), a condition known as **polyphyodonty**. Mammals, by contrast, typically have a maximum of two sets of teeth, although the wave-like pattern of replacement persists in modified form – the **milk teeth** and the **permanent teeth** (**diphyodonty**). As the names imply, the milk teeth are usually fairly short-lived (deciduous) and their replacements (if any) must last for the rest of an individual's life. It should also be noted that mammalian teeth are characterized by prismatic enamel (Boyde, 1971), which is typically lacking from reptilian teeth. Possession of prismatic enamel may have been necessary for the evolution of diphyodonty, as such enamel may permit teeth to last longer (Grine, Vrba and Cruickshank, 1979; Grine and Vrba, 1980).

Teeth in the permanent dentition of modern mammals are allocated to four separate categories on the basis of their shape and their sequence from the front to the back of the jaw: **incisors**, **canines**, **premolars** and **molars** (Fig. 4.3). Of these, the molars can be simply defined as teeth that appear only in the permanent set, having no precursors among the deciduous milk teeth. Alternatively, they can be seen as teeth actually belonging to the first (milk) set that have no replacements. The common picture of mammalian tooth replacement is that the milk dentition consists of incisors, canines and premolars (also referred to as 'milk molars' or 'deciduous molars'), while the permanent dentition is constituted by the replacements of these teeth together with the molars. However, the situation is more complex than this and it has been shown, for instance, that in living marsupials the only teeth to be replaced are the last lower premolars (Archer, 1978), with the remaining teeth – incisors, canines and anterior premolars – retained throughout life. Nevertheless, the basic distinction between incisors, canines, premolars and molars remains a useful one, particularly as they generally represent functionally separate categories.

The numbers of the different kinds of teeth in the upper and lower jaws are customarily presented as a **dental formula**, prescribing the teeth present on one side of the skull (the other side being essentially a mirror image). For instance, the maximum numbers of the different categories of teeth found among living adult placental mammals (with the exception of the aberrant dolphins and their relatives) are generally regarded as: three incisors, one canine, four premolars and three molars on each side of the upper and lower jaws. This maximal dentition (total number of teeth = 44) is represented by the formula $I_3^3 C_1^1 P_4^4 M_3^3$. Until relatively recently (see Chapter 6), it was almost universally accepted that this is the ancestral formula for the placental mammals (e.g. Weber, 1927; Young, 1957; Colbert, 1969). Among living placentals, it still occurs in pigs and certain species of insectivore. It is also found in some extinct placentals, such as the early Oligocene *Anagale*, once believed to be allied to tree-shrews. The extant marsupials have a different maximal dental formula: $I_4^5 C_1^1 P_3^3 M_4^4$ (total number of teeth = 50). Note that living marsupials can have up to four molars in their upper and lower jaws; living placentals never have more than three. The premolars and molars of these two major groups also have different morphology (Clemens, 1979), which suggests some fundamental divergence between marsupials and placentals in terms of dental development and function early in their evolution.

The premolars and molars – **cheek teeth** – perform the function of mastication, whereas the anterior teeth (incisors and canines) typically have a grasping function. Accordingly, the cheek teeth are characterized by varying degrees of complexity of cusp pattern and they commonly have two or even three roots to ensure better anchorage in the jaws. Although the cusp patterns of the cheek teeth of modern marsupials and placentals show considerable variation, it is now widely accepted that the basic pattern in the upper and lower molars is a triangular array of cusps with an accessory heel on each lower molar. It is generally assumed that the marsupials and placentals, at least, can be traced back to a common ancestor in which such molars (**tribosphenic molars**) were already present (Patterson, 1956; see Chapter 6). The acquisition of a complex cusp pattern on the cheek teeth during marsupial and placental evolution permitted the development of increasingly refined mechanisms for mastication. It also required far more precise juxtaposition (**occlusion**) of the working surfaces of the upper and lower tooth rows during mastication; occlusal relationships between upper and lower teeth are therefore of particular significance. Details of cheek tooth form, masticatory mechanisms and occlusal relationships are given in Chapter 6 and a particularly instructive review is provided by Lumsden and Osborn (1977).

The occurrence of limited tooth replacement (diphyodonty) among mammals generally implies some connection with growth and development. In fact, as has been pointed out by several authors (e.g. Tanner, 1963; Colbert, 1969; Pond, 1977), growth in reptiles is typically a relatively continuous process throughout life, whereas in both mammals and birds there is an initial period of quite rapid growth followed by an extensive period of clearly recognizable adulthood in which body size remains largely unchanged. As Tanner (1963) has put it, growth in mammals is 'target-seeking'. In reptiles, increasing body size as an individual grows is typically matched by a progressive switch in diet (e.g. from small animal prey to large animals or even to plant food), but in mammals there is instead a major switch from the mother's milk to the typical adult diet, and this may be either abrupt or somewhat drawn out (see Pond, 1977). The milk dentition of placental mammals can thus be regarded as a temporary arrangement to bridge the gap between initial independent feeding and the attainment of adulthood, serving its function during the main period of postnatal growth. During this period, the skull is growing and the adult jaws can consequently accommodate longer tooth rows (i.e. including the molars) in the permanent dentition (Ewer, 1963). Seen in this light, the odd condition in modern marsupials, with only the last lower premolars replaced (see Archer, 1978), is particularly significant, as it suggests a major difference in developmental strategy between marsupials and placentals that probably also accounts for the difference in number of molar teeth (four against three). It is reasonable to conclude, therefore, that the evolution of diphyodonty might have proceeded hand in hand with the evolution of mammalian suckling behaviour (Ewer, 1963; Hopson, 1973). Mammals are born with disproportionately large heads and the head remains relatively large throughout development. Although the large head size at birth is to some extent associated with the precocious development of the brain compared with other organ systems (see Chapter 8), it also ensures the presence of relatively large jaws early on to accommodate the milk set of teeth (Pond, 1977).

Living mammals also differ from living reptiles in other features of skull structure, notably in the anterior region of the skull (see Fig. 4.3). Reptiles (with the exception of the secondarily specialized, aquatic crocodilians) do not have a horizontal bony partition between the nasal cavity and the oral cavity; the mouth serves as a common passage for both food and air and the nostrils are small and widely separated. Mammals, however, have (in parallel with the crocodilians) developed a bony **secondary palate**. This horizontal partition, together with the soft palate behind it, maintains the separation between air and food up to the region of the pharynx, where the air must cross the food channel on its way to and from the ventrally situated trachea (windpipe). In mammals, the bony secondary palate is formed primarily by the maxilla and palatine bones, with a small anterior contribution from the premaxilla (Fig. 4.3). Mammals also differ from reptiles in that the nasal openings of the skull (nostrils) are large and are combined into a single opening at the front of the skull, divided only by a thin nasal septum.

The development of a secondary palate in mammals has several implications. First, it permits an adult mammal to continue breathing while masticating food, and, of course, uninterrupted, relatively rapid respiration is required for homeothermy/endothermy (there is naturally a direct connection between breathing rate and energy requirements). A very similar argument applies to suckling, in that a secondary palate permits a young mammal to breathe during suckling bouts. This is a particularly acute requirement for certain marsupials, in which an infant maintains a firm grasp on a teat during much of its life in the pouch. It is also likely that the bony palate reinforces the upper jaw. Last, but not least, separation of the nasal cavity from the oral cavity has permitted certain additional developments in mammalian evolution. The effectiveness of respiration can be increased by filtering, heating and moistening air before its arrival in the lungs. In the nasal cavity of most mammals are **turbinal bones**, which are extensively folded bony scrolls bearing moist nasal epithelium. The increased surface area provided by these bones and the tortuous pathway imposed on inhaled air together ensure the necessary pretreatment of the air. Once again, increased respiratory efficiency doubtless enhanced the development of homeothermy, at least at some stage of mammalian evolution. In fact, most of the turbinal bones of mammals are **ethmoturbinals**, formed from outgrowths of the ethmoid bone (Fig. 4.3), which is a universal element in mammalian skulls but which is completely lacking from reptiles. The ethmoturbinals are supplemented by a **maxilloturbinal** on each side of the skull (see also Chapter 7).

Respiration

Efficient respiration is further enhanced in mammals by several other special adaptations that distinguish them from reptiles. The most important of these concern the heart and the lungs. In mammals and birds, the heart has become completely divided into four chambers (two auricles and two ventricles) such that the circulation to the lung is effectively separated from that to the rest of the body, with venous blood passing to the lung for oxygenation and oxygenated blood supplying the remaining tissues (Romer, 1955). All living mammals also have a muscular **diaphragm**, separating the heart and lungs in a thoracic cavity from the other viscera in the general abdominal cavity. Reptiles lack a diaphragm and consequently have less effective control over the inhalation and exhalation of air. (Birds also lack a diaphragm, but they have developed a totally different mechanism for increasing respiratory efficiency; see Romer, 1955.) The operation of the mammalian diaphragm in breathing is consolidated by muscular action of the thoracic rib-cage, which is strengthened by reinforcement of the ribs and attachment of the cartilaginous ventral ends of most of them to a prominent **sternum**. In association with this special development of the thoracic region, living mammals normally lack proper ribs on the neck (cervical) vertebrae and on the lumbar vertebrae, although ribs occur in these areas in modern reptiles. These modifications of the mammalian thorax, combined with refinements of the nasal passage and of the circulatory system, mean that mammals can breathe more rapidly and more effectively than reptiles, to support a markedly higher level of energy metabolism. In fact, recent experimental work with rats (Ruben, Bennett and Hisaw, 1987) has indicated that the diaphragm in mammals is required not for maintenance of basal metabolic turnover or of a constant body temperature, but for increase in energy metabolism during activity.

Smell

The development of a separate nasal cavity and proliferation of turbinal bones in mammals has also been linked to increasing emphasis on the olfactory sense. Among other things, the turbinal bones provide a very large surface area for olfactory epithelium. Within the nasal cavity of all mammals, at least during embryonic development, there are also two elongated, cigar-shaped organs (**Jacobson's organs**) containing olfactory epithelium and lying in scrolls on either side of the midline nasal septum (see Fig. 7.14). These organs are present in amphibians and in most reptiles as well, but they are restricted to relatively small pouches in the roof of the mouth. In many mammals they are considerably larger and serve as important accessory olfactory organs in a wide variety of species. In short, mammals are typically extremely well equipped with olfactory structures in comparison to modern reptiles and birds, and it is likely that the increased emphasis on olfaction in mammalian evolution generally has influenced the evolution of the central nervous system (see Chapter 8). Unfortunately, there has not yet been a proper quantitative comparison to show whether the olfactory components of the mammalian brain are larger, relative to body size, than in reptiles, but Jerison (1973) has suggested that an increase in relative size has taken place in at least some lines of mammalian evolution during the Cenozoic era.

It is also worth recording that the skin glands developed in association with mammalian hair have also provided the basis for the development of a wide variety of localized **marking glands** that permit many mammalian species to perform scent-marking in their home ranges (Schaffer, 1940; Ralls, 1971; von Holst, 1974; Müller-Schwarze, 1983). The secretions of such glands, commonly in conjunction with urine, faeces and other body secretions, provide for a rich repertoire of olfactory communication in numerous mammalian species.

Locomotion

Another suite of characters typifying modern mammals is related to locomotor developments. Living reptiles, in so far as they have retained limbs at all, have a sprawling posture with the limbs spread out to the sides. By contrast, the limbs of mammals are essentially placed vertically beneath the body, with the elbows directed more or less backwards and the knees pointing forwards. This constitutes yet another adaptation for active movement, which is a hallmark of mammalian evolution generally. In association with the more vertical orientation of mammalian limbs, the three component bones on each side of the pelvis (**ilium, ischium** and **pubis**) have become fully fused in all adults and the ilium (the blade-like extension above the socket containing the articular head of the femur) is inclined forwards rather than backwards. In addition, the **scapula** has become the major bone in the mammalian shoulder girdle and in living species it typically bears a prominent spine running across its dorsal surface. Other major diagnostic features of the mammalian post-cranial skeleton are particularly related to the emergence of structural consistency. The basic components have become considerably more standardized than those in reptiles. For instance, the vertebral column has an essentially fixed pattern; living mammals characteristically have 7 cervical vertebrae, 12–14 thoracic vertebrae, 5–7 lumbar vertebrae and 2–5 (usually 3) sacral vertebrae. It is only in the number of tail (caudal) vertebrae that substantial variation is found; but, even so, the mammalian tail tends to be shorter than the reptilian tail.

There is also a very characteristic pattern of bones in the mammalian hand and foot. The numbers of terminal bones (**phalanges**) in the digits of both hand and foot are less than in reptiles and there is a distinctive mammalian pattern. The number of phalanges in each digit from the innermost (thumb or big toe) to the outermost is expressed by a **phalangeal formula**. The typical reptilian phalangeal formulae of 2.3.4.5.3 and 2.3.4.5.4 for the hand and foot, respectively, contrast with the uniform formula of 2.3.3.3.3 for both hand and foot in generalized mammals (Fig. 4.4). Further, in the ankle of marsupials and placentals two of the tarsal bones have assumed a distinctive and very consistent arrangement for articulation with the lower leg bone, the **tibia**. The **talus** (= astragalus) has a well-marked articular surface and lies on top of the **calcaneus**, which has a backwardly directed heel section for attachment of the tendon of the calf muscles that provide the power for foot movement during locomotion (see Chapter 10). In a series of papers, Lewis (1980a, b, c, 1983) has provided a very effective review of major features in the evolution of the mammalian foot (see Chapter 10).

Mammals also have a distinctive and consistent mechanism for movement of the head relative to the vertebral column. There are two rounded articular surfaces (**occipital condyles**) at the rear end of the skull, located on either side of the lower margin of the foramen magnum. These condyles articulate with the first cervical vertebra, the **atlas**, and permit movement of the head in an essentially vertical plane. The atlas, in turn, articulates with a specialized second cervical vertebra, the **axis**, which bears a prominent peg (the **odontoid process**) permitting rotation of the head and atlas around the long axis of the body. This arrangement provides for a combination of strength and versatility in head movements. By contrast, reptiles typically have only a single occipital condyle and the atlas/axis complex, though present, lacks complete fusion (e.g. of the peg to the axis). One further striking feature of the mammalian locomotor skeleton meets requirements for increased stability of the component bones while allowing for growth to the 'target' body size for each species (Tanner, 1963). This is, namely, the possession of cartilaginous growth zones separating the main shafts (**diaphyses**) of numerous bones from their articular ends (**epiphyses**). Distinct cartilaginous growth zones permit a bone to grow in size while having reinforced (ossified) articular ends for functional, weight-bearing joint relationships with other bones. Reptiles and birds typically lack epiphyses; the articular ends of their bones are themselves cartilaginous and must combine growth requirements with the function of articulation. Such an arrangement is understandable for continuously growing reptiles with a sprawling gait, but it is difficult to explain why birds lack epiphyses. In mammals, the epiphyses can fuse with the diaphyses once the target body size has been attained and the articular surfaces are throughout better suited for more vertical limb orientation combined with enhanced locomotor activity.

Brain

Finally, there are numerous differences between modern mammals and reptiles in the size and organization of the central nervous system. Jerison (1973) has clearly shown that brain size, relative to body size, is typically far smaller in reptiles than in mammals. In fact, reptiles and amphibians seem to be rather similar in this respect, whereas both birds and mammals have notably larger brains, relative to body size (see later). As both birds and mammals are endothermic, in contrast to reptiles and amphibians, it seems likely that some link exists between increased energy turnover and relative

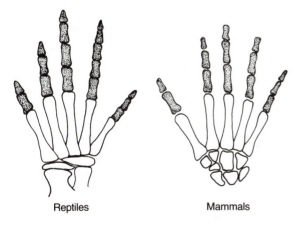

Reptiles   Mammals

Figure 4.4. The basic hand patterns of mammals and reptiles illustrating the standard phalangeal formulae. The phalanges are stippled. (After Wood Jones, 1941.)

brain size. One suggestion, for which there is increasing supporting evidence, is that this link involves the investment of parental energy in the offspring (Martin, 1981c; see also Chapter 9). This would, among other things, explain why parental care is widespread among birds and universal among mammals (birds and mammals having relatively large brains), whereas it is typically poorly developed among reptiles and amphibians.

The difference between mammals and reptiles in brain size relative to body size involves several areas of the brain, but particularly the cerebral hemispheres and the cerebellum (see Fig. 8.2). The larger **cerebellum** of mammals provides a further indication of the importance of locomotor developments in mammalian evolution. It is known that the cerebellum is concerned with the control of movement in mammals, especially with respect to equilibrium and fine control of body movements. It therefore seems quite likely that expansion of the mammalian cerebellum was related, among other things, to the development of an increased vertical orientation of the limbs and an overall increase in activity. There are also qualitative differences between the mammalian and reptilian brains (see also Chapter 8). Expansion of the mammalian **cerebral hemispheres** has centred around the development of a new area of cortex (**neocortex** or **neopallium**) over and above the original cortical area (**palaeocortex** or **palaeopallium**) found in reptiles (see Fig. 8.2). In addition, there are conspicuous fibre tracts linking the cortex to the cerebellum in mammals (once again underlining the importance of locomotor adaptations), and the cerebellum itself is characterized by subdivision into convoluted 'hemispheres' and flocculi. In reptiles, the two cerebral hemispheres of the brain are connected by a single major transverse tract of fibres forming the **anterior commissure**. In placental mammals (but not in monotremes or marsupials), there is an entirely new, and typically prominent, tract called the **corpus callosum** that supplements and replaces the anterior commissure to various degrees. This unique feature reflects the importance of the cerebral hemispheres, and particularly of the cerebral cortex, in the evolution of placental mammals (see also Chapter 8).

This brief, and necessarily superficial, account of the major distinctions between living reptiles and mammals clearly demonstrates that the modern mammals as a group are characterized by a number of interrelated developments that can be gathered under the following headings:

1. Elaboration of parental provision for the offspring, involving adaptations for suckling and (in marsupials and placentals) vivipary.
2. Development of metabolic mechanisms for internal generation of body heat (endothermy), associated with metabolic and other adaptations for the maintenance of a constant, elevated core body temperature (homeothermy). The development of a highly efficient kidney in mammals provides further evidence of the importance of enhanced metabolic turnover in mammalian evolution (Kemp, 1982).
3. An overall increase in the level of locomotor activity, fuelled by more rapid energy turnover resulting from improved food-processing techniques (notably efficient mastication) and enhanced respiratory performance, together with modifications in cranial, dental and postcranial morphology. (Carrier (1987) argues that many changes, including increased lateral stability of the vertebral column, were necessary to enable mammals to breathe while moving.)
4. Determination of a target body size that is characteristic of each species, combined with associated growth mechanisms.
5. Increased elaboration of the central nervous system, with emphasis on the cortex, the cerebellum, (probably) the olfactory structures and more effective exchange of information between the two halves of the brain. The larger brain of mammals is directly linked to their higher energy turnover.
6. Development of a body covering of hair, with specialized glands present in the skin for sweating, scent production and other functions.

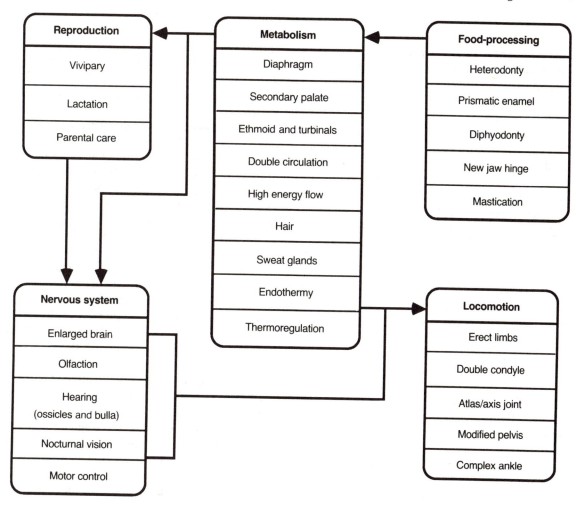

Figure 4.5. Functional interrelationships of mammalian characteristics. (Adapted and expanded from Kemp, 1982.)

One of the main points to emerge from this survey is the complex pattern of interrelationships between the various distinguishing features of mammals, which suggests that they have evolved slowly as an integrated complex (Fig. 4.5; see also Kemp, 1982). However, comparisons thus far have been confined to living mammals and reptiles. It is now necessary to turn to the fossil record to determine how mammalian features emerged over time and, if possible, to establish when the adaptive radiation of the placental mammals (including primates) began.

THE FOSSIL RECORD OF MAMMALIAN ORIGINS

The most startling aspect of the divergence of mammals from an ancestral reptilian stock is that it apparently began at an extremely early stage of vertebrate evolution. It is now generally accepted that all modern mammals are descended from a distinct branch of early reptiles derived from the Palaeozoic cotylosaurs ('stem reptiles'). This early group of reptiles, commonly referred to as 'mammal-like reptiles' (synapsids) became established during the Carboniferous period over 290 Ma ago. Synapsids persisted through the Permian and Triassic periods and finally disappeared from the fossil record during the early Jurassic period some 175 Ma ago, their only modern descendants being the mammals. The transition from these mammal-like reptiles to the mammals approximately coincided with the Triassic/Jurassic boundary, over 200 Ma ago. Modern reptiles and mammals have thus evolved completely separately for over 300 Ma and the various distinctions between them that have been outlined above could therefore have arisen at any point between the mid-Carboniferous and the estimated time of divergence between the monotremes and viviparous mammals (marsupials + placentals) at the Triassic/Jurassic boundary (or even thereafter, if parallel evolution of modern mammalian characters has occurred). Given this long history of separation between mammals and reptiles, one must be very wary of regarding the characters of modern reptiles as necessarily primitive merely because the common ancestor of the reptiles and mammals is classified as a 'reptile' (see also Romer, 1968). After all, the various lines of living reptiles are the end-products of evolution over a period of some 300 Ma during which time many independent specializations have surely emerged.

The mammal-like (synapsid) reptiles are quite well known, despite their antiquity, probably because the known forms were relatively large-bodied in general (see Kemp, 1982 for a comprehensive review). They can be sub-divided into two largely successive groups: the Carboniferous/Permian **pelycosaurs** and the Permian/Triassic/early Jurassic **therapsids** (Romer, 1956) – see Fig. 4.6. The pelycosaurs comprised three major groups (ophiacodonts, edaphosaurs and sphenacodonts) and it is likely that the later therapsids were derived from carnivorous pelycosaurs of the sphenacodont group, which include the well-known *Dimetrodon*. The therapsids comprised at least seven distinct groups (gorgonopsians, therocephalians, bauriamorphs, anomodonts, ictidosaurs, tritylodonts and cynodonts; see

Fig. 4.6) and it seems that advanced cynodonts of presumed carnivorous habits gave rise to the earliest true mammals. Several well-preserved skeletons of an early Triassic cynodont called *Thrinaxodon* are known (e.g. see Brink, 1958); it was not quite as advanced in certain key factors as the presumed ancestors of the mammals. Therapsids finally disappeared from the fossil record about 180 Ma ago.

The next phase of mammalian evolutionary history (for reviews see Lillegraven, Kielan-Jaworowska and Clemens, 1979; and Kermack and Kermack, 1984) covers the timespan from the beginning of the Jurassic to the end of the Cretaceous periods, from 210 to 66 Ma ago (see Fig. 4.6), which, as noted by Lillegraven *et al.*, represents the first two-thirds of mammalian history. It was not until the end of the Cretaceous that the large-scale evolutionary radiation of marsupials and placentals took place, as far as we can tell from available fossil evidence. It is particularly important to establish the details of mammalian evolution during the Jurassic and Cretaceous periods but, unfortunately, the fossil record for this stage of mammalian evolution is particularly poor, several key groups being known only from teeth and jaw fragments, as will be seen later. The Jurassic/Cretaceous record of mammals has thus been aptly named the 'dark ages'. The scarcity of available information is undoubtedly partly attributable to the uniformly very small body size of mammals at this point in their evolutionary history (Lillegraven *et al.*, 1979). However, mammals were also rare during this time and it has been said that they could easily have become extinct.

In view of their diminutive body sizes, it seems very likely that the Jurassic/Cretaceous mammals required a relatively high-energy diet and the morphology of their teeth suggests that they relied heavily on arthropods (e.g. see Romer, 1966; Jenkins and Parrington, 1976). Thus, the mainstream of mammalian evolution, from the pelycosaurs onwards (carnivorous sphena-codonts to carnivorous cynodonts to 'insec-tivorous' early mammals), seems to have shown a consistent emphasis on consumption of animal prey. However, there was repeated independent

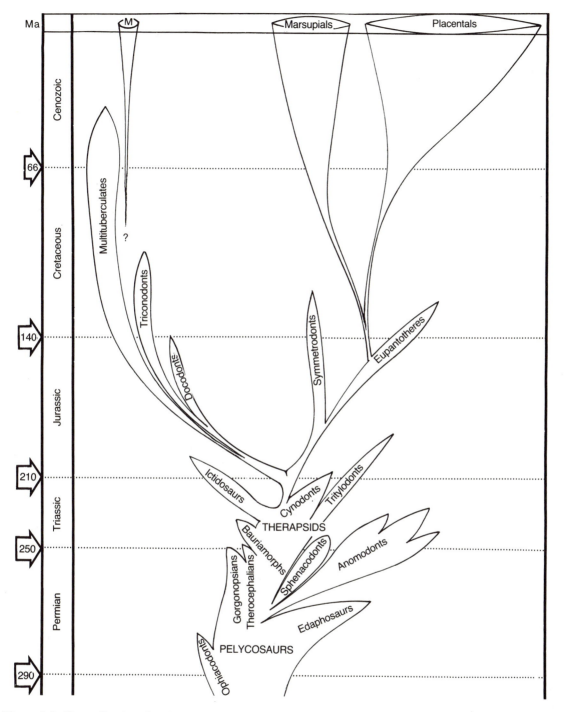

Figure 4.6. Generalized outline diagram of one widespread interpretation of evolutionary relationships between modern mammals – placentals, marsupials and monotremes (M) – and their reptilian predecessors. The boundary between the Triassic and Jurassic periods approximately marks the presently documented transition from mammal-like reptiles to true mammals, some 210 Ma ago. The most recent evidence indicates that monotremes may be more closely related to marsupials and placentals than is indicated here. (Based on Thenius and Hofer, 1960; Romer, 1966; Colbert, 1969; Crompton and Jenkins, 1973; Gingerich, 1977a; Crompton and Jenkins, 1979; Kemp, 1982; Kermack and Kermack, 1984.)

evolution of side groups of relatively large-bodied herbivorous forms at successive stages of mammalian phylogeny (e.g. edaphosaurs, anomodonts, herbivorous marsupials and placentals).

There is a broad consensus, based both on a comparison of living forms and on interpretation of fossil evidence, that the monotremes diverged relatively early in the evolution of mammals, although the exact timing of the separation is controversial. Indeed, many authors have proposed that the monotremes diverged at or very soon after the reptile/mammal transition and this widely published interpretation (e.g. see Crompton and Jenkins, 1979; Kermack and Kermack, 1984) is illustrated in Fig. 4.6. There has been considerable support for the view that a dichotomy is identifiable among the earliest known mammals at the Triassic/Jurassic boundary. Some have been linked to a branch leading to the monotremes, multituberculates and docodonts (i.e. the morganucodonts: *Morganucodon* (= *Eozostrodon*), *Megazostrodon* and *Erythrotherium*), whereas the marsupials, eupantotheres and symmetrodonts have been linked to the contemporary late Triassic *Kuehneotherium*. *Megazostrodon* and *Erythrotherium* are known from quite well-preserved skulls and postcranial elements, but *Kuehneotherium* is known only from partial jaws and teeth.

However, there has been mounting evidence from comparative morphological studies (e.g. Kemp, 1982, 1983; Lewis, 1983) that the divergence of the monotremes from viviparous mammals (marsupials + placentals) occurred at a significantly later stage than is indicated in Fig. 4.6. Particularly important in this respect has been the discovery of a partial lower jaw of an early Cretaceous monotreme in Australia (Archer, Flannery *et al.*, 1985), bearing teeth indicating some similarities to the hypothetical ancestral condition for marsupials and placentals. In the light of this evidence, it has now been accepted by several previous supporters of the phylogenetic pattern indicated in Fig. 4.6 that the monotremes did not diverge from the line leading to marsupials and placentals until the

later part of the Jurassic (Kielan-Jaworowska, Crompton and Jenkins, 1987). Further, the discovery of a typical mammalian set of three ear ossicles in the multituberculate *Lambdopsalis* from the Palaeocene of China indicates that this extinct group of mammals (see Fig. 4.6) may also be more closely related to the origin of marsupials and placentals than is commonly supposed (Miao and Lillegraven, 1986).

By any account, however, the marsupials and placentals clearly share a more recent common ancestry, postdating the divergence of the monotremes. An origin of marsupials and placentals from a stock such as the eupantotheres close to the Jurassic/Cretaceous boundary seems likely on present evidence (Lillegraven, Kielan-Jaworowska and Clemens, 1979). It is therefore clearly important to draw a distinction between those characters that are shared by all living mammals and those shared only by marsupials and placentals. For instance, all mammals share the possession of mammary glands and suckling behaviour, but vivipary is restricted to marsupials and placentals (Long, 1969; Pond, 1977). It is therefore reasonable to suggest that suckling adaptations were present in the common ancestors of the mammals at a relatively early stage, but that vivipary did not appear until later, in the stock that gave rise to marsupials and placentals. This underlines the fact that suckling is more fundamental to mammalian evolution than vivipary (which has evolved sporadically and independently among fish, amphibians and reptiles and is not necessarily associated with intensive parental care). As vivipary is universal to marsupials and placentals, they have been termed 'therian mammals' and their common ancestor has been referred to as the 'ancestral therian'. However, this is somewhat misleading, because the formal taxonomic category Theria includes certain fossil groups that probably predated the common ancestral stock of marsupials and placentals and it is therefore preferable to refer to the 'ancestral stock of viviparous mammals' when referring specifically to the origins of marsupials and placentals (see also Kemp, 1982).

Subsequent evolution of the mammals is

particularly poorly documented in the fossil record up to the late Cretaceous. Virtually nothing is known for the first half of the Jurassic, and for the second half of this period most species are known only from teeth and partial jaws. Things are much the same for the Cetaceous, with virtually no fossil evidence reported for the middle of that period (some 80–100 Ma ago). Apart from the first recognizable 'therians' from the late Cretaceous (some possible identifiable as marsupials or placentals; Clemens, 1971), five major groups of Jurassic/Cretaceous mammals are currently recognized: early multituberculates, triconodonts (= eutriconodonts), docodonts, symmetrodonts and eupantotheres (Fig. 4.6).

The multituberculates are unusual early mammals with peculiarly specialized dentitions, documented from the late Cretaceous through to the early Cenozoic (i.e. to the end of the Eocene epoch). They are probably not directly related to any extant mammals. Clemens and Kielan-Joworowska (1979) and Kermack and Kermack (1984), largely on the basis of similarities in the structure of the brain case, have suggested that multituberculates were linked to the origins of the monotremes, although until recently only the Miocene form *Obduradon* from Australia could be directly allocated to the monotreme group. As noted above, a partial lower jaw from the early Cretaceous *Steropodon* of Australia has now been allocated to the monotremes (Archer, Flannery *et al.*, 1985) and indicates that monotremes are more closely related to marsupials and placentals than was once believed.

Unfortunately, before the late Cretaceous there is no documented postcranial evidence for multituberculates. Similarly, the docodonts and triconodonts from the late Jurassic/early Cretaceous were originally known only from teeth and jaws. However, some docodont skulls, along with an incomplete skeleton, have now been reported (Henkel and Krusat, 1980) and postcranial skeletons of early Cretaceous tricondonts have also been discovered in the Cloverly formation of Montana (Jenkins and Crompton, 1979). To date, the symmetrodonts and eupantotheres are still documented almost exclusively by teeth and jaws, although a brief

report (Henkel and Krebs, 1977) has announced the discovery of an almost complete eupantothere skeleton from late Jurassic deposits of Portugal. Unfortunately, most of this late Jurassic/early Cretaceous cranial and postcranial material has yet to be described in detail. As will be seen later, even with the new finds, our restricted knowledge of the first two-thirds of mammalian evolution is a major handicap not only in terms of the remarkably small number of substantial fossil species known but also with respect to the unrepresentative geographical distribution of known fossil sites. The present state of affairs has been aptly summarized by Lillegraven *et al.* (1979): 'Paleontological enquiry to date, however, has documented only a tiny fraction of these evolutionary radiations; we have much to learn.'

The gradual emergence of the mammals from stem reptiles can, with reference to the general scheme of inferred phylogenetic relationships in Fig. 4.6, be considered in terms of four successive stages:

1. Early mammal-like reptiles: sphenacodont pelycosaurs (e.g. *Dimetrodon*).
2. Advanced mammal-like reptiles: cynodont therapsids (e.g. *Thrinaxodon* and its later relatives, such as *Cynognathus*, *Diademodon* and *Probainognathus*).
3. The earliest mammals (e.g. *Megazostrodon* and *Erythrotherium; Kuehneotherium*).
4. Jurassic/early Cretaceous transitional mammals (eupantotheres, symmetrodonts, docodonts, triconodonts and early multituberculates). Because no skeletal reconstructions have yet been published for any of these forms, a late Cretaceous therian mammal (*Zalambdalestes*) must be taken to illustrate the completion of this stage.

(Note: This phylogenetic sequence is considered in much greater detail by Kemp (1982), who discusses fossil forms representing 21 successive stages of mammalian evolution, starting with the stem reptiles of 300 Ma ago.)

The four stages defined above can be examined in turn to trace the gradual emergence of mammalian characteristics. As with the survey of fossil primates (Chapter 2), emphasis will be placed on substantial fossils known from dental, cranial and postcranial remains.

Early mammal-like reptiles (sphenacodont pelycosaurs)

The sphenacodont pelycosaurs were still close to the general reptilian pattern in many respects, but they also showed some clear signs of the mammalian condition. Unlike the earliest true mammals, however, these pelycosaurs were all quite large in body size: their body weights were in the range 3–330 kg, with a weight of about 85 kg being representative for the group (Romer and Price, 1940). Their limbs were still oriented in an essentially reptilian fashion, despite the large body size, with the elbow and knee joints roughly on a level with the shoulder and hip joints, respectively. However, the limb bones had become longer and, although a reptilian phalangeal formula of 2.3.4.5.4/3 was retained, in some species the digits had become shorter to approximate to the mammalian condition without actual loss of phalanges. Ribs were present on the neck vertebrae and on all trunk vertebrae, so there was no indication of the

formation of a distinct 'chest' (thorax) with a diaphragm. The tail was typically long and the pelvic girdle was basically reptilian in form, with a backward-pointing ilium. In the pectoral girdle, however, the scapula was quite prominent, indicating a trend towards the mammalian form.

Much has been made of the fact that some (but by no means all) sphenacodont pelycosaurs had a dorsal 'sail' based on large, bony projections (neural spines) from the trunk vertebrae, as in *Dimetrodon* (Fig. 4.7). Similar sails were also present in some edaphosaurs (e.g. *Edaphosaurus*), and it has been suggested that these structures represented early devices for behavioural thermoregulation, used for absorbing or radiating heat by appropriate orientation as and when required (Romer, 1966; Kemp, 1982). Some support for this notion is provided by the fact that, among pelycosaur species with sails, the area of the structure increases in a fairly predictable fashion with body size, being disproportionately larger in large-bodied species to match the geometric increase in body volume (Romer and Price, 1940). On the other hand, it is puzzling that certain edaphosaurs and sphenacodont pelycosaurs lacked sails altogether, and reasonable alterna-

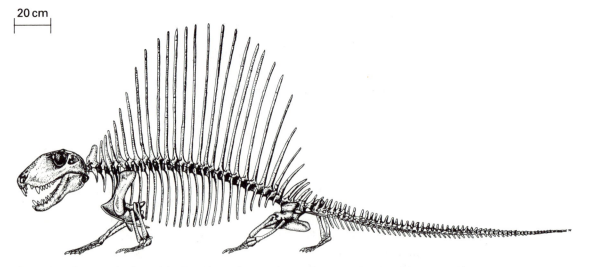

20 cm

Figure 4.7. Reconstruction of the Permian sphenacodont pelycosaur *Dimetrodon*. Note the large 'sail' composed of vertical processes from the vertebrae. This has been interpreted by some authors as an early adaptation permitting a degree of thermoregulation. (After Romer and Price, 1940.)

tive hypotheses can be advanced (e.g. the sails might have been used in intraspecific display behaviour, on a parallel with numerous modern lizard species that have special display organs and associated postures). Nevertheless, the possibility remains that some pelycosaurs, at least, were beginning to develop devices for thermoregulation as an early step towards the origin of homeothermy/endothermy in mammals.

The skull of sphenacodont pelycosaurs also followed a basically reptilian pattern with a few modifications in the direction of mammals. There was a large pineal foramen on the roof of the skull in virtually all species (indicating the presence of a 'third eye' – a primitive feature) and the nostrils were still separate, although perhaps a little closer together than in the typical reptilian condition. There was no secondary palate and absence of this feature, combined with the lack of a well-defined 'thorax', suggests that respiration was still essentially reptilian. Similarly, at the rear of the skull there was a single occipital condyle and the atlas/axis complex retained a reptilian form. The teeth remained relatively simple, although there was some initial differentiation into incisors, canines and cheek teeth. The teeth lacked proper cusps and the cheek teeth were accordingly little differentiated from the anterior dentition, but a few large, canine-like teeth were present in the premaxilla and maxilla (above) and towards the front of the dentary (below). The teeth were quite numerous. *Dimetrodon*, for instance, had a dental formula of $I_2^3 C_1^{2/3} P + M_{16}^{11-14}$ with a maximum total of almost 80 teeth. (As it is not possible to distinguish premolars and molars at this stage of mammalian evolution, the cheek teeth are lumped together in the formula.) Tooth replacement was polyphyodont and resembled the pattern in modern crocodiles, with replacement teeth emerging inside their predecessors and then moving outwards to replace them. There were several waves of tooth replacement, passing from the back of the jaw forwards. The jaw hinge was of the reptilian articular/quadrate type and there were typically eight bones on each side of the lower jaw

(dentary, two coronoids, surangular, angular, prearticular, articular and splenial in *Dimetrodon*). However, there were a few signs of the later mammalian condition in that the dentary was somewhat enlarged relative to the other bones in the lower jaw and it possessed a small 'ascending process'. Further, the angular bone bore a small flange (**reflected lamina**) separated from the main body of the jaw by a notch, resembling the condition found in the later therapsids and connected perhaps with the gradual conversion of the angular to the ecto-tympanic during mammalian evolution (Allin, 1975). The stapes was a large, thick bone on the ventral aspect of the skull, even more massive than in typical reptiles.

Advanced mammal-like reptiles (cynodont therapsids)

Cynodont therapsids, such as *Thrinaxodon* (Fig. 4.8), had developed further than the sphenacodont pelycosaurs in the direction of mammals. Body size was generally moderate in comparison both to pelycosaurs and to modern mammals, with estimated body weights of 500 g for *Thrinaxodon*, 1.25 kg for *Probainognathus* and 7 kg for *Diademodon* (Jerison, 1973; Crompton and Jenkins, 1978). In contrast to the pelycosaurs, the limbs were brought in beneath the body in cynodonts such as *Thrinaxodon*, with the elbows bent backwards and the knees bent forwards. The limb bones were more gracile in form and the femur, in association with its new position beneath the body, had its articular head at the side as in modern mammals, rather than in a terminal position, as in reptiles. The digits were somewhat reduced to an almost uniform length, but this was achieved by shortening rather than losing bony elements and *Cynognathus*, for instance, still had a phalangeal formula of 2.3.4.4.3, close to the standard reptilian formula. Nevertheless, some therapsids did exhibit development of the mammalian phalangeal formula (e.g. therocephalians; Kemp, 1982). Ribs were still present on most vertebrae, but differentiation of a thorax was well under way, with noticeably longer ribs on the first 13 trunk

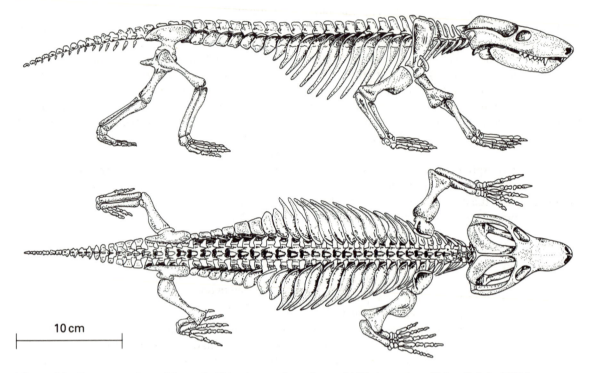

Figure 4.8. Reconstruction of the early Triassic cynodont therapsid *Thrinaxodon*. (After Brink, 1958.)

vertebrae (Fig. 4.8). Ribs on the neck vertebrae were reduced and quite a sharp transition occurred in the trunk after the thirteenth vertebra, suggesting that differentiation of a diaphragm might have begun (see Fig. 4.8 and Brink, 1958), although Gregory (1951) doubted that a complete diaphragm was present. An abrupt change in the ribs of the advanced cynodont *Procynosuchus*, indicating possible separation into thoracic and lumbar regions, provides evidence of the development of a diaphragm in this genus (Kemp, 1982). The tail was relatively short. Both limb girdles showed premammalian developments; the scapula was prominent in the pectoral girdle and had the beginnings of a scapular spine, and the ilium sloped forwards in the pelvic girdle. The heel on the calcaneus also makes its first appearance among therapsids (Kemp, 1982).

The skull of cynodont therapsids also has several modifications towards the mammalian condition. Although a pineal foramen was still present in most cases, it was reduced in size and in some cases had disappeared altogether (e.g. in bauriamorphs). The nostrils were closer together, although still not united in the typical mammalian fashion. A quite substantial secondary palate had developed, too. However, there was some variation in this feature between genera and its posterior margin was markedly further forwards than in modern mammals. At the rear of the skull, the occipital condyle had become a double structure in many cynodonts (e.g. *Thrinaxodon* and *Cynognathus*), with matching parallel changes in the atlas. The teeth were more differentiated than in sphenacodont pelycosaurs, with advanced cynodonts exhibiting typical heterodonty (Colbert, 1969). Molars with cusps were developed independently in several later therapsid groups (Bown and Kraus, 1979), including certain cynodonts that developed cheek teeth rather like those of the later triconodonts (early mammals; see later). Most cynodonts had developed a single canine in the lower jaw and at the anterior margin of the maxilla in the upper jaw on each side, although some species had two such teeth in that position. The dental formula was, however, little different from that of sphenacodont pelycosaurs, with most cynodonts approximating the general dental formula of $I_{3/4}^{4} C_{1/2}^{1/2} P + M_{14(max)}^{14(max)}$ corresponding to a maximum total of about 80 teeth. Nevertheless, some species had markedly fewer teeth than this, largely as a result of reduction in the number of cheek teeth. *Cynognathus platyceps*, for instance, had a dental formula of $I_4^4 C_1^1 P + M_{11}^{10}$, corresponding to a total of only 60 teeth. No cynodont species had fewer than seven cheek teeth. It is not clear whether there was any definite trend towards the mammalian pattern of tooth replacement (diphyodonty). The advanced cynodont *Diademodon* had the most mammal-like dentition among the therapsids, with an incipient distinction between premolar-like and molar-like cheek teeth, and several authors have suggested that there were only two sets of teeth (Broom, 1913; Brink, 1957; Tarlo, 1964; Ziegler, 1969). But Hopson (1971) established from a detailed study that there were several waves of tooth replacement in *Diademodon*, though there was a shift to a 'sequential', rather than 'alternating' pattern of tooth replacement (see also Kemp, 1982). Hopson explains this shift as a necessary feature of deeply rooted cheek teeth that exhibit occlusal relationships between upper and lower sets (lacking in earlier, more primitive cynodonts such as *Thrinaxodon*). Thus, although *Diademodon* apparently had a polyphyodont pattern of tooth replacement, this genus shows some developments that are relevant to the eventual emergence of mammalian diphyodonty. Further, Crompton and Jenkins (1979) report a reduced frequency of tooth replacement in the advanced cynodont *Probainognathus*, suggesting some approximation to the mammalian pattern.

The jaw hinge in cynodont therapsids is still of the reptilian articular/quadrate type, although *Probainognathus* possibly had an incipient dentary/squamosal hinge as well (Romer, 1969; Kemp, 1982). There are still quite a large number of bones in the lower jaw (usually seven: dentary, just one coronoid, surangular, angular, pre-articular, articular and splenial); but the dentary is conspicuously enlarged relative to the other bones and its coronoid process is well developed. In addition, the squamosal is enlarged relative to the condition in pelycosaurs. The tympanic membrane was apparently borne in a reflected lamina of the angular in therapsids such as *Thrinaxodon*, and it has been suggested that the articular and quadrate bones functioned in sound transmission (along with the stapes) as well as acting as elements of the jaw hinge (Allin, 1975; Crompton and Jenkins, 1979). The stapes was still quite a substantial bone, however, and this has led to suggestions that it served to transmit ground-borne vibrations rather than air-borne sound (Allin, 1975; Kermack and Kermack, 1984).

The cynodonts also had some other features of special relevance to mammalian evolution. Watson (1913) noted that the cerebellum was relatively large in the brain of *Diademodon*, although there was no evidence of development of a neocortex. Olson (1944) confirmed this observation for other therapsids. Watson also reported, on the basis of natural endocasts of the nasal region, that the cynodonts *Diademodon* and *Nythrosaurus* had cartilaginous ethmo-turbinals (see also Kermack and Kermack, 1984). In the case of *Nythrosaurus*, the turbinals were associated with well-developed olfactory bulbs in the brain (see also Simpson, 1927). Kemp (1982) has also inferred the presence of sheets of cartilage, possibly supporting olfactory epithelium, in the nasal cavity of gorgonopsian and therocephalian therapsids. Finally, Watson (1931) inferred from the presence of pits on the facial region of the skulls of certain cynodonts that **vibrissae** (whiskers) might have been developed by this stage of mammalian evolution. The presence of whiskers would imply at least a capacity to develop hair on the body surface, and it is accordingly just possible that one of the key mammalian features (hair) was already developed in cynodont therapsids. Overall, the evidence suggests that although cynodonts had developed certain mammal-like features –

Table 4.2 Data on brain size, body size and degree of encephalization in cynodont therapsids

Genus	Brain weight (g)	Body weight (kg)	EQ_R*	EQ_M*
Thrinaxodon[†]	0.45	0.50	0.94	0.07
Diademodon[†]	2.6	7.0	1.31	0.05
Probainognathus[‡]	1.5	1.25	1.91	0.12

* Reptilian encephalization quotient (EQ_R) and mammalian encephalization quotient (EQ_M) values calculated according to formulae provided by Martin (1981c).
[†] Data on brain and body size from Jerison (1973).
[‡] Data on brain and body size from Crompton and Jenkins (1978). (Quiroga, 1980, using different figures, obtained an EQ_M value of about 0.14.)

notably in certain aspects of dentition and in adaptation for more active locomotion and more efficient respiration – they were still reptilian in many respects.

Limited data on relative brain size (Quiroga, 1980) indicate that cynodonts were at the boundary between reptilian and mammalian grades of organization, and Kemp (1982) suggested on qualitative grounds that some brain expansion had taken place in cynodonts. Information on brain and body size is available for three well-known cynodont genera (Table 4.2) and it is possible to express the relative size of the brain as an **encephalization quotient** value calculated with respect to standard relationships for modern reptiles and mammals (Jerison, 1973; Martin, 1981c; see Chapter 8 for a detailed discussion of encephalization quotients). In each case, an encephalization quotient value of unity would correspond to the 'average' condition for reptiles and mammals, respectively. As can be seen from Table 4.2, the three cynodonts have average or above average quotient values compared with modern reptiles (though lying within the natural range of variation), whereas in comparison to modern mammals the average quotient value is 0.08, indicating a relative brain size inferior to that of any living mammal species and representing only 8% of the brain size expected for a modern mammal of comparable body size. (Note that a typical modern mammal weighing 1 kg has a brain about 16 times larger than a typical modern reptile of the same body

weight and that the overall range of encephalization quotient values for modern mammals is about 0.30–5.00.)

The earliest mammals

The emergence of 'true' mammals at the Triassic/Jurassic boundary was marked by certain changes that were of considerable significance for the later adaptive radiation of the mammals. One point that is immediately striking is that these early mammals were all very small; known late Triassic mammals had body weights in the range 20–30 g only (Crompton and Jenkins, 1979). The best-known forms are the morganucodonts *Megazostrodon* and *Erythrotherium*, for both of which fairly complete postcranial material is available (Fig. 4.9). Jenkins and Parrington (1976) provided a detailed treatment of these two genera, commenting particularly on locomotor adaptations. Continuing the trend seen in the cynodonts, the limbs in these morganucodonts were tucked in beneath the body, with the elbows bent backwards and the knees bent forwards. Indeed, the posture seems to have been very similar to that of a modern mammal of similar body size. The head of the humerus was lateral to the shaft of the femur and spherical in shape, as in modern mammals. The phalangeal formula was mammalian (2.3.3.3.3), the hallux was probably somewhat divergent and it seems likely that the digits bore narrow, pointed claws. The pelvis had a typical mammalian form, with a forward-

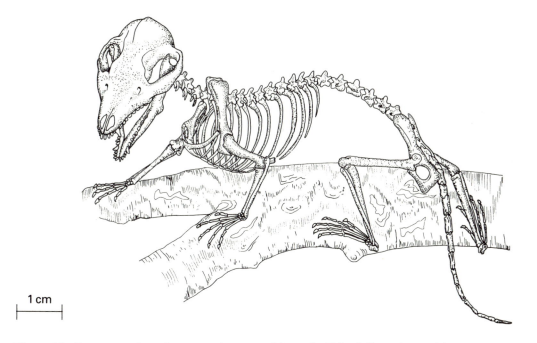

1 cm

Figure 4.9. Reconstruction of a very early mammal from the Triassic/Jurassic transition zone, based on skeletal material from *Megazostrodon* and *Erythrotherium*. (After Jenkins and Parrington, 1976.)

sloping ilium, although the pectoral girdle was still essentially like that of cynodonts. The vertebral column was also much more similar to a modern mammal's, with a definite thoracic ribcage (see Fig. 4.9) and with major structural differentiation between thoracic and lumbar vertebrae. Dorsoventral flexure of the vertebral column was thus possible. In short, these early morganucodonts had a postcranial skeleton well adapted for active locomotion and Jenkins and Parrington (1976) suggest that they were able to move about in trees as well as on the ground, although they were not specially adapted for arboreal locomotion as such.

With the advent of the first true mammals came marked changes to the skull. The pineal foramen was apparently absent, in so far as suitable fossil evidence is available to confirm this point. A well-developed bony secondary palate was present, although its posterior edge was still somewhat farther forwards than in modern mammals, and the nostrils were close together.

Morganucodonts, at least, had double occipital condyles and the atlas/axis joint incorporated a relatively large, protuberant peg.

The teeth of early mammals already had the typical mammalian heterodont pattern (i.e. the presence of different kinds of teeth) and, according to Parrington (1971), those of morganucodonts at least were diphyodont. Kemp (1982) regards diphyodonty (restriction to two sets of teeth) as probable for both morganucodonts and *Kuehneotherium*, although the replaced postcanine teeth were apparently resorbed rather than shed. The numbers of cheek teeth were generally reduced in comparison to cynodonts. Morganucodonts had a dental formula of $I_{3/4}^{3/4}$ C_1^1 $P_{3/5}^{4/5}$ $M_{4/5}^{3/4}$ with a total of about 50 teeth, and *Kuehneotherium* had a formula of $I_4^?$ $C_1^?$ $P_{4/6}^{5/6}$ $M_{4/5}^4$. The cheek teeth exhibited proper occlusal relationships and there was a clear differentiation between premolars and molars. Both in morganucodonts (e.g. *Megazostrodon* and *Erythrotherium*) and in

Kuehneotherium, there was contact between the dentary and squamosal, but the jaw hinge was transitional in form. Most or all of the original reptilian postdentary bones were still retained as a compound rod housed in a groove in the lower jaw (dentary). *Kuehneotherium* apparently had the basic complement of six postdentary bones (coronoid, surangular, angular, prearticular, articular and splenial), while *Morganucodon* may have lost the splenial. But in all cases, the dentary was by far the largest bone in the lower jaw and *Kuehneotherium* exhibited a well-developed dentary condyle. Similar developments in the lower jaw and jaw hinge were present in the contemporary ictidosaurs (e.g. *Diarthrognathus*) and, to a lesser degree, in tritylodonts (e.g. *Oligokyphus*). In fact, *Diarthrognathus* exhibited a more advanced condition than either the morganucodonts or *Kuehneotherium*. However, ictidosaurs and tritylodonts are so specialized in certain respects (e.g. dental morphology) that they are unlikely to lie close to the ancestry of the mammals.

Given the transitional nature of the jaw hinge in all of these later Triassic/early Jurassic forms, the ear ossicles had clearly still not acquired their definitive mammalian form and the articular and quadrate still served a dual function as jaw hinge components and as sound conductors. However, the cochlear part of the ear region of the morganucodont skull was large in comparison to cynodonts, indicating an increased emphasis on the sense of hearing. Unfortunately, little is known about relative brain size of the earliest mammals, so it is difficult to assess whether they exhibited any significant advance over the cynodont condition, which was still essentially reptilian (see Fig. 4.2). Nevertheless, Crompton and Jenkins (1978) point out that the braincase of *Megazostrodon* was apparently very similar in design to that of the late Jurassic *Triconodon*, which is itself quite close to modern mammals in terms of relative brain size (see later). Also, Kemp (1982) has inferred an increase in brain size from a change in composition of the braincase wall. Hence, there is at least some circumstantial evidence that the transition to the earliest mammals at the boundary between the Triassic and Jurassic may have been accompanied by a marked increase in relative brain size.

Jurassic/Cretaceous transitional mammals

The known Jurassic/Cretaceous transitional mammals (eupantotheres, symmetrodonts, docodonts, triconodonts and early multituberculates) were also typically quite small. Most species had a body size ranging from that of modern shrews to that of rats (Lillegraven, 1979), although one of the triconodont specimens from the early Cretaceous Cloverly deposits in Montana was larger, with a head and body length of 35 cm, indicating a body weight of 1–2 kg. Because the fossil evidence for this level of mammalian evolution has only recently been expanded to include postcranial material (which is still largely undescribed), there are numerous morphological points that remain to be settled. As has been noted above, the first postcranial evidence for multituberculates comes from late Cretaceous deposits, and the symmetrodonts are still known only from jaws and teeth. Our knowledge of postcranial anatomy, and indeed much of the skull anatomy, of late Jurassic/early Cretaceous transitional mammals therefore depends heavily on the Cloverly triconodont specimens and the late Jurassic docodont and eupantothere skeletons, all of which have yet to be fully described.

Available information indicates that, at this transitional stage of mammalian evolution, locomotor adaptations were quite similar to those reported for late Triassic morganucodonts, although some further advances may have been present. The 'mouse-sized' late Jurassic eupantothere skeleton from Portugal had strong claws and a well-developed tail with elongated vertebrae, perhaps indicating active arboreal habits (Henkel and Krebs, 1977), while the early Cretaceous triconodonts apparently had a quite advanced scapula (Jenkins and Crompton, 1979). For the time being, however, the rest of our information is limited to cranial and dental features.

The pineal foramen seemingly vanished altogether from the mammalian skull at the

Triassic/Jurassic boundary and it certainly has not been reported in any later mammals. Similarly, the basic mammalian pattern of nostrils brought close together and formation of a bony secondary palate was laid down early in mammalian evolution and persisted thereafter. As in the later Triassic morganucodonts, the occipital condyle was a double structure in those late Jurassic/early Cretaceous mammals for which this region of the skull has been described (e.g. triconodonts). Presumably, the atlas/axis complex was also of a recognizable mammalian type at this point.

Obviously, much more is known about the teeth and jaws of eupantotheres, symmetrodonts, triconodonts, docodonts and early multituberculates. They all showed typical mammalian heterodonty, but critical evidence regarding the presence or absence of diphyodonty is lacking (Bown and Kraus, 1979). Cheek teeth were quite complex, but none of the species concerned possessed fully developed tribosphenic molars of the kind that were presumably present in the common ancestors of marsupials and placentals (see Chapter 6). There was considerable variation in dental formulae from group to group and even within groups. The various formulae of the triconodonts can be summarized as $I_{1/2}^{1/2} C_1^1 P_{2/4}^{2/4} M_{3/5}^{3/5}$, and the total number of teeth probably varied between 36 and 52. The docodont *Haldanodon* had a dental formula of $I_{2+}^5 C_1^1 P_3^3 M_5^5$ (total: 50+ teeth), whereas early multituberculates (late Jurassic/ early Cretaceous plagiaulacoids) had a formula of $I_1^3 C_0^0 P_{3/4}^{4/6} M_2^2$ (total: around 34 teeth). The possession of only two molars in the upper and lower jaws is a consistent feature of multituberculates. The best-known symmetrodonts (spalacotheriids) had a dental formula of $I_3^? C_1^1 P_3^3 M_7^7$ (total: probably 56 teeth), although in some other symmetrodonts the cheek teeth apparently consisted of four premolars and four molars on each side of the upper and lower jaws. Finally, the eupantotheres apparently showed considerable variation in dental formulae, but the anterior dentitions are unfortunately relatively poorly known. For instance, *Peramus* had only eight cheek teeth, $P_3^3 M_5^5$ or $P_4^4 M_4^4$, (the

latter interpretation being favoured by Kermack and Kermack, 1984), whereas *Crusafontia* and some other eupantotheres had 12 cheek teeth ($P_4^4 M_8^8$). The lower jaw of *Crusafontia* had a dental formula of $I_4^? C_1^? P_4^? M_8^?$. Overall, then, these late Jurassic/early Cretaceous mammals exhibited a great diversity of dental formulae, although there was a general trend towards reduction in the total number of teeth, with certain species (e.g. plagiaulacoid multituberculates) having quite specialized dentitions in this respect.

In all of these transitional mammals, the jaw hinge had become confined to the typical mammalian dentary/squamosal articulation, with the possible exception of the docodonts. Although by now the dentary largely made up the lower jaw, some of the supplementary reptilian jaw bones were still present as vestigial splints. Earlier forms probably had a small splenial and coronoid (e.g. early eupantotheres), whereas later forms seem to have lost the splenial, with the coronoid retained but fusing with the dentary early in development. The dentary uniformly had a well-developed coronoid process, a distinct condyle and a clearly identifiable angle on its lower posterior border. Little is known of the ear region, but multituberculates, like modern monotremes, had a straight (rather than coiled) cochlea and lacked any kind of ossified ventral floor (bulla) in the middle-ear region.

Because of the relative paucity of well-preserved cranial material of transitional mammals from the late Jurassic/early Cretaceous, relatively little can be said about the brain at this stage of mammalian evolution. However, Simpson (1927) described the skull of the late Jurassic triconodont *Triconodon* as having large nasal chambers that were probably occupied by cartilaginous turbinals. Simpson also described a natural cast of the inside of the braincase (endocast) of *Triconodon*, which, among other things, revealed that the parts of the brain associated with the sense of smell (the olfactory bulbs) were relatively large. Hence, the olfactory sense seems to have been well developed in this early triconodont. Jerison (1973) has provided estimates of brain and body size for *Triconodon* (cranial capacity, 0.73 cm³; body weight, 100 g). Calculation of an encephalization quotient, as for the cynodont therapsids listed in Table 4.2, shows that relative to modern reptiles *Triconodon* had a quotient value of 3.6, whereas relative to modern mammals it had a quotient value of 0.38. Hence, *Triconodon* had a relatively large brain even in comparison to modern reptiles, but it just falls into the lower end of the modern mammalian range of relative brain sizes (see Chapter 8). It would seem, therefore, that by the late Jurassic/early Cretaceous (if not earlier) transitional mammals were well on the way towards developing brains of a typically mammalian grade. For reasons given above (see also Martin, 1981c), it is likely that increased relative brain size in early mammals would imply increased energy turnover and presumably some development of homeothermy/endothermy.

As no reconstructions of the skeleton are as yet available for Jurassic/Cretaceous transitional mammals, it is only possible to illustrate the condition that had been attained by placental mammals in late Cretaceous times. A suitable representative, albeit somewhat specialized in certain respects, is *Zalambdalestes* (Fig. 4.10; see also Chapter 5).

This cursory review of the fossil evidence of mammalian evolution is sufficient to demonstrate several important points of direct relevance to any attempt to reconstruct the evolution of a group of modern placental mammals such as the Primates. It is, for instance, extremely significant that the divergence of mammals from reptiles can be traced back so far in time, to over 300 Ma ago. Unfortunately, we still know relatively little about the first two-thirds of mammalian history, starting with the first true mammals (as defined by the presence of a dentary/squamosal jaw hinge) about 200 Ma ago and ending in the late Cretaceous, about 70 Ma ago. Particular caution is therefore required in interpretations of the early origins of the placental mammals and certainly in any attempts to assign dates to the origins of particular mammalian groups. There is now widespread support for the view that the common ancestral stock of the marsupials and placentals was derived from the eupantotheres at some stage towards the close of the Jurassic period (Fig. 4.6; see Gingerich, 1977a; Kielan-Jaworowska, Bown and Lillegraven, 1979; Kraus, 1979; Kermack and Kermack, 1984). There is also common agreement that the most suitable candidate for the ancestry of marsupials and placentals is *Aegialodon* from early Cretaceous deposits of Sussex, England. However, this genus is known only from a single lower molar tooth, a fact that starkly underlines both the imperfections of the mammalian fossil

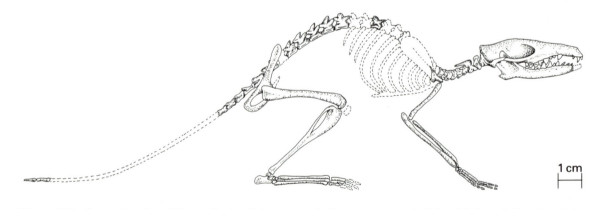

1 cm

Figure 4.10. Reconstruction of the earliest well-documented placental mammal, *Zalambdalestes lechei*, from late Cretaceous deposits of Mongolia. (After Kielan-Jaworowska, 1978.)

record and the extremely tenuous nature of many current hypotheses regarding the early stages of mammalian evolution.

The fossil record of mammal-like reptiles and early mammals also establishes the point that the evolution of mammalian characteristics was a drawn-out process, with changes occurring in mosaic fashion. As we have seen, the earlier mammal-like reptiles, the pelycosaurs, were still essentially reptilian with respect to characters preserved in the fossil record, although there was some evidence of initial changes in the skeleton and jaw apparatus suggestive of increased activity and improvement of the jaw mechanism. In the later mammal-like reptiles, the therapsids, such modifications were increasingly evident, notably among the cynodonts. The limbs and limb girdles progressively became more mammal-like in form and orientation, and the double occipital condyle made its appearance, indicating greater emphasis on control of head movement relative to the trunk. The secondary palate was also quite well developed in the later therapsids, confirming the impression from the postcranial skeleton that these mammal-like reptiles were more active creatures, requiring more efficient respiration. Further confirmation of this is provided by evidence of cartilaginous turbinals in certain therapsids, and by apparent differentiation of a thoracic ribcage. Certain restricted changes in the jaw apparatus were present, with the dentary becoming increasingly prominent among the lower jaw bones, leading to the apparent presence of a double jaw hinge (articular/quadrate + dentary/squamosal) in the cynodont *Probainognathus*. In parallel with this, there was increasing involvement of the articular and quadrate bones in the conduction of sound to the inner ear, despite retention of their participation in the jaw-hinge arrangement. However, in certain respects even the cynodont therapsids remained reptilian, and the limited evidence available on relative brain size (e.g. Table 4.2) indicates that the central nervous system was still more or less reptilian grade.

The emergence of the earliest known true mammals at the end of the Triassic involved a fairly sharp transition in some respects, whereas in other respects there was continuation of modifications that had already begun among the mammal-like reptiles. One extremely important change indicated by the known fossil record is that the earliest mammals were markedly smaller than the mammal-like reptiles, although some of the cynodonts (e.g. *Thrinaxodon* and *Probainognathus*) were intermediate in size between the typical weight of 85 kg for the sphenacodont pelycosaurs and an average body weight of 25 g for the late Triassic mammals. The postcranial skeleton of these first known mammals was in many respects much closer to that of modern mammals, particularly in exhibiting a well-defined thoracic region and several adaptations for greater mobility, allied to the earliest known appearance of the mammalian phalangeal formula. The brain had apparently reached a mammalian grade, in terms of relative size, by the end of the Jurassic if not in the earliest mammals. The dental apparatus, on the other hand, was still undergoing progressive changes that continued throughout the Jurassic and Cretaceous periods before the emergence of the typical dental patterns of marsupials and placentals. The earliest mammals still had a double jaw hinge and some of the postdentary bones (other than the articular, which became integrated into the middle-ear mechanism along with the quadrate and stapes) persisted in the jaw at least until the end of the Jurassic. Whereas marked heterodonty had appeared among the cynodont therapsids and diphyodonty apparently emerged with the first mammals, reduction of the dental formula occurred progressively throughout the Jurassic and Cretaceous. Similarly, the origin of tribosphenic molar teeth, heralding the emergence of therian mammals, can be traced back only as far as the early Cretaceous, although occlusal mechanisms of a less-refined kind had begun to emerge among advanced cynodonts some 65 Ma previously.

The first clearly recognizable placental mammals are known from the late Cretaceous (Kielan-Jaworowska, Bown and Lillegraven, 1979; see Chapter 5). However, the predominant feature of the fossil record of the 'dark ages' of mammalian evolution is its extremely fragmentary nature and we should therefore be particularly wary of concluding that the first known placental mammals are the first placentals that ever existed. This is of crucial importance, as inferred dates of divergence of placental mammals, including the primates, have been directly based on this assumption (see Chapter 12). Gregory (1951, p. 353) originally summarized the situation as follows:

> Indeed all the known Jurassic and Cretaceous mammalian fossils in the world could be put into a small box, although, owing to the mouse-like size of most of these broken jaws and teeth, a great many different kinds would be crowded together in that box.

In the 30 years since Gregory made this statement, matters have progressively improved through the discovery of new fossil material, particularly associated postcranial elements; but the fact remains that there are still vast gaps to be filled in our knowledge of early mammalian evolution. In addition, we should not lose sight of the fact that many of the most fundamental adaptations of modern marsupials and placentals (e.g. vivipary, suckling, presence of hair and homeothermy/endothermy) are not directly documented in the fossil record, so for a comprehensive understanding of mammalian evolution we must resort to inference from living forms and from fossilizable characters that provide indirect evidence of reproductive and energetic features.

CONTINENTAL DRIFT AND MAMMALIAN EVOLUTION

Patterns of geographical distribution obviously must have been important in the evolution of mammals, particularly as speciation typically seems to involve some form of geographical separation between animal populations (see Chapter 3). It has already been shown (Chapters 1 and 2) that the subdivision of living and fossil primates into 'natural groups' has a strong geographical component and there is a well-established literature dealing with the relationships between animal evolution and distribution patterns (zoogeography). Over the past 20 years or so, however, there has been a veritable revolution in our understanding of the Earth's history, with the establishment of the theory of **continental drift** as an acceptable explanation of several related aspects of the dynamics of continental landmasses. Accordingly, no discussion of mammalian adaptive radiation can be complete without reference to the far-reaching implications of continental drift.

The theory of continental drift was first advanced as an articulated body of arguments by Wegener (1915), although individual proposals involving continental movement of some kind had been put forward earlier (see Tarling and Tarling, 1971, for a discussion). Wegener, who is now generally regarded as the founder of continental drift theory, was motivated particularly by the need to account for a series of puzzling indications of ancient climates in the geological record. He later received support from du Toit (1937), who in fact presented some of the most convincing geological evidence for large-scale continental movements. The theory aroused considerable interest and discussion at the outset, but thereafter the balance of scientific opinion increasingly shifted against it, to the point where it was widely discredited by the mid-1940s. At that time, Simpson (1943) specifically considered the evolution of the mammals in relation to three competing hypotheses regarding relationships between continents: (1) continental drift; (2) transoceanic continents; and (3) stable continents. Having examined the evidence for the distributions of modern and fossil mammals, Simpson concluded quite emphatically (1943, p. 29):

> The known past and present distribution of land mammals cannot be explained by the hypothesis of drifting continents. . . . The distribution of mammals definitely supports the hypothesis that continents were essentially stable throughout the whole time involved in mammalian history.

The most telling objection to Wegener's continental drift hypothesis was that no clearly formulated **mechanism** was known that could explain such large-scale movements of the major landmasses. The geological evidence cited was almost entirely circumstantial and accessory arguments regarding the distribution patterns of plants and animals had the drawback that alternative scenarios could easily be constructed (see later). Eventually, the tide began to turn in favour of the theory of continental drift when studies of **palaeomagnetism** began to burgeon in the 1950s. With both molten rock and certain fine sedimentary deposits, the prevailing magnetic field becomes imprinted when the rock is consolidated and this palaeomagnetic imprint remains as an indicator of the original orientation of the consolidating rock relative to the Earth's magnetic poles. Studies of palaeomagnetism led to the recognition of **polar wandering curves** for individual continents that could be explained only by assuming either that the poles had changed position in a bizarre pattern over time or that the continents themselves had drifted in relation to the fixed poles. A major breakthrough came when it emerged that the polar wandering curves for individual continents closely fitted the pattern predicted by Wegener's original hypothesis (Blackett, 1961; Bullard, Everett and Smith, 1965; Runcorn, 1962; Tarling and Tarling, 1971). Over the past two decades, there has been remarkably rapid development in the theory of continental drift and a plausible geophysical mechanism has been provided in the form of **plate tectonics** (McElhinny, 1973; A.L. Graham, 1981). There are several lines of evidence indicating that the continents move apart because of **seafloor spreading**, with new crust material emerging at mid-oceanic ridges and migrating away on either side. Conversely, continents may move together when intervening oceanic crust is driven downwards in a **subduction trench**, thus reducing the expanse of the ocean floor separating the landmasses. There is therefore an established body of theory indicating that the pattern of the continents on the Earth's surface has been constantly changing throughout geological time, with new oceans appearing and old ones disappearing as the continental plates have changed their relationships with one another (van Andel, 1985).

As the geological and geophysical evidence in favour of continental drift is now so convincing, it is instructive to establish why numerous authors, including such eminent palaeontologists as Simpson (1943), were so adamant that the zoogeographical data were in conflict with drift theory. The explanation would seem to lie in a combination of negative evidence (the relative paucity of fossil material from southern continents) and the inherent reversibility of many zoogeographical arguments. Certain groups of animals and plants occur as disconnected isolates in southern continents (e.g. living marsupials, which are largely confined to South America and Australia, with a few species penetrating into Central and North America). Given such a pattern, it is possible to argue either that the ancestral stock originally occurred in the Northern Hemisphere and subsequently declined to leave isolated southern remnants, or that the original stock was in the southern continents and was directly subdivided by the break-up of Gondwanaland.

The case of the marsupials is particularly informative in that fossil mammals identified as marsupials had been identified in late Cretaceous deposits of South America and North America and in early Tertiary deposits of Europe and South America. Simpson (1943) postulated a 'little-differentiated Holarctic marsupial stock' (presumably linked by the Bering landbridge as the only available northern intercontinental connection) that gave rise to modern South American and Australian marsupials through two separate developments. This proposal suffered both from the fact that no fossil marsupials had been reported from Asia or South-East Asia and from the problem of the considerable sea barrier now separating Australia from the South-East Asian mainland. Followers of the theory of continental drift, on the other hand, were able to argue that the ancestral marsupials originally existed in some combined southern continental landmass (e.g. South America + Antarctica + Australia) and that the few marsupials known from other areas of the world were secondary derivatives. The subsequent discovery of fragmentary fossil evidence of marsupials in Eocene deposits of Antarctica (Woodburne and Zinsmeister, 1982) has, of course, removed one of the apparent obstacles to this latter hypothesis. More recently, the picture has become even more complicated with the discovery of jaw fragments attributed to marsupials in Africa from the early Eocene (Mahboubi, Ameur *et al.*, 1983) and Oligocene (Bown and Simons, 1984a) and of a single tooth identified as marsupial from central Asian Oligocene deposits (Benton, 1985). Despite the addition of new fossil material, it is still possible to interpret the biogeographic history of marsupials in different ways (Bown and Simons, 1984a,b; Jaeger and Martin, 1984; Benton, 1985). The reversibility of zoogeographical arguments is here seen very clearly: as a legacy from the stable continents hypothesis, numerous northern continental fossils are seen as representing the ancestral stocks from which isolated southern continental species have been derived, whereas an alternative view, rendered equally likely by continental drift theory, is that those northern continental forms may themselves be no more than secondary derivative outliers. The main lesson to be drawn from all this is that biogeographical arguments do not carry much weight either for or against acceptance of the theory of continental drift. Instead, now that drift theory has been independently validated on the grounds of geological and geophysical evidence, biogeographers should take reconstructions of past continental configurations as a reliable foundation on which to construct plausible hypotheses of animal and plant evolution in relation to geographical distribution.

Somewhat surprisingly, mammalian palaeontologists have been rather slow to explore the implications of continental drift for the evolution of mammals, although there are a few notable exceptions (e.g. Cox, 1970, 1974; Kurtén, 1973; Lillegraven, Kielan-Jaworowska and Clemens, 1979). If anything, primate palaeontologists have been even slower to exploit this new possibility. Among recent texts on the primate fossil record, some (e.g. Szalay and Delson, 1979) make no explicit mention of continental drift, while others (e.g. Simons, 1972) mention it only in passing.

Simons (1972) discusses continental drift merely in relation to the probability that primates could have migrated between Europe and North America in the early Tertiary, proceeding to the following conclusion regarding primate origins:

> Since the primates of latest Cretaceous and Paleocene times represent one of the oldest identifiable orders of placental mammals, it seems likely that their occurrence then in Western North America and almost as early in Europe, documents the origin of the order in the Northern Hemisphere. There is only a very low probability that there was time enough in the early stages of the placental radiation for the first primates to have arisen in the Southern Hemisphere and later migrated north.

This statement encapsulates three widespread ideas with respect to the evolution of placental mammals:

1. The adaptive radiation of the placental mammals only began during the latter part of the Cretaceous.
2. Continental drift is therefore largely irrelevant to the evolution of this group; hypotheses originally developed according to the stable continents model can be simply carried over without major modification.
3. The preponderance of fossil evidence of early placental mammals from the Northern Hemisphere reflects actual distribution patterns at the time rather than the preponderance of fossil-hunting activity (and fossil discovery) in that part of the world.

Because these beliefs derive essentially from negative evidence (the relative lack of early fossil mammals from the southern continents), they deserve careful scrutiny in the light of modern evidence on palaeocontinental positions.

Discussion of mammalian evolution in relation to past continental configurations has been considerably simplified by the publication of a series of palaeocontinental maps by Smith and Briden (1977) (see also Smith, Hurley and Briden, 1981). There are, of course, still considerable uncertainties surrounding such palaecontinental reconstructions, particularly with respect to earlier periods of the Earth's history. Further, it is important to remember that extensive **epicontinental seas** have covered large expanses of the continental landmasses at various times in the past. These must be included in any assessment of the existence of landbridges or barriers in relation to mammalian adaptive radiation. Ideally, the mammalian palaeontologist requires a series of accurate palaeocontinental maps incorporating the most recent evidence on the presence and extent of any epicontinental seas (e.g. see Lillegraven, Kielan-Jaworowska and Clemens, 1979). Broad information on the occurrence and approximate extent of epicontinental seas at various times in the past was provided by Termier and Termier (1960) – unfortunately in relation to a stable continents model – and there have been several more recent attempts to unite epicontinental seas and palaeocontinental positions in overall maps (e.g. Hallam, 1973; Barron, Harrison *et al.*, 1981; Cocks, 1981; Howarth, 1981). Although the outlines provided are still extremely tentative, particularly as continuous expanses of epicontinental sea must be inferred from geographically patchy evidence, they do at least establish some basic points in relation to mammalian evolution (see also Lillegraven *et al.*, 1979). It is important to assess the fossil record of mammalian evolution against a series of maps combining inferred palaeocontinental configurations with reconstructions of epicontinental seas, however tentative they may be, in order to demonstrate the principles involved.

Throughout most of the known fossil record of the mammal-like reptiles (from about 300 to 200 Ma ago; see Fig. 4.6), all of the world's continental landmasses were united in a single supercontinent, Pangaea. For part of this period (i.e. during the early part of the Permian), the supercontinent was divided into a northern region (**Laurasia**: North America + Eurasia) and a southern region (**Gondwanaland**: South America + Africa + Madagascar + India + Antarctica + Australasia) by an epicontinental sea lying north of the Equator (the Tethys Sea). However, throughout the Triassic and the earlier

part of the Jurassic, it would seem that Pangaea was essentially a unified supercontinent without major intervening sea barriers (Fig. 4.11). During the Triassic, there were considerable faunal and floral similarities between the constituent continents of Pangaea and close similarities persisted, at least in some cases, into the Cretaceous. Cox (1973) analysed the distribution patterns of Triassic tetrapods, which included both mammal-like reptiles (representing the adaptive radiation of the therapsids) and the earliest known mammals (see Fig. 4.6). As Cox (1973) stated: 'The Triassic is the only Period during which terrestrial vertebrates show clearly that land connections existed between every one of today's continents.' Indeed, Cox found a minimum of 41% similarity at the family level between any two continents and a maximum value of 89% similarity for two quite widely separated continental areas, Africa and Asia.

One of the most widely cited examples of a disconnected southern distribution best explained by the former existence of a single super-continent is that of a mammal-like reptile, the anomodont therapsid *Lystrosaurus*, from late Permian or early Triassic deposits. *Lystrosaurus* was originally reported from Africa, India and Asia and it was particularly exciting for proponents of continental drift when remains attributable to this genus were discovered in Antarctica (Elliot, Colbert *et al.*, 1970). Similarly, the cynodont therapsid *Cynognathus* has been reported from early/middle Triassic deposits of both South America and South Africa. Perhaps the best example of a widely distributed terrestrial tetrapod genus for that period of the Earth's history, however, is provided by the labyrinthodont amphibian *Parotosaurus*, which has been reported from early/middle Triassic deposits of North America, Europe, Asia, Africa, India and Australia (Cox, 1973). It therefore seems likely that during the Triassic and up until the break-up of Pangaea there were no major barriers to migration of therapsids or early mammals. As support for this, the earliest known mammals also had wide geographical distributions. For instance, morganucodonts have been reported from late

Triassic deposits of South Africa, Europe and China, and they possibly occurred in North America as well. Similarly, remains of late-surviving therapsids (tritylodonts) have been found in South Africa, Europe and China. Hence it is likely that the latest therapsids and the earliest mammals had an essentially worldwide distribution (see Fig. 4.11). Nevertheless, it is interesting to note that fossils of late Permian mammal-like reptiles are known predominantly from South Africa and that this region has yielded the most substantial evidence of Triassic therapsids and early mammals. Further, Cox (1973) points out that there seems to be a bias in fossilization of Triassic terrestrial tetrapods in relation to the likely palaeolatitudes involved (cf. Fig. 4.11). There are no known fossil sites for the inferred Triassic equatorial zone (10° N to 10° S) and there is a marked concentration of fossil sites at intermediate palaeolatitudes (30° N to 50° N and 30° S to 50° S). Cox attributes this bias to latitudinal variation in conditions favouring fossilization.

The Mesozic era as a whole seems to have been characterized by relatively stable climatic conditions and several lines of evidence indicate that relatively warm, dry climates prevailed. Mild temperatures may have been present as far as 50° N and S of the Equator during the Triassic and on into the Jurassic (Robinson, 1971; Hallam, 1975). The existence of widespread warm climatic conditions and relative climatic stability during the Triassic confirms the interpretation that the earliest mammals originated within a general Pangaean fauna lacking any major regional subdivisions.

Pangaea did not begin to fragment until Jurassic times, approximately 160 Ma ago, with Gondwanaland gradually becoming separated from Laurasia through a complex pattern of continental movements combined with establishment of the Tethys seaway. By mid-Cretaceous times, some 100 Ma ago (Fig. 4.12), Gondwanaland was completely separated from Laurasia and exhibited some internal subdivision after seafloor spreading began in the south about 130 Ma ago. At the same time, there was a general northward drift of both Laurasia

and Gondwanaland (Tarling and Tarling, 1971).

As we have seen, the fossil record of mammalian evolution from 200 Ma ago (with the continents still in the positions indicated in Fig. 4.11) to 100 Ma ago (Fig. 4.12) is particularly poor, especially when geographical representation is taken into account. Haramyids, symmetrodonts and triconodonts have been reported from early Jurassic deposits of India (Datta, Yadagira and Rao, 1978; Datta, 1981; Savage and Russell, 1983). Middle Jurassic fossil mammals are known only from the United Kingdom (Kermack and Kermack, 1984). As has been pointed out by Lillegraven, Kielan-Jaworowska and Clemens (1979), these fossil mammals came from an insular fauna that included morganucodonts, docodonts, triconodonts, multituberculates, symmetrodonts and eupantotheres, so one can only imagine what mammals might have been present in the rest of Pangaea 160 Ma ago when continental drift began to get under way. The situation is only a little better in the late Jurassic (160–140 Ma ago), with simultaneous representation in Europe and North America of five groups of early mammals – docodonts, triconodonts, multituberculates, symmetrodonts and eupantotheres. For the southern continents, the only evidence relating to mammalian evolution during the middle and late Jurassic consists of a jaw lacking tooth crowns from Tanzania provisionally classified as belonging to a eupantothere (*Brancatherulum*) and a set of footprints in Argentina dubiously attributed to an early mammal (*Ameghenichnus*). Early Cretaceous fossil evidence of mammalian evolution is also confined to the nothern continents, with multituberculates, symmetrodonts, eupantotheres and early therians present in Europe, North America and Asia. We are therefore faced with the unusual situation that therapsids and the earliest mammals are best known from South Africa, whereas there is virtually no further fossil evidence of mammalian evolution in the entire continent of Africa until Palaeocene times! Given the central position occupied by Africa in Pangaea and the apparent worldwide distribution of therapsids and early mammals in the late

Triassic, this gap of 120 Ma surely represents an enormous omission in our knowledge of mammalian evolution rather than a real absence of mammals from Africa throughout the Jurassic and Cretaceous periods. (Despite its uncertain affinities, the essentially toothless lower jaw of *Brancatherulum* at least documents the presence of mammals in Africa during the late Jurassic.) We must therefore exercise particular caution in inferring when the evolution of placental mammals began and where the initial radiation was located.

Lillegraven *et al.* (1979) state with respect to continental positions and the incidence of epicontinental seas 100 Ma ago (Fig. 4.12): 'Probably the greatest barriers to overland dispersal for all of the Mesozoic existed at that time.' The Tethys Sea was particularly wide and represented a major barrier separating northern and southern continents. In addition, there are other barriers that are relevant to the diversification of mammals. South America was probably separated from Africa by a narrow seaway by mid-Cretaceous times, but this separation may well have been overshadowed for a while by the opening of the trans-Saharan epicontinental seaway from about this time. The trans-Saharan seaway apparently existed intermittently during the latter half of the Cretaceous period and during the Palaeocene epoch (Figs 4.12 and 4.13) and it is likely that it constituted a partial barrier to faunal migration between West Africa and the rest of the continent at this critical stage of mammalian evolution. There was also a major epicontinental seaway that developed across the mid-western region of North America. This continent was divided into western and eastern regions by a widening epicontinental seaway (the 'Skull Creek Seaway'; Lillegraven *et al.*, 1979) that occurred in the mid-Cretaceous (Fig. 4.12) and persisted until the end of the Cretaceous, perhaps with intermittent interruptions. Hence, migration of terrestrial vertebrates between eastern and western North America was probably greatly limited during the second half of the Cretaceous. Finally, a major epicontinental seaway (combining the Obik Sea with the Turgai Strait) existed across Russia, perhaps intermittently, from late Jurassic times until the Oligocene and separated Europe and Asia. Given the simultaneous presence of the Skull Creek Seaway and the Obik Sea during the latter half of the Cretaceous, it is clear that there would have been considerable barriers to holarctic dispersal of terrestrial mammals at that time even with stable continents (cf. Simpson's proposal of a holarctic ancestral marsupial stock).

As far as climatic factors are concerned, there is some evidence that there was a general cooling of world temperatures during the latter half of the Cretaceous period, accompanied by an increase in humidity. In addition, it would seem that the temperature gradient between the Equator and the poles was greater then than during the Triassic and Jurassic (Lillegraven *et al.*, 1979). Accordingly, it is likely that the effect of increased geographical separation of continental areas produced by a combination of continental drift and extensive development of continental seas during the Cretaceous (see Fig. 4.12) was enhanced by increased climatic differentiation according to latitude. All in all, the conditions for adaptive radiation of the mammals through geographical subdivision exhibited maximal development during the Cretaceous. The particular significance of the Cretaceous for the evolution of mammals is underlined by the fact that the known fossil record documents mammals of uncertain affinities at the outset (docodonts, triconodonts, multituberculates, symmetrodonts and eupantotheres) and dentally recognizable marsupials and placentals by the end of the period.

It is generally maintained, on the basis of a strict interpretation of the known fossil record, that the spectacular radiation of marsupial and placental mammals was largely confined to the Cenozoic era (the 'Age of Mammals'). The earliest recognizable placental fossils are known from the late Cretaceous of North America, South America and Asia (with a single lower molar tooth from Europe provisionally identified as placental), and the earliest reliably identifiable marsupials have been reported from late Cretaceous deposits of North and South

174

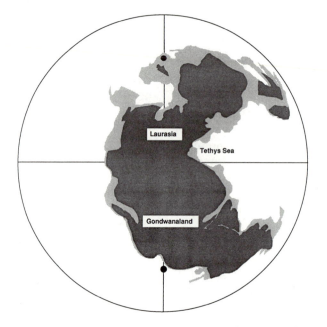

Figure 4.11. Inferred continental positions in the late Triassic, about 220 Ma ago. The configuration of continents is based on the Lambert equal-area map provided by Smith and Briden (1977); the distinction between exposed land areas (heavy stippling) and epicontinental seas (light stippling) is based on Howarth (1981). The positions of the North and South poles are shown by black circles. At this time, all of the major continental landmasses were united into the single supercontinent Pangaea, combining Laurasia in the north with Gondwanaland in the south.

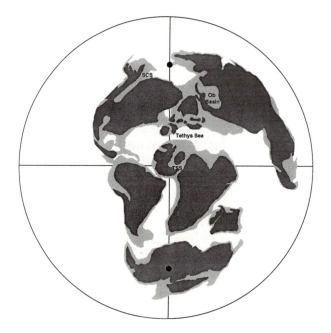

Figure 4.12. Inferred continental positions and epicontinental seas in the mid-Cretaceous, about 100 Ma ago, based on Smith and Briden (1977) and Barron, Harrison *et al.* (1981). Conventions as in Fig. 4.11. At this time, Africa was subdivided by the Trans-Saharan Seaway (TSS), the Skull Creek Seaway (SCS) at least partially subdivided North America, and Europe was more or less separated from Asia by the Ob Basin (Obik Sea) and the adjacent Turgai Strait.

Figure 4.13. Inferred continental positions and epicontinental seas in the mid-Palaeocene, about 60 Ma ago, based on Smith and Briden (1977) and Barron, Harrison *et al.* (1981). Conventions as in Fig. 4.11. It is not certain whether the Trans-Saharan Seaway (TSS) or the Ob Basin were connected with the Tethys Sea at this time.

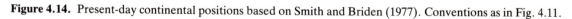

Figure 4.14. Present-day continental positions based on Smith and Briden (1977). Conventions as in Fig. 4.11.

America. Taken at face value, the fossil evidence suggests a central role of North America in the origin of both placentals and marsupials, and this is indeed the interpretation that has been made by numerous authors. As has been mentioned, all of the known late Cretaceous placentals and marsupials were relatively small-bodied species, as were the contemporary multituberculates of North America and Asia, and it is not until the start of the Cenozoic during the Palaeocene epoch that considerable diversity in both form and size of mammals is documented in the fossil record. The multituberculates apparently exhibited their greatest diversity during the Palaeocene and relatively rich therian mammal faunas (marsupials and/or placentals) are known from Palaeocene deposits of South America, North America, Europe and Asia. Kurtén (1973) has conducted an analysis of the late Palaeocene mammalian faunas of these geographical regions, the results of which accord quite well with the map of continental seas for the mid-Palaeocene (see Fig. 4.13). On the basis of faunal similarities between continental areas, he recognizes three distinct zoogeographical regions: (1) North America + Europe; (2) Asia; and (3) South America. The close faunal similarity between North America and Europe is particularly striking; all 16 families of terrestrial mammals represented in Europe are also found in North America and, of 28 European genera, 11 are documented in North America (including the 'archaic primate' *Plesiadapis*; see Chapter 2). This suggests that the North Atlantic barrier was scarcely developed at the beginning of the Cenozoic.

On the other hand, the fact that the mammalian fauna of Europe was apparently quite distinct from that of Asia in the late Palaeocene indicates that the Obik seaway did constitute a major barrier to faunal migration between Europe and Asia during the latter part of _ the Cretaceous and on through the Palaeocene. There was a markedly greater degree of faunal similarity between North America and Asia during the late Palaeocene, although this did not approach the level of similarity between North America and Europe.

The late Palaeocene mammalian fauna of South America (known only from restricted sites in Patagonia and Brazil) was essentially isolated from those of North America, Europe and Asia, with 11 of 13 families represented exclusively in South America and all 32 genera known only from that continent. Hence, faunal exchange between North America and South America during the late Cretaceous and Palaeocene was probably greatly restricted (see Figs 4.12 and 4.13). Unfortunately, virtually nothing is known of Africa, Madagascar, India, Antarctica or Australia for the Palaeocene, so Kurtén was unable to assess (for instance) the degree of faunal similarity between South America and Africa or between Africa and Europe for the late Palaeocene. Late Eocene mammalian faunas from West and North Africa indicate that large numbers of species were restricted to Africa by that stage, but by the late Eocene the pattern in the northern continents had changed considerably, with faunal barriers between Europe and Asia breaking down and with Europe becoming more isolated from North America. The disappearance of the trans-Saharan seaway by the end of the Palaeocene complicates the picture in Africa, because no Palaeocene fossil sites have been reported for Africa east or south of the seaway. Palaeocene fossil mammals have recently been discovered in North Africa (Cappetta, Jaeger *et al.*, 1978), but these are fragmentary and do not greatly clarify the question of the early evolution of placental mammals in Africa.

Overall, our understanding of the distribution and evolution of mammals during the Cretaceous and the Palaeocene is notable more for the vast gaps in the fossil record in the southern continents than for the reliability of any conclusions that have been drawn. The cooler temperatures that characterized the latter part of the Cretaceous (in contrast to the Triassic and Jurassic periods) persisted through the Palaeocene, although there was a gradual increase in temperatures from the late Palaeocene to the late Eocene or mid-Oligocene, followed by another decline that has continued to the present day (Wolfe, 1978; Hambrey and

Harland, 1981). The latitudinal bias against fossil discovery for tropical and subtropical latitudes identified by Cox (1973) for Triassic terrestrial tetrapods is essentially maintained throughout the Jurassic, Cretaceous and Palaeocene, and the cooler temperatures prevailing during the late Cretaceous and Palaeocene render this bias even more unfortunate. Apart from the few fossil sites that reliably document the presence of mammals in South America during the late Cretaceous and Palaeocene and the questionable late Jurassic evidence of *Ameghenichnus* (South America) and *Brancatherulum* (Africa), we have no fossil evidence of mammals in the southern continents from the early Jurassic right through to the Palaeocene. Among other things, this means that we have virtually no evidence of mammalian evolution in tropical and subtropical latitudes over a critical timespan of some 120 Ma.

We can now return to the widespread assumptions about the evolution of placental mammals that have been made on the basis of the known fossil record, particularly with respect to primates. Firstly, it can be seen that only flimsy evidence is available to support the notion that the adaptive radiation of the placental mammals did not begin until the latter part of the Cretaceous. We really have no indication of the progress of mammalian evolution in the southern continents for most of the Jurassic and the Cretaceous. Yet it would seem likely that the divergence between marsupials and placentals took place in the early Cretaceous (Slaughter, 1981; Savage and Russell, 1983). Further, there are a number of worrying features with respect to the likely continental configuration in the mid-Cretaceous (see Fig. 4.12). If the adaptive radiation of placentals and marsupials only began after that point in time, when South America was probably effectively separated from North America by an epicontinental seaway and Australia + Antarctica were apparently well isolated from South America and Africa, how can we explain the occurrence of a variety of marsupials in North America and South America during the late Palaeocene despite the relatively low degree of faunal similarity between those two continents? Kurtén (1973) suggests that the late

Palaeocene mammalian fauna of South America 'bears the stamp of earlier resemblance to North America and even Asia and Europe' and in fact concludes that the faunal similarity between South America and North America was *decreasing* during the Palaeocene and Eocene. A possible solution would be that marsupials were present in the southern continents considerably earlier than is generally assumed and that marsupials were common to Australia, Antarctica, South America and North America when those continents were closer together. Riek (1970) reported fossil fleas from early Cretaceous deposits in Australia that exhibit adaptations for clinging to hair, suggesting the presence of mammals in Australia at that time. An early Cretaceous 'monotreme' has now been reported from Australia (Archer, Flannery *et al.*, 1985) and it is quite possible that marsupials were present at that time. The possibility that marsupials were present in Africa at such an early stage (i.e. between the times represented by Figs 4.11 and 4.12) cannot be ruled out as there is simply no mammalian fossil evidence.

The presence of placental mammals in South America, North America, Asia and possibly Europe in the late Cretaceous similarly presents problems with respect to the continental configuration indicated for the mid-Cretaceous (Fig. 4.12). How could placental mammals arise during the latter part of the Cretaceous and spread so widely, despite the considerable sea barriers that were apparently present? Again, a possible solution would be provided by a considerably earlier origin for the placental mammals; again, the lack of suitable fossil sites, particularly in Africa, rules out any direct test of this possibility. Therefore, although the available fossil evidence indicates that the origin and initial diversification of marsupials and placentals took place after the mid-Cretaceous, there remains a strong possibility that fossil evidence from suitable sites in the southern continents (especially in Africa) could document a much earlier date somewhere between the early Jurassic and mid-Cretaceous. Whereas it is perfectly understandable that palaeontologists should rely on available fossil evidence, rather than on hypotheses based on gaps in the fossil record, it is nevertheless true that our present knowledge of Jurassic/Cretaceous mammals is so biased and limited that we cannot yet draw any lasting conclusions. There has been a common tendency to imply that the almost exclusive limitation of the fossil record of Jurassic/ Cretaceous mammals to the northern continents reflects zoogeographical reality, but we must bear in mind the alternative possibility that the (as yet largely undocumented) adaptive radiation of mammals in the southern continents was in fact of central importance and that northern continental mammals represent peripheral offshoots.

The biogeographical evidence is particularly relevant to discussion of primate evolution, as modern primates are essentially restricted to tropical and subtropical regions of the world. As has been shown in Chapter 2, early fossils attributed to the order Primates can be fairly sharply divided into 'archaic primates' ranging from the late Cretaceous through to the Eocene and 'primates of modern aspect' ranging from Eocene to Recent times. All of the fossils concerned, from the late Cretaceous to the end of the Eocene, are essentially confined to the Northern Hemisphere, and this has been taken to indicate a northern continental origin for primates. Further, since the earliest known 'archaic primate' (*Purgatorius*) is known from North America, some authors have concluded that primates originated in that continent. However, the 'archaic primates' have only a tenuous link with primates of modern aspect (suggested predominantly by cheek-tooth morphology; see Chapter 6) and their geographical distribution during the cooler times of the late Cretaceous and Palaeocene is surely peripheral with respect to tropical and subtropical environments of the

time. Taking the reasonable assumption that primates have always been adapted particularly to tropical and subtropical forest conditions, it seems more likely that they would have evolved in Africa and that the Eocene 'primates of modern aspect' of Europe and North America represent peripheral populations that spread northward as world temperatures increased during the Eocene, only to decline as temperatures decreased later in the Cenozoic (see also Gingerich, 1986a). The 'archaic primates' doubtless represent an earlier northern radiation from a stock of placental mammals that branched off either from the earliest primates or from some completely independent subdivision of primitive placental mammals; they could not have been directly ancestral to any 'primates of modern aspect'. This alternative interpretation of primate evolutionary radiation is supported by some geographical considerations (e.g. the Eocene 'primates of modern aspect' and the ancestors of the modern lemurs of Madagascar must surely be linked by an African stock if the reconstruction in Fig. 4.13 is basically correct), but it remains extremely tentative until the fossil history of Africa is better documented. For the time being, therefore, present interpretations of primate origins in the northern continents are heavily dependent on negative evidence and on the clinging legacy of reconstructions based on the view that modern continental positions (Fig. 4.14) have remained essentially unchanged throughout mammalian evolution. The implications of continental drift for mammalian evolution remain to be comprehensively tested by fossil discoveries in the southern continents and we must bear in mind the distinct possibility that placental mammals (and hence primates) began to diversify far earlier than is commonly supposed.

APPENDIX: THE IMPORTANCE OF BODY SIZE

It has already emerged that body size was a significant feature in the early origin and subsequent evolutionary radiation of the mammals. Whereas the mammal-like reptiles (pelycosaurs and therapsids), from which the mammals emerged, were typically quite large-bodied, the known fossil mammals from the Triassic/Jurassic boundary right through to the end of the Cretaceous were all quite small, with a range of body sizes comparable to that of modern shrews, mice and rats. It would therefore seem that small body size was an integral feature of early mammalian adaptation that was subsequently maintained for the first two-thirds of mammalian history. Among other things, this suggests that the earliest primates were probably relatively small-bodied and that (as in other placental mammal groups) there has been a general tendency for body size to increase throughout the Cenozoic era, although some exceptions to this trend clearly exist.

In broad ecological terms, body size is evidently closely connected with overall evolutionary adaptation and it is reasonable to suggest from comparisons with living mammals that the consistently small size of early mammals was specifically linked with nocturnal habits and heavy reliance on small animal prey (particularly arthropods) in the diet (see also Jerison, 1973; Kemp, 1982; Kermack and Kermack, 1984). Correspondingly, the spectacular radiation of mammals documented during the Cenozoic was characterized by the emergence of a variety of diurnal forms and the adoption of a wider range of dietary habits as diversification in body size took place. Certainly, among modern primates there is a reasonably close correlation between small body size and nocturnality (Fig. 4.15). The modal weight of nocturnal primates is about 500 g, whereas that of diurnal primates is about 5 kg, some 10 times greater. In fact, among modern mammals small-bodied species tend to

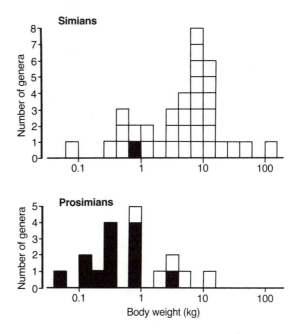

Figure 4.15. Average body weights for primate genera: simians ($N = 35$) and prosimians ($N = 18$). In both primate groups, nocturnal forms (black boxes) tend to be relatively small.

be nocturnal whereas diurnal habits are largely confined to large mammals. There are a few exceptions in that some relatively small mammals are typically diurnal (e.g. squirrels, tree-shrews, marmosets and tamarins), and there are some relatively large mammals (e.g. aardvark, hyena) that are active at night; but most mammals comply with the general rule.

In a perceptive discussion of distinctions between nocturnal and diurnal habits in mammals, Charles-Dominique (1975) reached the interesting conclusion that competition with diurnal birds has generally restricted mammals to nocturnal lifestyles. An analysis of the avian and mammalian faunas of two modern tropical forest ecosystems (Gabon and Panama) revealed basically similar patterns with the following characteristics: (1) the great majority of bird species are diurnal, whereas most mammalian species are nocturnal; and (2) the average body weight of diurnal mammals is markedly greater than that of nocturnal mammals. In fact, diurnal mammal species become more frequent than nocturnal mammal species only when the body weight exceeds 5 kg, which is close to the approximate upper limit for bird flight. (The absolute maximum weight for modern flying birds is 15 kg, whereas flightless birds weigh as much as 90 kg.) It is therefore reasonable to suggest that birds can out-compete mammals for the occupation of diurnal ecological niches over the body-size range where winged flight is possible. (In this context, it is interesting to note that bats, the only winged mammals, are without exception nocturnal in habits.) As Charles-Dominique (1975) notes: 'As an overall impression, one can conclude that birds occupy primarily the diurnal ecological niches and that mammals occupy the nocturnal ecological niches.' Given the additional strong association between small body size and nocturnal habits in mammals and the fossil evidence that the earliest mammals were consistently small in size, it seems likely that mammals have typically remained nocturnal throughout most of their evolution. This may well explain why the olfactory and tactile senses became important early in mammalian evolution whereas in birds olfaction is poorly developed and vision is predominant. Accordingly, one might expect mammals that have increased their body size and become diurnal in habits to show significant shifts in sensory adaptation and in other features. Indeed, the shift from nocturnal to diurnal habits can be regarded as a significant transition into a new 'adaptive zone' (Jerison, 1973).

In some areas of the world (Madagascar, Africa, Asia and South America), both nocturnal and diurnal primate species now occupy the same general habitat areas, and there is typically a fairly sharp distinction between the two types. In Africa and Asia, the nocturnal primates are prosimians (members of the loris group and tarsiers), whereas the diurnal non-human primates are simians (Old World monkeys and apes). The Afro-Asian nocturnal prosimian species are uniformly small, whereas the diurnal

simian species are markedly larger. This particular contrast is interesting in that a similar size distribution, but with quite different origins, is found among Eocene fossil primates in North America. Fleagle (1978) has shown, on the basis of molar tooth dimensions (which serve as an approximate indication of body size for the fossil species), that North American Eocene 'tarsioids' were uniformly quite small, whereas Eocene 'lemuroids' from the same region were generally larger in size (Fig. 4.16). And it seems likely on other grounds (e.g. relative size of the orbits; see Chapter 7) that Eocene 'lemuroids' were typically diurnal whereas 'tarsioids' were probably nocturnal. Hence, we have the unusual situation in which, among Eocene primates of North America, 'lemuroids' (possibly related to modern lemurs and lorises) presumably occupied the diurnal niches and 'tarsioids' (possibly more closely related to modern simians) presumably occupied nocturnal niches. Later in primate evolution this situation was reversed, with the simian relatives of the 'tarsioids' occupying daytime niches and the lorises confined to nocturnal niches. Furthermore, the data on North American fossil primates (Fig. 4.16) also indicate that distinct 'adaptive zones' were occupied by individual groups of primates, just as with the modern representatives.

Apart from its general relevance to nocturnal/diurnal distinctions and differential occupation of 'adaptive zones', body size is important in that it can influence morphological and other characters in several broadly predictable ways. Such size effects must clearly be taken into account when mammalian species of different body sizes are being compared. The basic question to be asked in conducting comparisons between several species of different body sizes is: 'To what extent are the observed similarities and differences attributable to body size alone, and to what extent can they be traced to fundamental similarities or differences in bodily organization?' Arguably, similarities between two species that are attributable merely to similarity in body size should be given very little weight or discounted altogether, whereas

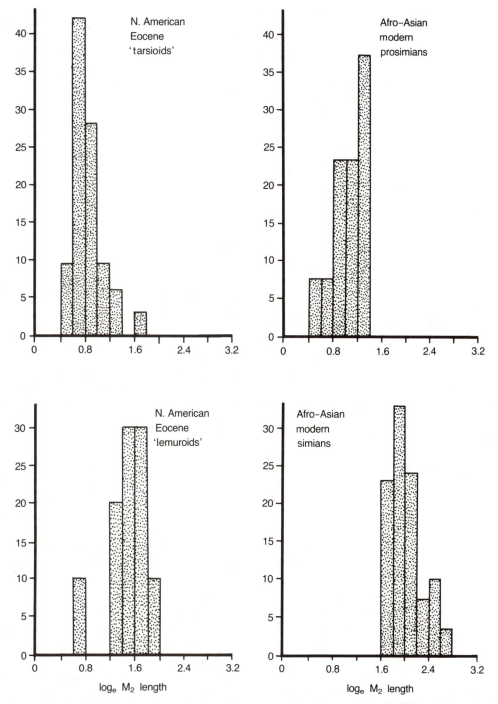

Figure 4.16. Size distributions based on lengths of lower second molar teeth for (left) North American Eocene primates ('tarsioids' and 'lemuroids') and (right) modern Afro-Asian primates (prosimians and simians). Vertical scale indicates per cent representation for each size category. (After Fleagle, 1978.)

similarities between species of different body sizes that are revealed after body size has been taken into account are likely to reflect some fundamental similarity in organizational principles.

We know from simple geometric considerations that if body size increases without any change in shape the surface:volume ratio must decrease, because surface area increases according to the square of a linear dimension while volume increases according to the cube. Thus, regardless of their particular phylogenetic relationships, large mammals will tend to share common features distinguishing them from small mammals in any functional systems influenced by surface:volume ratios, or by other comparable size-dependent relationships. This situation can be simply illustrated by taking a series of cubes of increasing volume and comparing them, in terms of surface:volume relationships, with a similar series of spheres. If surface area is plotted against volume for these two sets of objects, the resulting curves (Fig. 4.17) exhibit two consistent features. Firstly, the surface:volume ratios of both cubes and spheres decrease as volume increases. Secondly, at any given volume a cube always has greater surface area than a sphere. One interesting corollary of these two rules is that, for every sphere with a given surface:volume ratio, one can find a matching cube with the same ratio but with a greater volume (in fact, just less than twice the volume of the sphere). This can be regarded as a special case of 'convergence', with a similarity in one respect (surface:volume ratio) being achieved by different combinations of shape and volume. Now, if one were to consider, instead of sets of spheres and cubes, two series of organisms with different organizational patterns and covering the same range of body sizes, similar relationships might be expected. For example, one might wish to consider some surface-dependent functional system such as respiration or digestion in a comparison of the two groups in the attempt to define some systematic organizational difference between them. This, however, would involve empirical data for the surface-dependent system (e.g. lung

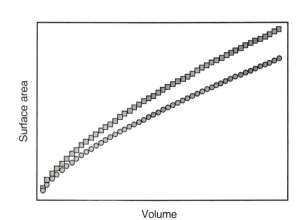

Figure 4.17. Plot of surface area against volume for a series of spheres and a series of cubes, expressed in arbitrary units.

architecture or morphology of the digestive tract) and body size (weight or volume), rather than values calculated from standard formulae for ideal geometric bodies, as in Fig. 4.17. Thus it is necessary to arrive at some standard form of analysis of such empirical data that will take body-size differences into account and reveal the systematic difference between the two groups.

It must be emphasized that comparisons between animals of different body sizes become complicated only when the relationship between a given feature (e.g. surface:volume ratio) and body size changes in a **non-linear** fashion. In the relatively rare cases in biology where there is a simple proportional relationship between the dimensions of a given functional system and body size (**isometry**), no chance convergence would be expected between members of groups with different organizational principles; a simple ratio between a particular dimension and body size expresses the relationship quite adequately. For example, in mammals generally the volume of the lungs and the weight of the heart increase isometrically with body weight (Schmidt-Nielsen, 1984) and mammals with similar lung:body ratios or heart:body ratios can be taken as organizationally comparable.

In most instances, however, there is a **curvilinear** relationship between the dimensions of functional systems and body size (**allometry**). Where the dimension concerned increases more slowly than body size, as is the case with surface area in relation to volume (Fig. 4.17), the relationship is said to be one of **negative allometry**. Where the dimension increases more rapidly than body size, as is often said to be the case with the weight of the skeleton in relation to overall body weight, the relationship is one of **positive allometry**. When comparing organisms with respect to characters that vary allometrically with body size and therefore show curvilinear relationships, the empirical relationships must be converted to a linear form to permit more direct comparison. Here, the standard approach is to use logarithmic coordinates for the plotting of the allometric data. When the data illustrated in Fig. 4.17 for surface:volume relationships of spheres and cubes are converted to logarithmic form, the two original curves are transformed into two parallel straight lines (Fig. 4.18). The similar slopes of the two lines reflect the common **scaling principle** involved in the increase of surface area with increasing volume, and the constant interval between the two lines reflects the systematic difference in shape between spheres and cubes. This result is achieved because for both spheres and cubes the curvilinear relationship between surface area and volume can be expressed by the common formula:

$$S = kV^{0.67}$$

(where S is the surface area, V is the volume and k is the constant). In its logarithmic form, this equation becomes linear:

$$\log S = 0.67 \log V + \log k$$

The difference between spheres and cubes, in terms of surface area:volume relationships, resides in the different values of the constant k. For cubes, the constant takes the value 6; for spheres, it is 4.84. In other words, at any given volume, the surface area of a cube is bigger than that of a sphere by a factor of 6/4.84, or 1.24 (representing an increase of 24%). In Fig. 4.18, the vertical distance between the lines is equivalent to the difference between the

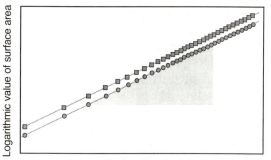

Figure 4.18. Logarithmic plot of the data shown in Fig. 4.17. The two original diverging curves have been converted into parallel straight lines. The slope of the lines (indicated by the shaded triangle) is 0.67, reflecting the ratio between the power functions involved (surface is proportional to the square of a linear dimension; volume is proportional to the cube). The vertical separation between the two lines (= difference in intercept values on the ordinate) reflects the systematic difference in shape (and hence surface: volume relationships at any given size) between spheres and cubes.

logarithmic values of k (i.e. log 6 − log 4.84) and directly indicates the factor by which the surface area of a cube exceeds that of a sphere of the same volume (because log (6/4.84) = log 6 − log 4.84).

The formula given above for the relationship between surface area and volume is, in fact, just a special case of the general **allometric formula**, which has been found to provide a useful empirical approximation to numerous sets of data for size relationships in biology (Gould, 1966; Martin, 1980; McMahon and Bonner, 1983; Calder, 1984; Schmidt-Nielsen, 1984; Jungers, 1985). The basic allometric formula is commonly written:

$$Y = kX^{\alpha}$$

giving:

$$\log Y = \alpha \log X + \log k$$

(where X is some measure of body size, Y is the dimension of some character under investigation, k is the allometric coefficient and α is the allometric exponent.

One great advantage of the logarithmic bivariate data plot (e.g. Fig. 4.18) is that the value of the allometric exponent (α) is directly indicated by the slope of the line, and that of the allometric coefficient (k) can be determined from the intercept on the vertical axis or ordinate (intercept = log k). In Fig. 4.18, the slope of both lines is 0.67.

In any discussion of the concept of allometry, it is particularly important to distinguish **intra-specific** from **interspecific** relationships. As an individual grows, an allometric relationship may be operative between, say, some measure of a particular organ and the body weight attained at different ages. **Ontogenetic allometry** of this kind was specifically studied by Huxley (1932), who found that the allometric formula gave a good empirical fit for developmental data from a variety of different sources. One can also conduct intraspecific comparisons by investigating the dimensions of given features in relation to body weight in a sample of adults, thus quantifying intraspecific variation in the adult condition. Again, the allometric formula gives a good empirical fit in many cases, although charac-teristically the slope is relatively low (probably reflecting adaptation of bodily characteristics to match an optimum 'target' body size for the species).

These situations are quite distinct from that of **interspecific allometry**, where average adult conditions for a range of different species are compared so as to investigate broad evolutionary relationships. In interspecific comparisons of this kind, there is no direct continuity between the points on the allometric plots, and any empirical relationships determined can be taken to reflect general scaling principles governing the adaptation of individual species, rather than the expression of developmental processes *per se*. Where the primary concern is with phylogenetic relationships, as in this survey of primate evolution, it is largely interspecific allometry that will be tackled. Identification of common organizational principles using allometric analysis provides one particularly useful criterion

for suggesting common ancestry in certain cases. Further, it should be remembered that the constraints imposed by body size alone are so pervasive, because of the multiple allometric relationships involved, that they probably constitute one of the major selective forces leading to convergent evolution (see p. 136). It is therefore important to investigate allometric effects both to identify fundamental scaling principles and to highlight convergent responses to constraints imposed by body size (see Fig. 3.12).

The special case of sets of cubes and spheres (Figs 4.17 and 4.18) can be generalized to a broad biological context, as illustrated in Fig. 4.19, where an allometric relationship for two sets of hypothetical organisms is portrayed in a bivariate logarithmic plot. Within each set of organisms, the size of the character concerned is 'scaled' to body size in a reasonably systematic fashion, but each set of organisms follows the scaling principle in a particular way, resulting in a vertical separation of the points on the plot. Such vertical separation (reflecting differences in the value of the allometric coefficient k) can be said to define **allometric grades** of organization: the two sets of organisms obey the same scaling principle, but fundamental differences in organization generate differences in detail (Martin, 1980). It should be noted that the logarithmic plot of surface area:volume relationships of cubes and spheres in Fig. 4.18 yields the same basic pattern as that shown in Fig. 4.19, and that the geometrical difference between a cube and a sphere is thus comparable to a biological difference in organizational grade. Taking the basic model illustrated in Fig. 4.19, one can imagine that general scaling principles (such as the surface:volume law) have prevailed throughout mammalian evolution, but that the emergence of new organizational features has permitted 'grade shifts' to occur occasionally.

Naturally, when comparing sets of living mammals in allometric terms, one is likely to encounter established grades with relatively few intermediate cases. Nevertheless, the dynamic aspect of evolutionary shifts between grades is important and it is therefore necessary to proceed

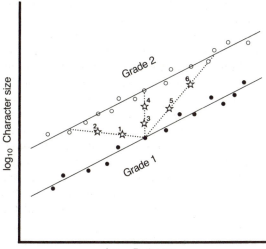

Figure 4.19. General logarithmic model for inter-specific allometric analysis. In this example, two series of living organisms (●; ○) both obey the same 'scaling principle', which is reflected by the common slope of the lines. However, at any given body size it is found that a species belonging to the second group (○) consistently has a higher value for the character dimension concerned than a species belonging to the first group (●). Thus, two separate grades of organization can be identified. Assuming that, in this case, grade 1 represents the primitive condition, it is likely that intermediate forms (☆) existed at some time, as grade 2 became established. Fitting of lines to successive intermediate fossils (e.g. 1–2, 3–4, 5–6), in an attempt to express 'phylogenetic allometry', actually confuses change in body size over time with the grade shift. It is theoretically possible for the slope of the (dotted) line to take any value between negative and positive infinity. (Figure reprinted from Martin (1983) with kind permission of the American Museum of Natural History, New York.)

with particular caution when attempting to trace an evolutionary transition from one grade to another. In certain cases, it may be possible to find a series of fossil forms that approximate to an evolutionary lineage involved in a grade shift, as indicated in Fig. 4.19. Although straight lines can be fitted to data points for such a series of fossil forms, it should be noted that '**phylogenetic**

allometry' of this kind is very difficult to interpret. The parameters for the line fitted to the series of data points represent a combination of the grade shift and the direction of body-size change. Hence, the allometric exponent (α) can, in theory, exhibit any value from negative to positive infinity. Accordingly, scaling principles governing allometric relationships can only be determined for sets of organisms that are broadly contemporary and occupy the same allometric grade. The problem of objectively defining allometric grades remains one of the fundamental problems involved in the analysis of scaling relationships.

Assuming that a reliable allometric interpretation is attainable, the empirically determined allometric equation allows one to take the effects of body size into account when making comparisons between species. This is particularly important where one is comparing two groups of species that are suspected to be fundamentally different in organizational grade. The empirical definition of two distinct allometric relationships, with approximately the same slope values but with differing intercept values, is in itself of value in demonstrating a grade difference between the two groups. However, one can take the procedure further by calculating, for each species, a value that reflects its position relative to all others in the comparison. Taking the empirical values determined for the slope and intercept of the allometric relationship, one can relate each data point to some standard and thus calculate an **allometric index** value. The index value (I) for any given species can be determined from the following formula:

$$I = Y_O/Y_E$$

where Y_O is the observed value of the character dimension for the species concerned and Y_E is the 'expected' value for its body size). Two different strategies can be adopted for calculating Y_E from an established allometric relationship of the form:

$$Y = kX^\alpha$$

Once this formula has been determined for all of the species in a comparison, it represents the typical condition for any given body size. Thus, Y_E is simply the value of Y indicated for any given species when the value for its body size (X) is entered in the equation. Alternatively, one can determine some minimal condition for a given set of organisms either by fitting a line to the lowest grade represented or by taking the slope value for the overall best-fit line (α) and determining the appropriate value of the allometric coefficient (k') to fit the lowest point in the distribution. Y_E is then determined from the formula:

$$Y_E = k'X^\alpha$$

The important point is that the use of such allometric indices permits direct, meaningful comparison between species of different body sizes, provided that the empirically determined value for α is reliable. The existence of any more-or-less distinct grades should then be reflected by the distribution of the values of the index calculated for the species concerned.

One further issue fundamental to the successful analysis of allometric relationships involves the selection of an appropriate technique for determining a best-fit line in relation to a logarithmic data set and hence for empirical determination of the allometric power function (α) and the allometric coefficient (k). The simplest approach in conceptual terms, and one that has been widely used in allometric studies, is the calculation of a **least-squares regression**. However, as was pointed out by Kermack and Haldane (1950), there are underlying assumptions in regression analysis that are frequently not compatible with biological data (see also Harvey and Mace, 1982). Regression analysis is designed for situations in which one variable (Y) is clearly dependent on the other (X), and where the latter (the independent variable) is measured 'without error'. With allometric relationships, it is by no means obvious that body size (usually taken as the X variable) is the independent variable, and any error entailed in measuring other dimensional characters (the Y variable) are generally comparable both in kind and in magnitude to those present in measurements of body size. Therefore, it may

be more appropriate to use a line-fitting procedure that does not involve such a distinction between the two variables. In other words, one really needs a procedure that treats the two variables symmetrically, rather than asymmetrically as is the case with regression. For this reason, Kermack and Haldane (1950) proposed the **reduced major axis** as an appropriate best-fit line for biological data and other authors (e.g. Teissier, 1948) have favoured use of the **major axis**. Both the reduced major axis and the major axis represent symmetrical approaches to the question of determining a best-fit line, but they differ in the mode of calculation and the major axis is probably to be preferred (see Harvey and Mace, 1982 for a discussion of the essential points).

Regardless of the statistical procedure actually used for calculating a best-fit line for allometric data, the strength of the relationship between two variables can be expressed (in a very general sense) by the **correlation coefficient** (r). This can take any value between $+1$ (representing a perfect positive correlation) and -1 (representing a perfect negative correlation). The squared value of the correlation coefficient (r^2, the **coefficient of determination**) indicates the proportion of the total variation that is 'explained' by the relationship between the two variables. Where a very high value is obtained for the correlation coefficient (i.e. where r is greater than 0.97 or less than -0.97), there is little problem involved in determining a best-fit line for an allometric relationship, because the regression, reduced major axis or major axis all yield essentially similar results. However, as the correlation coefficient value approaches zero, the discrepancies between the best-fit lines obtained with these three statistical procedures increase. In particular, the regression usually becomes markedly divergent from the reduced major axis and major axis (except in unusual cases where the slope of the line is very shallow, when the regression and major axis differ markedly from the reduced major axis). The slope of the regression (α_r) is related to the slope of the reduced major axis (α_m) by the following simple formula:

$$\alpha_r = r\alpha_m$$

(with α_m always taking the same sign as α_r). In many allometric analyses, the slope value determined with a regression is markedly lower than that determined with a reduced major axis or major axis and the difference may be quite substantial even with correlation coefficient values that might otherwise be regarded as quite high (e.g. $r = 0.90$).

With a correlation coefficient value of $r = 0.90$, a regression slope of 0.67 is equivalent to a reduced major axis slope of 0.74. This, of course, has important implications for the identification of scaling principles and the determination of values of allometric indices (quotients) that supposedly cancel out the effects of body-size differences. If the value adopted for the allometric slope is (for example) too low, calculated values for an allometric index will be too low for small-bodied species and too high for large-bodied species, and the scaling principle involved will be misinterpreted because of the erroneous slope value. It is important to understand that allometric interpretations depend on the line-fitting technique used, even though there is a continuing, as yet unresolved, debate over the most appropriate approach. In the text that follows, allometric analyses have been based primarily on the major axis (although other line-fitting procedures may have been used where results are reported from prior publications), on the grounds that a symmetrical best-fit line is preferable to an asymmetrical one where biological data are concerned. In cases where correlation coefficient values between -0.97 and $+0.97$ are involved, it should be remembered that regressions would generally yield a lower value for the allometric exponent (α) and a higher value for the allometric coefficient (k).

SCALING OF ENERGY REQUIREMENTS

Specific applications of allometric analysis in the clarification of particular aspects of primate evolution will be described in the relevant chapters to follow (e.g. for dentitions in Chapter 6, brain size in Chapter 8 and locomotor adaptations in Chapter 10). There is, however, one allometric relationship that is so fundamental to the general body functions of mammals that it deserves special mention at the outset – the relationship between metabolic turnover and body weight (see also Chapter 6). Obviously, the availability of energy for life processes must have profound implications for all other aspects of mammalian biology.

Because of the necessity to measure metabolic requirements according to precise and standardized criteria, the energy relationships of organisms are usually expressed in terms of **basal metabolic rate** (BMR). BMR, which can be measured either in terms of oxygen consumed or in terms of calories generated over a standard period of time (e.g. 24 h), is ideally established with resting animals that are 'postabsorptive' (i.e. are not utilizing energy for digestive processes) and are maintained under conditions of thermoneutrality (i.e. at an ambient temperature that does not require special energy expenditure for heating or cooling the body). Numerous studies have been conducted on BMR in mammals and other animals and there is now an extensive literature on the subject (e.g. see Schmidt-Nielsen, 1972, 1984). One major generalization to emerge from these studies is that, for various groups of organisms, basal metabolic rate scales according to body weight in the following fashion:

$$BMR = kW^{0.75}$$

(where W is body weight and k is the allometric coefficient for any group of organisms). Because recognition of this widespread scaling principle is in large part due to the investigations conducted by Kleiber (1947, 1961), this allometric formula is commonly referred to as Kleiber's Law (but see also Brody, 1945). For the time being, however, the scaling principle for basal metabolic rate remains empirical in nature. Although several possible explanations have been advanced to account for the exponent value ($\alpha = 0.75$) in the allometric formula, the theoretical basis for the scaling of basal metabolism has yet to be established unequivocally. Nevertheless, there is abundant evidence that the scaling principle recognized by Kleiber is widely applicable. In their famous 'mouse-to-elephant curve', Brody, Proctor and Ashworth (1934) showed quite clearly for mammals generally that the scaling exponent value was close to 0.75 and that there was an extremely good fit of individual points to the straight line determined for a logarithmic plot of BMR against body weight. (Incidentlly, this provides a good example of a case where the correlation is extremely high and where regression, reduced major axis and major axis

all yield closely similar results.) Kleiber (1961) reported the following allometric relationship between BMR (in kilocalories per day) and body weight (in kilograms) for a sample of 26 data points:

$$\log_{10} \text{BMR} = 0.756 \log_{10} W + 1.83$$

The allometric relationship for these data is, in fact, so consistent that the correlation coefficient for the logarithmic plot is $r = 0.998$. Conversion from the logarithmic form yields the following formula:

$$\text{BMR} = 67.6\ W^{0.756}$$

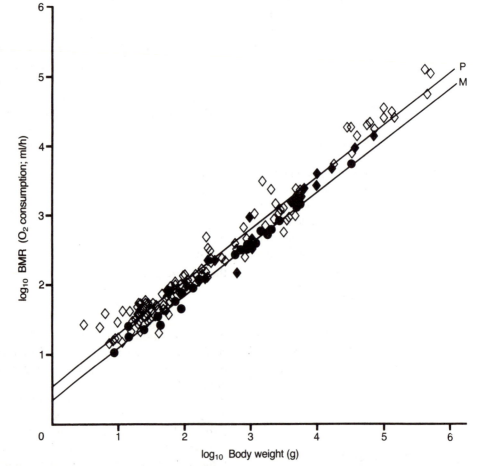

Figure 4.20. Logarithmic plot of basal metabolic rate (BMR) against body weight for a large sample of mammalian species ($N = 149$). Key: $\diamondsuit$, placental mammals including primates ($N = 126$); $\blacklozenge$, primates alone ($N = 19$); $\bullet$, marsupials ($N = 23$); P, major axis for placental mammals; M, major axis for marsupials. Note: basal metabolic rate can be expressed either in kilocalories, as in Kleiber's equation, or in terms of oxygen consumption, as in this graph.

(in kilocalories per day). This is commonly rounded off to give the standard formula for placental mammals:

$$BMR = 70\ W^{0.75}$$

In the words of Kleiber (1961, p. 179), this standardized equation expresses the fact that: 'Under standard conditions, fasting homeotherms produce daily an average of about 70 kcal of heat per $kg^{3/4}$ per hour.'

Despite the close fit of Kleiber's data to a straight line (on logarithmic coordinates), there is some scatter of individual mammalian species around the line when the sample is expanded to include many more species (Fig. 4.20). Some of this scatter doubtless occurs because the values indicated (collected from diverse sources in the literature) were not obtained under such rigorous conditions as those specified by Kleiber (1961). However, there are also significant departures from the best-fit line that reflect special metabolic adaptations of individual species. A particularly clear example of such significant departures is provided by the contrast between marsupials (see Dawson and Hulbert, 1970) and placentals. Marsupials and placentals both obey 'Kleiber's Law' in that the value of the allometric exponent is in both cases close to 0.75, but the allometric coefficient value is lower in marsupials than in placentals. The equations for the best-fit lines (major axes) in Fig. 4.20 are as follows:

Marsupials ($N = 23$):
$\log_{10} BMR = 0.75 \log_{10} W + 0.37$ ($r = 0.995$)
Placentals ($N = 126$):
$\log_{10} BMR = 0.76 \log_{10} W + 0.53$ ($r = 0.98$)

(where BMR is the basal metabolic rate in millilitres of oxygen consumed per hour and W is body weight in grams). Conversion of these logarithmic equations yields the following standard formulae:

Marsupials: $BMR = 47.4\ W^{0.75}$ (in kcal/day)
Placentals: $BMR = 75.7\ W^{0.76}$ (in kcal/day)

As can be seen, the implication of these formulae is that, at any given body weight, marsupials utilize about 35% less energy for basal metabolism than typical placentals do. Figure 4.20 shows a very good example of a difference in 'allometric grade' between marsupials and placentals.

The equation for primates alone does not differ significantly from that for placental mammals generally (Fig. 4.20), but there are some primate species that have significantly lower basal metabolic rates than would be expected for typical placental mammals of the same body weight. For example, it is known that all members of the slow-moving loris subgroup have low basal metabolic rates relative to body size (see also Chapter 6), and the same is true of the only nocturnal simian primate, the owl monkey (*Aotus trivirgatus*). Similar exceptions exist among the non-primate placental mammals, the most obvious example being provided by the sloths. Lorises and sloths have basal metabolic rates close to 50% of the values expected from Kleiber's standard equation (i.e. even lower than in typical marsupials). McNab (1978b, 1980, 1986) has discussed in detail the deviations of individual mammalian species from basal metabolic rate levels expected from Kleiber's equation and he has identified some common factors. In particular, mammals that are specialized for feeding on leaves (folivory) exhibit depressed basal metabolic rates, relative to body size, and the same is true of mammals that are specialized for feeding on ants and termites (myrmecophagy). In both of these cases, McNab has suggested that toxins present in food items may require a slowing of body metabolism in the mammals feeding on them. However, Elgar and Harvey (1987) have shown that basal metabolic rates mainly vary according to taxonomic group among mammals and may not be directly linked to diet.

Allometric grades in basal metabolic rates are even more marked when comparisons are widened to include all vertebrates. Hemmingsen (1950, 1960) has shown that the homeothermic birds and mammals together constitute a grade that is quite distinct from that constituted by poikilotherms such as reptiles and amphibians. The comparison here is somewhat complicated by the fact that poikilotherms do not maintain a constant body temperature, in contrast to the homeothermic birds and mammals, so the basal metabolic rates of poikilotherms must be determined for a standard temperature. A temperature of 20 °C is commonly taken as the reference level for basal metabolic rate in poikilothermic vertebrates, as this represents a realistic average figure for ambient temperatures under natural conditions. For homeotherms, on the other hand, the body temperature is maintained relatively constant by endothermic mechanisms at about 37 °C. When the allometric grades identified by Hemmingsen (1950, 1960) for poikilothermic and homeothermic vertebrates are examined in detail, it emerges that the basal metabolic rate of a homeotherm (e.g. a mammal) with a body temperature of 37 °C is approximately 25 times higher than that of a poikilotherm (such as a reptile) of the same body size at an ambient temperature of 20 °C! Clearly, there is a considerable energetic cost involved in endothermic/homeothermic mechanisms of birds and mammals and there must

have been a considerable selective advantage of such mechanisms to explain their emergence in the evolution of these two vertebrate groups. Since basal metabolic rates per unit body weight are highest for small vertebrates – because of the negative allometric scaling of basal metabolism to body weight – small mammals obviously have relatively high energy demands. It is therefore hardly surprising that the earliest mammals may have required a considerable period of time to establish and refine mechanisms for endothermy/ homeothermy before they were in a position to compete effectively with reptiles. Energy relationships are therefore directly relevant to the emergence of mammalian characteristics and hence to the earliest beginnings of primate evolution. Indeed, the linked features of body size and energy turnover represent two major themes that will be encountered repeatedly in the chapters that follow. In attempting to trace the adaptive radiation of the primates, their humble origins among small-bodied nocturnal mammals of the Mesozoic must be borne firmly in mind.

Chapter five

Are tree-shrews primates?

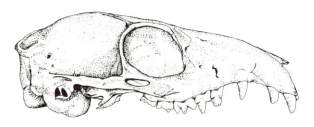

The possibility of a phylogenetic relationship between tree-shrews (family Tupaiidae) and primates has been a dominant theme in discussions of primate evolution since virtually the beginning of this century. Following a variety of early reports of similarities between modern tree-shrews and primates (e.g. Parker, 1885; Kaudern, 1910, 1911; Gregory, 1913), Carlsson (1922) concluded after a fairly comprehensive survey of morphological features of tree-shrews that a direct phylogenetic relationship with primates existed. This conclusion was consolidated by Le Gros Clark in a series of publications dealing with tree-shrew anatomy, notably concerning the brain (Le Gros Clark, 1924b,c, 1925, 1926, 1934b). An official seal of approval was then given to the proposal of a link with primates when Simpson (1945) included the family Tupaiidae in the order Primates in his influential classification of the mammals. Le Gros Clark's conclusions were subsequently further supported by Saban (1956/57). However, there was some opposition to this interpretation (e.g. Lorenz, 1927; Henckel, 1928; Roux, 1947) and over the past two decades doubts about a phylogenetic link between tree-shrews and primates have been expressed with increasing regularity (see particularly: Campbell,

1966a,b, 1974; Goodman, 1976; Lehmann, Romero-Herrera et al., 1974; Luckett, 1969, 1974a,b, 1980a; Martin, 1967, 1968a,b, 1973, 1975; McKenna, 1966; Romero-Herrera, Lehmann et al., 1976b; Tattersall, 1984; Van Valen, 1965; and the useful collection of review papers in Luckett, 1980b). As a result, many recent authoritative texts (e.g. Simons, 1972; Szalay and Delson, 1979) have once again excluded tree-shrews from the primates.

In itself, the question of whether to regard tree-shrews as primates may not seem to be of great significance, but from the point of view of phylogenetic reconstruction it impinges directly on the fundamental subject of the evolutionary origin of the primates as a distinct group of mammals. Assessment of the evidence for and against a phylogenetic relationship between tree-shrews and primates actually provides us with a particularly valuable test case in the search for a genuinely scientific understanding of primate origins.

LIVING TREE-SHREWS

As a preliminary step in assessing the phylogenetic significance of tree-shrews, it is essential to establish an accurate overall picture of the entire family Tupaiidae. This is particularly necessary because attempts to investigate the evolutionary relationships of tree-shrews have often been based on comparisons of a single species ('the tree-shrew') with primates and other mammals, rather than on consideration of the family Tupaiidae as a whole. Although the modern tree-shrews represent only a limited adaptive radiation, comprising about 18 extant species (Table 5.1), there is nevertheless appreciable diversity in morphology and behaviour. Furthermore, the modern species have a fairly substantial geographical distribution in South and South-East Asia (Fig. 5.1).

The tree-shrews are generally recognized as falling into two distinct subgroups – the subfamilies Ptilocercinae and Tupaiinae (Lyon, 1913; Steele, 1973). The first subfamily contains a single nocturnal species (the pen-tailed tree-shrew, *Ptilocercus lowii*; Fig. 5.2) and the second contains the five diurnal genera *Tupaia, Anagale,*

Lyonogale, Urogale and *Dendrogale*. (It was suggested by Davis (1938) that *Dendrogale* is intermediate between other tupaiines and *Ptilocercus*, but so little is known about this genus of tree-shrew that this possibility cannot be assessed as yet.) More recent studies by Steele (1973), Butler (1980) and Zeller (1986a,b) confirm the basic distinction between Tupaiinae and Ptilocercinae on cranial and dental grounds. Relatively little is known about the pen-tailed tree-shrew, particularly its natural habits (see Lyon, 1913; Le Gros Clark, 1926; Davis, 1962; Lim, 1967; Gould, 1978), but the species generally seems to be more primitive than species in the subfamily Tupaiinae – for instance, in being nocturnal (Gould, 1978), in retaining certain inferred primitive dental features (Butler, 1980) and in several other respects (Le Gros Clark, 1926; Campbell, 1974). For this reason, *Ptilocercus* could shed special light on the evolutionary radiation of tree-shrews. It is important to

Table 5.1 Living species of tree-shrews revised from Lyon (1913) and Napier and Napier (1967)

Scientific name	Common name
Subfamily Ptilocercinae	
Ptilocercus lowii	Pen-tailed tree-shrew
Subfamily Tupaiinae	
Tupaia belangeri	Belanger's tree-shrew
Tupaia glis	Common tree-shrew
Tupaia longipes	Long-footed tree-shrew
Tupaia montana	Montane tree-shrew
Tupaia nicobarica	Nicobar tree-shrew
Tupaia picta	Painted tree-shrew
Tupaia palawanensis	Palawan tree-shrew
Tupaia splendidula	Rufous-tailed tree-shrew
*Tupaia minor**	Pygmy tree-shrew
*Tupaia javanica**	Indonesian tree-shrew
*Tupaia gracilis**	Slender tree-shrew
Anathana ellioti	Indian tree-shrew
Lyonogale tana	Terrestrial tree-shrew
Lyonogale dorsalis	Striped tree-shrew
Urogale everetti	Philippine tree-shrew
Dendrogale melanura	Southern smooth-tailed tree-shrew
Dendrogale murina	Northern smooth-tailed tree-shrew

* Small-bodied, relatively arboreal *Tupaia* species.

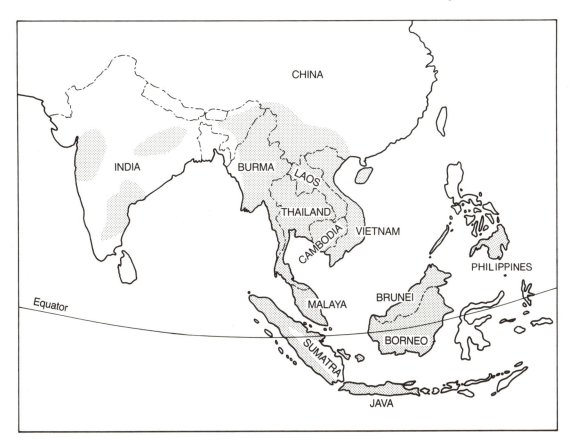

Figure 5.1. Geographical distribution of modern tree-shrews. (Adapted from Lyon, 1913; see also Fiedler, 1956.)

recognize the point that assessment of the likely phylogenetic relationships of tree-shrews must involve inference of the hypothetical ancestral condition from which all living tree-shrew species diverged. It is obvious that the common ancestral stock of modern tree-shrews must have been at least as primitive as the most primitive surviving species and was probably more primitive than any of them. Thus, assessment of the phylogenetic relationships of tree-shrews on the basis of a single surviving species (particularly if it happens to be one of the more advanced among them) is likely to confuse rather than clarify the question of actual ancestral relationships.

Although there are some significant differences between tree-shrew species, they share a basic common pattern that can be described with reference to the relatively well-known Belanger's tree-shrew, *Tupaia belangeri* (Fig. 5.3). All tree-shrews are relatively small (body weight 45–350 g, depending on the species), and they are generally omnivorous, with a predilection for small animal prey (mainly arthropods) and small fruits. The skull and postcranial skeleton follow a fairly unspecialized placental mammal pattern and there are claws on all the digits of both hands and feet (see Fig. 10.2). There is no evidence that the hands or feet are adapted in any particular way for grasping, although the big toe (hallux) is somewhat divergent in *Ptilocercus* (see Fig. 5.2). In spite of their name, tree-shrews are not uniformly arboreal, and herein resides the basis for much of the divergence between the species. Some tree-shrews (e.g. *Ptilocercus*, *Dendrogale* and *Tupaia minor*) spend a considerable amount of time in

Figure 5.2. The small pen-tailed tree-shrew, *Ptilocercus lowii* (body weight 45 g). Note the divergent big toe on the hindfoot. (After Lim, 1967; see also Gould, 1978.)

trees, whereas others (e.g. *Lyonogale* and *Urogale*) are predominantly or exclusively terrestrial. But most tree-shrew species, like Belanger's tree-shrew and the common tree-shrew, are semi-arboreal and feed extensively on the ground, rather in the manner of European squirrels (e.g. see Langham, 1982). Indeed, tree-shrews in general are so similar to squirrels in external appearance and habits that the Malay word 'tupai' (from which the generic name *Tupaia* is derived) is reportedly used indiscriminately for tree-shrews and squirrels.

The overall spectrum of habits, from predominantly terrestrial to predominantly arboreal seen among tree-shrews, is accompanied, as in squirrels, by a gradation in anatomical characters (Martin, 1968a; D'Souza, 1974). Terrestrial tree-shrews are typically larger and have longer snouts (Fig. 5.4) and relatively shorter tails than the more arboreal species. In addition, *Ptilocercus lowii*, which is probably the most arboreal tree-shrew species, has more forward-facing eye sockets (orbits) than other tree-shrews (Fig. 5.5). All tree-shrews seem to use nests of some kind, both for occupation by adults when sleeping and for rearing of the offspring. (See Chapter 9 for an account of the

Figure 5.3. Belanger's tree-shrew, *Tupaia belangeri* (body weight 170 g).

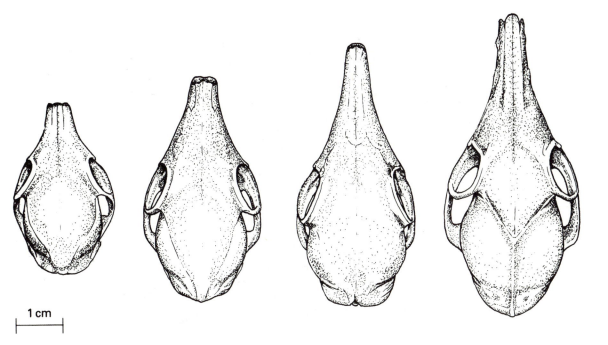

Figure 5.4. Dorsal views of the skulls of (from left to right): pygmy tree-shrew, *Tupaia minor* (relatively arboreal); common tree-shrew, *Tupaia glis* (semiterrestrial); terrestrial tree-shrew, *Lyonogale tana* (terrestrial); and Philippine tree-shrew, *Urogale everetti* (terrestrial).

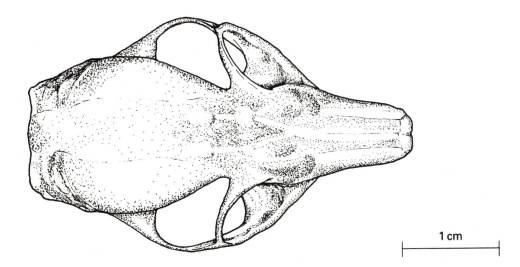

Figure 5.5. Dorsal view of the skull of the pen-tailed tree-shrew, *Ptilocercus lowii*. Note the more forward-facing orbits compared to the other tree-shrews (Fig. 5.4).

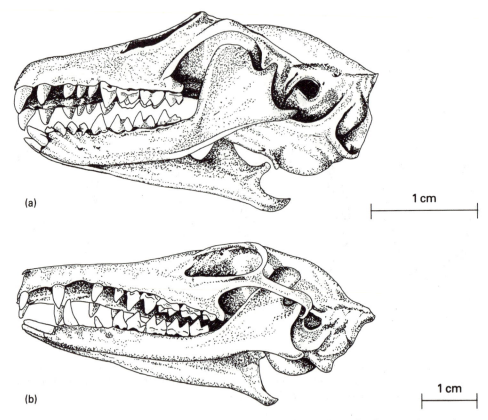

(a)

1 cm

(b)

1 cm

Figure 5.6. Oblique lateral views of the skull and lower jaw of the pen-tailed tree-shrew (*Ptilocercus lowii*, a) and the Philippine tree-shrew (*Urogale everetti*, b), showing marked elongation of the snout and associated wider spacing of the anterior teeth in the terrestrial form (*Urogale*). (After James, 1960.)

unusual maternal behaviour of tree-shrews.) The nests are commonly located in trees, although the predominantly terrestrial *Lyonogale* and *Urogale* may well use nests at ground level (e.g. in fallen tree trunks). Even those tree-shrews that spend part or all of their time in the trees (e.g. *Tupaia minor* and *T. glis*) typically use broad branches as supports, and these are negotiated with their sharp claws acting as grapples. Tree-shrews generally avoid fine branches and they do not usually leap actively within or between trees (D'Souza, 1974; see also Jenkins, 1974b). In other words, tree-shrews tend to make use of trees largely as vertical extensions of the terrestrial substrate, rather than as a three-dimensional environment of a fundamentally different nature, and it is perhaps more appropriate to describe their locomotion as 'scansorial'. Hence, tree-shrews are 'arboreal' only in a very limited sense of the term, even in the case of species that spend a considerable proportion of time in trees. Most tree-shrew species have relatively long snouts (see Fig. 5.4) adapted for foraging along broad trunks or for rooting in litter on the forest floor, and this adaptation is taken to an extreme in *Urogale everetti* (Fig. 5.6).

TREE-SHREWS AND SQUIRRELS

The resemblance noted above between tree-shrews and squirrels goes well beyond the simple sharing of features, such as (1) a generalized pattern of quadrupedal locomotion associated with clawed extremities and (2) a limited spectrum of body sizes linked with a gradient from arboreal to terrestrial habits. In South-East Asia, there seem to be pairs of tree-shrew and squirrel species, each closely matched in body size, external appearance and general habits (e.g. degree of arboreality and diet). In Borneo, according to Shelford (1916), the following five pairs of species are matched on grounds of external appearance and geographical distribution: *Tupaia glis ferruginea/'Sciurus notatus'* (= *Callosciurus notatus*); *Tupaia minor/'Sciurus jentinki'* (= *Sundasciurus jentinki*); *Tupaia gracilis/'Sciurus tenuis'* (= *Sundasciurus tenuis*); *Tupaia montana/'Sciurus everetti'* (= *Dremomys everetti*); and *Tupaia tana/'Funambulus insignis diversus'* (= *Lariscus insignis*). Shelford went on to state: 'This segregation in given localities of similarly coloured species, belonging to two very different orders of mammals, is evidence enough that the resemblances are not fortuitous, but it is by no means easy to explain them satisfactorily.'

Two alternative hypotheses were considered by Shelford, the first being that tree-shrews feed more on animal prey and profit from their similarity to squirrels when approaching their prey ('aggressive mimicry'), and the second being that squirrels profit from the resemblance because tree-shrews are unpalatable (a fact personally tested by Shelford; see also Langham, 1982) and are hence avoided by predators. Shelford opts for the latter hypothesis, but admits that it does not accord with his own observation that the squirrels are far more abundant than the tree-shrews. (In such cases of mimicry, it is expected that the mimic should be less common than the model.) Cases of mimicry seem to be generally rare among mammals, and explanation of the remarkable resemblance between tree-shrews and certain squirrel species requires a detailed field investigation. Whatever the explanation, however, it is significant that tree-shrews and squirrels can exhibit such close morphological behavioural similarity. Consequently, whenever an apparently primate-

like feature is identified in tree-shrews, it is particularly instructive to enquire whether that same feature is also found in comparable squirrel species. Where that is so, there is a high probability that the resemblance between tree-shrews and primates can be ascribed either to convergence or to retention of characteristics of ancestral placental mammals.

It is also worth commenting on the close resemblance that is seen in skull structure between the shrew-faced squirrel (*Rhinosciurus laticaudatus*) of Borneo, Sumatra and the southern part of the Malay peninsula, and terrestrial tree-shrews of the genus *Lyonogale* (Fig. 5.7). (One of the squirrel subspecies actually bears the name *Rhinosciurus laticaudatus tupaioides*.) The squirrel's snout is elongated like the terrestrial tree-shrew's, and the dentition has been modified from the typical squirrel pattern to resemble the tooth pattern of the tree-shrews, notably through considerable reduction of the upper incisors. According to E.P. Walker (1968), *Rhinosciurus* has an unusual diet for a squirrel, consisting of large ants, termites, beetles and fruit (see also Harrison, 1951) and Medway (1978) recorded a stomach containing insects and earthworms. But such a diet is quite close to that of *Lyonogale* and this probably explains the broad agreement in skull structure, essentially reflecting specialization for feeding on arthropods (see also Eisenberg, 1981). Once again, this special case of very close similarity between a squirrel and a tree-shrew underlines the general resemblance between these two groups of mammals, which was specifically noted by Weber (1928) in his review of mammals and was later echoed by Fiedler (1956).

In spite of the close resemblances between tree-shrews and squirrels, discussions of the possible phylogenetic relationships of tree-shrews have rarely included any reference to squirrels, doubtless because rodents in general have such specialized dental features that they are not regarded as relevant to primate evolution. For quite understandable historical reasons, assessments of the likely evolutionary relationships of tree-shrews have been confined

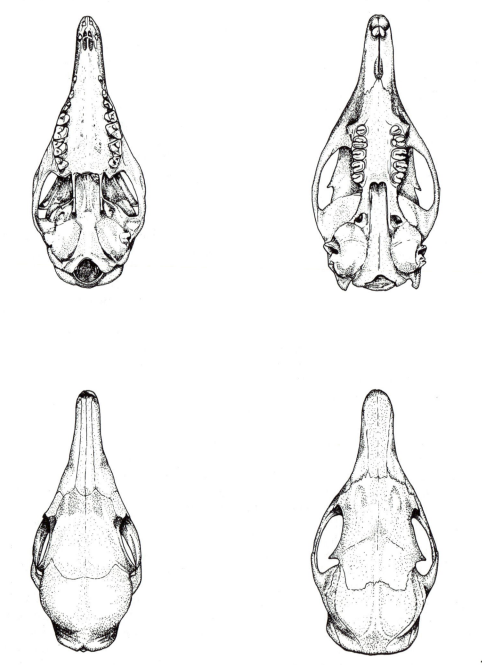

1 cm

Figure 5.7. Ventral and dorsal views of the skulls of (left) the terrestrial tree-shrew (*Lyonogale tana*) and (right) the shrew-faced squirrel (*Rhinosciurus laticaudatus*). Note the general resemblance in skull form, particularly in the elongation of the snout and in the orientation of the orbits. The teeth of the shrew-faced squirrel are reduced compared to those of most other squirrel species.

largely to comparisons between insectivores, tree-shrews and primates.

TREE-SHREWS AS INSECTIVORES OR PRIMATES

Tree-shrews were initially regarded as relatively unspecialized placental mammals and were accordingly classified in the order Insectivora (e.g. Haeckel, 1866; Weber, 1928). Following comments made by Parker (1885) on similarities between tree-shrews and lemurs, a possible phylogenetic link with primates was postulated by Gregory (1910, 1913), who agreed with Haeckel and Weber in dividing the Insectivora into the two subgroups Menotyphla (tree-shrews and elephant-shrews) and Lipotyphla (remaining insectivores, such as shrews, hedgehogs, tenrecs and moles). Gregory further suggested that the Menotyphla should be combined in the 'superorder' Archonta together with the Primates, the Chiroptera (bats) and the Dermoptera ('flying lemurs'). This suggestion carried the implication that tree-shrews, in common with the other mammals included in the Archonta, bore a particular phylogenetic relationship to primates. As noted above, subsequent work by Carlsson and Le Gros Clark, in particular, emphasized the possibility of a specific phylogenetic relationship between tree-shrews and primates, and they were included in the order Primates in Simpson's classification of mammals (1945), which was produced as a successor to Gregory's earlier version (1910).

At present, there is still a divergence of opinion in the literature over the relationship between tree-shrews and primates. Some authors of synthetic works have included tree-shrews within the order Primates (e.g. Le Gros Clark, 1959; Buettner-Janusch, 1966; Napier and Napier, 1967), whereas others have excluded them (e.g. Hill, 1953; Simons, 1972; Szalay and Delson, 1979). But virtually all authors seem to agree that tree-shrews are either insectivores or primates and that they lie close to what is now commonly termed the 'insectivore–primate boundary'. In essence, modern authors have generally upheld Simpson's (1945) view that 'all recent students agree that the tupaioids are *either* the most primate-like insectivores *or* the most insectivore-like primates' (author's emphasis). As a result, it is easy to slip into the allied view that a detailed comparison between insectivores, tree-shrews and primates is adequate both to decide on the relationships of tree-shrews and to explore the early origins of primates (e.g. see Saban, 1956/57; Le Gros Clark, 1959). Further, if we agree that tree-shrews are intermediate between insectivores and primates, it may seem to be simply a matter of classificatory convention whether they are included in the order Primates or not. This point deserves closer examination.

The question 'Are the tree-shrews primates?' actually combines two logically distinct elements that have rarely been examined separately in the literature. The first element relates to the scientific undertaking of *phylogenetic reconstruction* and the second concerns the more subjective process of *classification*. The really significant step in examining the possibility of an evolutionary relationship between tree-shrews and primates is that of generating a hypothetical pattern of ancestral branching points for the evolutionary tree of placental mammals. This relates to the first question, which is open to scientific investigation: 'Within the adaptive radiation of the placental mammals, did tree-shrews and primates share an ancestral stock exclusive of all other mammals?' Assuming that a scientifically valid process of phylogenetic reconstruction has generated a reliable outline of placental mammal evolution, thus answering this first question, it is possible to seek a classificatory arrangement that will be compatible with the inferred phylogenetic relationships. This relates to the second question, which depends on the definition of arbitrary classificatory boundaries: 'Should tree-shrews be classified as primates?' Among other things, a response to this question requires selection of a particular branching point in the inferred phylogeny for definition as the 'ancestral primate stock'. That selection, in turn, depends on a general assessment of what is the most meaningful assemblage of living and fossil species deserving the collective label 'Primates'.

Now it is probably correct to state that most authors who have considered the phylogenetic relationships of tree-shrews have either concluded from comparative investigations, or simply assumed, that the tree-shrews are indeed the closest relatives, among living mammals, of the living and fossil primates listed in Chapters 1 and 2. If this conclusion is correct, then it is certainly the case that inclusion of tree-shrews within the order Primates, or their exclusion therefrom, amounts to a matter of somewhat arbitrary definition. This would, in fact, appear to constitute the only real difference between Hill's view (1953) that the order Primates is best defined with the tree-shrews excluded (see also Hill, 1972) and Le Gros Clark's view (1959) that the order should include tree-shrews. If this were the only issue concerned in the question 'Are the tree-shrews primates?', it would scarcely be worth all the ink that has been expended on it. Yet, Simpson (1965) summarized and favoured the widely accepted interpretation of tree-shrew relationships and its implications for classification in his dismissal of what he chose to call 'long abandoned views':

> The tupaioids arose, and still stand, somewhere between the earliest placental (nominally insectivore) stem and that of the Primates. Their reference to one group or the other is in part arbitrary or semantic. Use of them to represent the earliest primate or latest preprimate stage of evolution is as valid and useful, and subject to as much caution, as is any use of living animals to represent earlier phylogenetic stages.

Accepting, for the sake of argument, the view that tree-shrews are 'either primate-like insectivores or insectivore-like primates' as a valid reflection of actual phylogenetic relationships, the problem of classificatory definition of the order Primates is simply explained in relation to Fig. 5.8. Given the branching pattern shown, the order Primates can be defined either with the tree-shrews included (solution 2: Simpson, 1945; Le Gros Clark, 1959) or to their exclusion (solution 3: Simpson, 1931a; Hill, 1953; Simons, 1972). In fact, this does not

exhaust the range of possible solutions. The order Primates could, for example, be defined to the exclusion of lemurs and lorises (solution 4), as suggested by Mivart (1873) and by Wood Jones (1929). (Incidentally, note that Wood Jones specifically stated his belief that tree-shrews are the closest living relatives of lemurs and lorises, while denying the latter any specific link to tarsiers, monkeys, apes and humans, which together constituted the Primates by his definition.) Going to the other extreme, the assumed phylogenetic relationships depicted in Fig. 5.8 would also permit an even broader definition of the order than generally accepted, to include elephant-shrews as well (solution 1: Evans, 1942). All of these various solutions to definition of the ancestral primate stock are connected with the concept of a diffuse 'insectivore–primate boundary' (see Szalay, 1968b). The relative merits of the different solutions depend essentially on practical utility and the degree to which any resulting grouping of living and fossil species as 'Primates' is meaningful in a general diagnostic sense. In all cases, essentially the same basic pattern of phylogenetic relationships has been accepted; but if the inferred phylogeny is itself in error, the situation can of course be radically altered (see later). As a final note on classificatory aspects, it should be pointed out that the problem of definition illustrated in Fig. 5.8 exists regardless of the classificatory strategy used (cladistic or classical; see Chapter 3).

Some comment is also required on Gregory's concept of a superordinal group Archonta containing primates, tree-shrews, bats and 'flying lemurs' (colugos). Although such a grouping was firmly rejected by Simpson (1945), it has recently been revived by several authors (e.g. McKenna, 1975; Szalay, 1975b, 1977; Cronin and Sarich, 1980), although usually to the exclusion of elephant-shrews and bats. If it is accepted that primates, tree-shrews and 'flying lemurs' are more closely related to one another than to other placental mammals, it is once again a matter of arbitrary definition whether tree-shrews and/or 'flying lemurs' are included in the order Primates or not.

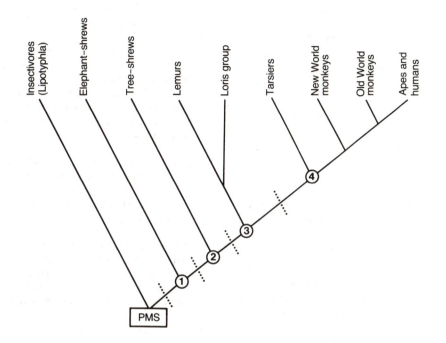

Figure 5.8. Summary of one commonly accepted interpretation of the phylogenetic relationships between insectivores, tree-shrews and primates, showing a variety of permissible definitions of the 'ancestral primate stock' (PMS, placental mammal stock):

1. Inclusion of elephant-shrews and tree-shrews in the order Primates.
2. Inclusion of tree-shrews in the order Primates.
3. Primates as defined in this book.
4. Exclusion of lemurs and the loris group from the order Primates.

DEFINITION OF THE ORDER PRIMATES

Definition of the order Primates in accordance with inferred phylogenetic relationships relates to the question of diagnosis. Obviously, an all-embracing diagnosis of all living and fossil primates is difficult to formulate, since the earliest primates (by any definition) were only marginally different from the earliest representatives of other orders of placental mammals. Diagnoses of primates have hence been directed primarily at the living forms, particularly as many features shared by living primates are of considerable significance (e.g. placentation; see Chapter 9) yet do not fossilize.

The question of the far narrower diagnosis of primates for purposes of classification of early fossil representatives is a special issue dependent on fossilizable 'hard parts' of the skull and skeleton and is heavily biased towards dental evidence. This will be considered in the relevant chapters on the skull, dentition and postcranial skeleton (Chapters 6, 7 and 10).

A widely quoted basic diagnosis of living primates was provided by Mivart (1873), whose list of defining features has had a considerable influence on subsequent discussions of primate evolution, particularly because tree-shrews have been widely quoted (rightly or wrongly) as sharing at least some of the characters involved:

> Unguiculate claviculate placental mammals with orbits encircled by bone; three kinds of teeth, at least at one time of life; brain always with a posterior lobe and calcarine fissure; the innermost digits of at least one pair of extremities opposable; hallux with a flat nail or none; a well-developed caecum; penis pendulous; testes scrotal; always two pectoral mammae.

Mivart's definition has been expanded by Le Gros Clark (1959) and by Napier and Napier (1967) to give a composite diagnosis (Table 5.2) that certainly applies to the living primates surveyed in Chapter 1. These supposedly diagnostic features of living primates have figured prominently in assessments of phylogenetic relationships between insectivores, tree-shrews and primates. However, to be of value with respect to phylogenetic reconstruction, it is not sufficient that a diagnosis of living primates should accurately summarize shared features of living primates. In the first place, it should also effectively distinguish primates from other placental mammals, and, secondly, it should ideally emphasize derived characteristics of primates rather than primitive characteristics of placental mammals that have been retained by primates but lost by other mammals (Martin, 1968a, 1986a). Wood Jones (1929) neatly summarized the limitations of Mivart's definition and, by implication, the composite list given in Table 5.2:

> *There is no single character in this definition which constitutes a peculiarity of the Primates: for a primate animal may only be diagnosed by possessing the aggregate of them all.* The Primates, therefore, constitute a defined group of animals which possess no single distinguishing feature; but they all retain a surprisingly larger number of primitive and common mammalian characters collected together in one animal type.

This sentiment has been repeated by numerous subsequent authors, including Cain (1954), Le

Table 5.2 Composite diagnosis of living primates, based on Mivart (1873), Le Gros Clark (1959) and Napier and Napier (1967); see also Martin (1968a, 1986a)

1. Preservation of a generalized postcranial skeleton, including retention of a pentadactyl (five digit) pattern of the limbs and of certain skeletal elements (e.g. clavicle) lost in some other mammalian groups. Development of an upright orientation of the trunk.

2. Free mobility of the digits, associated with a grasping action of the thumb (pollex) and big toe (hallux) and with possession of nails rather than claws, at least on the hallux. Sensitive tactile pads present on the digits.

3. Possession of a fairly large brain, associated with elaboration of the visual apparatus and reduction of the olfactory apparatus. Linked, in the skull, with presence of a postorbital bar and abbreviation of the snout. Calcarine sulcus present in the brain.

4. Preservation of a relatively primitive pattern of dentition, although with some reduction and modification from the ancestral placental mammal condition.

5. Possession of a well-developed caecum.

6. Penis pendulous. Testes scrotal. One pair of pectoral mammae.* Increased efficiency of processes in gestation. Prolongation of postnatal life periods.

---

* This is, in fact, an error; see text.

Gros Clark (1959) and Napier and Napier (1967), and has generally reinforced the notion that there is a very diffuse boundary between insectivores and primates. Indeed, Simpson (1955) specifically denied the possibility of any sharp distinction between insectivores and primates, contrasting this with the examples of the bats (which possess wings) and of the artiodactyl ungulates (which have a 'double-pulley' mechanism in the ankle region). Cartmill (1982) has provided a thoughtful discussion of the origins of this view, which has exerted a profound influence on discussions of primate evolution. The fact of the matter is that a definition which does not clearly exclude other mammals and does not clearly identify probable derived features of primates is of very little use for any attempt to define the early origins of the order Primates. Indeed, careful reading of Mivart's (1873) discussion of primate characteristics reveals that he did not regard the order Primates as monophyletic. In other words, he believed that strepsirhine and haplorhine primates diverged separately from ancestral placental mammals. Accordingly, one would scarcely expect to find shared derived features of primates in his definition. The list in Table 5.2 therefore requires close scrutiny with this in mind, and it is particularly important to determine whether primates do share any derived characters that distinguish them from other mammals, including insectivores.

One point that must be cleared up straight away is that Mivart (because of limited information available at the time) was in error in stating that living primates are characterized by 'two pectoral mammae'. Living primates can have up to three pairs of teats (see Chapter 9) and in any case there are other placental mammals characterized by small numbers of mammae. The remaining drawbacks of Table 5.2 result mainly from the inclusion of many characters that have been developed in other groups of placental mammal (e.g. relatively large brain, see Chapter 8; reduction in dental formula; and descent of the testes) and indeed from inclusion of several features that were probably present in the common ancestral stock of the placental mammals (e.g. generalized skeleton; free mobility of the digits; generalized dentition). A particularly good example of the latter is provided by the caecum (a blind sidebranch of the alimentary tract, located at the junction of the small and large intestines; see Fig. 6.2). The caecum is widely represented among both placental and marsupial mammals (Hill and Rewell, 1954; Le Gros Clark, 1959) and it is even present in various amphibians and reptiles (Romer, 1955). It seems quite likely that this organ was present in the common ancestral stock of marsupials and placentals, if not at some earlier stage in the evolution of terrestrial

vertebrates. Inclusion of possession of a caecum, or any other primitive mammalian feature, in a diagnosis of the living primates therefore serves little purpose. Thus, assessment of tree-shrew relationships on the basis of the composite diagnosis given in Table 5.2 does not meet the criteria for phylogenetic reconstruction set out in Chapter 3.

As stated above, with respect to Fig. 5.8, the question of inclusion or exclusion of the tree-shrews in the definition of the order Primates is only a matter of classificatory convention *provided that it has been demonstrated that tree-shrews are indeed more closely related to primates than to any other group of placental mammals.* It is often simply assumed that this latter requirement has been met, but one must at least consider the possibility that tree-shrews represent an entirely separate evolutionary lineage that diverged from the ancestral stock of placental mammals quite independently of the origin of primates. In other words, it is necessary to test the hypothesis that the last common ancestor shared by tree-shrews and primates was, in fact, identical with or very close to the ancestral stock of placental mammals. Three hypotheses concerning the phylogenetic relationships between tree-shrews, insectivores (= Lipotyphla) and primates are shown in Fig. 5.9. The customary view, as depicted in Fig. 5.8, is shown in Tree 1, but it is possible that tree-shrews actually form an integral part of the insectivore radiation (Tree 2) or even branched off before the divergence between Lipotyphla and primates (Tree 3; see Martin, 1967, 1968a). If either Tree 2 or Tree 3 in Fig. 5.9 were to emerge as the most likely reconstruction of the actual phylogenetic relationships involved, then there would be little or no justification for classifying tree-shrews with primates. Instead, it would be more appropriate to place the tree-shrews in the separate order Scandentia, as proposed by Butler (1972). Thus everything depends ultimately on establishment of the probability that tree-shrews and primates exhibit shared derived characters indicative of the presence of a specific tree-shrew/primate ancestral stock during the evolutionary radiation of the placental mammals.

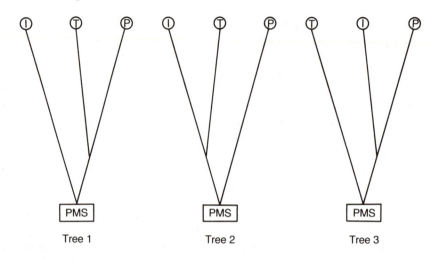

Tree 1   Tree 2   Tree 3

Figure 5.9. Three hypotheses concerning the phylogenetic relationships between tree-shrews (T), true insectivores (= Lipotyphla, I) and primates (P). PMS, placental mammal stock. Note: Possible relationships to other groups of placental mammals have been ignored for the sake of simplicity.

1. Usual conception of tree-shrews as an early offshoot from the primate lineage.
2. Tree-shrews as an integral part of the insectivore radiation.
3. Tree-shrews branched off before the divergence between Lipotyphla and primates.

At this point, it is particularly revealing to review in more detail some historical aspects of assessments of the relationships of tree-shrews to insectivores and primates. It has already been mentioned that tree-shrews were originally classified with the Insectivora as relatively generalized mammals with no striking specializations (e.g. see Gray, 1825). Following Haeckel (1866), however, it became common practice to subdivide the order Insectivora into two major subgroups. The tree-shrews (Tupaiidae) were grouped with the elephant-shrews (Macroscelididae) in the suborder Menotyphla, while the remaining insectivore families (the Chrysochloridae, Erinaceidae, Potamogalidae, Solenodontidae, Soricidae, Talpidae and Tenrecidae as recognized by Simpson, 1945) were allocated to the suborder Lipotyphla (see also McDowell, 1958). The subdivision, and the names of the two suborders, were based on the possession of a caecum by tree-shrews and elephant-shrews and its absence in lipotyphlan insectivores. Lyon (1913) followed Haeckel in recognizing the suborder Meno-typhla.

Eventually, partly or largely because of Gregory's (1910) inclusion of the Menotyphla in the superorder Archonta along with primates, bats and 'flying lemurs' (colugos), it became customary to regard the Menotyphla as 'advanced' insectivores and the Lipotyphla as 'primitive' insectivores. This was not, however, the original interpretation stated by Haeckel (1866), who regarded the Menotyphla as the basic representatives of the order Insectivora (see also McDowell, 1958). Indeed, the possession of a caecum, which distinguishes the Menotyphla from the Lipotyphla, is (as noted above) a probable ancestral feature of the marsupial/placental mammal stock. It is the Lipotyphla that seem to be specialized in this respect, having lost an original characteristic of ancestral placental mammals.

Loss of the caecum has also occurred in certain other placental mammals, such as many carnivores and dolphins, and it seems to be associated with particularly heavy concentration on animal food in the diet, whereas retention of the caecum generally characterizes mammals that include at least some plant food in their diet (see Chapter 6). Accordingly, there is good reason to suspect that in the evolution of the lipotyphlan insectivores there may have been a shift to heavier concentration on animal prey. Although modern lipotyphlans have a variety of diets, most species concentrate heavily on animal prey (e.g. arthropods and earthworms) and loss of the caecum implies the existence of an ancestral stage in which bacterial fermentation of plant food in a special gut chamber was of negligible importance. It follows from this that ancestral placental mammals were not 'insectivores' (a loose term indicating dietary concentration on small animal prey) in the same specialized sense as the ancestors of the Lipotyphla. Ancestral placentals apparently had a caecum and probably ate a significant proportion of plant food. Incidentally, note that the terrestrial tree-shrew (*Lyonogale tana*), which seems to concentrate on animal food, has apparently lost the caecum as well.

TREE-SHREWS AS PRIMITIVE MAMMALS

Repeated mention has been made in the literature of the fact that, regardless of any similarities to primates, tree-shrews have numerous features other than the caecum that would appear to be primitive for placental mammals generally. Carlsson (1922), while emphasizing similarities between tree-shrews and prosimian primates, made it clear in her discussion of individual features that tree-shrews retain many characteristics that can be regarded fairly reliably as primitive for placental mammals generally. Although Carlsson concluded that tree-shrews should be regarded as related to prosimians, this conclusion was plainly based on overall similarity rather than on the possession of shared derived character states by tree-shrews and primates. Indeed, Carlsson explicitly states in several places that retention of primitive features by tree-shrews and primates is indicative of 'genetic relationship'. Particular attention was devoted to the musculature, notably of the lower limbs and extremities, and whereas Carlsson emphasized features shared by tree-shrews and primates, it actually emerges that the greatest degree of shared similarity in her comparisons was between tree-shrews and marsupials (Table 5.3). Moreover, in a more recent review of the limb morphology of tree-shrews, George (1977) concluded that the primate-like features may be a retention from the 'basal mammalian heritage'.

In a similar vein, Simpson (1945), while noting 'unmistakable and significant resemblances to primates' recorded the fact that tree-shrews 'are less specialized in their average or conjoint

Table 5.3 Quantitative summary of shared similarities in the limb musculature of tree-shrews, *Didelphis*,[*] insectivores,[†] *Sciurus*[‡] and prosimians[§] (from Carlsson, 1922). The numbers indicate shared characters out of a total of 32

	Prosimians	*Tupaia*	*Sciurus*	*Erinaceus*
Didelphis	12	21	15	13
Erinaceus	6	16	13	
Sciurus	10	13		
Tupaia	15			

[*] The American opossum (a marsupial).
[†] Heavily based on the hedgehog (*Erinaceus*).
[‡] A squirrel (rodent).
[§] Mainly based on *Lemur*.

anatomy than are any of the (other or true) living insectivores, and they share with the didelphids the distinction of being the most nearly generalized of surviving Theria'.

The possibility that the Menotyphla may be more primitive, in some respects at least, than the Lipotyphla was given considerable support in Romer's (1966) classification of the mammals, which was part of a comprehensive review of the vertebrate group as a whole. Following common practice, Romer included the order Insectivora as the least specialized assemblage in his category of placental mammals (infraclass Eutheria). But he divided the Insectivora into four suborders (Protoeutheria, Macroscelidea, Dermoptera and Lipotyphla), with the tree-shrews and certain fossil insectivore families (see later) included in the Protoeutheria. As the name implies, the suborder Protoeutheria is based on Romer's interpretation that tree-shrews are 'persistently primitive in many regards'. In a later commentary on this classification, Romer (1968) succinctly summarized his views on the relatively primitive nature of the tree-shrews. After noting that the tree-shrews are 'certainly very primitive placentals' he went on to identify three different categories into which the observations of various authors may be placed:

1. Tree-shrews have many primitive placental characters from which primate features may have developed.
2. Tree-shrews and primates show some 'common positive characters' absent in placentals generally.
3. In many respects, tree-shrews and primates clearly differ.

From this, Romer went on to conclude:

> The conflict of testimony in categories (2) and (3) mirrors the fact that, after all, modern tree-shrews have had many tens of millions of years to evolve since eutherian history began and although remaining primitive in many regards, they may well have, during that time, made modifications of one sort or another either converging toward or diverging from primate descendants . . . The tupaiids may be

primate ancestors. But not improbably their importance is far greater. It may well be that in the tree-shrews we see the most primitive of living placentals – forms not too distant from the common base of all eutherian stocks.

It is therefore by no means a foregone conclusion that tree-shrews can be regarded as 'advanced' insectivores intermediate (in a phylogenetic sense) between other insectivores and primates, as is commonly assumed. The balanced assessment made by Romer instead suggests that Haeckel's original concept of the Menotyphla as basal insectivores is one 'long-abandoned view' that deserves serious re-examination.

Because of the historical shift in prevalent interpretations of the tree-shrews (initially as primitive insectivores, then as advanced insectivores, and eventually as primitive primates), there has been a growing tendency to regard any shared features of tree-shrews and primates as necessarily 'advanced' in contrast to the 'primitive' features of insectivores. It is common to see references to ancestral placental mammals as 'ancestral insectivores'. This can probably be traced back to Huxley (1880), who stated that the Insectivora 'represent more than any other order, the stock from which the various groups of placental mammals have descended'. Simpson (1945) later supported the widespread view 'that the insectivores are the most primitive of placentals and stand near the origin of all other groups'. But Simpson went on to give the following warning:

> It is, however, also true that each group of living Insectivora and most known fossil forms are strongly specialized in some peculiar direction and that they are hence not generalized placentals despite their many primitive characters.

In spite of this warning, the insectivores, treeshrews and primates together have come to be regarded as constituting a classical example of the *Scala naturae* (see Chapter 3), with living insectivores providing a model for the ancestral placental mammal condition and tree-shrews serving as a model for the ancestral primate condition. Le Gros Clark (1959), for instance, based his discussion of the phylogenetic relationships of tree-shrews heavily on a limited comparison between living insectivores (often only a single genus, typically the hedgehog *Erinaceus*), living tree-shrews (usually a single species, notably *Tupaia minor*) and primates. As noted in Chapter 3, Simpson (1950) implied that tree-shrews represent virtually unchanged survivors from the ancestral primate stock:

> Even primate re-radiation would be entirely possible if present primates were wiped out and their empty ecological niches continued to exist, because living tree-shrews have no specializations that would clearly exclude approximate repetition of such a radiation.

The net result of thinking within the framework of the *Scala naturae* and hence labelling insectivores and tree-shrews as 'living fossils', virtually frozen at early stages in mammalian evolution, has been to oversimplify vastly the complex problems involved in the reconstruction of primate evolution. It is not sufficient simply to assume that living insectivores (Lipotyphla) are primitive in every respect compared to tree-shrews and primates. The primitiveness, or otherwise, or living lipotyphlans must be established for each character through an overall analysis of placental mammals, with due reference to marsupials and to the fossil record, following the principles set out in Chapter 3. McKenna (1975) specifically noted that living lipotyphlan insectivores share several derived features.

FOSSIL INSECTIVORES

As mentioned above, it is customary to refer to early placental mammals as 'insectivores' and to classify them in the order Insectivora, and this has led to the oversimplified view that *all* insectivores occupy the bottom rung of the mammalian *Scala naturae*. The earliest known fossils that can be regarded reliably as placental mammals come from late Cretaceous deposits. The only substantial fossil remains – virtually complete skulls of four genera (*Asioryctes,*

Barunlestes, Kennalestes and *Zalambdalestes*) – come from Mongolian sites; quite extensive associated postcranial remains are also known for all of these except *Kennalestes* (Kielan-Jaworowska, Bown and Lillegraven, 1979; see Fig. 4.10). Remains of a fifth genus, *Deltatheridium*, have also been found in Mongolia, but it is now questioned whether this mammal is a placental (e.g. see Butler and Kielan-Jaworowska, 1973). Similar late Cretaceous mammals are known from other areas, notably North America, but they are represented so far only by jaws and isolated teeth. All of the known forms are small, ranging from the size of a modern shrew to that of a marmot (Kielan-Jaworowska *et al.*, 1979), and their cheek teeth show some resemblance to those of modern lipotyphlan insectivores, particularly in bearing relatively high, pointed cusps. The combination of their small body size (requiring a relatively high energy diet; see Chapter 6) with sharp-cusped teeth certainly seems to justify the term 'insectivore' as a broad indication of likely dietary habits (Fig. 5.10). But resemblance in molar morphology and associated dietary adaptations does not necessarily mean that the earliest placental mammals resembled modern lipotyphlan insectivores in all respects. In addition, it must be remembered that chance plays a great part in shaping our knowledge of the fossil record of mammals (see Chapter 2) and it does not of course follow that the earliest known placental mammals from the late Cretaceous are the earliest placentals to have existed.

In fact, Kielan-Jaworowska *et al.* (1979) noted that there is quite marked diversity in dental morphology among the known genera of Mongolian late Cretaceous placental mammals. The molar teeth of *Kennalestes* (awkwardly the only genus not known from postcranial remains) seem to be relatively primitive and to accord well with the hypothetical ancestral type for placentals and marsupials (Crompton and Kielan-Jaworowska, 1978), whereas the molars of other Mongolian (and North American) genera are more specialized. Accordingly, *Asioryctes, Barunlestes* and *Zalambdalestes* (see

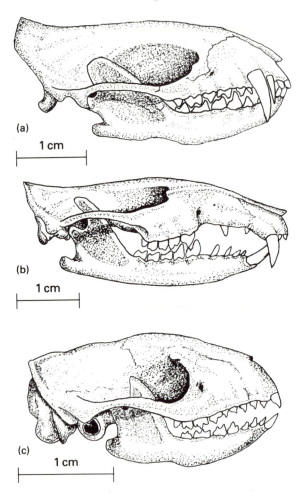

Figure 5.10. The skulls of two late Cretaceous mammals (a, *Deltatheridium*; b, *Zalambdalestes*) compared with that of a modern hedgehog, *Erinaceus* (c). The phylogenetic affinities of *Deltatheridium* are debatable, but *Zalambdalestes* is widely believed to be an early placental. Note the specialized anterior dentition of *Zalambdalestes*. (Fossil skulls after Gregory and Simpson, 1926 and Romer, 1966; hedgehog skull after Kühn, 1957.)

teeth and jaws. Early representatives of the families Leptictidae (e.g. *Procerberus*) and Zalambdalestidae (e.g. *Zalambdalestes*) were included in his suborder Protoeutheria along with the modern tree-shrews, while early members of the family Adapisoricidae (e.g. *Gypsonictops*) were included in the suborder Lipotyphla. Cretaceous representatives of the family Deltatheridiidae (e.g. *Deltatheridium* and *Cimolestes*) were classified along with the later members of the family as primitive members of the order Creodonta. Finally, *Protungulatum* was classified in the family Arctocyonidae in the order Condylarthra. Subsequent discoveries, notably in Mongolia, have only served to emphasize the considerable diversity in both cranial and postcranial features that was already present in late Cretaceous placental mammals. Thus, even the earliest known placental mammals cannot be simply equated with the ancestral placental condition; each character must be examined on its merits.

The fossil family Leptictidae, which (as defined by Romer, 1966) includes a variety of forms ranging from late Cretaceous *Procerberus* to late Oligocene *Leptictis* (= *Ictops*), has come to be regarded by many authorities as a relatively primitive assemblage of insectivore species in terms of both dental and cranial characters (see also Novacek, 1977b, 1986a). Romer (1966) included the Leptictidae in his suborder Protoeutheria together with the tree-shrews, thus underlining his interpretation of these fossils as closely approximating the ancestral placental condition. McKenna (1975), while departing quite radically from conventional views of the evolution of placental mammals (e.g. Simpson, 1945; Romer, 1966), also shows the leptictids (= his 'grandorder' Ictopsia) as a very early branch in his hypothetical phylogenetic tree of the placentals. Novacek (1986a) indicates that skull morphology in *Leptictis* is relatively primitive, although he infers a phylogenetic link to lipotyphlan insectivores. In general skull form and dental patterns, the Leptictidae are quite similar to modern tree-shrews, and in some respects different from modern lipotyphlan insectivores, thus suggesting that in certain

Fig. 4.10) probably do not directly reflect the condition of the ancestral placental mammals, particulary as the postcranial elements of the latter two genera seem to be rather specialized. Romer (1966) recognized five quite distinct groups among known late Cretaceous placental mammals in his classification, which included several additional genera known only from

character states at least the tree-shrews may be more primitive than the Lipotyphla.

Obviously, clarification of the relationships between tree-shrews, insectivores and primates must await more satisfactory resolution of the early adaptive radiation of the marsupials and placentals; but there are already definite grounds for suspecting that the tree-shrews are not necessarily intermediate (in a phylogenetic sense; cf. Figs 5.8 and 5.9) between lipotyphlan insectivores and primates. The alternative possibilities shown in Fig. 5.9 therefore remain as real alternatives that must be borne in mind in assessment of similarities shared by tree-shrews and primates.

FOSSIL TREE-SHREWS

Attempts to trace the phylogenetic relationships of tree-shrews have been particularly hindered by the absence of any convincing early fossil representatives. The only substantial fossil of any antiquity that has been seriously proposed as a relative of tree-shrews is *Anagale gobiensis* (Fig. 5.11) from early Oligocene deposits of Mongolia (Simpson, 1931b). This species was originally described on the basis of a well-preserved skull and a partial postcranial skeleton and Simpson noted several apparent resemblances to tree-shrews in the relative size and structure of the orbits, in the structure of the ear region and in other details of skull morphology. He also noted that in some respects *Anagale* seemed to be more similar to primates than the tree-shrews themselves, particularly in terms of dental morphology, shortening of the anterior part of the snout (premaxilla region) and certain details of limb morphology. The hindlimb of *Anagale* is unusual in that the terminal bones of the toes are flattened in a manner analogous to that found in most modern primates, suggesting the presence of nail-like structures rather than claws. By contrast, however, the terminal bones of the fingers are extremely unusual, being elongated, curved and laterally compressed bones with a marked central fissure – clearly adapted for the support of particularly powerful claws. Simpson

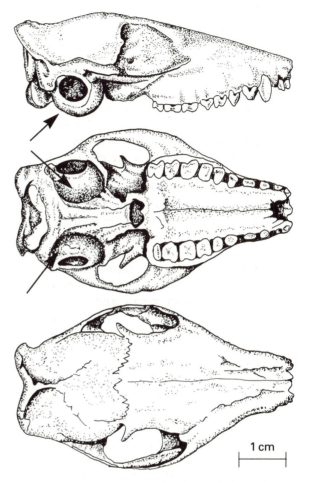

Figure 5.11. Lateral, ventral and dorsal views of the skull of *Anagale gobiensis* from the early Oligocene of Mongolia. Note the large orbits, which lack a complete postorbital bar (in contrast to living tree-shrews and primates), and the inflated auditory bulla (arrowed). (After McKenna, 1963.)

(1945) described *Anagale* as the 'only known fossil that is definitely, unquestionably tupaioid' and went on to state (p. 183):

In some respects this ancient tupaioid was even more lemur-like than are the living forms, and it strongly supports the inference that the tupaioids arose from primitive Lemuriformes and slightly diverged from the lemuroids proper while retaining most of their ancestral characters.

This interpretation of *Anagale* as a definite relative of tree-shrews, consolidating the proposed link with primates, was taken up by Le Gros Clark (1959) and integrated into his overall scheme of primate evolution.

Subsequent studies have, however, led to widespread rejection of the hypothesis that *Anagale* is related to tree-shrews and to primates. The first step in this direction came with Bohlin's description (1951) of dental and cranial material from a closely related species *Anagalopsis kansuensis* from deposits in northwestern China, tentatively regarded as of late Oligocene age. Bohlin, on the basis of this material, questioned Simpson's (1931b) interpretation of the auditory region of the skull of *Anagale* and suggested that *Anagalopsis* and *Anagale* together should be regarded as placental mammals of unknown affinities. At a later stage, McKenna (1963) reconsidered all of the available evidence, including a recently prepared second specimen of *Anagale* collected with the skull originally described by Simpson. This second specimen, consisting of a skull fragment and jaws, had better preserved teeth than the other specimens of *Anagale* and *Anagalopsis*, in which details of cheek tooth morphology had been obscured by extremely heavy wear. Further, McKenna was able to describe the ear region of the original skull of *Anagale* in more satisfactory detail, after expert preparation. On closer examination it emerged that the dental details were far less primate-like than was originally thought and that the ear region of *Anagale* was in fact quite different from that of modern tree-shrews or primates, but similar to that of *Anagalopsis*. McKenna emphasized the peculiar limb morphology of *Anagale*, involving a combination of large, fissured claws on the hand and spatulate 'nails' on the foot, and pointed out that the only mammals known to him that even approached this condition were late Oligocene creodonts from eastern Asia belonging to the family Didymoconidae. McKenna concluded by supporting Bohlin (1951) in his contention that *Anagale* and *Anagalopsis* (united into the single family Anagalidae by McKenna) should be regarded as placental mammals of uncertain affinities but having no connection with either tree-shrews or primates. Le Gros Clark later accepted this view (1971 revision of Le Gros Clark, 1959). *Anagale*'s combination of primitive placental mammal features (such as its dental formula; see Chapter 6) with unusual specializations (such as the structure of the ear region and the peculiar limb morphology) certainly seems to indicate an early, discrete origin and hence no clear connection with any living placental mammals. The very heavy wear found on the cheek teeth suggests a rather specialized, herbivorous diet.

Evans (1942) had noted a possible link between *Anagale* and elephant-shrews, whereas Van Valen (1964) subsequently suggested that the Anagalidae might be phylogenetically related to the order Lagomorpha (rabbits, pikas and hares). Eventually, McKenna (1975) followed both Evans and Van Valen in proposing a close phylogenetic association between the Anagalidae and elphant-shrews, within a larger grouping also containing the Lagomorpha and the fossil 'insectivore' family Zalambdalestidae. Romer (1966), on the other hand, included the Anagalidae in the suborder Protoeutheria, uniting them in the superfamily Tupaioidea along with the tree-shrews and a variety of fossil 'insectivore' families including the Leptictidae and the Zalambdalestidae, while noting that he did not regard *Anagale* as a true tupaiid. All of these authors agree, however, on the two fundamental points that the Anagalidae are not direct fossil relatives of the tree-shrews and that they are not related to primates. Simpson's (1945) interpretation has thus been convincingly rejected.

A variety of other early Tertiary fossil genera have, at one time or another, been suggested as possible relatives of tree-shrews, but they are all fragmentary fossils confined to teeth and partial jaws. Such proposed fossil relatives of tree-shrews include *Adapisorex*, *Adapisoriculus*, *Entomolestes*, *Leipsanolestes*, *Litolestes*, *Messelina* and *Tupaiodon* (Carlsson, 1922; Matthew and Granger, 1924; Simpson, 1931b;

Van Valen, 1965; Romer, 1966; Szalay, 1968b, 1977). However, none of these fossil genera has been widely accepted as a serious contender for relationship with tree-shrews. To a large extent, this state of affairs can be attributed to the fact that living tree-shrews have relatively primitive cheek teeth, with the result that any un-specialized early placental mammal is likely to share dental similarities with tree-shrews. Indeed, all of the above fossil genera have been allocated at one time or another to 'insectivore' families such as the Leptictidae for precisely the same reason. Until more substantial fossil remains are available, any allocation of early Tertiary fossil 'insectivores' to the tree-shrews must accordingly be regarded as extremely tentative.

More recently, acceptable evidence of fossil tree-shrews has been described from much younger fossil deposits, namely the middle Miocene Siwalik deposits of India (Chopra, Kaul and Vasishat, 1979; Chopra and Vasishat, 1979) and Pakistan (Jacobs, 1980), dating back to about 10 Ma ago. The most complete specimen is a partial skull, confined to the snout and anterior orbital region, allocated to the species *Palaeotupaia sivalicus* (Chopra and Vasishat, 1979). A smaller fragment of the snout lacking tooth crowns has been described without attribution of a specific name by Jacobs (1980). The remaining specimens are confined to isolated teeth and a jaw fragment, although a ribcage from upper Siwalik deposits of India has been tentatively identified as that of a tree-shrew (Dutta, 1975). If the allocation of the ribcage to the tree-shrews were to be confirmed, it would extend the range of known fossil tree-shrews up to the Pliocene; but the specimen has yet to be fully described and it will in any case yield little further information of value regarding the evolution of tree-shrews.

The partial skull of *Palaeotupaia* from India exhibits so many similarities to modern tree-shrews, particularly in the dental formula, morphology of the cheek teeth and preserved details of skull anatomy, that its allocation to the Tupaiidae can be regarded as fairly safe. It is comparable in dimensions to the skull of modern *Tupaia minor*, although the snout was relatively longer. The other skull fragment from Pakistan, described by Jacobs (1980), seems to have come from a somewhat larger skull, again with a moderately elongated snout. In so far as diagnostic features are preserved, all of the known specimens seem to be related to the Tupaiinae rather than to the Ptilocercinae and conform with the interpretation that modern tupaiine tree-shrews are derived from a semiterrestrial ancestral form of moderate body size and with a mildly elongated snout. There is also a certain amount of dental diversity among the known fossil tree-shrew specimens from the Siwaliks, suggesting that the evolutionary radiation of the Tupaiidae began some time earlier than 10 Ma ago. For the present, however, we have no fossil evidence to document the early origins of tree-shrews and for most of the Tertiary the evolutionary history of this group remains a complete mystery.

For the time being, then, the question of the likely phylogenetic relationships of tree-shrews must be settled almost exclusively on the basis of what is known about the biology of living tree-shrew species. Table 5.4 lists the major similarities shared by tree-shrews and primates that have been cited in the literature as evidence of a phylogenetic connection between them (e.g. Kaudern, 1910, 1911; Gregory, 1913; Carlsson, 1922; Saban, 1956/57; Fiedler, 1956; Le Gros Clark, 1959; Goodman, 1962a; Buettner-Janusch, 1966; Hafleigh and Williams, 1966; Napier and Napier, 1967). In the chapters that follow, the characters concerned will be scrutinized in detail to assess their relevance to the early origins of the Primates, bearing in mind the principles of phylogenetic reconstruction outlined in Chapter 3 and the particular points that have emerged from the above discussion, which can be listed as follows:

1. Comparisons should be as broad as possible and should include due reference to the marsupials, as the 'sister group' of the placentals.

Table 5.4 The main similarities between tree-shrews and primates widely cited as indicating phylogenetic affinity between them (compiled from Doran, 1879; Gregory, 1910, 1913; Kaudern, 1910, 1911; Carlsson, 1922; Portmann, 1952; Fiedler, 1956; Saban, 1956/57; Le Gros Clark, 1959; Biegert, 1961; Goodman, 1962a; Buettner-Janusch, 1966; Hafleigh and Williams, 1966)

Context	Shared similarities
Skull	1. Snout relatively short 2. Simplified set of turbinal bones 3. Enlarged, forward-facing orbits 4. Postorbital bar present 5. Pattern of bones in medial orbital wall 6. Well-developed jugal bone with foramen 7. Enlarged braincase 8. Inflated auditory bulla containing 'free' ectotympanic ring 9. Internal carotid pattern (bony tubes) 10. 'Advanced' form of auditory ossicles
Dentition	1. Tooth-comb present at front of lower jaw, linked with a specialized, serrated sublingua 2. Reduced dental formula 3. Similarities in cheek teeth between tree-shrews and certain primates with relatively primitive cheek teeth (e.g. *Tarsius*)
Postcranial morphology	1. Limbs and digits highly mobile 2. Numerous details of limb musculature 3. Osteological similarities in both forelimbs and hindlimbs (e.g. presence of entepicondylar foramen on humerus; presence of free centrale in hand) 4. Ridged skin on palms and soles
Brain and sense organs	1. Olfactory apparatus reduced 2. Visual apparatus enhanced 3. Central, avascular area of retina 4. Neocortex expanded; brain size increased 5. Calcarine sulcus present
Reproductive biology	1. Penis pendulous; testes scrotal 2. Discoidal placenta, as in tarsiers and simians 3. Small litter size; small number of teats
Miscellaneous	1. Caecum present 2. Molecular affinities (e.g. albumins)

2. Variability within the family Tupaiidae must be taken into account and a hypothetical reconstruction of ancestral tree-shrew characteristics should be conducted as part of the overall exercise.
3. Wherever possible, information on the earliest known fossil placental mammals should be included.
4. Particular attention should be paid to similarities between tree-shrews and squirrels, which may reflect either retention of characteristics of ancestral placental mammals or convergent adaptations for scansorial locomotion in trees.

It is particularly important to recognize the possibility that tree-shrews may, in fact, have no specific phylogenetic relationship to primates. If this possibility should be confirmed, many previous discussions of tree-shrews as intermediate in a phylogenetic sense between insectivores and primates will require thorough re-examination. As Romer (1966, 1968) suggested, it may well be that the tree-shrews, rather than providing a 'model' for ancestral primates, actually provide an indication of the ancestral placental mammal condition instead (see also George, 1977). A corollary to this, of course, is that the widespread interpretation of ancestral primates as rather like modern tree-shrews must be well wide of the mark. Rejection of tree-shrews as early offshoots from the primate stock necessarily requires a radical revision of concepts of the earliest primates.

Chapter six

Primate diets and dentitions

Feeding behaviour is clearly an important basic feature of any animal and associated adaptations are typically prominent components of body design. In the case of mammals, as has been explained in Chapter 4, the energy requirements of the body have been considerably increased (in comparison to reptiles) because of the maintenance of a comparatively constant, elevated body temperature (homeothermy). Accompanying the emergence of these increased energy needs during mammalian evolution, there has been considerable remodelling of the jaws and teeth along with modification of numerous other morphological features. In particular, emergence of the new dentary/squamosal jaw hinge and differentiation of the teeth into subsets with separate functions (heterodonty) occurred at an early stage in the emergence of mammals (see Chapter 4). Because teeth and jaws are also the most durable parts of the mammalian skeleton, they have obviously played a predominant part in interpretations of the fossil record. Such fossil evidence of teeth and jaws has permitted hypothetical reconstruction of the general outlines of mammalian evolution (see Chapter 4) and current ideas of the adaptive radiation of mammals have been heavily

influenced by available dental evidence. On the one hand, broad dental patterns seem to have remained relatively conservative throughout mammalian evolution, because of the need for the teeth and jaws to operate together as components of an interacting complex. On the other hand, finer details of the jaws and teeth may reflect quite closely the dietary habits of individual species, such that it is becoming increasingly possible to make well-founded inferences about the diets of fossil primates.

Body size is unquestionably a key factor in dietary adaptations, both because energy requirements scale in a fairly predictable fashion to body weight and because the different stages of food-processing depend on size in various ways. For this reason, allometric analysis is essential for the interpretation of many aspects of mammalian dietary adaptation. In the following discussion of primate diets and dentition, therefore, there will be frequent reference to the effects of body size.

DIETS

The diet of any primate species obviously contains an enormous variety of individual substances and the study of primate nutrition is therefore an extremely complex, potentially endless undertaking. Simplistically, however, it is possible to consider primate diets in terms of three major categories of foodstuffs – proteins, fats and carbohydrates – supplemented by certain essential components, such as vitamins and minerals. In a crude sense, it is possible to regard proteins (and their basic subunits, the amino acids) as being required for growth and repair, and fats and carbohydrates can be seen primarily as energy sources. (Naturally, this is a considerable oversimplification, as carbohydrates and fats can also act as important structural components, especially for cell walls and membranes, and proteins can act as a source of energy, notably under conditions of starvation.) Overall, the diet of any individual primate species must obviously be balanced to provide for all the needs of growth, repair, expenditure of locomotor energy and reproduction.

Energy requirements

It has already been shown in Chapter 4 that one of the most fundamental scaling relationships in mammals generally is Kleiber's Law, which governs the relationship between **basal metabolic rate** (standard energy requirements at rest) and body weight (Kleiber, 1947, 1961). Kleiber's Law states that basal metabolic rate (BMR) is related to body weight (W) by an allometric relationship of the following form for the placental mammals as a group:

$$BMR = kW^{0.75}$$

(where k is the allometric coefficient). The exponent in this equation (i.e. 0.75) is < 1, and the relationship is therefore **negatively allometric**. In other words, as body size increases among mammals, the increase in basal energy requirements does not keep pace. As a result, the amount of energy required per unit body weight declines progressively with increasing body weight. Because there is little change in the size of individual cells with increasing body weight among mammals (Munro, 1969; Calder, 1984)

this means that large mammals require less energy per cell than do small mammals. This phenomenon can be expressed in terms of **mass-specific basal metabolic rate** (MSR), which is the basal energy requirement per unit body weight:

$$MSR = BMR/W = kW^{-0.25} \quad (\text{i.e. } k/W^{0.25})$$

This second equation reflects the fact that, whereas total energy requirements increase in a predictable fashion with increasing body weight, there is an accompanying steady decline in the amount of energy required for a given mass of body tissue. The effects of this decline are quite dramatic over a large range of body sizes. Other things being equal, a primate weighing 100 kg (e.g. a male gorilla) would require less than one-sixth of the energy required, per unit body weight, of a primate weighing 60 g (e.g. a lesser mouse lemur).

This pronounced difference in basal energy requirements (i.e. per cell) obviously has considerable implications for dietary intake in relation to body size. Whereas small-bodied primates, in common with other small mammals, must eat food with a relatively high content of energy in readily accessible form, large-bodied primates and other mammals can consume food with relatively less energy, requiring quite extensive digestion. Sailer, Gaulin, *et al.* (1985) have found that a crude measure of 'dietary quality' indicates a progressive decline with increasing body size among primates generally, reflecting decreasing MSR. It should be emphasized, however, that large-bodied mammals are not obliged to consume low-energy food for this reason (although they may have to do so because of the comparative rarity of high-energy foods or because of competition with other species). If large-bodied mammals can obtain high-energy food (as, for example, large-bodied carnivores do), then they can benefit from the relatively low basal energy requirements of their cells. Thus, whereas small-bodied mammals are generally obliged to seek out high-energy food that is readily digestible, large-bodied mammals may show a range of dietary habits, in many cases involving the consumption of at least some low-energy food requiring lengthy digestion. Accordingly, small-bodied primates (60 g–1.5 kg) typically rely extensively on animal food (especially arthropods), and regular leaf-eating is generally confined to large-bodied primate species (6 kg or more). Primates of intermediate body size characteristically feed primarily on fruits, with a limited supplement of either animal food or leaves (see Kay and Covert, 1984). The only obvious exceptions to this general rule are the sportive lemur (*Lepilemur*) of Madagascar, which weighs only 600–800 g (according to species) and yet feeds extensively on leaves and flowers (Charles-Dominique and Hladik, 1971; Hladik, Charles-Dominique and Petter, 1980), and the little-studied *Avahi*, which has a similar body weight (ca. 1 kg) and apparently also depends heavily on leaves in its diet.

Although Kleiber's Law constitutes a useful and valid generalization with respect to the scaling of basal energy needs to body size in mammals generally, there are individual deviations from the general scaling trend. In some cases, such deviations are considerable. Perhaps the most striking examples among placental mammals are provided by sloths, which need less than 50% of the energy for resting metabolism that would be predicted for their body weight (about 4 kg). As has been shown by McNab (1978b, 1986), there is a strong correlation among mammals generally between the proportion of leaves contained in the diet and reduction in basal metabolic rate relative to the standard level expected from Kleiber's Law. Sloths feed predominantly on leaves, which make up at least 70% of the diet, and they represent an extreme case among placental mammals for lowering of BMR below the expected level. McNab has also demonstrated that mammals with low metabolic rates, relative to the expectation from Kleiber's Law, are subject to certain limitations. For instance, the proportion of body weight represented by muscle (normally about 40%) is smaller for species with low basal metabolic rates (McNab, 1978b). Further, the capacity for increasing metabolic turnover during periods of intense activity ('metabolic scope') is severely constrained in such species. Whereas mammals with 'normal' BMR (i.e. conforming

with the expectation from Kleiber's Law) can increase their basal rates of metabolism by a factor of about 6 at sustained peak activity, mammals with relatively low BMR, such as sloths, may show no more than a doubling of basal rates during peak activity. Hence, mammals in the latter category are generally rather sluggish and it is easy to understand how sloths earned their name.

Overall, then, mammalian species that include a significant proportion of leaves in their diets generally require relatively low MSRs. Large-bodied mammals already have low MSRs by virtue of their size, but small-bodied mammals can exhibit such low rates only if they undergo some reduction of metabolic rate relative to the level predicted by Kleiber's Law. In fact, a sloth weighing 4 kg, with a BMR only 40% of that predicted for its body size from Kleiber's Law, has the same MSR as a typical placental mammal weighing about 160 kg. Hence, lowering of BMR relative to the expected level can considerably modify the tolerance of a given species for low-energy food. For all of the reasons given above, it can be predicted with some confidence that the sportive lemur (*Lepilemur*) has a BMR markedly lower than predicted for its body size. No data are yet available on the BMR of this leaf-eating lemur, but its relatively sluggish behaviour under natural conditions (e.g. see Charles-Dominique and Hladik, 1971) indicates that its energy expenditure is severely limited. In southern Madagascar, white-footed sportive lemurs (*Lepilemur leucopus*) apparently have very small home ranges and travel only a short distance each night. Further, *Lepilemur* can reingest faecal material from the enlarged caecum (**caecotrophy**), an adaptation that presumably permits extraction of additional nutrients after bacteria have broken down the leaf material (Charles-Dominique and Hladik, 1971; Hladik, Charles-Dominique and Petter, 1980). Thus, *Lepilemur* would seem to be specially adapted in several ways for subsisting on a diet consisting mostly of leaves, despite its small body size. It is highly likely that *Avahi* has similar special adaptations, but this lemur has yet to be studied in detail.

Among primates, reduction of BMR below the level predicted by Kleiber's Law is not confined to species that depend heavily on leaves in their diet. For example, prosimian BMR values are generally low (Müller, 1983, 1985). In particular, members of the loris subgroup (subfamily Lorisinae) have unusually low BMRs. This has been shown for the angwantibo (*Arctocebus*; Hildwein, 1972), for the potto (*Perodicticus*; Hildwein and Goffart, 1975), for the slender loris (*Loris*; Müller, Nieschalk and Meier, 1985) and for slow lorises (*Nycticebus*; Müller, 1979). However, these primates typically feed on a mixture of animal prey (predominantly arthropods) and fruits, so, in this case, the low BMR cannot be explained in terms of the low energy content of the diet. Charles-Dominique (1977a) has noted that *Arctocebus* and *Perodicticus*, in contrast to bushbabies (*Galago* spp.), tend to feed on 'noxious' arthropod species and it is possible that their reduced BMR permits them to tolerate toxins in their prey. Whatever the reason, lorisines are renowned for their sluggish behaviour (see also Chapter 10), as is apparent from the common name 'slow loris' for *Nycticebus*. Their limited energy turnover is doubtless related to their tendency to depend on concealment (e.g. to escape from predators) rather than relying on fast movement like their close relatives the bushbabies (Charles-Dominique, 1977).

The owl monkey (*Aotus*), which has a diet dominated by fruit, supplemented by leaves and a small quantity of animal food (Wright, 1981), also has an unusually low BMR. In its natural environment, *Aotus* has relatively restricted activities, as would be predicted from its lowered BMR. However, it is not clear why *Aotus* (the only nocturnal simian species) should have a low basal metabolic rate, as there is no obvious special feature of its diet that would seem to require this. It is, of course, possible that lowering of the BMR may also occur as a response to competition between mammalian species. This would have the general benefit that any species with a lowered metabolic turnover requires a smaller energy intake and the more specific benefit that it can tolerate foodstuffs that

either contain toxins or require extensive digestion. It is also possible that nocturnal habits favour low BMR in primates.

Dietary categories

With the exception of these special cases, primate diets show a fairly predictable relationship to body size (Kay, 1984; Kay and Covert, 1984). It is now customary to recognize three major dietary groups among living primates according to the major component of the diet: **insectivores** (arthropod-eaters), **frugivores** (fruit-eaters) and **folivores** (leaf-eaters). Although this subdivision is necessarily crude and omits some special cases (e.g. heavy dependence on gums and other plant exudates in some prosimians and marmosets; see later), it does permit certain useful generalizations. For instance, there is a fairly sharp boundary between predominant insectivores and predominant folivores among primates, at a body weight of about 600 g. No insectivore so far studied exceeds this body weight and no folivore weighs less than 600 g (Kay, 1984). As already noted, small-bodied primates cannot subsist on leaves because of their need for rapid energy turnover; and large-bodied primates cannot depend predominantly on insects because capture rates do not normally increase significantly with body size (Charles-Dominique, 1977; Kay, 1984). Among mammals generally, the only large-bodied species that feed predominantly on arthropods are those that seek out special categories of prey, notably colonies of social species such as termites (e.g. anteaters and pangolins). It is possible, in fact, that the aye-aye (*Daubentonia*) is an exception among primates to the rule stated by Kay; with a body weight of about 3 kg, this species is known to feed extensively on wood-boring arthropod larvae (see also Richard, 1985). However, there are as yet no field data to indicate whether arthropod food makes up the greatest proportion of the aye-aye's diet, thus justifying the label 'insectivore'.

Among the predominantly frugivorous primate species, the smaller-bodied forms typically round off their diet primarily with arthropod food, whereas the larger-bodied forms typically eat a supplement of leaves (Kay, 1984; see also Hladik, 1981). Hence, with increasing body size, the general trend among primates is to pass through the following sequence: insectivory, insectivory–frugivory, frugivory, frugivory–folivory and folivory. One important feature of this general sequence is that no primate species has a diet combining large amounts of insects and leaves (see also Richard, 1985). These two components of primate diets seem to be more or less mutually exclusive, as is shown by a ternary diagram of the proportions of arthropods, fruits and leaves in the diets of different primate species (Fig. 6.1). This diagram also reflects the significant fact that most primate species include at least some fruit in their diet.

Harding (1981) found that in a sample of 121 primate species (excluding tree-shrews), 114 species (94%) included fruit in their diets; 78 (64%) consumed animal food (arthropods with or without other kinds of animal prey); and 103 (85%) ate at least some mature leaves. The relative importance of fruit in primate diets is revealed even more clearly when primate species are examined in terms of major dietary components. Harding's analysis shows that 48% of primate species eat fruit (including seeds) as a major component of the diet, whereas 23% concentrate on arthropods and 29% eat leaf material of various kinds in large quantities. As noted by Richard (1985), 'there is at least one frugivore in every primate family'. From all of this, it is reasonably clear that primates generally depend on fruits, to some extent at least, and that fruit is usually supplemented with animal food (mainly arthropods) or with leaf material, but not with significant amounts of both in any individual case. Because small-bodied primates generally supplement their diets with arthropods rather than with leaves, and the earliest primates were probably relatively small in body size, it can be inferred that the ancestral primates probably fed on a mixture of fruit and arthropods, perhaps with small quantities of other kinds of food. This kind of diet has often been described as 'omnivorous' (although this is an extremely vague term), and Harding has concluded that the primates generally are best regarded as

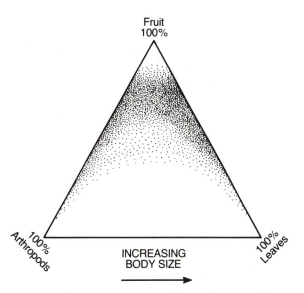

Figure 6.1. Ternary diagram (three-dimensional graph) showing the range of compositions of the diets of primate species. The three corners of the triangle (unoccupied in practice) represent 100% leaves, 100% arthropods and 100% fruit, respectively, and the position of any species on the plot indicates the relative contributions of these three components to the diet. Degree of shading indicates the probability that a particular dietary combination will occur. Note that there is general avoidance of diets combining significant proportions of both arthropods and leaves. It should also be noted that, whereas a species may occupy a particular position with respect to its average diet overall, there can be considerable variation over the annual cycle. The trend with increasing body size is no more than a general tendency. (After Chivers and Hladik, 1980 and Kay, 1984.)

'omnivores'. Certainly, a mixed diet seems to have been a dominant feature throughout primate evolution.

Although it is reasonable to consider primate diets, in the first instance, in terms of the relative contributions of arthropods, fruits and leaves, such an approach does tend to obscure the great diversity of items eaten by various species. In some cases, there are problems of classification of particular dietary items, notably with respect to seeds, flowers, buds and rhizomes. In addition, there are some dietary items that are totally excluded from consideration in the tripartite approach illustrated in Fig. 6.1. This is the case for plant exudates, particularly gums. Certain strepsirhine primates – notably, the fork-crowned lemur (*Phaner furcifer*) and the needle-clawed bushbaby (*Galago elegantulus*) – depend on gums to such an extent that they dominate the diet (Petter, Schilling and Periente, 1971; Charles-Dominique, 1977). Other strepsirhine species eat less gum but this food still represents a major source of nutrients, especially during the dry season, as, for example, in the lesser bushbaby *Galago senegalensis moholi* (Bearder and Martin, 1980a). It has also become apparent that marmosets and tamarins commonly feed on plant exudates, including both sap and gums, and marmosets are specially adapted for this (Coimbra-Filho and Mittermeier, 1976, 1977; see later). Gums generally seem to consist of polymerized sugars, along with certain minerals and perhaps traces of protein. In their sugar content, they are closest to fruits, and they are clearly different from both arthropods and leaves because they contain negligible protein. Yet, because the sugars contained in gums are polymerized, digestion of gums is not as straightforward as the digestion of fruit pulp. It would seem that symbiotic bacteria are usually required for the efficient digestion of gums, and primate species that feed extensively on these plant exudates have an enlarged caecum to house these bacteria in large numbers (see later). From this point of view, then, gums impose similar requirements to leaves; they cannot be readily digested and assimilated. Strepsirhine primates that feed extensively on gum typically supplement their diets with arthropods and thus constitute a rather distinct category of 'gummivory–insectivory'. On the basis of similar considerations of other special dietary items, Richard (1985) has identified seven different dietary groups of primates: insectivores, gummivores, frugivores (plus insects), frugivores (plus leaves), graminivores (seed-eaters), folivores and herb-eaters.

As was emphasized at the outset, detailed analysis of primate diets is an exceedingly complex matter and the above discussion deals only with some of the most general aspects. A great deal of work remains to be done on the relationships of natural primate diets to requirements for vitamins, minerals and other essential nutrients. Furthermore, it is now clear that the presence of metabolic inhibitors and toxins of various kinds in plant and animal foods is a potent factor affecting the choice of dietary items and their processing. Particular attention is now being focused on the 'secondary compounds' of plants (e.g. tannins and alkaloids), which have evolved as a defensive response to the feeding activities of animals (Freeland and Janzen, 1974). Such compounds are commonly present in mature leaves and thus represent an additional burden for the mammalian digestive system, over and above the intrinsic resistance of leaf material to digestion. This subject has been succinctly reviewed by Glander (1982) and by Waterman (1984). In this context, it should not be forgotten that arthropods may pose similar problems for digestion. Many arthropod species are known to contain noxious substances as a defence against predation, and animals that feed extensively on arthropods face the additional problem of digesting large amounts of chitin. The enzyme required for this purpose (chitinase) may be produced by the predator's own digestive system, but often this need is fulfilled by the symbiotic microbes in the digestive tract. For primates generally, very little is known as yet about the availability of chitinase for coping with arthropod food in those species that feed predominantly on arthropods.

Finally, before passing on from the topic of nutrient requirements and food digestibility, it is worth noting that long-term field studies have revealed that a surprisingly large number of primate species regularly include small quantities of soil in their diets. This is particularly unusual in that such behaviour (geophagy) has been observed in many species that are otherwise typically arboreal and must descend to the ground in order to feed on soil. Although the primate species concerned are usually selective with respect to the soil consumed and often concentrate on a few sites only, it is not obvious why soil is consumed. Analyses of soil samples (Hladik and Guegen, 1974; Eudey, 1978) have not revealed any consistent pattern across primate species and it is not clear whether the soil serves a direct nutritional function (e.g. in providing essential minerals) or whether it serves as a physical aid to digestion (e.g. in absorption or ion exchange).

THE DIGESTIVE TRACT

The digestive tract in primates follows the typical mammalian pattern (Fig. 6.2), although various kinds of specialization are shown by individual species with particular dietary habits. All living primates have a **caecum** – a blind pouch at the junction of the small intestine with the large intestine (colon). This feature was noted by

Mivart (1873) in his much-quoted definition of the order Primates, but the possession of a caecum is probably a primitive feature of placental mammals generally rather than a derived feature of primates (see Chapter 5). In fact, most living placental mammals have a caecum and it has been reduced or completely lost as a secondary development in various groups of mammals that feed exclusively, or almost exclusively, on animal food. The caecum is completely lacking in lipotyphlan insectivores (e.g. shrews, hedgehogs, tenrecs and moles), in cetaceans (whales and dolphins) and in pinnipeds (seals and sealions). It is also lacking in most members of the order Carnivora, although a much reduced caecum has been retained by

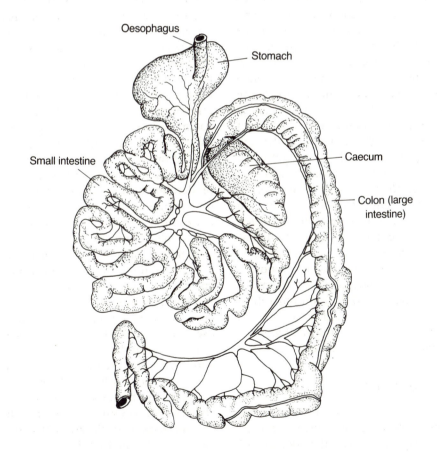

Figure 6.2. The digestive tract of a macaque (*Macaca mulatta*), showing the basic primate pattern. All living primates have a caecum. (Drawing after Chivers and Hladik, 1980.)

certain species, such as canids (dogs and their relatives) and felids (cats). It would seem that the caecum, wherever it is present in mammals, serves to house symbiotic bacteria that assist in the digestion of plant material and that it has been lost only in mammalian groups in which such material has been completely eliminated from the diet (see later). This being the case, it would appear to be misleading to refer to the ancestral placental mammals as 'insectivores', as is common practice (see also Chapter 5). Hence, early primates may have differed from ancestral placental mammals only in terms of a shift of emphasis to more consumption of plant food (fruits).

Broadly speaking, the separate main compartments of the mammalian digestive tract have clearly identifiable functions. The stomach is where the primary digestion of food occurs, whereas the small intestine serves mainly for the absorption of the breakdown products of such digestion. Additional digestion takes place in the caecum and, although it is sometimes maintained that the colon is involved only in the absorption of water before the expulsion of the faecal pellets, some additional absorption of nutrients does take place in the colon and perhaps through the walls of the caecum itself. Specialization of the caecum for extensive bacterial digestion of plant material is characteristic of lagomorphs (rabbits and hares) and of many rodents, and numerous species of both groups reingest faeces derived from the caecum. Among primates, however, such reingestion of faeces of caecal origin (caecotrophy) has hitherto been reported only for the sportive lemur, *Lepilemur* (Charles-Dominique and Hladik, 1971), so presumably in other primate species the products of bacterial breakdown in the caecum are absorbed at some stage before the faeces are expelled.

Allometric scaling of the digestive tract

Given these basic functions of the main compartments of the mammalian digestive tract, it is to be expected that the degree of development of each compartment should reflect the nature of the typical diet for each species.

However, the dimensions of these compartments must obviously be influenced in a rather complex way by differences in body size, so allometric analysis is necessary to clarify the relationships involved (see Appendix to Chapter 4). Until recently, most discussions of the dimensions of the mammalian digestive tract failed to take account of the scaling effects of body size. Although ratios were occasionally used, in attempts to take account of body-size effects, their value was severely limited. This is partly because of the complex allometric relationships involved and partly because some ratios (e.g. those involving two compartments of the tract) bear no relation to overall body size and hence to energy needs.

The first effective approach to the allometric scaling of compartments of the mammalian digestive tract was made by Chivers and Hladik (1980). They collected standardized data on dimensions of the stomach, small intestine, caecum and colon for a large sample of mammalian species (including primates). A logarithmic plot of the total volume of potential fermenting chambers (stomach + caecum + colon) against body size (represented by the cube of body length) produced a fairly clear separation of three dietary groups: folivores, frugivores and 'faunivores' (i.e. species feeding predominantly on animal food, including both insectivores and carnivores). These three categories appeared as typical 'grades' on the allometric plot, with the greatest total volume of fermenting chambers, relative to body size, in folivores and the smallest total volume in faunivores. A similar analysis of total potential surface area for absorption (small intestine area + half of the combined area of stomach, caecum and colon) led to identification of the same three grades. Relative to body size, folivores have the greatest potential surface area for absorption, whereas faunivores have the smallest. On the basis of these analyses, Chivers and Hladik (1980) calculated **indices of gut specialization** that closely matched the dietary habits of individual species.

The data collected by Chivers and Hladik were subsequently re-analysed in order to relate dimensions of the digestive tract to actual body weight and to examine in more detail the overall adaptations of the tract in individual species (Martin, Chivers *et al.*, 1985; MacLarnon, Martin *et al.*, 1986; MacLarnon, Chivers and Martin, 1986). A sample of 73 mammalian species, including 42 primate species, was covered by the analysis. This began with logarithmic plots of the surface areas of individual compartments of the digestive tract against body size. In each case, it was decided to take a best-fit line of fixed slope of 0.75 for each plot. This was done because it had emerged that, with the possible exception of the

colon, scaling of the gut compartments generally seems to conform with the expectation that the surface area of the gut would scale in the same way as basal metabolic rate (Chivers and Hladik, 1980; Martin *et al.*, 1985).

Stomach

A plot of surface area of the stomach against body weight (Fig. 6.3) shows that primates generally conform to the pattern for placental mammals, although there is some scatter around the best-fit line. However, it is notable that Old

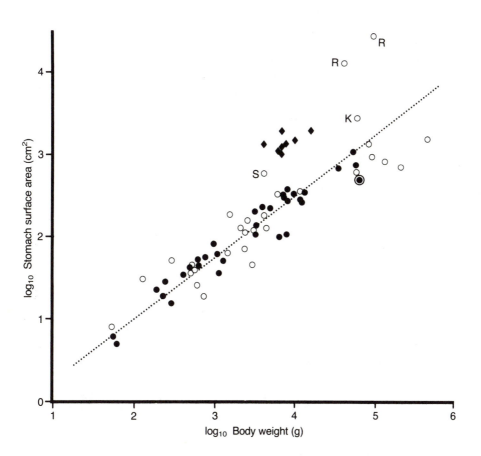

Figure 6.3. Logarithmic plot of surface area of the stomach against body weight for 73 mammalian species (42 primates; 31 non-primates): ●, primates (point for *Homo sapiens* is circled); O, non-primates. Old World leaf-monkeys (Colobinae) are separately indicated (◆) to distinguish them from other primates. The best-fit line has a fixed slope of 0.75. Non-primate mammals with relatively large stomachs are indicated by letters (R, ruminants; K, kangaroo; S, sloth). (After Martin, Chivers *et al.*, 1985.)

World leaf-monkeys (Colobinae) lie well above the best-fit line, indicating that they have much larger stomachs than expected for their body size. In fact, their position relative to the line indicates that the surface areas of their stomachs are about five times larger than expected for a typical mammal of the same body size. Enlargement and specialization of the stomach are so pronounced in Old World leaf-monkeys that qualitative mention of this feature is often made in the literature. In addition to its relatively large size, the stomach of these primates is characteristically subdivided into recognizable subcompartments (Kuhn, 1964; Bauchop and Martucci, 1968; Suzuki, Nagai *et al.*, 1985). In both respects, leaf-monkeys parallel the ruminants (a subgroup of the artiodactyls that includes cattle, sheep, deer, antelopes and giraffes) and certain other mammals, such as kangaroos and sloths (Fig. 6.3). In all cases where such marked enlargement of the stomach, relative to body size, is identifiable, large populations of symbiotic bacteria are housed in special chambers. Because these bacteria are present, bulky plant food (e.g. leaf material) can be fermented to yield volatile fatty acids and gases. In addition, the bacteria can utilize non-protein nitrogen (e.g. in the form of urea recycled through the stomach by the host) to produce protein that is harvested, along with other nutrients such as vitamins, by the host from any bacteria subsequently digested (Bauchop, 1978).

The stomach of Old World leaf-monkeys is, in fact, divided into four distinct subcompartments: cardiac pouch (presaccus), gastric sac (saccus gastricus), gastric tube (tubus gastricus) and pyloric chamber (pars pylorica). The symbiotic bacteria are housed in the first two chambers, where neutral conditions are maintained, and these chambers together constitute the major part of the stomach. There is a marked shift to acid conditions in the gastric tube (Kuhn, 1964), presumably indicating a shift from fermentation to digestion. In fact, there is an oesophageal groove (*Magenstrasse*) leading from the oesophagus to the gastric tube, which may serve either to transport certain dietary components (e.g. liquids) directly to a digestive stage (Bauchop and Martucci, 1968) or to release gases

produced by fermentation and digestion (Kuhn, 1964), if not both.

For certain ruminants, the volatile fatty acids produced by bacterial fermentation in the stomach supply about 70% of the total energy needs of the host. As yet, no detailed studies of the energy yield from bacterial fermentation have been carried out on Old World leaf-monkeys. However, Bauchop and Martucci (1968) made a simple calculation for langurs (*Presbytis*) in the leaf-monkey group indicating that energy derived from volatile fatty acids of bacterial origin probably constitutes a major proportion of their total energy needs. Hence, the special adaptation of the stomach in leaf-monkeys is of crucial importance for their ability to subsist to a large extent on mature leaves.

Small intestine

In contrast to the plot of the stomach, there are no conspicuous outliers in the scaling of the surface area of the small intestine against body weight (Fig. 6.4). Overall, there is a fairly consistent relationship between the two variables, as might be expected from the fact that the small intestine is largely concerned with absorption and might therefore be expected to reflect the energy needs of the body rather than any adaptations for a particular diet. It is noteworthy that the empirical best-fit line (major axis) in this case has a slope value of 0.75, possibly reflecting a correspondence between the scaling of the small intestine and the scaling of basal energy requirements (cf. Kleiber's Law). In this respect, it should be noted that scaling of the surface area of the small intestine in primates is essentially the same as for non-primates. This is to be expected from the fact that in primates and non-primates the scaling of basal metabolic requirements is basically the same (see Fig. 4.20).

Caecum

In direct contrast to the picture with the small intestine, a logarithmic plot of surface area of the caecum against body weight shows considerable scatter (Fig. 6.5). At one extreme are species, including both primates and non-primates, with a

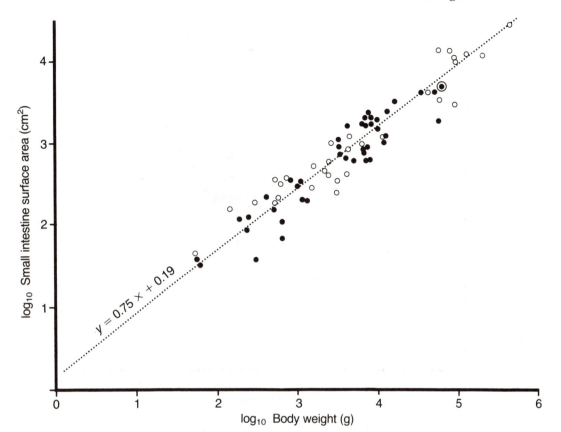

Figure 6.4. Logarithmic plot of surface area of the small intestine against body weight for 74 mammalian species (symbols as for Fig. 6.3). For this figure, the best-fit line was determined empirically from the major axis and the slope is 0.75 ($r = 0.95$). There are no conspicuous outliers and the primates are not distinguished from the non-primates. (Reprinted from Martin, Chivers *et al.*, 1985 with kind permission from Plenum Press, New York.)

greatly enlarged caecum; at the other extreme are species, mainly non-primates, with a very small caecum. Indeed, there are several mammalian species in the sample analysed that lack a caecum altogether (lipotyphlan insectivores, some carnivores, pinnipeds and cetaceans). These cannot appear in Fig. 6.5 because a zero value has no logarithmic equivalent. However, there are also many species, particularly primates, that lie quite close to the line of fixed slope, in agreement with the inference that possession of a caecum of moderate size was the ancestral condition for mammals. As a general rule, it can be stated that mammalian species with a relatively large caecum are secondarily specialized for a diet including large amounts of plant material

requiring extensive digestion (e.g. leaves), and species with a reduced caecum, or lacking a caecum entirely, are usually specialized for heavy reliance on animal food. Complete loss of the caecum, as noted above, is a feature of mammals with almost exclusively insectivorous or carnivorous habits.

The primates showing the greatest enlargement of the caecum, relative to body size, are lemur species that feed heavily on leaves (*Lepilemur* and *Avahi*). In these prosimians, the surface area of the caecum is about 10 times larger than expected (from the best-fit line) for a typical mammal of the same body size. There is a similar degree of enlargement of the caecum in, for example, the rabbit, the tree-hyrax (*Dendrohyrax*) and the horse (Fig. 6.5). In all

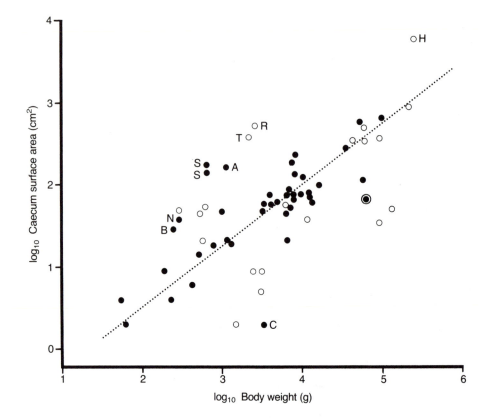

Figure 6.5. Logarithmic plot of surface area of caecum against body weight for 61 mammalian species (symbols as for Fig. 6.3). The best-fit line has a fixed slope of 0.75. In this plot, the number of non-primates was reduced from 31 to 19 because 12 species (lipotyphlan insectivores, some carnivores, pinnipeds and cetaceans) lack a caecum altogether and therefore cannot appear in a logarithmic presentation. Letters indicate species with leaf-eating habits and particularly pronounced enlargement of the caecum: A, *Avahi*; H, horse; R, rabbit; S, sportive lemur (*Lepilemur*); T, tree-hyrax (*Dendrohyrax*). The gum-feeding needle-clawed bushbaby (*Galago elegantulus*, N) and the fruit-eating Allen's bushbaby (*G. alleni*, B) also have markedly enlarged caeca. Reduction in size of the caecum is rare among primates but is found in the capuchin monkey (*Cebus capucinus*, C) and in humans. (After Martin, Chivers *et al.*, 1985.)

cases, symbiotic bacteria in the caecum generate volatile fatty acids and other nutrients from plant material in a similar fashion to the stomach flora of ruminants and of Old World leaf-monkeys. However, absorption of the products of such fermentation is more problematic for mammals in which the caecum has been enlarged as a fermentation chamber, because absorption cannot take place via the small intestine in the normal way.

There is also relative enlargement of the caecum in the needle-clawed bushbaby (*Galago*

elegantulus), which is known to be a specialized gum-feeder (Charles-Dominique, 1977). In this prosimian species, the surface area of the caecum is about five times greater than expected for a typical mammal of the same body size, and this tends to confirm suggestions that digestion of the polymerized sugars in gums requires some special adaptation in the digestive system. However, it should be noted that Allen's bushbaby (*Galago alleni*) also has a relatively enlarged caecum, with a surface area almost four times greater than expected, even though in Gabon the major part of its diet consists of fruit (Charles-Dominique, 1977).

Among the New World monkeys, in contrast to the Old World monkeys, specialization on leaf-eating is very limited. In fact, the only genera that can be described as folivorous are *Brachyteles* (woolly spider monkey) and *Alouatta* (howler monkeys); both species generally eat more leaves than fruit. Other New World monkeys are best described as frugivores or (for the smaller species) as insectivore–frugivores. It is therefore only to be expected that no New World monkey species shows significant enlargement of the stomach, relative to body size. However, there is mild enlargement of the caecum in a few species – in *Alouatta*, as might be expected, in the woolly monkey (*Lagothrix*) and in the owl monkey (*Aotus*).

As a general rule, primates have moderately sized or enlarged caeca; reduction in size of the caecum is quite rare. There are, however, two notable exceptions: in capuchin monkeys (*Cebus*) and in humans the caecum is markedly reduced in size, relative to the overall best-fit line (Fig. 6.5). Although this could be interpreted as an indication of increased emphasis on animal prey in the diet, it is more likely that it reflects a more general emphasis on easily digestible food with a relatively high energy content (see also Fig. 6.31).

Large intestine (colon)

The last major compartment of the gastro-intestinal tract, the colon, shows a rather different pattern (Fig. 6.6). Among mammals

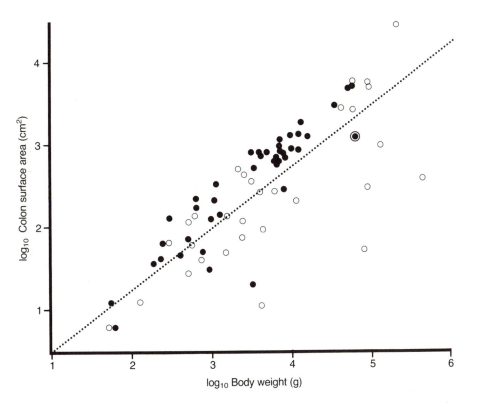

Figure 6.6. Logarithmic plot of surface area of colon against body weight for 73 mammalian species (symbols as for Fig. 6.3). The best-fit line has a fixed slope of 0.75. Note that the colons of primates generally have relatively large surface areas (i.e. lie above the best-fit line), although there are several obvious exceptions. (Reprinted from Martin, Chivers *et al.*, 1985 with kind permission from Plenum Press, New York.)

generally, there seem to be relatively few cases of conspicuous enlargement of the colon. The most extreme case is that of the horse, which has a colonic surface area approximately five times greater than expected for a typical mammal of the same body size. In primates, the colon is typically somewhat enlarged; on average, it has a surface area about twice as big as that of other mammals, after taking body size into account. As one of the main functions of the colon is to resorb water before the elimination of the faeces, one possibility is that the typically arboreal primates have a mildly enlarged colon as an adaptation to reduce their requirements for drinking water. Nevertheless, it is obvious from Fig. 6.6 that a few primate species have a rather smaller colonic surface area than expected from the best-fit line, although only one of the species concerned (*Homo sapiens*) is terrestrial in habits. There are several cases of conspicuous reduction in size of

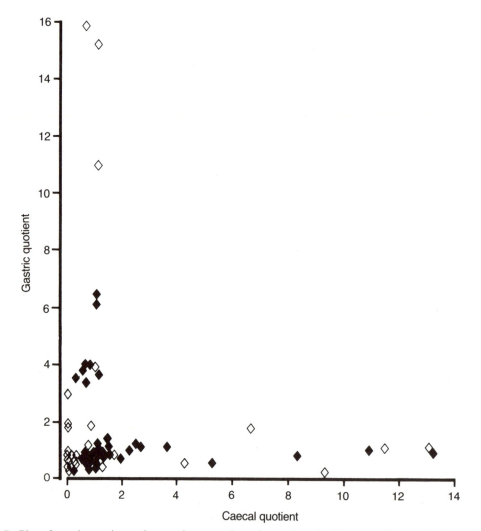

Figure 6.7. Plot of gastric quotient values against caecal quotient values for 73 mammalian species (species lacking the caecum altogether have a caecal quotient value of zero): ◆, primates; ◇, non-primates. Note that mammals may have either enlarged stomachs (i.e. elevated gastric quotient values) or enlarged caeca (i.e. elevated caecal quotient values), but not both together. (Reprinted from Martin, Chivers *et al.*, 1985, with kind permission from Plenum Press, New York.)

the colon in other mammals, notably groups with meat-eating habits (carnivores, pinnipeds and cetaceans), and this would suggest that a small colonic surface area is one further reflection of the relatively rapid passage of the food through the gastrointestinal tract among carnivorous mammals. Non-primates with a diet composed predominantly of plant food tend to lie above the best-fit line in Fig. 6.6, along with most primates, and it is therefore quite possible that the relatively large surface areas of primate colons simply reflect a general emphasis on plant food, rather than on animal food.

Additional information about the adaptations of the gastrointestinal tract for particular dietary regimes can be obtained by calculating quotient values for the individual compartments from the plots presented in Figs 6.3–6.6 (Martin, Chivers *et al.*, 1985). It is possible to calculate for each mammalian species concerned four quotient values (gastric quotient, intestinal quotient, caecal quotient and colonic quotient) reflecting the extent to which each compartment is larger or smaller than expected. For example, a gastric

quotient of 1 would indicate that the surface area of the stomach is exactly the size expected for a typical mammal, whereas a value of 5 would indicate that the surface area is 5 times greater than expected. On this basis, it is possible to make comparisons between compartments of the gastrointestinal tract. For instance, a plot of gastric quotient values against caecal quotient values (Fig. 6.7) shows that, among mammals generally, enlargement may occur either in the stomach or in the caecum; in no mammalian species is there marked enlargement of both compartments. This supports the general distinction made in the literature between **foregut fermentation** and **caecal fermentation** as two alternative strategies for coping with food requiring extensive digestion. The same distinction applies among primates: leaf-monkeys (Colobinae) are foregut fermenters, whereas sportive lemurs (*Lepilemur*) and *Avahi* (for example) are caecal fermenters.

It is also obvious from Fig. 6.7 that enlargement of the stomach does not usually entail any reduction in size of the caecum and vice versa, although in a few species caecal enlargement

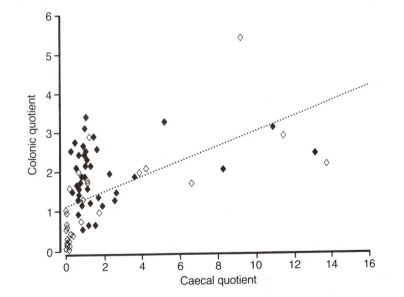

Figure 6.8. Plot of colonic quotient values against caecal quotient values for 73 mammalian species (symbols as for Fig. 6.7). The dotted line is an empirical best-fit line (major axis) reflecting a weak positive correlation ($r = 0.51$) between the two quotients. (Reprinted from Martin, Chivers *et al.*, 1985, with kind permission from Plenum Press, New York.)

does seem to be associated with some reduction in stomach size. A similar plot of colonic quotient values against caecal quotient values (Fig. 6.8), on the other hand, reveals a weak positive correlation. In other words, there is some tendency for enlargement of the caecum to be accompanied by enlargement of the colon. In fact, almost all species that have an enlarged caecum (i.e. caecal quotient value greater than 1) also have an enlarged colon (i.e. colonic quotient value exceeding 1). However, there are quite a number of species (mainly primates) in which there is an enlargement of the colon without any significant enlargement of the caecum.

Such pairwise comparisons of quotient values are somewhat limited and in order to exploit all of the information available on surface areas of the compartments of the gastrointestinal tract, it is necessary to conduct a multivariate analysis of some kind. A particularly useful technique has proved to be that of multidimensional scaling, which generates a two-dimensional plot taking into account residual values (quotients) for three or more parameters (Martin, Chivers *et al.*, 1985; MacLarnon, Martin *et al.* 1986; MacLarnon, Chivers and Martin, 1986). Multidimensional scaling plots generated using quotient values for all four gastrointestinal compartments derived from data presented in Figs 6.3–6.6 have revealed general clustering tendencies that reflect broad dietary categories, such as foregut-fermenting herbivores, caecum-fermenting herbivores, frugivores, insectivore-frugivores and carnivores (see later; Fig. 6.31).

DENTAL PATTERNS

As explained in Chapter 4, one of the striking characteristics that distinguishes mammals from other vertebrates is the possession of morphologically and functionally distinct categories of teeth (**heterodonty**). In a typical mammal, the teeth in each jaw can be identified as **incisors**, **canines** and **cheek teeth** (**premolars** and **molars**) in sequence from front to back. The incisors generally serve a prehensile (grasping) function, although in some cases they have been modified (e.g. for nipping or slicing of food). The canine teeth, usually restricted to one in each jaw, are sharp, stabbing teeth and their crowns usually project well beyond the crowns of the adjacent incisors and cheek teeth. Finally, the cheek teeth are typically adapted for mastication of food, to reduce it to relatively small fragments that can be more easily digested.

Identification of the teeth belonging to each category in mammalian jaws provides one of the basic items of information necessary for examining likely phylogenetic relationships (see Fig. 6.9). For this purpose, some simple ground rules are usually followed (e.g. see Simons, 1972). In the first place, the upper incisors can be defined as teeth located on the premaxilla, the most anterior bone of the upper jaw (see Fig. 4.3). As a rule, incisors are relatively small, peg-like or shovel-like teeth, but in some cases they are very large and adapted for continuous growth throughout life, in association with gnawing habits. Such enlarged incisors (commonly limited to one in each jaw) are typical of rodents and lagomorphs, among placental mammals, and they are also found in one living primate (the aye-aye, *Daubentonia*; Fig. 6.9) and in some 'archaic primates' (e.g. *Plesiadapis*; Fig. 2.4). In the upper jaw, the canine tooth (where present) can usually be identified both by its typical dagger shape and by its characteristic location at the anterior border of the maxilla (Fig. 4.3). By definition, all teeth posterior to the upper canine are cheek teeth. Identification of the teeth in the lower jaw depends on another generally accepted ground rule that the teeth in the upper and lower jaws are typically matching sets, with each tooth in the lower jaw closing in front of its counterpart in the upper jaw. Thus, the lower canine can usually be recognized as a tall, sharp-pointed tooth closing in front of the upper canine and any teeth anterior to the lower canine are, by definition, lower incisors. As with the upper jaw, any teeth posterior to the lower canine are defined as cheek teeth.

The distinction made between premolars and molars in the row of cheek teeth is somewhat more complex in that ideally it involves developmental evidence. By definition, molar teeth have no precursors in the milk dentition (see Chapter 4), whereas premolars are typically represented in both milk and permanent sets.

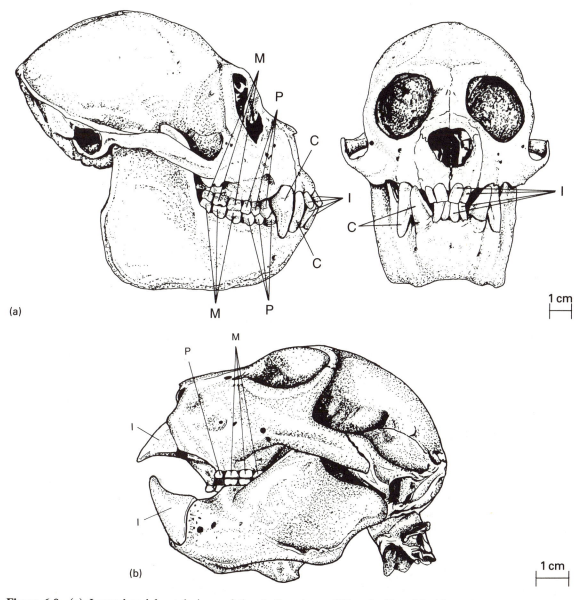

Figure 6.9. (a) Lateral and frontal views of the skull and mandible of a New World monkey, the monk saki (*Pithecia monachus*), showing a little-modified dentition containing all four categories of teeth (I, incisors; C, canines; P, premolars; M, molars. (b) Lateral view of the skull of the aye-aye (*Daubentonia madagascariensis*). This species has the most specialized dentition of any living primate, with continuously growing anterior teeth adapted for gnawing and a reduced array of cheek teeth. ((a) After Elliot, 1913; (b) after James, 1960.)

(There are some exceptions to this; teeth that appear to be premolars in terms of their position in the cheek-tooth row may not be replaced in certain mammalian species. For example, Archer (1978) concluded that only the last lower premolars are replaced in modern marsupials.)

In many adult mammals, there is a distinct increase in complexity from the last premolar to the first molar in both upper and lower jaws and it is therefore possible to infer which teeth are premolars and which are molars in the absence of developmental evidence. In fact, it is commonly implied that the primitive condition for mammals involves a sharp morphological contrast between the last premolar and the first molar. Accordingly, mammals lacking such a sharp distinction are said to have 'molarized premolars', suggesting that this is a derived condition. However, for placental mammals at least, there is a definite possibility that a relatively smooth transition in form from premolars to molars was the primitive condition (see also Clemens, 1979) and that 'molarization' of premolars is therefore a primitive, not a derived, feature (see later). Whatever the case may be, in mammals with a smooth transition from premolars to molars it is impossible to separate these two types of teeth reliably in the absence of developmental evidence.

As has been explained in Chapter 4, the numbers of teeth in each category in the upper and lower jaws are summarized in the **dental formula**. It is widely accepted at present that the common ancestor of modern placental mammals had three incisors, one canine, four premolars and three molars on each side of both upper and lower jaws, giving a dental formula of: $I_3^3 C_1^1 P_4^4 M_3^3$. (In principle, the milk dentition corresponding to this adult condition should have the formula $I_3^3 C_1^1 P_4^4$, following the rule that all teeth other than the molars typically have deciduous precursors.) Interpretation of the evolution of dental formulae among mammals is facilitated by the empirical rule that there have been two associated trends in most lineages: there has been a general trend towards reduction in numbers of teeth, accompanied by an equally general trend towards increasing complexity of those teeth that remain. There are only a few exceptions to this rule. For instance, in dolphins there has been a secondary return to a reptile-like condition, with an increase in numbers of teeth associated with a decrease in their complexity (the simple, conical teeth in the jaws of a dolphin are superficially similar to those of a crocodile). Otherwise, however, it is reasonable to assume that the trend among mammals has been towards a reduction in the numbers of teeth. Accordingly, in inferring a hypothetical ancestral condition for a group of mammals, it is reasonable to take the maximum dental formula represented within the

group as closest to the ancestral condition, if not actually identical with it. For instance, among modern placental mammals (excluding obviously aberrant cases such as the dolphins) the maximum dental formula is generally taken to be $I_3^3 C_1^1 P_4^4 M_3^3$ and this provides one of the reasons for taking this as the ancestral formula for placentals. Simons (1961a) succinctly summarized this principle as follows:

> Once a tooth is lost from the series, it cannot be reproduced again as such. From this latter, it follows that species postulated to lie on or near the line of ancestry of a given form must have the same or a greater number of each kind of tooth than does a supposed descendant.

There is a corollary to this inference of a trend towards a reduction in numbers of teeth among mammals, in that one would expect numbers of teeth to increase on average as one goes back through the fossil record (see Chapter 3). This is, indeed, the case, although the rule does not necessarily apply to individual species at any point in the fossil record, and the generalization applies right back to the origins of the mammals among their reptilian forebears. In dental terms, then, mammals with many teeth are likely to be more primitive than mammals with fewer teeth, barring a few unusual exceptions such as the dolphins.

Length of the tooth row and jaws

Before turning to a detailed account of primate dental formulae, it is worth considering some of the overall dimensions related to the length of the tooth row. It is to be expected that the numbers of teeth are broadly related to the length of the snout in mammals and that any reduction in the dental formula is accompanied by a reduction in the length of the snout. It has often been suggested that both primates and tree-shrews are characterized by shortening of the snout, the usual explanation being that such shortening is associated with a reduction in the importance of the sense of smell (see Table 5.4 and Chapter 7). However, quantitative evidence relating to brain morphology presented in Chapter 8 indicates that the olfactory apparatus has not been reduced either in tree-shrews or in many prosimians (most lemurs and lorises). Further, any shortening of the snout in primates should not be considered merely in relation to olfaction, as the upper jaw is an integral part of the snout and hence influences its length. Thus, even if shortening of the snout could be demonstrated to have taken place during the evolution of primates and tree-shrews, it would be necessary to relate such a change to the functional morphology of the jaws and teeth.

An indirect approach to the question of snout length in tree-shrews and primates can be made by plotting overall skull length against body weight and making a comparison with insectivores (Fig. 6.10). This plot, incidentally, shows that the length of the skull is related to body weight in a negatively allometric fashion, scaling with an exponent value of 0.28 rather than 0.33. That is to say, the size of the skull increases at a slower pace than overall body size. This is to be expected if skull size reflects jaw size, since Kleiber's Law implies that food requirements (and hence food-processing requirements) should scale in a negatively allometric fashion to body size. Assuming that skull length does reflect jaw length, reduction of the snout during primate evolution should be revealed by a shorter overall skull length, relative to body size. Somewhat surprisingly, comparatively little difference is found between insectivores and primates in terms of relative skull length in Fig. 6.10. However, it turns out that tree-shrews have particularly long skulls, relative to body size, even in comparison to lipotyphlan insectivores. Thus, there is some suggestion that skull length may have increased during the evolution of the tree-shrews. Examination of tree-shrew skulls (see Fig. 5.4) indicates that any such increase in relative skull length is probably attributable to lengthening of the snout and hence of the tooth row. In other words, allometric analysis in this case reveals that tree-shrews have undergone evolution in the opposite sense to that commonly reported in the literature (see Table 5.4) and that some special dietary adaptation may be involved. In this context, it should be noted that the snout of tree-shrews generally curves down at its tip, most

noticeably in the larger-bodied species. Relative elongation of the snout combined with downward curvature occurs in an extreme form in the anteaters of South America and it is perhaps significant that all tree-shrew species depend heavily on arthropod food. It is therefore possible that mild elongation of the snout has occurred during their evolution – as a specialization for feeding on arthropods among leaves and under stones.

Although Fig. 6.10 shows that there is no clear distinction between primates and insectivores in the relative length of the skull, simian primates have somewhat shorter skulls overall. This is indicated by the separate best-fit line for simians, which suggests that a limited 'grade shift' has taken place. In fact, it is possible that shortening of the snout in primates generally – and particularly in simian primates – has been masked to some extent by secondary lengthening of the braincase occurring as a result of relative enlargement of the brain (see Chapter 8). It seems likely that all primates (but not tree-shrews) have undergone some reorganization of the upper jaw, given the empirical observation that the premaxilla is generally quite short in primates (see later). This point can be investigated further by analysing dimensions of the upper jaw and teeth in relation to overall length of the skull.

A logarithmic plot of the length of the palate against skull length provides one indication of the length of the upper jaw (Fig. 6.11). This plot has the advantage that it is possible to include fossil species that are known from relatively well-preserved skulls (although distortion during fossilization introduces an element of uncertainty into the measurements). It can at once be seen that (relative to skull length) living primates do, indeed, have shorter palates than lipotyphlan insectivores, whereas tree-shrews and elephant-shrews do not. Tree-shrews lie very close to the best-fit line for lipotyphlan insectivores. As it has already been shown (Fig. 6.10) that they have relatively long skulls for their body size, it follows that they also have relatively long palates. Saban (1956/57) noted that the palate extends further back in tree-shrews than in prosimians. Prosimians generally have palates about 30%

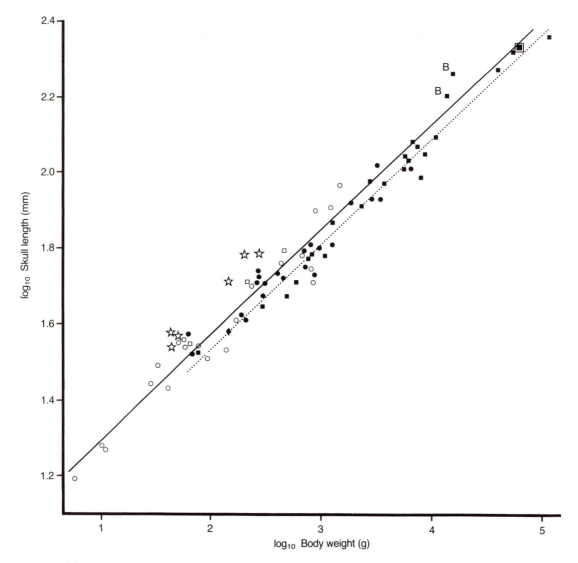

Figure 6.10. Logarithmic plot of skull length (*SL*) against body weight (*W*) for a sample of insectivores, tree-shrews and primates. The major axis for lipotyphlan insectivores (continuous line) has the following formula:

$$\log_{10} SL = 0.28 \log_{10} W + 1.01 \qquad (r = 0.97)$$

Many strepsirhine primates lie close to this line, whereas the tarsier, some strepsirhines and most primates lie below the line. Two baboon species (B) lie above the insectivore line, indicating that there has been secondary elongation of the snout in these simian primates. The major axis (dotted line) yields the following formula for simian primates (excluding baboons):

$$\log_{10} SL = 0.28 \log_{10} W + 0.97 \qquad (r = 0.99)$$

This indicates that, relative to body size, the skull is generally about 9% shorter in simian primates, as compared with lipotyphlan insectivores. By contrast, in tree-shrews the skull is generally about 26% longer than in lipotyphlan insectivores. (Data for this plot and those in Figs 6.11 and 6.12 were collected as part of the overall study of primate skull dimensions described in Chapter 8; see Fig. 8.3.)

Key to symbols: ○, lipotyphlan insectivores (*N* = 20); □, elephant-shrews (*N* = 4); ☆, tree-shrews (*N* = 6); ●, strepsirhine primates (*N* = 22); ◆, tarsier; ■, simian primates (*N* = 25; point for *Homo sapiens* is outlined).

shorter, relative to skull length, than lipotyphlan insectivores, whereas simian palates are about 40% shorter. In *Homo sapiens*, shortening of the palate is particularly pronounced, with a further reduction of about 40% relative to other simians. In other words, modern humans have palates 65% shorter, relative to skull length, than expected from the line for lipotyphlan insectivores in Fig. 6.11. It can be seen that most of the Tertiary fossil primates of modern aspect fall between the modern insectivores and the modern primates, although it is noteworthy that Oligocene *Aegyptopithecus* (the earliest known simian skull) lies quite close to the modern simian line. *Necrolemur* seems to be somewhat aberrant, in that it lies very close to the line for lipotyphlan insectivores. The 'archaic primate' *Plesiadapis* also falls quite close to the lipotyphlan insectivores, although there might be some mild shortening of its palate. Overall, Fig. 6.11 shows that modern primates are characterized by relatively short palates and that a trend towards a reduction in the length of the palate was probably under way among early Tertiary primates (although *Necrolemur* seems to be an exception). As the premaxilla contributes to the length of the palate, this shortening of the palate among primates can doubtless be linked to a marked shortening of the premaxilla, associated with a reduction in the number of incisors (see later).

It is also instructive to consider the total length of the cheek-tooth row in relation to skull length, in order to see whether there are any distinctive features associated with a reduction of palate length (Fig. 6.12). In fact, it emerges that in prosimian primates generally there is the same relationship between the length of the cheek-tooth row and skull length as in lipotyphlan insectivores, elephant-shrews and tree-shrews. The aye-aye (*Daubentonia*) is an obvious exception, in that there has clearly been marked reduction in the length of the cheek-tooth row, to less than half that expected for a typical strepsirhine primate. Compared with prosimian primates, simians constitute a separate grade, reflecting a reduction of about 25% in the cheek-tooth row relative to skull length. As with palate length, *Homo sapiens* is a conspicuous outlier,

with a marked further reduction in the length of the cheek-tooth row. Tertiary fossil prosimians, like modern prosimians, generally fit the line for lipotyphlan insectivores, although the 'lemuroid' *Adapis magnus* possibly shows some mild reduction in the length of the cheek-tooth row. In the 'archaic primate' *Plesiadapis*, perhaps reflecting a similar adaptation to the aye-aye's, the length of the cheek-tooth row is relatively reduced. *Aegyptopithecus* once again falls into the modern simian range. Overall, as insectivores, tree-shrews, modern prosimians and most early Tertiary primates conform quite closely to the upper best-fit line in Fig. 6.12 (i.e. the insectivore line), it seems likely that this represents the primitive condition. Accordingly, reduction in the length of the cheek-tooth row has doubtless occurred in the evolution of simian primates, an evolutionary development carried to an extreme in *Homo sapiens*.

Similar results from an analysis of the area of the lower second molar tooth in relation to body size in insectivores and primates were obtained by Gingerich and Smith (1985). As with the length of the cheek-tooth row, insectivores constitute a distinct grade relative to frugivorous and folivorous primates (mainly simians), with a larger area of the lower second molar at any given body size. *Tarsius*, like insectivores, has large lower second molars – in accordance with its exclusive dependence on animal food (mainly arthropods) – whereas the dwarf bushbaby (*Galago demidovii*), the grey lesser mouse lemur (*Microcebus murinus*) and the slender loris (*Loris tardigradus*) are intermediate.

Against this background of overall skull size and the relative size of the palate and cheek teeth, it is now possible to consider the general evolutionary trend towards reduction in numbers of teeth during primate evolution. Consideration of living forms (Gregory, 1920b–e, 1921, 1922; James, 1960; Swindler, 1976) permits identification of the maximal dental formula – as a provisional indication of the possible common ancestral condition – and also allows inference of probable shared derived conditions characterizing individual subgroups. These inferences can then be tested by examining the available fossil evidence.

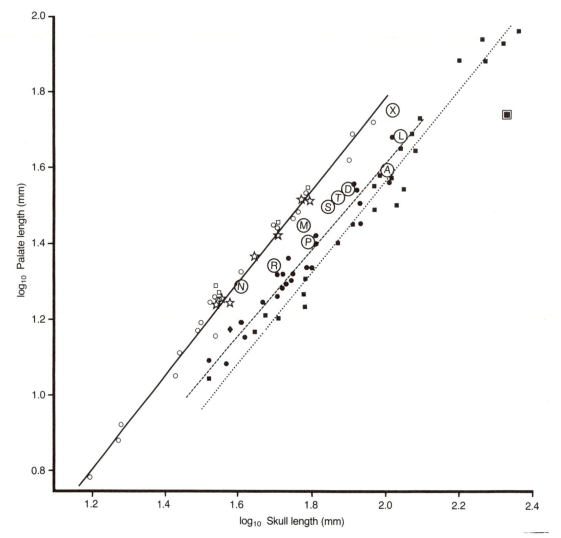

Figure 6.11. Logarithmic plot of palate length (*PAL*) against skull length (*SL*) for a sample of insectivores, tree-shrews and primates. The major axis for lipotyphlan insectivores (continuous line) yields the following formula:

$$\log_{10} PAL = 1.22 \log_{10} SL - 0.66 \qquad (r = 0.995)$$

Tree-shrews and elephant-shrews lie very close to this line. Modern strepsirhine primates fit a major axis (dashed line) lying well below the insectivore line, with the following formula:

$$\log_{10} PAL = 1.14 \log_{10} SL - 0.66 \qquad (r = 0.97)$$

The tarsier lies slightly above this line. Finally, simian primates (excluding *Homo sapiens*, as an obvious outlier) fit a major axis with the formula:

$$\log_{10} PAL = 1.22 \log_{10} SL - 0.87 \qquad (r = 0.98)$$

Insectivores, prosimians and simians thus represent three grades. Relative to lipotyphlan insectivores, the palate of prosimians is 71% of the expected length and that of simians, 62%. The palate of *Homo sapiens* is only 60% of the length expected for a typical simian with the same skull length.

Key: living species as for Fig. 6.10 (including one additional tree-shrew species and one additional strepsirhine primate species). Fossils (circled letters): A, *Aegyptopithecus*; D, *Adapis parisiensis*; L, *Adapis magnus*; M, *Mioeuoticus*; N, *Necrolemur*; P, *Pronycticebus*; R, *Rooneyia*; S, *Smilodectes*; T, *Notharctus*; X, *Plesiadapis*.

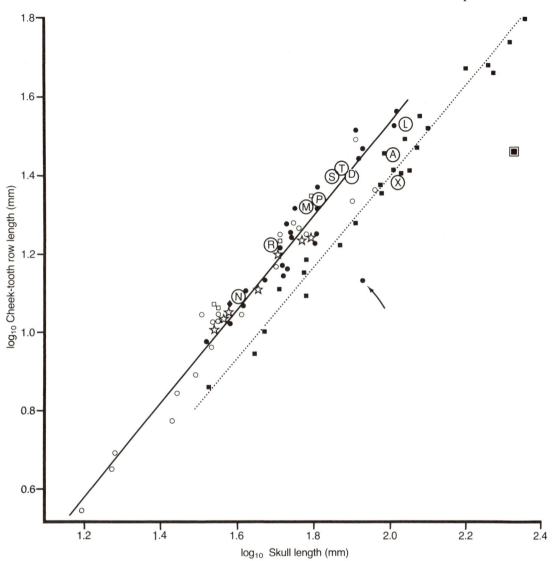

Figure 6.12. Logarithmic plot of the length of the cheek-tooth row (*CTL*) against skull length (*SL*) for a sample of insectivores, tree-shrews and primates (symbols as for Fig. 6.11). The major axis for lipotyphlan insectivores (continuous line) yields the following formula:

$$\log_{10} CTL = 1.20 \log_{10} SL - 0.87 \qquad (r = 0.98)$$

Elephant-shrews lie somewhat above this line, whereas tree-shrews cluster closely around it. The major axis for strepsirhines primates almost coincides with that for lipotyphlan insectivores – if the aye-aye (arrowed) is excluded as an obvious outlier – and tarsiers also fit the insectivore line. Hence, the length of the cheek-tooth row has an approximately uniform relationship to overall skull length in insectivores, tree-shrews and prosimian primates. By contrast, simian primates represent a separate grade. The major axis for simians, excluding *Homo sapiens* as an obvious outlier, yields the following formula:

$$\log_{10} CTL = 1.16 \log_{10} SL - 0.92 \qquad (r = 0.99)$$

This corresponds to a reduction of about 25% in the length of the cheek-tooth row, relative to skull length, in simian primates generally. In *Homo sapiens*, however, the length of the cheek-tooth row is only one-third of that expected for an insectivore of the same skull length and about half of that expected for a typical simian.

Dental formulae of living primates

Comparison of living primates reveals that the maximum dental formula is $I_2^2 C_1^1 P_3^3 M_3^3$. This reflects the loss of one incisor and one premolar in both upper and lower jaws relative to the ancestral dental formula of $I_3^3 I_1^1 P_4^4 M_3^3$, commonly inferred for living placental mammals generally. This maximum dental formula of $I_2^2 C_1^1 P_3^3 M_3^3$ for living primates is found in both strepsirhines (lemurs and lorises) and haplorhines (tarsiers, monkeys, apes and humans). As a provisional hypothesis, therefore, it is reasonable to infer that ancestral primates may have already had a reduced dental formula with only two incisors and three premolars on each side of the upper and lower jaws. In this connection, it is particularly important to note that, in all living primates, the possession of a maximum of two incisors in the upper jaw is

associated with a marked reduction in the length of the premaxilla (Fig. 6.13), and this doubtless accounts for at least some of the reduction in palate length (Fig. 6.11). As a result, the incisors are arranged more or less transversely, rather than longitudinally, in the upper jaw.

The primitive condition for placental mammals in this respect remains to be established. It seems likely, however, that the incisors of early mammals were arranged longitudinally along the premaxilla and that this bone was at least moderately well developed. Kermack and Kermack (1984) illustrate the early mammal *Morganucodon* as having a moderately developed premaxilla bearing four incisors in a longitudinal array on each side. A similar condition is found in the known skulls of the late Cretaceous placental mammals *Asioryctes*, *Barunlestes*, *Kennalestes* and *Zalambdalestes* (Kielan-Jaworowska, Bown and Lillegraven,

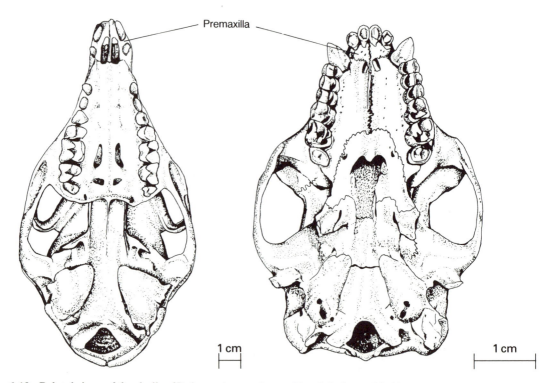

Figure 6.13. Palatal views of the skulls of Belanger's tree-shrew, *Tupaia belangeri* (left) and of the howler monkey, *Alouatta* sp. (right) to show the relatively short premaxilla of the latter (the difference is even more pronounced in a dorsal view). Although both skulls have two incisors on each side, the premaxilla is noticeably shorter in the howler monkey and the incisors are arranged transversely rather than longitudinally, as in modern primates generally. (Tree-shrew skull after Martin, 1967; howler monkey skull after Gregory, 1920a.)

1979). Further, a relatively well-developed premaxilla is found in all marsupials and all insectivores that have kept a relatively large number of teeth, and the incisors are typically arranged longitudinally rather than transversely. Tree-shrews, in contrast to living primates, have a well-developed premaxilla with longitudinally arranged incisors, even though these are reduced to two on each side (Fig. 6.13). In primates, reduction in the number of incisors to two on each side of the upper and lower jaws is probably allied to reduction in the length of the premaxilla and transverse orientation of the upper incisors, as an overall shared, derived feature. However, reduction in the length of the premaxilla and transverse arrangement of the incisors is not unique to primates; it also occurs in other placental mammals, notably in bats and in carnivores belonging to the cat family (Felidae). Indeed, Weber (1927) has noted that the premaxilla may be lost altogether in certain bats. Conversely, there are doubtless some placental mammals in which the premaxilla has undergone a secondary increase in size. This is true of the aye-aye and of many other mammals with continuously growing incisors adapted for gnawing (e.g. rodents and lagomorphs) and it has been taken to an extreme in certain mammals with elongated snouts adapted for feeding on ants and termites (e.g. the South American anteaters *Myrmecophaga* and *Tamandua*). It is, in fact, possible that the premaxilla has undergone secondary elongation in tree-shrews, thus accounting for their relatively long skulls (Fig. 6.10) and enhancing the contrast with primates (Fig. 6.13).

Among modern strepsirhine primates, there are many cases involving reduction from the maximum dental formula of $I_2^2 C_1^1 P_3^3 M_3^3$, but the full formula is found in all members of the loris group (lorises, pottos and bushbabies) and in several genera among the extant Malagasy lemurs (*Allocebus, Cheirogaleus, Microcebus, Phaner, Hapalemur, Lemur* and *Varecia*). The sportive lemur (*Lepilemur*) has departed from this pattern in that the permanent upper incisors have been lost, reducing the adult dental formula to $I_2^0 C_1^1 P_3^3 M_3^3$. This may be associated with the predominantly leaf-eating habits of *Lepilemur*,

because the typical peg-like upper incisors of modern lemurs doubtless became redundant with such a diet. There is reduction of the dental formula in another direction in members of the family Indriidae (*Avahi, Indri* and *Propithecus*). In these lemurs, an additional premolar has been lost in the upper and lower jaws and there has been loss of one of the anterior teeth in the lower jaw. There has been some doubt about the identity of the tooth lost from the lower jaw, because of the unusual arrangement of the anterior lower teeth in strepsirhines (see later), but it seems most likely that it is the canine (Gingerich, 1977d). Hence, the dental formula of the extant Indriidae is probably $I_2^2 C_0^1 P_2^2 M_3^3$. (An alternative interpretation is given by James, 1960 and Schwartz, 1974a). Even greater reduction has taken place in the aye-aye (*Daubentonia*), which has a rodent-like dentition, as noted above (see also Fig. 6.9). There is a single, enlarged tooth at the anterior end of each jaw and the cheek teeth have been reduced both in number and in size (see later). It is customary to write the dental formula of *Daubentonia* as $I_1^1 C_0^0 P_0^1 M_3^3$, but Gregory (1922) has identified the anterior lower tooth as a canine and Schwartz (1975a) has suggested that the enlarged front teeth in both upper and lower jaws are canines. Incidentally, these anterior teeth are not replaced during development, so it is possible that they are technically milk teeth (but see later). Whatever the correct identification of the anterior teeth may be, the reduced formula of the aye-aye can be derived from the maximum identified for strepsirhines generally and could also be derived from the indriid condition.

As noted in Chapter 2, the subfossil lemurs of Madagascar should really be regarded as members of the extant lemur fauna, as their extinction took place only in very recent times. It is therefore appropriate to consider their dental formulae in the context of modern lemurs. A useful review of the dentitions of subfossil lemurs is provided by Tattersall (1982). As with extant lemurs, there are some species (subfossil representatives of certain modern genera) with the full strepsirhine dental formula of $I_2^2 C_1^1 P_3^3 M_3^3$, but in most subfossil lemurs there has been reduction of some kind. Subfossil

Megaladapis species resemble modern *Lepilemur* in that they have lost the upper incisors from the adult dentition, and Tattersall has inferred that there was a horny pad at the anterior end of the upper jaw, as is found in some hoofed mammals (all ruminants except members of the camel family). *Archaeolemur* and *Hadropithecus* have lost one lower anterior tooth from the basic strepsirhine formula, presumably an incisor, and *Mesopropithecus*, *Palaeopropithecus* and *Archaeoindris* all seem to have the same dental formula as modern indriids. There is, however, some doubt about the identity of the lower anterior teeth in all of the subfossil lemurs that show a reduced number of teeth in the lower jaw.

Turning to the haplorhine primates, the tarsier has a generally primitive dentition, but it is unusual in that there is a difference in numbers of teeth between upper and lower jaws, associated with modification of the anterior lower dentition. Because a tooth has been lost from the lower tooth row, in comparison to the upper tooth row, and because the upper and lower anterior teeth do not show a clear interlocking pattern, there is continuing uncertainty about the dental formula of *Tarsius*. The formula is customarily written as $I_1^2 C_1^1 P_3^3 M_3^3$, indicating the loss of a lower incisor (James 1960); but alternative interpretations are possible (e.g. Schwartz, 1974b). However, it is reasonable to derive the condition of the dentition in *Tarsius* from an ancestral formula of $I_2^2 C_1^1 P_3^3 M_3^3$, at least in the first instance.

Modern simian primates, in contrast to prosimian primates, exhibit general stability of dental formulae. Among the New World monkeys, all members of the family Cebidae (the 'true' monkeys of the New World) have a formula of $I_2^2 C_1^1 P_3^3 M_3^3$. By contrast, all members of the family Callitrichidae (marmosets and tamarins) have lost a molar from each tooth row and consequently have a formula of $I_2^2 C_1^1 P_3^3 M_2^2$. This is very unusual in comparison to other living primates, because reduction in the number of cheek teeth otherwise involves loss of premolars. The fact that the cheek-tooth row has been reduced in this distinctive fashion in marmosets and tamarins may well lend support to the interpretation that these forms have undergone a secondary reduction in body size ('dwarfism')

during their evolutionary divergence from the common ancestral stock of the New World monkeys (see Chapter 12). Marmosets and tamarins show the same relationship between overall length of the cheek-tooth row and skull length as other simian primates (Fig. 6.12), so loss of a molar tooth from each tooth row could be interpreted as an adaptation to preserve this relationship following a reduction in body size. Goeldi's monkey (*Callimico goeldii*), which is intermediate between typical cebids and typical callitrichids in several respects, is also intermediate with respect to its dental formula. Although it has the full cebid dental formula of $I_2^2 C_1^1 P_3^3 M_3^3$, the third molar is quite small in both upper and lower jaws in comparison to other cebids. Thus, *Callimico* represents an intermediate stage in the evolution of the reduced dental formula of callitrichids.

Old World simian primates (monkeys, apes and humans) all share the same dental formula $I_2^2 C_1^1 P_2^2 M_3^3$, thus differing from all New World simians in the loss of a premolar from both upper and lower tooth rows. The consistency of the dental formula among Old World simians, despite other differences in dental adaptation, is a striking feature that suggests a relatively close evolutionary relationship between members of this group of primates.

At first sight, then, the distribution of dental formulae among living primates is consistent with the hypothesis that there was a common ancestral dental formula of $I_2^2 C_1^1 P_3^3 M_3^3$, with further reduction occurring in different lineages. This preliminary hypothesis can be tested by examining the dental formulae of fossil primates, following the principle that the maximum formula observed is likely to be the closest to the common ancestral condition.

Dental formulae of fossil primates

'Archaic primates'

The most striking feature of the 'archaic primates' (Plesiadapiformes) is the marked degree of reduction in numbers of teeth observed in many genera, primarily affecting the anterior dentition. However, such dental reduction is not found in all genera of the Plesiadapiformes, so dental formulae have doubtless been reduced as a parallel development within the group. The earliest known 'archaic primate' is *Purgatorius*. The late Cretaceous species *P. ceratops* is at present documented only by a single lower molar, but the early Palaeocene form *P. unio* is quite well known in dental terms and had a dental formula of $I_3^3 \, C_1^1 \, P_4^4 \, M_3^3$. This is, of course, the dental formula widely attributed to the ancestral placental mammals and *Purgatorius* is therefore generally primitive rather than primate-like in this respect. Incidentally, it should be noted that *Anagale*, which was once thought to link tree-shrews to primates, also had the formula of the hypothetical ancestral placental mammal (see Chapter 5).

The remaining members of the Plesiadapiformes (as defined by Szalay and Delson, 1979) had a maximum dental formula of $I_2^2 \, C_1^1 \, P_3^3 \, M_3^3$ and further reduction apparently took place in several separate lineages, as every family contains at least one genus with this formula. As Gingerich (1976b) has pointed out, the dental formula was apparently reduced in parallel in two separate lineages even within the single family Plesiadapidae. In all cases, dental reduction has accompanied specialization of the anterior dentition, with extreme enlargement of the anterior incisors evident in the most specialized forms such as *Phenacolemur*, *Plesiadapis* and *Chiromyoides*. However, as noted above, all 'archaic primates', with the exception of *Purgatorius*, could be derived from an ancestral form with a relatively unspecialized dentition (formula $I_2^2 \, C_1^1 \, P_3^3 \, M_3^3$).

The position of the Microsyopidae is still uncertain. Szalay and Delson (1979) exclude this family from the Plesiadapiformes, whereas other authors (e.g. Gingerich, 1976b) regard microsyopids as an integral member of the plesiadapiform radiation. The dental formula of *Microsyops* was apparently $I_1^1 \, C_0^1 \, P_3^3 \, M_3^3$ (Szalay, 1969), so inclusion of the Microsyopidae as part of the plesiadapiform group would rule out a dental formula of $I_2^2 \, C_1^1 \, P_3^3 \, M_3^3$ as a maximum for forms postdating the early Palaeocene *Purgatorius* and would tend to confirm $I_3^3 \, C_1^1 \, P_4^4 \, M_3^3$ as the maximum for the plesi-adapiforms generally, at least with respect to numbers of cheek teeth.

Regardless of the ancestral condition for the 'archaic primates', variation in dental formula among these early forms was as great as among all of the primates of modern aspect. This suggests that the 'archaic primates' were subjected to relatively intense selective pressure favouring dental specialization and this, in turn, indicates that their dietary habits were generally very different from those of most primates of modern aspect. However, as many authors have noted, there is a striking degree of resemblance in overall dental pattern between the aye-aye (*Daubentonia*) and some of the later 'archaic primates' such as *Plesiadapis*. Furthermore, Gingerich (1976b) suggested a resemblance between early plesiadapiforms (e.g. *Pronothodectes* and *Palaechthon*) and 'tarsioids' with respect to modifications of the lower anterior dentition.

Early Tertiary primates of modern aspect

The expectation that living primates of modern aspect might be derived from a common ancestor with a dental formula of $I_2^2 \, C_1^1 \, P_3^3 \, M_3^3$ is partially contradicted when information on Eocene 'lemuroids' and 'tarsioids' is taken into account. Virtually all of the Eocene 'lemuroids' (family Adapidae) had a formula of $I_2^2 \, I_1^1 \, P_4^4 \, M_3^3$. Among the North American forms, this formula was present in *Pelycodus*, *Notharctus* and *Smilodectes*, although in the late Eocene genus *Mahgarita* (Wilson and Szalay, 1976; Wilson and Stevens 1986) the dental formula had been reduced to $I_2^2 \, C_1^1 \, P_3^3 \, M_3^3$ with only tiny anterior premolars. Among the European forms, a formula of $I_2^2 \, C_1^1 \, P_4^4 \, M_3^3$ was similarly the rule (e.g. in *Protoadapis*, *Pronycticebus* and *Adapis*), although *Europolemur* had only three premolars in the adult dentition (Franzen, 1987). In the late-surviving adapid *Sivaladapis* of the Indian Miocene (Gingerich and Sahni, 1979, 1984), the dental formula had also been reduced. The early Tertiary 'lemuroids' thus had a maximum dental formula of $I_2^2 \, C_1^1 \, P_4^4 \, M_3^3$, with reduction from this maximum exhibited only by relatively few genera.

Examination of early Tertiary 'tarsioids' indicates the same maximum dental formula, although in most known species there was some reduction. In fact, the European Eocene 'tarsioid' *Teilhardina* is the only form reliably reported to have had four premolars in both upper and lower jaws. However, both Bown (1976) and Gingerich (1976b) have suggested that, although some specimens had a formula of $I_2^2 C_1^1 P_4^4 M_3^3$, this feature might have been variable. All other European 'tarsioids' have a reduced formula, with 2.1.3.3 in the upper jaw and only eight teeth in the lower jaw. As with modern *Tarsius*, identification of the anterior teeth in the lower jaw is problematic because the upper and lower teeth do not interlock in a simple fashion. It was originally suggested by Gregory (1915) that the dental formula of the Eocene 'tarsioid' *Necrolemur* was $I_1^2 C_1^1 P_3^3 M_3^3$, as in *Tarsius*; but Stehlin (1916) interpreted the formula of *Necrolemur* as $I_0^2 C_1^1 P_4^3 M_3^3$. Stehlin's interpretation of the dental formula of *Necrolemur* was supported by Simons (1961a), who extended that interpretation to cover the other European Eocene 'tarsioids' *Microchoerus*, *Nannopithex* and *Pseudoloris*. By contrast, Szalay and Delson (1979) 'tentatively' interpreted the dental formula of all four of these Eocene 'tarsioid' genera as $I_2^2 C_1^1 P_2^3 M_3^3$. This latter interpretation, which is supported by Schmid (1983), is problematic. The tooth identified as the anterior lower premolar by other authors clearly interlocks in front of the anterior upper premolar, as would be expected (see Simons and Russell, 1960; see also Fig. 2.16). Gingerich (1976b) has recently presented cogent arguments favouring the formula originally suggested by Gregory (1915), not only for *Necrolemur* but for *Microchoerus*, *Nannopithex* and *Pseudoloris* as well. This interpretation is accepted here as the most convincing in the light of present evidence. Thus, the European Eocene 'tarsioids', with the exception of *Teilhardina*, probably had the same dental formula as modern *Tarsius*, although there were differences in the relative sizes of the anterior teeth in the lower jaw, as noted by Gingerich (1976b). It should be noted that in *Teilhardina*, which was doubtless more primitive than other European Eocene

'tarsioids' in having a greater number of teeth, the canines were of typically mammalian form and the anterior lower dentition was not as greatly modified as in other European 'tarsioids'. If, as is widely accepted, *Teilhardina* is quite closely related to other fossil 'tarsioids' and possibly linked to their common ancestry (Bown, 1976), all or most of any resemblance in the anterior lower dentition between the remaining Eocene 'tarsioids' of Europe and certain 'archaic primates' must surely have emerged as a result of convergent evolution.

Among the Eocene 'tarsioids' of North America, which are known largely from fragmentary fossil evidence, the maximum dental formula so far identified is $I_2^2 C_1^1 P_3^3 M_3^3$. This is the case for *Hemiacodon* and *Tetonius*, which are the only Eocene genera that can be regarded as 'substantial' fossils of this group. In several genera, there was a reduction of the dental formula to $I_2^2 C_1^1 P_2^2 M_3^3$, as is the case, for example, with the fragmentary fossil *Anaptomorphus*. The Oligocene 'tarsioid' *Rooneyia* is also believed to have had this dental formula (Szalay and Delson, 1979), although in the original description of the skull (Wilson, 1966), the upper dental formula was interpreted as including three incisors because of uncertainty about the location of the suture between the premaxilla and the maxilla.

Because both Eocene 'lemuroids' and Eocene 'tarsioids' had a maximum dental formula of $I_2^2 C_1^1 P_4^4 M_3^3$ we must reject the hypothesis – based on a comparison of living primates – that the maximum dental formula of all primates of modern aspect (and hence their ancestral formula) is $I_2^2 C_1^1 P_3^3 M_3^3$. This hypothetical formula could be retained for the ancestry of living primates only if it were accepted that the Eocene 'lemuroids' and 'tarsioids' branched off before the modern primates began to diverge from their common ancestral stock. This seems unlikely and the most satisfactory explanation overall is that all primates of modern aspect (including Eocene 'lemuroids' as well as Eocene 'tarsioids') are derived from a common ancestor with a dental formula of $I_2^2 C_1^1 P_4^4 M_3^3$. This formula differs from the hypothetical ancestral formula for placental mammals generally only in the loss of an incisor from both upper and lower tooth rows. Accordingly, with respect to the dental formula, the definitive shift in the evolution of primates of modern aspect would seem to be in the reduction of the anterior dentition rather than in the modification of the number of cheek teeth. As with modern primates, all of the earliest known fossil primates of modern aspect ('lemuroids' and 'tarsioids') have a short premaxilla, as illustrated in Fig. 6.13 for the howler monkey (*Alouatta*). This accords with the picture presented in Fig. 6.11, which shows that in all living primates and all fossil primates of modern aspect (with the possible exception of *Necrolemur*) the palate is shorter relative to total skull length than in modern insectivores, tree-shrews and elephant-shrews.

In principle, it would be possible to derive most 'archaic primates' (e.g. *Palaechthon* and *Plesiadapis*) from a hypothetical stock with a dental formula of $I_2^2 C_1^1 P_4^4 M_3^3$ that also gave rise to primates of modern aspect. However, if *Purgatorius* is a definite member of the evolutionary radiation of the 'archaic primates', rather than a primitive form that branched off before the emergence of a common ancestor of the later plesiadapiforms and primates of modern aspect, then the ancestral dental formula of Plesiadapiformes was $I_3^3 C_1^1 P_4^4 M_3^3$. That is to say, reduction of the dental formula in later 'archaic primates' must have been a convergent development (repeated in the evolution of numerous other groups of placental mammals), rather than a shared derived feature linking them to the ancestry of primates of modern aspect. This possibility is strengthened by examination of the premaxilla in skulls of 'archaic primates'. Although the reconstructed skull of *Palaechthon nacimienti* (Kay and Cartmill, 1977; see Fig. 2.5) indicates a relatively short premaxilla, the skull of *Plesiadapis tricuspidens* (see Fig. 2.3) clearly had a large premaxilla with the two upper incisors arranged longitudinally, rather than transversely. Thus, in contrast to primates of modern aspect, 'archaic primates' do not consistently have a short premaxilla with transversely arranged incisors and they therefore cannot plausibly be regarded as derivatives of an ancestral stock showing this feature. In short, any

similarities in dental formula between certain 'archaic primates' and primates of modern aspect are best explained as the product of convergent evolution.

Miocene 'lorisoids'

Subsequent to the Eocene, primate dental formulae followed a fairly predictable pattern in which the maximum formula of $I_2^2 C_1^1 P_3^3 M_3^3$ found among living primates was never exceeded. On the strepsirhine side of the primate tree, fossil evidence is extremely limited. Apart from the subfossil lemurs discussed above, there are only the Miocene 'lorisoids' of East Africa and the Indian subcontinent. All three documented genera from East Africa (*Komba*, *Mioeuoticus* and *Progalago*) had the same dental formula as modern members of the loris group – $I_2^2 C_1^1 P_3^3 M_3^3$ – and this formula has also been inferred for *Nycticeboides* from Pakistan (L.L. Jacobs, 1981), thus confirming it as the maximum formula for strepsirhines (lemurs and lorises).

Haplorhines

On the haplorhine side of the primate tree, considerably more fossil evidence is available. For the 'tarsioids', apart from the very fragmentary specimen of *Afrotarsius* recently reported from the Oligocene deposits of the Fayum (Simons and Bown, 1985), there is no fossil record of dental evolution between the known Eocene 'tarsioids' and the modern *Tarsius*. Accordingly, the post-Eocene haplorhine evidence is essentially limited to simians. In the New World, the earliest known simian fossil is *Branisella* from the early Oligocene. Although the fossil evidence is limited to a few fragmentary specimens, it seems likely that the dental formula was $I_2^2 C_1^1 P_3^3 M_3^3$. This formula, which is characteristic of modern cebids, occurred in all of the adequately documented fossil New World monkeys from the late Oligocene and Miocene of South America (i.e. *Cebupithecia*, *Dolichocebus*, *Homunculus*, *Neosaimiri*, *Soriacebus*, *Stirtonia* and, probably, *Tremacebus*). Only in the subfossil *Xenothrix* from Jamaica was there an obvious departure from the typical cebid formula. Apparently, *Xenothrix* had lost a molar from both upper and lower tooth rows and the only known specimen has a reduced formula of $I_2^? C_1^? P_3^? M_2^?$. However, the superficial similarity between the large-bodied *Xenothrix* and the small-bodied callitrichids in having this reduced dental formula is doubtless attributable to convergent evolution.

The earliest known simians from the Old World are those from the Oligocene deposits of the Fayum; their dental formulae are interesting because two distinct types are identifiable. In *Apidium* and *Parapithecus*, there were six cheek teeth on each side of the upper and lower jaws, whereas in *Aegyptopithecus*, *Propliopithecus* and (presumably) *Oligopithecus*, the dental formula was $I_2^2 C_1^1 P_2^2 M_3^3$. The dental formula of *Apidium* and *Parapithecus* was for some time uncertain, but new material has permitted Kay and Simons (1983b) to confirm that in *Apidium* it was $I_2^2 C_1^1 P_3^3 M_3^3$, whereas one lower incisor had been lost in *Parapithecus fraasi* and both had been lost in *P. grangeri*. In the retention of three premolars, *Apidium* and *Parapithecus* resemble New World monkeys (although only in a primitive feature), and the loss of lower incisors in *Parapithecus* is a highly unusual development for any primate. In *Aegyptopithecus*, *Oligopithecus* and *Propliopithecus*, on the other hand, there was the standard dental formula found in all Old World simians from the early Miocene until the present day. This formula is found in all fossil forms that can be regarded as definite Old World monkeys (i.e. allied to the modern Cercopithecoidea), in *Oreopithecus*, in *Pliopithecus* and in all other fossils that may be regarded as 'apes' (e.g. dryopithecines). Not surprisingly, this formula is also found in *Australopithecus* species and in all fossil forms of *Homo*. Thus, once established, the standard dental formula of Old World simians remained stable for at least 20 million years.

Although it has been suggested that Old World monkeys may have developed a dental formula of $I_2^2 C_1^1 P_2^2 M_3^3$ independently of apes and humans, following separate derivation of the former from the *Apidium/Parapithecus* group of the Fayum (e.g. see Simons, 1972), it is at present more parsimonious to assume that all modern Old World simians were derived from a common ancestor in which that formula was already present.

DEVELOPMENT OF PRIMATE DENTITIONS

The outline of the evolution of dental formulae in primates presented above is cast in the traditional mould and is based on the assumption that homologous teeth occupy the same positions in different species. For example, upper incisors are customarily defined as teeth located in the premaxilla. However, there are several reasons for questioning this view. Leche (1907), a pioneer of studies of the development of mammalian teeth, specifically noted that teeth may be independent of the skeletal components of the skull. Weber (1927), following Leche, recorded that in closely related species a given tooth may be in either the premaxilla or the maxilla. He therefore suggested that position in the tooth row and the particular form of a tooth are important additional criteria for assessing dental homologies. For instance, in the European mole (*Talpa*), which has a dental formula of $I_3^3 C_1^1 P_4^4 M_3^3$ (as in the hypothetical ancestral placental mammal), the permanent canine (recognized on the basis of those additional criteria) is in the premaxilla although, as expected, the milk canine is in the maxilla. Conversely, for the desman (*Desmana moschata*), which has the same dental formula, Weber reported that the second and third upper incisors are in the maxilla. (Osborn (1978) has confirmed Weber's observations, at least for some specimens.) Both of these species belong to the insectivore family Talpidae in which the full dental formula of $I_3^3 C_1^1 P_4^4 M_3^3$ is commonly present, with the upper incisors usually in the premaxilla and the canine at the front end of the maxilla, as is typical among mammals.

Obvious exceptional cases such as those of *Talpa* and *Desmana* are rare, however, and the greatest challenge to conventional interpretations of mammalian dental formulae comes from developmental evidence. Such evidence has been relatively neglected in discussions of the evolution of tooth patterns, but it has much to offer. In the first place, as with other aspects of morphology, developmental evidence can provide valuable tests of hypotheses regarding the course of evolution. In addition, such evidence can provide entirely new sources of information, as is the case with sequences of development and eruption of teeth.

A simple example of the kind of corroboration of evolutionary hypotheses that can be obtained from developmental evidence is given by the milk dentition of species in which the adult formula has been reduced. Instances of this are rife among lemurs. In members of the family Indriidae, the milk-tooth formula has been reported as $I_2^2 C_1^1 P_3^3$, whereas the adult formula is $I_2^2 C_0^1 P_2^2 M_3^3$ or $I_2^2 C_1^1 P_2^2 M_3^3$ (James, 1960). This supports the inference that all living strepsirhine primates may be derived from a common ancestor with an adult dental formula of $I_2^2 C_1^1 P_3^3 M_3^3$ (the maximum observed).

Similarly, Friant (1947) reported that the milk-tooth formula of the sportive lemur (*Lepilemur*) is $I_2^1 C_1^1 P_3^3$, indicating that loss of the upper incisors was incomplete in the deciduous dentition. The single upper incisor is sometimes found in the adult (M. Cartmill, personal communication). James (1960) also reported that the milk dentition of the aye-aye (*Daubentonia*) has the formula $I_2^2 C_0^1 P_2^2$ (or possibly $I_1^2 C_1^1 P_2^2$). This confirms the interpretation that the aye-aye, with a formula of $I_1^1 C_0^0 P_0^1 M_3^3$, has a drastically reduced adult dentition, relative to the ancestral strepsirhine condition. In fact, the young aye-aye with a full set of milk teeth has the same total number of teeth as the adult! (Incidentally, the milk teeth of *Daubentonia* clearly demonstrate that the rodent-like nature of the adult dentition has been attained by convergent evolution and it rules out any direct link with the condition found in the 'archaic primate' *Plesiadapis*.) Franzen (1987) has reported that the milk dentition of the Eocene 'lemuroid' *Europolemur* included four premolars although the adults had only three premolars. This evidence further confirms that ancestral adapids had four premolars. Another example of the retention of original components of the dentition in immature stages is provided by the Oligocene simian *Parapithecus grangeri*. The lower dental formula of this species is $I_1 C_1 P_3$ for the milk dentition, as opposed to $I_0 C_1 P_3 M_3$ for the permanent dentition (Kay and Simons, 1983b), again providing developmental confirmation of evolutionary loss of teeth.

A full understanding of dental evolution in mammals, however, depends on detailed study of the development of teeth from the earliest embryological stages. A major impetus for this kind of work was given by Leche's pioneering studies (e.g. 1892, 1895, 1897, 1907, 1915), yet developmental evidence has been strangely neglected since that time, although there have been some notable exceptions (e.g. Butler, 1939, 1952a,b, 1956, 1978, 1982; Osborn, 1970, 1978). Discussions of primate dental evolution have commonly been based exclusively on adult dentitions, one reason for this being the general scarcity of immature dental stages in the fossil record. Although the available evidence is too scanty and dispersed to permit a comprehensive comparative review of primates, it is sufficient to show that the developmental perspective is of crucial importance for further advances in our understanding of primate dental evolution. A review of some key features of mammalian dental development will serve to illustrate this.

Pattern formation

In each jaw, teeth begin their development as buds produced by the **dental lamina**. Subsequently, the hard tissues of the tooth are laid down as the **tooth germ** develops within a **follicle**, which controls the shape of the developing tooth. In a typical mammalian tooth, the hard tissues are: a core of **dentine** (produced by odontoblasts), a cap of **enamel** (produced by ameloblasts) and **cementum**. Although there is room within the jaw bone for initial development of teeth, the tooth follicles are eventually only able to grow in size through the gradual removal of the surrounding bone. Following a relatively long period of such development, the teeth eventually erupt to occupy their appropriate places in the jaws (barring developmental accidents). Hence, the individual teeth of a completed dentition are, in principle, recognizable as soon as the buds form on the dental lamina and their developmental history can be traced from that point onwards. (For a fuller discussion of the ontogeny of mammalian teeth, see Butler, 1956.)

On the basis of developmental studies, Butler (1939) made the important observation that teeth evolve as part of a system and not as individual units. This is entirely to be expected, because upper and lower dentitions must interact in a complex fashion, with precise occlusal relationships between their functional surfaces (see later). Butler further noted that there is typically a **gradation of form** through a series of teeth, notably with the cheek teeth (Fig. 6.14).

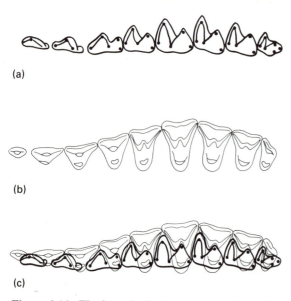

(a)

(b)

(c)

Figure 6.14. The hypothetical ancestral condition for the upper and lower cheek teeth in placental mammals, demonstrating the principle of gradation. Lower cheek teeth (a) are shown in heavy lines; upper cheek teeth (b) in faint lines. In (a), the cheek-tooth row is from the right side of the jaw and is shown as seen from above. In (b), the cheek-tooth row is also from the right side of the jaw, but is shown as if the teeth were transparent and seen from above to reveal the crown structure. In (c), the upper and lower tooth rows are superimposed to show their occlusal relationships. (After Gregory, 1922.)

As pointed out above, such a smooth gradation through premolars and molars was probably the primitive condition for placental mammals (i.e. the posterior premolars were primitively 'molarized') and morphological discontinuities between premolars and molars represent a secondary condition in certain mammalian groups. By contrast, in marsupials there is never a smooth continuity between premolars and molars (Kermack and Kermack, 1984). Although gradation of form along the cheek-tooth series is typical among placental mammals, the details of such gradation can vary quite noticeably even among closely related species (e.g. among the insectivores studied by Butler, 1937, 1939). It was this variation in the exact configuration of graded cheek teeth that led Butler to propose the existence of **morphogenetic fields** in the jaw, governing the development of incisors, canines and cheek teeth, respectively (Fig. 6.15). One implication of this model is that the completed form of the teeth depends not just on the position of tooth buds relative to skeletal landmarks (e.g. the premaxilla/maxilla boundary), but also on the configurations of the morphogenetic fields.

It is now widely accepted that some model of this kind must be invoked to explain the coordinated development of the teeth in mammals and special features such as gradation in completed dentitions. However, an alternative model, the **clone model**, has recently been proposed by Osborn (1978). As Osborn has observed, Butler's **field model** involves the inference that pattern formation depends on some influence external to the developing dentition (i.e. a **field substance** characterized by a diffusion gradient). In the clone model, on the other hand, it is proposed that there is some kind of precursor (a **clone**), which grows longitudinally from an original position, generating buds (each surrounded by a zone of inhibition) that give rise to all members of a tooth class, such as the set of cheek teeth. Gradation of form through a series of teeth derived from a clone is postulated to be an outcome of a gradient of cell ancestry. Hence, the clone model differs from the field model in that the gradient of form is an intrinsic property of the tooth buds, not a product of an external

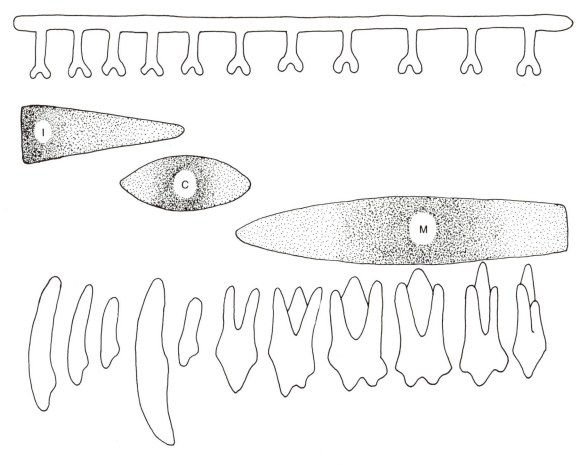

Figure 6.15. The concept of morphogenetic fields governing the development of the mammalian dentition. Top: The dental lamina with rudimentary tooth buds. Centre: The three morphogenetic fields (I, incisor field; C, canine field; M, cheek-tooth field). Bottom: The resulting completed dentition. (After Butler, 1939.)

field. Osborn proposed this alternative model of tooth development to account for certain findings that would seem to be incompatible with Butler's field model. Most notable among these findings is the observation that a developing tooth bud of the first lower molar of the mouse, when transplanted into the anterior chamber of the eye, will produce perfectly developed crowns of the complete set of three molar teeth found in adult mice (Lumsden, 1979).

Butler (1982) subsequently considered Osborn's clone model, but cites reasons for preferring the field model (see also Westergaard,

1980, 1983). It remains unclear which model provides the best explanation for observed patterns of development, particularly because the underlying processes remain hypothetical. To a large extent, the differences between the two models are immaterial with respect to interpretations of mammalian dental evolution and it is more important for present purposes to note the large degree of agreement between Butler and Osborn, especially with respect to the emphasis on overall patterns of dental development rather than on the properties of each individual tooth. Osborn, like Butler, accepts the existence of three categories of teeth (incisors, canines and cheek teeth) in developmental terms, although these are based on clones rather than on fields, with the clone for the cheek teeth originating centrally and developing both forwards and backwards.

Nevertheless, there is one significant difference between the field model (as espoused by Butler, 1939, 1982) and the clone model (as proposed by Osborn, 1978) that affects interpretations of dental evolution. For Osborn, the initial tooth bud in a given class (the 'stem progenitor') is likely to be homologous between species. The stem progenitor of a tooth class – for example, that of the cheek teeth – is seen as the first element to begin development as a separate bud and it follows that it therefore becomes identifiable as a landmark in the graded series of completed teeth. Osborn states that the posterior deciduous premolar is the stem progenitor of the cheek teeth in most mammals, whereas the penultimate deciduous premolar is the stem progenitor in *Homo* and in certain other mammals (e.g. *Tupaia* and *Felis*). One result of this interpretation is that it is necessary to invoke a dental formula for an ancestral placental mammal with five premolars and three molars in order to homologize the posterior premolar of most mammals with the penultimate premolar in *Homo* and certain other placental mammals (see Osborn, 1978). Another result is that it is necessary to reinterpret the anterior dentitions of certain mammals, including some primates, because the 'stem progenitor' concept requires that premolars should decrease in size towards the front of the jaw. Accordingly, the tooth generally recognized as an enlarged (caniniform) anterior lower premolar in certain lemurs (e.g. *Varecia* and *Indri*) must be interpreted as the canine in the lower jaw, giving a dental formula of $I_3^2 C_1^1 P_2^3 M_3^3$ for *Varecia*. Further, in the slow loris (*Nycticebus coucang*) it would follow that there are two canines on each side in the upper jaw accompanying a lemur-like condition in the lower jaw, yielding a dental formula of $I_3^2 C_1^2 P_2^2 M_3^3$. Similarly (according to Osborn, 1978), two canines would be identified on each side of the upper and lower jaws of *Tarsius*,

yielding yet another potential formula for this genus of $I_1^2 C_2^2 P_2^2 M_3^3$, although the relative sizes of the anterior teeth do not clearly indicate this even if the clone model is accepted. As Butler (1982) has commented, acceptance of Osborn's concept of the 'stem progenitor' for cheek teeth 'involves the writing of new dental formulas that are very difficult to reconcile with palaeontology'. Although the clone model should not be rejected merely because it might cause considerable headaches for the interpretation of mammalian dental formulae, it must be said that a considerable body of evidence would be necessary to justify such an upheaval. At present, the clone model remains essentially theoretical, as is indeed the case with the field model, and considerable caution is required in the interpretation of dental patterns according to either model. Nevertheless, identification of homologies of teeth must ultimately be decided on developmental grounds, as has been pointed out quite rightly by Osborn (1978) and thus the fate of mammalian dental formulae depends in large measure on the eventual outcome of the conflict between clone theory and field theory.

Whatever view one might take of the developmental homologies of teeth in mammals, there is clearly some overall limitation on the number of teeth in the adult dentition, which doubtless reflects a limitation on the number of tooth buds originally formed by the dental lamina. The fact remains that the maximum number of teeth found in any living placental mammal is typically 44 (i.e. 11 on each side of the upper and lower jaws). Whether or not one accepts the view that this full component of teeth conforms to a dental formula of $I_3^3 C_1^1 P_4^4 M_3^3$, with direct homology of teeth between species with this complete formula, it is significant that this maximum has been found to apply quite consistently. Nevertheless, although it is still widely accepted that the ancestral placental mammals had a dental formula of $I_3^3 C_1^1 P_4^4 M_3^3$, there have been suggestions that this was not the case. In particular, McKenna (1975) has proposed that the ancestral placental mammals had five premolars and three molars, citing evidence that five premolars were present in the Cretaceous mammals *Gypsonictops* and *Kennalestes* (see also Novacek, 1986b). In both cases, however, it would seem that fully adult individuals of both genera (at least in some cases) had only four premolars. Luckett and Maier (1982) have noted that it is difficult to exclude the possibility that a milk premolar had been retained in those jaws that seemed to indicate the presence of five premolars, and Westergaard (1983) proposed a reasonable model of dental evolution requiring only seven cheek teeth.

It must also be remembered that such Cretaceous forms as *Gypsonictops* and *Kennalestes* have been identified as placental mammals on dental grounds, so it is circular and dangerous to use other dental evidence derived from those fragmentary specimens to question the validity of the proposed ancestral dental formula of $I_3^3 C_1^1 P_4^4 M_3^3$ for placental mammals. If *Gypsonictops* and *Kennalestes* did indeed have five premolars, should this not instead provide grounds for questioning their status as placental mammals? Certainly, the limited and inconclusive dental evidence from these two Cretaceous mammals cannot be regarded as providing convincing support for Osborn's inference, based on the clone model, that ancestral placental mammals had five premolars rather than four.

Eruption sequences

Apart from the possibility of eventually resolving problems of dental homology on the basis of early ontogenetic events, developmental evidence can also provide useful information with respect to sequences of development and eruption of teeth. As noted by Tattersall and Schwartz (1974), this is an area that has been relatively neglected, although Schultz (1935) provided basic data on eruption of the permanent teeth in primates and Schwartz (1974c, 1975a,b, 1978, 1980) has made explicit use of eruption sequences, primarily for permanent teeth, to provide an additional perspective on the evolutionary relationships among primates (especially prosimians). However, little comparative work has been done on eruption sequences of milk teeth in primates and

it is important to emphasize that sequences of initial development, eruption of milk teeth eruption and of permanent teeth are not necessarily the same. In addition, the sequences determined for permanent teeth may differ according to whether eruption is defined on the basis of initial emergence of the crown of the tooth from the jaw bone (**alveolar eruption**) or from the gum (**gingival eruption**). Thus, whereas eruption sequences for permanent teeth may provide useful phylogenetic information in their own right, they do not faithfully reflect sequences of initial development of teeth (i.e. from tooth buds) and therefore they do not in themselves provide reliable criteria for the assessment of dental homologies.

Eaglen (1985) published some comparative information on the eruption sequences of the milk dentition in several lemur species, indicating that there is considerable variability between species. In all cases, however, the lower incisors and canines always erupt together, reflecting the fact that they constitute a functional unit (see later). By contrast, the upper canines consistently erupt before the upper incisors and the incisors may erupt at different times. In most species, the premolars erupt in sequence from front to back in both upper and lower jaws (i.e. in *Lemur* species, in *Varecia* and in *Propithecus*), although in the dwarf lemur (*Cheirogaleus medius*) the middle lower premolar may erupt after the posterior premolar (see also Schwartz, 1974c). Tattersall and Schwartz (1974) noted that eruption of the posterior deciduous premolar before the middle premolar is characteristic of lorisids, *Lepilemur* and *Microcebus*. In fact, in the neonates of both the grey lesser mouse lemur (*Microcebus murinus*) and the dwarf bushbaby (*Galago demidovii*) (Martin, 1975) the milk incisors and canines have already erupted, along with the anterior lower premolars, so in these species postnatal eruption of milk teeth is confined to the premolar region.

In the owl monkey (*Aotus*), as in some lemurs, the milk premolars erupt in sequence from front to back, but the sequence is reversed for the eruption of the permanent molars (Hall, Beattie and Wyckoff, 1977). Eaglen's study also clearly demonstrates that eruption sequences for milk teeth and for permanent teeth are not identical for the species of lemur investigated. For instance, whereas the milk canines erupt before the milk incisors in the upper jaw, the upper permanent canines consistently erupt after the upper permanent incisors, as is the case in lemurs generally (Schwartz, 1975a,b). In addition, in prosimians generally at least some replacement of milk teeth takes place before all of the molars have erupted. This contrasts with the suggestion that the primitive condition for placental mammals generally and also for primates involves initial eruption of the milk teeth, followed by eruption of the molars and then by replacement of the milk teeth. (This view is allied to the reasonable proposal that the molars are in fact part of the milk-tooth sequence and really represent unreplaced milk teeth that are retained into adult life.) Gregory (1922), for instance, clearly regarded eruption of the molars before the replacement of the milk cheek teeth as a primitive mammalian feature retained by modern lemurs. The data provided by Schwartz (1975b) and by Eaglen (1985) demonstrate that in lemurs and other prosimians replacement of the milk teeth begins while the molar series is still incomplete. Unfortunately, there is insufficient evidence for a more general analysis of milk-tooth eruption in primates, and this remains one of the priorities for future research in this area.

However, there is a more complete comparative picture of the eruption of permanent teeth in primates largely as the result of Schwartz's work (e.g. 1975b). Schwartz also attempted to review the development of the permanent teeth before eruption, but information was derived from radiographs of dry museum specimens, rather than from histological studies, so his conclusions regarding development are very provisional. (Radiographs can only reveal teeth at a fairly late stage in their development, so the sequence of origination of actual tooth buds cannot be inferred with confidence on that basis.) Schwartz (1975a,b) surveyed information on the eruption of permanent teeth in prosimian primates and made certain inferences with respect to phylogenetic

relationships. First, it seems that among prosimians the permanent premolars usually erupt in the sequence anterior–posterior–middle, and this is taken as the ancestral primate sequence (see also Tattersall and Schwartz, 1974). In some lemurs (e.g. *Lemur catta, Lepilemur, Archaeolemur* and *Hadropithecus*), there is a modified eruption sequence of posterior–middle–anterior, which would accordingly be a derived condition that has arisen in parallel several times. In any case, it is noteworthy that in no living prosimian is there the straightforward eruption sequence of anterior–middle–posterior that would be expected from a very simple model of dental eruption in regular sequence from the front to the back of the jaw.

Schwartz (1975b) also specifically discussed the phylogenetic significance of the eruption of the molars relative to the eruption of permanent replacements of incisors, canines and premolars. He proposed that there was replacement of at least some of the anterior dentition before the eruption of the upper and lower first premolars in the ancestral primates. From this hypothetical condition, he then postulated two directions ('morphoclines') of evolution of eruption sequences. The first of these involves progressively earlier appearance of the molars, with at least the first molars erupting before any permanent replacements of anterior teeth. This would apply to various prosimian species. Schwartz (1975b), citing Johnston, Dreizen and Levy (1970), also stated that it applies to marmosets and tamarins (Callitrichidae); but Byrd (1981) has since shown that in all callitrichids at least one incisor erupts before the second molar. In any case, eruption of molars in marmosets and tamarins is complicated by the fact that the third molar has been lost, uniquely among living primates. In Goeldi's monkey (*Callimico goeldii*), the reduced third molar is the last permanent tooth to erupt. Although it has been claimed that in the owl monkey (*Aotus*) all three molars erupt before any other permanent teeth (Byrd, 1981), it would seem that the first upper incisor is replaced before the third molar erupts (Hall, Beattie and Wyckoff, 1977).

The second direction of evolution of eruption sequences postulated by Schwartz involves early eruption of the first molars combined with increasing delay in the eruption of the second and third molars. This would apply to Old World simians, with the most extreme form occurring in hominids.

On the basis of his postulated first direction, Schwartz (1975b) proposed that the eruption sequence in bushbabies (Galaginae) and mouse and dwarf lemurs (Cheirogaleidae), with the first molars erupting before any permanent replacements of anterior teeth, is a shared derived condition. This is taken as an additional indication, supporting certain interpretations based on basicranial evidence (see Chapter 7), that the mouse and dwarf lemurs were not an integral part of the evolutionary radiation of the Malagasy lemurs, but shared a specific common ancestry with the loris group (see also Tattersall and Schwartz, 1974). This conclusion is, however, problematic because in members of the subfamily Lorisinae there is replacement of some anterior teeth before the first molars have erupted. Accordingly, the characteristic shared by the Galaginae and Cheirogaleidae (i.e. initial eruption of the first molars) could be a shared derived condition only if these two groups were more closely related to one another than the Galaginae are to the Lorisinae. This seems most unlikely on the grounds of virtually all other available evidence.

It is, in fact, difficult to see why Schwartz (1974b) regarded eruption of some permanent anterior teeth before the first molars as the primitive condition for primates. Among living primates, only *Tarsius* and the Lorisinae show this characteristic, although Schwartz's radiographic evidence indicated that the lower incisors and canines of the variegated lemur (*Varecia*) might develop (but not erupt) before the first molars, and that the anterior upper incisors of *Avahi* develop and erupt before the first molars. Because eruption of the first molars before the permanent replacement of any of the anterior teeth is far more common among living primates, this appears to be a more likely ancestral condition for primates. This eruption sequence seems to have occurred in the Eocene 'lemuroids' *Adapis* (Stehlin, 1912) and *Notharctus* (Gregory, 1920a). Indeed, contrary to the claim made by Schwartz (1975b), Stehlin's interpretation was that all three molars of *Adapis* erupted before the permanent replacement of any of the other teeth still present in modern primates. (As noted above, *Adapis* and *Notharctus* had four premolars on each side of upper and lower jaws. The anterior premolar of the series, which has been lost in all living primates, apparently erupted after the first molar but before the second.)

The most parsimonious interpretation of the available data for primates is that the first molars, at least, erupted before any permanent replacement teeth in the ancestral primates. Gingerich and Sahni (1984) demonstrated that this was the case in the Miocene adapid *Sivaladapis*, and Conroy, Schwartz and Simons (1975) reported that the teeth of the Oligocene simian fossil *Apidium phiomense* also erupted in this pattern. In fact, Kay and Simons (1983b) later showed that in the lower jaw both the first and second molars erupted before the permanent replacement teeth in *Apidium* (and possibly in *Parapithecus* as well). This tends to confirm the picture presented by *Adapis* and *Notharctus*, with molars erupting earlier (relative to permanent replacement teeth) than in living primates. It seems quite possible, in fact, that in ancestral primates all of the molars erupted before the replacement of any milk teeth, as originally suggested by Gregory (1920a, 1922), even though this condition has not been preserved in any living primate.

Schwartz (1975b) reported that the first molars erupt before the permanent anterior teeth in tree-shrews (*Tupaia*) and in at least one insectivore (*Chrysochloris*), whereas in most insectivores (e.g. *Suncus*, *Talpa*, *Erinaceus* and *Elephantulus*) part or all of the anterior dentition is replaced before the eruption of the first molars. Whereas Schwartz interpreted the condition in *Tupaia* as derived, it must be emphasized that insectivores tend to have peculiar features of dental eruption compared with other mammals. (The hedgehog and shrews have vestigial teeth representing either the aborted permanent successors of milk teeth or aborted milk teeth and thus either retain milk teeth without permanent replacement or possess precociously developed permanent teeth in the adult dentition; see Moss-Salentijn, 1978.) Once again, there is every reason to believe that modern insectivores are specialized in various ways and do not provide a reliable model for the ancestral condition of placental mammals.

In fact, tree-shrews might be a much better model for the ancestral placental condition, as in numerous other respects. Butler (1980) and Shigehara (1980) reported that in the Tupaiinae generally all three molars usually erupt before any anterior replacement teeth (although in some individual specimens there is a slight departure from this condition). In *Ptilocercus* (Ptilocercinae), by contrast, the permanent premolars erupt much earlier, with some (at least) erupting before the first molars (Butler, 1980). One possible interpretation of this is that the Tupaiinae have retained the primitive mammalian condition postulated by Gregory (1922), and essentially retained in the Eocene 'lemuroids' *Adapis* and *Notharctus*, with all molars erupting early in the sequence, whereas *Ptilocercus* (like many insectivores and a few predominantly insectivorous prosimian species) has undergone secondary specialization with a relative advance in the timing of eruption of anterior permanent teeth. Butler (1980) also stated that in all tree-shrews the typical eruption sequence for permanent premolars is anterior–posterior–middle, apparently confirming Schwartz's inference (1975b) that this sequence is primitive for primates and probably for placental mammals generally. Shigehara (1980), however, found that only the lower premolars erupted in the sequence indicated by Butler; the upper premolars erupted in the sequence posterior–middle–anterior. Accordingly, any inferences about the evolution of dental eruption sequences in mammals must remain tentative until a thorough comparative review has been conducted.

Gingerich (1976b) reported that in the 'archaic primate' *Plesiadapis* eruption of the first and second molars was followed by eruption of the permanent anterior premolar and only then by the third molar. This resembles the condition in some primates, but is presumably a convergent development in view of the evidence for Eocene 'lemuroids'.

An interesting test case of the use of eruption sequences and inference of developmental patterns to establish dental homologies is provided by the dentition of *Tarsius*. It has already been noted that this genus presents special problems for identifying the dental formula using traditional criteria, and Osborn (1978) has proposed one possible revision of that formula on the basis of his clone model. Schwartz (1980) has proposed an alternative radical reinterpretation of the dental formula of *Tarsius*, starting from the view that the position of the teeth (e.g. relative to the premaxilla/maxilla suture) is not a reliable criterion for assessing dental homologies. He relies instead on the observed shape of teeth and on sequences of development and eruption of permanent teeth. His first point is that the first tooth in the upper jaw of *Tarsius* is 'strikingly caniniform', and therefore likely to be a canine rather than an incisor, and the subsequent teeth are 'premolariform'. His second point is based on the interpretation (tentatively accepted above) that in ancestral primates the eruption sequence for permanent premolars was anterior–posterior–middle. If the dental formula of *Tarsius* is reformulated as $I_0^0 C_0^1 P_5^5 M_3^3$, and if it is assumed that the first and third teeth in the series of five premolars are unreplaced milk teeth (see later), then a similar eruption sequence can be recognized for the remaining three teeth identified as premolars (i.e. anterior–posterior–middle).

Among other things, this proposed dental formula conflicts with the generally accepted view that the dental formula of ancestral placental mammals was $I_3^3 C_1^1 P_4^4 M_3^3$, subsequently reduced to $I_2^2 C_1^1 P_4^4 M_3^3$ in ancestral primates of modern aspect. However, as noted above, McKenna (1975), Osborn (1978) and Novacek (1986b) have proposed that ancestral placental mammals had five premolars on either side of the upper and lower jaws, so Schwartz's dental formula for *Tarsius* is technically compatible with that minority interpretation.

Luckett and Maier (1982) specifically set out to test Schwartz's hypothesis with respect to the dental formula of *Tarsius*, using embryological evidence. They found that the tooth germs of the anterior two teeth on either side of the upper jaw are clearly associated with the premaxilla. These two teeth are hence identifiable as deciduous incisors, as the embryological evidence is compatible with the traditional approach to dental homologies in this case. In addition, the tooth traditionally identified as the upper deciduous canine develops from a tooth bud at the front end of the maxilla, as would be expected. As Schwartz had noted, there are peculiarities in the development of some of the teeth, which led him to propose that certain milk teeth remain unreplaced in the adult dentition. Taking the traditional dental formula of $I_2^2 C_1^1 P_3^3 M_3^3$, Luckett and Maier found that the second upper incisor and the anterior premolars of the milk dentition show abnormal development, becoming 'vestigial'. Moss-Salentijn (1978) has identified such vestigial teeth as aborted permanent teeth in other mammals (e.g. rodents and lagomorphs), but certain embryological evidence indicates that they may be milk teeth in *Tarsius*, at least. Deciduous teeth arise as buds from the primary dental lamina, whereas each permanent tooth arises from a successional lamina, originating on the inner side of the developing germ of its deciduous precursor. In *Tarsius*, the 'vestigial' teeth are derived from the primary dental lamina. Hence, the teeth identified by Schwartz as unreplaced milk teeth in the adult dentition of *Tarsius* were identified by Luckett and Maier (1982) as permanent teeth that replace vestigial predecessors. This being the case, Schwartz's argument (1980) based on sequences of permanent premolar eruption is undermined.

There is an additional problem in that, although the tooth traditionally identified as the upper canine (and seen by Schwartz as the second in a series of five premolars) erupts before the last two premolars, the embryological evidence shows that the canine begins to develop after the last premolar but before the penultimate one (Luckett and Maier, 1982). So Schwartz's presumed developmental sequence involving these three teeth is not correct. Indeed, there is a far simpler hypothesis concerning premolar development and eruption sequences for *Tarsius* and it is consistent with the traditionally accepted dental formula of $I_1^2 C_1^1 P_3^3 M_3^3$. In terms of both development (Luckett and Maier, 1982) and eruption (Schwartz, 1975b), the three premolars recognized in this formula for *Tarsius* appear in the sequence anterior–posterior–middle, thus complying with the pattern that may have been present in ancestral primates.

It is particularly encouraging that embryological evidence supports the traditional interpretation of the dental formula of *Tarsius*, as this suggests that current interpretations of the evolution of dental formulae in primates (customarily based on the formula of $I_3^3 C_1^1 P_4^4 M_3^3$ in ancestral placental mammals) may not require extensive revision. Nevertheless, it must be remembered that, although it is convenient to adhere to the traditional interpretation, worrying questions remain about the identification of dental homologies between species. With further studies of dental eruption sequences, following on from Schwartz's pioneering work, and with clarification of the processes involved in early dental development, following on from the work of Butler and Osborn, it is quite likely that present conceptions of dental homology will have to change in some respects.

FUNCTION OF PRIMATE DENTITIONS

The discussion thus far has been concerned with teeth as structural entities without any real reference to their functions. It is now necessary to turn to the question of dental function in order to obtain further insights into dental evolution in primates. To do this it is convenient to consider the dentition as consisting of two functional components – the anterior teeth (concerned with prehension of food in most cases) and the cheek teeth (essentially concerned with mastication). As before, living primates will be considered first and fossil primates will be examined subsequently to test and refine hypotheses regarding evolutionary change.

Anterior teeth (incisors and canines)

Primitively, the primary function of the anterior dentition (incisors and canines) was undoubtedly that of prehension. Given that ancestral placental mammals were probably omnivorous, with a mixed diet including both arthropods and easily digested plant foods (e.g. fruit), the anterior dentition was presumably at that stage adapted for the grasping and killing of small animal prey and perhaps for seizing plant food items such as small fruits. The incisors and canines of ancestral placental mammals were probably tapered with the incisors being rather small, peg-like teeth and the canines having the typical dagger-like form found in many modern mammals. This pattern has been retained without extensive change in the upper anterior dentition of most primates, but there have been some striking modifications of the lower dentition in many prosimians. Further, as has been established above, all primates of modern aspect are characterized by a reduction in the number of incisors to a maximum of two in each tooth row associated with shortening of the premaxilla. This development was very probably associated with a shift of prehensile functions from the anterior dentition to the hands (see Chapter 10), which are typically adapted for grasping in primates of modern aspect. One result of this shift has been to emancipate the remaining anterior dentition to serve new functions.

The most unusual modification of the anterior dentition among modern primates is seen in the so-called **tooth-comb** or **tooth-scraper** in strepsirhines (Fig. 6.16). This structure, which typically consists of six procumbent anterior teeth (two incisors and a canine on each side of the lower jaw), is present in all members of the loris group and in most modern lemur species. Among modern lemurs, the Indriidae are distinctive in that the tooth-comb has been reduced to a total of four teeth through loss of one tooth (probably the canine) on both sides,

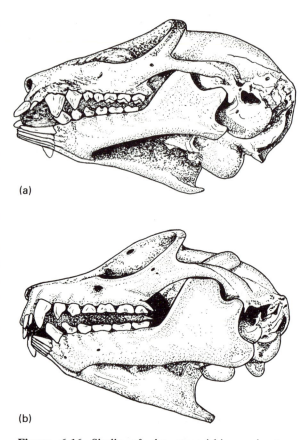

(a)

(b)

Figure 6.16. Skulls of the strepsirhine primates *Phaner* (a) and *Cheirogaleus* (b), showing the tooth-comb in the lower jaw. The tooth-comb in both cases consists of two incisors and one canine on each side, giving a total of six procumbent teeth. In *Phaner*, which is a specialized gum-feeder, the tooth-comb is particularly well developed in relation to overall skull length. (After James, 1960.)

whereas in the aye-aye (*Daubentonia*) there is only one continuously growing tooth on either side of the lower jaw (either an incisor or a canine; see Fig. 6.9b). The fact that in both cases reduction has taken place from an originally complete tooth-comb is clearly indicated (see above) by the presence of a full set of six teeth in the deciduous tooth-comb of indriids and by the presence of four teeth in the anterior milk dentition of the aye-aye. In addition, the existence of the tooth-comb as a functional unit is indicated by the simultaneous eruption of the constituent teeth in many species, as reported above.

Among subfossil lemurs, the picture is less clear-cut. Although a full tooth-comb is found in some cases (e.g. in subfossil *Varecia* and in *Megaladapis*), in several genera (e.g. *Archaeolemur, Hadropithecus, Palaeopropithecus* and *Archaeoindris*), both upper and lower incisors have an essentially shovel-like (spatulate) form and there is no marked procumbency of the lower anterior teeth. In this respect, these genera resemble simian primates rather than other strepsirhines. Again, it is likely that this represents a secondary condition involving specialization away from an ancestor with a typical strepsirhine tooth-comb, as – unlike simian primates – all subfossil lemurs lack a dagger-like canine in the lower jaw (see also Martin, 1972b).

Tree-shrews also have a procumbent tooth-comb in the lower jaw. Saban (1956/57) and Le Gros Clark (1959) specifically suggested that this indicates an affinity with the strepsirhine primates. However, in tree-shrews the primitive mammalian complement of three incisors on each side is still present in the lower jaw and the typical tooth-comb consists of six incisors, not four incisors and two canines as in strepsirhines. Le Gros Clark noted that in *Anathana* the canines are also procumbent and rather incisor-like, suggesting 'a phase of evolutionary development which surely must have preceded the more highly specialized dental comb characteristic of Recent lemuroids'. But the condition in *Anathana* must be a convergent development for two reasons. Firstly, any phylogenetic relationship would have to link ancestral primates to ancestral tree-shrews, not to the single modern genus *Anathana*. Secondly, the most parsimonious interpretation of dental formulae in primates of modern aspect (see above) is that the number of lower incisors had already been reduced to two on each side before the evolution of the strepsirhine tooth-comb. Thus, the presence of a tooth-comb in tree-shrews provides an interesting case of convergence rather than suggesting a specific phylogenetic relationship between tree-shrews and primates. An even

more striking example of convergence is provided by the Eocene mammal *Thryptacodon* (a condylarth), which had a tooth-comb very similar to that of lemurs and lorises but formed by six lower incisors without participation of the canine (Gingerich and Rose, 1979).

The question of the function of the tooth-comb in strepsirhine primates and in tree-shrews has been much debated. It is commonly accepted that this dental structure primarily serves for grooming of the fur (e.g. see Buettner-Janusch and Andrew, 1962). There is no doubt that the tooth-comb is used for grooming by tree-shrews and by strepsirhine primates, as numerous behavioural observations have now been corroborated by studies showing wear marks produced on the lower anterior teeth by the repeated passage of hair between them (Rose, Walker and Jacobs, 1981). However, it is also known that the lower anterior teeth are used for feeding by both tree-shrews and strepsirhine primates. Originally, it was suggested by Avis (1961) that the lower anterior teeth of strepsirhines serve a cropping role in feeding, but it is unlikely that the tooth-comb operates in this way, except perhaps in the case of *Lepilemur* (and, by implication, *Megaladapis*). On the other hand, there is substantial evidence that the tooth-comb is used as a scoop for the ingestion of soft plant food. In captivity, tree-shrews and a wide variety of strepsirhine primates scoop out the pulp of soft fruit (e.g. banana and peaches) using the tooth-comb, leaving characteristic furrows in the remaining pulp. In the wild, both *Indri* and *Propithecus* have been observed using the tooth-comb when feeding on plant foods (see Martin, 1979).

Use of the tooth-comb has also been implicated in the collection of a particular type of plant food that has only been properly investigated in recent years, namely exudates (such as gum). Field observations of small-bodied strepsirhine primates have now shown that members of the loris group (Charles-Dominique, 1977a; Bearder and Martin, 1980a) and nocturnal lemurs (Petter, Schilling and Pariente, 1971; Martin, 1972b; Charles-Dominique, Cooper *et al.*, 1980) commonly feed on plant exudates. Indeed, it has emerged that gums make up a major proportion of the diet in the needle-clawed bushbaby (*Galago elegantulus*) and in the fork-crowned lemur (*Phaner furcifer*). In both of these species, the area of the tooth-comb is quite large relative to overall skull length (Martin, 1979) and especially so in *Phaner* (see Fig. 6.16). This tends to support the suggestion made by Martin (1972b) that the tooth-comb is particularly important for the collection of plant exudates.

As has been pointed out by Szalay and Seligsohn (1977), the tooth-comb in nocturnal strepsirhine primates is too fragile for penetrating the bark of exudate-producing trees. In fact, field observations show that the exudates eaten by nocturnal strepsirhines are commonly produced by trees as a response to damage caused by wood-boring arthropods; the exudates are then scooped up from the surface of the trunk by strepsirhine consumers. This is, for example, the case with *Acacia* gums collected by the lesser bushbaby (*Galago senegalensis*) (Bearder and Martin, 1980a). At the end of a 2-year field study, Bearder and Martin found that the base-boards of all the traps in regular use had been scored with the tooth-comb by bushbabies taking the bait (the initially liquid bait had many of the properties of gum and became hard, like gum, if left *in situ*).

There is therefore no doubt that the tooth-comb can serve a dual function in both tree-shrews and lemurs, allowing for combing of the fur as well as the collection of soft plant foods. Given the evidence that the lower anterior teeth were used for grooming the fur in certain early Tertiary mammals (Rose, Walker and Jacobs, 1981), the most likely explanation would seem to be that grooming and feeding functions of the tooth-comb evolved in close association (see also Martin, 1981b). There is certainly no justification for claims that the tooth-comb has little or no function in feeding behaviour, and it seems inherently unlikely that grooming is the primary function in the sense that evolution of the tooth-comb was conditioned predominantly by the need to clean fur. There is no evidence that strepsirhine primates have been subject to

unusual selection pressures favouring the modification of the anterior lower dentition (uniquely including the canines) for grooming. If the grooming function is so important, it is difficult to see why *Tarsius* (for example) has no comparable tooth-comb or why in certain subfossil lemurs (but not *Megaladapis*) there should be secondary loss of the tooth-comb. As is so often the case, it seems that a correlation between possession of a tooth-comb and its use in grooming has been too readily assumed to reflect causation. Given that teeth typically serve feeding functions in vertebrates generally, it seems far more likely that tooth-combs developed as scoops in association with feeding behaviour and that the grooming function evolved in conjunction with this (perhaps determining the particular comb-like structure). Rosenberger and Strasser (1985) have proposed a modified version of the 'grooming hypothesis' to explain the origin of the tooth-comb in strepsirhines. They postulate that the upper dentition was reduced in connection with the olfactory role of the Jacobson's organ, which is associated with the rhinarium (see Chapter 7), and that this led to emancipation of the lower teeth for a non-dietary function. However, strepsirhines are not the only mammals with a well-developed rhinarium associated with Jacobson's organ, and this modified hypothesis still does not explain why strepsirhines should need to groom their fur so effectively in comparison with many other mammals.

The primate fossil record is poor with respect to documentation of the evolution of strepsirhine primates, but it is nevertheless revealing with respect to the evolution of the tooth-comb. In the first place, there is no tooth-comb comparable to that in strepsirhines in any of the Eocene 'lemuroid' species (Adapidae). Gregory (1920a) demonstrated that *Notharctus* had a typical mammalian dentition with simple peg-like incisors in both upper and lower jaws, accompanied by quite prominent, dagger-like canines (see also Rosenberger and Strasser, 1985). As far as is known, the dentition of all of the North American Eocene 'lemuroids' followed this pattern. The anterior dentition is poorly documented for European Eocene 'lemuroids' other than *Adapis parisiensis* and *A. (Leptadapis) magnus*. *Adapis magnus* had well-developed canines in the lower jaw. By contrast, *A. parisiensis* had relatively small lower canines that appear to have formed a functional unit with the lower incisors. Both upper and lower incisors were spatulate and, whereas the upper canine was a moderately well-developed conical tooth, the lower canines were spatulate and their crowns formed a continuous cutting edge with the lower incisors (Gingerich, 1975b). As Gingerich observed, this condition is unique among Eocene 'lemuroids' and it was largely for this reason that Szalay favoured the use of a separate generic name, *Leptadapis*, for *Adapis magnus* (see Szalay and Delson, 1979). This step has not been followed here, however, because of the overriding need to maintain stability of nomenclature wherever possible (see Chapter 3).

The close association between the lower incisors and canines in *Adapis parisiensis* led Gingerich to suggest that this species is 'the best candidate yet known for the common ancestor of lemurs and lorises'. However, it seems more likely that this modification of the lower anterior dentition in *A. parisiensis* represents a parallel development (see also Rosenberger, Strasser and Delson, 1985), as *Adapis* species all have several specializations that were presumably absent in the common ancestor of the strepsirhine primates (e.g. fusion of the two halves of the lower jaw; see later). On the other hand, the condition of the anterior lower teeth in *A. parisiensis* does illustrate evolutionary modification of these teeth for dietary reasons, as there is no evidence that they were specifically used for grooming and they do not form a 'comb'. On present evidence, it would seem that the tooth-comb of strepsirhine primates represents an unusual evolutionary development lacking in Eocene 'lemuroids' generally.

Because of the unusual nature of the strepsirhine tooth-comb, involving the lower canines as well as incisors, it seems likely that this feature was present in the ancestral strepsirhine stock, being retained by all extant members of the loris group and by many modern and subfossil

lemurs, but having undergone secondary modification in the remaining lemurs. However, this hypothesis was originally challenged in two different ways. Le Gros Clark and Thomas (1952) suggested that the tooth-comb was only weakly developed, at best, in Miocene lorisids of East Africa, whereas Simons (1962c), on dental grounds derived the lorises and the lemurs separately from Eocene 'lemuroids'. Both of these proposals have the typical six-tooth 'comb' developing separately in lorises and lemurs. Le Gros Clark and Thomas based their suggestion on the observation that the sockets (alveoli) for the incisors and canines in the lower jaw of the Miocene 'lorisoid' *Progalago* were more vertical than horizontal in orientation. However, as has been noted by Walker (1969b, 1974b), modern strepsirhines such as *Galago* are no different from *Progalago* in this respect. In fact, the procumbency of the tooth-comb is in large part due to the fact that the bilaterally flattened crown of each tooth is set at an angle to the root (Fig. 6.17). Walker (1978) has since reported on additional specimens of East African Miocene lorisids with closely packed, bilaterally flattened roots for the lower anterior dentition, thus confirming that a fully developed tooth-comb was present in these strepsirhines. A similar condition has been reported for the Miocene lorisid *Nycticeboides* from Pakistan by Jacobs (1981), who has also observed grooves produced by the passage of hair between the teeth. In view of these findings, it would seem highly unlikely that lemurs and lorises were separately derived from any known group of Eocene 'lemuroids'. On the contrary, the possession of a six-tooth comb including the canines is one of the most convincing examples of likely shared derived characters linking modern lorises and lemurs, marking out these strepsirhine primates as a monophyletic group (Martin, 1972b: Gingerich, 1975b).

Although the anterior lower dentition of modern tarsiers does have certain peculiarities, as has been noted above, there is little sign of the special adaptation of these teeth that characterizes lorises and lemurs. The anterior lower teeth are not bilaterally flattened and there

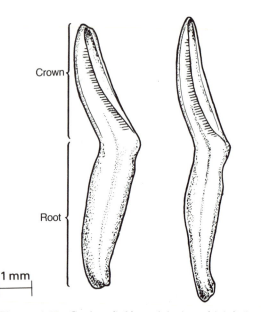

Crown

Root

1 mm

Figure 6.17. Canine (left) and incisor (right) from the tooth-comb of the lesser bushbaby (*Galago senegalensis*). Note the bilaterally flattened form of the crown, with a scoop-shaped dorsal surface. The crown is set at an angle to the root, so the marked procumbency of the crowns of the teeth in the tooth-comb is not identifiable from the angle of the alveoli in the lower jaw. The hatching indicates the location of grooves produced by hairs during grooming.

is no significant procumbency of their crowns. The same applies to known Eocene 'tarsioids', so the 'tarsioids' as a group are distinct from strepsirhine primates in this respect. Schmid (1983) reported that the Eocene 'tarsioid' *Necrolemur* had grooves on the anterior lower teeth indicative of a grooming function. This was used as an argument to link Eocene 'tarsioids' (Omomyidae) to strepsirhine primates. However, Musser and Dagosto (1987) have found similar grooves on the anterior lower teeth of *Tarsius pumilus*, so this feature alone does not warrant a link between omomyids and strepsirhines. The unusual features of the anterior dentition in modern tarsiers, at least, are probably related to the purely insectivorous/carnivorous habits of these primates (see also Luckett and Maier, 1982). This possibility is

supported by the observation that the relative dimensions of the cheek teeth also reflect special adaptation for such a diet (Gingerich and Smith, 1985).

The modern simian primates are distinguished from all recent and fossil prosimian primates (with the exception of some of the large-bodied subfossil lemurs identified above) by having spatulate incisors adapted for biting and cutting. As this is an apparently derived feature shared by all living simians, it is reasonable to suggest that it was a characteristic of the ancestral simian stock that gave rise to both New World monkeys (platyrrhines) and Old World monkeys and apes (catarrhines). Presumably, there must have been some significant dietary shift involving increased emphasis on incisal biting in the evolution of the simian primates.

An allometric study by Hylander (1975) of incisor size in Old World simians indicated that there is an overall isometric relationship between the total width of the upper incisors and body weight. Incisor size is greater (relative to body size) in frugivores than in folivores. This finding was explained on the grounds that fruits require more incisal biting than leaves. A similar investigation of incisor size in New World monkeys, which included an analysis of lower incisors as well as upper incisors, was subsequently carried out by Eaglen (1984). His results suggested that, in contrast to the Old World simians, there is a negatively allometric relationship between incisor size and body weight for both upper and lower jaws. As with Hylander's study of Old World simians, however, the few New World monkey species that eat significant quantities of leaves (*Alouatta* species and *Brachyteles arachnoides*) were found to lie below the best-fit line, although the separation from the other (predominantly frugivorous) species was not clear-cut. When Eaglen (1984) compared New World with Old World simians over the range of comparable body weights (5–10 kg), he further found that New World monkeys tend to have smaller incisors than Old World monkeys and apes, even though the former tend to be more frugivorous in habits. Thus, incisor size relative to body size cannot be used as a reliable criterion for inferring diet for a mixed sample of New World and Old World simians. In fact, Eaglen's bivariate plots of incisor size against body weight indicate that the incisors may be relatively narrow in the squirrel monkey (*Saimiri*) and in marmosets and tamarins (Callitrichidae), although these species are frugivore–insectivores rather than folivores. The overall width of the incisors would therefore seem to be a somewhat unreliable criterion for inferring dietary habits in simians generally, although within simian subgroups it may be of some value.

The dimensions of the incisors do seem to have a special dietary significance with respect to members of the family Callitrichidae. As noted by Napier and Napier (1967), marmosets have relatively long lower incisors and the lower canines project little above the crowns of adjacent teeth ('short-tusked' condition), whereas tamarins have relatively small lower incisors and quite prominent canines ('long-tusked' condition). There is a sharp distinction between all marmosets (*Cebuella* and *Callithrix* species) and all tamarins (*Leontopithecus* and *Saguinus* species). Coimbra-Filho and Mittermeier (1976, 1977) observed that several marmoset species use their lower anterior dentition to perforate tree bark and thus actively stimulate the flow of exudates; tamarin species do not do this, although some species will eat exudates on occasions when these are naturally available. It therefore seems that the 'short-tusked' condition in marmosets represents a special adaptation for feeding on exudates by active gouging of holes in trees.

The available fossil evidence indicates that ancestral simians were characterized by the possession of spatulate incisors and prominent, dagger-shaped canines. This is the condition identifiable in the first adequately preserved fossil simians from Miocene deposits of the New World and it is the condition generally characterizing the Oligocene simians of the Fayum, which are the earliest documented fossil simians of the Old World (see Szalay and Delson, 1979 for illustrations). Hence, the shift from peg-like or conical incisors to spatulate incisors with a cutting edge is confirmed as a significant shift in the emergence of simian primates. It should also be noted that Old World simian primates (catarrhines) are characterized by a special 'honing' mechanism in which the upper canine is sharpened against the anterior lower premolar. This mechanism is particularly well developed in certain Old world monkeys (Zingeser, 1969). A parallel development is also found in some New World monkeys.

Cheek teeth (premolars and molars)

Most of the discussion of the evolution of primate dentitions has understandably centred on the cheek teeth and the part that they play in the mastication of food. Indeed, discussions of cheek-tooth morphology have generally overshadowed any consideration of the anterior dentition, partly because cheek teeth (especially the molars) are more complex in form than incisors or canines and partly because the anterior teeth are far less well represented in the fossil record. The complex structure of the molars in particular has provided a rich basis for phylogenetic reconstructions concerning mammals and it is a general rule that molar complexity has tended to increase in most lines of mammalian evolution. It is now necessary to consider the form and function of the cheek teeth in some detail.

Molar teeth of early mammals

Initial comparative work on the form of mammalian molar teeth concentrated almost exclusively on the shape of the molars, rather than on the functional dimension. Much discussion was devoted to the question of homologies between species. Such work eventually led to the tritubercular theory of Cope and Osborn, which was based on the concept that both upper and lower molars were originally based on a triangular pattern, with three main cusps in each case (Fig. 6.18). According to this scheme, the upper molar was a fairly simple triangle (**trigon**), whereas the lower molar had both a main triangle (**trigonid**) and a heel

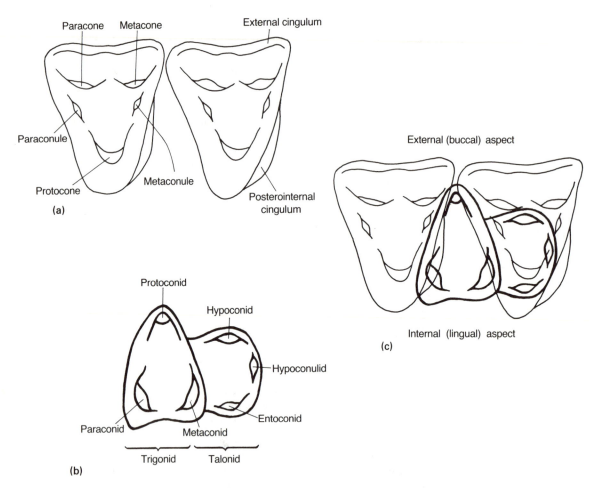

Figure 6.18. Schematic diagram of the ancestral features of tritubercular molars in viviparous mammals (marsupials and placentals). The molars shown here are from the right jaws and are oriented such that the anterior–posterior axis runs from left to right. The lower molar (b) is shown in heavy outline, presenting the cusps as seen from above. Two upper molars (a) are shown in faint outline, presenting the cusps as they would be seen 'by transparency' in a view from above. On the right (c), the upper and lower molars are superimposed to show their occlusal relationships. Note that the protocone of the upper molar fits into the talonid basin of the lower molar in occlusion. (See text for further details.)

(talonid). In lateral view, the cusps on the trigonid of the lower molar were markedly higher than the cusps on the weakly developed talonid. The main cusps on the upper and lower molars are traditionally named according to the system originally proposed by Osborn (1888, 1907), although alternative schemes have been suggested at various times (see Butler, 1978). In the upper molar, the apex of the triangle (trigon) is located internally (lingually) and bears the **protocone**. The two external (buccal) cusps are the **paracone** (anterior) and the **metacone** (posterior). In the lower molar, the apex of the main triangle (trigonid) is located externally (buccally) and bears the **protoconid**. The two internal (lingual) cusps are the **paraconid** (anterior) and the **metaconid** (posterior). (Note that the names of cusps on the lower molar end in

'-id' to distinguish them from cusps on the upper molar.) The heel (talonid) of the lower molar probably bore three cusps in the original condition: **entoconid** (internal), **hypoconid** (external) and **hypoconulid** (posterior).

Osborn's original nomenclature was based on the assumption that, in the derivation of mammalian molars from simple conical teeth, the primary cusp was the protocone on the upper molars and the protoconid on the lower molars. It is now clear from a combination of embryological and palaeontological evidence that this is not the case. In fact, the paracone was the original primary cusp of the upper molar, followed by the metacone and then by the protocone. Osborn was, however, correct in identifying the protoconid as the primary cusp on the lower molar. Because Osborn based his names for the cusps of the upper molar on a false premise, some authors have suggested alternative schemes of nomenclature as being 'more logical' (e.g. Vandebroek, 1961; Hershkovitz, 1971). However, as is the case with systems of nomenclature used in the classification of species (see Chapter 3), there is a strong case for retaining Osborn's terms, as the confusion generated by his original error of interpretation is negligible compared with the confusion generated by altering names that have been very widely used. Butler (1978) has aptly commented: 'Let those who are contemplating the introduction of new names pause to consider whether in so doing they are advancing the subject or making it more difficult to understand. Language is for communication.' The important thing, after all, is that Osborn was almost certainly right about the basic triangular pattern of mammalian molars.

Three-cusped molar teeth were already present in certain advanced mammal-like reptiles (cynodonts; see Chapter 4), but the cusps were originally arranged linearly rather than in a triangular pattern. Subsequently, in early mammals (symmetrodonts) the cusps were arranged in a V-shape, with the apex in the upper molar formed by the homologue of the paracone in mammals. This **primary trigon** of the upper molar in early mammals was replaced by the **secondary trigon** of later mammals when the tribosphenic pattern (see below) emerged with the addition of the protocone and suppression of part of the original trigon, probably at some time during the Cretaceous period. The secondary trigon (as illustrated in Fig. 6.18) was developed in close functional association with the talonid of the lower molar (Butler, 1961). Indeed, the presence of a protocone on the upper molar, interacting with a three-cusped heel (talonid) on the lower molar, may be cited as a shared derived feature of marsupials and placentals (Kemp, 1982). In the lower molar, the three main cusps on the trigonid (protoconid, paraconid and metaconid) have been present since the Triassic period. This sequence of evolutionary events is reflected by the embryological development of molar teeth in modern mammals. In upper molars, the first cusp to develop is the paracone, whereas in lower molars the first to develop is the protoconid and the talonid develops after the trigonid. Thus, a cohesive picture of the evolution of molar morphology in mammals has now been established.

Additional insights into molar evolution have been obtained from studies of functional aspects of molar morphology, particularly with respect to the **occlusal relationships** involved in the interaction between the working surfaces of upper and lower cheek teeth (e.g. see Gregory, 1920b,c, 1921, 1922 for an early contribution to this topic). This functional perspective led Simpson (1936) to coin the term 'tribosphenic' for the molars of mammals, which literally means 'grinding wedges'. In a useful review, Lumsden and Osborn (1977) examined in detail the way in which upper and lower molars interact to break food morsels down into small, easily digestible particles. They noted, in particular, that the evolution of complex cheek teeth in mammals has been associated with the emergence of *lateral* jaw movements, which were virtually absent from mammal-like reptiles. In addition, early mammal dentitions were characterized by contact between adjacent teeth (mesiodistal contact), which enabled the cheek teeth to act as a stable unit rather than as an assembly of isolated teeth, and the teeth were able to migrate progressively along the jaws to compensate for wear. Butler (1978) notes that tribosphenic lower molars are typically locked together by virtue of the fact that the hypoconulid of one molar fits into the next molar between a ridge on the anterior surface of the paraconid and a crested ledge (**cingulum**) on the internal (lingual) side.

In early mammalian dentitions, the primary function of the molars was that of shearing and the earliest emergence of the basic triangular pattern of cusps (seen, for example, in some early Jurassic mammals such as the pantothere *Kuehneotherium* – see Chapter 4) was significant more for the development of worn shearing edges than for the cusps themselves. The basic functional principle of simple triangular upper and lower molars (i.e. before the development of the talonid on the lower molar) is that each tooth has a straight edge and a concave edge. In upper molars, the anterior edge of the triangle is straight and the posterior edge concave, whereas in the lower molar the reverse is the case. As a result, lens-shaped gaps are initially present between the blades of opposing upper and lower teeth, trapping pieces of food that are sliced as the jaws close and the gaps disappear. This special design of the triangular upper and lower molars can only work effectively because the mammalian dentary/squamosal jaw joint permits lateral, as well as vertical, movement of the lower jaw.

The talonid (heel) of mammalian molars first emerged during the Jurassic period, being partially developed in pantotheres, and it permitted a change in the length of the shearing edge on the lower molar. The fully three-cusped talonid is, however, confined to viviparous mammals (marsupials and placentals) and, as has been indicated above, the emergence of the full talonid defined the origin of the true tribosphenic molar in their common ancestral stock. The development of the true protocone increased the length of the shearing surface available on the upper molar to interact with the talonid on the lower molar. More significantly, the basin of the talonid developed as a relatively wide valley to accommodate the protocone, and the external cusp of the talonid (i.e. the hypoconid) fitted into

the basin of the trigon of the upper molar (see Fig. 6.18). This arrangement allowed the talonid to serve the function of reciprocal crushing in interaction with the upper trigon. Hence, the molar teeth had shifted from an alternating pattern serving essentially for shearing to a pattern of direct surface-to-surface occlusion between the upper and lower teeth. In other words, true tribosphenic molars permit puncturing and crushing as well as shearing (see Lumsden and Osborn, 1977).

The modern New World opposum *Didelphis* has molar teeth quite similar to the hypothetical ancestral condition for marsupials and placentals. There are six shearing edges on each upper molar and five on each lower molar and a puncture–crushing interaction between the trigon and the talonid, as described above. Because the shift from the simple triangular molars of certain early mammals to the true tribosphenic molars of viviparous mammals (as exemplified by *Didelphis*) involved the addition of the puncture–crushing action to the original shearing function, some dietary explanation is required to account for this shift. Whereas it is conceivable that the earliest mammals were insectivorous in the strict sense of the term, the diet of ancestral viviparous mammals clearly included some components, presumably derived from plants, that required puncture–crushing. Inclusion of plant food in the diet, even as a relatively minor constituent, is further suggested by the inference that the digestive tract of ancestral viviparous mammals most probably incorporated a caecum (see pp. 221–2). It is also important to note that exclusive dependence on arthropod food poses a number of nutritional problems, notably with respect to calcium availability and to maintenance of an appropriate calcium:phosphate ratio (e.g. see Martin, Rivers and Cowgill, 1976). Diversification of the diet is therefore advantageous for insectivorous mammals.

Basic changes in mammalian molars

Among modern placental mammals, the cusp pattern of the molar teeth has in most cases been modified in various ways from the common ancestral pattern indicated in Fig. 6.18. In some mammals, such as insectivores, tree-shrews and primates, these modifications have been relatively mild; but in others, such as most rodents and ungulates (hoofed animals), they have been so extensive that it is difficult to recognize the original cusp pattern. As a general rule, modifications of the molar teeth have been related to a need for increased masticatory efficiency associated with heavy dependence on resistant plant food. However, in carnivorous species (for example) there has generally been a secondary return to marked adaptations for shearing to deal with meat-eating. Primates lack any such pronounced specializations and the most striking feature of their molars is that they differ so little from those of the earliest known placental mammals. Indeed, in the modern tarsier, which has the most primitive molar teeth found in any living primate, only minor changes from the likely ancestral placental mammal condition have taken place (Fig. 6.19). Although there may have been secondary simplification of the molars, given the fact that *Tarsius* is the only living primate genus to feed exclusively on animal food, it is more parsimonious to assume that the molars have retained primitive features. Accordingly, among living primates, *Tarsius* probably provides us with the best guide to the ancestral condition of the molars for primates generally.

Evolutionary modification of the molar teeth in mammals to permit improved mastication of plant food typically involves a simple initial transformation that has clearly occurred in parallel in numerous different lineages (Fig. 6.20). The original three primary cusps (protocone, metacone and paracone) on the upper molar are supplemented by a fourth cusp (**hypocone**), which typically develops on a small heel (talon) on the posterointernal margin of the tooth. The lower molar, by contrast, loses one of the original three primary cusps on the talonid (i.e. the paraconid), leaving only the protoconid and the metaconid. As a result, both the upper and lower molars assume a more rectangular appearance and the original shearing action of the trigonid against the posterior margin of the corresponding upper molar (Fig. 6.18) gives way to a grinding action of the reduced trigonid against the hypocone (Fig. 6.20). The working surfaces of the molars are also eventually levelled off. The talon and the talonid, originally lying well below the planes of the trigon and of the trigonid, respectively, are often raised

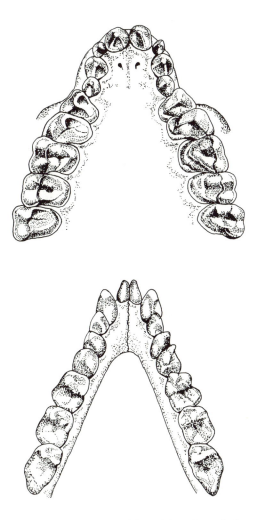

Figure 6.19. Upper dentition (above) and lower dentition (below) of *Tarsius*. The primitive three-cusped condition is essentially preserved on the trigon of the upper molars and on the trigonid of the lower molars. Note that the two halves of the lower jaw remain unfused. (Adapted from James, 1960 and Le Gros Clark, 1959.)

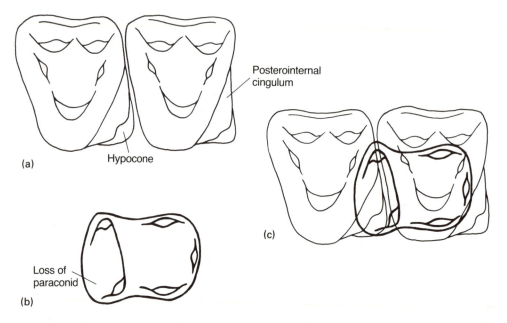

Figure 6.20. Schematic diagram of modified tribosphenic molars, showing a common trend among placental mammals (conventions as in Fig. 6.18). In the upper molar (a), a fourth cusp (hypocone) develops as an extension of the posterointernal cingulum. In the lower molar (b), one of the original three cusps of the trigonid – the paraconid – disappears. In both cases, the net effect is that the molars take on a more rectangular form. Note that the new cusp on the upper molar, the hypocone, occludes with the trigonid basin of the lower molar (c).

progressively as evolution of the molars progresses. Similarly, the anterior part of the talonid becomes wider to match the width of the trigonid. As pointed out above, many modern placental mammals have progressed well beyond this stage, with the development of additional molar complexity. In primates, however, even the most specialized species have evolved little beyond the condition illustrated in Fig. 6.20, except that there is a general tendency for small accessory cusps (e.g. the metaconule and paraconule shown in Fig. 6.18) to disappear. Nevertheless, there are subtle distinctions in molar morphology among primate species that are broadly indicative of phylogenetic relationships, and it is also possible (for example) to consider their affinities with tree-shrews on this basis. As a guiding principle in comparisons between species, it is reasonable to assume that the condition illustrated in Fig. 6.18 is primitive for placental mammals generally and that

departures from that condition represent secondary specializations (see also Szalay, 1968b; Crompton and Kielan-Jaworowska, 1978).

Molars of tree-shrews

It has already been mentioned in Chapter 5 that, somewhat surprisingly, molar morphology has not figured prominently in discussions of possible affinities between tree-shrews and primates, even though reconstructions of mammalian phylogenetic relationships have generally depended heavily on dental evidence. This point has also been made by Butler (1980), in a useful review of tree-shrew dentitions. An important first step in assessment of tree-shrew relationships is consideration of variability within the family Tupaiidae, as it is essential to establish the ancestral condition for tree-shrews generally before attempting to infer their affinities with other mammals. Steele (1973) and Butler (1980) have shown that there is considerable variation in molar morphology among tree-shrews. In particular, there are marked differences between pen-tailed tree-shrews (Ptilocercinae) and other members of the family (Tupaiinae). In *Ptilocercus*, the molar cusps are relatively blunt and there is a moderately developed, distinct hypocone. Features such as these led Gregory (1910) to infer that *Ptilocercus* is more omnivorous than other tree-shrews, but the limited data on the natural diet of the pen-tailed tree-shrew (see Butler, 1980) do not support this inference. In all species of the subfamily Tupaiinae, the molars bear relatively tall, sharp cusps and there is considerable variation in the degree of development of the hypocone. In some cases (e.g. *Dendrogale* and *Tupaia minor*) there is no trace of a hypocone, whereas at the other extreme the hypocone is quite well developed in *Anathana* and in *Urogale*. Butler (1980) inferred that a small hypocone was present in the ancestral stock of the tree-shrews, with secondary loss occuring in some members of the Tupaiinae; but it is also possible that ancestral tree-shrews lacked a hypocone and that this accessory cusp was developed independently in *Ptilocercus* and in some tupaiine tree-shrews. Steele (1973), for example, states that the hypocone of *Ptilocercus* is markedly different from that found in various members of the Tupaiinae.

All tree-shrews retain the paraconid in their lower molars, so they are more primitive than the living primates except *Tarsius* in this respect and certainly do not exhibit the advanced condition illustrated in Fig. 6.20. In fact, it is now widely agreed that tree-shrews generally have quite primitive molar teeth, not very different from those of ancestral placental mammals. Hence, Gregory's (1910) suggestion that tree-shrews (and particularly *Ptilocercus*) might provide a good model for the origins of primate molar adaptations may be valid only in the sense that primates are, of course, derived from ancestral placentals.

In tree-shrews other than *Ptilocercus*, both Gregory (1910) and Butler (1980) identified secondary specializations of the molars. In the upper molars, the paracones and metacones form

V-shaped ridges and the outer margins of the teeth bear small accessory cuspules (**styles**) that are functionally associated with the ends of the ridges. On each upper molar, the paracone, metacone and styles form a W-shaped (dilambdodont) array that has also been developed – presumably independently – in several other groups of placental mammals, including some lipotyphlan insectivores, insectivorous bats, 'flying lemurs' (Dermoptera) and some fossil and living primates (see Butler, 1980). Hiiemae and Kay (1973) point out that tree-shrews generally depart from the probable ancestral primate condition in that the external cingulum (**stylar shelf**) of the upper molars is conspicuously developed, and in the lower molars there is a marked interlocking relationship between the paraconid and the preceding hypoconulid. Thus, the molar teeth of tree-shrews show a combination of numerous primitive placental features with a variety of secondary specializations that do not indicate any specific connection with the origin of primates. Butler concludes from his comparative survey of tree-shrew dentitions that early primates cannot be derived from the condition represented by the Tupaiidae and that tree-shrews are therefore best allocated to their own order, Scandentia.

Molars of living primates

As might be expected, the molars of prosimian primates generally are more primitive than those of simians, and it has already been noted that among living primates the most primitive condition is seen in *Tarsius* (Fig. 6.19). Nevertheless, there is a considerable range of variation among living strepsirhine primates (lemurs and lorises) and some species exhibit quite specialized molar morphology. In particular, some of the subfossil lemurs, which are considered here as constituting an integral part of the modern lemur fauna, have molar specializations paralleling those of simian primates. In fact, in all modern strepsirhine primates there is virtual or complete loss of the paraconid from their lower molars, so in this respect they are uniformly more advanced than *Tarsius*.

Mouse and dwarf lemurs (Cheirogaleidae) have generally retained a relatively primitive pattern of molar morphology. The basic tribosphenic pattern of ancestral placental mammals is still clearly recognizable, except that both the paraconid and the hypoconulid have disappeared from the lower molars and that the hypocone is at least partially developed in the upper molars (weakly developed in *Cheirogaleus*; more obvious in *Microcebus* and *Phaner*). In the Lemuridae, the molars generally exhibit a more rectangular shape and the lower molars consistently lack both the paraconid and the hypoconulid, but there is still considerable variation. For instance, the hypocone is quite small in *Lemur* and virtually absent in *Varecia*. Yet in both *Varecia* and *Lemur* species there is variable development of a different accessory cusp, which, unlike the hypocone, occurs on the anterior internal border of the upper molar and is referred to as the **pericone** (or **protostyle**). In the remaining two genera, *Hapalemur* and *Lepilemur*, a small hypocone is present. *Lepilemur* is unusual, however, in that there is a posterior, blade-like extension of the protocone in addition to the weakly developed hypocone. Further, in the lower molar the four remaining cusps are linked by two oblique ridges (protoconid–metaconid and entoconid–hypoconid) arranged in a very distinctive pattern. Despite its considerably larger body size, the subfossil lemur *Megaladapis* has molar teeth that are very similar to those of *Lepilemur*. This, along with the fact that *Lepilemur* and *Megaladapis* have both lost the upper incisors, is evidence of a specific evolutionary relationship between these two genera (see also Tattersall and Schwartz, 1974). Although *Hapalemur* also has ridges linking the four remaining cusps on its lower molars, these are oriented more-or-less transversely, rather than obliquely as in *Lepilemur* and *Megaladapis*, to produce a characteristic **bilophodont pattern** (James, 1960).

In members of the family Indriidae, the tendency to develop a more quadratic shape of the molars has progressed even further. The hypocone is consistently well developed on the upper molars, such that they exhibit a clear quadritubercular appearance. On the lower molars, there are typically four cusps as in other lemurs. But the indriids are unusual in that the paraconid has been retained on the anterior lower molar (M_1), indicating that suppression of the paraconid was not complete in the common ancestral stock of the strepsirhines. In the indriids as a group there is a general tendency for both upper and lower molars to develop a bilophodont appearance. This trend is weakly recognizable in *Avahi*, moderately evident in *Propithecus* and most apparent in *Indri*. Bilophodont upper and lower molars are even better developed in the subfossil indriid *Archaeolemur*. Indeed, in *Archaeolemur* the bilophodont condition of the molars is so conspicuous that there is a remarkable resemblance to the condition characteristic of Old World monkeys (Maier, 1977; see later).

In addition to exhibiting radical modification of its anterior dentition, the aye-aye (*Daubentonia*) is also unusual in that its molar teeth have very little surface detail. However, the molars in both upper and lower jaws are rectangular in form, doubtless reflecting derivation from an ancestral stock in which there were four full cusps on both the upper molars (i.e. through the complete development of a hypocone) and the lower molars (i.e. through the loss of the paraconid and the hypoconulid).

Members of the loris group, in contrast to lemurs, show considerable uniformity in molar morphology. In all genera except *Perodicticus* a well-developed hypocone is present on the upper molars and this cusp is distinctive in that it is displaced backwards to produce a characteristic rhomboid shape of the upper crown. The lower molars consistently lack both the paraconid and the hypoconulid. At the same time, the molars of all lorisids have clearly retained features of the original tribosphenic pattern and the cusps are typically high and sharp. It has often been noted that bushbabies (galagines) differ from the true lorises (lorisines) in that the last premolar in both upper and lower jaws is clearly molariform in the former. This is a consistent difference and one possibility was that it has arisen through late retention of the last milk premolar in adult bushbabies (the last milk premolar typically being more complex in morphology than its permanent replacement among primates). This explanation was proposed by Bennejeant (1953) but has been ruled out by several other authors (e.g. Schwartz, 1974c). In other respects, however, galagines and lorisines are very similar in dental features and this suggests that they together make up a relatively cohesive group of species derived from a comparatively recent common ancestral stock.

Because the molars of *Tarsius* are seemingly very primitive (Fig. 6.19), it is difficult to propose a common ancestral condition for the haplorhine primates that would be significantly different from the ancestral condition for primates generally. The evidence for a phylogenetic link between tarsiers and simian primates is derived from non-dental features, such as the conformation of the nasal region and placentation. On the basis of molar morphology alone it is impossible to confirm or deny the existence of a haplorhine stock ancestral to tarsiers and simians. Hence, early fragmentary fossil primate species documented only by dental remains cannot be identified confidently as 'haplorhines', although they may in some cases be identified either as 'tarsioids' or as simians. This provides one of the most compelling reasons for retaining the grade-based classificatory distinction between prosimians and simians (see Chapter 3).

The simians as a group share several features of molar morphology suggestive of a specific common ancestry. For example, the lower molars in all living simians lack an identifiable paraconid and their shape is rectangular. The upper molars are also rectangular in most cases, with suppression of small accessory cusps (styles and conules), but usually with a well-developed hypocone. However, the marmosets and tamarins (Callitrichidae) are exceptional in that the hypocone is essentially lacking from the upper molars, which therefore have a more triangular shape than in other modern simians. Hershkovitz (1977) and Kinzey (1974) saw this as a primitive retention in callitrichids, with the implication that ancestral simians (and hence

ancestral New World monkeys) lacked a hypocone. However, Gregory (1920d) inferred that the common ancestor of New World monkeys in fact had tritubercular molars with small but clearly recognizable hypocones, and numerous authors have followed Gregory's inference that the hypocone in marmosets and tamarins has been secondarily reduced (e.g. Remane, 1960; Hoffstetter, 1974; Rosenberger, 1977; Ford, 1980). The callitrichids are, after all, unique among modern primates in having lost the third molars from their upper and lower jaws, and many of their apparently 'primitive' features are convincingly attributable to a secondary reduction in body size during their evolution (Ford, 1980; see also Chapter 12). Gregory (1920d) regarded owl monkeys (*Aotus*) and titi monkeys (*Callicebus*) as closest to the ancestral condition for New World monkeys, with respect to molar morphology, and both of these genera have four-cusped, rectangular upper and lower molars. Gregory proposed that the hypoconulid was still present in ancestral New World monkeys, but this cusp has not usually been retained in modern descendants and it may have been already lacking in the ancestral stock. As in other features, Goeldi's monkey (*Callimico*) is intermediate between typical cebids and typical callitrichids in molar morphology; all three molars are still present, although the third is reduced in size, and the upper molars are somewhat squarer in shape, even though the hypocone is barely developed.

All modern Old World simians have the hypocone as a well-developed cusp on the upper molars, which are consistently rectangular in shape. As the hypoconulid is retained on all lower molars of modern apes, this cusp was doubtless present in the ancestral stock of the Old World simians, even though it is typically lacking from at least the first and second lower molars of all modern Old World monkeys (Cercopithecoidea). Nevertheless, like the upper molars, the lower molars are consistently rectangular in shape in all Old World simians.

As a result of widespread reference to the *Scala naturae* (see Chapter 3) in discussions of primate evolution, there is a general expectation that apes, being more closely related to humans, should be more advanced than monkeys. With respect to molar morphology, however, this expectation is not fulfilled; it is in the molars of Old World monkeys that the most marked specializations occur. In all modern cercopithecoids, the upper and lower molars have four more-or-less equally developed cusps (i.e. the hypoconulid has commonly been lost from the lower molars), and the teeth are clearly divided into anterior and posterior sections, each bearing two cusps. These paired cusps are linked to varying degrees (according to species) by transverse ridges, such that some degree of bilophodonty is found in all Old World monkeys. Although the transverse ridges are only moderately developed in some genera (e.g. *Cercopithecus* species), bilophodonty is a consistent identifying feature of the Cercopithecoidea. The most extreme development of bilophodonty is to be found in leaf-monkeys (Colobinae), in which the transverse ridges are particularly high and trenchant (Fig. 6.21). It has already been noted that varying degrees of bilophodonty occur in *Hapalemur* and living

and subfossil indriids of Madagascar, parallel developments also occur in a few New World monkey genera such as *Ateles* and, especially, *Cebus* (Remane, 1960). However, it is only in the Cercopithecoidea that marked bilophodonty occurs as a universal feature characterizing an entire 'natural group' of living primates. The presence of bilophodonty is hence an important diagnostic feature of the Old World monkeys, which at the same time clearly marks them out as a specialized offshoot in the evolutionary radiation of the Old World simians. As Gregory (1920e) has aptly noted in relation to the Cercopithecoidea:

> The early pairing of the molar cusps in groups of two, together with the relative unimportance of the hypoconulid, are conspicuous characters which definitely rule all cercopithecoid monkeys out of the line of ascent leading to the anthropoid apes and man.

In comparison to the Old World monkeys, apes and humans (Hominoidea) have relatively unspecialized molar teeth. The upper molar teeth have four cusps, with a well-developed hypocone, but the cusps are not linked by transverse ridges. Indeed, there is generally an oblique crest (**crista obliqua** or **postprotocrista**) linking the protocone and metacone in hominoids, such that the hypocone remains isolated from the original trigon. (The crista obliqua is somewhat modified in modern *Homo sapiens*, but it is clearly present in all modern apes.) Although the hypoconulid is typically present, it tends to disappear in the latter stages of human evolution, such that its expression is variable in the lower molars of modern humans. As with the upper molars, there is no appreciable tendency for transverse ridges to develop in the lower molars of hominoids.

Gregory (1920e) states that, among modern hominoids, the gibbons and siamang (Hylobatidae) have remained the most primitive. They have, for instance, retained a fairly distinct trigon in the upper molars, with the hypocone less developed and more distinctly separate than in great apes and humans. The upper molars of

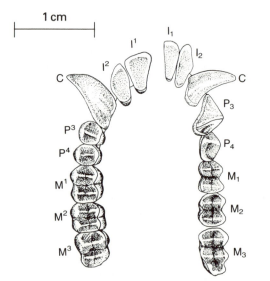

Figure 6.21. Dentition of a male silvered leaf-monkey (*Presbytis cristatus*), showing the typical cercopithecoid condition of bilophodonty. Left: upper jaw (right side) seen from below. Right: lower jaw (right side) seen from above. (After Swindler, 1976.)

hylobatids also lack the wrinkling of the enamel surface that other hominoids commonly have. In the lower molar of the gibbons and siamang, the hypoconulid is located centrally at the posterior end of the tooth. This corresponds to the primitive condition inferred both for primates in particular (Fig. 6.20) and for viviparous mammals in general (Fig. 6.18). This latter feature is particularly important because in the great apes and in humans (in those cases where the cusp has been retained) the hypoconulid is located externally and the metaconid is in contact with the hypoconid. This modification of the crown of the lower molar produces a special pattern known as **Y5** (Hellman, 1928) because a Y-shaped set of furrows runs between the five cusps (Figs. 6.22 and 6.23). As lower molars of this kind are characteristic of fossils of the *Dryopithecus* group (see later), Gregory (1916) originally coined the term '*Dryopithecus* pattern' for the Y5 arrangement of cusps. Some confusion has arisen in the literature, because Simons (1964) and others illustrated the Y of the *Dryopithecus* pattern as lying lengthwise along the lower molar, whereas by definition it runs

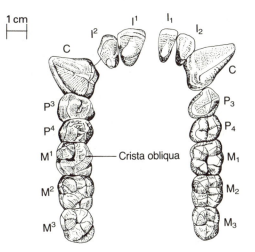

Figure 6.23. Dentition of a male gorilla (*Gorilla gorilla*), showing the crista obliqua in the four-cusped upper molars and the Y5 pattern in the lower molars. Left: upper jaw (right side) seen from below. Right: lower jaw (right side) seen from above. Note that, as is typical of simian primates, the two halves of the lower jaw are fused together. (After Swindler, 1976.)

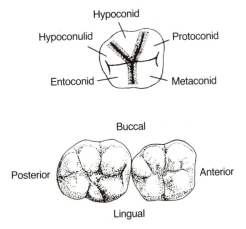

Figure 6.22. The characteristic Y5 (*Dryopithecus*) pattern in the lower molars of great apes. Above: diagram of the Y5 pattern, showing the arrangement of cusps and valleys. Note that the hypoconulid has shifted towards the external (buccal) margin. Below: M_2 and M_3 of *Gorilla* (after Remane, 1960).

across the molar with the base of the Y located on the inside (lingual aspect), as shown in Fig. 6.22 (see also Robinson and Allin, 1966). This characteristic Y5 pattern of the modern great apes is a distinctive feature almost equivalent, in terms of diagnostic value, to the bilophodonty of Old World monkeys, although the latter is more specialized. During human evolution, the Y5 pattern of the lower molars has been modified and in modern *Homo sapiens* there is considerable variation, with the most extreme form being a four-cusped '+pattern' that arises through complete suppression of the hypoconulid (Gregory, 1921).

This cursory survey of the major features of molar morphology in living primates has demonstrated that the order as a whole is characterized by relatively limited departure from the likely ancestral condition for viviparous mammals (Fig. 6.18). Nevertheless, there are diagnostic features that characterize certain stages of primate evolution and these can guide our interpretation of the fossil record (see later).

As yet, however, only the morphological features of primate molars have been considered and it is necessary to consider functional aspects of mastication (chewing) before attempting to interpret the fossil evidence. Whereas certain inferences about functional properties of molar teeth can be made purely on the basis of morphological features (e.g. see Gregory, 1920a–e, 1921, 1922), direct evidence of masticatory activity in living primates is ideally required in order to test such inferences and to provide a full description of mastication.

Mastication

The basic pattern of muscular activity governing the movement of the jaws during mastication is essentially the same in all mammals studied to date (e.g. see Hiiemae, 1978). Mastication is carried out primarily by movement of the lower jaw relative to the skull, although cranial flexion and extension (i.e. movement of the skull relative to the vertebral column) also contribute (Hiiemae, 1976, 1978). The lower jaw is raised by three main elevator muscles: the **temporalis** (which is attached to the lateral wall of the braincase), the **masseter** (which is typically attached predominantly to the zygomatic arch) and the '**medial pterygoid**' (which is attached to the base of the skull). Lowering of the jaw is brought about by two main depressor muscles: the **lateral pterygoid** and the **digastric**. In mammals, the elevators are well developed, whereas the depressors (especially the digastric) are small, reflecting the fact that muscular energy is predominantly devoted to the masticatory interaction between upper and lower teeth. In fact, as was noted above in the discussion of the origin of tribosphenic molars in viviparous mammals, it is a significant feature of jaw movement in marsupials and placentals that some lateral movement takes place between upper and lower jaws in addition to the vertical movement required to open and close the jaws. Because the basic features of the jaw muscles are the same in all viviparous mammals, it follows that differences in patterns of movement of the jaws between species must largely reflect

differences in tooth morphology and skull architecture, notably with respect to the jaw joint (i.e. the **temporomandibular joint**). This is particularly true of differences between species involving distinctions in the movement of the lower jaw in the horizontal plane.

There is, in fact, a major difference between living simians and living prosimians with respect to the structure of the lower jaw. As explained in Chapter 4, the mammalian lower jaw consists of a single bone on each side, the **dentary**. Primitively, the two sides of the lower jaw remained unfused in the adult (e.g. see Moore, 1981) and this condition is retained, for example, in modern lipotyphlan insectivores and tree-shrews. In some mammals, however, the two dentaries fuse together at their anterior junction (**symphysis**), such that the lower jaw becomes a single bony element in the adult (Beecher, 1977, 1979a). The evolutionary sequence from unfused symphysis to **symphyseal fusion** is reflected in the early development of such mammal species, because in all mammals the two dentaries initially develop as separate elements. In all modern simian primates, the dentaries are firmly fused together in the adult; indeed, symphyseal fusion takes place very early in life. By contrast, in all extant prosimian primates the dentaries typically remain unfused in the adult, although there are some cases of partial fusion late in life (Beecher, 1983). As symphyseal fusion is probably a derived feature, the universal presence of this feature as an early development in all simians clearly suggests that it was already present in the ancestral simian stock.

Symphyseal fusion is, in fact, found in some subfossil lemurs (*Archaeolemur, Hadropithecus, Archaeoindris, Palaeopropithecus* and *Megaladapis*). It is noteworthy that these are all of the largest known forms and that symphyseal fusion has probably developed in parallel at least three times in the lemurs, because the subfossil lemurs with this feature fall into three distinct groups, separately related to living lemurs that lack symphyseal fusion (i.e. *Archaeolemur* and *Hadropithecus* are arguably early derivatives of the indriid stock; *Archaeoindris* and *Palaeopropithecus* are probably more closely related to modern idriids; and *Megaladapis* is probably closely related to *Lepilemur* – e.g. see Martin, 1972b; Tattersall and Schwartz, 1974). These two facts taken together suggest that, among lemurs at least, symphyseal fusion is related to increased body size (see also Beecher, 1983). It has been noted in Chapter 4 that living simian primates are typically about 10 times heavier than living prosimian primates, so it is reasonable to suggest that the development of symphyseal fusion in ancestral simians was linked to a substantial increase in body size, relative to the earliest primates. However, body size alone cannot account for the universal presence of symphyseal fusion among modern simians, because the relatively small marmosets and tamarins (including the pygmy marmoset, *Cebuella*) have a fused symphysis, whereas the largest living prosimian (*Indri*), which is comparable in body size to typical monkeys and lesser apes, still has an unfused symphysis. Also, it is only in the subfossil Archaeolemurinae that the symphysis fused early in life, as in simians (Tattersall, 1973).

Hiiemae and Kay (1973) suggested that the origin of symphyseal fusion in simian primates was linked to the evolution of spatulate incisors and the associated phenomenon of incisal biting (see also Beecher, 1983). This would also account for the presence of symphyseal fusion in most of the large-bodied subfossil lemurs, which have separately developed spatulate incisors (i.e. *Archaeolemur, Hadropithecus, Archaeoindris* and *Palaeopropithecus*); but it does not account for fusion of the dentaries in *Megaladapis*, which has a typical strepsirhine tooth-comb in the lower jaw and completely lacks incisors in the upper jaw.

The hypothesis linking symphyseal fusion to incisal biting has also been rejected by Hylander (1979a,b), who used miniature strain gauges to measure bone stress *in vivo* in *Galago* (a typical prosimian lacking symphyseal fusion) and in *Macaca* (a typical simian with fused dentaries). Hylander's studies have shown that fusion of the symphysis permits more effective coordination of the muscles of the two sides of the jaws during mastication. During mastication on one side (the

working side), it was found that the ratio of bone strain on the working side to bone strain on the other side (balancing side) was 3.3:1 in *Galago*, but only 1.5:1 in *Macaca* (Hylander, 1979a; see also Hylander and Johnson, 1985). This suggests that muscular force is applied predominantly on the working side during chewing by *Galago*, whereas in *Macaca* the muscular force is more evenly distributed between the working and balancing sides. During the power stroke of mastication (see later), the balancing side of the lower jaw is subjected to bending in a vertical plane (sagittal bending), whereas the working side is predominantly twisted about its longitudinal axis. As a result, stress occurs in the symphyseal region. Hylander (1979a) notes that, in addition to having fused dentaries, simians also differ from living prosimians in having markedly deeper lower jaws. A deep lower jaw bone provides greater resistance against sagittal bending and fusion of the symphysis serves a similar function. Hylander therefore concludes that these two modifications are probably components of a single adaptation related to mastication, permitting the lower jaw to tolerate increased bite forces. Because incisal biting also generates bending and twisting stresses in the lower jaw, it is possible that symphyseal fusion and increased jaw depth may be related to the use of well-developed spatulate incisors in some simian primate species, but this is unlikely to be the primary context for the evolutionary development of these universal simian features. As is pointed out by Hylander (1979a), colobine monkeys have relatively less developed incisors than cercopithecine monkeys (see above), yet the lower jaw is deeper in colobines.

Hylander's hypothesis linking symphyseal fusion and increased depth of the lower jaw primarily to increased masticatory requirements in simians would therefore seem to be valid. However, because these two developments seem to have such obvious advantages with respect to providing increased protection against masticatory stresses in the lower jaw, it is difficult to see why the lower jaw remains unfused even in large-bodied modern lemurs such as *Propithecus* and *Indri*. Presumably, there is some selective advantage of maintaining an unfused symphysis in prosimians, perhaps permitting the working side of the lower jaw to rotate relative to the balancing side. It should be remembered, incidentally, that simian primates have a markedly shorter cheek-tooth row, relative to skull length, than prosimian primates (Fig. 6.12). Taken together with the derived features of the lower jaw discussed above, this suggests a fundamental remodelling of the jaws in ancestral simians that involved shortening of the cheek-tooth row as well as increasing the resistance of the lower jaw to masticatory stresses. As noted by Hiiemae and Kay (1973), relative shortening of the tooth row in simians is further associated with raising of the jaw hinge (temporo-mandibular joint) above the plane of occlusion between the upper and lower molars. Thus, the simians seems to constitute a well-defined phylogenetic unit with respect to masticatory function.

The process of mastication clearly depends heavily on the morphology of the cheek teeth, especially the molars. Hence, as in other areas of mammalian morphology, a major advance in our understanding of molar characteristics came with explicit consideration of functional aspects. Initially, functional interpretations were based on inferences from fine details of the molar teeth, but more recently the process of mastication has been studied directly with the aid of techniques such as cineradiography in attempts to relate the dynamics of jaw movement to the observed properties of the molar teeth. In all cases, however, the central feature of interest has been the manner in which upper and lower cheek teeth interact when they are brought together during chewing, that is to say their **occlusal relationships**. Patterns of occlusion between upper and lower teeth were emphasized by Gregory (1920a–e, 1921, 1922) in his extensive review of the evolution of primate dentitions and this functional perspective considerably improved our understanding of the significance of the fine morphological details of molar cusps. More importantly, in his monograph on the Eocene 'lemuroid' *Notharctus*, Gregory (1920a) made reference to the **wear facets** produced by

interaction between upper and lower molars (see later). It was through detailed analysis of these facets and inference of the jaw movements required to produce them (i.e. occlusal analysis) that Butler (1952a,b) was able to include a new functional dimension in his study of the dentition of horses. Soon afterwards, Butler's approach was extended to the study of primate molar teeth by Mills (1955, 1963), who thus introduced the concept of **dynamic occlusion** (i.e. changing relationships between teeth during chewing) into the interpretation of dental evolution in primates. Butler and Mills together (1959) also extended occlusal analysis to include fossil primates.

A key point of reference in the path of movement of the lower molars against the uppers is **centric occlusion**, at which point the closest fit between opposing teeth is achieved (hence the alternative term **maximum intercuspation**; Hiiemae, 1978). In centric occlusion, each cusp on a molar tooth fits into a corresponding valley between cusps in the opposing tooth row. In particular (as noted above), the protocone of the upper molar fits into the talonid basin of the lower molar, while the main trigon basin of the upper molar accommodates the hypoconid (see Fig. 6.18). With reference to the landmark position of centric occlusion, Mills (1955, 1963, 1978) recognized two phases in the dynamic occlusal relationships between upper and lower teeth. During the **buccal phase**, the lower jaw on one side moves inwards and upwards from a laterally displaced position into centric occlusion. As this movement takes place, each cusp on a lower molar produces a shearing action along a groove between two opposing upper cusps ('cusp-in-groove slicing action'). During the subsequent **lingual phase**, the lower jaw on that side continues to move towards the tongue (i.e. medially), thus passing out of centric occlusion. In this phase, the outside (buccal) cusps of the lower molars slide inwards and somewhat downwards across the internal (lingual) cusps of the upper molars, with a consequent grinding action. Clearly, when one side of the lower jaw is moving medially in this fashion, the other side must be moving laterally.

Mills proposed that when teeth on one side of the jaw are in the buccal phase, the teeth on the other side are simultaneously in the lingual phase, producing a condition known as **balanced occlusion**. It has since emerged, however, that dynamic occlusion is usually confined to one side of the jaw at a time (the working side), while the teeth on the other side are not in proper occlusal contact (the balancing side). Mills himself noted that occlusion is unilateral at the beginning of the buccal phase in *Indri*, and Hiiemae (1978) has pointed out that genuine bilateral chewing is rare among mammals (see also Butler, 1973). Occlusion on one side of the jaw at a time was, in fact, already present in early mammals of the late Triassic, namely in morganucodontids and in *Kuehneotherium* (Kermack and Kermack, 1984). Indeed, in the case of modern *Didelphis* (the opossum), even centric occlusion is unilateral, although it is typically bilateral in the various placental mammals (including primates) so far examined.

Because balanced occlusion does not seem to occur consistently as envisaged by Mills (1955, 1963), Hiiemae prefers to use the terms phase I and phase II instead of buccal phase and lingual phase (e.g. see Hiiemae, 1978). However, as in many other cases, it is not really essential to change previously established terminology because of a shift in interpretation, especially as centric occlusion remains the accepted reference point for the chewing cycle on the working side. Mills (1978) maintains that balanced occlusion is found in all 'higher primates' and in certain other groups, while accepting that chewing typically takes place only on one side of the mouth at a time. It remains to be seen whether wear facets on mammalian molars are typically produced by alternating use of the cheek teeth on each side in chewing, or whether wear facets can in some cases be produced simultaneously on both sides of the jaws during interaction between upper and lower teeth.

Occlusal analysis, involving the interpretation of jaw movement from wear facets on molar teeth, is limited to phases when the teeth are in contact and has been greatly reinforced by a series of cineradiographic studies by Ardran,

Crompton, Hiiemae, Kay and others on the dynamics of mastication in certain mammalian species. Such work was initiated with *Homo sapiens* (Klatsky, 1940; Ardran and Kemp, 1960), but has been extended to other mammals, including *Didelphis marsupialis* (Crompton and Hiiemae, 1970; Crompton, Thexton *et al.*, 1977), *Oryctolagus cuniculus* (Ardran, Kemp and Ride, 1958), *Rattus norvegicus* (Hiiemae and Ardran, 1968), *Tupaia glis* (Hiiemae and Kay, 1973), *Galago crassicaudatus* (Kay and Hiiemae, 1974), *Saimiri sciureus* and *Ateles belzebuth* (Hiiemae and Kay, 1973). These studies, which have been effectively reviewed by Hiiemae (1978) and by Moore (1981), broadly confirmed the inferences based on molar wear facets, notably in identifying two phases of jaw movement (the buccal and lingual phases of jaw movement of Mills, 1955) during occlusion. In addition, they have permitted identification of a basic chewing cycle that is common to all viviparous mammals (Fig. 6.24).

In the first place, it should be noted that when a morsel of food is introduced between the upper

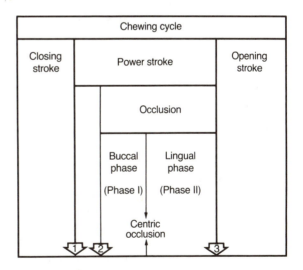

Figure 6.24. Characterization of the typical chewing cycle of viviparous mammals (see text for details). Numbered arrows indicate the following points in the cycle: 1, tooth–food–tooth contact is achieved; 2, tooth-to-tooth contact occurs; 3, tooth-to-tooth contact ceases. (After Hiiemae, 1978.)

and lower molars, actual chewing is preceded by a series of movements (**puncture–crushing cycles**) that ensure preparatory pulping of the food. Although the profile of jaw movement in puncture–crushing and **chewing cycles** is similar, the motion is more restricted to the vertical plane in puncture–crushing and there is no tooth-to-tooth contact, i.e. no occlusion (Hiiemae, 1978; Osborn and Lumsden, 1978). Subsequent chewing cycles exhibit the characteristic pattern indicated in Fig. 6.24. As has been noted above, chewing typically occurs only on one side of the mouth at any one time. There is an initial **closing stroke**, which acts to trap the food between upper and lower molar teeth. Once tooth–food–tooth contact has been achieved, with the buccal surfaces of upper and lower molars on the working side approximately aligned, the **power stroke** begins. Initially, the molar teeth on the working side are not actually in contact and tooth-to-tooth contact therefore occurs part of the way through the power stroke. From that point on, the molar teeth remain in occlusion until the power stroke is terminated. Loss of tooth-to-tooth contact introduces the third and final part of the chewing cycle, the **opening stroke**, following which **maximum gape** is achieved and another cycle can begin. During the power stroke, there is a buccal phase (Hiiemae's 'phase I') lasting from initial tooth-to-tooth contact until the attainment of centric occlusion, followed by a lingual phase ('phase II'), which represents the remainder of the power stroke (see Fig. 6.24). Whereas the power stroke is a relatively slow movement, the closing stroke and the latter part of the opening stroke take place quite rapidly.

Cineradiographic studies of chewing have also confirmed the inference from molar wear facets that the general direction of the power stroke is always buccal–lingual. During the buccal phase ('phase I'), there is an additional upward and forward motion of the lower jaw on the working side that brings the molars into centric occlusion. Slight downward movement of the lower jaw typically occurs in the lingual phase (although apparently not in the opossum *Didelphis*; Hiiemae, 1978), but this phase ('phase II') is generally more variable in mammals, becoming more horizontal and representing a greater proportion of the power stroke in species with an increased emphasis on grinding (e.g. *Saimiri*, *Ateles* and *Homo sapiens* as contrasted with *Didelphis* and *Tupaia*). Direct studies of mammalian mastication using cineradiography have also demonstrated that there is a smooth transition from the buccal phase to the lingual phase. Investigations of mammalian jaw movement using cineradiography have been supplemented by direct measurement of the activity of jaw muscles by electromyography (e.g. see Crompton, Thexton *et al.*, 1977 for *Didelphis*).

It has already been mentioned that the basic pattern of activity of the jaw muscles during chewing is relatively constant in mammals. However, there are differences between mammalian species in the relative development of the three major jaw muscles involved in the power stroke (temporalis, masseter and medial pterygoid). In primitive mammals the temporalis was probably the dominant muscle in mastication and this condition has been retained in *Didelphis*, insectivores, tree-shrews and primates. In carnivores, the dominance of the temporalis has been somewhat increased in many cases, whereas in specialized herbivores such as rodents and ungulates the situation has been radically altered such that the temporalis is relatively less developed and the masseter is the dominant muscle (see Moore, 1981). This pattern of relative development of the temporalis and masseter muscles among mammals indicates that increased emphasis on the lingual phase of jaw movement, which typifies forms with pronounced herbivorous habits, is associated with a shift in dominance from the temporalis to the masseter.

Thus far, the discussion of studies of jaw movement has followed the generally accepted assumption that wear facets on molar teeth are produced during normal chewing activity. As has been stated by Hiiemae and Kay (1973), the conventional view is that the action of food on teeth produces **abrasion**, whereas the mechanical interaction between opposing teeth produces

attrition. Abrasion by food is thought to result in randomly oriented scratching and pitting of the tooth surface, whereas attrition between teeth is thought to be responsible for the clearly recognizable polished enamel areas bearing fine, parallel striations that are now widely known as wear facets. As already noted, Butler and Mills used directional information from the striations of such facets to infer jaw movements during occlusion. They concluded that buccal and lingual phases are characterized by separate facets with different directions of wear. Direct studies of mastication – for example, using cineradiography – confirmed this inference from wear facets, at least to the extent that there are two directionally distinct phases of jaw movement during occlusion. This, however, does not demonstrate conclusively that the striated wear facets are produced during mastication and this is, in fact, very difficult to establish.

An alternative suggestion has been made to account for the striated wear facets (Every, 1970, 1974; Every and Kühne, 1970, 1971). It is proposed that occlusal interaction between opposing teeth during mastication can only blunt the cheek teeth and that striated wear facets are instead produced by a distinct set of jaw movements specifically adapted for sharpening the teeth (**thegosis**). Such specific sharpening ('honing') has been widely accepted in the case of canine teeth, which can exhibit comparable striated wear along their blade-like edges – for example, in Old World monkeys (Zingeser, 1969). But Every's hypothesis invoking thegosis as the agency responsible for striated wear facets on mammalian molars has not received general acceptance. One of the few supporters is Mills (1978), who regards the hypothesis as 'not improbable' and has suggested that balanced occlusion may in fact operate to project the jaw during thegosis, rather than serving a particular function during mastication. On the other hand, Osborn and Lumsden (1978) conclude from a detailed review of the operation of mammalian molars that striated wear facets are probably produced during normal mastication. For present purposes, however, it is not essential to decide how wear facets on mammalian molars are

produced. The important thing is that comparisons between species do indicate, both from wear facets and from evidence of jaw movement during mastication, systematic features that reflect both dietary adaptation and phylogenetic affinities.

The evidence available to date clearly indicates that interactions between upper and lower molar teeth were restricted to the buccal phase movement in early viviparous mammals. The opossum (*Didelphis*) shows no signs of a lingual phase (Crompton and Hiiemae, 1970); there are only six pairs of matching wear facets on its upper and lower molars (Crompton, 1971). By contrast, the molars of typical placental mammals with a buccal phase have 10 pairs of facets (Butler, 1952a, 1973). (Incidentally, Crompton (1971) did not follow Butler's numbering system for molar wear facets, but introduced a different system. This provides yet another example of differences in classification leading to potential confusion.) Lingual-phase facets are similarly missing from the molars of certain early Tertiary groups of mammals, including the Lepictidae and Palaeoryctidae, and from the molars of modern lipotyphlan insectivores (Butler, 1973). Tree-shrews (*Tupaia*) have a very restricted lingual phase and the upper and lower molars have only nine pairs of matching wear facets. In all living primates and fossil primates of modern aspect so far examined, a lingual phase has been identified (although its extent varies from group to group) and it is therefore reasonable to infer that ancestral primates of modern aspect had already developed some degree of lingual-phase movement. However, as noted by Butler (1973), there have been at least two independent origins of lingual-phase occlusion among placental mammals, so the presence of a very limited lingual phase in *Tupaia* provides no reliable evidence of an affinity with primates. Indeed, lingual-phase movement is found in some marsupials (e.g. certain modern phalagerids), so a multiple origin for this phase of occlusion is beyond doubt. On the other hand, there is evidence from cineradiographic studies of jaw movement that *Tupaia* is distinctly different from primates in at least one respect. Whereas in the primates so far examined (*Galago*, *Saimiri*, *Ateles* and *Homo*) the duration of puncture–crushing cycles is comparable to that of actual chewing cycles, in *Tupaia* the puncture–crushing cycles are significantly longer than the chewing cycles (see Hiiemae, 1978). Hence, evidence from molar wear facets and jaw movements indicates that *Tupaia* is distinct from primates rather than allied to them.

Studies of the occlusal relationships of molar teeth have also yielded new insights into functional aspects, two of which are of particular interest in terms of primate evolution. First, in some simian primates there is a novel adaptation called **compartmentalizing shear** (Hiiemae and Kay, 1973). In this, the cutting edges involved in the buccal phase serve to surround basins in such a way that food is cut into fragments that are trapped and then ground during the lingual phase. Compartmentalizing shear is found, for example, in spider monkeys (*Ateles*) and seems to have been present also in the Oligocene simian *Aegyptopithecus*. Second, Mills (1963) has noted that a bilophodont form of the molars, as in Old World monkeys, ensures that shearing is produced during the lingual phase of mastication as well as during the buccal phase. Hence, bilophodonty is clearly linked with increased cutting of food, which would apply (for instance) to specialized leaf-eating species.

Another feature of primate molar teeth that has recently attracted attention is **enamel thickness**. The importance of this feature was first clearly stated by Simons and Pilbeam (1972), who emphasized its significance for hominid evolution. Enamel thickness in a variety of simian primates was discussed by Molnar and Gantt (1977) and the topic was further explored by Kay (1981) and L.B. Martin (1985) with respect to Old World simians (cercopithecoids and hominoids). Initial expectations that thick enamel on the molar teeth would be of a diagnostic value in reconstructing human evolution have been confounded by evidence that this feature is sporadically distributed among living simian primates. (The distribution among prosimian primates is not yet known.) Apart from occurring in the hominids

Homo and *Australopithecus* and in the fossil hominoids *Ramapithecus*, *Sivapithecus* and *Gigantopithecus*, thick enamel is also found in the orang-utan (*Pongo*), in mangabeys (*Cercocebus* spp.) and, particularly, in New World capuchin monkeys (*Cebus* spp.). In the context of hominid evolution, it was initially proposed that thick enamel might be related to increased grinding capacity of the molars. An alternative interpretation is that enhanced enamel thickness is related to the cracking of hard fruits, nuts and seeds (but see Teaford and Walker, 1984). Kay (1981) found that there is an inverse relationship between relative thickening of molar enamel and relative development of shearing crests on the molars. The leaf-monkeys (Colobinae), for example, have relatively thin enamel and well-developed shearing crests. Here, the thin enamel is advantageous in that wear rapidly creates sharp shearing edges on the molars. By contrast, species such as capuchins and mangabeys have only poorly developed shearing crests on their molars.

Much work remains to be done on the topic of enamel thickness, especially with prosimian primates, but it is clear that this is an important feature of molar adaptation for specific dietary requirements. Valuable additional information on diets of individual mammalian species is also becoming available from studies of microscopic wear marks on teeth, particularly on molars (e.g. see Walker, Hoeck and Perez, 1978; Covert and Kay, 1980; Teaford and Walker, 1984). For example, Teaford and Walker (1984) found that microwear patterns may differ consistently between folivores, frugivores and consumers of hard items.

Fossil evidence for primate molar evolution

Having considered the main lines of evidence derived from studies of the cheek teeth of living primates, we can now turn to the fossil evidence. Two main objectives can be identified here: the first is that of reconstructing the likely ancestral primate condition of the molar teeth; the second is that of tracing the main lines of primate molar specialization in order to provide further information on the likely branching relationships within the evolutionary tree of primates. These two objectives are best tackled by starting with the earliest known primates of modern aspect from the Eocene – the 'lemuroids' (Adapidae) and the 'tarsioids' (Omomyidae).

The molars of Eocene primates

Eocene primates of modern aspect generally fit the pattern expected from the study of living primates, in that the molar teeth have departed little from the ancestral pattern inferred for viviparous mammals (Fig. 6.18), but they nevertheless show some characteristic developments. In particular, the molar cusps are relatively low and rounded and the talonid of the lower molars is somewhat raised to approach the level of the trigonid (see also Szalay, 1968b). In fact, in most known genera of Eocene primates of modern aspect there is some development of a fourth cusp on the upper molars and in this respect they are more advanced than certain living prosimian primates (some lemurs and tarsiers). However, some Eocene primates show virtually no trace of the development of a fourth cusp, as is the case with the adapids *Protoadapis recticuspidens* and *Europolemur koenigswaldi* (Franzen, 1987). Further, in the earliest known adapids not only is there no fourth cusp on the upper molars, but the paraconid cusp is also still clearly present on the lower molars (i.e. the otherwise typical primate developments illustrated in Fig. 6.20 have not yet begun). This is the case with *Pelycodus* (including *Cantius*), which is the stem genus of the North American adapids (see Gregory 1916, 1920a) and which is now recognized as an early adapid occurring in Europe as well (*Pelycodus* (*Cantius*) *eppsi*; Russell, Louis and Savage, 1967). In principle, it is reasonable to derive all Eocene 'lemuroids' from an ancestor with relatively primitive molar morphology close to that shown by early *Pelycodus*, and this means that many of the more specialized developments of later primates (e.g. full development of a fourth cusp on the upper molar) must have occurred as secondary changes and cannot be cited as ancestral primate features.

In fact, development of a fourth cusp (hypocone) is considerably more advanced in certain modern tree-shrews than in the earliest *Pelycodus* (see also Gingerich, 1986b).

One interesting feature of Eocene 'lemuroids' is that they have fused dentaries in the lower jaw (**symphyseal fusion**), paralleling that present in the larger subfossil lemurs of Madagascar and in all living simian primates. The presence of symphyseal fusion in adapids and in simians has sometimes been cited as indicating a specific phylogenetic link between these two primate groups (e.g. see Gidley, 1923; Gingerich and Schoeninger, 1977). However, it is important to note that symphyseal fusion was generally lacking in *Pelycodus*, *Smilodectes* and several other genera (Beecher, 1979b, 1983) and that fusion of the two dentaries in the lower jaw among adapids has probably occurred in separate lineages (Gingerich, 1980b). Further, in contrast to simians, the North American adapids (notharctines) show symphyseal fusion late in life, if at all, and it is only in adapines (e.g. *Adapis*) that fusion early in life occurs. Early fusion is also found in the Miocene *Sivaladapis* (Gingerich and Sahni, 1984). Thus, simians could be linked to adapids by symphyseal fusion only if the simian primates diverged from a particular adapid lineage in which symphyseal fusion had already occurred. Conversely, however, most known adapid species are too specialized to have given rise directly to any modern prosimians, as the former have symphyseal fusion and the latter do not. Only a few adapids such as *Pelycodus* have remained close to the hypothetical ancestral condition for modern prosimians in retaining unfused dentaries in the lower jaw.

The known adapids are a relatively diverse group in terms of dental features and this is particularly true of molar morphology. As already noted, many adapids have a prominent fourth cusp on the upper molars, but there is a major distinction between most North American forms (notharctines) and most European forms (adapines) in the manner of formation of that additional cusp (Gregory, 1920a,c). The typical pattern in mammals is for a fourth cusp to arise as an enlargement on the posterointernal cingulum

(Figs 6.18 and 6.19), in which case it is called a **hypocone** (Fig. 6.20). This is the pattern typically found in Adapinae (e.g. *Adapis*). By contrast, Gregory showed that in the Notharctinae studied by him a fourth cusp arises in later genera (*Notharctus* and *Smilodectes*) by budding off from the protocone to produce a **pseudohypocone**. One immediate indication of this alternative pathway to the evolution of a fourth cusp in the upper molars is that in *Smilodectes* and *Notharctus* the posterointernal cingulum is still present (see Fig. 6.25). In addition, Gregory listed other lines of evidence favouring a special origin of the fourth cusp in *Notharctus*. Following his 'premolar analogy theory' (which he successfully used to make the correct inference that the paracone is the primary cusp on the upper molar), one can trace the likely sequence of formation of the fourth cusp along the tooth row, beginning with the posterior premolars and progressing to the second molar

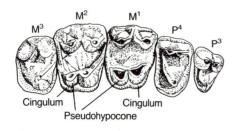

Cingulum   Cingulum
        Pseudohypocone

POSTERIOR   ANTERIOR

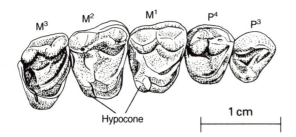

                                        1 cm
        Hypocone |————————————|

Figure 6.25. Comparison of the upper molars of the Eocene 'lemuroids' *Notharctus robustior* (above) and *Adapis magnus* (below). The fourth cusp of the notharctine is a pseudohypocone, formed independently of the cingulum, whereas the fourth cusp of the adapine is a true hypocone. (After Gregory, 1920a.)

(Fig. 6.25). The evolutionary sequence thus inferred is neatly confirmed by the actual sequence of fossil forms: early *Pelycodus*–late *Pelycodus*–early *Notharctus*–late *Notharctus* (Gregory, 1920a).

However, subsequent authors have often questioned the validity of Gregory's distinction between the pseudohypocone and the hypocone. Simpson (1955) described this as 'a distinction without a difference' – a point of view later echoed by Simons (1972), who described the evidence for differential derivation of the fourth cusp as 'overrated'. Butler (1963) also stated that the pseudohypocone and hypocone intergrade, such that use of the term 'pseudohypocone' is pointless. Yet Gregory (1920a,c) provided a detailed demonstration that the pseudohypocone and the hypocone are distinctly different not only in their origins but also in their functional attributes. Whereas the true hypocone of other mammals has an occlusal relationship with the trigonid basin of the subsequent lower molar (Fig. 6.20), the pseudohypocone of *Notharctus* occludes between the entoconid of the opposing lower molar and the residual paraconid (see later) of the subsequent molar. This specific occlusal relationship is doubtless of some importance, as in *Notharctus robustior* the pseudohypocone on M_1 is bigger than the protocone. Gregory linked this special relationship betwen the pseudohypocone and the entoconid in *Notharctus* to an increased emphasis on transverse movement of the lower jaw, which he believed to be lacking in *Adapis*. This difference in jaw movement is indicated by a difference between *Notharctus* and *Adapis* in the shape of the articular condyle (Fig. 6.26). When seen from above, the condyle of *Notharctus* is bean-shaped and the inner end of the condyle is extended backwards and downwards so that a large part of the articular surface can be seen from the rear. By contrast, in *Adapis* the articular surface of the condyle is flatter above and shows an essentially transverse orientation. This suggests that the lower jaw of *Adapis* was primarily adapted for vertical movement, with only a limited capacity for lateral translocation, whereas the lower jaw of *Notharctus* was suited to

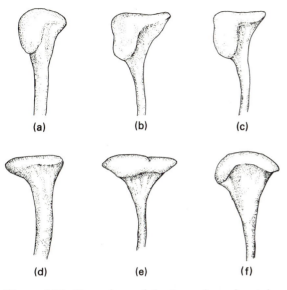

(a) (b) (c)

(d) (e) (f)

Figure 6.26. Rear views of the lower jaw of certain Eocene 'lemuroids' and modern lemurs. Drawings are not to scale. Top row: (a) *Notharctus robustior*; (b) *Megaladapis edwardsi*; (c) *Lepilemur mustelinus*. Bottom row: (d) *Adapis parisiensis*; (e) *Lemur fulvus*; (f) *Propithecus verreauxi*. Note that the lower jaws illustrated in the top row all have a downward and backward extension of the articular surface of the condyle, whereas those in the bottom row do not. ((a) and (d) after Gregory, 1920a; (b),(c),(e) and (f) after Tattersall and Schwartz, 1974.)

relatively free movement, including pronounced lateral and forward motion. Interestingly, comparable variation in the form of the condyle is sometimes found among modern lemurs (including subfossils). As shown in Fig. 6.26, *Lepilemur* and *Megaladapis* resemble *Notharctus* in having a downward and backward extension of the articular surface of the condyle, whereas *Lemur* and *Propithecus* (for example) resemble *Adapis* in lacking such an extension. As has been noted above, *Lepilemur* and *Megaladapis* have only weakly developed hypocones on their upper molars and possess instead a backward ridge-like extension of the protocone (Fig. 6.27). This ridge may be functionally equivalent to the pseudohypocone of *Notharctus*, interacting in a similar way with the entoconid of the lower molar.

Although Butler (1963) initially doubted the importance of the distinction between the pseudohypocone and the hypocone, he subsequently (1973) radically revised this view as a result of studying molar wear facets in the Notharctinae. He noted that the hypocone of early primates is too small to contact the entoconid, whereas in certain *Notharctus* specimens there is a small wear facet on the anterointernal surface of the pseudohypocone, corresponding to a reciprocal facet on the external face of the entoconid. This confirms Gregory's inference of an occlusal relationship between the pseudohypocone and the entoconid. Butler (1973) further observed that *Pelycodus* has a small facet on the posterior surface of the protocone and this apparently occludes with the anterior surface of the paraconid of the next lower molar. In *Notharctus*, the corresponding facet occurs on the posterior face of the pseudohypocone, rather than on the protocone, thus producing the relationship between the pseudohypocone and the residual paraconid identified by Gregory. Hence, the pseudo-hypocone cannot be regarded as homologous with the hypocone; it is both morphologically and functionally distinct. In fact, Butler suggested that the true hypocone, rather than interacting with the entoconid and paraconid, may have initially served the distinct function of guiding the protoconid of the lower molar along the internal cingulum of the upper molar, after the height of the trigonid had been reduced.

Given the valid distinction between the pseudohypocone and the hypocone in Eocene 'lemuroids', it is reasonable to infer that the common ancestor of the Adapidae must have lacked any marked development of either type of fourth cusp. In other words, the upper molars of the ancestral adapid were doubtless essentially tritubercular with, at the most, very weak development of the posterior slope of the protocone (incipient pseudohypocone) or of the internal corner of the cingulum (incipient hypocone). This condition is, in fact, more or less documented by the earliest species of *Pelycodus* (Gingerich, 1986b) and by other Eocene adapids such as *Protoadapis reticuspidens* and

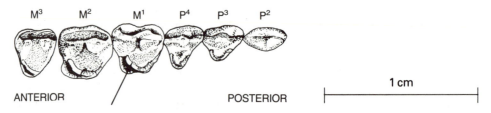

Figure 6.27. Upper left cheek-tooth row of the sportive lemur (*Lepilemur mustelinus*). Note the backward, ridge-like extension of the protocone (arrowed), which is also present in the upper molars of *Megaladapis*. This ridge may be functionally equivalent to a *Nannopithex*-fold (see text for details). (After Gregory, 1920 and Remane, 1960.)

Europolemur koenigswaldi (Franzen, 1987). The Miocene adapid *Sivaladapis* also lacks a hypocone of any kind on its upper molars (Gingerich and Sahni, 1984). Further, parallel development of the fourth cusp on the upper molars in different adapid lineages may be confirmed by the observation that the later Eocene form *Mahgarita stevensi* from North America, unlike *Notharctus* and *Smilodectes*, has a true hypocone developed as a distinct, sharp cusp on the cingulum (see Wilson and Szalay, 1976; Szalay and Delson, 1979). Conversely, Godinot (1984) has reported fragmentary dental evidence of a European adapine (*Cryptadapis*) with a pseudohypocone linked to the protocone. Although it could be argued that *Mahgarita* is an adapine that migrated to North America, and *Cryptadapis* is a notharctine that migrated to Europe, convergent development of the fourth cusp in different lineages seems more likely. Indeed, Franzen (1987) has proposed that a hypocone developed independently within the genus *Europolemur*, being absent in *E. koenigswaldi* but present in *E. klatti*.

The distinction between the pseudohypocone and the true hypocone also applies to Eocene 'tarsioids', although this terminology has not been applied in the same fashion to these early primates of modern aspect. Stehlin (1916) noticed that in the European Eocene 'tarsioid' *Nannopithex* there is a distinct ridge running backwards from the protocone. Hürzeler (1948) later coined the term '*Nannopithex*-fold' for this ridge, and this term has been quite widely used in the literature. A *Nannopithex*-fold is found quite widely among Eocene 'tarsioids' and is present,

for example, in the North American genus *Tetonius* (Fig. 6.28). Simpson (1955) then suggested a connection between the *Nannopithex*-fold and the pseudohypocone. In his discussion of the 'archaic primate' genus *Phenacolemur*, Simpson noted that there are three crests descending from the protocone of the upper molar. One runs to the middle of the anterior border of the tooth, one runs obliquely back towards the metacone and one runs backwards to the inner rear corner. It is the last crest that is identified as the *Nannopithex*-fold and it obviously occurs among 'archaic primates' (e.g. in *Phenacolemur*) as well as among Eocene 'tarsioids'. Simpson noted that the cusp identified as a pseudohypocone arises from the ridge termed the *Nannopithex*-fold, thus identifying a parallel between developments among Eocene Adapidae and developments among Eocene Omomyidae. Recent authors have also recognized the importance of this ridge, but have suggested alternative names, including **postprotocingulum** (Gingerich, 1974) and **postprotocone fold** (Szalay and Delson, 1979).

The *Nannopithex*-fold does not occur universally among Eocene 'tarsioids'. Among North American forms, in addition to *Tetonius*, it is found (along with a small hypocone) in *Hemiacodon*, for example, whereas in *Omomys* there is a distinct true hypocone and no trace of a *Nannopithex*-fold (Szalay and Delson, 1979). The later surviving Oligocene 'tarsioid' *Rooneyia* has a well-developed hypocone that appears to be a true hypocone rather than a derivative from the posterior slope of the protocone. Among European forms, the fold is, of course,

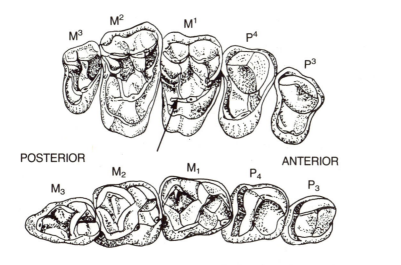

Figure 6.28. Cheek teeth of the North American Eocene 'tarsioid' *Tetonius homunculus*. Above: last two premolars and the three molars from the upper dentition. Below: last two premolars and the three molars from the lower dentition. Note the distinct *Nannopithex*-fold (arrowed) on the first and second upper molars. (After Szalay and Delson, 1979.)

prominent in *Nannopithex*, whereas *Microchoerus*, *Pseudoloris* and *Necrolemur* all have a well-developed hypocone that is essentially distinct from the protocone. The distribution of the *Nannopithex*-fold and of true hypocones among early 'tarsioids' thus suggests that both of these features have developed in parallel in several lineages and were, at most, only weakly developed in the ancestral 'tarsioid'. This inference is supported by the fact that the modern *Tarsius* (Fig. 6.19) lacks any real development of either the *Nannopithex*-fold or the hypocone.

It is important to note that the paraconid cusp on the lower molars, which has been essentially suppressed in all living primates except *Tarsius*, is still present in many Eocene primates. It is uniformly present in Eocene 'tarsioids' (Omomyidae), regardless of whether a well-developed hypocone is present. (Unfortunately, *Rooneyia* is known only from the single skull illustrated in Fig. 2.17, so it cannot yet be stated whether the paraconid was still present in this Oligocene genus.) In fact, minor variations in the disposition of the paraconid provide one of the main bases for taxonomic distinctions among Eocene 'tarsioids' (Simons and Bown, 1985).

Among Eocene 'lemuroids' (Adapidae), the paraconid is still present in earlier forms (e.g. *Pelycodus*), but it generally tends to disappear in later forms both in North America and in Europe, although a paraconid is reportedly still present in European *Pronycticebus*.

The manner of disappearance of the paraconid differs according to whether a true hypocone develops on the upper molar (e.g. *Adapis*) or the fourth cusp is a pseudohypocone (e.g. *Notharctus*). In the lineage leading from *Pelycodus* to *Notharctus*, the trigonid basin of the lower molar becomes increasingly compressed in an anteroposterior direction and this eventually leads to virtual loss of the paraconid (Gregory, 1920a). A similar development would seem to account for the loss of the paraconid in *Smilodectes*. In European adapids, on the other hand, loss of the paraconid (e.g. in *Adapis* species) has taken place without prior compression of the trigonid. The difference in the process of paraconid suppression between *Notharctus* and *Adapis* is correlated with the functional distinction between the pseudohypocone and the hypocone (Butler, 1973). As might be expected, therefore, the North

American Eocene 'lemuroid' *Mahgarita* – which has a true hypocone, rather than a pseudo-hypocone – has lost the paraconid without obvious compression of the trigonid basin. All of the evidence taken together clearly indicates that the paraconid was still present in ancestral primates of modern aspect (as is the case in modern *Tarsius*) and that loss of this cusp has occurred in parallel on numerous occasions during primate evolution.

The molars of 'archaic primates'

It is now possible to consider the question of the phylogenetic relationship between 'archaic primates' (i.e. Plesiadapiformes) and the primates of modern aspect. The primary reason for proposing a link between plesadapiforms and primates was the resemblance in dental morphology, notably in the molar teeth. Indeed, all of the 'archaic primate' genera were first documented from teeth alone and postcranial material is extremely scarce. The Plesi-adapiformes have hence been aptly termed 'dental primates' in the literature. It must be said at once that there are striking similarities in molar morphology between many 'archaic primates' and certain fossil primates of modern aspect. This was clearly stated by Simpson (1935a):

> As regards molar pattern, *Plesiadapis* resembles the primitive Notharctinae more closely than any other group . . . The resemblance to *Pelycodus*, most primitive known notharctine, is really amazing and extends to the apparently most insignificant details. The upper molars are of almost identical structure throughout, differing only in details of the cingula and proportions such as may characterize species of one genus. In the lower molars, *Pelycodus* has the paraconids slightly more distinct, but the resemblance is equally striking and includes even such features as the minute grooving of the trigonid face of the metaconid and the exact structure of the complex grooving of the talonid face of the hypoconid and of the whole heel of M_3.

This assessment was repeated by Gingerich (1986a), who regards the similarity in molar morphology between *Plesiadapis* and *Pelycodus* as providing convincing evidence of a phylogenetic relationship between Plesi-adapiformes and primates of modern aspect.

There are undeniable similarities in molar morphology between 'archaic primates' and certain Eocene primates of modern aspect (e.g. notharctines), as can be seen from the cheek teeth of *Palaechthon* (Fig. 6.29). In addition to comparable general features of the molars, such as the low, bulbous form of the cusps, there is a specific resemblance in that a *Nannopithex*-fold is clearly present on the upper molars. This feature is widespread among Palaeocene 'archaic primates', including *Plesiadapis*, and (as noted above) is found in some Eocene 'tarsioids' (e.g. *Tetonius*; Fig. 6.28) as well as in certain Eocene 'lemuroids'. This has led some authors to make the explicit proposal that the presence of a *Nannopithex*-fold on the upper molars was an ancestral primate characteristic, linking 'archaic primates' to primates of modern aspect. Kay and Hiiemae (1974) suggested that *Palenochtha*, a Palaeocene plesiadapiform with a *Nannopithex*-fold, had a condition 'closest to the protoprimate dentition' and Gingerich (1974, 1975a) emphasized the apparent importance of this feature as a defining characteristic of ancestral primates. Szalay and Delson (1979) also state that presence of a *Nannopithex*-fold, leading to development of a pseudohypocone, is 'almost certainly primitive for primates'.

There are, however, several problems involved in the proposal that a well-defined *Nannopithex*-fold was already present in a common ancestral stock that purportedly gave rise to plesiadapiforms and primates of modern aspect. In the first place, this feature is not consistently present among Eocene 'lemuroids' and 'tarsioids' and it is not found in any living primates, with the possible exception of *Lepilemur* (Fig. 6.27). Hence, in order to accept that ancestral primates had a *Nannopithex*-fold, one must also accept that this defining feature was suppressed in several early lineages and in virtually all lineages leading to modern primates.

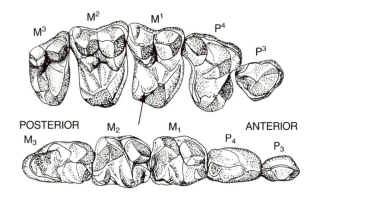

Figure 6.29. Cheek teeth of the Palaeocene 'archaic primate' *Palaechthon alticuspis*. Above: last two premolars and the three molars from the upper dentition. Below: last two premolars and the three molars from the lower dentition. Note the *Nannopithex-fold* (arrowed) on the first and second upper molars. (After Szalay and Delson, 1979.)

There is also a serious problem in that there is no evidence of a *Nannopithex*-fold in the Cretaceous/Palaeocene *Purgatorius*, which is supposedly close to the ancestry of both 'archaic primates' and primates of modern aspect (Szalay and Delson, 1979). *Purgatorius* in fact has a weakly developed true hypocone, located on the internal cingulum as in most living and many fossil primates of modern aspect. Therefore, there are no reliable grounds for postulating that a *Nannopithex*-fold was present either in the ancestral stock of primates of modern aspect or in a hypothetical stock linking them with 'archaic primates'.

Although it is possible that more subtle features of molar morphology, combined with a potential to develop a *Nannopithex*-fold, might provide evidence of a phylogenetic link between plesiadapiforms and primates of modern aspect, the detailed similarity between *Plesiadapis* and *Pelycodus* that was given such prominence by Simpson (1935a) must surely be attributable largely to convergent evolution. In fact, it is a logical conclusion both from a comparison of living primates and from a study of early Tertiary primates of modern aspect that the upper molars of ancestral primates must have been quite primitive, such that later forms could develop either a pseudohypocone or a true hypocone.

As it has not even been established that development of a pseudohypocone occurs only among primates, it is difficult to specify discrete specializations of the ancestral primate molars that would distinguish them from the molars of ancestral placental mammals. In this context, it is interesting to note that Gregory (1920a) specifically pointed out that development of the pseudohypocone in notharctines was paralleled by a very similar development in Eocene 'titanotheres' (now known as brontotheres), early relatives of the perissodactyls. This being the case, one surely cannot rule out the possibility that the dental resemblance between plesiadapiforms and certain early primates of modern aspect has arisen through convergent evolution. For this reason, it is indispensable to consider other evidence (cranial and postcranial) in assessing the validity of the hypothesis that 'archaic primates' shared a specific common ancestry with primates of modern aspect.

Study of molar wear facets in 'archaic primates' has thrown further light on the evolution of the pseudohypocone. Kay and Hiiemae (1974) showed that in *Palenochtha*, which possesses a small hypocone in addition to a *Nannopithex*-fold, there is a wear facet on the latter corresponding to a wear facet on the entoconid of the lower molar, whereas the hypocone does not have an occlusal relationship with the entoconid. In the more specialized *Plesiadapis rex* (Gingerich, 1974), the occlusal relationship between the *Nannopithex*-fold and the entoconid is even more pronounced and a definite pseudohypocone is present. Further, it would seem that in plesiadapiforms, as in typical notharctines, suppression of the paraconid took place through compression of the trigonid basin. (The paraconid is still clearly present in *Palenochtha*, as in *Palaechthon* (Fig. 6.29) and *Purgatorius*, but it is partially or completely suppressed in later plesiadapiforms such as *Plesiadapis* and *Phenacolemur*.)

Dental similarities have also been invoked to support the hypothesis that there is a specific phylogenetic relationship between 'archaic primates' and 'tarsioids'. An early proponent of this interpretation was Gidley (1923) and the hypothesis was elaborated (although subsequently rejected) by Gingerich (1976b). Gingerich proposed that primates should be classified into the two suborders Plesitarsiiformes (plesiadapoids and 'tarsioids') and Simio-lemuriformes (adapids and strepsirhines and simians) to reflect an inferred early dichotomy in their evolution. Although both Gidley and Gingerich relied heavily on features of the anterior dentition in proposing a phylogenetic link between plesiadapiforms and 'tarsioids', Gingerich (1976b) also suggested that there were similarities in molar morphology, notably in the constriction of the trigonid and expansion of the talonid on the lower molars. However, the arguments set out above concerning the possible phylogenetic relationship between 'archaic primates' and primates of modern aspect apply even more forcefully to the proposal that 'tarsioids' are more closely related to plesiadapiforms than they are to other, undoubted primates. Given the overwhelming evidence from other functional systems that primates of modern aspect, including 'tarsioids', form a monophyletic assemblage, it seems far more likely that any resemblances between 'archaic primates' and 'tarsioids' in the anterior dentition and in molar morphology must be the result of convergent evolution. It is, in any case,

inconsistent to observe that the molars of the 'lemuroids' *Pelycodus* and *Plesiadapis* are 'amazingly' similar (Simpson, 1935), yet to suggest that *Plesiadapis* is more closely related to 'tarsioids' (Gingerich, 1976b, 1986a).

Origins of modern primate molars

As it is clear that the ancestral primates must have had molar teeth that were still quite primitive and as certain living primates have retained essentially primitive molars (e.g. *Microcebus* and *Tarsius*), it follows that most or all of the known Eocene primates of modern aspect are unlikely to be directly ancestral to extant primates. For instance, specialization of the molars and fusion of the symphysis of the lower jaw in many later Adapidae (i.e. most forms other than *Pelycodus*) almost certainly excludes them from direct relationship with any modern strepsirhines, although on dental grounds the earliest known *Pelycodus* could fit the bill (Gingerich, 1986b). It is therefore more reasonable to postulate the existence of hypothetical ancestral stocks for the earliest stages of the evolution of primates of modern aspect. Subsequent to the Eocene, however, the molar morphology of fossil primates fits a fairly predictable pattern.

It has already been noted that there is little fossil evidence to document the evolutionary history of the strepsirhine primates (lemurs and lorises). Whereas it is possible that the Adapidae are specifically related to strepsirhines, rather than to haplorhines (e.g. Rosenberger, Strasser and Delson, 1985), it can be argued that adapids share with strepsirhines only a set of characters that would be ancestral for primates of modern aspect overall. Indeed, Gingerich (1976b, 1980b) believes that adapids may be more closely related to simians than to strepsirhines. However, a radically different interpretation has been advanced by Schwartz and Tattersall (1985; see also Tattersall, 1982). These authors have proposed, on the basis of certain shared similarities (interpreted as shared derived features) in the jaws and teeth, that individual species among the European Eocene adapids are specifically related to discrete groups among the Malagasy lemurs. For instance, it is suggested that *Adapis* (including *Leptadapis*) might be specifically related to the indri family (Indriidae), while the fragmentary fossils *Anchomomys* and *Huerzeleris* could be related to the mouse and dwarf lemurs (Cheirogaleidae). Such an interpretation, however, conflicts with other evidence (e.g. symphyseal fusion) indicating that *Adapis* species, at least, are specialized forms with no direct relationship to modern strepsirhines. It should also be noted that post-cranial bones attributed to *Adapis* indicate locomotor adaptations quite unlike those in any modern Malagasy lemurs (see Chapter 10). Thus, it seems far more probable that any similarities between individual adapid species and particular groups of living strepsirhines in terms of dental features provide additional examples of convergent evolution, a phenomenon that is clearly rife in primate dental evolution.

The only other fossils relevant to the evolution of modern strepsirhines are members of the Miocene *Progalago* group (*Komba*, *Mioeuoticus* and *Progalago*). In all of these forms, as in most living members of the loris group (excluding *Perodicticus*), there is a well-developed true hypocone on the upper molars and the paraconid has been lost from the lower molars. Walker (1978), partly on grounds of molar morphology, links *Mioeuoticus* to the modern, slow-moving lorises (Lorisinae), while suggesting that *Komba* and *Progalago* are linked to bushbabies (Galaginae). If this is accepted, it follows that the divergence of Lorisinae from Galaginae must have taken place at some time before 20 Ma ago. However, it should be noted that in *Komba* and *Progalago* the last upper premolar is not molariform, so these Miocene forms lack an important distinguishing feature of modern *Galago* species (probably not due to late retention of the last milk premolar without replacement, as indicated earlier in this chapter).

On the haplorhine side of the primate tree, there is a similar paucity of fossil evidence with respect to the modern *Tarsius*. Although the poorly known Oligocene form *Afrotarsius* may throw some light on the earliest origins of tarsiers, as discussed above, there is no more recent evidence to document the evolutionary history of this small but significant group of living primates. In any case, as has been emphasized above, *Tarsius* has an extremely primitive dentition (Fig. 6.19), so it would be difficult to document the evolutionary history of tarsiers on the basis of molar morphology alone.

The first known simian primates are derived from Oligocene deposits in both the New World and the Old World. In the New World, the earliest recorded form is *Branisella*, a genus from the late Oligocene documented only by a few fragments bearing cheek teeth. The upper molars are rather narrow but nevertheless rectangular in form, with a fairly prominent true hypocone on the internal cingulum (Hoffstetter, 1969; Wolff, 1984). The lower molars attributed to this genus are also 'squared off' and appear to lack a paraconid (see Szalay and Delson, 1979; Wolff, 1984). Other fossil simians from the New World are considerably more recent, dating from the later Oligocene and the Miocene, and they are typically somewhat more advanced in dental terms. It is significant, however, that the new Patagonian fossil platyrrhine *Soriacebus* reported by Fleagle, Powers *et al.* (1987) has a *Nannopithex*-fold and a small hypocone on the upper molars. Like *Branisella*, all other genera possess rectangular upper and lower molars, with a well-developed hypocone on the upper molars and with the paraconid suppressed on the lower molars. In addition, in so far as is known, in all of the late Oligocene/Miocene forms the dentaries were fused. (The relevant part of the lower jaw is not preserved in the *Branisella* material.) The fossil evidence from the New World therefore provides some support for the inference that ancestral simians were already characterized by essentially rectangular upper and lower molars and by symphyseal fusion, while secondary simplification has occurred in the molars of modern marmosets and tamarins, leading (for example) to suppression of the original small hypocone.

In terms of overall jaw morphology and details of molar structure, individual fossil genera from the Miocene of South America show specific resemblances to particular modern forms among the New World Monkeys (Cebidae only). For instance, *Neosaimiri* resembles *Saimiri*, *Stirtonia* resembles *Alouatta*, *Kondous* resembles *Ateles* and *Cebupithecia* resembles *Pithecia*. If these resemblances reflect actual phylogenetic affinities, this suggests that the adaptive radiation of the New World monkeys was well under way by the Miocene. This is dramatically indicated by the discovery of Miocene jaws and teeth attributed to the modern genus *Aotus* (Setoguchi and Rosenberger, 1987). However, none of the fossils discovered to date clearly documents the evolutionary origins of the marmosets and tamarins (Callitrichidae), so there is no direct evidence of molar simplification in these forms. Setoguchi and Rosenberger (1985) have allocated isolated Miocene teeth (incisor, premolar amd molar) to Callitrichidae on size grounds, but the evidence is very tenuous. The isolated upper molar has a true hypocone and thus differs from that of modern callitrichids.

In the Old World, the earliest known simians come from the Oligocene Fayum deposits of Egypt. Largely on dental grounds, Simons (1967) traced the origins of all of the main modern groups of Old World simians (monkeys, apes and hominids) back to the various forms represented in the Fayum deposits. In spite of the variation in morphological details among the Fayum simians, however, they all had rectangular molars (hypocone present in upper molars, paraconid greatly reduced or suppressed in lower molars) and (as far as is known) symphyseal fusion, thus providing further confirmation of the hypothesis that these are ancestral simian characteristics. It has already been noted that, in terms of dental formulae, the Fayum simians fall into two groups: *Apidium* and *Parapithecus* with three upper and lower premolars and *Aegyptopithecus*, *Propliopithecus* and (presumably) *Oligopithecus* with only two. It is now commonly accepted that *Aegyptopithecus*, *Propliopithecus* and (perhaps) *Oligopithecus* are related to the Old World hominoids (apes and humans), as was originally suggested by Simons (see Simons, Andrews and Pilbeam, 1978).

On the other hand, opinions vary about the phylogenetic relationships of *Apidium* and *Parapithecus*. Szalay and Delson (1979) inferred that these two genera represent early offshoots in the Old World simian radiation, diverging before the split between modern Old World monkeys and apes: but Simons (1972) saw *Apidium* and *Parapithecus* as directly related to the origins of the Old World monkeys (Cercopithecoidea). According to Szalay and Delson (1979), *Apidium* and *Parapithecus* have some derived features that separate them from other Old World simians, combined with several primitive features. As an example of the latter, the hypoconulid is still present on the lower molars and it is located in the midline rather than on the outside (buccal side) of the tooth. Hence, these two genera lack any significant development of bilophodonty, which is the most reliable defining feature of Old World monkeys. Thus, it would seem that Szalay and Delson (1979) are right in suggesting that the relatively minor similarities in molar morphology that led Simons to propose a phylogenetic link between *Apidium* + *Parapithecus* and Old World monkeys are best explained as convergent developments.

Incipient bilophodonty is not documented until the early Miocene (*Prohylobates*) and is already fully developed in the early/middle Miocene genus *Victoriapithecus*. Thereafter, bilophodonty is consistently present in all fossil genera allocated to the Old World monkeys, from late Miocene *Mesopithecus* through to the present day. The actual origin of the Old World monkeys is therefore somewhat obscure and the evolutionary origin of bilophodonty in this group remains to be documented in detail.

The Oligocene 'apes' from the Fayum deposits (*Oligopithecus*, *Aegyptopithecus* and *Propliopithecus*) resemble modern apes primarily in primitive features (Harrison, 1982). As in *Apidium* and *Parapithecus*, a hypoconulid cusp is present in the lower molars and it occupies a similar midline position. There is no evidence of the buccal shift of the hypoconulid that produced the distinctive Y5 pattern of the great apes (see later). In fact, *Oligopithecus*, is somewhat unusual in still having a paraconid on the first lower molar, while wear facets on the lower molars indicate that the hypocone was only moderately developed (if at all) on the upper molars. This and other features led Szalay and Delson (1979) and several other authors to suggest that *Oligopithecus* may occupy a very peripheral position in the evolutionary radiation of the Old World simians, with no definite affinities either with monkeys or with apes. In fact, because the ear region is more primitive in Fayum simians (e.g. *Apidium* and *Aegyptopithecus*) than in any modern Old World simians (see Chapter 7), one must consider the possibility that none of the Oligocene fossil forms is specifically related to the modern Cercopithecoidea or Hominoidea.

All modern Old World simians have a dental formula of $I_2^2 C_1^1 P_2^2 M_3^3$ and a well-developed bony ectotympanic tube on the auditory bulla. It is reasonable to assume that these constant features were present in the common ancestral stock that gave rise to extant Old World monkeys and apes. Yet *Apidium* and *Parapithecus* have a (doubtless more primitive) dental formula including three upper and lower premolars and there is no evidence that a bony ectotympanic tube had developed in any of the Fayum simian genera. Thus, although it is common practice to refer to *Aegyptopithecus* and *Propliopithecus* as 'apes', they may fit this term only in so far as they share relatively primitive dental features with modern apes (gibbons, siamang and great apes). Overall, as Andrews (1981) has aptly stated with respect to the Fayum simians: 'There is no good evidence linking these fossil taxa with any of the extant families of the Catarrhini, and it is possible that their occurrence pre-dates the split between the Cercopithecoidea and the Hominoidea.'

Miocene Old World simian fossils regarded as 'apes', all sharing the dental formula of $I_2^2 C_1^1 P_2^2 M_3^3$, can be divided into two groups (Andrews, 1981; Szalay and Delson, 1979). The first group contains *Pliopithecus*, *Dendropithecus* and *Micropithecus*, which are genera of uncertain relationships to modern apes, while the second group contains the dryopithecines (e.g. *Dryopithecus*, *Proconsul*, *Gigantopithecus*, *Ramapithecus* and *Sivapithecus*), which are probably directly linked to the evolutionary radiation that gave rise to the modern great apes and humans. The group containing *Pliopithecus*, *Dendropithecus* and *Micropithecus* has often been seen as having a possible phylogenetic relationship to modern gibbons and the siamang (lesser apes). However, it would seem that the only dental features shared by these Miocene 'apes' and modern lesser apes were probably primitive for the Old World simians in general. In the lower molars, for instance, the hypoconulid of the Miocene 'apes' and modern lesser apes occupies a central position and is not displaced towards the external (buccal) margin. With respect to dental morphology, therefore, the origins of the modern lesser apes remain obscure and it is best to regard forms such as *Pliopithecus* and *Dendropithecus* as generalized Old World simians with no proven link with the ape radiation (see also Harrison, 1982).

As has been described above, the dryopithecines (i.e. members of the subfamily Dryopithecinae) are characterized by the distinctive Y5 pattern on the lower molars (Figs 6.22 and 6.23). In addition to the basic arrangement of furrows, involving the buccal shift of the hypoconulid, this pattern is characterized by other features such as the loss of prominent shearing edges, associated with a lowering and expansion of the five main cusps (Gregory and Hellman, 1927). It should also be noted that formation of the distinctive Y5 pattern requires complete suppression of the paraconid on all lower molars. Once established, the Y5 pattern ('*Dryopithecus*' pattern) apparently remained very conservative. Although there are some minor modifications of the pattern in modern great apes – as with the pronounced wrinkling of the molar surface in the orang-utan – it is only during the latter stages of human evolution that the Y5 arrangement of cusps on the lower molars underwent secondary suppression (Gregory and Hellman, 1927). Thus, in terms of molar morphology, dryopithecines, modern great apes and hominids seem to constitute a cohesive phylogenetic unit.

The late Miocene Old World simian fossil *Oreopithecus* remains an enigma. On dental grounds, many authors (e.g. Gregory 1920e;

Simons, 1960; Szalay and Delson, 1979) have linked *Oreopithecus* to the Old World monkeys (Cercopithecoidea). Indeed, Simons (following Gregory) suggested that *Oreopithecus* may be specifically related to *Apidium* from the Fayum Oligocene (see also Simons 1964, 1972). On the other hand, Hürzeler (1949, 1958) concluded that *Oreopithecus* might be related to the hominid line, especially as *Oreopithecus* and hominids share a number of features related to reduction in size of the canine teeth (see later). Certainly, Hürzeler was convinced that *Oreopithecus* was a hominoid, rather than a cercopithecoid. This diversity of interpretations has doubtless arisen because *Oreopithecus* has remained primitive in many features of its dentition (Butler and Mills, 1959). For example, there is no real evidence of a tendency to develop bilophodonty in *Oreopithecus*, nor is there any sign of a Y5 pattern in the lower molars. The upper molars have a well-defined crista obliqua and there are distinct protoconules and metaconules (all features suppressed in the bilophodont upper molars of Old World monkeys). In the lower molars the hypoconulid is only moderately

developed and occupies a central position. All molars of *Oreopithecus* have sharp, high cusps – probably a primitive retention – and the first lower molar is unusual in that it has a paraconid. *Oreopithecus* does share with Old World monkeys an apparently derived condition in that the molars are rather elongated, but Butler and Mills (1959) reasonably suggest that this is likely to be a result of convergent evolution. As *Oreopithecus* shares with modern hominoids several distinctive features associated with suspensory locomotion, notably loss of the tail and development of curved phalanges (see Chapter 10), the most likely explanation would seem to be that this late Miocene simian is not related to Old World monkeys at all but may be linked to the earliest origins of the modern hominoids (apes and humans), before the development of the characteristic Y5 pattern.

HUMAN DENTITION AND DIET

Along with the enlarged brain and the special locomotor pattern of bipedalism, the unusually modified human dentition represents one of the three major characteristics distinguishing modern humans from the great apes. If modern *Homo sapiens* is compared with the great apes, there are some striking differences (Gregory, 1921; Le Gros Clark, 1967; Pilbeam, 1972):

1. Human jaws are considerably shorter than those of great apes. (This is documented both by reduction of palate length relative to skull length (Fig. 6.11) and by shortening of the length of the cheek-tooth row relative to skull length (Fig. 6.12) in modern *Homo sapiens*, compared to all other primates.) Human molars are much smaller than those of great apes, both absolutely and relatively.

2. The dental arcade in humans has a smoothly rounded parabolic shape, in contrast to the U-shaped dental arcade of great apes (Fig. 6.30; contrast with Fig. 6.22).

3. The canine teeth are small and non-interlocking in *Homo sapiens*. The tips of the canines are no higher than those of the adjacent teeth and they are very unusual in

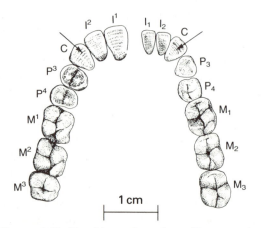

1 cm

Figure 6.30. Dentition of modern *Homo sapiens* (male), showing the typical rounded (parabolic) shape of the dental arcade. Left: upper jaw (right side) seen from below. Right: lower jaw (right side) seen from above. Note the small, spatulate canines (arrowed) and the bicuspid form of the anterior lower premolar. The lower molars show a somewhat modified Y5 pattern. (After Swindler, 1976.)

that they have a spatulate cutting edge, rather than a sharp point. Contrary to the condition typical of the great apes, there is no marked sexual dimorphism in size or shape of the human canines. The permanent canines also erupt earlier, relative to the molars, in humans.

4. In association with the reduced size of the human canines, there is no gap (diastema) between the upper incisors and canine or between the lower canine and the anterior molar. Further, the lower anterior premolar has very little occlusal interaction with the upper canine and it is accordingly bicuspid in humans, whereas it is sectorial (adapted for honing) in great apes.

5. The incisors are relatively narrow in the human dentition; they are also more erect than in great apes.

6. Whereas the lower jaw of the great apes has a transverse buttress (simian shelf) reinforcing the symphysis, this structure is lacking from the human mandible. By contrast, the modern human mandible has a chin.

7. The Y5 pattern of the lower molars is typically somewhat modified in the human dentition. The hypoconulid is usually much less well developed than in great apes and the pattern of furrows between cusps is accordingly different. The third upper molar tends to be three-cusped, rather than four-cusped, in humans.

8. The cusps on human molars are more rounded and more compacted together. This is associated with the fact that human molars bear thick enamel. There is also a marked differential wear gradient in human molars, such that the first molar can be quite worn before the third erupts. The wear gradient could arise either because of heavier wear or because of an increase in the time taken for the molars to erupt in sequence, or perhaps because of both influences combined. There is also marked 'interstitial wear' between human molars, which are crowded together in the jaw and rub together during use.

9. The mandible has a high, vertical ascending ramus in humans, associated with the fact that the jaws are set well back beneath the skull. Shortening of the cheek-tooth row combined with this backward shift, relative to the condyles, increases mechanical efficiency by shortening the load arm of the jaw. The temporal musculature is best developed anteriorly in humans, ensuring a more vertical pull on the lower jaw. On the other hand, the zygomatic arch is relatively weakly developed, suggesting a decreased emphasis on the masseter muscles.

The phylogenetic interpretation of these differences between humans and great apes has been dogged by the tendency, arising from the concept of the *Scala naturae*, to assume that the human condition is necessarily advanced when it differs from the condition typical of great apes. This assumption is undoubtedly unjustified with respect to some of the human dental features listed above. Le Gros Clark (1967), for instance, recognized that great apes are probably specialized in having relatively broad incisors and a simian shelf. Accordingly, the presence of narrow incisors and the absence of a simian shelf in an early hominoid would not represent derived features indicating a specific phylogenetic relationship to the human lineage. With other dental distinctions between great apes and humans, however, the primitive condition for hominoids generally is not known with certainty. It is commonly assumed that ancestral hominoids had large canines and that the anterior lower premolar was sectorial to provide for a honing action against the upper canine. According to this view, small canines and a bicuspid anterior lower premolar in modern humans necessarily represent secondary specializations. However, it is conceivable that ancestral hominoids had relatively small canines and that the anterior lower premolar was still bicuspid – as is the case in *Oreopithecus*. Nevertheless, it is undoubtedly true that the modern human dentition is extremely specialized in many respects and that there must have been a significant functional shift in the early evolution of the hominid lineage. It has been claimed that this shift was related to the development of a more rotary motion of the jaws in hominids, with an increased emphasis on powerful chewing.

Virtually all of the features listed above as distinguishing modern *Homo sapiens* from great apes are also found in fossil forms of *Homo* (including *H. erectus* and *H. habilis*) and in both gracile and robust australopithecines, and this provided the most convincing evidence at the outset for regarding the australopithecines as definite hominids (see Le Gros Clark, 1967). However, miniaturization of the teeth (essentially affecting the cheek teeth, rather than the anterior dentition) has taken place progressively in the evolution of hominids (Pilbeam and Gould, 1974). In australopithecines, the cheek teeth were, if anything, larger relative to body size than in great apes, and reduction in size of the cheek teeth has been confined to the genus *Homo*, taking place progressively in the sequence *Homo habilis–H. erectus–H. sapiens*. This reduction in size of the teeth, which has been associated with decreasing development of the zygomatic arch, is therefore distinct from the establishment of a basic hominid dental pattern in an ancestral stock that gave rise to *Australopithecus* and to *Homo*. Formation of a recognizable chin on the mandible is also a relatively recent acquisition of *Homo sapiens*.

Numerous attempts have been made to trace the hominid lineage further back through the fossil record, and these have been based primarily on dental evidence, because of the relative paucity of cranial and postcranial remains. That is to say that, prior to the australopithecine level, any proposed early relatives of the human lineage are as yet largely 'dental hominids'. Without doubt, the most favoured candidate for early hominid affinities has been the Miocene form *Ramapithecus*, now included in *Sivapithecus* (e.g. Pilbeam, 1972; Simons, 1972; Szalay and Delson, 1979). It was initially maintained that, apart from small differences of detail, *Ramapithecus* shared with *Australopithecus* and *Homo* almost the entire set of distinctions from great apes listed above, excluding features related to miniaturization of the cheek teeth (Simons and Pilbeam, 1965). But

reanalysis of the evidence, incorporating new specimens, demonstrated that the dental arcade of *Ramapithecus* is not, in fact, parabolic and that reduction of the canines was less pronounced than originally thought (Walker and Andrews, 1973). In addition, it has now become apparent that several of the dental features that distinguish *Homo* and *Australopithecus* from modern great apes and from Miocene species of the genera *Dryopithecus* and *Proconsul* occur in a group of hominoids including *Sivapithecus* and *Gigantopithecus*. Indeed, interpretations of the fossil evidence have changed so greatly that, whereas *Sivapithecus* was originally regarded as a typical ape-like dryopithecine contrasting with the apparently hominid-like *Ramapithecus* (Simons and Pilbeam, 1965), it is now widely accepted that the latter may not be generically distinct from *Sivapithecus* (e.g. Kay, 1982).

Andrews (1978b) specifically recognized *Gigantopithecus*, *Ramapithecus* and *Sivapithecus* as constituting a group separate from other dryopithecines and emphasized the fact that the group is characterized by thick enamel on the cheek teeth, which bear low, crowded cusps. Further, specimens attributed to these genera are said to be characterized by relatively small canines, transverse thickening of the lower jaw and the presence of a pronounced wear gradient on the molars. There are, however, some significant differences between members of the group. In particular, the canines of *Gigantopithecus* are flattened by wear of the crowns, so that they seem to have been functionally incorporated into the cheek-tooth row (Simons, 1972; Szalay and Delson, 1979). This special feature, which is not found in *Ramapithecus/Sivapithecus*, contrasts with the modern human condition in which the canines appear to be functionally incorporated into the incisor row, but it is found to some extent in robust australopithecines. Nevertheless, in *Gigantopithecus*, *Ramapithecus* and *Sivapithecus* the incisors are relatively small and tend to be vertically implanted, so all three genera resemble *Australopithecus* and *Homo* in these respects. In addition, *Gigantopithecus*, *Ramapithecus* and *Sivapithecus* all have bicuspid anterior lower premolars. Although the anterior lower premolar is to some degree adapted for honing against the upper canine, this function is limited in comparison to its development in modern great apes. Thus, on dental grounds it might seem reasonable to trace the origins of the human lineage to an ancestral form located somewhere within the adaptive radiation of the group of hominoids characterized by thick-enamelled cheek teeth – *Gigantopithecus*, *Ramapithecus* and *Sivapithecus* (Andrews, 1978b). As has been pointed out above, many authors have traced hominid ancestry back to *Ramapithecus* – for example, on the grounds that this form apparently exhibits an incisor-like conversion of the canines (Szalay and Delson, 1979). There are, however, some authors who have argued that *Gigantopithecus* might be specifically related to human origins (e.g. Eckhardt, 1975; Frayer, 1973).

Discussions of early hominid origins have increasingly emphasized the importance of thick enamel on the cheek teeth as a defining characteristic of hominids. The potential significance of this feature was first clearly recognized by Simons and Pilbeam (1972), who regarded the presence of thin enamel on the cheek teeth as the primitive condition for hominoids generally. This interpretation has been followed by several other authors, such as Kay (1982). As was noted by Pilbeam (1972), the presence of thick enamel on the cheek teeth may allow a species to cope with particularly tough food items or it may allow for a greater resistance of the teeth over an extended lifespan (or perhaps permit both functions combined). Comparative studies of enamel thickness in primates (Molnar and Gantt, 1977; Kay, 1981) have indicated that only relatively few species have thick-enamelled cheek teeth. On the one hand, this suggests that thin enamel was indeed the primitive condition for primates generally. On the other hand, the occurrence of thick enamel in a number of unrelated modern primates, such as *Cebus* (New World capuchin monkeys), *Cercocebus* (Old World mangabeys) and *Homo*, indicates that this feature has been developed convergently in several lineages. As noted above, Kay (1981) has

linked the development of thick-enamelled cheek teeth to special requirements for dealing with hard food items (nut-cracking). Thus, the occurrence of thick enamel in the hominoids *Gigantopithecus*, *Ramapithecus* and *Sivapithecus* could be attributed to similar dietary adaptation, without the need to invoke a specific relationship to the hominid lineage.

In fact, however, a detailed study of enamel thickness in hominoids (L.B. Martin, 1985) has now suggested an entirely different interpretation. In this study, considerable care was devoted to accurate measurement of enamel thickness, since previous studies (e.g. Molnar and Gantt, 1977; Kay, 1981) had relied on measurements of worn teeth or naturally fractured teeth. In addition, the fine structure of the enamel layer in hominoid teeth was investigated in detail. It emerged that *Homo* and *Sivapithecus* (including *Ramapithecus*) are characterized by thick enamel, whereas *Hylobates*, *Pan* and *Gorilla* have thin enamel. The orang-utan (*Pongo*) is intermediate with respect to enamel thickness. However, it also emerged that, whereas thin enamel seems to be a primary condition in *Hylobates*, a substantial outer layer of enamel in great apes (*Pongo*, *Pan* and *Gorilla*) is structurally different, reflecting a phase of slower growth following an initial phase of fast enamel deposition of the kind that produces virtually all of the enamel layer in *Hylobates* and *Homo*. Accordingly, Martin (1985) proposed that the common ancestor of modern great apes and humans already had thick enamel, which has simply been retained in *Homo*, and that in great apes secondary thinning of the enamel has occurred since their divergence from that ancestor (moderate thinning in *Pongo*; pronounced thinning in *Pan* and *Gorilla*). A corollary of this interpretation is that *Gigantopithecus* and *Sivapithecus*/*Ramapithecus* share only a primitive hominoid feature with humans in having thick enamel and need not, therefore, be linked specifically to hominid origins on that count.

Recently discovered specimens of *Sivapithecus* revealing details of facial morphology (Andrews and Tekkaya, 1980; Ward and Pilbeam, 1983) indicate, contrary to previous interpretations based on dental evidence alone, that this genus may be specifically related to the orang-utan, rather than to hominids. Before the description of those specimens, the 'thick-enamelled hominoids' (i.e. *Gigantopithecus* and *Sivapithecus/Ramapithecus*) were known almost exclusively from partial jaws and teeth. As has been pointed out in Chapter 2, a drastic transformation has taken place in assessments of the affinities of *Ramapithecus*. It has quickly passed from protohominid status to inclusion within the genus *Sivapithecus* and hence to alliance with the orang-utan, providing a graphic illustration of the dangers of basing anything other than the most provisional conclusions on 'fragmentary fossils'. The postcranial adaptations of *Sivapithecus/Ramapithecus* and *Gigantopithecus* are still virtually unknown and it is quite possible that discovery of relevant fossil material, which would transform the status of these forms into that of 'substantial fossils', will lead to further radical revision of hypotheses regarding their phylogenetic relationships. For the present, the case for proposing a specific phylogenetic link between *Sivapithecus/Ramapithecus* or *Gigantopithecus* and the hominid lineage (*Australopithecus* and *Homo*) has been considerably weakened. It now seems quite likely that the ancestral stock that gave rise to modern great apes and humans already had some supposedly distinctive hominid characteristics, such as thick-enamelled cheek teeth and limited development of the canines, and that the early origins of the human lineage within the adaptive radiation of the great apes remain obscure (see L.B. Martin, 1985).

The fact remains that modern great apes are generally distinguished from *Australopithecus* and *Homo* by numerous features of the jaws and teeth, even if there are certain exceptions from the list given at the beginning of this section (e.g. orang-utans have enamel of intermediate thickness). Although some hominid features may be primitive retentions from the ancestral stock giving rise to great apes and humans (e.g. thick enamel; relatively small incisors; weak development of the canines and associated honing mechanism), there are others that are

probably derived features (e.g. genuinely parabolic dental arcade; incisor-like conversion of the canines; shortening of the lower jaws associated with more vertical arrangement of the temporal musculature). This suggests that there was a significant dietary shift associated with the divergence of the hominid lineage from the hominoid stock. Several hypotheses have been advanced in the literature to identify the particular dietary conditions involved. In many cases, however, these rest on the assumption that there was a simultaneous shift from forest-living to terrestrial (usually open-country) habits. Szalay and Delson (1979), for instance, refer to 'thick-enamelled "ground" apes' in discussing the relationship between *Ramapithecus* and hominid origins, while the 'savanna hypothesis' has in general had a great influence upon interpretations of early human evolution. Szalay (1972b), among others, has proposed a hunting and scavenging model for hominid origins, which clearly depends on a transition to open country.

An alternative, very imaginative, proposal made by C.J. Jolly (1970b) similarly depends on a shift to terrestrial habits in early hominids and has gained wide currency as the 'seed-eating hypothesis'. Jolly set out from a comparison of *Homo* with modern great apes and noted that in a comparison of gelada baboons (*Theropithecus*) with other baboons (*Papio* species) numerous parallels emerge. Of 48 points of comparison, he found that in 23 the differences between *Theropithecus* and *Papio* paralleled those between *Homo* and great apes. (In fact, in an accompanying paper Jolly (1970a) pointed out that differences between the two subfossil lemurs *Hadropithecus* and *Archaeolemur* provided a third example of such parallelism.) Many of the shared distinctions of *Homo* and *Theropithecus* were linked to terrestrial, open-country adaptation; but a significant number were also connected with adaptations of the teeth and jaws, including the following:

1. Origins of the temporal muscle set forward on the skull.
2. Ascending ramus of the lower jaw oriented vertically.
3. Lower jaw robust in the region of the molars.

4. Premaxilla reduced.
5. Dental arcade narrows anteriorly.
6. Incisors relatively small and narrow.
7. Canine relatively small and erupting early relative to molars.
8. Molar crowns more parallel-sided, with cusps set near the edge.
9. Cheek teeth markedly crowded along the long axis of the jaw.
10. Rapid wear on the molars, producing a marked wear gradient.

Jolly accounted for these shared features of *Theropithecus* and hominids by proposing a parallel switch in their early ancestral forms to 'small-object feeding', with increased emphasis on crushing and grinding. There are also some marked differences between *Theropithecus* and *Homo*, however, notably in that the molar teeth of the gelada baboon have a complex system of crown ridges rather than the low-cusped, thick-enamelled condition found in hominids. Jolly explains this by proposing that, whereas gelada baboons include grass in their diet, early hominids were more specifically adapted for eating solid, hard and spherical objects, particularly cereal grains ('seed-eating' adaptation).

By extension from the parallels between *Theropithecus* and *Homo*, Jolly went on to propose a two-phase model for the evolution of hominids. Phase 1 (the 'seed-eating' phase) involved a transition from forest to grassland, with the adoption of grain-feeding habits and development of the refined hand-and-eye coordination required for seed-collection. Phase 2 (the 'hunting transition' phase) involved a shift from seed-eating to hunting and gathering, with sharply increased emphasis on meat-eating.

Although this model has been widely cited and, to varying degrees, quite widely accepted, there are several problems with it. For instance, drawing on the parallel with gelada baboons, Jolly proposed that during phase 1 hominids engaged in 'sitting-while-foraging'. It is difficult to see how this unusual kind of activity could have led on, in phase 2, to the unique human locomotor pattern of 'bipedal striding'. There is also a problem with Jolly's claim that the canine

teeth are reduced in size in *Theropithecus*. Although it is possible that the canines were reduced in size in larger-bodied, extinct forms of *Theropithecus* (for which the diet is, of course, unknown), there is little evidence that canine reduction has taken place in modern gelada baboons. Jolly (1970b) provided only qualitative statements in his paper, whereas subsequently published quantitative data indicate that, in females at least, canines are not significantly smaller in *Theropithecus* than in *Papio*. Data provided by Smith (1981) show that, when body size is taken into account through allometric analysis, the upper canine of female *Theropithecus gelada* has a somewhat greater relative height than that of female *Papio anubis* or *Mandrillus leucophaeus*, although its base is relatively more slender than in either of these two species. Finally, as Jolly himself noted, there is a major discrepancy in that *Theropithecus* does not have thick enamel on its cheek teeth. Kay's (1981) provisional measurements of enamel thickness based on worn teeth indicate that *Theropithecus* has somewhat thinner enamel than expected (compared to molar length) from an allometric comparison of 37 Old World simian species, whereas *Papio* species have mildly thicker enamel than expected. Thus, the *Theropithecus:Papio* difference is in the opposite direction to that of the human:great ape difference in this case.

Kay (1981), following Kinzey (1974), has suggested that thick enamel in simian primates, along with numerous other features of the teeth and jaws that distinguish hominids from great apes, is associated with dietary concentration on hard nuts and seeds, which are more likely to occur in the trees than on the ground. Although Jolly (1970b) has undoubtedly identified some interesting parallels between *Theropithecus* and hominids and has rightly emphasized the importance of increased grinding and crushing capacity in the evolution of hominid jaws and teeth, it would seem that early human origins are best sought among essentially tree-living apes adapted for feeding on hard nuts and seeds, rather than among ground-living forms. It now seems certain from Martin's study (1985) of enamel thickness that this particular feature was not restricted to the hominid lineage in the adaptive radiation of the hominoids and it is likely that many of the dental features that now distinguish *Homo* from great apes were present in forms that had no direct phylogenetic relationship to human origins. Given the fact that most of the additional peculiarities of the human dentition were already present in *Australopithecus* (see above), the dietary conditions that produced the characteristic human pattern (other than miniaturization of the cheek teeth) are to be sought in the earlier stages of adaptive radiation of forest-living hominoids, rather than in the more recent open-country adaptation of hominids.

Further insights into the origins of human dietary adaptation can be obtained by comparing the digestive tract of modern *Homo sapiens* with tracts of other primates (Martin, Chivers *et al.*, 1985; MacLarnon, Martin *et al.*, 1986; MacLarnon, Chivers and Martin, 1986). Following the procedures explained earlier in this chapter, one can examine the allometric relationships between the major components of the digestive tract and body size in order to infer a general pattern of adaptation of the tract for particular dietary conditions (Figs 6.3–6.6). Of course, there is a problem in attempting to do this for humans, in that the dimensions of the digestive tract may be modified by the individual's diet, such that the pattern shown by any modern human being may reflect individual adjustment to the requirements of an artificial diet, rather than general adaptation for a particular natural range of dietary conditions. However, although the digestive tract is more variable in its dimensions than most other organ systems of the primate body and can indeed be modified to some extent by artificial diets, there are limits to the degree of plasticity involved, such that relatively extreme natural dietary adaptations can still be recognized (Martin, Chivers *et al.*, 1985).

It can be seen from Figs 6.3 to 6.6 that in modern *Homo sapiens* all four major components of the digestive tract (stomach, small intestine, caecum and colon) are smaller than predicted for a 'typical mammal' from the best-fit

lines in bivariate plots. In particular, the caecum (Fig. 6.5) in *Homo sapiens* is markedly smaller than expected (a degree of reduction unlikely to be due to the direct effect of a particular diet on the individuals concerned). As explained earlier in this chapter, it is possible to express departures of individual species from the best-line in each plot as 'quotient' values (gastric, intestinal, caecal and colonic quotients). Using multivariate methods of analysis, it is then possible to combine the various quotient values in order to examine the overall pattern of differentiation of the digestive tract. One suitable procedure is to calculate a matrix of Euclidean distances between pairs of species, taking all four quotient values into account, and then to present the results in a two-dimensional diagram that effectively summarizes those distances (multidimensional scaling). When this is done (Fig. 6.31), it is possible to recognize fairly obvious clusters that reflect broad dietary adaptations (MacLarnon, Martin *et al.*, 1986). Primates and other mammals lacking obvious specializations of the digestive tract and with broadly omnivorous habits are located in the centre of the plot, whereas mammals with notable specialization of the tract lie in three outlying clusters – two containing herbivores and one containing 'carnivores'. One herbivore cluster consists of species with enlarged stomachs in which extensive foregut fermentation occurs and includes the Old World leaf-monkeys (Colobinae). The other herbivore cluster contains species in which caecal fermentation takes place, and includes several primates such as the sportive lemur (*Lepilemur*). The 'carnivore' cluster includes all of the species belonging to the mammalian order Carnivora and also contains *Homo sapiens* and *Cebus capucinus* (the New World capuchin monkey). Members of this cluster are characterized, in particular, by marked reduction of the caecum. Although this cluster consists mainly of 'carnivores', it should be emphasized that many species in the order Carnivora eat a very varied diet and that the common factor, rather than heavy reliance on animal food, may be adaptation for a relatively energy-rich diet associated with a rapid transit of

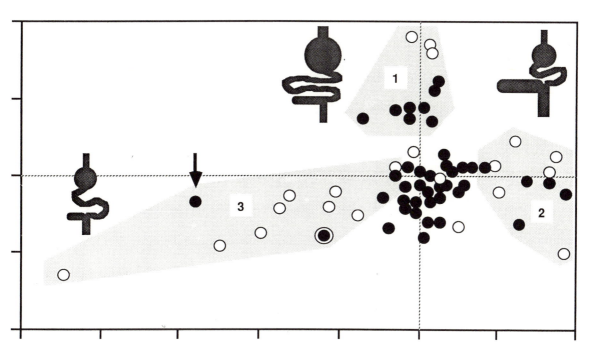

Figure 6.31. Multidimensional scaling plot for the digestive tract dimensions of 61 mammalian species, based on four sets of quotient values (gastric, intestinal, caecal and colonic quotients) derived from the bivariate plots in Figs 6.3–6.6. The quotient values have been kept in logarithmic form, as this preserves the symmetry of data derived from the original plots (MacLarnon, Martin *et al.*, 1986; MacLarnon, Chivers and Martin, 1986b); but this means that 12 mammalian species lacking a caecum are omitted, as there is no logarithmic equivalent of a zero value. Key: ○, Non-primates ($N = 21$); ●, primates ($N = 42$; point for *Homo sapiens* is outlined).

Omnivorous species lacking obvious specialization of the digestive tract are in the centre of the plot. There are three outlying clusters with specialized dietary adaptations: 1, foregut fermenters; 2, caecal fermenters; 3, 'carnivores' (including *Homo* and *Cebus*, the latter indicated by the arrow).

food through the digestive tract (hence the reduction in size of all compartments).

Thus, the overall dimensions of the digestive tract suggest that *Homo* and *Cebus* are unusually adapted compared to other primates, for a relatively high-energy diet. The parallel with *Cebus* is particularly interesting in that this is one of the few primates other than orang-utans and humans to have thick-enamelled cheek teeth (Kay, 1981), and there are other resemblances between *Cebus* and hominids in the morphology of teeth and jaws. The parallel is also interesting because progressive expansion of the human

brain can be linked to the need for a relatively energy-rich diet (see Chapter 8) and *Cebus* in fact has the largest relative brain size of all non-human primates (Table 8.1). Accordingly, detailed comparisons between *Cebus* and *Homo* may throw further light on the dietary factors related to the origin of hominid dental patterns. In any event, the parameters of the modern human digestive tract, relative to body size, confirm the inference made by Jolly (1970b) that a switch to a high-energy diet was one of the key factors in hominid evolution.

Chapter seven

The skull and major sense organs

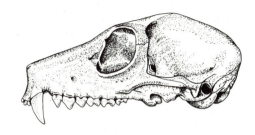

The mammalian skull must meet numerous functional requirements and its overall form thus represents a compromise solution. The jaws and teeth, already discussed in the previous chapter, obviously constitute a major functional component of the skull. Many features, notably in the facial region of the skull, are related to the mechanical properties of the jaws. The skull also houses the major sense organs – the eyes, ears and olfactory apparatus – and provides protection for the body's control centre, the brain (see Chapter 8). Parts of the skull serve additional important functions, such as the channelling of air to and from the lungs through the nasal cavity. An understanding of the skull as a whole therefore ideally requires integrated study of a variety of different functional systems, and skull structure alone requires book-length treatment for satisfactory coverage (e.g. Moore, 1981). Here, however, it is justifiable to concentrate on the major sense organs and their relationship to the skull, as the characters involved are particularly informative with respect to primate evolution.

When considering the evolution of the primate skull it is necessary to take into account not only features of the adult skull but also the **development** (or **ontogeny**) of those features. As will be seen, inclusion of information on developmental aspects does much to clarify certain aspects of the evolution of the primate skull that would otherwise remain obscure. Of particular importance in this respect is the identification of component bones of the skull. There are numerous separately identifiable components that eventually combine to form the bony skull of adult mammals (see Fig. 4.3), but their relationships are often unclear in the adult skull. Component bones of the skull initially have clear **sutures** at their junctions with one another. These sutures, which represent important sites of bone growth during skull development, provide an objective basis for the recognition of boundaries of individual bones of the skull. However, in the skulls of adult mammals it is quite common to find that some sutures, at least, have completely disappeared. Closure of sutures in adult skulls is quite common among primates. Indeed, as Moore (1981) has noted, the sequence of suture closure in simian primates is relatively regular and can be used to infer the approximate age of individuals (although there is a large degree of variation in timing between individuals).

Because of suture closure in adult skulls, it is often difficult to identify the exact bony pattern involved and erroneous conclusions have sometimes been drawn, notably with respect to the eye (orbital) region and to the ear (otic) region of the primate skull (see below). This problem is particularly acute with respect to fossil specimens. It is relatively rare to find immature skulls in the mammalian fossil record, especially for small-bodied species in which the skull is particularly fragile. As a result, interpretation of skull structure for fossil primates is subject to a great deal of uncertainty, although this is often overlooked.

THE EYES AND THE ORBITS

Central nervous processing of sensory information from the eyes will be considered in Chapter 8, so the discussion here will be restricted to the eyes themselves and to their relationship to skull structure.

The eye and vision

The eye has the same basic pattern in all mammals, with only minor variations in detail (Fig. 7.1). Light sensitivity of the eye depends on the inner lining of the optic cup, the **retina**, which contains the light-sensitive photoreceptors. The constitution of the retina is, in fact, somewhat unusual in that the light receptors are directed away from the centre of the eye rather than lining the internal surface. Incoming light must therefore typically pass through a superficial network of blood capillaries, a layer of nerve fibres and a substantial layer of nerve cells before reaching the receptors themselves. The outer tips of the receptors are inserted in a layer of pigment cells, which is overlain by a vascular **choroid**. It is in this latter region of the eye that an important distinction is to be found between nocturnal and diurnal mammals.

Tapetum

In nocturnal mammals, between the choroid and the retina, there is commonly a reflecting zone

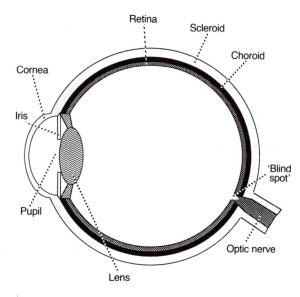

Figure 7.1. Schematic illustration of the basic structure of the mammalian eye. When a reflecting tapetum is present, it is located at the interface between the retina and the choroid.

called the **tapetum lucidum**, which reflects light back through the retina (Tansley, 1965). It seems likely that the tapetum serves to increase the sensitivity of the retina in dim light conditions, although there is probably some reduction in the sharpness of the retinal image brought about by the reflected light.

The tapetum was originally described in some detail by Johnson (1901), who proposed that it had a fibrous structure in ungulates (hoofed animals) and a cellular structure in carnivores and certain primates (see also Pirie, 1959; Pedler, 1963). The validity of this overall distinction has, however, been questioned (Wolin and Massopust, 1970) and there is still considerable uncertainty both about the distribution of the tapetum among mammals generally and about the occurrence of different types of tapetum. Mammals that have a well-developed tapetum are characterized by bright eye reflections ('eyeshine') when observed with a spotlight at night and incidental field sightings by the author showed that such reflections are by no means confined to ungulates, carnivores and primates.

They are characteristic of certain marsupials (e.g. the opossum, *Didelphis*), certain rodents (e.g. the springhaas, *Pedetes*) and at least some fruit-bats (e.g. *Pteropus*). There are significant differences between species in the nature of the tapetum: eye reflections from ungulates, carnivores and springhaas are greenish-blue in colour, whereas those from primates and marsupials are golden yellow or yellow/red. Pedler (1963) and Pirie (1966), have commented on the different kinds of tapetum in mammals, but as yet there has not been a comprehensive survey to define the tapetum clearly in the various mammalian groups. Pedler states that there is a fibrous tapetum in marsupials, rodents and ungulates. Pirie also reports that the tapetum of ungulates is fibrous, whereas that of carnivores and seals has a complex of zinc and cisteine as the active principle. In fruit-bats, a fatty material seems to be involved.

The pattern of distribution of the tapetum among primates is of particular significance. The reflecting structure is found in all prosimian species with the exception of certain diurnal or crepuscular *Lemur* species and of *Tarsius* (Pariente, 1979; see also Wolin and Massopust, 1970). Among diurnal simian primates the tapetum is universally lacking. The nocturnal owl monkey (*Aotus*) is said to have a fibrous tapetum (Rodieck, 1988), but there is no obvious eyeshine (author's observations). Hence, a tapetum producing conspicuous eyeshine characterizes virtually all of the strepsirhine primates (lemurs and lorises), whereas it is absent from all haplorhine primates (tarsiers, monkeys, apes and humans). Why some *Lemur* species lack a tapetum is puzzling because the structure is well developed in the typically diurnal ringtail lemur (*Lemur catta*) and some other clearly diurnal forms (e.g. sifakas, *Propithecus*). But it is striking that all nocturnal species of lemurs and lorises have a tapetum producing conspicuous eyeshine, whereas the two nocturnal haplorhine genera – *Tarsius* and *Aotus* – lack this feature. This suggests some marked separation between nocturnal strepsirhines and nocturnal haplorhines within the evolutionary radiation of primates.

Although it is common practice to interpret the tapetum as an adaptation for vision in dim light conditions, the structure is not confined to nocturnal mammals. Some ungulate species with a tapetum are essentially diurnal in habits and (as noted above) a tapetum is present in some diurnal lemurs such as the ringtail and sifakas. The presence of a tapetum in such diurnal forms may represent a simple retention of a feature originally developed in a nocturnal ancestor. If this is so, however, it is difficult to explain why the tapetum has apparently been lost in some diurnal *Lemur* species (e.g. the black lemur *L. macaco*), but retained in others (e.g. the ring-tail, *L. catta*). It would seem that the structure may still serve some residual function in certain diurnal forms even if it was originally developed as an adaptation for nocturnal life. The role of the tapetum in diurnal lemurs and ungulates is as yet unexplained and it is in fact difficult to understand why the structure should have been retained, as reflection from the tapetum in bright light would be expected to reduce the clarity of the retinal image. Possibly, retention of a tapetum in diurnally active forms is related to a continued need to avoid nocturnal predators, although one would surely expect all *Lemur* species to be subject to broadly similar predation risks.

Additional evidence regarding the evolution of the tapetum has come from studies of the composition of the structure in lemurs and lorises. In the thick-tailed bushbaby (*Galago crassicaudatus*), the tapetum contains flat crystals of riboflavin (vitamin B_2) as the active principle (Pirie, 1959). The riboflavin crystals may act in two ways to increase the sensitivity of the retina in dim light conditions: in addition to reflecting light back through the retina, the crystals are fluorescent and it has been suggested that they thus shift the wavelength of any incident short-wave light towards the optimum for the light receptors. (Tansley (1965) questioned this latter possibility, because short-wave ultraviolet light may be absorbed by the media of the eye, but Afieri, Pariente and Solé (1976) showed that ultraviolet light is perceived by the retina of *Lemur mongoz*, probably through fluorescence of the tapetum.) It has been shown that riboflavin is very probably the active principle in the tapetum of at least some other strepsirhine primates: *Microcebus murinus, Hapalemur griseus, Lemur catta* and *Perodicticus potto* (Alfieri, Pariente and Solé, 1976; A.E. Williams, personal communication). Although it has yet to be shown that riboflavin is the active principle in the tapetum of the remaining lemurs and lorises, indirect evidence (e.g. uniformity of eyeshine coloration) suggests that this is likely to be a universal feature of the tapetum in strepsirhine primates. As far as is known, the massive presence of flat riboflavin crystals in the tapetum is unique to lemurs and lorises, although some riboflavin occurs in the tapetum of the cat and certain fish (Pirie, 1966). This would therefore seem to be a derived feature of the ancestral strepsirhines. There is no evidence that such a tapetum existed at any stage in the evolution of haplorhine primates and the absence of a tapetum in the nocturnal *Tarsius* provides further evidence of a divergence between strepsirhines and haplorhines at an early stage of primate evolution.

Given the facts (1) that a tapetum is found only in nocturnal mammals with relatively large eyes (e.g. it is lacking from microchiropteran bats), (2) that it typically develops postnatally and (3) that its composition seems to differ significantly from one order of mammals to another, it seems likely that the tapetum has evolved separately in different groups of mammals (e.g. ungulates, carnivores, strepsirhine primates, rodents and fruit-bats). However, it remains to be seen whether the tapetum of marsupials or that of rodents differs from that of strepsirhine primates. The eyeshine of marsupials resembles that of lemurs and lorises in its general colour, but the active principle is apparently unknown. Although diurnal tree-shrews (subfamily Tupaiinae) lack a tapetum, it is not known whether the nocturnal pen-tailed tree-shrew, *Ptilocercus lowii* (Ptilocercinae) has one. If it does, then identification of the active principle could provide valuable evidence regarding the evolution of the tapetum in mammals. At present, there are no grounds for proposing a link between tree-shrews and primates in this respect.

Retina

The general morphology of the primate retina itself has been reviewed by Wolin and Massopust (1970). These authors have identified an important universal feature of primates in the pattern of blood vessels lining the surface of the retina inside the eye. It is a characteristic feature of the vertebrate eye that the nerves and blood vessels associated with the retina pass through a single aperture (i.e. the site of exit of the optic nerve), which is necessarily free of receptors and is therefore known as the 'blind spot' (see Fig. 7.1). Blood vessels supplying the retina radiate outwards from this point. In the eyes of all primate species examined, Wolin and Massopust found that the distribution of the retinal blood vessels is asymmetrical, with prominent vessels passing laterally (see also Johnson, 1901). In all cases, the main blood vessels pass around a **central visual area** located in a lateral (temporal) position only a small distance away from the blind spot. The central visual area is thus left free of major superficial blood vessels, permitting greater precision of the retinal image in that area. However, Samorajski, Ordy and Keefe (1966) and Wolin and Massopust (1970) found that the retinal blood vessels in *Tupaia* radiate out from the blind spot in a relatively symmetrical pattern, like the spokes of a wheel. A zone free of blood vessels was located in an extreme lateral position, some distance away from the blind spot. (Wolin and Massopust linked this latter feature to the more lateral position of the eyes in *Tupaia*; see later.) Hence, the arrangement of the retinal blood vessels in tree-shrews seems to be sharply different from the pattern found in primates. In fact, the pattern found for tree-shrews resembles that in some carnivores and rodents (Johnson, 1901). However, it is not clear whether primates are unique among mammals generally in having a relatively avascular central visual area close to the 'blind spot'.

Rohen and Castenholz (1967) examined the concept of 'centralization' of the primate retina in considerable quantitative detail. On the basis of gross morphology of the retina combined with reference to ratios between photoreceptor cells and nerve cells (see later), these authors identified three different levels of centralization among primates. At a gross morphological level, most nocturnal strepsirhines (*Microcebus*, *Cheirogaleus* and *Avahi*) and some diurnal lemurs (*Lemur fulvus* and *Varecia*) show no clear sign of centralization, whereas other nocturnal strepsirhines (*Galago*) and diurnal lemurs (*Lemur catta*, *Propithecus* and *Indri*) have a thickening of the retina in the temporal region. The third group of primates, including *Tarsius* and all diurnal simian primates (New World monkeys, Old World monkeys, apes and humans), have a small pit (**fovea**) in the temporal region of the retina, surrounded by a ring-shaped, raised area (**parafoveal** region). The foveal pit of *Tarsius* can be detected even by simple visual examination (Wolin and Massopust, 1970) and, as reported by Polyak (1957), it is clearly evident in histological sections (see Fig. 7.2). Contrary to statements made by Polyak (1957) and Le Gros Clark (1959), the nocturnal owl monkey *Aotus* has a thickening of the retina in the temporal region but a foveal pit is usually lacking (see Jones, 1965; Wolin and Massopust, 1970). However, there may be

variability in the formation of a fovea (J. Allman, personal communication).

Because a fovea is a universal feature of haplorhine primates apart from *Aotus*, it seems quite likely that a fovea was present in the common ancestor of haplorhines and has, for some reason, been secondarily lost in the evolution of the owl monkey. Although the loss of the fovea from the retina of *Aotus* could have occurred as a result of a shift from diurnal to nocturnal habits (Rohen, 1962; Rohen and Castenholz, 1967), it is difficult to understand why a fovea is still present in the retina of *Tarsius*, which is also nocturnal and may have experienced a similar shift from original diurnal habits (see later). In any event, haplorhine primates seem to be unique among mammals in having a fovea (Tansley, 1965) and this greatly increases the probability that the fovea is a derived feature of the haplorhine primates. It should be noted, however, that a fovea is characteristically found in birds as well as in some reptiles (e.g. lizards and chameleons) and in some teleost fish, presumably as a result of convergent evolution. As diurnal habits are generally typical for all vertebrates with a fovea, the presence of this feature in *Tarsius* is clearly best explained as a retention from an ancestral haplorhine adapted for diurnal activity.

The fovea of diurnal simians and possibly that of *Tarsius* is also associated with the presence of deposits of yellow pigment (carotenoids) in the inner layers of the retina in the region of the central visual area. The retinal area immediately surrounding the fovea is therefore called the **macula lutea** (yellow spot) in these primates. *Aotus* lacks a yellow spot as well as the fovea and is once again distinct from all other simians. The diurnal simians and *Tarsius* (if it does indeed have a yellow spot) are apparently unique among mammals in having a macula lutea and the occurrence of this in combination with a fovea in these haplorhine primates suggests a close functional association between the two features (see Weale, 1966). The avascular central visual area, found in all primates, is generally regarded as a region of enhanced visual acuity in the retina. The additional occurrence of a fovea and macula lutea within this central visual area in haplorhine

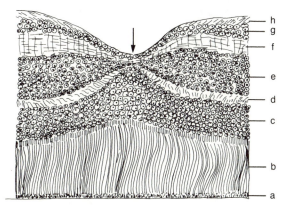

Figure 7.2. Diagram of a histological section through the central area of the retina of *Tarsius*, showing the distinct foveal pit (arrowed). Abbreviations: a, pigment epithelium; b, photoreceptors; c, outer nuclear layer (photoreceptor nuclei); d, outer plexiform layer (synapses + fibres); e, inner nuclear layer (bipolar cell nuclei); f, inner plexiform layer (synapses + fibres); g, ganglion cell layer (ganglion cell nuclei); h, nerve fibres (passing to optic nerve). (After Polyak, 1957.)

primates (excluding *Aotus*) appears to be related to a further enhancement of visual acuity. Indeed, there would seem to be a direct link with diurnal habits. The macula lutea of haplorhine primates in fact owes its yellow colour to the presence of carotenoid xanthophyll pigments, which filter out blue-violet light; such filtering is achieved in different ways in a variety of vertebrates adapted for diurnal colour vision (Tansley, 1965).

The photoreceptors of the retina are, of course, of primary importance for vision. Two main categories of photoreceptors are recognized in the retina of typical vertebrates: **rods** and **cones** (Fig. 7.3). Both kinds of receptor consist of an outer segment, an inner segment, a nucleus, and a synapse terminal located in the outer plexiform layer (Fig. 7.2). Rods are typically relatively thin and the outer segment has a diameter similar to that of the inner segment. Cones, by contrast, are typically quite thick, but the outer and inner segment are commonly tapered and the outer segment is thinner than the inner segment. However, such differences in shape between rods and cones are not consistently observed in different mammalian species (Jacobs, 1981). Where rods and cones occur together in the retina, there is usually little difficulty in distinguishing one from other; but where the retina is predominantly or exclusively composed of one kind of photoreceptor (rods or cones), problems of interpretation can arise. For this reason, it is important to bear in mind a crucial distinction between rods and cones noted by Boycott and Dowling (1969). In mammals with both rods and cones, the synaptic terminals of these two types of photoreceptor are quite different – a cone has a large, conical swelling (**pedicle**) with basal filaments, whereas a rod has a small swelling (**spherule**) lacking any filaments. Because of this difference, it is possible to identify rods and cones more reliably even when the outer and inner segments lack clear distinguishing features. Rods also typically differ from cones in the pattern of the layered membranes in the outer segment (G.H. Jacobs, 1981).

Reliable identification of rods and cones is essential because of the relationship between

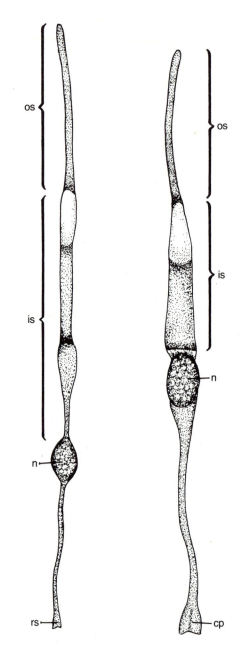

Figure 7.3. Schematic drawings of a typical rod and a typical cone from the human retina, illustrating their distinctive features. The rod (left) has a relatively thin inner segment (is) comparable in width to the outer segment (os) and its synaptic termination is a small swelling called a spherule (rs). The cone (right) has a thicker, somewhat conical inner segment and its synaptic termination is a large swelling called a pedicle (cp). n, Nucleus. (After Rodieck, 1973.)

photoreceptor type and activity pattern (nocturnal versus diurnal). Rods are typically adapted for visual sensitivity in dim light (**scotopic** conditions) and become saturated in bright light, whereas cones are adapted for visual acuity in bright light (**photopic** conditions). This distinction depends partly on the constitution of the photoreceptors themselves and partly on the relationship between the photoreceptors and the nerve fibres leaving the retina to conduct impulses to the brain. Each individual cone typically connects up with a single nerve ganglion cell, but several rods are normally connected to one ganglion cell, thus permitting summation of the responses of a cluster of rods (Tansley, 1965). Given this fundamental distinction between rods and cones, with the former providing sensitivity at the cost of acuity, it is not surprising to find rods predominating in the retina of nocturnal species of mammals and cones predominating in diurnal species. Most (if not all) mammals in fact have both kinds of receptor and it seems likely that a mixed rod and cone retina is generally advantageous. It is commonly stated, however, that in some nocturnal primates, at least, the retina contains only rods (e.g. Detwiler, 1939, 1940; Rohen, 1962; Dartnall, Arden *et al.*, 1965; Jones, 1965; Rohen and Castenholz, 1967), with the implication that cones are not required in species with exclusively nocturnal habits. This conflicts with an early report by Kolmer (1930) that cones do occur in the retina of various nocturnal lemurs and lorises (e.g. *Cheirogaleus* and *Nycticebus*), although they are very sparsely distributed. Kolmer also reported the presence of rudimentary cones in the retina of *Aotus*. Jones (1965) subsequently claimed that the retina of *Aotus* consists exclusively of rods, but Kolmer's original report has now been substantiated by several separate studies (Hamasaki, 1967; Ferraz de Oliveira and Ripps, 1968; Ogden, 1974, 1975), the most recent of which provides electrophysiological evidence that the owl monkey's retina has a duplex (scotopic/photopic) function. *Aotus* also has some capacity for colour vision (G.H. Jacobs, 1977), comparable to that of the diurnal posimian *Lemur catta* (Blakeslee and Jacobs, 1985).

The reason for confusion about the occurrence of cones in nocturnal primates would seem to be that cones do occur in the retina of these primates, but that they are present in such low proportions compared to rods that they are identifiable only in very thin sections (i.e. 5 μm or less in thickness). This interpretation is strengthened by a study showing that thin sections of the retina of the lesser mouse lemur (*Microcebus murinus*) reveal sparsely distributed cones. In thicker sections, the cones are obscured, but their presence in the retina is nevertheless indicated by the occurrence of clearly identifiable cone pedicles alongside rod spherules (A.E. Williams, personal communication). Comparable results have been reported for the potto (*Perodicticus potto*), along with electrophysiological evidence for specific perception of red light (Goffart, Missotten *et al.*, 1976).

Because cones are commonly linked to diurnal colour vision, Rohen (1962) suggested that the scattered cones reported for the retina of various nocturnal primate species by Kolmer (1930) have little functional importance. Although this might seem to be the case with respect to colour vision, it is possible that the sparse cones in these nocturnal primates do in fact serve a special function. In the human eye the cones cease to function in dim light (scotopic conditions), whereas the rods will continue to function, permitting residual black-and-white vision. The transition from photopic vision to scotopic vision, dependent exclusively on rods, is associated with a pronounced shift in spectral sensitivity (Purkinje shift). The point at which colour vision ceases as light intensity decreases at dusk is clearly noticeable. During field observations of the nocturnal lesser bushbaby (*Galago senegalensis*), Bearder and Martin (1980b) noted that the onset of activity in this species was regularly related to the switch between photopic and scotopic vision of the human eye. It is accordingly possible that the presence of a relatively small number of cones in the retina of nocturnal primate species may provide a basis for timing activity rhythms in relation to sunset and, perhaps, to sunrise.

Diurnal simians all have a mixed retina with numerous cones, but the position in diurnal lemurs is not quite so clear. Rohen and Castenholz (1967) did not find clearly identifiable cones in the retina of the brown lemur (*Lemur fulvus*) or variegated lemur (*Varecia variegata*), but Kolmer (1930) noted both rods and cones in the retina of *Lemur catta*, with about one cone for every five rods. Castenholz (1965) also reported cone-like receptors in *Lemur catta* but expressed some doubt about their identification as true cones. As this lemur species has some capacity to distinguish colours (see Wolin and Massopust, 1970; Blakeslee and Jacobs, 1985), it seems quite likely that functional cones are present in the retina. Further, Alfieri, Pariente and Solé (1976) report that *Lemur mongoz* has both rods and cones. Identifiable cones are present in the retina of *Propithecus* and *Indri* but are considerably outnumbered by rods (Rohen and Castenholz, 1967).

In diurnal primate species that clearly have both cones and rods, the cones generally seem to become more numerous, relative to the rods, towards the central area of the retina. Indeed, in humans and possibly in diurnal simian primates generally the central visual area is distinctive in that the fovea is composed exclusively of cones (which are usually more slender than elsewhere). This distinction further underlines the importance of the fovea as a retinal area specially adapted for enhanced visual acuity. Because of the location of the fovea in the temporal area of the retina, such enhancement applies to the area of binocular overlap within the visual field (see later). Given this interpretation, the presence of a fovea in the retina of *Tarsius* appears all the more enigmatic, in that only rods seem to be present in the central visual area, as elsewhere in the retina (Polyak, 1957; Castenholz, 1965; Rohen and Castenholz, 1967).

As is well known, diurnal simian primates have well-developed colour vision and at least some diurnal lemurs such as *Lemur catta* have a rather limited capacity for colour vision (Pariente, 1979; Blakeslee and Jacobs, 1985), so there is a broad correlation between the degree of representation of cones in the retina and a capacity for colour vision.

A further complication is introduced by the composition of the tree-shrew retina. Castenholz (1965) and Ordy and Keefe (1965) reported that the retina of the diurnal *Tupaia* contains only cones. Samorajski, Ordy and Keefe (1966) essentially confirmed this, although some rod-like photoreceptors were identified at the periphery of the retina. Rohen and Castenholz (1967) similarly reported that in the diurnal *Urogale* the photoreceptor layer of the retina is of the 'pure cone' type, although Polyak (1957) stated that in this tree-shrew the retina is composed 'almost exclusively of cones' and that rod nuclei were identifiable at a ratio of 1:5–10 cone nuclei. In Polyak's case, identification of the cones was also supported by recognition of typical cone pedicles in the outer plexiform layer of the retina. As the retina of diurnal tree-shrews is composed predominantly of cones (perhaps exclusively of cones in the central zone), one would expect these animals to have well-developed colour vision (Polyak, 1957). But although *Tupaia* performs moderately well in colour discrimination tests (Tigges, 1963; Shriver and Noback, 1967), this tree-shrew experiences difficulties in discriminating red from yellow. Further, Cai, Kuang *et al.* (1984) have found that *Tupaia* cannot see well in dim light. Thus, the retina of diurnal tree-shrews seems to have no advantage over that of diurnal simians and may actually suffer disadvantages, so the reasons for evolution of a 'pure cone' retina are unclear. It is noteworthy, however, that the visual acuity of tree-shrews is comparable to that of simian primates (Ordy and Samorajski, 1968). 'Pure cone' retinas were also originally reported for certain diurnal squirrel species (Tansley, 1965), once again underlining the close resemblance between tree-shrews and squirrels (see Chapter 5); but, as with tree-shrews, more recent research has demonstrated the presence of rods in the retina of squirrels, although cones are dominant (Jacobs, 1981). However, the advantage of a cone-rich retina has yet to be demonstrated in the case of squirrels as well.

Because research into the visual capacities of tree-shrews has been confined to the diurnal genera *Tupaia* and *Urogale*, the impression is often given that all tree-shrews have a cone-rich retina. But the pen-tailed tree-shrew (*Ptilocercus*) is exclusively nocturnal in habits and it has been stated (Le Gros Clark, 1959) that its retina is of the pure rod type. If this is the case, there is a stark contrast between diurnal and nocturnal tree-shrews in the composition of the retina. Further, if the common ancestor of tree-shrews was nocturnal in habits, colour vision in diurnal tree-shrews must surely have evolved independently of colour vision in diurnal primates. This conclusion is in any case suggested by the finding that *Tupaia* has a dichromatic system of colour vision that is apparently unique among mammals (Jacobs, 1981).

In a study of centralization in the primate retina, Rohen and Castenholz (1967) also provided quantitative data on the relationships between the photoreceptors, the underlying bipolar nerve cells and the ganglion cells. By counting the numbers of nuclei in the outer nuclear layer (receptor nuclei), in the inner nuclear layer (bipolar cell nuclei) and in the ganglion layer (ganglion cell nuclei) (Fig. 7.2), these authors were able to make very general inferences about the degree of summation in central and peripheral parts of the retina. Clearly, such inferences are very approximate as (for example) cross-connections occur among the neurones of the retina and as simple counting of receptor nuclei does not permit the separation of rods from cones. Nevertheless, some useful information can be extracted from counts of nuclei. Rohen and Castenholz found that in all of the primate species investigated there was a noticeable decline in the degree of summation from the periphery of the retina to the central visual area in the temporal zone. In other words, the triple ratio of receptor nuclei to bipolar nuclei to ganglion nuclei (r:b:g ratio) was higher at the periphery than in the visual centre of the retina. This gradation from the periphery to the centre was least marked in lemurs lacking retinal thickening in the central visual area (e.g.

Microcebus and *Cheirogaleus*) and most marked in diurnal simians, which (as noted above) have a fovea. An intermediate level of gradation was found in prosimians (e.g. *Galago* and *Indri*) with retinal thickening in the central visual area and in *Tarsius*, which has a fovea but is nocturnal in habits. Prosimians generally had higher r:b:g ratios than diurnal simians, reflecting the predominance of rods in the former and substantial predominance of cones in the latter.

Since rods are generally thinner than cones, more of them can be packed into a given unit area of the retina. As a result, the outer nuclear layer (composed of the receptor nuclei) is typically quite thick in a rod-rich retina. However, the rod-rich retina typically has a greater degree of summation, so the inner nuclear layer (composed of the bipolar cell nuclei) and – particularly – the ganglion cell layer are characteristically thin. By contrast, the cone-rich retina typically has a lower density of receptors. Figures provided by Rohen and Castenholz (1967) show that diurnal simians have only half the number of receptors per unit area of retina that is characteristic of most prosimians and of *Aotus*. (Among the prosimians, *Lemur catta* was the only exception in having a receptor density as low as that found in diurnal simians.) Accordingly, the outer nuclear layer is relatively thin in the retina of diurnal simians, although it usually increases markedly at the fovea, where the receptors are packed together more tightly. On the other hand, the inner nuclear layer and the ganglion cell layer are typically rather thick in diurnal simians, reflecting their lower r:b:g ratios. Low r:b:g ratios, particularly at the fovea, provide a direct neural basis for enhanced visual acuity.

Once again, tree-shrews are quite distinctive. In the first place, the density of receptors in the retina of diurnal tree-shrews (*Tupaia* and *Urogale*) is even lower than in diurnal simians, by around 50%. Second, the figures given by Rohen and Castenholz (1967) for numbers of nuclei in the retina of *Tupaia* and *Urogale* show that there is very little summation of photoreceptors. In fact, in the temporal area of the retina they found the reverse of the usual situation, with two ganglion nuclei for every receptor nucleus. In all areas of the retina, the numbers of bipolar cell nuclei exceed the numbers of receptor nuclei, indicating diffusion, rather than summation, of receptor response. Finally, Rohen and Castenholz found relatively little difference between the periphery and the centre of the retina of the diurnal tree-shrews. All of these observations taken together indicate that the 'pure cone' retina of diurnal tree-shrews is organized on a totally different basis from the mixed retina of diurnal primates. It has been aptly said that, in neurological terms, the entire retina of a diurnal tree-shrew seems to be organized like the central visual area of a diurnal primate retina (Samorajski, Ordy and Keefe, 1966); but the exact significance of the unusual retinal organization of tree-shrews remains to be determined. Certainly, there is little justification for suggesting any phylogenetic link between tree-shrews and primates on the basis of retinal structure.

Evolution of the composition of the retinal photoreceptors in primates must obviously be examined within the context of the evolution of the retina in vertebrates generally. As has been pointed out by Tansley (1965), ancestral vertebrates were most probably diurnal in habit and it seems likely that (if anything) the cone would have evolved before the rod (Kermack and Kermack, 1984). It is highly probable that stem reptilians were characterized by a retina incorporating cones, whether or not rods had appeared by that stage. Rods may have appeared as a single adaptation early in vertebrate evolution, or they may have emerged convergently in several different lines of vertebrates. Whatever the case may be, there is convincing evidence that mammals were essentially nocturnal for a long period of their evolution, perhaps as much as the first 70% thereof (see Chapter 4). Hence, the eye of early mammals was undoubtedly adapted for nocturnal vision, with a predominance of rods. The most parsimonious interpretation is that nocturnal lemurs and lorises have retained this condition to the present day and that those

primates that are now diurnal in habits (some lemurs and all the simian primates except *Aotus*) have secondarily reverted towards the ancestral vertebrate condition, with a re-emphasis on cones in the retina. The fact that sparsely distributed cones seem to be present in the retina of nocturnal primates generally supports this interpretation, in that a secondary return to a diurnal retina could take place through a shift in the relative frequencies of rods and cones, rather than through the redevelopment of cones *de novo*. It will be interesting, therefore, to discover whether various nocturnal mammals other than primates and the domestic cat (Jacobs, 1981) – for example, the pen-tailed tree-shrew *Ptilocercus* and nocturnal rodents – retain small numbers of cones in the retina. Otherwise, the cones of diurnal primates are unlikely to be homologous with those of other diurnal mammals (e.g. diurnal tree-shrews). Indeed, the cones of diurnal lemurs and simian primates can be homologous only if the (presumably nocturnal) ancestral primates had at least some cones.

Tarsius and *Aotus* occupy a special place among primates with respect to the development of nocturnal habits. The presence of a fovea and (apparently) a macula lutea in *Tarsius* clearly suggests secondary derivation from a diurnal haplorhine stock, as these two features are unique to haplorhine primates among mammals and probably represent shared derived characters. For *Aotus*, the evidence is less clear, because earlier reports of a fovea and macula lutea in this nocturnal simian have been contradicted by subsequent studies. However, *Aotus* resembles *Tarsius* in numerous other features suggestive of a secondary readaptation for nocturnal activity (e.g. in lacking a reflective tapetum and in features of the orbit discussed below), so it is still reasonable to suggest that *Aotus* is secondarily nocturnal as well. In fact, *Aotus* apparently has trichromatic colour vision, as is typical for simian primates, although its capacity for such vision is severely limited (Jacobs, 1981). Overall, therefore, it is possible to identify tentatively the following stages in the evolution of the primate retina:

1. Ancestral vertebrate stage: pure cone retina, or mixed cone-and-rod retina.
2. Early reptilian stage: mixed cone-and-rod retina. (Condition retained in later reptiles and birds.)
3. Ancestral mammalian stage: the emergence of nocturnal habits associated with a marked increase in frequency of rods in the retina. Cones reduced in frequency, but not lost altogether. (Condition retained in nocturnal lemurs and lorises, with visual sensitivity enhanced by the evolution in the ancestral strepsirhine stock of a tapetum incorporating riboflavin crystals.)
4. Ancestral haplorhine stage: a shift back to a cone-dominated retina with the emergence of diurnal habits. (Condition retained in diurnal simians.)
5. Emergence of trichromatic colour vision among simian primates.
6. Secondary return in *Tarsius* and *Aotus* to a rod-dominated retina with readaptation to nocturnal life.

Added to this, the retina of ancestral primates probably showed some degree of development of a central visual area, at least with respect to relative absence of superficial blood vessels.

The orbit

The bony orbit that houses the eye in primates is of particular interest for two reasons. In the first place, the structural composition of the orbit provides useful information for the reconstruction of phylogenetic relationships. Second, study of relationships between the eye and the orbit in living primates can enable us to make inferences regarding functional adaptations in fossil primates, for which only the bony structure of the orbit is documented. Attention has so far been focused particularly on the external margin of the orbit, on the structure of the internal (medial) wall, and on the size and orientation of the orbits with respect to the skull.

Postorbital bar

One significant structural feature of the eye region of the primate skull concerns the degree to which support and/or protection are provided by the surrounding bone. The typical mammalian condition is for the lower margin of the eye to be supported laterally by the cheek bone, or **zygomatic arch**, which is formed from the jugal anteriorly and from the squamosal posteriorly. The zygomatic arch marks off a cavity (the **orbitotemporal fossa**) on the side of the skull. The eye is located above the front end of the arch and the temporomandibular jaw muscles pass through behind to splay out on the side of the braincase. It seems likely that in primitive mammals, as in many relatively unspecialized modern mammals (e.g. lipotyphlan insectivores, most marsupials, most rodents and most carnivores), the orbit was not isolated from the temporomandibular muscles (Weber, 1927). Instead, there was a wide orbitotemporal fossa and the contents of the orbit were separated from the temporal musculature only by a membranous sheet of tissue. All living primates, however, have a bony **postorbital bar** formed by the union of an ascending process of the jugal with a descending process of the frontal. This postorbital bar provides support for the outer margin of the eyeball, although it does not in itself separate the contents of the orbit from the temporal musculature (Fig. 7.4).

In the literature on primate evolution, much has been made of the fact that living tree-shrews, in common with living primates, have a postorbital bar, whereas this structure is lacking from lipotyphlan insectivores and elephant-shrews. Possession of a postorbital bar was one of the features cited in Mivart's oft-quoted definition of primates (1873; see Chapter 5) and several authors have interpreted this as a shared, derived feature of tree-shrews and primates (e.g. Napier and Napier, 1967; Le Gros Clark, 1959). However, as was noted by Wood Jones (1929), possession of a postorbital bar is not restricted to tree-shrews and primates among living mammals. Wood Jones specifically discussed the

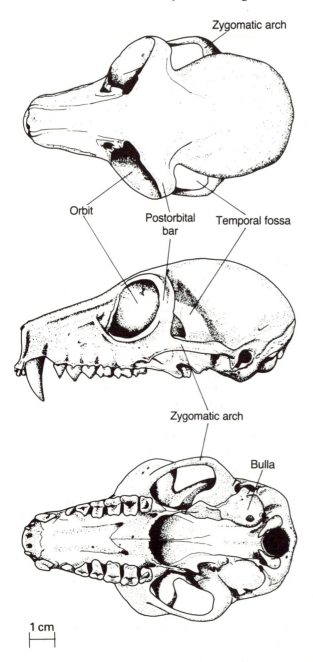

Zygomatic arch

Orbit

Postorbital
bar

Temporal fossa

Zygomatic arch

Bulla

1 cm

Figure 7.4. Dorsal, lateral and ventral views of the skull of a prosimian primate – the variegated lemur (*Varecia variegata*). Note the relatively forward-facing eyes, with a lateral postorbital bar. In living strepsirhine primates (lemurs and lorises), the orbital cavity is confluent with the temporal fossa, whereas in living haplorhine primates (tarsiers and simians) there is a postorbital plate marking off the rear of the orbit.

separate modes of formation of the postorbital bar in the horse (a perissodactyl) and in the cow (an artiodactyl) and this structure is a widespread, although not universal, feature of hoofed mammals (ungulates). In fact, the pattern of distribution of the postorbital bar among modern ungulates, when viewed in combination with fossil evidence, clearly suggests that this structure has emerged as a convergent development in numerous separate lineages. The gradual development of a postorbital bar is, for instance, documented by the well-known fossil record of the evolution of the horse, extending from Eocene forms such as *Hyracotherium* (which lacked a postorbital bar) to the modern *Equus* (which has a well-developed postorbital bar).

Complete postorbital bars are also found in representatives of the placental orders Hyracoidea (*Dendrohyrax* and at least one fossil form), Carnivora (species of the subfamily Herpestinae), Edentata (in the fossil glyptodont *Panochthus   tuberculatus*) and Sirenia (*Trichechus senegalensis*), as well as occurring in certain living and fossil marsupials (*Myrmecobius* and *Thylacosmilus*). Prominent processes of the frontal and jugal bones are found in a variety of other mammals, including some rodents (e.g. arboreal squirrels), some elephant-shrews and the 'flying lemur' (*Cynocephalus*). As all mammals have at least a postorbital ligament, linking the zygomatic arch to the frontal bone and thus demarcating the orbit from the temporal musculature, it is easy to imagine that multiple convergence could have occurred in the evolution of the postorbital bar through simple ossification of the postorbital ligament. Accordingly, the presence of a postorbital bar in tree-shrews does not in itself provide convincing evidence of a phylogenetic relationship with primates. This was, in fact, the conclusion reached by Gregory (1913), who stated: 'It seems quite likely that many of the lemur-like features of *Tupaia*, such as the large brain-case, ringed orbits . . . may have developed independently after the separation of the Primates took place.'

Interpretation of the significance of the postorbital bar in primate evolution really requires some understanding of the functional role of this structure. It has often been suggested that development of a postorbital bar is in some way associated with arboreal life, perhaps because of a requirement for protection of the outer margin of the eye (see Prince, 1953). This would possibly account for the development of the postorbital bar in primates and in tree-shrews, and it is notable that it is the arboreal forms among rodents (e.g. living tree-squirrels) and among hyraxes (i.e. the tree-hyrax, *Dendrohyrax*) that show the greatest degree of development of postorbital processes. However, the terrestrial ungulates with a postorbital bar are clearly an exception in this respect. An alternative suggestion has linked the postorbital bar to the transmission of forces during mastication, but once again the scattered occurrence of the postorbital bar among mammals with a wide variety of dietary adaptations presents an obstacle. Greaves (1985) has specifically proposed that the postorbital bar acts as a strut to resist compressive forces during mastication. Because mammals typically chew on one side at a time (see Chapter 6), the skull is asymmetrically loaded. The anterior part of the skull, containing the upper jaw, tends to twist against the posterior part of the skull, largely consisting of the braincase. There is, therefore, a potential point of weakness at the junction between the two parts of the skull, at the rear of the orbits. In species such as carnivores, in which the temporal musculature is prominent, the twisting effect is not pronounced. However, in species with well-developed masseter and pterygoid musculature (e.g. many ungulates and primates), potential twisting would be pronounced. Greaves suggested that, in ungulates and primates, postorbital bars are orientated in line with a 45° helix representing the line of maximal compressive and shear stresses.

Cartmill (1972) proposed convincingly that the postorbital bar may prevent deformation of the orbital margin during mastication and it is possible that this function supplements that identified by Greaves (1985). In fact, mammals with relatively large eyes that depend on very accurate vision (e.g. leaping arboreal forms and

fast-running terrestrial forms) may well require extra stabilization of the orbital margin, not only during mastication but during general activity. It is also possible that enlargement of the eye may itself necessitate some supplementary reinforcement of the skull, with the postorbital bar serving as a flying buttress. Hence, there may be several factors involved in the formation of the postorbital bar in various mammalian groups. Whatever the particular functional relationships may be in any given case, it is highly likely that development of a postorbital bar in mammals is typically linked both to masticatory forces and to enhancement of the visual sense.

Although possession of a postorbital bar is not restricted to primates, and therefore cannot be used as a reliable indicator of affinities with other groups of mammals (e.g. tree-shrews), the universal presence of this feature among living primates does suggest that their common ancestor had already developed this feature. Indeed, so far as is known all fossil primates of modern aspect have a postorbital bar. In many cases, the bar itself has been preserved more or less intact, as in the Eocene 'lemuroids' *Adapis*, *Notharcus* and *Smilodectes* (Figs 2.6, 2.8 and 2.10) and in the Eocene 'tarsioids' *Tetonius* and *Necrolemur* (Figs 2.15 and 2.16). An intact postorbital bar has also been reported for the Eocene 'lemuroid' *Europolemur* (Franzen, 1987). In other cases, although the skull has not been preserved intact, the original presence of a postorbital bar is indicated by irregular broken surfaces on the frontal and/or jugal. This is, for instance, true of the Eocene 'lemuroid' *Pronycticebus* (Fig. 2.7), of the Oligocene 'tarsioid' *Rooneyia* (Fig. 2.17) and of the Miocene 'lorisoid' *Mioeuoticus* (Fig. 2.11). There is also evidence that the Miocene 'lorisoid' *Nycticeboides* had a postorbital bar (MacPhee and Jacobs, 1986). It is therefore very likely that a postorbital bar was present in the common ancestral stock of primates of modern aspect.

In stark contrast, the 'archaic primates' (Plesiadapiformes) apparently lacked any trace of a postorbital bar. This has long been known for the skull of *Plesiadapis* (Fig. 2.3) and it has been more recently inferred for the skull of

Palaechthon (Kay and Cartmill, 1974, 1977; see Fig. 2.5). It is somewhat surprising that these presumed early relatives of the primates of modern aspect should lack this key feature, because it doubtless to some extent reflects the special part that the visual sense has played throughout the evolution of primates of modern aspect.

In living haplorhine primates (tarsiers and simians), there is a further development in that the postorbital bar is joined to the braincase by a bony **postorbital septum**. As noted by Cartmill (1980), this is a feature found in no other vertebrates and it is undoubtedly a derived condition. The postorbital septum effectively separates the orbit from the temporal musculature, such that in all modern haplorhines the eye is housed in an almost complete bony cup. In both tarsiers and simians, the postorbital septum consists primarily of bony flanges of the frontal, the zygomatic and the alisphenoid. It might therefore seem reasonable to propose that the postorbital plate is a shared derived feature of haplorhines, traceable to their common ancestor. However, several objections have been made to this interpretation, firstly with respect to details of the bony composition of the postorbital septum in living haplorhines and secondly in relation to the fossil evidence. A valuable review of the evidence has been provided by Cartmill (1980).

With respect to living haplorhines, it has been argued that the postorbital septum of *Tarsius* represents a convergent development, independent of that found in simians (e.g. Schwartz, Tattersall and Eldredge, 1978), and it has even been suggested that the septum emerged independently in New World and Old World simians (see Ashley Montagu, 1933; Simons, 1962c, 1972; Rosenberger and Szalay, 1980). The latter proposal is, in fact, a necessary corollary of the hypothesis that these two groups of simians originated separately from 'tarsioid' (omomyid) stocks with only a postorbital bar. In all cases, assessments of the likelihood of convergence have depended on the presence of apertures (notably the inferior orbital fissure) in the septum of various haplorhine species and on

differences in the pattern of bones in the septum. It has often been concluded that differences in the composition of the postorbital septum indicate that the septum was relatively incomplete in the common ancestral condition, with different patterns emerging as the septum became more complete in separate lineages. This argument, however, is not particularly convincing. As Cartmill (1980) has aptly noted, there are differences in the bony composition of the braincase wall among living haplorhine primates, but this does not indicate that the braincase was incompletely ossified in the ancestral haplorhines and in the ancestral simians! In fact, Cartmill presents a convincing case for the presence of an ossified postorbital septum, incorporating flanges of the frontal, zygomatic and alisphenoid, in the common ancestor of simians and possibly in the ancestral stock of living haplorhines.

The fossil evidence presents more of a problem for interpretation of the evolution of the postorbital septum in haplorhines. As far as simians are concerned, there is nothing to conflict with the view that a postorbital septum was present in the common ancestral stock. The earliest known simian skull, that of the Oligocene *Aegyptopithecus*, indicates that a well-developed septum was already present. Although it has been suggested that the late Oligocene New World monkey *Tremacebus* had a very large inferior orbital fissure (Hershkovitz, 1974b), it would appear that the size of this fissure has been exaggerated by breakage (Rose and Fleagle, 1981). All fossil simian skulls from the Miocene and from more recent fossil deposits have a well-formed postorbital septum.

By contrast, it seems very likely that early Tertiary 'tarsioids' lacked a postorbital septum. Simons and Russell (1960) found no evidence of a septum in the Eocene *Necrolemur*, which is the best known of the fossil 'tarsioids'. They did note, however, that the postorbital aperture was smaller than in any of the Eocene 'lemuroids' (Adapidae) or in any of the extant lemurs and lorises. Further, they noted that the adult *Necrolemur* skull resembles that of a juvenile *Tarsius* in having flanges on the postorbital bar without formation of a septum. A postorbital septum was also apparently absent from the skull of the Eocene *Tetonius*. The only other relatively intact fossil 'tarsioid' skull, that of *Rooneyia* from the Oligocene, similarly had a postorbital aperture. Nevertheless, there is a definite postorbital flange of the frontal, covering the gap between the braincase and the upper end of the postorbital bar, that may represent an initial development of postorbital closure (Wilson, 1966).

Overall, the evidence indicates that the known 'tarsioids' from the early Tertiary lacked any substantial development of the postorbital septum. Two reasonable interpretations can be made in the light of this evidence. The first is the common suggestion that the postorbital septum emerged as a convergent development in tarsiers and simians (Rosenberger and Szalay, 1980). A second possibility, however, is that the early Tertiary 'tarsioids' branched off before *Tarsius* and simians diverged from a common ancestor that already had a postorbital septum (Cartmill, 1980). Choice between these two alternative interpretations depends on the strength of the evidence that early Tertiary 'tarsioids' known from cranial remains are specifically related to modern tarsiers (e.g. see Schmid, 1983).

As with the postorbital bar, elucidation of the function of the postorbital septum should throw some light on its evolutionary background. Several different suggestions have been made with respect to possible functions of the septum, and these have been effectively reviewed by Cartmill (1980). After a discussion of the competing hypotheses, Cartmill concludes that the postorbital septum of living haplorhines probably developed to insulate the eye from movements of the temporal musculature. It certainly seems unlikely that the postorbital septum evolved either to support the eye or to protect it from external impact, as both of these functions are adequately served by the postorbital bar alone. There are serious flaws in the arguments put forward by Ehara (1969) and by Cachel (1979) that the postorbital septum developed to provide an increased surface for attachment of the anterior temporal muscle (see

Cartmill, 1980), and it also seems unlikely that the postorbital septum serves to transmit masticatory stresses. On the other hand, if the postorbital bar acts as a flying buttress (as suggested above), it is possible that development of a septum further reinforces this role and that the septum, like the bar, may serve a number of functions.

Internal (medial) wall

Discussions of the evolution of the orbit in primates have also emphasized the structure of the internal (medial) wall of the eye socket. The bony surface of the skull adjacent to the inner side of the eye incorporates a mosaic of bony elements that may include all of the following: frontal, maxilla, lacrimal, orbitosphenoid, palatine, alisphenoid and ethmoid. Le Gros Clark (1959) and Saban (1956/57) placed particular stress on the phylogenetic significance of certain resemblances between tree-shrews and some lemurs in the composition of the medial orbital mosaic. According to Le Gros Clark, the primitive mammalian condition is for the maxilla to form a broad suture with the frontal within the orbit (Fig. 7.5(a)). This pattern, he stated, undergoes modification in one of two ways in association with orbital enlargement: either the palatine passes between the frontal and maxilla to contact the lacrimal (Fig. 7.5(b)), or the ethmoid produces within the orbit a flat plate (os planum) that essentially separates frontal, maxilla, lacrimal and palatine from one another (Fig. 7.5(c)). Tree-shrews and lemurs were said to show pattern 7.5(b) (palatine:lacrimal contact), whereas all other primates were said to show pattern 7.5(c) (exposure of the ethmoid). However, Le Gros Clark did note some exceptions to this general rule. Among the living lemurs, the aye-aye (*Daubentonia*) is distinctive in having pattern 7.5(a) (i.e. broad contact between the frontal and the maxilla), and Le Gros Clark included a footnote indicating that mouse lemurs (*Microcebus*) show pattern 7.5(c), with ethmoid exposure. It was further noted that pattern 7.5(a) (frontal:maxilla contact) was characteristic of Eocene 'lemuroids' (Adapidae),

The skull and major sense organs

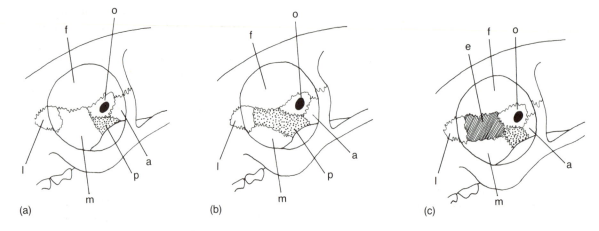

Figure 7.5. Composition of the medial orbital mosaic in primates. Abbreviations: a, alisphenoid; e, ethmoid (hatched); f, frontal; l, lacrimal; m, maxilla; o, orbitosphenoid; p, palatine (stippled).
(a) Condition in some lemurs (e.g. *Daubentonia*, *Archaeolemur* and *Palaeopropithecus*); (b) condition in tree-shrews and some lemurs (e.g. *Lemur* and *Varecia*); (c) condition in most primates (*Cheirogaleus*, *Microcebus*, *Allocebus*, 'lorisoids', *Tarsius* and simians). (Modified from Le Gros Clark, 1959 and Cartmill, 1975.)
Le Gros Clark interpreted condition (a) as primitive for mammals, but it is more likely that condition (b) was the primitive pattern, and that condition (c) was the ancestral primate pattern.

and Le Gros Clark interpreted the presence of frontal:maxilla contact in the orbital mosaic as a primitive retention in adapids and in *Daubentonia*. The logical conclusion from this, as pointed out by Le Gros Clark (1959), is that lemurs and lorises must be derived from a common ancestor in which the inferred primitive condition of frontal:maxilla contact was still present. However, there is an equally logical conclusion that the presence of palatine:lacrimal contact in tree-shrews and some lemurs must accordingly be a convergent development and that the orbital mosaic (as interpreted by Le Gros Clark) can hence provide no direct evidence for a shared common ancestry of tree-shrews and lemurs!

The constitution of the medical orbital mosaic in primates has been reviewed in detail by Cartmill (1975, 1978). Two main points have emerged quite clearly. First, in all of the major groups of living primates except Malagasy lemurs the ethmoid is exposed in the orbit (Fig. 7.5(c)). It is only among the lemurs that there is

substantial variation in the orbital mosaic. (Indeed, Cartmill observed that among the lemurs there is frequently variation in the mosaic within species.) Second, ethmoid exposure in the orbit is actually quite common even in lemurs. Cartmill (1978) reported that ethmoid exposure is typically present in *Microcebus* and *Allocebus* (but apparently not in *Cheirogaleus* or *Phaner*), predominantly present in *Lepilemur* (about 60% of cases), and occasionally present in *Hapalemur*, *Avahi* and *Indri*. (Cartmill and Gingerich (1978) reported separately that apparent ethmoid exposure is present in 35% of adult *Indri* skulls in which the sutures are still recognizable.) According to Cartmill, only in *Lemur*, *Varecia*, *Propithecus* and *Daubentonia* is ethmoid exposure in the orbit typically absent, although he found a very small ethmoid component in the orbit of one specimen of *Lemur mongoz*. Frontal:maxilla contact, in addition to being present in *Daubentonia*, also occurs as a variant in *Phaner*, *Hapalemur* and *Lepilemur* as well as occurring in the subfossil genera *Archaeolemur* and *Palaeopropithecus*. Palatine:lacrimal contact is the typical condition for *Lemur* and *Varecia* and occurs as a variant in *Phaner, Hapalemur* and *Lepilemur*. Wood Jones (1917) was able to show quite clearly that in newborn *Lemur* the palatine contacts the lacrimal. There is therefore no 'typical' condition of the orbital mosaic for lemurs; but the most widespread pattern would seem to be orbital exposure of the ethmoid, which (according to Cartmill, 1978) occurs in three of the four main groups of lemurs.

Inference of the likely ancestral condition for placental mammals constitutes an essential step in reconstructing the evolution of the orbital mosaic in primates. Le Gros Clark (1959), as noted above, inferred that broad contact between frontal and maxilla (Fig. 7.5(a)) was the ancestral placental condition. However, he seems to have based this inference primarily on the observation that some modern lipotyphlan insectivores (e.g. hedgehogs and shrews) show this condition. Instead, it now seems more likely that palatine:lacrimal contact (Fig. 7.5(b)) was the ancestral mammalian pattern as originally suggested by Muller (1935). This pattern was present in the early fossil mammal *Morganucodon* and also in the fossil 'insectivore' family Lepictidae (Novacek, 1986a). Among living mammals, palatine:lacrimal contact is the commonest pattern in marsupials and it also occurs in some carnivores and ungulates as well as being present in both tree-shrews and elephant-shrews (see Cartmill, 1975). Further, as there is partial or complete loss of the lacrimal and of the zygomatic arches in most modern lipotyphlan insectivores, some caution is needed in suggesting that they have retained a primitive pattern in the medial orbital mosaic (i.e. frontal:maxilla contact). Therefore, it may be the case that modern tree-shrews have simply retained a primitive pattern of palatine:lacrimal contact in the orbital mosaic from the ancestral placental stock (Martin, 1967).

The distribution of orbital mosaic patterns among living primates suggests that exposure of the ethmoid was present in their common ancestral stock. This is, after all, the pattern found exclusively in five out of the six major

groups of living primates (lorises; tarsiers; New World monkeys; Old World monkeys; apes and humans) and it is also found in several lemurs, either as the typical condition or as a variant. The apparent occurrence of occasional ethmoid exposure in certain lemur genera (e.g. *Indri*) might therefore be interpreted as a partial retention of the primitive condition. In fact, occurrence of an os planum in the medial orbital mosaic is a very rare phenomenon among placental mammals. Apart from the presence of this feature in most living primates, it has been reliably reported elsewhere only for the domestic cat (perhaps as a variable trait) and it may also be present in 'flying lemurs' (M. Cartmill, personal communication). Some authors have claimed that an os planum is present in the orbits of other mammals, but closer examination has revealed this to be an artifact. Carlsson (1922) reported an os planum in tree-shrews and Saban (1953, 1956/57) referred to its presence in the hedgehog (*Erinaceus*) and in elephant-shrews (Macroscelididae). However, Martin (1967) showed that the apparently separated bony element in the orbital mosaic of tree-shrews is in fact the anterior portion of the orbital wing of the palatine. The sphenopalatine foramen passes through the orbital wing of the palatine and the anterior part of that wing forms a suture with the posterior part of the palatine above the foramen. (A similar error was made by Forsyth Major (1901) in his identification of an 'os planum' in certain Malagasy lemurs (e.g. *Lemur*); Kollmann (1925) and Cartmill (1975) showed that there is a suture formed between anterior and posterior portions of the palatine of *Lemur*, just as in tree-shrews and elephant-shrews.) Thus, a true os planum is almost confined to primates and is therefore most likely to be a derived feature, if not a shared derived feature, of the primates that have it.

At first sight, the early fossil evidence of primate evolution seems to present a totally different picture of the medial wall of the orbit from that found among living primates. It has been widely accepted that Eocene primates of modern aspect ('lemuroids' and 'tarsioids') had frontal:maxilla contact in the orbit (Fig. 7.5(a)), and this would seem to suggest that this was the primitive pattern for primates (Simons and Russell, 1960; Le Gros Clark, 1959; Cartmill, 1975; see also Gingerich and Martin, 1981). There is, however, a specific difficulty involved in identification of the medial orbital pattern in fossil forms in that fusion of the relevant sutures is known to occur in many living primate species once the adult condition has been attained (Kollmann, 1920). Because such fusion of sutures can occur differentially, it does not necessarily follow that the orbital mosaic of a fossil primate can be directly identified simply because some sutures are identifiable. For instance, Simons and Russell (1960) describe frontal:maxilla contact in the orbit of *Necrolemur* and rule out any possibility of ethmoid exposure. But Cartmill (1971) points out that the medial orbital wall of *Necrolemur* is very similar to that of adult *Microcebus* after closure of the suture between the frontal and the ethmoid, and concludes: 'a distinct os planum may well have been present in juvenile *Necrolemur*'. Further, in one particularly well-preserved skull of *Adapis parisiensis*, Gingerich and Martin (1981) identified a small ethmoid exposure. A similar degree of ethmoid exposure apparently occurs as an occasional feature in the orbit of adult *Indri* (Cartmill and Gingerich, 1978). It is possible that such ethmoid exposure is a minor variant from the more typical pattern of frontal:maxilla contact in both *Adapis* and *Indri*. On the other hand, it is also possible that in Eocene primates of modern aspect there was generally some degree of ethmoid exposure during development of the orbit and that this became obliterated in the fully adult skull. A thorough survey of ontogenetic aspects of the orbital mosaic in living primates (especially lemurs) is hence necessary before the patterns found in fossil skulls can be interpreted with any confidence (but see later).

At present, it is not possible to draw any firm conclusions regarding the presence of a particular medial orbital mosaic in ancestral primates, but it does seem very likely that the palatine:lacrimal contact in tree-shrews and some lemurs cannot be a shared derived feature indicative of specific

common ancestry. Comparison of living primates suggests that ethmoid exposure (Fig. 7.5(c)) may have been the condition in ancestral primates of modern aspect and that both frontal:maxilla contact (Fig. 7.5(a)) and palatine:lacrimal contact represent secondary specializations.

According to Russell (1964), the 'archaic primate' *Plesiadapis* shows frontal:maxilla contact within the orbital mosaic. Assuming that there had been no fusion of sutures and that the pattern was correctly identified, the orbital mosaic of *Plesiadapis* provides no evidence of a specific relationship to primates. Frontal:maxilla contact is common among non-primates and would not reliably indicate an affinity with primates even if it were accepted that this was the ancestral primate condition. For the time being, the nature of the medial orbital mosaic in ancestral primates must remain in some doubt, pending precise information on the ontogenetic processes involved.

Size

Another important aspect of the primate orbit is its overall size. This feature is significant not only because it is generally held that relatively large eyes set the primates apart from other mammals (at least to some extent), but also because nocturnal primates have particularly large eyes. However, as it is also obvious from superficial examination alone that eye size does not increase at the same pace as body size, some kind of allometric analysis is necessary to take account of negative scaling effects. One simple way of approaching the problem is to plot orbit diameter against skull length on logarithmic coordinates (Fig. 7.6). When this is done it becomes apparent that different grades can be recognized with respect to orbit size relative to skull size (see also Kay and Cartmill, 1977). Tree-shrews have small orbits relative to skull length, whereas diurnal lemurs and simians have relatively enlarged orbits and nocturnal primates fairly consistently have the largest orbits of all. When best-fit lines (major axes) are determined for these three groups, the following empirical formulae are

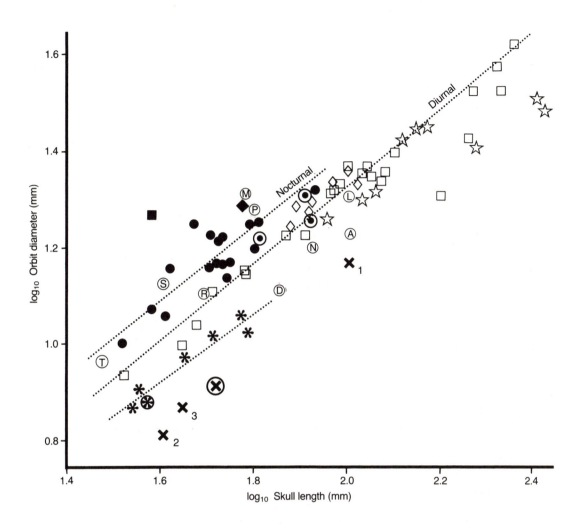

Figure 7.6. Logarithmic plot of orbit diameter against skull length for primates and tree-shrews. The best-fit lines indicated are major axes. Key:

Black symbols = nocturnal primate species: ●, nocturnal lemurs and lorises ($N = 17$); ■, tarsier; ◆, owl-monkey (*Aotus*).

Circled black dots = crepuscular (?) lemurs.

Open symbols = diurnal primate species: ◇, diurnal lemurs ($N = 7$); ☆, subfossil lemurs ($N = 9$); □, diurnal simians ($N = 24$).

Circled letters = fossil primates of modern aspect ($N = 9$): A, *Aegyptopithecus zeuxis*; D, *Adapis parisiensis*; L, *Adapis magnus*; M, *Mioeuoticus* sp.; N, *Notharctus tenebrosus*; P, *Pronycticebus gaudryi*; R, *Rooneyia viejaensis*; S, *Necrolemur antiquus*; T, *Tetonius homunculus*.

Crosses = fossil 'archaic primates' ($N = 3$): 1, *Plesiadapis tricuspidens*; 2, *Palaechthon nacimienti*; 3, *Phenacolemur jepseni*; circled cross = fossil 'insectivore' *Leptictis* (= *Ictops*).

Asterisks = tree-shrews ($N = 7$); circled asterisk, nocturnal *Ptilocercus*.

(Original data for living tree-shrews and primates supplemented with data on relevant fossils from Gingerich and Martin, 1981, and from Kay and Cartmill, 1977.)

obtained for the relationship between orbit diameter (*D*) and skull length (*SL*):

Tree-shrews (*N* = 7):

$$\log_{10} D = 0.73 \log_{10} SL - 0.25 \qquad (r = 0.96)$$

Diurnal lemurs + simians (excluding long-snouted baboons) (*N* = 29):

$$\log_{10} D = 0.80 \log_{10} SL - 0.27 \qquad (r = 0.99)$$

Nocturnal strepsirhines (*N* = 17):

$$\log_{10} D = 0.79 \log_{10} SL - 0.17 \qquad (r = 0.85)$$

As can be seen from Fig. 7.6, the slopes of the best-fit lines are approximately parallel, notably for the diurnal and nocturnal primates, which are represented by relatively large samples. In all cases, the scaling component is less than unity (i.e. 0.73–0.80), in accordance with the negative allometry of eye size relative to skull size. Clearly, there is some degree of scatter about the best-fit lines and this can be portrayed by calculating an **orbital index** for each species relative to the best-fit line for diurnal lemurs and simians. The index value for any species is calculated by dividing the observed orbital diameter of that species by the diameter predicted from the best-fit line for a diurnal lemur or simian of the same body size. The values obtained can then be presented in histogram form (Fig. 7.7). Orbital index values for fossil species plotted in Fig. 7.6 are included in Fig. 7.7.

From these two presentations, it is possible to draw some interesting conclusions. It is notable, first of all, that tree-shrews have markedly smaller orbits than diurnal lemurs or simians, relative to skull size. Further, it is striking that the nocturnal pen-tailed tree-shrew (*Ptilocercus*) does not have larger orbits, relative to skull size, than other tree-shrews. This contrasts with living primates, where nocturnal forms characteristically have relatively larger orbits than diurnal forms. This difference alone suggests that there is some radical difference in visual adaptation between tree-shrews and primates.

Among living primates, an orbital index value of 1.10 provides a fairly clear dividing line

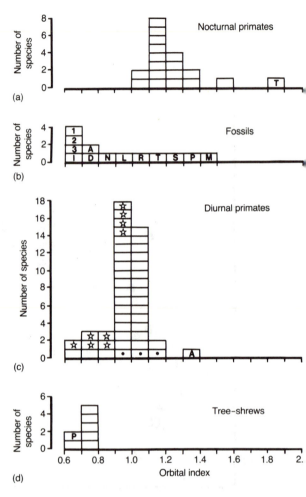

Figure 7.7. Histograms of orbital index values derived from the data presented in Fig. 7.6, based on the best-fit line for diurnal lemurs and simians. (See text for explanation.)

(a) Nocturnal primates (excluding *Aotus*): index values typically exceed 1.10. *Tarsius* (T) has the highest value of any primate (1.87).

(b) Fossils: letters indicate fossil primates of modern aspect and numbers indicate fossil 'archaic primates', following the key in Fig. 7.6. I, *Leptictis* (= *Ictops*).

(c) Diurnal primates and *Aotus*: ●, lemurs with crepuscular (?) habits; ☆, subfossil lemurs. *Aotus* (A) has the highest index value of any simian primate (1.38), falling well within the range of nocturnal lemurs and lorises.

(d) Tree-shrews: the nocturnal *Ptilocercus* (P) does not have a higher orbital index value than the other, diurnal, tree-shrews.

between diurnal forms (which typically have lower index values) and nocturnal forms. Only two diurnal forms, both lemurs, have index values greater than 1.10 (i.e. *Hapalemur simus* and *Lemur rubriventer*) and at least one of these species is probably crepuscular in habits. Among nocturnal primates, only two relatively slow-moving forms, the angwantibo (*Arctocebus calabarensis*) and the potto (*Perodicticus potto*) have index values lower than 1.10. (Goffart, Missotten *et al.*, 1976, specifically commented on the relatively small eyes of the potto.) It is particularly striking that the only nocturnal simian, the owl monkey (*Aotus*), has an index value of 1.38, in the upper end of the range of nocturnal lemurs and lorises. In fact, *Aotus* has quite a high index value even in comparison to nocturnal strepsirhines. *Tarsius* has the highest index value of all (1.87). As both of these nocturnal haplorhines lack a reflective tapetum, it seems quite likely that the eyes have become particularly large to compensate for the absence of this aid to vision in dim light conditions.

Having established a good empirical distinction between nocturnal and diurnal species among living primates, it is possible to turn to the interpretation of fossil skulls. Using the cut-off point of an orbital index value of 1.10, it would seem likely that, among early Tertiary primates, both *Adapis* species (*A. magnus* and *A. parisiensis*), *Notharctus* and *Aegyptopithecus* had diurnal habits. The same probably holds for *Rooneyia* as well, but in this case the orbital index value of 1.05 is sufficiently close to the borderline to leave a large element of doubt. By contrast, it is probable that *Mioeuoticus*, *Pronycticebus*, *Necrolemur* and *Tetonius* were all nocturnal (see also Kay and Cartmill, 1977), although this interpretation is least secure for *Tetonius*, which has an index value of only 1.13. It should also be noted that, whereas *Adapis magnus* has an orbital index value close to that of typical diurnal lemurs and lorises, *A. parisiensis* has an appreciably smaller value (0.79 versus 0.94). This is quite a large difference and presumably reflects some distinction in visual adaptations between the two *Adapis* species. *Aegyptopithecus* also has quite a low orbital index value (0.78) compared

to modern simians, indicating a significant difference in this case, too.

The smallest orbits of all, relative to skull length, are found in the fossil 'archaic primates' (*Plesiadapis*, *Palaechthon* and *Phenacolemur*) and in the early Tertiary insectivore *Leptictis* (= *Ictops*), as pointed out by Kay and Cartmill (1977). As 'archaic primates' have even smaller orbits than the modern tree-shrews, they clearly do not show any sign of the emphasis on vision that has been one of the key features of primate evolution (see also Chapter 8). This observation adds particular significance to the fact that 'archaic primates' lack the postorbital bar that is universally present among primates of modern aspect.

Figures 7.6 and 7.7 also include some subfossil lemur species, all of which have moderate-sized or quite small orbits relative to skull size. Walker (1967b) commented on this fact and inferred that these large-bodied lemurs were probably all diurnal in habits. This would seem to be borne out by the histograms in Fig. 7.7, which show the subfossil lemurs lying within the lower end of the range of orbital index values for diurnal lemurs and simians. However, Kay and Cartmill have suggested, on the basis of comparisons with marsupials, carnivores and squirrels, that interpretations based on orbit size relative to skull size become less reliable as body size increases. At the larger body sizes, they found considerable overlap in orbit size relative to skull size for nocturnal marsupials, nocturnal carnivores and diurnal monkeys, whereas such overlap was far more limited for smaller-bodied species. Obviously, some caution is needed in interpreting the habits of large-bodied mammals on the basis of relative size of the orbits. But the problem may not be as acute as indicated by Kay and Cartmill. It can be seen from Fig. 7.6 that two diurnal simian species (the long-snouted baboons *Papio cynocephalus* and *Therapithecus gelada*) lie well below the others, and they were in fact excluded as conspicuous outliners when the best-fit line for diurnal primates was calculated. A similar problem is encountered with some long-snouted subfossil lemurs (*Archaeoindris*, *Megaladapis* and *Palaeopropithecus*) that also fall well below the best-fit line. Naturally, if a particularly long snout is developed for some reason (e.g. in association with elongation of the tooth row), the effect will be to render the orbital index smaller. In this context, it is worth pointing out that *Aegyptopithecus* has a relatively long snout and it may be primarily for this reason that this early fossil simian appears to have relatively small orbits in Fig. 7.6. It should also be emphasized that relative orbit size may be of value for indicating nocturnal versus diurnal habits only among primates and that the appropriate procedure is therefore to compare subfossil lemurs with other primates, rather than with mammals generally, in the first instance. As noted above, relative size of the orbits does not distinguish the nocturnal *Ptilocercus* from other tree-shrews, to take just one example from non-primates.

Nevertheless, Kay and Cartmill (1977) raised an important point with respect to the interpretation of orbit size in larger-bodied mammals, as they specifically noted that the size of the orbit is less closely related to the size of the eye in the larger primate species (e.g. great apes). In order to reach a reliable interpretation of orbit size, it is necessary to consider the size relationship between the orbit and the eye itself, taking allometric factors into account. Unfortunately, only limited data are available from the literature on the diameter of the eyeball, but enough information is available from two sources (Schultz, 1940; Rohen, 1962) to permit a preliminary plot of eyeball diameter against orbit diameter for a sample of primate species (Fig. 7.8). This plot shows that in primates generally the diameter of the eye usually represents 75–100% of the diameter of the orbit. However, it can be seen that great apes fall below this level, with the eye diameter representing only 50–75% of the diameter of the orbit. Thus, for great apes, orbit diameter does not provide a good indication of eyeball diameter and even for other primates there is some variation in the eyeball:orbit relationship. In the absence of a detailed analysis of a really satisfactory sample, it is difficult to say more at this stage, although it is clear that orbit size may indeed provide a better indication of

eyeball size in smaller-bodied species, as suggested by Kay and Cartmill (1977).

There is, however, one intriguing feature of Fig. 7.8 that deserves special mention. The data currently available suggest that for marmosets and tamarins (arrows in Fig. 7.8) the diameter of the eyeball exceeds that of the orbit by some margin. As this is not true even of the large-eyed *Tarsius* and *Aotus*, some special explanation is required. One possible explanation is that marmosets and tamarins (Callitrichidae) are in fact dwarf forms, as has been suggested on dental grounds (see Chapter 6). They could have retained oversized eyes because of their derivation from a larger-bodied ancestral form with a well-developed visual system requiring a certain minimum eye size for effective operation. But more data are required to confirm the provisional picture presented in Fig. 7.8.

Orientation of primate eyes

Another distinctive feature of living primates that is commonly cited is the forward rotation of the eyes such that the two ocular scanning areas overlap to produce an extensive **binocular field**. As discussed in Chapter 8, the existence of binocular overlap is a prerequisite for **stereoscopic vision**, which depends on central nervous processing of information derived from simultaneous observation of objects with both eyes. It is therefore of interest to examine the degree of forward rotation ('frontality' of Simons, 1962c) shown by the orbits of primates and other mammals, a subject that has been examined in some detail by Cartmill (1970, 1971, 1972, 1974b). Once again, measurements on the bony orbit do not necessarily provide a direct indication of the relationships of the eye itself (see Cartmill, 1972); but they do give a general indication of overall visual adaptation. As Cartmill has pointed out, forward rotation of the eyes in primates is a complex phenomenon and actually involves three main components:

1. **Convergence**: forward rotation of the planes of the orbits (measured as the angle formed between the plane of either orbit and the midsagittal plane of the skull).

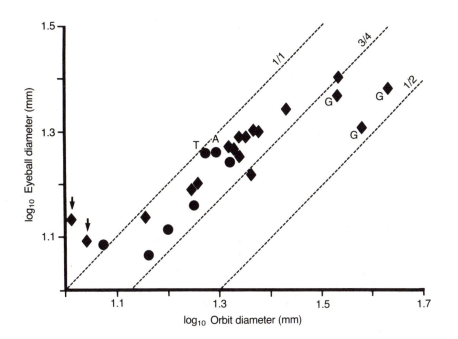

Figure 7.8. Logarithmic plot of eyeball diameter against orbit diameter for a sample of 25 primate species. Key: ●, nocturnal primates; ◆, diurnal primates; T, *Tarsius*; A, *Aotus*; G, great apes; arrows, callitrichids (*Callithrix* and *Saguinus*). The dashed lines indicate isometric scaling of eye to orbit diameter in the proportions of 1 : 1, 3 : 4 and 1 : 2.

2. **Frontation**: tilting of the eyes relative to the horizontal plane (measured as the angle between the longitudinal axis of the skull and the line of intersection between orbital and mid-sagittal planes).

3. **Relative interorbital breadth**: the degree of separation of the inner margins of the two orbits (measured as a ratio of interorbital distance to orbital diameter).

The interaction between the first two of these components is illustrated by certain primate species with upward-tilted eyes. Although the degree of orbital convergence in such primates may seem to be only moderate, when the head is held with the snout sloping downwards the effective degree of binocular overlap is increased. Upward-tilted eyes combined with a downward-tilted head posture are characteristic of the slow-moving lorises (e.g. *Loris tardigradus*; Fig. 7.9). Similar upward tilting of

the orbits is found in some of the subfossil lemurs (e.g. *Palaeopropithecus*; see Simons, 1962c) and also in the Eocene 'lemuroid' *Adapis parisiensis* (see Fig. 2.6), and it is likely that these primates also increased their effective binocular overlap by tilting their heads downwards during activity. Nevertheless, simple measurements of orbital convergence do provide some useful information about binocular overlap, provided that allometric effects are taken into account.

It has already been demonstrated above that orbit size does not increase at the same pace as skull size (i.e. there is a negative allometric relationship). Cartmill (1972) has shown that small-bodied mammals with relatively large eyes must therefore be subject to a limitation on orbital convergence, whereas this constraint is lifted in larger mammals in which the orbit is relatively smaller. Orbital convergence should therefore be plotted against some measure of skull size for meaningful interpretation of the

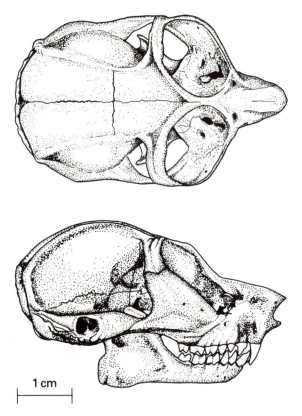

1 cm

Figure 7.9. Dorsal and lateral views of the skull of the slender loris (*Loris tardigradus*). In all members of the loris subgroup, the head is typically held with the snout inclined downwards, such that effective convergence of the orbits is increased.

data (Cartmill, 1971, 1974b). A plot of convergence against skull length for living primates and tree-shrews (Fig. 7.10) is, in fact, quite instructive. In the first place, it is seen that Cartmill's postulation of an overall increase in convergence with increase in skull size is borne out for primates, in that the two variables are positively correlated ($r = +0.59$). On the other hand, tree-shrews show a negative correlation between the two variables ($r = -0.77$), indicating that there has been far less emphasis on orbital convergence in the evolution of tree-shrews than in primate evolution (see also Fig. 5.4). In line with this, tree-shrews all show low values for orbital convergence (mean: 28°; range: 25–33°).

Kay and Cartmill (1974) estimated that the Palaeocene 'archaic primate' *Palaechthon* had a maximum value of 35° for orbital convergence, thus resembling the extant tree-shrews quite closely. By contrast, living primates, typically have markedly higher values for orbital convergence. With the sole exception of *Phaner*, which has an unexpectedly low value of 30°, the primate species surveyed (see Cartmill, 1971) uniformly have quite high degrees of orbital convergence (mean: 55°; range: 40–75°, excluding *Phaner*), markedly exceeding the values found among tree-shrews. This is of considerable importance, as the relatively large amount of binocular overlap thus ensured provides the necessary basis for the special pattern of stereoscopic vision found in modern primates (see Chapter 8). In addition, as was noted by Cartmill (1971), it emerges that the presence of ethmoid exposure in the orbit is typically found in those primate species that have the highest degrees of convergence for any given skull length (see Fig. 7.10).

The discovery of ethmoid exposure in the orbits of some *Indri* by Cartmill and Gingerich (1978; see above) does not accord with this general pattern and led these authors to reject Cartmill's original suggestion that exposure of the ethmoid in the medial orbital wall is a direct concomitant of a high degree of orbital convergence. For its skull size, *Indri* has a relatively low degree of orbital convergence (i.e. 57°), so encroachment of the orbits on the nasal cavity would not seem to explain the occasional occurrence of ethmoid exposure in this case. However, Cartmill's original hypothesis does fit all other living primates with ethmoid exposure and, if *Indri* is regarded as an odd exception, convergence of the orbits provides a plausible explanation for the emergence of this feature during primate evolution.

If increasing convergence of the orbits leads to encroachment on the intervening nasal cavity, this should be reflected in reduction of the interorbital breadth (i.e. the minimum distance between the inner margins of the two orbits). Hence, further information can be obtained by plotting interorbital breadth against skull length

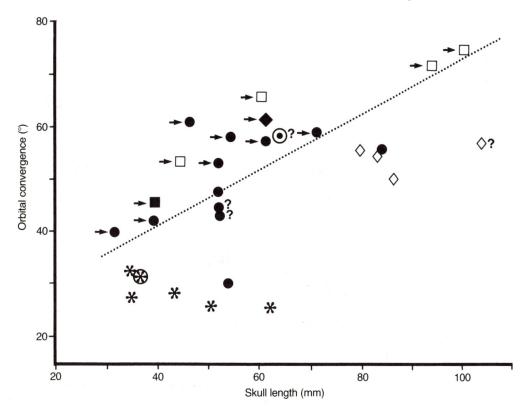

Figure 7.10. Plot of orbital convergence against skull length for primates and tree-shrews (symbols as for Fig. 7.6). Species known to have ethmoid exposure in the orbit are marked with an arrow; those with occasional ethmoid exposure (cf. Cartmill, 1978) are marked with a question mark. (*Indri* is plotted on the extreme right.) The dotted line indicates the lower limit for species with consistent ethmoid exposure. Note that complete forward rotation of the orbits would produce a convergence value of 90°, whereas a value of 0° would indicate complete lateral orientation of the eyes. (Data from Cartmill, 1970, 1971.)

for primates and tree-shrews (Fig. 7.11). In this case, as a fairly simple measurement on a comparatively resistant part of the skull is involved, it is possible to include data from fossil skulls without too much fear of error arising through distortion during fossilization (see also Kay and Cartmill, 1977). As expected, tree-shrews, which have less orbital convergence than primates (Fig. 7.10), have greater interorbital breadths relative to skull length. In this case, however, *Ptilocercus* differs from the other tree-shrews and approximates the primate condition, even though its orbits are no larger than those of other tree-shews (Fig. 7.6) and do not show more convergence (Fig. 7.10) relative to skull length. Primates typically have quite low values for

interorbital breadth relative to skull length, and especially low values are found for certain simians (e.g. *Cebus* and *Cercopithecus*) in the sample studied by Cartmill (1970). Verheyen (1962) and Kay and Cartmill (1977) have shown that colobine monkeys (leaf-monkeys) have rather higher values for interorbital breadth, although these are still well within the primate range (relative to skull length). However, certain prosimian primates have relatively wide separation of the inner margins of the orbits. This is particularly true of the aye-aye (*Daubentonia*), which has a condition comparable to that found in tree-shrews, and to a lesser extent it is true of various diurnal lemurs (*Lemur, Propithecus* and *Indri*). It is noteworthy that primates known to

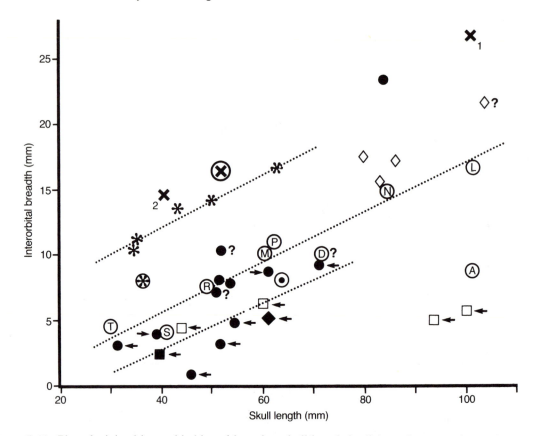

Figure 7.11. Plot of minimal interorbital breadth against skull length for living primates and tree-shrews, with inclusion of relevant fossil forms (symbols as for Fig. 7.6). Species known to have ethmoid exposure in the orbit are marked with an arrow; those with occasional ethmoid exposure are marked with a question mark. (*Indri* is plotted on the extreme right.) The three best-fit lines (major axes) have been determined for (from top to bottom): diurnal tree-shrews; early Tertiary primates of modern aspect; and modern nocturnal prosimians with ethmoid exposure. (Data for living primates and tree-shrews from Cartmill, 1970, and for fossil forms from Kay and Cartmill, 1978.)

show consistent ethmoid exposure in the orbit all have small interorbital breadths, relative to skull size, as predicted by Cartmill's hypothesis (1971) that ethmoid exposure results from orbital encroachment on the nasal capsule. Once again, however, the apparent occurrence of ethmoid exposure as an occasional feature in *Indri* does not fit the general pattern.

Kay and Cartmill (1977) suggested that there is a positive allometric relationship between interorbital breadth and skull length, with an exponent value of approximately 1.25. This does not apply to tree-shrews, however, and a perfectly good isometric fit can be obtained for

the relationship between interorbital breadth (*B*) and skull length (*SL*) for diurnal tree-shrews without converting the data to logarithmic form:

$$B = 0.21 \, SL + 3.57 \qquad (r = 0.98; N = 6)$$

In fact, fitting a best-fit line to the logarithmically transformed data in this case indicates a negatively allometric relationship, with an exponent value of 0.75, but no improvement in the correlation coefficient is obtained through logarithmic conversion (as before, *r* = 0.98).

A best-fit line of similar slope can be determined without logarithmic transformation of the data for nocturnal prosimian species in which ethmoid exposure is known to occur (see Fig. 7.11):

$$B = 0.19\,SL - 4.60 \qquad (r = 0.79; N = 8)$$

In this case, logarithmic transformation indicates a positively allometric relationship between interorbital breadth and skull length (exponent = 1.57), but as the value of the correlation coefficient falls dramatically ($r = 0.59$), this is probably a spurious result. It would seem reasonable to assume that interorbital breadth scales isometrically with skull length within groups of mammals that contain functionally comparable species. This view is borne out when fossil forms are included in the comparisons. All of the early Tertiary (Eocene/Oligocene) primates of modern aspect have interorbital breadths that, relative to skull length, are well below those found in modern tree-shrews. On the other hand, the values for these fossil primates tend to lie within the upper region of the overall primate distribution, suggesting that there was generally less encroachment on the nasal capsule by the orbits than is the case with many modern primates.

A best-fit line determined for the seven early Tertiary fossil primates (*Adapis magnus*, *A. parisiensis*, *Pronycticebus*, *Notharctus tenebrosus*, *Necrolemur*, *Rooneyia* and *Tetonius*) has a very similar slope to that of the lines determined for diurnal tree-shrews and for nocturnal prosimians with ethmoid exposure:

$$B = 0.19\,SL - 1.88 \qquad (r = 0.97)$$

This best-fit line for fossil primates of modern aspect lies between the other two lines (Fig. 7.11) and above all the modern primate species that show ethmoid exposure (although *Necrolemur* and *Adapis parisiensis* lie below the fossil primate line and overlap with such primates). The Miocene 'lorisoid' *Mioeuoticus* lies just above the line for the earlier fossil primates and has a larger value for interorbital breadth, relative to skull length, than any of the extant loris group species. The possibility therefore arises that exposure of the ethmoid was not present in ancestral primates, but has emerged as a convergent phenomenon in several primate lineages as a concomitant of a general trend for the inner margins of the orbits to be brought closer together. This could explain why ethmoid exposure has rarely been reported for early Tertiary primates of modern aspect, although Fig. 7.11 suggests that ethmoid exposure was probably present in *Necrolemur* (at least), as hypothesized by Cartmill (1971), as well as in *Adapis parisiensis*.

Figure 7.11 shows that the fossil 'archaic primates' *Palaechthon* and *Plesiadapis* have quite high values for interorbital breath, relative to skull length, as does the fossil 'insectivore' *Leptictis* (= *Ictops*). Thus, none of these fossil forms has the approximation of the inner margins of the orbits that characterizes primates of modern aspect (with the exclusion of the highly specialized *Daubentonia*). Kay and Cartmill (1977) have shown that modern tree-shrews, *Palaechthon*, *Plesiadapis* and *Leptictis* overlap with squirrels in terms of the relationship between interorbital breadth and skull length. Carnivores and marsupials lie between tree-shrews and primates, overlapping the upper end of the prosimian distribution. In other words, modern marsupials and carnivores are more like primates of modern aspect than are tree-shrews or 'archaic primates' in terms of interorbital breadth relative to skull length. In this feature, *Aegyptopithecus* is closer to modern diurnal simians than to diurnal lemurs, in apparent confirmation of its position close to the stem of the simian radiation. But it should also be noted that the subfossil lemur *Archaeolemur* approximates to the condition found in leaf-monkeys (colobines) in terms of relative interorbital breadth (Kay and Cartmill, 1977), emphasizing the convergent monkey-like form of the skull in this genus.

THE NASAL CAVITY AND THE SENSE OF SMELL

It has often been claimed that in primates generally increased emphasis on vision, as reflected in the modifications of the orbits discussed above, has been closely accompanied by reduction in emphasis on the sense of smell (olfaction). The latter process should necessarily affect structures in the nasal cavity, and it might therefore be expected that primates should be uniformly characterized by a reduction of such structures. The reciprocal relationship between increased visual capacities and decreased olfactory capacities was stressed by Elliot Smith (1927), who wrote in terms of a 'struggle for dominance between smell and vision'. Elliot Smith saw reliance on olfaction as a primitive feature of mammals, such that any relative enhancement of vision was necessarily an advanced feature. This theme was taken up by Le Gros Clark (1959), who in particular argued that tree-shrews shared with lemurs and lorises an increase in visual capacity allied to a decrease in olfactory capacity. Part of the evidence for this was derived from the bony structures located within the general nasal cavity (**nasal fossa**) of the skull.

Nasal fossa

The most obvious structures within the nasal fossa of the mammalian skull are the thin scrolls of bone called **turbinal bones**. In a typical mammalian skull, the nasal cavity is separated along the midline by a nasal septum. If the skull is sectioned longitudinally just adjacent to that septum, the turbinal bones can be seen in the exposed nasal fossa of one side, with their scroll-like surfaces directed towards the midline (Fig. 7.12). On each side of the skull, all mammals have one **maxilloturbinal**, which arises from the maxilla and may be very complex in structure. This turbinal bone differs from all of the others in that it is a separate dermal bone and has a purely respiratory function. It is covered with respiratory tract epithelium and has no connection with the olfactory bulb of the brain (see Chapter 8). In fact, behind the maxilloturbinal there is typically a horizontal shelf of bone (the transverse lamina) that separates a ventral air-channel, which leads directly to the lungs, from the dorsal part of the nasal fossa (Fig. 7.12).

The remaining turbinal bones, which are at least partially covered with specialized olfactory epithelium, are thus usually partitioned off in a cul-de-sac that serves the sense of smell. All of these turbinal bones are connected to the ethmoid bone and they are accordingly called **ethmoturbinals**. In fact, the uppermost of these is also attached to the nasal bone and is commonly termed the **nasoturbinal** (e.g. Le Gros Clark, 1959; Cave, 1967, 1973). However, Paulli (1900), in the classical review of the nasal fossa of mammals, treats the nasoturbinal as the first of the ethmoturbinals. This procedure has been followed in an excellent recent summary by Moore (1981) and will also be followed here. Beneath the ethmoturbinals there may be accessory scrolls called **ectoturbinals**. At the rear of the ethmoid is a flat plate termed the **cribriform plate**. This is typically perforated to allow the passage of individual nerves passing to the (normally adjacent) olfactory bulb. The ethmoid, which is a novel element in the mammalian skull, is hence a key component of the nasal fossa with respect to olfactory processes.

Although all mammals (except the highly specialized cetaceans) have a basic complement of a maxilloturbinal and a set of ethmoturbinals on each side of the skull, there is considerable variation in the number and arrangement of the ethmoturbinals. It was on this basis that Le Gros Clark (1959) suggested that there is a specific resemblance between tree-shrews and most lemurs and lorises in that there are only five ethmoturbinals (i.e. including the nasoturbinal) and two accessory ectoturbinals. He interpreted this as a 'reduced' set of turbinal bones, reflecting a shift in emphasis from olfaction to vision in a common ancestor (see also Fiedler, 1956; Saban, 1956/57). Even if this were the case, however, the resemblance between tree-shrews and lemurs/lorises could not represent a shared derived condition in terms of Le Gros Clark's own interpretation, as he went on to note that the aye-aye (*Daubentonia*) has six ethmoturbinals. This was interpreted as a more primitive, less-reduced condition correlated with a 'relatively larger olfactory bulb' in *Daubentonia*. However, allometric analysis (Fig. 8.16) has subsequently shown that the aye-aye

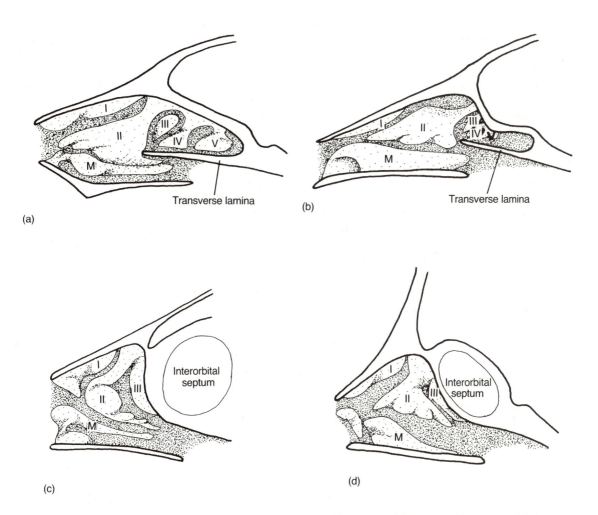

Figure 7.12. Structure of the nasal fossa in various primates: (a) *Galago*; (b) *Lemur*; (c) *Tarsius*; (d) *Saimiri*. *Galago* and *Lemur* have the basic complement of turbinal bones: five ethmoturbinals (I–V; I = nasoturbinal) and one maxilloturbinal (M). Further, the transverse lamina isolates the ethmoturbinals in a separate olfactory chamber, above the respiratory passage. In *Tarsius* and *Saimiri*, the number of ethmoturbinals has been reduced to three and the transverse lamina has disappeared. (After Cave, 1973.)

does not in fact have a larger olfactory bulb than expected for a nocturnal lemur of that body size.

It emerges, however, there was no basis for Le Gros Clark's assumption that five ethmoturbinals constitute a 'reduced' set in comparison to a more numerous set in a hypothetical ancestral condition for placental mammals. Paulli's review (1900) of the nasal fossa in mammals demonstrated quite clearly that most mammals have a basic set of five ethmoturbinals (including the nasoturbinal) and a few ectoturbinals (see also Cave, 1973; Moore, 1981). This basic complement of olfactory turbinal bones is found in most marsupials, insectivores, rodents, many carnivores, hyraxes and 'flying lemurs', as well as occurring in tree-shrews and most lemurs and lorises. Certain mammalian groups, such as the edentates (sloths, anteaters and armadillos), some carnivore families and ungulates, (perissodactyls and artiodactyls) have more than five ethmoturbinals; but this is probably a specialized condition rather than primitive. It therefore seems likely that, as in several other contexts, tree-shrews merely resemble lemurs and lorises in the retention of a primitive feature of the placental mammalian stock instead of sharing a derived condition.

Apparently, Le Gros Clark was working on the assumption that development of the visual system is necessarily associated with a reduction of the olfactory apparatus, but this does not seem to be true of tree-shrews or of lemurs and lorises generally (see also Chapter 8). In fact, it seems likely that certain groups of mammals with quite well-developed vision, notably the ungulates, have undergone a further increase in olfactory capacity since diverging from the ancestral stock of placental mammals. Elliot Smith (1927) erroneously stated that 'behaviour is controlled mainly by the sense of smell' in primitive vertebrates, in support of his notion that emphasis on olfaction is a primitive feature. To the contrary, mammals as a group are distinguished from other vertebrates by their generally increased emphasis on the sense of smell, doubtless as a result of a long period of adaptation to nocturnal habits during the first two-thirds of mammalian evolution (see Chapter 4). It is therefore only to be expected that certain placental mammals may have undergone further expansion and specialization of the olfactory apparatus since diverging from the common ancestral stock.

Whereas lemurs and lorises do not exhibit any significant reduction in the olfactory structures within the nasal fossa, such reduction is clearly evident in the nasal fossa of tarsiers and simians (Paulli, 1900; Le Gros Clark, 1959; Cave, 1973; Moore, 1981). As Cave has pointed out, the degree of reduction observed is quite unusual among mammals and it is exceeded only in cetaceans (whales and dolphins), in which adaptation for a fully aquatic existence has led to complete remodelling of the nasal fossa. In all simians and in *Tarsius* the number of ethmoturbinals has been decreased from five to three (including the nasoturbinal), although some simians do show a vestige of an additional ethmoturbinal. As Le Gros Clark (1959) has pointed out, vestiges of the suppressed ethmoturbinals may occasionally be observed in adult humans and it is possible to recognize five ethmoturbinals during human embryonic development; so there is good evidence for secondary loss of ethmoturbinals in the evolution of tarsiers and simians (Fig. 7.12). In these haplorhine primates, the first ethmoturbinal has been considerably reduced and in Old World monkeys, apes and humans it is no more than a simple ridge (the **agger nasi**). Further, the transverse lamina has been lost (see Fig. 7.12), such that there is no longer a horizontal partition between the respiratory passage and the olfactory chamber of the nasal fossa. This extensive reduction of the nasal fossa in tarsiers and simians is so unusual among mammals that it is quite likely to represent a shared derived condition. Indeed, Cave (1973) went so far as to suggest that the primate order should be redefined to include only the tarsier and simians, as lemurs and lorises have a typical mammalian nasal fossa.

It has also been pointed out (Cave, 1967; Cartmill, 1972) that reduction in the volume of the nasal fossa has accompanied the partial suppression of the ethmoturbinal system. This has permitted another unique development in *Tarsius* and some of the smaller-bodied simian primates (e.g. marmosets and tamarins, *Aotus*, *Saimiri*, *Pithecia* and *Miopithecus*). In these forms, the orbits actually meet in the midline of the skull and a thin, bony partition (**interorbital septum**; Fig. 7.12) is formed between them (Cave, 1967; Cartmill, 1972; Starck, 1975). As a result, the cribiform plate is displaced upwards and the olfactory bulbs are separated from the plate, such that the olfactory nerves must pass through an olfactory tube on their way from the nasal cavity to the brain. That olfactory tube lies above the interorbital septum. As has been explained above, eye size is related to skull size in a negatively allometric fashion. For this reason, the eyes do not encroach on the midline of the skull to such an extent in large-bodied simians (e.g. *Alouatta*; most Old World monkeys, apes and humans), and no interorbital septum is formed. Nevertheless, the cribiform plate shows the same vertical displacement in all simian primates, with the result that the ethmoturbinals have a predominantly vertical rather than horizontal arrangement. It seems likely that an interorbital septum might have been present in the common ancestor of the simians (provided that the body weight did not exceed about 5 kg). Cartmill (1972) suggested that an interorbital septum might have been present in at least some of the simians represented in the Fayum Oligocene deposits, and also in certain Eocene 'tarsioids' (e.g. *Necrolemur* and *Pseudoloris*). The latter proposal, incidentally, reinforces the likelihood that the ethmoid was exposed in the orbit of *Necrolemur* (see Cartmill, 1975). Hence, if olfactory reduction, as reflected by the structures of the nasal fossa, is a shared derived feature of tarsiers and simians it is also possible that the presence of an interorbital septum in the common ancestral stock was associated with such reduction.

It has often been suggested (e.g. Spatz, 1968; Le Gros Clark, 1959) that formation of an interorbital septum, and associated suppression of part of the ethmoturbinal system, developed as a convergent feature in *Tarsius* because of the unusually large eyes of this genus (Fig. 7.6). However, it seems far more likely that formation of an interorbital septum could only arise as a consequence of a reduction in the olfactory apparatus. After all, *Tarsius* has no more orbital convergence than nocturnal lemurs or lorises of the same skull size (Fig. 7.10) and its interorbital breadth is actually less reduced than that of one of the lorises (*Loris*) in relation to skull size (Fig. 7.11). Cartmill (1972) has in fact shown that *Loris* has convergently developed an interorbital septum, probably in association with the extreme degree of approximation of its eyes; but that the septum lies above the olfactory bulbs and does not coincide with any obvious reduction in the ethmoturbinal system. Accordingly, it seems likely that reduction of the olfactory apparatus developed as the primary feature of both tarsiers and simians (quite probably as a shared derived

characteristic of a common ancestral stock). Closer approximation of the orbits – with formation of an interorbital septum in smaller-bodied forms – was therefore able to develop as part of the continuing refinement of the visual system.

Rhinarium

The sharp distinction in terms of the morphology of the nasal fossa between lemurs and lorises on the one hand and tarsiers and simians on the other is matched by an equally sharp distinction with respect to the external appearance of the snout. This distinction concerns the **rhinarium** (Fig. 7.13), an area of moist, glandular and essentially hairless skin surrounding the nostrils. Lemurs and lorises, in common with most mammals, have a rhinarium, whereas tarsiers and simians do not. This led Pocock (1918) to divide the order Primates into two suborders: Strepsirhini (primates with a rhinarium – lemurs and lorises) and Haplorhini (primates without a rhinarium – tarsiers and simians). Wood Jones (1929) went a step further and proposed that the term 'Primates' should be applied only to the tarsiers and simians (monkeys, apes and humans), as these forms share a whole suite of characters not found in lemurs and lorises. More recently, Hill (1953) used the terms Strepsirhini and Haplorhini for the two major subdivisions ('grades') of his classification of the Primates. As has been discussed in Chapter 3, the wisdom of devising a classification to reflect this apparent sharp division among two groups of living primates is open to some doubt and these formal taxonomic terms will not be used here. Nevertheless, the common names 'strepsirhine' and 'haplorhine' do provide a useful shorthand device for referring to living lemurs and lorises as a group apart from living tarsiers and simians, and this procedure has been followed throughout this book. (Hofer (1980) has questioned the use of the terms 'strepsirhine' and 'haplorhine' because they originally referred to the shape of the nostrils, but they are useful general labels for the complex of features discussed below.)

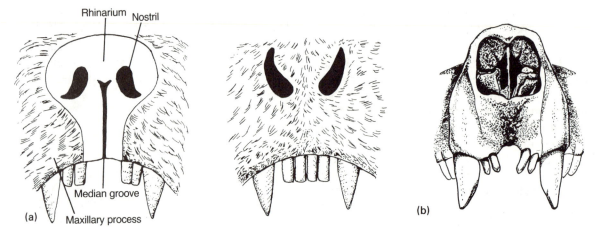

Figure 7.13. (a) Comparison of the strepsirhine condition of the snout in lemurs and lorises (left) with the haplorhine condition in tarsiers and simians (right). In strepsirhines, the primitive rhinarium is still present surrounding the nostrils. The rhinarium has a marked median groove and is attached to the gum at the anterior end of the upper jaw. In haplorhines, the hairy maxillary processes have completely obliterated the rhinarium. (b) Anterior view of the snout region of the skull of *Loris tardigradus*, showing the gap between the upper incisors.

In a major review of the form of the rhinarium among living mammals, Boyd (1932) defined five separate classes (with some intermediate cases). One important point to emerge from this review was that the tarsiers and simians are unique among living mammals in lacking all trace of a rhinarium in the adult. There can therefore be little doubt that this is a derived condition of tarsiers and simians (haplorhines) and that lemurs and lorises have a rhinarium as a primitive retention from the ancestral stock of placental mammals. As Boyd demonstrated, the anterior end of the snout in mammals is formed during development by a median **frontonasal process**, which gives rise to the rhinarium, and by two lateral **maxillary processes**. In the primitive condition – exemplified by various marsupials, insectivores, carnivores and tree-shrews as well as by lemurs and lorises – there is a clear labial extension of the rhinarium descending between the hair-covered lateral maxillary processes. This extension of the rhinarium, which is marked by a pronounced median groove (see later), fuses with the gum of the anterior part of the upper jaw, thus preventing any real mobility of the upper lip.

Other mammalian groups show varying degrees of departure from this basic pattern, beginning with the disappearance of the median groove on the rhinarium (e.g. sloths) and passing through varying degrees of coverage of the rhinarium by the maxillary processes. In lagomorphs, camels and kangaroos, the maxillary processes meet in the midline but do not actually fuse, whereas in the aardvark, pangolins and various carnivores (Canidae, Felidae and Mustelidae) there is definite fusion and the exposed area of the rhinarium is limited to a patch around the nostrils. However, fusion of the two maxillary processes is recognizable even in the adult condition, as a median groove persists along the junction between them. Finally, in some carnivores and in ungulates, the two maxillary processes fuse over the frontonasal process without leaving a groove, but the rhinarium still persists around the nostrils. The haplorhine primates therefore exhibit the greatest degree of obliteration of the rhinarium in that it has disappeared even from the area surrounding the nostrils. As Boyd (1932) has noted, obliteration of the labial extension of the rhinarium from its original attachment to the gum of the upper jaw permits a much greater degree of mobility of the upper lip in haplorhine primates.

Among the haplorhines themselves, a distinction has often been drawn between the New World simians and the Old World simians, with respect to the location and orientation of the nostrils. In New World monkeys typically (and, incidentally, in *Tarsius*), the nostrils are separated by a wide partition (septum) and they are forward-facing. These monkeys are therefore called **platyrrhines**, and the formal taxonomic term Platyrrhini has often been used in place of the term Ceboidea for New World monkeys. In Old World monkeys and apes, on the other hand, the nostrils are typically outward-facing and separated by a relatively narrow septum. The Old World simians are accordingly called **catarrhines** and the taxonomic term Catarrhini has been used (see Table 3.1) to group together the Cercopithecoidea and Hominoidea of Simpson's classification (1945). However, although the distinction between platyrrhines and catarrhines is generally valid (Hofer, 1980), it should be noted that exceptions apparently exist. Weber (1928) specifically commented that among New World monkeys the woolly spider monkey (*Brachyteles*) has catarrhine-like nostrils, whereas among Old World monkeys langurs (*Presbytis*) have somewhat platyrrhine-like nostrils. Certainly, the distinction between platyrrhine and catarrhine conditions does not seem to be as clear-cut or as fundamental as the distinction between strepsirhines and haplorhines as defined here.

Jacobson's organ

Detailed study of the olfactory system in the lesser mouse lemur (*Microcebus murinus*) has shed new light on the significance of the rhinarium in strepsirhine primates (Schilling, 1970). It would seem that there may be a direct connection between the rhinarium and a specific olfactory structure called the **Jacobson's organ** (or, alternatively, the **vomeronasal organ**). The organ of Jacobson is commonly overlooked in discussions of primate olfactory structures, perhaps because it is virtually absent in human beings; but it has proved to be of considerable interest with respect to the distinction between strepsirhines and haplorhines. It is an ancient accessory olfactory organ of vertebrates that is present in amphibians and is typically quite well developed in reptiles. The usual condition is for a pair of Jacobson's organs to occur in the nasal region of the skull, with one organ on each side of the midline. According to Romer (1955), the original function of the Jacobson's organ was connected with detection of olfactory sensations from food in the mouth cavity. In modern snakes, in fact, the paired Jacobson's organs are present as pits at the front end of the roof of the mouth and the tips of the forked tongue are inserted into the pits in connection with olfaction during feeding. In mammals, as pointed out in Chapter 4, a secondary palate separates the nasal cavity from the oral cavity and the Jacobson's organ has therefore come to adopt a quite specific location (Fig. 7.14). In a typical mammal, the paired Jacobson's organs lie just above the palate and beneath the mucous membrane of the nasal floor. The two organs lie on either side of the nasal septum that passes down the centre of the nasal cavity, separating the latter into right and left chambers. Each Jacobson's organ takes the form of a hollow tube, closed at both ends, and rests in a gutter-shaped cartilaginous capsule (the **paraseptal cartilage**) arising from the nasal septum. Both the paraseptal cartilage and the enclosed Jacobson's organ are supported by a bony **palatine process** of the premaxilla (Fig. 7.14). The Jacobson's organ itself is typically kidney-shaped in cross-section; its inner wall is lined with recognizable olfactory epithelium, and its outer wall consists of thinner, cuboidal epithelium.

Because the nasal and oral cavities are separated by the bony secondary palate in mammals, the Jacobson's organs are not directly accessible to the oral cavity. Accordingly, the opening of each Jacobson's organ typically bears a special relationship to both oral and nasal cavities in mammals. In fact, these two cavities are usually linked at their anterior ends by a pair of **nasopalatine canals** that pass through the premaxilla a short distance behind the upper incisor teeth, via the **incisive foramina**. The incisive foramina are normally present in the

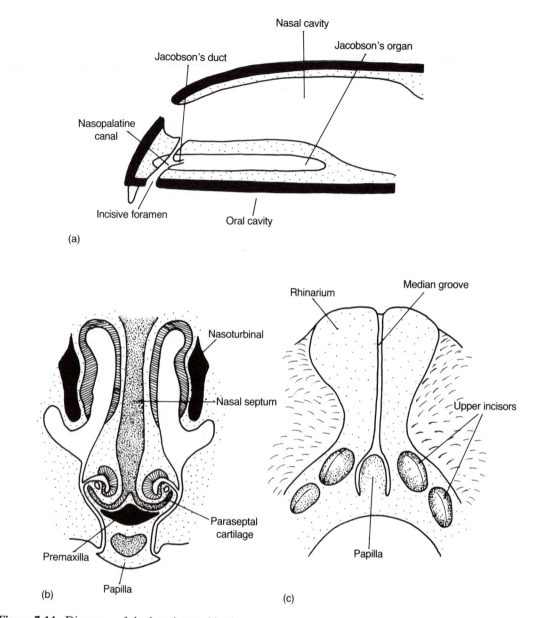

Figure 7.14. Diagrams of the location and basic organization of the Jacobson's organ in a typical mammal (e.g. strepsirhine primate).

(a) Longitudinal section through snout, just adjacent to midline. The nasopalatine canal links the nasal cavity to the oral cavity. The Jacobson's organ is linked to the nasopalatine canal by the Jacobson's duct. Note that the nasopalatine canal passes through the palate via the incisive foramen.

(b) Cross-section through the anterior end of the snout of a typical mammal, showing the two Jacobson's organs in section (cartilage is shown by heavy stippling, bone in black). Olfactory epithelium (hatching) is found in the general nasal cavity – in this section adjacent to ethmoturbinal I (= nasoturbinal) – and lining the Jacobson's organ. The Jacobson's organs are supported by the paraseptal cartilages, which are outgrowths from the nasal septum. The paraseptal cartilages themselves rest on the palatine process of the premaxilla. The papilla at the oral end of the nasopalatine canals contains a small element of cartilage.

(c) Ventral view of the rhinarium in the lesser mouse lemur (*Microcebus murinus*). The median groove passes back between the upper incisors and bifurcates to join the oral openings of the nasopalatine canals adjacent to the papilla. (After Schilling, 1970.)

mammalian skull even if the nasopalatine canals have been lost (see later), but they are not found in the skulls of cetaceans (whales and dolphins), which completely lack Jacobson's organs. In a typical mammal, the Jacobson's organ (which always opens close to its anterior end) has a short **Jacobson's duct** that opens into the nasopalatine canal. Accordingly, the Jacobson's organ can potentially receive olfactory inputs from both the oral cavity and the nasal cavity in mammals that have not undergone any reduction (Estes, 1972).

Weber (1927) noted that the internal cavity (lumen) of the Jacobson's organ is always filled with fluid in mammals and it is significant in this connection that there are numerous glands opening into the lumen in addition to the presence of typical olfactory epithelium (Moore, 1981). In all mammals that have Jacobson's organs, there are longitudinal blood vessels (often forming a plexus) alongside the organs. Through regulation of the supply of blood to these vessels, it is possible to compress the Jacobson's organs (thus expelling their contents) or to allow them to inflate (thus drawing material in), and this presumably permits olfactory sampling of particular substances. There has been relatively little research on the functioning of mammalian Jacobson's organs and it is therefore not entirely clear why mammals require these special olfactory structures in addition to the general olfactory epithelium of the nasal cavity. However, there is some indication that the Jacobson's organs may serve a special function in permitting recognition of species-specific odours, such as those commonly present in the urine of female mammals as indicators of reproductive condition (Estes, 1972). It is highly likely that the Jacobson's organ plays a specific part in olfaction, as it has its own nerve supply connected directly to the **accessory olfactory bulb**, rather than to the main olfactory bulb (see Chapter 8), which receives inputs from the olfactory epithelium lining the nasal cavity.

The link between the Jacobson's organs and the rhinarium demonstrated for *Microcebus murinus* by Schilling (1970) is illustrated in Fig. 7.14c. It can be seen that the median groove of the rhinarium connects directly with the oral openings of the nasopalatine canals. This

suggests that fluid on the rhinarium may be channelled directly into the Jacobson's organs and that the rhinarium may therefore be directly linked to the olfactory functions of these organs. It should also be noted that the continuation of the rhinarium back to the openings of the nasopalatine canals in strepsirhine primates such as *Microcebus* requires the presence of a median gap between the upper incisors (see Fig. 7.13b). This raises the possibility that the strepsirhine condition might be recognizable in fossil primate skulls on the basis of such a gap between the upper incisors (Martin, 1973). Certainly, in the absence of a gap it can be concluded confidently that there was no link between the rhinarium (if present) and the oral openings of the nasopalatine canals (if present). Among extant lemurs and lorises, a pronounced median gap between the upper incisors is present in all nocturnal forms with the exception of the aye-aye (*Daubentonia*), which has a specialized rodent-like dentition. This median gap tends to be less obvious in the skulls of diurnal lemurs, notably in *Indri* and *Propithecus*, and it has been almost completely suppressed in many (presumably diurnal) subfossil lemurs, especially in the monkey-like forms with spatulate upper incisors (e.g. *Archaeolemur* and *Hadropithecus*). Assuming that there is, indeed, some connection between the rhinarium and the Jacobson's organ, this suggests that there has been some reduction of the olfactory functions of this organ in diurnal lemurs. A pronounced gap was present between the upper incisors in some Eocene 'lemuroids', such as *Europolemur* and *Notharctus* (Franzen, 1987), although it had been reduced in *Adapis parisiensis* (Gingerich and Martin, 1981). By contrast with most strepsirhines, in extant haplorhine primates (tarsiers and simians) there is no median gap between the upper incisors, even in the two nocturnal genera (*Tarsius* and *Aotus*). In simian primates, the incisors are spatulate in both upper and lower jaws, whereas in tarsiers the simple, conical upper incisors are borne on a narrow rostrum that provides insufficient space for any separation of the upper incisors.

Because extant haplorhines lack a rhinarium and a median gap between the upper incisors, it was suggested (Martin, 1973) that this is indicative of a reduction of olfactory function at a very early stage of haplorhine evolution. Assuming that the rhinarium was, in fact, lost as an ancestral derived feature of tarsiers and simians, it is in any case obvious that the oral openings of the Jacobson's organs must have differed from the condition found in modern strepsirhines such as *Microcebus*. Nevertheless, Starck (1975) has since reported that in adult *Tarsius* the nasopalatine canal does open into the oral cavity alongside an identifiable papilla and that the Jacobson's organ is quite well developed. Starck also reported that, contrary to many previous statements, the Jacobson's organ is also well developed in adult New World monkeys and that the opening of the nasopalatine canal into the oral cavity is present, as in *Tarsius*. By contrast, in Old World simians the Jacobson's organ is virtually absent in adults. Vestigial components of the Jacobson's organ can be found in early developmental stages of Old World simians, as is the case with human beings, but these later disappear. Hence, it is only in Old World simians (catarrhines) that Jacobson's organ has been suppressed entirely. In New World simians (platyrrhines) and in tarsiers, the Jacobson's organ remains apparently functional in the adult, but the original connection with the rhinarium has been lost. Contrary to previous statements (e.g. Martin, 1973, 1979), it now seems likely that Eocene 'tarsioids' (e.g. *Necrolemur* and *Pseudoloris*) could have had a rhinarium as a median gap was present between the upper incisors (Schmid, 1982; Aiello, 1986). It is therefore possible that tarsiers lost the median gap independently of simians, as a convergent development, although it is also possible that omomyids are less closely related to tarsiers than generally believed and that loss of the rhinarium was an ancestral feature of tarsiers and simians only (Cartmill, 1980; Schmid, 1982; Aiello, 1986). By contrast, the earliest known simian primate skull (that of the Oligocene *Aegyptopithecus*) had spatulate upper incisors with no apparent median gap, so loss of the median gap was probably an ancestral feature of simian primates at least.

The loss of any connection between the oral openings of the nasopalatine canals and a rhinarium in all modern haplorhines is suggestive of some reduction in the importance of the Jacobson's organs in these primates. Unfortunately, however, there are no quantitative data on the overall size of the Jacobson's organ in insectivores and primates, so no effective comparison can be made. It is possible to approach this question indirectly, though, because of the specific connection between the Jacobson's organs and the accessory olfactory bulbs of the brain. Data on the size of the accessory olfactory bulbs of insectivores and primates are available (Stephan, Frahm and Baron, 1981) and comparisons can be made with due allowance for the effects of body size. In Fig. 7.15, the volume of the accessory olfactory bulb is plotted against body weight for a sample of treeshrews ($N = 3$) and primates ($N = 31$). Because Old World simian primates in fact do not have accessory olfactory lobes, reflecting the complete absence of a functional Jacobson's organ in adults, they do not appear on this logarithmic plot. It is clear from Fig. 7.15 that nocturnal lemurs and lorises typically have larger accessory olfactory blubs at any given body size than do diurnal New World monkeys. *Tarsius* occupies an intermediate position, as do some of the diurnal lemurs, indicating that the Jacobson's organs are less important than in nocturnal strepsirhines. Hence, Fig. 7.15 does provide evidence that, although the adult tarsier has a quite well-developed Jacobson's organ, this is not as fully developed as in nocturnal lemurs and lorises with a rhinarium that links up with the oral apertures of the nasopalatine canals. Interestingly, the owl-monkey (*Aotus*) has very small accessory olfactory bulbs even in comparison to other New World monkeys, despite its nocturnal habits. This shows that the Jacobson's organ is of reduced importance (compared to nocturnal strepsirhines) in all simian primates, although this reduction is only relative in New World monkeys. Uniquely, Old World simians have undergone complete suppression of the organ.

Figure 7.15 also shows that the size of the

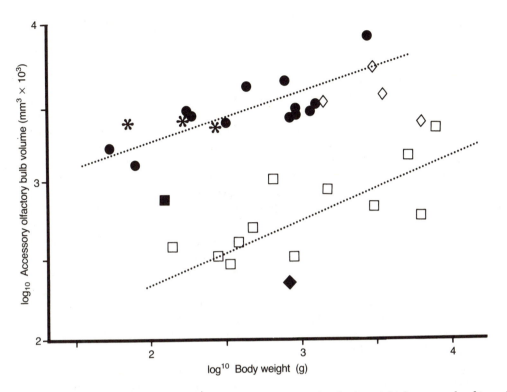

Figure 7.15. Logarithmic plot of accessory olfactory bulb volume against body weight for a sample of tree-shrew ($N = 3$) and primate ($N = 31$) species. Key: *, tree-shrews; ●, nocturnal lemurs and lorises; ◇, diurnal lemurs; ■, tarsier; ◆, owl-monkey; □, diurnal simians. Note that the best-fit line (major axis) for nocturnal lemurs and lorises lies well above that for simian primates. The tarsier occupies an intermediate position relative to these two groups, but the owl-monkey (in spite of its nocturnal habits) has very small accessory olfactory bulbs even in comparison to diurnal simians. Note: Old World simians lack accessory olfactory bulbs altogether, so they do not appear on this logarithmic plot. (Data from Stephen, Frahm and Baron, 1981.)

accessory olfactory bulbs, relative to body size, is about the same in tree-shrews as in nocturnal strepsirhines. The upper best-fit line in Fig. 7.15 might therefore be regarded as reflecting the primitive condition for primates and perhaps for placental mammals generally. However, the data provided by Stephan *et al.* (1981) show that in lipotyphlan insectivores the relative volume of the accessory olfactory bulbs is not as large as in tree-shrews and in nocturnal strepsirhines (see also Stephan, 1972). From this one must conclude either that the Jacobson's organ has undergone some expansion (at least with respect to size of the accessory olfactory bulb) in the evolution of tree-shrews and nocturnal

strepsirhines, or that this accessory olfactory system has become reduced in size during the evolution of lipotyphlan insectivores. The choice between these two alternatives is important because *Tarsius* lies within the range of the lipotyphlan insectivores and therefore there may or may not have been some evolutionary reduction of the Jacobson's organ in this primate. At present, it is not clear whether the lipotyphlan insectivores have a primitive or a reduced system and detailed comparative studies of the Jacobson's organ in insectivores and prosimian primates are needed to provide the necessary information. It is, however, clear that in all simians (especially Old World simians) the relative size of the accessory olfactory bulbs is reduced.

Comparative information on the development of the Jacobson's organ in mammals was presented in a series of papers by Broom (1898, 1902, 1915a, 1915b; see also Weber, 1927). From these studies, it emerged that the Jacobson's organ is well developed in many marsupials, carnivores, rodents, lagomorphs, elephant-shrews and ungulates, as well as in tree-shrews and in strepsirhine primates. There are certain variations of detail among these mammals with prominent Jacobson's organs. Whereas the Jacobson's duct opens at some point along a nasopalatine canal possessing both oral and nasal openings in some marsupials, insectivores and carnivores – just as it does in tree-shrews and in strepsirhine primates – there is a direct connection between the Jacobson's organ and the nasal cavity in certain marsupials and bats and in all rodents, lagomorphs and ungulates (see also Moore, 1981). This latter condition suggests that the nasal opening into the Jacobson's organ is more important (perhaps exclusively so) in these mammals. In the case of rodents and lagomorphs (rabbits, hares and pikas), it is possible that the development of continuously growing incisors has ruled out the use of the oral openings of the nasopalatine canals as a major supply route for the Jacobson's organs.

Simian primates are not the only mammals to have undergone reduction or loss of the Jacobson's organs (Broom, 1898; Weber, 1927; Moore, 1981). In most bats the organs are considerably reduced and fully aquatic mammals (i.e. cetaceans and pinnipeds) completely lack them. As noted above, cetaceans have even lost the incisive foramina, although these have been retained in the Old World simians despite complete suppression of the Jacobson's organs.

As a final note on the occurrence of the Jacobson's organ in mammals, it should be recorded that Broom (1898) regarded this accessory olfactory system as being of particular interest with respect to mammalian evolution:

From the small tendency to variation in the organ and its cartilages, we have in them a

factor of considerable value in the classification of the Eutherian orders, probably of more value than either dentition or placentation.

Broom noted that marsupials differed from most placentals in two particular features of the Jacobson's organ. Firstly, marsupials have an additional strip of cartilage (the outer bar) that passes along the outer margin of the Jacobson's organ, dorsal to the outer rim of the paraseptal cartilage. Secondly, marsupials lack a complex scroll-like extension of the nasal floor cartilage surrounding the nasopalatine canal that is typical of most placentals. It was originally pointed out by Broom (1898) that lagomorphs, like marsupials, lack the scroll-like extension of the nasal floor cartilage and also have a rudimentary outer bar. In subsequent papers, he reported that elephant-shrews, tree-shrews and at least some edentates (*Dasypus* and *Orycteropus*) similarly resemble marsupials rather than other placentals in these two respects (Broom, 1902, 1915a; see also Christie-Linde, 1914). As he regarded the possession of an outer bar and the lack of a scroll-like extension of the nasal floor cartilages as primitive features, Broom divided the placental mammals into two major groups – the 'Archaeorhinata' (tree-shrews, elephant-shrews, edentates and lagomorphs) and the 'Caenorhinata' (lipotyphlan insectivores, bats, carnivores, ungulates, hyraxes, primates, sirenians and cetaceans). He did, however, note that the lipotyphlan *Chrysochloris* is intermediate between these two groups. It is particularly important to note that, on this basis, Broom stated that tree-shrews and elephant-shrews (the 'Menotyphla'; see Chapter 5) 'must be removed far from the Insectivora and placed in quite a different phylum and not far from the early marsupials'. With respect to the structure of the cartilages of Jacobson's organ, lipotyphlan insectivores are apparently closer to primates than are tree-shrews. The marsupial-like pattern found in tree-shrews provides yet another indication of the early divergence of this group from other placental mammals. As has already been mentioned, the Jacobson's organ and associated structures have been relatively neglected in recent discussions of mammalian evolution, but it is fairly obvious that much useful information can be obtained from study of this accessory olfactory system.

THE EAR REGION

As has been shown in Chapter 4, associated changes in the jaw mechanism and in the ear apparatus distinguished mammals from reptiles at a critical stage in mammalian evolution. This indicates that, in addition to significant changes in the functions of jaws and teeth, modifications in the apparatus of hearing were of considerable importance in the early evolution of mammals. If, as seems likely, the adoption of nocturnal habits was a particular and enduring characteristic of early mammals, it may be suggested that increased sensitivity of hearing was vital to this particular lifestyle. Further, as changes in the ear mechanism closely accompanied changes in the jaws and dentition early in the evolution of mammals, it is not surprising that characters of the ear region of the skull have played an important part in interpretations of the mammalian fossil record, second only to dental features (MacPhee, 1981).

The basic organization of the ear region of the mammalian skull is shown in Fig. 7.16. The hearing mechanism itself depends on the **cochlea**, which is coiled in all mammals, whereas in most reptiles the cochlea is very rudimentary and in birds and crocodilians it is essentially a linear

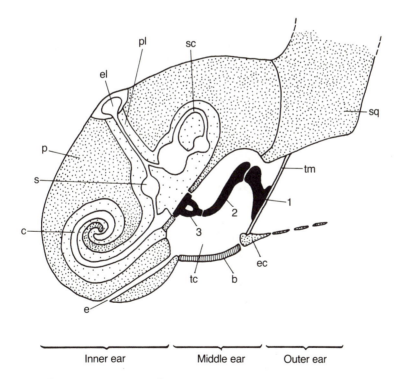

Inner ear   Middle ear   Outer ear

Figure 7.16. Diagrammatic transverse section of the typical mammalian ear region. Ear ossicles are shown in black (1, malleus; 2, incus; 3, stapes). Abbreviations: b, auditory bulla (ventral floor); c, cochlea; e, eustachian tube; ec, ectotympanic (tympanic ring); el, endolymph; p, petrosal (periotic); pl, perilymph; s, sacculus; sc, semicircular canal (one of three shown); sq, squamosal; tc, tympanic cavity; tm, tympanic membrane (eardrum).

tube (see Romer, 1955). The special development of the cochlea in mammals in fact provides an additional indicator of the increased importance of hearing in mammalian evolution. The cochlea is linked to the major organ of balance – three **semicircular canals** arranged in planes intersecting more or less at right angles – and the entire system is housed in the **petrosal** (periotic) bone. Adjacent to the inner ear lie the three **ear ossicles**, which form a sound-conducting chain linking the cochlea (via the **fenestra ovalis**) with the eardrum (**tympanic membrane**) in the **middle ear cavity** (Fig. 7.16). As explained in Chapter 5, the original single ear ossicle of terrestrial vertebrates has become the stirrup-shaped **stapes**, and it has been joined by two additional ossicles, the **incus** and **malleus** (derived from the reptilian quadrate and articular, respectively).

In all mammals, the tympanic membrane is supported by a special bony element, the **ectotympanic** (= tympanic), believed to be derived from a reptilian lower jaw bone, the angular. On the outside of the eardrum lies the external ear region, consisting of a tubular auditory meatus leading to the ear pinna. Sounds from the outside world enter the meatus and vibrations generated in the eardrum are transmitted through the ossicles to the cochlea. It is commonly suggested that the main function of the chain of three ossicles is to provide a finely controlled protective device for the ear mechanism, but (as explained in Chapter 4) recent evidence suggests that the ossicles represent a special adaptation to increase the sensitivity of the ear. Because air and body fluids have very different acoustical impedance values, a great deal of sound energy would be lost in the transference of airborne vibrations to the fluid-filled cochlea in the absence of the specially adapted chain of ossicles. These appear to operate together as an 'impedance transformer' (see Moore, 1981).

Bulla

Most living mammals possess some kind of bony structure in the ventral floor of the middle ear

region, isolating the tympanic cavity from the soft issues of the upper neck (see Fig. 7.16). This ventral floor may be either flat or inflated downwards to varying degrees. It will be termed the **auditory bulla** here (see Fig. 7.4), but some authors extend this term to cover all of the walls surrounding the tympanic cavity (see Moore, 1981). However, although most mammals possess a bulla of some kind, there is considerable variation from group to group in the manner of formation of the ventral floor of the middle ear cavity. Accordingly, characters of this region of the skull have provided a rich source of inferences regarding mammalian evolutionary relationships. Classic studies were conducted by van Kampen (1905) and by van der Klaauw (1931) and excellent modern reviews have been provided for placental mammals generally by Novacek (1977a) and Moore (1981), for tree-shrews and certain primates by Cartmill and MacPhee (1980), for prosimian primates by MacPhee (1981) and for primates generally by MacPhee and Cartmill (1986). The differences among mammals, primarily concern: (1) the contributions made by various skull components to the ventral floor of the middle ear cavity, and (2) the relationships of the ectotympanic to other components.

A few living mammals lack a bulla, in which case the ectotympanic is usually present as a relatively simple ring-shaped structure lying obliquely across the base of the middle ear, usually at an angle of less than 45° to the horizontal plane. This is true of monotremes, of certain lipotyphlan insectivores (shrews and some moles) and of certain other placental mammals (e.g. sirenians – sea cows). There are also a few mammals in which the ventral floor of the tympanic cavity is not ossified but cartilaginous. This condition is found in a few megachiropteran bats, in an edentate (the armadillo *Dasypus*) and in a carnivore (the viverrid *Nandinia*). All other living mammals have an ossified bulla involving one or more of the following bones as major components: alisphenoid, basisphenoid, ectotympanic, entotympanic(s), petrosal (Novacek, 1977a).

In living marsupials, the ventral floor of the

bulla is typically formed primarily by the alisphenoid, whereas this bone is only rarely involved as a major component of the bulla in placental mammals. There is a major contribution of the alisphenoid to the composite bulla of elephant-shrews, and Cartmill and MacPhee (1980) have stated that the alisphenoid contributes to the bulla of the pen-tailed tree-shrew (*Ptilocercus*). However, the alisphenoid is not involved in the ventral wall of the bulla of other tree-shrews (Tupaiinae) and Zeller (1986a,b) has produced evidence that there is no alisphenoid contribution to the bulla in *Ptilocercus*. According to Weber (1927), the alisphenoid is involved in the bulla of the edentate *Myrmecophaga*. By contrast, in most of those lipotyphlan insectivores that have a bulla (hedgehogs and tenrecs), the basisphenoid usually forms the major part of the ventral wall, although in certain cases (some tenrecs) there is also a significant contribution from the petrosal (see Novacek, 1977a). Perhaps the commonest condition among living mammals is for the ectotympanic bone itself to make a major contribution to the ventral bulla wall. This condition is found in ungulates (artiodactyls and perissodactyls), in cetaceans, in elephants, in rodents, in lagomorphs, in many carnivores and in the remaining lipotyphlan insectivores (some moles and golden moles).

Entotympanics are also widely found as components of the bulla in placental mammals. Cartmill and MacPhee (1980) note that 11 out of 20 orders of living placental mammals (as recognized in their classification) probably have some entotympanic contribution to the bulla. But it is only in just over half of these (bats, 'flying lemurs', edentates, hyraxes, elephant-shrews, tree-shrews, pangolins and some carnivores) that

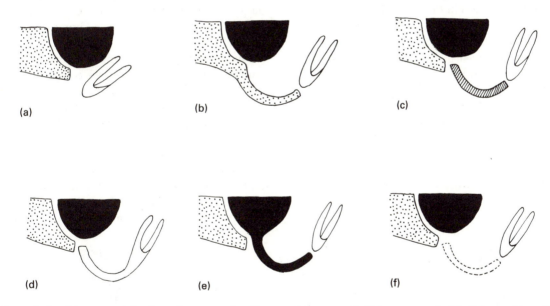

Figure 7.17. Diagrams of the six major types of auditory region present in living mammals. Medial aspect is on the left of each diagram; lateral aspect is on the right. Key: □, ectotympanic; ■, petrosal; ▓, basisphenoid or alisphenoid; ▨, entotympanic; ⌐⌐, cartilage. (Adapted from Novacek, 1977a.)

   Type a: No ossified ventral floor of the middle ear region (monotremes; some insectivores). Note the marked inclination of the ectotympanic.
   Type b: Bulla formed from the alisphenoid (marsupials) or basisphenoid (some insectivores).
   Type c: Bulla formed from the entotympanic (present in pure form only in tree-shrews).
   Type d: Bulla formed from the ectotympanic (present in pure form in rodents and ungulates).
   Type e: Bulla formed from the petrosal (primates).
   Type f: Cartilaginous bulla (rare among mammals, present in a few fruit-bats, one carnivore and one armadillo).

the bulla is thought to be primarily composed of one or more entotympanic elements. Further, elephant-shrews are unusual in having a bulla involving nine elements, with significant contributions from the alisphenoid, ectotympanic, two entotympanics and the petrosal, such that they have the most complex bullar composition among the modern placental mammals (van der Klaauw, 1929; MacPhee and Cartmill, 1986). By contrast, tree-shrews are unusual in that they have a very simple construction of the auditory bulla, with the ventral floor produced largely or exclusively from a single entotympanic element. In this, they are unique among living mammals. Finally, primates are outstanding in that the ventral floor of the bulla is formed exclusively from the **petrosal**. It is widely acknowledged that this is a distinctive feature that sets living primates apart from other mammals. The closest approximation to the primate condition in this respect is found among certain tenrecid insectivores, which have a composite bulla formed from the basisphenoid and the petrosal. The basic kinds of bulla composition among mammals are illustrated in Fig. 7.17.

In addition to these major distinctions among groups of mammals in the gross composition of the ventral floor of the middle ear region, there are numerous differences of detail within groups. For instance, although carnivores generally have a bulla in which there is a combination of the ectotympanic with entotympanic elements, the relative contributions of these components vary from one subgroup to another. On this basis, Hunt (1974) has produced a detailed reappraisal of evolutionary relationships among carnivores. Similarly, there are differences of detail among primates that broadly characterize individual subgroups (Fig. 7.18). In all primates, the petrosal forms the ventral floor of the bulla, but the relationship of the **ectotympanic** to the petrosal bulla can follow one of three basic patterns. In lemurs, the ectotympanic is present as a 'free', simple ring within the bulla, attached to the skull principally at its upper end and typically arranged in an oblique fashion. In lorises and in New World monkeys, the ectotympanic is fused to the outer margin of the bulla. Although the ectotympanic may be slightly extended outwards, thus departing from a simple ring-like shape, it does not form a definite bony tube. Finally, in tarsiers and in Old World simians the ectotympanic is fused to the outer margin of the petrosal bulla and extended into an obvious tube, constituting a bony auditory meatus.

Given these major distinctions among mammalian groups in the gross composition of the bulla (Fig. 7.17) and the finer distinctions that may be observed within groups, as in the case of the primates (Fig. 7.18), reconstruction of the evolutionary changes involved is obviously a complex matter. However, from a survey of living mammals alone it is possible to make a few important generalizations. In the first place, the considerable diversity found among modern placental mammals in terms of bullar structure

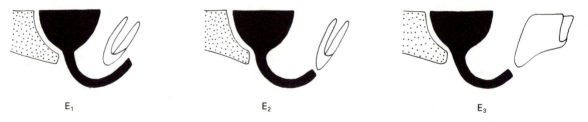

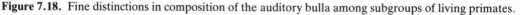

Figure 7.18. Fine distinctions in composition of the auditory bulla among subgroups of living primates.
   Type E_1: Ectotympanic enclosed as a simple ring within the petrosal bulla (lemurs, including subfossil forms).
   Type E_2: Ectotympanic fused to the rim of the petrosal bulla and perhaps slightly extended into a short tube (lorises; New World monkeys).
   Type E_3: Ectotympanic fused to the rim of the petrosal bulla and extended into a definite tube-like bony auditory meatus (tarsiers; Old World monkeys; apes and humans).

clearly suggests that there have been multiple independent developments (Presley, 1978). In other words, the ancestral placental mammals almost certainly had a condition that may have contained some or all of the contributory elements but was not committed to full formation of a bony bulla comparable to any of the major types identifiable today (see Fig. 7.17). The marked variation in bullar composition among modern insectivores (Lipotyphla) indicates that their common ancestor did not have a bony bulla showing any of the modern patterns, especially as some of them (shrews and some moles) completely lack a bony floor to the middle ear region. At the same time, this marked variation among living insectivores confirms the view that these mammals began to diverge very early in the evolution of placental mammals and that the Lipotyphla may indeed constitute a 'wastebasket' category, containing groups of mammals united largely by primitive retentions rather than by derived features. It is also clear, overall, that a basic composition of the auditory bulla became established at an early stage in many individual mammalian groups such that it became a reliable identifying feature of all later members. However, phylogenetic reconstruction requires careful attention to the development of the ear region in individual mammals as well as appropriate interpretation of available fossil evidence.

Ontogeny of the ear region

It has often been noted that, as in the case of the medial orbital mosaic (see above), interpretation of the composition of the bulla in adult mammals is hindered by fusion of sutures during maturation. As a result, ontogenetic information is essential for any confident assessment of bullar formation. In the absence of such information, erroneous conclusions may be drawn. A case in point is provided by confusion in the literature over the constitution of the bulla in tree-shrews. Originally, the bulla was said to be derived from the ectotympanic. Although van Kempen (1905) then correctly stated that the tree-shrew bulla is

formed from an independent entotympanic element, Saban (1956/57, 1963) firmly concluded from a study of adult skulls that the bulla is an outgrowth of the petrosal bone. Detailed study of the ontogeny of the tree-shrew bulla by Spatz (1966) subsequently clearly demonstrated that the bony ventral floor of the middle ear region is produced by ossification of an independent entotympanic element. By contrast, ontogenetic studies of the primate skull (e.g. lemurs and lorises; Forsyth Major, 1899) reveal equally clearly that the primate bulla arises as a direct outgrowth of the petrosal (see also Van Valen, 1965; McKenna, 1966; Cartmill and MacPhee, 1980; MacPhee and Cartmill, 1986).

Apart from permitting reliable identification of the component bones of the bulla, such ontogenetic studies also yield valuable accessory information that permits insights into evolutionary origins. For example, the ectotympanic element of the middle ear region is shown by developmental studies to be a particularly important component in all mammals, as is to be expected from its primary function of providing support for the tympanic membrane. The ectotympanic is always the first component of the bulla to appear and to ossify during ontogeny, and this alone suggests that it was the first element (in addition to the ossicles) to emerge during evolution of the typical mammalian middle ear. Regardless of its ultimate shape and orientation, the ectotympanic in all mammals always passes through a developmental stage in which it has a ring-like (or horseshoe-like) form and is oriented almost horizontally beneath the middle ear. The ventral margin of the ectotympanic ring then swings outwards to varying degrees and the ectotympanic may expand to contribute to the bony bulla and/or a bony external auditory meatus. It is the outward rotation of the lower margin of the ectotympanic ring that creates a ventral gap that is subsequently or simultaneously filled by a bony bullar floor in most mammals.

Ontogenetic studies have also shown that components of the bulla, other than the ectotympanic, fall into two distinct classes

(MacPhee, 1981). Most components arise as **tympanic processes** of the constant bones of the basicranium (i.e. alisphenoid, basisphenoid or petrosal) and thus typically show periosteal growth. Two separate tympanic processes of the petrosal (rostral and caudal) can in fact be recognized. By contrast, **entotympanics** arise as developmentally independent components that typically show endochondral development (i.e. originate by ossification of pre-existing cartilaginous structures). Hence, entotympanics would seem to be novel components that have arisen during mammalian evolution. In fact, no entotympanic elements have been identified in monotremes and it is now believed that marsupials also lack entotympanics, or at least lack entotympanic elements homologous with those of placental mammals, so it is quite likely that entotympanics are novel developments confined to the adaptive radiation of placental mammals.

It is now recognized from ontogenetic studies that there are at least two different kinds of entotympanic element among placental mammals (van der Klaauw, 1931; Novacek, 1977a; MacPhee, 1981), which may occur together in the bullae of certain mammals (e.g. elephant-shrews and various carnivores). The name rostral entotympanic has been given to an anterior element associated with the cartilage of the eustachian tube, and a more posterior element associated with a cranial component known as the tympanohyal is called the caudal entotympanic. In most cases, it is plausible to derive mammalian bullar components from one or both of these elements, although it is possible that there may occasionally be as many as three separate entotympanic elements involved in the bulla. Hunt (1974) has identified three entotympanics – one rostral and two caudal – in the bullae of certain carnivores (e.g. bears). It should be noted that tree-shrews have a single entotympanic element and this can be clearly identified as a rostral entotympanic (Van Valen, 1965; Spatz, 1966; Cartmill and MacPhee, 1980).

Developmental evidence therefore demonstrates beyond doubt that the bulla in tree-shrews is formed by ossification of a distinct rostral entotympanic element, whereas in living primates the bulla typically arises as an outgrowth of the petrosal. Nevertheless, certain authors have suggested that the condition in tree-shrews could easily have given rise to the primate condition through very early fusion of the entotympanic with the petrosal, such that it secondarily lost its developmental independence (van Kampen, 1905; McDowell, 1958; Spatz, 1966; Presley, 1978; Moore, 1981). This hypothetical process of 'primordial fusion' (de Beer, 1937) provides a beguiling means of reconciling tree-shrews and primates with respect to development of the bulla, but it has never been verified (see MacPhee and Cartmill, 1986). In fact, several features of bullar ontogeny in tree-shrews conflict with this interpretation. For instance, Spatz (1966) has shown that the entotympanic of tree-shrews arises as a rostral component, associated with the cartilage of the eustachian tube. The entotympanic then expands in a posterior direction to fuse with the caudal tympanic process of the petrosal, before expanding laterally to form the bulla. As noted by Cartmill and MacPhee (1980), the entotympanic in tree-shrews does not fuse with the rostral tympanic process of the petrosal even in adults, whereas in lemurs and lorises (at least) it is this latter process that grows out in an essentially lateral direction to constitute most of the ventral wall of the bulla. It is significant that there is no evidence for any placental mammal of primordial fusion between an entotympanic and the rostral process of the petrosal.

Le Gros Clark (1959) actually reversed the suggested process of primordial fusion to propose that in tree-shrews the entotympanic element is 'an extension of the petrous bone which has secondarily acquired a developmental independence'. He indicated in a footnote that support for this reverse hypothesis was provided by Saban's investigation of the tree-shrew skull (1956/57). However, it is difficult to see why a tympanic process of the petrosal should become independent in tree-shrews to form an entotympanic, only to rejoin the caudal tympanic process of the petrosal so closely that Saban describes the tree-shrew bulla as an integral part of the petrosal! In any event, the development of the bulla in tree-shrews is so different from that of the primate bulla in several significant respects that the most reasonable interpretation is that the bulla had quite separate evolutionary origins in tree-shrews and primates (Cartmill and MacPhee, 1980).

The fact remains that tree-shrews resemble lemurs, but not other living primates, in the enclosure of a 'free' ectotympanic ring within the bulla (Fig. 7.18E$_1$). It is this striking similarity, unique among living placental mammals, that has provided the most persuasive argument for suggesting a close phylogenetic relationship between tree-shrews and primates. Nevertheless, it should not be forgotten that enclosure of the ectotympanic ring within the bulla is also found in the marsupial *Dasycercus*, although the bulla in this case is formed primarily by a tympanic process of the alisphenoid in the typical marsupial fashion (Wood Jones and Lambert, 1939). *Dasycercus* also differs from tree-shrews and extant lemurs in that the ectotympanic grows outwards to form a bony auditory meatus within the alisphenoid bulla. This difference is not so great as it may seem, however, as lemurs typically possess an annulus membrane extending from the ectotympanic ring to the bullar margin and merging into the cartilage of the external auditory meatus (Cartmill, 1975). Further, in the largest subfossil lemurs (*Archaeoindris, Palaeopropithecus* and *Megaladapis*), the intrabullar meatus formed by the annulus membrane becomes surrounded by bone, as in *Dasycercus* (Lamberton, 1941; Saban, 1963, 1975). The occurrence of an intrabullar ectotympanic in *Dasycercus* in any case shows that this condition is not a unique feature linking tree-shrews to primates. Further, if it is to be argued that the presence of an intrabullar, 'free' ectotympanic ring in tree-shrews and lemurs is a shared derived condition indicative of a phylogenetic relationship, it follows that their common ancestor (i.e. the common ancestor of tree-shrews and primates) must have had this feature. In order to pursue this point further, it is necessary to consider the relevant fossil evidence.

The bulla in fossil mammals

Investigation of the composition of the ear region in fossil mammals is beset by two particular difficulties. First, as has been mentioned above, fusion of sutures in the adult skull can alter the apparent composition of the bulla. Because immature stages are rarely preserved in the mammalian fossil record, the palaeontologist is usually obliged to resort to speculative inference, and there is the additional complication that fossil specimens may be distorted and cracked. The second difficulty is, quite simply, that the bulla may be completely lost from fossil specimens. Of course, this is most likely to happen when the bulla is formed not from one or more tympanic processes of the constant bones of the basicranium but from the ectotympanic and/ or entotympanic elements. Novacek (1977a), for instance, found that the bony bulla was present in only 5 out of 66 well-preserved early Tertiary 'insectivore' skulls from species of the family Leptictidae. Members of this family (e.g. the Eocene/Oligocene *Leptictis*) probably had an entotympanic bulla (McDowell, 1958), but the apparent entotympanic element has been lost during fossilization in most cases. Because of these two major limitations (i.e. fusion of sutures and loss of the bulla during fossilization), considerable caution is required in the interpretation of bullar morphology in fossil mammals.

Loss of the bulla during fossilization has been a major source of confusion in discussions of the early evolution of the auditory bulla among mammals. It was originally believed that the bulla was generally absent among early placental mammals (see Novacek, 1977a). However, van der Klaauw (1931) pointed out that a cartilaginous bulla or a bony bulla consisting of an entotympanic could easily be lost during fossilization and various studies have now reported the presence of a bony bulla in several early Tertiary mammalian genera previously believed to lack one (see Novacek, 1977a). There is still no evidence for the presence of a bulla in certain early placental mammals, notably for the Palaeocene *Palaeoryctes* (McDowell, 1958) or the late Cretaceous *Asioryctes* (Kielan-Jawarowska, Bown and Lillegraven, 1979); but even for these forms one can conclude only very provisionally that the bulla was really lacking in life. Kielan-Jawarowska *et al.* (1979) cite the absence of a bulla as a general characteristic of Cretaceous placental mammals, and this conclusion gains some support from the fact that in the well-preserved skulls of *Asioryctes* the ectotympanic is present as a large ring-like structure inclined at an angle of about 45° to the horizontal plane. However, we have as yet only a very small sample of Cretaceous mammals (see Chapter 4), so caution must still be exercised in accepting the provisional conclusion that the ossified bulla was developed quite independently in different mammalian lineages after the close of the Cretaceous.

The auditory bulla of the fossil 'insectivore' family Leptictidae is of particular interest. In the Eocene/Oligocene form *Leptictis* (= *Ictops*), the ectotympanic was a ring-like structure more or less enclosed by the entotympanic bulla (McDowell, 1958; McKenna, 1966; Novacek, 1977a). Although the bulla was probably incomplete, in that it did not cover the entire middle ear cavity, inclusion of a ring-like ectotympanic within an entotympanic bulla represents a noteworthy similarity between leptictids, tree-shrews and lemurs. It is marginally possible that enclosure of the ectotympanic ring within the bulla was a common ancestral feature of leptictids, tree-shrews and primates (being lost in all descendants except lemurs); but the most likely explanation is that this condition arose independently in these three groups. Whatever the case may be, identification of an intrabullar ectotympanic ring in *Leptictis* considerably undermines the argument, based on living placental mammals, that tree-shrews and lemurs share a unique derived feature in this respect. Novacek (1977a) reports that the late Palaeocene leptictid *Prodiacodon* had an ossified bulla similar to that of *Leptictis*, but it is not yet known whether the possession of an entotympanic bulla with an enclosed ecto-tympanic ring was an ancestral feature of the Leptictidae.

Like all living primates, fossil primates of modern aspect for which the ear region is known had a complete, ossified bulla. As far as is known, the ventral wall of the bulla was formed from the petrosal, as no suture has ever been identified between the petrosal and the bulla. However, the conclusion that all primates of modern aspect, including fossil forms, have or had a petrosal bulla is subject to some element of doubt because of the problem of fusion of sutures in adult skulls.

The bulla of Eocene 'lemuroids' (Adapidae) was very similar to that of modern Malagasy lemurs (e.g. see Gregory, 1920a). The ectotympanic was enclosed as a 'free' ring within the bulla, which was relatively inflated. This condition is seen in *Adapis* (Stehlin, 1912; Simons, 1961a; Saban, 1963; see Fig. 7.19), in *Pronycticebus* (Le Gros Clark, 1934a; Saban, 1963), in *Europolemur* (Franzen, 1987), in *Smilodectes* (MacPhee and Cartmill, 1986), and, by inference, in *Notharctus* (Gregory, 1920a). Eocene 'tarsioids' also had well-developed auditory bullae, in so far as suitably preserved specimens are known. The auditory

Figure 7.19. Ventral view of the auditory bulla of the Eocene 'lemuroid' *Adapis parisiensis*, with part of the bulla floor removed to reveal the enclosed ectotympanic ring (er). (After Simons, 1961a.)

region is best known for *Necrolemur*, which had an inflated bulla with an ossified, tube-like external auditory meatus. Simons (1961a) described the bulla of *Necrolemur*, showing that the ectotympanic ring was actually enclosed within the bulla, but was co-ossified with the lateral margin of the bullar wall and hence continuous with the bony auditory meatus (see Fig. 7.20). Simons concluded that the auditory meatus was therefore a tubular extension of the ectotympanic (see also Saban, 1963). However, because no suture is visible to indicate the division between petrosal and ectotympanic components of the bulla, this conclusion is insecure (Conroy, 1980). The subfossil lemurs *Archaeoindris*, *Palaeopropithecus* and *Megaladapis* indicate a comparable organization of the bulla, but in these forms the ectotympanic ring was relatively small and located so far within the bulla (see Saban, 1963, 1975) that formation of the external bony auditory meatus by the ectotympanic seems most unlikely (see also Szalay, 1975a). (Once again, no suture is visible in the preserved skulls, so the relative contributions of the petrosal and the ectotympanic can only be inferred). As with the marsupial *Dasycercus* (Wood Jones and Lambert, 1939), it seems likely that there has been ossification around the annulus membrane within the bulla to form an enclosed, short bony tube. Whether or not the tube-like meatus extending beyond the lateral margin of the bulla in *Necrolemur* and subfossil lemurs such as *Megaladapis* is formed from the ectotympanic, rather than from the petrosal, is at present a matter of conjecture (Conroy, 1980).

In the Oligocene 'tarsioid' *Rooneyia*, the ectotympanic ring had a similar disposition to that in *Necrolemur*, but with the ring located deep within the bulla (Szalay, 1975a, 1976). Like *Necrolemur*, *Rooneyia* had a short but clearly identifiable bony auditory meatus adjacent to the bulla. It is similarly uncertain whether the external bony meatus was formed from the ectotympanic or from the petrosal, as no suture between these bones is visible.

Subsequent to the Eocene, patterns of bullar formation among fossil primates of modern

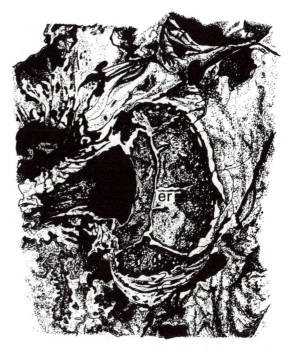

Figure 7.20. Ventral view of the auditory bulla of the Eocene 'tarsioid' *Necrolemur antiquus*, with part of the bulla floor removed to reveal the enclosed ectotympanic ring (er), which is joined to the outer wall of the bulla by an ossified tube. (After Simons, 1961a.)

aspect essentially conform to expectations derived from comparisons of living primates (see Fig. 7.18), although there has been some controversy with respect to early Old World simians (see later). Among strepsirhine primates, the only relevant fossil forms are the Miocene 'lorisoids'. The ear region is documented by skulls of two East African forms: *Komba robustus* (Le Gros Clark and Thomas, 1952) and *Mioeuoticus* sp. (Le Gros Clark, 1956) and by a partial skull of the Siwalik form *Nycticeboides* (MacPhee and Jacobs, 1986). Walker (1974b) has shown that in both African forms the ear region resembles that of modern lorises in several respects, notably in the relationship between the ectotympanic ring and the petrosal and the same is true of *Nycticeboides* (MacPhee and Jacobs, 1986). However, Walker does note that in *Mioeuoticus*, although the bulla is fairly similar to that of the modern *Galago crassicaudatus*, the ectotympanic ring is 'thicker and less closely moulded to the bulla wall'. Overall, it would seem that the essential details of the bullar pattern of modern lorises had been established by the Miocene. Similarly, available fossil evidence (actually quite limited) suggests that the typical bullar pattern in New World monkeys was established at least by the Miocene. In New World monkeys, as in the lorises, the ectotympanic ring is fused to the lateral margin of the petrosal bulla and there is no significant development of a bony external auditory meatus (see Fig. 7.18). This is the pattern found in the Miocene cebids *Cebupithecia* (Stirton, 1951) and *Tremacebus* (Hershkovitz, 1970).

The situation is somewhat different with respect to fossil evidence of Old World simians. Surprisingly, the earliest known skulls from the Oligocene Fayum site had an external conformation of the bulla similar to that in New World monkeys, with the ectotympanic ring fused to the outer margin of the bulla but not extended to form a tubular meatus. This pattern is found in the known skulls of *Aegyptopithecus* (e.g. see Simons, 1972), and Gingerich (1973) has inferred from isolated skull fragments attributed to *Apidium* that this genus also lacked an ectotympanic tube. Indeed, Gingerich even suggested, on the basis of indirect evidence, that *Apidium* might have had an intrabullar 'free' ectotympanic tube. Indeed, Gingerich even contrast to later fossil Old World simians, which conform to the modern pattern, with the ectotympanic applied to the outside of the bulla and usually extended outwards to form a prominent bony meatus (Fig. 7.18). However, subsequent re-examination indicated that the key cranial specimen attributed to *Apidium* by Gingerich was probably derived from a creodont (Cartmill, MacPhee and Simons, 1981). *Apidium* in fact resembles *Aegyptopithecus* in possessing a simple ectotympanic fused to the external margin of the bulla (MacPhee and Cartmill, 1986).

In the earliest Old World monkey skulls (i.e. those of *Mesopithecus*) there is an external, tubular ectotympanic, and this feature is found in all more recent skulls of this group (see illustrations in Szalay and Delson, 1979). Miocene fossil apes and Plio-Pleistocene hominids also typically had a tubular bony external meatus, presumably formed from the ectotympanic, although relevant fossil evidence for the apes is actually quite rare. *Proconsul africanus* had a well-developed ectotympanic tube (Davis and Napier, 1963) and so probably did *Oreopithecus* (Hürzeler, 1968); all known fossil skulls of *Australopithecus* and of *Homo* also have a tubular ectotympanic extension. Nevertheless, it should be noted that the Miocene ape *Pliopithecus* apparently had only partial development of an ectotympanic tube (Zapfe, 1960) and would therefore seem to be intermediate between the Oligocene simians of the Fayum and the modern Old World monkeys, apes and humans. In consequence, there is some suggestion from the fossil evidence that the tubular extension of the ectotympanic to form a bony external auditory meatus may have arisen independently in different lineages of living Old World simians (unless the Fayum simians and *Pliopithecus* diverged before the evolutionary radiation of the remaining known Old World simians).

As with fossil primates of modern aspect, possession of an ossified bulla seems to have been characteristic of *Plesiadapis* and *Phenacolemur* (Szalay, 1976), although this may not have been a universal feature of the 'archaic primates' (plesiadapiforms). The skull is as yet unknown for the earliest representative thereof, *Purgatorius*, and two entire families of plesiadapiforms (i.e. Picrodontidae and Carpolestidae) are still known exclusively from teeth and jaw fragments. The partial skull of *Palaechthon nacimienti* (Kay and Cartmill, 1974, 1977) has traces of ossified bullae, but the specimen is not sufficiently complete to reveal any details of bullar morphology. Nevertheless, there is fairly close similarity in bullar structure between the plesiadapid *Plesiadapis* and the paromomyid *Phenacolemur*, so it is possible that these two genera reflect the general condition for plesiadapiforms. In both cases, the ventral bulla wall appears to be continuous with the petrosal and there is a bony external auditory meatus. It has been concluded by several authors (e.g. see Szalay, 1976) that the plesiadapiform bulla was therefore formed by the petrosal, as in modern primates, and that there was a tube-like extension of the ectotympanic. For the reasons given above, however, this conclusion cannot be regarded as fully reliable. No relevant ontogenetic evidence is available (see also MacPhee, Cartmill and Gingerich, 1983; MacPhee and Cartmill, 1986).

A detailed description of the bullae of *Plesiadapis tricuspidens* was provided by Russell (1964), who noted that they were large and probably completely ossified. Russell suggested that the bulla was formed uniquely from the ectotympanic and an 'entotympanic', although he also pointed out that it was impossible to determine whether the bulla was formed independently of the petrosal or not. The ectotympanic ring was enclosed within the bulla, being completely fixed and continuous with a long, bony auditory meatus (see also illustration in Cartmill, 1975). In this, *Plesiadapis* exhibits a condition superficially resembling that in *Necrolemur* (Gingerich, 1976b) and in *Megaladapis*. Szalay (1972a) suggested that the bulla in *Phenacolemur* was similar to that of *Plesiadapis*, with an ectotympanic ring extended laterally to form a bony auditory meatus lateral to the bulla. However, he noted that the ectotympanic of *Phenacolemur* was largely extrabullar, thus resembling *Tarsius* and Old World simians rather than *Necrolemur*. Szalay has consistently inferred that the bulla of plesiadapiforms was formed from the petrosal and regards this as a derived condition linking them to primates of modern aspect (see also Szalay and Delson, 1979); but this assumption may not be justified (MacPhee, Cartmill and Gingerich, 1983). In fact, the bulla of *Plesiadapis* is superficially very similar to that of the hystricomorph rodent *Lagostomus*, in which the ectotympanic is not only present as a ring-like structure in the bulla, but also forms the ventral bulla wall. As *Plesiadapis* shows no suture

between the ectotympanic ring and the bulla, it cannot be ruled out that the bulla is similarly formed by the ectotympanic (MacPhee and Cartmill, 1986).

The question of bullar composition in plesiadapiforms is complicated by differing views regarding the evolutionary relationships of the Microsyopidae. Szalay (1969) originally linked the Microsyopidae to the origin of primates, but subsequently excluded this family from the order Primates (e.g. see Szalay and Delson, 1979). Bown and Gingerich (1973), on the other hand, regard the Microsyopidae as members of the Plesiadapiformes, as do MacPhee and Cartmill (1986). This difference of interpretation is further complicated by the fact that certain genera included in the Microsyopidae by Bown and Gingerich (e.g. *Palaechthon*) are placed in the Paromomyidae by Szalay and Delson. If the Plesiadapiformes, as defined by Szalay and Delson (1979), represent a monophyletic assemblage, it is possible to retain the provisional hypothesis that possession of a petrosal bulla is a shared feature of plesiadapiforms and primates of modern aspect. However, if the plesiadapiform radiation includes all of the Microsyopidae, there is a major problem in that the Eocene microsyopid *Cynodontomys* (for example) lacked any continuity of the bulla with the petrosal and possibly had an entotympanic bulla (McKenna, 1966; Szalay, 1969). As noted by MacPhee, Cartmill and Gingerich (1983), this would eliminate possession of a petrosal bulla as a consistent shared, derived feature linking plesiadapiforms and primates of modern aspect.

Evolution of the bulla

Both ontogenetic evidence from living mammals and evidence from the mammalian fossil record indicate that the ancestral placental mammals lacked a fully formed auditory bulla (van Kampen, 1905; Gregory, 1910; Novacek, 1977a; MacPhee, 1981). It can be concluded provisionally that the earliest placental mammals known from the fossil record (e.g. *Asioryctes*) lacked an ossified ventral floor to the middle ear region and that there was a fairly large, ring-like ectotympanic inclined beneath the middle ear cavity. This condition has been preserved in certain extant insectivores (e.g. shrews) and it is present in all placental mammals during their development. Although various authors have suggested that early placental mammals may have had a cartilaginous bulla that did not survive fossilization (McKenna, 1966; Martin, 1968a; Szalay, 1975a), this now seems unlikely. As Novacek (1977a) has noted, cartilaginous bullae are extremely rare among modern placental mammals and can best be explained as secondary developments, arising through suppression of ossification.

The great variety of patterns of formation of the auditory bulla among living placental mammals itself suggests multiple origins for this structure during the adaptive radiation of marsupials and placentals. As noted above, among lipotyphylan insectivores there is considerable diversity in this feature, as one would expect from a group that began to subdivide very early from a hypothetical common ancestor lacking an ossified bulla. It is also noteworthy that tree-shrews and elephant-shrews are utterly different with respect to bullar composition. This indicates that the 'Menotyphla' is an artificial grouping based only on the retention of certain primitive placental characteristics (e.g. possession of a caecum) and that tree-shrews and elephant-shrews similarly diverged at a very early stage, before the development of definitive ossified bullae. It has been proposed (Cartmill and MacPhee, 1980; MacPhee, 1981) that ancestral placental mammals had tympanic processes of the alisphenoid and of the petrosal (i.e. a caudal tympanic process of the petrosal), but that all other components of the bulla of various placental mammals developed independently. Significantly, in elephant-shrews there is involvement of both an alisphenoid tympanic process and a caudal tympanic process of the petrosal in the bulla and this can accordingly be explained on the basis of primitive retention from the ancestral placental stock.

Although entotympanics are quite common in the bullae of placental mammals, it is not clear whether there were any precursors to these components in ancestral placental mammals. It seems very likely that rostral and caudal entotympanics arose as novel features of the ossified bullae of placental mammals, but their time of origin remains obscure. Perhaps the most plausible explanation is that ancestral placentals had small cartilaginous elements that later gave rise to ossified caudal and/or rostral entotympanics in several different lineages. All other developments in the evolution of ossified bullae of placental mammals (involvement of the basisphenoid; medial expansion of the ectotympanic into the ventral floor of the middle ear; formation of a rostral tympanic process of the petrosal) would seem to be secondary, arising after the initial divergence from the placental stock. Formation of the bulla predominantly from a rostral process of the petrosal would hence appear to be a unique, shared derived feature of primates, possibly (but not yet demonstrably) including plesiadapiform primates as defined in the narrow sense by Szalay and Delson (1979).

As has been pointed out several times (e.g. Novacek, 1977a; MacPhee, 1981; Moore, 1981), interpretations of the evolution of the auditory bulla in mammals have been hampered by our restricted understanding of functional aspects. Without some grasp of the functional implications of particular features of bullar morphology, it is difficult to comprehend the diversity of patterns found among placental mammals. Nevertheless, it is fairly clear that there is some general functional advantage of an enlarged, enclosed middle ear chamber, perhaps in relation to reception of low-frequency sounds (Moore, 1981). In this connection, MacPhee (1981) has made the significant observation that there is a close correlation between the degree of development of the bulla overall and inflation (**pneumatization**) to produce large air spaces in the ear region. Such pneumatization may involve the bulla itself and it may also affect surrounding parts of the skull. As with the composition of the bulla, there is considerable variation among mammals generally and the lipotyphlan insectivores show a particularly high degree of variability.

The middle ear of all modern primates shows

pneumatization but the site differs from one group to another. For instance, inflation in lemurs primarily takes place ventrally, affecting the bony floor of the middle ear, whereas in lorises it is initiated dorsally and then spreads through the petrosal. This indicates that in the common ancestor of strepsirhine primates there was relatively little pneumatization of the middle ear and that divergent specialization took place subsequently. In tarsiers, pneumatization primarily involves the anterior part of the petrosal, whereas in simians there is an extensive system of air cells in the mastoid region of the petrosal and squamosal as well as in the anteromedial part of the bulla. Pneumatization to form an anterior cavity may have been developed in the common ancestor of living haplorhines (MacPhee and Cartmill, 1986), but divergent developments obviously took place in 'tarsioids' and simians subsequent to their establishment as separate groups. Since pneumatization seems to be closely linked with the overall development of the bulla, the variability among modern primates with respect to patterns of pneumatization provides further evidence that the bulla was relatively poorly developed in ancestral primates and that divergent specialization took place subsequently. Additional support for this interpretation is provided by Szalay's observation (1975a) that three different patterns of pneumatization can be recognized: two in early Tertiary 'tarsioids' (*Necrolemur* and *Rooneyia*) and one in modern tarsiers. Hence, reconstructions of the evolution of the auditory bulla should start from the assumption that the ancestral primates probably had a condition more primitive than in any known living or fossil primate species, rather than from the viewpoint that an essentially primitive condition may be found among known primates.

This point can be illustrated with respect to the difference between extant lemurs and lorises in terms of middle ear structure. As has been discussed above, all living lemurs (and all known subfossil forms) have an intrabullar ectotympanic ring, whereas in all living lorises (and in the Miocene fossils *Komba* and *Mioeuoticus*) the ectotympanic is extrabullar and fused to the outer

margin of the petrosal bulla. Because a condition resembling that in modern lemurs is also found in Eocene 'lemuroids' (e.g. *Adapis* – see Fig. 7.19), there has been a common tendency to regard the condition in modern lemurs as primitive and that in modern lorises as derived. This is possible, particularly as early Tertiary 'tarsioids' (e.g. *Necrolemur* – see Fig. 7.20) had an intrabullar ectotympanic. However, developmental evidence suggests an alternative interpretation (Fig. 7.21).

Cartmill (1975), has demonstrated that the typical adult condition in lemurs is attained following an initial developmental stage in which the ectotympanic ring is adjacent to the developing petrosal floor of the bulla and not yet enclosed within it (Fig. 7.21a,2). Subsequently, the petrosal grows past the ectotympanic ring such that it comes to occupy its eventual intrabullar position (Fig. 7.21a,3). It has also been shown by MacPhee (1977) that pneumatization of the bulla takes place postnatally in lemurs and lorises, following divergent courses in the two groups. In lemurs, a **hypotympanic sinus** is produced through ventral expansion of the middle ear cavity (Fig. 7.21b,II), whereas in lorises (including the Miocene fossil forms *Komba*, *Mioeuoticus* and *Nycticeboides*) a **petrosal sinus** system is produced by pneumatization of the petrosal, beginning at the dorsal limit of the middle ear cavity (Fig. 7.21(b),III). Given that lemurs and lorises diverge not only in the relationship between the ectotympanic and the petrosal bulla but also in patterns of pneumatization, the most likely common ancestral condition for these strepsirhine primates would probably be quite close to the neonatal or early postnatal condition for both lemurs and lorises (i.e. close to the condition shown in Fig. 7.21a,2 and b,I). Further, as haplorhines constitute the sister group of strepsirhines and as in all haplorhines the ectotympanic is fused outside the bulla, it follows that the comparable condition in lorisids is likely to be more primitive than that in lemurs, in which the ectotympanic ring is enclosed within the bulla. It would be appropriate to derive all primates of modern aspect from an ancestor exhibiting the condition of the auditory bulla shown in Fig. 7.21b,I, given the considerable

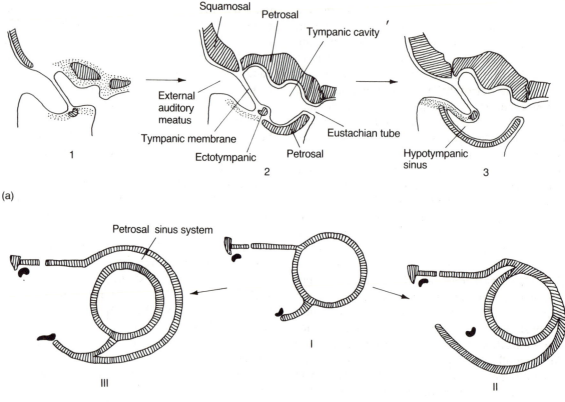

Figure 7.21. (a) Development of the middle ear cavity in a typical lemur: 1, fetal stage; 2, neonatal stage; 3, adult condition (, bone; , cartilage.) (After Cartmill, 1975.) (b) Development of pneumatization, illustrated by schematic diagrams of transverse sections through the cochlea region: I, neonatal lemur or loris (equivalent to stage 2 in part (a); II, adult lemur condition (equivalent to stage 3 in part (a); III, adult loris condition. (After MacPhee, 1977.)

divergence found both in terms of relationships between the ectotympanic and the petrosal bulla and in terms of pneumatization. According to this hypothesis, however, all primates of modern aspect would at least be derived from an ancestor already committed to formation of a ventral floor to the bulla from the petrosal bone.

It should also be noted with respect to pneumatization that all extant simian primates have a common pattern of extensive formation of small air cells in the mastoid area of the petrosal and squamosal as well as in the anteromedial region of the petrosal (Szalay, 1975a; MacPhee,

1981). A similar pattern of pneumatization is apparently found in simian fossils, from the Oligocene Fayum forms onwards. Hence, pneumatization of the petrosal and squamosal with an extensive system of cellules would seem to be a shared derived feature of simians generally.

The internal carotid system

An additional feature of the mammalian auditory bulla that has attracted considerable attention is its relationship to branches of the internal carotid artery. The internal carotid system is of great functional significance since, in concert with the vertebral artery (which enters the skull through the foramen magnum), it ensures the supply of blood to the brain in primates. As with the bulla itself, there is considerable variation among mammals in the patterns shown by the internal carotid system and for this reason it is reasonable to expect that they should yield comparable insights into mammalian evolution.

There has been extensive discussion of the various known patterns of internal carotid circulation, but there is still considerable confusion over nomenclature, largely because of doubts about the underlying homologies. It is generally accepted that in mammals the **common carotid artery** on each side divides into two branches, the **external carotid** and the **internal carotid**. The intrabullar circulation, where it is present, is derived from the latter and it is common to find two distinct branches of the internal carotid artery within the bulla – a **stapedial branch** (which for some reason passes through the aperture of the stapes, hence accounting for its typical stirrup shape in mammals) and a **promontory branch** (which passes over the promontory of the petrosal). Two other vessels have been recognized in connection with the internal carotid, both running along the base of the skull medial to the bulla, rather than passing through the bulla. The first is called the **medial internal carotid** and the second has been termed the **ascending pharyngeal artery**. Both vessels, whether present alone or together, enter the skull through a foramen close to the antero-medial border of the bulla ('middle lacerate foramen'). Confusion has arisen, however, over other vessels and over the identity of the artery involved when there is just a single vessel running medially to the bulla.

A satisfactory resolution to the problem seems to have emerged as a result of careful consideration of criteria for determining homology (Cartmill, 1975; Cartmill and MacPhee, 1980) and of developmental evidence (Presley, 1979). It would seem that the promontory artery and the medial internal carotid artery (which are never found together)

are in fact homologous derivatives of the dorsal aorta. The only significant difference between the two is that the promontory artery passes through the bulla, whereas the medial internal carotid artery passes along the medial side of the bulla. Accordingly, there are only three major vessels to be considered in interpreting the internal carotid circulation of any mammal: a stapedial artery, a promontory or medial internal carotid artery (depending on the relationship between this single vessel and the bulla) and an ascending pharyngeal artery. Typically, modifications to the internal carotid system in mammals seem to have taken place partially through reduction from this original complement of three vessels.

The characteristic shape of the stapes in mammals suggests that a substantial stapedial artery was present in the earliest mammals. In at least some advanced mammal-like reptiles (i.e. therapsids; see Chapter 4) the stapes was perforated (Kermack and Kermack, 1984) and this was the condition in the early mammal *Morganudon* (Kermack, Mussett and Rigney, 1981), so the stapedial branch of the internal carotid circulation would seem to be an ancient characteristic. Novacek and Wyss (1986) have questioned this interpretation on the grounds that the foramen in the stapes is small or lacking in monotremes, most marsupials and some eutherians. However, secondary reduction of the foramen in these mammals would seem to be a more likely explanation in view of its early appearance in the fossil record and the low probability of convergent evolution. The Palaeocene multituberculate *Lambdopsalis* also lacks a foramen in its stapes (Miao and Lillegraven, 1986), but again secondary reduction seems to be most likely. Packer (1987) has shown that noise from the stapedial artery may in some cases interfere with hearing, so it is easy to understand why the stapedial artery and the foramen in the stages may have been lost in some mammals. Most mammals have either a promontory artery or a medial internal carotid artery and the ascending pharyngeal artery is almost universally found in placental mammals (Cartmill and MacPhee, 1980), so it is reasonable to assume that ancestral placentals had all three basic vessels defined above. At present, it is unclear whether the second vessel originally ran medially (as a medial internal carotid) or through the bulla (as a promontory artery). However, Presley (1979) has noted that a medial internal carotid is found in monotremes and marsupials as well as in a variety of placental groups (e.g. rodents, carnivores, 'flying lemurs', edentates, pangolins and leptictids; see also Novacek, 1980; Cartmill and MacPhee, 1980; Wible, 1983). A medial internal carotid was apparently also present in the late Cretaceous placental mammals *Asioryctes* and *Kennalestes* (Kielan-Jaworowska, 1981). It therefore seems likely that the vessel concerned had an extrabullar course in ancestral placentals, but that an intrabullar course developed secondarily in several separate lineages during the adaptive radiation of placental mammals. A promontory artery is found in lipotyphlan insectivores, artiodactyls, cetaceans, most primates, elephant-shrews, tree-shrews and microchiropteran bats (Novacek, 1980; MacPhee, Cartmill and Gingerich, 1983; Wible, 1983).

Following these considerations, it is possible to define a hypothetical ancestral primate condition in which stapedial, promontory and ascending pharyngeal arteries were all present (Fig. 7.22), although the last may not have been connected with the internal carotid (Wible, 1983). This provides a basis for interpreting the various patterns found among living primates, all of which can be seen primarily in terms of reductions from the original pattern of vessels. It is particularly noteworthy that, whereas most living Malagasy lemurs lack any marked development of the ascending pharyngeal artery and have an enlarged stapedial branch of the internal carotid (Fig. 7.22D), mouse and dwarf lemurs (Cheirogaleidae) resemble the lorises in that the intrabullar circulation has been largely or completely suppressed and the ascending pharyngeal artery has been emphasized instead (Fig. 7.22C). The ascending pharyngeal artery enters the skull through the 'middle lacerate foramen', which is located at the anteromedial margin of the bulla.

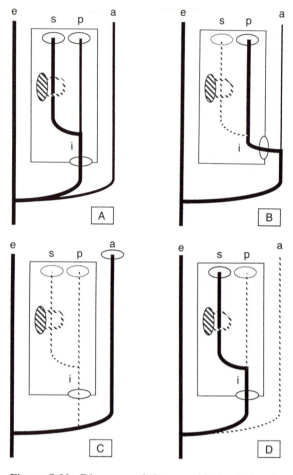

Figure 7.22. Diagrams of the carotid circulation in living primates:

A. Hypothetical ancestral condition.
B. Condition in modern haplorhines.
C. Condition in lorises and in mouse and dwarf lemurs (Cheirogaleidae).
D. Condition in the remaining living lemurs (Lemuridae, Indriidae and Daubentoniidae).

Abbreviations: a, ascending pharyngeal artery; e, external carotid artery; i, internal carotid artery; p, promontorial branch of the internal carotid; s, stapedial branch of the internal carotid (passing through the stapes). The dashed lines indicate reduction or loss of a particular vessel; the central shaded rectangle indicates the location of the auditory bulla and the ovals represent foramina. Note the medial entry of the internal carotid artery into the bulla in the haplorhine condition (2). (Adapted from Cartmill, 1975 and incorporating more recent information.)

This rather striking similarity between cheirogaleids and lorisids has attracted a considerable amount of attention (Cartmill, 1975; Groves, 1974; Hoffstetter, 1974; Szalay, 1975a; Szalay and Katz, 1973; Tattersall and Schwartz, 1974) and it has been widely concluded that a shared derived condition is involved. However, in some cases the reasoning behind this conclusion was based on a misidentification of the ascending pharyngeal artery, either as a medial internal carotid or as an entirely new vessel, and there has been a common assumption that the condition shown by other lemurs (Fig. 7.22D) was the primitive condition for primates. As can be seen from Fig. 7.22, it may be suggested that the condition found in cheirogaleids and lorisids arose primarily by reduction from a common ancestral pattern. A novel development is, however, necessary in that the ascending pharngeal artery must enter the braincase through a special foramen and link up with the brain. It is nevertheless quite reasonable to suggest that convergent evolution might have taken place (see also Cartmill, 1975). Similarity in the carotid circulation between cheirogaleids and lorisids has provided the main basis in the literature for the interpretation that the lemurs are not a monophyletic group completely separated from the lorises. This, in turn, has led to speculation about repeated invasion of Madagascar, or even back-migration of lemurs from Madagascar to the African mainland (Szalay and Katz, 1973). Given the low probability of rafting across the Mozambique Channel and the relatively simple transformation required to derive cheirogaleids and lorisids separately from the hypothetical ancestral condition shown in Fig. 7.22, it would be more parsimonious overall to retain the hypothesis that the Malagasy lemurs are a monophyletic group (Martin, 1972b; see also Chapter 12). Incidentally, it should be noted that the stapedial artery had been greatly reduced in large-bodied subfossil lemurs (*Archaeoindris, Palaeopropithecus* and *Megaladapis*), so the blood supply to the brain must have been ensured by some alternative to the internal carotid system (MacPhee and Cartmill, 1986).

The situation with respect to haplorhines (tarsiers and simians) is relatively straight-forward (Fig. 7.22B). In all haplorhines, the promontory artery is the main vessel in the intrabullar circulation. Tarsiers and simians also share a derived feature in that the promontory artery is walled into the partition separating the anterior cavity of the bulla from the true tympanic cavity (MacPhee and Cartmill, 1986). In tarsiers, the stapedial artery is still present as a small vessel, but in simians this branch of the internal carotid has apparently been completely suppressed. The ascending pharyngeal artery is not present as a major vessel in any living haplorhine, although it may be present as a minor feature of the carotid circulation (e.g. in humans; see Cartmill, 1975).

Overall, it would appear that divergent specialization has occurred in primates, with one of the three original components of the internal carotid system being emphasized in each case. The strepsirhines have emphasized either the stapedial artery or the ascending pharyngeal artery, whereas the haplorhines have emphasized the promontory artery. However, there is at present no obvious and convincing functional explanation for this evolutionary divergence.

Another factor that has often been mentioned with respect to the intrabullar carotid circulation is the presence of bony tubes surrounding the internal carotid and perhaps the stapedial and promontory branches thereof (Bugge, 1974). Such bony tubes are found in tree-shrews and in primates, and this has sometimes been seen as providing evidence of a phylogenetic relationship between these two placental groups (Saban, 1956/57; Szalay and Delson, 1979). However, such bony tubes are also found in a variety of lipotyphlan insectivores (talpids, chrysochlorids and at least some erinaceids – Van Valen, 1965; McKenna, 1966; Novacek, 1980) and they are also partially developed in elephant-shrews (see also Szalay, 1972a). Thus, although it is probably true to suggest that bony tubes enclosing the intrabullar branches of the carotid system were lacking in early placentals (as this is the condition found in most modern mammals), possession of these tubes does not specifically link tree-shrews to primates (see also the later discussion of fossil evidence). Indeed, it seems most likely that tree-shrews and primates developed intrabullar bony tubes separately (Van Valen, 1965; Novacek, 1977a; Cartmill and MacPhee, 1980). Packer (1987) has produced evidence suggesting that arteries within the bulla are enclosed in bony tubes as a device to attenuate noise in mammals adapted for low-frequency hearing. It is for this reason, presumably, that large-bodied mammals typically lack a stapedial artery altogether.

In tree-shrews, as with certain other placental mammals noted above, the intrabullar carotid circulation has both stapedial and promontory branches, with the former being somewhat more prominent than the latter (Cartmill and MacPhee, 1980). There is no substantial vessel running medial to the bulla in tree-shrews, although it is possible that a vestigial ascending pharyngeal artery is present. Hence, the tree-shrew pattern may also be derived from a hypothetical ancestral pattern comparable to that shown in Fig. 7.22A for primates (which would also be the likely ancestral pattern for lipotyphlan insectivores, elephant-shrews and microchiropteran bats). In moderately emphasizing the stapedial branch of the internal carotid, tree-shrews resemble Malagasy lemurs (other than cheirogaleids) to a certain degree; but the great reduction or loss of the ascending pharyngeal artery in both cases could have occurred independently. There is therefore no convincing evidence from the pattern of the internal carotid circulation to suggest a specific phylogenetic relationship between tree-shrews and primates.

It is also possible to obtain evidence regarding the evolution of the internal carotid system from the fossil record, insofar as the pattern of vessels can be inferred from the presence of bony tubes or of grooves within the bulla. However, great care must be exercised; Conroy and Wible (1975) found that a bony tube passing across the promontory within the bulla of the variegated lemur (*Varecia variegata*), alleged to carry a promontory branch of the internal carotid, actually serves as a canal for the internal carotid nerve and the tympanic nerve. These authors also report that several other strepsirhine species (*Microcebus murinus, Lemur catta, Galago senegalensis* and *Nycticebus coucang*) have a bony promontory canal while lacking a promontory artery. Accordingly, interpretations of the internal carotid system of living species should ideally be based on specimens in which the blood vessels have been specially prepared so as to permit direct observation (e.g. Bugge, 1972, 1974; Cartmill, 1975; Cartmill and MacPhee, 1980). Further, any interpretations based on bony patterns alone must inevitably be rather tentative, particularly where the size of a blood vessel is inferred from the size of a bony canal.

It has been claimed that the intrabullar carotid patterns in Eocene 'lemuroids' (e.g. *Notharctus* and *Adapis*) were essentially similar to those in lemurs other than cheirogaleids, with the stapedial artery emphasized, whereas in early Tertiary 'tarsioids' (*Necrolemur* and *Rooneyia*) the promontory artery may have been emphasized, as in modern haplorhines (Szalay, 1975a). In addition, Szalay has suggested that early Tertiary 'tarsioids' share with modern haplorhines a condition in which the internal carotid enters the bulla medially rather than posteriorly (Fig. 7.22B), and that this may represent a derived feature of haplorhine primates (see also Rosenberger and Szalay, 1980). However, the relative degree of development of the stapedial and promontory arteries was apparently variable among Eocene 'lemuroids'. In some specimens, the promontory artery was apparently markedly larger than the stapedial (MacPhee and Cartmill, 1986). Further, in *Necrolemur* at least the stapedial and promontory canals were both large, and the basicranium of this Eocene 'tarsioid' was more like that of contemporary 'lemuroids' than that of modern haplorhines. Approximately equal development of the stapedial and promontory arteries, in Eocene primates generally, fits well with the hypothetical ancestral condition proposed in Fig. 7.22. Further, a medical entry of the internal carotid into the bulla does not in itself provide convincing evidence of a link between omomyids and haplorhines. Medial entry is quite widespread among placental mammals and may, in fact, be the primitive condition (Schmid, 1981; MacPhee and Cartmill, 1986). Omomyids lack the anterior chamber of the bulla found in tarsiers and simians and the promontory artery is hence not walled into a transverse partition. In short, there is no character of the ear region in omomyids that would clearly link them to modern haplorhines.

No fossil primates of modern aspect other than the known Miocene lorisids (e.g. *Mioeuoticus*) had a foramen close to the anteromedial boundary of the bulla indicating the presence of a prominent ascending pharyngeal artery as in modern lorises and cheirogaleids. Hence, it would seem that the ascending pharyngeal artery was either relatively small in ancestral primates or underwent reduction in most lineages early in primate evolution. The fact that the intrabullar internal carotid pattern of Eocene 'lemuroids' seems to resemble that of modern lemurs other than cheirogaleids has tended to add weight to the interpretation that it was the ancestral strepsirhine pattern (e.g. see Szalay, 1975a). However, it is not parsimonious to propose that the ascending pharyngeal artery, which is apparently lacking in these forms, has arisen on different occasions during mammalian evolution, with the same basic associations with other morphological features in each case.

Little is known of the internal carotid circulation in 'archaic' primates (plesiadapiforms). According to Russel (1964), the initial course of the internal carotid artery in *Plesiadapis* was enclosed in a bony tube, but there were no bony tubes for the promontory or stapedial branches. Grooves on the promontory were taken to indicate the presence of a promontory artery and a smaller, rather variable stapedial artery. Subsequent examination of a different specimen by Gingerich (1976b) indicated that the stapedial artery may have been completely lacking and that the internal carotid artery was rather small, suggesting a general reduction of the intrabullar carotid circulation. However, there is apparently no evidence that *Plesiadapis* had a substantial medial internal carotid artery or ascending pharyngeal artery and no 'middle lacerate foramen' has been reported,

so it remains to be established how arterial blood was supplied to the brain in adequate quantities. Szalay (1972a, 1975a) reported that the paromomyid *Phenacolemur* may have had a bony canal for a relatively small promontory artery, but this was later ruled out by MacPhee and Cartmill (1986) and no bony canal for a stapedial artery has been reported for this genus. It now seems to be clear that *Plesiadapis* and *Phenacolemur* did not have a set of bony tubes enclosing an internal carotid system within the bulla, so the suggestion by Szalay and Delson (1979) that ancestral primates had such bony tubes as a derived feature is not logically consistent if plesiadapiforms are included (see also MacPhee, Cartmill and Gingerich, 1983). The virtual absence of bony tubes in the bulla of *Plesiadapis* and *Phenacolemur* also conflicts with the suggestion made by Szalay and Delson that possession of intrabullar bony tubes is a shared derived feature linking tree-shrews with primates (including plesiadapiforms).

In a reconsideration of the morphology of the basicranial region in plesiadapiforms, MacPhee *et al.* (1983) noted that the paromomyid 'archaic primate' *Ignacius* had a foramen, identified as the 'middle lacerate foramen', close to the anteromedial border of the bulla, clearly suggesting the presence of a major vessel passing medial to the bulla. This vessel would have been either a medial internal carotid artery or an ascending pharyngeal artery (as found in Cheirogaleidae and Lorisidae). Presence of an ascending pharyngeal artery in *Ignacius* would be entirely consistent with the view, favoured here, that this vessel was present in ancestral mammals and retained in ancestral primates (Fig. 7.22).

Because of the confusion surrounding the status of the Microsyopidae, which are regarded as 'archaic primates' by some authors, it is difficult to assess the relevance of evidence concerning the carotid system in these forms. Szalay (1969) has described in some detail the basicranial region of *Microsyops* (including *Cynodontomys*, which is recognized as a distinct genus by certain other authors; e.g. McKenna, 1966; MacPhee, Cartmill and Gingerich, 1983). In the region presumably contained within an entotympanic bulla (see above), Szalay identified clear grooves on the promontory for the internal carotid and for its promontory and stapedial branches. In addition, there is a well-defined groove running along the petrosal, medial to the location of the auditory bulla. Anterior to this groove lies a large foramen identified as a 'middle lacerate foramen'. Szalay (1969) interpreted this as evidence for the presence of a medial internal carotid artery in *Microsyops*, but this conflicts with the view (supported here) that placental mammals cannot possess both a promontory artery and a medial internal carotid artery. Instead, it can be argued that *Microsyops*, perhaps in common with the paromomyid *Ignacius*, had an ascending pharyngeal artery that entered the skull through the 'middle lacerate foramen'. In fact, this would mean that *Microsyops* had the same basic set of three branches of the internal carotid artery as proposed here for the ancestral primates (Fig. 7.22A) and for certain other groups of placental mammals.

Whether or not *Microsyops* and its close allies are excluded from the order Primates, as proposed by Szalay and Delson (1979), there clearly was considerable variability among 'archaic primates' with respect to the internal carotid circulation. No clear ancestral pattern emerges for the Plesiadapiformes in relation to the degrees of development of the different components of the internal carotid system, although bony tubes were probably very limited. Overall, there is no convincing evidence from carotid patterns of a direct phylogenetic relationship between plesiadapiforms and primates of modern aspect (see also MacPhee *et al.*, 1983).

THE COMBINED EVIDENCE OF THE SENSE ORGANS

Evidence from the major sense organs clearly has a great deal to contribute to our understanding of the phylogenetic relationships of primates. In the first place, such evidence permits us to recognize living and fossil primates of modern aspect as a group characterized by a number of distinctive features that are generally lacking from 'archaic primates' and from tree-shrews. Secondly, there are numerous indications from the major sense organs that there is a sharp dichotomy among primates of modern aspect – between lemurs and lorises (strepsirhines) on the one hand and tarsiers and simians (haplorhines) on the other. Indeed, as has been shown, the terms 'strepsirhine' and 'haplorhine' relate to a fundamental division with respect to the organization of the olfactory system in these two groups of primates.

Overall, it can be stated that primates of modern aspect have undergone certain specializations of the visual and auditory systems that set them apart from other mammals, whereas the olfactory system has generally retained a primitive configuration or has been subject to reduction. The visual system is particularly important with respect to primate evolution. All primates of modern aspect have at least a postorbital bar, but this structure is uniformly lacking from all known 'archaic primates'. Although it seems likely that ancestral primates of modern aspect had a postorbital bar, the possession of this feature by other mammals (e.g. tree-shrews) cannot be regarded as a reliable indicator of phylogenetic affinity, as a postorbital bar has clearly developed independently in a wide variety of mammalian lineages. Primates of modern aspect also have relatively large, forward-facing orbits, permitting a pronounced degree of binocular overlap. Although certain other mammals (e.g. arboreal marsupials, rodents and carnivores) may resemble primates of modern aspect in terms of relative size and/or forward rotation of the orbits, there is an almost universal distinction in both respects between primates of modern aspect on the one hand and 'archaic primates' and tree-shrews on the other. The orbits are relatively small and essentially laterally oriented in 'archaic primates' and tree-shrews. In living primates the characteristic pronounced degree of binocular overlap is associated with a special arrangement of the visual pathways in the brain (see Chapter 8), which contrasts with the typical vertebrate pattern found in the tree-shrews and in most other mammals. Living primates are further characterized by varying degrees of development of a central area of the retina, involving at least a special arrangement of retinal blood vessels that leaves a clear area adjacent to the blind spot. Tree-shrews lack any such development and have a radial arrangement of retinal blood vessels. The tree-shrew retina also has some unusual features, not found among primates, notably the cone-rich arrangement in *Tupaia*, combined with diffusion rather than summation of the inputs from the photoreceptors.

By contrast with the visual system, the olfactory system of primates of modern aspect does not seem to have any unique specializations that can be interpreted as distinguishing all primates from other placental mammals. It can therefore be inferred that ancestral primates had a primitive olfactory system. With respect to the ear region, there is considerable variation among primates of modern aspect and it is difficult to identify with confidence a common ancestral pattern. For the same reason, there are few clear differences from other placental mammals. However, primates of modern aspect (including fossil forms that have the ear region preserved) uniformly have an auditory bulla formed from the petrosal and they also have bony tubes within the bulla to house branches of the internal carotid artery. Although this latter feature is not confined to primates, it is absent in most placental mammals and was probably not present in the ancestral placental stock. Accordingly, development of bony tubes to house the intrabullar carotid circulation is probably a shared, derived feature of primates of modern aspect.

Strepsirhine primates differ sharply from haplorhine primates in numerous features associated with all of the major sense organs. All living haplorhines have, in addition to the postorbital bar, a postorbital plate that isolates the orbit completely from the temporal musculature, although some evidence suggests that the postorbital plate of modern tarsiers evolved independently of the equivalent structure in simians. Whereas strepsirhine primates typically have a tapetum behind the retina, no haplorhine primate shows any sign of this reflective layer. However, the retina of haplorhines typically shows a greater degree of centralization than that of strepsirhines and all haplorhines except the owl monkey (*Aotus*) have a macula lutea containing a fovea. The distinction between haplorhines and strepsirhines was originally based on the fact that the former (uniquely among mammals) have lost all trace of the rhinarium, whereas the latter have retained it. Examination of the nasal system generally reveals that loss of the rhinarium in tarsiers and simians is associated with at least some degree of reduction of the Jacobson's organ, with reduction in the number of turbinal bones (from the original five to only three) and with loss of the transverse lamina originally separating the posterior end of the nasal cavity into a ventral air passage and a dorsal olfactory chamber. Such reduction within the nasal cavity is associated with closer approximation of the orbits, relative to skull size, with the result that an interorbital septum is formed in the smallest forms.

It therefore emerges that strepsirhines and haplorhines differ in numerous features of the visual and olfactory systems. All of these features may be linked to the fact that strepsirhine primates are predominantly nocturnal, whereas haplorhine primates are typically diurnal (the only exceptions being *Tarsius* and *Aotus*). It is reasonable to suggest that the ancestral primates, like early mammals generally, were nocturnal in habits and that those strepsirhines that have remained nocturnal have generally maintained a primitive pattern of the olfactory system and have undergone only mild specialization of the visual system, involving relative enlargement of the eyes. Haplorhines, on the other hand, can be traced to a diurnal ancestor and the proposed switch from nocturnal to diurnal habits at the time of origin of the haplorhines would account for the reduced emphasis on olfaction and enhanced emphasis on vision. This interpretation is supported by evidence on the relative size of the olfactory bulbs in strepsirhines and haplorhines (see Chapter 8). Additional support for this interpretation comes from the fact that in lemur species that have become diurnal in Madagascar certain changes to the major sense organs have taken place that to some extent parallel those found in haplorhines.

As yet, there is little to suggest that differences between nocturnal and diurnal habits are reflected in differences in organization of the ear region between strepsirhines and haplorhines. However, there is a systematic difference between these two groups in that the internal carotid artery enters the auditory bulla in a medial position in living haplorhines, rather than posteriorly as in living strepsirhines, and that the promontory artery is predominant in haplorhines, whereas in strepsirhines this vessel is reduced and emphasis is placed on either the stapedial artery or the ascending pharyngeal artery. The functional significance of these distinctions between strepsirhines and haplorhines remains obscure, however, and it is true to say that a proper understanding of the evolution of the ear region in primates must await clarification of the functional implications of specific features of construction of the auditory bulla. Functional interpretation of the different patterns of the internal carotid system is also required, although it may be that changes to the vessels to some extent passively follow other changes in the ear region.

As haplorhine primates, humans show all of the distinctive features of the major sense organs that separate haplorhines from strepsirhines. Some other features, notably the complete loss of the accessory olfactory bulb and the configuration of the auditory bulla, distinguish Old World simians from other haplorhines, and we share these features with Old World simians. However, there seem to be no unique features of the major sense organs that set humans apart from all other primates, as is the case (for example) with locomotor adaptations (see Chapter 10). In terms of major sensory adaptations, we are typical Old World simian primates and the most important developments of all clearly took place a long time ago early in the evolutionary history of the primates of modern aspect.

Chapter eight

Evolution of the primate central nervous system

The central nervous system (brain and spinal cord) has been a prime focus of attention in discussions of primate evolution, especially as the relatively large size of the human brain is one of the major hallmarks distinguishing human beings from virtually all other mammals. It is widely, though somewhat uncritically, accepted that the relative degree of development of the brain is particularly prominent among the primates as a group and that the unusually great development of the human brain constitutes a further advance beyond the typical primate pattern. Napier and Napier (1985), for instance, have expressed a common viewpoint in stating: 'The most distinctive characteristic of primates is the size and complexity of the brain'. It is therefore to be expected that evidence from the central nervous system (CNS) should play a major part in any comprehensive attempt to reconstruct primate and human origins. Correspondingly, information regarding the CNS has commonly been at the forefront in discussions of the likely evolutionary relationships of tree-shrews with respect to insectivores on the one hand and to primates on the other (see Chapter 5). In short, a proper understanding of primate evolution requires both a grasp of those features of the CNS

that are of significance throughout the order Primates and an assessment of the extent to which such features are uniquely characteristic of the primates.

In considering the evolution of the primate CNS through comparison of living forms, valuable evidence can be obtained from two distinct sources. The first is provided by comparative quantitative analysis of the overall size of the system or of its component parts. The second involves largely qualitative comparison of particular structural features. For the former, allometric analysis (see Chapter 4, Appendix) is absolutely essential, because the complex relationship between brain size and body size can be expressed satisfactorily only in terms of an allometric equation. If absolute brain size is considered in isolation, then the largest-bodied mammals (e.g. the Indian elephant, with an average brain weight of 5400 g) are clearly superior to modern humans (with an average brain weight of about 1300 g). On the other hand, because of the negative allometric scaling of brain size to body size, use of simple ratios of brain size to body size is equally misleading, this time favouring the smallest-bodied mammals. For example, the relatively primitive lesser mouse lemur (*Microcebus murinus*), with an average brain weight of 1.8 g and an average body weight of 60 g, has a brain:body ratio of 1:33, whereas modern humans, with an average brain weight of 1300 g and an average body weight of 60 kg, have a ratio of only 1:46.

Only after a comprehensive allometric analysis of an adequate sample of species (Bauchot and Stephan, 1966, 1969; Jerison, 1973; Martin, 1973, 1983; Passingham, 1975, 1981; Stephan, 1972) can meaningful comparisons be made between species of different body sizes. Failure to comply with this requirement has, in the past, led to suspect conclusions, as with Le Gros Clark's (1959) and Hill's (1972) discussion of brain size in tree-shrews in comparison with insectivores and primates. The same applies to the common tendency (still prevalent in the literature) to discuss brain size in human evolution without taking into account the probable body size of our fossil relatives. Allometric factors must be taken into account for the pre-eminence of human brain size, relative to all other primates and relative to mammals in general, to become clearly recognizable (see later).

Turning to the question of the structure of the CNS in living primates, there is the advantage that detailed morphological and functional comparisons are possible, at least in some instances, and that these can throw additional light on the evolutionary background to the primate CNS. Studies of individual structural features can be supplemented by allometric analysis of the size relationships of separate components of the CNS, ranging from major subdivisions of the brain (e.g. Stephan, Bauchot and Andy, 1970; Stephan, 1972) down to discrete populations of nerve cells (Armstrong, 1982a). In any case, it is important as a general rule to bear in mind the pervasive influence that body size may exert on central nervous adaptation, even when qualitative structural features of the CNS are being considered.

In several respects – namely, for the external morphology of the brain and for the size of the brain and its major components relative to body size – comparative studies of living species can be extended back in time to include fossil evidence.

In some cases, natural fossil **endocasts** have been produced by infilling of the cranial cavity with matrix during fossilization. Subsequent removal of the surrounding skull bones, either by natural erosion or through careful dissection of the fossil in the laboratory, then reveals a natural replica of the internal contours of the braincase. Further, where fairly complete fossil skulls are available and where it is possible to clean the cranial cavity thoroughly, the cranial capacity can be measured and it is also feasible to produce an artificial endocast of the braincase (Bauchot and Stephan, 1967; Radinsky, 1968). Thus, although the internal morphological features of the brains of fossil forms are completely inaccessible it is possible to trace the evolution of the primate brain through the fossil record quite effectively – in so far as suitable fossil specimens are available. In this way, various inferences based on the studies of modern species alone can be tested.

BASIC STRUCTURE OF THE MAMMALIAN BRAIN

Before embarking on a detailed analysis of evidence relating to the evolution of the primate CNS, it is necessary to consider the basic features typical of placental mammals in general (Fig. 8.1). It is customary to divide the CNS into two main parts – the **brain** and the **spinal cord** – although it essentially functions as an integrated system. Viewed in the simplest terms, the brain is an expanded anterior aggregation of nervous tissue directly linked to the major sense organs (eyes, ears and olfactory apparatus). It also acts as an integration centre for sensory inputs and motor outputs serving the body generally. The more-or-less tubular spinal cord, which runs along most of the length of the vertebral column, is concerned with local reflex responses and with lower-level nervous integration, as well as with conveying nerve fibres to and from the brain. One important distinction between the brain and the spinal cord concerns the distribution of the bodies of nerve cells, or **neurones**, constituting the 'grey matter' relative to the tracts of nerve fibres, or **axons**, constituting the 'white matter'.

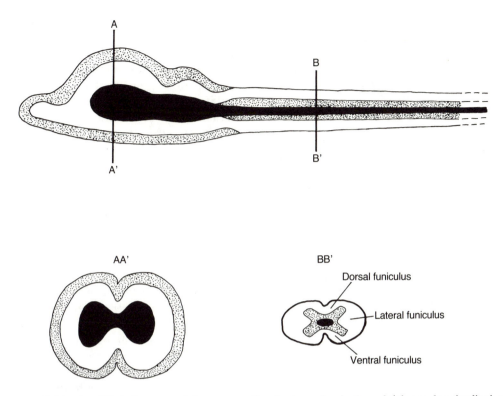

Figure 8.1. Highly simplified diagram of the mammalian brain and spinal cord (above: longitudinal section; below: transverse sections). Note that in the brain (section AA′) the grey matter forms the surface layer (cortex), whereas in the spinal cord (section BB′) the grey matter occupies a central position and has a cross-like configuration with four 'horns' (cornua). Key: stippling, grey matter (neurones + glial cells); white, white matter (nerve fibres); black, cerebrospinal fluid (ventricles in the brain; spinal canal in the cord).

In the brain itself, the grey matter represents the surface layer (**cortex**), whereas in the spinal cord it is located in the core (see Fig. 8.1). With both the brain and the spinal cord, cross-sections reveal a centrally located, interconnected system of cavities (**ventricles** in the brain; **spinal canal** in the cord) containing the cerebrospinal fluid. The cross-section of the spinal cord also shows that the grey matter is arranged in the form of a cross, with two ventral and two dorsal horns (**cornua**). The dorsal horns are concerned essentially with sensory inputs arriving via the spinal nerves, and the ventral horns are involved mainly with the transmission of motor impulses to appropriate regions of the body. The horns divide the white matter into dorsal, lateral and ventral sectors (**funiculi**).

On the basis of gross external morphology (Fig. 8.2), the brain of a typical placental mammal can be subdivided into various regions that reflect broad functional categories. The system is essentially bilaterally symmetrical, with the left half of the brain virtually constituting a mirror image of the right half. The **olfactory bulbs**, the most anterior parts of the brain, are concerned exclusively with the receipt of sensory inputs from the olfactory organs housed in the nasal region of the skull. They are clearly demarcated from the **cerebral hemispheres**, which are concerned with other sensory inputs and with the general integration of sensory information with motor outputs. In any mammalian species with relatively limited expansion of the cerebral cortex (e.g. any

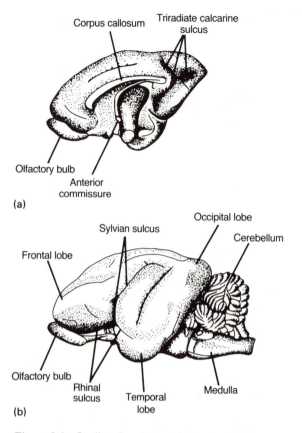

Corpus callosum   Triradiate calcarine
                         sulcus

Olfactory bulb
Anterior
commissure
(a)

                                    Occipital lobe
Sylvian sulcus
                                       Cerebellum
Frontal lobe

Olfactory bulb
   Rhinal   Medulla
   sulcus   Temporal
(b)   lobe

Figure 8.2. Outline drawings of the brain of the brown lemur (*Lemur fulvus*), illustrating the main features of the typical placental brain: (a) cerebral hemisphere and olfactory bulb of the right side of brain separated to expose the internal surface; (b) external view of the left side of the brain. (After Elliot Smith, 1902.)

modern insectivore and some relatively small-brained primates, such as the lemur shown in Fig. 8.2), the lateral aspect of the cerebral hemisphere has an approximately horizontal cleft or fissure – the **rhinal sulcus**. The position of this sulcus marks the separation between the ventral **palaeocortex** (= archipallium or pyriform lobe) and the dorsal **neocortex** (= neopallium). The neocortex owes its name to the observation that it seems to be a new region of the brain developed exclusively in mammals (Elliot Smith, 1901, 1910). A neocortex, as such, is lacking from the brains of reptiles and birds, although in these two vertebrate groups there has been a functionally

convergent development of the brain (Webster, 1973). Thus, the degree of development of the neocortex with respect to the palaeocortex provides one good indicator of the degree of advance of any mammalian brain beyond the condition present in the ancestral placental mammals, and this is reflected by the position of the rhinal sulcus (typically high up in modern insectivore brains; typically low down in modern primate brains; see later, Fig. 8.17).

The sensory inputs from the eyes to the brain are recognizable as the ventrally placed **optic nerves**, which cross over partially or completely before entering the brain, forming the **optic chiasma**. (See later for a discussion of functional aspects of this crossing of the optic tracts.) Corresponding to the relatively posterior location of the ear on the mammalian skull, the **vestibulocochlear nerve** enters the brain posteriorly, close to the rear end of the cerebral hemisphere. Visual and auditory sensory inputs are integrated in the neocortical regions of the brain, in contrast to the olfactory inputs, which are primarily associated with the palaeocortex. The other major regions recognizable from the external surface of the mammalian brain are the **cerebellum**, which is concerned with the integration of motor functions, and, ventral to this structure, the **medulla**, which leads on to the spinal cord approximately at the point where the latter emerges from the skull through the foramen magnum (see Fig. 8.3). The size of the foramen magnum accordingly provides some indirect indication of the dimensions of the spinal cord at its point of junction with the medulla (see later).

The cerebral hemispheres are traditionally described as having three separate lobes – the anterior **frontal lobe**, adjacent to the olfactory lobes; the posterior **occipital lobe**, adjacent to the cerebellum; and the **temporal lobe**, in a ventrolateral position. In the skull, the temporal lobe is located at the rear of the orbits, so its position and relative degree of development depend, at least in part, on the location and size of the eyes. The occipital lobe is primarily concerned with visual functions, the frontal lobe probably with attention-focusing functions and the temporal lobe predominantly with integrative functions involving a variety of sensory inputs and motor outputs (Le Gros Clark, 1959).

A sagittal section through the brain of a typical placental mammal reveals other areas of major topographical and functional importance. Of special significance are two transverse tracts, seen in cross-section (Fig. 8.2), concerned with transmission of information between the two cerebral hemispheres (which otherwise would be obliged to operate as independent integrative units). The most ventral and anterior of these two transverse tracts is appropriately known as the **anterior commissure**, and in reptiles, birds and even some mammals (monotremes; most or all marsupials) this represents the only channel of intercommunication between the left and right cerebral hemispheres. The second (dorsal) commissure shown in Fig. 8.2, which is unique to placental mammals, is called the **corpus callosum**. The corpus callosum is apparently a novel development within placental mammals and, as a universal characteristic, is doubtless derived from an original development in ancestral placental mammals following their divergence from marsupials. Its degree of development relative to that of the anterior commissure therefore provides another indication of the relative degree of advance of the brain of any living placental mammal beyond the ancestral condition. This development of a major second channel of communication between the two halves of the brain is just one indication of the greater overall degree of central nervous coordination characterizing this group of terrestrial vertebrates.

Having established these major features of the CNS of placental mammals, it is now possible to proceed to an analysis of those features that have been particularly important in the evolution of primates.

RELATIVE BRAIN SIZE IN PRIMATES

The first step in the analysis of changes in relative brain size during primate evolution naturally involves the comparative investigation of living primates with respect to other extant mammalian groups. In order to accomplish this first step, it is necessary to decide on standard measures that can subsequently be used for the study of fossil forms, so that they can be compared with hypothetical ancestral stages. For this reason, it is preferable to use **cranial capacity** as a measure of brain size, at least in the first instance, and to investigate the relationship of this measure to body weight. (The problem of establishing body weight for fossil mammals, where complete skeletons are not usually available for the necessary calculations, will be considered in due course.) The systematic use of cranial capacity as a standard measure of brain size is also preferable because it permits the inclusion of mammalian species for which skulls are available in museum collections, but for which no data are available on actual brain weight. Further, cranial capacities can be measured on a relatively large sample of skulls for each modern species, thus allowing individual variation to be taken into account. The analyses discussed below are therefore based primarily on measurements of cranial capacity from a sample of 47 primate species (22 strepsirhines; 25 haplorhines including humans), 5 tree-shrew species and 23 insectivore species. Eight intact skulls were measured for all except the rarest species – such as the pen-tailed tree-shrew (*Ptilocercus lowii*) and the aye-aye (*Daubentonia madagascariensis*), which are poorly represented in museum collections. Cranial capacities were measured using sintered glass beads (see Martin, 1980) to obtain the average values listed in Tables 8.1 and 8.2. Selected dimensions of the skulls were also recorded (Fig. 8.3) to assess possibilities for indirect estimation of both body size and brain size (see later).

It is, of course, possible that the volume of the cranial cavity does not reflect brain size sufficiently accurately to permit the use of cranial capacities for a reliable analysis of the evolution of brain size. In addition, some systematic decrease in the accuracy of cranial capacity, as a measure of brain volume, might accompany increase in body size, as it has been suggested that the external morphology of the brain is least accurately reflected by the internal surface of the

Table 8.1 Body weights, cranial capacities and indices of cranial capacity for (a) strepsirhine and (b) haplorhine primates. Body weights mainly from Stephan, Bauchot and Andy (1970), Eisenberg (1981) and B.C.C. Rudder (personal communication)

(a)

Species	N	W	C	ICC
Lemurs				
Microcebus murinus	7	60	1.8	3.4
Microcebus coquereli	2	385	5.8	3.1
Cheirogaleus medius	8	178	2.6	2.4
Cheirogaleus major	8	403	5.6	2.9
Phaner furcifer	5	440	6.7	3.3
Lepilemur mustelinus	8	630	8.1	3.1
Lemur mongoz	8	1 669	23.0	4.5
Varecia variegata	8	3 388	31.8	3.9
Avahi laniger	8	1 071	9.8	2.6
Propithecus verreauxi	6	3 384	30.6	3.7
Indri indri	8	6 250	33.4	2.7
Hapalemur griseus	8	830	14.6	4.6
Daubentonia madagascariensis	3	2 800	45.2	6.3
Loris group				
Galago demidovii	8	63	2.6	4.8
Galago senegalensis	8	229	3.7	2.8
Galago alleni	8	246	3.7	2.7
Galago crassicaudatus	9	1 340	10.2	2.3
Euoticus elegantulus	8	287	5.1	3.3
Arctocebus calabarensis	8	203	7.6	6.3
Perodicticus potto	8	1 053	13.1	3.5
Nycticebus coucang	8	1 110	11.9	3.1
Loris tardigradus	8	271	6.4	4.4

(b)

Species	N	W	C	ICC
Tarsier				
Tarsius spp.	8	112	3.0	3.7
New World monkeys				
Aotus trivirgatus	7	985	16.9	4.8
Callicebus moloch	8	1 078	18.3	4.9
Saimiri sciureus	8	914	23.6	7.0
Cebus apella	8	2 437	76.2	11.7
Ateles spp.	8	8 200	108.8	7.3
Lagothrix lagotricha	8	6 248	97.2	7.8
Alouatta seniculus	8	6 556	60.3	4.7
Callimico goeldii	1	471	11.1	5.2
Cebuella pygmaea	3	72	6.1	10.2
Callithrix jacchus	8	203	7.2	6.0
Saguinus spp.	8	534	9.9	4.3
Old World monkeys				
Miopithecus talapoin	7	1 250	39.0	9.4
Cercopithecus ascanius	7	3 605	63.4	7.4
Cercocebus albigena	7	7 758	96.9	6.7
Macaca mulatta	8	4 600	83.0	8.2
Papio anubis	8	16 650	177.0	7.3
Theropithecus gelada	6	21 500	133.0	4.6
Colobus badius	8	8 617	61.6	4.0
Apes and humans				
Hylobates lar	8	5 442	99.9	8.8
Hylobates syndactylus	8	10 725	123.7	6.9
Pongo pygmaeus	8	55 000	418.0	7.7
Pan troglodytes	8	45 290	393.0	8.2
Gorilla gorilla	8	114 450	465.0	5.2
Homo sapiens	8	65 000	1409.0	23.0

Abbreviations: N, number of skulls measured; W, body weight (g); C, cranial capacity (cm³, converted to mm³ for calculation of index); ICC, index of cranial capacity.

braincase in large-brained mammals (Radinsky, 1972). Jerison (1973) specifically noted that cranial capacity considerably overestimates brain size in cetaceans (whales and dolphins). These possibilities must be examined as a first priority by analysing the relationship between cranial capacity (C) and some measure of actual brain size (e.g. brain weight, E). Data on cranial

capacities from Tables 8.1 and 8.2 have been plotted, on logarithmic coordinates, against available data on brain weights (Stephan, Bauchet and Andy, 1970) for 33 primate species including *Homo sapiens* (Fig. 8.4). There is a very consistent relationship, with a high correlation coefficient value of $r = 0.996$. (Some departures from an ideal relationship are to be

Table 8.2 Body weights, cranial capacities and indices of cranial capacity for insectivores and tree-shrews. For sources of body weights and abbreviations see Table 8.1

Species	N	W	C	ICC
'Basal' insectivores				
Sorex minutus	8	5.2	0.10	1.0
Sorex araneus	8	10.2	0.15	0.96
Crocidura russula	8	11.0	0.15	0.91
Crocidura occidentalis	8	28.2	0.35	1.1
Suncus murinus	8	35.5	0.31	0.84
Echinops telfairi	8	87.0	0.60	0.88
Hemicentetes spp.	8	110.0	0.70	0.88
Setifer setosus	8	245.0	1.5	1.1
Tenrec ecaudatus	8	1 995.0	3.1	0.54
Microgale talazaci	8	55.0	0.64	1.3
Limnogale mergulus	3	91.0	0.90	1.3
Erinaceus europaeus	8	776.0	3.10	1.0
Echinosorex spp.	7	1 202.0	6.4	1.6
'Advanced' insectivores				
Solenodon paradoxus	4	890.0	5.1	1.6
Potamogale velox	8	660.0	3.7	1.4
Talpa europaea	8	76.0	0.85	1.4
Chrysochloris stuhlmanni	7	39.0	0.75	1.9
Chrysochloris asiatica	8	49.0	0.50	1.1
Desmana moschata	1	437.0	3.4	1.7
Galemys pyrenaicus	8	58.0	1.2	2.3
Elephantulus rufescens	8	38.0	1.3	3.4
Petrodromus tetradactylus	8	186.0	3.0	2.6
Rhynchocyon cirnei	8	427.0	4.9	2.5
Tree-shrews				
Ptilocercus lowii	4	43.0	1.5	3.6
Tupaia glis	8	150.0	3.4	3.5
Tupaia minor	8	45.0	2.0	4.6
Lyonogale tana	8	209.0	3.8	3.1
Urogale everetti	4	347.0	4.3	2.5

expected, as brain weights and cranial capacities were derived from different samples.) The best-fit line (major axis) yields the formula:

$$C = 0.94 \, E^{1.02}$$

This relationship is virtually isometric (see Chapter 4, Appendix) and it can therefore be assumed that cranial capacity provides a fairly reliable indicator of brain weight for primates generally. Indeed, the implication of the empirical formula is that one can regard cranial capacity as approximately equivalent to brain weight for simple comparative purposes for the sample of primates covered (and presumably for fossil primates and other mammals of similar body size). However, it must be remembered that over a large range of body sizes there will be a systematic deviation of C values from E values.

Grades in brain size

Plotting of cranial capacities against body weights on logarithmic coordinates for the species listed in Tables 8.1 and 8.2 broadly confirms the picture originally reported by Bauchot and Stephan (1969) on the basis of actual brain weights. Following Bauchot and Stephan, the overall sample has been subdivided into four categories: 'basal' insectivores, 'advanced' insectivores (here including tree-shrews), strepsirhines (lemurs and lorises) and haplorhines (tarsiers and simians). As Fig. 8.5 shows, the four separate major axes have approximately parallel slopes. All slope values are reasonably close to 0.67 (i.e. 2/3), as is indicated by the following formulae based on the major axes:

1. 'Basal' insectivores ($N = 13$):

   $$\log_{10} C = 0.68 \log_{10} W + 1.51 \quad (r = 0.98)$$

2. 'Advanced' insectivores, including tree-shrews ($N = 15$):

   $$\log_{10} C = 0.62 \log_{10} W + 1.99 \quad (r = 0.94)$$

3. Strepsirhine primates ($N = 22$):

   $$\log_{10} C = 0.68 \log_{10} W + 2.07 \quad (r = 0.94)$$

4. Haplorhine primates, excluding humans ($N = 24$):

   $$\log_{10} C = 0.75 \log_{10} W + 2.07 \quad (r = 0.97)$$

(where C is the cranial capacity in cubic millimetres and W is body weight in grams). Thus, these four groups seem to represent classic allometric 'grades' (see Chapter 4, Appendix) with brain size increasing in a predictable fashion

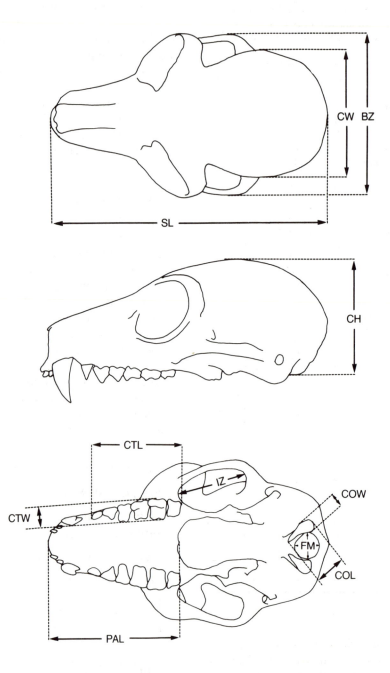

Figure 8.3. A representative prosimian primate skull (*Varecia variegata*), showing standard cranial dimensions used in scaling analyses (top: dorsal view; centre: left lateral view; bottom: ventral view). Finer details of the skull are illustrated in Fig. 7.4. Abbreviations: BZ, bizygomatic width; CH, vertical height of braincase; COL, maximum length of occipital condyle; COW, maximum width of occipital condyle; CTL, length of cheek-tooth row; CTW, maximum width of cheek teeth; CW, maximum width of braincase base; FM, height and width of foramen magnum; IZ, internal length of zygomatic aperture; PAL, length of palate; SL, maximal skull length.

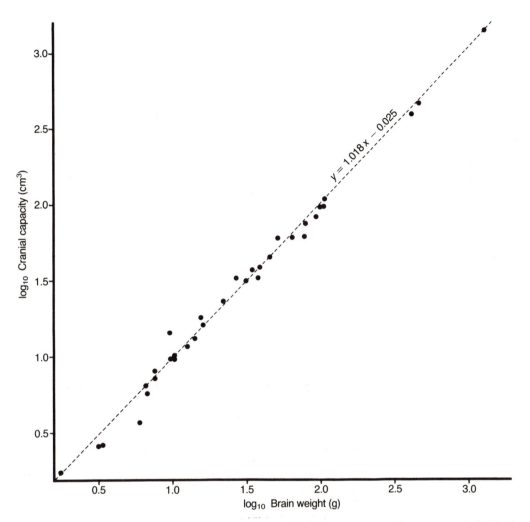

Figure 8.4. Logarithmic plot of cranial capacity against brain weight for primates ($N = 33$ species). The best-fit line is the major axis, the equation for which is indicated. (Brain weights from Stephan, Bauchot and Andy, 1970; cranial capacities from Table 8.1.)

relative to body size within each grade. Each grade follows a common scaling principle, but grade shifts produce progressively larger brain sizes relative to any given body size in the series 'basal' insectivores–'advanced' insectivores–strepsirhine primates–haplorhine primates.

At this level, then, allometric analysis of cranial capacities replicates the conclusions drawn from actual brain weights by Bauchot and Stephan (1969). Therefore, as an overall generalization, the relative size of the brain (or the 'degree of encephalization') can be said to increase progressively through a graded series passing from the 'basal' insectivores through the 'advanced' insectivores and then the strepsirhines on to the haplorhines. Allometric analysis thus makes it possible to quantify the overall differences between the four groups, which are reflected in the vertical distances between the lines. Therefore, in comparison to a

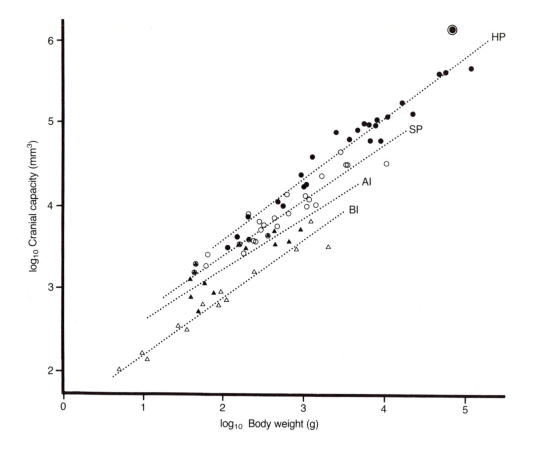

Figure 8.5. Logarithmic plot of cranial capacity against body weight for primates, tree-shrews and insectivores (data from Table 8.1). The best-fit lines are the major axes (equations given in text). Key: BI = 'basal' insectivores ($\triangle$; $N = 13$); AI = 'advanced' insectivores combined with tree-shrews ($\blacktriangle$ and $\oplus$, respectively; total $N = 15$); SP = strepsirhine primates – lemurs + lorises ($\bigcirc$; $N = 22$); HP = non-human haplorhine primates – tarsiers + simians ($\bullet$, $N = 25$; point for *Homo sapiens* is outlined). (Reprinted from Martin, 1983 with the kind permission of the American Museum of Natural History, New York.)

typical 'basal' insectivore of any given body size, a typical 'advanced' insectivore of the same size will generally have a brain about twice as big, a typical strepsirhine will have a brain about four times as big and a typical haplorhine will have a brain about seven times as big. Compared to the typical condition established for the simian primates, humans have a brain some three times bigger when body size has been taken into account. In other words, the human brain (relative to body size) is more than 20 times bigger than a typical 'basal' insectivore's brain.

A standard for comparison

Hence, allometric analysis can be used to make meaningful comparisons between species of different body size that avoids the problems involved in taking either absolute brain size or simple ratios of brain to body size as a basis for the comparison. But it is also evident from Fig. 8.5 that, within each of the four categories examined, there is considerable variation around the best-fit line and, indeed, some degree of overlap between the categories – for example, between 'advanced' insectivores (including tree-shrews) and strepsirhine primates (i.e. lemurs and lorises). In other words, although it is possible to recognize different grades of relative brain size among insectivores and primates, comparisons of individual species require a more detailed examination of the allometric relationships involved. This can be achieved by calculating an index of the vertical deviations of individual points from some standard reference line that expresses the overall scaling of brain to body size. Various calculations of this kind have been made, but all rely on the following:

1. The basic assumption that some standard value of the allometric exponent applies to all species compared.
2. Selection of an appropriate baseline equation against which to compare all species concerned.

Obviously, the value of the index used depends on the biological relevance of the standard equation used in the comparison, and it is here that some fairly complex problems arise.

The scaling equation for mammals

For a long period, there was widespread acceptance of the view that an allometric exponent value of 0.67 is systematically encountered in allometric plots of brain and body size for a wide range of vertebrate groups (Bauchot and Stephan, 1969; Stephan, 1972; Jerison, 1973). This was taken as representing a 'biological law' for vertebrates generally (Gould, 1975). It has been explicitly proposed that an allometric slope value of 0.67 reflects a surface:volume relationship (Brandt, 1867; Jerison, 1973), with the brain increasing in size across a range of species to keep pace with some correlate of body surface area, rather than with body volume. However, the consistent occurrence of an allometric slope value of 0.67 for brain:body size relationships has been questioned (e.g. Portmann, 1962; Szarski, 1980; Martin, 1981c), and two major difficulties can be be identified with the use of a standard exponent value of 0.67 for calculating indices of relative

brain size. First, re-analysis of the scaling of brain size to body size in placental mammals generally indicates that the exponent value is closer to 0.75 than to 0.67 (Bauchot, 1978; Eisenberg, 1981; Martin, 1981c; Hofman, 1982). Second, it has been observed repeatedly that exponent values tend to decrease with decreasing taxonomic rank of the group of species examined (e.g. exponent values determined for entire orders of mammals are typically higher than exponent values determined for individual families; see Martin and Harvey, 1985).

For the overall scaling of brain weight to body weight, analysis of a large sample of placental species ($N = 309$) from 10 different orders produced the following empirical equation (see Fig. 8.6):

$$\log_{10} E = 0.76 \log_{10} W + 1.77 \qquad (r = 0.96)$$

(where E is brain weight in milligrams and W is body weight in grams). The 95% confidence limits on the slope of the major axis (0.73–0.78) indicate that the commonly accepted value of 0.67 for the allometric exponent is definitely too low for placental mammals generally, and an exponent value of about 0.75 has been confirmed for large samples of placental mammal species (Bauchot, 1978; Eisenberg, 1981; Armstrong, 1982b; Hofman, 1982). Figure 8.6 shows that the primates as a group lie almost parallel to the best-fit line and that only a few primate species lie below the line, confirming the view that primates are relatively large-brained mammals. Further, this graph shows that *Homo sapiens* does indeed have the largest brain size relative to body size among placental mammals, but that the closest species in terms of large relative brain size is a cetacean species (dolphin), not another primate. In fact, all of the points close to *Homo* in the graph are cetaceans; the great apes are noticeably lower. Moreover, there is considerable overlap between primates and other placental mammals, so it is clear that primates are not as sharply distinguished from other mammals, in terms of brain size, as is often maintained.

However, the objection could be raised that the slope of the best-fit line in Fig. 8.6 (and hence the inferred exponent value of 0.76) may be

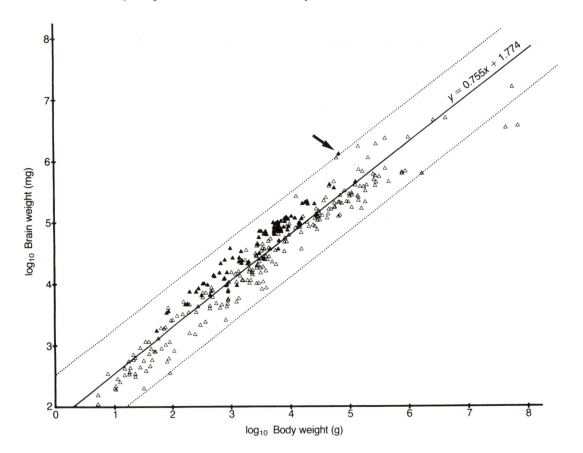

Figure 8.6. Logarithmic plot of brain weight against body weight for 309 extant placental mammal species: △, non-primates; ▲, primates; arrowed point, *Homo sapiens*. The best-fit line (solid line) is the major axis for the entire sample. The dotted lines indicate fivefold variation on either side of the major axis. (Reprinted by permission from Martin, 1981c; *Nature* vol. 293, pp. 220–3. Copyright (C) Macmillan Journals Limited.)

biased by the range of species represented. There is, in fact, a fairly large proportion of small-bodied mammals (notably rodents) with relatively small brains for their body sizes, which might tend to increase the slope of the line. On the other hand, the largest mammals included (whales) also have relatively small brains for their body sizes, which would tend to decrease the slope of the line. In order to counter this potential problem of bias, average logarithmic values of brain and body weight have been calculated for the individual orders of placental mammals such

that the best-fit line is not affected by the number of species at any given body size (Armstrong, 1982b; Martin, 1983). When these average values for individual orders are examined (Fig. 8.7) it emerges that the value of 0.67 for the exponent is ruled out even more emphatically. The best-fit line (major axis) yields the following empirical relationship:

$$\log_{10} E = 0.78 \log_{10} W + 1.61 \qquad (r = 0.993)$$

The 95% confidence limits on the slope are quite narrow (0.72–0.84), again excluding the value of

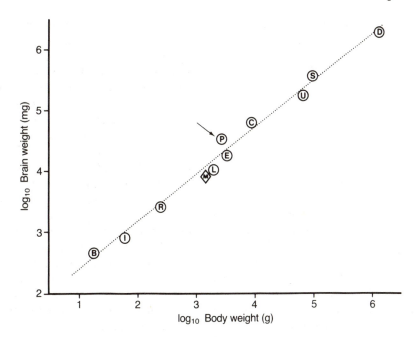

Figure 8.7. Plot of average logarithmic values for brain and body weights for 10 orders of placental mammals (circles) and for marsupials (diamond). The best-fit line (major axis) has been fitted to data for placental mammals only, and it is noteworthy that marsupials nevertheless lie as close to the line as rabbits and hares, contrary to the general belief that marsupials are 'primitive' mammals. Key: B, bats (Chiroptera); C, carnivores (Carnivora); D, dolphins and whales (Cetacea); E*, edentates (Edentata); I*, insectivores + tree-shrews (Insectivora + Scandentia); L, rabbits + hares (Lagomorpha); M*, marsupials; P, primates; R, rodents (Rodentia); S, seals and sea-lions (Pinnipedia); U*, hoofed mammals (Artiodactyla + Perissodactyla). (* Treated as single orders for purposes of analysis.) Note that the point for primates (arrowed) is the most prominent outlier above the line. (Reprinted from Martin, 1983 with the kind permission of the American Museum of Natural History, New York.)

0.67 as being too low. This analysis also shows (Fig. 8.7) that the Primates as an order are distinctive in having a relatively high average brain weight compared with the general trend, a feature that only becomes clear once the scatter of individual species seen in Fig. 8.6 has been condensed into a single value.

Even this approach is open to the objection that biased distribution of species within orders might influence the average values used to determine the best-fit line. For this reason, Martin and Harvey (1985) repeated the analysis by calculating logarithmic averages for brain and body weights at successively higher taxonomic levels to arrive at weighted average values for 15 orders of mammals (including marsupials). The

best-fit line (major axis) for this analysis yields an empirical exponent value of 0.72 (95% confidence limits: 0.68–0.77), once again excluding the value of 0.67 as being too low but permitting a value of about 0.75. Thus it can be concluded that all analyses involving large samples of mammalian species agree in ruling out a scaling exponent value of 0.67 as too low and in yielding exponent values of around 0.75.

Encephalization quotient

This conclusion is important both for the calculation of indices of relative brain size and (see later) for the interpretation of brain-size scaling itself. Jerison (1973), for instance, defined an index of relative brain size, termed the **encephalization quotient** (EQ), based on an overall scaling equation for living mammals (mostly placental mammals). The equation was derived by assuming an exponent value of 0.67 and applying it to brain and body sizes of a range of mammalian species:

$$E = 0.12 \, W^{0.67}$$

(where E is brain weight and W is body weight, both in grams). For each species, the EQ value is calculated by first determining the brain size expected (E_E) for its body size from the equation and then relating it to actual brain size (E_A) as follows: $EQ = E_A/E_E$. In other words, EQ expresses the ratio between the actual brain size of a given species and the brain size expected for a hypothetical 'typical' mammal of the same body size. In principle, this is a very useful basis for comparing mammalian species, but in practice its value is now questionable because the assumed exponent value of 0.67 is not compatible with the results of subsequent analyses, as indicated above. If, as now seems likely, the exponent value for placental mammals generally is actually close to 0.75, the EQ values determined from Jerison's formula will tend to underestimate relative brain size in small mammals and to overestimate it in large mammals. Hence, Eisenberg (1981) has redefined mammalian EQs on the basis of the following equation (converted to correspond in form to Jerison's equation above):

$$E = 0.05 \, W^{0.74}$$

Given that the scaling exponent value of 0.74 is compatible with numerous analyses of large samples of mammalian species, this equation is probably more suitable than Jerison's for comparisons of brain size among mammals generally.

Changes with taxonomic level

There remains, however, the second major problem, posed by the variation of scaling exponents with the taxonomic level at which allometric analysis is conducted (Bauchot, 1978;

Gould, 1975). Although there is now good evidence that a scaling exponent close to 0.75 applies for analyses of placental mammals and possibly for the class Mammalia as a whole, exponent values determined at lower taxonomic levels are typically lower than this (i.e. for species within individual orders, suborders, families, subfamilies or genera). As has been pointed out by Holloway and Post (1982), the conclusions drawn from allometric analysis may differ markedly according to which taxonomic level is selected for comparison. For primates, the following average exponent values are determined at different taxonomic levels (Martin and Harvey, 1985): infraorder ($N = 4$): 0.76; subfamily ($N = 11$): 0.59; genus ($N = 9$): 0.45. A detailed analysis of variation in scaling exponent values with taxonomic level (Martin and Harvey, 1985) showed that there is a general trend for these to decrease with decreasing taxonomic level in six orders of mammals (Primates, Rodentia, Carnivora, Insectivora, Chiroptera and Artiodactyla). However, the pattern of change varies considerably from one order to another. Nevertheless, an exponent value of 0.67 does not emerge as particularly prominent at any taxonomic level. At the ordinal level for Insectivora and Primates values significantly above 0.67 are found, although in no case is there an exponent value significantly greater than 0.75. Thus, exponent values for brain-size scaling may increase up to 0.75, but do not significantly exceed this level, suggesting that this may be a biologically important limiting value for mammals generally (Martin and Harvey, 1985). Martin (1981c) has suggested that a scaling exponent of 0.75 is found because the basal metabolic rate of the mother, which scales with this exponent according to Kleiber's Law (Chapter 4, Appendix), acts as a constraint on fetal brain development.

Exponent values below 0.75, and even below 0.67, that have been recorded for low-level taxonomic groups may be partly explained with respect to the process of brain development. Very low scaling exponent values are recorded in intraspecific comparisons, where the individual points in a bivariate plot represent adult members of the same species rather than average values for separate species. It is commonly stated that intraspecific scaling exponent values lie in the range of 0.2–0.4, but in some cases the comparisons have not been clearly restricted to mature adults and primates typically have intraspecific exponent values indistinguishable from zero (Martin and Harvey, 1985). Re-analysis of intraspecific data for primates and other mammalian species (Martin and Harvey, 1985) indicates that the relationship between brain weight and body weight during development can be broken down into three separate phases (Fig. 8.8). During the earliest phase of brain development (phase 1), brain weight increases almost in proportion to body

size and the slope of the line is therefore fairly close to unity (Martin, 1983). In mammalian species such as primates, which produce relatively well-developed neonates (**precocial** type; see Fig. 9.12), the end of this phase coincides approximately with the time of birth, after a relatively long gestation period. In other mammalian species, such as most insectivores, which produce poorly developed neonates (**altricial** type; see Fig. 9.12) after a relatively short gestation period, this initial 'fast' phase of brain development continues for some days or weeks after birth.

During the ensuing phase 2, brain growth is much less rapid than body growth generally and the slope of the line falls into the range 0.2–0.4. This phase typically persists until the time of attainment of sexual maturity. The final phase, phase 3, corresponds to differences in brain size among adults of different body sizes and in this case it is not so easy to equate body size with age. In phase 3, scaling exponents tend to be even lower than for phase 2, commonly falling below 0.2. In the case of primates, as noted above, phase 3 scaling exponents for brain:body size relationships among adults are generally close to or equal to zero when males and females are examined separately to eliminate the effects of dimorphism in body size (Martin and Harvey, 1985).

From a phylogenetic point of view, it is really phase 2 that is of most interest here. In the divergence between closely related species, differences in body size are common. If such closely related species share similar growth trajectories for brain and body size (see Shea, 1985), and if differences in body size between species are achieved largely by altering the timing of sexual maturity in phase 2 (Fig. 8.8), it follows that brain:body size scaling among such species (e.g. several species belonging to a single genus) would have a low exponent value in the range 0.2–0.4. Shea (1985) has specifically suggested that various morphological differences among the extant great apes may be explained in terms of scaling along common growth trajectories. It can further be suggested (see Lande, 1979a) that in the early stages of species diversification,

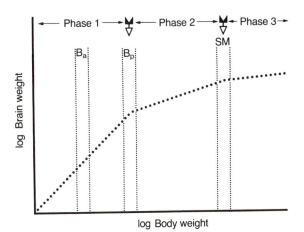

Figure 8.8. Schematic breakdown of the relationship between brain size and body size within a single mammalian species (intraspecific allometry). Phase 1 represents rapid early brain growth in development. Birth in altricial mammals (B_a) occurs during this phase, so that rapid brain growth continues for a period after birth, whereas birth in typical precocial mammals (B_p) approximately coincides with the end of phase 1. During the subsequent phase 2, which terminates approximately at the time of attainment of sexual maturity (SM), brain growth continues much more slowly than body growth and the slope of the line is significantly decreased. Phase 3 represents variation in brain weight with body weight among adults. Note that body weight gives an approximate indication of age, but particularly in the case of adults it should be remembered that time is not directly represented in this diagram.

selection may act largely to favour different body sizes, with brain size simply following (with a low scaling exponent) as a correlated character. Subsequently, however, selection may act directly on brain size itself, such that in a broad comparison of mammalian species brain size will be found to scale to body size in a manner reflecting optimization of brain:body relationships (i.e. with an exponent of approximately 0.75, as indicated above). This could provide a general explanatory basis for the observation that exponent values vary according to the taxonomic level of comparison, ranging from the species level up to the level of the entire class Mammalia.

Variation in exponent values with taxonomic level may also be partly attributable to statistical problems arising from fluctuations in body weight and from declining sample sizes (e.g. see Pagel and Harvey, 1988). It is noteworthy in this respect that among orders of placental mammals empirically determined exponent values vary according to the sample size of the species examined. Taking Bauchot's (1978) figures, it is found that for the six largest orders ($N = 34$–147 species per order) the average exponent value is 0.71 ± 0.09, whereas for the six smallest ($N = 4$–25 species per order) the average exponent value is 0.48 ± 0.12. This might be explained on the basis that the largest placental orders represent cases where there has been greater selection for 'optimal' brain:body size relationships (e.g. because of more ancient divergence or because selection on competing species belonging to large orders has been more intense). Alternatively, it might simply reflect statistical aberrations.

Because of variation in scaling exponents with taxonomic level, and because of continuing uncertainty about their significance, it is difficult to provide binding criteria for analysis of relative brain size (e.g. by prescribing standard exponent values for each taxonomic level). Considerable variation among mammalian orders in the patterns of change with changing taxonomic level makes the problem even more difficult. The best that one can do in the circumstances is to adopt pragmatic guidelines that take account of the

observed variation in exponent values with taxonomic level:

1. Analyses should ideally be conducted at several different taxonomic levels, to reveal any variation in exponent values.
2. If a single taxonomic level is selected for analysis, it should be chosen according to the biological relevance of that level for the specific aims of the study.
3. It should be remembered that decreasing taxonomic level entails decreasing sample size, such that at the lowest levels (e.g. comparison of species within a single genus) it may be difficult, for purely statistical reasons, to determine scaling exponents that are really meaningful.

Indices of cranial capacity

Returning now to the four grades of relative brain size identified by Bauchot and Stephan (1969; see also Stephan, 1972) and illustrated in Fig. 8.5, it can be seen that the analysis meets the above criteria for a restricted comparison between insectivores, tree-shrews and primates. If we are concerned only with assessing the relative brain size of insectivores, tree-shrews and primates, rather than with considering placental mammals overall, it is reasonable to use the equations for the best-fit lines in Fig. 8.5, as a basis for quantitative comparison.

Bauchot and Stephan (1966, 1969), Stephan and Andy (1969) and Stephan (1972), considering data on actual brain weight in relation to body size, selected their equation for 'basal' insectivores as a baseline against which to assess relative brain size in their other three grades ('advanced' insectivores, prosimian primates and simian primates):

$$\log_{10} E = 0.63 \log_{10} W + 1.632$$

In effect, they took the 'basal' insectivore line as indicative of a minimal condition for all species involved in the comparison. An 'index of progression' (IP) was then calculated as the ratio between the actual brain weight of any species (E_A) and the brain weight predicted for the body size of that species from the 'basal' insectivore line (E_P): IP = E_A/E_P. This contrasts with the approach taken by Jerison (1973) and by Eisenberg (1981), who calculated relative brain size (EQ) in relation to the average condition established for placental mammals generally. For the specific purposes of comparison of primates, tree-shrews and insectivores, the 'index of progression' defined by Bauchot and Stephan would nevertheless seem to be perfectly justifiable.

By the same token, it would seem to be quite reasonable to take the best-fit lines illustrated in Fig. 8.5 as a basis for deriving an **index of cranial capacity** (ICC) for a limited comparison between primates and insectivores. The average exponent value for the four best-fit lines in Fig. 8.5 is 0.68

and this happens to coincide with the exponent value in the empirical formula for 'basal' insectivores. That formula can therefore be used as a basis for calculating an index of cranial capacity as the ratio between the actual cranial capacity (C_A) of any species and the cranial capacity 'expected' (C_E) for that body size from the 'basal' insectivore equation: ICC = C_A/C_E. Indices of cranial capacity have accordingly been calculated for all of the primate and insectivore species listed in Tables 8.1 and 8.2. The index values obtained are presented in histogram form in Fig. 8.9 and the cranial capacity indices for individual groups of species are summarized in Table 8.3. These index values now provide an acceptable basis for comparison of brain size between species (following the pattern

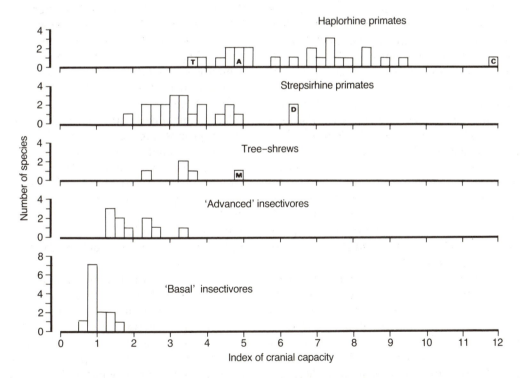

Figure 8.9. Histograms of cranial capacity index (ICC) values for 'basal' insectivores, 'advanced' insectivores, tree-shrews, strepsirhine primates and haplorhine primates, compiled from the data in Table 8.1. *Homo sapiens*, with an index value of 23.0, is not shown so as to avoid undue compression of the non-human primate distribution. Species of particular interest are indicated as follows: A, owl monkey (*Aotus trivirgatus*); C, capuchin monkey (*Cebus apella*); D, aye-aye (*Daubentonia madagascariensis*); M, pygmy tree-shrew (*Tupaia minor*); T, tarsier (*Tarsius* spp.).

Table 8.3 Group-by-group summary of indices of cranial capacity (ICC values given in Tables 8.1 and 8.2)

Group	N	Average ICC $\pm$ s.d.*	95% range[†]
'Basal' insectivores	13	1.0 ± 0.3	0.4–1.6
'Advanced' insectivores (excluding tree-shrews)	10	2.0 ± 0.7	0.6–3.4
Tree-shrews	5	3.5 ± 0.8	1.8–5.0
Strepsirhine primates	22	3.6 ± 1.1	1.4–5.8
Haplorhine primates (excluding *Homo sapiens*)	24	6.7 ± 2.1	2.5–10.9
Simian primates (excluding *Homo sapiens*)	23	6.9 ± 2.0	2.9–10.9

* s.d., standard deviation; † 95% range = mean $\pm$ (2 × s.d.). Note: Use of standard deviation to define 95% confidence limits requires normal distribution of the data. This requirement may not be met – see Fig. 8.9.

established by Bauchot, Stephan and others) in that appropriate allowance for the effects of body size has been made. It is now possible to turn to specific questions regarding the evolution of brain size in primates and, in particular, to consider the status of the tree-shrews in this respect.

Although it is possible to accept the existence of four different 'grades' on the basis of Figs. 8.5 and 8.9, it is also clear (especially from Fig. 8.9) that there is considerable overlap between them. In other words, relative brain size (as expressed by the index of cranial capacity) provides a useful criterion for distinguishing between entire groups of primates and insectivores, but it is not in itself adequate for assessing the position of individual species. Tree-shrews, for instance, can be regarded either as large-brained insectivores or as small-brained primates, and tarsiers can be regarded either as prosimians with moderate brain size or as small-brained haplorhines. On grounds of relative brain size alone, there is no more justification for postulating a link between tree-shrews (average ICC = 3.5) and primates than for suggesting a similar link between primates and certain species of elephant-shrew such as *Elephantulus rufescens* (ICC = 3.4) or *Petrodromus tetradactylus* (ICC = 2.7). This conclusion is amply reinforced when, following the ground-rules set out in Chapter 3, the

comparison is extended to cover placental mammals generally. As is clear from Fig. 8.6 (see also Jerison, 1973; Passingham, 1981), strepsirhine primates are actually far from outstanding in terms of relative brain size in comparison to other placental mammals. Among simian primates the relative size of the brain does indeed tend to exceed that in many other placental mammals, but even here other mammalian groups (e.g. seals and dolphins) overlap the simian range. Only in humans is relative brain size clearly exceptional among mammals; yet, as has been noted, the values closest to that of *Homo sapiens* are found not among simian primates but among the cetaceans (i.e. in dolphins). For similar reasons, the moderately sized brain of *Tarsius* cannot be cited as a reliable criterion for either inferring or refuting a particular link with simians. In sum, one must conclude that relative brain size, in itself, is not an adequate criterion for inferring phylogenetic relationships among mammals and that the primates as a group are not as outstanding among the mammals in this respect as a limited comparison with insectivores might suggest. This is but one example of the more reliable perspective that can be achieved by assessing primate evolution within the broader framework of the adaptive radiation of the placental mammals generally.

Evolution of brain size

Returning once again to the graphical comparison between insectivores and primates presented in Fig. 8.5, the apparent step-wise progression from one subgroup to another ('basal' insectivores to 'advanced' insectivores, to strepsirhines, to haplorhines and on to humans) deserves some special comment. This provides one of the most striking illustrations of an apparent *Scala naturae* available in the literature (see Chapter 3), and also demonstrates clearly how this appearance is only superficial and can act as a beguiling source of error in phylogenetic analysis. It is, for instance, tempting to regard the typical condition established for relative brain size in modern 'basal' insectivores as equivalent to the ancestral placental mammalian condition. Similarly, one might easily be led to equate the typical condition established for living strepsirhine primates as equivalent to the ancestral primate condition, and so on. In both cases, these tempting parallels involve the hidden assumption that expansion of relative brain size has been completely arrested in entire groups of mammals (e.g. 'basal' insectivores and strepsirhines) at different times and subsequently maintained unchanged for tens of millions of years. In short, although it is both valid and instructive to compare relative brain sizes in living primates and insectivores, when it comes to tracing the likely course of brain evolution through the phylogenetic history of the order Primates it is necessary to conduct careful reconstruction with reference to the fossil record.

Cranial capacity of fossil mammals

The first question that must be asked with respect to the fossil record relates to the likely relative brain size of the earliest mammals. Since the placental mammals apparently arose at some time during the Cretaceous period (see Chapter 4), it would ideally be necessary to establish relative brain sizes (i.e. indices of cranial capacity or encephalization quotients) for a sample of Cretaceous placental mammals. Unfortunately, however, the known fossil remains of fossil

mammals for this period are very limited and there are as yet no published accounts of relative brain size. The closest that one can come to determining an appropriate basic value, at present, is through the examination of a sample of early Tertiary (Palaeocene/Eocene) placental mammals for which data on both brain size and estimated body size are available. Jerison (1973) analysed just such a sample, which he referred to as 'archaic mammals', and found that all of the values lay near the lower margin of the polygon on a graph containing all the available values for modern mammals. Although Jerison concluded that the values for these 'archaic' mammals are included within the range of modern mammals, his figure indicates that some species, at least, lie below the modern range.

Taking Jerison's figures (1973) for the cranial capacities and inferred body weights of the 'archaic' mammals, it is possible to calculate indices of cranial capacity (ICC values) using the 'basal' insectivore formula given above (Table 8.4). It emerges that some of these early mammals had very low index values compared to modern 'basal' insectivores (compare Table 8.4 with Tables 8.1–8.3), whereas others were already quite advanced in the early Tertiary (Palaeocene/ Eocene). Given the fact that at least 30 million years had elapsed between the presumed origin of the placental mammals and the beginning of the Palaeocene, it can be fairly confidently concluded that the 'archaic' mammals examined had already developed a modest degree of diversity in relative brain size. The ancestral placentals doubtless had quite small brains even in comparison with modern 'basal' insectivores. Conversely, Novacek (1982b) has reported an estimated cranial capacity of 6 cm^3 and an estimated body weight of 750 g for the Oligocene leptictid *Leptictis dakotensis*. These values yield an ICC value of 2.1, lying well within the range of modern 'advanced' insectivores and overlapping with the range for tree-shrews. The neocortex was clearly better developed than in some modern lipotyphlan insectivores. It is therefore misleading to treat either known fossil or modern insectivores as 'frozen ancestors' with respect to the relative size of their brains and to equate

Table 8.4 Indices of cranial capacity (ICC) for Palaeocene/Eocene (PAL/EO) 'archaic' mammals (data from Jerison, 1973)

Species	Epoch	C (ml)*	W (kg)*	ICC
Multituberculate				
Ptilodus montanus	PAL	1.09	0.20	0.92
Condylarths				
Arctocyon primaevus	PAL	38	86	0.52
Arctocyonoides arenae	PAL	8.3	11	0.46
Pleuraspidotherium aumonieri	PAL	6.0	3.3	0.75
Phenacodus primaevus	EO	31	56	0.57
Meniscotherium robustum	EO	15	6.2	1.22
Hyopsodus miticulus	EO	3.2	0.63	1.23
Amblypods				
Pantolambda bathmodon	PAL	19	30	0.53
Leptolambda schmidti	PAL	69	205	0.52
Barylambda or *Haplolambda* sp.	PAL	102	620	0.36
Coryphodon hamatus	EO	93	270	0.58
Coryphodon elephantopus	EO	90	540	0.35
Uintatherium anceps	EO	250	1400	0.56
Tetheopsis ingens	EO	350	2500	0.48
Creodonts				
Thinocyon velox	EO	5.7	0.80	1.87
Cynohyaenodon cayluxi	EO	8.3	3.0	1.11
Pterodon dasyuroides	EO	62	42	1.38

* Converted into units of mm^3 for cranial capacity (C) and g for body weight (W) to calculate 'expected' cranial capacities from the formula:

$$\log_{10} C = 0.68 \log_{10} W + 1.51$$

them with the ancestral placental mammal condition as has been done (implicitly or explicitly) in many discussions of the evolution of the primate brain.

Reference to fossil evidence in tracing the evolution of relative brain size in mammals raises two important issues concerning the collection of data:

1. The application of suitable methods for the assessment of brain size in fossil species.
2. Establishment of reliable techniques for inferring the body weight of fossil species.

The first issue has already been broached above with respect to natural and artificial endocasts produced by moulding of the internal contours of the braincase. Such endocasts have the advantage, as discussed later, that in addition

to direct measurement of endocast volume (e.g. by water displacement) it is possible to identify major external morphological features of the brain. As with the skulls of modern primate and insectivore species, it is also possible to determine cranial capacity by packing the braincase of a suitably prepared fossil with small particles (e.g. lead shot, mustard seed or sintered glass beads) and then measuring their volume. However, it should be emphasized that consistent measurement of cranial capacity in this way requires a standardized procedure because of variability arising from the special packing properties of small solid objects (e.g. see Broca, 1861). This point is now frequently overlooked, yet it may lead to difficulties in comparisons of cranial capacities determined by different authors using different techniques.

It is also possible to obtain an approximate calculation of endocranial volume from standard outlines of the cranial cavity (e.g. derived from X-ray pictures), one taken in the horizontal plane and one taken in the sagittal plane. Jerison (1973) has described the method of **double graphic integration** based on determination of the average height (h) and the average width (w) of the cranial cavity from a series of parallel, equidistant lines superimposed on tracings of the internal contours of the brain case. The approximate cranial capacity (C_{EST}) can then be calculated using the following formula:

$$C_{EST} = (\pi/4 \times w \times h) \times l$$

(where l is the total length of the cranial cavity). One caveat that must be attached to the use of this method of double graphic integration is that it is only reliable if there are no major discontinuities in the contours of the cranial cavity. When, as is usually the case with fossil mammals, there are distinctly separate olfactory bulbs on stalks (peduncles) extending some way in front of the cerebral hemispheres, the method leads to overestimation of brain volume. In such cases, the volume of the olfactory bulbs must be determined separately from the volume of the rest of the cranial cavity by applying the above formula separately to the two parts. With certain well-prepared fossil skulls (e.g. the Eocene 'lemuroid' *Adapis parisiensis*; Martin, 1973) it is possible to use X-rays to trace standard outlines of the cranial capacity and to apply the method of double graphic integration (with separate calculation for the olfactory bulbs) without the need either to dissect the skull or to prepare an endocast. Table 8.5 presents the results of the determination of the endocranial volume in two well-prepared skulls of *Adapis parisiensis*, using all of the methods discussed above.

Body weight of fossil mammals

Estimation of body weight for species of fossil mammals poses a very different kind of problem, as there is no possibility of direct measurement of any kind. Where fairly complete specimens of fossil mammals are available, consisting of both

Table 8.5 Determination of the endocranial volume of two skulls of the Eocene 'lemuroid' *Adapis parisiensis*, using different methods

Method	Skulls	
	BMNH M.1345*	Cambridge M.538*[†]
1. Packing of cranial cavity with sintered glass beads or mustard seed	8.8 cm³	8.8 cm³
2. Volume of artificial endocast (latex mould of contours of internal braincase)	8.0 cm³	8.1 cm³
3. Double graphic integration (following Jerison, 1973) using X-rays	7.2 cm³	7.1 cm³

* BMNH, British Museum (Natural History); Cambridge, Zoological Museum, Cambridge, UK.
† See description by Gingerich and Martin (1981).

skulls and skeletons, it is possible to estimate body weight by direct comparison with living mammalian species of comparable overall dimensions. Alternatively, one can establish some standard relationship between a given parameter (e.g. head and body length) and body weight in living mammals and then apply this relationship to estimate body weight in fossil mammals. Both of these techniques have been used by Jerison (1973) and can be taken as generally reliable. But skulls of fossil primates and other mammals that are sufficiently well preserved to permit estimation of brain size often lack accompanying postcranial material. This rules out any direct comparison of overall body dimensions with those of living mammals. One simple approach that has been used in this situation is that of making direct comparisons of overall skull size between fossil forms and individual modern mammalian species. For instance, Jerison (1973) and Martin (1973) independently selected the skull of the living lemur *Lemur mongoz* as broadly comparable in overall size to that of the Eocene 'lemuroid'

Adapis parisiensis and concluded that the latter probably had a body weight near the average value of 1700 g established for *L. mongoz*. But Radinsky (1977) has rightly criticized such single-species comparisons as not being systematic enough, and it is certainly preferable to seek general relationships permitting reliable estimation of body weight that are applicable to a large sample of mammals, or at least to a representative sample of a particular subgroup of mammals (e.g. primates).

In an attempt to respond to the need for a reliable systematic indicator of body weight that can be determined from the skull alone, Radinsky (1967, 1970) suggested the use of the cross-sectional area of the foramen magnum. Given the height (h) and the width (w) of the foramen magnum, its approximate area (F) can be determined with the following formula:

$$F = (\pi/4) \times w \times h$$

(In the following text, the constant $\pi/4$ will be ignored and F will be simply indicated by $w \times h$.) A very simple interpretation is that a measure of foramen magnum area directly reflects the cross-sectional area of the spinal cord at the point of exit from the skull and that the calibre of the cord is essentially determined by the number of incoming (afferent) and outgoing (efferent) nerve fibres. It might be expected that, at any given body size, a standard number (and hence cross-sectional area) of afferent and efferent fibres might be required by a typical mammal. Hence, the foramen magnum area could be taken as a useful and consistent indicator of body weight for mammals generally. At first sight, it would seem that these expectations are fulfilled. Radinsky (1967) showed that within selected groups of placental mammals (insectivores, rodents, carnivores, artiodactyls, prosimian primates and simian primates), there is a very close correlation between foramen magnum area and body weight on a standard bivariate logarithmic plot. Further, Radinsky stated that there is relatively little distinction between these mammalian groups in the allometric constants (slope; intercept) determined. In other words, a single overall best-fit line might conceivably be

used as an indicator of body weight for all mammals covered by the analysis. If it is assumed, on this basis, that foramen magnum area is a reliable indicator of overall body size in modern mammals and that this is also true for fossil mammals, it follows that relative brain size can be investigated using bivariate plots of brain size against foramen magnum area, rather than against body size itself. The advantage of this procedure, if it is indeed reliable, is that relative brain size can be assessed for fossil species represented only by skulls, without resorting to indirect estimation of body weight.

The foramen magnum and brain size

Radinsky (1970) accordingly produced a graph in which cranial capacity was plotted against foramen magnum area on logarithmic coordinates for a comprehensive series of modern prosimian species ($N = 15$). Included on the graph were measurements for three subfossil lemur species (*Archaeolemur majori*, *Megaladapis edwardsi* and *M. madagascariensis*) and for four early Tertiary (Eocene/Oligocene) fossil primate species (*Adapis parisiensis*, *Necrolemur antiquus*, *Rooneyia viejaensis* and *Smilodectes gracilis*). The points for the subfossil lemurs and the early Tertiary fossil primates all clustered quite tightly around the best-fit line (regression) determined by Radinsky for the 15 modern prosimian species. This result is essentially replicated by an analysis of a larger sample of modern, subfossil and fossil primate species, as shown in Fig. 8.10. Now, it would seem justifiable to conclude from these relationships that early Tertiary primates had the same relative brain size as modern prosimians. This, in turn, might be taken to indicate that the relative brain sizes found in modern prosimians were already achieved in the Eocene, some 50 million years ago. Indeed, Radinsky (1970) specifically concluded that *Adapis* falls 'within the range of modern prosimians' with respect to relative brain size assessed according to foramen magnum area. Taken at face value, such an interpretation implies that relative brain size in modern prosimians can, after all, be regarded as a 'frozen ancestral condition' and that the *Scala naturae* evident among living primates with respect to relative brain size (see Figs 8.5 and 8.9) can be equated with the actual course of evolutionary history.

In the first place, there are in fact identifiable grade distinctions between groups in a plot of foramen magnum area against body size (Fig. 8.11). The distribution of points follows a pattern very similar to that found with the plot of cranial capacity against body weight (Fig. 8.5), although there are some differences in detail. Once again,

it is possible to identify four main grades defined by best-fit lines of similar slope and *Homo sapiens* is found to be a marked positive outlier relative to all other species in the comparison. However, the insectivores show more scatter for foramen magnum area than for cranial capacity and the New World monkeys show an unusual distribution, with unexpectedly small foramen sizes in most cases. (This latter feature is to be expected from the fact that New World monkeys are almost without exception positive outliers in the plot of cranial capacity against foramen magnum area in Fig. 8.10. Clearly, New World monkeys are somewhat different from Old World simians with respect to the size of the foramen magnum in relation both to body weight and to brain size.) For this reason, the fourth grade in Fig. 8.11 has been confined to Old World simians, rather than including all haplorhines as in Fig. 8.5. Similarly, tree-shrews have not been included in the calculation of a best-fit line for 'advanced' insectivores, as their inclusion renders the line indistinguishable from that for strepsirhine primates. With respect to the relationships between foramen magnum dimensions and body size, at least, tree-shrews

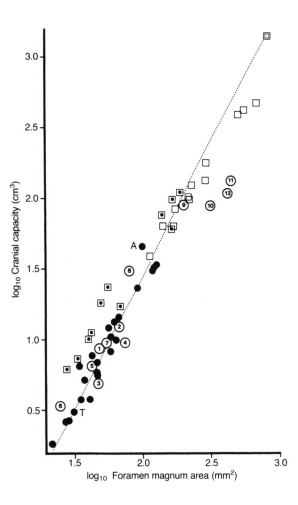

Figure 8.10. Logarithmic plot of cranial capacity against foramen magnum area for modern and fossil primate species. The best-fit line (major axis) is for modern prosimian species and has the following formula:

$$\log_{10} C = 1.85 \log_{10} F - 2.26$$

(where C is the cranial capacity in cubic centimetres and F is the inferred foramen magnum area in square millimetres).

Key: Black circles = modern prosimians (T, tarsier; A, aye-aye). Squares = modern simians (▣, New World simians; □, Old World monkeys + apes; point for *Homo sapiens* is outlined). Numbered circles = fossil primates: 1, *Adapis parisiensis*; 2, *Adapis magnus*; 3, *Pronycticebus gaudryi*; 4, *Smilodectes gracilis*; 5, *Mioeuoticus* sp.; 6, *Necrolemur antiquus*; 7, *Rooneyia viajaensis*; 8, *Aegyptopithecus zeuxis*; 9, *Archaeolemur edwardsi*; 10, *Megaladapis madagascariensis*; 11, *Megaladapis edwardsi*; 12, *Palaeopropithecus maximus*.

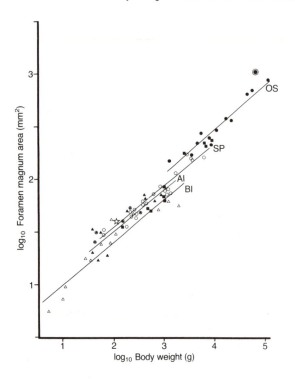

Figure 8.11. Logarithmic plot of foramen magnum area against body weight for primates, tree-shrews and insectivores (body weight data from Table 8.1). The best-fit lines are the major axes.

Key: BI = 'basal' insectivores   ($\triangle$; $N = 13$); AI = 'advanced' insectivores, excluding tree-shrews ($\blacktriangle$; $N = 10$); SP = strepsirhine primates : lemurs + lorises ($\bigcirc$; $N = 22$); OS = Old World non-human simian primates : apes + monkeys ($\bullet$; $N = 12$); tree-shrews ($\blacktriangle$; $N = 5$); *Tarsius* ($\star$); New World monkeys ($\blacksquare$; $N = 11$); *Homo sapiens* ($\circledcirc$).

are clearly closely similar to strepsirhine primates. The following formulae for the relationships between foramen magnum area (F) and body weight (W) are indicated by the best-fit lines (major axes) for the four grades in Fig. 8.11:

1. 'Basal' insectivores:
   $\log_{10} F = 0.41 \log_{10} W + 0.58$ ($r = 0.94$)

2. 'Advanced' insectivores:
   $\log_{10} F = 0.43 \log_{10} W + 0.66$ ($r = 0.89$)

3. Strepsirhjne primates:
   $\log_{10} F = 0.40 \log_{10} W + 0.74$ ($r = 0.96$)

4. Old world monkeys + apes:
   $\log_{10} F = 0.43 \log_{10} W + 0.76$ ($r = 0.97$)

The fact that the grade pattern in Fig. 8.11 is broadly similar to that in Fig. 8.5 indicates that there may be some connection between scaling of brain size to body size and scaling of foramen magnum area to body size. However, because the grade lines are not so clearly separated in the case of foramen magnum area, there seems to have been rather less evolutionary divergence in foramen magnum size than in brain size. Nevertheless, the distances between grade lines in Fig. 8.11 translate into quite appreciable differences when predictions of body size are made from dimensions of the foramen magnum. Taking the best-fit line for 'basal' insectivores, a foramen magnum area of 65 mm^2 would correspond to a body weight of 1 kg, whereas a foramen magnum area of this size would correspond to a body weight of only 280 g for the formula derived for Old World simians (monkeys and apes). Or, to take another example, extrapolation from the dimensions of the human foramen magnum using the best-fit line for Old World simians would indicate a body weight of 175 kg. Indeed, prediction from foramen magnum area using the formula of the best-fit line for 'basal' insectivores would indicate a human body weight of 869 kg! For this reason alone, caution must be exercised in the estimation of mammalian body weights from foramen magnum area. Accordingly, it cannot simply be assumed that the best-fit line determined for a group of modern primates can be applied for the inference of body weights for fossil primate species from (say) the early Tertiary.

A far more fundamental problem attaching to the use of foramen magnum area as an indicator of overall body size is suggested by the grade distinctions identifiable in Fig. 8.11 and was independently noted by Gould (1975), Jerison (1973) and Martin (1973). It is possible, at least in theory, that foramen magnum area is actually associated with brain size rather than (or as well

Table 8.6 Correlation coefficients for the relationship between foramen magnum area and body weight or cranial capacity in insectivores and primates

Group	N	Bivariate comparison			
		$\log_{10} FA$ vs $\log_{10} W$		$\log_{10} FA$ vs $\log_{10} C$	
		r	MA	r	MA
Primates (including humans)	47	0.977	0.485	0.979	0.584
Insectivores (including tree-shrews)	27	0.923	0.433	0.951	0.591

Abbreviations: N, number of species included in analysis; FA, foramen magnum area; W, body weight; C, cranial capacity; r, correlation coefficient; MA, slope of major axis. Note: The direct comparison of correlation coefficients is not really justified as FA and C were measured on the same skulls, whereas W was taken from the literature. However, the point is that FA is highly correlated with both body weight and cranial capacity.

as) with body size. Given that brain size is highly correlated with body size within major groups of mammals because of the general scaling trend (see Fig. 8.5), a close correlation between any dimension of the brain and body size would be entirely predictable. Consequently, if the foramen magnum area actually represents a cross-sectional area of the brain (e.g. approximating to the cross-section of the medulla), even to some degree, the close correlation between foramen magnum area and body size recorded for various mammalian groups by Radinsky (1967) would not, in itself, demonstrate that one measure can be taken as a reliable indicator of the other. In other words, there is a possibility that foramen magnum area may be scarcely more useful as an indicator of body size than brain size itself would be. This point is borne out in Table 8.6, where it is shown that correlation coefficients for the relationship between cranial capacity and foramen magnum area (on logarithmic coordinates) are as high as those for the relationship between foramen magnum area and body size, for both living insectivores and primates. The same conclusion has been reached by Jerison (1973) through the use of partial correlation coefficients: foramen magnum area is closely correlated with body weight, but is also closely correlated with brain size.

Use of foramen magnum area as an indicator of body size throughout primate evolution, as well as just for living primates, is based on an inherent assumption illustrated in Fig. 8.12. For foramen magnum area to serve as a consistent measure of body size over time, there must be no change in spinal cord dimensions for a given body size. However, this is a somewhat unlikely assumption; if the brain can increase in size (relative to body size) over time, there is no obvious reason why the spinal cord should not do the same. In fact, use of the foramen magnum area as an indicator of body size for fossil as well as living primates involves a modified version of the 'frozen ancestor' concept, although in this case it is implied that the spinal cord remains in a frozen ancestral condition throughout primate evolution while the rest of the CNS – the brain – undergoes varying degrees of increase in size. In this context, Passingham (1975, 1978) has analysed brain weight in relation to medulla weight. In so far as internal relationships of the brain are concerned, this is a useful approach; but the size of the medulla should not be seen as a surrogate measure for body size.

In the graph of cranial capacity against foramen magnum area (Fig. 8.10), it can be seen that the points for certain early Tertiary primate fossils such as the Eocene 'lemuroid' *Adapis* fall towards the lower end of the line. For *Adapis*

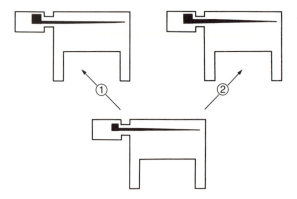

Figure 8.12. Alternative hypotheses regarding the relationship between body size and area of foramen magnum through an evolutionary sequence involving no change in body size. (Reprinted from Martin, 1982, courtesy of Plenum Press, New York.)

Hypothesis 1: Increase in brain size is not associated with any increase in the dimensions of the spinal cord. Hence, foramen magnum area, as an indirect indicator of the cross-sectional area of the spinal cord at the point of exit from the skull, does not change over time and can be used as a consistent guide to body size.

Hypothesis 2: Increase in brain size is associated with an increase in the dimensions of the spinal cord. Hence, foramen magnum area increases over time and because of this cannot be used as a consistent guide to body size.

parisiensis, a best-fit line for modern primates determined for the relationship between foramen magnum area would in fact indicate a body weight of about 440 g (see also Table 8.8). It has already been mentioned that with *Adapis parisiensis*, where insufficient postcranial material is available for a reconstruction of the skeleton (see Chapter 10), simple comparison of skull size with modern prosimians suggested a body weight in the region of 1700 g(Jerison, 1973; Martin, 1973). There is, therefore, a considerable discrepancy between the body size indicated for this Eocene fossil primate by foramen magnum area and that suggested by a simple assessment of likely body size based on comparison with a single modern primate species. Radinsky himself (1977) provided a revised estimate of 1600 g for the body weight of *A. parisiensis*. It would therefore seem that the relationship between foramen magnum area and

body size in *Adapis* was in fact markedly different from that in modern prosimians and that some other procedure is required for reliable estimation of body weight for fossil primate species lacking postcranial material. By the same token, there seems to have been some increase in foramen magnum area (and therefore, by inference, in the girth of the spinal cord) during primate evolution, paralleling the increase in relative brain size. The possible functional significance of this will be discussed later in this chapter. MacLarnon (1987) has shown that the Eocene 'lemuroids' *Notharctus* and *Smilodectes* had smaller vertebral canals than modern primates, indicating that the spinal cord was, indeed, smaller relative to body size. Like *Adapis*, *Notharctus* had a very small foramen magnum for its body size, whereas the foramen of *Smilodectes* (curiously) was comparable in size to that of some modern prosimians. This shows that the size of the foramen magnum may not necessarily provide a reliable indication of the girth of the spinal cord.

Inferring body size from skulls

One must therefore look elsewhere for dimensions that can be measured on skulls of primates (or other mammals) to obtain estimates of body size. Linear dimensions include the maximum length, the maximum width across the zygomatic arches, the length of the palate, or the maximum length of the horizontal aperture defined by the zygomatic arch (see Fig. 8.3). In addition, one can measure the maximum length and width of one of the occipital condyles, to establish a crude measure of condylar surface area, and the maximum length and width of the cheek-tooth row (i.e. premolars and molars), to establish an equally crude measure of the surface area of the cheek teeth. Taking the sample of modern primate skulls already mentioned regarding the determination of cranial capacities (see Table 8.1), but with the exclusion of some species (including *Homo sapiens*) to leave a total sample of 36 species, standard allometric relationships have been established for each of these measures with respect to body weight

Table 8.7 Empirical allometric formulae relating various skull measures to body weight for a sample of modern primate species ($N = 36$)

Dimension	Empirical formula*	$r^{\dagger}$
Skull length (*SL*)	log *P* = 3.89 log *SL* − 4.09	0.98
Bizygomatic width (*BZ*)	log *P* = 3.77 log *BZ* − 3.19	0.98
Internal zygomatic length (*IZ*)	log *P* = 3.26 log *IZ* − 0.96	0.96
Palate length (*PAL*)	log *P* = 3.68 log *PAL* − 2.08	0.96
Cheek-tooth row length × width (*CTL* × *CTW*)	log *P* = 2.06 log (*CTL* × *CTW*) − 1.00	0.95
Condyle length × width (*COL* × *COW*)	log *P* = 2.16 log (*COL* × *COW*)	0.98
Foramen magnum height × width (*FMH* × *FMW*)	log *P* = 2.15 log (*FMH* × *FMW*) − 1.20	0.98

* Formula is based on the major axis; logarithms are all to base 10. Skull dimensions are in millimetres (or square millimetres); body weight is in grams.
† *r*, correlation coefficient. Note: With these relatively high values, the choice of best-fit line has little influence on the empirical formula.

(Table 8.7). It can be seen that in all cases there are quite high correlations between the cranial/dental measures and body weight for the entire non-human primate sample. The lowest correlations are found with the measurements of the internal zygomatic length, the palate length and the surface area of the cheek-teeth, probably because of the existence of separate grades (see Chapter 4, Appendix for a discussion of allometric grades, and Chapter 6 for specific reference to grades in cheek-tooth area).

It is also interesting to note that body weight (*W*) does not scale isometrically against any of the linear skull measurements (skull length, palate length, internal zygomatic length or bizygomatic width). The average exponent value is 3.65, rather than the value of 3 which would be expected on simple geometric grounds ($W \propto L^3$, where *L* is a linear measure). Inversion of the allometric formulae (Table 8.7) suggests instead that skull volume scales to body weight with an exponent value of about 0.82 instead of 1.0. This is perhaps understandable, since food collection is one of the major functions of the skull, and food requirements increase with body size along with metabolic rate, the basal level of which is related to the 3/4 power of body size (see Chapter 4, Appendix). Hence, as body size increases in primates (or in other mammals and other

terrestrial vertebrates generally), the head becomes progressively smaller in proportion to body size. In line with this, one must be wary of treating skull size as a direct isometric indicator of body size (Martin, 1982). To 'predict' body weights from skull parameters, it is necessary to make use of allometric relationships of the kind listed in Table 8.7.

The allometric relationships given in Table 8.7 were established using the list of primate body weights published by Stephan, Bauchot and Andy (1970), which left out several species in the skull sample. It was therefore possible to 'predict' body weights for these additional species and then to compare them with actual body weights obtainable from other sources (Table 8.8). Fairly close agreement was achieved between the average 'predicted' values and the observed values in most cases, although quite substantial discrepancies are seen with certain primate species, particularly those with long snouts and/or extreme sexual dimorphism (e.g. baboons and orangutan). When the average body weights estimated from the various skull dimensions are compared to the actual body weights for the primate species in Table 8.8, the average prediction error is found to be ± 32.3% (range: 0.5–79.3%). In some cases, the errors involved are trivial, but in others they are far from negligible.

384

Table 8.8 Predictions of body weight for 12 modern primate species based on allometric formulae for various skull dimensions (see Table 8.7). Table is modified and expanded from Martin (1980); some slight differences arise from using the major axis instead of the reduced major axis

Species	Body weight (g) predicted from relevant empirical allometric formula							Mean predicted body weight (g)	Actual body weight (g)	Ratio of predicted to actual body weight*
	SL	BZ	IZ	PAL	CTL×CTW	COL×COW	FMH×FMW			
Phaner furcifer	**399**†	354	481	621	233	589	398	439	440	1.00 (0.91)
Hapalemur simus	2 150	**2 850**	2 460	4 130	6 830	2 980	2 000	3 340	2 370‡	1.41 (1.20)
Lemur mongoz	2 460	1 430	**1 910**	3 960	4 360	1 740	1 720	2 510	1 670	1.50 (1.14)
Euoticus elegantulus	260	392	**280**	306	294	263	255	293	287	1.02 (0.98)
Galago alleni	351	379	324	**368**	714	430	303	410	246	1.67 (1.50)
Cebuella pygmaea	**69**	78	75	58	35	39	130	69	72	0.96 (0.96)
Callimico goeldii	371	339	297	224	392	290	**316**	318	471	0.68 (0.67)
Papio anubis	49 400	**25 500**	14 200	115 600	42 200	26 900	21 700	43 600	16 700	2.61 (2.13)
Theropithecus gelada	**29 200**	29 301	13 600	71 800	34 000	19 700	14 300	30 300	21 500	1.41 (1.36)
Hylobates lar	5 290	5 560	5 860	**5 790**	4 420	6 500	11 200	6 370	5 440	1.17 (1.06)
Hylobates syndactylus	**11 500**	11 200	13 200	19 900	11 400	9 040	12 600	12 700	10 700	1.19 (1.07)
Pongo pygmaeus	84 700	107 700	52 300	108 000	**86 400**	145 000	82 800	95 300	55 000	1.73 (1.57)

* Ratio in parentheses was calculated by taking the modal value rather than the mean.
† Values in bold type are the modal values for every species.
‡ Data from a single male (Meier, Albignac *et al.* 1987).

Table 8.9 Predictions of body weight for fossil primate species based on allometric formulae for modern primate species (see Table 8.7 for explanation of abbreviations)

Species	Body weight (g) predicted from relevant empirical allometric formula							Mean predicted body weight (g)*	Ratio to FM prediction†
	SL	BZ	IZ	PAL	CTL×CTW	COL×COW	FMH×FMW		
Eocene 'lemuroids'									
Adapis parisiensis (N = 2)	2550	2350	3050	–	2150	1650	440	2350	5.4
Adapis magnus (N = 4)	6800	9700	12000	12500	10500	6000	880	9600	10.9
Pronycticebus gaudryi	840	1500	950	1200	1950	880	410	1200	2.9
Notharctus tenebrosus	1400	(1400)	1400	2500	4500	760	350	2000	5.7
Smilodectes gracilis	1400	1750	1500	2200	3400	1450	1100	1950	1.8
Miocene 'lorisoid'									
Mioeuoticus sp.	690	–	–	1800	1400	–	290	1300	4.5
Subfossil lemurs									
Archaeolemur edwardsi (N = 2)	16500	23500	–	–	39000	–	9000	26000	2.9
Megaladapis edwardsi (N = 2)	390000	85500	120000	610000	735000	–	54000	390000	7.2
Palaeopropithecus maximus	61000	44000	–	–	166000	–	46000	90000	2.0
Early Tertiary 'tarsioids'									
Tetonius homunculus	70	–	–	–	–	–	–	70	–
Necrolemur antiquus (N = 4)	150	180	–	460	260	120	90	230	2.6
Rooneyia viejaensis	380	(820)	560	780	1200	950	470	780	1.7
Oligocene simian									
Aegyptopithecus zeuxis	5100	4400	–	–	11000	–	1300	6800	5.2
'Archaic primate'									
Plesiadapis tricuspidens	6000	6300	15500	21000	6000	7950	630‡	10500	16.7

* Mean excludes prediction from dimensions of foramen magnum.
† 'FM prediction', prediction from dimensions of foramen magnum.
‡ *FMW* × *FMW* taken, as *FMH* is distorted by crushing.

For the plains baboon (*Papio anubis*), for instance, the estimated body weight is over twice the actual average body weight for males and females, whereas for Goeldi's monkey (*Callimico goeldii*) the estimated body weight is about two-thirds the actual figure.

Thus, despite the high correlation coefficients recorded in Table 8.7, use of skull measures to 'predict' body weight for any primate species is clearly a very approximate solution. However, any 'prediction' of primate body weight in the absence of a complete skeleton is likely to be equally prone to error. Further, even the substantial errors recorded in Table 8.8 are considerably less than the difference between an estimate based on foramen magnum area and other estimates for the Eocene 'lemuroid' *Adapis parisiensis* (see below). It remains to be established, however, that the allometric relationships determined for living primates (Table 8.7) and tested on other living primates (Table 8.8) can also be applied to fossil primates. One test of this is to estimate body weights for the 'lemuroids' *Adapis parisiensis* and *Smilodectes gracilis*, by applying the allometric formulae given in Table 8.7 to their cranial and dental dimensions (see Table 8.9). It emerges that there is very close correspondence between these estimates and those derived from other sources, with the exception of estimates based on foramen magnum area. As a provisional approach, then, it

would seem justifiable to utilize the panel of dental and cranial measures included in Table 8.7 to estimate body weights of fossil primate species for which postcranial skeletal evidence is lacking. Several authors have estimated body weights for fossil primate species using allometric formulae based on individual tooth dimensions, notably length and width of molars (Gingerich, Smith and Rosenberg, 1982; Gingerich and Smith, 1985; Fleagle and Kay, 1985). The results obtained are generally similar to those reported here, although dental measures are less well correlated with body weight than cranial measures because of grade distinctions (Martin, 1982). Insectivores (including tree-shrews), for instance, have relatively larger molars than primates that eat mainly fruit or leaves (Gingerich and Smith, 1985). *Tarsius* is unusual among living primates in having relatively large molars like those of insectivores (see also Chapter 6).

While considering the possibility of estimating body size from cranial and dental measures, it is also worthwhile to ask whether measures taken on the skull can also be used to estimate brain size. There are, after all, numerous fossil primate species for which endocasts are lacking because the skulls in collections have not been dissected or cleaned to expose the interior of the braincase. Use of skull parameters to estimate brain size would greatly expand the sample size of fossil primate cranial capacities available for allometric

Table 8.10 Empirical allometric formulae for inferring cranial capacity from linear dimensions of the braincase for a sample of modern primate species

Dimension*	Empirical formula[†]	r[‡]
Braincase width (CW)	$\log C = 3.24 \log CW - 3.75$	0.99
Braincase height (CH)	$\log C = 2.91 \log CH - 2.91$	0.98
Braincase length (CL)	$\log C = 3.28 \log CL - 4.37$	0.98
$MO = CW + CH + CL$[§]	$\log C = 3.12 \log MO - 5.18$	0.995
$PR = CW \times CH \times CL$[‖]	$\log C = 1.02 \log PR - 3.54$	0.995

* Dimensions of CW and CH are shown in Fig. 8.3; CL, chord of midline suture through occipitals, parietals and frontals.
† Formula is based on the major axis; logarithms are all to base 10.
‡ r, Correlation coefficient. Note: With these relatively high values, the choice of best-fit line has little influence on the empirical formula.
§ MO is the modulus of the linear dimensions.
‖ PR is the product of the linear dimensions.

Table 8.11 Predictions of cranial capacities (cm^3) of fossil primates derived from linear dimensions of the braincase, compared with other measures of cranial capacity (see Table 8.10 for explanation of abbreviations)

Species	Prediction from *CW*	Prediction from *CH*	Prediction from *MO*	Other measures
Eocene 'lemuroids'				
Adapis parisiensis	6.5	7.2	–	8.8*; 8.3[†]
Adapis magnus	25.8	17.6	–	–
Pronycticebus gaudryi	7.8	4.8	4.8	–
Notharctus tenebrosus	12.4	–	–	10.4[†]
Smilodectes gracilis	13.8	6.5	10.1	9.5[‡]; 9.6[§]; 9.5[†]
Miocene 'lorisoid'				
Mioeuoticus sp.	8.8	4.8	7.8	5.6[‖]
Subfossil lemurs				
Archaeolemur edwardsi	150	108	132	88.3*
Megaladapis edwardsi	176	217	460	132*
Palaeopropithecus maximus	196	169	242	107*
Early Tertiary 'tarsioids'				
Necrolemur antiquus	4.4	2.6	3.2	4.4[‡]; 4.2[§]; 2.7[†]
Rooneyia viejaensis	10.5	9.9	10.2	7.5[‡]; 6.4[§]; 7.4[†]
Oligocene simian				
Aegyptopithecus zeuxis	41.9	20.4	34.4	27.8[‖]; 30[‡]
'Archaic primate'				
Plesiadapis tricuspidens	16.6	(3.3)	(9.4)	14[¶]

* Direct measurement of cranial capacity with mustard seed/sintered glass beads (see Gingerich and Martin, 1981).
[†] Measured volume of restored endocast cited by Gurche (1982).
[‡] Value reported by Radinsky (1977).
[§] Value determined by double graphic integration by Jerison (1979).
[‖] Determined by double graphic integration.
[¶] Selected by Jerison (1979) from the range of 12–17 cm^3 suggested by Radinsky (1977). Note: Values predicted from *CH* and *MO* for *Plesiadapis* are misleading because of marked flattening of the skull during fossilization.

analysis. Three standard measures of the braincase (two of which are illustrated in Fig. 8.3) were taken on the large sample of modern primate skulls examined, and the allometric relationships between these measures and the cranial capacities (determined by packing of the braincase with sintered glass particles) were determined (Table 8.10). It emerges that all of these measures – one (*CH*) indicating the height of the brain case, another (*CW*) indicating its width, and the third (*CL*) indicating its length – are very closely correlated with cranial capacity across the primate sample as a whole. The 'prediction' of primate brain sizes from cranial measures is therefore possible, although it is by no means obvious that the relationships established for living primates should apply uniformly to fossil species. In order to test this, cranial capacities 'predicted' from skull dimensions for a variety of fossil primate species have been compared to those established by other techniques (Table 8.11). The agreement obtained is surprisingly good, and it can be concluded that this technique is reliable for the assessment of cranial capacities in fossil primates generally.

Relative brain size in fossil primates

By applying the above methods (other than inference from foramen magnum area) for estimating body weights of fossil primate species and for inferring cranial capacities in cases where direct measurement is precluded, it is possible to draw up a reasonable comprehensive list of these parameters, and hence of cranial capacity indices (ICC values), for fossil primates other than hominids (Table 8.12). The ICC values permit a meaningful direct comparison with modern insectivore and primate species, as illustrated in Fig. 8.13. The first important conclusion to emerge from this comparison is that fossil primates generally have lower ICC values than their nearest apparent relatives among the living primates (see also Gurche, 1982). For example, the 'archaic primate' *Plesiadapis* (which has the lowest value) and the Eocene 'lemuroids' (*Adapis*, *Notharctus*, *Pronycticebus* and *Smilodectes*) have ICC values well below the average value for living strepsirhine primates and uniformly below the minimum value for a modern strepsirhine (2.4 for *Cheirogaleus*; see Table 8.3), although in some cases just within the 95% confidence limits. Similarly, the Miocene

Table 8.12 Inferred body weights (*W*) and cranial capacities (*C*) for fossil primate species, together with calculated indices of cranial capacity (ICC values)

Species	Estimated *W* (g)	Estimated *C* (cm^3)	ICC
'Archaic primate'			
Plesiadapis tricuspidens	10 500	16.6	0.95
Eocene 'lemuroids'			
Adapis parisiensis	2 350	8.8	1.39
Adapis magnus	9 500	21.7	1.32
Pronycticebus gaudryi	1 220	4.8	1.18
Notharctus tenebrosus	1 990	10.4	1.84
Smilodectes gracilis	1 960	9.5	1.69
Miocene 'lorisoid'			
Mioeuoticus sp.	1 280	7.8	1.86
Early Tertiary 'tarsioids'			
Necrolemur antiquus	233	3.8	2.88
Tetonius homunculus*	74	1.5	2.48
Rooneyia viejaensis	782	7.4	2.47
Fossil simians			
Aegyptopithecus zeuxis	6 710	34.4	2.66
Proconsul africanus[†]	10 500	167	9.51
Australopithecus africanus[‡]	30 000	442	12.33
Homo habilis[‡]	40 000	642	14.73
Homo erectus[‡]	50 000	941	18.55

* Estimates from Jerison (1979).
[†] Estimated figures from Walker, Falk *et al.* (1983).
[‡] Estimated figures from Martin (1983).

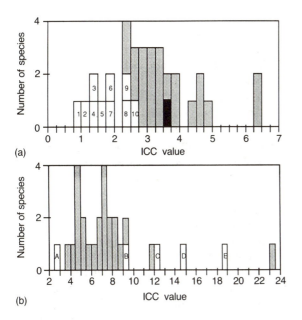

Figure 8.13. Indices of cranial capacity (ICC) for living and fossil primates, taking the data listed in Table 8.1 and adding information for various fossil primate species from Table 8.11 (black box = *Tarsius*).

(a) Prosimians: 1, *Plesiadapis tricuspidens*; 2, *Pronycticebus gaudryi*; 3, *Adapis magnus*; 4, *Adapis parisiensis*; 5, *Smilodectes gracilis*; 6, *Notharctus tenebrosus*; 7, *Mioeuoticus* sp.; 8, *Rooneyia viajaensis*; 9, *Tetonius homunculus*; 10, *Necrolemur antiquus*.

(b) Simians: A, *Aegyptopithecus zeuxis*; B, *Proconsul africanus*; C, *Australopithecus africanus*; D, *Homo habilis*; E, *Homo erectus*.

'lorisoid' *Mioeuoticus* has an index value only 50% of that found for modern lorisids (lorises and bushbabies), and the cranial capacity index value for the Oligocene simian *Aegyptopithecus zeuxis* is only 40% of the average for modern simians and well below the minimum value of 4.0 found for *Colobus badius* (see Table 8.3), although just outside the 95% confidence limits.

The only exception to this general trend among fossil prosimians is that Eocene 'tarsioids' have ICC values within the range of modern prosimians although somewhat below that found with the modern *Tarsius*. The ICC values for *Necrolemur*, *Rooneyia* and *Tetonius* are markedly greater than for the contemporary

Eocene adapids and the evidence therefore suggests that among 'tarsioids' the brain began increasing in size quite rapidly early in primate evolution. However, assuming that modern tarsiers are related to early Tertiary 'tarsioids', brain expansion apparently slowed markedly, such that they show relatively little advance. This may, in fact, constitute one of the very rare cases in primate evolutionary history where it is justifiable to think almost in terms of a 'frozen ancestral condition'. It must be emphasized at once, however, that this only applies to the tarsier's brain; in other respects (e.g. locomotor adaptation; see Chapter 10) the tarsier is clearly quite different from the Eocene 'tarsioids'.

As a general rule, then, it can be stated that there has been gradual expansion of the brain (relative to body size) throughout most of primate evolution, although a slower overall rate of expansion among strepsirhines resulted in restriction of the modern species to a lower grade than that now occupied by haplorhine primates (Fig. 8.5). The haplorhines were clearly characterized by a faster expansion of relative brain size at a comparatively early stage of primate evolution, as both early Tertiary 'tarsioids' and the Oligocene *Aegyptopithecus* have cranial capacity indices lying within the range of modern strepsirhine primates. Divergence in the degree of expansion of brain size between strepsirhines and haplorhines must consequently have taken place during the early Tertiary (if not earlier). This divergence was subsequently maintained in spite of continued expansion of the brain, relative to body size, in both primate groups beyond the level found in adapids on the one hand and in early Tertiary 'tarsioids' and the Oligocene *Aegyptopithecus* on the other. Hence, although the modern tarsier is intermediate, in terms of relative brain size, between strepsirhines and simians (Fig. 8.9), and can therefore be described as a 'moderately large-brained' prosimian, the fossil evidence suggests a distinct difference in the evolutionary pathways leading to the present overlap between *Tarsius* and the larger-brained strepsirhines: in strepsirhines, the pace of brain size expansion seems to have been relatively consistent over time, whereas in 'tarsioids' an initial rapid phase of expansion was followed by a long period of comparative stagnation continuing to the present day.

With reference to the order Primates as a whole, it is significant that the Eocene Adapidae – in several respects the most primitive of the known fossil primates 'of modern aspect' – should uniformly be characterized by cranial capacity indices inferior to those found in all living primate species (compare Table 8.12 with Table 8.3). As the Adapidae themselves were presumably somewhat advanced with respect to the considerably earlier (Cretaceous) ancestral primate stock, it can be concluded with a fair degree of certainty that all living primates exhibit a greater degree of encephalization than was present in the ancestral primate stock. It is therefore erroneous to equate the apparent steps of a *Scala naturae* in Figs 8.5 and 8.9 with actual phylogenetic stages in the evolution of brain size in primates. No living primate species has a brain as primitive as that of ancestral primates, just as no living placental mammal has a brain as primitive as that possessed by ancestral placentals.

Relative brain size in hominids

Assessment of brain size in relation to body size through allometric analysis and the elimination of 'frozen ancestor' concepts are particularly relevant to discussion of the evolution of human brain size (Jerison, 1973; Martin, 1983). There is a reasonably good fossil record for human evolution from about 3.5 million years ago to the present, so that it is possible to determine cranial capacities and estimate body weights for species of *Australopithecus* (*A. afarensis*, *A. africanus* and *A. robustus*) and of *Homo* (*H. habilis*, *H. erectus* and *H. sapiens*). However, discussion of the evolution of human brain size, in relation to our closest zoological relatives the great apes, is subject to two potential sources of error:

1. Body size has often been overlooked as a factor relating to brain size and this is particularly important as the lightly built ('gracile') australopithecines (*A. afarensis* and *A. africanus*), at least, were clearly significantly smaller than recent species of *Homo* (*H. erectus* and the average modern *H. sapiens*).

2. Comparison of modern humans and fossil hominid species with modern great apes involves the inherent assumption that the latter have remained unchanged, in terms of relative brain size, since the divergence of the hominid lineage from the common ancestral stock of great apes and humans. Until recently, no cranial capacities were available for relevant fossil species reasonably close to that ancestral stock (i.e. dryopithecines, see Chapter 2), so this assumption was previously untestable. However, estimates of cranial capacity and body weight have been provided by Walker, Falk *et al.* (1983) for *Proconsul africanus*. These indicate a cranial capacity index value of 9.5, which is higher than in any modern ape and at the upper end of the range for modern non-human simians generally. If the estimates of brain and body size for *Proconsul africanus* are correct, this means that this species is exceptional among fossil primates in having a larger brain than most of its modern relatives. It is conceivable that relative brain size has remained stable, or has even undergone reduction, during the evolution of the great apes. Alternatively, *Proconsul africanus* might have been unusual among contemporary simians in having a very large brain. Only further skull material for other Miocene simians can resolve this issue. Thus, the relative brain size of the common ancestor of great apes and humans remains unknown.

For purposes of discussion, evolution of the human brain can be related to the approximate series *A. africanus–H. habilis–H. erectus–H. sapiens*. Even though this series probably does not represent an actual evolutionary sequence, it does represent a succession of stages through time illustrating the broad trend of expansion in relative brain size in hominids. Data on average cranial capacities and inferred body weights for

Table 8.13 Inferred body weights and cranial capacities (averages) for four hominid species (from Martin, 1983)

Hominid species	Inferred body weight (kg)	Cranial capacity (cm^3)
Australopithecus africanus	30	442
Homo habilis	40	642
Homo erectus	50	941
Homo sapiens	57	1230

these four hominid species are presented in Table 8.13 (see also Martin, 1983). In order to trace the increase in relative brain size over time through this series, it is necessary to calculate appropriate indices of cranial capacity and to make an appropriate inference regarding the common ancestral condition from which humans and the great apes diverged.

Various different approaches to the calculation of indices of relative brain size for fossil hominids have been adopted (e.g. Jerison, 1973; Pilbeam and Gould, 1974; Passingham, 1975). Because allometric exponent values vary with taxonomic level, it is necessary to decide on an appropriate comparison for calculating relative brain size in fossil hominids. The broadest approach is to compare modern humans and fossil hominids with placental mammals in general, following the reasoning behind Jerison's encephalization quotient (EQ). Taking an empirical formula for the relationship between brain weight and body weight in placental mammals (Martin, 1981c; see above), it is possible to calculate such EQ values for the various hominid species and to plot these values against time (see also Passingham, 1975). Alternatively, it might be argued that a more realistic comparison would be restricted to Old World simians (Old World monkeys, apes and hominids) because of their closer degree of relationship. Following the general trend discussed above, an empirical allometric formula for the relationship between brain weight and body weight for Old World simian genera ($N = 18$) has a lower exponent value than that

found for placental mammals generally (Martin, 1983):

$$\log_{10} E = 0.60 \log_{10} W + 2.68 \qquad (r = 0.97)$$

Using this formula, it is possible to calculate 'Old World simian encephalization quotient' values (EQ_{OW}) for hominid species and to plot these against time instead. There remains the problem of inferring the likely relative brain size of the common ancestor of great apes and modern humans, in the absence of any reliable fossil evidence. As an interim solution, a reasonable assumption is that the common ancestor would not have had a relative brain size exceeding the minimum found among modern great apes. One can therefore compare encephalization quotient values for hominids (EQ and EQ_{OW}) to that minimum value on the grounds that any advance beyond the common ancestral condition found in hominids is likely, if anything, to be underestimated. This procedure has been followed in the plots of EQ and EQ_{OW} against time for hominid species presented in Fig. 8.14.

Two points are clear from Fig. 8.14, regardless of which encephalization quotient is most appropriate. First, the brain of *Australopithecus africanus* was already significantly larger relative to that of the likely common ancestor, so that it can be inferred that the hominid brain began enlarging relative to body size at least 4 million years ago. Second, this increase in brain size has been a fairly continuous phenomenon throughout hominid evolution and present evidence does not indicate any sharp change of tempo at any given point. There is therefore no basis for hypotheses that invoke a recent abrupt onset of expansion of the hominid brain only 2 million years ago (approximately coinciding with the appearance of the first stone tools in the archaeological record) or postulate an accelerated enlargement of the brain in association with particular events in human evolutionary history.

Note that the calculations of encephalization quotients illustrated in Fig. 8.14 follow common practice in relating cranial capacity measures for fossil hominids to formulae based on actual brain

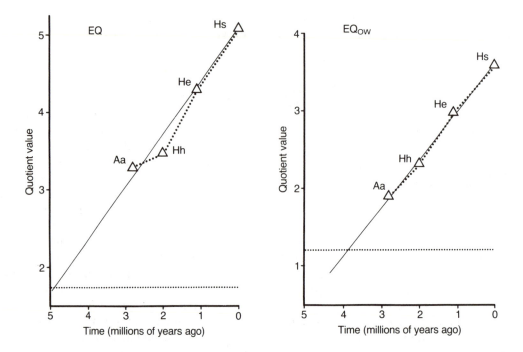

Figure 8.14. Plots of EQ values (encephalization quotients relative to placental mammals generally) and EQ_{OW} values (encephalization quotients relative to Old World non-human simian primates only) against time for hominid species, using the data provided in Table 8.13. The horizontal dotted line represents the minimal condition for modern great apes, and the oblique continuous line represents a best-fit line through the hominid points. The heavy dotted lines joining the hominid points indicate a general trend only and are not intended to imply strict evolutionary continuity. Abbreviations: Aa, *Australopithecus africanus*: Hh, *Homo habilis*; He, *Homo erectus*; Hs, *Homo sapiens*. (Reprinted from Martin, 1983, with kind permission from the American Museum of Natural History, New York.)

weights for the other mammals used in the comparison. As explained above, cranial capacity is approximately equivalent to actual brain weight, but there is a systematic effect of the equation presented in that cranial capacity increasingly exceeds brain weight with increasing brain size. Assessment of hominid cranial capacities should therefore ideally be conducted in comparison with formulae based on the cranial capacities (rather than brain weights) of other mammals. However, examination of the data for primates indicates that reassessment of the evidence on this basis will not modify the picture presented in Fig. 8.14 to any significant degree and the conclusions reached above remain valid.

Scaling of the olfactory bulbs

Another very good illustration of the potential dangers of discussing primate brain evolution in the absence of quantitative assessment of scaling effects of body size and only with reference to modern species is provided by analysis of the relationship between the olfactory system and the rest of the brain. Through the writings of Elliot Smith (1902, 1927), followed by those of Le Gros Clark (1934b, 1959), a general consensus view of primate brain evolution emerged. According to this, the presumed transition from a typical terrestrial insectivore condition to the typical arboreal primate condition was particularly marked by a switch in emphasis from olfactory functions to visual functions (see also Chapter 7). This view is encapsulated in the following quotation from Le Gros Clark (1959):

> With the increasing perfection of the visual sense, therefore, and also because the sense of smell does not serve the immediate practical use in arboreal life for which it is required in terrestrial life, the olfactory organ and its centres undergo a progressive atrophy in the evolution of Primates.

Le Gros Clark, in company with several other authors, specifically suggested that tree-shrews resemble primates in having a reduced olfactory system and that this is evidence of a phylogenetic relationship between them. It is certainly true that in living primates, as compared to living insectivores, vision plays a far more important part as a sensory system. This observation has led to the conclusion, almost universally cited in standard works on primate evolution (e.g. Buettner-Janusch, 1966; Hill, 1972; Le Gros Clark, 1959; Napier and Napier, 1967; Shultz, 1969), that the olfactory system has been physically reduced in the evolution of all living primates and of tree-shrews as well. Perhaps the most explicit statement of this widespread view is given by Buettner-Janusch (1966): 'The olfactory apparatus of *Tupaia* is reduced absolutely and is small relative to other bony structures and to the rest of the brain.'

Yet such a conclusion really requires a comparison of the brains of living primates with the hypothetical ancestral placental condition (not with the condition present in living insectivores). It should also be based on a genuinely quantitative comparison in which the scaling effects of body size are taken into account. It has already been inferred above, on the basis of allometric analysis, that the ancestral placental mammals probably had smaller relative brain sizes than any living mammals, including 'basal' insectivores. There is no a priori reason why enlargement of the brain, which has generally characterized the evolution of all extant species of mammals, should not also have affected the olfactory system in particular

mammals such as the insectivores. Thus, inferences based on a modern insectivore model as a 'frozen ancestor' in the reconstruction of primate evolution may not be borne out in a more realistic reconstruction based on derivation from a hypothetical ancestral placental condition (Fig. 8.15). Further, it is possible that development of other areas of the brain may simply overshadow the olfactory system in primate evolution, without involving any reduction in the latter. That is to say, it is quite possible that, although the main olfactory components of the primate brain (olfactory bulbs and palaeocortex) appear to be small relative to the rest of the brain, they have not actually undergone reduction ('atrophy') in relation to body size. Only a quantitative comparison of the size of olfactory components of the brain relative to body size can clarify this somewhat complex situation.

Stephan, Bauchot and Andy (1970) measured the volume of the olfactory bulb in a large series of modern primate and insectivore species (see also Stephan, 1972). From a logarithmic plot of these volumes against body size (Fig. 8.16), it can be seen that, relative to body size, the olfactory bulb volumes of tree-shrews and of most nocturnal lemurs and lorises fall within the range shown by modern insectivores (both 'basal' and 'advanced'), although the average level for the strepsirhines is lower than for the insectivores. If the allometric relationship between olfactory bulb size and body size shown by living insectivores is taken, for the moment, as representative of the ancestral condition for placental mammals, it can be concluded that there is in fact no evidence of any reduction of the olfactory bulbs in the evolution of tree-shrews and that there is at most very mild reduction in most nocturnal lemurs and lorises (see also Gurche, 1982; Baron, Frahm *et al.*, 1983). However, it is clear from Fig. 8.16 that the olfactory bulbs are markedly smaller, in relation to body size, among living haplorhines (tarsier and simians) than among modern insectivores, tree-shrews and most nocturnal prosimians. The same applies to the diurnal lemurs and to two nocturnal lemur species that are largely or exclusively folivorous in habits (*Lepilemur* and *Avahi*). It is

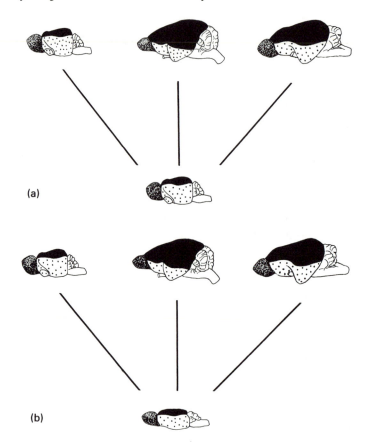

Figure 8.15. Two alternative approaches to the interpretation of brain evolution in primates: olfactory bulb (heavy stippling); palaeocortex (spaced dots); neocortex (black). (Reprinted from Martin, 1968a, with kind permission from the Royal Anthropological Institute of Great Britain and Ireland, publishers of *Man*.)

Hypothesis a. The brain in the ancestral placental mammal was comparable to that in a modern species of insectivore (hedgehog). In the evolution of tree-shrews (top centre) and primates (e.g. mouse lemur, top right), the size of the olfactory bulb and the volume of the palaeocortex were physically reduced.

Hypothesis b. The brain in the ancestral placental mammal was smaller than in any modern mammal. Hence, the size of the olfactory bulb and the volume of the palaeocortex may have undergone some secondary increase in certain lineages (e.g. insectivores).

apparent, then, that reduction of the olfactory bulbs relative to body size may have occurred independently in the evolution of (1) certain lemur species that have adopted diurnal and/or folivorous habits, and of (2) haplorhines, even in the two genera that are nocturnal (*Tarsius* and *Aotus*).

Note that haplorhines, with the exception of *Tarsius* and *Aotus*, are diurnal and that they all have small olfactory bulbs in comparison to insectivores. By contrast, most strepsirhines (like insectivores) are nocturnal and the olfactory bulbs are reduced in size (relative to insectivores) in all genera that are now diurnal (*Lemur*, *Hapalemur*, *Varecia*, *Propithecus* and *Indri*). This suggests a definite link between diurnal habits in primates and relatively small olfactory bulbs (implying, in turn, less reliance on the olfactory sense). It is therefore remarkable that *Tarsius* and *Aotus*, despite being nocturnal,

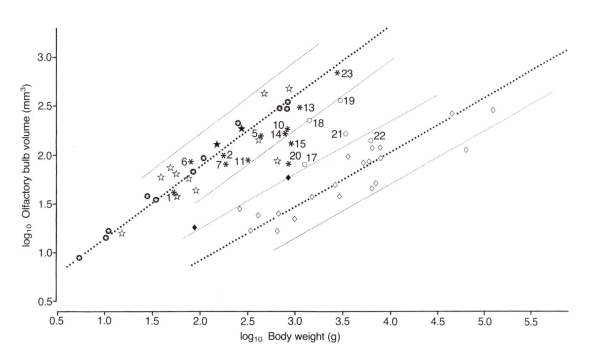

Figure 8.16. Logarithmic plot of olfactory bulb volume against body weight for insectivores, tree-shrews and primates. Heavy dotted lines represent reduced major axes for insectivore species (above; $N = 22$) and for non-human simian primate species (below; $N = 20$). Lighter dotted lines approximately indicate 95% confidence limits for each of the reduced major axes.

Key to symbols: 'basal' insectivores (●); 'advanced' insectivores (☆); tree-shrews (★); nocturnal lemurs + lorises (∗); diurnal lemurs (○); *Tarsius + Aotus* (◆); simians (◇). Key to strepsirhine primate species: 1, *Microcebus murinus*; 2, *Cheirogaleus medius*; 5, *C. major*; 6, *Galago demidovii*; 7, *G. senegalensis*; 10, *G. crassicaudatus*; 11, *Loris tardigradus*; 13, *Perodicticus potto*; 14, *Nycticebus coucang*; 15, *Lepilemur mustelinus*; 17, *Hapalemur griseus*; 18, *Lemur fulvus*; 19, *Varecia variegata*; 20, *Avahi laniger*; 21, *Propithecus verreauxi*; 22, *Indri indri*; 23, *Daubentonia madagascariensis*. (Data from Stephen, Bauchot and Andy, 1970. Reprinted from Martin, 1979 by permission of Academic Press Inc., New York.)

appear to be like other haplorhine primates in having reduced olfactory bulbs relative to body size. If reduction in size of olfactory bulbs in primates is primarily related to a switch from noctural to diurnal habits, this provides additional evidence supporting the suggestion that both *Tarsius* and *Aotus* have become secondarily adapted for nocturnal habits, following divergence from a diurnal ancestry (see Chapter 7). As ancestral placental mammals were very probably nocturnal (Chapter 4), it seems likely that ancestral primates were still

nocturnal and that extant nocturnal lemurs and lorises have generally remained primitive in this respect. This might explain why the size of the olfactory bulbs relative to body size of these primate genera (with the exception of *Lepilemur* and *Avahi*) falls within the range for modern insectivores.

There is, however, an enigma surrounding the position of the two tree-shrew species – the common tree-shrew (*Tupaia glis*) and the Philippine tree-shrew (*Urogale everetti*) – in the plot of olfactory bulb volume against body size

(Fig. 8.16). Both of these tree-shrew species are completely diurnal in habits, yet on the figure they lie just above the best-fit line for modern (nocturnal) insectivore species. Hence, these tree-shrews do not follow the trend towards smaller olfactory bulb shown by all the diurnal modern primates. In other words, not only is there no evidence to support the widespread interpretation that tree-shrews have a reduced olfactory system relative to insectivores, there is also the surprising fact that they fail to show such a reduction in spite of being diurnal. This suggests that tree-shrews are, in fact, significantly different from primates in terms of their reliance on olfaction, in that well-developed olfactory bulbs are retained even in diurnally active forms. Particular reliance on the olfactory sense in tree-shrews is further indicated by the presence of a very well-developed Jacobson's organ in these mammals (see Chapter 7). Hence, evidence from the olfactory system indicates a gulf between tree shrews and primates rather than any phylogenetic affinity.

One problem remains here in that it may not even be correct to assume that living insectivores are primitive with respect to the relationship between olfactory bulb size and body size. It is theoretically possible that, under the influence of continued nocturnal habits, the olfactory system has been expanded during the evolution of modern insectivores (Fig. 8.15), of tree-shrews and of most extant nocturnal lemurs and lorises. This possibility can be rejected or confirmed only by an analysis of the relationship between olfactory bulb size and body size in fossil placental mammals generally. Unfortunately, although such an analysis would be possible if currently available endocasts for fossil non-primates were appropriately measured, few data are available as yet for this test to be done. Gurche (1982) conducted an allometrical analysis of olfactory bulb volume, determined from endocasts, for the early Tertiary primates *Adapis*, *Notharctus*, *Necrolemur* and *Tetonius*, concluding that they fall within the modern prosimian range. In fact, *Adapis* and *Notharctus* lie just within the 95% limits for the insectivore distribution in Fig. 8.16, whereas *Necrolemur*

and *Tetonius* lie closer to *Tarsius*. As *Adapis* and *Notharctus* were probably diurnal in habits (e.g. see Fig. 7.6) and as *Necrolemur* and *Tetonius* may be related to modern haplorhines, the olfactory bulbs may have been reduced in all of these fossil primates. Notwithstanding the absence of data on olfactory bulb size in fossil insectivores, it seems reasonable to conclude provisionally that the general allometric relationship shown by modern insectivores (although not the condition in many individual species) is close to the ancestral condition for placental mammals on the following grounds:

1. The 'basal' and 'advanced' insectivores are not separated overall on the basis of olfactory bulb size relative to body size (Fig. 8.16), in spite of marked differences in relative brain size and considerable variation in behavioural adaptation.

2. Tree-shrews show the same relationships as insectivores.

3. Measurements on two endocasts of *Adapis parisiensis* by the author indicate rather larger olfactory bulbs than concluded by Gurche (1982), giving approximately the same olfactory bulb size: body size relationships as for modern nocturnal lemurs and lorises, which also overlap with the modern insectivore condition.

4. Accessory evidence on the morphology of the sensory apparatus for olfaction (Chapter 7) confirms the presence of a basically primitive pattern in insectivores, tree-shrews and most modern lemurs and lorises.

In summary, therefore, it can be said that quantitative investigation of the relative size of the brain in general, and of the olfactory bulbs in particular, provides only very limited support for the view that the condition for any living mammalian species has remained unchanged since divergence from the ancestral placentals. Although there is some evidence that the olfactory components of the CNS may have undergone relatively little change in certain lineages, the general picture is one of a consistent trend throughout mammalian evolution towards increase in relative brain size. Tree-shrews, it turns out, merely resemble many other mammalian species in having relative brain sizes intermediate between those of insectivores and primates and they are actually rather distinct from the latter in terms of their reliance on olfaction in spite of adoption of diurnal habits. Within the order Primates, strepsirhines generally have smaller brains than haplorhines and clear evidence of reduced olfactory bulbs in the latter would seem to be linked with a shift from nocturnal to diurnal habits.

It is now necessary to turn from gross size relationships of the brain to qualitative features of brain morphology.

EXTERNAL MORPHOLOGY OF THE PRIMATE BRAIN

The major details of the surface morphology of the primate brain are illustrated in Fig. 8.2, taking a relatively primitive modern primate species of moderate body size (*Lemur fulvus*). Apart from the typical mammalian features already mentioned above, the following characteristics have generally received attention as being of special interest in primate evolution:

1. All three major lobes of the cerebral hemispheres (frontal, occipital and temporal) are well developed, in association with the moderate to large size of the brain relative to body size. In all living primates, the frontal lobes of the two hemispheres overlap the olfactory bulbs, and the occipital lobes overlap the cerebellum.

   Extensive development of the occipital lobes of the cerebral hemispheres in living primates is a reflection of the particular importance of vision in primate behaviour (see Chapter 7). As mentioned above, vision may also be indirectly related to the development of the temporal lobes of the cerebral hemispheres because of their typical position in primates just posterior to the orbit. Further, Allman (1977, 1982) has noted that a large proportion of the cortex of the temporal lobe in primates is devoted to vision (see also Fig. 8.22). Development of the temporal

lobes, for whatever reason, is particularly conspicuous among living primates in comparison to other mammals.

2. In the brain of living primate species, at least, the surface of the cortex has certain infoldings (**sulci**) that are relatively constant in position and characteristic of each species. Where numerous sulci are present, the surface of the brain is subdivided into convolutions (**gyri**). The common sulcus of placental mammals dividing the palaeocortex from the neocortex, the **rhinal sulcus**, has already been mentioned. In addition to this basic mammalian sulcus, particular mention has been made of two other sulci in primate evolution: the **Sylvian sulcus** and the **calcarine sulcus**. As shown in Fig. 8.2, the former is located just anterior to the temporal lobe on the dorsolateral aspect of the brain, and the latter is located on the inside surface of the occipital wing of each cerebral hemisphere. Elliot Smith (1902) reported the presence of both of these sulci in all the brains of living primate species examined by him.

Taking these specific features of the brain surface as seen in living primates, one can assess their phylogenetic significance by dealing with a number of questions in turn. First, it is necessary to establish to what extent these features are both common to living primates and exclusive to them. Second, one can examine the brains of tree-shrews to establish whether they show any specifically primate features. Finally, one can turn to the primate fossil record in the attempt to trace, as far as fossilizable details will allow, phylogenetic developments in the morphology of the brain within the order Primates.

Sulcal patterns

It must be pointed out at once that, as with so many other morphological features, certain characters identifiable in the mammalian brain depend heavily on body size. A particularly obvious example is provided by the influence of brain size on the formation of sulci and gyri, regardless of phylogenetic affinities (Le Gros

Clark, 1945a; Elias and Schwartz, 1971; Jerison, 1977, 1982; Prothero and Sundsten, 1984). This dependency exists because the neurones (grey matter) of the mammalian brain are distributed as a surface layer (cortex) over the central mass of fibres (white matter; see Fig. 8.1). As the volume of the brain increases, the volume of the grey matter can only keep pace with the volume of the

white matter either by increasing in depth or by undergoing folding to generate an increased surface area. In fact, the thickness of the grey matter increases only marginally as brain size increases (Jerison, 1977, 1982; Hofman, 1982), perhaps because of the basic mammalian pattern of cortical organization involving columnar units of cells (Hubel and Wiesel, 1967; Palay, 1967;

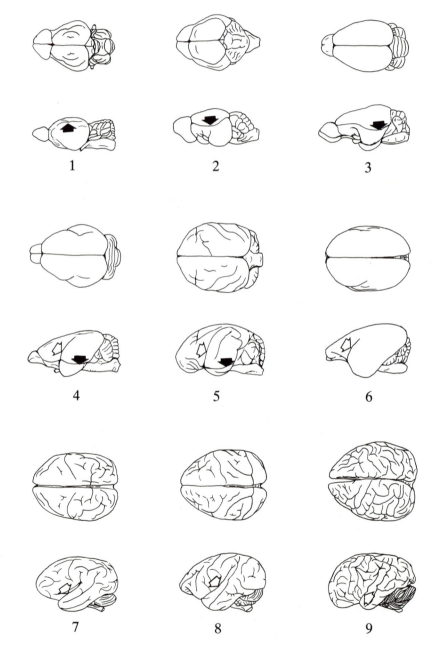

Towe, 1975; Mountcastle, 1978; Jerison, 1982; Changeux, 1983). As a result, the formation of sulci ('fissurization' as defined by Jerison, 1982) is a regular outcome of increase in brain size in mammals. Indeed, Jerison (1982, p.82) has concluded: 'The result is, essentially, that the increase in gyri and sulci that can be seen in larger brains is almost entirely a result of the enlargement.'

Two corollaries emerge from this observation, otherwise known as Baillarger's Law of Folding Compensation (Noback and Moskowitz, 1963). First, within each grade of brain size relative to body size (e.g. see Fig. 8.5), larger-bodied species will necessarily have more sulci than smaller-bodied species. Second, as brain size increases at a given body size (i.e. as there is a shift from one grade to another in relative brain size), more sulci will form. The first corollary is important because comparisons between species within a grade must involve a consideration of the effects of body size. The second is even more important, because it means that parallel expansion in relative brain size in a multiplicity of mammalian lineages must have been accompanied by a general increase in formation of sulci. Both of these effects are involved in the observation that the brains of simians typically have more sulci than those of prosimians (Starck, 1962). On the one hand, simians overall have relatively larger brains than prosimians (Figs 8.5 and 8.9); on the other, the average body size of simians is greater than that of prosimians, so simians would in any case be expected to have larger brains with more surface folding.

As is only to be expected from the above considerations, the brains of the smallest-bodied primates – both prosimians and simians – have virtually smooth cortical surfaces (lissencephalic condition). This is true not only of *Microcebus*, *Cheirogaleus*, *Galago demidovii* and *Tarsius* spp., which are all small-bodied prosimians, but also of the smallest simian species *Cebuella pygmaea* (see Fig. 8.17). As brain size increases, regardless of any other considerations, the complexity of cortical folding patterns increases in a very predictable fashion. In considering the evolutionary relevance of sulci, then, it is important to look for the minimal condition among living primates, on the grounds that the ancestral primates were probably small-bodied and undoubtedly had quite small brains. A

Figure 8.17. Outline drawings of the brains of selected insectivore, tree-shrew and primate species of differing body sizes, showing the general relationship between brain size and degree of cortical folding. For each species, a dorsal view (above) and a left lateral view (below) are shown. Key to species:

1. Solenodon, *Solenodon* (medium-sized insectivore with very small absolute brain size).
2. Hedgehog, *Erinaceus* (medium-sized insectivore with very small absolute brain size).
3. Pygmy tree-shrew, *Tupaia minor* (small-bodied tree-shrew with small absolute brain size).
4. Dwarf lemur, *Cheirogaleus* (small-bodied lemur with small absolute brain size).
5. Sifaka, *Propithecus* (large lemur with moderate absolute brain size).
6. Pygmy marmoset, *Cebuella* (smallest New World monkey with small absolute brain size).
7. Woolly monkey, *Lagothrix* (large New World monkey with moderately large absolute brain size).
8. Gelada baboon, *Theropithecus* (large Old World monkey with moderately large absolute brain size).
9. Gorilla, *Gorilla* (very large Old World great ape with substantial absolute brain size).

In insectivores and in tree-shrews, the only well-defined sulcus visible externally is the rhinal sulcus (black arrow), separating the neocortex (above) from the palaeocortex (below). All of the primates exhibit a true Sylvian sulcus (open arrow). The rhinal sulcus is only visible on the lateral surface of the brain in the lemurs *Cheirogaleus* and *Propithecus*, which have only moderate relative brain sizes. In the other primates shown, expansion of the neocortex has displaced the rhinal sulcus ventrally. Among primates, increasing complexity of sulcal patterns is generally associated with increasing brain size. Hence, *Cebuella* (the smallest simian primate) has an almost completely smooth (lissencephalic) brain in spite of having a relatively large brain for its body size. (1, 2, 6, 7, 8, 9 after Starck, 1962; 5 after Stephan, Frahm and Bauchot,1977; 3 and 4 from specimens studied by the author.)

modern primate species such as the lesser mouse lemur (*Microcebus murinus*), which counts among the smallest living primates and also belongs to the lowest grade of relative brain size (Fig. 8.5), provides a useful limiting case for assessing the phylogenetic significance of sulci. In spite of its small size, *Microcebus murinus* has a well-defined Sylvian sulcus (the only sulcus clearly identifiable on the dorsolateral surface of the brain), as shown for the related form *Cheirogaleus* in Fig. 8.17, and there is also a calcarine sulcus present on the internal wall of the cerebral hemisphere (Elliot Smith, 1902). Thus, these two sulci, recognized as characteristic of the primates (Elliot Smith, 1902; Martin, 1973), are present in the least developed brain found among living primates.

It is important to emphasize that Elliot Smith (1902) not only recognized the presence of these two sulci as characteristic for primates but also defined them in a particular way. He pointed out that the primate Sylvian sulcus arises from a junction with the rhinal sulcus, thus distinguishing it from the 'pseudosylvian sulcus' in a similar position in the brains of certain other mammals. Further, he noted that the calcarine sulcus in the primate brain typically has a unique three-branched (triradiate) shape (Martin, 1973). Le Gros Clark (1959) only laid emphasis on the rear branch of this sulcus system, the 'retrocalcarine sulcus', as being a particular feature of primates. These special features of the Sylvian and calcarine sulci are present in the brain of *Microcebus murinus*, just as they are in the brains of *Galago*, *Tarsius* and all other primates examined by Elliot Smith. In fact, the brains of certain living primates – notably marmosets and tamarins – apparently do not have a triradiate calcarine sulcus (L.J. Garey, personal communication). As marmosets and tamarins have probably undergone secondary dwarfing, it is possible that the calcarine sulcus system has been modified as a result. However, detailed study of this feature of the primate brain is required for reconstruction of the likely ancestral condition.

For the reason explained above, the sulcal patterns of the brains of larger primates show increasing degrees of complexity. In principle, therefore, the additional sulci may provide information about their phylogenetic relationships (Connolly, 1950; Falk, 1978). However, there is a major problem of potential convergence here. There may be only a limited number of possible solutions for folding of the cortex as brain size increases within a skull of a particular shape and similarity in sulcal pattern may therefore tell us relatively little about phylogenetic relationships. Indeed, the obligatory folding of the cortex that accompanies increase in brain size provides one of the clearest examples of a general trend that is likely to lead to numerous cases of convergent evolution (Fig. 8.17; see also Chapter 3).

Nevertheless, it is possible that specific sulci may provide evidence of phylogenetic affinity between particular primate species, although any assessment of this possibility should allow for the high probability of convergent evolution. For example, Radinsky (1974b) reports that a postlateral sulcus is present in the occipital region of the brains in all lorisine species (except apparently in the small-bodied *Nycticebus pygmaeus*), whereas this sulcus is absent from the brains of all bushbabies and lemurs of broadly comparable body and brain size. In this case, the lorisine sulcal pattern cannot be simply attributed to size effects, in comparison to other strepsirhines, so it is reasonable to regard the pattern as indicative of some special phylogenetic affinity. Conversely, it is instructive to consider the case of the subfossil lemurs of Madagascar, several species of which had quite large bodies. These subfossil lemurs almost certainly increased in size independently of other primates, as an integral part of the adaptive radiation of lemurs within Madagascar (Martin, 1972b). Because of brain:body size scaling effects, they also developed quite large brains. As Radinsky (1970) has shown in a very effective survey of subfossil lemur endocasts, all of the large-bodied genera so far examined (*Archaeolemur*, *Hadropithecus*, *Megaladapis* and *Palaeopropithecus*) showed some degree of fissurization of the cortex. Indeed, the sulcal pattern in *Archaeolemur* was quite complex. After studying an *Archaeolemur* endocast, Le Gros Clark (1945b) concluded that the brain of this prosimian was generally similar to that of simians. However, Radinsky (1970) correctly recognized the considerable probability of convergence between this large-bodied lemur and simians. He concluded that the sulcal pattern in *Archaeolemur* actually resembles that of indriids rather than that of simians in fine detail, when allowance is made for differences in body size.

Given the general size-dependency of sulcal patterns, it is also hardly surprising that the greatest complexity of sulci is found in the largest-bodied simian species. The large brain sizes and the sulcal complexity of great apes are only to be expected from the scaling of both of these features to brain size and hence to body size among the simians (see Figs 8.5 and 8.17). Only in *Homo sapiens* is there a major departure from the typical simian condition, with a brain size (relative to body size) some three times bigger than expected from the general trend (Figs 8.5 and 8.9). Accompanying this marked increase in brain size, there is an entirely predictable, concomitant increase in sulcal complexity.

The brain of tree-shrews

At this point it is instructive to return to the question of the relationship between tree-shrews and primates, as it was evidence from the size and morphology of the brain that particularly led Le Gros Clark (1924c, 1926, 1931, 1959) to postulate a specific phylogenetic link. Le Gros Clark listed the following external morphological features of the tree-shrew brain as indicating an affinity with primates and as distinguishing them from insectivores:

1. The relative size of the brain as a whole.
2. Expansion of the neocortex, accompanied by downward displacement of the rhinal sulcus.
3. Formation of a distinct temporal lobe of the neocortex.
4. Backward projection of the occipital lobe of the neocortex.
5. The presence of a calcarine sulcus.

The first four of these features are likely to be developed in any mammal showing an increase in relative brain size above the minimal level found in modern 'basal' insectivores. As discussed earlier, the tree-shrews are by no means unique among placental mammals in showing a modest expansion of the brain relative to body size. It is also significant that tree-shrews, like most primate species, are adapted (at least to some extent) for arboreal life, whereas modern insectivores are essentially terrestrial. We must therefore be careful to take account of possible convergence between tree-shrews and primates arising because both groups are adapted for life in the trees and because of a general tendency for relative brain size to increase in mammalian evolution. In short, we must seek specific features of the brain linking tree-shrews to primates, rather than relying on general characteristics that may be developed by any arboreal mammal with a moderately expanded brain.

In point of fact, all or almost all of the supposedly primate-like features of the tree-shrew brain listed above are commonly found in other placental mammals with arboreal tendencies, such as squirrels (*Sciurus* spp.), cats (*Felis* spp.) and even certain marsupials (e.g. the brush-tailed possum, *Trichosurus vulpecula*), as noted by Campbell (1974). In terms of gross brain morphology, there are only two features of primate brains that can be regarded as derived, in that they are both unique to modern primates and universal (or almost universal) to them. These are the triradiate calcarine sulcus and the Sylvian sulcus arising from the rhinal sulcus (connected with the formation of 'a distinct temporal lobe of the neocortex', as listed above). Le Gros Clark (1959) mentions only a 'shallow Sylvian fossa' for *Tupaia minor* and the calcarine sulcus illustrated is barely detectable. Indeed, Le Gros Clark himself notes that the retrocalcarine sulcus (the rear branch of the primate triradiate sulcus) is absent from *Tupaia*, in contrast to all living primates. The fact of the matter is that tree-shrews do not have a clear dorsolateral sulcus anterior to the temporal lobe and even if a small depression (fossa) can be detected in that region there is certainly no sign of a true Sylvian sulcus joining the rhinal sulcus in the manner defined by Elliot Smith (1902). Similarly, there is no evidence of a triradiate calcarine sulcus on the internal wall of the cerebral hemisphere, even if a small depression may be identifiable there. Calcarine sulci of various kinds are found in several non-primate mammalian groups (Elliot Smith, 1902), and only the presence of an obvious triradiate sulcus in tree-shrews would suggest a definite phylogenetic link with primates. In short, living tree-shrew species such as *Tupaia minor* do not have any clear derived characters in external brain morphology that are shared with living primates and the list of features compiled by Le Gros Clark does not provide a valid guide to phylogenetic affinities.

It is also revealing to examine the manner in which Le Gros Clark (1959) evaluated the evidence that led him to suggest a phylogenetic relationship between tree-shrews and primates. The living tree-shrews do, after all, have some broad resemblances to living prosimians in terms of relative brain size, general morphology of the brain and enhancement of the visual system. The question that must be asked, however, is: 'What is the likelihood that a hypothetical ancestral tree-shrew and a hypothetical ancestral primate would have shared specific characteristics in the development of the central nervous system indicative of common ancestry?' (see Chapter 3). It emerges that this question was not really considered by Le Gros Clark from the outset (1934b). Nor was it raised by numerous subsequent workers who have similarly concluded that the brain morphology of tree-shrews is indicative of a common ancestry with primates (e.g. Starck, 1962; Stephan, 1972; Buettner-Janusch, 1966; Napier and Napier, 1967). All of these authors confined their attention to demonstration of similarities between living tree-shrews and living primates rather than attempting to reconstruct the possible phylogenetic history underlying such present-day similarities. Indeed, Le Gros Clark confined his approach even more by emphasizing similarities between primates and a particular tree-shrew species with the most advanced degree of development of the central nervous system:

There is a marked difference between the brain of the Ptilocercinae (represented by *Ptilocercus*) and that of other tree-shrews, the Tupaiinae. In most of its anatomical systems *Ptilocercus* is a much more primitive creature and, in contrast to the diurnal habits of tree-shrews in general, it is crepuscular. These differences are reflected in the general features of the brain. We shall here confine our attention to *Tupaia*, partly because its structure has been studied in some detail, and partly because it displays a number of features which clearly indicate affinities with the lemuroid brain.

(Note: the pen-tailed tree-shrew, *Ptilocercus*, is in fact nocturnal as described in Chapter 4.) Le Gros Clark actually limited his attention to a single tree-shrew species, *Tupaia minor*, which has the largest relative brain size for the family (Fig. 8.9).

This single quotation reveals a logical flaw that has characterized much of the discussion about possible phylogenetic affinity between tree-shrews and primates as indicated by evidence from the central nervous system. This flaw arises because of a failure to follow implications consistently through a hypothetical phylogenetic scheme. As noted above, the primary question that should be asked is: 'Do tree-shrews and primates share characteristics of the central nervous system that are based on some property present in a common ancestral stock?' Answering this question requires, among other things, reconstruction of a hypothetical common ancestral condition for tree-shrews and a hypothetical common ancestral condition for primates. Obviously, if the most primitive of living tree-shrews (i.e. *Ptilocercus*) does not share with living primates specific features of the brain that characterize the order Primates, there can be no case for postulating retention of shared, derived character states from a common ancestral stock. On this criterion alone, the lack of specifically primate-like features in the brain of *Ptilocercus lowii* recorded by Le Gros Clark (1926, 1959) surely indicates that any resemblance between the brains of other tree-shrew species and those of primates must have arisen through independent (i.e. convergent) evolution.

Brain morphology of fossil primates

It is, then, a logical requirement for any reconstruction of primate phylogeny that an attempt should be made to trace the evolution of the external morphology of the central nervous system through the fossil record with the aid of available endocasts. Particularly valuable research in this direction has been conducted by Edinger (1929), by Radinsky (1968, 1970, 1972, 1974b) and by Jerison (1970, 1973). A good starting-point for such reconstruction is provided by consideration of the morphology of the brain of the Eocene 'lemuroid' *Adapis parisiensis*, for which two particularly informative artificial endocasts are available (Le Gros Clark, 1945b; Martin, 1973; Gingerich and Martin, 1981). Figure 8.18 shows that this Eocene primate species lacked the overlapping of the olfactory bulbs by the frontal lobes and of the cerebellum by the occipital lobes of the cerebral hemispheres that characterizes all living primate species. This feature of living primates, which (as explained above) is a direct consequence of expansion of the brain relative to body size, may therefore have arisen as a convergent similarity among the major groups of modern primates. Lack of such overlapping in the brain of *Adapis parisiensis* directly supports the interpretation (above) that this Eocene species had a smaller relative brain size than any living primate. By contrast, the temporal lobe was already well developed in *Adapis parisiensis* and examination of the endocast indicates the presence of a true Sylvian sulcus arising from the rhinal sulcus (Radinsky, 1970; Martin, 1973; Gingerich and Martin, 1981).

Naturally, the internal face of the cerebral hemisphere is not reflected in the endocast, so it is impossible to determine whether or not a triradiate calcarine sulcus was present. Although it is reasonable to infer that, because most living primates (regardless of body size) have a triradiate calcarine sulcus, this feature might have been present in the common ancestor of the

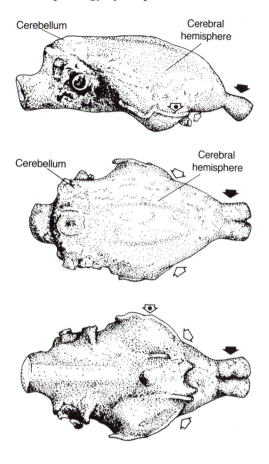

Figure 8.18. Outline drawings of the (partially reconstructed) endocast of the Eocene 'lemuroid' *Adapis parisiensis*. Note the pedunculate olfactory bulbs (black arrow) and the weakly indicated Sylvian sulcus (open arrow). The position of the rhinal sulcus, which is indicated by the impression of a superficial venous sinus (open arrow with dot), is relatively low down on the temporal pole of the brain, indicating quite marked expansion of the neocortex in this 'lemuroid'. (After Gingerich and Martin, 1981.)

primates, this inference must unfortunately remain speculative. In observable features of its overall morphology, the brain of *Adapis parisiensis* was (as expected) more primitive than in any living primate species, but it also had certain features judged to be characteristic of the evolution of the primate brain. Precocious development of the temporal lobe in the brain of

Adapis parisiensis, along with the formation of a true Sylvian sulcus, was perhaps physically associated with the fairly prominent development of the orbits, given that the temporal lobe is located just behind the orbit, as noted above. However, it may also indicate early emphasis on expansion of the visual cortex. Whatever the case may be, this evidence indicates that the development of the temporal lobe of the cerebral hemisphere was an early and significant feature of primate evolution, as suggested by Le Gros Clark (1959).

Examination of other early Tertiary primate endocasts, from *Smilodectes*, *Tetonius* and *Necrolemur*, confirms the general absence of overlap of the olfactory bulbs by the frontal lobes of the cerebral hemispheres (Radinsky, 1970). Pedunculate (stalked) olfactory bulbs were therefore undoubtedly characteristic in early primates (see also Gurche, 1982). As in *Adapis*, the cerebellum of the 'lemuroid' *Smilodectes* is not overlapped by the occipital lobes of the hemispheres. The 'tarsioids' *Tetonius* and *Necrolemur*, on the other hand, do show a small degree of overlapping of the cerebellum, and this suggests that the visual system (as indicated by the relative extent of the occipital cortex) was developed to a greater degree in Eocene 'tarsioids' than in Eocene 'lemuriods'. This prominent development of the occipital lobes of the cerebral hemispheres is also seen in the partially exposed endocast of the North American 'tarsioid' *Rooneyia* from the Oligocene (Wilson, 1966; Hofer and Wilson, 1967; Radinsky, 1970). Overall greater development of the neocortex, reflected by partial overlapping of the cerebellum, hence characterizes early Tertiary 'tarsioids' generally in contrast to Eocene 'lemuroids', and also correlates with the fact that the former had larger brains relative to body size (see Table 8.12; Fig. 8.13).

Both *Tetonius* and *Necrolemur* had a true Sylvian sulcus (Radinsky, 1970), but the endocast of *Smilodectes*, somewhat surprisingly, indicates that this 'lemuroid' did not (Gazin, 1965; Radinsky, 1970). This latter observation can be interpreted in three ways. One possibility is that

the ancestral primates lacked a Sylvian sulcus and that this was developed independently in some 'lemuroids' (e.g. *Adapis*) and in one or more other lineages leading to the extant primates. The second possible interpretation is that a Sylvian sulcus was developed in a common ancestor of the European Adapinae and all living primates, while the North American Notharctinae (including *Smilodectes*) diverged at an earlier stage of evolution. Finally, it is always possible that the brain of *Smilodectes* did in fact have a Sylvian sulcus but this did not leave an impression inside the braincase. Gurche (1982) has suggested that the Sylvian sulcus could have been masked by a superficial blood vessel in *Smilodectes*. In any event, this single character cannot be granted much weight in the reconstruction of early primate phylogeny and it remains to be seen whether a true Sylvian sulcus was an ancestral primate feature or arose separately in different primate lineages. The brain of *Smilodectes* does, however, fit the general primate pattern in having a fairly well-developed temporal lobe of the cerebral hemisphere and this development is even more pronounced in the Eocene 'tarsioids' *Tetonius* and *Necrolemur* (Radinsky, 1970).

INTERNAL ORGANIZATION OF THE PRIMATE BRAIN

A great deal is now known about the internal organization of the primate brain. However, it is beyond the scope of the book to consider this evidence in any detail, particularly because the range of primate species covered is generally too limited for an effective inference of possible phylogenetic relationships. Instead, discussion here will focus on certain specific components of the brain concerned with the visual system. For one thing, visual developments have clearly played a central part in primate evolution (see Chapter 7) and corresponding features of the brain have accordingly been highlighted in the literature, notably with respect to the affinities of tree-shrews. Further, the major features of the visual system have been studied in a sufficient

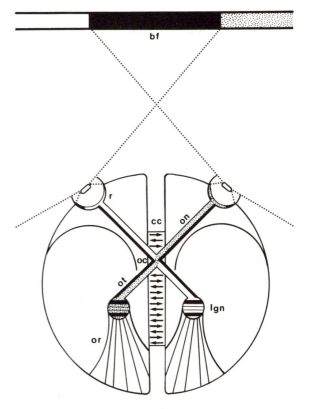

number of primate species to permit a reasonable evolutionary interpretation.

In addition to defining characters observable externally on the intact brain, Le Gros Clark (1959) listed the following internal features of the brain as indicative of affinities between tree-shrews and primates, on the grounds that they are lacking from insectivores:

1. Well-marked lamination and cellular richness of the neocortex in general, and a pronounced degree of differentiation of the separate cortical areas.
2. Pronounced elaboration of the visual apparatus of the brain, particularly at the cortical level.
3. Well-defined cellular lamination of the lateral geniculate nucleus.

All of these features are directly or indirectly connected with the visual system (e.g. well-marked lamination is a particular feature of the primary visual cortex in primates). It is necessary to examine them more closely to see whether tree-shrews and primates do, indeed, share special features suggestive of common ancestry, or whether there are significant differences of detail that indicate convergent evolution, perhaps in response to general selection pressures favouring enhanced processing of visual information in an arboreal environment.

It should be noted at once that there are two major visual systems in the mammalian brain that seem to be largely independent of one another. The primary visual system (Fig. 8.19) is the **retinogeniculostriate system** which involves projection of retinal inputs to the visual cortex via a relay centre (**dorsal lateral geniculate nucleus**) on each side of the brain. The secondary visual system is the **retinotectal system**, in which retinal inputs project to the **optic tectum** (also called the superior colliculus) in the midbrain. (The location of the optic tectum is shown in the illustration of the brain of *Aotus* in Fig. 8.22.) The retinotectal system is of special significance for primate evolution, but it was until recently generally omitted from discussions of primate evolution and will therefore be discussed separately below. There is, however, one

Figure 8.19. Basic diagram of the primary visual system (retinogeniculostriate system) in placental mammals. Non-primates typically have a restricted field of binocular overlap (bf). Each retina (r) receives inputs from the binocular field and from the relatively large monocular field viewed by that eye. Optic nerve fibres from the retina carry this information along the optic nerve (on) to the optic chiasma (oc). At the chiasma, fibres relating to the monocular field and to the adjacent half of the binocular field cross over, while fibres from the opposite half of the binocular field remain uncrossed (ipsilateral). The fibres on each side of the brain then pass via the optic tract (ot) to the dorsal lateral geniculate nucleus (lgn) and on through the optic radiation (or) to reach the visual cortex. Information can be exchanged between the two cerebral hemispheres via the corpus callosum (cc).

common feature relating to both visual systems in that optic nerve fibres from the two eyes cross over to varying degrees at the **optic chiasma** before passing to one of the two major visual systems. In vertebrates other than mammals (i.e.

in fish, amphibians, reptiles and birds), crossing-over (decussation) is complete, such that all visual information from the left eye initially passes to the right side of the brain and vice versa. (This follows the basic vertebrate pattern in which most sensory inputs from one side of the body pass to the opposite side of the brain.) By contrast, in mammals there is a significant difference in that at least some optic nerve fibres do not cross over at the chiasma (e.g. see Allman, 1977). As a result, each side of the brain receives a proportion of fibres from the eye on the same side of the body (**ipsilateral fibres**) along with those originating from the eye on the opposite side of the body (**contralateral fibres**). This arrangement provides a direct neural basis for integration of information from both eyes on each side of the brain. In fact, the primary visual system is organized such that the left half of the visual field is represented in the right visual cortex and vice versa (Allman, 1977). This is achieved by virtue of the fact that retinal fibres carrying information from one half of the total visual field pass to the opposite side of the brain (these being the contralateral fibres), while the ipsilateral fibres are those carrying information from the remaining half of the binocular field. In other words, on each side of the brain contralateral fibres representing the opposite half of the total visual field (opposite monocular field plus opposite half of the binocular field) are brought together with ipsilateral fibres also representing the opposite half of the binocular field. Hence, in the right side of the brain (for instance) contralateral and ipsilateral inputs from the left half of the binocular field can be brought together to provide a basis for three-dimensional interpretation of that sector of the visual field.

Primary visual system

In the mammalian primary visual system, the contralateral and ipsilateral fibres constituting the **optic tract** on each side of the brain pass first to the dorsal lateral geniculate nucleus, from which numerous fibres forming the **optic radiation** spread out to reach the **visual cortex** in the occipital region of the brain (Fig. 8.19).

Subsequently, communication between the two halves of the brain can take place through transverse connections, thus bringing together the two halves of the total visual field. In vertebrates other than placental mammals, as noted above, such communication between the two halves of the brain is limited to the anterior commissure. In the placental mammals, however, there is an additional transverse commissure, the **corpus callosum**, which is more substantial and largely supersedes the anterior commissure (e.g. see Figs 8.2 and 8.21). Before communication between the right and left of the brain takes place, however, there is a complex process of coordination of visual information in each half of the brain.

Returning now to Le Gros Clark's list (above) of internal features of the brain shared by tree-shrews and primates, but not by insectivores, the first important observation to be made is that these features are also shared by several other mammalian groups. Well-defined lamination of the cortex, particularly of the primary visual cortex, and of the dorsal lateral geniculate nucleus are found in a variety of mammals, such as certain rodents (notably squirrels), carnivores, artiodactyls and even marsupials (specifically members of the family Phalangeridae; e.g. the brush-tailed possum, *Trichosurus vulpecula*). Hence, a reasonable interpretation would seem to be that tree-shrews and primates share only generalized features of the primary visual system that are likely to be enhanced in any mammal that relies heavily on vision as a major sense. Active arboreal life would obviously require a well-developed visual system and it may be simply for this reason that certain similarities between tree-shrews and primates are found. Accordingly, Campbell (1975) reported that the visual cortex of the arboreal grey squirrel (*Sciurus carolinensis*) is 'almost identical' to that of *Tupaia*, whereas it is less well developed in terrestrial rodents. Giolli and Tigges (1970) concluded that the primary visual system of *Tupaia* quite closely resembles that of two non-arboreal mammals, the rat and the rabbit, in overall features, so any resemblance between tree-shrews and primates is clearly of a fairly general nature. Although such observations do not rule out a phylogenetic relationship between tree-shrews and primates, they do indicate that the probability of primitive retention and/or convergent evolution is high, thus considerably weakening the case.

Taking a different tack, it is pertinent to ask whether the visual system of primates has other features that are both unique and universal to the group, suggesting a derived condition traceable to the hypothetical ancestral primate stock. Le Gros Clark's view, relating to the CNS in general, was that primates are not defined by any such universal derived features but by general trends, with the tree-shrews having a condition intermediate between that of insectivores and that of primates. However, in the light of recent evidence, this view may now be effectively challenged on numerous grounds.

Mention has already been made of the behaviour of optic nerve fibres at the optic chiasma. Comparison of a tree-shrew with various primate species, including both prosimians and simians, shows that there is a striking difference (Table 8.14). In the tree-shrew there is almost complete crossing-over, with only a small percentage of ipsilateral fibres, whereas all primates investigated show an approximate balance between contralateral and ipsilateral fibres passing to each side of the brain (Giolli and Tigges, 1970). Predominant crossing-

Table 8.14 Degree of crossing-over (decussation) of the optic nerve at the chiasma in tree-shrews and primates (data from Giolli and Tigges, 1970)

Species	Decussation (%)	No. of studies
Tupaia glis	90–97	5
Galago crassicaudatus	50–60	3
Nycticebus coucang	50–60	2
Perodicticus potto	50–60	1
Macaca mulatta	~ 60	1
Cercocebus atys	~ 60	1
Saimiri sciureus	50–60	2
Ateles geoffroyi	50–60	1
Homo sapiens	53	1

over at the optic chiasma is characteristic of mammals with relatively laterally oriented eyes, whereas a high proportion of ipsilateral fibre projection is associated with frontally directed eyes and a concomitantly large binocular field of vision. Tree-shrews, including *Ptilocercus*, have relatively small, laterally directed eyes (Figs 5.4 and 5.5) with less convergence than those of primates and, in common with most non-primate mammals, they have a more restricted field of binocular overlap. Very low percentages of ipsilateral fibres are also found, for instance, in lagomorphs and rodents. A high proportion of ipsilateral fibres (40–50%) in the primary visual system would therefore seem to be a shared derived feature of primates. However, this is not a unique shared derived feature of primates as certain other mammals with forward-facing eyes, such as cats, also have a fairly high proportion of ipsilateral fibres.

Secondary visual system

In fact, Allman (1977, 1982) identified a far more fundamental distinction between non-primates and primates with respect to visual projection to the optic tectum in the midbrain. In the first place, although some non-primate mammals (e.g. cats) do have a fairly high proportion of uncrossed (ipsilateral) fibres at the chiasma, primates seemed to be unique in that there is a high proportion (one-third to one-half) of ipsilateral fibres among the population of fibres passing specifically to the optic tectum. In all of the other mammals investigated by Allman, including cats and tree-shrews, only a very low proportion of the fibres passing to the optic tectum are ipsilateral. This difference in the ratio between ipsilateral and contralateral fibres passing to the optic tectum is associated with a marked divergence in the organization of the tectum itself (Allman, 1977, 1982). In all primates so far examined, including both prosimians and simians but excluding tree-shrews, the retinal projection to the optic tectum follows the same principle as the projection to the visual cortex: each optic tectum contains a representation of the opposite half of the total visual field. By contrast, a large sample of non-primate mammals was found to resemble other vertebrates (fish, amphibians, reptiles and birds) in that each optic tectum contains a representation of the visual field seen by the opposite retina. This distinction is illustrated in Fig. 8.20. As yet, the functional basis for the special pattern in primates is poorly understood. However, Allman (1977) has cited evidence suggesting that the optic tectum links conjugate eye movements with the retinal projection, such that the eyes could automatically focus together on any point of interest in the binocular field. This would indicate that there is a direct link between the primate pattern of tectal representation and emphasis on binocular vision associated with frontally directed eyes. Possession of a high proportion of ipsilateral fibres passing to the optic tectum, combined with representation of the opposite half of the visual field in each tectum, therefore, seemed to constitute a unique feature of primates among

vertebrates, Allman reasonably concluded that this represents a derived development in the ancestral stock that gave rise to living primates, excluding tree-shrews (see also Martin, 1986a).

Subsequent to Allman's discussion of the unusual tectal projection of primates (1977, 1982), it emerged that this pattern is not, after all, unique to primates. Fruit-bats (Megachiroptera) share the same features as primates – namely, a relatively high proportion of ipsilateral fibres passing to the optic tectum and representation of the opposite half of the visual field in each tectum (Pettigrew, 1986). The other bats (Microchiroptera) do not have these features at all and therefore resemble other vertebrates, including most other mammals. On this basis, Pettigrew suggested that the two main groups of bats may have evolved independently, with the fruit-bats sharing a specific common ancestry with primates. Among other things, this hypothesis requires that wings must have evolved independently in the two groups of bats,

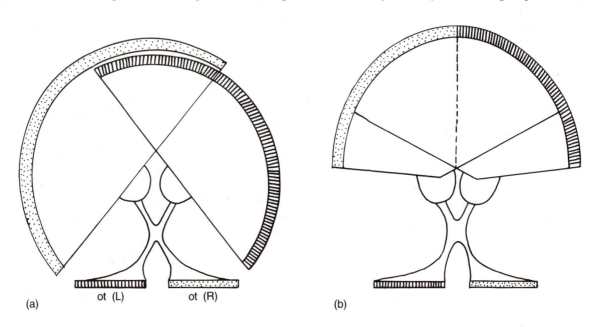

(a)   ot (L)   ot (R)

(b)

Figure 8.20. Diagrammatic comparison of the visual projections to the optic tectum (ot) in virtually all non-primate vertebrates (a) and in primates (b). In typical non-primates, the entire visual field of one eye is projected to the tectum in the opposite side of the brain. In primates, the total visual field is bisected and each half is projected to the tectum in the opposite side of the brain. (Reprinted from Martin, 1986a, with kind permission from Cambridge University Press.)

and it also conflicts with conclusions drawn from serological comparisons (e.g. Cronin and Sarich, 1980). Further, there is a striking resemblance between fruit-bats (but not other bats) and primates in the degree of forward rotation of the orbits and it is conceivable that a primate-like pattern of organization of the optic tectum has emerged in fruit-bats as a convergent development associated with adaptation of the visual system for an extensive binocular field (Martin, 1986b). Nevertheless, the identification of a primate-like tectal projection pattern in fruit-bats indicates that these mammals should be specifically taken into account in future discussions of the early origins of primates (see Chapter 12). Regardless of the outcome of such discussions, however, it is important to note that, on present evidence, tree-shrews definitely lack the primate-like pattern and are therefore markedly more distinct from primates in this respect than are fruit bats.

Allman (1982) also listed three additional features of the visual system that are universal to primate species studied to date and may well indicate some common ancestral condition:

1. A high concentration of ganglion cells in the central retina, associated with greatly expanded representation of the central retina in neural maps of the visual field.
2. Distinct lamination of the dorsal lateral geniculate nucleus, in which inputs from the two eyes are 'brought into precise visutopic register' before passing on to the primary visual cortex.
3. Marked expansion of the visual cortex, which contains a number of separate 'maps' of the visual field.

In fact, whereas all of the above features are characteristic of primates, they are by no means unique to primates. For example, retinal centralization is not unique to primates, and both lamination of the dorsal lateral geniculate nucleus and development of numerous visual 'maps' are found quite widely among other mammals.

Dorsal lateral geniculate nucleus

It has already been pointed out that the dorsal lateral geniculate nucleus is laminated in many mammalian species, and not just in primates and tree-shrews. Campbell (1975) reported that at least eight mammalian orders include species with laminated nuclei, with more than half of the mammalian orders remaining to be investigated. He summed up the evidence as follows: 'the conclusion is inescapable that lamination has arisen independently many times in animals which must rapidly evaluate spatial relations. These are animals that fly, are arboreal, glide, or move rapidly on the ground.' Kaas, Huerta *et al.* (1978) have, in fact, reported that lamination of the dorsal lateral geniculate nucleus is probably the rule among mammals, but that such lamination is not revealed by simple histological examination in some species and can be demonstrated only with special techniques. It would seem that overt lamination as in primates and tree-shrews has arisen as a common convergent feature among mammals, and it is therefore necessary to consider the organization of the dorsal lateral geniculate nucleus in more detail to identify indicators of phylogenetic relationships.

As Noback (1975) has pointed out, the primary visual pathways from each eye remain separate, even in mammals, until the cortex is reached and integration of information from the left and right eyes is therefore a property of the visual cortex. Hence, the dorsal lateral geniculate nucleus seems to act essentially as a relay station of some kind. Lamination of the dorsal lateral geniculate nucleus in primates, tree-shrews and other mammals has received detailed consideration from several authors (e.g. Noback and Moskowitz, 1963; Hassler, 1966; Noback, 1975; Noback and Laemle, 1970; Giolli and Tigges, 1970; Kaas, Guillery and Allman, 1972; Kaas, Huerta *et al.*, 1978). As a rule, each discrete layer of the nucleus is associated exclusively with fibres from just one eye (i.e. it receives either contralateral or ipsilateral inputs). In principle, then, it is possible to compare layers of the dorsal lateral geniculate nucleus between species, taking into account the type of input (contralateral or ipsilateral). In practice, however, this task has proved to be rather difficult because of uncertainties about homologies between species. The common practice has been to number the layers of the nucleus, usually starting with the layer closest to the optic tract and ending with the layer closest to the optic radiation (e.g. see Noback, 1975). In interspecific comparisons, problems of interpretation have arisen because layers may become subdivided and because new layers may emerge between pre-existing layers. Hence, a layer given a particular number in one species may not be homologous with a layer given the same number in another species. For this reason, Kaas *et al.* (1972, 1978) have adopted a different approach, using a classificatory scheme based on features of the cells composing the layers. Their scheme avoids many of the pitfalls of numbering systems and will be followed here (Fig. 8.21).

In all primates, it is possible to identify two large-celled (magnocellular) layers close to the optic tract. The external magnocellular layer (ME) consistently receives contralateral inputs while the internal layer (MI) consistently receives ipsilateral inputs, so it is reasonable to infer that these two layers are homologous in all primates and were present in their common ancestor. Similarly, all primates have two small-celled (parvocellular) zones lying further away from the optic tract. In this case, the situation is complicated by the fact that in many simian primates each parvocellular zone may be subdivided into leaflets, which can interdigitate. Nevertheless, as the external zone (PE) consistently receives contralateral inputs while the internal zone (PI) consistently receives ipsilateral inputs, it is reasonable to conclude that the ancestral primate had two parvocellular layers. The hypothetical basic four-layered pattern is present in *Tarsius*, which therefore seems to show the most primitive condition found among living primates (Kaas, Huerta *et al.*, 1978; McGuinness and Allman, 1985). Many simian primates (e.g. New World monkeys, gibbon, orang-utan – Tigges and Tigges, 1987) have

departed relatively little from this basic pattern. However, as noted above, simians commonly have subdivisions of the parvocellular layers and they also generally have one or two superficial layers (SE and/or SI) that receive ipsilateral and contralateral inputs, respectively (see Fig. 8.21). By contrast, in strepsirhine primates (lemurs and lorises) it would seem that two new layers have developed between the external and internal parvocellular layers. These new layers are characterized by very small cells and have therefore been termed 'koniocellular' (Kaas, Huerta *et al.*, 1978). The external koniocellular layer (KE) consistently receives contralateral inputs, while the internal (KI) consistently receives ipsilateral inputs, so the layers seem to be homologous among strepsirhines generally. It would therefore seem that the development of the koniocellular layers is a shared derived feature of lemurs and lorises, traceable to their common ancestor. Accordingly, the structure of the dorsal lateral geniculate nucleus does provide additional evidence that the lemurs and lorises constitute a monophyletic group. On the other hand, the lateral geniculate nucleus provides no evidence for a specific affinity between tarsiers and simians as tarsiers merely have a set of layers that would seem to be primitive for primates generally.

Noback (1975) noted that the organization of the layers of the dorsal lateral geniculate nucleus of *Tupaia* is quite distinct from that of all primates. This conclusion has been confirmed by Kaas, Huerta *et al.* (1978) on the basis of their revised scheme of homologies. In the tree-shrew, it is possible to identify six layers in the nucleus. The first of these may be homologous with the external superficial layer (SE) found in various primates, but it differs in that it receives contralateral, rather than ipsilateral, inputs. The second and third layers are magnocellular layers and may be homologous with the two magnocellular layers of primates. However, the inputs to the external magnocellular layer (= ME?) are ipsilateral and the inputs to the internal layer (= MI?) are contralateral, which is the reverse of the universal primate condition. The fourth layer of the tree-shrew nucleus is

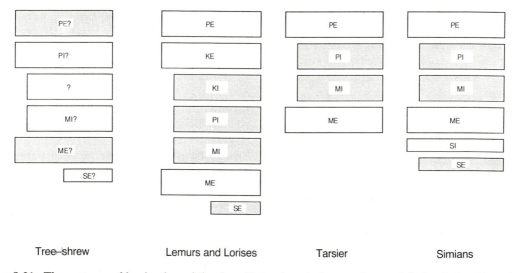

Tree–shrew   Lemurs and Lorises   Tarsier   Simians

Figure 8.21. The patterns of lamination of the dorsal lateral geniculate nucleus and their relationship to inputs from the two eyes in tree-shrews (*Tupaia*), strepsirhine primates (lemurs and lorises) and haplorhine primates (tarsiers and simians). Key: white, contralateral inputs; stippled, ipsilateral inputs; KE, external koniocellular layer; KI, internal koniocellular layer; ME, external magnocellular layer; MI, internal magnocellular layer; PE, external parvocellular layer; PI, internal parvocellular layer; SE, external superficial layer; SI, internal superficial layer.

Note that strepsirhines differ from haplorhines primarily in the presence of two additional layers (KE; KI) between the two parvocellular layers (PE; PI). The parvocellular layers can show quite complex subdivisions in simian primates (not shown here). There is also some apparent variability in the superficial layers. The tree-shrew differs from all primates in the inputs to layers that seem to be comparable in terms of relative position and cell size. (Diagram based on information provided by Kaas, Huerta, *et al.*, 1978.)

unusual in that it does not seem to have a counterpart, contrary to the situation in primates, in which all of the main layers can be classified in pairs. Finally, the fifth and sixth layers of the tree-shrew nucleus are parvocellular and may be equated with the two parvocellular layers of primates. Once again, however, the tree-shrew shows reversal of the inputs, with PE(?) receiving ipsilateral inputs and PI(?) receiving contralateral inputs. Overall, therefore, the pattern of lamination of the dorsal lateral geniculate nucleus in the tree-shrew is very different from that of primates and the most reasonable interpretation is that it has developed largely or completely independently from a very simple ancestral condition.

In view of the observation, discussed above, that the optic tectum of fruit-bats is organized in a primate-like pattern, it is important to note that these mammals also have a six-layered dorsal

lateral geniculate nucleus. However, whereas the two magnocellular layers close to the optic tract in fruit-bats are very similar to those of primates in both morphology and connections, the remaining four layers are different in that the inputs alternate in the sequence contralateral–ipsilateral–contralateral–ipsilateral away from the magnocellular layers. Thus, the nucleus of fruit-bats is more similar to that of primates than is the nucleus of the tree-shrew, but the condition in fruit-bats cannot be derived directly from the hypothetical ancestral condition for primates outlined above.

Kass, Huerta *et al.* (1978) suggested that the ancestral condition for placental mammals, as exemplified by modern insectivores, was a relatively simple dorsal lateral geniculate nucleus with concealed lamination – a central cluster of cells with ipsilateral inputs sandwiched between two outer clusters of cells with contralateral inputs. Following McKenna's (1975) proposal that tree-shrews, bats, 'flying lemurs' and primates constitute a distinct phylogenetic assemblage (Archonta), Kaas, Huerta *et al.* also proposed a common ancestral stage ('visually specialized, semi-arboreal insectivores') with overt lamination of the nucleus. However, given the differences in inputs into the successive layers reported for fruit-bats and tree-shrews, it would seem that they must have developed independently from an ancestral stage in which the pattern of lamination was still relatively simple. It should also be remembered that the practice of taking a group of living mammals (insectivores) as representatives of the ancestral placental condition is as suspect here as anywhere else. There is always the possibility that the dorsal lateral geniculate nucleus of insectivores has undergone secondary regression and hence reduction of lamination as a result of evolutionary decrease in the degree of dependence on vision.

As with any other morphological characters, interpretation of the evolution of the dorsal lateral geniculate nucleus would benefit greatly from an understanding of functional aspects. Unfortunately, the function of the nucleus remains obscure, although it is known that there is precise 'mapping' such that ipsilateral and contralateral inputs from corresponding parts of the visual field are brought together (Allman, 1977). Noback (1975) has suggested a number of possible roles for this relay station, all concerned with refinement of the information transmitted to the visual cortex. Further, Kaas, Huerta *et al.* (1978) have pointed out that the different layers in the nucleus seem to be concerned with functionally different types of visual information. However, it remains to be determined why particular patterns of lamination have developed in individual mammalian groups. Whatever its function, the dorsal lateral geniculate nucleus is obviously an important component of the primary visual system and even basic features of its organization are clearly of value with respect to preliminary inference of phylogenetic relationships (Fig. 8.21).

Maps in the visual cortex

The question of 'mapping' of visual information on the cortex is one that has become increasingly important, particularly because of the observation that inputs from the eyes are mapped in several different ways in discrete areas of the visual cortex. The visual cortex has long been known to be subdivided into separate areas that were first clearly defined by Brodmann (1909) and have been intensively investigated in recent years (e.g. see Allman, 1977, 1982). The posteriorly located **primary visual cortex**, corresponding to Brodmann's area 17, is particularly well laminated, as has been noted above, and is hence also called the **striate cortex**. Bordering this is the **second visual area** or **parastriate cortex**, corresponding to Brodmann's area 18, and anterior to that is the **peristriate cortex**, corresponding to Brodmann's area 19. However, the number of separate visual maps varies greatly among mammals. Allman (1982) emphasized that there are only two such maps in hedgehogs, whereas cats have at least 12. The owl monkey (*Aotus trivirgatus*), which is the primate species whose visual cortex has been most completely mapped to date, was initially shown to have nine discrete areas (Fig. 8.22), and 13 have now been identified (J. Allman, personal communication). Hence, the number of separate cortical maps of the visual field is not in itself indicative of affinity with primates.

Allman (1977, 1982) has suggested that primates probably share a basic set of visual sensory domains ('maps') in the visual cortex, including the primary visual cortex (V-I) the second visual area (V-II) and two subregions of the peristriate cortex – the middle temporal visual area (MT) and the dorsolateral visual area (DL) – all of which are present in *Aotus* (Fig. 8.22). However, some of the additional visual areas found in the cortex of individual primate species apparently represent independent specializations and the number of cortical maps is not a reliable guide to phylogenetic relationships even among primates. Nevertheless, Allman has noted that primates (both simians and

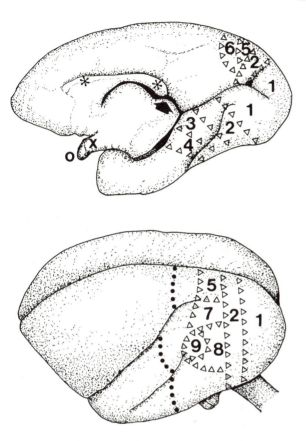

Figure 8.22. Representation of the visual sensory domains in the cerebral cortex of the owl monkey (*Aotus trivirgatus*): above, ventromedial view; below, dorsolateral view.

The row of black dots indicates the approximate anterior limit of the visually sensitive cortex. Rows of open triangles indicate the boundaries of discrete visual areas ('maps'). The black arrow indicates the location of the optic tectum, the asterisks indicate the corpus callosum, the cross shows the position of the optic chiasma and the optic nerve is indicated by O. Key to visual areas: 1, first visual area (V-I); 2, second visual area (V-II); 3, ventral posterior visual area (VP); 4, ventral anterior visual area (VA); 5, dorsomedial visual area (DM); 6, medial visual area (M); 7, dorsointermediate visual area (DT); 8, dorsolateral visual area (DL); 9, middle temporal visual area (V-II).

Note that the triradiate calcarine sulcus on the medial aspect of the cerebral hemisphere (upper drawing) is intimately associated with the boundaries of the first and second visual areas. (Adapted, with simplifications, from Allman, 1977 and 1982.)

prosimians) seem to be unusual among the mammals investigated to date in that the organization of area V-II is very different from that of V-I, whereas in other mammals V-II is essentially just a mirror image of V-I. In this respect, tree-shrews once again resemble non-primates but cats resemble primates, emphasizing the fact that in several respects the cat's visual system is more primate-like than that of the tree-shrew. It is also important to note that the triradiate calcarine sulcus of the primate brain, which has been identified above as a special feature of most living primates, is intimately associated with the boundaries between visual maps. The posterior, deeply penetrating branch of the calcarine sulcus (retrocalcarine sulcus) is particularly significant in this respect (see Fig. 8.22). Allman (1977) has noted that the cortex bordering the calcarine sulcus is mainly concerned with representation of the peripheral visual field. Thus, there would seem to be a link between the distinctive, typically triradiate form of the calcarine sulcus in primates and the organization of neural maps in the visual cortex.

As a final note on the visual cortex of primates, it should be mentioned that Horton (1984) reported the occurrence of regularly spaced patches of cytochrome oxidase activity in the striate cortex in several primates (*Galago, Aotus, Saimiri, Papio, Macaca* and *Homo*). These patches, which are probably linked to the columnar organization of the cortex (Hubel and Wiesel, 1967), were not found in a variety of other mammals, including the tree-shrew (*Tupaia*). However, only one prosimian (*Galago*) was investigated, so it is unclear whether these patches are a shared derived feature of primates.

EVOLUTION OF THE SPINAL CORD IN PRIMATES

As an integral part of the primate CNS, the spinal cord has been relatively neglected in discussions of primate evolution. Indeed, it is often implied that the spinal cord has remained relatively unchanged throughout mammalian evolution.

As has already been explained above, use of the foramen magnum area as an indicator of body size for determining relative brain size in fossil primates is based on the implicit assumption that, at any given body size, foramen magnum area (and, by implication, spinal cord cross-sectional area in the cervical region) has remained constant (Fig. 8.12, hypothesis 1). However, this implied view of the spinal cord as an unchanging, passive channel for ingoing (afferent) sensory impulses and outgoing (efferent) motor commands is both simplistic and unjustified. First, a certain amount of integration takes place within the spinal cord itself, and there is no obvious reason why such integrative functions should not have been enhanced during mammalian evolution. Second, it is known that new sensory and motor pathways have been laid down along the spinal cord during mammalian evolution. Both of these developments would have increased the average girth of the spinal cord during mammalian radiation, as has now been shown by an investigation of the scaling of the spinal cord and vertebral column in living and fossil primates (MacLarnon, 1987).

Noback and Moskowitz (1963) have broadly divided the ascending (sensory) pathways of the spinal cord into the **reticular system** and the **lemniscal system**, and they have similarly divided the descending (motor) pathways into the **extrapyramidal system** and the **pyramidal system**. The reticular system and the extrapyramidal system together seem to constitute a phylogenetically ancient combination, characterized by a somewhat diffuse nature and by numerous intervening synapses (which are reflected by slower transmission, as synaptic delay is the major determinant of nervous transmission times). By contrast, the lemniscal system and the pyramidal system together constitute a phylogenetically recent combination with relatively direct, more rapid pathways. Whereas the reticular system and the extrapyramidal system are present in all living vertebrates, the lemniscal system is more prominent in mammals and the pyramidal system is largely confined to mammals. (Noback and Moskowitz (1963) stated that the pyramidal system is unique to mammals, but Campbell (1975) has reported the presence of a similar system in at least some birds.) It is the combined, more recent development of the faster lemniscal and pyramidal systems that most probably accounts for the increase in girth of the spinal cord during evolution that has been demonstrated for primates by MacLarnon (1987).

The pyramidal system is of particular interest with respect to the adaptive radiation of mammals, notably in connection with assessments of a possible phylogenetic link between tree-shrews and primates. The system is actually quite complex, but there is a major pyramidal tract (also called the **corticospinal tract**) that is identifiable in all mammals, running down each side of the spinal cord. Within the pyramidal system, fibres come together from widely separate areas of the cortex to pass through the internal capsule of each cerebral hemisphere and on through the 'pyramids' in the medulla (hence the name of the tracts). They terminate in the grey matter of the spinal cord. The corticospinal tract thus represents a direct connection between the cortex and motor neurones in the grey matter of the spinal cord (see Fig. 8.1). The left and right corticospinal tracts typically cross over at some stage, normally at the junction between the brain stem and the spinal cord (Campbell, 1975), and they then pass down the spinal cord along the dorsal, lateral or ventral funiculi (Fig. 8.1). The location of the major pyramidal tract within the spinal cord actually provides a very useful indicator of phylogenetic relationships, as it seems to be consistent (virtually without exception) for each order of mammals. In all primates examined to date (at least 14 species; Campbell, 1975), the major crossed pyramidal tract is found in the lateral funiculus of the spinal cord. In the common tree-shrew (*Tupaia glis*), on the other hand, the major tract is found in the dorsal funiculus (Fig. 8.23).

A review of mammalian orders shows that the major pyramidal tract is found in the dorsal funiculus in monotremes, marsupials, edentates, rodents and tree-shrews, in the lateral funiculus

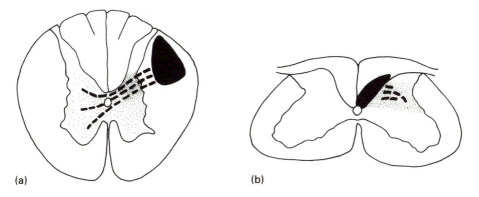

(a)   (b)

Figure 8.23. Diagrams of cross-sections of the spinal cord (neck regions) in the slow loris, *Nycticebus coucang* (a) and in the common tree-shrew, *Tupaia glis* (b), showing the position of the major pyramidal tract (black) on one side of the cord. Studies of fibre degeneration show that fibres from the major pyramidal tract of *Nycticebus* pass to the contralateral side of the grey matter and synapse with cells over a wide area, whereas in *Tupaia* fibres are confined to the iplsilateral grey matter and synapse only in the dorsal region of the grey matter. Key: dashed lines, fibres passing from the pyramidal tract; stippling, area of synapses with neurones. (After Jane, Campbell and Yashon, 1965.)

in ungulates, carnivores, lagomorphs and primates, and in the ventral funiculus in insectivores and elephants (Jane, Campbell and Yashon, 1965). Incidentally, the major pyramidal tracts of insectivores are also distinctive in that they are uncrossed; individual fibres cross to the opposite side of the grey matter at the point where they terminate in the spinal cord (Campbell, 1975). *Tupaia glis* is distinguished from the primates so far examined not only by the position of the major pyramidal tract in the spinal cord, but also by finer details of its connections. In the tree-shrew, the major pyramidal tract rapidly peters out below the cervical level, very few fibres reach as far as the mid-thoracic level and none penetrates as far as the lumbar and sacral regions of the spinal cord. In primates such as the slow loris (*Nycticebus*), by contrast, the major pyramidal tract runs the entire length of the spinal cord down to the sacral level (Jane *et al.*, 1965). Limitation of the major pyramidal tract to the cervical area of the cord is characteristic of marsupials, insectivores, edentates, chiropterans, elephants and artiodactyls as well as of tree-shrews, while extension of the tract down to the sacral level is found in rodents, lagomorphs, hyracoids,

carnivores and possibly monotremes as well as in primates (Towe, 1973). Jane *et al.* (1965) also showed a marked contrast between *Tupaia* and *Nycticebus* in the termination of the pyramidal tract fibres in the grey matter of the spinal cord. In *Tupaia*, fibres are confined to the same side of the spinal cord as that containing the tract, and the fibres terminate only in the dorsal part of the grey matter. In *Nycticebus*, on the other hand, fibres terminate in the dorsal, intermediate and ventral parts of both sides of the grey matter, albeit with differential frequencies (Fig. 8.23). Verhaart (1970) noted that in most mammals the pyramidal tract fibres terminate in the dorsal horn of the spinal cord and in the intermediate grey matter, whereas in all primates examined there are large numbers of terminations within the groups of motor neurones located in the ventral horn. In agreement with this distinction, Phillips (1971) classified corticospinal pyramidal fibres into two groups:

1. those terminating in the dorsal half of the spinal grey matter and controlling transmission of reflexes by spinal inter-neurones;
2. those terminating in the ventral half of the

grey matter; fast-conducting corticospinal fibres that are concerned with the innervation of the distal musculature by motoneurones.

Phillips pointed out that the second kind of termination is also found, to some degree at least, in certain non-primate mammals (e.g. in a carnivore, the raccoon), but he nevertheless identified this cortico-motoneuronal connection as 'a major characteristic of primates'. Clearly, the tree-shrew lacks this characteristic (Fig. 8.23).

As with the dorsal lateral geniculate nucleus, variation in the location of the pyramidal tracts within the spinal cord, from one order of mammals to another, suggests that the major pyramidal tract system was barely developed (or completely absent) in ancestral placentals and that independent evolution of the system within different mammalian lines has been associated with chance differences in the eventual arrangement of the tracts. On this criterion, tree-shrews can be seen as clearly divergent from all primates examined to date, and it would appear that any common ancestor of tree-shrews and primates must have lacked any significant development of the pyramidal system. In other words, the common ancestor of tree-shrews and primates would have been indistinguishable, in this respect, from the ancestral placental mammal. It has been suggested that the location of the major pyramidal tract in the dorsal funiculus might be primitive, as it is present in monotremes and marsupials (for example) and that in the evolution of the primates there was a shift towards location of the tract in the lateral funiculus. However, this suggestion overlooks the fact that the major pyramidal tract is in the ventral funiculus in insectivores, and it seems unlikely that that tract would change its location once established in a particular funiculus of the spinal cord. Noback and Shriver (1966) and Campbell (1975) prefer the hypothesis that the corticospinal pyramidal tracts evolved independently in the different orders of mammals, hence adopting different locations in the funiculi. This interpretation is further supported by the fact that the major pyramidal

tract system develops quite late in the ontogeny of individual primate species, probably reflecting a relatively late phylogenetic origin. It manifests itself only at about 6 months after birth in the rhesus monkey (Kuypers, 1968) and at about 9 months after birth in the human infant.

Because of its rapid operation and the direct link between motor neurones and the cortex, the major pyramidal tract system permits very fine coordination of individual movements. In human beings, for example, it is by virtue of the pyramidal system that the fine finger movements necessary for the playing of a musical instrument are guaranteed. Accidental damage to the pyramidal system obliterates this ability for fine coordination of the fingers. The same observation seems to apply to other simian primates, to the extent that the control of individual digits seems to be a function of the pyramidal system. Surgical disruption of the pyramidal system will obliterate independent control of the digits in a rhesus monkey trained to use a single finger for performing a standard task, and the experimental animal will revert to whole-hand patterns (Passingham, 1981). Experimental work on baboons (Phillips, 1971; Phillips and Porter, 1977) has shown that the pyramidal system operates as a sophisticated feedback network guaranteeing versatility of muscular activity in the limbs. Thus, it can be concluded that the evolution of the pyramidal system has been associated with the development of much finer control over the body musculature and of individual extremities in primates, where some of the corticospinal fibres terminate on motor neurones in the ventral horn of the spinal grey matter. The features shared by all primate species so far investigated (location of the corticospinal tract in the lateral funiculus; termination of fibres on the motor neurones of the ventral horn on both sides of the spinal cord; penetration of the tract down to sacral levels) suggest a single origin for the system in the ancestral primate stock. Tree-shrews, on the other hand, would seem to be totally different (if *Tupaia glis* can be taken as broadly representative of the family Tupaiidae) both in terms of a poorer development of the corticospinal tract, with no direct synapses between its fibres and the motor neurones of the ventral horn, and in terms of location of the tract in the dorsal funiculus.

With respect to the pyramidal system, at least, there is thus considerable evidence that the spinal cord has undergone progressive modification during the evolutionary radiation of placental mammals. As the pyramidal system seems to be a new system of nerve fibres that emerges late in individual development in addition to pre-existing systems, it would also seem that the spinal cord must have increased in girth during the evolution of placental mammals. This increase in girth would doubtless have applied at all levels of the spinal cord containing the pyramidal system (i.e. at least at cervical levels). Accordingly, it would seem likely that the size of the foramen magnum relative to body size has generally increased during the adaptive radiation of mammals. This observation suggests one possible interpretation of the fact that at least some Eocene 'lemuroids' (e.g. *Adapis*) had a smaller foramen magnum in relation to overall body size than any living primate. Evidently, any comparatively new system such as the pyramidal system would have been at a relatively early stage of differentiation in the Eocene, and it is only to be expected that the foramen magnum might have been smaller relative to body size. At the same time, it seems almost certain that Eocene adapids such as *Adapis* would have had poorer muscular control than in any living primate, and this must necessarily have reflected a lesser degree of locomotor sophistication. This must be borne in mind when considering the evolution of locomotion in the primates (Chapter 10), specifically in relation to *Adapis*. As has already been mentioned, MacLarnon (1987) has now shown that the spinal cord was probably smaller, relative to body size, in the Eocene 'lemuroids' *Notharctus* and *Smilodectes*, than in modern primates, although *Smilodectes* differed from *Notharctus* and *Adapis* in having a foramen magnum almost comparable in size to that of a modern primate of the same body weight (see Table 8.9).

In summary, then, firm evidence indicates that both the brain and the spinal cord have increased in size and become further elaborated during the evolutionary radiation of placental mammals. As part of this radiation, primates have shown the same general trends, such that no living primate has a brain or a spinal cord as primitive as those possessed by the ancestral primate. Further, the combined expansion of the brain and spinal cord has been closely associated with increasing sophistication of neuromuscular control. Given these links, a very close relationship between evolution of the CNS and evolution of locomotion seems likely, and this point will be raised again in Chapter 10.

DEVELOPMENT OF THE PRIMATE BRAIN

As with the analysis of any other characteristics of living primates, consideration of the **ontogeny** (development) of the brain can be expected to yield supplementary information of particular value for the reconstruction of phylogenetic relationships. In fact, the ontogeny of the brain is likely to be especially revealing because the CNS is unusual among the organ systems of the mammalian body in that it reaches its definitive size and form very early in development. Indeed, the CNS has only a very limited capacity for modification or repair of its basic structure once it has been laid down.

Because of its early maturation, the central nervous system has a particularly close link with reproduction (Chapter 9). The major part of CNS development occurs during fetal life and during the early postnatal period, such that the definitive size of the brain has been virtually attained by the age of weaning in mammals generally. This implies that resources supplied by the mammalian mother to her offspring are crucial in determining the eventual size of the brain. It has already been noted that brain size in adult mammals scales to adult body size in the same way as basal metabolic rate (i.e. with an allometric exponent value of approximately 0.75) and it has been specifically suggested (Martin, 1981c) that it is the metabolic turnover of the mother that constrains brain size in mammals. This is probably an indirect constraint in that, other things being equal, mammalian mothers are likely to devote a standard proportion of their own resources to the development of their offspring. If the total resources available to mammalian mothers scale overall in accordance with Kleiber's Law (i.e. with an exponent value close to 0.75), the resources devoted to brain and body development in the fetus and early postnatal stages may be expected to scale in the

same way (Martin and MacLarnon, 1988). Of course, at any given body size mammalian mothers may differ in their reproductive 'strategies' and therefore in the proportion of available resources provided for individual offspring, thus accounting for scatter of points around any allometric best-fit line. But when mammals are considered overall, it is reasonable to expect that the basic constraint exerted by

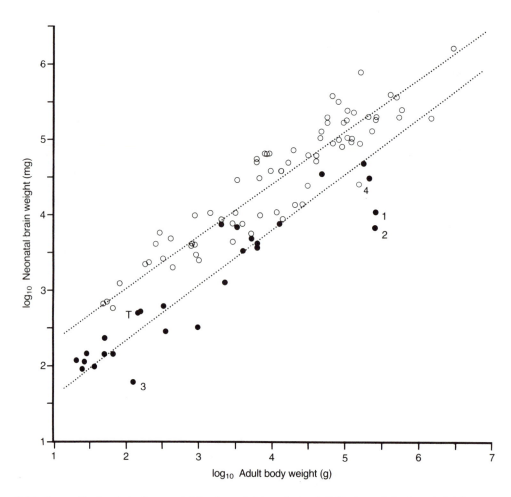

Figure 8.24. Logarithmic plot of neonatal brain weight against adult body weight for altricial mammals (●, N = 27; T, *Tupaia*) and for precocial mammals (○, N = 73). The best-fit lines (dotted) are the major axes, excluding four aberrant species: 1, polar bear; 2, Alaskan brown bear; 3, golden hamster; 4, domestic pig. The two bears are unusual among altricial mammals in that the mothers hibernate or at least become lethargic during pregnancy, while the pig is unusual among precocial mammals in producing quite large multiple litters. Both of these conditions can be expected to reduce maternal resources available for fetal brain development. The hamster is unusual because it has the shortest gestation period, relative to body size, of any placental mammal. (Reprinted from Martin, 1983, courtesy of the American Museum of Natural History, New York.)

metabolic scaling will be revealed in the overall trend of the data.

With respect to the development of the CNS, the basic distinction between altricial and precocial mammals (see Chapter 9; Fig. 9.12) is particularly important. Altricial neonates are born in a relatively helpless condition after a short gestation period (relative to the mother's body size; see Fig. 9.18). The implication of this is that fetal brain development is relatively restricted and that more brain development is likely to take place after birth than in precocial mammals. In the latter, the neonates are born in a quite advanced state after a relatively long gestation period, so quite a large proportion of CNS development can take place within the relatively sheltered confines of the mother's uterus. It seems likely that development of the offspring's CNS within the mother's body is a more efficient process than postnatal development, where the mother must supply the nutrients for brain development less directly (through her milk) and where the infant's metabolism must carry out the necessary transformations without the sheltered conditions provided by the maternal uterus. It is therefore to be expected that, as a general rule, altricial mammals will be born with smaller neonatal brain weights than precocial mammals (for a given maternal body weight) and that the brain will grow more during postnatal life in altricial mammals than in precocial mammals. Both of these expectations are confirmed by analysis of available data (Martin, 1983). Figure 8.24 shows that neonatal brain weight is typically considerably larger in precocial mammals than in altricial mammals, once adult (maternal) body weight has been taken into account, although there is some scatter and a mild degree of overlap between the two distributions. The size of the brain at birth in a typical precocial mammal is about 4.5 times greater than for a typical altricial mammal born to a mother of the same body weight, so the overall difference is quite considerable.

There is, in fact, a hidden assumption behind the above hypothesis relating scaling of brain size in adult mammals to constraints operating during fetal development of the brain within the mother's body. It has already been established (Figs 8.6 and 8.7) that scaling of adult brain size to adult body size in placental mammals has an empirical exponent value close to 0.75. It is also true that neonatal brain weight scales with similar exponent values with respect to adult (i.e. maternal) body weight. The best-fit lines in Fig. 8.24 actually show empirical exponent values of 0.70 for precocial mammals and 0.74 for altricial mammals. However, for the link between neonatal brain weight and adult brain weight to be associated with a common exponent value of 0.75, it must also follow that scaling of adult brain size to neonatal body size will have an exponent value of unity (i.e. the relationship must be isometric). Obviously, this requirement applies to altricial and precocial mammals separately, because the former have much shorter gestation periods (Martin and MacLarnon, 1985) and start off from a much earlier stage of development at birth. In other words, for altricial and precocial mammals separately, the following set of formulae must apply:

$$E_N = k \, W_M{}^{0.75} \qquad (1)$$

$$E_A = k' \, E_N \qquad (2)$$

Hence:

$$E_A = k'' \, W_M{}^{0.75} \qquad (3)$$

(where E_N is neonatal brain weight, E_A is adult brain weight, W_M is maternal body weight and k, k' and k'' are the allometric coefficients). Equation (1) is confirmed in Fig. 8.24 and equation (3) is confirmed in Figs 8.6 and 8.7. It remains to be established that equation (2) applies to altricial and precocial mammals separately. Calculation of best-fit lines (Fig. 8.25) demonstrates that the relationship is indeed isometric; the empirically determined exponent values are 0.99 for precocial mammals and 1.01 for altricial mammals (Martin, 1981c). The equations for the best-fit lines reveal that in altricial mammals the adult brain is typically about 7.5 times as big as in the neonate, whereas in precocial mammals the adult brain is typically only 2.5 times as big as in the neonate (Martin, 1983). Hence, the brain of an altricial mammal

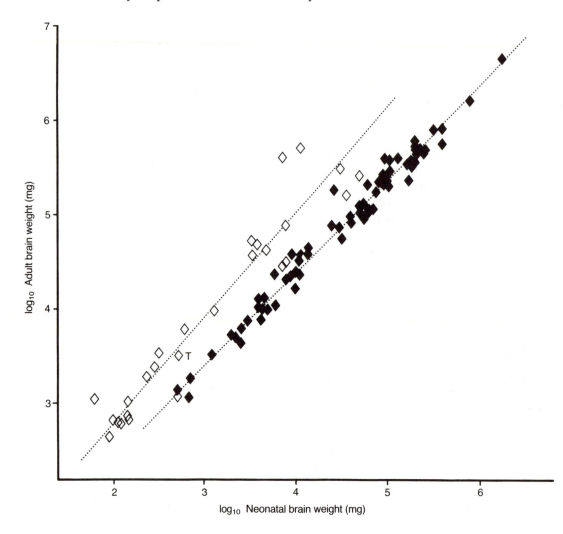

Figure 8.25. Logarithmic plot of adult brain weight against neonatal brain weight for altricial mammals ($\diamondsuit$, $N = 27$; T, *Tupaia*) and for precocial mammals ($\blacklozenge$, $N = 71$). The best-fit lines (dotted) are the major axes.

usually increases in size markedly more than that of a precocial mammal during postnatal development, the increase being greater by a factor of about 3. It should be noted that the tree-shrew (*Tupaia glis*) conforms with other altricial mammals, rather than with the precocial primates, in this respect (Figs 8.24 and 8.25).

The consistency of the isometric scaling relationship between adult brain weight and neonatal brain weight is clearly shown by an analysis of primates alone (Fig. 8.26). Although there is some scatter about the best-fit line, the correlation coefficient is very high ($r = 0.992$) and the allometric exponent is exactly 1.00, indicating perfect isometry overall. The empirical equation indicates that in the typical primate the adult brain is about 2.3 times bigger than the neonatal brain (thus agreeing closely with the relationship for precocial mammals generally, illustrated in Fig. 8.25).

With respect to the relationships established thus far, tree-shrews are totally different from

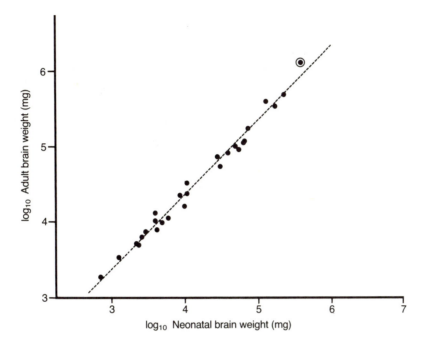

Figure 8.26. Logarithmic plot of adult brain weight against neonatal brain weight for primates ($N = 27$ species; point for *Homo sapiens* is ringed). The best-fit line (dashed) is the major axis. (Reprinted from Martin, 1983, courtesy of the American Museum of Natural History, New York.)

primates by virtue of the fact that they produce altricial offspring (Fig. 9.13) rather than precocial offspring like primates. As a direct result, neonatal brain weight in tree-shrews is smaller than in any primate species (relative to maternal body weight) and there is considerable postnatal brain growth. In short, tree-shrews lie close to the best-fit lines for altricial mammals shown in Figs 8.24 and 8.25. However, it could be argued that tree-shrews merely show a primitive (altricial) condition that gave way to a more advanced (precocial) condition seen in the later common ancestral stock of primates. After all, the switch from a primitive altricial state to a derived precocial state must have taken place at some stage during primate evolution (if not several times independently in separate primate lineages), and various groups of placental mammals have developed the precocial condition.

Nevertheless, there is one feature of fetal brain development in primates that is apparently universal to this group of mammals but is not found in tree-shrews, insectivores or any other placental mammals. This feature concerns the relationship between brain and body size of the fetus throughout fetal development. A logarithmic plot of fetal brain weight against fetal body weight for primates and other mammals (Fig. 8.27) reveals that all primate fetuses (regardless of their age) fit one line whereas virtually all non-primate mammalian fetuses (regardless of age) fit another line, with no overlap between the two distributions. The only exception to this rule, which has been recognized independently by several different authors (Holt, Cheek *et al.*, 1975; Holt, Renfree and Cheek, 1981; Sacher, 1982; Martin, 1983), is found among the cetaceans (whales and dolphins). Toothed cetaceans seem to lie on a line

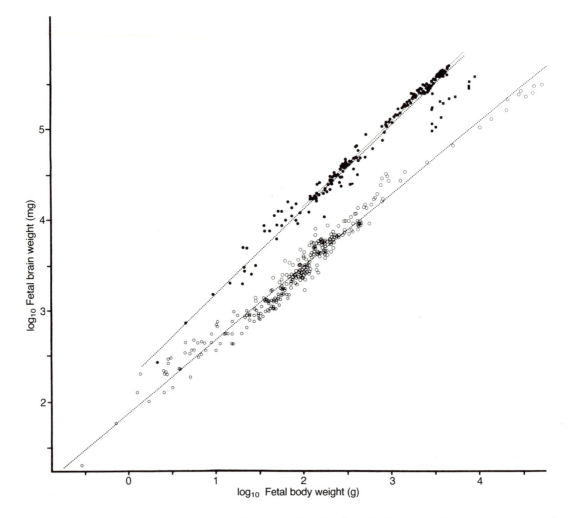

Figure 8.27. Logarithmic plot of fetal brain weight against fetal body weight for a sample of mammalian species. Each point represents a single fetal specimen: ○, non-primate mammals other than toothed cetaceans ($N = 291$ points for 10 species); ■, toothed cetaceans ($N = 14$ points for 3 species); ●, primates ($N = 185$ points for 6 species). The two main best-fit lines (dotted) are the major axes for non-primates (excluding toothed cetaceans) and for primates, respectively. The fine dotted line above the primate major axis is for fetal stages of *Homo sapiens* alone, showing that human fetuses closely conform to the fundamental primate pattern. (Reprinted from Martin, 1983, courtesy of the American Museum of Natural History, New York.)

intermediate between the primate line and the line for other non-primate mammals (Fig. 8.27). The scaling relationships for primates and non-primates indicated by the best-fit lines in Fig. 8.27 are, unfortunately, not parallel, so it is impossible to give a simple interpretation of the scaling principle involved. But Sacher (1982) has

simplified matters by pointing out that up to a body weight of 1 kg the primate line roughly corresponds to a typical value of 12% for the brain weight in relation to body weight for any primate fetus, while the non-primate line roughly corresponds to a typical value of 6% brain weight in relation to body weight for any non-primate fetus. In other words, at a given fetal body weight, a primate fetus will have twice as much brain tissue as a non-primate fetus (excluding toothed cetaceans). As Sacher (1982) has pointed out, this distinction cuts right across the altricial/precocial divide.

Now Fig. 8.27 is, for obvious reasons, based on a limited number of species for which data are available on fetal brain and body weights (i.e. six primate species, including *Homo sapiens*, and 10 non-primate species). In particular, these primates do not include any prosimian species. It

is therefore possible that the apparent distinction shown between primates and non-primates may be an unrepresentative effect, resulting from a limited sample size. Fortunately, it is possible to test this possibility (Sacher, 1982; Martin, 1983). If the fetal trajectories shown in Fig. 8.27 apply throughout fetal life, as they appear to do, then the distinction should still be apparent in neonatal brain:body weight relationships. Figure 8.28 shows that this is indeed the case. The primates ($N = 25$ species, including both prosimians and simians) are still clearly separated

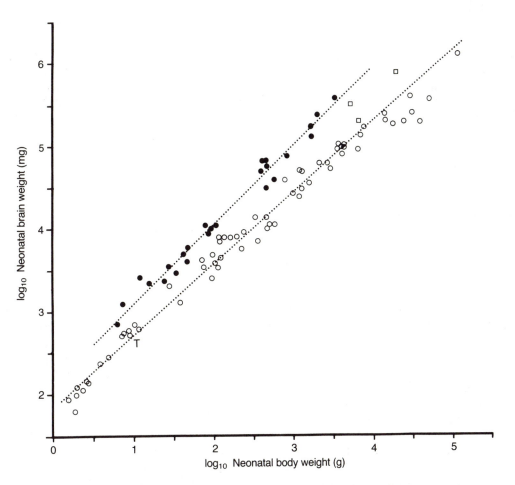

Figure 8.28. Logarithmic plot of neonatal brain weight against neonatal body weight for a sample of mammalian species. Each point represents the average condition for a single species: ○, non-primate mammals other than toothed cetaceans ($N = 65$ species; T, *Tupaia*); □, toothed cetaceans ($N = 3$ species); ●, primates ($N = 25$ species). The best-fit lines (dotted) are the major axes. (Reprinted from Martin, 1983, courtesy of the American Museum of Natural History, New York.

from most non-primate mammals ($N = 65$ species), with the toothed cetaceans occupying an intermediate position. The empirical formulae determined for the primate and non-primate best-fit lines in Fig. 8.28 are virtually identical to those determined for the fetal trajectories shown in Fig. 8.27 (Martin, 1983). Hence, it can be confidently concluded that the difference in fetal brain:body weight relationships between primates and non-primate mammals suggested in Fig. 8.27 is a consistent difference which relates to a fundamental fetal growth pattern in primates unlike that found in other mammals. This conclusion has now been confirmed by data for fetal brain and body weights in *Galago crassicaudatus*, the first prosimian species to be documented (Newman and Hendrickx, 1984). Four values for brain weights at different stages of gestation for this bushbaby species lie closest to the best-fit line for primates shown in Fig. 8.27. Hence, at any stage of fetal development, primates (prosimians and simians) devote a greater proportion of available resources to brain tissue than do any other mammals.

The differences between primates and non-primates shown in Figs 8.27 and 8.28 are so striking that Sacher (1982) regards the distinctive pattern shared by strepsirhine and haplorhine primates as a derived feature of primates of modern aspect. He traces this derived feature to an 'extraordinary evolutionary event' that took place at some time before the adaptive radiation of the primates: 'The schedule of primate fetal development was modified by reducing by half the amount of non-neural somatic tissue associated with a given amount of neural tissue throughout the greater part of fetal life.' As Sacher notes, this evolutionary modification was not directly connected with the development of relatively large brain size, as early primates, particularly Eocene adapids (Table 8.11), did not have very large brains. It can be confidently concluded (see above) that the common ancestral stock that gave rise to primates of modern aspect had relative brain sizes indistinguishable from those of modern 'advanced' insectivores. Further, if the alteration of fetal brain growth patterns in primates were connected with relative

brain size as such, one would expect to find a similar alteration among other mammalian groups in which there have clearly been increases in relative brain size since their divergence from the ancestral placental stock (e.g. carnivores, hoofed mammals). It seems, in fact, that the shift in fetal brain:body weight relationships that characterizes primates as a group was not achieved by an increase in resources devoted to fetal brain development but by a decrease in resources devoted to fetal body development. It is now well established (Martin and MacLarnon, 1988; Chapter 9) that primates are characterized by the lowest rates of fetal body development among the mammals generally. Hence, the distinctive primate trajectory of fetal brain development in relation to fetal body development (Fig. 8.27) is almost certainly a by-product of the typical primate reproductive 'strategy' of reduced investment in total fetal tissue per unit time.

To turn now to the question of the tree-shrews, neonatal brain and body weights for *Tupaia glis* show that this species fits the non-primate trajectory rather than the primate trajectory (Fig. 8.28). Thus, as was recorded by Sacher (1982), *Tupaia* resembles other mammals such as insectivores rather than primates in terms of its pattern of fetal brain development. This is particularly interesting because, taken together with the fact that tree-shrews produce altricial offspring, it means that the prosimian-like relative brain sizes of adult tree-shrews (Figs 8.5 and 8.9) are attained by utterly different ontogenetic pathways. Once again, it is possible that tree-shrews are related to primates but branched off before a typical primate pattern of brain development was acquired by the common ancestral stock of strepsirhines and haplorhines. Yet the fact remains that tree-shrews have not been shown to share any derived features of brain ontogeny with primates.

It should also be noted that data for the fruit-bat *Pteropus* (kindly provided by Dr C. West) show that the relationship between neonatal brain weight and body weight follows the non-primate pattern in Fig. 8.28. There is therefore no support here for the suggestion that fruit-bats are specifically related to primates (Pettigrew, 1986).

HUMAN BRAIN DEVELOPMENT: A SPECIAL CASE

As discussed above, primates produce precocial infants that are typically well developed and fairly active soon after birth. Human infants, however, are somewhat different. Although the human neonate does have many precocial features at birth – such as completed opening of the eyes and ears – it is nevertheless somewhat helpless, relatively immobile and heavily dependent on parental care. Thus it is that some authors (e.g. Portmann, 1941) refer to the human neonate as 'secondarily altricial'. The clue to this special condition would seem to reside in the state of the brain at birth.

It is first of all necessary to establish what happens in terms of brain development after birth in mammals. In precocial mammals, be they primates or non-primates, there is a fairly rapid departure after birth from the fetal brain:body growth trajectories shown in Fig. 8.27. From soon after birth onwards, a different trajectory is found, with the brain growing far less rapidly in relation to body size (Fig. 8.29; see also Fig. 8.8). However, in altricial mammals the fetal trajectory of brain:body growth continues for days or weeks after birth. Thereafter, there is a similar switch to a different trajectory of slower brain growth in relation to general body growth (see Fig. 8.8). It is essentially for this reason that altricial mammals can 'catch up' with precocial mammals by means of greater postnatal brain growth. However, the human neonate differs from all other placental mammals in that the

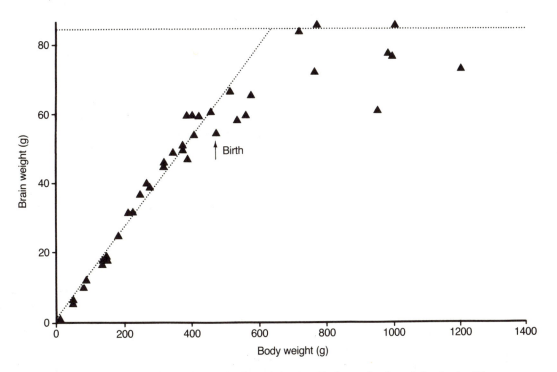

Figure 8.29. Fetal and postnatal increase in brain weight, in relation to body weight, in the Rhesus macaque (*Macaca mulatta*). The fetal trajectory for brain:body weight relationships is indicated by the inclined dotted line (major axis), and the average adult brain weight is indicated by the horizontal dotted line. From the moment of birth onwards, brain weight increases less rapidly in relation to body weight. Note: As the relationship between fetal brain weight and fetal body weight is almost isometric, logarithmic transformation was unnecessary in this case. (Reprinted from Martin, 1983, courtesy of the American Museum of Natural History, New York.)

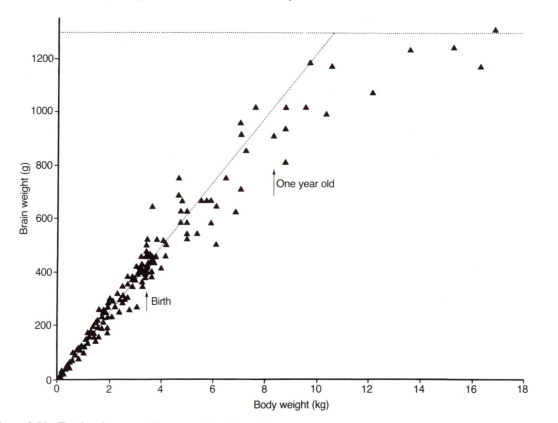

Figure 8.30. Fetal and postnatal increase in brain weight, in relation to body weight, in *Homo sapiens*. The fetal trajectory for brain:body weight relationships, indicated by the inclined dotted line (major axis), is followed for about 12 months after birth before the switch to a typical mammalian postnatal trajectory occurs. The average adult brain weight is indicated by the horizontal, dotted line. As for Fig. 8.29, logarithmic transformation was unnecessary. (Reprinted from Martin, 1983, courtesy of the American Museum of Natural History, New York.)

typical primate fetal trajectory of brain:body growth (Fig. 8.27) is followed for approximately 12 months after birth (Fig. 8.30). Only at about 1 year of age does the switch take place to a typical postnatal trajectory, with brain weight increasing more slowly in relation to body weight. Because of this continuation of the fetal brain:body growth trajectory for the first 12 months of postnatal life, the brain increases in size by a factor of almost 3.5 between birth and the completion of brain growth in humans, compared to a factor of 2.3 for typical primates (as noted above; see also Fig. 8.26). It would seem that the reason for this special pattern in *Homo sapiens* is that the head of the neonate is already quite big at birth and that attainment of the large adult brain

size found in modern human beings could only be possible by continuing a fetal pattern of brain growth into the first year of postnatal life. Portmann is therefore essentially correct (1941) when he states that the human gestation period is really 21 months – 9 months intrauterine and 12 months extrauterine. It is the prolongation of a fetal pattern of brain growth during the first year of postnatal life that renders the human infant 'secondarily altricial' and, incidentally, introduces an important new element of flexibility into human brain development. Thus, the comparatively recent great expansion of the human brain, relative to body size, is based on a very special development from the typical primate pattern.

Chapter nine

Primate reproductive biology

Reproductive functions are of central importance in the evolution of organisms, in terms of both energetic costs to the parents and survival of the offspring. The fundamental evolutionary process of natural selection, after all, depends on differential reproductive success among the members of a species. Of course, reproductive characteristics are themselves subject to natural selection just like any other biological features. It is to be expected that they would respond particularly sensitively to environmental pressures because of their more intimate connection with the basic process of evolutionary change. Natural selection influences not only the primary reproductive processes of adult mammals, but also any of their adaptations that enhance the survival of offspring, most notably parental behaviour. Any consideration of the evolution of primate reproduction must therefore include parental care in addition to the direct reproductive features of the adults themselves. Indeed, it is now clear that full appreciation of the evolution of reproduction in mammals ultimately depends on an understanding of 'reproductive strategies' (Pianka, 1970; McNab, 1980; Eisenberg, 1981). In other words, we must seek an understanding of

the set of reproductive characteristics that constitutes the typical life-history pattern of each species.

In tracing the evolution of individual reproductive characteristics, and of integrated 'reproductive strategies', in primates, it is particularly instructive to examine the way in which the tree-shrews have been cited to support different interpretations. Reproductive characteristics counted among the features originally cited in support of a phylogenetic link between tree-shrews and primates (Le Gros Clark, 1934b, 1959; see Chapter 5) and it was particularly on grounds of reproductive biology that certain authors eventually questioned this supposed link (J.P. Hill, 1965; Luckett, 1969; Martin, 1968b, 1969). In fact, tree-shrews turn out to be sharply distinct from the primates in their reproductive biology, and the differences observed are extremely revealing both with respect to the general evolutionary radiation of the mammals and in relation to the specific evolutionary history of the order Primates.

Any reconstruction of the evolution of primate reproduction must, of course, be based almost exclusively on inference from a study of living forms, because there is little of relevance to be sought in the fossil record. The following discussion hence depends on comparative study of living primates and other mammals, leading to an assessment of the implications of overall 'reproductive strategies'. Placental mammals as a group are characterized by internal fertilization and extensive development of the offspring within the mother (vivipary with placentation) followed by suckling (lactation) for a period after birth. For this reason, it is primarily the female that is of interest with respect to the evolution of reproductive characteristics among mammals in general and among primates in particular.

THE REPRODUCTIVE ORGANS

Before considering the details of the male and female reproductive organs, it is important to emphasize two conclusions from studies of vertebrate embryological development. First, there is a close association between the reproductive and urinary systems in both male and female vertebrates, which has persisted as a general rule in mammals. Second, in both male and female vertebrates the reproductive system has a basic bilaterally symmetrical (mirror-image) pattern, which is also found in the urinary system. In the course of embryonic development, the intimacy of association between the urinary and reproductive systems is typically reduced to varying degrees and midline fusion of paired ducts within the female reproductive tract can take place to some extent. It can therefore be assumed as a guiding principle that in mammalian evolution there has been a progressive decrease in the association between the reproductive and urinary systems, with midline fusion in the female reproductive tract becoming increasingly prevalent. Embryological evidence and comparative studies of the vertebrates together indicate that the reproductive and urinary outlets were originally located very close to the anus in a recognizable cloaca in primitive vertebrates. This association with the anus has, however, been virtually lost in all living placental mammals; only in the monotremes has a cloaca been retained, and marsupials show no more than vestigial traces (Fig. 9.1).

Male reproductive organs

In mammals, as with all other terrestrial vertebrates, internal fertilization is the rule and the male has an erectile penis that is inserted into the female tract during mating. One peculiarity that distinguishes most marsupials and most placental mammals from other terrestrial vertebrates, however, is **descent of the testes**. Rather than remaining in their original developmental location close to the kidneys, the testes of most viviparous mammals migrate towards the tail and often descend to varying degrees to lie closer to the ventral abdominal wall. Frequently, they emerge into special external compartments of the abdominal cavity, the **scrotal sacs**. In all normal adult male primates, the testes are descended into a clearly separated **scrotum** incorporating the two scrotal sacs. As in all other placental mammals with testicular descent, but not in marsupials, the vas deferens from the testis in male primates loops over the ureter on each side of the body (Fig. 9.1).

Testicular descent has been the subject of much discussion and it is of particular interest here as a universal characteristic of all adult male primates that was specifically noted in Mivart's definition (Mivart, 1873; see Chapter 5). Because descent of the testes is unique to marsupials and placentals, but does not occur in all representatives of these two mammalian groups, two alternative hypotheses are possible. The first is that the common viviparous ancestor of all marsupials and placentals lacked descent and that this condition emerged on numerous separate occasions in the evolution of these two major mammalian groups. The second is that testicular descent was, in fact, present in that common viviparous ancestor, or at least in the separate ancestral stocks of marsupials and placentals, but became secondarily suppressed in some cases during the subsequent evolution of marsupials and placentals. This latter hypothesis is supported to some extent by the very wide distribution of full testicular descent among marsupials and placentals (Weber, 1927; Eckstein and Zuckerman, 1956; Eisenberg,

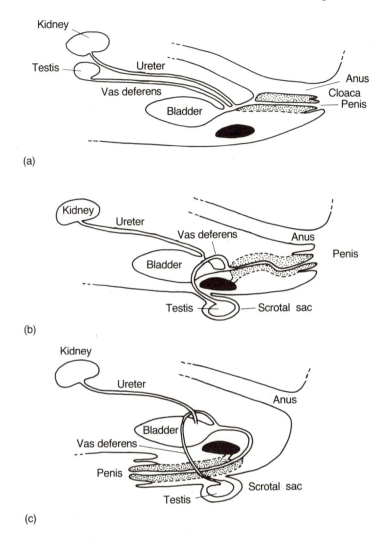

Figure 9.1. Simplified diagrams of one side of the adult male reproductive system in monotremes (a), marsupials (b) and placentals (c). Note the location of the testis in its original position close to the kidney in the monotreme condition. In most marsupials and placentals, the testis descends into a scrotal sac by passing ventrally past the pubic bone (black). In placentals, the vas deferens loops over the ureter passing from the kidney to the bladder, whereas this is not the case in marsupials. In all cases, the erectile penis is shown by stippling; in monotremes it is contained in a clearly defined cloaca along with the anus. The extreme backward direction of the penis shown for marsupials is schematic only. (After Weber, 1927.)

1981), as shown in Table 9.1. Further support comes from the fact that in some mammals, such as whales and dolphins (Cetacea), some insectivores and armadillos (Dasypodidae), the testes do in fact migrate to some extent, although they remain dorsal within the abdomen. It is also significant that testicular descent is absent or relatively restricted (e.g. with the testes remaining subintegumental rather than emerging into a proper scrotum) in marsupials and placental mammals with obvious secondary specializations. This is true of viviparous

Table 9.1 Distribution of testicular descent among mammals, based on Pocock (1926), Weber (1927), Carrick and Setchell (1977), Eisenberg (1981) and author's own observations

Order	Position of testes	Comments
Monotremata	Undescended	
Marsupialia	Descended	Testes remain subintegumental in Notoryctidae and Phascolomidae
Insectivora	Variable degree of descent	Testes undescended in Centetidae and Chrysochloridae; testes remain intra-abdominal in Solenodontidae; subintegumental testes in Talpidae, Soricidae and Erinaceidae. Seasonal variation can occur
Dermoptera	Descended	
Chiroptera	Descended	Subintegumental testes in Microchiroptera. Seasonal variation can occur
Macroscelidea	Undescended	
Scandentia	Descended	
Primates	Descended	
Edentata	Variable degree of descent	Testes undescended in Bradypodidae; partial descent to intra-abdominal pelvic position in Myrmecophagidae and Dasypodidae
Pholidota	Descended	No scrotum, but testes descend at sexual maturity; testes remain subintegumental
Tubulidentata	Descended	Testes may remain subintegumental
Lagomorpha	Descended	
Rodentia	Descended	Testes may remain subintegumental or emerge into a true scrotum. Seasonal variation can occur
Cetacea	Descended	Testes remain intra-abdominal
Carnivora	Descended	Testes remain subintegumental in some cases
Pinnipedia	Descended	Testes remain subintegumental in Phocidae
Proboscidea	Undescended	
Hyracoidea	Undescended	
Sirenia	Undescended	
Perissodactyla	Descended	Testes remain subintegumental in Rhinocerotidae and Tapiridae
Artiodactyla	Descended	Testes remain subintegumental in Hippopotamidae

mammals with a burrowing habit (certain marsupials, insectivores and rodents), of those with aquatic adaptations (certain marsupials, insectivores and rodents; seals; hippopotamus; sea-cows; whales and dolphins) and of large-bodied pachyderms (rhinoceros and elephant).

It has long been apparent that descent of the testes is in some way connected with the elevated core body temperature of mammals (Cowles, 1958). It is equally striking, however, that birds, which share the characteristic of a high body temperature, do not show any sign of testicular descent. Thus, any explanation of the evolution of testicular descent among mammals should ideally account for the minority of mammals without full descent and also explain the absence of descent in birds. For this reason, unless special cooling mechanisms exist within the abdomen, we must discard the commonly accepted view (e.g. Romer, 1955, p. 421) that testicular descent evolved because 'the delicate process of sperm maturation cannot go on at high temperatures'. Mammals lacking testicular descent do not have lower core body temperatures than mammals with descended testes (Carrick and Setchell, 1977), yet they show the same pattern of normal development of sperm (spermatogenesis). However, there is good evidence that in mammalian species with testicular descent the temperature of the testis is significantly lower (by as much as 7 °C) than the core abdominal temperature (typically maintained in the region of 37 °C). Moreover, in mammalian species that normally show testicular descent, spermatogenesis is arrested by abnormal retention of the testes within the abdomen (cryptorchidism), by surgical return of the testes to the abdominal cavity or by artificial prolonged heating of the testis. In human beings, where unilateral or bilateral cryptochidism occurs as an occasional developmental abnormality, testes retained in the abdominal cavity into adult life always lack spermatogenesis. Further, in adult males of certain mammalian species the testes may be retracted into the abdominal cavity. This may occur either as a relatively short-term response to stress (e.g. in tree-shrews; Martin, 1968b; von Holst, 1969) or as a longer-term phenomenon occurring during the non-breeding period of seasonally breeding species (e.g. certain marsupials, insectivores, bats and rodents). In both types of natural testis retraction, return of the testes to the abdominal cavity leads to the arrest of spermatogenesis.

Nevertheless, as has been pointed out by Johnson and Everitt (1980), the requirement of the scrotal testis for a lower ambient temperature 'may be a secondary *consequence* of its scrotal position rather than the original evolutionary *cause* of its migration'. As the elevated core body temperature does not suppress spermatogenesis in birds or mammals that lack testicular descent (e.g. monotremes, tenrecs, elephant-shrews, hyraxes, elephants and sea-cows), it seems more likely that the evolutionary reason for descent must lie elsewhere. There is now good evidence to suggest that sperm storage, rather than spermatogenesis, is the key factor (Martin, 1968b; Bedford, 1978, 1979). In at least some mammalian species lacking testicular descent (e.g. hyraxes – Glover and Sale, 1968; elephant-shrews –Tripp, 1970) the tail of the **epididymis**,

which is the site of sperm storage (see Johnson and Everitt, 1980), is displaced to occupy a position close to the ventral abdominal wall. In these species, sperm production evidently takes place at the core body temperature, whereas sperm storage occurs at a somewhat lower temperature. In species where both the testis and the epididymis descend to occupy a scrotum, the epididymis descends first and progresses farthest of all (involving a realignment relative to the testis) and the area of scrotal skin covering the epididymis (rather than the testis) is commonly devoid of hair (Bedford, 1978). Furthermore, in the common tree-shrew (*Tupaia belangeri*) short-term retraction of the testis into the abdomen in stressed adult males is not accompanied by retraction of the epididymis (Martin, 1968b).

Considerable support for a link between sperm storage and testicular descent in mammals is provided by Wolfson's observation (1954) that in certain passerine birds, at least, the 'seminal vesicles' (the functional equivalents of the mammalian epididymes) are located in sacs close to the root of the penis with temperatures some 7 °C lower than the core body temperature. Passerines generally have higher metabolic rates, relative to body size, than other birds (Lasiewski and Dawson, 1967). It is therefore possible that sperm storage in passerines takes place at a lower temperature for reasons paralleling those responsible for testicular descent in mammals.

However, even if descent of the testis and/or epididymis in most mammals can be explained in terms of sperm storage at lower temperatures, there remains the problem that most birds are apparently able to produce and store sperm at a normal core body temperature exceeding 37 °C. Hence, some additional explanation is necessary to account for the particular susceptibility of stored sperm to adverse effects of the elevated core body temperature in mammals. Bedford (1978, 1979) suggests that sperm storage at a lower temperature in viviparous mammals may be related to their general polygynous habits. It is true that most birds are pair-living, and there is some evidence that testicular (i.e. epididymal) descent in mammals is particularly well developed in markedly polygynous species.

However, this is not the whole story, because numerous mammalian species with intra-abdominal testes are polygynous, whereas some mammals with descended testes (e.g. certain primates) are monogamous. Further, passerines do not include significantly more polygynous species than other bird groups. One notable universal feature of viviparous mammals, however, is the very small size of the egg (microlecithic ovum) compared with the eggs of birds. It is conceivable that fertilization of such a small egg requires special characteristics of the sperm that are enhanced by storage at a reduced ambient temperature.

Whatever the reason for the evolution of testicular descent in most mammals, the evolutionary history of this characteristic is of special relevance here because, as noted above, Mivart's widely quoted definition of the order Primates (see Chapter 5) lists scrotal testes as a distinctive primate feature. It is true that in all male primates there is descent of the testes. But, if this feature was already present in the common ancestors of marsupials and placentals, possession of testicular descent yields no information with respect to evolutionary relationships among placental mammals. For example, the occurrence of testicular descent in tree-shrews has long been cited (Gregory, 1910; Kaudern, 1911; Carlsson, 1922; Le Gros Clark, 1959) as an 'advanced' feature indicative of a specific ancestral relationship to primates (see Table 5.4), whereas this may indicate nothing more than common retention of a primitive characteristic of the ancestral stock of placentals. Even if the common ancestors of marsupials and placentals had undescended testes, this condition must have evolved so many times in parallel during mammalian evolutionary history that it is, by itself, of little value as an indicator of phylogenetic relationships (Martin, 1968b; Luckett, 1980c).

It is therefore important to consider in more detail the question of testicular descent in primates. Closer examination shows that there is one significant difference separating all tree-shrews from all primates in the degree to which the testes descend along the longitudinal axis of the body. In tree-shrews, the testes descend only to the anterior margin of the root of the penis (Martin, 1968b) and the poorly developed scrotum is accordingly **prepenial** (not parapenial, as suggested by Le Gros Clark, 1959). In primates, by contrast, the testes typically descend past the root of the penis to occupy a well-developed **postpenial** scrotum. The prepenial scrotum is, in simple developmental and evolutionary terms, a logical antecedent to the postpenial scrotum and it is, in fact, extremely rare to find the prepenial condition among placental mammals. Apart from tree-shrews, this characteristic apparently occurs only in the Lagomorpha (rabbits, hares and pikas) among placentals. On the other hand, in confirmation of the presumed primitive nature of this feature, a prepenial scrotum is characteristic for marsupials with descended testes. Thus, the position of the scrotum of tree-shrews is primitive for placentals and suggests that this group has long been separated from primates. (As postpenial testes are a characteristic of living primates, it is reasonable to infer that this condition was present in the ancestral primate stock.)

Primates are also sharply demarcated from tree-shrews by the irreversible nature of testicular descent. Whereas, as mentioned above, the testes may be retracted back into the abdominal cavity during the adult life of male tree-shrews (e.g. in response to stress), this is not the case in any primate. Once the testes have descended in male primates, the canal connecting the scrotal cavity with the general abdominal cavity (i.e. the inguinal canal) becomes constricted by the inguinal ring and descent is rendered permanent (Harms, 1956). In the Malagasy lemurs, which reproduce seasonally (see later), the testes and the scrotum are reduced considerably in size during the non-breeding period. It is sometimes stated that the testes are 'retracted' at this time, but they are withdrawn only to the external orifice of the inguinal canal and do not pass through it to re-enter the abdominal cavity. Hence, the clear distinction between tree-shrews and primates with respect to the permanency of testicular descent holds good.

Primates may also be distinctive with respect to the timing of testicular descent. In many mammalian groups (e.g. some insectivores, some rodents, most carnivores and tree-shrews), the testes descend only when the male approaches sexual maturity, although such mammals typically reach sexual maturity relatively early in life (see later). By contrast, although attainment of sexual maturity in primates is relatively late, testicular descent frequently (if not universally) occurs very early in life. Unfortunately, this feature has not yet been surveyed systematically. However, it would seem that in most, if not all, primates the testes have already migrated posteriorly from their original position close to the kidneys by the time of birth so that they are already in the scrotum or very close to passage through the inguinal canal. This state of affairs is, at first sight, paradoxical in that late attainment of sexual maturity (i.e. onset of spermatogenesis) is associated with early testicular descent, and vice versa. In view of this observation, it would be interesting to examine the early endocrine activity of the testis in male primates generally. A postnatal peak of testosterone production has been reported for humans (Forrest, Caithard and Bertrand, 1973), chimpanzees (Winter, Faiman *et al.*, 1980), macaques (Robinson and Bridson,

1978) and common marmosets (*Callithrix jacchus* – Hearn, 1977; Dixson, 1986) and it has been suggested that it is related to sexual differentiation of the brain. It is possible that in male primates early descent of the testes might be associated with hormonal effects leading to a significant distinction between males and females at an early age. Such a distinction might be absent in those mammalian species (e.g. tree-shrews) lacking testicular descent until the age at which spermatogenesis is initiated. If this interpretation should prove to be correct, it could have important implications for the development of social integration within primate social groups.

Female reproductive organs

As with the typical condition for the male mammal, the reproductive tract of female mammals is basically bilaterally symmetrical, though with varying degrees of distal fusion, and there is an intimate association between the reproductive and urinary systems during development. In adult female primates, as in males, the cloaca is no longer developed as such. The typical mammalian female reproductive tract (Fig. 9.2) is bilaterally symmetrical in having paired ovaries and oviducts, which lead to

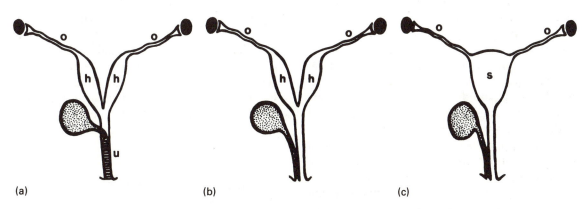

(a)   (b)   (c)

Figure 9.2. Simplified diagrams of the adult female reproductive system in tree-shrews (a), prosimian primates (b) and simian primates (c). In all cases, paired ovaries (black) shed their eggs into paired oviducts (o), where fertilization takes place. Tree-shrews are distinguished from all primates by retention of a long urogenital sinus (u) in which the urinary tract leading from the bladder (stippled) is combined with the genital tract. Simian primates differ from both prosimians (including tarsiers) and tree-shrews in the fusion of the two uterine horns (h: bicornuate condition) to form a single uterine chamber (s: simplex condition).

two separate uterine horns. Below the level of the uteri, midline fusion in placental mammals leads to the formation of a single **vagina** emerging through a vulval aperture, contrasting with the retention of a double vaginal tract in marsupials (associated with a bifid penis in the male). Primitively in mammals, the embryological association between the reproductive and urinary tracts of the female is retained into adult life, in that the ureter joins the vagina anterior to the vulva to form a common urogenital tract (**urogenital sinus**; Fig. 9.2). This urogenital sinus is typically retained in marsupials and also persists in some placental mammals; but in certain placental groups – such as primates (apparently without exception) – the urethra opens through an external aperture separate from the vulva. In embryological terms, this is undoubtedly the more advanced condition and the retention of a relatively long urogenital sinus in female tree-shrews (Martin, 1968b) once again represents a primitive condition distinguishing the Tupaiidae from the Primates. As the loss of the urogenital sinus would appear to be a universal characteristic of living primates, it seems likely that the sinus had already been suppressed in their common ancestor.

It is possible that loss of the urogenital sinus could be associated with the phenomenon of vulval closure, which is widespread among small-bodied mammals that have seasonal breeding and/or relatively long gestation periods (e.g. the European mole, various rodents, various small-bodied lemurs and bushbabies). Intermittent sealing of the vulva during pregnancy or during non-breeding periods of the year may have some selective advantage (e.g. in the exclusion of pathogens from the female tract) and would necessarily be associated with the development of a separate urethral aperture (i.e. loss of the urogenital sinus). Hence the absence of a urogenital sinus in female primates might indicate that the common (small-bodied) ancestral form had a relatively long gestation period, a strictly seasonal pattern of breeding, or perhaps both features combined. On the other hand, mammals such as tree-shrews, which have relatively short gestation periods and can

potentially breed at any time of the year (Lyon, 1913; Zuckerman, 1932; Harrison, 1955; Martin, 1968b), have doubtless remained primitive in these respects, in association with the retention of the urogenital sinus. Of course, once the urogenital sinus has been lost in a given ancestral form (as may have been true of the ancestral primate), there is no reason why this condition should be reversed even if the original causative influence were to disappear (e.g. secondary return to a non-seasonal pattern of breeding).

In other features, the female reproductive tract has remained relatively primitive in certain primate groups. For example, retention of twin uterine horns (**bicornuate uterus**) characterizes all of the prosimian primates including tarsiers (Fig. 9.2) and contrasts with the midline fusion to produce a single uterine cavity (**simplex uterus**) present in all simian primates. This latter condition is extremely rare among the mammals generally, apparently occurring only in certain edentates (sloths and armadillos) in addition to the simian primates (Weber, 1927). As this condition is an advanced characteristic of all simians, and as it is so rare among mammals generally, special conditions must have led to its evolution, presumably in the common simian ancestor of monkeys, apes and humans. It is likely that such a condition would only occur in association with reduction of the litter size to a single offspring (as is the case with nearly all living simians and with sloths and some armadillos). Therefore, one might infer that the common ancestor of monkeys, apes and humans typically produced only a single infant at a time. As will be seen, such a reduction in litter size would have had far-reaching implications for overall reproductive strategies.

THE FEMALE CYCLE AND SEASONALITY OF BREEDING

In mammals typically, the female has a characteristic short period of 'heat' (**oestrus**) during which she is receptive to mating by the male. Given that the common ancestor of marsupials and placentals was most probably nocturnal in habits, and therefore probably 'solitary' in its lifestyle, mating between male and female would doubtless have required special signals indicating the receptive state of the female, inciting the male to approach and mount. Olfactory cues (pheromones) indicating female receptivity would probably have been the most effective means of communication. Mating between early mammals would therefore have depended on two important features: the frequency with which a female would have been receptive (i.e. the duration of the oestrous cycle) and coordination between male and female behaviour at oestrus.

As a reflection of detailed laboratory investigations into the ovarian functions of female mammals, there is a widespread tendency to think in terms of a natural **oestrous cycle**, with each female having oestrus at regular intervals (in the absence of any seasonal suppression of reproduction). However, it is far more relevant for most purposes to think in terms of a **pregnancy cycle** under natural conditions, or at least some alternation of pregnancy and short periods of repeated oestrus, as long periods of oestrous cycling would imply the failure of reproductive coordination between males and females. By the same token, it is unlikely that natural selection would have operated directly on the oestrous cycle in isolation, and it may well be misleading to seek evolutionary explanations for the cycle as an isolated feature. Instead, it is more productive to consider the evolution of oestrous cyclicity and pregnancy as intimately linked phenomena.

If natural selection were to favour high reproductive output, then one would expect to find short periods between oestrous states, short gestation periods associated with relatively large litter sizes, and reproduction over an extended period each year. Given selection for high reproductive output, it is conceivable that the intervals between oestrous states would be determined simply by the minimum period required for development of the eggs (ova) to a level suitable for eruption from the ovary (ovulation). In principle, female mammals could maintain continuously overlapping cycles of ovum development, such that ripe ova would be present at all times in the ovaries. In practice, however, in all female mammals there seem to be waves of ovum development leading to the cyclical presence of batches of ripe ova at regular intervals. Therefore, it might be the case that the shortest oestrous cycles (lasting only a few days) are determined by the basic requirements of ovum development combined with a need for maximal reproductive potential (i.e. the greatest possible frequency of female receptivity). Conversely, an oestrous cycle of longer duration could be interpreted as an adaptation associated with a reduced breeding potential; but it is also possible that long oestrous cycles are a secondary feature resulting from some other aspect similarly associated with reduced reproductive turnover (e.g. increased gestation period).

Cycles of ovulation

In order to examine the significance of the duration of the oestrous cycle in more detail, it is necessary to consider ovum development and ovulation in relation to fertilization and subsequent embryonic development (Fig. 9.3). It is known from laboratory investigations that ovulation in some mammalian species depends on the act of mating (**induced ovulation**). In the absence of mating, ripe ova simply degenerate without further development (i.e. they undergo **atresia**). By contrast, in other mammalian species ovulation takes place regardless of whether the female is mated (**spontaneous ovulation**). In both cases, when ovulation has taken place the residual cellular mass of the ovarian (Graafian) follicle typically forms a **corpus luteum** (yellow body) which produces the hormone progesterone. This hormone acts to maintain the pregnancy that normally ensues, ensuring the survival of the embryo at least during the early stages of pregnancy (see Fig. 9.3). When mating occurs, followed by pregnancy, there is little practical difference between these two systems, although in one case the ovulatory stimulus (OS in Fig. 9.3) is provided by mating, whereas in the other it is directly provided by the female's own internal system of hormonal control.

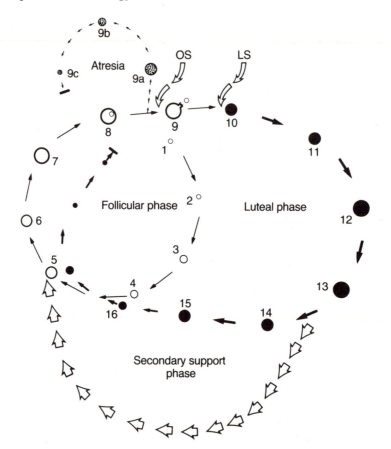

Figure 9.3. Schematic diagram of female reproductive cycles in mammals. Groups of ovarian follicles mature in cyclical fashion (1–9). In the absence of an ovulatory stimulus of some kind (OS), ripe follicles subsequently degenerate (atresia: 9a–c). If ovulation is stimulated and followed by luteinization, a luteal phase follows, characterized by the presence of a corpus luteum (black circles, 10–16).

A separate stimulus (LS) may, in some cases at least, be required to ensure that luteinization follows ovulation. The corpus luteum is required for the hormonal support of pregnancy in mammals generally, although in species with long gestation periods a secondary support phase (involving production of secondary corpora lutea or the production of hormones by the placenta itself) is typically necessary during the latter stages of pregnancy. (Reprinted from Martin, 1975, courtesy of Plenum Press, New York.)

In fact, the classical division between induced and spontaneous ovulators is not clear cut; for example, in some mammalian species (e.g. murine rodents – mice, rats and hamsters; Perry, 1971) ovulation takes place spontaneously, but the stimulation provided by mating is required for the formation of a functional corpus luteum (LS in Fig. 9.3). It is therefore more meaningful to draw a distinction between species that will not normally form a functional corpus luteum in the absence of mating (**induced luteinizers**) and those in which ovulation and the subsequent formation of a functional corpus luteum occur regardless of mating (**spontaneous luteinizers**). The practical implication of this distinction is that in the absence of mating the former will have a short oestrous cycle reflecting only the period involved in follicular development (follicular phase), whereas the latter will have a long oestrous cycle incorporating both this period and the period over which the spontaneously formed corpus luteum remains active in the absence of ensuing pregnancy (follicular phase plus luteal phase; see Fig. 9.3). This division into short and long oestrous cycles is to some extent obscured by intergradations (Conaway, 1971) and by variations in detail from one mammalian group to another, but as a basic generalization it is both valid and useful. Certainly, the association between long female cycles and spontaneous ovulation/luteinization is generally found in primates. The female cycle of primates is typically quite long (Table 9.2); it lasts about a lunar month in most cases (e.g. Lang, 1967) and even exceeds this in some prosimian species.

The only exceptions to the rule regarding cycle length in primates seem to be found, somewhat surprisingly, among New World monkeys rather than in the generally more primitive prosimians. The most extreme case seems to be that of the female squirrel monkey (*Saimiri sciureus*), which probably has a cycle lasting only about 9 days. Several other New World monkeys, such as the owl monkey (*Aotus trivirgatus*), the howler monkey (*Alouatta caraya*) and the capuchin monkey (*Cebus capucinus*), have cycle lengths intermediate between this minimal value and the approximately monthly durations of most other primates (Table 9.2). All of the prosimians so far investigated have long oestrous cycles, associated (as far as is known) with spontaneous ovulation and obligatory formation of a corpus luteum. As this is also true of Old World monkeys, apes and humans, it seems likely that some New World monkeys have become secondarily specialized away from this common pattern. It is now well known that the reproductive endocrinology of female New World monkeys has some distinctive features, notably in the maintenance of circulating steroid levels far higher than those found in other mammals, so secondary specialization is to be expected.

The typical primate pattern of a long oestrous cycle with spontaneous ovulation contrasts markedly with the pattern in tree-shrews. Female tree-shrews have short-phase oestrous cycles lasting 6–12 days and various lines of evidence

suggest that ovulation is induced rather than spontaneous (Conaway and Sorenson, 1966; F. D'Souza, personal communication). Thus, tree-shrews differ fundamentally from primates, in all of which there is a complex of reproductive characteristics indicative of restricted breeding potential (see later). It should also be noted that spontaneous ovulation, followed by automatic formation of a functional corpus luteum and hence by an automatic delay before conception becomes possible once again, implies particularly close coordination between males and females. The behaviour of males must be finely adapted for accurate recognition of oestrus in any available females, as failure to mate at the appropriate time would inevitably lead to a loss of breeding potential. By contrast, in species with induced luteinization, the female will rapidly return to oestrus if mating does not take place, so such fine coordination between males and females becomes unnecessary.

Here we need to comment on the use of the term oestrous cycle in relation to female primates and on the related concept of the **menstrual cycle**. The term menstruation, first used for the human female and then extended to certain other primates, refers to the discharge of blood from the uterus at approximately monthly intervals (Latin *mensis* = month). It is often implied that oestrous cycles and menstrual cycles in primates are necessarily equivalent. Further, as Keverne (1981) has pointed out, if the term oestrus (heat) is used in relation to primate behaviour, it implies a 'recurrent, restricted period of sexual receptivity . . . marked by intense sexual urge'. The term also implies a well-defined underlying physiological basis involving hormonal activity of the ovary. But for monkeys, apes and humans, at least, it is generally inappropriate to refer to 'oestrus' or to an 'oestrous cycle', as female receptivity, though often maximal around the time of ovulation, is not tightly restricted to a specific stage of the non-pregnant cycle and is not so heavily dependent on hormonal factors. Partly for this reason, many authors refer to the 'menstrual cycle' of primates, but this introduces a different kind of complication. The occurrence of menstruation is a variable feature among

Table 9.2 Characteristic lengths of female cycles determined for a variety of primate species. See surveys by Asdell (1964), Lang (1967), Butler (1974), Van Horn and Eaton (1979) and Robinson and Goy (1986) in addition to the sources quoted

Species	No. of females	No. of cycles	Cycle length (days) ± s.d.	Range	Source
Lemurs					
Cheirogaleus medius	3	9	19.7 ± 1.6	18–23	Foerg (1982)
Microcebus murinus	–	20	50.5 ± 6.4	36–63	Petter-Rousseaux (1964)
Lemur catta	16	40	39.3 ± 2.7	32–45	Evans and Goy (1968)
Lemur fulvus	8	68	29.5 ± 4.1	–	Boskoff (1978)
Lorises					
Galago crassicaudatus	–	10	44.0 ± 8.0	35–49	Eaton, Slob and Resko (1973)
Galago senegalensis	11	168	32.9 ± 7.8	–	Darney and Franklin (1982)
Arctocebus calabarensis	1	4	39.0 ± 4.1	36–45	Manley (1966)
Loris tardigradus	3	7	33.6 ± 4.5	29–40	Izard and Rasmussen (1985)
Nycticebus coucang	2	17	42.3	37–54	Manley (1966)
Perodicticus potto	2	39	37.9	–	Ioannou (1966)
Tarsiers					
Tarsius bancanus	3	15	24.0 ± 3.2	–	Izard, Wright and Simons (1985)
Tarsius syrichta	1	11	23.2 ± 3.0	19–29	Catchpole and Fulton (1943)
New World monkeys					
Aotus trivirgatus	3	8	15.6	–	Dixson (1982)
Cebus apella	–	108	20.0 ± 3.0	13–28	Nagle and Denari (1983)
Saimiri sciureus	–	–	9.1	–	Wolf, O'Connor and Robinson (1977)
Alouatta caraya	5	10	20.4 ± 2.6	17–24	Colillas and Coppo (1978)
Ateles geoffroyi	–	–	25.5	24–27	Goodman and Wislocki (1935)
Callithrix jacchus	–	19	28.6 ± 4.4	–	Hearn (1983)
Saguinus oedipus	4	10	22.7 ± 1.7	19–25	Brand (1981)
Old World monkeys					
Cercopithecus aethiops	5	32	33 (median)	25–46	Rowell (1970)
Cercopithecus mitis	6	72	30 (median)	19–57	Rowell (1970)
Erythrocebus patas	18	64	30.6 ± 2.6	24–38	Sly, Harbaugh *et al.* (1983)
Macaca fascicularis	–	–	30.8 ± 4.6	24–42	Jewett and Dukelow (1972)
Macaca mulatta	22	392	26.6 ± 4.4	14–42	Hartman (1932)
Cercocebus albigena	3	43	30.2 ± 3.1	22–36	Rowell and Chalmers (1970)
Papio anubis	17	63	33.2 ± 4.0	19–43	Hendrickx and Kraemer (1969)
Papio ursinus	–	404	35.6 ± 4.0	29–42	Gilman and Gilbert (1946)
Presbytis entellus	10	18	27.6 ± 3.1	23–35	Jay (1965)
Apes and humans					
Hylobates lar	2	17	29.8 ± 2.5	21–43	Carpenter (1941)
Pongo pygmaeus	6	18	29.6	24–32	Graham (1981)
Pan troglodytes	22	653	37.3 ± 3.6	22–187	Young and Yerkes (1943)
Gorilla gorilla	7	56	31.1 ± 7.5	13–55	Nadler (1975)
Homo sapiens	17	523	27.6 ± 3.5	18–53	King (1926)

primates. It is not found at all in lemurs or lorises (probably because of their non-invasive form of placentation; see later) and it is only weakly identifiable in tarsiers and New World monkeys, for which microscopic examination is often necessary to confirm the incidence of menstrual bleeding. In Old World monkeys and apes, menstruation is typically quite easily detected by externally visible signs of bleeding, but in no species is the menstrual flow as copious and as easily recognizable as with the human female. (This apparent *Scala naturae* in the occurrence of menstrual bleeding broadly parallels the gradation found in the invasiveness of the processes associated with placentation, as recorded by Hill (1932), and this parallelism suggests some functional connection between the two phenomena.) It is therefore quite misleading to use the term 'menstrual cycle' to characterize the non-pregnant cycle of all female primates, as is often done to avoid using 'oestrous cycle'. Indeed, lemurs, lorises and tarsiers do seem to show 'oestrus' of the classical type, as defined by Keverne (1981), and for these prosimian primates it is perfectly justifiable to use oestrous cycle, whereas reference to a menstrual cycle would be totally misleading for all except possibly tarsiers. Nevertheless, all primate species do seem to share a basic set of characteristics common to non-pregnant female cycles and, to avoid problems of terminology, these will be referred to here simply as **female cycles**.

Some comment is also needed on the phenomenon of **pseudopregnancy** in mammals, which occurs when a female fails to return to an expected non-pregnant cycle even though conception has not taken place. In mammalian species with induced luteinization (e.g. rabbits), pseudopregnancy can be reliably produced in the laboratory by mating females with males that have been rendered infertile without impairment of their copulatory activity, either through stress or by bilateral section of the vas deferens (vasectomy). The ensuing pseudopregnant condition may have many of the features of normal pregnancy (e.g. increase in body weight; development of the mammary glands; occurrence of nesting behaviour) and usually

persists for at least half of the normal gestation period. In such species, mating is normally required for the formation of a functional corpus luteum, and under natural conditions matings would usually be fertile, so there is probably little scope for selection against the persistence of a corpus luteum following infertile mating. It is therefore not surprising that a relatively long pseudopregnancy should follow artificially contrived infertile matings in captivity. Likewise, the tree-shrews *Tupaia longipes*, *T. belangeri* and *Lyonogale tana* have shown pseudo-pregnancy lasting about 25 days (Conaway and Sorenson, 1966; F. D'Souza, personal communication) in cages containing both intact and vasectomized males. By contrast, in mammalian species with spontaneous ovulation and luteinization, such as primates, pseudopregnancy is relatively rare and is not specifically brought about when vasectomized males are mated with females in captivity. (This, indeed, is just as well in view of the part played by vasectomy in modern large-scale programmes of human birth control!). It would therefore seem that species with spontaneous ovulation/luteinization have particularly well-developed mechanisms for the 'recognition' of pregnancy within the female's body, such that there is an immediate return to normal non-pregnant cycling if mating is not followed by conception.

Conaway and Sorenson (1966), among others, recorded uterine bleeding in several tree-shrew species, notably at the end of pseudopregnancy. Such bleeding has been equated with the menstruation of simian primates, with the implication that there is some phylogenetic connection between tree-shrew pseudo-pregnancy and primate menstrual cycles. But such a connection is most unlikely for several reasons. First, uterine bleeding in tree-shrews does not occur in association with their biologically relevant short oestrous cycle, but in association with pseudopregnancy. The latter is probably extremely rare under natural conditions and is therefore not open to selection towards a simian-like condition. Second, tree-shrews have oestrus and will ovulate at the end of pseudopregnancy, such that the preceding follicular phase must be 'telescoped' into the end of pseudopregnancy rather than following uterine bleeding as in the case of the menstrual cycle of simian primates (see Perry, 1971). Finally, lemurs and lorises – as noted above – do not menstruate, so any similarity between tree-shrews and simian primates must surely be the product of convergent evolution unless the strepsirhine primates (lemurs and lorises) have secondarily lost menstruation.

Returning to the basic relationship between pregnant and non-pregnant cycles in females, it is noteworthy that mammalian species with long oestrous cycles typically have long gestation periods as well (see later). The function of the original corpus luteum in maintaining pregnancy (through the production of progesterone) is characteristically taken over by some secondary support mechanism part of the way through gestation (see Fig. 9.3). In some cases, this secondary support is ensured by the formation of secondary corpus lutea (e.g. in horses and elephants); in others it is ensured by production of hormones by the placenta itself (e.g. in at least some primates). There seems to be some limit to the period over which a single corpus luteum can function to maintain a pregnancy, with the result that evolution of a long gestation period necessarily entails the development of a secondary support mechanism to maintain the latter part of pregnancy.

There has been some discussion of the question as to whether induced ovulation/luteinization or spontaneous ovulation/luteinization represents the primitive condition for mammals (e.g. Conaway, 1971; Weir and Rowlands, 1973; Martin, 1975). This is a classical case in which either condition can be taken as primitive, with the other derived from it in accordance with a perfectly reasonable set of explanations. However, it is clear that the induced condition is associated with relatively short gestation periods and a high reproductive potential, whereas the spontaneous condition is combined with relatively long pregnancies and a lower reproductive potential. Thus, resolution of the primitive condition depends on analysis of other

aspects of reproduction and this point will be reconsidered further on.

Seasonality of reproduction

The relationship between length of the female cycle and reproductive potential obviously also depends on the degree of **seasonality** of reproduction. It is therefore important to give some attention to the question of seasonality, particularly as Zuckerman (1932) suggested that a seasonal pattern of reproduction was the primitive condition for mammals generally, with a non-seasonal pattern emerging as an unusual feature among primates. Before examining the annual breeding patterns of primates, however, it is necessary to define what is meant by 'seasonality'. Virtually all mammalian species are seasonal at least to the extent that birth frequency varies in a predictable fashion over the year, with recognizable peak and trough levels. There are extreme cases – even amongst the primates – as with all of the Malagasy lemurs (Petter-Rousseaux, 1964; Martin, 1972; van Horn, 1980) – where births are confined strictly to just one or two relatively brief periods in the year. Conversely, there are some cases where births seem to occur at the same level throughout the year, with no more than random fluctuations from month to month. One can recognize four different annual breeding patterns as arbitrary subdivisions on the continuous spectrum ranging from year-round, non-seasonal breeding through seasonal breeding peaks to strictly seasonal breeding (Fig. 9.4).

In fact, as is shown by Lindburg (1987), there is no regular distribution of annual reproductive patterns among primates and hence no simple indication of the evolutionary background to those patterns. For example, it would seem that three of the four patterns shown in Fig. 9.4 occur in the five prosimian species studied in Gabon by Charles-Dominique (1977a). The potto (*Perodicticus potto*) seems to be a strictly seasonal breeder (condition d), while the three bushbaby species (*Galago demidovii, G. alleni* and *G. (Euoticus) elegantulus*) are year-round

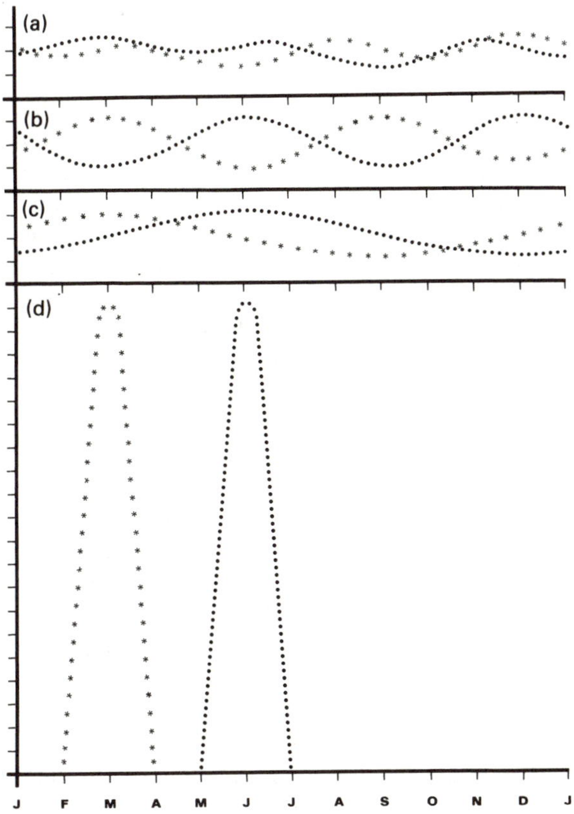

Figure 9.4. Hypothetical diagram illustrating varying degrees of seasonality of breeding, taking a standard number of successful matings (∗) leading to a similar number of births (●) and allowing a uniform gestation period of 3 months. (a), no seasonality, just irregular fluctuations; (b) two regular annual birth peaks; (c) one regular annual birth peak; (d) strict seasonality; matings and births confined to very restricted periods of the year. Note the substantial numbers of births concentrated into a single period of the year in condition D. (Reprinted from Martin, 1975, courtesy of Plenum Press, New York.)

breeders with a seasonal peak (condition c) and the angwantibo (*Arctocebus calabarensis*) apparently breeds year-round with no obvious seasonal fluctuation (condition a). Nor is it true that all simian primate species show the supposedly advanced condition of year-round, non-seasonal breeding (e.g. van Horn, 1980;

Lindburg, 1987). It is true that some simian species show relatively little restriction of breeding to a particular period during the year, as is the case with (for example) the two chimpanzees, the gorilla and various Old World monkeys such as the hanuman langur, *Presbytis entellus* (Lancaster and Lee, 1965). However, several cases of fairly strict breeding seasonality have been reported among simian primates, notably in various macaque species such as the barbary macaque (*Macaca sylvanus*; MacRoberts and MacRoberts, 1966), in the golden lion tamarin (*Leontopithecus rosalia*; Coimbra-Filho and Maia, 1979), in the cotton-top tamarin (*Saguinus oedipus*; Brand, 1980) and in the squirrel monkey (*Saimiri sciureus*; Baldwin and Baldwin, 1981). With the squirrel monkey, seasonal limitation of mating and births is so pronounced that both females and males show clear differences between breeding and non-breeding states; in females ovarian cyclicity is absent during the non-breeding season, and in males there is a marked increase in body weight at the onset of the breeding season in association with specific hormonal changes – the 'fatted male' condition (see Baldwin and Baldwin, 1981).

As a general rule, seasonality of reproduction is doubtless related ultimately to seasonal patterns of food availability. Nevertheless, this need not always be the case exclusively. One advantage of strict seasonal breeding in any primate species, regardless of food availability, is that the level of any predator population must be adapted to year-round food supplies. Restriction of births to a short period of the year might reduce predation by 'saturating' the local predator population. At the same time, however, synchronization of births in a strictly seasonal primate species necessitates a relatively abundant supply of food during the breeding season to cope with the abrupt increase in population size. (Note the relative sizes of the peaks in the different annual breeding patterns illustrated in Fig. 9.4.) An explanation in terms of saturation of predators may account for the occurrence of strict seasonal breeding among pottos and squirrel monkeys in rainforest environments alongside other primate species lacking such obvious seasonality.

In other cases of strict breeding seasonality among primates, however, marked seasonal fluctuations in food availability may well be directly responsible for the restriction of births to a particular season. Two bushbabies that inhabit relatively dry forest regions of Africa with a shortage of insect food and fruits for a given period each year are strictly seasonal breeders: there are two birth periods a year in *Galago senegalensis* and one in *G. crassicaudatus* (see Van Horn and Eaton, 1979). Similarly, among the lemurs of Madagascar strict breeding seasonality is generally associated with clear seasonality in food availability, although the eastern coastal rainforest where some species live (e.g. *Propithecus diadema*, *Indri indri* and *Daubentonia madagascariensis*) does not have a well-defined annual dry season. It is possible that strict breeding seasonality in the rainforest lemur species may be associated more with saturation of local predators than with variations in food availability. Among simian primates, relatively marked seasonality in breeding is most notable in macaque species living in strictly seasonal environments on the fringes of the zone of geographical distribution of Old World primates (e.g. *Macaca sylvanus* and *M. fuscata*). Overall, the presence or absence of breeding seasonality in primates generally seems to be explained in terms of a combination of saturation of local predators and seasonality of food availability. There is no obvious case for postulating a particular annual pattern of reproduction for the ancestral primate in the absence of other evidence. Certainly, there is no justification for regarding seasonal breeding as necessarily primitive in all those mammalian species in which it now occurs. Year-round breeding without a peak may perhaps be viewed as a derived feature wherever it happens, as this is a relatively uncommon condition among mammals under natural conditions.

Where seasonal patterns of reproduction are associated with clear annual variation in availability of suitable foods, as is typical in habitats other than tropical rainforest, the timing of the birth season seems to be related primarily to the requirements of the offspring as they develop. Among Malagasy lemurs, for example, it is possible for copulation, gestation and even part of the lactation period to occur during the dry season (the time when food is limited for most lemur species). As body size of the species concerned increases, birth takes place earlier and earlier with respect to the ensuing wet season (Martin, 1972b). The crux of the matter seems to be that birth is timed such that the developing infant has time to grow to independence and to accumulate adequate reserves when food is in greatest supply, such that it will survive the subsequent season of poor availability. This interpretation also applies to the two mating and birth seasons of *Galago senegalensis moholi* in South Africa, where births are confined to October/early November and late January/early February, respectively. Both sets of infants are able to reach almost full adult size by the end of the wet season in March.

Seasonality of breeding in certain simian primate species also seems to fit this explanation. Lancaster and Lee (1965) reported that the onset of copulations in rhesus macaques (*Macaca mulatta*) and Japanese macaques (*M. fuscata*) typically coincides with 'the time when food is abundant'. These authors suggested a possible link between food availability and 'triggering mechanisms of reproductive cycles', but there is an alternative interpretation. Given a gestation period of about 6 months for macaques and the beginning of independent feeding by infants at about 3 months of age (with decreasing reliance on lactation thereafter), there would be broad coincidence between the period of maximum food availability one year after conception and the period when young infants depend on independent feeding to build up their body weight and reserves. Similarly, Dawson (1977) reports for Geoffroy's tamarin (*Saguinus geoffroyi*) in Panama a birth peak in the period March–June. Because these tamarins are weaned by 4 months of age (i.e. during the period July–October) they have from 2 to 5 months in which to build up reserves before the onset of the dry season (mid-December–mid-April). This apparent association between seasonal breeding patterns and the post-weaning survival prospects of developing offspring is just one facet of the overall reproductive strategy that relates the breeding of any primate species to its natural environmental setting.

CONCEPTION, GESTATION AND BIRTH

The life history of any individual primate begins when the freshly discharged ovum is fertilized, at a site relatively high in the oviduct, by a single sperm (spermatozoon). Even before this point, in connection with the process of ovulation itself, there is an interesting difference between lemurs and lorises (strepsirhine primates) on the one hand and tarsiers, monkeys, apes and humans (haplorhine primates) on the other. In female mammals generally, the ovary on each side is commonly enveloped by a membraneous pocket (**ovarian bursa**). This bursa links up with the general abdominal cavity only through a relatively small opening and presumably ensures that the ovum safely enters the oviduct funnel to begin its passage down the female tract. The possession of a well-developed ovarian bursa is extremely widespread among mammals and is therefore likely to represent the primitive mammalian condition. The ovarian bursa has been retained by strepsirhine primates, but in haplorhines it has been virtually lost as an unusual specialization. Presumably, the loss of the bursa might increase the danger that an ovum could escape into the abdominal cavity, rather than descending the oviduct as required, in the absence of some special mechanism to take over the enclosing function of the bursa. Direct laparoscopic observation of the process of ovulation in rhesus monkeys and in women has, in fact, shown that the oviduct funnel is closely applied to the ovarian surface and envelops the ovary by active movement. Given the absence of the ovarian bursa in all haplorhine primates, it seems likely that this special function of the oviduct funnel must be similarly well developed throughout the entire group. It is not immediately clear why the specialization should

have emerged in haplorhines, unless it ensures greater efficiency in the transfer of the ovum to the oviduct. In humans, however, an ovum may on rare occasions fail to enter the oviduct and, if it is fertilized, the fetus will develop in the general abdominal cavity (ectopic pregnancy) – a condition that is fatal if untreated. There must obviously be strong selection pressures favouring efficient mechanisms for the transfer of eggs from ovary to oviduct in mammals generally, and it remains to be seen why haplorhine primates should have developed a rare specialization involving the loss of the ovarian bursa. Whatever the functional background to this evolutionary development, its presence in tarsiers as well as in simian primates (monkeys, apes and humans) certainly provides a clear example of a probable shared derived character indicating a specific ancestral relationship linking haplorhine primates.

Once the ovum has safely arrived in the oviduct and has begun its journey towards the uterus, normal fertilization can take place. Following fertilization, the resulting zygote undergoes cell division while still within the oviduct. By the time it reaches the uterus, it has become a hollow ball of cells called the **blastocyst**, and it is at this stage that implantation on the wall of the uterus can take place – to provide the initial direct link between the mother's tissues and the developing embryo. While the zygote is descending the oviduct (a process that takes several days), the wall of the uterus undergoes preparatory changes (e.g. modification of the internal lining of epithelial cells and proliferation of the vascular system of the uterine wall) that are required for successful implantation. This preparation of the uterine wall following ovulation will take place regardless of whether the ovum has, in fact, been fertilized and it is possibly for this reason that certain primate females (notably Old World monkeys, apes and humans) menstruate after any ovulation not accompanied by fertilization (see later).

After implantation of a blastocyst, certain physiological changes ensure that the corpus luteum continues to function in hormonal support of the ensuing pregnancy, at least during the vital initial stages before the placenta is sufficiently developed to take over that role. In this way, normal ovarian cyclicity is arrested and the progress of pregnancy is ensured. The blastocyst is, at the time of implantation, two-layered (bilaminar) and the outer layer of cells (called the **trophoblast**) is responsible for the initial process of attachment to the uterine wall and for any invasion of the maternal tissues. Meanwhile, within the blastocyst, the future embryo begins to develop as the **embryonic disc**. Following a complex sequence of developmental changes, the developing embryo of any mammal eventually achieves a form in which the embryo or fetus (the named used when the organs have developed a recognizable form) is enveloped by a series of membranes (Fig. 9.5; see also Steven and Morriss, 1975). The outermost membrane, the **chorion**, provides a continuous boundary layer for the embryonic system, and the embryo

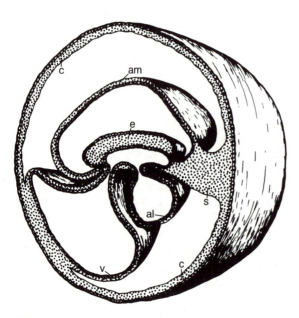

Figure 9.5. Diagram of the embryonic/fetal membranes in a typical placental mammal. Abbreviations: c, chorion; v, vitelline membrane (yolk sac); al, allantoic membrane (allantois); am, amnion; e, embryo; s, connecting stalk. (After Amoroso, 1952 and Le Gros Clark, 1959; reprinted from Martin, 1976, with kind permission from the Zoological Society of London and Academic Press, London.)

itself is eventually surrounded by a protective, fluid-filled membrane called the **amnion**. (It is the rupture of this membrane, followed by the release of the amniotic fluid, which produces the flow of 'waters' heralding the imminent birth of a human infant.) Within the chorion, but outside the amnion, there are typically two other membranes – the **vitelline membrane** (forming the yolk sac) and the **allantoic membrane** (forming the allantois) – that play a major part in the nutrient supply of the developing embryo. In fact, as pointed out by Hill (1932) and Luckett (1975), there is a significant difference between prosimian primates (lemurs, lorises and tarsiers) and simian primates in the mode of formation of the amnion. In prosimians, the amnion is formed by fusion of two folds of tissue over the embryo, whereas in simians the amnion is formed by a cavitation process. Because formation of the amnion from fusion of folds is likely to be the primitive condition for mammals (Mossman, 1953; Luckett, 1975), it is fairly obvious that simian primates share a derived character of some significance.

The nutrient supply of the developing embryo, based on the inward transfer of material of maternal origin, primarily depends on: (1) the relationship established between the chorion and the uterine wall; and (2) the pattern of vascularization of the internal wall of the chorion associated with the yolk sac and/or allantois, which provides the blood supply to the embryo through the umbilical cord. These two aspects will be considered in some detail, as they provide crucial evidence relating to the phylogenetic subdivision of the primates.

Nutrients utilized by the developing embryo or fetus within the uterine cavity are derived from three main sources. The first, and the least important overall, is the breakdown and consumption of maternal cells (phagocytosis) carried out during any invasion of the maternal tissues. The second involves the transfer of nutrients by a combination of diffusion and active transport from maternal blood vessels to embryonic blood vessels. (It should be noted that there is an equally important transfer of waste products in the opposite direction, and this dual role of the functional association between maternal and embryonic circulatory systems is clearly a central feature of any form of mammalian placentation.) The third source of embryonic nutrition is provided by coiled uterine glands in the wall of the uterus; these open into the uterine cavity. This third source of embryonic nutrition is often overshadowed by the central importance of exchange between maternal and embryonic blood vessels, but it is nonetheless important and deserves more attention than is generally given.

Types of placentation

Following Grosser (1909, 1927), it is possible to classify types of **placentation** according to the relationship established – in the definitive placenta – between the chorion of the embryo and the uterine wall with its underlying maternal circulatory system (Fig. 9.6; see also Steven, 1975a; Johnson and Everitt, 1980; King, 1986). In the simplest system, the uterine wall is not invaded and as a result the internal epithelial lining of the uterus remains intact, providing the surface of contact with the chorion – the **epitheliochorial** type of placenta. This non-invasive placental system always involves a large area of contact between the chorion and the uterus (in many cases yielding a **diffuse** type of placentation) and there is conspicuous development of the uterine glands. Indeed, the relationship with the uterine glands is so well developed in the typical epitheliochorial placenta that there are special absorptive areas of the chorion (e.g. **chorionic vesicles**; see Fig. 9.6) developed at points of contact with multiple outlets of the uterine gland. These absorptive areas doubtless serve a special function in enhancing the transfer of particular nutrients; such as iron and lipid, from the maternal uterus to the embryo (King, 1984).

The remaining types of placentation, following Grosser's simple scheme of classification, involve varying degrees of invasion of the uterine wall by the embryo. A system with the uterine epithelium eroded, leaving the chorion in contact with the maternal connective tissue layer, would be the next logical stage in the relationship between the embryonic system and the uterine wall. However, increasing use of electron microscopy has generally led to exclusion of this

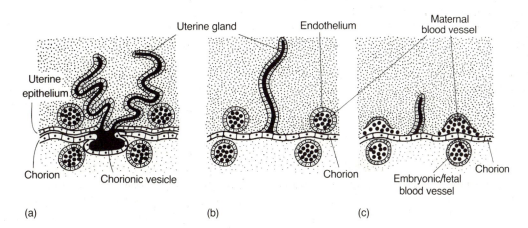

Figure 9.6. Schematic illustrations of Grosser's basic classification (1909) of types of placenta: (a) epitheliochorial; (b) endotheliochorial; (c) haemochorial. Embryonic or fetal tissues (light stippling) and maternal tissues (heavy stippling) are typically separated by the chorion, which constitutes a barrier to diffusion between maternal and embryonic/fetal blood vessels (seen in section). Uterine glands, producing uterine milk, are best developed in species with the epitheliochorial type of placenta, in which special areas of the chorion (e.g. chorionic vesicles) are related to grouped outlets of such glands. (After Amoroso, 1952 and Martin, 1976.)

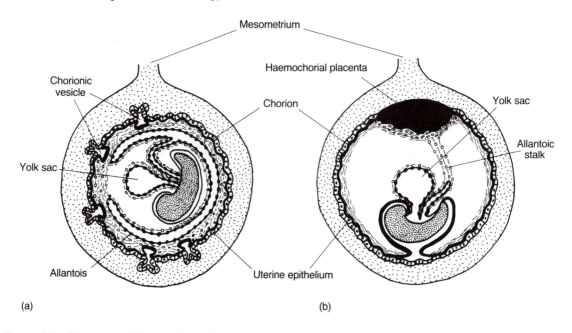

Figure 9.7. Diagrams of the fetal membranes and placentation in strepsirhine primates (a) and haplorhine primates (b). Note the diffuse epitheliochorial form of placentation in the strepsirhine, associated with the development of chorionic vesicles, and the contrasting discoidal, haemochorial form of placentation in the haplorhine. In the haplorhine, the vessels of the yolk sac are not involved in placentation and the allantois develops only as a vestigial stalk-like structure. (After Luckett, 1975.)

intermediate category of placentation, known as the **syndesmochorial** type (Steven, 1975b). Given a further degree of invasion of the maternal tissue, the maternal blood vessels themselves become closely applied to the chorion (Fig. 9.6). This type of placentation is referred to as **endotheliochorial**, because the cellular sheath of blood vessels is called endothelium. Finally, the most invasive kind of placentation occurs where the endothelium of the maternal blood vessels has itself been broken down so that the chorion is directly bathed by maternal blood – the **haemochorial** type of placenta (Fig. 9.6). Both the endotheliochorial and the haemochorial types of placentation differ from the epitheliochorial type in that there are one or more localized areas of invasive contact (**discoid** or **zonary** placentation) between the chorion and the uterine wall.

It was noted at a very early stage by the founder of comparative embryology, Hubrecht (1908),

that the primates show a remarkable dichotomy with respect to placentation types – all strepsirhine primates (lemurs and lorises) have the least invasive, epitheliochorial type of placentation, whereas all haplorhine primates (tarsiers and simians) have the most invasive, haemochorial type (Fig. 9.7). This marked difference in placentation has remained at the core of discussions of the phylogenetic relationship between tarsiers and the other primates since Hubrecht first pointed it out. It should be noted here that tree-shrews have a quite distinctive form of endotheliochorial placentation (Fig. 9.8). This involves twin discoidal attachments to special placental pads developed on the lining of the uterus (Hubrecht, 1898; van der Horst, 1949; Martin, 1968b; Luckett, 1969; Hill, 1965). There is therefore nothing in the basic type of placentation that would indicate a specific relationship between tree-shrews and any group of living primates.

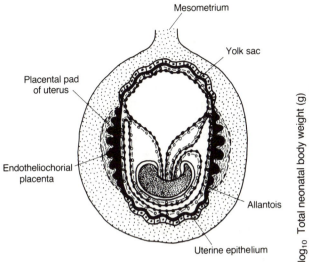

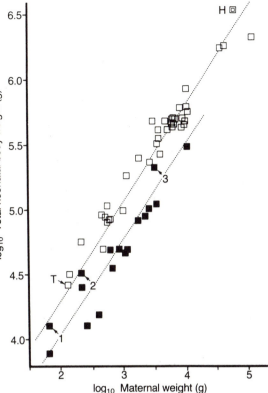

Figure 9.8. Diagrammatic cross-section of the uterus of a tree-shrew during mid-pregnancy. Note the replacement of the yolk sac by the allantois over the placentae at this stage of fetal development. Although the blood vessels of the yolk sac are no longer involved in placentation at this stage, they continue to absorb the uterine milk produced by the uterine glands. (After Luckett, 1969, and Martin, 1969.)

Figure 9.9. Logarithmic plot of total neonatal body weight against maternal body weight for primates. Key: ■, strepsirhine primates (1, *Microcebus murinus*; 2, *Galago senegalensis*; 3, *Varecia variegata*); □, haplorhine primates (T, *Tarsius*; H, *Homo sapiens*). The lines of fixed slope (= 0.75) have been fitted to strepsirhines and haplorhines separately. Note that the strepsirhines with multiple litters (points 1, 2 and 3) approach the haplorhine line for total neonatal mass but not for individual neonate weight. (Data collected from the literature and from additional zoo specimens – see Sacher and Staffeldt, 1974; Rudder, 1979; and Martin and MacLarnon, 1988.)

(Preliminary reports by Meister and Davis (1956, 1958) – cited by Le Gros Clark (1959) – that tree-shrews have haemochorial placentation were based on a very limited range of embryos and have since been effectively discounted.)

At one time, it was maintained that complex aspects of placentation in mammals provide particularly valuable indicators of evolutionary relationships because the lack of any clear functional significance ensures that they behave as conservative characters (Mossman, 1937, 1953). It is undoubtedly correct that complex aspects of placentation provide valuable clues to mammalian evolutionary relationships, but at the same time it is unlikely that differences in the organization of placentation between mammalian groups are completely devoid of functional significance. An interesting case in point is provided by the sharp dichotomy between epitheliochorial placentation in strepsirhine primates and haemochorial

placentation in haplorhines. In spite of the long history of discussion of the potential evolutionary significance of this distinction (Hubrecht, 1908; Wood Jones, 1929; Mossman, 1953; Le Gros Clark, 1959; Luckett, 1975), the question of its functional significance was barely raised until very recently. This situation was apparently radically modified by the observation of Leutenegger (1973) that there is a similar sharp dichotomy between strepsirhines and haplorhines with respect to the relationship between the weight of the neonate and the body weight of the mother. A logarithmic plot of total neonate weight against maternal weight shows that there are two quite separate linear relationships ('grades') for strepsirhines and haplorhines, respectively (Fig. 9.9). The allometric exponent is more or less the same in both cases, approximating to the value of 0.75. However, the vertical separation between the two lines indicates that, at any given maternal weight, a haplorhine primate will produce a neonate about three times heavier than that produced by a strepsirhine primate.

There is, of course, the possible objection that the striking difference in relative size of the neonate arises because strepsirhines have considerably shorter gestation periods, relative to maternal body weight, than do haplorhines. When gestation period is plotted against maternal body weight on logarithmic co-ordinates, however, it emerges that there is no such clear dichotomy between strepsirhines and haplorhines (Fig. 9.10); although certain lemur species do have notably brief gestation periods compared to haplorhines (i.e. *Microcebus murinus*, *Microcebus coquereli* and *Varecia variegata*), most strepsirhine species overlap with haplorhines in this relationship. In short,

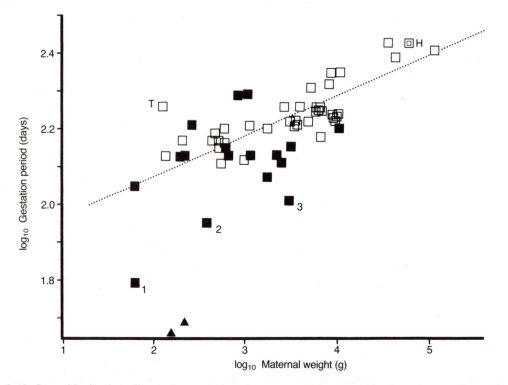

Figure 9.10. Logarithmic plot of gestation period against maternal body weight for primates. Key: ▲, tree-shrews; ■, strepsirhine primates (1, *Microcebus murinus*; 2, *Microcebus coquereli*; 3, *Varecia variegata*); □, haplorhine primates (T, *Tarsius*; outlined square, *Homo sapiens*). The best fit-line is the major axis for haplorhine primates. (Data collected from the literature and from additional zoo specimens – see Sacher and Staffeldt, 1974; Rudder, 1979; and Martin and MacLarnon, 1988. The unusually long gestation period for *Tarsius* – 178 days – has been confirmed by Izard, Wright and Simons, 1985.)

haplorhines typically produce considerably heavier neonates than do most strepsirhines when the effects of both maternal body weight and gestation period are taken into account (Martin and MacLarnon, 1988).

As the epitheliochorial placenta of strepsirhines is far less invasive than the haemochorial placenta of haplorhines, it may seem reasonable to suggest a connection between the typically heavier neonates of haplorhines and their more invasive form of placentation (Leutenegger, 1973; Martin, 1975). This interpretation comes readily to mind because of the widespread assumptions that non-invasive epitheliochorial placentation is primitive and that it is associated with lower rates of exchange of material between mother and fetus. However, it must be remembered that scaling analyses (Figs 9.9 and 9.10) merely establish a correlation between relative size of the neonate and the type of placenta in primates. In addition, several authors (e.g. Steven, 1975b; King, 1986) have warned against the assumption that there is any simple relationship between Grosser's classification of placentae and efficiency of transfer of nutrients. Before accepting the functional inference that haemochorial placentation permits more rapid transfer of nutrients from the mother's circulation to that of the developing fetus (see Fig. 9.6), some additional checking is advisable (Martin, 1984). A suitable test is provided by considering neonate weight in relation to maternal body weight across placental mammals generally (Fig. 9.11). The comparison is complicated by the fact that some placental mammals produce relatively well-developed neonates after a long gestation period, whereas others give birth to relatively poorly developed neonates after a short gestation period (precocial versus altricial mammals; see Fig. 9.12). Nevertheless, one can directly compare primates on the one hand and hoofed mammals (ungulates) and dolphins on the other, as the latter all resemble primates in producing well-developed precocial offspring after a long gestation period. (When maternal body weight is taken into account, it is found that primates, ungulates and dolphins have very similar

gestation periods (Martin and MacLarnon, 1988). If anything, ungulates and dolphins have slightly shorter gestation periods, relative to body size, than primates and should be expected to produce relatively smaller neonates.)

In fact, Fig. 9.11 shows that ungulates and dolphins produce neonates that are heavier relative to maternal body weight, than those of haplorhine primates. Yet both ungulates and dolphins have non-invasive epitheliochorial placentation similar to that of strepsirhine primates. Accordingly, one cannot accept the conclusion (Leutenegger, 1973) that a more invasive form of placentation can be directly linked to more rapid fetal development. Conversely, given the fact that ungulates (in particular) can clearly have very high rates of fetal development with non-invasive epitheliochorial placentation, there is no sound basis for arguing that this type of placenta in lemurs and lorises in some way limits the rate of fetal development. It must be emphasized that certain strepsirhine primates, such as *Microcebus* and *Varecia*, have relatively short gestation periods (Fig. 9.10) and yet produce total neonatal masses comparable to those of haplorhines (Fig. 9.9). When the effects of maternal body weight and gestation period are taken into account simultaneously, it emerges that inferred rates of individual fetal development are higher in these lemurs than in many haplorhine primates. In fact, detailed scaling analyses reveal that the typically lower rates of fetal growth of many strepsirhines may be linked to the relatively lower basal metabolic rates of their mothers (Martin and MacLarnon, 1988).

Thus, it may be concluded that there is no evidence from placental mammals generally that the epitheliochorial type of placentation necessarily limits rates of fetal development. The difference between strepsirhines and haplorhines in terms of neonatal body weight can be attributed to other factors (notably maternal basal metabolic rate), rather than to the striking difference in placentation between these two groups of primates. Accordingly, we have yet to find a convincing functional explanation for this difference in placentation. It should be noted,

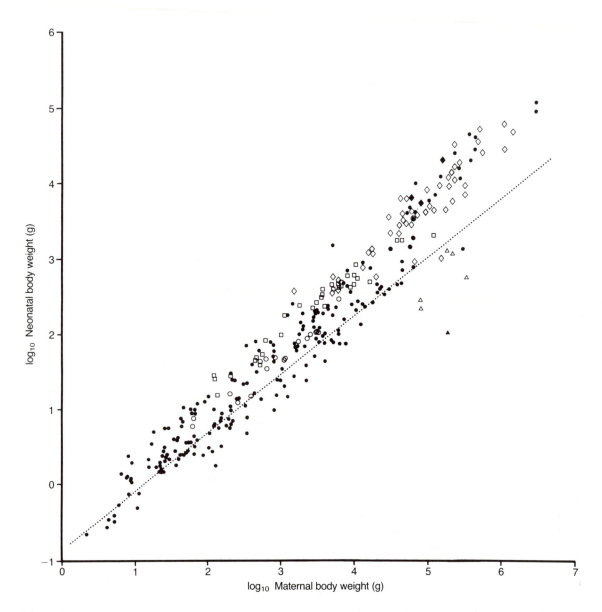

Figure 9.11. Logarithmic plot of neonatal body weight against maternal body weight for 201 species of placental mammals. Key: ●, mammals with short gestation periods and poorly developed neonates (altricial type; see Fig. 9.12); ○, strepsirhine primates; □, haplorhine primates; ◇, ungulates (hoofed mammals); ◆, dolphins; △, bears; ▲, giant panda. The best fit-line shown is the major axis fitted to altricial mammals excluding bears and the giant panda. (Reprinted from Martin, 1984, with kind permission of the Zoological Society of London and Academic Press, London.)

however, that invasive haemochorial placentation tends to be associated with earlier implantation among placental mammals generally (Johnson and Everitt, 1980) and this may be significant. Further, the mother is presumably more likely to produce antibodies against embryonic or fetal proteins if the placenta is of an invasive kind, and this factor must also be borne in mind when seeking a functional explanation for the fundamental dichotomy between strepsirhines and haplorhines in their placentation.

Turning now to the question of the evolution of placentation among placental mammals generally, the problem of establishing the primitive condition is encountered in an acute form. The difficulties involved in phylogenetic reconstruction are underlined by the fact that different authors have proposed all three basic types of placentation (Fig. 9.6) as ancestral for placental mammals and, by implication, for primates. Many authors have regarded epitheliochorial placentation as primitive (Grosser, 1909; Hill, 1932; Luckett, 1969, 1975), but others have inferred the primitive condition to be either endotheliochorial (Mossman, 1937; Martin, 1975) or haemochorial (Wislocki, 1929; Starck, 1956; Portmann, 1965; Potts, 1965). The main contenders for the ancestral condition in the literature have, in fact, been the two extremes, namely epitheliochorial versus haemochorial. Potts (1965) was led to comment: 'This was a unique situation in the history of comparative anatomy because nowhere else do opposing theories involve the complete reversal of the phylogenetic system.' The present author, by contrast, has concluded from the comparative evidence that the endotheliochorial type of placenta is ancestral for placental mammals.

The arguments in favour of the epithliochorial type of placenta as the ancestral condition for placental mammals are, at first sight, compelling. In the evolution of the basic placental pattern of embryonic development, it seems obvious that the earliest stage of retention of the developing embryo within the uterus must have involved simple apposition of embryonic and maternal tissues. This would hence have been a precursor of a non-invasive form of placentation. A modern parallel to this hypothetical stage in the evolution of placentation can be found among modern marsupials. In all marsupials, there is some degree of fetal development within the uterus, although a **shell membrane** is consistently present over the chorion throughout at least the major part of pregnancy. Despite the presence of this shell membrane, however, a rudimentary form of placentation is found in some marsupials. This is accompanied by a certain degree of invasion of the uterine lining and involvement of the allantois in addition to the yolk sac in the vascular system of the rudimentary placenta (e.g. in the bandicoot, *Perameles*). Nevertheless, it is common practice to cite an apparent series ranging from the egg-laying monotremes through the 'ovoviviparous' marsupials to the placentals as illustrative of the actual evolutionary sequence in the development of placentation. Once again, this is a case of arrangement of living species on a *Scala naturae* and it should be remembered that this approach can be very misleading (see Chapter 3). The monotremes and marsupials have, after all, undergone independent evolution over very long periods of time since their divergence from a common mammal-like reptilian stock. Particularly in view of the diversity of reproductive characteristics among modern marsupials (Tyndale-Biscoe, 1973), it is highly questionable to assume that all modern marsupials have retained a basic pattern of maternal–fetal relationships that is not only consistently more primitive than that in all modern placentals (including insectivores), but has not even changed significantly during more than 100 million years of evolution.

There are, in fact, several indications that marsupial reproduction is highly specialized and does not directly reflect the condition present in the common ancestral stock of marsupials and placentals. For instance, in marsupials, almost without exception, the oestrous cycle is longer than the gestation period, whereas this is never the case among placentals. Marsupials generally seem to have a long-phase oestrous cycle with spontaneous ovulation and an obligatory luteal phase during which progesterone is produced by a corpus luteum (Tyndale-Biscoe, 1973). If, following the logic outlined in Fig. 9.3, the luteal phase represents a period of preparation for and maintenance of an eventual pregnancy, it does not really make sense if the pregnancy itself is very short in comparison to the luteal phase. This might, then, suggest that gestation periods have actually been reduced during marsupial evolution. As noted above, female marsupials differ from placentals in having paired vaginal tracts rather than a single vagina in the midline. (This is associated with the fact that marsupials have a different alignment of the ureters. In males, this is responsible for the typical marsupial condition shown in Fig. 9.1, with the vas deferens not looping over the ureter as in placentals. In females, the ureters pass through the gap between the lateral vaginae and thus preclude the midline fusion found in placentals.) This unusual condition may well have favoured the evolution of increasing emphasis on lactation, rather than intra-uterine sustenance, to supply the needs of the developing offspring of marsupials. Increasing refinement of this adaptation would have permitted birth to take place at an earlier stage, as the sheltering function of the uterus was replaced by a sheltering function of the pouch (marsupium).

Contrary to the implication of the term 'marsupial', and contrary to popular supposition, not all marsupial species have pouches and the variety of patterns of formation and orientation of pouches suggests a multiple origin. Clearly, evolution of pouches and evolution of birth of offspring at a very vulnerable stage of development must have taken place in parallel and it seems reasonable to conclude that ancestral marsupials probably produced rather better developed young after somewhat longer gestation periods. Therefore, the type of placentation in modern marsupials should not be accepted unhesitatingly as a model for tracing the evolutionary development of placentation in the ancestral placental mammals. It is also necessary to consider the possibility, raised by Tyndale-Biscoe (1973), that vivipary evolved separately in marsupials and placentals and that

any similarities in adaptations for intrauterine development of offspring are accordingly the product of convergence. Whatever the case may be, it is simply not justifiable to equate placentation in ancestral placentals with that in modern marsupials.

A potentially far more convincing argument for the identification of epitheliochorial placentation as primitive for mammals generally was proposed by Luckett (1975). He has pointed out that placental mammals with the epitheliochorial type of placenta all share a series of other features of fetal membrane development. This, he states, is true of strepsirhine primates, hoofed mammals (artiodactyls and perissodactyls), dolphins and whales (cetaceans) and pangolins (Pholidota). The features concerned are, most notably, the orientation of the embryonic disc at the time of implantation and the relationships of the yolk sac and allantois. In fact, Luckett is not quite correct in stating that all mammals with epitheliochorial placentation share the same basic set of characteristics in fetal development. Most mammals, including hoofed mammals and pangolins, have an **antimesometrial** orientation of the embryonic disc of the blastocyst at implantation (i.e. the embryonic disc is located on the opposite side of the uterus to the mesometrium from which the uterus is suspended within the abdominal cavity). By contrast, strepsirhine primates (like haplorhines) have an **orthomesometrial** orientation of the embryonic disc (i.e. the embryonic disc is in a lateral position with respect to the mesometrium). There are also other differences of detail between epitheliochorial placentae in different mammalian groups (Steven, 1975b). However, it remains true that in many features strepsirhine primates are similar to other placental mammals with epitheliochorial placentation: the blastocyst attaches by means of a paraembryonic trophoblast (i.e. a trophoblast that lies laterally with respect to the embryonic disc); the amnion develops by folding; the yolk sac is prominent and well developed during the initial stages of pregnancy (choriovitelline placentation); and the allantois develops as a proper sac to produce the definitive placenta (chorioallantoic placenta). Luckett argues that such a set of shared similarities in fetal membrane development indicates that the epitheliochorial type of placenta is the primitive form for placental mammals, as multiple convergence would not have generated such uniformity. By contrast, placental mammals with haemochorial placentation have a variety of apparent specializations that differ from group to group, suggesting independent evolution of the haemochorial placenta type several times during the adaptive radiation of mammals. Luckett (1975) argues that the same applies to placental mammals with endotheliochorial placentation; there is no uniform shared pattern suggestive of retention of the ancestral condition. He therefore concludes that endotheliochorial and haemochorial types of placentation have arisen in parallel numerous times in the evolution of placental mammals, developing from an ancestral epitheliochorial condition.

Several objections can be raised against Luckett's interpretation. There is, first of all, the general point that no living mammalian species is likely to possess placentation as primitive as that present in ancestral placentals. Yet Luckett (1975) tends to imply that the essential features of placentation in strepsirhine primates, hoofed mammals, cetaceans, and pangolins have remained unchanged during the adaptive radiation of mammals. His cladogram of primate relationships based on features of fetal development indicates that lemurs and lorises are primitive in every single feature of their placentation. Yet the possibility of mosaic evolution, with some features remaining primitive and other become specialized, is present even with mammals that have epitheliochorial placentation, despite the presence of numerous shared characteristics. Further, the possibility of convergent evolution should not be underestimated. Certain mammals with endotheliochorial placentation, such as carnivores and tree-shrews, in fact share with strepsirhines, hoofed mammals, cetaceans and

pangolins all of Luckett's inferred primitive features of fetal development other than the character of non-invasiveness (i.e. paraembryonic blastocyst attachment: formation of the amion by folding; choriovitelline stage of placentation; and formation of a large, vesicular allantois). Evidently, then, it is possible for this set of apparently primitive features to exist regardless of whether the placentation is invasive (endotheliochorial) or non-invasive (epitheliochorial).

Further clarification of this issue has been provided by Kihlström (1972), who found that gestation periods – relative to maternal body weight – are consistently larger in mammals with epitheliochorial or highly invasive (villous) haemochorial placentation than in those with an intermediate degree of invasiveness of the placenta (endotheliochorial or labyrinthine haemochorial). This is crucial, because there is good evidence (see later) that mammals with long gestation periods (relative to body size) are generally less primitive than those with short gestation periods. Relatively short gestation periods are associated with altricial neonates (Fig. 9.12), which seem to be primitive for placental mammals. Further, the largest relative brain sizes among living placental mammals occur without exception in precocial species with relatively long gestation periods. The dolphins, which are the mammals closest to humans in terms of relative brain size (Fig. 8.6), in fact have long gestations, precocial offspring and non-invasive, epitheliochorial placentation. It therefore seems inherently more likely that a relatively primitive type of placentation among modern mammals would be associated with a relatively short gestation period and primitive altricial neonates. Given that epitheliochorial placentation is almost entirely restricted to mammals with precocial offspring, it is difficult to accept that this type of placentation is primitive. (The only apparent exception is the American mole, *Scalopus*, which has epitheliochorial placentation, altricial offspring and a relatively short gestation period.) Mammals with short gestation periods and altricial offspring typically have placentae with an intermediate degree of invasiveness (endotheliochorial or labyrinthine haemochorial), which are more likely to resemble the ancestral condition for placental mammals.

If epitheliochorial and haemochorial placentae have been derived from an ancestral endotheliochorial type during the evolution of placental mammals, multiple convergence has obviously taken place for both types of placenta. Although such widespread convergence may initially seem improbable, examination of Fig. 9.7 reveals two important facts that are often overlooked. First, mammalian species – such as haplorhine primates – that have a localized invasive placenta also have a large area of 'free' chorion. As a general rule among placental mammals, the more invasive the placenta, the smaller the proportion of the chorion directly involved in the placenta proper. Second, the degree of development of uterine glands during pregnancy increases with decreasing invasiveness of the placenta. Hence, there would seem to be some kind of trade-off between close approximation to maternal blood in a localized, invasive placenta and absorption of uterine milk over a large area of the chorion. Although the 'free' area of chorion is not usually directly involved in a recognizable placental attachment in forms with invasive placentation, it is inevitably brought close to the uterine wall at least during the later stages of pregnancy and (provided that it is vascularized) can potentially absorb material produced by the uterus. Accordingly, it is possible to imagine an ancestral condition for placental mammals in which the special functions of a localized, moderately invasive placenta were combined with a general absorptive function of a vascularized 'free' area of chorion, which may or may not have been attached to the uterine wall. The free chorion would have been particularly concerned with the absorption of uterine milk, discharged by the uterine glands into the uterine cavity, and may have borne specialized areas that were forerunners of the chorionic vesicles that modern lemurs and lorises have. An approximate model

for this condition is perhaps provided by the modern European mole (*Talpa*), which has endotheliochorial placentation and yet has chorionic vesicles, showing that these two features can occur together.

Hill (1965) reported that part of the chorion not involved in the twin discoidal placentae of tree-shrews (*Tupaia*) bears a few small local depressions (crypts). This part of the chorion is located opposite the mesometrium and is still vascularized by the yolk sac when the latter has been displaced from the placentae by the allantois (see Fig. 9.8). It should be noted that in *Tupaia* no uterine glands open on the placental sites themselves, so uterine milk is necessarily absorbed by non-placental areas of chorion. Hill interpreted the chorionic crypts of *Tupaia* as vestigial structures, retained from an ancestor with epitheliochorial placentation and fully developed chorionic vesicles; but they could equally well be forerunners of such vesicles. As Luckett himself has noted (1980c), tree-shrews have a set of features that may be primitive for placental mammals generally (antimesometrial orientation of the embryonic disc at implantation; blastocyst attachment at the paraembryonic pole; large yolk sac with choriovitelline stage of placentation; and large allantoic vesicle). It may therefore be that in their placentation, as in many other features, tree-shrews are relatively close to the ancestral condition for placental mammals. Tree-shrews have short gestation periods (Fig. 9.9) and altricial offspring (see Fig. 9.13) and it is therefore likely, following the arguments above, that their placentation is generally more primitive than in living primates. Although tree-shrews do have some obvious specializations (e.g. development of double placental sites and placentae), they would fit an ancestral condition for placental mammals in which a localized, endotheliochorial placental area was supported by a general absorptive function of the free chorion. Divergent specialization from such a condition, in numerous lines of mammalian evolution, could then have led either to emphasis on direct diffusion between maternal and fetal blood (haemochorial placenta) or to emphasis of absorption of uterine milk (epitheliochorial placenta).

Luckett (1975) has argued that evolution of an epitheliochorial placenta from a more invasive (e.g. endotheliochorial) ancestral condition violates the principle of the 'logical antecedent' (Chapter 3). However, this is not the case if epitheliochorial placentation involves the development of specialized features such as chorionic vesicles (which are not identical between mammalian groups and may have evolved convergently). Further, Hubrecht (1908) convincingly argued that the logical antecedent of simple apposition of the chorion against the uterine lining occurred well before the origin of the ancestral placental mammals.

There is actually some embryological evidence to suggest that the strepsirhine primates evolved from an ancestral stage with a greater degree of invasiveness of the placenta. The presence of invasive characteristics of the placenta at an early stage of pregnancy was first reported for the dwarf bushbaby (*Galago demidovii*) by Gérard (1932). Although questioned by Hill (1965), this was subsequently confirmed by Luckett (1974a). A temporary invasive giant cell trophoblast has also been reported for the lesser bushbaby (*Galago senegalensis*) by Butler (1964). But the isolated occurrence of such transitory invasiveness of the placenta in certain *Galago* species could well be a secondary specialization and has been interpreted as such by Hill (1965) and by Luckett (1974a, 1975). Were this condition to represent a primitive condition, however, one would reasonably expect to find it in other strepsirhine primates, particularly among the more primitive representatives such as the mouse and dwarf lemurs (Cheirogaleidae). This expectation was initially discounted by Luckett's report (1974b) that the lesser mouse lemur (*Microcebus murinus*) lacked any sign of invasiveness in its placentation, thus reinforcing his interpretation of the condition in the two *Galago* species as a secondary specialization.

More recently, however, study of a new sample of gravid uteri of *Microcebus murinus*

collected by the author in Madagascar, together with material originally collected by H. Bluntschli, has demonstrated that the lesser mouse lemur not only has an initial invasive attachment comparable to that of *G. demidovii* but actually retains an endotheliochorial ('syndesmochorial') zone in its definitive placenta (Reng, 1977; Strauss, 1978a, b). This new evidence neatly complies with the prediction that residual endotheliochorial placentation would be found among the more primitive strepsirhines. It can also be predicted that similar conditions are present in the remaining cheirogaleid species and, perhaps, in the variegated lemur (*Varecia variegata*), which has some apparently primitive features in other aspects of its reproduction (e.g. in having a relatively short gestation period, Fig. 9.10, and in producing multiple litters in association with nest-use – see later). In fact, Benirschke and Miller (1982), in a valuable survey of the mature placenta in primates, reported that in *Varecia* there is an intimate involvement of maternal tissue in the placenta and perhaps even some decidual property.

It has, on occasions, been suggested that a highly invasive (haemochorial) placenta is required for the development of a large brain relative to body size (e.g. Leutenegger, 1973; Goodman, 1961), because of the dependency of the developing brain on a rich supply of oxygen, and is hence more advanced. (During human birth, fetal deprivation of oxygen through obstetric accident can rapidly lead to degeneration of the brain tissue.) Among primates, this suggestion is apparently supported by the fact that haplorhines typically have relatively larger brains than strepsirhines (see Fig. 8.5), which could conceivably relate to the marked difference in placentation. However, outside the order Primates, the largest brain size relative to body size known among placental mammals is that of the dolphins (exceeding the relative size typical of monkeys and apes and approaching that of *Homo sapiens*; see Fig. 8.6). As noted above, cetaceans such as the dolphin have epitheliochorial placentation. Hence, there is no obvious conflict between the possession of epitheliochorial placentation and the development of a relatively large brain. This conclusion is borne out by the existence in certain strepsirhine primates (e.g. the aye-aye, *Daubentonia madagascariensis*) of relative brain sizes falling within the simian range (Chapter 8). In addition, Sacher and Staffeldt (1974) demonstrated that neonatal brain size in placental mammals is broadly related to the gestation period without reference to the type of placentation. Therefore, an explanation for the generally smaller relative brain sizes of strepsirhine primates, compared to haplorhines, must be sought elsewhere and is probably attributable to lower basal metabolic rates – as is the case with their smaller neonate sizes.

The manner in which the embryonic membranes become supplied with blood vessels (vascularized), thus ensuring the flow of nutrients to the embryo, provides a further source of information about the evolutionary radiation of primates (see Fig. 9.3). In the enclosed macrolecithic (large-yolked) egg of modern reptiles and birds, the yolk material inside the yolk sac provides the unique source of nutrients for the embryo. This was most probably the ancestral condition for the terrestrial tetrapod vertebrates in general. The retention of the egg within the female tract in the evolution of viviparous mammals was presumably followed by the progressive reduction of the yolk and its replacement as a nutrient source by maternal uterine products. It is therefore likely that the function of the blood vessels of the yolk sac was gradually converted from that of transporting material from inside the sac to that of carrying material passing through the chorion from the outside. Accordingly, in most marsupials, where the developing egg enclosed in its shell membrane is simply retained in the uterine cavity without formation of a true placenta, it is the yolk-sac vessels that typically perform the function of taking up nutrients emanating from the uterus wall. In placental mammals generally, however, the function of the yolk-sac blood vessels is supplemented or replaced at some stage by vascularization of the allantois. In reptiles and

birds (and by implication in their common ancestor), the allantois functions as a reservoir for accumulated nitrogenous waste products. In placental mammals, such waste products diffuse into the maternal blood system and are thus discharged through the mother's excretory system. As a result, the allantois must have become available for an alternative function – that of supplying the embryo with nutrients – during the evolution of placental mammals.

In fact, the allantois is also involved in the embryonic circulatory system in certain marsupials (e.g. bandicoots, *Perameles*). It is a moot point whether this rare occurrence of chorioallantoic circulation in marsupials represents an independent development parallel to that in the ancestral placental, or whether in the earlier common ancestor of marsupials and placentals the allantois was already involved in the nutrient supply of the embryo. If, as suggested above, there has been a general evolutionary trend towards reduction of the marsupial gestation period, the absence of a chorioallantoic circulation in most marsupials could be attributable to this influence. Be that as it may, it is a general rule for placental mammals that vascularization of the yolk sac (if it occurs at all) typically precedes vascularization of the allantois (or some secondary development thereof). An apt illustration of this is provided by the living tree-shrews (Hill, 1965; Luckett, 1969, 1980c), in which the two discoidal placentae of each embryo are served initially by choriovitelline blood vessels, but these are replaced by chorioallantoic vessels about half way through pregnancy (Fig. 9.8). As noted above, the yolk sac does not entirely lose its role in supplying the tree-shrew embryo with nutrients, despite this development, because its blood vessels apparently act thereafter to transfer nutrients passing from the uterine wall to regions of the chorion that are not involved in the discoidal endotheliochorial placental attachments.

In strepsirhine primates, vascularization of the yolk sac and (subsequently) of the allantois takes place according to the basic pattern hypothesized for the ancestral placental mammal, with the exception that the vascular role of the yolk sac itself disappears part of the way through pregnancy (in contrast to the condition in tree-shrews). In this respect, then, strepsirhines can be regarded as relatively primitive (Luckett, 1975). With haplorhine primates (including tarsiers), in direct contrast, the role of the yolk sac in placentation is completely suppressed and the allantois develops only as a rudimentary stalk (Fig. 9.7). (The spider monkey, *Ateles*, appears to be exceptional in that the allantois does develop as a proper vesicle – Benirschke and Miller, 1982). It would seem that vascularization of the chorion in haplorhines is usually achieved through precocious development of allantoic blood vessels in the absence of any choriovitelline vessels. In effect, haplorhine primates pass directly to a chorio-allantoic form of embryonic circulation at a very early stage of pregnancy. In this feature, haplorhines are unusual among placental mammals in general and there is consequently little doubt that this represents a significant specialization peculiar to this group of primates.

Naturally, there are many finer details of primate placentation that render the overall picture extremely complex; particularly valuable surveys have been provided by Hill (1932), Mossman (1937) and Luckett (1974a, 1975). Hill, in particular, made use of these fine details to establish a phylogenetic scale among living primates, passing from strepsirhines through tarsiers to New World monkeys and on through Old World monkeys and apes to humans. In this sequence, he saw a gradual increase in the invasiveness of placentation (essentially based on a shift from epitheliochorial to haemochorial placentation, with increasing degrees of invasiveness occurring in the latter among haplorhines), accompanied by progressively earlier development of the chorioallantoic circulation. In human beings, for instance, the blastocyst embeds in a small crypt in the uterine wall (interstitial implantation) and the chorioallantoic circulation is established with particular rapidity.

But analysis of developmental biology in general, as outlined above, indicates that this apparent phylogenetic scale may be largely illusory. First, it would appear that the development of epitheliochorial placentation in strepsirhine primates can be explained as a divergent specialization away from a hypothetical endotheliochorial type in the ancestral primates, rather than as a purely primitive condition. Second, it should be emphasized that Hill's sequence, traced from one group of living primates to another, is accompanied by an increase in average body size from strepsirhines to haplorhines and among haplorhines. It has now been demonstrated by Rudder (1979) that as body size increases in primates basic geometrical influences render it necessary for the placenta to become more effective as a collector of maternal nutrients – for example, by increase in surface complexity. This is to be expected on theoretical grounds as the metabolic turnover of the embryo overall is likely to be matched to the basal metabolic rate of the mother (i.e. to the three-quarters power of maternal body weight; see Martin amd MacLarnon, 1988) whereas the surface area of the chorion is likely to increase according to the normal surface relationship (i.e. with the two-thirds power of maternal body weight). As maternal body weight increases from one species to another, the increasing deficit between fetal metabolic turnover and chorionic surface area must be met by increasing effectiveness of the maternal–fetal nutrient exchange system. Hence, the progressively earlier establishment of the chorioallantoic circulation and the more invasive character of the haemochorial placentation of larger-bodied haplorhines could be explained in terms of fetal metabolic requirements without recourse to the anthropocentric concept of the *Scala naturae*. In sum, there is no good reason for regarding the placentation of any living primate species as representative of an ancestral stage.

Special features of pregnancy and birth

Before passing on to the related questions of postnatal development and overall reproductive strategies in primates, there are some special features of primate pregnancy and birth that deserve mention. The first concerns an unusual form of placentation that occurs in the marmosets and tamarins (family Callitrichidae), in association with the characteristic occurrence of twins. Although the twins in each case are normally developed from separate fertilized eggs (i.e. dyzygotic twins), there is fusion of the placental membranes to the extent that the twins share a pair of discoid placentae and a common placental circulation. In fact, Benirschke and Layton (1969) reported fusion at the blastocyst stage for the golden lion tamarin (*Leontopithecus rosalia*). This unexpected specialization, which is apparently unique among placental mammals, requires some explanation. The Callitrichidae are, in any event, unusual among simian primates in giving birth to twins as a rule, even though they have a simplex uterus, which is indicative of an adaptation for production of a single neonate at a time. As marmosets and tamarins also resemble other simians in having a single pair of teats (see later), there is a distinct suggestion that their ancestor (i.e. that giving rise to all simians) produced only a single infant. Following Leutenegger's suggestion (1973), one can imagine that there has been a reduction in body size during the evolution of the Callitrichidae generally and that this has been accompanied by an increase in litter size as would be expected within the reproductive strategy of a small-bodied mammal (see later). Perhaps the pre-existence of a simplex uterus, derived from the common ancestor, in some way favoured the development of a single placental system in spite of the presence of twins in the uterus. Certainly, this unique specialization must have been favoured by some considerable selection pressure, as the sharing of a placental circulation by one male and one female embryo (as occurs in half the cases) must present peculiar problems in the hormonal control of sexual differentiation, especially in brain development (A.F. Dixson, personal communication).

It is noteworthy that the unusual Goeldi's monkey (*Callimico goeldii*), which is generally intermediate between typical New World monkeys (Cebidae) and the marmosets and tamarins in numerous features, is larger in body size than most true marmosets and tamarins and gives birth to only a single infant, like the Cebidae. It is hence reasonable to attribute the evolution of reproductive peculiarities among the Callitrichidae to relatively recent reduction in body size following divergence from the common ancestor of the South American monkeys.

Another important characteristic with respect to gestation in mammals generally is the rapidity with which a female may return to reproductive condition after giving birth. In many relatively unspecialized mammals, such as tree-shrews, the female may come into oestrus immediately (postpartum oestrus) and thus initiate gestation of the next litter while suckling the previous one (Martin, 1968b). As a general rule, postpartum oestrus of this kind is lacking among primates. In most strictly seasonally breeding species (e.g. most Malagasy lemurs, the potto, squirrel monkeys and the Barbary macaque), a female may produce only one infant a year. Postpartum oestrus in these species is thus absent, as all primate gestation periods are less than a year in duration. In primate species that breed all the year round, there is usually a delay of several weeks at least before a female returns to breeding condition even when the infant fails to survive. If the infant survives, so that lactation progresses normally, there is typically a fairly long interval before the mother returns to breeding condition. However, small-bodied primates that produce several infants each year through repeated breeding (e.g. *Microcebus murinus*, *Galago senegalensis* and at least some callitrichid species), have a postpartum oestrus. In the lesser mouse lemur (*M. murinus*), postpartum oestrus is observed in captivity when the neonates fail to survive, but not when the infants are suckled to weaning by the mother (Martin, 1972c). However, in the lesser bushbaby (*G. senegalensis moholi*) a postpartum oestrus seems to be typical under natural conditions even when the infants survive, as the two breeding peaks that span the wet season (Bearder and Doyle, 1974) are separated by a period equivalent to the gestation period (about 4 months). A similar situation has been observed in captive female common marmosets (*Callithrix jacchus*), which will typically show oestrus about 10 days after giving birth and can conceive while suckling the twins from that birth (Hearn, 1977). In all of these cases, postpartum oestrus (whether or not it depends on the loss of the litter) can be interpreted as an adaptation maximizing the reproductive potential of females. It remains to be seen whether marmosets and tamarins under natural conditions have a regular postpartum oestrus despite the establishment of lactation, but the observation of this capacity in captive *C. jacchus* certainly indicates that this callitrichid species, at least, is adapted for relatively rapid breeding.

Finally, a brief note should be added on the **timing of birth** in primates. Primates typically give birth during the inactive phase of the daily cycle (A. Jolly, 1972b, 1973). In other words, nocturnal primates usually give birth during the daytime, whereas diurnal primates typically give birth at night. Such timing of birth may well occur in mammals generally. But it is particularly necessary in a largely arboreal group of mammals in which the mother will usually give birth in the trees and must do so discreetly to avoid attracting predators. Interestingly, this restriction of birth timing to the inactive phase of the daily cycle has been retained in *Homo sapiens*; natural human births typically occur during the night, most frequently during the early hours of the morning.

POSTNATAL DEVELOPMENT AND PARENTAL CARE

The state of the infant at birth itself provides vital clues to the evolution of reproduction in the order Primates. A major contribution to this topic was made by Portmann (1939, 1952, 1962), who recognized two distinct categories of mammalian neonates, based on their degree of development at birth. Some mammalian infants are relatively poorly developed at birth (**altricial**) whereas others are quite well developed (**precocial**); Table 9.3. Marsupial neonates generally can be taken as an extreme case of the altricial type; placental neonates may be

of either type, depending on the taxonomic group involved. As Portmann pointed out, the distinction between altricial and precocial neonates would seem to be particularly significant because, as a general rule, each mammalian order or at least suborder is characterized by one type of neonate. For example, lipotyphlan insectivores, carnivores and most rodents (myomorphs and sciuromorphs) typically produce altricial infants, while ungulates (hoofed mammals), cetaceans (whales and dolphins), hystricomorph rodents, elephant-shrews and primates typically produce precocial infants. The overall indication is that neonate type represents a characteristic of great importance in evolution (see also Eisenberg, 1981). This is not at all surprising, as neonate type is associated with numerous other important features such as gestation period (relative to maternal body size), litter size, relative brain size in the adult, and the duration of several life-history periods (see Table 9.3). Mammals with altricial infants tend to be small-bodied, small-brained, fast-breeding species, while the opposite is generally true of mammals with precocial infants. Portmann also noted a general relationship with nest-use, in that mammals with altricial neonates tend to make use of nests (*Nesthocker*; i.e. **nidicolous** species), while those with precocial neonates typically do not do so (*Nestflüchter*; i.e. **nidifugous** species). Although this is a useful generalization for placental mammals as a group, it does not apply consistently to primates. Among strepsirhine primates (mouse and dwarf lemurs, sportive lemurs, gentle lemurs, variegated lemurs, the aye-aye and bushbabies), nest-use is common, despite the uniform presence of precocial neonates. Indeed, mouse lemurs, some of the bushbabies and the aye-aye can construct quite complex leaf-nests, rather than simply making use of available tree hollows or tangles of dense vegetation. Conversely, some mammalian species with altricial young (e.g. certain microchiropteran bats) do not make use of nests and cannot therefore be classified as nidicolous. The question of the presence or absence of nest-use should not, therefore, be allowed to obscure

Table 9.3 Contrasting characteristics of mammalian species giving birth to altricial versus precocial offspring (after Martin, 1975)

Altricial type	Precocial type
1. Adult body size tends to be small	1. Adult body size tends to be large
2. Adults tend to be nocturnal in habits	2. Adults tend to be diurnal in habits
3. Adults typically use nests, at least when relatively young offspring are present	3. Adults do not usually use nests at any stage
4. Infants are born virtually naked. Ears and eyes are sealed by membranes for some time after birth. Offspring often have reduced homeothermic capacity initially, compared to adults	4. Infants are born with at least moderate covering of hair. Ears and eyes open at birth or soon afterwards. Offspring usually show well-developed homeothermic capacity, compared to adults, from birth onwards
5. The jaws and jaw-hinge are incompletely developed at birth and the middle ear is incompletely formed as a result. Teeth erupt some time after birth	5. The jaws and jaw-hinge are quite well developed at birth and the middle ear is accordingly quite advanced in formation. Teeth often erupt quite soon after birth, particularly in relatively small-bodied species
6. Infants typically have low mobility at birth	6. Infants are typically quite mobile at birth
7. The gestation period is short (relative to body size); litter size and number of teats are quite large	7. The gestation period is long (relative to body size); litter size and number of teats are small (often only one infant and one pair of teats)
8. Infants are born with very small brains, which grow considerably after birth	8. Infants are born with fairly large brains, which grow only moderately after birth
9. Lactation period, time taken to reach sexual maturity and longevity are all short (relative to body size)	9. Lactation period, time taken to reach sexual maturity, and longevity are all long (relative to body size)

the clear distinction between the separate character complexes of altricial and precocial neonates.

As the ancestral viviparous mammals were undoubtedly small-bodied (Chapter 4) the features listed in Table 9.3 suggest that the altricial complex would have been primitive, implying relatively short gestation periods and multiple litters. Weber (1950) and Portmann (1962) have provided support for this conclusion with the observation that in mammals with precocial neonates the eyelids fuse over the eyes of the embryo and the ears become sealed, as if in preparation for emergence into a nest, only to re-open again at about the time of birth (Fig. 9.12). In the precocial neonate, the events usually associated with a period of nest development seem to have been incorporated into the phase of uterine development. This contrasts with the situation in birds, where the altricial condition apparently arose as a secondary condition from a precocial ancestral type and where there is no closure of the eyelids during the embryonic life of precocial species (Portmann, 1962). The altricial complex can, accordingly, be reliably taken as the ancestral condition for the viviparous mammals generally. Nevertheless, one should not necessarily regard the altricial neonates of modern placental mammals (e.g. insectivores, rodents or carnivores) as necessarily reflecting the condition of ancestral viviparous mammals in every detail. It is quite possible that selection may have favoured an increase in reproductive potential in the evolution of certain mammalian groups, leading to a combination of reduced gestation periods and enhanced litter sizes. This

Birth

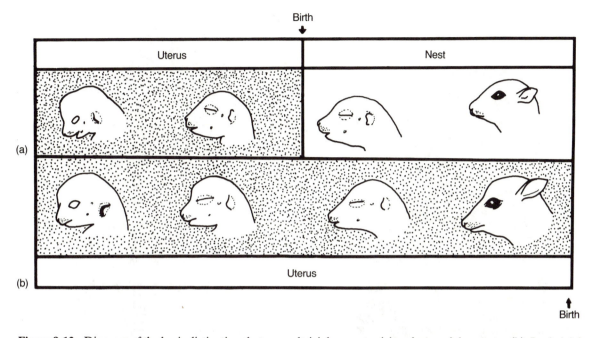

Birth

Figure 9.12. Diagram of the basic distinctions between altricial neonates (a) and precocial neonates (b). In altricial neonates, birth takes place after a relatively short gestation period (stippling) and the eyes and ears are still sealed at birth. Altricial neonates typically develop for some time in a nest prior to emergence into the outside world. In precocial neonates, the gestation period is relatively long and the nest-phase of altricial offspring is, so to speak, compressed into uterine development such that the eyes and ears are open at birth. Other distinctions between altricial and precocial neonates are listed in Table 9.3. (After Portmann (1962), reprinted from Martin (1983) courtesy of the American Museum of Natural History, New York.)

may well have been the case in the evolution of the murine rodents (mice and rats), which give birth to neonates at a remarkably early stage of development. In addition if there has been evolutionary reduction of gestation periods in marsupials (as suggested above), the extreme altricial condition of their neonates may well represent a secondary specialization. Certainly, one should not unhesitatingly assume that the placental mammals passed through an evolutionary stage in which extremely altricial infants were produced after a gestation period of only 2 weeks or so. After all, the reptilian ancestors of the mammals undoubtedly produced precocial infants that had to make their own way after emerging from the egg and the origin of altricial mammals must have depended on the parallel evolution of lactation to support the helpless neonates.

Given the far-reaching consequences of the distinction between altricial and precocial neonates, it is a fact of fundamental importance that the tree-shrews (without exception as far as is known) produce altricial offspring after a gestation period which is conspicuously short relative to maternal body weight (Fig. 9.10). Tree-shrew infants are born hairless and with the eyes and ears closed, in direct contrast to the situation found with typical primate neonates (Fig. 9.13). In this respect, they share a primitive feature present in most insectivores and they lack a distinct secondary specialization characterizing primate evolution. Once again, tree-shrews provide a better illustration of the hypothetical ancestral placental mammal than of the hypothetical ancestral primate.

The state of the human infant at birth deserves a special mention, because it might at first sight

(a)

(b)

Figure 9.13. Comparison of an altricial tree-shrew
neonate (*Tupaia belangeri* (a)) with a typical precocial
primate neonate (*Microcebus murinus* (b)). Note the
swollen belly of the tree shrew neonate, which has
been provided with a 48-h supply of milk.

appear that the human infant is of the altricial
type in terms of its relatively helpless state at
birth (compared, for example, with neonatal
chimpanzees). Close examination shows that the
condition of the human neonate is, in fact, a
specialized development from the precocial type
characteristic of primates (see also Chapter 8).
As with precocial neonates generally, the eyes
and ears of the human neonate are open by birth
and the typical covering of hair is actually present
just before birth, but is lost as a secondary

specialization *in utero*. (Premature human
infants are sometimes born with this covering of
hair, the **lanugo**, still intact – somewhat to the
consternation of any unprepared mother – and
there is a rare hereditary condition with which
this hair is retained into adult life.) The helpless
state of the human neonate resides mainly in its
poor locomotor development at birth and the
impression created by the secondary loss of body
hair. The explanation for this probably lies in the
rapid expansion of the human brain over a

relatively short period of evolution (see Fig. 8.14 and Martin, 1983). The human infant is unique among primates in that its brain continues to expand at the comparatively fast rate typical of fetal development for about 1 year after birth, rather than growing at a slower pace from birth onwards. Because of the great development of the human brain, the large size of the head of the human neonate has probably ruled out further extension of the gestation period (which would normally be expected to occur with increase in adult brain size; Chapter 8). Consequently, the human infant is, in terms of development of the central nervous system, still 'embryonic' during the first year of extra-uterine life and not surprisingly gives the impression of an altricial stage of development generally. Portmann (1941) has aptly referred to this as a 'secondarily altricial' condition.

Parental care

Parental (particularly maternal) care of the neonate after birth is particularly pronounced among primates. This is to be expected in any mammalian group characterized by precocial infants, as litter sizes are small and loss of a single infant represents a proportionately greater detriment to reproductive capacity. Most primate species produce only one infant at birth and multiple litters (of two or three infants) occur only in some mouse and dwarf lemurs, in the variegated lemur (*Varecia variegata*), in some bushbabies, and in marmosets and tamarins. In line with this, most primate mothers maintain extremely close contact with their infants throughout the first weeks or months of development. Most of the larger Malagasy lemurs, true lorises and all of the simian primates carry the infant, which clings independently to the adult's fur with its hands and feet. Usually, it is the mother that carries the infant, at least during the early period of postnatal life, but the marmosets and tamarins once again present a clear exception. In these primates, the infants are typically carried by a male (the father or a subadult male in the family group) virtually from

birth onwards, with the mother accepting the infants only for short periods of suckling from time to time. There is similar involvement of males in the transport of infants, but not until several days to a few weeks after birth, in other New World monkeys, such as the owl monkey (*Aotus trivirgatus*; Dixson and Fleming, 1981) and Goeldi's monkey (*Callimico goeldii*; Heltne, Wojcik and Pook, 1981). Male Old World siamangs (*Hylobates syndactylus*) start carrying infants at an even later stage (Chivers, 1974).

A variety of different relationships between mother and infant are shown by the prosimian primates, although in all cases the resting period (i.e. the daytime in most cases) is spent in close contact. In those prosimians that use nests of some kind (e.g. mouse and dwarf lemurs, variegated lemurs and bushbabies), the infant is characteristically carried in the mother's mouth when transported during the active period, as is the case with most mammals with altricial infants (insectivores, rodents and carnivores). Apparently, tarsiers also transport their infants in this way, at least occasionally (Le Gros Clark, 1924a). The size of the adult probably determines how the infant will be carried, mouth transport being favoured by small-bodied species. This suggestion is supported by the observation that the infant of the largest bushbaby species, *Galago crassicaudatus*, is carried on the mother's fur as well as in the mouth (Buettner-Janusch, 1964). Among the members of the loris group and in certain Malagasy lemurs (e.g. *Microcebus coquereli* and *Lepilemur mustelinus*), an interesting behaviour pattern known as 'baby-parking' has been observed. During nocturnal activity, the mother will transport her infant(s) either in her mouth or on her fur to the feeding area and will then leave the infant(s) clinging to a suitable branch in dense vegetation while moving around to feed. The infant(s) will remain immobile until collected by the mother and may be shifted to a new 'parking' site during the night. Obviously, this behaviour – like the continuous carriage of the infant on the fur shown by most primate species – depends on the infant's ability to cling independently from birth onwards and hence on the precocial state of the infant.

Lactation

Suckling of the infant by the primate mother (with the possible exception of the marmosets and tamarins) follows a pattern of feeding 'on demand', rather than 'on schedule' (Ben Shaul, 1962). In other words, while the mother is close to the infant (as is typical during the lactation period), the infant itself may determine the times of suckling by moving to the mother's teat. It does not usually depend on suckling visits that the mother makes according to some schedule determined by her own behaviour. As a result, suckling of primate infants tends to occur frequently over each 24-hour period. Incidentally, in mammals generally the number of teats that the mother has (in bilaterally symmetrical pairs) is broadly related to the number of infants in the typical litter (Schultz, 1948; Gilbert, 1986). Among the smaller-bodied prosimians (mouse and dwarf lemurs, bushbabies and *Tarsius*), several of which have litter sizes of two or three, it is common to find females with two or three pairs of teats. But the larger-bodied Malagasy lemurs (except *Varecia*, which has multiple births and multiple teats), the true lorises and the simians (where single births are usual) normally only have one pair of teats. The presence of only a single pair of teats in marmosets and tamarins (Callitrichidae) therefore provides additional support for the interpretation that multiple litters (twinning) represent a secondary specialization in this group of simians. With this exception, in primates generally the number of pairs of teats is broadly equivalent to the typical litter size. A similar relationship occurs in rodents (Gilbert, 1986).

Tree-shrews differ from primates in that the mother does not carry her infant(s) (usually one to three in each litter, corresponding to the presence of up to three pairs of teats). Further, those tree-shrew species that have been properly investigated to date (i.e. *Tupaia belangeri, T. minor* and *Lyonogale tana*) all show an extreme and rather peculiar case of suckling 'on schedule', with the infants kept in a separate nest and visited by the mother for suckling only once every 48 hours (Martin, 1968b; D'Souza and Martin,

1974; see also Fig. 9.13). As each suckling visit lasts only 5–10 minutes, any tree-shrew infant will accordingly spend a total of less than 2 hours in contact with its mother during the month-long nest phase, following which the infant is virtually independent of its mother for feeding purposes. Therefore, tree-shrews show the lowest level of mother–infant contact yet known among viviparous mammals. The closest parallel is, in fact, found in rabbits. A mother rabbit makes suckling visits at 24-hour intervals to offspring kept in a separate side-chamber of the burrow (called a 'stop') and mother–infant contact is strictly limited, just as in tree-shrews.

Tree-shrews thus provide a stark contrast to primates, in which mother–infant interaction is so pronounced as to be one of their most conspicuous behavioural features (Martin, 1968b). The developmental period from birth to weaning, the **lactation period** is also generally quite long in primates, relative to body size, whereas in tree-shrews the lactation period is comparatively short (Fig. 9.14), in common with other placental mammals that produce altricial offspring.

Composition of milk

The contrasting types of maternal behaviour seen in tree-shrews and primates are associated with equivalent contrasts in the composition of the milk. The three major nutrient components of mammalian milk – fat, protein and sugar – show characteristic patterns of concentration in direct association with maternal behaviour (Ben Shaul, 1962; Martin, 1968b). The differences between mammals in this respect depend essentially on the mother's suckling behaviour and on the rate of postnatal growth. In mammals that suckle 'on schedule', the milk tends to be high in fat. This is presumably because the offspring of such mammals are usually of the poorly developed, altricial type and are confined to a nest in which they must maintain their body temperatures during the mother's absence. By contrast, suckling 'on demand' is typically found in precocial mammals that are relatively active from birth onwards. In this case, maintenance of the

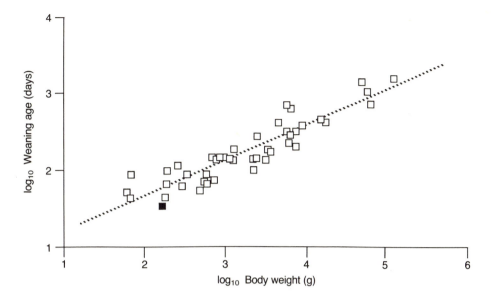

Figure 9.14. Logarithmic plot of the duration of lactation (birth to weaning) against body weight for 43 primate species (□) and for the tree-shrew *Tupaia belangeri* (■). The best-fit line is the major axis ($r = 0.92$). At least part of the scatter for primates is caused by problems in determining the precise ages of weaning, whereas for the tree-shrew the time of weaning is reliably known because of the unusual suckling behaviour of this species (Martin, 1968b). Although tree-shrews, in contrast to primates, produce poorly developed altricial offspring that must grow more extensively after birth, the lactation period of *Tupaia belangeri* is one of the shortest, relative to body size. (Data from: B.C.C. Rudder, personal communication; Doyle, 1979; Eisenberg, 1981; Harvey, Martin and Clutton-Brock, 1987; Izard, 1987; and author's own data.)

body temperature is not such a problem for the comparatively large-bodied, active young (particularly when they are carried on an adult's body, as in the case of most primates); but they do require sugar (rather than fat) to meet immediate energy demands. As a result, the milk of precocial mammals is usually rich in sugar and low in fat, although aquatic species (e.g. pinnipeds and cetaceans) exceptionally require a fat-rich milk in order to provide insulation for their offspring in the marine environment. In fact, there is a reasonably consistent inverse relationship between the contents of fat and sugar in mammalian milks generally (Fig. 9.15). Tree-shrews and primates differ markedly in this respect. Whereas tree-shrew milk is rich in fat and low in sugar, as would be expected for a mammal with altricial offspring that are suckled 'on schedule' at very long intervals, all primate

milks are characterized by high amounts of sugar and relatively low amounts of fat. As can be seen from Fig. 9.15, the mammals that are closest to primates with respect to the fat and sugar content of their milk are the perissodactyls (the horse group of hoofed mammals).

The rate of postnatal growth is fairly directly matched by the concentration of protein in the milk. Mammals that give birth to large litters of fast-growing altricial offspring produce high-protein milk, while mammals that give birth to individual, slowly growing precocial offspring produce low-protein milk. Once again, there is a stark contrast between tree-shrews and primates in that the former have a high-protein milk permitting very rapid postnatal growth of offspring, whereas the latter produce milks that are very low in protein. The protein content of the milk shows a consistent positive correlation

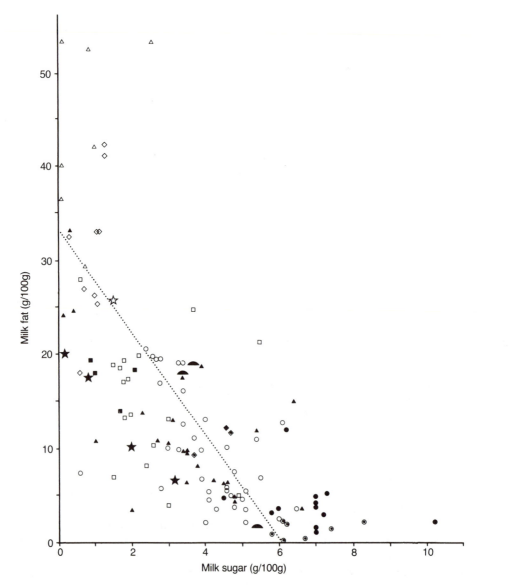

Figure 9.15. Plot of milk fat concentration against milk sugar concentration for mammals. The best-fit line is the reduced major axis. Key: ◗, bats; ★, insectivores; ☆, tree-shrew; ■, lagomorphs; □, rodents; ▲, carnivores; △, pinnipeds; ●, primates; ○, artiodactyls; ◉, perissodactyls; ◇, cetaceans; ◈, elephants; ◆, aardvark. (Reprinted from Martin, 1984, with kind permission of the Zoological Society of London and Academic Press, London.)

with the fat content of the milk across mammals generally (Fig. 9.16). The only major exception to the general trend is provided by aquatic mammals with precocial offspring (pinnipeds and cetaceans). These mammals, as noted above, have milks with an unusually high concentration

of fat because of the particular thermoregulatory problems associated with living in an aquatic environment, but the protein content of the milk is typically only moderate in association with the moderate rates of postnatal growth.

Primates are also somewhat unusual among

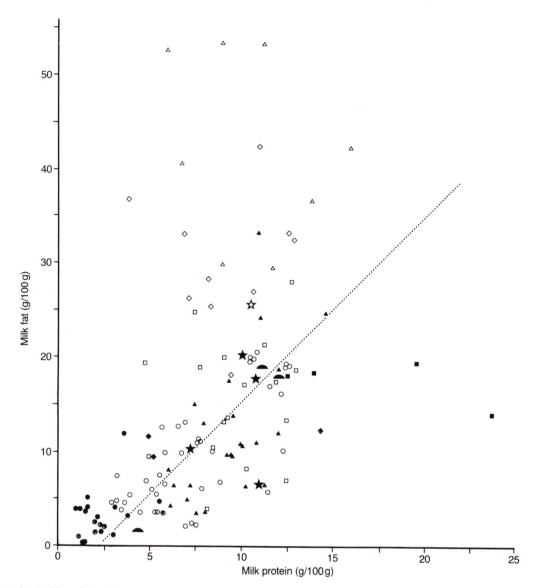

Figure 9.16. Plot of milk fat concentration against milk protein concentration for mammals (symbols as for Figure 9.15). The best-fit line is the reduced major axis. (Reprinted from Martin, 1984, with kind permission of the Zoological Society of London and Academic Press, London.)

placental mammals generally in terms of the energy content of milk produced per day. Because measurement of milk energy requires collection of the complete milk output per day, information on this aspect of lactation is naturally limited. However, an allometric plot (Fig. 9.17) shows quite clearly that the two primate species

that have been examined to date (plains baboon and humans) have a noticeably low milk energy yield per day in comparison to the other placental mammals that have been sampled. The only value that is smaller, relative to body size, than that of the baboon and humans, is that determined for the echidna (an egg-laying

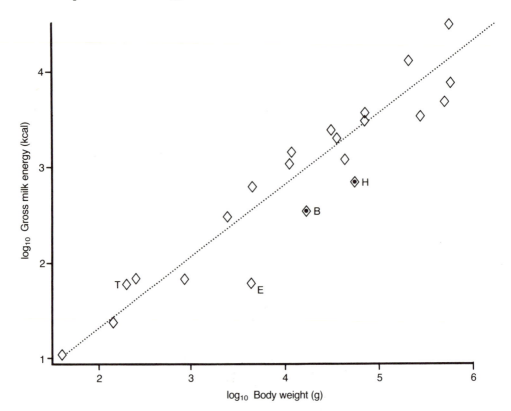

Figure 9.17. Logarithmic plot of gross yield of milk energy per day against body weight for a sample of 22 mammalian species. The slope of the best-fit line (major axis), determined with the exclusion of primates (negative outliers) and mammals domesticated for milk production (positive outliers), is 0.70. Key to labelled points: E, echidna (monotreme); B, plains baboon; H, human; T, common tree-shrew. (After Martin, 1984.)

monotreme), which is less well adapted for lactation than are placental mammals. The milk energy of the tree-shrew, in contrast to the two primates, lies in the upper range for mammals generally, relative to body size. The best-fit line in Fig. 9.17 has a slope of 0.70, which is compatible with the expectation that milk production might scale in a similar way to the basal metabolic turnover of the mother and thus conform to Kleiber's Law (see Chapter 4, Appendix), with an exponent value close to 0.75. In fact, in a more detailed analysis, Oftedal (1984) has shown that three separate 'grades' may be recognized among placental mammals, with an average exponent value of 0.74. In decreasing order of milk energy relative to maternal body size, the first grade consists of placental mammals with multiple litters (e.g. carnivores, rodents and lagomorphs), the second consists of hoofed mammals (ungulates) with single offspring and the third grade consists just of the plains baboon and humans. In other words, relative to her metabolic capacity, a primate mother devotes a smaller proportion of available energy to lactation than any other placental mammals so far investigated. This, in turn, is reflected by the fact that primates show slower postnatal growth in comparison to other mammals, once correction has been made for the effects of body size (Payne and Wheeler, 1968).

REPRODUCTIVE STRATEGIES

It has emerged progressively from this survey of

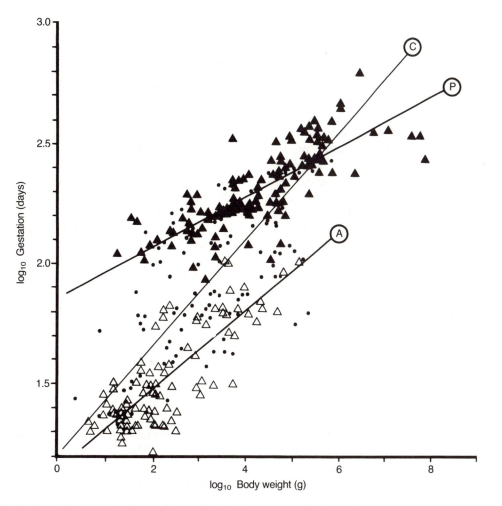

Figure 9.18. Logarithmic plot of gestation period against maternal body weight for a sample of 394 species of placental mammal. Key to best-fit lines: A, altricial mammals; P, precocial mammals; C, total sample. Mammals regarded as 'intermediate' (see text) are indicated by black dots. (From Martin, R.D. and MacLarnon, A.M., 1985. Reprinted by permission from *Nature* vol. 313 (no. 5999), pp. 220–3. Copyright © Macmillan Magazines Limited.)

primate reproductive characteristics, in the context of mammalian reproduction generally, that it is possible to recognize two broad complexes of mammalian reproductive parameters. The **altricial complex** is associated with a high reproductive potential, whereas the **precocial complex** is associated with a much more limited capacity for reproductive turnover. The tree-shrews fit into the first category, whereas primates belong in the second. Among primates,

the following features (most of which vary systematically with body size) can be listed as interrelated components of an overall pattern of low reproductive potential combined with intensified maternal care: (1) precocial neonates; (2) long gestation periods; (3) long oestrous cycles (typically associated with spontaneous ovulation); (4) the common absence of postpartum oestrus; (5) slow postnatal development; and (6) long lactation period. To these can be added the linked empirical observations that primates, in comparison to other mammals, reach sexual maturity relatively late and also have relatively long lifespans (Western, 1979).

The question of the length of the gestation period, relative to maternal body size, is of particular importance with respect to the distinction between altricial and precocial mammals. Indeed, many studies of the scaling of mammalian gestation periods have failed to take this distinction into account and have thus produced misleading results because of the problem of 'grade confusion'. In an analysis of the gestation periods of placental mammals, Martin and MacLarnon (1985) defined three separate categories. Mammals producing litters of three or more offspring in which the eyes do not open until at least 5 days after birth were regarded as clearly altricial. Mammals with an average litter size of less than 1.5, producing neonates in which the eyes are already open, were regarded as clearly precocial. Any remaining mammals, with litter sizes between 1.5 and 3 and/or with the eyes opening between birth and 5 days of age, were regarded as intermediates. As can be seen from Fig. 9.18, the slope for the best-fit line determined for the whole sample is considerably steeper (0.22) than the slopes of the individual lines determined for clearly altricial mammals (0.17) and clearly precocial mammals (0.11), respectively. An almost complete separation of clearly precocial from clearly altricial mammals in terms of gestation periods emerges once body size is taken into account as in Fig. 9.18.

Further analysis shows that some grade confusion is still present in Fig. 9.18, particularly among altricial mammals, and an appropriate slope for an overall best-fit line was found to be 0.10 (Martin and MacLarnon, 1985). Using this exponent value, it is possible to generate a histogram of residuals showing the relationship of individual points to the line. Deviations of residual values from zero indicate the degree to which the gestation period of any given mammalian species is shorter than (negative value) or greater than (positive value) expected

from the overall best-fit line (Fig. 9.19). The distribution of residual values in Fig. 9.19 is clearly bimodal, indicating a relatively sharp separation between altricial mammals, with relatively short gestation periods, and precocial mammals, with relatively long gestation periods (about four times longer, on average). Indeed, the 'intermediate' mammals defined above also conform to the bimodal distribution, as is also apparent from Fig. 9.18. Primates, almost all of which are clearly precocial by the definition given above, consistently show positive residual values for gestation period, indicating that all have long gestation periods compared to other placental mammals. Tree-shrews, which are altricial in most respects but are 'intermediate' by the definition given above because of their relatively small litters, have negative residuals and therefore have relatively short gestation periods. An order-by-order analysis of residual values indicates that primates, most artiodactyls (i.e. excluding pigs), perissodactyls, elephants, hyraxes, cetaceans and pinnipeds all have relatively long gestation periods. On the other hand, relatively short gestation periods are found in most insectivores, tree-shrews, carnivores, most rodents (sciuromorphs and myomorphs) and lagomorphs (Martin and MacLarnon, 1988). Thus, the relative lengths of mammalian gestation periods, as indicated in Figs 9.18 and 9.19, closely follow the accepted division between precocial and altricial mammals. It is hence possible to state unequivocally that there is a virtual dichotomy between altricial mammals adapted for the relatively rapid production of multiple litters and precocial mammals typically adapted for the slow production of individual, well-developed offspring.

Given that the precocial complex, with all of its ramifications, is a universal characteristic of living primates, it is highly likely that the common ancestor of primates had a similar complex of adaptations. There are equally good reasons for believing that the ancestral placental mammals had altricial characteristics. For instance, in precocial mammals there is transitory closure of the eyes and ears during gestation, suggesting that they are derived from an altricial

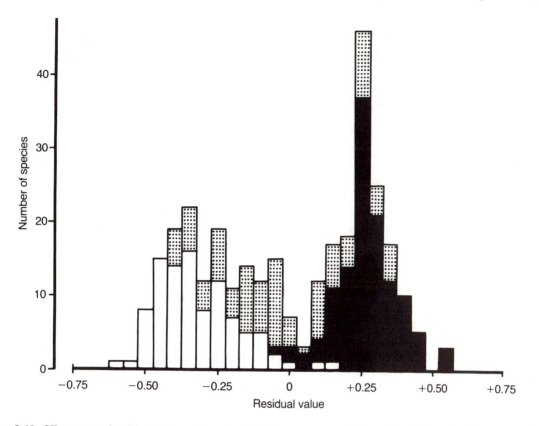

Figure 9.19. Histogram of residual values determined relative to an overall line of fixed slope 0.10 for the species shown in Fig. 9.18. There is an almost complete separation between clearly altricial mammals (white boxes) and clearly precocial mammals (black boxes); the 'intermediate' species (stippled boxes) also conform to the bimodal pattern. (From Martin, R.D. and MacLarnon, A.M., 1985. Reprinted by permission from *Nature* vol. 313 (no. 5999), pp. 220–3. Copyright © Macmillan Magazines Limited.)

ancestor. Moreover, the smallest relative brain sizes among living mammals are found in species that have altricial offspring, whereas the largest relative brain sizes are found in species with precocial offspring. Hence, one can view a hypothetical ancestral precocial state for the primates as a derived feature defining the group. Tree-shrews, which have retained the essentially primitive altricial complex, are excluded from primates by this definition. As noted above, there are several mammalian groups other than primates that have precocial characteristics, so the primates are by no means unique in being precocial. However, as Shea (1987) has pointed out, primates are relatively unusual in that they developed precocial features at a relatively small body size. It is in any case a significant feature of primate evolution that there must have been an early shift from altricial to precocial characteristics. It is therefore important to enquire why primates should have shifted to the precocial complex rather than retaining the (apparently primitive) altricial condition.

In general terms, one can refer to the overall combination of reproductive functions within the life history of an individual mammalian species as constituting its **reproductive strategy**. (Although the term 'strategy' has a somewhat anthropomorphic and teleological connotation, it has now become widely established in the literature on evolutionary biology and it does provide a very convenient short-hand expression for referring to the overall complex of reproductive parameters.) Obviously, for any mammalian species, some balance must be achieved in the allocation of energy between the parent's own maintenance requirements and the requirements for production and rearing of offspring. Taking the basic Darwinian paradigm of evolution through natural selection (see Chapter 3), it can be stated that, in the evolution of the reproductive strategy of any species, selection will favour a balance between individual maintenance and reproduction that allows for maximization of reproductive success relative to other individuals of the same species. Hence, selection will favour a high reproductive potential only in situations where production of a relatively large number of offspring in a short space of time might be accomplished without detriment to the parents (and hence to their overall reproductive capacity). Further, those situations must allow at least the possibility that individuals can thereby produce large numbers of surviving offspring.

It should be noted that, even in the relatively slow-breeding primates, each female has a total reproductive potential of 10–40 offspring, according to species (i.e. the maximal output of offspring per year, allowing for any limitation associated with lactation, multiplied by the reproductive lifespan in years). However, only two surviving offspring per female would be required (on average) for maintenance of a stable population. Thus, even primates generally have a considerable surplus reproductive capacity that is presumably partly geared towards offsetting the effects of predation and other causes of mortality, but is also doubtless favoured to some extent by the possibility that any individual in a population may contribute more than the average of two surviving offspring permitted under stable conditions. Obviously, mammals with an even greater total reproductive potential than that found in primates must live under environmental conditions favouring a greater allocation of energy resources to reproduction. In captivity, a female common tree-shrew (*Tupaia belangeri*) has the potential of producing around 100 offspring in a lifetime, and this greatly

exceeds the maximum attainable by even the most prolifically breeding primate species. More significantly, a female common tree-shrew can theoretically produce up to 15 infants in a year, whereas the theoretical maximum for rapidly breeding primates (e.g. *Microcebus murinus*, *Galago senegalensis* and *Callithrix jacchus*) is only four or five per female. In fact, most female primates can produce only one infant per year. Naturally, these theoretical maximum levels are rarely, if ever, reached in the wild; but it is important to bear in mind the total available reproductive capacity should environmental conditions allow reproductive output to increase.

Understanding of the evolutionary background to the distinction between slow-breeding and fast-breeding 'strategies' in mammals has been greatly enhanced by development of the model of **r- and K-selection** (MacArthur and Wilson, 1967; Pianka, 1970). This evolutionary model, which is applicable to organisms in general, can be simply explained with reference to the standard growth curve for an animal population (Fig. 9.20). When an animal population is in an early phase of expansion (i.e. during colonization or recolonization of an 'open' area of suitable habitat), food – and other density-dependent resources – will be superabundant and the population will increase in size geometrically. If population growth is unchecked by special factors (e.g. predation or parasitism), the form of the growth curve will be determined simply by the **intrinsic rate of population increase** (r), which depends on the standard reproductive parameters of the species concerned (e.g. litter size and interbirth interval). During this geometric phase of population growth, selection will favour maximal reproductive output, as resources are superabundant and individuals of a species will be selected on the basis of their capacity to breed rapidly. However, any expanding population will eventually attain a maximal level permitted by available resources in a given environment. This level is the **carrying capacity** (K) of the habitat concerned. Under conditions where a population exists at or near carrying capacity for any length of time, intraspecific competition for resources

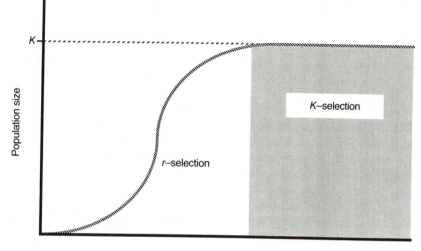

Figure 9.20. Simplified illustration of the principle of *r*- and *K*-selection. The hatched curve represents the growth in size of a given population, beginning with a phase of geometric increase and then levelling off as the carrying capacity (*K*) of the environment is reached. During the phase of geometric increase, the rate of growth of the population is primarily determined by the intrinsic rate of increase (*r*) of the species, which depends on basic reproductive parameters such as age at first breeding, litter size and interbirth interval. Species that commonly exist under conditions allowing population expansion will be subject to *r*-selection (promoting high reproductive potential), whereas species that commonly exist at or near carrying capacity will be subject to *K*-selection (promoting increased efficiency in the use of food resources).

will be intense and selection will favour those individuals that utilize resources most efficiently, establishing a finely tuned balance between allocation of energy for maintenance and energy expenditure on the production and rearing of offspring. Selection will also favour mechanisms that enhance the survival of the relatively small numbers of offspring that are produced.

MacArthur and Wilson (1967) coined the terms *r*-selection and *K*-selection to refer to these two different kinds of selection, depending on the predominant phase of selection, depending on the predominant phase of population growth shown by the population of any species (i.e. geometric growth or relative stability at carrying capacity). It is to be expected that species typically living under relatively stable environmental conditions would be *K*-selected, resulting in relatively low reproductive turnover and increased efficiency of use of food resources. By contrast, species in

relatively unstable environments (e.g. those with unpredictable climatic conditions or with harsh conditions at one time of the year predisposing to high seasonal mortality) or acting as colonizers would be *r*-selected, resulting in high reproductive turnover. One would therefore expect to find a predominance of *K*-selected species in environments such as tropical rainforest and a predominance of *r*-selected species in open-country habitats (with relatively low and unpredictable rainfall) or in high-latitude environments (with pronounced seasonal changes). It should also be noted that large-bodied animals, because of their greater degree of insulation from environmental fluctations, are more likely to be *K*-selected than small-bodied animals.

The concept of *r*- and *K*-selection has not passed unchallenged, especially because it has largely been invoked in a descriptive sense and has yet to be rigorously tested. An alternative model based on the concept of 'bet-hedging' has been proposed by Schaffer (1974). This model emphasizes the importance of age-specific patterns of mortality and predicts that fluctuating environments producing high mortality among juveniles will favour long-lived, slow-breeding individuals, whereas the occurrence of high mortality among adults will favour short-lived, rapidly breeding individuals. In some respects, then, the bet-hedging model stands in opposition to that invoking *r*- and *K*-selection. In an unpredictable environment, the outcome will depend on the pattern of age-specific mortality in the population concerned. In practice, it seems that individual cases can be cited in support of both models and it is undoubtedly true that the situation is not as simple as is indicated by the model based on *r*- and *K*-selection (see also Stearns, 1976, 1977; Richard, 1985; Ross, 1988). Nevertheless, this model does seem to fit the data on reproductive parameters for mammals quite well and it does provide a convincing basis for explanation of the overall reproductive strategy of primates. Available field data suggest that mortality is typically highest among young individuals in many mammalian species and there is therefore no reason to invoke the bet-hedging model in most cases.

There is, in fact, a remarkable degree of correspondence between the life-history parameters predicted by the *r*- and *K*-selection model and the divergent reproductive characteristics found in the altricial and precocial complexes (see Table 9.3). This, in itself, suggests that there is likely to be a connection between the universal presence of the precocial complex among primates and *K*-selection. As Pianka (1970) emphasized, *r*-selection and *K*-selection are relative terms and can be used only in a comparative sense. For example, the arthropods as a group would generally appear to be *r*-selected in comparison with the vertebrates as a group. Yet, among vertebrates, some types are apparently *r*-selected or *K*-selected in comparison with others. The mammals generally occur at the *K*-selected end of the spectrum for

vertebrates, but there is also a substantial range of variation among the mammals themselves. Overall, mammals with the precocial complex probably show some of the most extreme consequences of *K*-selection to be found among vertebrates and primates are a case in point. On the other hand, mammals with the altricial complex can be regarded as the products of *r*-selection, relative to precocial mammals.

It has already been established that primates probably originated as arboreal inhabitants of a tropical rainforest environment. It is therefore possible that a long period of adaptation to such relatively stable habitats was associated with *K*-selection, which established a basic pattern of low reproductive turnover combined with intensive parental care. This had a profound influence on all subsequent phases of primate evolution. By contrast, tree-shrews are obviously *r*-selected in comparison with the primates. Although they also inhabit tropical rainforests, there must be some fundamental distinction separating them from primates, reflecting a considerable period of evolutionary adaptation for high reproductive turnover. It would seem that tree-shrews, as a group, are primarily adapted for occupation of secondary forest areas and act as colonizers that can rapidly occupy new areas of habitat.

As a final point, it should be noted that *Homo sapiens* has the most extreme form of the slow-breeding strategy among primates. Odd as it may seem, in the light of modern world population problems, human beings have drawn-out life-history phases and low reproductive potential in comparison to all other primates. Indeed, modern population problems are to some extent the result of artificial intervention leading to shortening of natural life-history phases. For instance, bottle-feeding of infants reduces the natural period of infertility that is usually associated with lactation. Taking the natural length of life-history phases, once the effects of body size have been taken into account it emerges that human beings reach sexual maturity the latest and have the longest lifespan among primates generally. In so far as the *r*- and *K*-selection model is valid, it therefore follows that human beings are the most *K*-selected of all living primates (Rudder, 1979). This is an important point to be taken into account in framing hypotheses regarding the environmental conditions that were associated with human evolution. Whatever those conditions were, they also favoured adaptation for a very low reproductive potential.

Chapter ten

Primate locomotor patterns

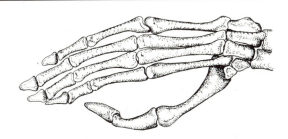

Locomotion, the sum of movements through the environment, is clearly a major component of the overall adaptation of any mammal. The postcranial skeleton is subject to selection pressures primarily related to locomotor behaviour, particularly as the energy costs of locomotion represent a significant proportion of a mammal's total energy budget. Locomotor energy is expended predominantly in the context of food-gathering; in simple terms, such expenditure must yield a 'profit' of food energy that will support the essential processes of maintenance, growth and reproduction. Locomotion is also important for other vital activities such as the avoidance of predators. Every mammalian species has what might be termed a locomotor 'strategy' that fulfils all of these basic functions and profoundly influences the structure of the postcranial skeleton. The central aim of this chapter is to identify a common pattern in the structure of the primate postcranial skeleton and to infer an ancestral locomotor 'strategy' that has influenced the adaptive radiation of primates.

Active locomotion is, of course, part of a continuum. Even when relatively immobile – for example, during feeding or resting – any primate species shows **postural behaviour** requiring appropriate muscular and skeletal adaptations. Prost (1965) refers to the overall combination of locomotor and postural behaviour as **positional behaviour,** whereas Ripley (1967) uses the term **total locomotor pattern** in the same sense. In principle, the skeletal morphology of any primate species must be adapted for its complete range of positional behaviour, rather than just for active locomotion, and any approach to the evolution of primate locomotion should take this into account (e.g. see Rose, 1973). At present, however, relatively little is known about primate postural behaviour and an integrated view of the evolution of primate positional behaviour remains an elusive prospect.

It is widely accepted that both locomotor behaviour and postural behaviour in primates are closely related to a general theme of adaptation to arboreal life (Jenkins, 1974a). Indeed, many authors have suggested that numerous features of primates, in addition to postcranial adaptation *per se*, owe their origin to a shift to arboreal habits in ancestral primates. As the earliest origins of primate adaptations represent the main concern of this book, special attention will be devoted to this 'arboreal theory' of primate evolution after a survey of the relevant evidence.

CLASSIFICATION OF PRIMATE LOCOMOTION

A useful step in the investigation of relationships between behaviour and morphology in primate locomotion is to construct a meaningful **classification** of locomotor behaviour. One of the most widely cited classifications of primate locomotor patterns, which built on earlier studies by Petter (1962) and by Ashton and Oxnard (1964), is Napier and Walker's (1967). In this classification, four major categories of locomotion are defined (Fig. 10.1): (1) vertical-clinging-and-leaping; (2) quadrupedalism; (3) brachiation; and (4) bipedalism.

In recognizing the **vertical-clinging-and-leaping** pattern, Napier and Walker made a particularly valuable contribution to our understanding of primate locomotion (see also Napier, 1967a). Drawing on field studies, particularly that by Petter (1962) of Malagasy lemurs, they noted that the locomotion of various prosimian primates is characterized by a pattern of clinging to predominantly vertical tree trunks and leaping between them. Such active leaping emphasizes the hindlimbs both for the provision of power and, typically, for the absorption of shock on landing (Fig. 10.1a). Vertical-clingers-and-leapers thus have especially well-developed hindlimbs and they commonly show bipedal hopping, rather than quadrupedal running, when moving along broad horizontal branches or across the ground. Vertical-clinging-and-leaping of this overall form is seen in representatives of all three modern prosimian groups. It occurs in sportive lemurs (*Lepilemur*), gentle lemurs (*Hapalemur*) and all members of the indri family among the Malagasy lemurs, in most bushbabies (*Galago* spp.) and in tarsiers (*Tarsius* spp.).

Quadrupedalism is, of course, the basic locomotor pattern of all terrestrial vertebrates, involving approximately equal emphasis on the forelimbs and hindlimbs. The type of locomotion is found in various forms in living primates and it

is noteworthy that it occurs in predominantly terrestrial species (e.g. baboons, *Papio* spp.; Fig. 10.1b) as well as in predominantly arboreal primates (e.g. many New World and Old World monkeys). The ringtail lemur (*Lemur catta*), the only modern prosimian to spend a significant proportion of its time on the ground, also has a predominantly quadrupedal form of locomotion.

In the pattern of arboreal locomotion termed **brachiation** (Fig. 10.1c), the forelimbs are emphasized as opposed to the hindlimbs in vertical-clinging-and-leaping. There is, unfortunately, some confusion over use of the word 'brachiation'. Originally, the term was used largely or exclusively for the ricochetal form of arm-swinging shown by members of the gibbon family (Hylobatidae), with the body passing through a free flight phase between successive hand-holds (see Trevor, 1963). However, following Keith (1899), many authors have equated brachiation with any form of suspensory locomotion involving an element of arm-swinging. Napier (1963) dealt with this problem by referring to the agile, ricochetal arm-swinging of gibbons as 'true brachiation' and to the more deliberate suspensory locomotion of great apes as 'modified brachiation'.

The last major category of primate locomotor behaviour is **striding bipedalism** (Fig. 10.1d), which is unique to *Homo sapiens* among modern primates and, indeed, among mammals generally. As was pointed out by Napier (1967b, 1971), bipedal locomotion in the broad sense (i.e. any form of movement involving just the hindlimbs) is by no means restricted to the human species, but habitual striding bipedalism is a distinctive human feature.

Napier and Walker (1967) divided their major categories of primate locomotor behaviour into several subtypes, and further refinements were introduced by Rose (1973). Table 10.1 summarizes the resulting overall classification of primate locomotion, with the incorporation of a few minor modifications (e.g. see Rollinson and Martin, 1981). As can be seen, most of the subdivisions relate to the category of 'quadrupedalism', which includes several varieties in addition to the basic distinction

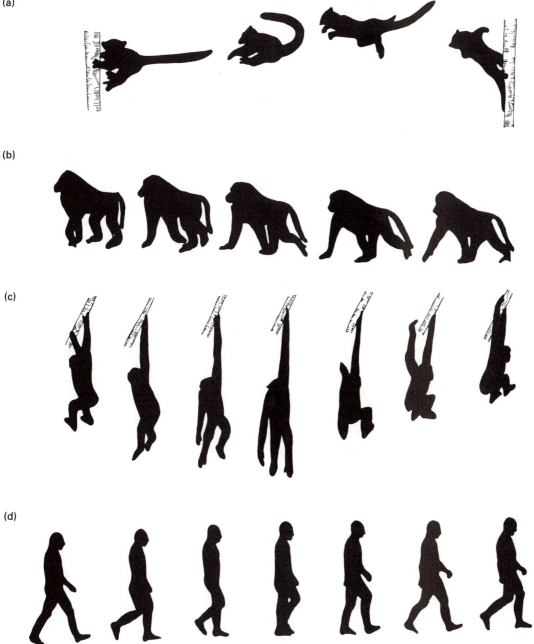

(a)

(b)

(c)

(d)

Figure 10.1. The four main types of locomotion of living primates:

(a) Vertical-clinging-and-leaping: Allen's bushbaby, *Galago alleni*. (After Charles-Dominique, 1977.)
(b) Terrestrial quadrupedalism: plains baboon, *Papio anubis*. (After Rollinson and Martin, 1981.)
(c) Brachiation: siamang, *Hylobates syndactylus*. (After Fleagle, 1974.)
(d) Bipedalism: modern humans, *Homo sapiens*. (After Napier, 1967a.)

Table 10.1 Outline classification of locomotor categories among living primates (revised from Rollinson and Martin, 1981)

Major category	Size group	Subcategory	Representative*	
Vertical-clinging-and-leaping	Small		(A)	*Galago†, Tarsius*
	Medium		(A)	*Avahi, Hapalemur, Indri, Lepilemur, Propithecus*
Arboreal quadrupedalism	Small	(a) Clawed	(B)	*Callimico, Callithrix, Cebuella, Leontopithecus, Saguinus*
		(b) Non-clawed, agile	(C)	*Cheirogaleus, Microcebus, Phaner*
		(c) Non-clawed, slow climbing	(D)	*Arctocebus, Loris, Nycticebus, Perodicticus*
	Medium		(E₁)	*Daubentonia, Lemur, Varecia*
			(E₂)	*Aotus, Callicebus, Cebus, Chiropotes, Pithecia, Saimiri*
	Large	(a) Branch sitting and walking	(F)	*Cercocebus‡, Cercopithecus, Macaca‡, Mandrillus*
		(b) Old World suspensory type	(G)	*Colobus, Nasalis, Presbytis, Pygathrix, Rhinopithecus*
		(c) New World suspensory type	(H)	*Alouatta, Ateles, Brachyteles, Lagothrix*
Terrestrial quadrupedalism	Medium	(a) Ground standing and walking (with digitigrady)	(J)	*Cercocebus‡, Erythrocebus, Macaca‡, Mandrillus, Papio, Theropithecys,*
	Large	(b) Knuckle-walking		*Gorilla, Pan*
Arboreal arm-swinging	Medium	(a) True brachiation	(K)	*Hylobates*
	Large	(b) Quadrumanous climbing	(L)	*Gorilla, Pan, Pongo*
Terrestrial striding bipedalism	Large		(M)	*Homo*

* Letters in parentheses refer to groups indicated in the plot of intermembral index values against body weight (Fig. 10.5).
† *Galago crassicaudatus* differs from other bushbabies in both size and locomotor habits and is not included here (see text).
‡ These genera contain some species that are predominantly arboreal and others that are predominantly terrestrial.

between predominantly arboreal and predominantly terrestrial forms. Among the smaller arboreal quadrupeds, it is essential to recognize separate subcategories for the slow-climbing lorisines (*Arctocebus, Loris, Nycticebus* and *Perodicticus*) and for the claw-bearing marmosets and tamarins (family Callitrichidae), as in both cases there is a marked departure from the typical primate pattern of relatively agile arboreal locomotion using grasping extremities (see later). Similarly, among the larger arboreal

quadrupeds there is some justification for distinguishing certain New World and Old World monkey species that perform suspensory locomotor activity in trees a significant proportion of time. There has been considerable discussion regarding suitable terminology for this distinctive monkey type of locomotion, particularly following a seminal paper by Erikson (1963) on 'brachiation' in New World monkeys. Napier (1961) coined the term 'semibrachiation' for the suspensory locomotor activity of the New

World and Old World monkey species involved and later (1963) distinguished the two types 'New World monkey semibrachiation' and 'Old World monkey semibrachiation' (see also Napier and Walker, 1967).

Extension of the term brachiation in this way to include suspensory behaviour in certain monkey species was criticized by Stern and Oxnard (1973), who favoured tight restriction of the term to the pattern of locomotion shown by members of the gibbon family. Stern and Oxnard suggested, quite reasonably, that it is more important to emphasize the fact that limbs are subjected to tension during suspensory motion than to attempt to define shades of meaning in the term 'brachiation'. In a similar vein, Mittermeier and Fleagle (1976) reported that arm-swinging activities – as opposed to suspensory postures overall – are relatively uncommon in the locomotion of *Ateles* and *Colobus* under natural conditions. Further, recent research has suggested that climbing may be more important than arm-swinging as a major component in the locomotor repertoire of many primates that have been termed brachiators or semibrachiators (Fleagle, Stern *et al.*, 1981; Aiello and Day, 1982). For all of these reasons, the term 'semibrachiation' has been eliminated in favour of the term 'suspensory' in Table 10.1, as a modification from the classification outlined by Rollinson and Martin (1981). Finally, it should be noted that terrestrial locomotion of the gorilla and chimpanzee is included as a subcategory (knuckle-walking) of terrestrial quadrupedalism in Table 10.1, in agreement with Rose (1973), but departing from the classification of Napier and Walker (1967), which listed all of the great apes as 'modified brachiators'. Incidentally, it should be noted that the predominantly arboreal orang-utan uses a distinctive pattern of forelimb support known as 'fist-walking' during terrestrial locomotion, approximating to the knuckle-walking pattern of chimpanzees and gorillas, but differing in that the hand rests on its outer margin rather than on the knuckles.

Of course, any classification of primate locomotor behaviour will have limitations, and intermediate cases or straightforward exceptions are only to be expected. The essential point is that classification, locomotor or otherwise, should group data into meaningful categories that facilitate analysis and interpretation. Hence, any classification of primate locomotion should be evaluated not on the basis of difficulties involved in classifying particular species (which must inevitably arise with any classification), but in terms of its contribution to our understanding of the subject. From this viewpoint, the classification originally proposed by Napier and Walker (1967) has undoubtedly served a useful purpose, despite several justifiable criticisms and the understandable need for subsequent modifications (as in Table 10.1).

Criticisms of the kind of locomotor classification given in Table 10.1 fall into two main groups. Many objections concern difficulties encountered in the allocation of individual species to locomotor categories or the integrity of certain categories. For instance, Cartmill (1972) regarded the vertical-clinging-and-leaping category as artificial, because it included 'two quite different kinds of adaptation' represented by *Galago/Tarsius* on the one hand and by *Lepilemur/Avahi/Propithecus/Indri* on the other. Stern and Oxnard (1973) similarly questioned the integrity of the vertical-clinging-and-leaping category. They cited results of multivariate studies of the scapula, forelimb, pelvis and hindlimb indicating a clear separation between *Avahi*, *Indri* and *Propithecus* in one cluster and *Galago* species together with *Tarsius* in another. Even within the genus *Galago*, there is considerable differentiation between species in locomotor habits (e.g. see Charles-Dominique, 1977a) and only two species – Allen's bushbaby (*Galago alleni*) and the lesser bushbaby (*G. senegalensis*) – show well-defined habitual vertical-clinging-and-leaping. By contrast, it has been suggested by other authors that the category of vertical-clinging-and-leaping should in fact be enlarged to include at least some of the marmosets and tamarins. This has been proposed for the red-handed tamarin, *Saguinus midas* (Thorington, 1968b) and the pygmy marmoset, *Cebuella pygmaea* (Kinzey, Rosenberger and Ramirez, 1975), on the grounds that these species have been observed clinging to and leaping between vertical supports. However, marmosets and tamarins lack the conspicuous emphasis on the hindlimbs found in all vertical-clinging-and-leaping prosimians and therefore lack correlated features such as bipedal hopping.

A basic problem is that the locomotor repertoire of any individual primate species includes a variety of different locomotor patterns. Hence, labelling the locomotion of each species with a simple category name, such as 'vertical-clinging-and-leaping', cannot possibly do justice to the complex repertoire involved. For this reason, Walker (1979) suggested that quantification of locomotor behaviour for individual primate species might yield a meaningful basis for comparisons. His detailed study of the dwarf bushbaby (*Galago demidovii*) in captivity revealed that leaping between vertical supports accounted for 14.2%, bipedal hopping along the ground for 3.4% and various forms of quadrupedalism (walking or running) for about 24% of locomotor activity. Hence, although the locomotion of this species does have features typical for vertical-clinging-and-leaping (Napier and Walker, 1967), they represent a minor part of the locomotor repertoire, as is also indicated by Charles-Dominique's report (1977a) on locomotion in *G. demidovii* under natural conditions. R.H. Crompton (1984) has since provided a detailed account of locomotion in *G. senegalensis* and *G. crassicaudatus* under natural conditions, showing both variability within species and overall differences between species (see also Crompton, Lieberman and Oxnard, 1987). Further, Gebo (1987c) has extended Walker's approach to 17 species of lemurs and lorises in captivity. He showed, among other things, that suspensory locomotion is particularly prevalent in lorises and that there is good agreement between field and laboratory data for several species.

Similar considerations are obviously involved in the continuing controversy over the categories of 'brachiation' and 'semibrachiation'. Several authors have implicitly or explicitly proposed that use of these terms should depend on a quantitative assessment of the importance of arm-swinging (with or without a free-flight phase) in the locomotor repertoires of the species concerned. Thus, as with Walker's analysis of locomotion in *Galago demidovii*, there is perhaps a case for classifying primate species according to their most frequent mode of locomotion. But such a procedure would beg a number of questions concerning the underlying rationale of primate locomotor classifications. Interpretation of a link between locomotor behaviour and (for example) skeletal morphology presupposes identification of the main behavioural features associated with the selection of particular skeletal adaptations. However, it could be argued that postcranial morphology is determined not by the frequency with which individual locomotor patterns occur but by the severity of stresses imposed by those patterns. Locomotor activities shown during rapid and precarious movement, notably during escape from predators, might be crucial in influencing postcranial morphology, even if such activities represent a relatively minor proportion of the locomotor repertoire in terms of frequency of occurrence. Similarly, if energy expenditure is at a premium, selection may favour those features of locomotor morphology that are associated with significant energy savings. Given general uncertainty about the operation of natural selection to shape the locomotor morphology of individual species, the most satisfactory interim measure is to use a classification that seems to group living primates into meaningful categories as far as overall behavioural similarities between species are concerned. This criterion is met adequately by the classification in Table 10.1.

The second kind of criticism that has been made of locomotor classification, although relying to some extent on the arguments mentioned above, is far more fundamental in that it questions their validity altogether. Stern and Oxnard (1973), possibly the sternest critics of such classifications, have stated: 'as more data on behaviour and structure are accumulated, it becomes increasingly clear that such classifications no longer serve a useful purpose'. Their main reason for drawing this conclusion is that locomotor categories coincide closely with taxonomic categories and that a locomotor classification therefore conveys no information additional to that already contained in a standard classification of living primates (see Table 3.1). Indeed, it is clear from Table 10.1 that many of the subcategories of locomotor behaviour listed coincide with established taxonomic groupings. For instance, the small-bodied, clawed group of arboreal quadrupeds is constituted by the family Callitrichidae (marmosets and tamarins) and the large-bodied Old World suspensory group consists of the subfamily Colobinae (leaf-monkeys). Similarly, true brachiation is confined to the family Hylobatidae (gibbons and siamang). However, this certainly does not apply in all cases. The small-bodied vertical-clinging-and-leaping group includes representatives of two widely separated taxonomic categories among prosimians (Galaginae and Tarsiidae), and the medium-sized, non-clawed arboreal quadrupeds include one subfamily of Malagasy lemurs (Lemurinae) along with a number of New World monkey subfamilies (Aotinae, Cebinae and Pithecinae). In these two instances, it is surely of some importance that representatives of quite distinct taxonomic groupings have come to occupy the same subcategory of locomotor adaptation and this alone should throw some light on the evolutionary processes that have shaped primate locomotion. Moreover, the close coincidence between locomotor classification and general taxonomic groupings is in itself extremely revealing. Far from simply illustrating the redundancy of locomotor classification, as claimed by Stern and Oxnard (1973), this coincidence actually yields the vital piece of information that locomotor adaptation was a central feature of primate adaptive radiation. Further, it is clear that within individual taxonomic groups locomotor patterns have remained relatively conservative. In any case, in spite of the unavoidable shortcomings of any individual locomotor classification, it is an inescapable fact that investigation of links between locomotor behaviour and skeletal

morphology depends on some meaningful system for organizing data. This is particularly true when the locomotion of fossil primates is inferred from morphological evidence alone (Gebo, 1987c).

LOCOMOTOR MORPHOLOGY

An important point to recognize at the outset is that the locomotor morphology of primates is, on the whole, characterized by retention of a relatively primitive skeletal framework. As a group, primates typically lack obvious specializations of the postcranial skeleton associated with restricted locomotor repertoires such as rapid movement across the ground (ungulates) or burrowing (various insectivores, lagomorphs and rodents), let alone flight (bats) or aquatic locomotion (e.g. seals, dolphins and manatees). This is because primates in general show considerable flexibility in locomotor behaviour, depending on a combination of relatively unspecialized skeletal morphology with a fairly well-developed central nervous system (see Chapter 8). Indeed, primates are unique among mammals with relatively large brains in having a relatively primitive postcranial skeleton.

Broad comparisons of vertebrates in general (Romer, 1966) and of mammals in particular (Weber, 1927) have led to the identification of a **basic postcranial skeleton** for mammals. As was noted in Chapter 4, mammals as a group are characterized by relative uniformity in the composition of the skeleton, although some mammals have obviously become specialized in various ways. Typically, such specializations have involved fusion or actual loss or basic skeletal elements. With relatively few exceptions, however, the basic pattern has been retained by comparatively unspecialized, small-bodied and small-brained mammals such as living insectivores, rodents and carnivores, as well as persisting in primates.

The basic mammalian pattern can be illustrated conveniently with the skeleton of the common tree-shrew, *Tupaia glis* (Fig. 10.2). The tree-shrew skeleton has remained relatively

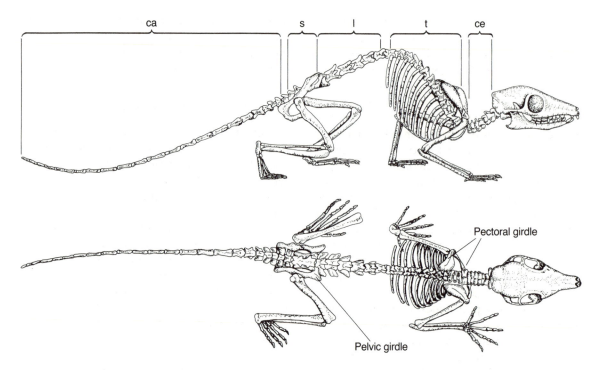

Figure 10.2. Outline drawing of the skeleton of the common tree-shrew *Tupaia glis*; top, lateral view; bottom, dorsal view. Abbreviations: ca, caudal vertebrae; ce, cervical vertebrae; l, lumbar vertebrae; s, sacral vertebrae; t, thoracic vertebrae. (After Jenkins, 1974b).

primitive, with little fusion or loss of component elements. Both forelimbs and hindlimbs conform to the basic five-rayed (pentadactyl) pattern (see later) and little modification has taken place in the shoulder (pectoral) or in the hip (pelvic) girdle. The pectoral girdle, for instance, retains the collar-bone (clavicle), whereas this has been greatly reduced or lost in various other groups of placental mammals (e.g. ungulates, carnivores, pinnipeds and cetaceans). The backbone (vertebral column) follows the standard mammalian arrangement (see Chapter 4) in being clearly divided into five functionally separate regions. The following set of vertebrae are typically found in *Tupaia glis* (Jenkins, 1974b): 7 neck vertebrae (cervicals), 13 chest vertebrae (thoracics), 6 loin vertebrae (lumbars), 3 fused sacral vertebrae (the first articulating with the hip girdle) and about 24 tail vertebrae (caudals). The possession of a relatively long tail

seems to be a primitive feature for mammals generally.

As noted above, primates have similarly retained a relatively primitive postcranial skeleton and for this reason alone there is quite close resemblance in skeletal morphology between them and tree-shrews. For instance, the clavicle has been retained in all living primates (see Mivart's definition in Chapter 5) and the numbers of vertebrae have generally remained close to the inferred primitive pattern. A relatively unspecialized prosimian primate such as the mongoose lemur (*Lemur mongoz*) has a vertebral column very similar in composition to that of *Tupaia glis*. Gregory (1920a) gives the following vertebral formula for *L. mongoz*: 7 cervicals, 12 thoracics, 6–8 lumbars, 3 sacrals and 28 caudals, noting that the Eocene 'lemuroid' *Notharctus* had a similar pattern. MacLarnon (1987) has since confirmed that *Notharctus tenebrosus* had a probable vertebral formula of 7 cervicals, 12 thoracics, 8 lumbars, 3 sacrals and 19(+) caudals, whereas *Smilodectes gracilis* apparently had 13 thoracics and 7 lumbars. Nevertheless, there are some distinct differences between tree-shrews and primates in the postcranial skeleton, notably in the extremities, and these will be examined in detail below.

In many modern mammals other than tree-shrews and primates the postcranial skeleton is considerably modified through fusion and/or loss of individual components. For example, some limb elements of hoofed mammals (ungulates) have been lost altogether or have become fused in association with the development of strong, gracile limbs adapted for fast terrestrial locomotion. Similarly, cetaceans (dolphins and whales) have become adapted for an aquatic existence through modification of the forelimbs to act as flippers (pectoral fins) and through complete atrophy of the hindlimb (pelvic) girdle, leaving only a pair of isolated bones embedded in the body wall. In other mammals the postcranial skeleton has become greatly modified without marked fusion or loss of elements. For instance, bats (Chiroptera) have hypertrophied forelimbs bearing the wings on elongated fingers and their hindlimbs have been reduced in size. All of these mammals with modified skeletons have become adapted for special modes of locomotion, so the retention of a relatively primitive skeleton in modern primates clearly suggests that primate locomotor patterns have not diverged markedly from the ancestral condition for placental mammals.

BODY SIZE AND LOCOMOTOR ADAPTATION

It has already been established (see Chapter 4) that **body size** has been a major factor in the adaptive radiation of mammals. Numerous physiological, morphological and other features are intimately connected with body size and for several reasons it is particularly necessary to take body size into account when assessing locomotor adaptations. In the first place, locomotion depends directly on energy relationships and body size influenzes both the **availability of**

energy for locomotor activity (e.g. see the discussion of basal metabolic rate in the appendix to Chapter 4) and the **energy costs** of such activity. In both contexts, the scaling of individual parameters to body size is commonly allometric (rather than isometric). Hence, scaling analyses should be included in any comparative study of primate locomotor adaptation. It is also to be expected, for various basic mechanical reasons, that there should be systematic modification of skeletal morphology with changing body size, and such modification is again usually allometric in nature.

A very simple example of systematic change with body size that has often been cited is the relationship between the total weight of the skeleton and overall body weight in mammals. It has been argued (see Schmidt-Nielsen, 1972) that this relationship should be positively allometric. This is because the weight of the body (W) increases according to the cube of linear dimensions, whereas (other things being equal), the capacity of the limb bones to support the body depends on their cross-section, that is to say on the square of certain linear dimensions. Using a very simple model, one might therefore expect the limb bones to increase disproportionately in girth with increasing body weight, with a consequent disproportionate increase in total weight of the postcranial skeleton, as was originally proposed by Galileo (1638). A crude relationship of the following kind between postcranial skeletal weight (S) and body weight (W) might therefore be expected:

$$S = kW^{3/2}$$

(where k is the allometric coefficient). This would be a positively allometric relationship, because the expected exponent value is greater than 1 (i.e. 1.5). In fact, the prediction was only weakly supported by scaling studies. A positive allometric relationship between skeletal weight and body weight has indeed been reported for both mammals and birds (Prange, Anderson and Rahn, 1979), and this has also been indicated for other animal groups (Anderson, Rahn and Prange, 1979). The reported exponent value was, however, considerably smaller than predicted

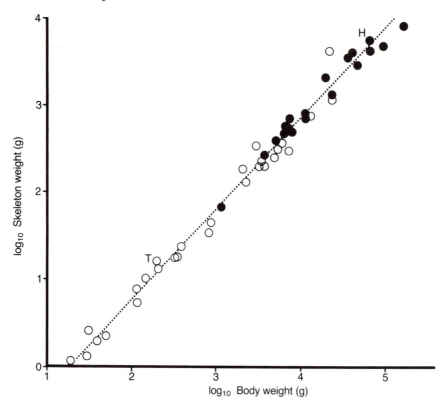

Figure 10.3. Logarithmic plot of the weight of the postcranial skeleton (*S*) against body weight (*W*) for a sample of placental mammal species (*N* =47). The empirical scaling formula derived from the best-fit line (major axis) is:

$$\log_{10} S = 1.05 \log_{10} W - 1.348 \qquad (r = 0.99)$$

This yields the relationship (with both *S* and *W* in grams):

$$S = 0.045\, W^{1.05}$$

Key: ●, primates (*N* = 19; H, *Homo sapiens*); ○, non-primates (*N* = 28; T, tree-shrew). (Data for primates and tree-shrews from Schultz, 1962; data on non-primates from Prange, Anderson and Rahn, 1979.)

from simplistic geometrical considerations (i.e. significantly smaller than 1.5). For a sample of a mammals ranging in size from 20 g to 160 kg, the empirically determined exponent value was only 1.05 (Fig. 10.3). Nevertheless, an exponent value of 1.05 would correspond to a noticeable increase in the weight of the skeleton as a proportion of body weight from 5.5 to 8.2% over the size range of modern primates (60 g to 150 kg). However, it has since emerged that skeletal weight actually varies isometrically within mammalian orders (i.e. with an exponent

value of 1) and the slope of 1.05 is attributable to minor 'grade' differences between orders (Potter, 1986).

In fact, it would seem that subtle changes occur in skeletal parameters other than simple dimensions, such as in overall bone shape and histological architecture, to counter the problem identified by Galileo (see Aiello, 1981a for discussion). There is also the point that there may be some overall constraint on calcium-containing skeletal tissue in mammals, with the result that the scaling requirement of increasing body weight identified by Galileo cannot be overcome by simply increasing the girth of supporting bones. Hence, it is possible that with increasing body size among mammals, the postcranial skeleton becomes increasingly vulnerable to stresses and strains, and that locomotor behaviour is increasingly restricted (e.g. through reduction in leaping) to protect the bones. This provides an additional reason for expecting body size to play a major part in primate locomotion and it should therefore always be taken into account in studies of primate locomotor evolution.

A potential source of confusion arises from the widespread use of simple indices (ratios) in quantitative studies of primate postcranial morphology. It is commonly implied that the use of simple osteometric indices 'eliminates' the effects of body size. Yet indices can only be said to eliminate the effects of body size if quite specific conditions are met. Consider two linear measurements of the skeleton, L_1 and L_2, that are allometrically related to body weight (W) as follows:

$$L_1 = k_1 W^{\alpha_1}$$
$$L_2 = k_2 W^{\alpha_2}$$

A ratio of L_1 to L_2, consituting a classical osteometric index, would therefore be governed by the following formula:

$$\text{Index } \frac{L_1}{L_2} = \frac{k_1}{k_2} W^{(\alpha_1 - \alpha_2)}$$

It is obvious from this relationship that the index (L_1/L_2) will itself behave allometrically with respect to body weight unless α_1 and α_2 happen to take the same value. At least in some cases, this latter requirement is not met. In practice, it is easy to determine whether a given index eliminates the effect of body weight by plotting values of that index against body size to verify the absence of any identifiable trend. Alternatively, the two component variables, L_1 and L_2, can be examined directly on a logarithmic plot to determine whether the relationship between them is isometric (slope value = 1), as would be the case with identical values for α_1 and α_2. For instance, Aiello (1981a) has plotted the length of

the shinbone (tibia) against the length of the thighbone (femur) for 51 simian primate species. The ratio between the lengths of these bones is widely used in the literature as the 'crural index' (see later), without any explicit reference to body weight. But Aiello's analysis shows that there is an overall negative allometric relationship between tibia length and femur length, with an empirical exponent value of 0.88 ($r = 0.99$). The 95% confidence limits on the exponent (0.85–0.91) clearly rule out isometry (exponent value = 1) in this case. The net result is that, regardless of any other considerations, the crural index must show a regular and progressive decline with increasing body weight among primates (Rollinson and Martin, 1981). Similar considerations apply to many other indices calculated for the primate skeleton and commonly cited in the literature as indicators of locomotor adaptation.

Given the dependence of various simple osteometric indices on body weight, it is difficult to interpret the results of multivariate studies that have utilized a set of such indices as a starting point for analysis without explicit consideration of the influence of body weight. For example, a multivariate study of various limb and limb girdle ratios (Oxnard, 1973; Stern and Oxnard, 1973) has been cited as providing evidence that the small-bodied vertical-clingers-and-leapers *Galago* and *Tarsius* are distinct from the large-bodied vertical-clingers-and-leapers *Avahi*, *Propithecus* and *Indri*. However, without a specific analysis of the effects of body size on individual index values, it remains a moot point whether these authors have merely separated small-bodied from large-bodied vertical-clingers-and-leapers or whether they have uncovered a more fundamental distinction in locomotor adaptation. This uncertainty is underlined by the fact that *Galago* and *Tarsius* differ fundamentally in hindlimb morphology (see later) even though they are grouped together by multivariate analysis. Until the results of multivariate studies can be interpreted in step-by-step fashion without losing sight of possible effects of body size, their implications for functional morphology will remain difficult to decipher.

LIMB PROPORTIONS

There is an abundant literature on the skeletal morphology and musculature of the limbs in primates and the wealth of information available should ideally make a major contribution to our understanding of primate locomotor evolution. However, broad comparative studies are relatively rare (e.g. Miller, 1932; Schultz, 1933, 1937, 1969; Jouffroy, 1960, 1962, 1975; Stern, 1971; Stern and Oxnard, 1973; McArdle, 1981; and references cited below). Further, it is exceedingly difficult to condense the available data and to integrate it with information on other mammals so that derived features of primates can be distinguished from primitive features retained from the ancestral stock of placental mammals. An effective review of this kind is sorely needed. However, interpretation of the evolution of primate locomotion has been advanced significantly through quite simple analyses of the relative proportions of forelimbs and hindlimbs (Mivart, 1867; Mollison, 1910; Schultz, 1933, 1937; Jouffroy, 1960, 1975; Napier and Walker, 1967; Walker, 1967a; Aiello, 1981a,b). Such analyses reveal little about the features distinguishing primates from other mammals, but they are particularly useful in throwing light on adaptive radiation in locomotion during primate evolution.

One index that has figured prominently in comparative studies of primate limbs is the **intermembral index**. This is the ratio of forelimb length (humerus + radius) to hindlimb length (femur + tibia), expressed as a percentage. Mollison (1910) identified this index as being of particular value in comparisons of primate species and Napier and Walker (1967), in their seminal paper defining vertical-clinging-and-leaping, stressed the significance of a low intermembral index value for this pattern of locomotion. Napier (1971) used the intermembral index as a key criterion for the recognition of locomotor categories in living primates. As a general rule, index values are low (less than 70%) in vertical-clinging-and-leaping primates (bushbabies, tarsiers and certain lemurs), moderate in quadrupedal primates (70–

100%) and high in primates showing a prevalence of suspensory locomotion (100–150%). Among primates, then, the intermembral index alone provides a crude guide to broad locomotor categories. In fact, a low intermembral index is found in postcranial skeletons of the Eocene 'lemuroids' *Notharctus* and *Smilodectes*. This provided one of the primary reasons for the inferences that these early primates were vertical-clingers-and-leapers and that this pattern of locomotion characterized the ancestral stock of primates of modern aspect (Napier and Walker, 1967;. Napier (1967a) went on to recognize three stages in the evolution of primate locomotion, reflecting the predominance of low intermembral indices among prosimians, of moderate indices among monkeys and of high indices among apes (Figure 10.4).

Although vertical-clinging-and-leaping is widespread among living prosimians, it is by no means universal. In the loris group, the bushbabies (with the conspicuous exception of the thick-tailed bushbaby, *Galago crassicaudatus*) broadly fit into that category, but the slow-climbing lorises and pottos are primarily quadrupedal and have relatively high intermembral indices to match. Among the Malagasy lemurs, all members of the indri group and *Lepilemur* are specialized vertical-clingers-and-leapers, but species of the genera *Phaner*, *Cheirogaleus*, *Daubentonia*, *Lemur* and *Varecia* are essentially quadrupedal; *Microcebus* and *Hapalemur* are intermediate. Indeed, when subfossil lemurs are taken into account, it emerges that two large-bodied genera, *Megaladapis* and *Palaeopropithecus*, had high intermembral indices comparable to those in modern great apes. This probably reflects an emphasis on suspensory behaviour (Walker, 1967a, 1974a; Rollinson and Martin, 1981). Although it can be argued that vertical-clinging-and-leaping is the most widespread locomotor pattern among modern prosimians, because it occurs in all three 'natural groups' (lemurs, lorises and tarsiers), it cannot be concluded with any confidence that it characterized the common ancestral stock of the ancestral primates. In fact, it seems more reasonable to suggest that the

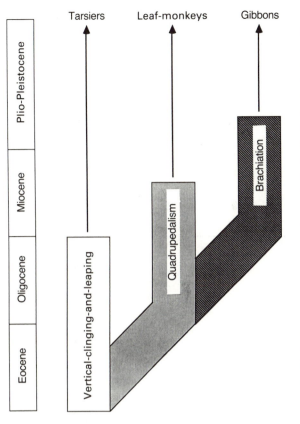

Figure 10.4. A hypothetical general trend in the evolution of primate locomotion – from vertical-clinging-and-leaping in prosimians, through quadrupedalism in monkeys and on to brachiation in apes (after Napier, 1967a). The living primate types indicated at the top of the diagram are 'specialists' identified by Napier for each category of locomotion.

ancestral primate condition was relatively unspecialized and that both vertical-clinging-and-leaping and slow quadrupedal climbing represent divergent specializations. This suggestion is supported by studies of the tarsal bones of the foot (see later) and by recent studies showing that the hindlimb of tarsiers differs in many morphological and metrical details from that of bushbabies (Jouffroy, Berge and Niemitz, 1984; Gebo, 1987a).

Intermembral index values for primates have often been considered without any explicit reference to body size, perhaps because of the mistaken view (see above) that calculation of

indices 'eliminates' the influence of body weight. However, it emerges that additional information can be gained from intermembral index values if they are plotted against body weight (Fig. 10.5; see also Rollinson and Martin, 1981). There is, for example, an overall trend for intermembral indices to increase with increasing body size, which is evident among both living and fossil primates. This trend is undoubtedly related to the progressive switch to suspensory locomotion with increasing body weight among arboreal primates, involving greater emphasis on the role of the forelimbs. Among prosimians, at the lower end of the body-size range, modern vertical-clingers-and-leapers have lower-than-expected index values, as do the Eocene 'lemuroids' *Notharctus* and *Smilodectes*, and the slow-climbing lorises and pottos have higher-than-expected values. It should be noted, however, that intermembral index values of less than 100 are generally typical of quadrupedal placental mammals, and values even lower than those of vertical-clinging-and-leaping primates occur in certain rodents (Howell, 1944). A wide-ranging survey of mammalian species led Howell to conclude that in the generalized mammalian condition the intermembral index is close to 75, well within the modern prosimian range. Hence, low intermembral index values cannot be regarded as diagnostic for the relatively small-bodied prosimian primates, although they are useful for distinguishing locomotor categories among primates (see also McArdle, 1981). No special significance can be attached to the observation that the intermembral index values of tree-shrews and of the 'archaic primate' *Plesiadapis* fall among those of quadrupedal primates of comparable body size (Fig. 10.5); this is a common feature of placental mammals and probably reflects the ancestral condition.

High intermembral index values are confined to primates at the upper end of the size range. Values in excess of 100 are relatively rare among mammals and the modern sloths (also adapted for suspensory arboreal locomotion) are the only placental mammals in Howell's survey (1944) that overlap the range of large-bodied arboreal primates to any significant extent. Gibbons and

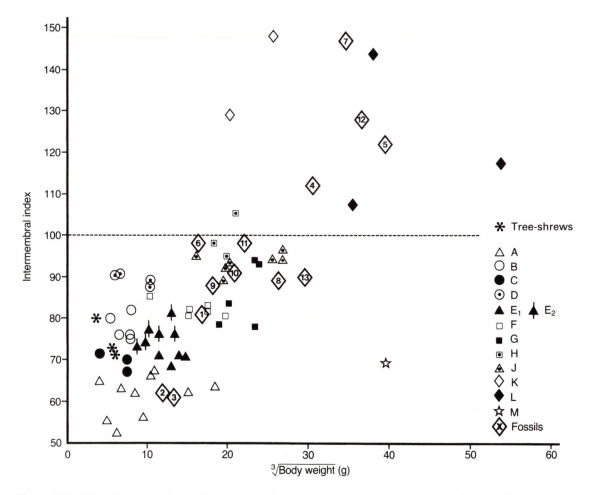

Figure 10.5. Plot of intermembral index values against the cube root of body weight for living and fossil primate species. Key:

Extant species (see Table 10.1): A, vertical-clinging-and-leaping prosimians; C, D, E₁, quadrupedal prosimians; B, E₂, F, J, quadrupedal monkeys; G, H, suspensory monkeys; K, gibbons; L, great apes; M, humans; *∗*, tree-shrews.

Fossil species (numbered diamonds): 1, *Plesiadapis tricuspidens*; 2, *Smilodectes gracilis*; 3, *Notharctus tenebrosus*; 4, *Megaladapis madagascariensis*; 5, *Megaladapis edwardsi*; 6, *Varecia insignis*; 7, *Palaeopropithecus maximus*; 8, *Archaeolemur edwardsi*; 9, *Mesopithecus penteleci*; 10, *Pliopithecus vindobonensis*, 11, *Dendropithecus macinnesi*; 12, *Oreopithecus bambolii*; 13, *Australopithecus afarensis*. (After Rollison and Martin, 1981.)

the siamang are somewhat unusual, in comparison to the overall primate trend, in having high intermembral index values at a moderate body weight of 5–11 kg (Fig. 10.5). This suggests that the active, ricochetal brachiation shown by gibbons is a specialized feature, rather than resembling the ancestral pattern for apes and humans (hominoids). It is also noteworthy that *Homo sapiens* is unusual compared to other relatively large-bodied primate species in having a relatively low intermembral index value. If body size is not taken into account, this important distinction is obscured because there are numerous small-bodied prosimian species with intermembral index values of around 70 (Fig. 10.5). The earliest known hominid skeleton for which an intermembral index value can be calculated, from *Australopithecus afarensis*, has a value that is consistent with the overall primate trend but below that of modern great apes. This provides confirmation for the view that there was no real 'brachiating' stage in human evolution and that the modern human skeleton is derived from a relatively generalized primate condition.

Similar observations can be made with respect to other standard indices based on limb dimensions, such as the **crural index** (ratio of tibia length to femur length in the hindlimb) and the **brachial index** (ratio of radius length to humerus length in the forelimb). Both of these indices show a progressive decrease with increasing body weight among primates, as is true of mammals generally, such that reference to body size is again necessary for meaningful interpretation. Howell (1944) suggested that in the generalized mammal both crural and brachial indices (expressed as percentages) would approximate to 100 (i.e. with distal and proximal segments roughly equal in length in both hindlimbs and forelimbs). This condition is seen in modern tree-shrews and in small quadrupedal lemurs such as *Cheirogaleus* and *Phaner* (Rollinson and Martin, 1981). The brachial index yields some interesting points in relation to body size. Vertical-clingers-and-leapers tend to have higher values than other primates of comparable body size, as noted by Napier and Walker (1967), although the distinction is not clear. Further, among medium-sized primates gibbons and the siamang have relatively high brachial indices whereas New World suspensory forms have relatively low brachial indices (see also Aiello, 1981b). The crural index, on the other hand, yields little information of value from a simple plot against body size, as the downward trend with increasing body size is more uniform (see Rollinson and Martin, 1981). However, the very consistency of the relationship between the crural index and body size, in contrast to that between the brachial index and body size, bears witness to the relative conservatism of the hindlimb in

primate evolution. As will be seen, this point is of considerable significance.

The question of skeletal allometry in relation to locomotor behaviour has been considered in more detail for prosimians, with special reference to the subfossil lemur *Megaladapis*, by Jungers (1977, 1978) and for simians by Aiello (1981a, b). These studies have further demonstrated the need for taking body size into account to gain a proper appreciation of the factors involved. Aiello's detailed analyses of lengths and diameters of the major limb bones have shown that considerable care must be exercised in the interpretation of indices derived from such data.

Apart from their systematic variation with body size, limb indices can also be potentially confusing for other reasons. There is a natural tendency to assume that mammals with the same index value at a given body weight are necessarily closely comparable, but this is not always true. For instance, among modern placental mammals primates tend to have longer limb bones than other mammals in relation to body weight (Alexander, Jayes *et al.*, 1979) and this is associated with longer stride lengths and lower stride frequencies (Alexander and Maloiy, 1984). Hence, although primates may have similar limb ratios to other mammals, their limbs are generally more elongated. This point is illustrated by the Eocene 'lemuroids' *Notharctus* and *Smilodectes*, in which the limbs were markedly shorter in relation to overall body size than in comparable modern prosimians (e.g. note the comparison between *Smilodectes* and the modern sifaka, *Propithecus* in Fig. 2.10). This suggests that there has been a progressive trend towards elongation of the limbs relative to body size among primates, and perhaps in other mammals as well. This underlines the need for caution in interpreting the superficial similarity between Eocene 'lemuroids' and modern vertical-clinging-and-leaping prosimians in intermembral index values. The intermembral index is a simple mathematical abstraction that can clearly be misinterpreted if considered without reference to primate species as entire, living organisms and this particular example highlights the less-obvious dangers of bivariate or multivariate abstractions. The evidence on limb size relative to overall body size in Eocene 'lemuroids' in fact suggests that these early primates of modern aspect were less agile leapers than modern lemurs with comparable intermembral index values (see also Chapter 8).

THE EXTREMITIES

The basic mammalian skeleton is characterized by possession of **pentadactyl limbs**. That is to say, the extremities (hands and feet) commonly have five digits apiece and each is attached to a limb with two main segments (Fig. 10.6). In the standard pentadactyl limb, the upper segment articulating with the limb girdle has a single long bone (**humerus** in the forelimb; **femur** in the hindlimb) and the more distal segment incorporates two parallel bones (**radius** and **ulna** in the forelimb; **tibia** and **fibula** in the hindlimb). The fundamental pentadactyl pattern, which is in fact characteristic of terrestrial vertebrates in general, is retained without significant modification in the overwhelming majority of primates. There are, however, several species in which one or more digits have been reduced. For example, in all members of the loris subgroup there has been marked reduction of the second digit of both hand and foot (Fig. 10.7). This specialization reflects the development of an enhanced pincer-like action of the hand and foot associated with the slow-climbing habit of these primates (see later). Another kind of specialization involving reduction of digits is found in primates with patterns of locomotion in which the hands are used more as suspensory and prehensile hooks than as grasping pincers. In such species, the thumb is considerably reduced (e.g. in langurs, *Presbytis*, or in gibbons, *Hylobates*) or lost altogether (e.g. in spider monkeys, *Ateles* and in guerezas, *Colobus*), as shown in Fig. 10.7.

The individual bones of the digits (phalanges; singular = phalanx) are also of interest. It was shown in Chapter 4 that the mammals as a group are characterized by a standard **phalangeal formula** of 2.3.3.3.3 in both the hand and the foot. With the exception of the special cases just

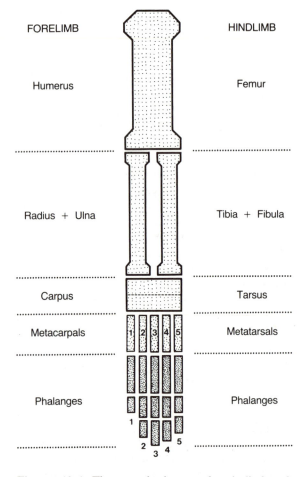

Figure 10.6. The standard pentadactyl limb of mammals. The digits of both hand and foot typically have the basic mammalian phalangeal formula 2.3.3.3.3 and the basic mammalian digital formula 3:4:2:5:1. Note that the radius and ulna are commonly crossed in mammals to permit rotation of the hand about a longitudinal axis (pronation/supination).

mentioned, in which digital reduction has taken place, this standard formula is retained in living primates generally (including humans) and it is present in all fossil primates of modern aspect for which the postcranial skeleton is adequately known (e.g. the Eocene 'lemuroid' *Notharctus*). Retention of all five digits and the standard phalangeal formula in most primates provides further confirmation of the generally primitive nature of the primate postcranial skeleton.

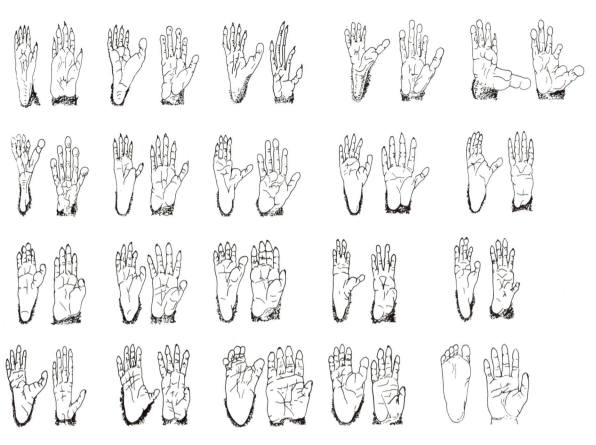

Figure 10.7. Ventral (palmar and plantar) views of the right hands and feet of a tree-shrew and of selected primate species. In each pair of extremities, the foot is shown on the left and the hand on the right. Key (from left to right): Top row: *Tupaia*, *Lemur*, *Daubentonia*, *Galago* and *Nycticebus*. Second row: *Tarsius*, *Leontopithecus*, *Aotus*, *Cebus* and *Ateles*. Third row: *Cercopithecus*, *Macaca*, *Papio*, *Colobus* and *Presbytis*. Bottom row: *Hylobates*, *Pongo*, *Gorilla*, *Pan* and *Homo*. Note the reduction of the second digit of the hand and foot in *Nycticebus* and the loss of the first digit (thumb) of the hand in *Ateles* and *Colobus*. (After Schultz, 1956.)

The overall form of the hand and foot in mammals can also be expressed in terms of the **digital formula** (Wood Jones, 1929). This formula simply expresses the relative lengths of the digits by listing them in order of decreasing length. In the typical mammalian condition, the longest digit of both hand and foot is the third and the digital formula is 3:4:2:5:1. This formula is present, for instance, in modern tree-shrews (Wood Jones, 1929), as shown in Fig. 10.7. Most simian primates also have it, although there are a few exceptions (e.g. gibbons, siamang and

humans). In prosimian primates, however, there is an interesting departure from the basal mammalian digital formula. In lemurs (including the aye-aye) and lorises, the fourth digit is the longest in both hand and foot, so the digital formula is 4:3:5:2:1. Thus, in this respect, strepsirhine primates are apparently specialized in comparison to simian primates, with the functional axis of the foot running through the fourth digit rather than the third. Tarsiers are intermediate in that the hand has the inferred primitive mammalian digital formula of

3:4:2:5:1, whereas the foot has the same specialized digital formula as that found in lemurs and lorises (i.e. 4:3:5:2:1).

The feet

The most significant feature of primate hands and feet is a general adaptation for **grasping** (prehensile action). This is most obvious in the case of the **hallux** (big toe), which is widely divergent from the other digits of the foot in all species except humans and provides for a pincer-like action of the foot (Fig. 10.7). As the possession of a widely divergent hallux with a pincer-like grasping capacity is virtually a universal characteristic among living primates, it is reasonable to infer that this condition was present in the common ancestor of primates of modern aspect (Cartmill, 1972; Conroy and Rose, 1983; Martin, 1986a). Although there is no pronounced divergence of the hallux during human fetal development (Wood Jones, 1949), an original grasping function of the hallux in human ancestry is indicated both by the retention of intrinsic muscles that serve that function in other primates and by the retention of torsion (twisting) along the long axis of the first metatarsal bone (Le Gros Clark, 1959). As striding bipedalism is a unique and obviously specialized form of locomotion among primates, and among mammals generally, it is almost certain that the grasping power of the human hallux was secondarily lost in this connection (see later).

The expectation that the ancestral primates would have had a divergent, grasping hallux is confirmed by postcranial evidence from fossil primates of modern aspect. Careful studies (usually involving reconstructions) of partial or complete foot skeletons have confirmed the presence of a divergent hallux in Eocene 'lemuroids' (e.g. *Pelycodus* (*Cantius*) – Rose and Walker, 1985; *Notharctus* – Gregory, 1920a: bones attributed to *Adapis* – Dagosto, 1983; the first Messel adapid – von Koenigswald, 1979), in Eocene 'tarsioids' (e.g. bones attributed to *Hemiacodon* – Simpson, 1940) and in Miocene/ Piiocene simians (e.g. *Mesopithecus* – Gaudry,

1862; *Pliopithecus* – Zapfe, 1960; *Proconsul* – Walker and Pickford, 1983; *Oreopithecus* – Szalay and Langdon, 1986). In both living and fossil primates, the divergent hallux has a characteristic saddle-shaped articulation at its base, between the metatarsal and the tarsus, which accounts for its rotary pincer-like action (see Fig. 10.8; see also Fig. 10.13). For this reason, it is comparatively easy to recognize the typical primate condition in the first metatarsal (e.g. in *Hemiacodon* – Simpson, 1940; *Adapis* – Dagosto, 1983; *Aegyptopithecus* – Conroy, 1976b).

Given the strong inference that primates of modern aspect are derived from a common ancestor with a grasping adaptation of the hallux, it is important to establish whether this characteristic is found either in tree-shrews or in 'archaic primates' such as *Plesiadapis*. In tree-shrews, the hallux does have a weakly defined saddle-shaped articulation and there is some evidence that this digit plays a special part in locomotion. In the common tree-shrew (*Tupaia glis*; Jenkins, 1974b), the hallux is somewhat divergent from the other toes; this divergence seems to be even more marked in the pen-tailed tree-shrew (*Ptilocercus*; see Fig. 5.2). Both Jenkins (1974b) and Le Gros Clark (1959) have commented on the particular mobility of the hallux in tree-shrews and both authors have suggested that the hallux may even be able to grasp during arboreal locomotion. However, it seems that the primary function of the mildly divergent hallux in tree-shrews is to prevent the foot slipping away down the side of a horizontal or oblique branch, and the foot is by no means as well adapted for grasping in tree-shrews as it is in primates (see Fig. 10.7). Hence, although it may be argued that the structure and operation of the tree-shrew hallux show some similarity to the primate condition, there is still a considerable gulf between tree-shrews and primates.

It has only recently become possible to make any reliable statement about the structure of the hallux in 'archaic primates'. Although the postcranial skeleton was reasonably well known for the genus *Plesiadapis* (Szalay, Tattersall and Decker, 1975), virtually no evidence was

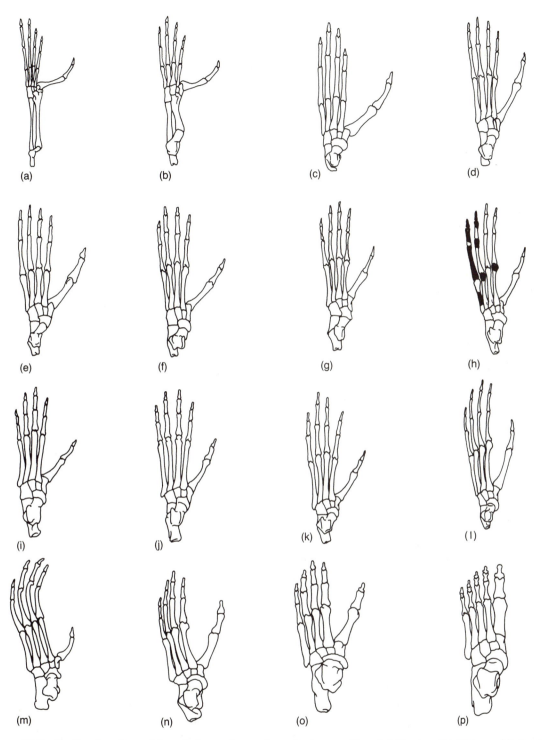

Figure 10.8. Outline drawings of foot skeletons from selected primates. Key: (a) *Tarsius*; (b) *Galago*; (c) *Loris*; (d) *Lemur*; (e) *Lepilemur*; (f) *Callicebus*; (g) *Cebus*; (h) *Notharctus* (fossil; black parts restored); (i) *Papio*; (j) *Macaca*; (k) *Presbytis*; (l) *Hylobates*; (m) *Pongo*; (n) *Pan*; (o) *Gorilla*; (p) *Homo*. Note the divergence of the hallux (except in *Homo*), associated with a saddle-shaped articulation between the first metatarsal and the tarsus. (After Morton, 1924.)

available initially concerning the morphology of the foot and no metatarsals or phalanges had been described. It was inferred that *Plesiadapis* had hands and feet with strongly developed claws (see Fig. 2.3), but satisfactory description of the foot had to await discovery of more complete fossil evidence. A suggestion by Szalay and Drawhorn (1980) that plesiadapiforms might have possessed a grasping hallux rested entirely on speculative interpretation from the morphology of the calcaneus (see later). In fact, Gingerich (1986a) has now reported on a specimen of *Plesiadapis* in which the foot bones are preserved along with the (separated) hallux. The present author unfortunately misreported Gingerich as believing the hallux to be absent in *Plesiadapis* (see Martin, 1986a), whereas a separated digit identified as the hallux is in fact present in the specimen. The digit concerned bore a robust claw and it is therefore highly unlikely that it served a grasping function as in primates of modern aspect.

The hands

The grasping adaptation of the foot in primates of modern aspect has been somewhat neglected in discussions of the evolution of primate extremities. Instead, many authors have concentrated on the structure and function of the **hand**. This, of course, is a reflection of the fact that the hand has been particularly important in human evolution. Indeed, a curious reversal took place during human evolution: the grasping action of the foot (retained in all other living primates) was suppressed in connection with adaptations for bipedal striding, while the hand was liberated from direct locomotor involvement and developed enhanced grasping and manipulatory capacity. As a result of a widespread anthropocentric approach to primate evolution, which has emphasized the primacy of the human hand over the human foot, the great importance of the grasping foot in primate evolution has been somewhat overlooked. In fact, in one early classification (Blumenbach, 1791), the non-human primates were placed in the order Quadrumana ('four-handed' forms),

and humans were placed in a separate order Bimana. This division clearly reflected the fact that the typical primate condition involves quadrumanous grasping. Any consideration of the evolution of the primate hand should therefore be seen in this context.

In primates generally, the thumb (**pollex**) is to some degree divergent from the other digits of the hand, but this divergence is usually far less pronounced than that of the hallux (see Fig. 10.7). Thus, the grasping action of the primate hand can be regarded, as a general rule, as subsidiary to that of the foot. As evidence of this, the pollex has been greatly reduced or completely lost in several primates adapted for suspensory locomotion (e.g. in *Ateles*, *Presbytis*, *Colobus* and *Hylobates*; Fig. 10.7), whereas the hallux has not undergone reduction or loss in any primate of modern aspect. The typical primate condition is exemplified by the Eocene 'lemuroid' *Notharctus* (see Fig. 2.9), which has a well-developed grasping foot but only a weakly grasping hand.

For the reason given above, considerable attention has been directed to the structure and function of the hand in primates generally (e.g. Napier, 1960, 1961; Bishop, 1962, 1964). However, the special concern with human hand-use has commonly led to an emphasis on those aspects that are significant for the human hand rather than on features that characterize primate hands in general. This is reflected, for instance, in concern with the term **opposition**, which was originally defined in relation to the human hand but was subsequently extended to cover other primates. Whereas it is true to say that primates generally have some degree of divergence of the pollex from the other digits of the hand (Fig. 10.7) and also tend to use the hand for grasping objects, it is not true to say that all primates show opposition of the thumb. Napier (1961) has defined **true opposability** of the pollex as requiring a special saddle-shaped joint between the metacarpal and the carpus at the base of the thumb, as is present in the human hand. By virtue of this special articulation, the thumb can rotate around its longitudinal axis as it is brought towards the other digits, such that the ventral surface of the thumb can be opposed to the ventral surfaces of the other fingers in grasping. Full opposability of this kind is present only in Old World simians and it is developed to the greatest degree in the human hand. Prosimian primates (lemurs, lorises and tarsiers), according to Napier, do not have a saddle-joint at the base of the thumb, so true opposability is ruled out. (This generalization is not entirely accurate, however, as Altner (1971) has shown that the lesser mouse lemur, *Microcebus murinus*, does have a simple saddle-shaped articulation at the base of the pollex.) Nevertheless, some degree of rotation of the pollex is possible at the joint between the metatarsal and the proximal phalanx in prosimians. Interestingly, New World monkey species resemble prosimians, rather than Old World simians in this feature, as rotation of the pollex is restricted to the metacarpophalangeal joint. The limited degree of rotation of the thumb relative to the other fingers that this arrangement allows was termed **pseudo-opposability** by Napier (1961). Incidentally, it should be noted that true opposability of the thumb in Old World monkeys, apes and humans involves not only rotation at the saddle-joint between the metacarpal and the carpus but also some degree of rotation at the metacarpophalangeal joint.

In tree-shrews, in contrast to all living primates, there is little special divergence of the pollex from the other fingers and there is no significant rotation either at the metacarpophalangeal joint or at the carpo-metacarpal joint. Haines (1958), followed by Napier (1961), aptly described the tree-shrew hand as convergent in that the tips of all the fingers are brought together by flexion at the metacarpophalangeal joints. Jenkins (1974b), on the basis of detailed laboratory observations, discounted various published claims that the tree-shrew pollex is to some extent opposable.

As far as fossil evidence is concerned, there is relatively little information available. Incomplete hand skeletons of the Eocene 'lemuroid' *Notharctus* indicate that the pollex was somewhat less divergent than in modern lemurs, although apparently adapted to some extent for a grasping action (Gregory, 1920a). Beard (1987) has since confirmed that the wrist in

Smilodectes gracilis, now documented by several partial skeletons, is relatively primitive in comparison to modern lemurs. In addition, a hand (complete except for the phalanges) attributed to *Adapis* has recently been described by Godinot and Jouffroy (1984). The reconstruction indicates that the thumb was moderately divergent and permitted some degree of grasping, although again not as much as in typical modern lemurs. Among 'archaic primates', the hand was until recently known only from a few elements in the case of *Plesiadapis*, and allocation of some of them may be questionable (Szalay, Tattersall and Decker, 1975). The only fairly reliable evidence came from bilaterally flattened, pointed terminal phalanges that indicate that the hand of *Plesiadapis* had powerful, curved claws that would scarcely be associated with grasping extremities of the kind typically found in primates of modern aspect. Gingerich (1986a) has now confirmed that the hand of *Plesiadapis* was characterized by digits bearing strong claws.

Bishop's comparative studies of hand-use in prosimian primates (1962, 1964) have added a detailed behavioural dimension to studies of primate hand morphology (e.g. Napier, 1960, 1961). For a variety of species in captivity, Bishop recorded both the range of orientations of their hands on arboreal supports (wooden dowels of different diameters) and the form of the hand during grasping actions (i.e. the **prehensive pattern**). The common tree-shrew (*Tupaia glis*) proved to be distinguishable from prosimian primates (*Microcebus*, *Lemur*, *Galago*, *Nycticebus* and *Loris*) in several ways. In particular, the tree-shrew showed the greatest variability in orientation of the hand relative to a support. Further, among all of the variable orientations that were recorded, the commonest axis lay across the thenar pad and between digits 4 and 5 (see later text and Fig. 10.9), whereas this orientation was very rare both in prosimians and in a sample of simian species (*Callimico*, *Aotus*, *Saimiri*, *Callicebus*, *Cercopithecus* and *Erythrocebus*). This predominant orientation of the hand relative to a support in *Tupaia* is linked to the unusual degree of development of the hypothenar pad (see later text and Fig. 10.9; Haines, 1955). With respect to the typical prehensive pattern of *Tupaia glis*, Bishop confirmed the convergent action of the hand defined by Naper (1961) and noted that the common tree-shrew rarely grasps an object with just one hand. In common with a variety of other generalized small mammals, such as rats and squirrels, it usually holds food items in both hands. As Napier has noted, the tree-shrew typically uses its two hands in opposition in this way, rather than using two parts of one hand to oppose one another.

Prosimian primates differ from *Tupaia* in that the orientation of the hand on a support is less variable and in that the support is more commonly oriented between digits 1 and 2 of the hand (i.e. between thumb and index finger). This suggests a greater emphasis on the role of the thumb in prosimians, and there is certainly no indication of the favoured tree-shrew pattern of gripping a support between the hypothenar pad and digits 1–3 of the hand. Nevertheless, there is great variation in the orientation of the hand on supports even in prosimian primates and the thumb is not sharply differentiated from the other digits in its action. Similar findings emerged from Bishop's study (1962, 1964) of prehensive patterns in prosimians. All prosimians investigated seem to be characterized by a whole-hand mode of control, although Bishop noted the possibility that the aye-aye (*Daubentonia*) may well have individual control of the middle digit of the hand in association with its unusual feeding behaviour. Prosimians do use a single hand in prehension – for example, in holding plant food items and (at least in the case of the small nocturnal forms such as bushbabies and mouse lemurs) in grasping arthropod prey. The hands are also used to grasp the fur of another individual during social grooming. Furthermore, the survival of a prosimian infant carried on its mother's fur (see Chapter 9) depends on the infant's ability to grasp with both hands and feet.

Bishop also points out that there is a consistent difference between lemurs and members of the loris subgroup in the prehensive pattern. Whereas lemurs reach out their hands with digits 2–5 essentially parallel and with the fingertips leading, members of the loris group (*Galago*, *Loris* and *Nycticebus*) reach out with the hand closed initially, opening the hand when it is near to the object such that the digits are spread out and the first contact with the object takes place with the interdigital pads (see later text and Fig. 10.9). Bishop was unable to include *Tarsius* in her laboratory study, but did note from photographic evidence that the prehensive pattern of the tarsier seems to resemble that of the loris group rather than that of lemurs. Overall, however, it emerges that prosimians generally have little fine control of their hands and they tend to grip objects between fingers and the palm rather than between the thumb and the other fingers.

Although Bishop's publications (1962, 1964) deal particularly with prosimians, they also include some valuable information on hand-use in simian primates. This information clearly confirms the distinction between New World and Old World simians drawn by Napier (1960, 1961). Somewhat surprisingly, all of the four New World monkey genera examined (*Callimico*, *Aotus*, *Saimiri* and *Callicebus*) orientate their hands variably on supports, with no significant emphasis on the axis between the thumb and the index finger. Hence, when these New World monkeys use arboreal supports the thumb is even less differentiated than in the prosimian primates studied by Bishop. Moreover, the prehensile use of the hand essentially follows the prosimian, whole-hand pattern, although, compared with prosimians, use of the hand in parting the fur of a partner during social grooming and in running finger-tips over novel objects is more sophisticated (Bishop, 1962).

These behavioural observations closely match Napier's conclusion that the hand of New World monkeys is psuedo-opposable rather than truly opposable. By contrast, the Old World monkeys observed by Bishop (1964) consistently moved their thumb independently against some part of the index finger or the side of the hand. This was true both of guenons (*Erythrocebus* and three *Cercopithecus* species) and of leaf-monkeys (*Presbytis* and *Colobus*), even though the thumb has been reduced to a small stump in *Colobus*. In addition, numerous Old World monkey species have progressed beyond simple use of the thumb in opposition and are able to use the index finger (digit 2) separately. Napier (1961) found that prehensile use of the hand is particularly well developed in Old World monkey species with terrestrial habits, and Bishop confirmed this in a comparison between arboreal guenon species (*Cercopithecus*) and the terrestrial patas monkey (*Erythrocebus patas*). The most refined prehensile use of the hand among non-human primates occurs in the gelada baboon (*Theropithecus*), which is highly specialized for feeding on the ground (Dunbar, 1984). Given the fact that prosimians and New World monkeys, in which prehensile use of the hand is relatively poorly developed, are almost exclusively arboreal, it would seem that life in trees actually inhibits the development of refined prehensile use of the hand. Hence, whereas the grasping power of the hallux is universal to non-human primates and undoubtedly plays a major part in arboreal locomotion, the grasping power of the pollex is only moderately developed in arboreal primates and requires terrestrial activity for further elaboration.

The fact that all Old World simians have a truly opposable thumb, regardless of their present lifestyle (arboreal or terrestrial), may suggest that their common ancestor was at least partly terrestrial. This would explain, for instance, why a predominantly arboreal leaf-monkey like *Colobus* shows both true opposition of the thumb (despite its reduction to a small stump) and independent use of the index finger, whereas broadly comparable New World monkey species that are arboreal in habits (e.g. the howler monkey, *Alouatta*) do not have such refined hand-use.

Pads on palms and soles

Patterns of use of the hand and foot in primates are also associated with the characteristics of the naked skin covering the ventral surface of the extremities (effectively reviewed by Biegert, 1961). Whipple (1904) concluded from a broad comparison of mammals that there is a basic pattern of **cutaneous pads** that can be identified both on the palm (palmar surface) of the hand and on the sole (plantar surface) of the foot (Fig. 10.9). In this basic pattern, each of the digits bears an **apical pad** (= terminal pad) and there are six pads on both the palm and the sole. On the palm, there are four pads at the junctions of the basal phalanges with the metacarpals (**interdigital pads**) and two other pads adjacent to the wrist articulations of the radius (**thenar pad**) and the ulna (**hypothenar pad**), respectively. Similarly, the sole bears four interdigital pads at the junctions of the basal phalanges and the metatarsals, a thenar pad corresponding to the articulation of the ankle with the tibia and a hypothenar pad corresponding to the position of the end of the fibula (Fig. 10.9). The basic pattern of six pads on the palm and sole is retained in a variety of relatively unspecialized small mammals, notably in arboreal marsupials and in most modern insectivores. The pentailed tree-shrew (*Ptilocercus*) also retains six pads on the palm and the sole, but in *Tupaia* the pattern has been modified in the foot by fusion of the first interdigital pad with the thenar pad, whereas the hand retains the primitive pattern.

In many mammalian groups, the original pattern of pads on the palm and sole has been modified to varying extents by fusion or obliteration. Of course, in cases where the limbs have been completely remodelled for particular patterns of locomotion (e.g. in ungulates and bats), the pads are no longer recognizable at all. In primates, however, it is generally possible to identify palmar and plantar pads, albeit with certain modifications in most cases. Prosimians generally have fairly distinct pads on the palms and soles and in the loris group (Lorisiformes) it is possible to identify all six of the original pads on both hand and foot. There has been some tendency for mild coalescence of the pads, however, and this is particularly true of the slow-climbing loris subgroup (Lorisinae). By contrast, the number of pads has been reduced to five on both hand and foot in all Malagasy lemurs through fusion of the first interdigital pad with the thenar pad, and this would seem to be a shared derived feature of the Lemuriformes. In tarsiers, there is a quite distinctive combination in that the hand retains the primitive pattern of six pads, as in the loris group (although there is some tendency for the first interdigital pad to be fused with the thenar pad), whereas in the foot there has been extensive fusion of pads to leave only three. These are represented by an enlarged first interdigital pad, a fused combination of second and third interdigital pads, and a large, curving structure formed by fusion of the fourth interdigital pad with both the hypothenar and the thenar pads (see Fig. 10.9).

In simian primates, it is still possible in many cases to recognize the basic array of six pads on the palm and sole, but coalescence of the originally separate cutaneous zones with the rest of the palmar and plantar skin has tended to obscure the pattern. As has been pointed out by Schlaginhaufen (1905), the conspicuously raised, demarcated pads of prosimian primates generally contrast with the diffuse spread of comparable areas of skin in simians, in which the entire ventral surfaces of the digits are included as an extension of the apical pads and of the palm or sole. Nevertheless, the original presence of clearly marked pads in primates is confirmed by their retention in the early human fetus (Johnson, 1899).

The cutaneous pads of the hands and feet of primates are of particular functional significance because they play a part both in gripping arboreal supports and in providing for tactile sensitivity. The frictional and tactile properties of the skin on the pads can be traced to the fine epidermal ridges (**dermatoglyphs**) that, among other things, are responsible for human fingerprints. Most mammals lack epidermal ridges, but they are present on the cutaneous pads of the hands and

feet in various arboreal marsupials, in tree-shrews and in all primates. There are, however, some significant differences in the orientation of ridges on the apical pads of the digits in these mammals. In marsupials, the typical pattern is a set of concentric loops on the apical pads. In diurnal tree-shrews (e.g. *Tupaia*), the ridges are oriented transversely across the apical pads, although *Ptilocercus* has looped ridges reminiscent of the marsupial condition (Carlsson, 1922; Haines, 1955). In primates, in contrast to tree-shrews, the predominant orientation of the ridges of the apical pads is clearly longitudinal. Schlaginhaufen (1904) showed that the apical pads of prosimian primates bear essentially longitudinal ridges, whereas the pattern in simian primates is generally complicated by whorls. Further, Schlaginhaufen observed that epidermal ridges are typically confined to the cutaneous pads in marsupials, tree-shrews and most prosimians, whereas in simian primates (and in members of the loris subgroup, such as *Loris* and *Nycticebus*) these ridges are also found on other parts of the palm and sole. Indeed, in simian primates epidermal ridges are found along the entire ventral surfaces of the digits rather than being confined to the apical pads, as in prosimians.

Whipple (1904) inferred that epidermal ridges on the cutaneous pads of the hands and feet of mammals primarily serve to increase friction between contact surfaces and this function has been explored in detail by Cartmill (1974c, 1979). Cartmill (1979), in fact, produced a theoretical explanation for the empirical observation that coalescence of pads tends to increase with increasing body size among primates generally. Schlaginhaufen (1905), however, proposed that

Figure 10.9. Ventral views of the hands (left column) and feet (right column) of (a) a tree-shrew (*Tupaia belangeri*), (b) a mouse lemur (*Microcebus murinus*) and (c) a tarsier (*Tarsius bancanus*), showing the raised cutaneous pads with their epidermal ridges. The first digit (pollex or hallux) is labelled 1.

The basic mammalian pattern is seen in the hands of *Tupaia* and *Tarsius* and in the foot of *Microcebus*, with apical pads on all five digits and six pads on the palm or sole. The tree-shrew bears sharp claws on all digits, whereas the mouse lemur and tarsier have flat nails on virtually all digits (see inset for *Microcebus* foot). 'Toilet claws' are seen standing away from the dorsal surface of the second and third digits in the foot of *Tarsius* and of the second digit in the foot of *Microcebus*. Note that the epidermal ridges on the apical pads are transverse in the tree-shrew and longitudinal in the mouse lemur and tarsier. Note also the presence of an accessory pad on the hallux of the mouse lemur, a specialization common among primates. (After original drawings made by the late N.A. Barnicot.)

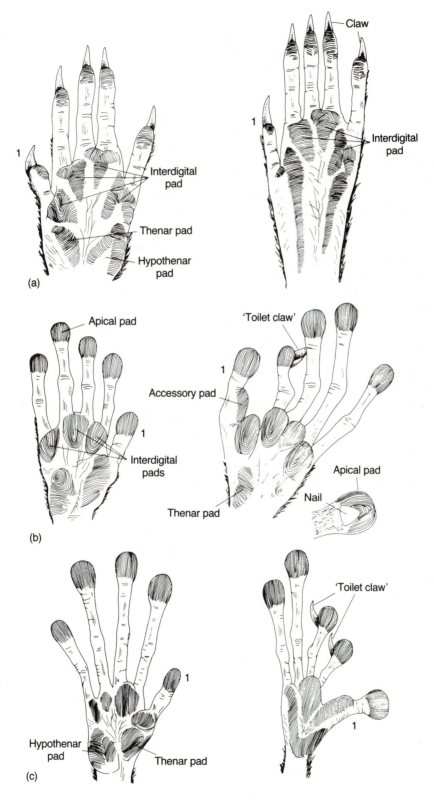

the ridges play a special part in tactile discrimination. He showed through experiments with human subjects that tactile discrimination of separate points is more successful when the points lie on a line running across a ridge rather than lying parallel to it. Further, skin bearing such epidermal ridges was found to be generally more capable of discriminating fine tactile sensations than skin elsewhere on the body. Schlaginhaufen's demonstration of a relationship between patterns of human finger ridges and tactile sensitivity has recently been supported by more sophisticated studies (Loesch and Martin, 1984).

Epidermal ridges doubtless serve a dual function, frictional and tactile, which is clearly best served when the ridges are oriented at right angles to the direction of motion of the digits along the substrate (Whipple, 1904). Hence, it can be inferred that the terminal pads on the digits of tree-shrews are primarily adapted for orientation of the axes of the digits along the axis of an arboreal support (i.e. with the essentially transverse epidermal ridges at right angles to that axis), whereas the terminal pads on the digits of prosimian primates are adapted for orientation of the axes of the digits perpendicular to the axis of an arboreal support (i.e. with the longitudinal epidermal ridges at right angles to that axis). In simian primates, which have whorls rather than straight ridges on their terminal pads, it would

seem that the digits are secondarily adapted in an all-purpose way to provide frictional resistance and tactile sensitivity in virtually any axis relative to an arboreal support.

Histological studies of the skin in primates have shown that there is, in fact, a well-defined pattern that is specifically adapted for the enhancement of tactile sensitivity (Cauna, 1954, 1956). The ridges on the outer surface of the epidermis, consisting of stratified horny (keratinized) epithelium, are found to have more extensive counterparts on the internal surface. Corresponding to each external epidermal ridge, there is a well-developed internal ridge (**intermediate ridge**) projecting down into the underlying layer (dermis) of the skin. In addition, these intermediate ridges alternate with other ridges (**limiting ridges**) that correspond in position to the grooves between external epidermal ridges (Fig. 10.10). There are cross-connections between all of these internal ridges, forming box-like compartments extending down into the dermis. Cauna (1954) has argued convincingly (with respect to human skin) that the intermediate ridges act like magnifying levers to increase sensitivity to pressure or friction applied to the external surface of the skin, thus permitting more efficient tactile perception.

The part played by the epidermal ridges of the cutaneous pads on primate hands and feet depends, of course, on the sense organs of the

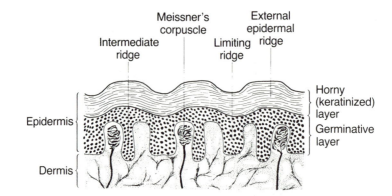

Figure 10.10. Diagram of a section through the ridged skin of a cutaneous pad (e.g. apical digital pad) of a typical primate. The epidermal ridges are linked to internal ridges that magnify any movement at the surface. Meissner's corpuscles, which are apparently unique to primates among placental mammals, but which have also been reported from a marsupial (opossum, *Didelphis*), are specialized tactile end organs directly associated with the internal ridge system (see Cauna, 1954; Winkelmann, 1963).

skin that provide for tactile sensitivity. Winkelmann (1963, 1965) has reviewed the evidence concerning tactile nerve endings in the skin of primates and other mammals, recognizing three distinct categories: (1) free nerve endings in dermal nerve networks; (2) networks of nerve endings surrounding hair follicles; and (3) specially organized nerve endings. Free nerve endings in dermal nerve networks occur throughout the skin of the body in mammals generally and seem to serve a general range of functions including coarse tactile perception. Wherever the skin is covered in hair, the nerve endings surrounding hair follicles are prominent and serve an essential tactile function. This function is enhanced with specialized long hairs (**vibrassae**), such as whiskers, that are located on particular sites of the body in many mammalian species. Finally, all mammals seem to have **organized nerve endings** of some kind that contribute to tactile perception. A well-known example is provided by the Vater–Pacini corpuscle, which is common along nerve trunks in the extremities of mammals and also occurs in the skin, although it is more often found in the connective tissue (panniculus) layer underlying the skin than in the dermis itself. The Vater–Pacini corpuscle, which is known to serve as a pressure receptor, consists of an inner bulb, containing the nerve fibre, surrounded (encapsulated) by a layered series of membranes.

In several mammalian groups other than primates, the dermis contains (just beneath the epidermis) special end organs, at least in particularly sensitive areas of skin, that seem to be equivalent both structurally and histochemically to the inner bulb and nerve fibre of the Vater–Pacini corpuscle (Winkelmann, 1963). These organized nerve endings, called **mammalian end organs** have so far been found in all non-primates studied, including a variety of carnivores, rodents, insectivores, lagomorphs and ungulates (Winkelmann, 1965).

Winkelmann (1963) found that all of the primate species contained in his sample, which included several prosimian species as well as a variety of simians, had a different kind of organized nerve ending in the dermis of the skin. This nerve ending, which appears to be confined largely or completely to naked (glabrous) areas of skin such as the cutaneous pads of the extremities, is known in humans as **Meissner's corpuscle**. Meissner's corpuscle differs from the mammalian end organ in that it is not encapsulated; instead, several nerve fibres with knoblike or reticular expanded endings are coiled and layered together without a surrounding membrane (Fig. 10.10). Meissner's corpuscles are located between the internal ridges of the epidermis and are clearly well placed to detect the magnified movements arising from friction or pressure at the external surface. There is therefore little doubt that the combination of the ridge system of the cutaneous pads with Meissner's corpuscles provides primates generally with an enhanced tactile sense. Interestingly, New World monkeys with a well-developed prehensile tail (e.g. spider monkeys, *Ateles*, and woolly monkeys, *Lagothrix*) have papillary ridges on the naked skin on the underside of the tail tip and Meissner's corpuscles are also present in that skin (Biegert, 1961). In tactile terms, then, the prehensile tail of these primates really acts like a fifth extremity.

Only one tree-shrew species, *Urogale everetti*, was studied by Winkelmann, but it is significant that no Meissner's corpuscles were found in the ventral skin of its extremities. In fact, Winkelmann found evidence only of dermal nerve networks and Vater–Pacini corpuscles and did not even find the mammalian end organs reported for other non-primates. Hence, it can be concluded that tree-shrews, with their transversely oriented ridges on the terminal pads of the digits and their apparently more primitive innervation of the cutaneous pads, differ quite sharply from primates and probably have a more limited tactile sense in their extremities. The universal occurrence of Meissner's corpuscles among living primates suggests that they were present as a feature of the ancestral primate stock (excluding tree-shrews). At first sight, it might seem that they could be a unique derived feature of primates, but Winkelmann (1964) has reported the presence of structurally identical Meissner's corpuscles in the skin of a marsupial (the common opossum, *Didelphis virginiana*). To explain this, one must either postulate the convergent evolution of Meissner's corpuscles in marsupials and primates or (as proposed by Winkelmann, 1965) accept the existence of these corpuscles in the common ancestral stock of marsupials and placentals, with subsequent loss occurring in many placental groups (including tree-shrews).

Nails and claws

Another significant feature of primate extremities is that the digits usually bear blunt, flat **nails** rather than the bilaterally compressed, sharp **claws** typical of relatively unspecialized mammals such as insectivores, rodents and carnivores. Indeed, Le Gros Clark (1936) described the possession of flat nails as 'one of the most distinctive features of the Primates'. However, although most living primates have nails on all digits, there are several exceptions. Among prosimians, the only clear exception is the aye-aye (*Daubentonia*), which has sharp, pointed claws on all digits except the hallux. The feet of prosimians generally are also somewhat unusual in having '**toilet claws**' ('grooming claws'). But these structures are not typical claws, as they are neither sharp nor bilaterally compressed; they are simply elongated, curved nails that stand at an angle to the dorsal surface of the digit rather than lying horizontally across that surface (Fig. 10.9). Lemurs and lorises have one 'toilet claw' on each foot, on the second toe, and tarsiers have two 'toilet claws' on each foot, on the second and third toes. It is a moot point whether ancestral primates already had a 'toilet claw' on the second toe, or whether convergent evolution occurred in the ancestor of lemurs and lorises and in ancestral tarsiers. As the name implies, these special structures seem to be adaptations for more effective grooming of the fur. Finally, among simians, the entire marmoset and tamarin family (Callitrichidae) – along with Goeldi's monkey (*Callimico goeldii*) – is characterized by the possession of sharp claws on all digits except the hallux. In fact, despite all of the exceptions noted, there is always a flat nail on the hallux in living primates. This feature – like the grasping adaptation of the big toe – is a universal characteristic of living primates, indicating that it was present in their common ancestral stock (see also Cartmill, 1974c). By contrast, tree-shrews have claws on all digits of the hand and foot including the hallux; so they are once again distinct from all modern primates.

The usual interpretation of the facts given above is that ancestral placental mammals, like modern tree-shrews, had claws on all digits and that in primates the claws have been replaced by nails to varying degrees in connection with the development of grasping functions of the extremities. According to this view, those primates that have sharp claws on most of their digits (the aye-aye and marmosets and tamarins) have retained them from an ancestral primate stock in which there were claws on all digits, except perhaps on the hallux (Le Gros Clark, 1959; Cartmill, 1974c). Support for this interpretation was apparently provided by Le Gros Clark's detailed study (1936) of primate 'claws'. As was shown by Le Gros Clark, tree-shrews (*Tupaia glis* and *Ptilocercus lowii*) have a typical mammalian claw (**falcula**) on all digits (Figs 10.9 and 10.11). The claw is strongly curved, markedly compressed from side to side and closely applied to the terminal phalanx of the digit, which is itself claw-shaped (Le Gros Clark, 1959). Histological examination shows that the horny substance of the claw has two distinct layers: a thin **superficial stratum** with almost horizontal fine lamellae and a far thicker **deep stratum** with the lamellae oriented obliquely upwards, pointing towards the tip of the claw. Beneath the horny substance of the claw lies the matrix, which can be divided into three successive zones. The first zone, close to the root of the claw, is the **basal matrix**, which gives rise to the superficial layer of the claw. This is followed by the **terminal matrix**, which produces the deep layer of the claw. Finally, there is the **sterile matrix**, which is the largest part of the matrix but plays no part in the formation of the claw. In comparison to this typical structure of a claw, the nail (**ungula**) of primates is a much simpler structure. The nail is only moderately curved from side to side, and the lack of bilateral flattening is reflected in the flat shape of the

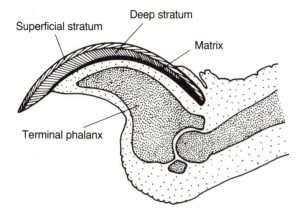

Superficial stratum

Deep stratum

Matrix

Terminal phalanx

Figure 10.11. Longitudinal section through the tip of a digit of the common tree-shrew (*Tupaia glis*), showing the basic structure of the claw. Note the claw-shaped terminal phalanx around which the bilaterally flattened claw fits closely. The claw itself has three recognizable layers: matrix, deep stratum and superficial stratum. In a typical primate digit, a flat nail lies above a dorsoventrally flattened phalanx and the deep stratum is absent from the nail substance. (After Le Gros Clark, 1936.)

terminal phalanx of the digit. Further, the nail has only a basal matrix and a superficial layer; the terminal matrix and deep layer characteristic of claws are both absent.

Le Gros Clark (1936) showed that the claws of the aye-aye and of marmosets and tamarins are intermediate between the typical sharp claws of other mammals (e.g. tree-shrews) and the typical nails of primates. In the claws of *Daubentonia*, all of the basic components of the mammalian claw system (as shown in Fig. 10.11) are present, but there are differences of degree. The basal matrix and the superficial stratum are relatively more prominent in a claw of the aye-aye and the terminal phalanx is not as curved as in tree-shrews. The claws of the marmosets and tamarins (e.g. *Callithrix* and *Leontopithecus*) are even further removed from the typical mammalian condition. The terminal matrix is not as distinct as in tree-shrews and the deep stratum represents little more that 15% of the thickness of the claw substance. Overall, lamination of the claw substance is less obvious than in tree-shrews and the terminal phalanx is only moderately claw-

shaped. Le Gros Clark also studied the 'toilet claws' of some prosimians. For example, in mouse lemurs (*Microcebus*) and bushbabies (*Galago*), he found no evidence of a terminal matrix or deep stratum, but he was able to recognize an ill-defined terminal matrix in the 'toilet claws' of *Tarsius*, along with an unlaminated deep stratum representing about a third of the thickness of the nail substance.

All of these findings are, of course, compatible with the hypothesis that the ancestral primates still had claws on all digits except the hallux and that any claw-like structures in living primates have been retained from that stage (Le Gros Clark, 1959; Cartmill, 1974c). However, they are also compatible with the alternative hypothesis that the ancestral primate already had nails on all digits and that in certain cases (e.g. the aye-aye, marmosets and tamarins, 'toilet claws' of prosimians) there have been special secondary developments to produce claw-like structures, none of which shows the full characteristics of the typical mammalian claw as exemplified by tree-shrews. Cartmill (1974c) cites as evidence against this alternative hypothesis Thorndike's report (1968) that there is a thin deep stratum in the nails on the foot of the white-fronted capuchin (*Cebus albifrons*) and Le Gros Clark's identification of a similar condition in the 'toilet claw' of *Tarsius* (as noted above).

In fact, Thorndike extended Le Gros Clark's study of the claws of marmosets and tamarins by examining the histological structure of the nail on the hallux in two tamarin species (*Saguinus oedipus* and *S. nigricollis*) as well as that of the claws. Surprisingly, a thin deep stratum is also identifiable in the nail of the hallux; it is more diffuse than in the claws borne by the other digits, but it is actually somewhat thicker. Although Thorndike interprets this unusual observation in terms of retention of claws as a primitive feature in marmosets and tamarins, it actually suggests that the distinction between claws and nails is rather less clear than is often assumed. For instance, if it seems reasonable that the ancestral primates had a flat nail on the hallux, and if marmosets and tamarins have nevertheless 'retained' a thin deep stratum in that nail, then it is surely possible that a thin deep stratum might have persisted in various stages of the evolution of nails elsewhere among primates. Gregory (1920a) noted that the terminal phalanges of the hands and feet of the Eocene 'lemuroid' *Notharctus* were intermediate in shape between those of the aye-aye and those of lemurs with typical flat nails (Lemuridae and Indriidae). As Gregory suggests, the digits probably bore quite narrow nails in *Notharctus*, and it therefore seems plausible that they might have been intermediate in histological structure between typical claws and nails of modern mammals. (Incidentally, Franzen (1987) has reported two flattened terminal phalanges of the Eocene 'lemuroid' *Europolemur koenigswaldi*, providing further confirmation that flat nails were present on the digits of Eocene 'lemuroids'.) In fact, fine gradations in nail structure can be found among modern primates such as hominoids (great apes and humans), as shown by Sprankel (1969).

Cartmill (1974c) also raised the objection that it is unparsimonious to propose that the ancestral primates had flat nails on all digits, as claws would then have to re-appear in the evolution of certain modern primates. It is, in fact, rather difficult to apply the criterion of parsimony because of the intermediate conditions noted above, and the differences between claws and nails (shape of terminal phalanx, shape of claw or nail and presence or absence of terminal matrix and deep stratum) are by no means so complex that evolutionary reversals would be inherently unlikely. Nevertheless, it is interesting to consider how many transitions would be required from obvious claws, excluding 'toilet claws', to obvious nails (or vice versa) within primate evolution (Fig. 10.12). It emerges that at least six transitions from claws to nails would be required within the framework of primate evolution if the common ancestor had claws on all digits except the hallux, whereas only three transitions (one from claws to nails at the beginning and two to secondary claws from nails later on) would be required if ancestral primates had nails on all digits. If 'toilet claws' are regarded as proper claws, and if it is assumed that they have been

developed secondarily in strepsirhines (lemurs and lorises) and in tarsiers separately, the total number of transitions rises to five. The criterion of parsimony therefore presents no obstacle to the proposition that ancestral primates – like most living primate species – had nails (perhaps with a weakly developed terminal matrix and deep striatum) on all digits. It should also be noted that it is now widely accepted (e.g. see Rosenberger, 1977; Ford, 1980) that claws were redeveloped in the common ancestor of marmosets and tamarins in association with secondary dwarfing (see Chapter 12).

ANKLE AND WRIST JOINTS

Use of the foot and hand in locomotion and/or prehension obviously depends heavily on the joints of the ankle (tarsus) and of the wrist (carpus). It is to be expected that characters identified in the wrist and ankle should be particularly revealing with respect to the adaptative radiation of mammals in general and of primates in particular. For example, it is very likely that characteristic features should distinguish arboreal from terrestrial mammals, since the former must cope with more complex substrates. To date, however, the postcranial skeleton has been relatively neglected in discussions of primate evolution, partly because of a tendency to overemphasize dental features (as noted by Szalay, 1977) and partly because of the sheer complexity of certain components such as the ankle and wrist.

The ankle

Despite considerable variation in detail, the ankle (**tarsus**) of mammals has remained relatively consistent in terms of the basic bony elements present (Fig. 10.13). In all modern viviparous mammals (marsupials and placentals), the **talus** (= **astragalus**) sits astride the **calcaneus** (= **calcaneum**) and articulates with the tibia of the lower leg to constitute all or most of the **primary ankle joint.** In addition, some degree of movement is usually possible between

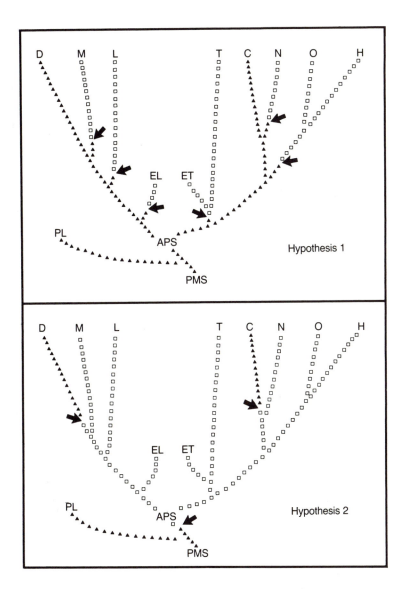

Figure 10.12. Alternative hypotheses to explain the evolution of nails versus claws in primates: (1) If it is assumed that the ancestral primates had claws on all digits (except the hallux), which were retained in the aye-aye and in marmosets and tamarins, six transitions from claws to nails (arrows) are required. (2) If it is assumed that the ancestral primates already had nails on all digits (involving one prior transition from claws to nails), only two transitions back to claws (arrows) on all digits except the hallux are required (arrows).

Abbreviations: D, aye-aye; M, other Malagasy lemurs; L, loris group; T, tarsiers; C, marmosets and tamarins; N, other New World monkeys; O, Old World monkeys; H, hominoids (apes and humans); PMS, placental mammal stock; PL, 'archaic primates' (plesiadapiforms); APS, ancestral stock of primates of modern aspect; EL, Eocene 'lemuroids'; ET, Eocene 'tarsioids'. Note that the number of transitions required obviously depends on the basic pattern of phylogenetic relationships assumed, but quite drastic alterations in the branching pattern are required to affect the basic point regarding parsimony.

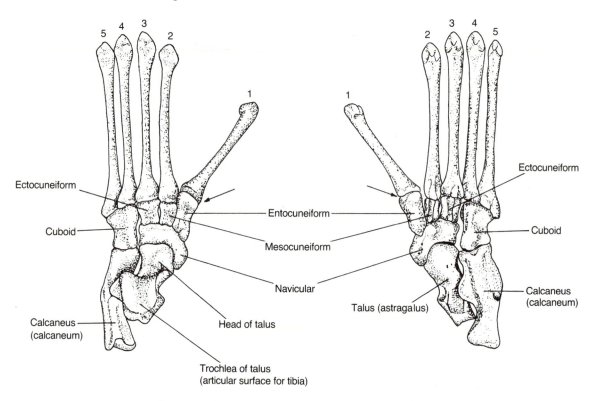

Figure 10.13. The pattern of tarsal and metatarsal bones in the foot of a simian primate (dusky leaf-monkey, *Presbytis obscura*): left, dorsal view; right, ventral view. The metatarsals are numbered 1–5. Note that there is a saddle-joint (arrowed) between metatarsal 1 and the entocuneiform, in association with the special grasping action of the hallux (big toe).

the talus and calcaneus; the articulation between these two bones is supplemented by articulations between talus, navicular, calcaneus and cuboid to constitute the complex **subtalar joint**. The talus and calcaneus together are hence of special importance for overall coordination of movements of the foot, and the calcaneus in particular has numerous diagnostic features (Stains, 1959).

The distal end of the calcaneus articulates primarily with the **cuboid**, and the talus articulates at its distal end primarily with the **navicular**. The navicular articulates in turn with a row of three **cuneiform bones** (Fig. 10.13). Metatarsals 4 and 5 articulate directly with the cuboid; the ectocuneiform articulates with metatarsal 3, the mesocuneiform with metatarsal 2 and the entocuneiform with metatarsal 1 (the

metatarsal of the hallux). Accordingly, when the hallux is adapted for pincer-like grasping, the first metatarsal and the entocuneiform are involved in a typical saddle-shaped joint and have distinctive facets. As all primates of modern aspect, excluding humans, have a grasping hallux, the saddle-shaped facet on the entocuneiform is a virtually universal feature of primates. A facet of this kind is, for instance, plainly recognizable in the Eocene 'lemuroid' *Notharctus* (Gregory, 1920a; Fig. 10.8) and in the Eocene 'tarsioid' *Hemiacodon* (Simpson, 1940). Although a number of entocuneiforms are apparently known for the 'archaic primate' *Plesiadapis*, they have yet to be adequately described and there is no evidence of a grasping adaptation of the foot in 'archaic primates'.

A considerable amount of further information

can be obtained by investigating the structure of the primary ankle joint and of the subtalar joint. In attempting to reconstruct the evolution of the tarsus in primates, a suitable procedure is to identify first of all features common to modern representatives (excluding tree-shrews) and then to relate the common primate pattern to an inferred ancestral pattern for placental mammals generally. This task is greatly facilitated by a fine series of papers by Lewis (1964a, 1964b, 1980a,b,c, 1983), based on detailed dissections of various primates and other mammals, supplemented by study of serial radiographs of ligamentous preparations fitted with metal markers. In conjunction with other detailed studies of the mammalian tarsus (e.g. Barnett, 1970; Decker and Szalay, 1974; Schaeffer, 1941, 1947; Szalay, 1977; Szalay and Drawhorn, 1980),

these papers yield a wealth of insights into functional morphology and evolution. A particularly effective review of the evolution of the primate foot, with special emphasis on the ankle, is provided by Conroy and Rose (1983).

One of the most striking features of the primate tarsus is that the compromise axis of the subtalar joint is typically oriented obliquely, rather than transversely, across the long axis of the foot (although *Tarsius* is a striking, secondarily specialized exception; see Jouffroy, Berge and Niemitz, 1984). The axis actually passes through the neck of the talus and on through (or parallel to) the hallux, thus defining the hallux as the reference component for foot movement. As noted by Lewis (1980b), this may be mechanically advantageous, because the hallux remains relatively stationary as the rest of

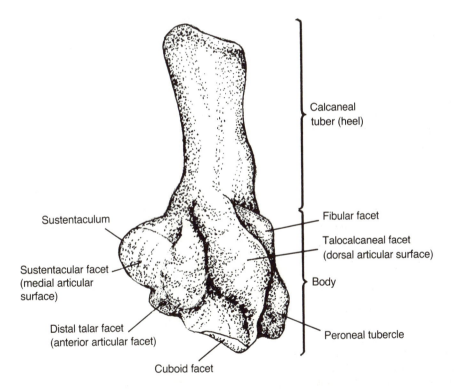

Figure 10.14. Dorsal view of a typical mammalian calcaneus, that of the Palaeocene condylarth *Claenodon*. Note the relatively distal position of the peroneal tubercle, quite close to the cuboid facet, and the somewhat oblique orientation of the latter, relative to the long axis of the calcaneus. Primates differ from this in the retracted position of the peroneal tubercle and in the transverse orientation of the cuboid facet. (After Schaeffer, 1947.)

the foot moves into a grasping posture. In fact, the subtalar joint has a characteristic screw-like (helical) form, associated with a relatively elongated, convex facet for the talus on the dorsal surface of the calcaneus. As a result, the calcaneus moves forwards and downwards relative to the talus when the sole of the foot is turned inwards (inverted) – for example, in grasping a branch. This movement is, of course, reversed when the sole of the foot is turned downwards (everted) for application to a relatively flat substrate (Fig. 10.14).

Lewis and others have recorded several additional features that typify the primate tarsus. On the talus (see Fig. 10.13), the lateral margin of the saddle-shaped trochlea (the surface articulating with the tibia) is more prominent than the inner (medial) margin. Further, the articular surface for the tibia extends forwards onto the neck of the talus, to end in an articular cup (termed the 'squatting facet' by Decker and Szalay, 1974). This cup acts as a kind of stop during the backward rotation of the talus and may therefore set a limit to upward rotation (dorsiflexion) of the foot. On the calcaneus (Fig. 10.14), the distal facet for articulation with the cuboid is oriented transversely, rather than obliquely, to the long axis of the calcaneus and the joint between the calcaneus and cuboid is of rotating (pivoting) form, with semicircular or kidney-shaped facets on both bones. The distal portion of the calcaneus is relatively elongated, such that it excludes articular contact between the talus and the cuboid and often forms a substantial joint with the navicular in addition to its primary articulation with the cuboid. Finally, a projection called the peroneal tubercle (Fig. 10.14), which is typically quite small or entirely absent in modern prosimian primates, is usually well removed from the distal end of the calcaneus when present. It should also be noted that Decker and Szalay (1974) and Dagosto (1983) have reported the presence of a deep groove beneath the sustentaculum (supporting shelf for the talus) of the calcaneus. This groove, which apparently serves as a channel for the flexor fibularis tendon, is regarded by these authors as a typical feature of the primate tarsus.

To proceed further with tracing the evolution of the primate tarsus, with due reference both to the fossil record and to the question of affinities between tree-shrew and primates, it is essential to establish the likely ancestral features of the placental mammals in general. Only on this basis is it possible to identify probable derived features of the primate tarsus and hence to assess relationships between primates and other mammals.

Evolution of the mammalian ankle

A series of inferences concerning the evolutionary history of the primate tarsus have been made by Szalay and his colleagues (Szalay, 1966, 1975b, 1977; Decker and Szalay, 1974; Szalay and Decker, 1974; Szalay, Tattersall and Decker, 1975; Szalay and Drawhorn, 1980) on the basis of detailed information on the morphology of the talus and calcaneus of living and fossil primates and other mammals. In a nutshell, these studies led to the conclusion that the tarsus of the 'archaic primate' *Plesiadapis* shares certain derived features with that of primates (including Eocene 'lemuroids'), indicating that the 'archaic primates' are the closest known relatives of primates of modern aspect. Szalay and Drawhorn have further proposed that apparent shared derived features in the tarsus of 'flying lemurs' (dermopterans), tree-shrews and primates (including 'archaic primates') justify recognition of a superorder Archonta including all of these mammals (and possibly bats), as originally proposed by Gregory (1910). Hence, although not actually including tree-shrews within the order Primates, Szalay and Drawhorn have lent credence to the idea that there is still a fairly close link between tree-shrews and primates (see also Szalay, 1977).

The key to the above interpretations resides in the procedure used to infer the ancestral condition for placental mammals. Szalay and Decker (1974) considered isolated late Cretaceous tarsal bones (calcanei and tali) attributed to *Protungulatum donnae* (an early condylarth) and to *Procerberus formicarum* (an early palaeoryctid 'insectivore') as broadly representative of the primitive condition. They listed the following inferred ancestral features for placental mammals (see Fig. 10.13 for the talus and Fig. 10.14 for the calcaneus):

1. Peroneal tubercle distal on the calcaneus.
2. Cuboid facet oriented obliquely across the long axis of the calcaneus.
3. Dorsal articular facet for the talus forming a relatively large angle with the long axis of the calcaneus.
4. Distal portion (body) of the calcaneus relatively short. Shallow, indistinct groove beneath the sustentaculum.
3. Broad, low trochlea with only moderate grooving on the talus.
6. Head of the talus wide and probably in contact with the cuboid.

In addition, Szalay and Decker (1974) noted that the fibula articulated with the calcaneus as well as with the talus in *Protungulatum* (although not in the contemporary *Procerberus*) and described this condition as 'undoubtedly primitive for the Eutheria and other Mammalia' (see Fig. 10.14).

In contrast to this inferred ancestral condition for placental mammals, Szalay and Decker (1974; see also Szalay, Tattersall and Decker, 1975) identified several features as apparent derived character states linking *Plesiadapis* with primates:

1. Peroneal tubercle somewhat more proximal.
2. Cuboid facet oriented less obliquely.
3. Dorsal articular facet for the talus formed a relatively smaller angle with the long axis of the calcaneus.
4. Deeply excavated groove beneath the sustentaculum of the calcaneus.
5. Trochlea of the talus longer than wide, although still rather low: lateral border high and sharply crested; medial border low and rounded.
6. Loss of articular contact between the fibula and the calcaneus.

Overall, Szalay and Decker (1974) concluded that the pattern inferred for ancestral placental mammals was associated with terrestrial habits

and that features typical of *Plesiadapis* and primates of modern aspect emerged as a common ancestral pattern associated with the initial adoption of arboreal habits. 'Archaic primates' such as *Plesiadapis* were seen as representing definite intermediates in the transition from terrestrial ancestral placental mammals to fully arboreal primates of modern aspect.

In subsequent studies (Szalay, 1977; Szalay and Drawhorn, 1980), most of the features shared by *Plesiadapis* and primates were described as characteristic of tree-shrews and dermopterans as well. The only feature of the tarsus shared by *Plesiadapis* and primates but lacking in tree-shrews and dermopterans was identified as the presence of a deep groove beneath the sustentaculum of the calcaneus, associated with a wide excavation on the rear of the trochlea of the talus to guide the flexor fibularis tendon. Hence, almost all of the characters originally listed as defining the tarsus of primates were later interpreted as common ancestral features of the Archonta.

This later development in interpretation of the evolution of the primate tarsus in fact involved a radical shift. Szalay and Decker (1974) specifically stated that tree-shrews did not seem to share homologous derived features with 'archaic primates' (i.e. *Plesiadapis*) with respect to tarsal morphology. Indeed, they indicated that the tree-shrew tarsus shares derived features with palaeoryctid 'insectivores', whereas Szalay (1977) later postulated a major evolutionary separation between palaeoryctids on the one hand and Archonta (tree-shrews, dermopterans and primates) on the other. This shift in interpretation appears to have occurred because further studies revealed a gradation in tarsal morphology among living tree-shrews, apparently related to the spectrum from predominantly arboreal to predominantly terrestrial habits (see Chapter 5). Szalay and Drawhorn (1980) concluded that the arboreal *Ptilocercus*, which does seem to be closer to primates in its tarsal morphology, is representative of the ancestral condition for tree-shrews and thus provides support for recognition of the Archonta as a phylogenetic unit.

The reliability of any such phylogenetic inferences depends both on the extent to which the ancestral placental condition can be correctly identified and on the degree to which dermopterans, tree-shrews, 'archaic primates' and primates of modern aspect can be shown to share a unique set of apparently derived character states. Particular importance must be attached here to comparison of placental mammals with their sister group, the marsupials, as a test of hypotheses regarding the ancestral placental condition. The distinction between terrestrial and arboreal habits is of central significance, as mammals included in the Archonta as defined above are characterized by arboreal habits (assuming that the flight adaptations of bats evolved in the trees). Tarsal features reflecting arboreal habits may have arisen by convergent evolution, if the ancestral placental mammals were in fact adapted for terrestrial life, and one must also consider the possibility that such features are retained primitive character states inherited from an arboreal common ancestral stock of placentals (see later).

As explained in Chapter 4, the fossil record for the earliest phases of mammalian evolution is both very poor and subject to considerable bias, so it is not very convincing to argue that the earliest known placental mammals were necessarily primitive in all respects. In fact, postcranial bones from the earliest known placental mammals of the late Cretaceous already reveal a considerable diversity of adaptations, as emphasized by Kielan-Jaworowska, Bown and Lillegraven (1979). This diversity demonstrates that ancestral adaptations of the tarsus of placental mammals must have emerged at an earlier stage, for which fossil evidence is completely lacking. Kielan-Jaworowska (1977, 1978) has provided valuable accounts of postcranial skeletons of the palaeoryctid 'insectivore' *Asioryctes* and of the zalambdalestid 'insectivores' *Zalambdalestes* and *Barunlestes*, all from late Cretaceous deposits in Mongolia. In contrast to the isolated tarsal bones attributed to the somewhat more recent *Protungulatum* and *Procerberus* of North America, the Mongolian skeletal remains are quite substantial and were found in direct association with cranial and dental material.

The postcranial skeleton of zalambdalestids had a combination of apparently primitive features with numerous likely specializations. Several postcranial character states identified by Szalay and Drawhorn (1980) as derived features of 'archontans' were already present in this, the earliest known skeleton of a placental mammal. For instance, the fibula apparently did not articulate with the calcaneus. Further, the distal cuboid facet of the calcaneus was oriented at right angles to the long axis of the bone, although the facet faced somewhat downwards rather than forwards. The talus had a deep trochlea on which the lateral ridge was more prominent than the medial. There were also some unusually specialized features. For instance, there was no peroneal tubercle on the calcaneus and the fibula was fused with the tibia for the distal two-thirds of its length. In fact, Kielan-Jaworowska suggested, on the basis of features such as the fusion of the tibia and fibula, that the essentially terrestrial elephant-shrews (Macroscelididae) provide the closest modern parallel to *Zalambdalestes* and *Barunlestes*. Hence, any features shared by zalambdalestids with the Archonta cannot be attributed to the acquisition of arboreal habits, although they might be explained by retention from an arboreal ancestor of placental mammals. The postcranial skeleton of the zalambdalestids was so specialized in many other respects (e.g. in the shortness of the tarsal region in relation to the metatarsals and in the presence on the axis vertebra of a long process passing horizontally backwards over the adjacent neck vertebrae) that these early 'insectivores' must surely have diverged considerably from the ancestral placental stock.

The tarsus of the contemporary early palaeoryctid *Asioryctes* (Kielan-Jaworowska, 1977) is totally different from that of the zalambdalestids. The peroneal tubercle was well developed on the calcaneus and located close to its articulation with the cuboid, which was arranged obliquely to the long axis of the calcaneus. The trochlea of the talus was low and poorly differentiated. Unfortunately, it is not known whether the fibula articulated with the calcaneus. The most striking feature reported for the tarsus of *Asioryctes*, however, is that the trochlea of the talus apparently did not lie directly above the calcaneus. All modern placental mammals, all modern marsupials and all other known fossil forms attributable to these two groups have a tarsus in which the trochlea sits clearly astride the calcaneus and it has usually been inferred (quite reasonably) that this feature has been retained from the common ancestral stock of placentals and marsupials (see Schaeffer, 1941; Lewis, 1980a, b, c, 1983; Kemp, 1982). Kielan-Jaworowska (1977; see also Kielan-Jaworowska, Bown and Lillegraven, 1979), concluded that, as *Asioryctes* appears on dental grounds to be a definite early placental mammal, superposition of the talus on the calcaneus evolved separately in placentals and marsupials, after the divergence of *Asioryctes* from other placentals. A well-developed sustentaculum was, in fact, present on the calcaneus of *Asioryctes*, so one must consider the possibility that the known postcranial skeleton was distorted during fossilization. It should be noted that in the (presumably related) palaeoryctid *Procerberus* the talus sat astride the calcaneus. Alternatively, it is possible that the condition in *Asioryctes* represents an unusual secondary specialization. There is also the (relatively unlikely) possibility that *Asioryctes* branched off before the origin of talar superposition on the calcaneus in the common ancestor of marsupials and placentals. All of these possibilities require examination, as the suggestion that superposition of the talus on the calcaneus and associated features arose separately in marsupials and placentals requires a remarkable degree of parallel evolution, violating the principle of parsimony.

Kielan-Jaworowska (1977, 1978) inferred terrestrial habits for *Asioryctes, Zalambdalestes* and *Barunlestes*, in keeping with indications that the fossil deposits were laid down under semi-desert conditions. She specifically noted that it would be naive to assume that there were no contemporary mammals with semi-arboreal habits. As the known late Cretaceous mammals already had divergent postcranial specializations and as the available evidence is extremely limited – particularly in zoogeographical terms – it is simply not possible to make any direct inference of the likely habits of ancestral placental mammals from the available fossils alone. Accordingly, the fact that the tarsus of *Protungulatum* seems to have had certain adaptations for terrestrial locomotion can in no way be taken as adequate evidence of a terrestrial origin for placental mammals (*contra* Szalay and Decker, 1974).

Inference of the ancestral condition of the tarsus of placental mammals has been particularly bedevilled by the widespread assumption that articulation of the fibula with the calcaneus (involving a clearly recognizable facet close to the calcaneal facet for the talus; see Fig. 10.14) is necessarily a primitive feature of placental mammals (Schaeffer, 1947; Szalay and Decker, 1974; Kielan-Jaworowska, 1977, 1978). As Schaeffer (1941) pointed out, articulation between the fibula and calcaneus was present in mammal-like reptiles, including the later therapsids. Evolution of the mammalian foot was characterized by superposition of the talus on the calcaneus, such that weight-bearing contact between the fibula and calcaneus was lost (see also Lewis, 1983). Hence, articulation between the fibula and calcaneus was indeed a primitive feature in the earliest origins of the mammals. The presence of such an articulation in Triassic morganucodonts (Jenkins and Parrington, 1976) and in the modern echidna (a monotreme) can be fairly confidently interpreted as due to primitive retention.

Among modern marsupials and placentals, some species have an articulation between the fibula and calcaneus, whereas others lack any sign of such an articulation. At first sight, it would seem reasonable to conclude that the former have retained a primitive feature from the ancestral stock of viviparous mammals, whereas the latter have lost it. Now, in very simple terms one can envisage a foot with a fibulocalcaneal articulation alongside a tibiotalar articulation as being of a hinge type, with little or no rotation along the long axis of the foot. By contrast, a foot

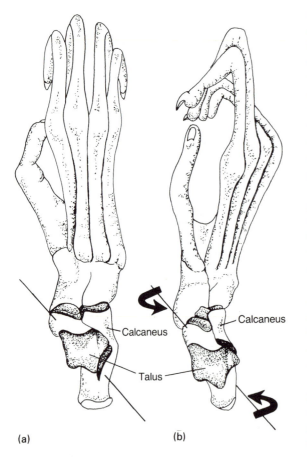

Figure 10.15. Dorsal view of the foot of the monk saki (*Pithecia monachus*), showing the everted position (a), with the sole of the foot essentially flat, and the inverted position (b), for grasping of a branch. The compromise axis of the subtalar joint is indicated by the oblique thin line and the arrows indicate the screwing motion involved in inversion of the foot. (After Lewis, 1980b.)

with only a tibiotalar articulation can show rotation of the calcaneus relative to the talus. This allows inversion and eversion (illustrated for the typical primate foot in Fig. 10.15) and it also permits the foot to swing backwards during head-first descent or hanging from the hindlimbs in trees (Cartmill, 1974c; Jenkins and McClearn, 1984). It follows from this that terrestrial locomotion is likely to be associated with the presence of an articulation between the fibula

and the calcaneus, whereas locomotion on complex three-dimensional substrates (notably in trees) is likely to be associated with the absence of such an articulation. This in turn indicates that if a fibulocalcaneal articulation is a primitive retention wherever it occurs among modern viviparous mammals then their common ancestor was characterized by terrestrial habits. This line of reasoning undoubtedly provides one of the main props for the hypothesis that the ancestral placental mammals were terrestrial in habits. After all, it would seem to violate the principle of parsimony to suggest that articulation between the fibula and calcaneus was present in mammal-like-reptiles, was then lost in the ancestral viviparous mammals and was subsequently regained in various descendants.

In fact, modern marsupials provide an excellent example of the way in which a shift from arboreal to terrestrial habits can be accompanied by secondary redevelopment of a fibulocalcaneal articulation. It is widely accepted (see later) that all modern marsupials are derived from an arboreal ancestral stock and that the most primitive modern survivors, in general morphological terms, are the Phalangeridae (Australian 'opossums') and the Didelphidae (New World opossums). Both groups are essentially arboreal and have several primate-like features, such as a grasping hallux with a flat nail and ridged skin on the tactile pads of hands and feet. In these arboreally adapted marsupials, articular contact between the fibula and the calcaneus is completely lacking. The compromise axis of the subtalar joint is oblique within the foot and a considerable range of movement is possible (Barnett, 1970; Lewis, 1980b). Some modern marsupials have, however, become secondarily adapted for running (cursorial) or leaping (saltatory) locomotion on the ground, the extreme case being that of the kangaroos. This transition to terrestrial locomotion has been accompanied by re-emergence of a well-developed articulation between the fibula and calcaneus. The axis of the subtalar joint has also been re-aligned to run transversely across the foot, coinciding approximately in its orientation with the transverse axis of the overlying primary ankle joint between the talus and the tibia. This 'drastic remodelling' of the tarsus in kangaroos (Lewis, 1980b) has produced an overall arrangement that is strikingly similar to that in modern artiodactyls among placental mammals (see Schaeffer, 1947).

Among modern placental mammals, contact between the fibula and the calcaneus occurs in a variety of forms adapted for terrestrial locomotion. The greatest degree of development of this joint, alongside the tibiotalar joint, is present in artiodactyls, in which inversion and eversion are completely ruled out and the subtalar joint is a true 'lower ankle joint' permitting only fore-and-aft rotation (Schaeffer, 1947). Clear articulation between fibula and calcaneus also occurs in, for example, certain insectivores (hedgehogs, shrews and moles, but not tenrecs), elephant-shrews, lagomorphs (rabbits and hares) and armadillos (Stains, 1959; Lewis, 1983). Interestingly, there is contact between the fibula and the calcaneus in the tarsus of microchiropteran bats (Stains, 1959), which no longer use their hindlimbs for arboreal locomotion as such. Thus, among modern placental mammals, articulation between fibula and calcaneus is confined to forms that do not require movements of inversion, eversion or backward rotation of the foot during locomotion. As Lewis (1983) has pointed out, in several groups of modern cursorial or saltatory terrestrial mammals with a fibulocalcaneal articulation there is fusion of the tibia and fibula, clearly suggesting that secondary specialization is involved.

Rather than accepting the talus and calcaneus of the late Palaeocene *Protungulatum* as representative of the condition in ancestral placental mammals (Szalay and Delson, 1974), because of the presence of fibulocalcaneal articulation, it is preferable to draw inferences from broad comparisons. Such comparisons should ideally include comprehensive studies of modern placentals (Lewis, 1980c, 1983; Novacek, 1980; Stains, 1959), consideration of additional early fossil evidence (e.g. early Palaeocene condylarths; Schaeffer, 1947) and appropriate reference to arboreal marsupials as relatively primitive representatives of the sister group of placentals (Lewis, 1980c, 1983). Typical features of the primate tarsus, as listed above, can then be interpreted more reliably. For instance, Novacek (1980), has examined individual tarsal features across several orders of placental mammals and has rightly questioned the diagnostic value of some features listed by Szalay and Drawhorn (1980) as typifying the Archonta.

There are several indications that the subtalar joint of ancestral placental mammals had an oblique compromise axis. This is the condition present in arboreal marsupials, it was the condition present in early Palaeocene condylarths before the origin of modern artiodactyls in the early Eocene, and it is found in several groups of modern placental mammals in addition to primates, tree-shrews and dermopterans. This evidence clearly suggests that ancestral placental mammals were adapted for lomomotion involving inversion and eversion movements of the foot, implying at least semi-arboreal habits.

It is by no means as easy to infer the orientation of the talar facet on the dorsal surface of the calcaneus in ancestral placental mammals. Arboreal marsupials have a rounded facet and it is only in placentals that it is elongated and hence has an identifiable orientation with respect to the long axis of the calcaneus. In modern arboreal placental mammals, the facet is oriented along the long axis of the calcaneus (e.g. in squirrels as well as in dermopterans, tree-shrews and primates; see Stains, 1959), whereas in terrestrial forms it is oriented obliquely or even transversely, as in the highly specialized artiodactyls. In the early Palaeocene condylarths the dorsal talar facet is oriented obliquely on the calcaneus, as it was in *Protungulatum* and possibly in ancestral placental mammals (see also Fig. 10.14). However, the orientation of this facet is obviously quite variable, as it changes from longitudinal to oblique according to the degree of arboreality within the otherwise relatively uniform tree-shrew family (see Novacek, 1980).

It seems likely that the calcaneus of ancestral

placental mammals had a well-developed peroneal tubercle (Fig. 10.14), as this feature is present in modern arboreal marsupials, in various early fossil placentals and in a variety of modern placentals. The position of the peroneal tubercle on the calcaneus is, however, directly related to the degree of arboreality. A retracted position of the peroneal tubercle is common to marsupials. Among placental mammals, it is found in those primates that still have the tubercle (e.g. Eocene 'lemuroids' and simians), in tree-shrews and in squirrels, which have the most retracted peroneal tubercle among all the mammals examined by Stains (1959). Although a distally located peroneal tubercle was present in *Protungulatum*, *Procerberus* and early Palaeocene condylarths (Fig. 10.14) and is seen in a variety of modern terrestrial placental mammals, it cannot be concluded with any confidence that this is an ancestral placental feature. Further, even if the distal location of the peroneal tubercle were to be recognized as a feature of ancestral placental mammals, the retracted position is clearly not a unique shared, derived feature of tree-shrews, primates and dermopterans.

There is similar uncertainty about the likely orientation of the distal calcaneal facet for articulation with the cuboid in the ancestral stock of placental mammals. The earliest known skeleton of a placental mammal, from a zalambdalestid 'insectivore', shows a transverse orientation of the facet, but in other early placental mammals there was an oblique orientation. As arboreal marsupials today have an oblique orientation of the calcaneocuboid articulation, it seems on balance likely that this was the condition present in ancestral placental mammals. Nevertheless, it is once again unjustifiable to regard the development of a transverse calcaneocuboid articulation as a unique shared, derived feature of primates, tree-shrews and dermopterans, as this character state is also found among rodents (notably in squirrels), in lagomorphs, in certain carnivores, in elephant-shrews and in bats (Stains, 1959; Novacek, 1980). This feature is hence not even restricted to arboreal mammals and its diagnostic

value is extremely limited. Returning to the features listed by Szalay and Delson (1974) as typical of primates, and later regarded as typical of the Archonta by Szalay and Drawhorn (1980), it can be seen that few, if any, are unique shared, derived features of primates, tree-shrews and dermopterans. Many are evident adaptations to arboreal life that have either evolved in parallel in several separate mammalian groups or have persisted as retained primitive character states from an arboreally adapted ancestral stock of placental mammals.

A restricted comparison of the tarsal bones of *Protungulatum*, *Plesiadapis*, and primates of modern aspect is bound to overemphasize any similarities between *Plesiadapis* and primates of modern aspect and to undervalue any distinctions between them. Characteristics of the talus and calcaneus of *Protungulatum* are definitely suggestive of terrestrial habits, so any degree of arboreal adaptation in *Plesiadapis* (either as a new development or as a retention from arboreal ancestral placentals) would, of course, be associated with some similarity to primates of modern aspect. It is therefore important to note that there are certain sharp distinctions between *Plesiadapis* and primates of modern aspect in tarsal morphology. In particular, the calcaneal facet for articulation with the cuboid was oriented quite obliquely in *Plesiadapis*, rather than transversely as in primates, and the calcaneocuboid joint had a gliding arrangement (Szalay and Decker, 1974) instead of the typical primate pivot form. Further, the peroneal tubercle was quite distally located on the calcaneus of *Plesiadapis*. It is therefore logically inconsistent to define the novel development of a transverse, pivotal calcaneocuboid junction and retraction of the peroneal tubercle as shared, derived features of Archonta including 'archaic primates' (Szalay, 1977; Szalay and Drawhorn, 1980).

Although *Plesiadapis* does appear to have had a 'squatting facet' on the neck of the talus, in common with many primates of modern aspect, the Eocene 'lemuroid' *Notharctus* lacked such an articular cup (Decker and Szalay, 1974) and no comparative information is available on the occurrence of this feature in other mammals. Szalay and Drawhorn (1980) have stated that the only feature shared by *Plesiadapis* and primates of modern aspect, but lacking from tree-shrews and dermopterans, is the presence of a deep groove for the flexor fibularis tendon, running beneath the sustentaculum of the calcaneus and linking up with a furrow on the back of the trochlea of the talus. The presence or absence of this feature in other placental mammals has yet to be properly documented, but a pronounced groove beneath the sustentaculum is certainly present in elephant-shrews and possibly in certain other groups of placental mammals (see Novacek, 1980), so it is not a unique shared, derived feature linking *Plesiadapis* to primates of modern aspect.

Elongation of the calcaneus

There is, in fact, one special feature of the tarsus in primates of modern aspect that does stand out as an apparently unique characteristic among placental mammals. As Lewis (1980c) has noted, the functioning of the typical primate foot depends on a switch from the primitive 'alternating tarsus' in which the talus articulates with the cuboid. The talocuboid articulation is eliminated in all modern primates and in many cases there is a new kind of alternating tarsus, with an articulation between the calcaneus and the navicular. This new development in all cases requires relative elongation of the distal portion of the calcaneus. Calcaneonavicular articulation is generally typical of modern prosimians and was already present in, for instance, the Eocene 'lemuroid' *Notharctus*. This special feature of the primate foot is reflected by an increased ratio of the distal portion of the calcaneus to the heel and is thus amenable to quantitative investigation.

Morton (1924) emphasized the elongation of the distal portion of the calcaneus in the primate foot, which is developed to an extreme in certain prosimians such as *Galago* and *Tarsius* (see Fig. 10.8), and linked it to arboreal locomotion involving a grasping hallux. As Morton observed, a grasping foot differs from a non-grasping, terrestrially adapted foot in that the principal fulcrum in the former lies at the distal end of the mid-tarsal bones (**tarsifulcrumation**), whereas that of the latter lies at the distal ends of the metatarsals (**metatarsifulcrumation**). Clearly, the arboreal grasping foot must differ in propulsion in that a branch will usually lie in the fork between digits 1 and 2, just beyond the cuneiforms, whereas in a terrestrial clawed foot the foot pushes against an essentially flat substrate, with the metatarsals and even the phalanges of the foot playing a part in propulsion. Accordingly, Morton concluded that elongation of the foot for leaping (saltation) would primarily involve the tarsal bones in arboreal forms with a grasping foot, whereas the metatarsals would be particularly affected in a terrestrially adapted foot lacking a grasping hallux. Elongation of the tarsal bones in arboreal primates that leap is especially noticeable in the distal portion of the calcaneus and in the navicular (Fig. 10.8; see also Fig. 10.16).

Gregory (1920a) had also noted this important feature of the grasping foot of primates in his outstanding monograph on *Notharctus*. He likened the tarsal region to a first-order lever system, with the fulcrum represented by the tibiotalar joint, with the heel portion of the calcaneus acting as the **lever arm**, and with the distal portion of the calcaneus together with more distal bones (e.g. the cuboid) acting as the **load arm**. Although it is a considerable oversimplification to treat the calcaneus and cuboid as a simple lever system, it was eventually shown (Hall-Craggs, 1965; Walker, 1967a) that the ratio of the lever arm to the load arm provides a very useful guide to primate locomotor adaptation. Ideally, calculation of a ratio of the lever-arm length to load-arm length should be carried out for the calcaneus and cuboid together, taking some easily recognizable fixed point on the calcaneus as equivalent to the midpoint of the articulation between the talus and the tibia. Hall-Craggs (1965) defined as the reference point on the calcaneus the junction between the distal end of the dorsal talar facet and the main body of the bone. This reference point was used to calculate a partial **foot-lever index** as the ratio between the heel portion of the calcaneus and the combined length of the remaining distal portion of the calcaneus and the cuboid. This index is, however, of very limited value for interpretation of fossil material. Although isolated calcanei are known for several fossil species of primate, associated cuboids are rare. Walker circumvented this problem by defining the **calcaneal index** as the ratio between the heel portion (tuber) and the distal portion (body) of the calcaneus alone. Even though this index omits the cuboid, and hence part of the effective load arm of the tarsus, it has proved to be both informative and practically useful in that it can be applied in broad comparisons of primates and other mammals (Walker, 1967; Martin, 1972b, 1979).

Examination of a broad range of mammalian calcanei suggested that a convenient reference point on the calcaneus for calculation of a lever arm:load arm index was provided by the junction of the proximal border of the sustentaculum with

the body of the calcaneus (Martin, 1972b; 1979). This junction corresponds approximately to the midway position of the centre of the tibiotalar joint during rotation of the calcaneus relative to the talus and it is a readily identifiable point that is accessible for measurement even on an articulated foot. In fact, with a *Galago* calcaneus, this reference point coincides quite closely with that selected by Hall-Craggs (i.e. the distal end of the dorsal talar facet on the calcaneus), as can be seen from Fig. 10.16. In other primates, however, the reference points do not coincide to the same degree, so calcaneal index values vary according to the method used for calculation. The situation is complicated because Berge and Jouffroy (1986) used the reference point defined by Hall-Craggs, whereas McArdle (1981) has used a third reference point for calculation of these values, namely the proximal end of the articular facet on the sustentaculum. However, provided that a single reference point is used for

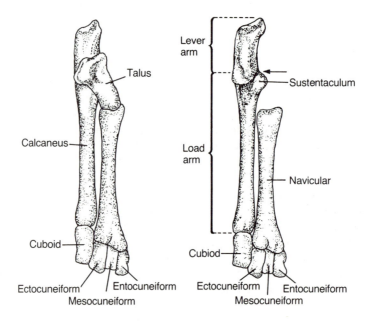

Figure 10.16. Tarsal bones of a typical bushbaby (*Galago*), showing the pronounced elongation of the distal segment of the calcaneus and of the navicular. In the illustration on the right, the talus has been removed to show how the calcaneal index is defined here, with respect to the junction between the proximal end of the sustentaculum and the body of the calcaneus (arrowed). The heel segment of the calcaneus is taken as the lever arm (Le) and the distal segment is taken as the load arm (Lo). The calcaneal index = (Le/Lo) × 100.

calculating calcaneal index values, in any comparison between species a consistent pattern is observed (e.g. compare Martin, 1979 with McArdle, 1981 and Berge and Jouffroy, 1986). For present purposes, all index values have been calculated as indicated in Fig. 10.16.

As explained at the beginning of this chapter, any osteometric index should be considered in relation to body size. Elongation of tarsal bones is clearly subject to scaling effects because body weight will increase faster than the cross-sectional area of a bone, other things being equal. Ideally, therefore, calcaneal index values should be plotted against some measure of body size. Once again, problems arise with isolated fossil calcanei, but it was found from a study of modern prosimian calcanei (Martin, 1972b) that there is a fairly consistent allometric relationship between lever-arm length and body size, whereas load-arm length can vary widely at any given body size. Hence, calcaneal index values can be plotted against lever-arm lengths for isolated primate calcanei as a means of examining the index in relation to a crude indicator of body size (Martin, 1972b, 1979; see also McArdle, 1981). Inclusion of other mammals in the overall comparison permits interpretation of primate calcaneal elongation, as defined here, in relation to the general mammalian pattern (Fig. 10.17).

As could be predicted from Morton's concept of tarsifulcrumation, it emerges that primates are almost unique in comparison to other placental mammals in that calcaneal index values typically lie below 100%. Among primates, there is a general trend for calcaneal index values to increase with body size (as indicated by lever-arm length), doubtless as a reflection of the fact that increasing body weight places a disproportionate strain on the cross-section of a bone such as the calcaneus. Extreme calcaneal elongation of the kind found in tarsiers or bushbabies (Fig. 10.16) is almost certainly ruled out for larger-bodied primates on mechanical grounds. Thus, at body weights exceeding 5 kg (approximately corresponding to a lever-arm length of 13 mm in Fig. 10.17), calcaneal index values may exceed 100% even in primates. Below this body weight, however, no modern primate species has a value

exceeding 100%. In stark contrast, virtually all other placental mammals have values greater than 100%, regardless of body size. Tree-shrews clearly fall among non-primate placentals in this respect, as is only to be expected as they lack the key primate characteristic of a grasping hallux.

The clear separation between primates and non-primates among placental mammals, at least at body weights below 5 kg, provides a valuable criterion for assessing fossil calcanei. In the case of the North American Eocene 'lemuroids' *Smilodectes* and *Notharctus*, almost complete skeletons associated with skulls are known and these two genera provide us with the most reliable evidence regarding the form of the tarsus in early primates. As noted by Gregory (1920a), these Eocene 'lemuroids' closely resemble modern Malagasy lemurs in calcaneal proportions and this is confirmed in Fig. 10.17. Isolated tarsal bones also indicate that this was true for the early Eocene 'lemuroid' *Pelycodus* (including *Cantius*) – Matthew and Granger (1915) and Covert (1987).

For other early primates, we are obliged to rely largely on isolated tarsal bones. In cases where a low calcaneal index value is found, there can be little doubt that the calcanei come from primates, as no other placentals are known to show this condition. This is true, for example, of calcanei attributed to a variety of Eocene 'tarsioids' (Szalay, 1976; Savage and Waters, 1978; Schmid, 1979). Several calcanei have been attributed to the North American Eocene 'tarsioid' *Hemiacodon gracilis* (e.g. see Simpson, 1940), and their calcaneal index values clearly lie within the range of modern Malagasy lemurs, though they are markedly higher than in the modern tarsier (Fig. 10.17). Calcanei attributed to the European Eocene 'tarsioid' genera *Microchoerus*, *Necrolemur*, *Nannopithex*, *Tetonius* and *Teilhardina* all seem to show a degree of elongation of the distal portion comparable to that in *Hemiacodon*, although no data on calcaneal index values are available for inclusion in Fig. 10.17. Szalay (1976) justifiably concluded that moderate elongation of the distal portion of the calcaneus was probably an ancestral feature of Eocene 'tarsioids'. In fact, it

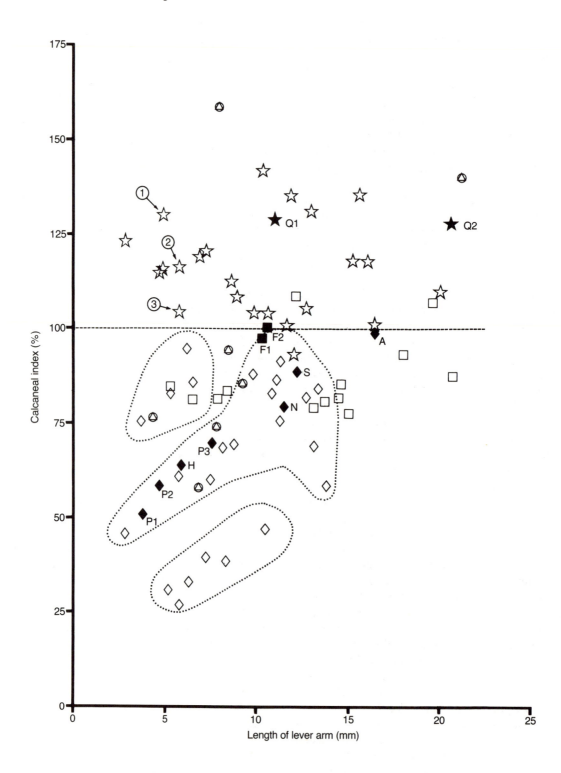

seems that *Necrolemur* – at least – might have had more pronounced elongation of the calcaneus (Gebo, 1987a).

Evidence from more recent prosimian fossil species is provided by a series of isolated calcanei from Miocene deposits of East Africa, allocated to the genera *Progalago* and *Komba* on grounds of overall size and morphology (Walker, 1970, 1974b; Gebo, 1986a). In all of these calcanei the distal portion is elongated and hence the calcaneal index values are quite low (Fig. 10.17), such that allocation to primates is clearly justified. Along with the calcanei from *Smilodectes*, *Notharctus* and *Hemiacodon*, the Miocene lorisid calcanei fall into the range of modern lemurs. It is particularly noteworthy that, in terms of calcaneal proportions, the calcanei of Miocene lorisids and of Eocene 'tarsioids' resemble those of the mouse and dwarf lemurs of Madagascar rather than those of modern bushbabies or tarsiers (Martin, 1972b; Fig. 10.17).

Several isolated calcanei from European Eocene deposits have in the past been allocated to the 'lemuroid' genus *Adapis*. Gregory (1920a) figured a calcaneus attributed (tentatively) to *Adapis parisiensis* and Decker and Szalay (1974) discussed calcanei and tali attributed to that species and to *A. (Leptadapis) magnus*. Gregory, followed by Decker and Szalay, noted that calcanei attributed to *Adapis* differ sharply from all other prosimian primates of modern aspect in that elongation of the distal portion is lacking. In fact, calcanei attributed to both *A. parisiensis* and *A. magnus* have index values well in excess of 100% (Fig. 10.17). Because of this unusual

feature, and because of the absence of convincing evidence for allocation of the calcanei to *Adapis*, it was suggested (Martin, 1979) that it was a moot point whether *Adapis* had an unusual adaptation of the foot or whether the isolated calcanei had in fact been misidentified. Subsequently, Dagosto (1983) conducted a detailed study of a large number of isolated postcranial bones attributed to *A. parisiensis* and *A. magnus*, concluding that allocation of the calcanei and other foot bones is unquestionably correct. The postcranial elements apparently fell into two fairly distinct size classes, providing some justification for allocating the smaller bones to *A. parisiensis* and the larger bones to *A. magnus*. Several first metatarsals have been attributed to each species and they are distinctive in that they have all of the typical features associated with the grasping hallux in the primate foot, notably a saddle-shaped articular surface at the proximal end and torsion of the shaft. There are therefore good grounds for identifying the metatarsals as primate foot bones. By implication, then, it is reasonable to accept the identification of calcanei and tali assigned to *Adapis* species, which are broadly comparable to the metatarsals on size grounds. It should be emphasized, however, that there are no known naviculars or entocuneiforms to link the first metatarsals to the tali. The composite partial feet figured by Dagosto (1983) involve bones that were not found in direct association, and we still lack fossil finds firmly linking any of the postcranial bones to cranial and dental evidence. Nevertheless, Dagosto has provided a very detailed treatment of the available material attributed to *Adapis* and

Figure 10.17. Calcaneal index values plotted against lever-arm lengths for primates and other mammals. Primates generally have index values less than 100%, whereas non-primates typically have values exceeding 100%. Dotted outlines indicate approximate overall ranges of variation for the loris subgroup (above), for extant Malagasy lemurs (centre) and for bushbabies and tarsiers (below). Numbered arrows indicate tree-shrews: 1, *Tupaia*; 2, *Urogale*; 3, *Lyonogale*. Key to symbols: ◇, modern prosimians; ◆, fossil prosimians; □, modern simians; ■, fossil simians; ☆, non-primate placentals; ⬠, marsupials. Key to fossil forms: A, *Archaeolemur edwardsi* (*N* = 2); F1, F2, Fayum simians (Oligocene); H, *Hemiacodon gracilis* (*N* = 6); N, *Notharctus tenebrosus* (*N* = 8); P1, P2, P3, East African Miocene lorisids; Q1, Q2, calcanei attributed to *Adapis parisiensis* and to *A. (Leptadapis) magnus*, respectively; S, *Smilodectes gracilis* (*N* = 2). (After Martin, 1979, with kind permission from Academic Press, New York.)

provides a convincing appraisal of postcranial morphology in that genus.

Dagosto explicitly recognized the distinctiveness of the calcanei attributed to *Adapis*, in terms of lever arm:load arm relationships. She accepts that the unusual characteristic of a high calcaneal index would not have been present in the common ancestral stock of primates of modern aspect. Indeed, her thorough analysis of available fossil material allocated to *Adapis* reveals many features indicative of a non-leaping, slow-climbing adaptation comparable in several respects to the pattern found in members of the modern loris subgroup (e.g. the potto, *Perodicticus*). It is therefore possible to argue that *Adapis* had diverged markedly from the typical ancestral primate condition, perhaps because of special adaptation for slow arboreal locomotion.

The unusual nature of the tarsal bones allocated to *Adapis* has been further underlined by the recent description of two remarkable specimens from middle Eocene deposits of Messel (Germany), one consisting of a completely articulated pelvic girdle and hindlimbs, and the second similar but less complete (von Koenigswald, 1979, 1985). Unfortunately, no cranial or dental remains were found with the specimens, although an incomplete skull and partial skeleton from the same site has been allocated to *Europolemur* (Franzen, 1987). The two posterior postcranial specimens have numerous primate-like features and von Koenigswald accordingly allocated them to the Adapidae without attempting to define the genus. In particular, the foot of the first specimen has a grasping hallux and the calcaneus has a relatively elongated distal portion, resembling the condition found in *Notharctus* and *Smilodectes* but sharply distinct from the condition found in calcanei attributed to *Adapis*. The articulated hindlimbs of the Messel 'lemuroid' are remarkably similar to those of modern lemurs as well as resembling those of North American Eocene 'lemuroids' (e.g. *Notharctus*). In fact, von Koenigswald (1979) noted a special similarity to the modern lemurs and lorises in that the terminal phalanx of the second digit in the first specimen is longer than

that on any of the other digits of the foot (excepting the hallux) and has a deeper base, which may indicate the presence of a 'toilet claw'. The terminal phalanx of the hallux is particularly broad, as in modern primates generally, and the structure of all of the terminal phalanges suggests the presence of nails rather than claws in the strict sense (von Koenigswald, 1979). We are hence presented with a strange situation in which isolated tarsal bones attributed to *Adapis* are clearly distinct from those of all other known primates of modern aspect, whereas a contemporary partial postcranial skeleton that neatly fits the general primate pattern cannot be allocated to any known European Eocene 'lemuroid' genus because of the absence of associated dental or cranial material! The lengths of the femur and tibia of the first Messel 'lemuroid' are intermediate between those of *Notharctus tenebrosus* (Gregory, 1920a) and those for material attributed to *Adapis parisiensis* (Dagosto, 1983), the latter being the shortest.

It is therefore reasonable to conclude that the Messel 'lemuroid' (which probably had a body weight in the region of 1 kg) fell within the size range of Eocene 'lemuroids' generally and was certainly larger than contemporary 'tarsioids', as noted by von Koenigswald. In fact, because of the marked distinctions between the Messel adapids and *Adapis* in numerous features of the postcranial skeleton, there is now an increasing tendency to regard *Adapis* (including *Leptadapis*) as an aberrant form (Franzen, 1987). European Eocene 'lemuroids' other than *Adapis* are generally similar to the broadly contemporary North American genera *Pelycodus*, *Notharctus* and *Smilodectes* in postcranial morphology and have the typical pattern for primates of modern aspect. It is therefore important to treat *Adapis* as a special case in discussions of postcranial adaptations in early primates of modern aspect.

Elongation of the distal portion of the calcaneus in virtually all primates of modern aspect corresponds to the pattern of tarsal elongation that Morton (1924) linked to tarsifulcrumation. It also corresponds to the shift from the primitive 'alternating tarsus', leading to elimination of the talocuboid articulation and widespread formation of a calcaneonavicular articulation noted by Lewis (1980c) for primate feet. As emphasized by Morton, all of these features depend on the presence of a grasping hallux. In this connection, it is noteworthy that although low calcaneal index values distinguish modern primates from virtually all other modern placental mammals, certain arboreal marsupials with grasping feet also have low values, confirming Morton's inferences regarding tarsifulcrumation (Martin, 1979). However, it is of central importance that Morton (1924) defined tarsifulcrumation with respect to saltatory arboreal locomotion in trees. Tarsal elongation would not necessarily be expected in an arboreal mammal with a grasping foot but lacking any tendency for arboreal leaping. (This point was not adequately emphasized by Martin, 1979.) Hence, the presence of a high calcaneal index value in calcanei attributed to *Adapis* does not necessarily preclude possession of a grasping hallux, as was noted by Dagosto (1983). However, it must be remembered that modern members of the slow-climbing loris subgroup still uniformly have calcaneal index values below 100% (Fig. 10.17), doubtless as a reflection of their derivation from an arboreal saltatory ancestor. Calcaneal index values considerably in excess of 100% for tarsal bones attributed to *Adapis* therefore suggest derivation from an ancestral form lacking arboreal saltatory locomotion, an unexpected feature for any primate of modern aspect.

The distinction between tarsifulcrumation and metatarsifulcrumation is of crucial importance in the interpretation of mammalian foot bones and analyses that ignore this can be quite misleading. This is, for example, the case with the 'foot-lever index' defined by Schultz (1963a,b) as the ratio of the lever arm (heel portion) of the calcaneus to the overall length of the distal portion of the calcaneus, the cuboid and metatarsal 3, taken together. The foot-lever index completely obscures the distinction between the grasping foot and the non-grasping, clawed foot and accordingly allows no meaningful interpretation. Schultz (1963a,b), for instance, showed that the vertical-clinging-and-leaping *Galago* has the same foot-lever index value as clawed,

quadrupedal marmosets and tamarins, whereas the vertical-clinging-and-leaping *Tarsius* differs completely from *Galago* in its foot-lever index value and resembles the slow-climbing *Arctocebus* instead! Hall-Craggs (1965) specifically noted that the use of an index including metatarsal length is inappropriate for primates that show tarsifulcrumation, such as *Galago* and *Tarsius*. It is therefore of no functional or evolutionary significance that treeshrews have foot-lever index values lying within the modern prosimian range.

The question of tarsifulcrumation in primates of modern aspect, allied to low values for the calcaneal index, is particularly relevant to the status of 'archaic primates', as exemplified by *Plesiadapis*. The aberrant high values for the calcaneal index recorded for tarsal bones attributed to *Adapis* have unfortunately tended to blur the otherwise sharp distinction between *Plesiadapis* and all other primates of modern aspect. The calcaneal index of *Plesiadapis* clearly exceeds 100%; a value of approximately 120% can be estimated for the calcaneus of *Plesiadapis tricuspidens* figured by Szalay and Decker (1974). As noted by Dagosto (1983), the high calcaneal index values reported for *Adapis* must necessarily reflect a derived condition relative to the common ancestor of primates of modern aspect, a point that has been confirmed in subsequent studies (e.g. Franzen, 1987). Hence, there cannot possibly be homology with the condition in the more distantly related *Plesiadapis*. In the absence of firm evidence of a grasping hallux in 'archaic primates', the high calcaneal index value for *Plesiadapis* merely reflects the condition typical of non-primate placental mammals. Given the other distinctive features of the tarsus of 'archaic primates', noted above, the overall adaptation of the primate tarsus can be defined most satisfactorily to the exclusion of *Plesiadapis* and its relatives.

The wrist

The primate wrist (**carpus**), in contrast to the ankle, has attracted relatively little detailed comparative study. As might be expected from studies of hand function (see above), the wrist is more variable in form among primates than is the ankle, and there is far less evidence of a clear ancestral pattern specifically related to saltatory arboreal locomotion. This is borne out by the fact that, in contrast to the tarsal bones, there is some variation in the basic pattern of carpal bones among primates, arising from fusion of originally separate elements.

The probable primitive mammalian pattern of **carpal bones** (see Steiner, 1942) is still retained in many modern primates, as exemplified by the hand of the variegated lemur (*Varecia*; Fig. 10.18). The proximal row of carpals consists of a **scaphoid** on the radial side, a **triquetral** on the

ulnar side and a **lunate** situated between them. The distal row consists of four bones, with the **trapezium** corresponding to metacarpal 1 (and hence to the pollex), the **trapezoid** corresponding to metacarpal 2, the **capitate** corresponding to metacarpal 3 and the **hamate** corresponding to metacarpals 4 and 5. Sandwiched between the proximal and distal carpal rows, there is an additional bone appropriately called the **centrale**. Adjacent to the ulna, and articulating with it, is an extra small bone that is not, strictly speaking, one of the carpals. This is the **pisiform** and it bears a special relationship to the hypothenar pad. Finally, alongside the trapezium there is a small bone called the **prepollex**, which has sometimes been interpreted as a remnant of an original sixth digit in vertebrate ancestry. The prepollex is present in most mammalian groups, but it shows considerable variation in its relationships and may articulate with the scaphoid and/or the trapezium.

Kielan-Jaworowska (1977, 1978) has described the structure of the carpus in the earliest known postcranial skeletons of placental mammals from late Cretaceous deposits of Mongolia. A tentative reconstruction of the carpus of the early palaeoryctid 'insectivore' *Asioryctes* indicates the presence of all 10 bones identified for *Varecia* in Fig. 10.18. Whereas these elements apparently showed no fusion in *Asioryctes*, the carpus of the zalambdalestid 'insectivore' *Barunlestes* shows fusion of the scaphoid and lunate in the proximal row. Thus, the zalambdalestids had some specialization in the carpus, just as they did in the tarsus (see above).

As a general rule, relatively primitive modern mammals that have retained a pentadactyl limb structure have also retained the basic pattern of carpal bones. Fusion of the scaphoid with the centrale and/or the lunate is, however, a relatively common feature. The centrale is fused with the scaphoid in monotremes, in some marsupials and in a variety of placental mammals. Fusion can take place at a very early (embryonic) stage, as is the case with certain insectivores, rodents and edentates. Among placental mammals, the scaphoid and lunate are fused together in carnivores, pinnipeds, most

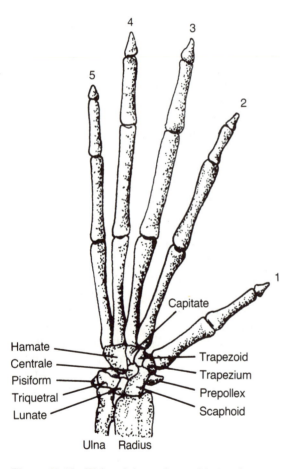

Figure 10.18. Wrist of the variegated lemur (*Varecia variegata*), showing the basic mammalian complement of carpal bones. (After Jouffroy, 1975.)

rodents, some insectivores, bats, 'flying lemurs', sirenians, pangolins and some primates (see below). Carlsson (1922) specifically noted that the scaphoid and lunate are fused in the hand of tree-shrews (*Tupaia*), hedgehogs (*Erinaceus*) and squirrels (*Sciurus*), whereas *Lemur* and American opossums (*Didelphis*) retain the primitive unfused condition. In modern carnivores, some living insectivores and bats, in fact, the scaphoid, lunate and centrale are all fused into a single bony mass. Fusion of the scaphoid with the lunate is common in marsupials but is significantly lacking from relatively primitive arboreal forms such as American opossums, gliders (*Petaurus*) and brushtail opossums (*Trichosurus*) – Weber (1927).

Modern primates, as noted above, generally have a relatively primitive pattern of carpal bones. Most prosimians, in common with *Varecia* (Fig. 10.18), show no fusion or loss of original elements, but the centrale and scaphoid are fused in *Hapalemur*, *Lepilemur*, *Indri* and *Avahi* (Jouffroy, 1975). Fusion of the scaphoid with the centrale is similarly rare among simian primates, although it occurs in the African great apes and in humans (Le Gros Clark, 1959; Lewis, 1974). It can be concluded fairly confidently, therefore, that ancestral primates still had the primitive mammalian pattern of eight carpal bones along with a pisiform and a prepollex. Unfortunately, however, there is little that can be tested against the fossil record. Even for the Eocene 'lemuroid' *Notharctus*, which is otherwise well represented by postcranial remains, the carpus is poorly known (Gregory, 1920a). However, Godinot and Jouffroy (1984) have recently been able to reconstruct a hand, lacking only the phalanges, attributed to the Eocene 'lemuroid' *Adapis*. The reconstruction of the carpus includes all of the elements indicated for *Varecia* in Fig. 10.18. Hence, any departures from the inferred basic pattern shown in Fig. 10.18 can be reasonably interpreted as a result of secondary specialization of wrist function, involving fusion or loss.

Altner (1971) made a detailed comparison of the hands of a tree-shrew (*Tupaia*) and a mouse lemur (*Microcebus*), with emphasis on carpal structure, in an attempt to elucidate further the evolution of hand function in primates. Both *Tupaia* and *Microcebus* have conserved the original set of carpal bones, although the scaphoid and lunate are fused in the adult hand of *Tupaia*, as reported by Carlsson (1922). Altner also noted a distinction in that in tree-shrews the centrale is functionally incorporated into the distal row of carpal bones whereas in mouse lemurs the centrale is incorporated into the proximal row. The effect of this is that the hand of *Microcebus* is divided into two functional units along a longitudinal axis passing through the trapezoid. The trapezium and metacarpal 1 thus form a relatively independent assemblage associated with a grasping function of the hand. Hence, although the hand of *Microcebus* lacks true opposability of the thumb, there is an important contrast between the grasping hand (*Greifhand*) of mouse lemurs and the convergent hand (*Spreizhand*) of tree-shrews. Incidentally, Carlsson (1922) cited the presence of a free centrale as a primate-like feature of *Tupaia*, whereas this is nothing more than a primitive mammalian feature.

Altner argued that the hand of *Tupaia* is, nevertheless, adapted to some degree in the direction of the primate hand (as exemplified by *Microcebus*) because the incorporation of the centrale into the distal carpal row is not as complete in *Tupaia* as it is in squirrels (*Sciurus* and *Xerus*). The condition in tree-shrews is hence seen as intermediate, in an evolutionary sense, between the 'primitive' condition present in squirrels and the 'advanced' condition in *Microcebus*. But this interpretation involves the assumption that the primate hand is necessarily advanced in comparison to the rodent hand, whereas careful phylogenetic reconstruction is necessary to reach such a conclusion. Altner himself reported that the centrale is included in the proximal carpal row in insectivores. Although the centrale, lunate and scaphoid are all fused together in most cases (e.g. hedgehogs and shrews), these bones are still separate in shrew-tenrecs (*Microgale*) and their disposition is quite similar to that in *Tupaia*. Given the widely accepted interpretation of the centrale as an interstitial element between the proximal and distal carpal rows (e.g. see Weber, 1927; Steiner, 1942), the condition in ancestral placental mammals was probably fairly close to that in *Tupaia* and *Microgale*. This view is confirmed by Kielan-Jaworowska's reconstructions of the carpus of *Asioryctes* and *Barunlestes* from the late Cretaceous, which show the centrale in an intermediate position between the proximal and distal carpal rows. Among primates, this intermediate position of the centrale is quite common and Le Gros Clark (1959) interpreted the proximal position of the centrale in modern lemurs as a secondary specialization, associated with development of contact between the centrale and the hamate. Once again, it emerges that an apparent affinity between tree-shrews and primates (Altner, 1971) is based on shared retention of character states that were probably present in ancestral placental mammals.

The primary wrist joint between the lower arm bones and the proximal row of carpals is essentially a hinge arrangement in its primitive form, as was the primary ankle joint in the earliest stage of mammalian evolution. In the wrist pattern of primitive mammals, which has been retained in most living primates, the radius articulates with the scaphoid and lunate, and the ulna articulates with the triquetral and pisiform, which together form a cup-like socket. However, in some primates this basic hinge arrangement has been modified to permit greater mobility of the wrist. Lewis (1969, 1971a,b, 1974), in particular, noted extensive remodelling of the wrist joint in great apes and humans (i.e. all hominoids). The essence of this remodelling is that the ulna has been withdrawn from direct articulation with the triquetral or with the (reduced) pisiform and an intra-articular meniscus of fibrocartilage has developed between the ulna and the triquetral. Lewis (1974) suggests that interpolation of a meniscus in this way represents a shared derived adaptation of hominoids, allowing extreme pronation and supination of the hand (i.e. rotation around the long axis of the arm), as the original hinge-joint has been partially dismantled to leave only a direct articulation between the radius and the carpus. Lewis linked this development

to increased emphasis on slow suspensory locomotion involving the forelimbs and suggested that remodelling of the hominoid wrist was therefore related to brachiation. It has been argued (e.g. Conroy and Fleagle, 1972) that this interpretation seems to conflict with the fact that wrist remodelling is least advanced in the specialized brachiating lesser apes (gibbons and the siamang). But Lewis (1974) countered this objection with the suggestion that the ricochetal brachiation of lesser apes may require less wrist mobility than slow, suspensory arm-swinging.

An alternative possibility is that remodelling of the wrist in hominoids might permit greater outward lateral movement of the hand (ulnar deviation), which could be of significance with respect to **knuckle-walking** during the terrestrial locomotion of African great apes (see Fig. 1.22). However, detailed studies of African great apes (Tuttle, 1967, 1969a,b, 1975) have indicated that knuckle-walking actually requires stabilization of the carpus and it has been observed that ulnar deviation is not a significant feature in the wrist during this form of locomotion (see Cartmill and Milton, 1977). Jenkins (1981) used cineradiography to analyse the suspensory behaviour of spider monkeys (*Ateles*) and woolly monkeys (*Lagothrix*), beneath a loop of rope. He showed that rotation at the mid-carpal joint may be an important feature permitting extensive supination in these New World semibrachiators. Studies of the hands of gibbons seemed to confirm that such mid-carpal rotation also occurs in lesser apes (but see Lewis, 1985a,b). Hence, suspensory locomotion clearly involves complex movements of the carpus, both at the primary wrist joint and at the mid-carpal joint, and it is not at present possible to link special features of the hominoid wrist specifically to individual locomotor activities such as brachiation. It is conceivable, as argued by Fleagle, Stern *et al.* (1981), that the remodelling of the hominoid wrist reported by Lewis is actually related more to adaptations for vertical climbing. Regardless of whether the remodelling represents an adaptation for climbing or for suspensory locomotion of some kind, the more extreme developments seen in great apes and humans may perhaps be related to the influence of their considerable body weights.

Cartmill and Milton (1977) reported that there are similarities in wrist morphology between the slow-climbing loris subgroup and hominoids, thus providing apparent support for the slow-climbing hypothesis. In all members of the loris subgroup, the articulation between the ulna and the triquetral has been somewhat reduced and the pisiform no longer articulates with the ulna. In radiographs there is, in fact, a noticeable gap between the slender distal articular process of the ulna and the triquetral, while the pisiform has been displaced distally. In some specimens of *Nycticebus*, an intra-articular fat pad was seen between the ulna and the triquetral. However, this fat pad is not really comparable to the specialized intra-articular meniscus found in hominoids and the gap between the ulna and triquetral seen on radiographs is actually bridged by cartilage articular surfaces (Lewis, 1985a,b). Remodelling of the wrist joint in slow-climbing lorises and pottos is markedly different from that seen in hominoids and involves an unusual specialization of the carpal joint that provides for the powerful pincer-like action of the lorisine hand (Lewis, 1985a). Although Mendel (1979) stated that sloths also have a hominoid-like adaptation of the wrist, citing this as additional support for the slow-climbing hypothesis, Lewis has shown that sloths and certain marsupials have an adaptation paralleling that of lorisines rather than that of hominoids. It is noteworthy, in fact, that most primates have retained a relatively primitive wrist structure and that divergent forms of remodelling of the wrist are only clearly evident in slow-climbing lorises and pottos on the one hand and in hominoids on the other.

HINDLIMB-DOMINATED LOCOMOTION

As Cartmill (1972) has observed, it is unlikely that the ancestors of primates of modern aspect were committed vertical-clingers-and-leapers like modern bushbabies, tarsiers or certain lemurs. Consideration of intermembral index values (Fig. 10.5) and of calcaneal index values (Fig. 10.17) suggests, instead, that modern

primates were derived from an ancestral form with a more generalized pattern of arboreal locomotion akin to that found in modern mouse and dwarf lemurs (Martin, 1972b; Gebo, 1987c). Modern vertical-clinging-and-leaping primates undoubtedly show specialized features in connection with this distinctive pattern of locomotion and some of the similarities they now share have probably arisen through convergent evolution. However, the fact remains that this locomotor pattern is widespread among prosimian primates, while being unique among placental mammals. Hence, it is highly probable that ancestral primates possessed certain distinctive features that permitted, and perhaps predisposed towards, the later development of vertical-clinging-and-leaping in several different lines of primate evolution. Those features are encapsulated by the concept of **hindlimb-dominated arboreal locomotion** (Walker, 1967a; Napier, 1967a; Martin, 1972a, 1979), which covers the following fundamental aspects of locomotor adaptation in primates:

1. A well-developed, grasping hallux.
2. Overall adaptation for active leaping in trees.
3. A low intermembral index, relative to body size.
4. A low calcaneal index, relative to body size.

As can be seen from Figs 10.5 and 10.17, modern prosimian vertical-clingers-and-leapers show extreme development of features 3 and 4, whereas in point 4 Eocene 'lemuroids' and Eocene 'tarsioids' generally resemble modern Malagasy lemurs that are not commited to such a specialized form of locomotion.

It has recently emerged that the typical primate characteristic of hindlimb domination is connected with overall patterns of limb coordination. The salient features of such coordination persist even in cases where the forelimbs have become specially developed for suspensory locomotion (thus increasing the intermembral index), as among modern apes. The evidence for this comes from a somewhat unexpected source – namely, a comparative consideration of mammalian gait patterns (Rollinson and Martin, 1981).

Because the arboreal environment is complex, the patterns of movement in primates are complex. However, a basic property of primate locomotor coordination can be recognized by considering the pattern of limb movement during walking over a simple substrate (flat ground or broad horizontal branch). This property can be illustrated by taking the right hindlimb as an arbitrary starting-point in the sequence (Fig. 10.19). In primates, the left forelimb is typically the next to contact the ground, followed by the left hindlimb and then by the right forelimb, to complete the cycle. This typical sequence in the primate symmetrical walking gait has been termed the **diagonal sequence** by Hildebrand (1967). Muybridge (1899) was apparently the first to recognize diagonal sequence walking in monkeys, referring to it as the 'pithecoid walk'. By contrast, non-primate placental mammals typically show a quite distinct sequence in the symmetrical walking gait (Fig. 10.19). In Hildebrand's terminology, this is the **lateral sequence**. (It should be noted that other authors (e.g. Howell, 1944) previously used the terms diagonal and lateral in the reverse sense to Hildebrand.) Although there are certain exceptions (see Rollinson and Martin, 1981), it is a general rule that primates have a diagonal sequence walking gait, whereas non-primate placentals – including tree-shrews (Jenkins, 1974b) – have a lateral sequence gait, using Hildebrand's terminology.

At first sight, it might seem that the distinction between the diagonal and lateral support sequences is of relatively little significance. This was, indeed, the interpretation made by Howell (1944), who stated: 'The only difference between the two is that one is a mirror image of the other'. However, although he predicted that the two support sequences should be used 'almost indiscriminately' by various mammals, he noted that the typical primate sequence (diagonal sequence as defined in Fig. 10.19) is actually quite rare among mammals. There is, in fact, a very good reason for this, which emerges when the forward motion of the body is taken into account along with the support sequence of the limbs.

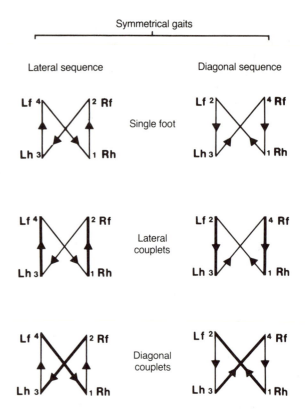

Figure 10.19. Limb-support sequences used in symmetrical gaits of mammals (in agreement with Hildebrand, 1967). Heavy lines indicate predominance of pairs of limbs (couplets) supporting the body during each cycle of limb movement. Primates typically show the diagonal sequence (right); non-primates typically have the lateral sequence (left). Arrows indicate the sequence of limb movement. Abbreviations: Rh, right hindlimb; Rf, right forelimb; Lh, left hindlimb; Lf, left forelimb. (Reprinted from Rollinson and Martin, 1981, with kind permission from the Zoological Society of London and Academic Press, London.)

In a theoretical consideration of slow quadrupedal gaits, McGhee and Frank (1968) concluded that the diagonal sequence should be extremely unstable, with extensive lateral rolling tendencies and that the lateral sequence should therefore be favoured exclusively (see also Gray, 1944). The fact that primates have a diagonal walking gait in spite of its apparent instability is explained when overall weight distribution within the body is taken into account. McGhee and Frank (1968) based their inferences on symmetrical models in which the centre of gravity of the body was implicitly assumed to be equidistant from all four limbs. Their conclusions are valid only if this is true, or if the centre of gravity is located closer to the forelimbs. By contrast, if the centre of gravity is closer to the hindlimbs than to the forelimbs, the diagonal sequence becomes a stable gait. This leads to the prediction that the centre of gravity should be located closer to the hindlimbs in primates, whereas in non-primate placental mammals the centre of gravity should be located midway between the two pairs of limbs or closer to the forelimbs. As yet, relevant data are available only for relatively few species, but this limited information broadly confirms the prediction (Table 10.2; see also Kimura, Okada and Ishida, 1979). It is interesting to note that the kinkajou

(*Potos flavus*), an arboreal carnivore with a prehensile tail, is an exception among non-primates in having its centre of gravity closer to the hindlimbs and a primate-like diagonal limb sequence during walking (Hildebrand, 1967).

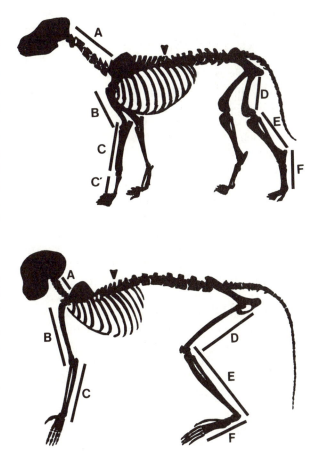

Table 10.2 Differential distribution of weight on forelimbs and hindlimbs in mammals (after Rollinson and Martin, 1981; see also Kimura, Okada and Ishida, 1979)

Species	Weight on forelimbs (%)	Weight on hindlimbs (%)
Terrestrial quadrupeds		
Dog (various breeds)	63	37
Cheetah	52	48
Horse	55	45
Ox	56	44
Dromedary	62	38
Camel	66	34
Llama	55	45
Elephant	55	45
Primates		
Spider monkey	29	71
Hamadryas baboon	44	56
Japanese macaque	46	54
White-collared mangabey	41	59
Atypical non-primates		
Kinkajou	43	57
Cuban hutia	45	55

Figure 10.20. The overall shape of the skeleton in a typical cursorial mammal, the dog (above), and in an arboreal monkey (below). The centre of gravity in non-primates is typically located closer to the forelimbs and, in association with this, the neck (A) is typically longer than in primates of comparable body size. Non-primates typically have shorter limb bones (B, C, D, E) than primates, relative to body size, but a tendency to walk on the digits (digitigrady) effectively lengthens the limbs through addition of proximal parts of the extremities (C′; F). In primates, the centre of gravity is typically closer to the hindlimbs as part of the overall phenomenon of hindlimb domination. (Illustration adapted from a drawing kindly supplied by Dr J.M.M. Rollinson; see Rollinson, 1975.)

Additional, striking evidence of a link between the position of the centre of gravity and the sequence of limb use during walking is provided by the results of a neat experiment conducted by Tomita (1967; cited by Kimura *et al.*, 1979). Tomita attached weights to the rear end of a dog and showed that the resulting backward displacement of the centre of gravity was accompanied by a switch from lateral-sequence to diagonal-sequence walking. Hence, it is reasonable to conclude that in primates there is a direct link between typical use of diagonal-sequence walking (Fig. 10.19) and a posterior location of the centre of gravity (Table 10.2) and that these two features constitute a significant component of hindlimb domination in primate locomotion. It can be argued that such emphasis on the hindlimbs is favoured in habitually arboreal mammals, whereas in terrestrial mammals emphasis on the forelimbs is favoured by requirements of fast locomotion over a relatively stable substrate (Rollinson and Martin, 1981; see Fig. 10.20). It seems quite probable that in ancestral primates of modern aspect, a posterior location of the centre of gravity of the body and diagonal-sequence walking were integral features of hindlimb domination. This would have provided a unique basis for such divergent specializations as habitual vertical-clinging-and-leaping among prosimians and striding bipedalism in humans.

ARBOREAL OR TERRESTRIAL ANCESTRY OF MAMMALS?

It is abundantly clear that the structure and function of the primate postcranial skeleton have been heavily influenced by adaptation for life in trees. This heritage is obvious even in the relatively few primates, such as humans, that have largely or completely abandoned arboreal habits, as a secondary development. Accordingly, it is possible to reconstruct a common ancestor of primates of modern aspect adapted for arboreal life and to follow the adaptive radiation in locomotor features that has taken place among the descendants of that ancestor. Any conclusions drawn in this respect

remain valid regardless of the nature of the ancestral stock of placental mammals from which the earliest primates were derived. However, in tracing the earliest origins of primates, notably with respect to the status of 'archaic primates' (early Tertiary plesiadapiforms) and with respect to the affinities of tree-shrews, it is essential to define the likely characteristics of the ancestral placental mammals.

It would be rash to assume that early primates had necessarily undergone advanced development relative to ancestral placentals in all aspects of their locomotion, merely because living primates are distinct from all or most other living placentals in certain features (e.g. in the almost universal possession of a grasping hallux, combined with hindlimb domination). Yet this is the assumption inherent in many comparisons aimed at identifying the phylogenetic relationships of 'archaic primates' and of tree-shrews. It is, for instance, commonly implied that comparisons of tree-shrews with terrestrial insectivores and arboreal prosimian primates provides an adequate basis for assessing the degree of relationship between tree-shrews and primates in locomotor terms (see Chapter 5). Similarly, it has often been accepted that any early fossil placental mammals with locomotor features resembling those of arboreally adapted primates of modern aspect must necessarily be related in some way to primate origins.

It has already been noted that the postcranial skeleton of primates is essentially primitive in many morphological features and that one of the hallmarks of primate locomotor evolution has been the development of relatively sophisticated central nervous control of locomotion, rather than the emergence of extreme skeletal specialization. The generally primitive nature of the postcranial skeleton, particularly among prosimian primates, is confirmed by retention of a little-modified set of tactile pads on the extremities. It is therefore necessary to proceed with caution in the identification of apparent specializations in primate locomotor morphology and function. Definition of the extent to which ancestral primates differed from ancestral placentals in their locomotor adaptations depends, above all else, on the question of whether the earliest placentals were terrestrial, semiterrestrial or arboreal in habits. If the ancestral primate condition was derived from a completely terrestrial placental ancestry, the modifications involved would certainly have been greater than if it originated from a placental ancestry already showing some degree of adaptation to arboreal life. At present, it would probably be true to say that the predominant view is that ancestral placental mammals were primarily terrestrial, but numerous objections have been raised against this view and the issue is by no means settled. Proper resolution of this question requires careful phylogenetic reconstruction rather than the *a priori* assumptions that have been all too prevalent in the literature.

One of the most influential papers on the arboreal or terrestrial origins of placental mammals was written by Haines (1958). As noted by Haines, the prevalent notion at that time was that the placental mammals were derived from arboreal ancestors. This view owed much to the writings of Matthew (1904 and later), who had concluded from studies of early Tertiary mammals that grasping adaptations of the hands and feet, along with a prehensile adaptation of the tail, were widespread. Gregory (1913, 1920a) initially concurred with this interpretation, notably because the earliest known placentals at that time shared many characters with 'primitive arboreal marsupials' and because postcranial specialization among mammals is particularly associated with terrestrial habits. An arboreal ancestry of the placental mammals was then accepted by such authorative palaeontologists as Simpson (1937) and Romer (1945). At the same time, a series of papers drawing on comparative anatomical investigations also supported the concept of arboreal adaptation in the earliest placentals (Böker, 1927; Böker and Pfaff, 1931; Panzer, 1932; Weidenreich, 1921). Steiner (1942) noted that all modern mammals share an unusual feature in the first digit of the hand and foot. The first distal carpal bone of the hand (trapezium – see Fig. 10.18) and the first distal tarsal bone of the foot (entocuneiform – see Fig. 10.13) become

elongated during embryonic development, thus displacing the base of the first digit from the others. Steiner saw this as a basic adaptation for grasping with the hand and foot. However, this prevalent view was not without its detractors. Gidley (1919) questioned Matthew's evidence for grasping adaptations of the hands and feet in early placentals and Wood Jones (1929) postulated a terrestrial ancestry for placental mammals. Gregory (1951) eventually somewhat modified his interpretation and took the modern terrestrial insectivore *Hylomys* (the lesser gymnure) as the most primitive of living placental mammals, thus giving some credence to the idea that modern insectivores could be regarded as models for placental ancestry. Haines (1958) therefore contributed to the debate at a time when the tide of opinion was beginning to shift against acceptance of an arboreal ancestry for placental mammals.

Haines attempted to resolve the issue regarding the locomotor habits of ancestral placentals largely through reference to hand structure. For reasons that are not clearly explained, he selected a modern carnivore, the Egyptian mongoose (*Herpestes ichneumon*), as a model for the primitive placental condition. The morphology of the hand of this species is described, in admirable detail, as exemplifying 'a typical terrestrial form, only slightly modified for cursorial locomotion'. However, it is curious that Haines should have selected for illustration of the condition of the inferred ancestral placental mammal the hand of a modern mammal with numerous obvious specializations:reduction of the first digit; fusion of the scaphoid, lunate and centrale into a single bony mass in the carpus (as in all other modern carnivores); and coalescence of all the interdigital pads to form a single 'interdigital eminence'. Modern prosimian primates typically lack these specializations, so derivation of a primate hand from a *Herpestes*-like hand involves multiple violation of the principle of parsimony.

The case made by Haines (1958) for a terrestrial ancestry of placental mammals rests mainly on a general discussion of hand musculature, based on the initial assumption that the condition in *Herpestes* is primitive, and on refutation of Matthew's claim that the pollex was opposable in early creodonts (e.g. *Claenodon*). Haines established quite convincingly that the pollex was not opposable in such creodonts. This weakens the force of some of Matthew's arguments, although it by no means demonstrates that creodonts were necessarily terrestrial. Indeed, Rose (1987) has identified climbing adaptations in the early Eocene arctocyonid creodont *Chriacus*, which may even have had a prehensile tail. Detailed treatment of mammalian hand musculature provides the strong point of Haines's paper, although once again most of his arguments merely indicate that the ancestral placentals were not necessarily arboreal. Evidence of actual terrestrial adaptation is largely restricted to interpretation of the contrahens muscles, used in closing the hand.

In generalized terrestrial placental mammals with 'convergent' hands, the ventrally located contrahens muscles are arranged as a simple fan of independent slips inserting on the basal phalanges. By contrast, in 'clasping' hands there is an approximately central **raphe** of connective tissue providing an attachment for the contrahens muscles, such that the muscle fibres are directed transversely across the palm. A median raphe of this kind is found in arboreal marsupials, in certain secondarily cursorial marsupials, in primates and in a variety of other placental mammals (see also Napier, 1961). Tree-shrews have a median raphe in the hand, in common with primates; but Haines, in a previous paper (1955), had specifically pointed out that it is located between contrahens muscles running to digits 2 and 5 in tree-shrews, and the contrahens muscle running to digit 1 (the pollex) is an independent slip. In the primate hand, however, the raphe is typically located between contrahens muscles running to digits 1 and 5. Haines (1955, 1958) interpreted this as evidence that tree-shrews have developed a separate adaptation of the hand for arboreal life. It is implicitly assumed that ancestral placentals lacked a median raphe; but this assumption seems to depend largely on the fact that the hand of *Herpestes* lacks a

raphe, rather than on actual phylogenetic reconstruction. Haines (1955) follows Forster (1916) in postulating that the hand of ancestral placentals had four contrahens muscles, running to digits 1, 2, 4 and 5. Among modern mammals, there has been a common tendency for one of the original set to be lost (either that to digit 2 or that to digit 4). It is therefore significant that both marsupials and the few placental mammals (notably primates) that still possess all four contrahens muscles also have a median raphe (see Napier, 1961). Tree-shrews have lost the contrahens muscle running to digit 4 and are thus specialized with respect to the hypothetical condition of ancestral placentals, but the presence of a median raphe may in fact be a retained primitive feature rather than a secondary specialization.

It is worth noting that the contrahens muscles have disappeared completely from the hand in certain insectivores (e.g. *Sorex*, *Crocidura* and *Talpa*), perhaps in association with burrowing habits (Haines, 1955). One must therefore be wary of regarding the hand of any terrestrial insectivore (e.g. *Hylomys*) as representative of the ancestral placental condition. Haines (1958) cites as evidence for the primitive nature of the fan-shaped arrangement of contrahens muscles in the hand of *Herpestes*, the fact that such an arrangement is also present in modern reptiles. However, *Herpestes* has lost the contrahens muscle running to digit 4 as yet another apparent specialization of its hand, paralleling the condition in tree-shrews. General similarity between *Herpestes* and reptiles does not in itself establish the ancestral placental condition. A broad comparative perspective, including marsupials, suggests instead that the ancestral placental hand was of a clasping kind, with four contrahens muscles and a median raphe, and that more specialized mammals (e.g. *Herpestes*) have undergone reduction from this original condition.

One can, in fact, take a view diametrically opposed to that of Haines, postulating that earlier authors were right after all and that habitual arboreal activity was characteristic of ancestral placentals (Lewis, 1964a; Martin,

1968a). This view links up with the generally accepted interpretation that all marsupials are derived from committed arboreal ancestors possessing a grasping foot and numerous other related adaptations. An arboreal ancestry for marsupials was originally proposed by Huxley (1880) and was consolidated by Bensley's detailed studies (1901a,b). Indeed, retention of certain arboreal characteristics in secondarily terrestrial marsupials was cited by Dollo (1899) as a prime example illustrating his renowned 'law of irreversibility of evolution' (see also Haines, 1958). Lewis (1964a) correctly maintained that evolution of locomotor adaptations in placental mammals could be considered effectively only in conjunction with studies of marsupials and noted several features common to the feet of placentals and marsupials, suggesting an ancestral condition with a grasping adaptation of the hallux. In particular, Lewis noted that in both marsupials and placentals the insertion of the peroneus longus tendon has migrated across the sole from the base of the fifth metatarsal to the base of the first, thus acting to swing the hallux towards the rest of the foot. Further, the function of the single-headed M. flexor accessorius was seen as that of correctly aligning the pull of the long flexor tendon attached to the widely divergent hallux. Lewis (1983) has since considerably extended his studies of the structure of the marsupial and placental foot and has somewhat modified his original interpretation, describing the earliest placentals as 'denizens of the spatially complex interface between arboreal and terrestrial habitats' rather than as fully committed arboreal forms comparable to modern opossums. However, the fact remains that modern maruspials and placentals (notably arboreal forms such as primates and tree-shrews) share numerous features of postcranial anatomy suggesting a common ancestral adaptation for locomotion on complex substrates. The foot of the ancestral viviparous mammal was clearly adapted for backward rotation, inversion and eversion, whereas this property is typically abolished in mammals adapted for exclusive terrestrial locomotion. Thus, whereas the exact degree of arboreal adaptation of ancestral

placental mammals remains undecided, it is highly unlikely that they were genuinely terrestrial in habits.

Cartmill (1974c) has reviewed some of the arguments for an arboreal ancestry of placental mammals, noting that various features of primates commonly seen as specializations would be revealed as primitive retentions if such an ancestry were demonstrated. He states that Matthew's arguments were convincingly answered by Gidley (1919) and by Haines (1958), but Cartmill's only real objection to Lewis's evidence (1964a) is that it is not confirmed by available fossil evidence. Indeed, Cartmill concludes judiciously: 'The notion that the ancestral mammals had didelphid-like opposable halluces remains tenable, but the evidence is not conclusive.' Arboreal marsupials and primates in fact share quite a number of features (e.g. primitive arrangement of tactile pads, ridged tactile skin, Meissner's corpuscles, grasping hallux and various features of the tarsal bones) that can be interpreted either as independent adaptations for life in trees or as retentions from an arboreal common ancestor of marsupials and placentals, and choice between these two possibilities depends on a balanced assessment of relative probabilities. It should be noted that Lewis (1964a) specifically compared the secondarily somewhat cursorial marsupial *Phascogale* (the brush-tailed marsupial mouse) with the tree-shrew *Ptilocercus*, suggesting that both show specializations away from an ancestral form adapted for arboreal locomotion involving grasping extremities. This underlines the fact that the condition in tree-shrews can be interpreted as the result of specialization away from an ancestral arboreal adaptation of placental mammals, rather than as the result of specialization towards the primate condition from ancestral terrestrial placental mammals.

Cartmill (1974c) correctly emphasized the importance of the fossil record as a test for hypotheses concerning the locomotor adaptations of ancestral placental mammals. Unfortunately, however, the fossil record is still simply inadequate to permit reliable testing in this way. As already noted, the earliest known postcranial remains of placental mammals come from late Cretaceous deposits of Mongolia and North America. The North American evidence is apparently confined to isolated tarsal bones attributed on size grounds to *Protungulatum* and *Procerberus* (Szalay and Decker, 1974), and the only substantial evidence is represented by Mongolian material from the 'insectivores' *Asioryctes*, *Zalambdalestes* and *Barunlestes* (Kielan-Jaworowska, 1977, 1978). The known postcranial evidence, taken as a whole, reveals that there was already considerable diversity by this stage, indicating the existence of ancestral placentals at some considerably earlier time for which no fossil evidence is available. Further, the late Cretaceous specimens described to date are extremely limited both in broad geographical terms and in terms of likely habitat conditions (e.g. inferred semidesert environment for the Mongolian deposits). Thus, although the known fossils of early placental mammals provide no direct support for an arboreal ancestry of placentals, it is certainly not justifiable to conclude that those fossils rule out such an ancestry. It should also be noted that the Palaeocene North American multituberculate *Ptilodus* had numerous postcranial adaptations for climbing, including great mobility in the ankle, a grasping hallux and a prehensile tail (Jenkins and Krause, 1983). Arboreal adaptations were prevalent in the multituberculates generally, indicating that tree-living habits may have been rife among early mammals.

For the time being, however, the locomotor adaptations of the ancestral placental mammals remain obscure and satisfactory resolution of this issue ideally requires a combination of additional broad comparisons of modern mammals (including marsupials) and more representative fossil evidence. Hypothetical identification of primitive versus derived character states has yet to be tested adequately in either sense. A possible compromise view for the present is that expressed by Jenkins (1974b), according to which modern tree-shrews provide a potential model for an ancestral placental mammal involving both terrestrial and arboreal locomotion with no firm commitment to either extreme. Nevertheless,

it should not be forgotten that tree-shrews themselves have some apparent specializations (e.g. in the fusion of the lunate and scaphoid in the wrist and in the arrangement of the contrahens muscles of the hand) and that a more specifically arboreal ancestry for placental mammals cannot be ruled out. Indeed, such an ancestry is perfectly plausible. There remains a distinct possibility that many locomotor features of modern primates involve primitive retentions from ancestral placentals, enhanced by greater sophistication of the central nervous system. This could explain why 'archaic primates', tree-shrews and 'flying lemurs' all share certain characteristics that may easily be misinterpreted as derived character states.

The question of the origins of primate arboreality is closely connected with the widely cited concept that numerous primate adaptations, such as emphasis on vision and marked increase in brain size and complexity, were developed as a result of an inferred shift from terrestrial to arboreal habits (Wood Jones, 1916; Elliot Smith, 1927). Cartmill (1972) reconsidered this 'arboreal theory of primate evolution' in some detail, identifying several logical shortcomings. As Cartmill observed, there are several other orders of placental mammals characterized by arboreal habits that lack the distinguishing features of primates of modern aspect (e.g. grasping foot), so arboreal activity *per se* does not account for primate origins. Elliot Smith particularly emphasized a possible link between developments of the visual apparatus and arboreal habits in primate evolution, but Cartmill demonstrated convincingly that additional factors must have been involved. He concluded instead that forward rotation of the eyes and other developments in the visual system of primates (see also Chapter 7) are associated with nocturnal, visually directed predation on insects in the terminal branches of trees. Whereas Cartmill (1972) rightly emphasized the importance of visually directed predation, however, it seems likely that a more comprehensive explanation for primate visual and locomotor adaptations can be based on original occupation of the 'fine-branch niche' by small-bodied, nocturnal and actively foraging ancestral primates (Martin, 1973, 1979; Charles-Dominique, 1977).

As Cartmill himself has demonstrated (1974c), non-primate placental mammals such as rodents and carnivores can move around with considerable agility in trees. However, such mammals tend to concentrate their activities in trees on broad trunks and branches. The same is generally true of tree-shrews (D'Souza, 1974). Indeed, for mammals in this body-size range, the possession of claws is almost certainly advantageous for negotiating broad arboreal supports, whereas grasping extremities are at a premium only among finer arboreal supports such as terminal branches and lianes.

It is significant that those primates with 'claws' of various kinds are commonly active on broad arboreal supports because of particular dietary habits. Marmosets and tamarins have been widely reported to feed on sap or gum from tree trunks and marmosets even have special dental adaptations for making holes in the bark of trees to induce the flow of both gum and sap (Coimbra-Filho and Mittermeier, 1977; see also Chapter 6). Two prosimian species that feed extensively on natural gum exudates, the needle-clawed bushbaby (*Galago elegantulus*) and the fork-marked lemur (*Phaner furcifer*), have sharp tips on the end of their nails ('needle-claws') that enable them to cling more securely to broad trunks while feeding head down on gum (see Charles-Dominique, 1977). It is, in fact, easy to imagine how 'needle-claws' could gradually evolve into full secondary claws, as in marmosets and tamarins, as a result of increased emphasis on movement along broad arboreal supports. The only other modern primate with well-developed claws, the aye-aye (*Daubentonia*), is well known for its habit of feeding on wood-boring insect larvae. There is a direct connection here with gum-feeding in that such larvae are known to be responsible for the production of natural gum-licks through damage to tree trunks and hence to contribute to the diet of primates such as the lesser bushbaby *Galago senegalensis* (Bearder and Martin, 1980a). It is not known whether the aye-aye eats gum, but one can postulate that gum-feeding would be a likely intermediate stage in the evolution of feeding upon the wood-boring larvae that generate natural gum-licks. In all cases where modern primates are found to possess claws, therefore, it is reasonable to postulate that there has been a secondary return from the fine-branch niche to locomotion on broad arboreal supports.

The fundamental locomotor adaptations of primates of modern aspect, involving grasping extremities and hindlimb domination combined with a posterior location of the centre of gravity, can be attributed fairly confidently to occupation of the fine-branch niche by a small- to medium-sized ancestor. This is neatly illustrated by the fact that small-bodied lemurs (Cheirogaleidae) and members of the loris group all show a cantilever posture in fine branches (Gebo, 1987c). In this posture, the feet grasp a fine branch and the body is held stretched out. Tarsiers also show the cantilever posture (Crompton and Andau, 1986). Hence, all small-bodied prosimians show this posture, probably as a reflection of ancestral adaptation to the fine-branch niche. As Jenkins (1974b) has aptly stated: 'The adaptive innovation of the ancestral primates was therefore not the invasion of the arboreal habitat, but their successful restriction to it.' 'Successful restriction' surely implies occupation of the finer branches. Arboreal marsupials are also occupants of the fine-branch niche and their many similarities to primates in locomotor adaptation may hence depend on this fact, although (as explained above) it remains to be seen whether ancestral marsupials and ancestral primates of modern aspect convergently occupied the fine-branch niche or whether they simply retained this feature from ancestral viviparous mammals.

ORIGINS OF HUMAN BIPEDALISM

The preceding survey of primate locomotor adaptations has yielded numerous indications of the likely ancestral pattern from which the diverse array of modern primate locomotor adaptations was derived. These indications can be combined to provide a framework for tracing in outline the origins of human bipedalism from the earliest beginnings of primates of modern aspect. For this purpose, it is convenient to consider three stages in the evolution of primate locomotor patterns: an ancestral primate stage, an ancestral simian stage and an ancestral hominoid stage.

It seems likely that the ancestral primates of modern aspect were relatively small, and this would have had specific implications for locomotor adaptations. The modal body weight for modern nocturnal prosimians is about 500 g, and it is reasonable to derive primates of modern aspect from nocturnal ancestors of broadly comparable size. Such small-bodied ancestral primates were probably 'quadrumanous' arboreal forms inhabiting the fine-branch niche, and grasping ability was particularly well developed in the foot by virtue of the presence of a powerful, widely divergent hallux. Arboreal activity of these early primates doubtless included leaping as well as climbing, and there was a characteristic combination of hindlimb domination, a diagonal-sequence walking gait and special adaptation of the tarsus (notably involving elongation of the distal portion of the calcaneus). Quadrumanous climbing in the fine branches of trees was also associated with the development of nails on all digits, linked with the presence of tactile pads incorporating Meissner's corpuscles on the ventral surfaces of hands and feet. Because of their role in the grasping action of the extremities (with respect to both frictional resistance and tactile sensitivity), the ridges on the terminal pads of the digits were probably arranged essentially longitudinally, as in all modern prosimians, to provide maximal effect.

The ancestral simian primates that emerged at a later stage during the radiation of early primates probably retained many of the original primate characteristics (i.e. 'quadrumanous', hindlimb-dominated arboreal climbing and leaping combined with enhanced tactile sensitivity), but they also showed certain innovations. Given that the modal body weight of modern simians (about 5 kg) is considerably greater than that of modern prosimians, it is

reasonable to conclude that ancestral simians had undergone a marked increase in size compared with ancestral primates (perhaps in association with a shift to diurnal habits; see Chapter 7). Accordingly, it is unlikely that the ancestral simians would have consistently occupied the finest terminal branches and twigs of trees. Instead, the body size of these primates would have been better suited for travel along trunks and relatively broad branches (see also Gebo, 1986b, 1987b). although feeding activity would still have been concentrated among the terminal branches. (For an illuminating discussion of terminal-branch feeding in simians, see Grand, 1972.) In all probability, in spite of retention of certain fundamental features of hindlimb domination (e.g. diagonal-sequence walking pattern associated with posterior location of centre of gravity), ancestral simians had a pronounced quadrupedal tendency that has persisted in modern monkeys of the New and Old Worlds. Emergence of this quadrupedal pattern was associated with the spread of ridged skin across the ventral surfaces of the hands and feet and the development of more complex ridge patterns (e.g. whorls). As in the ancestral primates and in most living prosimians, the tail was retained as an accessory balancing organ for arboreal locomotion. The ancestral simians were presumably still essentially arboreal in habits, as is the case with modern New World monkeys, but there is increasing evidence that the Old World monkeys at least underwent a shift to quite terrestrial habits that has left its mark on all modern representatives of this group (see Andrews and Aiello, 1984).

Evolution of the ancestral hominoids, from which apes and humans were derived, was probably accompanied by a further increase in body weight above 10 kg. This would have been correlated with a shift from above-branch quadrupedal locomotion to below-branch suspensory locomotion as the predominant mode of arboreal activity. (This trend was paralleled by large-bodied Malagasy lemurs such as *Palaeopropithecus* and *Megaladapis*.) The ancestral hominoids probably lost the tail and developed some other special features associated with suspensory locomotion and arboreal climbing, such as a high intermembral index and a barrel-shaped chest. Nevertheless, some degree of hindlimb domination probably remained, at least with respect to a relatively posterior location of the centre of gravity, such that bipedalism could still emerge at a later stage in the emergence of the human lineage.

It has been noted above that the lesser apes (gibbons and siamang) are unusual among primates in having very high intermembral indices in spite of their moderate body weights of 5–11 kg (see Fig. 10.5). This and other features suggest that the ricochetal brachiation of the lesser apes is a secondary specialization away from the hominoid condition, as is confirmed by the unusual dynamics involved in siamang brachiation (Fleagle, 1974). However, partly because of the common tendency to treat modern primates as constituting a 'phylogenetic scale' (see Chapter 3), with the gibbons intermediate between monkeys and great apes, there have been repeated suggestions in the literature that the active brachiation of modern lesser apes represents a possible stage in the origins of human bipedalism (Keith, 1923; Gregory, 1928; Washburn, 1950; Avis, 1962). As has been noted by several authors (e.g. Napier, 1963; Le Gros Clark, 1959), it is in fact highly unlikely that ancestral hominoids were as specialized as modern lesser apes. Discussion of this topic has been considerably hindered by variation in use of the term 'brachiation', however, and if this term is taken simply to mean suspensory arboreal locomotion, ancestral hominoids might well have been 'brachiators'. But even in this very loose sense, the term 'brachiator' may be misleading, as there is now increasing evidence that adaptations for **climbing**, rather than for suspensory locomotion alone may provide the key to various morphological developments in the relatively large-bodied ancestors of hominoids (e.g. see Fleagle, Stern *et al.*, 1981; Stern and Susman, 1983; Susman, Stern and Jungers, 1984). The special wrist adaptation of hominoids identified by Lewis (1971a,b, 1974) may actually be related to an ancestral combination of climbing and suspensory travel.

The development of habitual striding bipedalism as a unique feature of the hominoid lineage leading to modern humans has involved a suite of major readjustments in the morphology and function of the postcranial skeleton, literally from the neck downwards. These changes can be summarized by comparing the human skeleton with that of a great ape of about the same body size (Fig. 10.21). Although modern great apes also have certain specializations, the distinctive nature of bipedalism is such that virtually all differences represent derived human features. Starting at the foot, the human condition differs from that of all other primates in that the hallux is no longer divergent and actually projects forwards beyond the other toes. The hallux is physically bound in line with the other toes by the deep transverse metatarsal ligament, which attaches to the first metatarsal during fetal life (Wood Jones, 1949). As emphasized by Napier (1967b), this realignment of the hallux in the human foot is related to the fact that the final propulsive thrust during striding is provided by the terminal phalanx of the big toe. The human foot is also distinctive in that the usual transverse arch of the mammalian foot has been supplemented by longitudinal upward curving between the heel of the calcaneus and the heads of the metatarsals (Fig. 10.21c). This longitudinal curve serves as an elastic bow that allows the foot to absorb the effects of vertical forces, more or less like a spring (Wood Jones, 1949). As might be expected, the tarsal bones have been completely remodelled in connection with these developments and with the general requirements of terrestrial locomotion (Lewis, 1981).

Major changes have also occurred in the human leg, in addition to the secondary shift to a lower intermembral index (see Fig. 10.5). The human knee exhibits special features associated with full extension of the leg, with the femur and tibia in alignment, which occurs during standing and during bipedal striding. In all other primates, locomotion involving the use of the leg in support is characterized by more limited extension such that the leg is virtually always flexed at the knee. The human leg therefore requires special

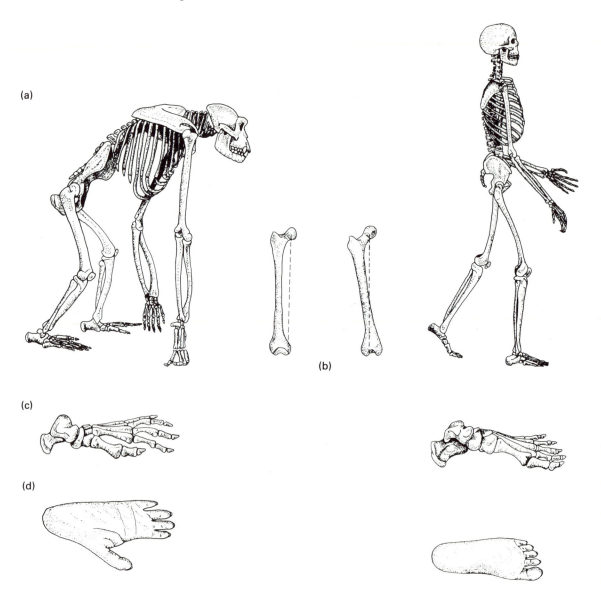

Figure 10.21. Comparison of gorilla and human to illustrate morphological changes associated with the evolution of striding bipedalism.

(a) Mounted skeletons of gorilla (left) and human (right). Note the shortening of the pelvis, the S-shape of the vertebral column, the reduction of the dorsal spines above the neck vertebrae and the balancing of the head on the vertebral column in the human skeleton. (After Napier, 1967b.)

(b) Femora of gorilla (left) and human (right). Note the 'carrying angle' of the human femur. (After Pilbeam, 1972.)

(c) Articulated bones of the gorilla foot (left) and the human foot (right, showing the longitudinal curve in the latter. (After Wood Jones, 1949.)

(d) Feet of gorilla (left) and humans (right). The big toe is no longer divergent in the human foot and it is the longest digit, in contrast to the big toe of great apes. (After Midlo, 1934.)

adaptations to prevent over-extension. There is also a modification to bring the knees in closer to the midline of the body, thus providing more direct support. This is achieved by orientation of the shaft of the femur at an angle outward from the knee joint (the 'carrying angle'). For the same reason, the outer articular condyle of the femur is the largest, as opposed to the inner in great apes, because weight transmission is concentrated on the outer side of the human knee.

The pelvic girdle has been extensively modified to meet the requirements of bipedal locomotion. In contrast to other primates, the human pelvis is broad and squat and the iliac blades have S-shaped rather than straight upper margins. The wedge shape of the sacrum is accentuated in the human skeleton and this contributes to the general broadening of the pelvic girdle. All of these changes serve to provide a broader base of support for the upper body on the hindlimbs, facilitating the maintenance of balance during bipedal walking. There has also been a significant change in the gluteus muscles connecting the femur and the pelvis. In non-human primates, the gluteus medius and gluteus minimus muscles serve a major role either as extensors of the leg or as rotators, whereas in humans they act as abductor muscles of the trunk, stabilizing the pelvis during bipedal striding. The gluteus maximus muscle, which is relatively less well developed in other primates, has been expanded in humans to take over the extensor role of the gluteus medius and gluteus minimus muscles (Napier, 1967b).

Above the pelvis, the vertebral column also has some special features connected with our erect body posture. In all other primates, the vertebral column curves dorsally between the hips and the shoulders and the ribcage is entirely ventral to the spine. In humans, by contrast, the vertebral column has an S-shaped profile when viewed from the side and the chest vertebrae are displaced inwards relative to the ribcage. Both of these human modifications are related to the essentially vertical forces that act downwards through the body. The sinuous form of the spine provides the equivalent of a vertical spring, absorbing downward-directed forces, while inward displacement of the spine relative to the ribcage brings it closer to the line of action of the body's centre of gravity. Further, in the neck (nuchal) region, the vertebrae lack the conspicuous dorsal spines present in great apes. These spines, together with a large nuchal area on the occipital region of the skull for attachment of powerful neck muscles, are required in great apes to maintain the skull in position relative to the obliquely slanted vertebral column. In human beings, by contrast, the head is quite well balanced on the vertebral column because the foramen magnum has been displaced forwards beneath the skull (Fig. 10.21a). Consequently, the nuchal area for neck muscle attachment on the human skull is relatively restricted and marked dorsal spines are no longer required on the neck vertebrae.

Because of these far-reaching changes in the human postcranial skeleton in connection with the emergence of striding bipedalism, many indicators are available for tracing the evolution of bipedalism through the fossil record. Unfortunately, however, there is still a major gap in the fossil record of human evolution before about 4 million years (Ma) ago and virtually nothing is known about the earliest phases in the evolution of bipedalism. The oldest known evidence comes from two sources: a fossilized footprint trail from the Laetoli site in Tanzania (Leakey and Hay, 1979; Leakey and Harris, 1987), dated at about 3.6 Ma, and a partial skeleton of a gracile australopithecine (*Australopithecus afarensis*), dated at about 3.1 Ma, from the Hader site in Ethiopia (Johanson and Taieb, 1976). Stern and Susman (1983) and Susman, Stern and Jungers (1984) have reviewed the material and have concluded, not surprisingly, that although there are some marked similarities to modern humans, the earliest evidence of hominid locomotor adaptations still reflects some features more reminiscent of modern great apes. The skeletal remains of *Australopithecus afarensis* show some evidence of suspensory locomotion and climbing and it is therefore unlikely that this species was as fully committed to bipedalism as modern humans are. The phalanges and metacarpals of the hand

and foot were curved, as in modern apes, and there is evidence that large wrist flexor muscles were present. As in modern apes, the shoulder joint was oriented upwards rather than directly out to the side as in modern humans. Similarly, the iliac blade of the pelvis, although squat, lacked the modern human S-shaped profile. Further, certain features of the ankle joint in *Australopithecus afarensis* indicate an ape-like adaptation for plantarflexion.

Although the femur of *A. afarensis* does have a carrying angle, other features of the knee-joint itself indicate that the knee had not yet become fully adapted for extension as in humans. Further, it has been observed by Prost (1980) that a carrying angle is identifiable in the femur of certain fully arboreal primates such as the orang-utan, and this may link initial development of a carrying angle of the femur with climbing rather than specifically with bipedalism. In this context, it should be noted that electromyographic studies conducted by Stern and Susman (1981) showed that in the chimpanzee there is the same pattern of muscle recruitment during climbing as during bipedal walking. All of this combines with the relatively high intermembral index value determined for *Australopithecus afarensis* (see Fig. 10.5) to indicate that this species had not made a complete transition from arboreal to terrestrial life. Incidentally, it should be noted that the fossil foot bones discovered in deposits dated at about 1.8 Ma ago in Olduvai Gorge (Tanzania) and attributed to *Homo habilis* may not be as similar to modern human foot bones as initially indicated by Day and Napier (1964). A detailed re-examination of the joints suggests, instead, a condition intermediate between modern great apes and man (Lewis, 1981). By contrast, more recent fossil material attributed to *Homo erectus* and to early *Homo sapiens* is generally much closer to the modern human condition. Nevertheless, an early, almost complete skeleton of *Homo erectus* (apparently from an adolescent male), dated at about 1.6 Ma, shows some australopithecine-like features (e.g. in the neck of the femur and in the shape of the ilium; see Brown, Harris *et al.*, 1985).

Detailed analysis of the fossil footprint trail from Laetoli (Day and Wickens, 1980) suggested that the footprints were made by a striding biped and that there was very little difference from modern human footprints. However, Stern and Susman (1983) and Susman, Stern and Jungers (1984) have questioned this interpretation, stating that there is little evidence of the presence of a proper ball on the sole of the foot and that there seems to be a deep heel impression more reminiscent of the terrestrial locomotion of a great ape. They suggest, in fact, that the hominid that made the prints may have curled the toes, thus modifying the form of the print to look more like a modern human print. This links up with the suggestion also made by Stern and Susman that *Australopithecus afarensis* may have retained some residual ability to abduct the hallux away from the rest of the toes. There is, of course, the problem that the Laetoli footprints may have been made by a hominid other than *A. afarensis*, and the situation is further complicated by the fact that examination of additional postcranial fragments from the Hadar site strongly suggests the presence of two distinct kinds of postcranial skeleton from hominids of different sizes. The partial skeleton of *A. afarensis* resembles fragments of the smaller size class, whereas fragments of the larger size class more closely resemble the modern human skeleton. Stern and Susman (1983), following a number of other authors, tentatively interpret the small and large individuals represented as females and males, respectively, of the single species *A. afarensis*; but there remains a distinct possibility that two separate hominid species were in fact present at Hadar some 3.5 Ma ago and that the larger species already had a relatively modern postcranial skeleton. In either case, the initial development of human bipedal locomotion still remains shrouded in mystery and only further fossil discoveries will provide the answers required. Nevertheless, it is clear that the origins of human bipedalism owe much to initial hindlimb-dominated adaptations for arboreal life that were developed in the earliest primates of modern aspect.

Chapter eleven

Evidence from chromosomes and proteins

The evidence relating to mammalian evolution discussed in previous chapters has been largely morphological in kind, with some mention of physiological and behavioural aspects where appropriate. In this chapter, we turn to a substantially different kind of evidence that has in recent years added an important new dimension to the undertaking of phylogenetic reconstruction. This evidence concerns the structure of the chromosomes carrying the fundamental genetic material (deoxyribonucleic acid or **DNA**) and, at a finer level of analysis, the structure of the DNA itself and of the proteins that are generated as its main products. Because the information derived from the comparative study of chromosomes, of DNA and of proteins is rather different in kind from the data discussed so far, it is necessary to consider some general background information before proceeding to an assessment of their potential contribution to phylogenetic reconstruction.

It is now firmly established that the basic genetic material of virtually all living organisms is the nucleic acid DNA. This material carries, in coded form, the genetic information transmitted from parent to offspring and it is essentially chance modifications (**mutations**) in DNA that

seem to provide the basic potential for evolutionary change. As has been explained in Chapter 2, Darwin's original theory of evolution through natural selection (1859) rested on the following interconnected tenets:

1. Populations of living organisms tend to produce more offspring than can be supported by available resources.
2. Numerous observed characters of any living organism have an hereditary basis.
3. Inherited differences in the expression of a given character can arise by chance in individuals of a species population.
4. In a given environment, a particular state of a character may bestow upon its possessor an increased probability of survival.
5. 'Natural selection' operates through differential survival of individuals exhibiting alternative states of a particular character.

Darwin was not aware of the actual properties of the material responsible for genetic transmission, although he did postulate the existence of particles ('gemmules') bearing hereditary information. In fact, at one stage he proposed a system of 'blending inheritance' that would have led inexorably to progressive dilution of any differences between individuals of a species population. Nevertheless, his two central tenets concerning the process of genetic transmission (the existence of hereditary material of some kind and chance modification thereof) were essentially correct and are fundamental to modern evolutionary theory.

Mendel's experiments on the inheritance of simple observable characters in plants (see Mendel, 1866) clearly indicated that the hereditary 'factors' responsible behave in a particulate fashion. To account for his experimental results, he was led to postulate that each organism has two factors operating together to determine the expression of any given characteristic. The two factors may be identical (**homozygous** condition), but they may also differ (**heterozygous** condition). In cases where the heterozygous condition was present, Mendel commonly found that one of the two factors was **dominant** to the other (the **recessive**), such that the expression of the latter was masked. Cross-breeding of plants with the heterozygous condition for a given characteristic showed that the two factors must be separable at some stage of reproduction, as it was possible to obtain offspring identifiably homozygous for the recessive condition (Mendel's **law of segregation**). Further, experiments involving the study of more than one characteristic at a time showed that the factors governing different characters could recombine at random in the offspring (Mendel's **law of independent assortment**), at least in those cases investigated. Thus, any mechanism subsequently proposed for genetic transmission required not only compatibility with Darwin's basic tenets, but also concordance with Mendel's observations of particulate inheritance involving paired 'factors'. (For a proper discussion of the historical background to modern genetical theory, the reader is referred to Dobzhansky, Ayala *et al.*, 1977).

As it turned out, Mendel's crucial observations were not incorporated into a general theory of inheritance for half a century, even though the foundations for such a theory were laid at about the same time through early microscopical studies of the cells of living organisms. Among other things, it was observed that, at cell division, it was possible to stain condensed rod-like or thread-like structures (the **chromosomes** – literally, 'staining bodies') in the cell nucleus. It had become obvious that cell division represents the primary process of growth and reproduction in all organisms and that some mechanism must exist to ensure that new daughter cells contain the requisite genetic information. The chromosomes, which occur in all living cells containing a nucleus (**eukaryotic** cells), were soon recognized as possible candidates for the role of genetic transmission.

Experimental evidence for the link between Mendel's laws of inheritance and the chromosomes was provided in the classical studies of the fruit fly (*Drosophila melanogaster*) conducted by Morgan (1910, 1911, 1926). The major advance made by Morgan resided in the discovery of deviations from Mendel's second law of independent assortment. He found that the hereditary factors (by that stage referred to as **genes**, following Johanssen, 1909) are present in linked groups and that true independent assortment occurs only between genes from different linkage groups. The number of linkage groups determined for the fruit fly through Morgan's experimental work was found to match the number of pairs of chromosomes. Morgan postulated that the degree of independent assortment found within linkage groups is a function of the linear distance separating genes on the same chromosome, and this eventually led to gene-mapping (e.g. see Bridges, 1935).

In a multicellular organism, such as a mammal, most cells in the body other than the sex cells (**gametes**) contain a standard number of chromosomes, constituting the **karyotype** of the species. With a very few exceptions involving slight numerical variation (see later), all members of a given species population will typically have a standard number of chromosomes in the non-reproductive (**somatic**) cells of the body. This, in itself, indicated to early workers that the chromosomes are likely to act as the agents of genetic transmission. In most cases, the chromosomes can be easily identified as matched pairs on the basis of shape and overall dimensions. The standard number of chromosomes for a species is hence referred to as the **diploid number (2n)**. One consistent departure from the matching of the chromosomes in pairs is, however, found with the **sex chromosomes** of sexually reproducing species. These chromosomes differ from the remaining chromosomes (**autosomes**) because of their specific role in sex determination. In mammals, the two sex chromosomes of the female are matched and are referred to as **X-chromosomes**. Males have only one X-chromosome in each diploid set of chromosomes and there is usually a distinctive, smaller **Y-chromosome** representing its partner.

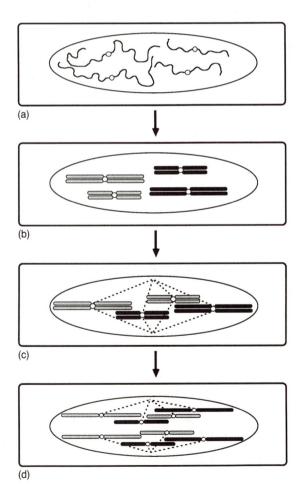

(a)

(b)

(c)

(d)

In routine cell division, as in the normal growth of a multicellular organism or in the asexual reproduction of certain plant and animal species, there is a standard process (**mitosis**) which ensures that (barring accidents) division of a parent cell will yield two daughter cells each containing a chromosome complement identical to that of the parent cell (Fig. 11.1). In fact, it is at the time of cell division that mutations typically seem to occur, so relatively minor differences between daughter cells may arise for this reason. Each chromosome has a specific narrowed region, the **centromere**, and in the process of cell division all of the chromosomes in the nucleus become attached to the so-called **spindle** by their centromeres.

Mitosis is divided into phases. The first one (**prophase**) involves the complete duplication (replication) of each chromosome, eventually including doubling of the centromere. The double centromere of each chromosome then becomes attached to the spindle during **metaphase**. During **anaphase**, the two halves of each replicated chromosome (**chromatids**) separate and begin to move to opposite poles of

Figure 11.1 Schematic illustration of the standard process of somatic cell divison (mitosis).

(a) In the resting state (interphase), the chromosomes are thin and filamentous and cannot be stained with conventional histological methods.

(b) At the beginning of mitosis (prophase), the chromosomes condense and undergo replication. At this stage, it is relatively easy to stain them. Note that it is now possible to recognize two matched sets of chromosomes (one set indicated by stippling, the other set indicated in black).

(c) The chromosomes subsequently become aligned in an equatorial plane (metaphase). Their centromeres, which by this stage have also undergone replication, are attached to the fibres of the spindle (dotted lines).

(d) One half of every chromosome migrates towards each pole of the spindle during anaphase, such that both daughter cells receive the species-specific diploid number of chromosomes. During telophase, the processes of division of the nucleus and of the cell are completed and the chromosomes then return to the resting state shown in (a).

the spindle. Finally, the two daughter sets of chromosomes reach the poles of the spindle and two separate nuclei form (**telophase**), thus concluding the process of cell division. In this way, the number and content of chromosomes are maintained essentially constant in all somatic cells of an organism produced by successive mitotic divison. Change over time can only occur

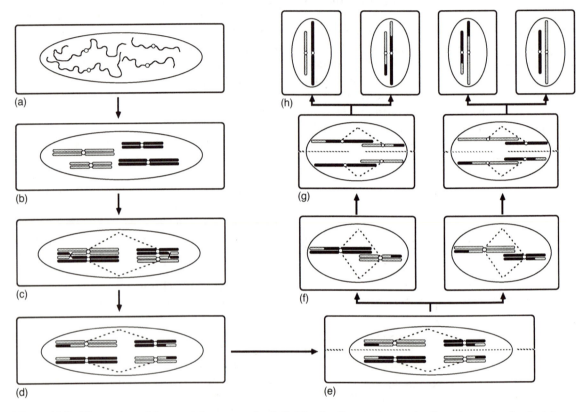

Figure 11.2. Illustration of the special process of cell division leading to the production of gametes (meiosis).

(a) Prior to cell division, the chromosomes are in the typical resting condition.

(b) At the beginning of meiosis, as with normal somatic cell divison (Fig. 11.1), the chromosomes condense and undergo replication, becoming relatively easy to stain. Two matched sets are recognizable (stippled and black); see also Fig. 11.5.

(c) The chromosomes subsequently become aligned in an equatorial plane. In contrast to mitosis, however, the chromosomes are assembled in matched pairs. Crossing-over can occur between the arms of the adjacent matched chromosomes.

(d) One chromosome from each pair migrates towards each pole of the spindle, with some arms modified by mutual exchange of chromosomal material brought about during crossing-over.

(e) As the first division of meiosis draws to a close, each daughter cell receives half of the normal complement of chromosomes (a haploid set).

(f) In the two daughter cells, the chromosomes once again become aligned on an equatorial plane. By this stage, the centromeres have been replicated, as in mitosis.

(g) One half of each chromosome migrates towards each pole of the spindle, such that all four daughter cells (gametes) receive half of the species-specific number of chromosomes. Following the completion of cell division to produce the gametes, the chromosomes return to the resting state shown in (a). Note that in the male mammal, all four daughter cells usually develop into viable sperm, whereas in the female three of the four daughter cells degenerate to produce 'polar bodies'.

through modification of the chromosomes by mutation.

In sexually reproducing organisms, however, the overall constancy of chromosome numbers per cell nucleus can be maintained from parent to offspring only by a special process of cell division that produces the gametes (**meiosis**; Fig. 11.2). Through this special process of division, the chromosome number is halved as the gametes are formed, so that fusion of the male gamete (the **sperm**) with the female gamete (the egg or **ovum**) restores the species-specific chromosome complement in the resulting **zygote**. Clearly, it is essential that in the process of gamete formation one member of each pair of matched chromosomes should pass to a gamete so that the zygote will have a reconstituted normal diploid set.

Meiosis actually involves two consecutive divisions, leading to the formation of four daughter cells. The first division depends directly on the fact that normal body cells, and hence the cells producing the gametes, have a diploid set of chromosomes matched in pairs. The chromosomes undergo replication, as in mitosis, but then the pairs of chromosomes become aligned together on the spindle during the first division, such that one chromosome from each pair passes to each daughter nucleus. While becoming aligned on the spindle, the arms of the matched chromosome pairs overlap and exchange genetic material (**crossing-over**). This direct exchange of material between paired chromosomes accounts for the limited degree of independent assortment that can occur with genes located on the same chromosome (i.e. within Morgan's linkage groups). The second meiotic division then takes place, with the half-complement of replicated chromosomes aligned on the spindle and with one duplicate chromatid passing to each of the daughter nuclei – as in the normal mitotic process. Each of the four daughter cells thus produced from an original parent cell accordingly has only a single, **haploid** set (*n* chromosomes).

The importance of sexual reproduction, involving the fusion of two gametes each containing a haploid set of chromosomes, lies in the introduction of variation through active **genetic recombination**, as distinct from the passive accumulation of random mutations that affects hereditary transmission with any form of cell division. Through the chance passage of one chromosome from each matched pair into every haploid nucleus, and through genetic exchange between chromosomes during crossing-over, the process of meiosis introduces two major sources of genetic recombination (see Dutrillaux, 1986). Subsequently, randomized fusion between gametes (egg and sperm) during sexual reproduction generates further variability among the reconstituted diploid chromosome sets of the offspring. All of these kinds of recombination together generate a considerably increased range of variation between individuals on which natural selection can act. This presumably explains why most living organisms (including all mammals) reproduce sexually at some stage of their life cycle, although the details of the evolution of sexual reproduction remain obscure.

The process of meiosis also ensures that only one of the two sex chromosomes passes to each gamete. In mammals, females are the **homogametic sex** and all mammalian ova normally contain one X-chromosome. Males are the **heterogametic sex** and each normal mammalian sperm will contain either one X-chromosome or one Y-chromosome. As a result, both male and female zygotes can be produced on fertilization, depending on the nature of the sex chromosome carried by the sperm. Indeed, unless special factors intervene, chance fertilization of eggs by sperm should produce equal numbers of male and female offspring.

The chromosomes themselves are composed of nucleic acids and several kinds of protein, forming a conjugate called **chromatin** or nucleoprotein. The protein fraction serves a dual role; some of the chromosomal proteins have a structural function, while the remainder play an important part in the regulation of gene action. Although there was initially some doubt about the identity of the hereditary material borne in the chromosomes, as some investigators believed that chromosomal proteins might be the carriers of genetic information, it was eventually established beyond doubt that the genes are carried in the DNA, which is of quite complex chemical structure. All nucleic acids are polymers of basic components termed **nucleotides**, each of which consists of a nucleotide base combined with a phosphate group and a sugar residue (Fig. 11.3). There are four kinds of nucleotide base in DNA: two purines (adenine and guanine) and two pyrimidines (cytosine and thymine). In DNA, every sugar residue is deoxyribose – hence the name of this particular nucleic acid. The nucleotides in each chain of nucleic acid are linked up in a linear sequence by bonds between adjacent phosphate and sugar groups. It was shown by Watson and Crick (1953) that in DNA there are two chains of nucleic acid combined to form a double helix. The cross-bridges between the two chains in the double helix are formed by hydrogen bonds linking nucleotide bases in specific combinations: adenosine links up with thymine and guanine links up with cytosine. It is this peculiarity in the specific bonding between nucleotide bases that gives DNA its vital property of consistent replication, because accurate reconstitution of the complementary chain can take place following separation of the two chains in the double helix.

DNA carries in coded form 'instructions' that govern protein synthesis in the cells of most living organisms. Proteins, which are also polymers, consisting of chains of **amino acids**, serve a variety of functions. Some are structural and others are regulatory, but the most important function is the part taken by many proteins as enzymes (catalysts). The concept of 'one gene – one enzyme', first formulated by Beadle and Tatum (1941), may be generalized to 'one gene – one protein'. This generalized version is still broadly accepted, although it is now emerging that genes do not necessarily consist of continuous linear segments of DNA (see later). The capacity of DNA to prescribe protein structure depends on the fact that a very limited set of amino acids is typically present in the vast array of different proteins possessed by living organisms. Twenty 'essential amino acids' are generally recognized (Table 11.1). The key to the 'instructions' for protein synthesis carried by

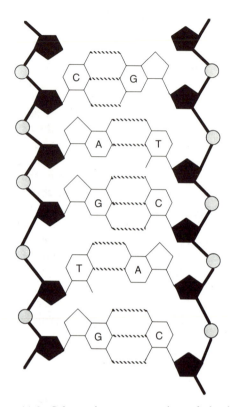

Figure 11.3. Schematic representation of the basic structure of DNA 'uncoiled' for the sake of clarity (adapted from Watson, Tooze and Kurtz, 1983). The two strands of the DNA molecule consist of alternating phosphate groups (stippled circles) and deoxyribose sugar groups (black pentagons), united by powerful chemical bonds (heavy black lines). Each sugar group is also attached to a single nucleotide base (A, adenine; C, cytosine; G, guanine; T, thymine). The two strands of DNA are held together by strong hydrogen bonds (hatched lines) between the nucleotide bases, which can only fit together in specific pairs (A–T or G–C). Because the G–C base pair involves three strong hydrogen bonds, whereas the A–T base pair involves only two, the stability of a DNA double helix increases as the proportion of G–C base pairs increases. The DNA molecule is coiled to form a double helix because of the specific packing properties of the component parts.

DNA resides in the **genetic code** which is based on triplet sequences of nucleotide bases (Figure 11.4). Most triplets code for a particular amino acid, although a small number represent 'stop' sequences signalling the end of a sequence to be

Table 11.1 The 20 'essential amino acids' that serve as the basic components of virtually all proteins present in living organisms (standard abbreviations shown in parentheses)

Alanine (Ala)	Leucine (Leu)
Arginine (Arg)	Lysine (Lys)
Asparagine (Asn)	Methionine (Met)
Aspartic acid (Asp)	Phenylalanine (Phe)
Cysteine (Cys)	Proline (Pro)
Glutamic acid (Glu)	Serine (Ser)
Glutamine (Gln)	Threonine (Thr)
Glycine (Gly)	Tryptophan (Trp)
Histidine (His)	Tyrosine (Tyr)
Isoleucine (Ile)	Valine (Val)

FIRST AND SECOND POSITIONS	THIRD POSITION			
	A	C	G	U
A A	Lys	Asn	Lys	Asn
A C	Thr	Thr	Thr	Thr
A G	Arg	Ser	Arg	Ser
A U	Ile	Ile	Met	Ile
C A	Gln	His	Gln	His
C C	Pro	Pro	Pro	Pro
C G	Arg	Arg	Arg	Arg
C U	Leu	Leu	Leu	Leu
G A	Glu	Asp	Glu	Asp
G C	Ala	Ala	Ala	Ala
G G	Gly	Gly	Gly	Gly
G U	Val	Val	Val	Val
U A	STOP	Tyr	STOP	Tyr
U C	Ser	Ser	Ser	Ser
U G	STOP	Cys	Trp	Cys
U U	Leu	Phe	Leu	Phe

transcribed. In principle, then, a linear sequence of nucleotide bases in a segment of a DNA chain (i.e. a 'gene') can prescribe a linear sequence of amino acids in a protein through the genetic code. The manner in which this is achieved is quite complex (e.g. see Watson, Tooze and Kurtz, 1983), but the basic elements of the process are straightforward.

The process of protein synthesis involves the participation of several different types of another, single-stranded nucleic acid – ribose nucleic acid (**RNA**). As the name implies, this nucleic acid, which also occurs in chromosomes, differs from DNA in that the sugar residues are ribose, rather than deoxyribose. There is a further difference in that one of the pyrimidine bases found in DNA, thymine, is replaced by another pyrimidine, uracil, in RNA. As the first main step in protein synthesis, a sequence of nucleotide bases in the DNA is transcribed to produce a matching, single-stranded RNA sequence called **messenger RNA**. The messenger RNA leaves the nucleus and attaches to a **ribosome**, containing a second class of RNA molecules (**ribosomal RNA**), in the cell cytoplasm. Individual amino acids in the cytoplasm become linked to specific varieties of a third form of RNA, **transfer RNA**, prior to assembly into a protein. By virtue of complementary triplet sequences, the transfer RNAs align themselves along the messenger RNA while it is on the ribosome and thus bring the appropriate amino acids into juxtaposition. In this way, the original triplet sequence of nucleotide bases in a stretch of DNA is eventually translated into a series of amino acids constituting a protein.

Figure 11.4. Tabular presentation of the genetic code (after Watson, Tooze and Kurtz, 1983). Note that the code is given for messenger RNA, rather than DNA itself, and therefore refers to uracil in place of thymine. The extreme left-hand column indicates the first and second position in each triplet sequence of nucleotide bases; the third position is indicated in subsequent columns. Abbreviations for amino acids corresponding to triplet sequences are given in Table 11.1. Abbreviations for nucleotide bases are as follows: A, adenine; C, cytosine; G, guanine; U, uracil.

CHROMOSOMAL EVOLUTION

The chromosomes are of special interest in evolutionary studies because they represent a level of organization intermediate between the gene and the whole organism (Marks, 1983). Yet the comparative study of chromosomes (**comparative karyology**) across the primates is a relatively recent undertaking, pioneered by a paper reporting diploid chromosome numbers for *Homo sapiens* and three species of Old World monkey (Darlington and Haque, 1955). Indeed, it was not until the following year that the normal diploid number of chromosomes for the human species was definitively established as $2n = 46$ (Ford and Hamerton, 1956; Tijo and Levan, 1956). Since that time, primate chromosomes have been fairly comprehensively investigated and there have been several synthetic reviews (e.g. Chu and Bender, 1961; Egozcue, 1969, 1974, 1975; Chiarelli, 1974; de Grouchy, Turleau and Finaz, 1978; Martin, 1978; Dutrillaux, 1979; Seuánez, 1979; Dutrillaux, Couturier and Viégas-Pequignot, 1981; Dutrillaux, Couturier *et al.*, 1982). The productivity of the various laboratories involved in chromosomal studies has been such that chromosomal evidence can now make a substantial contribution to the reconstruction of primate phylogenetic relationships.

As the diploid number of chromosomes ($2n$) is virtually constant for any mammalian species, it might be thought that it would provide a convenient, simple guide to phylogenetic relationships. However, this expectation is not fulfilled. Although there is indeed considerable variation in diploid number among mammals, from a minimum of $2n = 6$ for a hoofed mammal, the Indian muntjak, to a maximum of $2n = 92$ for a rodent, the Ecuador fish-eating rat *Anotomys leander* (Matthey, 1973a,b; Imai and Crozier, 1980), the diploid number may differ markedly even between closely related species belonging to the same genus. An extreme example of this among primates is provided by two bushbaby species: the thick-tailed bushbaby (*Galago crassicaudatus*) has a diploid number of $2n = 62$, whereas the lesser bushbaby (*G. senegalensis*)

typically has a diploid number of $2n = 38$ (e.g. see Egozcue, 1970; de Boer, 1973; Rumpler, Couturier *et al.*, 1983). The most extreme case so far reported among mammals is that of the Indian muntjac (*Muntiacus muntjak*), which has a diploid number of only $2n = 6$, whereas the closely related Chinese muntjak (*M. reevesi*) has a diploid number of $2n = 46$ (Wurster and Benirschke, 1970). Surprisingly, these two muntjak species are able to interbreed, in spite of the great disparity in diploid number between them. Such great deviations in diploid number between closely related species can occur because most chromosomal evolution seems to take place through rearrangement of genetic material rather than through substantial addition or loss of material. This is indicated by the fact that somatic cell nuclei of different species of placental mammals generally contain about the same quantity of DNA – about 8 picograms (pg, i.e. 8×10^{-12} g) of DNA per diploid nucleus (Mandel, Métais and Cuny, 1950; Atkin, Mattison *et al.*, 1965; Ohno, 1970; Bachmann, 1972) – although there is a two-fold range of variation overall. For a sample of 18 non-primate placental mammals, Bachmann (1972) indicates an average value of 8.2 pg, with extreme values of 6.1 pg for the bat *Tadarida* and 12.2 pg for the aardvark, *Orycteropus*. The average mammalian value of 8.2 pg corresponds to about 6×10^9 base pairs of DNA per diploid nucleus (Watson, Tooze and Kurtz 1983). For his sample of placental mammals, Bachmann found no correlation between diploid number and the amount of nuclear DNA, so an increase in the number of chromosomes does not appear to be associated with a systematic increase in nuclear DNA.

The quantity of DNA per nucleus for the human species is about 7.3 pg (Bachmann, 1972). A sample of 61 primate species was found to have an average value of 7.6 ± 1.2 pg of nuclear DNA, whereas the common tree-shrew (*Tupaia glis*) has a value of 7.0 (data from Pellicciari, Formenti *et al.*, 1982). Further, in a specific comparison between the bushbabies *Galago senegalensis* and *G. crassicaudatus*, Manfredi-Romanini, de Boer *et al.* (1972) found no significant difference in the quantity of DNA per somatic cell even though the two species have different diploid numbers. (Somewhat different nuclear DNA values were later reported for these species by Pellicciari, Formenti *et al.* (1982), but these results do not alter the conclusion that the greater number of chromosomes in *Galago crassicaudatus* is not associated with a higher content of nuclear DNA.) Nevertheless, there is a detectable range of variation in the quantity of nuclear DNA among primates (e.g. Pellicciari, Formenti *et al.*, 1982). The Philippine tarsier (*Tarsius syrichta*) and the moustached monkey (*Cercopithecus cephus*), which share a nuclear DNA content of 10.9 pg, have the highest recorded values; titi monkeys (*Callicebus* spp.), with a DNA content of about 4.7 pg, have the lowest values. This approximately two-fold range of variation among primates is comparable to that found for mammals generally by Bachmann (1972). Overall, therefore, it would seem that the quantity of DNA per somatic nucleus has tended to oscillate around a standard value of 7–8 pg both among placental mammals generally and among primates specifically. Therefore, although there are some significant differences in nuclear DNA content between species, most chromosomal differences among placental mammals must arise through redistribution of material within the karyotype.

Whereas it is generally true that the amount of DNA is the nuclei of placental mammals has remained relatively constant in spite of considerable chromosomal rearrangement, there are quite large differences between major groups of organisms. In bacteria, for instance, the quantity of DNA in the nucleus is only 0.007 pg, a thousandth of the amount in the mammalian nucleus. However, lest it be thought that the quantity of DNA in the nucleus increases in a regular fashion with the evolutionary complexity of organisms, it should be noted that birds have only half the amount of nuclear DNA of mammals (i.e. about 3–4 pg, as opposed to about 7–8 pg) and certain fish (dipnoans) and amphibians (urodeles) can have values of the order of 200 pg (Atkin, Mattison *et al.*, 1965; Szarski, 1976). In fact, Szarski (1970, 1976) has argued quite convincingly that the quantity of nuclear DNA in eukaryotic organisms is related both to the size of the nucleus and to the overall size of the cell. He hypothesized that selection acts primarily on the size of the cell, because of the relationship between cell dimensions and metabolic turnover (see Chapter 4), and that an increase or decrease in cell size leads to a corresponding increase or decrease in nuclear DNA. As it is known that mammalian cells increase in size only marginally with increasing body size (Munro, 1969; Calder, 1984), the relative constancy of DNA in the nuclei of placental mammals is in accordance with Szarski's hypothesis.

Structural rearrangement of chromosomes

The concept of chromosomal evolution through structural rearrangement of existing material was given strong empirical support by Matthey (1945, 1949). His observations were founded on a basic classification of chromosomes according to the position of the centromere (the site of attachment to the spindle during cell division; see Figs 11.1 and 11.2). Taking the form of the chromosome during the resting state of the cell (i.e. before replication), if the centromere is very close to one end the chromosome is essentially one-armed and can be referred to as **acrocentric** (see Fig. 11.5), whereas if the centromere is approximately in the middle (as indicated for all the chromosomes illustrated in Figs 11.1 and 11.2) the chromosome is two-armed or **metacentric**. Obviously, intermediate forms exist and additional categories are recognized (see Levan, Fredga and Sandberg, 1964). A chromosome with one long arm and one short arm is termed **submetacentric** or **subtelocentric**, depending on the relative length of the short arm, and a chromosome with the centromere in a genuinely terminal position is referred to as **telocentric**. It should be remembered, however, that White (1973) ruled out the existence of genuinely telocentric chromosomes on the grounds that they would be highly unstable. In any case, a simple bipartite division between acrocentric chromosomes (including both

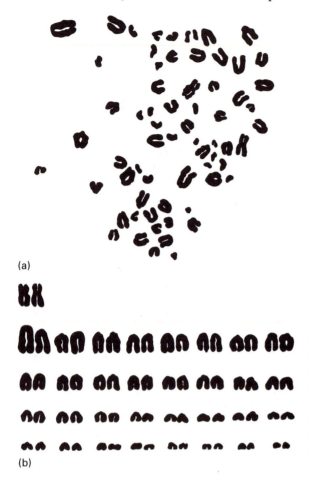

(a)

(b)

Figure 11.5. Normal diploid karyotype of a female lesser mouse lemur (*Microcebus murinus*): (a) the chromosomes as they are observed in a preparation made during the metaphase of mitotic cell division; (b) the chromosomes arranged in matching pairs. The diploid number of chromosomes (2*n*) is 66 in this species. All of the autosomes are acrocentric and only the X-chromosomes are metacentric. At metaphase, the chromosomes are double structures because one half of each replicated chromosome will go to each daughter cell. An acrocentric chromosome therefore appears to have two arms joined at the centromere and a metacentric chromosome appears to have four arms joined at the centromere. However, in the resting cell acrocentric chromosomes are one-armed and meta-centric chromosomes are two-armed. For calculation of the fundamental number of chromosomes, the arms are counted relative to the resting state. Therefore, in the karyotype of *Microcebus murinus*, there are 64 chromosome arms if only the autosomes are counted and 68 if the two X-chromosomes are included.

telocentrics and subtelocentrics) and metacentric chromosomes (including submetacentrics) is quite adequate for purposes of preliminary discussion and analysis (see also Mai, 1983). In fact, Imai and Crozier (1980) reported from an analysis of mammalian chromosomes that there is a fairly clear overall distinction between (1) chromosomes in which the weight of the shortest arm represents 0.6% or less of the total weight of the haploid set of autosomal chromosomes and (2) chromosomes in which the weight of the shortest arm exceeds this level. In the former case ('acrocentrics'), the short arm usually contains only genetically inactive DNA (**heterochromatin**), whereas in the latter case ('metacentrics') the short arm typically has at least some genetically active DNA (**euchromatin**).

It was on the basis of simple visual discrimination between acrocentrics and metacentrics that Matthey (1945) originally noted that the total number of chromosome arms provides a far more satisfactory guide to phylogenetic relationships than does the number of chromosomes (the diploid number, 2*n*). Matthey accordingly referred to the number of chromosome arms as the **fundamental number** (NF). (Note: in the calculation of the fundamental number, the sex chromosomes are often excluded because the Y-chromosome commonly differs in form from the X-chromosome and males and females could hence have different fundamental numbers.) As a general rule, the fundamental number tends to remain more constant between closely related mammalian species than does the diploid number, although there are a considerable number of exceptions (see later).

It is now widely recognized that the fundamental number is relatively stable because one of the major processes of chromosomal evolution in animals generally is **inter-chromosomal rearrangement** involving the reciprocal conversion of acrocentric and metacentric chromosomes (Matthey, 1949; White, 1973; Bender and Metler, 1958; Chu and Bender, 1961; Bender and Chu, 1963; Martin, 1978). Confusion has arisen in the literature because some authors have proposed that two acrocentric chromosomes typically 'fuse' to produce a metacentric chromosome (**centric fusion**), whereas others have suggested that the typical process involves 'fission' of a metacentric chromosome to yield two acrocentrics (**centric fission**). Clearly, phylogenetic arguments will be reversible according to the postulated direction of chromosomal change. (This is a prime example of the need to determine 'morphocline polarity' in phylogenetic reconstructions; see Chapter 3.) For instance, with respect to primate evolution some authors originally postulated that centric fusion predominantly or exclusively accounts for change from acrocentrics to metacentrics (e.g. Bender and Chu, 1963; Hamerton, 1963), whereas others claimed that the dominant process is centric fission (Staton, 1967; Todd, 1967; Giusto and Margulis, 1981). The first hypothesis would require an ancestor with a karyotype consisting largely or exclusively of acrocentric chromosomes whereas the second hypothesis would require an ancestral karyotype dominated by metacentrics. The problem with both hypotheses is that a unidirectional evolutionary trend should lead to progressive elimination of one type of chromosome from mammalian karyotypes. Centric fusion should eventually lead to karyotypes exclusively containing metacentrics, and centric fission should lead to karyotypes composed exclusively of acrocentrics. In fact, it seems likely that both processes can occur and that it is advisable to group them together under a single term such as **whole-arm transposition** (White, 1973) or **centric rearrangement** (Martin, 1978), except in cases where the direction of change is clearly identifiable. Part of the previous disagreement in the literature over the relative dominance of centric fusion and centric fission in chromosomal change stemmed from the fact that one or other process may be predominant in different groups of primates (see later). The general process of centric rearrangement is illustrated in Figure 11.6.

There is some divergence of opinion over the exact meaning of the terms 'centric fusion' and 'centric fission'. One view is that centric fusion

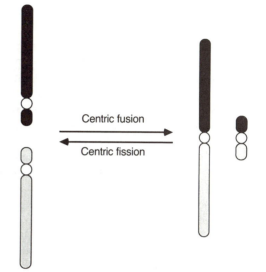

Figure 11.6. Reciprocal translocation (= Robertsonian translocation), one possible mechanism for the basic process of centric rearrangement. The centromere is indicated by a white circle and the chromosome arms are shaded (stippling; black). Note that 'fusion' of two acrocentric chromosomes through reciprocal translocation should theoretically produce a vestigial chromosome dominated by the centromere. Conversely, 'centric fission' of a metacentric chromosome to produce two acrocentrics through reciprocal translocation would require the prior existence of a vestigial chromosome as a source of the additional centromere. Alternative mechanisms of interconversion of metacentric and acrocentric chromosomes have been suggested, notably to account for 'fission' (see text).

takes place through **reciprocal translocation**, and that this perhaps applies to centric fission as well, as shown in Fig. 11.6 (White, 1973). The process of reciprocal translocation was first suggested by Robertson (1916) and is hence commonly labelled **Robertsonian translocation**. Robertson himself pointed out that the process might proceed in either direction. If a metacentric chromosome is generated by reciprocal translocation between two acrocentric chromosomes as portrayed in Fig. 11.6, the process should theoretically also produce a vestigial chromosome consisting of a centromere with two very short arms. Although the existence of such vestigial chromosomes is hence predicted, they have not actually been demonstrated in the karyotypes of species reasonably believed to have undergone 'centric fusion'. It is, of course, possible that these vestiges are rapidly lost in the process of cell division and remain undetected for this reason. The converse problem exists in an even more acute form with respect to 'centric fission', because derivation of two acrocentric chromosomes from a metacentric through reciprocal translocation would require some existing source for the additional centromere required. It is conceivable that the additional centromere might be acquired through reciprocal translocation involving a vestigial chromosome, as indicated in Fig. 11.6, but this would require the prior existence of 'spare' vestigial chromosomes in the nucleus for any process of chromosomal fission to take place.

An alternative view is that 'centric fission', rather than taking place through reciprocal translocation, may involve actual fission of a metacentric chromosome to produce two genuinely one-armed, telocentric chromosomes each bearing a half-centromere in a terminal position. Some authors (e.g. White, 1973) have claimed that telocentric chromosomes never occur in practice, but there are several reported cases suggesting their existence at least in certain circumstances. For example, Egozcue (1971) identified an apparent case of centric fission in one member of a chromosome pair in a leaf-monkey (*Presbytis entellus*) and Muleris, Dutrillaux and Chauvier (1983) have reported a male gelada baboon (*Theropithecus gelada*) in which a submetacentric chromosome had undergone centromeric fission to produce two telocentric chromosomes. There is the problem, of course, that persistence of chromosomal rearrangements produced through actual fission requires some mechanism for the regeneration of partial centromeres or for their formation *de novo* (see Imai and Crozier, 1980). In fact, it is also possible that 'centric fusion' may occur through end-to-end fusion of two telocentric or subtelocentric chromosomes to form a metacentric chromosome, rather than through

the process of reciprocal translocation illustrated in Fig. 11.6. If the two chromosomes actually fuse, there must be some mechanism for inactivation and perhaps eventual resorption of the second centromere and there is some evidence that this can occur (see Seuánez, 1979). It is almost certain that fusion of chromosomes has taken place in some instances. For example, Liming, Yingying and Xingsheng (1980) have demonstrated that the peculiar degree of chromosomal differentiation between the Indian and the Chinese muntjaks (*Muntiacus muntjak* and *M. reevesi*) has arisen through a combination of multiple fusions in tandem and Robertsonian translocations.

Regardless of the actual processes involved in centric rearrangement, it is abundantly clear from empirical evidence that this kind of chromosomal change is prevalent in mammals generally. It is an intrinsic feature of such rearrangement that the number of chromosome arms (NF) remains constant while the diploid number may vary considerably. There are numerous cases among primates where closely related species with different diploid numbers have the same, or very similar, NF values and where centric arrangement can account for all or virtually all of the difference. The lesser bushbaby (*Galago senegalensis*), with a diploid number of $2n = 38$, has 30 metacentrics and only 6 acrocentrics, excluding the sex chromosomes, whereas the thick-tailed bushbaby (*G. crassicaudatus*), with a diploid number of $2n = 62$, has 54 acrocentrics and only 6 metacentrics. The karyotype of *G. senegalensis* could be derived from that of *G. crassicaudatus* by the fusion of 24 pairs of acrocentrics to produce 24 metacentrics (see Rumpler, Couturier *et al.*, 1983). Alternatively, the karyotype of *G. crassicaudatus* could be derived from that of *G. senegalensis* by the fission of 24 metacentrics to produce 48 acrocentrics. The choice between these two alternative possibilities depends on the availability of other information indicating the direction of karyotypic evolution (see later).

Although centric rearrangement accounts for a great deal of chromosomal differentiation between species, other processes must obviously be involved as well. If centric rearrangement were the only process involved in chromosomal change, the fundamental number would remain constant and this is manifestly not the case either among mammals generally or among primates specifically. Therefore, there must be chromosomal rearrangements that alter the fundamental number. The commonest of these is almost certainly an intrachromosomal re-arrangement termed **pericentric inversion**, in which a central region of the chromosome including the centromere and parts of both arms becomes inverted and rejoins the remaining segments of the arms (Fig. 11.7). In extreme cases, pericentric inversion can lead to conversion of a metacentric chromosome to an acrocentric one, or vice versa (see Imai and Crozier, 1980). In the former case, the rearrangement will lead to a decrease in fundamental number, whereas in the latter it will lead to an increase. In either event, it is the fundamental number that changes while the diploid number remains constant. As with centric rearrangement, the direction of chromosomal

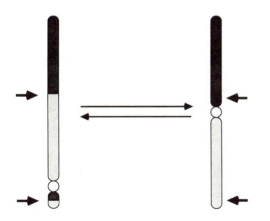

Figure 11.7. The process of pericentric inversion, in this case involving conversion between an acrocentric chromosome and a metacentric chromosome. In this intrachromosomal form of rearrangement, breaks occur in the chromosome (arrows) and the intervening region becomes inverted. As with centric rearrange-ment, the process can proceed in either direction. Conventions as in Fig. 11.6.

change through pericentric inversion is not immediately obvious from a comparison between the karyotypes of any two species.

Another mechanism that can potentially modify the karyotype, altering both the diploid number and the fundamental number, is **ploidy**. This can take two forms. In **aneuploidy**, one or more chromosomes may be added to or lost from the karyotype if chromosomes are unequally distributed to the gametes during meiosis. This leads to an unbalanced karyotype, except in the rare case where one or more pairs of matched chromosomes are added, so aneuploidy will usually have deleterious affects, as is the case with Down's syndrome in humans. By contrast, **polyploidy** involves straightforward multiplication of the haploid set of chromosomes beyond the normal diploid condition, again because of a departure from the normal processes of formation and combination of gametes to form a zygote. For example, it has been inferred fairly reliably for a variety of vertebrates (e.g. certain fish and amphibians) that the normal diploid set of chromosomes has doubled in evolution to produce a **tetraploid** condition. However, although it has been suggested that aneuploidy or polyploidy has occurred in certain areas of mammalian evolution, this is highly unlikely. Addition of one or more haploid sets of chromosomes to the normal karyotype should lead to a marked increase in the quantity of DNA in the nucleus and there is no evidence that this has happened in any abrupt fashion during mammalian evolution (see above). Addition or loss of individual chromosomes would not have any pronounced affect on the amount of nuclear DNA, but chromosomal evolution through aneuploidy is inherently unlikely because of the disruptive effect of imbalance in the karyotype. Whereas polyploidy would not necessarily lead to severe adverse effects for this reason, it would obviously disrupt the working of the universal sex-determination mechanism in mammals, which depends on the presence of either two X-chromosomes (female) or one Y-chromosome and one X-chromosome (male). Tetraploidy, for example, would double the number of sex chromosomes to four X-chromosomes in a female and two Y-chromosomes plus two X-chromosomes in a male. The gametes produced by such individuals would clearly not produce zygotes with normal sex determination.

It can therefore be concluded with some confidence that polyploidy has not occurred among mammals since the currently universal X/Y sex-determination mechanism was established (see also Seuánez, 1979). However, it is quite possible that tetraploidy occurred at some premammalian stage of vertebrate evolution (Ohno, 1970). This could explain why Comings (1972) was able to group pairs of chromosomes from the normal human diploid set into approximately matched 'quadruplets'. As Seuánez (1979) has noted, such matching cannot be expected to be complete, as the two pairs of chromosomes in each quadruplet have doubtless had a long time to undergo independent divergent evolution. Tetraploidy occurring at a premammalian stage of vertebrate evolution might also account for Bengtsson's observation (1975) that mammalian chromosomes of similar length are also similar in shape. Hence, it seems quite possible that the ancestral mammals were effectively tetraploid, but it seems highly unlikely that further polyploidy has occurred at any later stage in the evolution of the mammals.

Broad comparisons of primate karyotypes

With all of the types of chromosomal change discussed thus far, excluding the unlikely mechanisms of aneuploidy and polyploidy, it is relatively easy to determine the general nature of the change involved (e.g. centric rearrangement or pericentric inversion) from gross chromosomal morphology, but it is considerably more difficult to determine the direction of change (see also Imai and Crozier, 1980). Thus, there is a very good case for comparing primate karyotypes, in the first instance, without any a priori assumptions regarding the direction of chromosomal change.

One can begin with a straightforward examination of the distribution of **diploid numbers** ($2n$) among mammals. It can be seen from Fig. 11.8 that the distribution of diploid

numbers is essentially unimodal for placental mammals, with a peak at a diploid number of about $2n = 48$. (This is clearly confirmed by the much larger samples surveyed by Matthey, 1973a,b.) The distribution for primates broadly resembles that for placental mammals generally, and the same is true of certain other orders of placentals such as rodents, which are illustrated by the sample separately identified in Fig. 11.8. It is obvious from Fig. 11.8, however, that there is a radical difference between placental mammals and marsupials in their distributions of diploid numbers. In marsupials, diploid numbers are confined to the comparatively narrow range of $2n = 12$ to $2n = 24$ and there is relatively little overlap with the lower end of the distribution for placentals. This indicates that, in some cases at least, a major phylogenetic distinction based on morphological characters is matched by an equally striking distinction in terms of diploid number. It should be noted, however, that the average amount of nuclear DNA in marsupials is very similar to that in placental mammals, with an average value of 8.4 pg for 13 marsupial species when the data from Martin and Hayman (1967) are calibrated against the human karyotype (Bachmann, 1972; see also Pellicciari, Formenti et al., 1982). Hence, the striking difference between marsupials and placentals in terms of diploid number is presumably attributable essentially to chromosomal rearrangement. Incidentally, limited data also suggest that monotremes have about the same quantity of nuclear DNA as marsupials and placentals (Bick and Jackson, 1967; Bachmann, 1972), so it is reasonable to suggest that a nuclear DNA content of about 8 pg was established as an ancestral condition for the mammals generally and has been maintained as a modal value ever since.

If the diploid numbers of primates are examined in more detail (Fig. 11.9), it emerges that, although primates overall have the peak diploid number of $2n = 48$ characteristic of placental mammals generally, the strepsirhine primates (lemurs and lorises) show a distribution somewhat different from that of the haplorhines (tarsiers, monkeys, apes and humans). In both

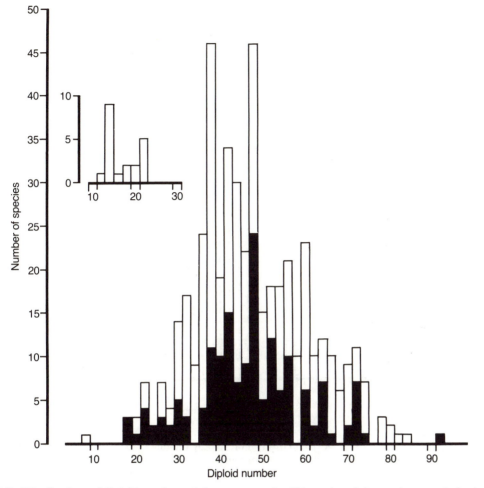

Figure 11.8. Distributions of diploid numbers of chromosomes for 466 species of placental mammals (main figure) and 20 species of marsupials (inset). Rodent species are indicated in black and show a similar distribution to placental mammals generally. (Figure reprinted from Martin, 1978, with the kind permission of Academic Press, London.)

cases, a skewed distribution is found and, although there is extensive overlap between the distributions, the main peaks are located in different places ($2n = 62$ for strepsirhines; $2n = 44$ for haplorhines). It should be noted that *Tarsius* stands out as having the highest diploid number of any primate examined to date ($2n = 80$) and that this number lies close to the upper end of the distribution for mammals overall (Fig. 11.8). This is interesting in view of the fact, noted above, that the amount of DNA in the genome of *Tarsius* is close to the upper limit for primates and for mammals generally.

The overall difference between strepsirhine and haplorhine primates is shown even more clearly when the distributions of fundamental numbers (NF values) are examined (Fig. 11.10). Strepsirhines have a modal value of NF = 66, whereas haplorhines have a modal NF of 84. In other words, a typical haplorhine will have 18 more chromosome arms than a typical strepsirhine. The distinction between strepsirhines and haplorhines in terms of their distributions of fundamental numbers is statistically highly significant ($P < 0.0003$) and it is far more pronounced than the distinction in

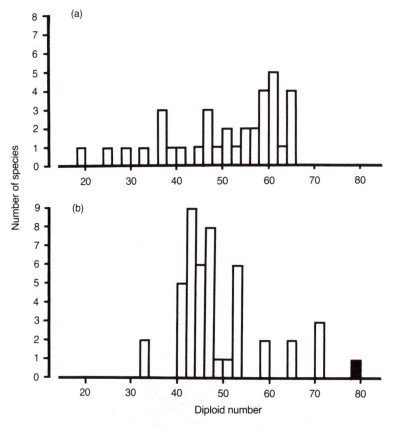

Figure 11.9. Distributions of diploid numbers of chromosomes for (a) strepsirhine primates (N = 36 species) and (b) haplorhine primates (N = 45 species). *Tarsius* is indicated by the black box. (Figure reprinted from Martin, 1978, with the kind permission of Academic Press, London.)

distributions of diploid numbers, which does not reach significance at the 0.05 level (P > 0.075; see Martin, 1978). It can be seen from Fig. 11.10 that *Tarsius* fits comfortably into the core of the haplorhine distribution in terms of its fundamental number, but would be a quite marked outlier within the strepsirhine distribution. Nevertheless, this does not provide a reliable basis for assessing the likely phylogenetic affinities of *Tarsius*, firstly because there are some strepsirhines (mainly members of the loris subfamily Lorisinae) with relatively high fundamental numbers similar to those found in tarsiers, and secondly because Fig. 11.10 does not provide any basis for discrimination between primitive and derived conditions.

When it comes to distinguishing primitive and derived conditions in the fundamental numbers of primates, at least three different hypotheses can be identified (Fig. 11.11; see also Martin, 1978). The first is that the modal fundamental number of strepsirhines (NF = 66) is primitive for primates generally and that an increase to NF = 84 occurred as a derived feature in the evolution of the ancestral stock of haplorhine primates. The second is that the ancestral fundamental number for primates was intermediate between the modern modal values for strepsirhines and haplorhines (e.g. NF = 74) and that divergent evolution led to a reduction in the fundamental number in ancestral strepsirhines and to an increase in ancestral

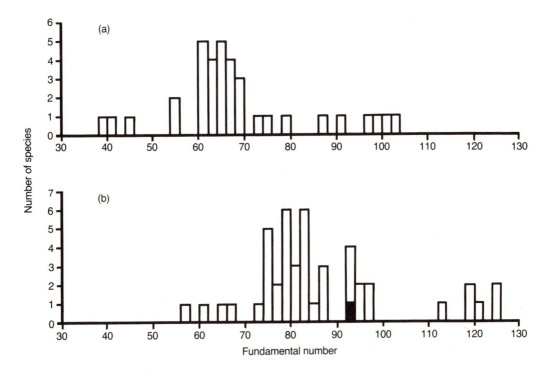

Figure 11.10. Distributions of fundamental numbers for (a) strepsirhine primates (*N* = 36 species) and (b) haplorhine primates (*N* = 45 species). *Tarsius* is indicated by the black box. (Figure reprinted from Martin, 1978, with the kind permission of Academic Press, London.)

haplorhines. Finally, it is possible that the haplorhines have remained primitive in having a relatively high modal fundamental number (NF = 84) and that a decrease to NF = 66 took place in the evolution of the strepsirhine ancestral stock.

One way of assessing the relative merits of these three competing hypotheses is to broaden the context of the comparison and to examine the distribution of fundamental numbers in mammals generally (Fig. 11.12). It emerges from this that the peak value for placental mammals is NF = 60) and the modal value is NF = 68, more or less coinciding with the modal value of NF = 66 determined for strepsirhine primates (see Fig. 11.11). Rodents show a pattern very similar to that of placental mammals generally. Hence, it would seem that hypothesis 1 in Fig.

11.11 is most likely to be correct and that strepsirhines have remained generally primitive in typically preserving fundamental numbers close to NF = 66. Haplorhines, by contrast, would seem to be specialized in having a modal fundamental number as high as NF = 84. Indeed, it can be seen from Fig. 11.12 that haplorhine primates must lie at the upper end of the distribution for placental mammals in this respect. One possibility is that the higher fundamental number of haplorhines could have arisen partly or entirely through the addition of new chromosomal material relative to the ancestral primate condition, presumably retained in strepsirhines generally. The data provided by Pellicciari, Formenti *et al.* (1982) on nuclear DNA suggest that this might be the case (Figure 11.13). The mean amount of nuclear

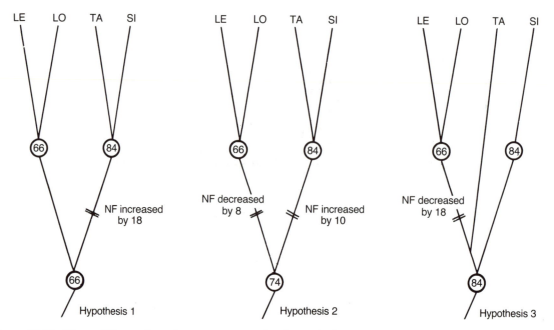

Figure 11.11. Three different hypotheses to account for the evolution of modal fundamental numbers (NF) in strepsirhine and haplorhine primates. The hypotheses differ according to the condition postulated as primitive for the ancestral primate stock, shown as the initial node in each tree. Abbreviations: LE, lemurs; LO, loris group; TA, tarsiers; SI, simians. (Figure reprinted from Martin, 1978, with the kind permission of Academic Press, London.)

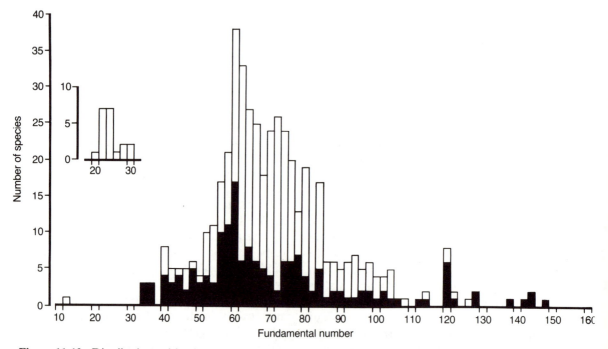

Figure 11.12. Distributions of fundamental numbers for 466 species of placental mammals (main figure) and 20 species of marsupials (inset). Rodent species are indicated in black. As in Fig. 11.8, the distribution for rodents matches quite closely the distribution for placental mammals generally. (Figure reprinted from Martin, 1978, with the kind permission of Academic Press, London.)

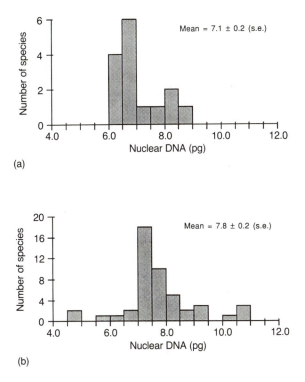

(a)

(b)

Figure 11.13. Distributions of values for nuclear DNA for (a) strepsirhine primates ($N = 15$) and (b) haplorhine primates ($N = 48$). (Data from Pellicciari, Formenti *et al.*, 1982.)

DNA determined for 15 strepsirhine species is 7.1 ± 0.2 (s.e.) pg, whereas the mean for 48 haplorhine species is 7.8 ± 0.2 (s.e.) pg, representing an average increase of some 9%. This difference is significant ($P < 0.01$).

In fact, the amounts of nuclear DNA in primates provide a further indication of a difference between strepsirhines and haplorhines. Pellicciari, Formenti *et al.* (1982) originally reported a significant correlation ($r = 0.39$; $P \leqslant 0.01$) between DNA content and diploid number for their entire primate sample, in contrast to Bachmann's finding (1972) for mammals generally. However, Pellicciari, Formenti *et al.* found that the correlation for the primate sample became insignificant when the

extreme cases of the yellow-handed titi monkey (*Callicebus torquatus*) with low DNA content and low diploid number and the Philippine tarsier (*Tarsius syrichta*) with high DNA content and high diploid number were removed from the analysis ($r = 0.20$; $P \geqslant 0.05$). If the same analysis is conducted on the relationship between DNA content and fundamental number, however, a consistent relationship is found. For the entire sample, there is a highly significant correlation ($r = 0.61$; $P < 0.001$) and this is barely affected by the removal of the two species *Callicebus torquatus* and *Tarsius syrichta* ($r = 0.55$; $P < 0.001$). But it turns out that this correlation is a property of haplorhine primates, which represent a disproportionate part of the sample. For haplorhines alone ($N = 47$) there is a highly significant correlation between the quantity of nuclear DNA and fundamental number ($r = 0.67$; $P < 0.001$) and this relationship accounts for 45% of the variation in nuclear DNA content among haplorhines. For strepsirhines alone ($N = 14$), however, the correlation between the amount of nuclear DNA and fundamental number is not significant ($r = 0.20$; $P > 0.52$). It therefore emerges that for haplorhines, but not for strepsirhines, an increase in fundamental number is associated with a detectable increase in nuclear DNA. In other words, it would appear that, among haplorhines, pericentric inversion producing metacentrics from acrocentrics in some way entails an addition to nuclear DNA. In this respect, *Tarsius* fits the haplorhine pattern in having a high content of nuclear DNA combined with a high fundamental number, whereas lorisines, which have the highest fundamental numbers among strepsirhines, do not have raised levels of nuclear DNA. In fact, placental mammals generally resemble the strepsirhines rather than the haplorhines. Bachmann's data (1972) indicate that there is no consistent relationship between fundamental number and content of nuclear DNA for placental mammals overall ($r = 0.03$; $N = 19$). Once again, therefore, strepsirhines fit the general mammalian pattern whereas haplorhines, including *Tarsius*, are distinctive.

It should also be noted from Fig. 11.12 that marsupials are even more sharply demarcated from placentals in terms of fundamental number than they are in terms of diploid number (Fig. 11.8). This provides yet another indication that the fundamental number generally serves as a more valuable guide to phylogenetic relationships among mammals than does the diploid number. Further, because (as noted above) marsupials do not differ from placentals in terms of the amount of DNA in their genomes, there obviously cannot be a correlation between fundamental number and DNA content applying to marsupials and placentals together. This provides further confirmation that an increase in number of chromosome arms is not consistently associated with an increase in nuclear DNA among mammals generally.

The direction of change

Thus far, discussion has been limited to the number of chromosomes in a somatic cell ($2n$) and to the total number of their constituent arms (NF). In themselves, these two numbers provide little information about the actual **direction of evolutionary change**; they merely provide a relatively crude guide to possible phylogenetic relationships. In particular, raw data on $2n$ and NF do not permit a distinction between primitive and derived conditions. Analysis of such numerical information can be taken a stage further, without needing to establish the direction of change in specific cases, by explicitly considering the effects of the two major processes of chromosomal rearrangement in mammals – centric rearrangement and pericentric inversion. As has been explained above, with centric rearrangement (Fig. 11.6) the fundamental number remains constant; it is related to the number of acrocentrics (A) and the number of metacentrics (M) by the following formula:

$$NF = A + 2M$$

Hence, in a graph in which the number of metacentrics is plotted against the number of acrocentrics, species differing from one another only through centric rearrangement would give a perfect fit to a straight line with the following formula:

$$M = (NF - A)/2$$

The slope of the line would therefore be -0.5 and the intercept with the abscissa (i.e. the horizontal axis indicating the number of acrocentrics) would reveal the fundamental number.

As has already been observed, however, chromosomal evolution among mammals is certainly not confined to the process of centric rearrangement. Other processes such as pericentric inversion have occurred, leading to interspecific variation in the fundamental number. In a plot of metacentrics against acrocentrics, pericentric inversions affecting the fundamental number (see Fig. 11.7) would be revealed through deviations from a straight-line relationship of the kind indicated by the above formulae. Pericentric inversion converting an acrocentric to a metacentric would displace a point above the line, whereas pericentric inversion converting a metacentric to an acrocentric would shift a point below the line. Following this reasoning, it is possible to propose a simple model of chromosomal evolution in which, beginning with an ancestral condition characterized by a given diploid number and a given fundamental number, most chromosomal changes have involved centric rearrangement, hence conforming to a linear equation of the form given above, while pericentric inversions and other changes affecting the fundamental number have generated scatter around the line. The implications of this model can be examined by fitting a line of slope -0.5 to a plot of metacentrics against acrocentrics for the sample of 466 placental mammals already mentioned with respect to diploid number (see Martin, 1978). It is clear from this plot (Fig. 11.14) that there is an overall inverse relationship between metacentrics and acrocentrics, as required by the model, and the idealized line indicates a fundamental number of 67, which coincides almost exactly with the modal value of 68 indicated for placental mammals by Fig. 11.12. As in other respects, rodents show the same

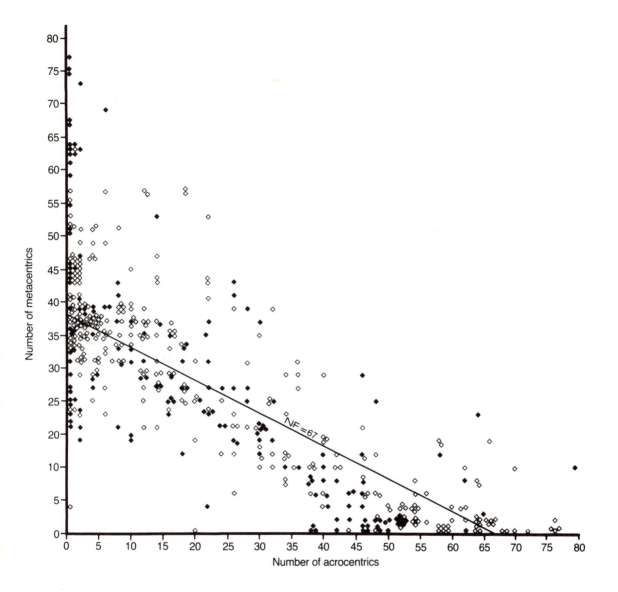

Figure 11.14. The relationship between numbers of metacentrics and numbers of acrocentrics for 466 species of placental mammals. Rodent species are indicated by black symbols. As in Figs 11.8 and 11.12, the distribution for rodents quite closely matches the distribution for placental mammals generally. A line of fixed slope −0.5 has been fitted to the data, yielding a corresponding fundamental number (NF) value of 67. (Figure reprinted from Martin, 1978, with the kind permission of Academic Press, London.)

overall pattern of distribution as placental mammals generally. It therefore seems reasonable to conclude that placental mammals may be derived from a common ancestor with a fundamental number of about 68 chromosome arms and that a substantial part of chromosomal evolution among placental mammals may be explained through centric rearrangement. Other processes, notably pericentric inversion, have also played a part and Fig. 11.14 indicates that there has probably been a bias towards the formation of metacentrics from acrocentrics through pericentric inversion, rather than vice versa. This bias is also reflected in Fig. 11.12, because the distribution of fundamental numbers is skewed towards the higher values.

The same kind of analysis can be applied to primate chromosomes alone, with the difference that strepsirhines and haplorhines must be analysed separately in view of the significant difference in modal NF values illustrated in Fig. 11.10. Fitting of separate lines of fixed slope − 0.5 to a plot of metacentrics against acrocentrics for strepsirhines and haplorhines (Fig. 11.15) produces the expected results. The line for strepsirhines yields an NF value of 66, whereas that for haplorhines yields an NF value of 84. These values coincide exactly with the modal values for the two primate groups indicated in Fig. 11.10. It can be seen that the point for *Tarsius* lies quite close to the haplorhine line (in fact, somewhat above it) but is a definite

positive outlier with respect to the strepsirhine line. This plot also reveals that *Tarsius* is quite unusual, in comparison to other primates, in that it combines a high fundamental number with a high proportion of acrocentrics. Other primates with more than 40 acrocentric chromosomes (all of which are strepsirhines) fit the line with NF = 66 very closely. As with placental mammals generally, there is an apparent bias in inferred pericentric inversion towards conversion of acrocentrics to metacentrics. This is reflected by the somewhat skewed distributions for NF values shown in Fig. 11.10. Although the plot for strepsirhine primates quite closely resembles that for placental mammals generally (Fig. 11.12), the plot for haplorhines is clearly somewhat divergent, reflecting an overall upward shift in fundamental number.

It therefore seems justifiable to conclude that haplorhine primates as a group may be distinguished from strepsirhines as a group in terms of the overall pattern of organization of the

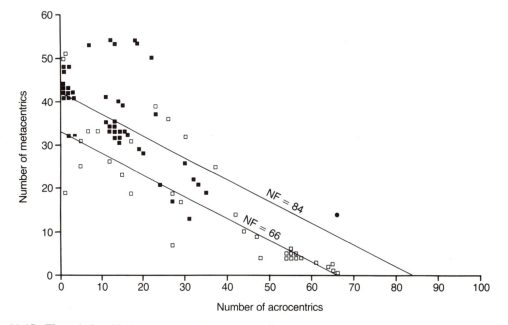

Figure 11.15. The relationship between numbers of metacentrics and numbers of acrocentrics for strepsirhine primates (□; *N* = 36) and for haplorhine primates (●, *Tarsius*; ■, simians; *N* = 44). Separate lines of fixed slope −0.5 have been fitted, yielding a corresponding fundamental number (NF) value of 66 for strepsirhines and of 84 for haplorhines. (Figure reprinted from Martin, 1978, with the kind permission of Academic Press, London.)

karyotype. It can be hypothesized that ancestral haplorhines underwent an underlying increase in fundamental number relative to ancestral primates and that this increase is at least partially reflected by an increase in the average quantity of nuclear DNA relative to strepsirhines. As Tarsius is closer to the typical simian condition than to the typical strepsirhine condition in several features, this provides supplementary (though by no means conclusive) evidence of the phylogenetic unity of the haplorhine primates.

Chromosome banding and detailed comparisons

Such broad comparative studies of chromosomes are necessarily superficial and permit only the most general conclusions. Their main advantage is that they can provide some evidence of the direction of change. In assessing evolutionary relationships on the basis of chromosomal features, considerably greater precision can be achieved by using techniques that provide information on the internal structure of individual chromosomes. The most useful techniques of this kind currently available are staining methods that generate **banding patterns** in chromosomes (Fig. 11.16). A variety of methods for generating banding patterns in chromosomes were introduced in the early 1970s. The production of these bands depends primarily on a distinction between two kinds of chromatin – **euchromatin** and **heterochromatin** (Heitz, 1928). Heterochromatin, in contrast to euchromatin, remains condensed throughout the cell cycle (except when replication is actually in progress) and it therefore shows different staining properties. In fact, it is now known that it is essentially the euchromatin that is genetically active, while the heterochromatin typically remains genetically inert. Some of the staining techniques widely used to produce banding patterns in chromosomes reveal the boundaries between euchromatic and heterochromatic regions. C-banding is widely recognized as a general indicator of the distribution of heterochromatin in chromosomes (Arrighi and Hsu, 1971), although this is not always the case (Dutrillaux, Couturier and Viégas-Pequignot,

1981). By contrast, R-banding and G-banding typically indicate regions containing euchromatin in chromosomes and therefore tend to produce a pattern complementary to that obtained with C-banding. However, R-banding will also stain some heterochromatin rich in guanine–cytosine links (Dutrillaux et al., 1981). A fourth technique, Q-banding, commonly indicates heterochromatic regions of chromosomes containing DNA rich in adenine–thymine links (see Comings, 1978). Using conventional banding techniques, 300 subregions can be recognized in the human karyotype and this can be increased to approximately 1000 by using the special technique of high-resolution banding (see Francke and Oliver, 1978; Dutrillaux et al., 1981; Harnden, Klinger et al., 1985).

Studies of primate chromosomes using banding techniques were initially restricted to a relatively narrow range of species, with an emphasis on the human karyotype. Comparative study of the banding patterns produced with different staining techniques revealed a high degree of correspondence between the chromosomes of great apes and humans (e.g. Turleau, de Grouchy and Klein, 1972; Dutrillaux, 1975; Dutrillaux, Rethoré and Lejeune, 1975; Miller, 1977; Seuánez, 1979; Mai, 1983; Marks, 1983). As a result, there is now a standard nomenclature for the chromosomes of these species, determined by a series of international conferences primarily concerned with the human karyotype (see: Stockholm Conference, 1978; Seuánez, 1979; Marks, 1983; Harnden, Klinger et al., 1985). Most of the differences between great apes and humans can be explained by localized rearrangements. Dutrillaux (1975) stated that 99% of the chromosome regions identifiable by G-banding or R-banding are common to all of these species and Seuánez (1979) noted that the X-chromosomes of great apes and humans in fact show identical patterns with C-, G-, Q- and R-banding. Naturally, the ability to recognize chromosome arms on the basis of specific banding patterns permits far more reliable interpretation than is possible merely from the study of gross chromosomal morphology. In

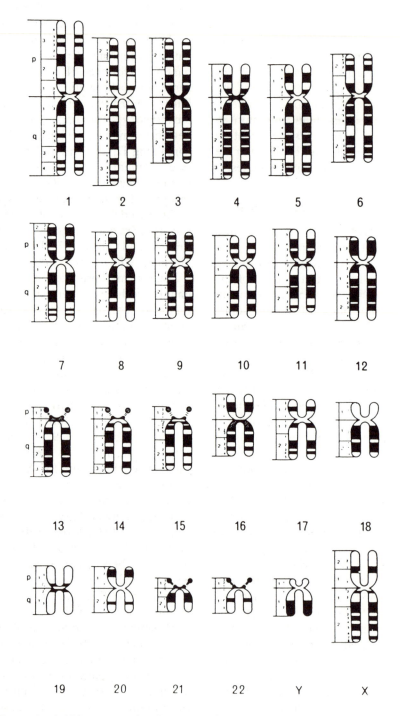

Figure 11.16. Human haploid karyotype showing composite banding patterns. (After Harnden, Klinger *et al.*, 1985. Reprinted with the kind permission of S. Karger AG, Basel.)

addition, banding patterns permit recognition of rearrangements that are not reflected in the overall shape of the chromosomes. For instance, certain pericentric inversions that have little overall effect on the shape of a chromosome could pass unnoticed without banding studies. Similarly, an inversion within a single chromosomal arm (**paracentric inversion**) should not significantly alter the length of the arm but it should alter the banding sequence. Banding studies also permit the recognition of translocations of material between chromosomes (i.e. in addition to Robertsonian translocations) and they reveal both additions and deletions affecting individual chromosomes.

Banding and the organization of chromosomes

Clearly, if the banding patterns of the chromosomes of related species show close similarity and if all or most observed differences between species can be explained through relatively simple rearrangements, it is reasonable to conclude that banding reflects fundamental features of the organization of chromosomes. This conclusion can be checked in several ways, the most direct being the localization of individual genes on chromosomes using a variety of special techniques (e.g. see Seuánez, 1979). The approach of **comparative gene-mapping** is now being increasingly applied to mammals (e.g. see Roderick, Lalley *et al.*, 1984), although, as in other instances, there has been a heavy concentration on humans and great apes among primates. For this reason, it is not yet possible to use data from actual gene-mapping for the reconstruction of phylogenetic relationships among primates generally. Instead, the available data for a limited sample of primate species can be used as a continuing test for inferences regarding the stability of chromosomal subregions across species.

There is now a substantial body of information on the location of individual genes on human chromosomes (McKusick, 1984) and for many of these genes comparisons with great apes have demonstrated almost without exception that widely accepted inferences based on banding

patterns are valid (see de Grouchy, Turleau and Finaz, 1978; Seuánez, 1979; Dutrillaux *et al.*, 1981; Marks, 1983). These comparisons have shown that, despite the occurrence of chromosomal rearrangements, linked groups (syntenic groups) of genes have been generally conserved. Restricted comparisons with other primates, including certain Old World monkeys, the New World capuchin monkey (*Cebus capucinus*) and the mouse lemur (*Microcebus murinus*) have shown that there has been quite marked conservation of linked groups of genes in identifiable chromosomal subregions even over long periods of primate evolutionary history (see Cochet, Creau-Goldberg *et al.*, 1982). Indeed, it has even been possible to recognize chromosomal subregions containing specific groups of genes in other mammals as well as in primates (Roderick, Lalley *et al.*, 1984). Overall, in spite of occasional reports of incongruities, the results of comparative gene-mapping have provided strong support for inferences derived from banding studies, not just with respect to human evolution but also with respect to the evolution of primates generally (Dutrillaux *et al.*, 1981).

In addition to gene-mapping, several supplementary techniques have permitted comparison of specific chromosomal subregions between species. Most of these techniques depend on the fact that different categories of DNA can be recognized. One of the major surprises that has emerged from studies of the DNA of eukaryotic organisms is that the genetic material is not confined to single copies of genes coding for individual proteins. As expected, there are numerous single-copy DNA sequences that do code for single proteins, but there are also several forms of **repetitive DNA**, consisting of multiple copies of the same basic sequence (Watson, Tooze and Kurtz, 1983). Some repetitive DNA consists of only moderately repetitive sequences, but some sequences exist in hundreds of thousands of copies (Britten and Kohne, 1968). It has been estimated that only 26% of the human genome consists of single-copy DNA sequences and it is possible that 'genes', in the classical sense, represent less than 2% of the genome (see Marks, 1983). In fact, chromosome banding seems to depend to a large extent on the distinction between repetitive and non-repetitive DNA, although the exact relationship has yet to be determined. The single-copy, genetically active DNA seems to be generally confined to euchromatin, whereas heterochromatin is typically genetically inert and contains a great deal of highly repetitive DNA. Hence, studies of the distribution of defined classes of repetitive DNA in chromosomes can provide valuable additional information on the evolutionary stability of chromosomal subregions. There are also differences between categories of DNA in the time of replication during cell division, and replication times can thus be used as an additional criterion for assessing homologies between chromosomal subregions of different species (see Dutrillaux *et al.*, 1981).

One major category of repetitive DNA that has been studied in humans and in some other primates is **satellite DNA**, which has received this name because it can be separated from other classes of DNA through centrifugation in a density gradient of caesium chloride. Satellite DNA represents about 4% of the human genome. It is the most highly repetitive class of DNA known for humans and it consists of relatively short sequences that are repeated at least 10^5 times (Britten and Kohne, 1968). Four different types of human satellite DNA (I, II, III and IV) have been recognized on the basis of specific physicochemical properties. It seems probable that satellite II DNA originated in the ancestral stock of Old World simians, that satellite I DNA and perhaps satellite IV DNA developed somewhat later in the ancestral hominoid stock that gave rise to gibbons, great apes and humans, and that satellite III DNA is unique to the human lineage (Mitchell, Gosden and Ryder, 1981). The patterns of distribution of satellite DNA in the chromosomes of humans and great apes show a fair degree of variability, but there are also some common features that provide further confirmation that inferences based on banding studies are generally sound (see Seuánez, 1979). However, although different kinds of satellite DNA have been reported for several mammalian species (e.g. for

the laboratory mouse), comparative studies of this class of repetitive DNA among primates have been largely confined to the Old World simians (especially great apes and human beings).

Another prominent category of repetitive DNA is **ribosomal DNA** (rDNA). This consists of multiple copies ('gene families') of sequences coding for three types of ribosomal RNA (5S, 18S and 28S sequences). The 18S and 28S DNA sequences are intimately associated because they jointly code for the 45S ribosomal RNA precursor that eventually produces the 18S and 28S RNA components of the ribosome. In fact, every 18S DNA sequence is linked to a 28S DNA sequence by two 'spacer' sequences, one of which codes for RNA and one of which does not. It is estimated that there are 600 copies of the combined 18S and 28S subunits in the human genome. There may be more than ten times as many copies of the 5S sequence of ribosomal RNA, but much less is known about this third type of ribosomal RNA, except that its distribution on the chromosomes of Old World simians seems to have remained very stable. The copies of each coding subunit of ribosomal DNA show a remarkable degree of sequence conservation throughout eukaryotic organisms generally (Seuánez, 1979), doubtless as a reflection of the fundamental part played by the ribosome in protein synthesis.

In human beings, the 18S and 28S sequences of ribosomal DNA are located on stalk-like appendages, confusingly known as 'satellites', on five pairs of acrocentric chromosomes (numbers 13, 14, 15, 21 and 22 in Fig. 11.16). As Marks (1983) has pointed out, these chromosomal 'satellites' containing 18S/28S ribosomal DNA have nothing to do with the definition of satellite DNA, which is based on physical properties of isolated DNA (see above). The ribosomal DNA components of individual chromosomes are also called **nucleolar organizer regions** because they appear to have a specific connection with a body termed the nucleolus in the cell nucleus. In great apes, as in humans, they are 18S/28S nucleolar organizer regions, commonly in satellites, on several pairs of chromosomes (five pairs in the chimpanzee, two pairs in the gorilla and eight to nine pairs in the orang-utan; see Marks, 1983). Two pairs of homologous acrocentric chromosomes consistently bear satellites containing 18S/28S ribosomal DNA in all great apes and humans, but distribution of the remaining ribosomal RNA varies from species to species. In gibbons and in Old World monkeys, by contrast, there is typically just a single pair of chromosomes bearing 18S/28S ribosomal DNA. A pair of submetacentric chromosomes is usually involved, but in the siamang (*Hylobates syndactylus*) it is a pair of acrocentrics. As nucleolar organizer regions can be specifically stained with silver, these chromosomes are called **marker chromosomes** (Chiarelli, 1971). Marks (1983) has inferred that the possession of a single pair of marker chromosomes is the ancestral condition for Old World simians and that redistribution of 18S/28S ribosomal RNA to occupy subregions (e.g. satellites) on several pairs of acrocentric chromosomes originated as a derived feature of the ancestral stock of great apes and humans, with further modifications occurring in individual descendant lineages.

Among New World monkeys, there is a single pair of metacentric marker chromosomes in some forms (e.g. owl monkey, spider monkey and squirrel monkey), but in others (e.g. marmosets and tamarins, sakis and capuchin monkeys) 18S/28S ribosomal DNA occurs on several chromosomes in a pattern that can differ from genus to genus. It is therefore possible that there was a single pair of marker chromosomes bearing 18S/28S ribosomal DNA in the common ancestral stock of all simian primates and that the involvement of additional chromosomes has evolved in parallel in certain New World monkeys and in great apes and humans. In the only prosimian species so far investigated, the brown lemur (*Lemur fulvus*), 18S/28S ribosomal DNA occurs on 10 pairs of very small chromosomes (microchromosomes), so it is difficult to infer the ancestral condition for primates generally. However, Henderson, Warburton *et al.* (1977) have reported that 18S/28S ribosomal DNA occurs on only two pairs of chromosomes in tree-shrews (*Tupaia*), suggesting that restriction of this category of repetitive DNA to a small number of chromosomes (one or two pairs) was probably the ancestral condition for mammals generally and for primates in particular. But far more comparative work is necessary before a convincing account of evolutionary changes in the chromosomal distribution of ribosomal DNA can be produced.

Several other categories of repetitive DNA have been recognized in primates and in other mammals, perhaps the most interesting of these being the category of 'alphoid' repetitive DNA. This occurs in humans and in at least some other primates and may be specific to primates (Musich, Brown and Maio, 1980). In the Old World vervet monkey (*Cercopithecus aethiops*), 'alphoid' DNA represents almost 25% of the genome, but in the other primate species that have been investigated it is much less abundant. The chromosomal distribution of 'alphoid' DNA is variable between primate species and it has yet to be established whether it is common to primates generally or confined to Old World simians. Another well-known category of repetitive DNA, the 'Alu' family of repeated sequences (Schmid and Jelenik, 1982), occurs in about 300 000 copies and constitutes 3–6% of the human genome. It is interspersed with single-copy DNA sequences and is distributed throughout the karyotype, so it may serve some regulatory function. This category of repetitive DNA is also present in other mammals (e.g. rodents), but each group of mammalian species seems to have characteristic Alu sequences. It is therefore possible that detailed comparative study of this type of repetitive DNA may throw additional light on phylogenetic relationships among mammals.

Although studies of different classes of repetitive DNA have provided some valuable accessory information on phylogenetic relationships, especially with respect to groups of closely related species such as great apes and humans (Seuánez, 1979; Marks, 1983), the main point to emerge from these comparisons is that the distribution of this type of DNA is far less stable than the distribution of linked groups of single-copy DNA. Repetitive DNA obviously behaves quite differently from single-copy DNA in several respects. For instance, aggregations of various types of repetitive DNA can apparently spread relatively quickly from one chromosome to another and can equally rapidly disappear from individual chromosomes. Because of this self-promoting property, repetitive DNA has been labelled 'selfish DNA' by some authors (e.g. Doolittle and Sapienza, 1980; Orgel and Crick, 1980). In fact, several lines of evidence suggest that there is some special mechanism for maintaining a consistent base sequence in individual classes of repetitive DNA within a species. For example, the chromosomal pattern of distribution of ribosomal DNA is very similar in humans and the common chimpanzee (*Pan troglodytes*), suggesting that most of the sites were established before the evolutionary divergence of the two species. But ribosomal DNA is more similar in sequence between different human chromosomes than between homologous chromosomes of humans and chimpanzees (see Marks, 1983). Further, it has been reported that the repetitive fractions of the DNA of humans and chimpanzees are virtually indistinguishable, whereas there is a detectable difference in the non-repetitive DNA (Deininger and Schmid, 1976).

Observations such as these have led to the proposal that there may be 'concerted evolution' of gene families distributed on different chromosomes (Arnheim, 1983). The mechanism of such coordination between gene copies is as yet obscure, but the existence of this phenomenon at least demonstrates that DNA does not always evolve in the straightforward fashion traditionally recognized for single-copy genes. On the one hand, this indicates that comparative studies of sequence information for repetitive DNA are unlikely to yield clear indications of phylogenetic relationships; on the other hand, it provides a warning that the evolution of single-copy genes may not be quite as straightforward as is often supposed. Unfortunately, very little is yet known about the functions of repetitive DNA. Although certain classes, such as ribosomal DNA, do obviously play a definite part in the cell and although some

types of repetitive DNA, such as the Alu family of sequences, may serve a regulatory function, most repetitive DNA has no obvious function and the label 'junk DNA' has often been applied. However, until the significance of repetitive DNA is fully understood, considerable uncertainty about classical interpretations of the evolution of DNA must remain.

Banding and phylogenetic reconstruction

Because of the unusual characteristics of repetitive DNA, attempts to trace evolutionary relationships on the basis of chromosomal banding patterns are most likely to be successful with techniques that emphasize the distribution of euchromatin (e.g. G-banding and R-banding). Using information primarily derived from such banding studies, it has been possible to produce convincing reconstructions of the major chromosomal events that have taken place during the evolutionary divergence of the great ape and human lineages. For example, humans differ from all of the great apes in having only 46 chromosomes rather than 48. At the gross morphological level, this difference reflects the fact that human diploid cells have one pair of metacentrics (the second pair of chromosomes labelled HSA2) in place of two pairs of acrocentrics or submetacentrics present in the diploid cells of great apes. Accordingly, it seemed quite likely that the difference between humans and great apes had arisen through some kind of conversion affecting these chromosomes. This hypothesis was confirmed by banding studies (Fig. 11.17), which showed that the arms of the human metacentric chromosome 2 correspond to the main arms of two separate acrocentric or submetacentric chromosomes in the karyotypes of great apes (see Seuánez, 1979). As great apes are generally fairly similar to one another in this respect and also quite closely resemble other primates, whereas humans are markedly different, it is reasonable to conclude that it is the human condition that is derived. Although this might at first sight appear to be a case of Robertsonian translocation, comparisons based on banding patterns suggest that the

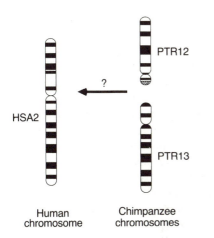

Human
chromosome

Chimpanzee
chromosomes

Figure 11.17. Comparison of the metacentric human chromosome 2 with two submetacentric chromosomes (nos 12 and 13) of the common chimpanzee (*Pan troglodytes*). The banding patterns strongly indicate that some form of centric rearrangement has occurred. (After Harnden, Klinger *et al.*, 1985.)

human metacentric chromosome is most similar to two submetacentric chromosomes of the chimpanzee. Dutrillaux (1979) has accordingly proposed that these submetacentrics fused end-to-end during human evolution (see also Seuánez, 1979). This, however, is the only case of 'fusion' of any kind inferred for the evolutionary divergence of the human and great ape lineages. Most of the chromosomal rearrangements inferred for these species by Dutrillaux (17 out of 28; i.e. 61%) are pericentric inversions, which are obviously best studied using banding techniques. For instance, banding studies have shown that the two subspecies of the orang-utan, the Bornean and the Sumatran (*Pongo pygmaeus pygmaeus* and *P. p. abelii*), differ from one another consistently by a pericentric inversion (Seuánez, Evans *et al.*, 1979).

Restricted comparisons of human and great ape chromosomes suffer from the disadvantage that the direction of evolutionary change is in many cases difficult to determine. Thus, in spite of considerable research, there is still disagreement over our relationship to the African great apes. Authors who have concentrated their attention on human evolution

(e.g. Miller, 1977; Seuánez, 1979; Mai, 1983; Marks, 1983; Yunis and Prakash, 1982) have variously reported that the two chimpanzees are our closest relatives, that the gorilla is our closest relative and that the chimpanzees and the gorilla are more closely related to one another than they are to us. Reliable determination of the direction of chromosomal change requires a broader perspective which, following the principles discussed in Chapter 3, is best provided by a comprehensive comparison of primates, with due reference to other placental mammals. Even then, a process of inferential reasoning is required and phylogenetic reconstructions have so far been generated on a trial-and-error basis. The principle of parsimony has been applied as the primary criterion for the selection of particular solutions, but, as has been explained in Chapter 3, problems can arise with this criterion if there are differences between lineages in rates of change (see later). It is therefore hardly surprising that there are disagreements between authors in the conclusions drawn and successive reconstructions by any given author have necessarily undergone refinements as more information has become available.

However, there may be a special advantage in basing phylogenetic reconstructions on chromosomal rearrangements in that possibilities for rearrangement are theoretically almost unlimited, with the result that convergence is relatively unlikely (Dutrillaux, 1979; Marks, 1983). This represents a major distinction both in comparison to the gross morphological level, where convergence is quite common because of adaptation to meet similar functional requirements, and in comparison to the level of the individual gene, where convergence is common because the possibilities for evolutionary change through point mutation of nucleotide bases are strictly limited. For most chromosomal rearrangements, fixation in evolution seems to be a unique event, so in many cases it may only be necessary to distinguish between primitive and derived conditions in reconstructing likely phylogenetic relationships. Nevertheless, it should be emphasized that particular types of chromosomal change seem to predominate in individual lineages, thus increasing the probability of convergence. Qumsiyen, Hamilton and Schlitter (1987) have, in fact, reported convincing evidence of convergence in the occurrence of particular reciprocal translocations in the rodent genera *Gerbillus* and *Tatera*. Further, Dutrillaux, Couturier *et al.* (1986b) have reported that pathological changes in human chromosomes often mimic changes known to have occurred in primate evolution.

Although a relatively broad comparison of primate karyotypes was initiated by de Grouchy, Turleau and Finaz (1978), humans and great apes have remained very much the focus, and the only genuinely comprehensive comparisons so far published are those of Dutrillaux and his colleagues (e.g. Dutrillaux, Viégas-Pequignot *et al.*, 1978; Dutrillaux, 1979; Dutrillaux, Couturier and Fosse, 1980; Dutrillaux and Rumpler, 1980; Dutrillaux and Couturier, 1981; Dutrillaux, Couturier *et al.*, 1982). Dutrillaux's group has studied an impressive number of different primate species, including both prosimians and simians, and they have also considered other mammalian groups, such as lagomorphs (rabbits; Dutrillaux, Viégas-Pequignot and Couturier, 1980) and carnivores (Dutrillaux and Couturier, 1983). Comparison of various primates with carnivores has revealed a striking degree of correspondence in chromosomal banding patterns (Nash and O'Brien, 1982; Dutrillaux and Couturier, 1983) and inferences regarding the conservation of homologous subregions have been confirmed by gene-mapping in the domestic cat (O'Brien and Nash, 1982). There is a similar degree of correspondence between the chromosomes of primates and those of the domestic rabbit (Dutrillaux, Viégas-Pequignot and Couturier, 1980) and confirmation of conservation of chromosomal subregions has been similarly forthcoming from gene-mapping (Soulié and de Grouchy, 1982). By including information on edentates, artiodactyls and rodents along with that for primates, carnivores and lagomorphs, Dutrillaux, Couturier *et al.* (1982) produced a hypothetical reconstruction of a common ancestral karyotype (i.e. the

karyotype of ancestral placental mammals). This hypothetical ancestral karyotype has a diploid number of 58, higher than the modal value for placental mammals indicated in Fig. 11.8, and a fundamental number of 66, coinciding closely with the modal value indicated in Fig. 11.12.

As a result of the extensive comparative work by Dutrillaux's group for mammals generally, it has been possible to compile an outline chromosomal phylogeny for primates based on the results of banding studies of more than 70 species (Fig. 11.18). The main features of this inferred phylogenetic pattern are summarized in Fig. 11.19. The most immediately striking aspect of Figs 11.18 and 11.19 is that there is very good general agreement between the pattern of phylogenetic relationships inferred from chromosomal rearrangements and the outline pattern inferred from morphological evidence (see Fig. 2.2). Unfortunately, however, evidence from chromosomal banding does not provide any clarification of the relationships between strepsirhines (lemurs and lorises) and haplorhines (tarsiers and simians). Poorman, Cartmill and MacPhee (1985) reported briefly on the G-banded karyotype of the western tarsier (*Tarsius bancanus*), stating that two-thirds of its chromosomes are visually and metrically indistinguishable from chromosomes of the thick-tailed bushbaby (*Galago crassicaudatus monteiri*). However, investigation of the karyotype of the Philippine tarsier (*Tarsius syrichta*) by Dutrillaux and Rumpler (1988), using several techniques, indicated that evolutionary change has been so extensive that a reconstruction of relationships is prohibited.

In the absence of directly comparable information on *Tarsius*, it can at least be concluded from Figs 11.18 and 11.19 that the chromosomal evidence indicates a basic division between strepsirhines and simians (monkeys, apes and humans). On the strepsirhine side, the chromosomal evidence clearly indicates a bifurcation between Malagasy lemurs and the loris group. The ancestral stock of the Malagasy lemurs is distinguished from the ancestral strepsirhine stock by a total of five shared derived rearrangements and the available data indicate

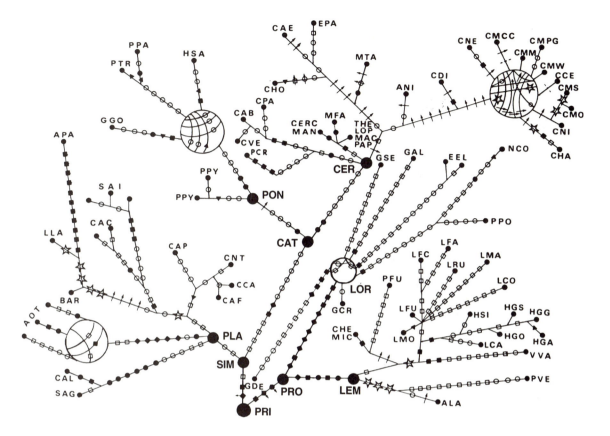

Figure 11.18. Outline phylogeny for primates based on inferred chromosomal rearrangements. Key to chromosomal rearrangements (symbols along lines): ○, pericentric inversion; ●, other inversion; □, Robertsonian translocation; ■, other translocation; →, fission; ☆, shift; ◆, complex rearrangement. (From Dutrillaux, Couturier *et al.*, 1986, with kind permission from Dr B. Dutrillaux and from *Mammalia*.)

1. Ancestral nodes (labelled large circles): CAT, ancestral Old World simian; CER, ancestral Old World monkey; LEM, ancestral lemur; LOR, ancestral lorisid; PLA, ancestral New World simian; PON, ancestral hominoid; PRI, ancestral primate; PRO, ancestral strepsirhine; SIM, ancestral simian.

2. Extant species (labelled terminal black circles): ALA, *Avahi*; ANI, *Allenopithecus*; AOT, *Aotus*; APA, *Ateles;* BAR, *Brachyteles*; CAB, *Colobus guereza*; CAC, *Cacajao*; CAE, *Cercopithecus aethiops*; CAF, *Cebus albifrons*; CAL, *Callithrix*; CAP, *Cebus apella*; CCA, *Cebus capucinus*; CCE, *Cercopithecus cephus*; CDI, *Cercopithecus diana*; CERC, *Cercocebus(Cercocebus)*; CHA, *Cercopithecus hamlyni*; CHE, *Cheirogaleus*; CHO, *Cercopithecus l'hoesti*; CMCC, *Cercopithecus campbelli*; CMM, *Cercopithecus mona*; CMO, CMS, *Cercopithecus mitis*; CMPG, *Cercopithecus grayi*; CMW, *Cercopithecus wolfi*; CNE, *Cercopithecus neglectus*; CNI, *Cercopithecus nictitans*; CNT, *Cebus nigrivittatus*; CPA, *Colobus angolensis*; CVE, *Colobus polykomos*; EEL, *Galago (Euoticus) elegantulus*; EPA, *Erythrocebus*; GAL, *Galago alleni*; GCR, *Galago crassicaudatus*; GDE, *Galago demidovii*; GGO, *Gorilla*; GSE, *Galago senegalensis;* HGA, HGG, HGO, HGS, *Hapalemur griseus*; HSA, *Homo sapiens*; HSI, *Hapalemur simus*; LCA, *Lemur catta*; LCO, *Lemur coronatus*; LFA, LFC, LFU, *Lemur fulvus*; LLA, *Lagothrix*; LMA, *Lemur macaco*; LMO, *Lemur mongoz*; LRU, *Lemur rubriventer*; LOP, *Cercocebus (Lophocebus)*; MAN, *Mandrillus*; MAC, MFA, *Macaca*; MIC, *Microcebus*; MTA, *Miopithecus*; NCO, *Nycticebus*; PAP, *Papio*; PCR, *Presbytis*; PFU, *Phaner*; PPA, *Pan paniscus*; PPO, *Perodicticus*; PPY, *Pongo*; PTR, *Pan troglodytes*; PVE, *Propithecus verreauxi*; SAG, *Saguinus*; SAI, *Saimiri*; THE, *Theropithecus*; VVA, *Varecia*.

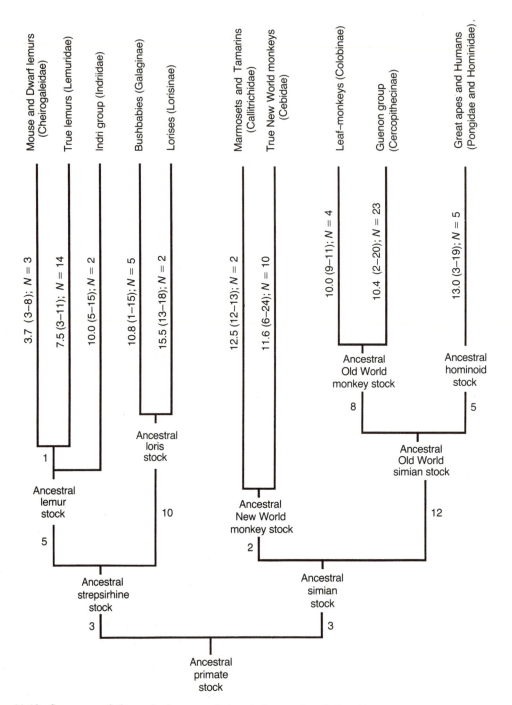

Figure 11.19. Summary of the main features of the phylogenetic relationships among primates inferred from chromosomal rearrangements by Dutrillaux, Couturier, *et al.* (1986). Numbers adjacent to lineages leading to ancestral stocks indicate numbers of shared derived rearrangements defining those stocks. Figures against lines leading to extant groups of primates indicate the mean number of chromosomal rearrangements separating each set of species from the ancestral node (range in parentheses) and the total number of species involved.

that the three families Cheirogaleidae, Lemuridae and Indriidae are all monophyletic groups, defined by a small number of shared rearrangements in each case. The loris group is defined by a total of 10 shared chromosomal arrangements and the two lorisines so far examined (*Nycticebus* and *Perodicticus*) are separated from the bushbabies by a further nine shared derived rearrangements. Incidentally, it should be noted that the phylogenetic reconstruction in Fig. 11.18 indicates that *Galago crassicaudatus* is relatively primitive in karyotypic terms and that repeated 'centric fusions' have indeed taken place in the evolution of the karyotype of *Galago senegalensis* (see Rumpler, Couturier *et al.*, 1983). On the simian side of the tree, a similar bifurcation is indicated between New World and Old World forms. The ancestral stock of the New World simians is defined by two shared derived rearrangements and the chromosomal evidence is consistent with a phylogenetic subdivision into marmosets and tamarins (Callitrichidae) versus 'true' New World monkeys (Cebidae), except that the owl monkey (*Aotus*) is not clearly linked to either of these two groups. Old World simians, in turn, are clearly subdivided into two groups. The first contains Old World monkeys, with an ancestral stock defined by eight shared derived chromosomal rearrangements, and the second contains great apes and humans, with an ancestral stock defined by five shared derived rearrangements. Among Old World monkeys, there are four shared derived rearrangements linking leaf-monkeys (Colobinae); but the remaining forms (Cercopithecinae) do not have a clearly defined ancestral node, probably because baboons, macaques, mandrills and mangabeys have remained chromosomally very conservative since diverging from the ancestral stock of Old World monkeys. For the group containing the great apes and humans, the inferred pattern of phylogenetic relationships is compatible with other evidence indicating that the orang-utan diverged first and that the relationship between African apes and humans is an unresolved trichotomy.

The attentive reader will have noticed that gibbons (family Hylobatidae) have been omitted from Figs 11.18 and 11.19. Although the karyotypes of these Old World simians have been studied, modification through chromosomal rearrangement has been so extensive that it has not yet proved possible to reconstruct their relationships to other primates (Dutrillaux, 1979). There is extensive karyotypic diversity even within the gibbon group and there is a particularly marked contrast between the six species of gibbons and the siamang, even though viable hybrids between gibbons and siamangs have been produced (Myers and Shafer, 1976; Shafer, Myers and Saltzman, 1984). In fact, comparison among different gibbon species suggests that chromosomal evolution in this group of primates has involved numerous non-Robertsonian reciprocal translocations, a mode of chromosomal rearrangement that is very rare among other primates (Couturier, Dutrillaux *et al.* 1982). Comparative gene-mapping has confirmed the interpretation that extensive karyotypic rearrangement has occurred in the gibbon group. It is possible to localize on gibbon chromosomes genes that have also been mapped for humans and great apes, but it is obvious that major dissociation and redistribution of linked groups has taken place (Turleau, Creau-Goldberg *et al.*, 1983).

The position of tree-shrews has yet to be clearly resolved with respect to chromosomal evidence. Dutrillaux, Couturier *et al.* (1986a) note that very little is yet known about the banding patterns of chromosomes in insectivores, so a direct comparison is ruled out. Tree-shrews have apparently retained a relatively primitive karyotype, so they show resemblances to various mammalian orders.

The pattern of chromosomal rearrangements in Fig. 11.18 reveals an interesting special feature, in that different types of chromosomal rearrangement seem to predominate in different groups of primates. Among strepsirhines, especially lemurs, chromosomal 'fusion' events (probably Robertsonian translocations) represent the dominant mode of inferred chromosomal rearrangement; pericentric inversions are relatively rare. By contrast, among simians, 'fusions' are comparatively uncommon and other types of chromosomal rearrangement predominate. Further, whereas (as noted above) pericentric inversions account for most of the chromosomal rearrangements within the group containing great apes and humans, the dominant mode of inferred chromosomal evolution within the guenon group of Old World monkeys (genera *Allenopithecus*, *Cercopithecus*, *Erythrocebus* and *Miopithecus*) has been chromosomal 'fission'. Indeed, it should be noted from Fig. 11.18 that the guenon group seems to have undergone an early dichotomy and that 'fissions' have apparently been the main mode of chromosomal evolution within both subgroups.

This dominance of a type of chromosomal rearrangement within a given group, which has also been emphasized by Dutrillaux, Couturier and Viégas-Pequignot (1981) and by Marks (1983), has been reported for a variety of other animal groups and has been referred to as 'karyotypic orthoselection' by White (1973). Dutrillaux, Couturier *et al.* (1986a) prefer the term 'orthomutation', as it more accurately reflects the tendency for chromosomes to show particular kinds of rearrangements in particular groups. No convincing explanation has yet been advanced to account for this tendency for a particular mode of chromosomal rearrangement to prevail in any one group, but (as suggested above), it has undoubtedly contributed to a common tendency for individual authors who have worked with a particular group of primates to emphasize one mode of chromosomal change at the expense of the others. In fact, it has been noted that chromosomal inversions observed as pathological aberrations in people today tend to resemble those that have been inferred in the evolution of humans and certain other primates. Further, irradiation of cells from guenons (*Cercopithecus* species) produce numerous fissions, whereas these have never been observed with irradiated human cells (Dutrillaux *et al.*, 1981). Thus, it would seem that the predominance of a particular mode of chromosomal rearrangement in individual primate groups reflects some natural propensity

of the chromosomes to undergo a particular kind of rearrangement.

Another point that Figs 11.18 and 11.19 make clear is that there has been quite marked variation in the **rates of chromosomal evolution**, in so far as the known forms of chromosomal rearrangement can be regarded as numerically equivalent. It has already been mentioned that the gibbons could not be included in these figures because, for some unknown reason, their karyotypes have undergone particularly dramatic reorganization. The average number of chromosomal rearrangements separating extant primate species from the inferred ancestral primate condition in Fig. 11.18 is 25.0 ± 9.5, but the individual values range from 11 to 43, which represents an approximately four-fold variation in the overall rate of change. In fact, Fig. 11.18 indicates that the smallest departure from the ancestral primate condition, in terms of the total number of inferred chromosomal re-arrangements, is found among New World capuchin monkeys (*Cebus*), closely followed by the mouse and dwarf lemurs (*Microcebus* and *Cheirogaleus*). Rates also vary along individual primate lineages. For instance, eight inferred shared derived rearrangements separate the ancestral lemur stock from the ancestral primate stock. Yet the genera *Cheirogaleus* and *Microcebus* show only four further rearrangements after diverging from the ancestral lemur stock, whereas 15 are found along the lineage leading to *Propithecus verreauxi*. On the simian side of the primate tree, 20 inferred shared derived rearrangements separate the ancestral stock of the Old World monkeys from the ancestral primate stock. Thereafter, baboons, macaques, mandrills and mangabeys show only two or three subsequent rearrangements whereas in the various species of the guenon group (*Allenopithecus*, *Cercopithecus*, *Erythrocebus* and *Miopithecus*) there were between 8 and 19 subsequent rearrangements over the same period of time. Hence, as with morphological evolution, reconstructions of chromosomal evolution must take account of the established fact that rates of change along individual lineages can vary quite substantially.

As can be seen from Fig. 11.19, when average numbers of chromosomal rearrangements are calculated for lineages leading to subgroups of primates, some of the variation in rates of chromosomal change is masked. The average number of chromosomal changes leading to the 10 specified groups of living primates is 23.3 and the range is from 12.7 to 33.4. This is less than half the range of variation found for individual species. Thus, there seems to be some tendency for rates of chromosomal evolution to balance out between subgroups over time and this, in turn, suggests that chance processes probably play at least some part. Nevertheless, it should be noted that quite low average rates of chromosomal change are found in the mouse lemur group and in true New World monkeys, whereas the highest average rates of change are found in the loris subgroup (Lorisinae) and in Old World simians.

The process of chromosomal change

In addition to considering the products of chromosomal evolution, it is also necessary to give some attention to the **process** through which a chromosomal arrangement can arise and become established in a population. In this respect, whole chromosomes or identifiable segments of chromosomes can be seen in the same light as individual genes. Clearly, rearrangements or 'mutations' of chromosomes (e.g. fissions, fusions or translocations) arise by chance rather like point mutations of individual genes, and chromosomal rearrangements must be similarly subject to selection and/or random changes in frequency. Further, as with single genes, it is to be expected that chromosomal **polymorphisms** might be found in natural populations, with individuals differing from one another in the occurrence of a particular chromosomal rearrangement. Just as with single gene loci, it should theoretically be possible to find both types of homozygote (one with both chromosomes of a pair showing the original pattern and one with both showing the rearranged pattern) and the heterozygous (mixed) form in a population.

This question of chromosomal polymorphism, however, introduces the problem of the **viability** and **fertility** of individuals with single chromosomal rearrangements in the heterozygous state. As a rule, although it is possible that viability might be adversely affected, one would not expect chromosomal rearrangement to be automatically detrimental to the possessor, because there could often be no significant loss or addition of genetic material. By contrast, there are good reasons to expect that chromosomal rearrangement might affect fertility. In principle, it is possible that the mismatching of a chromosome pair, because of rearrangement found in one chromosome but not shared by the other, could lead to problems when the chromosomes are aligned on the spindle in meiosis and would thus reduce the fertility of the possessor. There is, indeed, some evidence that certain chromosomal rearrangements in some species may cause problems at meiosis (e.g. see Tettenborn and Gropp, 1970, for mice; see also White, 1973).

If this were uniformly the case, there would obviously be consistent strong selection against chromosomal rearrangements and they could spread through a population only in an 'accidental' fashion (e.g. through isolation of a small inbred population in which a chromosomal rearrangement could become established as the norm by chance). In contrast to the situation for single gene loci, with chromosomal rearrangements it would then normally be expected that both homozygotes would consistently be at an advantage in comparison to the heterozygote, because only the latter would suffer from mismatching of chromosomes at meiosis. Hence, the chance local occurrence of a number of individuals sharing the same chromosomal rearrangement could lead to relatively rapid establishment of a subpopulation characterized by the rearrangement. Indeed, because selection would be expected to operate against any heterozygotes, an isolated subpopulation with a given chromosomal rearrangement could quite rapidly undergo full reproductive isolation from the main population and hence become a separate species. This **stasipatric** model of speciation through the chance establishment of a chromosomal rearrangement in an isolated subpopulation has been championed by White (1968, 1973, 1978) as an alternative to the commonly accepted standard **allopatric** model involving definite geographical separation (see Chapter 3, Fig. 3.6). As is discussed below, the stasipatric model has been specifically invoked for speciation of mammals by Bush (1975).

If it were uniformly true that new chromosomal rearrangements must lead to serious problems at meiosis, they should be relatively rare in any population. Overall, individual chromosomal rearrangements should be present at frequencies close to the natural mutation level and one would not expect to find **balanced polymorphisms** (i.e. apparent equilibrium states with both homozygous conditions and the heterozygous condition present at levels far exceeding the mutation rate). In the case of human beings, it would indeed seem to be the case that chromosomal rearrangements are frequently deleterious, commonly affecting viability as well as fertility. Hamerton, Canning *et al.* (1975) reported from a survey of a large sample of human neonates that chromosomal aberrations, both structural and numerical, occur at a frequency of about 2%. Most of the observed cases were associated with clinical abnormality and there was no evidence of polymorphisms substantially exceeding the mutation level. For the adult human population, three main types of 'centric fusion' or Robertsonian translocation involving acrocentric chromosomes have been reported (Hamerton, 1968). Two types are associated with an increased incidence of Down's syndrome, confirming the expectation that centric rearrangement may, in the heterozygous condition, lead to unbalanced segregation of chromosomes at meiosis. Nevertheless, it should be noted that the third type of 'centric fusion', which also happens to be the commonest (occurring at a frequency of up to one in a thousand individuals), is not associated with an increased incidence of Down's syndrome.

This indicates that some chromosomal rearrangements are more deleterious than others with respect to disruption of meiosis. Further, a number of chromosomal rearrangements not resulting in clinical abnormality have been traced through several generations in human families, including Robertsonian translocations, pericentric inversions, a small Y-chromosome and an inverted Y-chromosome (Seuánez, 1979). Thus, it is clear that certain human chromosomal rearrangements can be transmitted without major detriment, although their restriction to individual families might be seen as providing some indirect support for White's concept of speciation through chromosomal mismatching.

A survey of 110 adult yellow baboons (*Papio cynocephalus*) similarly revealed no sign of chromosomal polymorphism (Soulié and de Grouchy, 1981) and other investigators have failed to find chromosomal variants in this species. But, in contrast to the evidence available for human beings and yellow baboons, information derived from other primates and from other mammals indicates that established chromosomal polymorphisms are quite widespread. It has already been mentioned that the Bornean and Sumatran orang-utan populations differ consistently from one another with respect to a pericentric inversion identified through banding studies. In addition to this fixed chromosomal rearrangement distinguishing the subspecies, there is a complex rearrangement of another chromosome, explained as an inversion within an inversion, that is common to both orang-utan populations (Seuánez, Fletcher *et al.*, 1976). As this rearrangement is present in both orang-utan populations (usually regarded as subspecies), it seems likely that it was present in the ancestral population and has been maintained for some considerable time, perhaps as a balanced polymorphism. Seuánez (1979) has suggested a possible mechanism whereby this particular polymorphism could be maintained.

Chromosomal polymorphisms also occur in the owl monkey, *Aotus* (e.g. Brumback, Staton *et al.*, 1971; de Boer, 1974; Martin, 1978; Ma, Rossan *et al.*, 1978; Ma, 1981). This case is complicated by the fact that, although owl monkeys are customarily regarded as a single species (*Aotus trivirgatus*), significantly different karyotypes have been reported for specimens derived from different geographical areas ('chromosomal races'; see also Fig. 11.18). On the basis of the karyotypic and other evidence, it now seems highly likely that there are several species of owl monkey. (This is just one of several cases among primates where chromosomal evidence has suggested the existence of separate species that had not been clearly recognized on the basis of morphology alone.) In addition to the chromosomal distinctions between geographically separate populations of *Aotus*, however, there are polymorphisms within populations. Considerable confusion has arisen because a large number of different karyotypes have been reported for '*Aotus trivirgatus*' without a clear distinction between intrapopulation and interpopulation variation. Colombian owl monkeys, usually recognized as the subspecies *Aotus trivirgatus griseimembra*, have three different karyotypes with diploid numbers of 52, 53 and 54, respectively. The commonest homozygous karyotype found in wild-caught animals is that containing 54 chromosomes, of which 22 are metacentrics and 32 are acrocentrics. A karyotype of 52 chromosomes results from 'centric fusion' of two pairs of acrocentric chromosomes to produce a single pair of large metacentric chromosomes, giving a total of 24 metacentrics and only 28 acrocentrics. A karyotype with 53 chromosomes represents the heterozygous condition, with only one of each pair of acrocentrics fused to produce the large metacentric (see Martin, 1978). Hence, variation in the karotypes of Colombian owl monkeys represents a simple polymorphic condition. Ma, Rossan *et al.* (1978) give the relative frequencies of the diploid numbers 54, 53 and 52 as 41%, 46% and 13%, respectively, for a random sample of 363 individuals. These figures are consistent with the expectation for free interbreeding with no obvious selection against the heterozygous condition. Further, because all three karyotypes have been identified in several different batches of wild-caught owl monkeys from Colombia, it would seem that this is a genuine polymorphism

(perhaps a balanced polymorphism). The karyotype of Colombian owl monkeys is fundamentally different in numerous respects from the karyotypes of other geographical populations of owl monkeys and it is unlikely that fertile interbreeding between such populations occurs in the wild. Additional cases of polymorphism have been reported for other geographical populations of *Aotus* (see Ma, Rossan *et al.*, 1978; Ma, 1981).

Chromosomal polymorphisms involving 'centric fusions' (Robertsonian translocations) have been reported for other primates, such as the lesser bushbaby, *Galago senegalensis* (de Boer, 1973), and for a variety of both wild and domesticated mammals (see Martin, 1978 for a summary of some cases). White (1973) stated that Robertsonian translocation is the most frequent form of chromosomal polymorphism in mammals and this agrees well with the empirical evidence (e.g. see Fig. 11.14) that this is the predominant mode of chromosomal evolution. In fact, because it is necessary to study fairly large samples of individuals to detect chromosomal polymorphisms of any kind, it seems inevitable that further cases of 'centric fusion' and of other less-common rearrangements will be reported. For instance, Freitas and Seuánez (1982) have described apparent polymorphism in the size of heterochromatin regions on chromosomes of 20 capuchin monkeys (*Cebus apella*) collected from different parts of Brazil. The evidence already available hence convincingly shows that not all chromosomal rearrangements are detrimental to the extent that in large interbreeding populations selection will act to keep the frequency of heterozygotes close to the mutation level. Genuine polymorphisms obviously do exist in numerous species, providing the raw material for evolutionary change in the karyotype. As Lande (1979b) has noted, Robertsonian translocation seems to lead to problems at meiosis mainly when the chromosome arms concerned are of unequal length. Accordingly, it is not justifiable to propose a model of chromosomal evolution based on the assumption that a chromosomal rearrangement in the heterozygous condition necessarily disrupts meiotic division.

It has yet to be explained why chromosomal polymorphisms occur in some species and not in others and why some rearrangements are less deleterious than others within a given species. In this respect, however, chromosomal rearrangements are not very different from point mutations of genes; some are conspicuously deleterious whereas others are not. This touches directly on the question of possible selective advantages of certain chromosomal rearrangements, which is really the most fundamental of all issues concerning chromosomal organization. Ultimately, the most reliable approach to the question must be that of studying (preferably under natural conditions) the dynamics of a given chromosomal polymorphism. At present, however, there is no direct evidence of the presence or absence of a selective advantage for individual chromosomal rearrangements in any mammalian species. There is a major obstacle here in that presently documented cases of chromosomal polymorphism do not seem to be reflected by any gross morphological variation in the organism (Lande, 1979b; John, 1981). In the case of Colombian owl monkeys, for instance, individuals with different karyotypes have not yet been distinguished by any other criterion.

There are basically two schools of thought with respect to the fixation of chromosomal rearrangements in evolution. Most authors have adopted the view that chromosomal rearrangements do not themselves have specific adaptive significance, but that their fixation is directly related to the development of barriers to interbreeding between species populations (White, 1973, 1978). Marks (1983) has succinctly expressed this concept in stating that chromosomes are probably 'adaptively passive agents of reproductive isolation, important in speciation, but effectively neutral with respect to natural selection' (see also Lande, 1979b). The alternative interpretation is that chromosomal rearrangements can in fact be adaptively advantageous or disadvantageous in their own right, at least under certain conditions, and that they are therefore subject to natural selection just like other discrete features of organisms. Bickham and Baker (1979) proposed a 'canalization model' of chromosomal evolution with natural selection directly favouring rearrangements with advantageous effects. They suggested that chromosomal change would occur relatively rapidly during the initial adaptive radiation of a group of organisms and would subsequently decelerate after new adaptive arrangements of genes had become established. One prediction of this model (Bickham, 1981) is that chromosomal changes occurring during the early phase of adaptive radiation would be more fundamental in kind (e.g. inversions and translocations), whereas changes occurring during the later phase would be less dramatic (e.g. addition of heterochromatin). It follows from the 'canalization model' that there should be a general correspondence between degree of chromosomal evolution and degree of morphological evolution.

Bodmer (1975, 1981), among others, has suggested that change in the order of genes on chromosomes might lead to selective differences (the 'position effect'). The widespread conservation of linked groups of genes in chromosome subregions among placental mammals generally suggests that clustering of genes has some functional significance. Further, there are some specific cases in which alteration of the relative positions of the linear sequence of genes does seem to exert a significant influence on gene expression (e.g. Zieg, Silverman *et al.*, 1977). It has been suggested as a general rule that heterochromatin may have a suppressive influence on gene action and if this is the case then relative rearrangement of heterochromatic and euchromatic regions of chromosomes could exert general effects on gene expression. It has been mentioned above that dispersion of the Alu repeated sequences of DNA throughout the mammalian genome is indicative of some regulatory function (see Watson, Tooze and Kurtz, 1983), and this provides a further indication that the ordering of DNA sequences along chromosomes is probably of considerable functional significance. If the relative arrangement of individual genes and non-coding DNA sequences along chromosomes primarily affects levels of gene expression rather than exerting an all-or-nothing influence, this could explain why chromosomal polymorphism does not seem to be reflected in morphological polymorphism. Hence, it is by no means obvious that chromosomal rearrangements in themselves can be regarded as selectively neutral, although very little is as yet known about the actual conditions that might lead to replacement of one chromosomal pattern by another in a large, interbreeding population.

Dutrillaux (1986) has rightly emphasized another context in which chromosomal rearrangement may have long-term adaptive significance, namely in the level of recombination that can occur in the formation of gametes. The number of chromosomally distinct gametes that can be produced by any species is 2^n, where n is the haploid number, so species with a large diploid number have a vastly greater capacity for genetic recombination at sexual reproduction.

Chromosomal change and speciation

An attempt has been made to link mammalian chromosomal evolution to speciation patterns (Wilson, Bush *et al.*, 1975; Bush, Case *et al.*, 1977). The starting-point for this proposal was the observation that, whereas amphibians and mammals apparently have similar rates of molecular evolution (see later), mammalian chromosomes seem to have evolved at a markedly faster pace than the chromosomes of amphibians. No attempt was made to carry out an actual reconstruction of chromosomal evolution (e.g. on the basis of banding patterns); instead, an indirect method was used to estimate rate of change in chromosomes. Attention was confined to genera also known to occur as fossils and data on both diploid number ($2n$) and fundamental number (NF) were analysed. For each parameter ($2n$ or NF), an estimate of the average number of chromosomal changes per lineage within a genus (m) was initially obtained using the following formula (Wilson, Bush *et al.*, 1975):

$$m = (k - 1)/N$$

(where k is the number of different karyotypes

recorded for the genus and N is the number of extant species of that genus examined).

Having thus calculated values of m for each genus, it was possible to estimate the average rate of karyotypic evolution (r) for species within a genus as $r = m/t$, with t being the time of origin indicated by the earliest reported appearance of the genus in the fossil record. As has been pointed out in Chapter 2, it should be remembered that the time of first appearance of a taxon in the fossil record can only indicate a minimum value of t. The degree of underestimation of t will vary according to the effective sampling density of the fossil record for the group considered. Further, the equation given above is no more than a crude measure of present karyotypic diversity within a genus and it suffers from the particular disadvantage that it provides a tolerable estimate of the amount of change only if k is considerably smaller than N. For this reason, Bush, Case *et al.* (1977) subsequently used the following modified equation for estimating overall rate of karyotypic evolution (r'):

$$r' = \sum_{i+1}^{G} (a_i + b_i) \bigg/ \sum_{i+1}^{G} t_i$$

(where

G = number of genera examined karyotypically;

$$a = \frac{\text{highest } 2n \text{ in genus} - \text{lowest } 2n \text{ in genus}}{\text{number of species in genus}};$$

$$b = \frac{\text{highest NF in genus} - \text{lowest NF in genus}}{\text{number of species in genus}};$$

t = time of first reported appearance of genus in the fossil record).

Even this modified equation has a number of inherent problems. For instance, it is based on the assumption that every step in chromosomal evolution involves a change of ± 1 either in $2n$ or in NF. Accordingly, any pericentric inversions that do not change NF and all paracentric inversions will be ignored, as will balanced translocations and several other types of karyotypic change. In the absence of actual banding studies of chromosomes, many chromosomal changes will pass unrecognized (see Bickham, 1981). Further, because of the widespread occurrence of 'karyotypic orthoselection' (White, 1973), there will be a systematic bias against groups in which the dominant mode of chromosomal rearrangement does not modify 2n or NF. The equation also implies that all karyotypic evolution is divergent both with respect to 2n and with respect to NF. Nevertheless, the equation gives a very approximate estimation of overall degree of karyotypic change.

On the basis of estimates of rates of karyotypic change derived using this approach, Wilson, Bush *et al.* (1975) and Bush, Case *et al.* (1977) made several observations. The most important of these was that, in comparison to other vertebrates, placental mammals and marsupials seem to exhibit strikingly higher rates of chromosomal change than other vertebrates

(excluding birds, which were not included in the analysis because of difficulties in interpreting their karyotypes). Whereas the mammals examined exhibit an average combined change in 2n and NF of 0.43 modifications per lineage per million years, the other vertebrates (fish, amphibians and reptiles) exhibit average combined values of only 0.03 (Table 11.2). In other words, mammals generally show estimated rates of karyotypic change averaging about fourteen times greater than those for fish, amphibians and reptiles. However, there is considerable diversity among mammalian groups. Horses have the highest inferred rates of karyotypic change, bats and whales have the lowest (see Table 11.2). Whales, in fact, have inferred rates of karyotypic evolution indistinguishable from those of fish, amphibians and reptiles. Primates are characterized by quite high

Table 11.2 Inferred rates of chromosomal evolution in different groups of mammals (data from Bush, Case *et al.*, 1977)

Group	Number of genera	Average age of genera (Ma)	Karyotypic change* (2n + NF)	Corrected speciation rate†
Perissodactyla (horses)	1	3.5	1.40	2.8
Primates	13	3.8	0.75	2.8
Lagomorpha (rabbits, etc.)	3	5.0	0.63	1.6
Artiodactyla (even-toed hoofed mammals)	15	4.2	0.56	1.5
Rodentia	50	6.0	0.43	1.6
Insectivora (shrews, etc.)	7	8.1	0.19	1.4
Marsupialia	15	5.6	0.18	1.0
Carnivora	10	12.9	0.08	1.1
Chiroptera (bats)	15	9.0	0.06	0.7
Cetacea (dolphins and whales)	2	6.5	0.03	0.4
All mammals	131	6.5	0.43	1.5
Other vertebrates	82	22.1	0.03	0.3

* Karyotypic change per lineage per million years (Ma).
† Number of speciation 'events' per lineage per million years, allowing for estimated rate of extinctions.

rates of karyotypic evolution, ranking above all mammalian groups other than horses in the sample examined by Bush, Case *et al.* (1977). Although Wilson, Bush *et al.* (1975) claimed that for non-flying mammals the rate of karyotypic change was inversely correlated with body size (as indicated by overall body length), it can be seen from Table 11.2 that there is no consistent relationship of this kind. Horses and primates are generally significantly bigger than rodents and insectivores, yet they exhibit higher inferred rates of chromosomal change. Presumably, the analysis of Wilson, Bush *et al.* was biased by the fact that cetaceans are by far the largest-bodied mammals and also happen to have the lowest inferred rates of karyotypic change.

In an attempt to explain the high average rate of inferred karyotypic change in mammals generally, and to account for the differences among mammalian groups, Wilson, Bush *et al.* (1975) suggested that special features of population structure in mammals are involved, proposing that mammalian populations 'may be structured in such a way that chromosomal mutations have a better chance of surviving, becoming fixed, and spreading than is the case for other vertebrates'. They followed Bush (1975) in maintaining that patterns of social organization in mammals generally bring about subdivision of populations into social units ('demes'), which are effectively inbred (see also Lande, 1979b). Given small, inbred demes of 10 members or less, they argued, chromosomal rearrangements could become fixed in the homozygous state by chance. If, as has often been held, such fixation of chromosomal rearrangements presents a barrier to inbreeding because of selection against heterozygotes, it is accordingly possible that a new species could arise as a result of chance fixation of a chromosomal rearrangement in an inbred social unit. This relates directly to White's (1973, 1978) 'stasipatric' model of speciation, mentioned above, and provides one of the lines of apparent supporting evidence. Wilson, Bush *et al.* (1975) go so far as to suggest that this form of speciation might be the major mode of speciation for mammals, whereas other vertebrates have typically undergone speciation through geo-graphical isolation, in agreement with the standard allopatric model.

The fact that in certain mammalian groups, such as bats and whales, there are relatively low rates of inferred chromosomal change can, in principle, be explained on the grounds that in mammals with particularly high mobility (e.g. bats) or with a relatively homogeneous environment (e.g. aquatic mammals) the effective isolation of small inbred units ('demes') is far less likely. Arnason (1972, 1974) made a specific study of cetaceans (whales and dolphins) and of pinnipeds (seals and sea-lions) and found that in both of these groups there is very low karyotypic diversity. For 17 cetacean species (including both whales and dolphins), he recorded only two different diploid numbers ($2n = 42$; $2n = 44$) and for 18 pinniped species only three different diploid numbers ($2n = 32$; $2n = 34$; $2n = 36$). Similar consistency is found in fundamental numbers for both groups. Arnason suggested that low karyotypic diversity in cetaceans and pinnipeds is linked to three common features: low reproductive turnover (involving late sexual maturity and low annual birth rate), high mobility and the absence of clearly defined niches in the aquatic environment. He contrasted cetaceans and pinnipeds with rodents and insectivores, which are characterized by relatively high karyotypic diversity and show high reproductive turnover, restricted mobility and occupation of environments with well-defined niches. However, the apparent correlation between high karyotypic diversity and high reproductive turnover does not stand up to closer examination. The figures provided by Bush, Case *et al.* (1977) indicate that the highest inferred rates of karyotypic change among the mammals included in their sample were those for horses and for primates (see Table 11.2). As is explained in Chapter 9, one of the defining characteristics of the primates (excluding tree-shrews) is very low reproductive turnover, and horses are very similar in this respect. From Table 11.2 it would appear that prolifically breeding rodents and insectivores are characterized by only moderate rates of karyotypic change. Further, mouse and dwarf lemurs (Cheirogaleidae) have relatively high rates of reproductive turnover among primates, yet the rate of chromosomal evolution in these lemurs has apparently been one of the lowest for primates generally (Figs 11.18 and 11.19). Thus, one is left with the general possibilities that high karyotypic diversity might be related to low mobility and/or occupation of relatively patchy environments. Special factors such as rigid demarcation of social units and territoriality could reinforce low mobility and dispersion in patchy environments to produce small effective population sizes (Bush, Case *et al.* 1977).

Bush, Case *et al.* (1977) took the analysis of rates of chromosomal change further by including an assessment of **rates of speciation**. Given the hypothesis that mammals generally have high rates of karyotypic change because their patterns of social organization tend to subdivide their populations into small, inbred demes, it should follow that mammals also have higher rates of speciation than other vertebrates (fish, amphibians and reptiles). Successful inference of rates of speciation depends not only on reasonably accurate dating of the appearance of a genus but also on determination of the rate of extinction of species belonging to that genus. This again introduces the problem that the time of first appearance of a genus in the known fossil record cannot be simply equated with the real time of emergence of that genus in evolution. Nevertheless, by using a number of reasonable approximations, Bush, Case *et al.* were able to arrive at fairly acceptable estimates of corrected speciation rates (i.e. numbers of species per genus, including an estimate of the numbers of extinct species, generated per unit time). When inferred rates of karyotypic change were plotted against inferred rates of speciation, it was found that there was a highly significant correlation between the two (see also Lande, 1979b). For the 10 groups of mammals listed in Table 11.2, the correlation coefficient for the relationship between rate of chromosomal change and speciation rate is $r = 0.91$ ($P < 0.001$). There is, however, a statistical problem with this analysis because the correlation applies to two measurements of rates. Although there are some

differences in the actual mode of calculation, both karyotypic diversity and numbers of species per genus have been divided by the same time factor derived from the first appearance of each genus in the fossil record. If two sets of randomly paired numbers (A values and B values) are divided by another set of random numbers (t values), a spurious correlation will be found between A/t and B/t. Hence, one should really examine the strength of the relationship between the original data on karyotypic diversity and numbers of species per genus without dividing by time to infer rates. This has yet to be done, but it would in fact be unsurprising if there were some relationship between the number of species in a genus and the degree of karyotypic divergence among those species. Such a simple correlation would not require involvement of chromosomal change in the actual mechanism of speciation (see also Paterson, 1985).

Wilson, Bush *et al.* (1975) stated that there is no significant correlation between values for karyotypic diversity (m) and documented age of a genus (t) within a given vertebrate group. However, taking the data for different mammalian groups listed in Table 11.2, it emerges that there is a quite strong negative correlation between these two variables ($r = -0.70$; $N = 10$; $P < 0.025$). Taken at face value, this suggests that species belonging to genera first documented quite early in the fossil record show less karyotypic variation than species belonging to genera that appear more recently in the fossil record. Let us assume that practical recognition of genera across mammalian groups reflects a uniform level of morphological distinctiveness (a moot point indeed) and that rate of speciation is in some way related to rate of morphological change (see below). It would then be expected that long-lived but species-poor genera should show slower karyotypic change than short-lived but species-rich genera. This is clearly a point that deserves more detailed examination; but there is a fairly high probability that differential biases against fossilization between mammalian groups (e.g. in relation to body size and in relation to geographical occurrence) may lead to spurious differences in times of first occurrence of mammalian genera in the fossil record. There are also more subtle biases in that palaeontologists may more readily include fossil forms in certain extant genera (e.g. when those genera are currently species-poor or when reliable criteria for identification of fragmentary remains of fossil species are particularly difficult to set out).

Bush, Case *et al.* (1977) specifically mentioned various primates in support of their contention that rigidly structured social systems in mammals may produce small effective population sizes. For instance, they noted the high degree of karyotypic diversity in guenons (*Cercopithecus* spp.) and linked this to the prevailing tendency to live in harem groups. They contrasted the guenons with baboons, which show great karyotypic uniformity. However, although they state that baboons live in multi-male troops, some species (*Papio hamadryas* and *Theropithecus gelada*) live in troops subdivided into harem groups and bachelor-male groups and yet show no greater degree of karyotypic change than other baboons. Bush, Case *et al.* also noted that, whereas it is common to find several guenon species inhabiting the same forest area (i.e. living in sympatry), baboons generally show mutually exclusive geographic ranges with hybridization occurring on common boundaries. This could be taken as indicating that stasipatric speciation through chance establishment of chromosomal 'races' is particularly prevalent among guenons.

The overall hypothesis, that the social imposition of small deme sizes in mammals favours stasipatric speciation through the chance fixation of chromosomal rearrangements, is superficially appealing. There are, however, some problems in addition to those identified above. The first additional problem is that Wilson, Bush *et al.* (1975) and Bush, Case *et al.* (1977) have only established an apparent correlation between general social complexity, high rates of karyotypic change and high rates of speciation in mammals. As in many other contexts, one must be wary of concluding that a correlation necessarily indicates the existence of a causal relationship. This is particularly true in relation to the suggestion of stasipatric

speciation, because the authors concerned also suggest that there is some link between karyotypic diversity and morphological diversity. Although it might be thought that a high rate of speciation would necessarily be accompanied by a high rate of morphological change, there is no compelling reason why this should be so. Karyotypic changes in themselves may be selectively neutral and merely predispose towards speciation because of selection against heterozygotes. Therefore, it is theoretically perfectly possible for numerous species to arise from a common ancestor through karyotypic divergence with very little morphological change occurring. It has been noted above that chromosomal polymorphism within primate species has yet to be directly linked to morphological polymorphism of any kind. This can also apply across species. For instance, the gibbons show relatively little morphological diversity and yet, as has been noted above, they possess the most radically modified of all primate karyotypes and also show a great amount of within-group diversity. Because there is nevertheless an apparent overall correlation between karyotypic diversity and morphological diversity across primate species, there is an alternative possibility that these two features are separate byproducts of the process of evolutionary divergence. In other words, karyotypic change may be no more a 'cause' of speciation than is morphological change (see also Bickham, 1981).

The concept that social structuring in mammalian populations generates small inbred demes is also problematic in its own right. Although it is true that some primates, and some other mammals, tend to live in one-male breeding units, there is continuous reproductive interchange at each generation. Members of one sex (usually males) typically migrate out of the group in which they were born to become integrated in another group, perhaps moving to a number of groups in succession. Such migration characteristically occurs at or before the onset of sexual maturity, thus avoiding inbreeding by offspring of one sex in their natal group, and eventual entry into another group (often taking the form of a male takeover in harem-living primates) reduces the probability of inbreeding with maturing offspring of the other sex. The balance of available evidence suggests that localized inbreeding is not a consistent feature of natural mammalian populations and therefore cannot be invoked as a regular disposing factor in a model of stasipatric speciation. Hence there is no empirical justification for interpreting all mammalian social units as hermetically sealed demes. Further, Lande's calculations (1979b) reveal little real difference in deme size between mammals and lower vertebrates, even taking the figures of Bush, Case *et al.* (1977).

Bickham (1981) was able to test some of the ideas relating to chromosomal evolution on the basis of banded karyotypes for 48 turtle species belonging to eight different families. With due reference to the fossil record, he produced a parsimonious reconstruction of likely karyotypic changes in these turtles during their evolution over a period of at least 200 million years. His data in fact confirm the conclusion of Wilson, Bush *et al.* (1975) that overall karyotypic evolution has taken place more slowly in reptiles than in mammals. However, it emerged from Bickham's reconstruction that the average rate of chromosomal change seemed to be twice as high during the first half of the period covered (i.e. the period of initial adaptive radiation) than during the second half. Further, the predominant type of chromosomal change differed between the two periods. For the initial phase of evolution of the turtles, rearrangements such as centric fusions, pericentric inversions and translocations were identified, whereas for the second phase additions of heterochromatin predominated. Hence, the chromosomal banding data for turtles support the canalization model of karyotypic evolution rather than the proposal that rate of chromosomal change is related to the presence or absence of inbred demes.

The concept of acceleration of karyotypic evolution through the subdivision of a population into small inbred demes is also contradicted by the finding of Robbins, Moulton and Baker (1983) that there is a strong positive correlation between the extent of inferred chromosomal

change and overall geographical range for two rodent genera, *Peromyscus* and *Onychomys*. Bush, Case *et al.* (1977) explicitly linked their proposals regarding the rate of chromosomal evolution to the idea that speciation is most likely to occur in isolated peripheral subpopulations. Hence, one would be led to predict a negative correlation between inferred rate of karyotypic evolution and geographic range. Robbins *et al.* (1983) produced inferred chromosomal phylogenies for the genera *Peromyscus* and *Onychomys* based on banding studies in order to estimate the amount of karyological change that had taken place in each species. Apart from the fact that they found a positive correlation between numbers of inferred chromosomal changes and geographical range for the species within each genus, they also observed that chromosomal polymorphism (the raw material of chromosomal evolution) was most pronounced in the species with the largest geographical ranges. This is also contrary to the prediction that chromosomal evolution is likely to proceed most rapidly in small populations. Hence, Robbins *et al.* (1983), like Bickham (1981), concluded that chromosomal change is more likely to be linked with adaptive change in evolution rather than being a passive outcome of speciation. It should be noted, however, that no clear relationship between geographical range and degree of chromosomal change has been found for some other rodent genera (e.g. *Reithrodontomys*, *Oryzomys* and *Sigmodon*), so there is obviously some variation in patterns of chromosomal evolution from genus to genus. Hence, it would seem that there is no simple model of chromosomal change that will account for all of the observations made to date. As several authors have specifically invoked chromosomal change as a primary feature of rapid speciation within a model of 'punctuated equilibrium' (see Chapter 3), this conclusion throws further doubt on the general validity of that model as well.

GENETIC EVOLUTION

Comparisons between proteins on the basis of differences in amino acid sequences, as an aid to reconstruction of phylogenetic relationships, first became a practical possibility with the publication of sequence information for insulins (e.g. Brown, Sanger and Kitai, 1955). Once the amino acid sequence of a particular protein becomes known, it is possible to make some inferences about the nucleotide base sequences of the 'gene' (i.e. of a discrete region of DNA) coding for that protein, by extrapolating from the established genetic code (Fig. 11.4). In fact, techniques have recently become available for direct study and comparison of DNA sequences. It is the DNA that is the primary carrier of genetic information and therefore direct study of sequences of nucleotide bases provides the most straightforward information on evolutionary modification of the hereditary material. Point mutations replacing one nucleotide base by another in the DNA chain, of course, provide important raw material for evolutionary change. As some nucleotide base replacements do not lead to a change in coding for protein synthesis (Fig. 11.4; see later), determinations of amino acid sequences provide only a somewhat indirect perspective on the evolution of the hereditary material, and global comparisons of intact proteins obviously allow only very general inferences regarding evolutionary changes. However, it is appropriate to begin at the level of comparisons of intact proteins and to lead progressively towards the structure of the DNA itself, as this reflects the historical course of events. Two main classes of methods are used to compare intact proteins from different species – those based on their electrophoretic properties and those based on immunological properties.

Electrophoretic comparison of proteins

In **electrophoretic studies**, proteins are compared on the basis of their migration through an appropriate medium (e.g. a starch-gel) across which an electric current is passed. After a certain time has elapsed, the medium is stained with specific reagents to reveal the relative positions of individual proteins. The direction and rate of migration of a given protein through the gel will depend on a complex interaction

between the size of the protein molecule and its net electric charge (the balance between negative and positive charges). As the net electric charge of a protein depends on the pH of the surrounding medium, standardized buffered conditions must be used in any electrophoretic study. As can be imagined, this approach provides only a very crude basis for comparing corresponding proteins derived from different species, which may differ in net charge but will differ only marginally, if at all, in molecular weight. The value of electrophoretic studies is consequently confined to special areas, such as the demonstration and quantification of genetic polymorphism within species and allied calculations of **genetic 'distance'** between relatively closely related species.

An important finding from broad comparative electrophoretic studies, with respect to the interpretation of the genetic basis for evolutionary change, is that there is a considerably greater degree of genetic polymorphism in natural populations than was hitherto suspected (e.g. see Ayala, Tracy *et al.*, 1974). This has major theoretical implications (see later). Further, once the relative frequencies of alternative states (alleles) for several polymorphic genes in related species are known, it is possible to infer phylogenetic relationships among the species concerned. Calculations of genetic distance between populations can be made (Nei, 1972) and trees can be constructed on that basis. This approach has, for instance, been applied to hominoid primates (apes and humans). Bruce and Ayala (1979) concluded from a comparison of allele frequencies at 23 different gene loci in these primates that the gibbon and the siamang are definitely relatively distant from the great apes and humans, but that the relationship between the latter could not be clearly resolved. (In fact, there was a weak indication that the gorilla might have branched off first, with both the chimpanzee and the orang-utan being more closely related to *Homo*. This suggestion conflicts with a large body of morphological, chromosomal and biochemical evidence). The genetic distance between other relatively closely related primate species has been calculated in a similar way for a series of macaques (Weiss, Goodman *et al.*, 1973; Melnick and Kidd, 1985) and for the guenon group of Old World monkeys (Ruvolo, 1982; 1988).

The electrophoretic approach can be extended with two-stage separation; for example, with migration of the proteins in one direction achieved on filter paper and subsequent migration in a perpendicular direction achieved on a starch gel, to produce a two-dimensional scatter called a **fingerprint** (e.g. see Goodman, 1962a, 1963a). A study by Goldman, Giri and O'Brien (1987), using this more powerful technique of two-dimensional electrophoresis, has yielded a more reliable picture of relationships among hominoid primates. The gibbons and siamang were found to branch away first, followed by the orang-utan and then by the gorilla, with the chimpanzee being the closest relative of humans. This procedure can be carried further by digesting a given protein with a standard treatment and then examining the fingerprint pattern formed by the breakdown products (Zuckerkandl, 1963). Two-dimensional fingerprints of both kinds can provide valuable additional information on the similarities and differences between corresponding proteins (e.g. various serum enzymes) derived from different species. In spite of this refinement, electrophoretic studies have not permitted any overall reconstruction of evolutionary relationships among primates because the technique is best suited for comparisons between relatively closely related species.

Immunological comparison of proteins

Comparison of proteins on the basis of **immunological properties** has proved to be considerably more fruitful than the electrophoretic approach with respect to inference of phylogenetic relationships among primates in general. The technique relies on the fact that the immune system of a suitable host animal (e.g. a chicken or a rabbit) will produce **antibodies** in response to any foreign protein **antigen** that is introduced into the blood stream. These antibodies are relatively specific for the antigen that provoked their formation and can therefore be used to test the similarity of the original antigenic protein to corresponding proteins derived from other species. Once a host animal has developed an immunological response to an injected protein, it is possible to remove a sample of serum from the host and to use it for comparative studies. If a specific **antiserum** of this kind is mixed with a sample of the original protein under test conditions, it will bring about **precipitation** of that protein. If a corresponding protein from a different, fairly closely related, species is tested instead, precipitation will still occur, but the reaction will commonly be less intense because of differences in amino acid sequence between the test protein and the original protein. The strength of the cross-reaction between an antiserum to a given protein (e.g. serum albumin from a given species) and corresponding proteins (i.e. serum albumins from other species) can therefore be taken as an indirect measure of the degree of structural similarity between the corresponding proteins and the original protein. As yet, the details of the immunological response itself are still incompletely understood and hence the actual basis for cross-reactivity of an antiserum with a range of different proteins is uncertain. However, it is widely accepted that the immune system of the host organism responds to discrete antigenic sites (**determinants**) on the foreign protein in the blood stream and that a mixture of different antigenic substances responding to the individual sites is produced (Bauer, 1974). Atassi (1975) has, for example, identified five antigenic reactive regions, or determinants, on the myoglobin molecule from muscle.

Other things being equal, the levels of cross-reaction between an antiserum and a series of corresponding proteins from different species will therefore depend on the degree of matching between the determinant sites of the individual proteins. Presumably, the amino acid sequence of any particular determinant site is the primary feature governing the extent of reaction with a given antibody, although there is the prerequisite that the site must be located on the surface of the folded protein molecule to take part in such a reaction (Atassi, 1975). Indeed, there is evidence to suggest that all antigenic determinants of proteins may be discontinuous, resulting from the secondary juxtaposition of short combinations of amino acids through folding of protein chains (Barlow, Edwards and Thornton, 1986). Hence, immunological cross-reactions provide no more than an indirect means of sampling differences in amino acid sequences between corresponding proteins from a series of species. The main methods currently used assess the different degrees of precipitation of corresponding proteins from a series of species brought about by reaction with the antiserum under standardized conditions.

The technique of **comparative serology** as a tool for investigating phylogenetic relationships was pioneered by Nuttall (1904) and consolidated by Boyden (1953, 1958, 1959). Extensive application of the technique to the reconstruction of phylogenetic relationships among primates was initiated by Goodman (1961, 1962a, 1963a, 1966, 1967), whose research team has made the most comprehensive overall contribution to the subject now known as **molecular anthropology** (Zuckerkandl, 1963; Goodman and Tashian, 1976). Two other early papers on serological relationships among primates were those of Paluska and Korinek (1960) and of Hafleigh and Williams (1966). Since the mid-1960s, a considerable body of information on primates has also been published by Sarich and his colleagues (e.g. Cronin and Sarich, 1975, 1978a, 1980; Sarich, 1968, 1970a,b; Sarich and Cronin, 1976; Sarich and Wilson, 1966, 1967a,b; Wilson and Sarich, 1969). Altogether, this serological evidence now adds up to a valuable additional source of information for the reconstruction of phylogenetic relationships among primates.

Immunodiffusion

The procedure used by Goodman and his colleagues to assess the degree of cross-reactivity between an antiserum and corresponding proteins from a range of species is an **immunodiffusion technique** (see Goodman, 1962b; Goodman and Moore, 1971). In this procedure, an agar-gel plate (Ouchterlony, 1953) is used as a diffusing medium and the antiserum is tested simultaneously against two protein samples (e.g. the original protein sample and a corresponding protein sample from another species). In fact, Goodman and his colleagues used purified individual proteins only for humans. For individual non-human primate species, they used samples of unmodified serum to raise antisera in the host animals (chickens or rabbits). Hence, the antiserum used for a test of a non-human primate was a general one and it was applied to assess the cross-reactions of whole serum samples, rather than purified proteins, from the two species involved in any one test.

For the test, a central well is cut in a plate of agar gel to take the antiserum and two additional wells are cut equal distances away to take the two test samples of serum (see Fig. 11.20). Once the antiserum and the two serum samples have been introduced to the test plate, the system is left to develop for some time. Because different proteins migrate at different rates through the agar gel, a separate precipitin line is formed wherever a particular protein encounters an appropriate concentration of the antiserum. If a given protein in one serum sample differs from the corresponding protein in the serum sample from another species by one or more determinant sites involved in the immunological reaction, a **spur** will form on the otherwise cohesive precipitation front between the wells (Fig. 11.20b). The spur indicates a reaction of the antiserum with one or more determinant sites that are present on the protein derived from one of the wells but absent from the protein derived from the other well. The size of the spur reflects the overall degree of difference in reactive determinant sites between proteins and can be scored on a scale of 0–5. Because of certain complexities involved in comparisons of large numbers of species united by an asymmetrically branching phylogenetic tree, various adjustments are subsequently made to the raw data by Goodman and his colleagues to calculate **antigenic distances** between pairs of species (e.g. see Dene, Goodman and Prychodko, 1976a). A matrix of distances between all pairs of species in a comparison is then compiled and it is possible to apply a computer program to generate a best-fit divergence tree for the data. The unweighted pair-group method of Sokal and Michener (1958), originally developed for cluster analysis in numerical taxonomy, has been frequently used for this purpose (see also Sneath and Sokal, 1973).

Figure 11.21 provides an outline phylogenetic tree for living primates generated from data produced using the immunodiffusion procedure just described (Goodman, 1973; Dene *et al.*, 1976a). It can be seen at once that there is a good general agreement between that tree and an outline phylogenetic tree based on a preliminary assessment of morphological evidence (see Fig. 2.2). The inferred tree based on immunological evidence provides confirmation that there is a fundamental division of the primates into strepsirhines (lemurs and lorises) and haplorhines (tarsiers, monkeys, apes and humans). In the strepsirhine part of the tree, Malagasy lemurs and the loris group (bushbabies and lorises) can be recognized as separate monophyletic groups (Dene, Goodman and Prychodko, 1976b; Dene, Goodman and Prychodko, 1980). In the haplorhine part of the tree, the tarsier (as expected) branches off first (Goodman, Farris *et al.*, 1974), well before a common ancestral node for the simians (monkeys, apes and humans). Among the simians, New World monkeys form a clear-cut monophyletic group (Baba, Goodman *et al.*, 1975; Baba, Darga and Goodman, 1979), as do Old World monkeys, apes and humans (Goodman, 1968). Within the Old World simians, monkeys are clearly monophyletic (with a complete separation of leaf-monkeys from the remaining forms). In the hominoid sector of the tree, gibbons branch off first, followed by the orang-utan, and the African great apes (chimpanzee and gorilla) are shown to be our closest relatives.

There are, however, some unusual features of the tree in Fig. 11.21 that do not accord with traditional assessments of primate relationships. For instance, the ruffed lemur (*Varecia*) is shown as being closer than *Lemur catta* to *Lemur fulvus*, *L. macaco* and *L. mongoz*, contrary to the expectation that species allocated to the genus *Lemur* would form a monophyletic group with *Varecia* as the sister group. This latter expectation is confirmed by the chromosomal evidence (Fig. 11.18), so the antigenic distance

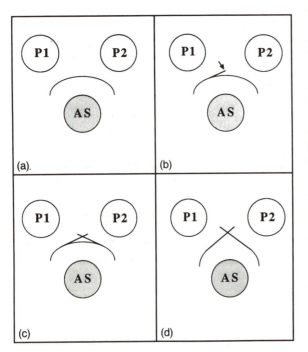

Figure 11.20. Diagrammatic representation of the comparison of proteins using the immunodiffusion technique. Antiserum (AS) to a particular protein (P) is placed in a central well in an agar-gel medium. Separate samples of protein from the original species or from other species are placed in the remaining two wells (P1, P2). Both the antiserum and the protein samples diffuse through the gel. Where the antiserum encounters the protein, a precipitin line is formed. (After Bauer, 1974.)

(a) If the proteins in the two wells are identical with respect to their determinants, a continuous precipitin line is formed.

(b) If protein P1 bears more determinants than protein P2, a spur (arrowed) will form on the line between P1 and AS.

(c) If P1 bears a determinant lacking in P2 and P2 bears a determinant lacking in P1, but the two proteins also share some common determinants, a complex pattern is obtained.

(d) If P1 and P2 bear no common determinants, a simple crossed pattern results.

results are definitely incompatible with other findings. Similarly, for relationships within the loris group Fig. 11.21 indicates that African lorises (*Arctocebus* and *Perodicticus*) are closer to bushbabies (*Galago* spp.) than to Asiatic lorises (*Loris, Nycticebus*). This runs counter to the usual interpretation that slow-moving lorises (subfamily Lorisinae) constitute a monophyletic group distinct from bushbabies (subfamily Galaginae). In this case, the chromosomal evidence (Fig. 11.18) clearly suggests that *Perodicticus* and *Nycticebus* share a considerable number of specific rearrangements sharply demarcating them from the bushbabies. Finally, although the relationships indicated on the haplorhine side of the tree broadly fit the pattern expected from assessments of the morphological evidence, there are some odd features in the branching pattern indicated for the New World monkeys. For example, *Aotus* is shown to be more closely related to the tamarin *Saguinus* and to Goeldi's monkey, *Callimico*, than to *Callicebus*, whereas there is much morphological evidence to suggest that *Aotus* and *Callicebus* are quite closely related to one another. Again, the chromosomal evidence (Fig. 11.18) supports the traditional interpretation based on morphological evidence, at least in ruling out any specific relationship between *Aotus* and marmosets and tamarins, and conflicts with the pattern inferred from immunological data.

It should be noted, however, that Dene *et al.* (1976a) provided an alternative evolutionary tree for primates based on antigenic distances determined with antisera raised in chickens. They found a general pattern of relationships very similar to that shown in Fig. 11.21, except that *Tarsius* was indicated as even more emphatically linked to simians. There were differences in branching detail, however, for two of the three groups singled out for comment above (i.e. for relationships within the loris group and among the New World monkeys). Unfortunately, the changes in detail do not resolve the problems, they merely produce patterns that are equally unexpected in comparison to morphological and chromosomal evidence. This, therefore, establishes two important points. First, differences in some branching patterns can be obtained by using antisera raised in different host animals (e.g. rabbit versus chicken). Second, it would appear that some parts of a phylogenetic tree based on immunological comparisons may be inherently unstable whereas certain main branching features remain relatively stable. It is also noteworthy that the tree based on results obtained with antisera raised in chickens indicates that *Gorilla* is closer to *Homo* than to *Pan*, whereas the tree derived from antisera raised in rabbits (Fig. 11.21) indicates that *Gorilla* is closer to *Pan* than to *Homo*.

Dene *et al.* (1976a) also compared primates immunologically with other mammals, although they did not provide an overall tree. They found that tree-shrews (Tupaiidae) and 'flying lemurs' (Dermoptera) show the smallest antigenic distances from primates, with rodents coming next. This work tends to confirm Goodman's early suggestion (1966) that immunological evidence provides support for a phylogenetic link between tree-shrews and primates. Other observations of Dene *et al.* were that the various groups of lipotyphlan insectivores show quite large antigenic distances from primates, as do bats and elephant-shrews (Macroscelididae). There is, in fact, one anomaly in the antigenic distances they reported, in that elephant-shrews are indicated as being more distant than marsupials from primates. As there can surely be no doubt that marsupials are the sister group of the entire set of placental mammals, including elephant-shrews, this demonstrates that the measure of antigenic distance cannot be entirely reliable as an indicator of phylogenetic affinity. Dene, Goodman and Prychodko (1978) and Dene, Goodman *et al.* (1980) have also reported specifically on relationships within the family Tupaiidae, based on immunological distances. Unfortunately, they did not have any material available for *Ptilocercus*, so they were unable to provide evidence concerning the relationship between the pen-tailed tree-shrew and other tree-shrews. However, their data for various representatives of the subfamily Tupaiinae confirm the general view that these tree-shrews

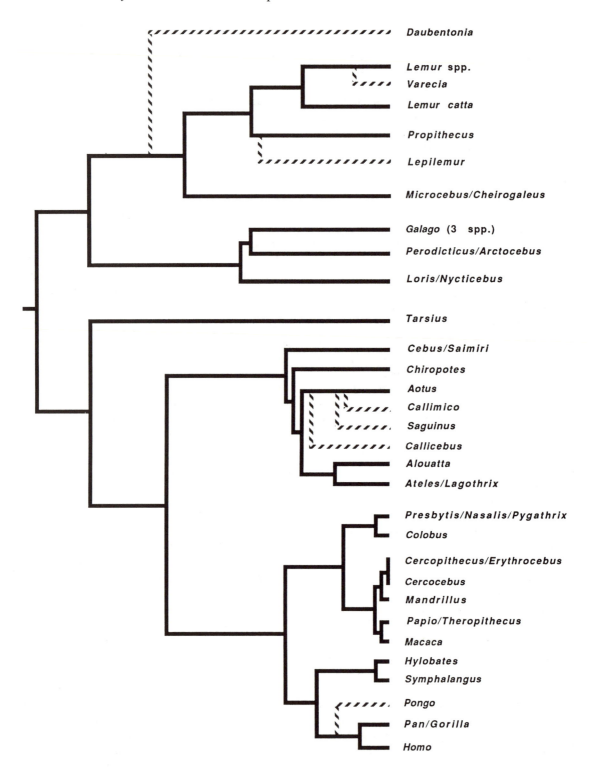

Daubentonia

Lemur spp.
Varecia
Lemur catta

Propithecus

Lepilemur

Microcebus/Cheirogaleus

Galago (3 spp.)

Perodicticus/Arctocebus

Loris/Nycticebus

Tarsius

Cebus/Saimiri

Chiropotes

Aotus

Callimico

Saguinus

Callicebus

Alouatta

Ateles/Lagothrix

Presbytis/Nasalis/Pygathrix

Colobus

Cercopithecus/Erythrocebus

Cercocebus

Mandrillus

Papio/Theropithecus

Macaca

Hylobates

Symphalangus

Pongo

Pan/Gorilla

Homo

form a fairly tight-knit group produced by a recent adaptive radiation. Among the Tupaiinae, the Philippine tree-shrew *Urogale* was found to branch off first, while the terrestrial tree-shrew *Lyonogale* was found to group with *Tupaia minor* and *T. montana*, leaving a residual group containing *Tupaia glis*, *T. belangeri* and *T. longipes*. The consequent separation of *Urogale* from *Lyonogale* conflicts with interpretations of dental evidence made by Steele (1973) and Butler (1980).

It is, in principle, possible to apply the immunodiffusion technique to individual, purified proteins. However, purification of individual proteins is a demanding and time-consuming procedure, so there have been very few studies of this kind for primates generally. An exception is provided by Bauer (1974), who raised antisera in rabbits to a series of proteins purified from human serum and used the immunodiffusion method to compare humans with other primate species. A minimum of 15 antisera to different serum proteins were tested against serum samples from three great apes (*Gorilla*, *Pan* and *Pongo*), four Old World monkeys (*Cercocebus*, *Cercopithecus*, *Erythrocebus* and *Macaca*), one New World monkey (*Cebus*) and one prosimian (*Galago*). As Bauer used purified proteins, he was able to infer from his results the number of antigenic determinant sites on each human protein molecule to which an antiserum had been raised. For 37 proteins examined, the number of determinant sites inferred per protein ranged from 1 (e.g. in cholinesterase) to 9 (in the γ_3-chain of gammaglobulin) and the average was 2.4. Using the inferred numbers of shared determinant sites as an indicator of phylogenetic relationship, Bauer generated an outline evolutionary tree for

primates showing the expected pattern. The bushbaby *Galago* branches off first followed by the New World monkey *Cebus* and then by the Old World monkeys in the sample, leaving a hominoid grouping (great apes and humans). Among the hominoids, the orang-utan branches off first, followed by the gorilla and the chimpanzee emerges as our closest relative.

Microcomplement fixation

An alternative approach to the immunodiffusion procedure is to assay the amount of precipitate produced by the reaction between standard amounts of an antiserum and of the antigenic protein concerned. The precipitate can, for instance, be assayed in a test vessel by virtue of its light-absorbing properties. Although it is a simple matter to standardize the quantity of antiserum used in a test reaction, it is impracticable to measure the amount of protein present in an unpurified serum sample. However, an alternative method of standardization is possible because of the form the dilution curve takes. If a sample of serum is progressively diluted and tested with the antiserum at each dilution level, a unimodal curve is consistently obtained. That is, there is always an optimal value for dilution of the serum, at which maximal precipitation occurs. Concentrations of serum higher or lower than this optimum will produce a reduced level of precipitation. Hence, dilution curves obtained with sera from several different species can be aligned on the basis of their peak values and dilutions can be expressed as fractions or multiples of the optimum level, taken as unity. Quantitative comparisons between species can then be based either on the peak precipitation

Figure 11.21. Outline reconstruction of phylogenetic relationships among primates based on immunodiffusion data obtained with antisera raised in rabbits. Heavy lines indicate species ($N = 29$) for which antisera were raised; hatched lines indicate species that were tested only against antisera raised for other species. The lengths of the branches provide an approximate indication of average antigenic distances. The common ancestral node represents an average antigenic distance from any living species of 10.3 units (see text for explanation). Tree partially condensed and adapted from Dene, Goodman and Prychodko, 1976a.)

levels themselves or on some other feature, such as the area enclosed between two specified dilution levels beneath each individual curve.

Using the approach outlined above, Hafleigh and Williams (1966) tested sera from various primates and from various other mammals including a tree-shrew (*Tupaia*), the hedgehog (*Erinaceus*) and the domestic pig, against an antiserum to human albumin raised in rabbits. The general pattern of relationships indicated was consistent with expectation. Cross-reactivity with the human antiserum decreased in the following order: great apes–Old World monkeys–New World monkeys–prosimians and tree-shrew–hedgehog–pig. In fact, the tree-shrew behaved like an average prosimian with respect to the level of the cross-reaction between its serum and the human antiserum. Because of this, Hafleigh and Williams concluded that their results could provide additional support for the then prevalent interpretation that tree-shrews are close relatives of primates. There were, however, some anomalies in their findings. In the first place, the comparisons indicated that some prosimian primates, notably sifakas (*Propithecus*), are more distant from humans than are tree-shrews. As there is scarcely any doubt that all prosimians, including *Propithecus*, must be phylogenetically closer than *Tupaia* to *Homo*, the recorded levels of antigenic cross-reaction cannot be taken as entirely reliable indicators of phylogenetic affinity. A similar anomaly was found with the owl monkey (*Aotus*), which appeared to be antigenically much closer to *Homo* than to any of the other New World monkeys examined and even to outrank a number of Old World monkeys in this respect. It seems most unlikely that *Aotus* can be as closely related to *Homo* as these results indicated, because there is overwhelming evidence that Old World simians are a monophyletic group, with New World monkeys constituting a well-defined sister group. Hafleigh and Williams suggest that the anomalous position of *Aotus* indicated in their tests may be due to the retention of certain primitive features in the albumin of *Aotus*. If this is true, however, the same argument could equally well apply to

Tupaia, which has a comparably anomalous position in relation to *Propithecus*. This possibility was explicitly recognized by Hafleigh and Williams.

At about the same time, Sarich and his colleagues also began to make quantitative immunological cross-comparisons on primates, assessing the degree of precipitation from the antigenic reaction with a technique known as **microcomplement fixation** (Sarich and Wilson, 1966). They began their work using pooled antisera raised in rabbits following injection of purified serum albumins from various primate species. At a later stage, they injected serum tranferrins in the same way and studied the antisera. The basic measure of antigenic difference between species was the **index of dissimilarity** (IND). This was determined in relation to a standard represented by a dilution curve for the interaction of fixed concentrations of the antiserum and the original protein used to raise the antiserum. With any sample of a corresponding protein from another species, the concentration of the antiserum was increased until a dilution curve was obtained with its peak matching the height of the peak of the curve obtained with the original protein. The factor by which the concentration of the antiserum had to be raised to achieve this result in each case yielded the IND value for that species. The number of amino acid sequence differences between proteins seems to be related to the logarithm of the IND value (Sarich and Wilson, 1967a; Sarich, 1968; Kirsch, 1969; Sarich and Cronin, 1976). Accordingly, the **immunological distance** (*ID*) between any two species was calculated from the following formula:

$$ID = \log \text{IND} \times 100$$

In practice, the *ID* values determined by Sarich's team ranged from 0 units (representing immunological identity) to 180 units (representing the maximum difference measurable between mammalian albumins using the technique of microcomplement fixation). Sarich (1970b) cited evidence showing that microcomplement fixation detects most amino acid differences in proteins of known sequence (e.g. human haemoglobins) and that there is a simple proportional relationship between empirical *ID* values and the number of amino acid differences between corresponding proteins (see also Sarich and Cronin, 1976). Hence, it was argued that *ID* values provide a convenient and reliable guide to differences in amino acid sequences between corresponding proteins from different species.

After numerous cross-comparisons, it is possible to establish a matrix of average *ID* values separating all pairs of species involved, just as with the antigenic distance values determined by the immunodiffusion technique used by Goodman and his colleagues (see above). This is illustrated in Fig. 11.22 for a hypothetical comparison of five species (A–E).

In case 1, the situation has been made as simple as possible by avoiding any discrepancies in immunological distances between species and by allowing no variation in rates of evolution per lineage. Given the matrix of *ID* values for pairs of species shown below the tree in case 1, it is possible to calculate branch lengths in the tree by solving sets of simultaneous equations (see also: Sarich, 1970b; Cook and Hewett-Emmett, 1974). For instance, just taking the three species A, B and C, it is possible to calculate the branch lengths a and b from the following equations derived from the matrix:

$$a + b = 20 \text{ units} \tag{1}$$
$$a + c + d = 40 \text{ units} \tag{2}$$
$$b + c + d = 40 \text{ units} \tag{3}$$

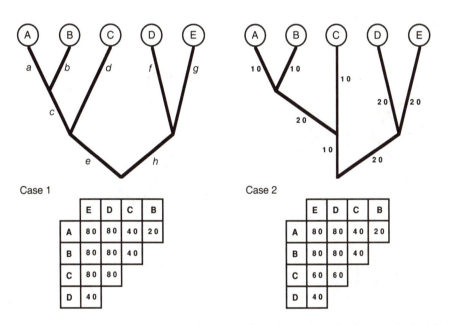

Figure 11.22. Diagrammatic explanation of the inference of branch lengths using immunological distances for a tree connecting five species (A–E).

Case 1: The matrix (below) indicates the immunological distances separating all pairs of species in the tree. Using a series of simultaneous equations, it is possible to calculate branch lengths a, b, c, d, f and g. The sum of the two basal branch lengths e and h can also be calculated, but cannot be reliably partitioned into two separate branch lengths without additional information from an outgroup (see text for details).

Case 2: As in case 1, the matrix indicates the immunological distances separating all pairs of species in the tree. However, in contrast to case 1, the lineage leading to species C is characterized by a conspicuously slow rate of evolution in case 2. Nevertheless, using the distance information provided in the matrix it is possible to infer the values of all branch lengths except for the two basal branches. As before, the sum of the two basal branch lengths cannot be partitioned into two separate branch lengths without additional information from an outgroup (see text for details).

From equations (2) and (3), it follows that:

$$a - b = 0 \text{ units} \qquad (4)$$

And from equations (1) and (4) it can then be determined that both of the branches a and b are 10 units in length. It can also be calculated from these equations that:

$$c + d = 30 \text{ units} \qquad (5)$$

However, it is impossible to ascribe separate values to the branches c and d without additional information in the form of some external reference point. In other words, the branch lengths for the relationship between species C and the two species A and B can be established only if the tree for A, B and C together can be rooted by consideration of at least one

additional, more distantly related species (i.e. an outgroup). In this instance, the species D and E together serve as an outgroup for this purpose. If the equations for all five species (A–E) in case 1 are solved, it is possible to infer the six branch lengths a, b, c, d, f and g. The combined total of the basal branch lengths $e + h$ can also be calculated, but in order to ascribe separate values to the two component branches it is necessary to root the overall tree with a further outgroup comparison. The application of the outgroup principle to achieve this end, by inclusion of an additional, more distantly related species F, is illustrated in Fig. 11.23. This example demonstrates the important point that in immunological tree-building, as in phylogenetic reconstruction on the basis of morphological data (see Chapter

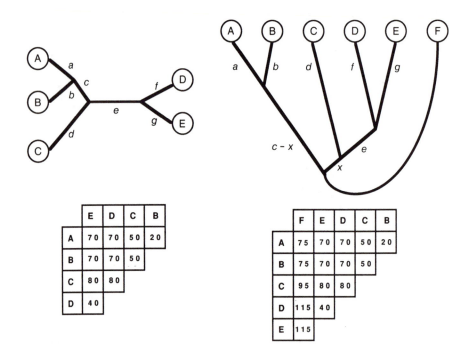

	E	D	C	B
A	70	70	50	20
B	70	70	50	
C	80	80		
D	40			

	F	E	D	C	B
A	75	70	70	50	20
B	75	70	70	50	
C	95	80	80		
D	115	40			
E	115				

Figure 11.23. The outgroup principle as applied to resolution of the basal branch lengths. In the absence of an external reference point, the tree cannot be reliably rooted and therefore must take the provisional unrooted form shown (top left). (Data matrix shown below left, as in Fig. 11.22.) By including immunological data for the more distantly related species F in the data matrix (below right), it is possible to root the tree as shown (above right). Contrary to the expectation that rates of evolution should be roughly constant, use of the outgroup F shows that species A and B have evolved rather slowly. The ancestral node is located on branch c, splitting it into two parts (x and c–x).

3), application of the outgroup principle is indispensable for the inference of ancestral nodes if rates of evolution are not constant.

As has been stated above, case 1 in Fig. 11.22 has been deliberately simplified by eliminating all of the discrepancies that are likely to arise with a real set of biological data. One such discrepancy is illustrated in case 2 of Fig. 11.22. Here, one species (C) has changed noticeably less than the others in comparison to the ancestral condition. Species A, B, D and E are all 40 units distant from the ancestral condition, whereas species C is only 20 units removed and has hence evolved at only half the rate overall. Nevertheless, given the matrix below the tree in case 2, it is possible to reconstruct the tree by solving simultaneous equations (except that, as before, the tree cannot be rooted reliably without additional information). In spite of the asymmetry in the data set brought about by the slow rate of change along the lineage leading to C, it is possible to infer the correct pattern of branch lengths among the five species A–E from the matrix. In other words, the fact that the lineage leading to C has been marked by a slow rate of evolutionary change can be determined by virtue of the internal consistency of the tree (see also Fig. 12.10). In any comparison with another species in the tree, C would show a smaller immunological distance than would be expected if evolutionary rates were constant.

Sarich (1970b) made much of the feature of internal consistency in making a case for the reliability of phylogenetic trees generated on the basis of immunological distances. One example that he gave involved a three-way comparison between *Homo*, slow lorises (*Nycticebus*) and Carnivora (the latter serving as an obvious outgroup in the comparison). Solution of the simultaneous equations for this comparison showed that the immunological distances of *Homo* and *Nycticebus* from their inferred common ancestor are distinctly different: the value for *Homo* (75 units) is more than twice that for *Nycticebus* (37 units). A similar, but less-extreme result was obtained with a three-way comparison between *Homo*, *Lemur* and Carnivora: *Homo* is separated from the ancestral

node shared with *Lemur* by an *ID* value of 76 units, whereas *Lemur* is separated only by 49 *ID* units. It was also found that *Nycticebus* and *Lemur* are separated by 86 units. The inference that these two prosimians have undergone slow evolution of their albumins in comparison to *Homo* therefore seemed to be neatly confirmed (but the figures conflict with the expectation from morphological comparisons that *Nycticebus* and *Lemur* should also be closer to one another than to *Homo* in terms of branching relationships). Sarich also found that the albumin of *Aotus* seems to have evolved very slowly in comparison to other New World monkey species examined. Hence, it is important to note that distinctly different rates of evolution of albumins, varying by an overall factor of about two, have been inferred from Sarich's immunological comparisons among primates and have in fact been specifically emphasized as demonstrating the principle of internal consistency within a tree generated from immunological distances. It should also be noted that all three primates found by Sarich (1970b) to be characterized by slow rates of evolution (*Nycticebus*, *Lemur* and *Aotus*) happen to be involved in suspect regions of the inferred primate tree (Fig. 11.21), based on the immunodiffusion data of Dene, Goodman and Prychodko (1976a), discussed above. Finally, Sarich's convincing demonstration of slow evolution in the albumin of *Aotus* indicates quite clearly that the anomalous results reported for the owl monkey by Hafleigh and Williams (1966) can be ascribed to shared retention of primitive features in *Aotus* and various Old World simian primates.

An overall phylogenetic tree for primates based on immunological distance data for serum albumins and transferrins has been produced by Sarich and Cronin (1976). In many respects, this tree (Fig. 11.24) resembles that of Dene *et al.* (1976a) derived from their general immunodiffusion studies of serum proteins (Fig. 11.21). However, there are some significant differences that deserve comment. Most importantly, although the strepsirhine primates (lemurs and lorises) and the simian primates (monkeys, apes and humans) emerge as two clear monophyletic

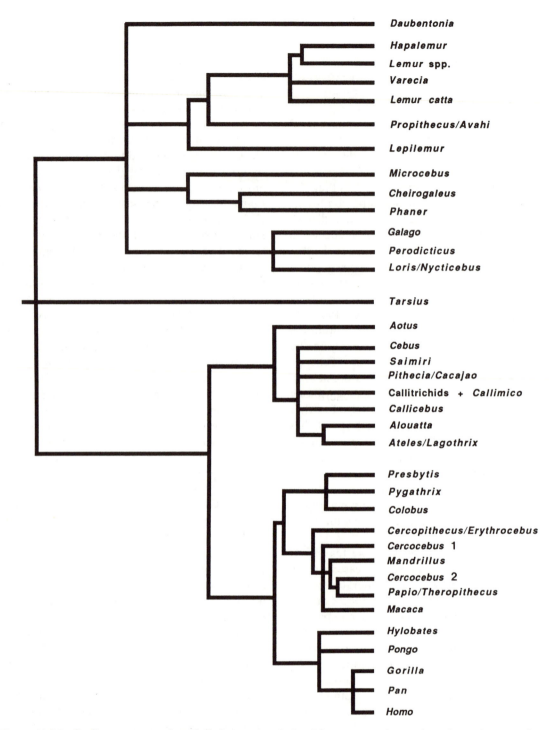

Figure 11.24. Outline reconstruction of phylogenetic relationships among primates based on microcomplement fixation data for albumins and transferrins obtained with antisera raised in rabbits (combined and adapted from Sarich and Cronin, 1976). The lengths of the branches provide an approximate indication of average immunological distances. The common ancestral node represents an average immunological distance from any living species of approximately 140 units. See text for explanation. Note: The part of the tree referring to the apes and humans is not based exclusively on immunological data but also includes information from DNA hybridization studies; see later.

groups, *Tarsius* is not specifically linked to the simians in the Sarich/Cronin tree. In other words, that tree provides no support for the hypothesis that *Tarsius* and simians are united in a monophyletic haplorhine group. There are also differences of detail in the branching patterns inferred for individual groups of primates. For instance, the Malagasy lemurs are not shown as clearly monophyletic. On the strepsirhine side of the tree, four separate lineages are seen to radiate from a common ancestral node, leading to: (1) the aye-aye (*Daubentonia*); (2) an assemblage containing the true lemurs (Lemuridae) and the indri group (Indriidae); (3) the mouse and dwarf lemurs (Cheirogaleidae); (4) the loris group (Lorisidae). However, Fig. 11.24 does agree with Fig. 11.21 in that there is no indication that the mouse and dwarf lemurs have any specific phylogenetic relationship with the loris group, contrary to suggestions based on the morphology of the ear region of the skull (see Chapter 7). Within the loris group, there is no suggestion of a specific link between *Galago* and *Perodicticus*, but, at the same time, lorisines (*Perodicticus*, *Loris* and *Nycticebus*) are not shown as clearly monophyletic, contrary to expectation. On the simian side of the tree in Fig. 11.24, the position of *Aotus* within the New World monkey radiation is shown to be quite different from that indicated in Fig. 11.21. Otherwise, the details of the radiation of New World monkeys are broadly similar between the two trees. Among the Old World simians, there are minor differences in the branching patterns indicated for baboons, macaques, drills and guenons, but the most striking discrepancy is that the tree provided by Sarich and Cronin (Fig. 11.24) indicates that the orang-utan is just as far removed from African apes and humans as are gibbons. This is certainly against expectation, since all of the morphological and chromosomal evidence clearly suggests that humans and all of the great apes (including the orang-utan) form a monophyletic group whose members did not begin to diverge until some time after the gibbons had branched off. Finally, it should be noted that the Sarich/Cronin tree indicates that *Gorilla*, *Pan* and *Homo* all diverged from a single common ancestral node.

Sarich and his colleagues have also produced evidence relating to the phylogenetic position of tree-shrews. Initially, Sarich (1970b) reported that the albumins of tree-shrews (*Tupaia* and *Urogale*) are anomalous with respect to their assessed immunological distances from the albumins of certain primates. Sarich stated: 'The problem is that *Tupaia* albumin is unreasonably more similar to that of *Homo* than can be logically defended on the basis of any phylogeny.' This, in fact, echoes the findings of Hafleigh and Williams (1966) discussed above. The explanation for this anomaly advanced by Sarich (1970b) was that considerable evolutionary change has occurred in the lineage leading to *Tupaia*, as in the evolutionary radiation of simian primates, and that the immunological distances involved therefore fall into a 'suspect range'. This explanation, which is diametrically opposed to that advanced by Hafleigh and Williams (i.e. retention of primitive features in tree-shrew albumin), raises the disturbing possibility that reconstruction of phylogenetic relationships on the basis of immunological distances suffers from a problem of resolving power and that early branching points in an inferred phylogenetic tree are accordingly likely to be subject to a large element of doubt. Nevertheless, Sarich went on to conclude: 'If the Lemuriformes, Lorisiformes, and *Tarsius* are primates, then phyletically *Tupaia* and *Urogale* are primates as well.'

Subsequently, Sarich and Cronin (1976) discussed the possible relationship between tree-shrews and primates in the context of comparisons of immunological distances determined for albumins and transferrins, including reference to certain other groups of placental mammals. They reported that a monophyletic group is formed by primates, tree-shrews and 'flying lemurs' (*Cynocephalus*), clearly demarcated from a composite outgroup consisting of bats, insectivores, edentates, carnivores and rodents. This conclusion was later underpinned by a very similar phylogenetic reconstruction based on a combined assessment of the immunological data and sequence data for haemoglobins (Cronin and Sarich, 1980). Hence, overall, Sarich and his group have concluded fairly firmly that immunological distance data support the hypothesis that tree-shrews have a specific phylogenetic relationship to primates. Cronin and Sarich also provide some limited information on relationships within the tree-shrew family (Tupaiidae) based on immunological distances. The three tree-shrew genera considered, *Ptilocercus*, *Tupaia* and *Urogale*, are all shown as fairly closely related together, confirming the morphological evidence indicating relatively recent divergence. *Ptilocercus* branches off before the division between *Tupaia* and *Urogale*, as is to be expected from the morphological distinctiveness of *Ptilocercus* in comparison to other tree-shrews.

General conclusions from immunological studies

The immunological evidence overall fits remarkably well with the inferences made from morphological characters and from chromosomal data, in spite of some variations in individual details of inferred phylogenetic trees. Although it is only natural to focus on those areas where the immunological evidence leads to differences in reconstruction, the general concordance in conclusions derived from such disparate sources of evidence should most definitely not be lost from sight. This concordance indicates that researchers in all fields are generally on approximately the right track when it comes to reconstructing the evolutionary history of primates.

The two overall reconstructions of phylogenetic relationships among primates discussed above (Figs 11.21 and 11.24), along with a number of less-comprehensive studies (e.g. Hafleigh and Williams, 1966), agree in several important respects. For example, the strepsirhine primates and the simian primates consistently emerge as monophyletic groups, although the inferred position of *Tarsius* is variable. Among strepsirhines, the loris group (family Lorisidae) is also consistently found to be monophyletic. Figures 11.21 and 11.24 further agree in identifying the three expected monophyletic groups of simian primates: (1) New World monkeys; (2) Old World monkeys; (3) apes and humans. As has been pointed out

above, the phylogenetic reconstruction produced by Dene, Goodman and Prychodko (1976a) is in some respects more concordant with the balance of inferences from morphological and chromosomal evidence than that produced by Sarich and Cronin (1976). In particular, the Dene *et al.* reconstruction (Fig. 11.21) shows the Malagasy lemurs as a monophyletic group and indicates a specific relationship between tarsiers and simians. It also shows the orang-utan as forming a monophyletic group with the African apes and humans. All of these features are lacking from the Sarich/Cronin reconstruction (Fig. 11.24). Throughout the studies cited, there are also some genera that are regularly found to produce odd results. The owl monkey (*Aotus*) is one obvious anomalous case and Sarich (1970b) demonstrated quite convincingly that confusing results reported for the immunological cross-reactions of this genus can be ascribed to an unusually slow rate of evolutionary change, at least in serum albumin. Similar problems are presented by tree-shrews such as *Tupaia*, which are commonly reported to be more similar to *Homo* than certain prosimians with respect to immunological cross-reactions, contrary to all of the evidence that tree-shrews must be an outgroup with respect to the primates as defined here. In spite of these anomalous results, it has been concluded both by Dene *et al.* (1976a) and by Cronin and Sarich (1980) that the immunological evidence suggests that tree-shrews seem to be allied to primates in some way. This conclusion must be treated with considerable caution until a convincing reason has been provided for the aberrant immunological distances determined for tree-shrews.

Thus far, the phylogenetic reconstructions generated on the basis of immunological difference (antigenic or immunological distances) have been accepted at face value. In principle, measurements of differences in immunological cross-reaction of a given type of protein from a series of different species have a definite advantage over classical morphological information or data on chromosomal rearrangements in that it is relatively easy to obtain quantitative data on a continuous scale. As differences in immunological response seem to be fairly closely related to changes in the amino acid sequence of the proteins tested (Sarich and Cronin, 1976), it would seem to be perfectly justifiable to take a simple quantitative assessment of immunological cross-reactivity as a measure of evolutionary distance. There are, however, numerous problems involved in such an approach and these must be borne in mind in examining any case where a phylogenetic reconstruction based on an assessment of immunological distances conflicts with a reconstruction based on one or more other kinds of evidence (e.g. morphological and/or chromosomal). A valuable discussion of the problems associated with interpretations of immunological comparisons is provided by Friday (1980).

The most severe limitation on the immunological approach is that it permits only an overall assessment of similarity and difference between proteins from different species. Similarity and difference cannot be broken down into individually identifiable character states and it is therefore impossible to draw a distinction between primitive, convergent and derived similarities (see also Friday, 1980). In other words, immunological comparisons of the kind described above are condemned to remain at the unsatisfactory level that was occupied by morphological comparisons before the need to identify shared derived features was clearly appreciated. Evidence on chromosomal re-arrangements, like morphological evidence, is more informative in that discrete changes can be recognized, permitting inferred distinctions between primitive, derived and convergent character states.

Sarich (1970b) in fact specifically claimed that data derived from proteins have a special advantage in that evolutionary change in protein structure is 'almost completely a divergent process with parallelism or convergence occurring at very near to statistical expectation'. But there are several immediately identifiable problems regarding possible convergence at the immunological level, which have generally not been seriously examined. In the first place, it is now known that convergence is rife at the level of the sequence of individual amino acids in a protein (see later). Such convergence would undoubtedly be reflected in the immunological cross-reactions of proteins. In addition to convergence occurring at the level of amino acid replacement in a protein, it is not beyond the bounds of possibility that different sequences of amino acids in a determinant site might in some cases show a similar degree of cross-reaction with an antibody, thus adding an additional source of convergence. However, this possibility pales into insignificance in comparison with the inevitable convergence that must occur in the loss of antigenic cross-reactivity through amino acid replacement. Given a particular amino acid sequence in a determinant site of a protein used to raise an antiserum, any evolutionary change at any amino acid position should lead to a reduction in cross-reactivity between the protein and the antiserum. In addition, repeated changes at the same amino acid position or at different amino acid positions within a determinant site will presumably pass undetected. Thus, numerous divergent evolutionary changes in the amino acid sequences of determinant sites will simply be lumped together as reflecting the same degree of immunological difference from the determinant of the protein used to raise the antiserum. The possible effects of these various forms of convergence in immunological comparisons of primate species have been largely ignored in the literature. Apparently, it has simply been assumed that any potentially confounding effects of convergence will be cancelled out by the tree-building approaches used. In view of the distortion caused by convergence at the level of morphological characters, this is undoubtedly a highly optimistic appraisal.

Equally importantly, however, techniques of phylogenetic reconstruction based on immuno-logical distance data take no account of the vital distinction between primitive retention and the development of derived features in specific ancestral stocks. Problems will predictably arise if there are differential rates of evolution (the existence of which at the immunological level has been reported above), because slowly evolving species will have reduced immunological distances from other species. Now it may be thought that this problem is easily eliminated because of the principle of 'internal consistency' discussed above (see Fig. 11.22, case 2). However, the discussion of that principle was set out in the context of a phylogenetic tree in which the branching pattern was taken as known at the outset and in which there was a perfect fit between the data matrix and inferred branch lengths. The problem is that the matrix of immunological distances between pairs of species must be used both for the inference of the branching pattern of the tree and for the calculation of branch lengths. There is, of course, a complex interaction between these two aspects. If the correct branching pattern is inferred, slow evolution in a given lineage should become obvious when simultaneous equations are solved in order to calculate branch lengths; but if the branching pattern is not correctly inferred, there will be a concomitant alteration in inferred rates of evolution. It has already been shown (Fig. 3.14) that, in at least some cases where evolution of discrete morphological characters is concerned, differences in rates of evolution between lineages can lead directly to erroneous reconstruction of the pattern of branching relationships.

In the case of tree-building based on immunological distances, any departure from a perfectly balanced matrix of values for pairs of species will raise serious problems of interpretation. The hypothetical examples provided in Figs 11.22 and 11.23 have been kept very simple, dealing with only five or six species. They also deviate from any real biological situation in that there is a perfect fit between the matrix and the inferred branch lengths. Given such a perfect fit, it is likely that the correct branching pattern, and therefore the correct branch lengths, would be inferred. However, with a matrix of real biological data deviations from a perfect fit are common. Part of the variation seems to be relatively random and may be ascribed to unavoidable errors in the techniques used to measure immunological distance, but certain systematic deviations are undoubtedly due to the effects of convergence combined with the mosaic retention of primitive features. Whatever the case may be, even relatively small deviations from a perfect fit in a given data matrix lead to practical problems both in reconstruction of the phylogenetic tree and in calculation of branch lengths. As with the analysis of discrete morphological characters, the severity of these problems can be mitigated, but not entirely eliminated, by taking a suitable outgroup and by increasing the numbers of species and proteins compared.

It is significant, therefore, that the question of the relationship between tree-shrews and primates suggested by immunological comparisons has not been explicitly discussed in relation to marsupials, which constitute the appropriate outgroup in this case. Cronin and Sarich (1980) noted the possibility that tree-shrew albumins might resemble primate albumins because of slow evolutionary change in the former, but they 'tested' (and rejected) this possibility by using bats as an outgroup. This is not an acceptable procedure, because the use of bats as an outgroup implies that it was already known that tree-shrews must be more closely related to primates than to bats. Similarly, although Dene, Goodman and Prychodko (1976a) do refer to immunological distances separating primates from various mammalian groups, including marsupials and monotremes, they do not provide information on the immunological distance between tree-shrews and marsupials and do not explicitly assess the position of tree-shrews in relation to marsupials as an outgroup. Whereas membership of the order Primates, in terms of the array of living species recognized in this volume, can be accepted with some confidence, relationships among orders of placental mammals, including the tree-shrew order Scandentia, can only be inferred by using marsupials as the outgroup. Indeed, given the occasional suggestions in the literature that the strepsirhine primates (lemurs and lorises) may not be directly related to the haplorhine primates (tarsiers and simians), it would be wise to include marsupials as an outgroup even in considerations of branching relationships among primates.

Another special problem involved in immunological approaches to the reconstruction of the evolutionary relationships of primates arises from the widespread practice of using a eutherian mammal (the rabbit) for raising antisera. The production of antibodies to a foreign protein by the rabbit's immune system depends on differences in determinant sites between the rabbit's own version of that protein and the injected version. Hence, rabbit antibodies to human albumin, for example, will be developed in response to differences in amino acid sequence, relative to rabbit albumin, occurring in exposed determinant sites of the albumin molecule. Testing of that antiserum against albumins from other primate species will then provide an assessment of the degree to which they share the same set of differences in amino acid positions compared to the rabbit albumin.

This immunological procedure may be sound in so far as comparisons between primate species are concerned. However, when it comes to comparing primates with other mammals of uncertain affinities (e.g. tree-shrews), the problem of outgroup relationships arises in a different form. Differences between rabbit albumin and, say, human albumin can have two distinct sources. One set of differences will have arisen because human albumin has changed relative to the albumin of ancestral placental mammals and the other set will have arisen because rabbit albumin has changed from the ancestral placental condition. Thus, any mammal that has retained the amino acids at positions that have undergone change in the rabbit but not in human will appear immunologically closer to humans and to any other mammals that have remained primitive in this respect. Whenever rabbits are used as the source of antisera, immunological distances between other mammalian species are in effect measured relative to the difference between the rabbit protein and the original protein injected as an antigen, rather than relative to the difference between the ancestral placental proteins and those of modern mammals. As it has been suggested both on morphological grounds (Martin, 1967) and on the basis of protein sequences (e.g. see Fig. 11.31 below) that rabbits and tree-shrews may be

phylogenetically related, there are further complications when antisera for immunological cross-comparisons are raised in rabbits to assess the phylogenetic relationships of tree-shrews.

Given the problems associated with the use of antisera raised in rabbits for the inference of phylogenetic relationships among placental mammals, it is particularly significant that Dene *et al.* (1976a), in contrast to Sarich and Cronin (1976), undertook parallel investigations using antisera raised in chickens. A second inferred phylogenetic tree for primates based on the results obtained with chicken antisera showed the same main monophyletic groups as shown in Fig. 11.21, which is based on data derived from antisera raised in rabbits. This correspondence indicates that the reconstruction shown in Fig. 11.21 may be more reliable than that of Sarich and Cronin (Fig. 11.24), which differs from both of the trees published by Dene *et al.* (1976a) in several important features, notably in not recognizing the Malagasy lemurs as a monophyletic group and in not linking tarsiers to simians.

A quite separate issue concerning the reliability of phylogenetic reconstructions based on immunological differences is that of reciprocity. The principle involved here is quite simple. If an antiserum to a specific protein of species A is raised in a suitable host animal and cross-reacts to a given degree with the corresponding protein of species B, then essentially the same result should be obtained if an antiserum is raised to the protein of species B and then tested against the protein of species A. More generally, it should also follow that inferred relationships between A, B and a series of other species should remain the same regardless of which species provides the protein for production of an antiserum. Hence, examination of the results obtained with different antisera provides a valuable means of testing the reliability of immunological differences as indicators of phylogenetic relationship. Ideally, if *N* species are involved in a comparison, separate antisera should be raised to the given protein (or whole serum) from every species and then systematically tested against the corresponding proteins from the remaining $N - 1$ species. In practice, however, this is very time-consuming (particularly if individual proteins must be purified to raise antisera) and reciprocity has usually been assessed on a piecemeal basis, if at all. Hafleigh and Williams (1966), for instance, reported results exclusively based on an antiserum to human albumin and Bauer (1974) has only used antisera to human proteins in his directed analyses of antigenic determinants. Reciprocal tests have, however, figured prominently in the work of both Goodman and colleagues (see Dene *et al.* 1976a) and of Sarich and colleagues (see Sarich and Cronin, 1976).

With respect to reciprocity, there is an important difference between studies involving immunodiffusion tests (e.g. Goodman, 1962a, 1963a; Bauer, 1974; Dene *et al.*, 1976a) and studies based on measurement of precipitation (e.g. Hafleigh and Williams, 1966; Sarich, 1970b; Sarich and Cronin, 1976). With the immuno-diffusion approach based on Ouchterlony plates (Fig. 11.20), it is possible to assess antigenic differences between all species by conducting successive tests on all possible pairs of serum samples. If an antiserum has been raised to the serum of species A, it is possible to make a direct assessment of the immunological difference between species B and species C by placing antiserum in the central well and then placing serum samples from B and C in the other two wells. There is the problem, though, that resolution will be poor for pairs of species that only show a weak cross-reaction with the antiserum, so reciprocal tests using a range of different antisera are required for this reason. By contrast, with any approach that measures the amount of precipitate formed in the antigenic reaction (e.g. the microcomplement fixation test of Sarich and Wilson, 1966) it is only possible to measure immunological distances with reference to a species for which an antiserum has been produced. Accordingly, a complete set of reciprocal tests is really required if a complete data matrix showing distances between all pairs of species is to be generated.

As a general rule, with reciprocal studies on immunological cross-reactivity between primate species, broadly comparable results have been obtained. It can therefore be stated that immunological distance data do seem to be relatively consistent. Nevertheless, reciprocal tests produce some differences and the inferred branching pattern for a given set of species will therefore vary according to the perspective that is used (i.e. according to which species is used to provide the original protein for raising an antiserum). It has also been noted that problems with reciprocity may arise with individual antisera (A. Friday, personal communication). The general approach to such differences between immunological distances measured in reciprocal tests has been to average the results to produce a compromise tree. Although this commonsense solution clearly has some merits, it should not be forgotten that the differences between reciprocal tests are of interest in their own right and could throw some light on the distinction between primitive, convergent and derived conditions in individual determinant sites. Present procedures do not explicitly take account of this three-fold distinction and averaging of results cannot be expected to provide a satisfactory solution to the fundamental problems posed by this distinction in any attempt to carry out phylogenetic reconstruction.

Amino acid sequences

In order to pursue the distinction between primitive, convergent and derived features of proteins, it is necessary to turn now to data on actual amino acid sequences. Such data clearly provide a very promising basis for the reconstruction of phylogenetic relationships (Zuckerkandl and Pauling, 1965). But it was until relatively recently an extremely laborious process to carry out the sequencing of individual proteins by stepwise chemical degradation. Accordingly, the data necessary for reconstruction of phylogenetic relationships on this basis have become available only quite gradually. Further, for many species data were obtained only on the **composition** of a protein in terms of the numbers of various kinds of amino acids incorporated in each fragment (peptide) and the **sequence** of those amino acids was inferred by comparing each peptide with its counterpart in a closely related species. Some 'sequences' that were used in initial tree-building attempts were therefore very provisional. However, now that the process of protein sequencing has become extensively automated (e.g. with the intro-duction of the sequenator), the pace of production of new, reliable data on amino acid sequences of proteins has increased markedly and it is now possible to carry out comparisons for several proteins across a quite impressive range of mammalian species, including a substantial proportion of primates.

Zuckerkandl (1963), who coined the term **molecular anthropology**, rightly predicted that data on protein structure and inferences regarding corresponding changes in the nucleotide sequences of DNA would come to play an important part in reconstructions of primate evolution. The first major steps towards the use of protein sequence data for the inference of phylogenetic relationships were made with the respiratory enzyme cytochrome *c* (Fitch and Margoliash, 1967; Margoliash and Fitch, 1968; Margoliash, Fitch and Dickerson, 1968) and with the major haemoglobin component of red blood cells (Zuckerkandl, 1965). Cytochrome *c* has clearly evolved relatively slowly compared to many other proteins and it therefore provides comparatively little evidence that is relevant to the evolutionary radiation of the primates, but studies of haemoglobins have been quite fruitful in this respect. Additional valuable information has been provided by studies of other proteins including minor haemoglobin components, myo-globin, fibrinopeptides, and eye lens proteins (notably α-crystallin A).

The amino acid sequences of a given protein from a range of different species can be directly compared. Given a protein containing a standard number of amino acids (e.g. individual haemo-globin chains contain about 140 amino acids), it is possible to treat each amino acid position as a character and to regard amino acid substitutions at any one position as alternative character states. Differences between species can

then be expressed in terms of **protein sequence distances**, and there is the considerable advantage that units of apparently equal value are used to quantify such distances (i.e. single amino acid substitutions at individual sites). There is one snag, however, in that proteins may occasionally evolve by processes other than substitution, with the addition or removal of amino acids occurring in some cases. As a general rule, though, this does not greatly affect protein sequence data used for reconstructions of primate evolution. In cases where proteins of different lengths are compared between species, it is usually relatively easy to infer from close similarity between flanking sequences where additions or removals have occurred, but this does introduce an element of circularity into the procedure of interspecific comparison.

Obviously, comparisons between species must be restricted to those regions of the protein that have been conserved at the same length. In fact, even after prior adjustments, direct comparisons of amino acid sequences are subject to reservations because not all changes at the level of the DNA sequence of the gene coding for the protein are reflected in changes in the amino acid sequence. This is because of the **redundancy** of the genetic code. As can be seen from Fig. 11.4, all amino acids apart from trytophan (Trp) correspond to more than one triplet of nucleotide bases. In some cases, as many as six different triplets code for the same amino acid. Hence, a substantial proportion of point mutations leading to nucleotide base substitutions, particularly those affecting bases in the third position of triplets, are 'silent' in the sense that they do not lead to amino acid substitutions in the protein produced. About 25% of nucleotide base substitutions are 'silent' in this way. In addition, if inferences are made about nucleotide base sequences, it may be found that more than one point mutation is required to account for a particular observed amino acid difference at a given site. Considerably greater accuracy can therefore be achieved by making inferences about the corresponding nucleotide base triplets and changes in them that are likely to account for a given pattern of amino acid sequence differences between the species in a comparison. This involves application of the principle of **parsimony** (see Chapter 3), in that the most direct route for transition from one amino acid to another is selected in proposing corresponding nucleotide base sequences.

One approach to the inference of phylogenetic relationships on the basis of amino acid sequences for a given protein is to examine every pair of species and to infer the smallest number of point mutations in the nucleotide base sequences required to account for the transitions between them. In this way, it is possible to calculate **minimum mutational distances** between pairs of species as a preliminary step towards reconstructing phylogenetic relationships. As with the distance information obtained from immunological studies, mutational distances are first used to compile a matrix showing the distances separating the members of all possible pairs of species (Fitch and Margoliash, 1967; Barnabas, Goodman and Moore, 1971, 1972). It is then possible to begin the process of tree-building using the method mentioned above with respect to the interpretation of immunological distances (see Cook and Hewett-Emmett, 1974). The standard approach to this is to begin with two species (A and B) that are separated by the shortest recorded mutational distance. All other species are lumped together as set C. Given the distance between A and B and the average of the distances separating A from C and B from C, it is possible to solve the resulting simultaneous equations, as explained above, to determine the three branch lengths of the unrooted tree joining A, B and C. This procedure can then be repeated in a stepwise fashion to generate an unrooted tree linking all species in the comparison.

As has been noted above in relation to tree-building using immunological distances, it is unlikely with a real set of biological data that there will be a perfect match between the matrix and an inferred tree, but a close approximation should be achieved. It is possible to assess the goodness of fit between the tree and the original matrix by deriving a reconstructed matrix from the branch lengths indicated on the tree and then

examining the degree of matching between the two matrices (Cook and Hewett-Emmett, 1974). In fact, it is unlikely that an optimal tree will be obtained directly by rigidly selecting each successive step on the basis of increasing distance between pairs of species. A major complicating influence would result from differential rates of evolution leading to small distances between some pairs of species because of retention of relatively primitive nucleotide base sequences, rather than because of the shared possession of derived changes from the ancestral sequence. Cluster analysis is in fact implicitly based on the assumption that there are even rates of evolution (Cook and Hewett-Emmett, 1974). The process of tree-building from a given matrix must therefore be repeated, slightly relaxing the criteria for successive selection of pairs of taxa, until no further improvement in the goodness of fit between the original matrix and the reconstructed matrix is achieved. Additional improvements can usually be obtained by randomly 'pruning' one species at a time from the tree and reinserting it at a series of different places on a trial-and-error basis, using a branch-swapping computer routine (e.g. see Goodman, Moore and Matsuda, 1975). Using these refinements, it is often possible to arrive at more parsimonious solutions than are permitted by straightforward cluster analysis.

Obviously, if there are major differences in the rates of evolution along individual lineages, this procedure for reconstructing phylogenetic relationships will remain suspect even after this process of refinement. As has been pointed out in Chapter 3 (Fig. 3.13), the number of possible trees increases alarmingly as the number of species is increased, such that it is impracticable for a computer to assess every tree except when very few species are involved in a comparison. It is for this reason that indirect methods must be used to approximate the most parsimonious tree. Hence, the investigator is left with a double problem: it is difficult to be sure that the most parsimonious tree has been found (the computer may become trapped in a localized 'rut' in terms of minimizing total genetic distance) and, if there is marked variation in evolutionary rates, application of the principle of parsimony may itself be in question.

The procedure of tree construction from a matrix of distances between species also suffers from some other limitations. Cook and Hewett-Emmett (1974) noted a 'major anomaly' in some early phylogenetic reconstructions based on inferred mutation distances, in that negative branch lengths were included. It is easy to understand how negative values for branch lengths might arise as a mathematical byproduct of the solution of a set of simultaneous equations, but obviously such negative values have no biological significance. Indeed, Cavalli-Sforza and Edwards (1967) suggest that generation of negative branch lengths provides a useful pointer to some incongruity in the initial clustering of species. Once recognized, the problem of negative branch lengths was eliminated with the aid of more sophisticated tree-building programmes. There is a remaining problem however, in that a bias is exerted by richly branching areas of the tree. For a symmetrically branching tree, the method is quite suitable, but with a phylogenetic tree showing a mosaic of species-rich and species-poor sectors (as is common in the real biological world) any tree-building method based simply on distances will produce questionable results because of 'hidden' changes in a long, undivided lineage. Successive substitutions may take place at a given locus, only the last being observable in the descendant species (see below). To counter this problem, it is possible to use a weighted method of cluster analysis (e.g. Fitch, Margoliash and Gould, 1969); but this approach suffers from other disadvantages. In particular, the lengths assigned by weighted cluster analysis to individual branches for a given tree will vary according to the sequence in which the branch lengths are determined, whereas with the unweighted method each branch length can take only one stable value (see Cook and Hewett-Emmett, 1974).

An alternative strategy for tree-building on the basis of protein sequence data is to take one amino acid site at a time. By inferring corresponding nucleotide base sequences, it is possible to identify the most parsimonious evolutionary pathway for that site across all of the species included in the comparison (Eck and Dayhoff, 1966; Dayhoff, 1969). As has been observed by Cook and Hewett-Emmett (1974), such **minimum path methods** are conceptually closer to the approach taken with morphological characters, in that each amino acid site is treated as a separate character. Further, it is at least theoretically possible to distinguish between primitive, derived and convergent character states with respect to the different nucleotide base sequences corresponding to alternative amino acids found at a given site in a series of species. Hence, one can apply precisely the same logic as is applied in the analysis of morphological character states (see Chapter 3). Unfortunately, the minimum path method has been relatively little used and the concept of minimum mutation distance has so far predominated in phylogenetic interpretations of protein sequence data for primates. Reconstructions based on character states rather than mere distances make for better use of the available information and distance should really only be used when there is no alternative, as in immunological comparisons.

Haemoglobins

Goodman (1975) provided a preliminary survey of phylogenetic trees generated from sequence data for individual proteins using a 'maximum parsimony' method based on a formulation of the matrix procedure described above (Moore, Barnabas and Goodman, 1973). At that time, partly through the work of Goodman's own research team, particular attention had been paid to **haemoglobins** in the investigation of evolutionary relationships among mammals. The haemoglobin component of mammalian red blood cells is a compound molecule incorporating four protein chains and an iron-containing haem group. In primates, the major haemoglobin component incorporates two α-haemoglobin chains and two β-haemoglobin chains ($\alpha_2\beta_2$)

surrounding the haem group. One or both types of haemoglobin chain have been sequenced in about 70 different mammalian species, including a few marsupials and monotremes (Goodman, Romero-Herrera *et al.*, 1982).

The primate region of a 'maximum parsimony' tree based on amino acid sequence data for β-haemoglobin chains is illustrated in Fig. 11.25. In this tree, a correction is indicated for nucleotide base replacements that are likely to

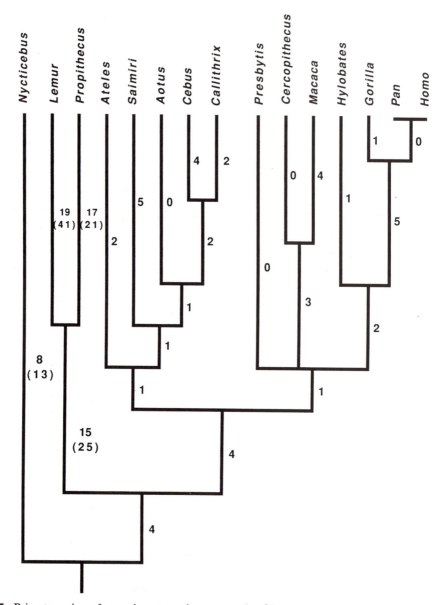

Figure 11.25. Primate region of a maximum parsimony tree for β-haemoglobin. The number to the right of each branch indicates the number of inferred nucleotide replacements allocated to that branch. Numbers in parentheses indicate augmented numbers taking account of 'hidden' nucleotide replacements occurring along species-poor lineages (see text). (Adapted extract from Goodman, 1975.)

remain undetected in lineages with relatively few branches because of superimposition of successive changes at a given site. As has been noted above, species-poor lineages can lead to bias in the application of the matrix method to tree-building. However, it is possible to estimate the likelihood that 'hidden' point mutations of a given nucleotide base have taken place in a long, unbranching lineage and to make an appropriate correction (Holmquist, 1972; Goodman, Moore *et al.*, 1974). Augmented numbers of inferred nucleotide base changes are accordingly indicated in parentheses for some lineages in Fig. 11.25. That figure, extracted from a larger tree given by Goodman (1975), indicates that the primates shown constitute a monophyletic group relative to the rabbit, various ungulates (hoofed mammals), the mouse and the dog. Within the primate cluster shown in Fig. 11.25, it can be seen that simian primates are a monophyletic group. By contrast, the slow loris (*Nycticebus*) does not form a monophyletic group with the two lemurs (*Lemur* and *Propithecus*), but instead branches off before the split between lemurs and simians. Among the simians, the New World monkeys and Old World simians (monkeys, apes and humans) emerge as two separate monophyletic groups, as expected. Within the New World monkey cluster, an unusual feature is that the marmoset (*Callithrix*) is illustrated as being specifically related to *Cebus*. The Old World simians show approximately the expected pattern of branching. However, the langur (*Presbytis*), which has apparently retained a relatively primitive β-globin chain, is not shown as having a monophyletic relationship with the guenon (*Cercopithecus*) and the macaque (*Macaca*). The gibbon (*Hylobates*) is shown as branching off between the Old World monkeys and the African apes and humans, while *Gorilla* is shown as slightly more divergent from *Homo* than the chimpanzee (*Pan*). In fact, it is noteworthy that the human β-haemoglobin chain is identical to that of the chimpanzee.

A similar tree, based on amino acid sequence data for α-haemoglobins, is shown in Fig. 11.26, with augmented figures for species-poor lineages indicated in parentheses, as before. This tree has the advantage that it incorporates data on a tree-shrew (*Tupaia*) and shows an inferred pattern of phylogenetic relationships connecting primates, tree-shrews and other mammals (see also Goodman, Barnabas *et al.*, 1971). Once again, primates are revealed as a monophyletic group in relation to the other mammals tested, and simian primates are shown as a monophyletic assemblage within the primate group. Yet again, however, strepsirhine primates (lemurs and lorises) are not shown to be monophyletic. In this case, it is the lemurs (*Lemur* and *Propithecus*) that branch off first, while the bushbaby (*Galago*) and the slow loris (*Nycticebus*) are depicted as being more closely linked to simian primates. Branching relationships for the relatively few simian primates included in the tree generally conform to the expected pattern. The two New World monkeys (*Ateles* and *Cebus*) form a group separate from the Old World simians. However, as with the β-haemoglobin tree (Fig. 11.25), the langur (*Presbytis*) is unusual in not forming a distinct monophyletic group with the other Old World monkeys (*Cercopithecus* and *Macaca*). Finally, it should be noted that the gorilla is somewhat divergent from *Pan* and *Homo*, which have identical amino acid sequences for α-haemoglobin, as for β-haemoglobin.

It can also be seen from Fig. 11.26 that *Tupaia* does not show a specific relationship to primates in terms of the amino acid sequence of its α-haemoglobin. In fact, the mouse emerges as the closest inferred relative of *Tupaia*, which also appears to be more closely related to the dog than to primates.

An important feature of both haemoglobin trees is that there is considerable variation in the inferred numbers of nucleotide base changes in different lineages derived from a common node, indicating marked variation in rates of fixation of point mutations. For instance, in Fig. 11.25 (β-haemoglobin) the total number of inferred nucleotide base changes accumulated by each primate varies from 9 in *Presbytis* to either 38 (taking the initial value) or 70 (taking the augmented value) in *Lemur*. In Fig. 11.26 (α-haemoglobin), the total number of inferred nucleotide base changes accumulated by each primate after divergence from the ancestral primate node varies from 8 in *Nycticebus*, *Pan* and *Homo* to 19 (taking the initial value) or 27 (taking the augmented value) in *Propithecus*. In *Tupaia*, there is also an accumulation of a very large number of inferred nucleotide base substitutions; it shows the maximum number (augmented value = 38) for any descendant from the ancestral node shared by primates, *Tupaia*, dog and mouse.

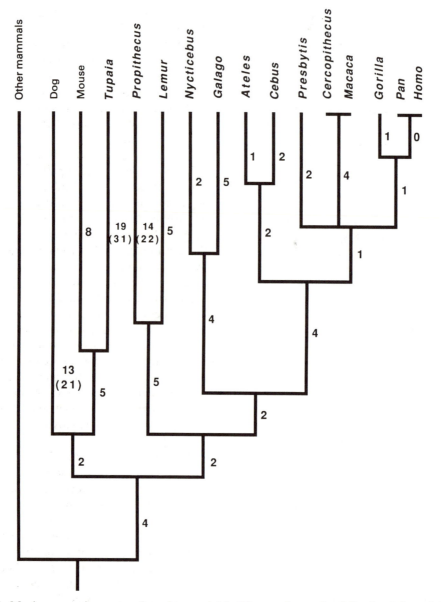

Figure 11.26. Maximum parsimony tree for α-haemoglobin. The number to the right of each branch indicates the number of inferred nucleotide replacements allocated to that branch. Numbers in parentheses indicate augmented numbers. The category 'other mammals' includes the rabbit and various ungulates (hoofed mammals). (Condensed and adapted from Goodman, 1975.)

The amino acid sequences of both the α-haemoglobin and the β-haemoglobin of *Tarsius* were eventually determined (Beard, Barnicot and Hewett-Emmett, 1976). Examination of individual amino acids at sites in the two proteins that are variable in the sample of primates investigated showed a somewhat equivocal picture. Relative to the hypothetical ancestral condition for the amino acid sequences of α-haemoglobin and β-haemoglobin proposed by

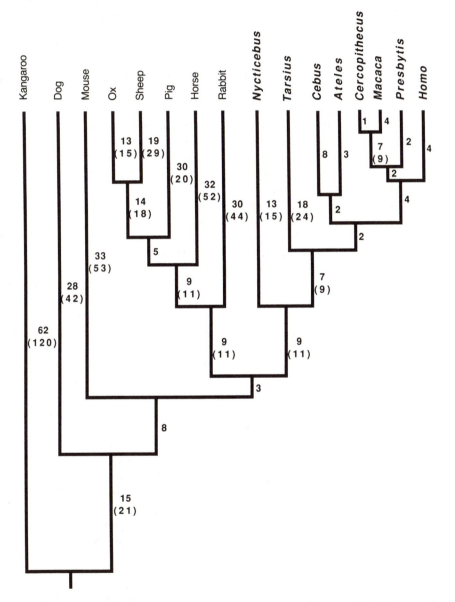

Figure 11.27. Combined maximum parsiony tree for α-haemoglobin and β-haemoglobin. The number to the right of each branch indicates the number of inferred nucleotide replacements allocated to that branch. Numbers in parentheses indicate augmented numbers, as in Fig. 11.25. Note: Two macaque species have been lumped together and the number of inferred nucleotide replacements given is the maximum for the two species. (Adapted extract from Beard, Barnicot and Hewett-Emmett, 1976.)

Goodman, Moore and Matsuda (1975), *Tarsius* was confusingly found to share an approximately equal number of apparently derived substitutions with *Nycticebus* and with the hypothetical ancestral condition for simian primates. *Tarsius* also has a substantial number of sites on both haemoglobin molecules where the amino acids present are unique among primates. However, it is possible to combine the amino acid sequence data for both types of haemoglobin, treating the two proteins together as if they were one double-length protein, and to produce a new maximum parsimony tree for primates and other mammals incorporating the information for *Tarsius*. When this was done for a sample of 20 vertebrate species – including nine primates, seven non-primate placentals, the kangaroo (a marsupial), the echidna (a monotreme), the chicken and the distantly related chordate *Amphioxus* – the most parsimonious solution grouped *Tarsius* with simians (Beard and Goodman, 1976). The relevant marsupial/placental part of this tree is shown in Fig. 11.27, which shows that *Tarsius* shares seven inferred nucleotide replacements (augmented value = 9) with simian primates. Figure 11.27 also shows that the lineage leading to *Tarsius* has accumulated 18 additional nucleotide replacements (augmented value = 24), underlining the fact that tarsiers have undergone a great deal of independent evolution relative to all other primates. As Beard *et al.* (1976) point out, the extensive accumulation of additional nucleotide replacements inferred for the tarsier lineage considerably exceeds that inferred for any of the simian lineages. This reconstruction therefore indicates that *Tarsius*, which has often been called a 'living fossil', is by no means primitive with respect to the amino acid sequences of its α-haemoglobin and β-haemoglobin chains. Unfortunately, *Nycticebus* is the only prosimian for which reliable data were available for the amino acid sequences of both α-haemoglobin and β-haemoglobin, so the reconstruction shown in Fig. 11.27 is necessarily provisional.

Lens crystallins

A large amount of information is now also available for **α-crystallin A**, one of the proteins incorporated in the mammalian eye lens (de Jong, 1982; de Jong and Goodman, 1988). Eye lens proteins have always figured prominently in molecular approaches to the reconstruction of phylogenetic relationships, as the eye lens is one of the most protein-rich organs of the body, and these proteins have been extensively used in both electrophoretic and serological studies. Cryst-allins are the characteristic proteins of the vertebrate eye lens and there are several different classes. The α-crystallins are the largest of the lens proteins and there are two kinds in mammals: α-crystallin A and α-crystallin B. In both cases about 175 amino acids are contained in the protein chain. Most research has concentrated on α-crystallin A, with the result that complete amino acid sequences are now known for 41 mammalian species (de Jong, 1982; de Jong and Goodman, 1988). In fact, there is a special advantage of the information now available for α-crystallins in that there has not been the customary bias towards humans and non-human primates. The amino acid sequence results have thus provided a more general perspective on mammalian evolution than is typical, which was de Jong's avowed goal. The tree shown in Fig. 11.28 shows the pattern of relationships among placental mammals inferred with the α-crystallin data, using the maximum parsimony tree-building approach developed by Goodman's group (de Jong, 1982). A constraint was, however, placed on the tree generated in that de Jong built in the assumption that generally accepted allocations of mammalian species to individual orders are correct (with tree-shrews placed in their own order Scandentia). Hence, construction of the tree permitted relatively free arrangements of species within orders and inference of the overall pattern of branching, but only for orders as defined at the outset. The two most parsimonious trees determined with that constraint each had a total of 157 inferred nucleotide replacements. In fact,

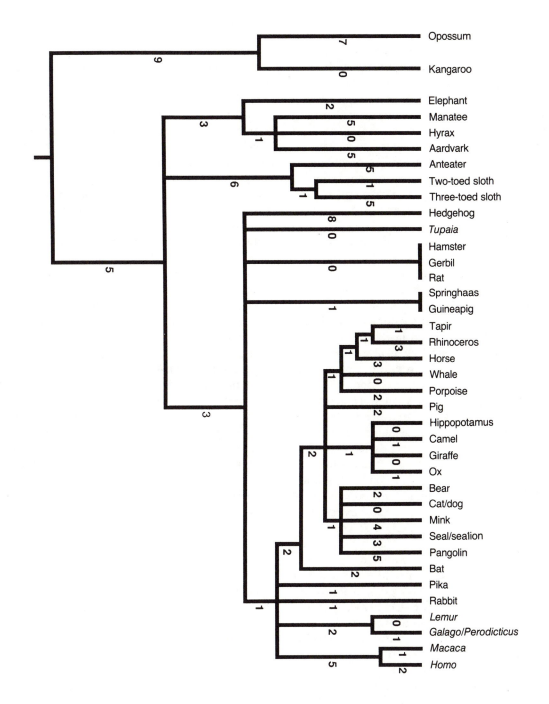

Figure 11.28. One of the two most parsimonious trees for mammalian α-crystallin A chains, generated on the assumption that widely accepted definitions of orders are correct. The number to the right of each branch indicates the number of inferred nucleotide replacements allocated to that branch and the tree requires an overall total of 157 nucleotide replacements. (After de Jong, 1982.)

these two trees differed significantly in a small number of branching features, indicating that certain relationships between orders are poorly defined by the data from α-crystallin A. (This is not too surprising because, as can be seen from Fig. 11.28, the average number of inferred nucleotide replacements allocated to each branch is quite small.)

Although there are differences between the two most parsimonious trees inferred from the α-crystallin A sequences, there are many points of agreement. For instance, marsupials and placentals consistently emerge as sister groups, each defined by a set of apparent shared derived features. Among the placentals, hoofed mammals (artiodactyls and perissodactyls) regularly group with cetaceans (whales and dolphins) and another regular grouping contains the elephant, hyrax and manatee. The three edentates in the sample (anteater and two sloths) are consistently associated in the parsimonious trees, forming a group that is always found to branch away quite early from other placentals. Even though primates were defined as a monophyletic group prior to analysis, neither of the most parsimonious trees indicated any derived nucleotide replacements shared by strepsirhines (lemurs and lorises) and simians. Indeed, in both trees primates were shown as being most closely related to lagomorphs (rabbit and pika) and there were no inferred nucleotide replacements indicating that the strepsirhines and simians are more closely related to one another than to lagomorphs. (This, again, may be an outcome of the small number of nucleotide replacements involved on individual branches of the tree representing long period of evolutionary time). Tree-shrews occupy an equivocal position in both trees, and this is undoubtedly associated with the fact that no inferred nucleotide replacements are allocated to the tree-shrew lineage. As de Jong (1982) has observed, the addition of only one nucleotide replacement to the tree could link tree-shrews to primates or lagomorphs. Indeed, tree-shrews could equally easily be linked to insectivores or to rodents, especially since the latter share with tree-shrews the smallest accumulation of inferred nucleotide

replacements since the ancestral node of the placental mammals.

After de Jong's parsimonious trees had been determined for mammals in general (Fig. 11.28), additional sequence information became available for tarsiers (*Tarsius syrichta*) and for owl monkeys (*Aotus trivirgatus*). This permitted refinement of the primate portion of the tree (de Jong and Goodman, 1988). *Aotus* was found to be unequivocally linked to *Macaca* and *Homo*, with three inferred shared mutations prior to divergence. The position of *Tarsius* proved to be problematic because of the preservation of a very primitive sequence overall, but one parsimonious solution links tarsiers to simians with a single shared mutation.

As can be seen from Fig. 11.28, there are considerable differences between lineages in apparent rates of evolution. The total number of inferred nucleotide replacements accumulated by each species of placental mammal since the common ancestral stock ranges from 3 to 12, representing a four-fold range of variation. There are marked differences between individual lineages (e.g. compare the hedgehog with *Tupaia* or with rodents). Another important aspect of de Jong's reconstruction based on the α-crystallin A sequences is that the trees reveal a substantial amount of convergent (or parallel) evolution. For both of the most parsimonious trees with totals of 157 inferred nucleotide replacements, de Jong (1982) found that 70 (45%) had occurred in parallel and a further 16 (10%) were reverse mutations. Hence, only 45% of the inferred nucleotide replacements represent uniquely derived transitions. Thus, convergent evolution, including secondary reversals, would appear to be rife at the molecular level, as at the morphological level.

Fibrinopeptides

Another group of proteins the **fibrinopeptides**, has attracted special attention with respect to reconstruction of phylogenetic relationships among mammals (particularly primates). Fibrinopeptides A and B are two chains of amino acids that become detached from the fibrinogen molecules of vertebrates during blood-clotting. Apparently, their function is essentially confined to prevention of spontaneous gel-formation by fibrinogen molecules and the exact sequence of amino acids contained in some regions of the fibrinopeptide chains is relatively immaterial. In any event, fibrinopeptides show a conspicuously high rate of change in mammalian evolution and are therefore well suited for comparative studies (Wooding and Doolittle, 1972; Doolittle, 1974). As some parts of the fibrinopeptide chains do remain relatively stable, it is possible to align A or B chains from different species to make comparisons between the variable regions. Doolittle (1974) reported that fibrinopeptides A and B had been sequenced for more than 45 mammalian species, including 12 primates.

In all mammals, the fibrinopeptide chains are quite short, ranging from 14 to 21 amino acids in length. In the primates examined fibrinopeptide A is typically 16 amino acids long (18 in *Nycticebus*), and fibrinopeptide B is typically 14 amino acids long (15 in *Nycticebus*). Wooding and Doolittle (1972) and Doolittle (1974) specifically discuss implications for phylogenetic relationships among primates. Unfortunately, only one prosimian (*Nycticebus*) is represented in the sample of 12 primates, so the conclusions largely concern relationships among simian primates. Doolittle presented separate parsimonious trees for fibrinopeptide A and fibrinopeptide B in primates, but these have been combined in Fig. 11.29 because compatible patterns emerged for the two chains. It can be seen that the branching pattern generally conforms with expectation. The slow loris (*Nycticebus*) is shown branching off first and a common ancestral node for simians is defined by at least two shared nucleotide base changes that are unusual among mammals generally. Within the simian group, there is a clear division between the two New World monkeys examined (*Ateles* and *Cebus*) and the Old World simians. The three Old World monkey genera (*Cercopithecus*, *Macaca* and *Mandrillus*) share a well-defined ancestral node. However, the gibbon (subgenus *Hylobates*) and the siamang (subgenus *Symphalangus*) are not shown as

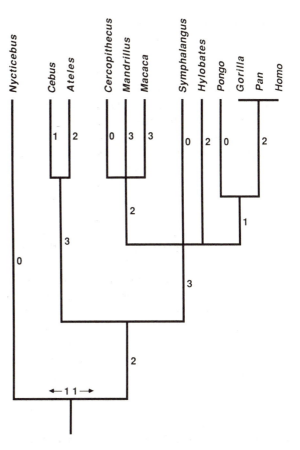

Figure 11.29. Parsimonious tree for fibrinopeptides A and B combined. The number to the right of each branch indicates the number of inferred nucleotide base changes, relative to the hypothetical ancestral condition, allocated to that branch. At the base of the tree, 11 amino acid changes have not been allocated because of the problem of 'rooting'. However, comparisons with other mammals suggest that most changes should be allocated to the lineage leading to simians, rather than to the *Nycticebus* lineage. Note: the siamang (*Hylobates syndactylus*) is here indicated as a separate genus (*Symphalangus*). (After Doolittle, 1974.)

having a specific relationship either to each other or to great apes and humans, whereas the gibbon differs from the siamang in two amino acid positions (one on each fibrinopeptide chain). Finally, in accordance with expectation, the orang-utan is shown as branching off before the divergence between *Gorilla*, *Pan* and *Homo*. In fact, these three genera have identical amino acid sequences both for fibrinopeptide A and for fibrinopeptide B.

Because there is only one prosimian genus, *Nycticebus*, in the sample, there is an acute rooting problem involved with the tree shown in Fig. 11.29. Although *Nycticebus* and the inferred ancestral stock of the simians differ by a total of 13 nucleotide replacements, only two of these can be allocated without much problem to the basal lineage leading to simians because of their unusual character in comparison with other mammals. The remaining 11 nucleotide substitutions might belong either to the lineage leading to *Nycticebus* or to that leading to simians. In fact, comparisons showed that the fibrinopeptide sequences of *Nycticebus* are very similar to those determined for the rabbit. Because a specific phylogenetic relationship between *Nycticebus* and the rabbit is highly unlikely, and because such a degree of convergence is also unlikely, the most reasonable explanation is that both the rabbit and *Nycticebus* have retained relatively primitive sequences. According to that interpretation, most of the 11 substitutions indicated at the base of the tree in Fig. 11.29 should be allocated to the simian lineage, which would suggest that fibrinopeptides have evolved remarkably slowly in the lineage leading to *Nycticebus*. Even among simian primates, it would appear that rates of change have varied by a factor of at least two, according to the reconstruction shown in Fig. 11.29. Wooding and Doolittle (1972) also compared primates with other mammals generally and concluded that both the rat and the rabbit may be linked to primates. This relationship is indicated to some extent by a few shared inferred nucleotide base substitutions, but the main evidence comes from a shared deletion of six to eight amino acids from the middle of the fibrinopeptide B chain, which could be interpreted as a derived feature indicating common ancestry.

Myoglobin

One of the most informative approaches to phylogenetic reconstruction using protein sequence data has involved **myoglobin**. This globin chain contains 153 amino acids in mammals and is a major component of muscle tissue (Joysey, 1978; Romero-Herrera, Lehmann *et al.*, 1978). Tree-building was carried out with myoglobin sequence data for a sample including 28 mammalian species (containing 13 primate species and *Tupaia*) and a small number of other vertebrates (chicken, penguin and lamprey). The main approach began with a partially resolved phylogenetic tree based on comparative morphological evidence and then used the myoglobin sequence data to test alternative refinements of that tree. Romero-Herrera, Lehmann *et al.* (1978) set out with an outline phylogenetic tree for placental mammals in which the relationships linking five main groups – primates, ungulates (hoofed mammals) and cetaceans (whales and dolphins), tree-shrew, hedgehog, carnivores and pinnipeds – had been left unresolved. Eight different arrangements of those five groups of placental mammals were then examined by taking the myoglobin sequence data and seeking the most parsimonious solution (in terms of total nucleotide substitutions required) in every case. One of the two most parsimonious solutions (involving a total of 281 nucleotide base changes) is illustrated in Fig. 11.30. It can be seen that the primate sector of the tree is shown to be monophyletic in terms of shared derived nucleotide base substitutions and has the expected internal branching pattern. The strepsirhines (*Lepilemur*, *Galago*, *Nycticebus* and *Perodicticus*) form a monophyletic cluster and the myoglobin sequence data quite clearly link the slow loris (*Nycticebus*) to the potto (*Perodicticus*). On the simian side, there is a basic division between New World monkeys and Old World simians (monkeys, apes and humans), and the latter group shows the expected pattern with the gibbons branching off first and an unresolved three-way split between *Gorilla*, *Pan* and *Homo*. Joysey (1978) provided additional information concerning relationships between apes and humans by including the siamang and the orang-utan. The myoglobin of the siamang (*Hylobates syndactylus*) is identical to that of gibbons. The myoglobin of the orang-utan (*Pongo*),

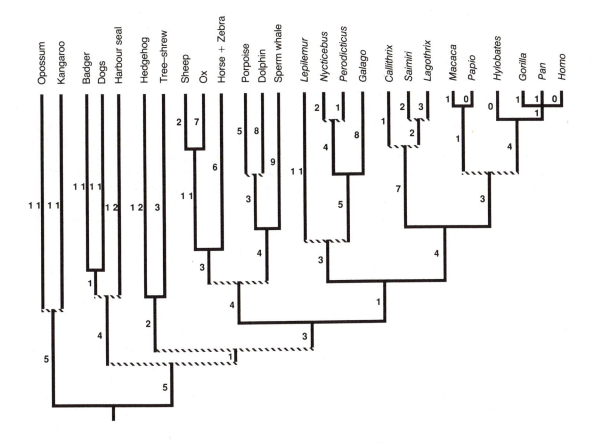

Figure 11.30. One of the two most parsimonious trees for myoglobin compatible with accepted views of morphological evolution in mammals. The number to the left of each branch indicates the number of inferred nucleotide replacements (point mutations). The complete tree required a total of 281 nucleotide replacements, but not all are shown here because of the omission of some species. Dashed horizontal lines indicate nodes where available results did not permit reliable inference of branching. Note: The domestic dog and the hunting dog have been lumped together under the heading 'dogs' and the number of inferred nucleotide replacements given is the maximum for the two species. (After Joysey, 1978; see also Romero-Herrera, Lehmann *et al.*, 1978.)

however, yielded rather surprising results (Romero-Herrera, Lehmann *et al.*, 1976a). When the orang-utan sequence was fitted into a most parsimonious solution for the Old World simians, *Pongo* was found to branch off before the divergence between *Hylobates*, African apes and humans. As there can surely be little doubt that *Pongo* is more closely related than gibbons to the African great apes and humans, this represents a clear anomaly in the inferences based on myoglobin sequence data for primates.

In Fig. 11.30, the tree-shrew (*Tupaia*) is linked to the hedgehog and is shown to be relatively distant from the primate assemblage. In fact, the ungulates and cetaceans are shown as being closer to primates than are the tree-shrew and hedgehog. Among the placental mammals investigated, only carnivores and pinnipeds (seal) are shown as further away from primates.

Romero-Herrera, Lehmann *et al.* (1978) also tried out several alternative approaches to the inference of branching relationships using the myoglobin sequence data. They applied the maximum parsimony procedure used by Goodman's group for haemoglobins, beginning with the matrix of differences between pairs of species and then generating the most economical tree using a branch-swapping programme. This procedure produced a tree requiring a total of only 242 nucleotide base substitutions, 39 (14%) less than for the tree shown in Fig. 11.30, in which the myoglobin data had been constrained to fit one of several possible trees judged to be compatible with morphological evidence. The unconstrained tree generated by the maximum parsimony procedure is thus considerably more economical in terms of overall evolutionary change required. However, it contains several strikingly anomalous features that are quite incompatible with widely accepted ideas regarding likely phylogenetic relationships among the mammals concerned: (1) The tree-shrew and the hedgehog are shown as intervening between simians and prosimians. (2) The sportive lemur is shown as grouping with the horse and zebra. (3) The harbour seal is grouped with cetaceans rather than with carnivores. (4) The badger is isolated from the dogs. In view of these anomalies, the reliability of the maximum parsimony tree as an indicator of phylogenetic relationships is in this case subject to a great deal of doubt.

Romero-Herrera, Lehmann *et al.* similarly generated parsimonious Wagner trees from the myoglobin data (see Kluge and Farris, 1969; Farris, 1972). Wagner trees have the advantage that they do not implicitly require constant rates of evolution along all lineages, in contrast to trees generated with the unweighted pair-group approach. However, a Wagner tree incorporating all of the mammals in the sample also contains numerous anomalies: (1) The opossum is shown as closer than the kangaroo to placental mammals, whereas it can surely be taken for granted that marsupials are monophyletic. (2) New World monkeys are shown as branching away at the beginning of the radiation of placental mammals, hence being widely separated from the remaining primates. (3) The hedgehog and the tree-shrew intervene between

Old World monkeys and prosimians. (4) The sportive lemur is shown as being most closely related to the horse and zebra (perissodactyls). (5) Artiodactyls (even-toed hoofed mammals) are shown to be closer to carnivores than to the horse and zebra. (6) The harbour seal is grouped with cetaceans rather than with carnivores. All of these anomalies, relative to widely accepted interpretations of the morphological evidence, suggest that Wagner trees are at least as unreliable as the maximum parsimony trees employed by Goodman and his colleagues in translating myoglobin sequence data into an inferred pattern of phylogenetic relationships among mammals.

Tandem alignment of amino acid sequences

As in the case of α-haemoglobin and β-haemoglobin discussed above, it is possible to combine information from several different proteins in a single tree using a procedure called **tandem alignment**. The amino acid sequences of the individual proteins are aligned in tandem and then analysed as if a single, large protein were involved. Goodman, Romero-Herrera *et al.* (1982) applied this procedure to generate an overall phylogenetic reconstruction for over 40 mammalian species with a tandem combination of seven proteins (α-haemoglobin, β-haemoglobin, myoglobin, α-crystallin A, fibrinopeptide A, fibrinopeptide B and cytochrome *c*). Unfortunately, not all proteins had been sequenced for all of the species included, so the reconstruction is somewhat uneven. Nevertheless, the most parsimonious trees produced (one of which is illustrated in Fig. 11.31) provide valuable summaries of the overall implications of the amino acid sequence data then available.

The primate sector of the tree in Fig. 11.31 agrees very well with the overall pattern indicated in the provisional tree based on morphological evidence (Fig. 2.2). The primates are indicated as a monophyletic group, branching fairly early to give rise to separate clusters of strepsirhines and haplorhines. The simians constitute a monophyletic group containing

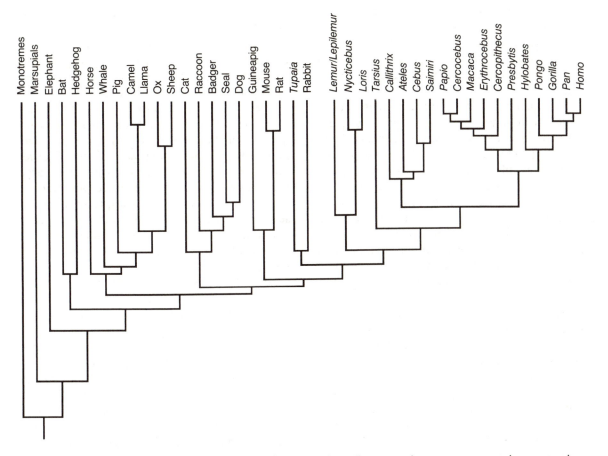

Figure 11.31. Parsimonious tree for mammals based on tandem alignment of two to seven proteins per species: α-haemoglobin, β-haemoglobin, myoglobin, α-crystallin A, fibrinopeptide A, fibrinopeptide B and cytochrome *c*. (Adapted extract from Goodman, Romero-Herrera *et al.*, 1982.)

distinct groups of New World monkeys and Old World simians. Among Old World simians, there is a basal divergence between monkeys and hominoids (apes and humans) and branching within the hominoid cluster follows the usual pattern, with the gibbon (*Hylobates*) branching off first, followed by the orang-utan, to leave a residual group containing the African apes and humans. Although the tree-shrew (*Tupaia*) is shown as being relatively close to primates within the overall tree for mammals, it is important to note that the rabbit is shown as the closest relative of the tree-shrew in this tree. Following the principle that increasing accuracy in phylogenetic reconstruction is likely to result

from increasing the number of characters considered, it is likely that a tandem alignment provides one of the most reliable guides to primate relationships yet available from amino acid sequence data. However, it should be noted that the tree shown in Fig. 11.31 was not, in fact, the most parsimonious tree determined for the data by Goodman, Romero-Herrera *et al.* (1982). That tree required a total of 2425 nucleotide substitutions, whereas the most parsimonious tree required marginally fewer (2415). The latter tree did not show any significant difference with respect to the inferred relationships of primates or *Tupaia*, but it did indicate that monotremes are closer to placentals and also showed some anomalies with respect to internal relationships of the cluster containing the hoofed mammals and the whale. Because of these clear anomalies, Goodman, Romero-Herrera *et al.* (1982) preferred the tree shown in Fig. 11.31.

This approach using tandem alignment of amino acid sequences of proteins was subsequently expanded by Goodman, Czelusniak and Beeber (1985). Thereafter, it proved possible to include information on an eighth type of polypeptide (ribonuclease), in addition to the original seven, and to include a far wider range of species. Thus, Miyamoto and Goodman (1986) were able to publish an analysis of protein sequences from 72 species of placental mammal and 18 other jawed vertebrate species. As with previous analyses, several most parsimonious trees were found, in this case all requiring the same number of inferred nucleotide substitutions (4062). Although details of the branching patterns differed between these trees, certain consistent features emerged. Among placental mammals, the edentates emerged as the sister group of all other species. Four consistent groupings were identified among the latter: (1) primates with rodents and lagomorphs; (2) bats with tree-shrews, lipotyphlan insectivores, carnivores and pangolins; (3) cetaceans with hoofed mammals (artiodactyls and perissodactyls); and (4) hyraxes, elephants, sea-cows and anteaters. In one of the most parsimonious trees, the expected general pattern of relationships among primates was found. There was an initial split between strepsirhines and haplorhines, followed by secondary divisions between lemurs and lorises and between tarsiers and simians. Among simians, New World monkeys branched away first and there was a subsequent division between Old World monkeys and hominoids (apes and humans). Further, it should be emphasized that none of the most parsimonious trees indicated a specific relationship between tree-shrews and primates.

The problem of choosing a tree

Unlike many other authors who have used amino acid sequence data to infer phylogenetic relationships, Romero-Herrera, Lehmann *et al.* (1978) went to great lengths to examine a variety of different methods of tree-building. As a result, they were able to show quite clearly the considerable discrepancies among reconstructions generated with different approaches from a given data set (see also Fitch, 1984). They also showed that many of the trees generated are quite incompatible with widely accepted interpretations of morphological evidence. Hence, there are sound reasons for suspicion that a large proportion of computer-generated phylogenetic trees based on amino acid sequence data are severely flawed. Once again, however, it is important to emphasize the broad level of agreement that is found, as well as recognizing the discrepancies.

Several details of the inferred evolutionary relationships of primates have received widespread confirmation from analyses of amino acid sequence data. For instance, there is overwhelming evidence that simian primates (monkeys, apes and humans) constitute a well-defined monophyletic group. New World monkeys are also generally separated quite clearly from Old World simians. Among Old World simians, there is usually a clear division between monkeys and hominoids (apes and humans), but several reconstructions have failed to link leaf-monkeys (as represented by *Presbytis*) definitely with the other Old World monkeys rather than with hominoids. Within the hominoid group, it is customary to find gibbons branching off first, followed by the orang-utan, leaving a cluster containing the African great apes and humans. As with the immunological evidence, there is uncertainty over which of the African apes – the chimpanzee or the gorilla – is our closest relative, although the balance of evidence (e.g. see Fig. 11.31) suggests that the gorilla branched away first. There is also fairly consistent, but by no means universal, evidence that the strepsirhine primates (lemurs and lorises) constitute a monophyletic group. The position of *Tarsius*, as indicated by amino acid sequence data, remains uncertain, largely because of the limited availability of relevant data. Of the proteins discussed above, only α-haemoglobin, β-haemoglobin and α-lens crystallin A have been sequenced to date, and these sequences provide a preliminary indication that tarsiers are linked to simians rather than to strepsirhines within the primate tree.

The primates as defined here, including lemurs, lorises, tarsiers, monkeys, apes and humans, are also commonly defined as a monophyletic group by analyses of protein sequence data, although it must be emphasized that monophyly of the primates has been assumed at the outset in numerous studies (e.g. those of α-crystallin A and myoglobin). Finally, in a pattern very reminiscent of that found in interpretations of the morphological evidence, tree-shrews are variously depicted as being part of the primate radiation, as being closely related to primates and as being quite distantly removed from primates. In short, uncertainty about the relationships of tree-shrews continues to reign at the molecular level as at the morphological level. This is not really surprising in that many studies have included a disproportionately large number of primate species and have not given adequate attention to the need to include appropriate outgroup comparisons, such that the earlier branches in any inferred tree for placental mammals generally must be subject to a large element of doubt.

As has become clear from the above discussion, the principle of **parsimony** has been explicitly invoked in all reconstructions of mammalian phylogenetic relationships based on amino acid sequence data. Indeed, parsimony has been invoked in at least two different contexts for the construction of trees based on inferred substitutions in nucleotide base sequences. The principle of parsimony is invoked in the inference of nucleotide base sequences corresponding to a given set of alternative amino acids at a given site in a protein and it is invoked again in using those inferred sequences to construct a tree based on some minimizing criterion. As has been noted in Chapter 3, there are grounds for concern about placing so much reliance on the principle of parsimony, as it is highly unlikely that the real biological world is consistently parsimonious.

It turns out, in fact, that there is another serious cause for concern about the use of the parsimony principle to select a particular phylogenetic reconstruction. This stems from the observation that it is generally possible to obtain several parsimonious trees that are virtually indistinguishable in terms of overall total branch lengths. To make matters worse, the branching patterns often differ quite radically between these parsimonious trees. Cook and Hewett-Emmett (1974), for example, generated 12 different relatively parsimonious trees using the amino acid sequence data for α-haemoglobin previously analysed by Barnabas, Goodman and Moore (1971). The four most parsimonious trees are shown in Fig. 11.32 and it can be seen that the sum of inferred branch lengths varies only slightly, from 124.8 to 125.1. (The remaining eight trees had summed branch lengths ranging from 126.2 to 129.3; but they all included negative branch lengths, so they were actually somewhat less economical than these sums indicate.) In spite of the narrow range in total branch lengths for the four trees shown in Fig. 11.32, radically different conclusions regarding phylogenetic relationships are indicated. The two slightly more 'costly' trees of the four (total branch lengths: 125.1; 125.2) show the primates as a monophyletic group and link the tree-shrew to the mouse. However, these two trees differ from one another with respect to the relationships indicated for lemurs and lorises. One shows three strepsirhine primates (*Propithecus*, *Lemur* and *Galago*) as constituting a monophyletic group, whereas the other indicates that *Galago* branched off separately from both lemurs and simians. The two trees sharing the minimal total branch length of 124.8 seriously conflict with interpretations of morphological evidence and also differ markedly from one another. In both cases, the lemurs are shown as being more closely related to the tree-shrew and the mouse than to simian primates, and the position of *Galago* differs radically between the two trees. Because the differences in total branch lengths between the four trees are so slight, there is no sound basis for choosing between them without additional information. It should be noted, however, that the tree that shows the expected pattern of relationships for the five primates, with monophyletic strepsirhine

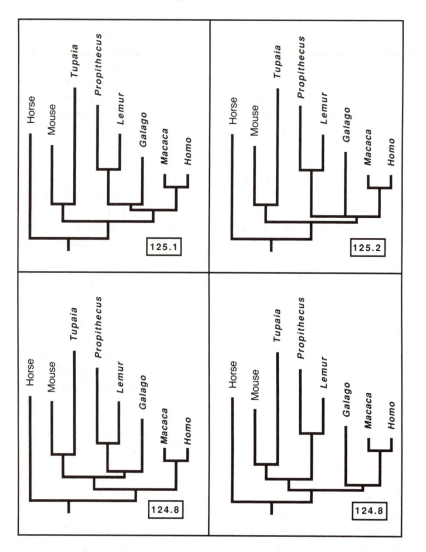

Figure 11.32. The four most parsimonious trees determined for relationships between eight mammalian species on the basis of nucleotide substitutions inferred from amino acid sequences of α-haemoglobins. The figure at the bottom right of each tree indicates the total number of substitutions required for the tree, as indicated by the sum of average values inferred for individual branches. The relative lengths of branches in each tree give an approximate indication of the numbers of substitutions allocated to each branch. It is clear that inferred rates of change differ substantially between lineages. (After Cook and Hewett-Emmett, 1974.)

and simian clusters, links the tree-shrew to the mouse. In fact, taking all 12 trees illustrated by Cook and Hewett-Emmett (1974), it is found that the tree-shrew shows the most variation in inferred relationships, confirming the observation (above) that there is something 'unstable'

about the relationships inferred for this group from sequence data for α-haemoglobin.

The problem of marginal differences in summed branch length between radically different parsimonious trees, as identified by Cook and Hewett-Emmett (1974), is only the tip of an enormous iceberg. Romero-Herrera, Lehmann *et al.* (1978), for instance, 'tested' eight alternative phylogenetic trees for mammals against the amino acid sequence data for myoglobin. The two most parsimonious reconstructions (one of which is shown in Fig. 11.30) required a total of 281 nucleotide substitutions each; but of the remainder one tree required 282 substitutions and five trees required 283 substitutions. Thus, the choice between eight radically different interpretations of mammalian evolutionary relationships in this case depends on a difference of only one or two substitutions out of an average of 282 (i.e. less than 1% of the total). This problem has been generalized by Fitch (1984), who took a number of relatively simple cases and calculated the number of nucleotide base substitutions required for all possible trees linking a small number of species. (As has been explained in Chapter 3, it is only feasible to examine all possible trees if the number of species involved is relatively small, because of the extremely rapid increase in numbers of alternative trees with increasing

sample size; see Fig. 3.13.) A good example is provided by the data for α-haemoglobin sequences for eight mammalian species originally analysed by Goodman, Czelusniak *et al.* (1979). Even taking the simplest model (unrooted, dichotomous branching pattern), there are 10 395 possible phylogenetic trees linking the eight species and Fitch showed that the distribution of total substitutions required for all these trees is continuous (Fig. 11.33). The range of total substitutions varies only from 133 to 154. As there is no discontinuity in the distribution to indicate which trees to choose as 'optimal' trees, an arbitrary decision must be made. One could, of course, decide to restrict the choice to the small set of most parsimonious trees with totals of 133 substitutions ($N = 6$). However, the example illustrated in Fig. 11.32 indicates that the trees thus selected might show serious anomalies in comparison with interpretations of the morphological evidence. Further, it must be remembered that a maximum parsimony tree generated for the mammalian myoglobin data discussed above requires far fewer substitutions

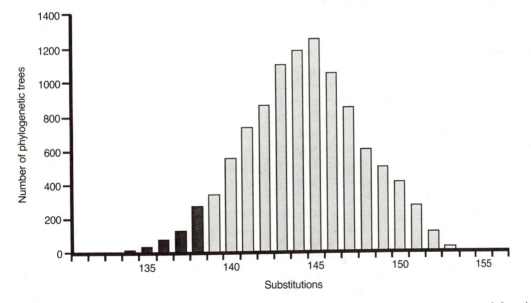

Figure 11.33. Total nucleotide substitution values inferred for all 10 395 possible trees generated for eight mammalian species on the basis of amino acid sequence data for α-haemoglobins. The 5% most parsimonious trees, indicated in black, cover a remarkably small range of values (133–8). (After Fitch, 1984.)

than any of the eight 'biologically reasonable' trees considered and yet shows major anomalies. Thus, even if the choice of suitable trees were to be restricted to the lowest 5% of a distribution such as that shown in Fig. 11.33, it is by no means certain that the real tree would be contained in that sample. Of the possible reconstructions based on the α-haemoglobin data, 5% amounts to 549 different trees and yet the difference in total substitutions involved is only five (range: 133–8), representing variation of less than 4%.

Fitch (1984) conducted a similar exercise with amino acid sequences of α-crystallin A for eight mammalian species. As with the haemoglobin results, the total substitutions per tree were distributed continuously but, in contrast to the pattern shown in Fig. 11.33, the distribution was highly skewed. A concentration of cases occurred at the higher end of the range of total substitutions. Fitch attributes this to the fact that the α-crystallin A data tend to generate highly asymmetrical trees with long branch lengths intervening between some nodes.

It is therefore clear that the criterion of parsimony is not adequate in itself for the selection of a particular phylogenetic reconstruction based on amino acid sequence data. This has, of course, been recognized in some of the studies cited above and various attempts have been made to cope with the problem. In most cases, the requirement of parsimony has been relaxed to varying degrees and a relatively parsimonious tree has been selected on the basis of some assessment of general compatibility with other evidence. One approach is to reject the most parsimonious tree generated in favour of one that is somewhat less parsimonious but more in tune with widely accepted interpretations of phylogenetic relationships based on morphological evidence (e.g. Goodman, Romero-Herrera *et al.*, 1982). An alternative approach is to place constraints on the trees generated by assuming the validity of certain phylogenetic relationships at the outset and to use the amino acid sequence data to 'test' remaining branching relationships (e.g. Romero-Herrera, Lehmann *et al.*, 1978; de Jong, 1982). However, any such approach inevitably introduces an element of circularity into the process of tree-building and there is the accompanying danger that the incorporation at some stage of assumptions about certain relationships will be forgotten when other authors cite the resulting trees as a source of supporting arguments for particular phylogenetic reconstructions. It is, perhaps, justifiable to use extraneous evidence for the selection of a particular tree generated on the basis of amino acid sequences if care is taken to avoid circularity. For example, if a tree for placental mammals is generated for the purpose of examining relationships among primates, it might be reasonable to select a tree that exhibits biologically 'reasonable' relationships for the other mammals included. To take just one point, it would be reasonable to reject any tree that does not show monotremes as the most distant mammalian relatives of placentals, with marsupials as the unequivocal sister group thereof. Nevertheless, even in this form, the introduction of evidence derived from analyses of morphological evidence weakens the case for subsequent discussion of the degree of concordance between trees based on morphological evidence and trees based on amino acid sequences.

With phylogenetic reconstructions based on assessments of morphological evidence, the existence of multiple alternative trees differing only slightly in perceived acceptability has not usually been seen as a problem. Investigators have, at least in some cases, considered alternative possible reconstructions, but the number of potential trees involved is usually quite small (not more than a dozen). This may, of course, be attributable to the fact that morphological information is not amenable to straightforward quantitative treatment comparable to that applied to amino acid sequence data and that morphologists have simply not been aware of the bewildering array of different possibilities for reconstruction. On the other hand, it must be recognized that reconstructions based on morphological evidence have a distinct advantage in that there are several accessory lines of evidence to support phylogenetic inference, such that it is not necessary to place such heavy reliance on slavish application of a simple mathematical version of the parsimony principle.

Supporting evidence for amino acid trees

In Chapter 3, a flow diagram (Fig. 3.12) was provided to show the various sources of evidence and argumentation that may be incorporated into an inferred phylogenetic reconstruction based on morphological features (Fig. 11.34). Many of these valuable guidelines (e.g. ontogenetic data and palaeontological information) are simply not applicable at the molecular level. However, some components of Fig. 11.34 are relevant to phylogenetic reconstruction using molecular data. It has been noted above that the **outgroup principle** is just as important at the molecular level as in the interpretation of morphological and other evidence. Widely used methods of tree-building using molecular data also exploit, at least implicitly, the relative **frequencies of character states** in inferring relationships (cf. 'universals' and 'common states' in Fig. 3.12). Further, reconstructions that take into account changes at individual amino acid sites in a protein can identify parsimonious **transformation series** as an aid to defining the direction of evolutionary change. For instance, if, for a particular protein sequenced for several species involved in a comparison, the three amino acids histidine (His), glutamine (Gln) and lysine (Lys) occur as alternatives at a given site, it is possible to infer the most probable sequence of change by examining the corresponding triplet codes in Fig. 11.4. The nucleotide base sequences CAC or CAT code for histidine, glutamine corresponds to the triplets CAA or CAG and lysine corresponds to the triplets AAA or AAG. From this, it can be inferred that the most likely transformation series for the three amino acids at that site in that protein is as follows:

Amino acid: His   ⇌   Gln   ⇌   Lys
Base sequence: (CAC or   (CAA or   (AAA or
                 CAT)   CAG)   AAG)

However, it must be remembered that the polarity of this sequence (i.e. the starting point in evolution) cannot be determined without additional evidence and that, as is shown above, it is not possible to choose between the alternative nucleotide base sequences corresponding to each amino acid without further comparative information.

One important point to have emerged from the above attempts at tree-building from amino acid sequences is that **convergence** (including 'parallelism' and reversed mutations) is particularly prevalent at the level of genes and proteins. Levels of about 50% convergence in amino acid substitutions at individual positions have been reported for inferred mammalian evolutionary trees based on two proteins (α-crystallin A and myoglobin) and such levels are found regardless of the actual branching sequence. This is not at all surprising, as the possibilities for change at the molecular level are so tightly constrained. Only four nucleotide bases are involved in the coding structure of DNA and, as a general rule, only 20 essential amino acids are incorporated into proteins produced by gene transcription. Hence, convergence is doubtless more of a problem at the molecular level than at the chromosomal level, where the possibilities for rearrangement are virtually boundless, and at the morphological level, where the complexity of many characters investigated makes independent acquisition of exactly the same character state relatively unlikely. In the absence of additional evidence to counter the confusing effects of convergence, it is only to be expected that the most parsimonious trees obtained for protein sequence data will show anomalous branching patterns.

At the morphological level, some cases of inferred convergence only come to light as an outcome of tree-building, as at the molecular level; but there are also cases where convergence can be recognized on independent grounds prior to tree-building. As has been noted in Chapter 3, ontogenetic evidence is valuable both in its own right and as a pointer to convergence (i.e. where character states that are similar in the adult condition differ in their developmental history). For amino acid sequence information, of course, there is no real equivalent to ontogenetic evidence. However, there is another kind of

independent evidence that is frequently invoked at the morphological level for the identification of convergence and that concerns **functional aspects**. Convergence in morphology is

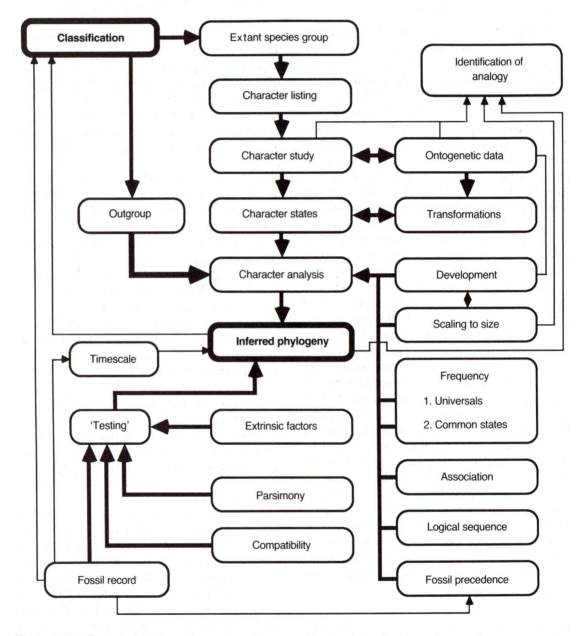

Figure 11.34. Repeat of the flow diagram from Chapter 3 (Fig. 3.12) showing the main procedures advocated for phylogenetic reconstruction using traditional biological evidence.

commonly defined as the development of similar adaptations to meet similar functional requirements. A parallel approach can be adopted at the molecular level, at least to the extent that the functional implications of changes in amino acid sequences of proteins are taken into account as far as possible. Now it must at once be stated that there is one well-represented school of thought which holds that most changes in DNA sequences and in protein sequences are selectively neutral. The arguments for this viewpoint will be discussed in a different context in Chapter 12, but it is obvious that convergence for functional reasons cannot be invoked for any changes in amino acid sequence that are neutral with respect to natural selection. Hence, the criterion of functional similarity as an aid to identification of convergence in amino acid substitutions can only be applied if it is accepted that those substitutions were subject to natural selection.

Romero-Herrera, Lehmann *et al.* (1978) specifically emphasized the need to consider the 'functional morphology' of a protein such as myoglobin. They noted, for instance, that two regions of the myoglobin molecule have remained remarkably conservative throughout all of the species investigated, indicating the preservation of important functional properties. Further, they observed that the myoglobins of both the diving mammals (cetaceans and pinnipeds) and the diving birds (penguins) in their sample are unusual in containing three to five arginine residues, in contrast to the one to two typical of land mammals. Myoglobin, like haemoglobin, has the essential property of combining reversibly with oxygen and it is highly likely that some amino acid substitutions, at least, must have direct implications for respiratory physiology. Diving vertebrates clearly have rather unusual requirements with respect to respiratory function and it is perfectly possible that convergent changes might have occurred in the structure of myoglobin to meet special physiological needs.

Another example of functional convergence is the sportive lemur (*Lepilemur*), which is similar to hoofed mammals, particularly perissodactyls (horse and zebra), in having a number of

dispersed amino acid sites of its myoglobin. In contrast to the 'biologically reasonable', but somewhat unparsimonious, myoglobin tree shown in Fig. 11.30, the most parsimonious trees determined without prior assumptions by Romero-Herrera, Lehmann *et al.* (1978) all grouped *Lepilemur* with the horse and zebra. If it is accepted that *Lepilemur* must be an integral member of a monophyletic primate radiation, it follows that any special similarities to the horse and zebra (none of which can be acceptably explained as a primitive retention) have necessarily arisen through convergence. It is marginally possible that such convergence (affecting at least five amino acid sites) has taken place by chance, but it seems more likely that some form of functional convergence is involved. It has been noted (see Chapter 6) that *Lepilemur* is a specialized folivore with an enlarged caecum and probably has an unusually low basal metabolic rate for its body size. The horse is also a folivore with an enlarged caecum and it has a relatively low mass-specific basal metabolic rate because of its large body size. Once again, therefore, it is possible that similarities in the amino acid sequence of myoglobins from different species are explicable in terms of metabolic demand.

Apart from the possible functional significance of individual amino acid substitutions, there can also be more diffuse reasons for convergence in DNA or protein sequences between species. For instance, it has been emphasized by Watson, Tooze and Kurtz (1983) that in the double-stranded DNA molecule the nucleotide bases cytosine and guanine are linked by three strong hydrogen bonds whereas adenine and thymine are linked by only two (see Fig. 11.3). As a result, the stability of the DNA chain increases as the proportion of cytosine–guanine base pairs increases. A similar overall effect has been identified by de Jong (1982) with respect to the amino acid composition of α-crystallin A in different mammalian species. de Jong noted that, of the inferred substitutions at individual amino acid sites, only 9% produced changes in the net amino acid side charge, whereas a figure of

32–44% would have been expected on the basis of studies of other proteins. (For this reason, in contrast to many other proteins, the electrophoretic mobility of α-crystallin A remains relatively constant across mammalian species in spite of differences in amino acid composition.) Thus, the α-crystallin A molecule seems to be under a selective constraint that minimizes changes in net charge accompanying amino acid substitutions. de Jong also reported a few cases in which two or more separate lineages had amino acid substitution at one specific site correlated with substitution at another specific site, indicating linked changes in coding triplets of nucleotide bases ('covarions'). Thus, there are several lines of evidence suggesting that functional convergence can occur at the molecular level. Accordingly, as in the case of morphological characters, an improved understanding of functional aspects should permit more reliable identification of cases of convergent evolution in DNA and proteins.

A further component of the flow diagram in Fig. 3.12 is also relevant at the molecular level: the criterion of **compatibility**. (This term is used in a special technical sense in connection with tree-building, but here it simply refers to general concordance.) As with morphological characters, there are two senses in which this concept is applicable. In the first place, any phylogenetic tree generated on the basis of amino acid sequence data for one type of protein should be consistent with a tree based on sequence data for another. One approach, already discussed above (see Fig. 11.31), is to combine evidence from several proteins in a tandem alignment and to generate an overall tree. Another approach is to compare the trees generated from comparisons of different proteins across a given range of species and to separate consistent features from idiosyncratic features of individual trees. This has been done very briefly and simplistically above, but there is considerable scope for the application of formal techniques for assessment of concordance between trees based on different sets of data and for selection of a consensus tree.

There is also a quite different sense in which

the criterion of compatibility may be applied. As at the morphological level, an attempt can be made to determine whether phylogenetic reconstructions based on molecular evidence are functionally feasible. Relatively little has been done in this direction as yet, but it is clear that much could be achieved. In particular, because phylogenetic trees based on amino acid sequences can lead to explicit reconstructions of hypothetical ancestral conditions, it is theoretically possible to examine the protein sequences suggested for particular ancestral stocks and to apply the criterion of functional compatibility. Obviously, the amino acid sequence suggested for a particular ancestral protein must be realistic in structural and functional terms and there is the possibility that hypothetical ancestral proteins may be synthesized so that their properties can be examined directly. Further, if several ancestral protein sequences have been inferred for a given common ancestor (e.g. the ancestral simian), they must be mutually compatible. In this respect, it should be particularly rewarding to investigate the metabolic properties of proteins such as haemoglobin, myoglobin and cytochrome *c*. For instance, if the reconstruction of the α-haemoglobin sequence for a particular ancestral stock is suggestive of a relatively low metabolic turnover, reconstruction of the myoglobin and cytochrome *c* sequences should lead to a matching conclusion. At present, however, investigation of such functional aspects of hypothetical ancestral molecules has hardly begun and in this respect phylogenetic reconstructions based on molecular data are lagging well behind those based on morphological characters. Without information on functional aspects, phylogenetic reconstructions lack an important source of testing.

Gene duplication

Quite apart from all of the usual problems involved in distinguishing between primitive, derived and convergent similarities, the interpretation of protein sequence data is beset by a special difficulty that has been realized only gradually. In the discussion so far, it has been taken for granted that it is a relatively simple matter to identify corresponding proteins in different species, such as albumins, individual haemoglobin chains or α-crystallins. In fact, care has been taken above to refer to 'corresponding' proteins, rather than to 'homologous' proteins, because the identification of directly homologous proteins in different species has turned out to be a much more complex matter than originally expected. (See Fitch (1970) for a discussion of criteria for the assessment of 'homology' among proteins.) Following an original suggestion by Bridges (1919), there is now abundant evidence that **gene duplication** is a regular phenomenon and several cases of 'gene families' containing several partially modified copies of an original gene sequence have been identified. Zuckerkandl (1963) alluded to the existence of such families of duplicated genes under the term 'isogenes' in his pioneering paper on the advent of molecular anthropology. As it is now known to be a common occurrence that some gene copies are **pseudogenes** that have been 'switched off' at some stage in evolution and are not expressed in protein synthesis, an investigator wishing to reconstruct the phylogenetic history of a particular protein is faced with an additional complexity. For a given type of protein to be strictly comparable between species, it must be the product of **orthologous DNA** sequences that have been transmitted to descendants in an unbroken series of straightforward replications from an original gene locus. Proteins produced by **paralogous DNA** sequences that have arisen through the intervention of gene duplication at some stage are not directly comparable.

If a particular gene locus is switched off, and hence not engaged in protein synthesis, it is obviously not directly subject to natural selection in the normal way. As a result, point mutations can presumably accumulate more rapidly in a gene copy that is switched off than in one that is operational and producing a protein whose properties are subject to natural selection. Nevertheless, it should not be assumed that pseudogenes necessarily serve no function. For

example, Hewett-Emmett, Venta and Tashian (1982) have suggested that pseudogenes may control the expression of adjacent genes that have not been switched off. Whether or not pseudogenes serve such a function, problems of interpretation can obviously arise when copies exist for a particular gene, such that the investigator may not be sure that it is the same (orthologous) copy that is switched on in all species. For instance, one can take the hypothetical case of an original gene A that undergoes duplication to produce a paralogous gene B. The original gene then becomes switched off and begins to accumulate point mutations relatively rapidly, while gene B is switched on and changes more slowly because it produces a protein that is exposed to natural selection. If, at some later stage in evolution, there is a reversal in a particular lineage, such that gene A is switched on again while gene B is switched off, discrepancies will be found in interspecific comparisons conducted in ignorance of this sequence of events. If proteins produced by gene A in one group of species are compared directly with proteins produced by gene B in another group, under the mistaken belief that the proteins are directly 'homologous' (i.e. orthologous), it will probably be concluded that the two groups are more divergent than they really are. Indeed, it is quite possible that the lineage in which there was the switch back to gene A actually arose somewhere within the group of species in which gene B is still operational. In other words, there will be the customary difficulty in distinguishing a primitive from a derived condition, compounded by the fact that the derived condition will almost certainly show marked divergence from the primitive.

Recognition of the phenomenon of gene duplication has introduced an entirely new dimension not only with respect to the reconstruction of phylogenetic relationships but also with respect to understanding of the evolutionary process itself. The classical view of evolutionary change was that it takes place by fixation of new versions of orthologous genes differing from one another only by point mutations. Now, however, the possibility exists that evolutionary change can take place both in this way and through the development of supplementary and/or alternative (paralogous) copies of a gene. Indeed, Ohno (1970) has argued persuasively that a major role has been played by gene duplication in evolution and Hewett-Emmett, Venta and Tashian (1982) have concluded that 'the creation of new genes is one of the principal means by which genes evolve' (see also Kohne, 1975). The implications of this concept for phylogenetic reconstructions have not yet to be fully appreciated and it is undoubtedly true that a better understanding of gene structure and function is currently leading to more, not less, uncertainty in the interpretation of molecular evidence for phylogenetic relationships. As has been noted by Hewett-Emmett *et al.* (1982), the need to distinguish orthologous from paralogous gene loci is the most persistent and challenging outstanding problem involved in tree-building using molecular data.

An excellent example of the phenomenon of gene duplication is provided by the mammalian globins. Thus far, discussion has been restricted to the major haemoglobin component of the blood of adult primates, which consists of $\alpha_2\beta_2$ chains bound to a haem group. There are, however, some other haemoglobins. In particular, adult hominoid primates (apes and humans) have a minor haemoglobin in their red blood cells, consisting of two α-chains and two δ-chains ($\alpha_2\delta_2$) bound to a haem group. Amino acid sequencing indicates an appreciable degree of overall similarity between the β-chain and the δ-chain, suggesting the possibility that the δ-chain arose as a duplication of the β-chain (Boyer, Crosby *et al.*, 1971). In other words, one can postulate an original condition in which only the major haemoglobin component was present, followed at some stage by emergence of the minor component after duplication of the β-gene to produce a δ-gene. However, an unusual distribution of the minor haemoglobin component exists among simian primates (Boyer, Crosby *et al.*, 1971; Cook and Hewett-Emmett, 1974; Hewett-Emmett *et al.*, 1982). Although New World monkeys resemble

hominoids in having a minor component incorporating apparent δ-chains, Old World monkeys lack a minor haemoglobin altogether (Barnicot and Wade, 1970). There are at least three possible explanations for this. One possibility is that the New World monkeys are more closely related to hominoids than the Old World monkeys are and that duplication of the β-gene to produce the δ-gene occurred in a specific ancestral stock that gave rise to New World monkeys and hominoids. This suggestion conflicts with an abundance of other evidence indicating that Old World simians constitute a well-defined monophyletic group to the exclusion of New World monkeys. The second possible explanation is that duplication of the β-gene to produce a δ-gene occurred independently in the ancestral stocks of New World monkeys and of hominoids and thus represents a convergent development. This seems unlikely because the δ-globins of New World monkeys and hominoids have certain similarities in sequence. The final and most promising explanation is that a duplication of the β-gene to produce a δ-gene was already present in the ancestral simians that gave rise to all the monkeys, apes and humans and that the δ-gene was secondarily switched off in the ancestral stock of the Old World monkeys. Indeed, it has been suggested that *Tarsius* has a minor haemoglobin component comparable to that found in New World monkeys and hominoids and that the δ-gene may have been present in the common ancestral stock of the haplorhine primates (Beard, Barnicot and Hewett-Emmett, 1976; Hewett-Emmett *et al.*, 1982).

In fact, it is now widely accepted that the α-chains and β-chains of the major haemoglobins of jawed vertebrates are themselves products of a very early duplication of a single ancestral haemoglobin gene. Further, it seems very likely that the genes for this ancestral haemoglobin and for the myoglobin of jawed vertebrates are derived from an even earlier duplication. Both of these early duplications of globin genes must have occurred successively during the initial radiation of the jawed vertebrates over 400 million years (Ma) ago (Goodman, Romero-

Herrera *et al.*, 1982; Hewett-Emmett *et al.*, 1982). Since that stage, both the α-gene and the β-gene have undergone duplication to produce clusters of related genes in mammals. Investigation of these two gene families has benefited both from tree-building based on amino acid sequences, which can lead to inference of the probable sequence of duplication events (Goodman, Romero-Herrera *et al.*, 1982), and from new techniques for direct examination of genes using DNA 'probes'. For example, in humans and some other simian primates, studies using gene probes have shown that there are two α-globin genes (α1 and α2), which code for identical amino acid sequences and cannot therefore be distinguished on the basis of their end-products. An extremely interesting point has emerged with respect to these two α-globin genes. The α1 and α2 genes of simian primates are very likely to be the products of a gene duplication already present in their common ancestor. Moreover, the chicken also has two α-globin genes and there is therefore a possibility that the duplication concerned took place before the evolutionary divergence between birds and mammals, over 300 Ma ago.

However, the α1 and α2 genes of any given simian primate species are more similar to one another than they are to their counterparts in related simian species. If the original duplication producing separate α1 and α2 genes was an ancient event, as the evidence cited suggests, it would seem that there must be some mechanism whereby the sequence correspondence between the α1 and α2 genes is maintained in any one species (**concerted evolution**; see Hewett-Emmett *et al.*, 1982). Evidently, any such corrective influence between gene copies presents additional complications for the reconstruction of phylogenetic relationships based on amino acid sequencing of a single, apparently 'homologous' protein from several species.

Studies using gene probes have now shown that the distinct clusters of α-globin and β-globin genes are in fact located on separate chromosomes, and this supports the idea that further duplication has taken place independently within each cluster. As would be expected from the

duplication hypothesis cited above to account for simian minor haemoglobins, the δ-globin gene of both New World monkeys and hominoids occurs on the same chromosome as the β-globin gene. Further, it has now been possible to demonstrate the presence of a 'switched-off' δ-globin gene, which is hence preserved as a pseudogene, on the corresponding chromosome of Old World monkeys (S.L. Martin, Zimmer *et al.*, 1980; Barrie, Jeffreys and Scott, 1981; Hewett-Emmett *et al.*, 1982). This provides one of the most convincing demonstrations of the predictive power of phylogenetic reconstructions. Before any studies of the duplication of globin genes, overwhelming evidence had already indicated that the simians (New World and Old World monkeys, apes and humans) constitute a monophyletic group. The presence of δ-haemoglobins with shared sequence similarities in New World monkeys and in Old World hominoids, but not in Old World monkeys, therefore led directly to the fairly confident prediction (as the most parsimonious hypothesis) that a δ-gene had been 'switched off' in Old World monkeys. Now independent evidence gathered with the aid of gene probes has confirmed the existence of a 'silent' δ-gene in at least some Old World monkey species.

The β-globin gene family of humans, and presumably of simian primates generally, also contains **embryonic haemoglobin** (ε-haemoglobin) and **fetal haemoglobins** (γ-haemoglobins). These haemoglobins are of special interest because they are actively produced at different stages of prenatal development, but subsequently give way to the normal adult haemoglobin(s). The existence of specially adapted haemoglobins during embryonic and fetal life is entirely to be expected, because metabolic turnover early in development is quite different from that in adult life and because oxygen exchange obviously takes place under quite different conditions while the developing individual depends on its mother's blood supply (containing adult haemoglobin) as a source of oxygen. Studies of embryonic and fetal haemoglobins are also still in their infancy. However, some unusual features have already come to light. For example, the mouse has two globins that appear from sequence information to be orthologous (i.e. 'homologous' in the strict sense) with human embryonic and fetal haemoglobin, respectively. Sequence information further indicates that the human ε-haemoglobin and γ-haemoglobin genes were generated through gene duplication from a single ancestral developmental haemoglobin, which was itself derived from an original duplication of the β-globin gene (Goodman, Romero-Herrera *et al.*, 1982). It would seem that duplication to produce distinct ε-globin and γ-globin genes preceded the evolutionary divergence among placental mammals, since the ε-globin gene of humans resembles that of the goat as well as that of the mouse (Hewett-Emmett *et al.*, 1982).

Sequence data also provide some interesting insights into the phylogenetic history of human fetal haemoglobin (γ-haemoglobin). Although both the mouse and the rabbit have globins that share sequence similarities with human fetal haemoglobin, the mouse and rat equivalents are not produced during fetal life. Further, it would appear that the original γ-globin locus was lost or suppressed at some stage in the evolution of the artiodactyl hoofed mammals, with a new 'fetal' haemoglobin emerging subsequently through a novel additional duplication of the β-globin gene. Further investigation of these globin genes of early development, particularly in marsupials, could eventually throw fresh light on the evolution of placentation in mammals and on the distinction between altricial and precocial mammals (see Chapter 9). Goodman, Romero-Herrera *et al.* (1982) have, for instance, suggested that evolutionary changes in the γ-haemoglobulins of simian primates have favoured an increased ability to compete with adult haemoglobin for oxygen, while the loss and subsequent redevelopment of a fetal haemoglobin in artiodactyls may indicate some secondary increase in gestation period (associated with the development of precocial infants) in that group.

Now that the phenomenon of gene duplication has been recognized as a prominent feature of molecular evolution, it is possible to account for some of the discrepancies encountered in comparative studies. If an abrupt shift in amino acid sequence of a given protein is found somewhere in a phylogenetic tree, there is obviously a strong possibility that gene duplication might be involved. Possible gene duplications are now being taken into account in the production of parsimonious trees based on amino acid sequencing (e.g. see Goodman, Romero-Herrera *et al.*, 1982). However, a great deal of effort is required to test whether such duplications have actually taken place and there is a danger that gene duplication may be invoked on an *ad hoc* basis to explain away inconvenient features of various parsimonious but apparently anomalous reconstructions. It should also be noted that, if a gene duplication has actually taken place and there has been a switch from an original gene to a new one, considerable doubt must be attached to the allocation of the species concerned to a place in a phylogenetic tree based essentially on proteins produced by the original gene. It has been noted above that the tree-shrew (*Tupaia*) stands out in the inferred phylogenetic tree based on α-haemoglobin sequence data (Fig. 11.26) because of an apparently very high rate of change from the inferred ancestral condition for placental mammals. For this reason, Goodman, Barnabas *et al.* (1971) suggested that there might have been an ancient duplication of the α-gene, with one copy being switched on in *Tupaia* and a different copy switched on in the primates included in the comparison. This, of course, would produce discrepancies in the assessment of the phylogenetic relationships of tree-shrews on the basis of α-haemoglobin sequence data and it is therefore possible that *Tupaia* is, in fact, related to primates despite the great degree of difference observed. On the other hand, as Cook and Hewett-Emmett (1974) have observed, the switching on of an ancient duplicate copy of an α-haemoglobin gene in *Tupaia* 'could equally well be used as further evidence that the tupaioids are not primates'.

It is apparent from all of the points raised above that the interpretation of amino acid sequence information for the inference of phylogenetic relationships between mammalian species is subject to several limitations. Rampant convergence at the molecular level combined with the phenomenon of gene duplication creates special difficulties that cannot really be resolved with the kind of data currently available. Hence, for the forseeable future it will be necessary to maintain a continuous process of cross-checking between phylogenetic trees based on traditional morphological evidence and trees based on molecular data in order to identify discrepancies and anomalies in both approaches. Moreover, given all of the problems that have been identified in reconstructing protein evolution in cases where the amino acid sequence has been determined, it necessarily follows that even greater uncertainties must be attached to the interpretation of immunological comparisons, where global reactivity of entire protein molecules is being assessed. The problems of convergence and gene duplication can only become worse at the immunological level, where it is in any case virtually impossible to recognize discrete character states, as opposed to diffuse overall similarity.

DNA hybridization

Although considerable progress is now being made with the actual sequencing of DNA, rather than depending on indirect inference from the amino acid sequences of proteins produced by individual genes, there is as yet not enough information of this kind to benefit the reconstruction of phylogenetic relationships among primates generally. However, a technique called **DNA hybridization** permits an overall comparison of coding DNA sequences between species (Britten and Kohne, 1968; Britten and Davidson, 1971; Kohne, 1970, 1975; Kohne, Chiscon and Hoyer, 1972). This technique depends on the fact that, under the right laboratory conditions, the double-stranded DNA molecule can be split into separate strands by heating. If the dissociated DNA strands are then slowly cooled, at a particular temperature they will join together in their original combinations because of the complementary pairing of

nucleotide bases (see Fig. 11.3). Now, if such heated, dissociated DNA strands from two different species (A and B) are mixed together, when they are allowed to cool some of them will rejoin to form hybrid DNA containing one strand from species A and one strand from species B. But, because the nucleotide base sequences of species A and species B will differ to a greater or lesser extent, there will be some mismatching in the hybrid DNA molecule, decreasing its thermal stability. The component strands of hybrid DNA will therefore dissociate at a lower temperature than the component strands of DNA from a single species. The amount by which the melting-point of the hybrid DNA is lowered relative to the original DNA provides a direct indication of the degree of mismatching in the nucleotide base sequence of the hydrid DNA. Kohne (1975) states that a lowering of 1 °C in melting-point corresponds to mismatching of approximately 1.5% between base pairs.

A major problem with the DNA hybridization technique stems from the fact, already discussed above, that a substantial proportion of DNA in the mammalian genome is repeated DNA consisting of multiple copies of particular sequences of nucleotide bases. Several recognizable 'families' of repeated DNA sequences contain hundreds of thousands of copies (Kohne, 1975). The remainder of the DNA consists largely of unique sequences, including the genes that code for protein synthesis. Because evolutionary change in repeated DNA seems to follow quite different principles from the evolution of unique sequences of DNA, it is necessary to separate the two kinds of DNA before attempting to use the thermal stability of hybrid DNA molecules as an indicator of phylogenetic affinity between species. Clearly, it should be more profitable to examine the unique sequences of DNA in this way, rather than the repeated DNA, as it is the former category that contains the traditionally recognized 'genes'. In fact, the separation of unique-sequence DNA from multiple-copy DNA turns out to be a relatively simple matter, because the same criterion of temperature of reassociation can be exploited. Because so many copies of repeated DNA sequences are available, this type of DNA can reassociate relatively quickly as dissociated DNA strands are cooled and can therefore be separated from single-copy DNA. It then remains to examine the thermal properties of hybrid molecules formed by the single-copy DNA derived from different pairs of species. To aid in the identification of different fractions of DNA with different thermal properties, the DNA of one species can be radioactively labelled by appropriate treatment of the tissue from which the DNA is extracted.

Kohne (1975) reported a preliminary attempt to reconstruct phylogenetic relationships among seven primate species with the aid of data derived from DNA hybridization, using labelled DNA from humans and the vervet monkey (*Cerco-pithecus aethiops*) as the standards for comparison. On the basis of thermal stability of hybrid DNAs formed between human or vervet DNA and the DNAs of other primates, it was possible to infer the overall percentage of nucleotide substitutions that had taken place since the divergence of each species. As with immunological comparisons and amino acid sequence information, it was then possible to calculate individual branch lengths in an inferred phylogenetic tree through the solution of simultaneous equations. An outline tree compatible with Kohne's data is shown in Fig. 11.35. It can be seen that the tree has the pattern predicted from phylogenetic reconstructions based on morphological evidence. The only prosimian in the sample, *Galago*, branches away first and is very distinct from the six simians in terms of degree of separation. The New World monkey, *Cebus*, branches off first among the simians investigated and there is then the expected dichotomy between the two Old World monkeys (*Cercopithecus* and *Macaca*) and the three Old World hominoids (*Hylobates*, *Pan* and *Homo*). Among the hominoids examined, the gibbon (*Hylobates*) is the most divergent, with *Pan* and *Homo* clearly quite closely related to one another. It should also be noted that there is no evidence of marked differences in rate of change along the different lineages in Fig. 11.35, although it is impossible to comment meaning-

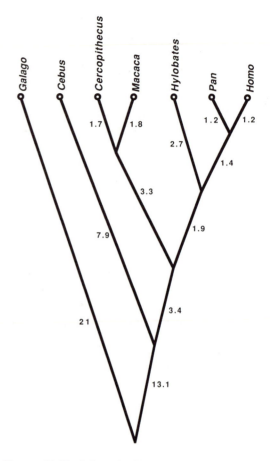

Figure 11.35. Inferred phylogenetic tree for seven primate species based on the DNA hybridization data for single-copy DNA provided by Kohne (1975). The figure alongside each branch indicates the percentage of nucleotide bases in the single-copy DNA believed to have undergone substitution along that branch. Although the tree has been 'rooted' by assuming that equal amounts of evolutionary change in the single-copy DNA have occurred along the lineage leading to *Galago* and along the lineages leading to individual simian species, the relative placing of the initial node of the tree is, of course, subject to doubt in the absence of data from a suitable non-primate outgroup.

fully on the separation between *Galago* and the simians because the tree has not been 'rooted' with data derived from an appropriate non-primate outgroup.

Benveniste and Todaro (1976) used the technique of DNA hybridization to examine the phylogenetic relationships among Old World simians. They compared samples from 10 monkey genera, 5 ape genera and humans. The resulting phylogenetic tree neatly fitted the expectation from morphological studies. Following a basic split between the monkeys and apes, there was a further division between leaf-monkeys (Colobinae) and guenons (Cercopithecinae). On the ape side of the tree, the gibbons and siamang branched away first, followed by the orang-utan, leaving the African apes as the closest relatives of humans.

A more wide-ranging study was later conducted by Bonner, Heinemann and Todaro (1980), who commendably included *Tarsius* along the primates examined (Fig. 11.36). They carried out DNA hybridization with DNA from nine primate species representing four main groups: lemurs (*Lemur catta* and *L. mongoz*), lorises (Galago, Loris and *Nycticebus*), tarsiers and simians (*Ateles*, *Papio* and *Homo*). Labelled DNA was prepared as a standard for comparison from one member of each group and hybridization was carried out essentially as described above. Bonner *et al.* (1980) differed from Kohne (1975) in equating a 1 °C difference in melting-point with a 1% divergence in nucleotide base sequence, rather than 1.5%, but otherwise their procedure was much the same. Indeed, after allowance has been made for the difference in 'calibration' of the thermal stability of hybrid DNA, the inferred degree of difference between lorises and simians is found to be reasonably similar between the two studies. Figure 11.36 also shows an important additional feature that neatly fits the prediction based on a provisional assessment of the morphological evidence (Fig. 2.2): the primate tree divided early on to yield a monophyletic strepsirhine group (lemurs and lorises) and a monophyletic haplorhine group (tarsier and simians). Further, although *Tupaia* is not included in Fig. 11.36, Bonner *et al.* (1980) did examine DNA from this genus in order to 'root' the tree. In fact, they found that the single-copy DNA of this tree-shrew is so divergent from that of all the primates investigated that direct comparison was hindered and an extrapolation had to be made. Following

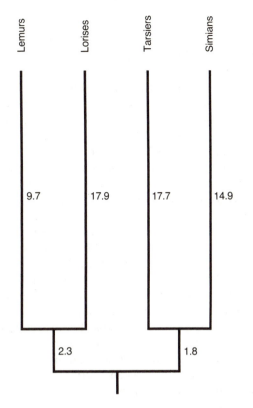

Figure 11.36. Inferred phylogenetic tree for four groups of living primates based on DNA hybridization data for single-copy DNA. As in Fig. 11.35, the figure alongside each branch indicates the percentage of nucleotide bases in the single-copy DNA believed to have undergone substitution along that branch. (After Bonner, Heinemann and Todaro, 1980.)

the necessary adjustment, the primate tree was then rooted as shown in Fig. 11.36. Bonner *et al.* also specifically noted that there are differences in apparent rate of change along the four main lineages in Fig. 11.36. The simian lineage seems to have accumulated nucleotide substitutions somewhat more slowly than either the loris lineage or the tarsier lineage, and the lemur lineage has a strikingly low level of accumulation of nucleotide substitutions in comparison to all of the other three lineages. Although no separate data are provided, single-copy DNAs from *Microcebus*, *Cheirogaleus* and *Propithecus* were also examined by hybridization and the results confirmed the general picture that there has been relatively slow evolutionary change among lemurs.

The technique of DNA hybridization has also been used for a specific examination of phylogenetic relationships among hominoid primates (Sibley and Ahlquist, 1984), again using the single-copy DNA fraction as a basis for comparison. Labelled single-copy DNAs were prepared for *Homo*, *Pan troglodytes*, *P. paniscus*, *Gorilla*, *Pongo* and *Hylobates lar*, thus providing a comprehensive basis for reciprocal comparisons among hominoids. Unlabelled DNAs from five Old World monkey genera (*Allenopithecus*, *Cercopithecus*, *Macaca*, *Papio* and *Pygathrix*) were also used for DNA hybridization in order to provide an appropriate outgroup to 'root' the phylogenetic tree determined for the hominoids. The overall result obtained was once again in accordance with expectation. Old World monkeys were approximately equidistant from all of the hominoids, which clearly formed a distinct monophyletic group. Among the hominoids, the gibbon branched away first, followed by the orang-utan, leaving a fairly close-knit cluster containing the African apes and humans. Of the African great apes, the two chimpanzees seemed to be somewhat more closely related to us than the gorilla.

Hence, phylogenetic relationships among primates inferred from the results of DNA hybridization show excellent general agreement with the pattern previously indicated by carefully conducted morphological and chromosomal comparisons. This is most encouraging, because the technique of DNA hybridization provides a means of comparing entire genomes between species. In terms of inferred branching patterns there has as yet been no recorded case of obvious incompatibility between the results of DNA hybridization studies and the results of morphological and chromosomal studies, as interpreted in previous chapters and in the first section of this chapter. Nevertheless, there has been considerable controversy in the recent literature over the degree of correspondence between morphological evidence and molecular evidence relating to primate evolution. However, this controversy has been concerned not so much with the branching patterns of inferred phylogenetic trees but with inferences about the timescale of evolution. This introduces a quite different topic that deserves some discussion in its own right. It will be examined as part of the overall review of primate phylogenetic relationships in Chapter 12.

Chapter twelve

A provisional synthesis

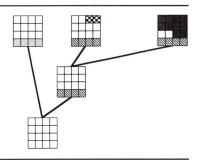

The central concern of previous chapters has been the integration of comparative information from several major areas of primate biology. Such wide-ranging coverage inevitably incurs the cost of superficial treatment of individual subject areas, but the view is taken here that the potential benefits far outweigh this price. As was emphasized in Chapter 3, the uncertainties involved in phylogenetic reconstruction call for assessment of primate evolution on the broadest possible basis. It cannot be emphasized enough that phylogenetic reconstruction requires **inference** of past ancestral conditions. As the ground rules for generating inferences are all of a probabilistic nature (see Chapter 3), every effort must be made to maximize the contribution made by hard information in phylogenetic analyses. This applies not only to the range of characters considered, which should ideally extend from morphological to molecular levels, but also to the sample of species covered. Thus, in addition to examining primate species, both living and fossil, it is necessary to incorporate information on other placental mammals and even on marsupials. An essential component of the reconstruction of primate evolution is the recognition of specific developments in the

earliest primates that set them apart during the initial phase of the adaptive radiation of the placental mammals. In other words, it is necessary to distinguish the derived features of early primates from the primitive features of ancestral placental mammals. Primate evolution can be understood effectively only within the broader framework of mammalian adaptive radiation. For this reason, the general background to mammalian evolution was specifically covered in Chapter 4 as an essential preliminary to discussion of the individual character systems of primates.

It is also worth repeating here the fundamental point established in Chapter 3, that classification (taxonomy) should be kept distinct from phylogenetic reconstruction as far as possible. There is, naturally, some overlap between these two subjects; but there are many important distinctions between them. In particular, a classification must serve various utilitarian functions, all requiring a large element of stability. Phylogenetic reconstructions, on the other hand, must necessarily change dynamically as we attempt to trace the evolutionary history of primates with ever greater precision. Baldly stated, classification is (or should be) an art, whereas phylogenetic reconstruction is (or should be) a science. In contrast to the application of scientific methods to reconstruct evolutionary history, classification is 'for the convenience of biologists and is in their minds and not in the universe of living things' (Kermack and Kermack, 1984).

Phylogenetic reconstruction, of course, requires some kind of hierarchial classification of species as a starting point, if only as a source of names for shorthand reference to groups of species. On one hand, the groups assembled in the classification should be 'meaningful', in some general evolutionary sense, so as to provide an appropriate set of names for phylogenetic analysis. On the other hand, it is frustrating both for individual researchers and for general readers if the use of group names is continually changing. It is difficult enough to cope with constantly changing interpretations of phylogenetic relationships without having to cope at the same time with major changes in classificatory schemes, often triggered by inferences based on surprisingly flimsy evidence. Far from contributing to a better scientific understanding of phylogenetic history, such classificatory changes act as stumbling blocks. As Kemp (1982) quite rightly observed:

> The construction of formal classifications based upon phylogenetic reconstruction has led to a great deal more controversy and misunderstanding than the method of phylogenetic reconstruction itself.

Worse, attempts to link classifications directly to individual hypotheses about phylogenetic relationships have diverted attention from the pressing need to achieve precise formulation and application of reliable methods for phylogenetic reconstruction. Far too often, great effort has been devoted to prescribing rules for the 'accurate' reflection of phylogenetic trees in classifications, while the accuracy of the trees themselves has been virtually taken for granted. For the reasons set out in Chapter 3, the cladistic approach to classification has been firmly rejected here in favour of the 'classical' approach. In a classical classification, relatively

stable, 'meaningful' grouping of species can be maintained by making minimal adjustments to ensure compatibility with consensus views of phylogenetic relationships. It is not necessary to reflect any particular hypothetical phylogenetic tree in detail.

By contrast, when it comes to phylogenetic reconstruction the more rigorous framework introduced through the cladistic school has been of enormous benefit. Provided that the individual ground rules for phylogenetic reconstruction outlined in Chapter 3 are taken as guidelines and not as infallible prescriptions, the more rigorous cladistic approach doubtless leads to greater precision. With the aid of those guidelines, it is possible to conduct more objective assessment of evidence from various areas of primate biology, such as those summarized in Chapters 6–11. The inclusion of additional evidence from comparative studies of chromosomal and molecular structure (Chapter 11) is especially important, as it not only significantly broadens the range of characters covered but also introduces some entirely new elements into the overall framework.

It is also worth re-emphasizing the proposition from Chapter 1 that the process of phylogenetic reconstruction should ideally begin with the comparative study of living forms. There are several reasons for this, but the most important is that vastly more characters are available for study with any living species. Following the ground rule that successful phylogenetic reconstruction requires the analysis of the maximum possible number of different characters, it can be argued that the integration of any fossil form into a hypothetical tree must always be somewhat tentative. Certainly, the distinction between substantial and fragmentary fossil species introduced in Chapter 2 is crucial here. With the former, sufficient characters are preserved to permit a reasonable assessment of probable phylogenetic affinities. With the latter, only a provisional label is justified. Indeed, the allocation of fragmentary fossil species is almost entirely a classificatory exercise. Such fossils are allocated to a group containing species showing the greatest degree of similarity for the few characters preserved. Without additional fossil material, and hence additional characters, it is impossible (for instance) to make a reasonable assessment of the probability of convergent evolution.

A cautionary tale is provided by the rise and fall of *Ramapithecus* as a candidate for early ancestry in the human lineage (see Chapter 2 and later in this chapter). When only fragmentary jaws bearing a few teeth were available, it seemed (to many authors) reasonable to conclude that *Ramapithecus* was a hominid. However, following the discovery of partial skulls, it now seems more reasonable to suggest that *Ramapithecus* is allied to the orang-utan lineage. This marked shift in interpretation is partly attributable to new hypotheses regarding the polarity of certain features (e.g. it has now been suggested that thick dental enamel, rather than thin enamel, was present in the common ancestor of the great apes and humans; L.B. Martin, 1985). However, it is undoubtedly the newly documented, extensive similarity in facial structure between *Ramapithecus* and the modern orang-utan that has led to ready acceptance of a major shift in phylogenetic inference. Further, certain features present in the jaws and teeth of *Ramapithecus*, but absent from those of the orang-utan, and once thought to link *Ramapithecus* specifically to later hominids, must presumably now be seen as products of convergent evolution. The revised interpretation, although based on several additional characters, is itself provisional. It cannot be ruled out that discovery of a complete skull and postcranial skeleton of *Ramapithecus* might lead to yet another fundamental shift in phylogenetic hypotheses, although this, admittedly less likely.

Hence, there is much to be said for an approach that excludes fragmentary fossils from phylogenetic reconstructions, at least in the first instance, and relies exclusively on substantial fossils. Even then, it is preferable to generate hypotheses regarding phylogenetic relationships initially on the basis of living species alone and to introduce information from substantial fossil species subsequently to test and refine those hypotheses. Of course, fossil evidence has a unique role in indicating a likely timescale for any given phylogenetic tree and there are other areas in which even fragmentary fossil evidence is especially valuable (e.g. in tracing micro-evolutionary changes at a single fossil site; see Gingerich, 1976a, 1977a). Nevertheless, within an overall reconstruction of primate evolutionary history, the interpretation of fossil evidence necessarily takes a subsidiary role. The formerly widespread practice of assigning directly ancestral positions to most fossil primate species, particularly (for obvious reasons) in the case of fragmentary forms, is now widely seen to be unacceptable. Once it is accepted that most known fossil primate species are likely to come from side branches of the tree, rather than from lineages leading directly to living forms, fossils are revealed in their true light as pale shadows of their living relatives – of equally uncertain affinities and documented only for a few features. We must naturally make the best use we can of the primate fossil record, but the great deficiencies in that record as a whole must always be borne in mind, The mammalian fossil record is typically deficient not only in terms of the material recovered for individual species but also in terms of grossly unequal representation of material from different geographical areas and from different geological ages. Kermack and Kermack (1984) aptly commented:

> The assumption that the collections that are available in museums give a representative sample of the fauna of a particular geological age is a highly dangerous one, although it can be difficult to resist.

DEFINITION OF THE ORDER PRIMATES

An essential aid in the reconstruction of primate evolutionary history is a working definition listing the significant features of primates. This is one example of the necessary overlap between pre-existing classifications of primates, reflecting consensus views of phylogenetic relationships, and phylogenetic reconstruction. In order to define primates, it is necessary to know which species, both living and fossil, are generally included in the order Primates. It is obvious from various widely cited classifications, and from accompanying discussions, that there are two main areas of contention regarding the boundaries of the Primates. As far as living forms are concerned, views have differed as to whether tree-shrews should be regarded as primates. Among fossil forms, there is uncertainty about the relationship between the 'archaic primates', now commonly included in the infraorder Plesiadapiformes (see Table 3.1), and the remaining fossil and living 'primates of modern aspect'. As a definition must provide a reliable basis for discussion, it is surely best to begin with species that are regarded by an overwhelming majority of experts as definite primates. So, as far as living primates are concerned, tree-shrews should be excluded from the order, at least initially, and the definition should be restricted to lemurs, lorises, tarsiers, monkeys, apes and humans.

Further, following the argument given above, it is preferable to exclude all fossil forms from the definition initially. Those characters in the definition that can be fossilized may then be used as a basis for deciding which fossil species can be regarded with some confidence as primates, because they have numerous features included in the definition based on living species, and which fossil species are of questionable status, because they possess only a few of those features. Naturally, if a given fossil form has only a few of the characters in the definition, there is the danger that a limited similarity to primates has arisen through convergent evolution. By contrast, if particular fossil forms show many of the features in the definition but do not exactly fit the expected pattern, this may lead to refinement of the definition itself, such that it will cover both living and fossil forms. In this respect, fossils serve as a test of hypothetical relationships inferred initially from living forms alone.

It must also be decided whether a definition of primates should summarize all features common to the group or only those than can be reasonably regarded as shared derived features of primates. For a general, all-purpose definition, it might be argued that all characteristic features should be included, regardless of their evolutionary origin. However, for a specific definition that will be used as a basis for phylogenetic reconstruction, inclusion of primitive features of the placental mammals is at best superfluous and at worst misleading. Simpson (1961) drew a distinction between definition, as the identification of the collective boundaries of a group of species, and diagnosis, as the identification of the features demarcating a group from other groups. However, this distinction is of little value with respect to phylogenetic reconstruction. Retained primitive features may be cited either in a definition (e.g. all living primates are stated to have a particular feature retained from the ancestral placental mammals) or in a diagnosis (e.g. that feature has been lost from most or all other living placental mammals). In either case, a given fossil form with the feature concerned might be identified as a 'primate', whereas it can in fact only be identified as a placental mammal. For this reason, the widely used term 'definition' is retained here, with the reservation that inclusion of primitive features of placental mammals in a definition of primates is not advisable if it is to be used as a basis for phylogenetic reconstruction.

As shown in Chapter 5, many current definitions of primates can be traced back to the brief list of characters given by Mivart (1873):

> unguiculate claviculate placental mammals with orbits encircled by bone; three kinds of teeth, at least at one time of life; brain always with a posterior lobe and calcarine fissure; the innermost digits of at least one pair of extremities opposable; hallux with a flat nail or none; a well-developed caecum; penis pendulous; testes scrotal; always two pectoral mammae.

In that definition, no attempt was made to distinguish between primitive features of placental mammals, novel features developed convergently in several mammalian groups and exclusive, derived features of primates. Indeed, a closer look at the article containing this widely quoted definition reveals a surprising fact: Mivart specifically pointed out that he did not regard the primates as a monophyletic group. He suggested that the strepsirhine primates (lemurs and lorises) and the haplorhine primates (tarsiers and simians) arose separately from the ancestral stock of the placental mammals. There has, therefore, never been any reason to expect that Mivart's definition would be of much value in reconstructing the evolutionary history of a monophyletic group including all living primates (strepsirhines and haplorhines). In fact, most features in Mivart's definition have been shown to be primitive retentions from the ancestral placental stock: presence of a clavicle; possession of three kinds of teeth; presence of a caecum; possession of a pendulous penis. The possession of scrotal testes is either a primitive feature for placental mammals or a feature that has evolved convergently in numerous lineages. Forgetting Mivart's error in specifying two pectoral teats (mammae) as a defining feature of primates (see Chapter 9), only two features may be retained from his definition as likely shared derived features of primates: (1) possession of a grasping big toe (hallux) combined with the presence of a flat nail at least on this digit (see Chapter 10); (2) presence of a calcarine sulcus in the brain (see Chapter 8). This provides a very slim foundation for the assessment of phylogenetic affinities, although living tree-shrews do lack both of these features, and 'archaic primates' apparently lacked the first (the second not being detectable in fossil forms).

The largely morphological features discussed in Chapters 6–10 can be used to construct a working definition of living primates. This can be achieved by taking features that are universal, or virtually universal, among living primates and, as far as possible, absent from other placental mammals. Some features, such as the postorbital bar and increased brain size, have been included despite relatively common occurrence among other mammals because all living primates have them. All universal features of primates, particularly when very rare or lacking among other mammals, are good candidates for interpretation as shared derived features. The resulting definition of living primates (with later refinements appended as notes [1]–[3]) is as follows (Martin, 1986a):

Primates are typically arboreal inhabitants of tropical and subtropical forest ecosystems. Their extremities are essentially adapted for prehension, rather than grappling of arboreal supports. A widely divergent hallux provides the basis for a powerful grasping action of the foot in all genera except *Homo*, while the hand uusually exhibits at least some prehensile capacity. The digits typically bear flat nails rather than bilaterally compressed claws; the hallux always bears a nail. The ventral surfaces of the extremities bear tactile pads with cutaneous ridges (dermatoglyphs) that reduce slippage on arboreal supports and provide for enhanced tactile sensitivity in association with dermal Meissner's corpuscles. Locomotion is hindlimb-dominated, with the centre of gravity of the body located closer to the hindlimbs, such that the typical walking gait follows a diagonal sequence (forefoot precedes hindfoot on each side). The foot is typically adapted for tarsi-fulcrimation, with at least some degree of relative elongation of the distal segment of the calcaneus, commonly resulting in reverse alternation of the tarsus (calcaneo-navicular articulation).

The visual sense is greatly emphasized. The eyes are relatively large and the orbits possess (at least) a postorbital bar. Forward rotation of the eyes ensures a large degree of binocular overlap. Ipsilateral and contralateral retino-fugal fibres are approximately balanced in numbers on each side of the brain and organised in such a way that the contralateral half of the visual field is represented [1]. Enlargement and medial approximation of the orbits is typically associated with ethmoid exposure in the medial orbital wall (though there are several exceptions among the lemurs). The ventral floor of the well-developed auditory bulla is formed predominantly by the petrosal. The olfactory system is unspecialized in most nocturnal forms and reduced in diurnal forms. Partly because of the increased emphasis on vision, the brain is typically moderately enlarged, relative to body size, in comparison to other living mammals. The brain of living primates always possesses a true Sylvian sulcus (confluent with the rhinal sulcus) and a triradiate [2] calcarine sulcus. Primates are unique among living mammals in that the brain constitutes a significantly larger proportion of body weight at all stages of gestation [3].

Male primates are characterized by permanent precocial descent of the testes into a postpenial scrotum; female primates are characterized by the absence of a urogenital sinus. In all primates, involvement of the yolk-sac in placentation is suppressed, at least during the latter half of gestation. Primates have long gestation periods, relative to maternal body size, and produce small litters of precocial neonates. Fetal growth and postnatal growth are characteristically slow in relation to maternal size. Sexual maturity is attained late and life-spans are correspondingly long relative to body size. Primates are, in short, adapted for slow reproductive turnover.

The dental formula exhibits a maximum of $I_2^2 \, C_1^1 \, P_3^3 \, M_3^3$. The size of the premaxilla is very limited, in association with the reduced number of incisors, which are arranged more transversely than longitudinally. The cheek teeth are typically relatively unspecialized, though cusps are generally low and rounded and the lower molars possess raised, enlarged talonids.

Since this definition was published, several refinements have become necessary:

[1] This distinction applies primarily to the subsidiary visual system involving the optic tectum. In all placental mammals, visual representation in the visual cortex of the primary visual system involves reversed representation of the left and right halves of the visual field. In a few non-primates (e.g. cats) there is an approximate balance between ipsilateral and contralateral retinofugal fibres. Primates are unusual in that the same pattern is found in the optic tectum as well, whereas in other mammals generally

the entire visual field seen by the contra-lateral eye is represented in each side of the optic tectum. Numbers of ipsilateral and contralateral retinofugal fibres passing to the optic tectum are also approximately balanced in primates, whereas in non-primates contralateral fibres typically predominate.

Further, it has recently been reported that fruit-bats (Megachiroptera), but not other bats (Microchiroptera), exhibit the same pattern of projection to the optic tectum as primates (Pettigrew, 1986). Hence, this feature is not unique to primates, unless fruit-bats are included in the order Primates. Nevertheless, the pattern of projection to the optic tectum in primates is unusual enough among placental mammals generally to warrant continued inclusion in the definition.

[2] There is, as noted in Chapter 8, at least one exception to this. In marmosets and tamarins (family Callitrichidae) the calcarine sulcus is simple, rather than triradiate (L. Garey, personal communication). Secondary dwarfing of marmosets and tamarins in evolution (see later) has perhaps led to a secondary modification of the original pattern.

[3] It should be added that the corticospinal tracts in primates have a characteristic pattern of organization – they are located in the lateral funiculi of the spinal cord and extend well down the cord (see Chapter 8, especially Fig. 8.23). Primates may also be distinctive in having cytochrome oxidase patches in the visual cortex (Horton, 1984).

Ideally, such a definition of primates should also refer to chromosomal and biochemical features. In theory, it is now possible to infer an ancestral set of chromosomes (karyotype) for primates and it is also possible to infer ancestral amino acid sequences for certain proteins and even nucleotide base sequences for the corresponding genes. This, however, touches on a fundamental problem of formulating a definition of primates. As has already been explained, in the definition given above emphasis has been given to features that are universal (or virtually universal) to living primates and rare or lacking in other living placental mammals. That is to say, the definition does not prescribe the ancestral condition for primates but a prevalent condition among living forms. With chromosomes and molecules, it is difficult to identify features of this kind common to all living primates, whereas it is now (at least in principle) relatively easy to identify likely ancestral conditions that are often no longer present among living forms. It is, of course, one of the major aims of this book to identify the derived features of the common ancestor of all living primates. It is quite likely for morphological features, as for chromosomes and proteins, that certain derived character states that were present in ancestral primates are no longer represented widely, if at all, among living primates, so those ancestral states must be inferred through a careful process of reconstruction. The above definition provides no more than an initial foundation from which to work towards eventual identification of the ancestral features of primates, including the ancestral karyotype, the amino acid sequences of ancestral proteins and the nucleotide base sequences of ancestral genes.

Unfortunately, only a reduced form of the above definition is applicable to fossil forms because many of the features listed are undetectable in fossils. It is, for instance, virtually impossible to discover anything about reproductive biology in fossil primates and only the most superficial details of brain morphology can be determined from a few, particularly well-preserved, fossil specimens. In order to test whether the definition given is applicable to fossil primates as well as to living primates, the following list of fossilizable features may be extracted:

1. The big toe (hallux) is well developed and divergent, reflecting its adaptation as a grasping organ. The presence of a flat nail on the hallux is revealed by the dorsoventrally flattened shape of the terminal phalanx of that digit.

2. The distal segment of the calcaneus is elongated, resulting in a low calcaneal index value (relative to body size).
3. The orbits are relatively large and convergent and the interorbital distance is correspondingly small. A postorbital bar is present. There may or may not be exposure of the ethmoid in the orbit, depending on the magnitude of the interorbital distance relative to skull size.
4. There is a relatively prominent auditory bulla, the ventral floor of which is formed from the petrosal.
5. The braincase is relatively large and an endocast, if available, should reveal the presence of a Sylvian sulcus.
6. The maximum dental formula is $I_2^2 C_1^1 P_3^3 M_3^3$. The premaxilla is relatively short and the upper incisors are arranged more transversely than longitudinally. The molar teeth have low, rounded cusps and the lower molars have raised, enlarged talonids.
7. It is also expected from the definition of living primates that related fossil forms were typically inhabitants of tropical or subtropical forest habitats.

The overall list can be taken as a basis for assessing the likelihood that individual groups of fossil species are related to modern primates. The same strategy can be adopted as with selecting living primates for formulating the full definition. In the first instance, at least, only undoubted fossil primates should be taken. Consideration of any forms of dubious status should be deferred. In practice, this means that the known post-Palaeocene fossil 'primates of modern aspect' should be considered initially, whereas the primarily Palaeocene 'archaic primates' should be assessed later. As shown in Chapters 6–10 (see also Martin, 1986a), most of the features cited in the restricted list above are identifiable in fossil primates of modern aspect back to the beginning of the Eocene, in so far as suitable fossil evidence is available. Indeed, the matching of this reduced definition of living primates with the comparatively well-known Eocene primates of modern aspect, both 'lemuroids' (Adapidae) and

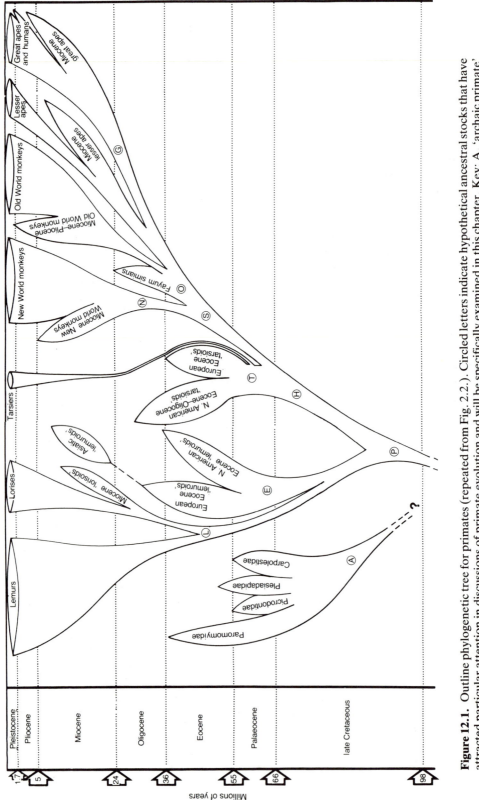

Figure 12.1. Outline phylogenetic tree for primates (repeated from Fig. 2.2.). Circled letters indicate hypothetical ancestral stocks that have attracted particular attention in discussions of primate evolution and will be specifically examined in this chapter. Key: A, 'archaic primate' stock; P, primate of modern aspect stock; E, Eocene 'lemuroid' stock; L, lemur/loris stock; H, 'tarsioid'/simian stock; T, 'tarsioid' stock; S, simian stock; N, New World simian stock; O, Old World simian stock; G, great ape stock.

'tarsioids' (Omomyidae), is so good that little doubt can exist about their status as members of a monophyletic radiation leading to modern primates. Further, given that most features fit the expectation very well, it is justifiable to modify the inferred ancestral primate condition in at least one specific point. Among both the Adapidae and the Omomyidae there are individual fossil species that still had four premolars in upper and lower dentitions. Hence, it can be concluded that the common ancestor of primates of modern aspect still had a dental formula of $I_2^2 C_1^1 P_4^4 M_3^3$. Further reduction of this dental formula to a maximum of $I_2^2 C_1^1 P_3^3 M_3^3$ through loss of a premolar in each upper and lower jaw presumably occurred subsequently – perhaps several times in parallel – during the evolution of living primates. There is also some doubt about the presence of the Sylvian sulcus in the common ancestor of primates of modern aspect. Although this sulcus is identifiable in endocasts of most Eocene primates of modern aspect, including both 'lemuroids' and 'tarsioids', the sulcus has not been identified on the endocast of the North American 'lemuroid' *Smilodectes*. It is a moot point whether the sulcus was genuinely absent from the brain of *Smilodectes*, or whether it was merely not reflected on the endocast. Even if the sulcus was, in fact, missing from the brain of *Smilodectes*, it would be reasonable to infer that secondary loss took place in this genus from some reason, as the sulcus was present in other Eocene representatives of the family Adapidae and was apparently uniformly present in the contemporary sister group Omomyidae. Apart from these reservations, it would seem that the full definition of living primates provides a useful first step towards the reconstruction of ancestral stages in the evolution of primates, at least back to the beginning of the Eocene. Having established this, it is now possible to return to the hypothetical evolutionary tree for primates originally given in Chapter 2 (Fig. 2.2) and to summarize provisional conclusions regarding some of the key ancestral stages (Fig. 12.1).

* Definition of primates reprinted with permission from Cambridge University Press.

THE MAJOR ISSUES

The first question that must be answered is that introduced in Chapter 5: Are tree-shrews primates? It is not the question of tree-shrew classification that is at stake here. The real issue, as far as this book is concerned, is: Did tree-shrews share a specific common ancestry with primates, to the exclusion of all other groups of living placental mammals? Whatever the answer to this question, assessment of the status of tree-shrews provides a valuable object-lesson in the reconstruction of the phylogenetic history of primates. It is essential to establish whether the widespread notion of a diffuse zone of transition between insectivores and primates is correct, with tree-shrews branching off somewhere within this zone, or whether it is possible to identify key features that were present in the common ancestor of modern primates (*sensu stricto*). If such specific features can be identified, their presence or absence in tree-shrews will effectively decide the question regarding a specific phylogenetic relationship between tree-shrews and primates. The issue of the phylogenetic relationships of tree-shrews also introduces the question of the validity – as a monophyletic assemblage – of the superordinal grouping Archonta as originally suggested by Gregory (1910). The possibility of a genuine phylogenetic association between primates, tree-shrews, bats and 'flying lemurs' has recently been revived by several authors (e.g. Szalay, 1977; Cronin and Sarich, 1980; Shoshani, 1986) and deserves serious examination. Having decided on the likely phylogenetic relationships of tree-shrews, it is possible to proceed to discussion of the most appropriate way of classifying them.

The question of the relationship of the living tree-shrews to primates is paralleled among fossil forms by the question: Are plesiadapiforms primates? Once again, the question is uniquely concerned with the presence or absence of a specific ancestral relationship between plesiadapiforms and the ancestral stock of primates of modern aspect. In addition, resolution of the status of the 'archaic primates' provides a further object-lesson, this time in the reliable identification of fossil primates, leading to further refinement of our understanding of the key developments in the earliest stages of primate evolution.

Once questions regarding the phylogenetic status of tree-shrews and plesiadapiforms have been resolved, it is possible to turn to a reconstruction of the ancestral primate of modern aspect. This requires identification of the key features that distinguished ancestral primates from ancestral placental mammals. In addition, it is possible to proceed to consider those characters as an integrated network, such that some inference of the lifestyle of ancestral primates may be made. This leads directly to the issue of the earliest phases of diversification of primates and hence to the question of the validity – in phylogenetic terms – of the distinction between strepsirhines (lemurs and lorises) and haplorhines (tarsiers and simians). The hypothesis that there was an initial divergence in primate evolution between strepsirhines and haplorhines must be examined in the light of all of the accumulated evidence. This is especially necessary because of competing hypotheses that suggest an ancestral link between tarsiers and 'archaic primates' and/or an ancestral link between strepsirhines and simians.

With respect to the later stages of primate evolution, certain topics have been given particular prominence in the literature. For strepsirhine primates, a major question is: Are the lemurs monophyletic? Numerous authors, particularly in recent years, have suggested that the lemurs of Madagascar are not derived from a single ancestral stock distinct from that which gave rise to the loris group. In particular, it has been claimed that the mouse and dwarf lemurs (family Cheirogaleidae) shared a specific ancestral stock with the loris group (family Lorisidae). For haplorhines, a parallel question is: Are the simians monophyletic? Although opinion has now swung towards a single origin for simians, there is a continuing minority view based on numerous suggestions in the earlier literature that New World monkeys and Old World simians

were separately derived from prosimian ancestors. In other words, it is still claimed by some that many of the similarities shared by modern simians were developed through convergent (parallel) evolution in the New and Old Worlds. In this case, the implications of continental drift for the phylogenetic separation between New World and Old World simians must be considered. Among the New World monkeys themselves, an important issue concerns the status of marmosets and tamarins. Some authors have interpreted these as the most primitive of living New World monkeys, whereas others have interpreted them as specialized products of a secondary reduction in body size ('dwarfing'). Among the Old World simians the traditional distinction between 'apes' and 'monkeys' deserves serious re-examination. Under the influence of the *Scala naturae*, there has been a general tendency to regard apes, which are more closely related to humans, as more advanced than the more distantly related monkeys. It is now emerging that Old World monkeys are in many ways quite specialized, and meaningful application of the terms 'ape' and 'monkey' to the fossil record has proved to be increasingly problematic.

A very general issue that arises with respect to the primate phylogenetic tree as a whole is that of determining an appropriate timescale. Here, fossil evidence comes into its own, but reliable dating of branching points in the primate tree is fraught with difficulty because of yawning gaps in the fossil record, most notably in the southern continents. Commonly accepted timescales for primate evolution must be regarded with a great deal of scepticism, particularly when the implications of continental drift for mammalian evolution in general are taken into account. The question of a timescale for primate evolution is specifically relevant to the special topic of human evolution. The implications of the present broad survey of primate evolutionary biology for the specific lineage leading to the modern human species must obviously be examined, particularly with respect to inferred dates of divergence. This is especially necessary because of controversy in the literature over conflicting interpretations from palaeontological and biochemical evidence. The specific topic of human evolution has received little special attention in the preceding chapters, in which the human species has figured simply as one of almost 200 living primate species. However, it should be recognized that a broader perspective from an examination of primate evolution overall can be more valuable than a more tightly focused, specific study of human evolution.

Finally, although it should be emphasized that the primary concern of this book is the reconstruction of primate evolutionary history, it is worthwhile considering the possible implications of the findings for primate classification. Every attempt has been made so far to avoid changes in classification wherever possible, but in some cases a point may be reached where an existing classification may be simply incompatible with highly probable phylogenetic relationships. It is instructive to consider the extent to which the classification given in Chapter 3 (Table 3.1) can absorb, without any need for modification, the specific hypotheses regarding phylogenetic relationships suggested here and to what extent limited changes might eventually become necessary.

ARE TREE-SHREWS PRIMATES?

As was emphasized in Chapters 3 and 5, resolution of the question of the phylogenetic affinities of tree-shrews has been generally hindered by confusion between the operations of classification and phylogenetic reconstruction. The balance of scientific opinion has now shifted against inclusion of tree-shrews in the order Primates. One can, for example, compare Napier and Napier (1985) with Napier and Napier (1967), or Buettner-Janusch (1973) with Buettner-Janusch (1966) (see also Luckett, 1980b). However, until relatively recently this shift owed more to changing perceptions of the optimal location of the 'insectivore–primate boundary' in classification than to a fundamental rethinking of primate origins based on step-by-step phylo-genetic reconstruction.

The concept of the 'phylogenetic scale' (see Fig. 3.5) has exerted a particularly strong influence on past discussions of primate evolution. Since living lipotyphlan insectivores are widely regarded as the most primitive surviving placental mammals, it follows that these mammals must occupy the lowest rung on the placental 'scale'. It also follows from such thinking that any 'missing link' in the supposed evolutionary chain between insectivores and primates should be intermediate in form between typical living insectivores (e.g. a hedgehog) and relatively primitive primates (e.g. certain lemurs).

Living tree-shrews conform with this expectation remarkably well in numerous features, notably because of various characteristics of the visual system, of the brain and of the postcranial musculoskeletal system, all of which are associated with their typical semi-arboreal habits. Given the inevitable conclusion that tree-shrews are morphologically intermediate between insectivores and primates, it is a seemingly obvious step to infer that they are phylogenetically intermediate. Consequently, their classification with either insectivores or primates seems to be a matter for relatively arbitrary choice. Seen in this light, the question 'Are tree-shrews primates?' is reduced to a comparatively uninteresting enquiry about arbitrary taxonomic boundaries. Indeed, many discussions of the affinities of tree-shrews, while purportedly dealing with the assessment of phylogenetic relationships, have actually been largely or exclusively concerned with the relatively trivial classificatory issue. It has been widely assumed, as a point of departure, that tree-shrews are either insectivores or primates in phylogenetic terms and that assessment of their status depends merely on evaluation of the undoubted similarities that tree-shrews share with primates. (Novacek, 1982b, referred to the 'highly publicized trio represented by primates, tupaiids and lipotyphlous insectivores'.)

A totally different picture emerges when the phylogenetic affinities of tree-shrews are assessed within an appropriate framework of phylogenetic reconstruction of mammalian evolution generally. In order to decide whether tree-shrews are really related to primates in a

Table 12.1 Summary of interpretations of major similarities shared by tree-shrew and primates, as listed in Table 5.4

Context	Shared similarities
Skull	1. Snout relatively short
	2. Simplified set of turbinal bones
	3. Enlarged, forward-facing orbits
	4. Postorbital bar present
	5. Pattern of bones in medial orbital wall
	6. Well-developed jugal bone with foramen
	7. Enlarged braincase
	8. Inflated auditory bulla containing 'free' ectotympanic ring
	9. Internal carotid pattern (bony tubes)
	10. 'Advanced' form of auditory ossicles
Dentition	1. Tooth-comb present at front of lower jaw, linked with a specialized, serrated sublingua
	2. Reduced dental formula
	3. Similarities in cheek teeth between tree-shrews and certain primates with relatively primitive cheek teeth (e.g. *Tarsius*)
Postcranial morphology	1. Limbs and digits highly mobile
	2. Numerous details of limb musculature
	3. Osteological similarities in both forelimbs and hindlimbs (e.g. presence of entepicondylar foramen on humerus; presence of free centrale in hand)
	4. Ridged skin on palms and soles
Brain and sense organs	1. Olfactory apparatus reduced
	2. Visual apparatus enhanced
	3. Central, avascular area of retina
	4. Neocortex expanded; brain size increased
	5. Calcarine sulcus present
Reproductive biology	1. Penis pendulous; testes scrotal
	2. Discoidal placenta, as in tarsiers and simians
	3. Small litter size; small number of teats
Miscellaneous	1. Caecum present
	2. Molecular affinities (e.g. albumins)

Results of analysis of characters

1. Snout in fact secondarily elongated in tree-shrews (Fig. 6.10)
2. Set of six turbinal bones probably primitive for placental mammals
3. Orbits relatively small and laterally facing (Figs 7.6 and 7.10)
4. Postorbital bar present as a convergent development in various mammals
5. Palatine/lacrimal contact in medial orbital wall probably primitive
6. Well-developed jugal bone with foramen probably primitive
7. Braincase has become enlarged convergently in various mammalian groups
8. Auditory bulla formed from entotympanic, not from petrosal; ecto-tympanic ring is primitively ring-shaped in placental mammals
9. Enclosure of internal carotid in bony tubes probably primitive
10. Auditory ossicles do not clearly share derived features with primates

1. Tooth-comb formed exclusively from incisors as a convergent feature; sublingua present in common ancestor of marsupials and placentals
2. Convergent reduction of dental formula in many mammalian groups
3. Limited similarities in cheek teeth between tree-shrews and certain primates undoubtedly due to primitive retention

1. Limbs and digits probably highly mobile in ancestral placental mammals
2. Limb musculature shares primitive retentions with prosimians (Table 5.3)
3. Osteological similarities in forelimbs and hindlimbs attributable to primitive retention from ancestral placental mammals

4. Ridged skin on palms and soles possibly a primitive feature for placental mammals; tree-shrews lack the characteristic Meissner's corpuscles of primates (Fig. 10.10)

1. Olfactory apparatus not reduced relative to body size in tree-shrews
2. Visual apparatus mildly enhanced; numerous primate features lacking
3. Unusual, spoke-like radiation of retinal vessels; unusual innervation
4. Expansion of neocortex and brain size found in many mammals
5. Calcarine sulcus not present in the brain of tree-shrews

1. Pendulous penis and scrotal testes present in many mammals
2. Discoidal placenta common in mammals; endotheliochorial in tree-shrews
3. Small litter size and small number of teats common in mammals; tree-shrew offspring are altrical, not precocial like those of primates

1. Caecum probably a primitive feature of marsupials and placentals
2. No convincing molecular affinities between tree-shrews and primates

strict phylogenetic sense, it is necessary to analyse all of the similarities between tree-shrews and primates within this broader context so as to distinguish as clearly as possible between the following: (1) convergent features, (2) primitive homologous character states retained from the ancestral stock of placental mammals and (3) derived homologous character states attributable to a hypothetical later stock uniquely ancestral to tree-shrews and primates. Only if it is possible to demonstrate convincingly the shared possession of a significant number of derived homologous character states in tree-shrews and primates is it permissible to conclude that the two groups are phylogenetically related. In fact, however, all of the features commonly cited as linking tree-shrews to primates can be reliably interpreted either as primitive character states retained from the ancestral placental mammals or as independent, convergent developments (Table 12.1). In this context, the extensive similarities between tree-shrews and squirrels, discussed in detail in Chapter 5, are particularly revealing, because most of the supposedly 'primate-like' features of tree-shrews are accordingly also found in squirrels.

A different perspective is obtained by assessing the characters of tree-shrews against the full definition of primates given above. This provides an answer to the enquiry as to whether tree-shrews share specific derived features with primates. In the locomotor sphere, it emerges from the discussion in Chapter 10 that resemblances between tree-shrews and primates are essentially limited to the bare facts that the former commonly show varying degrees of arboreal behaviour (although some species are almost completely terrestrial) and that they occur in tropical forest regions. By contrast, tree-shrews show no pronounced grasping adaptations of the extremities and certainly lack the clear-cut grasping adaptation of the big toe that is common to all non-human primates. Bilaterally compressed claws are present on all digits, including the big toe, and tree-shrews do not have any of the hallmarks of the typical primate pattern of hindlimb domination. Although the digits of tree-shrews have tactile

pads with cutaneous ridges, this may be a primitive feature for placental mammals and in any event the characteristic Meissner's corpuscles specifically associated with the ridges in primates are lacking.

A similar picture emerges from comparative study of the sense organs and of the central nervous system (Chapters 7 and 8). Although the visual system is quite well developed in tree-shrews, the eyes are markedly smaller than in primates (relative to body size). Further, although a postorbital bar is present, as in several other groups of non-primate mammals, there is little forward rotation of the orbits and there is no exposure of the ethmoid in the medial orbital wall. The essentially lateral orientation of the eyes is reflected in the projection of retino-fugal fibres to the visual cortex, which is over-whelmingly contralateral. Significantly, tree-shrews differ from all primates in the connections of the lateral geniculate nucleus (Kaas, Huerta *et al.*, 1978). Projection to the optic tectum also follows a typical non-primate pattern, with the visual field of the contralateral eye represented on each side. Finally, tree-shrews show a number of unusual features in the organization of the retina, which is of a 'pure-cone' type. There is a unique dichromatic system of colour vision, the density of receptors is generally low and summation of receptor inputs is restricted, being totally lacking in the temporal area. In the ear region, the ventral bulla is formed from an ento-tympanic, rather than from the petrosal as in primates, and in the nasal region tree-shrews seem to show a primitive mammalian pattern rather than any specific similarity to primates. Tree-shrews, like many other non-primate mammals, have moderately enlarged brains, but the Sylvian sulcus and prominent calcarine sulcus found in all primate brains are lacking. Further, the unique primate characteristic of emphasis of brain tissue during fetal development is absent from tree-shrews. Finally, in tree-shrews the corticospinal tracts have a significantly different pattern of organization. In contrast to the primate condition, the major tracts are located in the dorsal funiculi of the cord and do not pass back any further than the cervical (neck) level.

Furthermore, fibre connections are confined to the dorsal ipsilateral part of the cord.

As has emerged from Chapter 9, in the reproductive sphere tree-shrews again show only a superficial similarity to primates. Although in tree-shrews, in common with most marsupials and placental mammals, the testes descend into a scrotum, the scrotum is prepenial rather than postpenial as in primates, and descent does not occur until the approach of sexual maturity. Moreover, the testes may be retracted back into the abdominal cavity as a response to stress at any time. The reproductive system of female tree-shrews also differs significantly from that of primates in that an extensive urogenital sinus is present. Further, during gestation the yolk sac is well developed and apparently actively involved in the exchange of materials between mother and fetus throughout. Tree-shrews have exceptionally short gestation periods, relative to body size, in comparison with primates and, although litter size is relatively small (as is typical of arboreal placental mammals generally), the offspring are of the altricial type, in contrast to the typically precocial neonates of primates. Fetal and postnatal growth in tree-shrews is rapid in contrast to the slow development of all primates, and tree-shrews are clearly adapted for a markedly higher potential reproductive turnover than primates.

As far as teeth are concerned, there has apparently never been any serious suggestion that molar morphology indicates that tree-shrews are related to primates. The molar teeth of tree-shrews have relatively high, sharp cusps, reflecting a general emphasis on small animal prey in the diet, and there would seem to be no specific features indicating a common ancestral relationship with primates. Indeed, tree-shrews seem to have a unique combination of minor specializations with a basic molar pattern that is otherwise little different from that expected for the ancestral stock of placental mammals generally (Butler, 1980). Although tree-shrews have followed the common mammalian trend in showing some reduction in the dental formula from the hypothetical ancestral condition for placental mammals, all living tree-shrews have three incisors on each side of the lower jaw, whereas primates of modern aspect never have more than two. Correspondingly, the premaxilla in the upper jaw of tree-shrews is relatively long (strikingly so in terrestrially adapted species) and the two incisors on each side are arranged longitudinally, rather than essentially transversely as in the typical primate condition (see Fig. 6.13). Although the common possession of a tooth-comb by both tree-shrews and most strepsirhine primates has occasionally been cited as a feature linking tree-shrews to primates, closer examination reveals that any similarity must be attributable to convergent evolution (see Chapter 6).

As can be seen from Table 12.1, most of the similarities shared by tree-shrews and primates are attributable to retentions from the ancestral placental stock. It must be recognized that, contrary to widespread belief, the lipotyphlan insectivores are in many respects highly specialized in comparison both with tree-shrews and with the more primitive surviving primates. This has been increasingly accepted with respect to individual morphological features (e.g. see McKenna, 1975), but there has been only slow recognition of the logical corollary that lipotyphlan insectivores must nowadays have a lifestyle quite distinct from that of the ancestral mammals. It now seems likely that, far from providing a model for ancestral primates, living tree-shrews are in many respects the most primitive of surviving placental mammals and actually provide a better model for the ancestral placental mammals than do lipotyphlan insectivores. Accordingly, one may picture the ancestral placentals as adapted, at least to some extent, for arboreal life. Modern lipotyphlan insectivores, by contrast, have generally undergone secondary specialization for terrestrial activity, with consequent modification of the limbs and musculature (Chapter 10). There has also been a shift in emphasis to olfaction rather than vision, entailing reduction in size of the eyes, limitation on further expansion of the brain, and modification of associated skull structures. The loss of the caecum in modern lipotyphlan insectivores (see Chapter 6) provides one very good indication that their common ancestor underwent some dietary specialization with respect to the ancestral placental mammals. Hence, there is no basis for the general assumption that moderate adaptation for arboreal habits – in either living or fossil forms – provides significant evidence of an affinity with primates. In other words, there is no foundation for the common assumption (e.g. see Szalay, 1975c) that a gradient from terrestrial to arboreal adaptation among placental mammals can be automatically interpreted as an evolutionary sequence from primitive to derived conditions.

This conclusion leads directly on to the vexed issue of the superorder Archonta, a grouping originally suggested by Gregory (1910) to include Primates, Menotyphla (tree-shrews and elephant-shrews), Chiroptera (bats) and Dermoptera ('flying lemurs'). Simpson (1945), while advocating a definite phylogenetic relationship between tree-shrews and primates, in typical forceful style rejected the Archonta as an invalid assemblage:

> Gregory's 'Archonta' is almost surely an unnatural group. The 'menotyphlans' are certainly near the primates, in fact the forms that he considered typical, the tupaiids, are here classed as primates. The Dermoptera are perhaps allied to bats. But it is incredible to me, and to most recent students, that the primates are really more closely related to the bats than to the insectivores, and all recent research (including Gregory's own later work) opposes that opinion.

However, the issue of the possible validity of the Archonta as a monophyletic assemblage of placental mammals traceable to a specific common ancestry has recently been rekindled by several authors, following its acceptance by Butler (1956, 1972). McKenna (1975) proposed the revival of a modified concept of Archonta in an overview of mammalian phylogenetic relationships, and he was subsequently supported by Szalay (1977) in a survey of postcranial evidence and by Cronin and Sarich (1980) in a discussion of immunological evidence. As a first point, it should be noted that elephant-shrews (family Macroscelididae) have generally been quietly dropped from this assemblage. As the only exclusively terrestrial forms included in the Archonta as defined by Gregory, elephant-shrews were somewhat of an anomaly in certain other ways. Thus, McKenna (1975) divorced the elephant-shrews very distinctly from his 'grandorder' Archonta, which contained only the orders Scandentia (tree-shrews), Dermoptera, Chiroptera and Primates. McKenna cited Gregory (1910) as having identified certain shared derived character states linking these four groups of mammals. However, Gregory himself did not clearly recognize the vital importance of ensuring a sharp distinction between primitive and derived similarities and he did not make a convincing case for the Archonta as a genuine phylogenetic assemblage (Cartmill and MacPhee, 1980). Of course, it is absolutely crucial to apply reliable criteria for the inference of derived character states of an 'archontan' ancestral stock relative to an inferred ancestral condition for placental mammals generally.

It must be made absolutely clear that proposed recognition of a superorder Archonta is nothing other than an extension of the hypothesis that tree-shrews are phylogenetically related to primates. The very suggestion of this grouping, while distancing tree-shrews and primates from lipotyphlan insectivores, implies the existence of a specific common ancestry between tree-shrews and primates (albeit shared in some way with bats and 'flying lemurs'). If it is accepted that tree-shrews share any specific ancestry with primates among placental mammals, inclusion of tree-shrews in the order Primates once more boils down to a question of arbitrary recognition of taxonomic boundaries. The key question to be answered, therefore, is: Do tree-shrews, primates, bats and 'flying lemurs' have any shared derived character states indicating the existence of a specific common ancestor? As has been shown above, tree-shrews do not seem to share any of the major derived features that characterize primates of modern aspect and the same, so far as is known, applies to 'flying lemurs' and to bats (with one exception, to be discussed below). The conclusion reached above that tree-shrews share with primates only primitive features retained from the ancestral stock of placental mammals and convergent features associated with an arboreal way of life is in no way affected by the additional suggestion that bats and 'flying lemurs' may be related to primates. In fact, it is noteworthy that arboreal life is a major theme among 'archontans' (with the exclusion of elephant-shrews). Accordingly, convergent adaptation for arboreal habits is one probable source of similarities between tree-shrews, primates, bats and 'flying lemurs'.

Uncertainty about the lifestyle of the earliest placental mammals is a major issue here. If, as has been concluded by many authors, ancestral placental mammals were predominantly terrestrial, then similar arboreal adaptations in 'archontans' must be attributable either to convergent evolution or to the development of shared derived features in a specific ancestral stock. If, on the other hand, ancestral placental mammals were predominantly arboreal, many shared features of 'archontans' may be identifiable as primitive retentions. Hence, assessment of the validity of the Archonta as a monophyletic assemblage is inseparable from reconstruction of the probable lifestyle of ancestral placentals. At present, it is reasonable to conclude that ancestral placentals were at least partially arboreal and that similarities between 'archontans' are attributable to a combination of primitive retention and convergent adaptation, in some cases associated with more committed arboreal habits. As was shown in Chapter 10, this alternative interpretation can reasonably be applied to Szalay's (1977) list of possible shared derived features of the postcranial skeleton of 'archontans' (excluding bats).

Cronin and Sarich (1980) presented immunological evidence derived from albumins and transferrins suggesting a phylogenetic relationship between tree-shrews, primates and 'flying lemurs' – but bats were specifically excluded from this group. Thus, these authors did not in fact support the validity of the group Archonta as defined by McKenna (1975). In addition, there is a curious anomaly in that the analysis of albumin immunological distances suggests a closer relationship between 'flying lemurs' and primates than between tree-shrews and primates. This is in direct conflict with virtually all other interpretations of evolutionary relationships among 'archontans'. In fact, Cronin and Sarich (1980) themselves showed that the inferred rate of evolutionary change of albumin along the tree-shrew lineage was significantly slower than in primates, 'flying lemurs' or bats. It has already been noted in Chapter 11 that immunological results for tree-shrews are generally aberrant. Given the uncertainties attached to interpretation of 'immunological distances' between proteins, especially with differential rates of evolution manifest (see Chapter 11), the provisional conclusion of Cronin and Sarich regarding a specific relationship between tree-shrews, 'flying lemurs' and primates must be treated with caution. This is especially so because Dene, Goodman *et al.* (1980) concluded that their own earlier immunological results indicating such a relationship were not supported by subsequent (potentially far more reliable) comparative analyses of amino acid sequence data for primates, tree-shrews and bats. Later, far more comprehensive reviews of amino acid sequence data for a variety of mammalian proteins (Goodman, Romero-Herrera *et al.*, 1982; Miyamoto and Goodman, 1986) provided no evidence for specific links between tree-shrews, elephant-shrews, primates and bats. Thus the balance of evidence from comparative studies of proteins does not indicate that any combination of mammalian orders from Gregory's Archonta is valid in a phylogenetic sense.

Shoshani (1986) compared results from studies of proteins with those obtained from a broad investigation of non-dental morphological features in mammals. It was concluded that the evidence overall indicated a phylogenetic association between primates, tree-shrews, bats and 'flying lemurs'. However, Shoshani did not give any direct explanation of the morphological characters assessed, nor of the procedures used to determine derived conditions. Furthermore, it is noteworthy that the most parsimonious solutions for the morphological characters all linked carnivores to the above groups. This suggests a grouping based on primitive retentions rather than on shared derived features. Novacek (1982b), in a similar broad review of morphological features of mammals, found no evidence to support a phylogenetic association of primates, tree-shrews, bats and 'flying lemurs'. A similar conclusion was drawn from a study of basicranial features by MacPhee and Cartmill (1986). Overall, therefore, it can be concluded that there is at present no justification for recognizing the Archonta as a phylogenetic unit.

An additional complication to the problem of unravelling possible phylogenetic relationships among 'archontans' has arisen with Pettigrew's report (1986) that fruit-bats (Megachiroptera) share with primates a special pattern of projection of retinofugal fibres to the optic tectum. The inference made from this by Pettigrew is that fruit-bats shared a specific ancestry with primates and should perhaps be included in the order Primates. At first sight, this might seem to strengthen the case for a recognition of a superordinal grouping Archonta as a phylogenetic assemblage. However, it must be emphasized at once that Pettigrew demonstrated that the Microchiroptera do not show any trace of the primate pattern of projection to the optic tectum. The main weight of Pettigrew's evidence in fact bears on distinctions within the bat order between Microchiroptera and Megachiroptera, possibly indicating convergent evolution of winged flight. The grouping Archonta as proposed by various other authors requires the inclusion of all bats, not just fruit-bats, along with tree-shrews, primates and 'flying lemurs'. Furthermore, it has already been mentioned that tree-shrews also lack the typical primate pattern of projection to the optic tectum. Hence, possession of this pattern cannot be cited as a shared derived feature justifying the recognition of Archonta as a monophyletic assemblage. At the most, presence of the pattern may be cited as a possible shared derived feature linking fruit-bats to primates, but the possibility must also be considered that convergent evolution has occurred.

Fruit-bats have several other features included in the full definition of primates given above, but some unique, apparently shared derived features of primates are lacking, such as the petrosal bulla, the emphasis on the development of brain tissue during fetal development, and the entire suite of features associated with hindlimb domination. Of course, modification of the limbs for winged flight would have led to radical transformation of the postcranial skeleton and this could have obscured an original primate-like pattern. Nevertheless, it should be noted that a grasping big toe is in fact found in one bat, *Cheiromeles*; but this is a microchiropteran bat (a member of the family Molossidae), not a megachiropteran bat. It is curious that this typical feature of primates should have been developed (apparently through convergent evolution) in a microchiropteran bat, whereas the fruit-bats, specifically related to primates according to Pettigrew's hypothesis, show no trace of a grasping function of the hindlimb. Proper assessment of this hypothesis must await the accumulation of more evidence and a detailed analysis of individual characters permitting an assessment of the likelihood of convergence between fruit-bats and primates. At present, such an assessment is hindered by the virtual absence of fossil evidence for the evolution of megachiropteran (as opposed to microchiropteran) bats and by the extremely limited availability of relevant chromosomal or molecular data. Regardless of whether fruit-bats are eventually shown to be phylogenetically related to primates, this new element in the discussion provides no support whatsoever for a monophyletic grouping Archonta as originally conceived by Gregory (1910) or as recognized in a modified form by McKenna (1975).

A general obstacle to convincing resolution of the validity of the Archonta as a phylogenetic assemblage, as was noted by McKenna (1975), is the general lack of relevant fossil evidence, except for primates. It has been noted in Chapter 5 that there are no pre-Miocene fossil forms that can be allocated with any confidence to the tree-shrew group, so the origins of tree-shrews among early mammals are quite obscure. Entire skeletons of fully developed microchiropteran bats are known from Eocene deposits, but there are no earlier fossils to indicate the origins of these bats, during the key phase when the primitive mammalian forelimb was converted into a wing. There are no known fossil megachiropteran bats, apart from a few isolated teeth, and the fossil record for 'flying lemurs' is equally poor. Arguments concerning phylogenetic relationships among tree-shrews, primates, bats and 'flying lemurs' must, therefore, be based almost exclusively on the comparative study of living forms.

Accordingly, there is no convincing evidence from comparative studies to indicate the tree-shrews have any specific relationship to primates among placental mammals, either individually or as members of an 'archontan' assemblage of some kind. On the one hand, it may be said that equation of ancestral primates with modern tree-shrews has led to the biggest ever wild-goose chase in the literature on primate evolution. On the other hand, this very chase may ultimately lead to a more precise definition of primate origins.

ARE PLESIADAPIFORMS PRIMATES?

The 'archaic primates' allocated to the infra-order Plesiadapiformes have been included within the order Primates, primarily on dental grounds, for such a long period and by so many authors that it borders on heresy to suggest that they may not, in fact, be specifically related to primates of modern aspect. Once again, it is not a classificatory issue that it is of central importance here. There is already an increasing tendency in the literature to distinguish in one way or another between Plesiadapiformes and 'primates of modern aspect' (often referred to as Euprimates) and it would be but a short step to take the taxonomic decision to restrict the order Primates to primates of modern aspect. This could be done while acknowledging that Plesiadapiformes, among placental mammals, are the closest relatives of primates of modern aspect.

All of this, however, begs the question as to whether Plesiadapiformes are linked to primates

in any way in the phylogenetic tree of mammals. Satisfactory demonstration of such a linkage, through well-founded inference of shared derived features linking plesiadapiforms and primates of modern aspect, has yet to be achieved. The plesiadapiforms were firmly associated with primates by Simpson (1935a), primarily because of striking similarities in molar morphology between *Plesiadapis* and North American Eocene 'lemuroids' (Notharctinae), most notably *Pelycodus* (see also Gingerich, 1986a). Despite the eventual discovery of reasonably intact skulls for some genera (*Plesiadapis, Palaechthon* and *Phenacolemur*) and of specimens documenting most of the postcranial skeleton for *Plesiadapis*, discussion of the phylogenetic relationships of Plesi-adapiformes in the literature has typically started from the *a priori* view, derived from the original dental comparisons, that they are unquestionable primates. Relatively little attention has been given to the specification of probable shared derived features that would unequivocally link plesiadapiforms to primates.

When the known morphological features of plesiadapiforms are assessed against the reduced definition of primates applicable to the fossil record, the result is disappointing. *Plesiadapis* (the only genus known from associated postcranial material) lacks all of the major defining features of primates identified in the locomotor sphere. To begin with, there is no indication of primate-like hindlimb domination. Equally importantly, there is no evidence of a grasping adaptation in the foot. Indeed, the apparent presence of prominent, bilaterally flattened claws on all known digits of *Plesiadapis* indicates a style of locomotion fundamentally different from that of any primate of modern aspect. A separated digit identified as the hallux by Gingerich (1986a) apparently bore a sharp, pointed claw and this alone would rule out the presence of a primate-like grasping function of the hallux. Primates and all other modern mammals (including marsupials) with a properly prehensile foot consistently have a flat nail on the hallux. The absence of a primate-like grasping adaptation of the foot of *Plesiadapis* is further confirmed by the relatively short distal segment of the calcaneus (yielding a high value for the calcaneal index) and by other features of the ankle region (see Chapter 10).

Although Szalay and Decker (1974; see also Szalay, 1977) have hypothesized the presence of shared derived features in the ankle region of *Plesiadapis* and primates of modern aspect, this conclusion is suspect because their analysis was not based on well-supported inference of the ancestral placental condition. These authors assumed that the condition in isolated tarsal bones tentatively allocated to the late Cretaceous genera *Procerberus* and *Protungulatum* was relatively primitive and that any differences from this condition shared by *Plesiadapis* and primates were necessarily derived features indicative of common ancestry. However, this assumption was inextricably linked to the underlying, implicit assumption – firmly rejected here – that predominantly terrestrial adaptation was primitive for placental mammals (see also Lewis, 1983). Further, Szalay and Decker (1974) did not emphasize any of the specific features of primate locomotion that have been identified in the full definition of living primates given above and which were accordingly retained in the reduced definition applicable to the fossil record.

Plesiadapiforms also consistently lacked all of the characteristic features of the orbital region of the skull found in primates of modern aspect. There was no postorbital bar and there was little sign of forward rotation of the eyes. The orbits themselves were very small, relative to body size, indicating a rather minor role for vision in comparison to primates of modern aspect. As far as is known, there was no exposure of the ethmoid in the medial orbital wall, as is entirely to be expected from the comparatively massive interorbital pillar. The relative size of the brain was also apparently quite small in plesiadapiforms, although as yet there is no adequately preserved braincase either to permit proper confirmation of this point or to reveal the gross morphology of the brain. Evidence concerning the ear region is equivocal. Gingerich (1975a) reported on a well-preserved bulla of *Plesiadapis* indicating the presence of an ectotympanic ring within the bulla (see also

Russell, 1964). The bulla was completely ossified, but, as Gingerich himself noted, the existence of a separate entotympanic element during development of the bulla cannot be ruled out, as subsequent fusion and obliteration of a suture between the petrosal and the entotympanic would prohibit identification of the latter in a fully adult skull. Thus, although it is possible that in *Plesiadapis* the ventral floor of the bulla was formed from the petrosal, as in living primates and (it is presumed) in fossil primates of modern aspect, it is also possible that – as in tree-shrews – the bulla floor was derived from an entotympanic element that fused to the petrosal during development. Incidentally, it should be noted that Gingerich (1976b) went further than simply linking *Plesiadapis* to primates of modern aspect on grounds of the structure of the bulla. He specifically emphasized the overall similarity between the bulla of *Plesiadapis* and that of certain early Tertiary 'tarsioids' (*Necrolemur* and *Rooneyia*). He went on to suggest that this represented a shared derived character complex indicative of a specific phylogenetic relationship between Plesiadapiformes and Tarsiiformes, to the exclusion of other primates of modern aspect (see later).

Reduction of the dental formula from the hypothetical condition in ancestral placental mammals ($I_3^3 C_1^1 P_4^4 M_3^3$) is common among plesiadapiforms. Indeed, extensive reduction is common among later members of the group and the anterior dentition is very specialized in many later forms. The teeth of many plesiadapiforms are rodent-like, with markedly enlarged single upper and lower incisors on each side, associated with the loss of the canine teeth and the consequent presence of a gap (diastema) between the incisors and the cheek teeth. The incisors bear only a superficial resemblance to those of rodents, however, because in plesiadapiforms their crowns typically have a complex morphology (e.g. see Gingerich, 1976b). As in rodents, the numbers of cheek teeth in plesiadapiforms are also commonly reduced from the ancestral placental condition.

Because such far-reaching dental specialization is common among the Plesiadapiformes, there has been a tendency (once shared by the author) to operate on the assumption that all members of this group are very divergent from primates of modern aspect. Yet, in the earliest known representatives of the Plesiadapiformes there was relatively limited modification of the dentition. In particular, in the early Palaeocene form *Purgatorius* (also documented by a single lower molar from the latest Cretaceous) the full dental formula inferred for the ancestral placental mammals was still present ($I_3^3 C_1^1 P_4^4 M_3^3$). Hence, any reduction in dental formula shared by later plesiadapiforms and by primates of modern aspect must presumably have arisen by convergence, provided that it is correct to assume (e.g. Gingerich, 1976b; Szalay and Delson, 1979) that the Plesiadapiformes constitute an exclusive monophyletic group. It would, of course, be possible to postulate an origin for primates of modern aspect somewhere within the plesiadapiform radiation. In this case, a reduction of the dental formula to $I_2^2 C_1^1 P_4^4 M_3^3$ (as in the hypothetical ancestral condition for primates of modern aspect) could be seen as a shared derived feature. If this is proposed, however, various other implications must be accepted. For instance, this would probably imply an origin for primates of modern aspect at some time during the Palaeocene, with very rapid evolution of the suite of primate characters found in primates of modern aspect by the early Eocene. In fact, reduction of the number of incisors in the upper jaw in plesiadapiforms is not (or not consistently) associated with the marked reduction in length of the premaxilla that characterized the emergence of primates of modern aspect, so convergent evolution in incisor reduction seems far more likely.

The strongest evidence for a phylogenetic relationship between plesiadapiforms and primates of modern aspect remains the striking resemblance in molar morphology between *Plesiadapis* and *Pelycodus* originally emphasized by Simpson (1935a) and recently re-emphasized by Gingerich (1986a). The resemblance is, indeed, remarkable and has repeatedly convinced primate palaeontologists that there must be a phylogenetic connection between the plesiadapiforms and primates of modern aspect.

It is, of course, hypothetically possible that the primates of modern aspect originated from an ancestral form that possessed primate-like molar teeth and perhaps a petrosal bulla, but had not yet developed the characteristic primate features of the orbital region of the skull, of the brain and of the postcranial skeleton. However, once again the implications of linking primates of modern aspect to plesiadapiforms must be spelled out clearly. Let us assume that plesiadapiforms and primates of modern aspect constitute two monophyletic sister groups, as is currently accepted by many primate palaeontologists. In order to determine the likely characters of a hypothetical ancestral stock specific to those two groups, it is first necessary to infer the most primitive condition for plesiadapiforms and the most primitive condition for primates of modern aspect. It is not sufficient to emphasize similarities between one plesiadapiform (e.g. *Plesiadapis*) and one primate of modern aspect (e.g. *Pelycodus*), as these might have arisen by convergence. It must be shown that any inferred shared derived features linking plesiadapiforms and primates of modern aspect were present in the hypothetical ancestral stocks of both of these groups.

Among plesiadapiforms, the likely primitive condition of the molar teeth is found not in *Plesiadapis* but in the earlier forms, such as *Purgatorius*. The dental resemblance between *Purgatorius* and early primates of modern aspect such as *Pelycodus* is far less striking than that between *Plesiadapis* and *Pelycodus*. There is a comparable problem among primates of modern aspect in that certain living representatives, notably *Tarsius*, appear to have considerable more primitive molar teeth than those of *Pelycodus*. Hence, the common ancestor of primates of modern aspect presumably had more primitive teeth than those of *Pelycodus* originally considered by Simpson (1935a). It emerged in Chapter 6 that the molars of the hypothetical ancestral primate stock were probably very primitive indeed, differing only in relatively minor details (e.g. lowering and rounding of cusps; raising of the talonid of the lower molars) from the hypothetical ancestral condition for placental mammals generally. Indeed, there is a considerable difference between early and late species allocated to the genus *Pelycodus* (here including forms attributed by Gingerich to the genus *Cantius*). Gingerich's account (1986b) of the earliest known molars of *Pelycodus* matches quite closely the expected pattern suggested in Chapter 6 for the ancestral primate stock. The same applies to molars of *Protoadapis* and of early *Europolemur* (Franzen, 1987). Comparison between the molars of the earliest known *Pelycodus* (= *Cantius*) and those of *Purgatorius* suggests that much of the similarity between the molars of later *Pelycodus* and *Plesiadapis* must have arisen through convergent evolution. Although this does not rule out the possibility of a phylogenetic relationship between plesiadapiforms and primates, it considerably weakens the case put forward by Simpson (1935a).

An overall assessment of the evidence indicates that any phylogenetic relationship between plesiadapiforms and primates of modern aspect must be very tenuous. In fact, there has always been a close association between arguments for inclusion of tree-shrews in the order Primates and arguments for inclusion of the Plesiadapiformes. Remane (1956a), for instance, referred to tree-shrews and plesiadapiforms together as 'subprimates'. Recognition of the fact that tree-shrews do not share any unequivocal derived features with primates inevitably weakens the original case for linking plesiadapiforms to primates. Nevertheless, the case for such a link is certainly better than that for a link between tree-shrews and primates. The likelihood of a specific phylogenetic relationship between tree-shrews and primates has been assessed above as close to zero. By contrast, for present purposes, the plesiadapiforms have been kept in the order Primates in recognition of the possibility that there might have been a very early phylogenetic link. Continued use of the term 'archaic primates' for the Plesiadapiformes is therefore accepted here.

At the same time, however, the evidence linking 'archaic primates' to primates of modern aspect is of such a tenuous nature that one should be wary of basing arguments on the premise that plesiadapiforms are definite primates.

Incorporation of 'archaic primates' into arguments concerning other aspects of primate evolution is a highly dangerous procedure. For instance, arguments for the existence of a phylogenetic relationship between tree-shrews and primates, or for recognition of the Archonta as a genuine monophyletic assemblage, have often invoked plesiadapiforms (usually *Plesiadapis*) as well as primates of modern aspect. No credence should be given to reasoning of the following kind: '*Plesiadapis* shares this feature with X. *Plesiadapis* is a primate. Therefore, X is a primate.' If it ultimately emerges that plesiadapiforms are *not* related to primates, any conclusions based on such reasoning must collapse completely. For the time being, it may be accepted that plesiadapiforms possibly branched away at a very early stage from the lineage leading to primates of modern aspect, but the most appropriate verdict is: Not proven.

RECONSTRUCTION OF THE ANCESTRAL PRIMATE

A primary concern of this book has been the reconstruction of the ancestral primate. Such a reconstruction is envisaged not just as a relatively sterile listing of characters present in the common ancestor of primates, but in terms of a biological understanding of the lifestyle of that common ancestor. This task has been relatively neglected in the literature for the simple reason that most authors have accepted, explicitly or implicitly, that living tree-shrews provide an adequate model for the ancestral primate condition. If it is accepted that ancestral primates were rather like some kind of hybrid between a hedgehog and a lemur, then tree-shrews seem to fit the bill rather neatly. However, if the starting point is a survey of the living primates (excluding tree-shrews) using a definition based on them, an entirely difference picture emerges. That picture can be reached by taking the characters listed in the definition given earlier in this chapter and attempting to trace their origins back through time, with due reference to relevant fossil evidence.

Most living primates are associated with

subtropical and tropical forests and the predominant lifestyle is almost completely arboreal. Indeed, as noted by Richard (1985), all families of living primates include at least some representatives that live in tropical forests and 12 of 14 modern primate families are restricted to tropical biomes. It would also seem that most fossil primates of modern aspect were associated with tropical forest ecosystems (bearing in mind the fact that there would be selection against the fossilization of species living in such habitats). As is shown in the distribution map for living primates in Fig. 1.1, the generally more primitive living prosimians are even more tightly restricted to tropical and subtropical habits than living simians. Further, there is only one living prosimian primate, the ringtail lemur (*Lemur catta*), that shows any real tendency towards terrestrial activity, and, even in this case, about 75% of its time is spent in trees. It is therefore reasonable to infer that the common ancestor of modern primates was an arboreal inhabitant of tropical forests. This original arboreal lifestyle is reflected by the virtually universal presence among living primates of a grasping foot (secondarily transformed only in *Homo*) and a general prehensile adaptation of the extremities, combined with ridged digital pads and special tactile sensitivity. The general theme of hindlimb domination in primate locomotion is also best explained as an adaptation developed in association with grasping hindlimbs in an arboreal milieu. The same applies to the special developments of the ankle region identifiable in the primate hindlimb.

Of course, reconstruction of the original primate lifestyle is inextricably linked to inference of the ancestral body size, because so many features of locomotor adaptation are demonstrably dependent upon overall body weight. Many lines of evidence suggest that ancestral primates were relatively small in size, probably in the region of 500 g or less, and one must therefore concentrate on the smallest-bodied representatives among living primates to find clues to the ancestral condition. Given that modern prosimians are generally more primitive than simians, emphasis should obviously be placed on the smallest species among living prosimians. Thus, it is to be expected that members of the dwarf lemur family (Cheirogaleidae), members of the loris family (Lorisidae) and tarsiers (Tarsiidae) have remained closest to the ancestral primate condition. In other words, a hypothetical ancestor reconstructed as a kind of 'lowest common denominator' from dwarf lemurs (*Cheirogaleus*), mouse lemurs (*Microcebus*), bushbabies (*Galago*) and tarsiers (*Tarsius*) should provide a reasonable indication of the ancestral primate lifestyle. The picture that emerges bears no resemblance whatsoever to modern tree-shrews. Instead, one is confronted with a small-bodied, fully arboreal creature with grasping feet and a pattern of hindlimb domination, specifically adapted for active locomotion in the fine-branch niche. It is only among the fine, terminal branches of trees or on thin saplings that grasping hands and feet, rather than grappling claws, come into their own for such a small-bodied mammal.

Occupation of the fine-branch niche in trees would also explain numerous other features that are more or less universal to living primates, particularly in the context of vision. Active locomotion in a network of fine arboreal supports would account not only for the general emphasis on vision that characterizes primates of modern aspect, but also for forward rotation of the orbits and associated neural developments favouring enhanced stereoscopic vision. In a terrestrial environment where predators may approach from any direction, the potential advantages of enhanced stereoscopic vision are offset by the advantages of all-round vision and an essentially lateral orientation of the eyes is generally favoured. However, in the fine-branch niche of trees, where predators cannot easily approach undetected, marked forward rotation of the eyes permits improved three-dimensional vision without an increased risk of predation. Forward rotation of the eyes, accompanied by other features such as the development of a postorbital bar, has been a consistent theme throughout the evolution of the primates of modern aspect. Doubtless for this reason, enlargement and specialization of the brain, notably of those parts concerned with vision, have also been particularly prominent among primates.

The question of vision in primates is intimately linked to that of the distinction between nocturnal and diurnal habits. As explained in Chapter 4, there are very good reasons to believe that the ancestral placental mammals, and indeed early mammals in general, were nocturnal in habits. Nocturnal habits are predominant among living prosimians and it is only among some of the Malagasy lemurs that diurnal habits occur. By contrast, among simian primates diurnal activity is the rule, the only exception being the owl monkey (*Aotus*) – apparently because of a secondary transition to nocturnal life (see Chapter 7). The most parsimonious explanation would seem to be that ancestral primates remained nocturnal, like the common ancestor of placental mammals generally, and the diurnal habits emerged at a later stage, convergently in some lineages among the lemurs of Madagascar and in the common ancestor of simian primates (see later). It has been shown in Chapter 7 that nocturnal primates typically have larger orbits than diurnal primates, relative to body size. It may therefore be inferred that the nocturnal ancestral primates would have had prominent eyes and relatively large orbits. This, in itself, is of particular interest because nocturnal life among terrestrial mammals does not seem to be generally linked to an increase in the size of the eyes. Indeed, among lipotyphlan insectivores there is good evidence for a secondary reduction in the size of the eyes, in spite of the typical nocturnal habits of these mammals. Hence, the fact that nocturnal primates generally have larger eyes (and orbits) than diurnal primates can also be linked to the specific requirements of a fine-branch, arboreal habitat. Interestingly, the orbits in tree-shrews are generally quite small in comparison to those of primates, and the nocturnal pentailed tree-shrew (*Ptilocercus*) does not have larger orbits, relative to body size, than other, diurnal tree-shrews (see Fig. 7.6). This provides an additional indication that the arboreal adaptations of the smaller tree-shrew species (the larger species being predominantly

terrestrial) are significantly different from those of primates, probably differing in the lack of emphasis on the fine-branch niche.

Body size is also important for the inference of the dietary habits of ancestral primates. From broad correlations between dietary habits and body size among living primates (Chapter 6), it may be inferred that a small-bodied early primate would probably have fed on a mixture of energy-rich, easily digested plant parts (e.g. fruits) and small animal prey (especially arthropods). Universal retention of the caecum among the living primates indicates that plant food has typically played at least a contributory part in feeding adaptations throughout primate evolutionary history. Further, the implications of scaling of basal metabolic rate to body size (Chapter 6) are that small mammals must typically obtain their protein from small animal prey, rather than from leaves, which can be digested only with difficulty. Such inferences regarding the diet of the earliest primates of modern aspect agree with hypothetical reconstructions of the morphology of their molar teeth.

As has been shown in Chapter 6, the earliest primates probably still had relatively primitive molars that retained the basic tritubercular pattern. 'Squaring' of the molar teeth through formation of a fourth cusp (hypocone or pseudo-hypocone) on the upper molars and through reduction of the trigonid (involving loss of the paraconid) on the lower molars apparently occurred subsequently as parallel developments in numerous primate lineages. In so far as it is correct to infer that the ancestral placental mammals were 'insectivores', feeding predominantly on arthropod food, it is likely that ancestral primates still heavily depended on small animal prey. However, the molars of the ancestral primates were probably characterized by some degree of lowering and rounding of the principal cusps and by raising of the talonid in the lower molars, indicating increased emphasis on plant food in comparison with ancestral placental mammals. Hence, one can picture a small-bodied ancestral primate moving actively around among the fine branches of trees, foraging on small fruits and on small animal prey, rather like modern dwarf bushbabies and mouse lemurs (Charles-Dominique, 1972; Martin, 1972a). In the tropical forests inhabited by the earliest primates, small fruits would doubtless have been concentrated in the fine-branch zone of shrubs and trees and arthropod prey (for instance) could have been collected in an opportunistic fashion. This image of the ancestral primate fits quite closely with that of a small-bodied, visually oriented insect predator advanced by Cartmill (1972; 1974b).

Although there has been traditional concentration on the importance of molar morphology for inferring ancestral relationships among placental mammals, one of the most striking developments in the early development of primates apparently took place in the anterior dentition. Living and fossil primates of modern aspect never have more than two incisors on either side of the upper and lower jaws, as compared to the three incisors that were probably present in ancestral placental mammals. Further, the premaxilla in the upper jaw of primates is consistently short, such that the upper incisors are arranged more or less transversely. This characteristic feature of primates is probably linked to the marked prehensile capacity of the forelimb, rendering the original prehensile function of the incisors redundant. In the earliest placental mammals, as in modern lipotyphlan insectivores and tree-shrews, the incisors probably served a prehensile function for predation. The sharp, longitudinally arranged incisors were doubtless used as an efficient device for grasping and killing arthropods. By contrast, in the earliest primates it would seem that the hands were used, at least to some extent, for grasping prey, which were then brought to the mouth for the killing bite. This presumably permitted greater flexibility and versatility in patterns of predation on arthropods in the fine-branch niche, including the potential to seize prey in flight. The presence of hindlimb domination associated with a grasping foot would have allowed the ancestral primates to launch themselves at arthropod prey while retaining a firm grasp on arboreal supports with their hindlimbs. (Such a pattern of arboreal predation

on arthropods is prevalent among modern small-bodied, nocturnal prosimians.) A prehensile capacity of the forelimbs in the earliest primates would doubtless also have conferred advantages in foraging on small fruits dispersed on small twigs in the terminal branches. Of course, the combination of prehensile forelimbs with forward rotation of the eyes and enhancement of stereoscopic vision paved the way for the further refinement of the visual system in relation to object manipulation. Hence, one of the hallmarks of mankind – the manufacture and use of tools – can be traced back to the earliest primates.

The features of reproductive biology present in all living primates can also be traced back to the earliest primates, although in this case there is no relevant fossil evidence to test the inferences made. All living primates are characterized by the 'precocial complex'. They are adapted for slow reproductive turnover, with small litters of well-developed precocial offspring born after a relatively long gestation period and with conspicuous extension of other life-history phases (long period of lactation; late attainment of sexual maturity; long lifespan). Although there is continuing controversy over theoretical models advanced to explain 'life-history strategies' in mammals (Richard, 1985; Ross, 1988), the classical model of *r*- and *K*-selection (MacArthur and Wilson, 1967; Pianka, 1970) can in principle account quite well for the observed pattern in primates. According to this model, unstable and unpredictable environments favour high reproductive turnover (*r*-selection), whereas relatively stable, predictable environments favour efficient use of environmental resources, associated with low reproductive turnover (*K*-selection). It is reasonable to suggest that the earliest primates, as arboreal inhabitants of tropical rainforest, were generally subject to relatively stable and predictable environmental conditions and that this favoured the development of adaptations for low reproductive turnover, combined with relatively high total investment of resources in individual offspring.

The consistent primate pattern of low reproductive turnover combined with heavy total investment in individual offspring ('quality rather than quantity') is probably connected, at least indirectly, with the unique pattern of fetal brain growth found among primates. As has been shown in Chapter 8, at all stages of fetal growth, brain tissue represents a markedly greater proportion of total fetal body weight in primates than in all other placental mammals. This means that, at any given body weight, a neonatal primate has more brain tissue than any other placental mammal. This special emphasis on brain tissue in primate fetal development is doubtless a further reflection of emphasis on quality rather than quantity in the evolution of primate reproduction.

IS THE DIVISION BETWEEN STREPSIRHINES AND HAPLORHINES VALID?

Throughout the preceding chapters it has been necessary to distinguish repeatedly between two major groups of living primates: strepsirhines (lemurs and lorises) and haplorhines (tarsiers and simians). This distinction applies to a wide variety of features of living primates. This, in itself, indicates that there might be a major phylogenetic separation between these two groups of primates. It is important to note, however, that the features concerned are very largely non-fossilizable. Therefore, although the separation may be demonstrable for living primates, there is no guarantee that it will be either recognizable in the fossil record or testable on the basis of fossil evidence. Further, it is necessary to establish whether *both* groups of living primates, strepsirhines and haplorhines, are monophyletic. This can only be done if it can be demonstrated that each group is characterized by a distinctive set of derived features (Fig 12.2). For instance, in principle it is possible that all of the character states that separate haplorhines from strepsirhines have arisen as derived features in the evolution of haplorhines, while strepsirhines share only primitive character states retained from ancestral primates. If this were indeed found to be the case, it would only be justifiable to recognize the haplorhines as a monophyletic group.

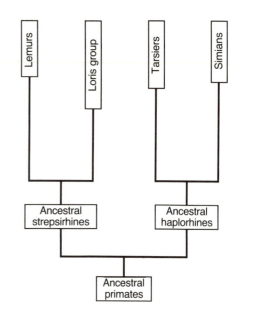

Figure 12.2. The suggested separation between strepsirhine primates (lemurs and lorises) and haplorhine primates (tarsiers and simians). To confirm that both groups are monophyletic, it is necessary to identify derived features for the ancestral strepsirhine stock as well as for the ancestral haplorhine stock.

In fact, with respect to characters of the teeth and jaws there is relatively little to indicate that the haplorhines constitute a monophyletic group. This is largely because living tarsiers apparently possess the most primitive cheek teeth of all extant primates (Chapter 6; see Fig. 6.19). The upper molars are essentially triangular and virtually tricuspid, and the lower molars retain the paraconid. In addition, the molar teeth are large relative to body size, as in living lipotyphlan insectivores and in tree-shrews, and this may be a retained primitive feature of tarsiers. It is correspondingly difficult to define an ancestral haplorhine stock on the basis of features of the cheek teeth, and it is equally difficult to recognize early Tertiary fossil 'tarsioids' as haplorhines on dental grounds. Living tarsiers do, admittedly, show some unusual specializations in the arrangement of the anterior dentition (incisors and canines) and in the overall shape of the upper dental arcade, and it should be remembered that tarsiers are unique among living primates in that they consume only animal food. As explained in Chapter 6, it seems likely that ancestral primates (and ancestral placental mammals) consumed at least some plant food, so exclusive concentration on animal food probably represents a specialization of modern tarsiers. It may be that the unusual features of their dentition are linked to this dietary peculiarity. Although tarsiers possess certain dental specializations, however, there is no convincing dental evidence to link tarsiers to simian primates, as would be required to define an ancestral stock for haplorhines generally. By contrast, the living simian primates share several apparently derived features of the teeth and jaws (see later), which are hence attributable to an ancestral simian stock, rather than to an ancestral haplorhine stock. Living strepsirhines also share at least one conspicuous derived dental feature in that they have a tooth-comb (or some secondary modification thereof) in the lower jaw.

The distinction between derived and primitive character states is fundamental to any discussion of the separation between strepsirhines and haplorhines and it is crucial for assessment of the relationships of tarsiers. Tarsiers are small-bodied, nocturnal primates that have remained relatively primitive in many aspects of their biology. It is for this reason that they have often been grouped with lemurs and lorises in primate classifications based on the grade concept, most notably within the suborder Prosimii in Simpson's classification (1945). It is undoubtedly the case that tarsiers share many features with lemurs and lorises in addition to sharing a suite of other features with simian primates. The question that must be asked is whether the tarsiers share derived features exclusively with simian primates, hence justifying recognition of the haplorhine primates as a monophyletic assemblage, or whether they share derived features only with lemurs and lorises, thus indicating that prosimian primates constitute a genuine monophyletic group rather than just a grade.

Characteristics from the sphere of reproductive biology in fact provided the first strong evidence of a specific phylogenetic relationship between tarsiers and simians, but some characters clearly illustrate the importance of distinguishing between primitive and derived character states. For instance, all prosimian primates (lemurs, lorises and tarsiers) show the typical mammalian condition in having two separate uterine horns (bicornuate condition), rather than the single, fused uterine chamber (simplex condition) present in all simian primates (Chapter 9; see Fig.9.2). This provides no evidence of a specific relationship between lemurs, lorises and tarsiers; it simply indicates that all prosimians have remained relatively primitive in this respect. By contrast, the possession of a simplex uterus suggests that simians belong to a monophyletic group. There are, however, numerous aspects of reproduction in which tarsiers and simians appear to share derived character states. Most importantly, as noted by Hubrecht (1898, 1908), both tarsiers and simians have an invasive, haemochorial type of placentation. There seems little doubt that this form of placentation represents a derived condition among placental mammals (see Chapter 9). Tarsiers and simians also share other special features, such as formation of the amnion by cavitation, general suppression of the contribution of the yolk sac to vascularization of the placenta and typical secondary modification of the allantois to form a stalk rather than a sac (see Fig. 9.7), associated with early establishment of the chorioallantoic circulation. Taken together, these features provide convincing evidence of a specific phylogenetic relationship between tarsiers and simians. Further support for this interpretation is provided by the observation that tarsiers and simians, apparently uniquely among mammals, have undergone secondary loss of the ovarian bursa, which normally envelops the ovary and virtually isolates it from the general abdominal cavity.

Strepsirhines differ sharply from haplorhines in that they typically have a non-invasive, epitheliochorial type of placentation. According to some authors (e.g. Hill, 1932; Luckett, 1975), this is a primitive feature among placental mammals and, if this view is correct, placentation accordingly provides no evidence that strepsirhine primates constitute a monophyletic group. There is, however, the alternative interpretation (Mossman, 1937; Martin, 1975) that the ancestral condition for placental mammals was a semi-invasive, endotheliochorial type of placentation associated with a relatively short gestation period and an altricial condition of the neonates (as is generally true of modern placental mammals with this type of placenta). According to this interpretation, both haemochorial placentation and epitheliochorial placentation represent derived character states and the latter would accordingly provide evidence that the strepsirhine primates constitute a monophyletic group. In other words, this alternative interpretation (favoured here) indicates that there has been divergent modification of the placenta during primate evolution, with the least invasive, epitheliochorial form developing in ancestral strepsirhines and the most invasive, haemochorial form emerging in ancestral haplorhines.

It has also been demonstrated (Leutenegger, 1973; see Fig. 9.9) that haplorhines typically produce markedly larger neonates, relative to maternal body weight, than do strepsirhines. Leutenegger linked this difference between haplorhines and strepsirhines to the difference in placentation between them, arguing that the invasive placentation of haplorhines permits more rapid transfer of nutrients from mother to fetus. Fundamental to this interpretation is the widespread notion that the non-invasive, epitheliochorial form of placentation represents a major barrier that limits diffusion of materials between mother and fetus. If it is inferred that the epitheliochorial type of placentation has remained essentially primitive, it may seem justifiable to conclude that the relatively small size of neonates produced by strepsirhine primates is also a primitive feature. This cannot be accepted as a simple retention of an ancestral feature of placental mammals, however, because ancestral placentals probably produced altricial neonates (Portmann, 1939) and because all living primates, including strepsirhines, produce

precocial neonates. Among modern mammals altricial neonates are (for obvious reasons) typically the smallest, relative to maternal body size. Accordingly, strepsirhine primates would be expected to produce relatively larger neonates than those originally produced by ancestral placental mammals. In fact, strepsirhine primates do produce relatively larger neonates than modern altricial mammals (Fig. 9.11), although their neonates are among the smallest (relative to maternal body weight) produced by modern precocial mammals. It is, of course, possible, that the ancestral primates produced neonates of intermediate size, as in modern strepsirhines. Nevertheless, one should at least consider the alternative possibility that there has been a secondary reduction in the relative size of the neonate in the evolution of strepsirhines from the ancestral primates, which would therefore constitute a shared derived feature of this group of primates. One possibility, suggested by Martin (1975), is that ancestral strepsirhines became secondarily adapted for reproduction in forest environments subject to relatively marked seasonal changes.

It has become increasingly obvious that tarsiers are linked to simians by a suite of (apparently) shared derived features from several functional contexts in addition to that of reproductive biology. In the ear region, tarsiers share with simians the development of an anterior chamber and the inclusion of the promontorial artery in a partition between this chamber and the main bullar cavity (MacPhee and Cartmill, 1986). In the visual and olfactory systems tarsiers also share numerous special features with simian primates. Although it has often been argued that some or all of these might have developed through convergent evolution, the evidence now available indicates that the overall degree of similarity between tarsiers and simians in apparently derived features of these systems is too great to be explained away by convergence. Given the fact that tarsiers are nocturnal whereas simians (with the sole exception of the owl monkey, *Aotus*) are typically diurnal, the initial expectation should surely be that divergent, rather than convergent, adaptation would have taken place in the evolution of the visual and olfactory systems of tarsiers and simians. Tarsiers share the following set of unique features of the visual apparatus with simians: there is a post-orbital plate and the eye is therefore enclosed in a bony socket separated from the jaw muscles; the retina has a fovea and a yellow spot (although these features are commonly lacking from *Aotus*); and there is a special pattern of organization of the lateral geniculate nucleus (Fig. 8.21). With respect to the latter feature, it is important to note that lemurs and lorises also share an apparently derived feature in the development of two new small-celled (koniocellular) layers (Kaas, Huerta *et al.*, 1978).

The list of resemblances between tarsiers and simians in the olfactory system is also striking: the rhinarium has been completely suppressed (Fig. 7.13); there has been extensive reduction of structures within the nasal cavity, with loss of two ethmoturbinals and of the transverse lamina (Fig. 7.12); and the olfactory bulbs are very small, relative to body size (Fig. 8.16). Indeed, it is the loss of the rhinarium in tarsiers and simians that led Pocock (1918) to propose the name 'Haplorhini' for this group of primates. Although some convergence may have taken place, it seems highly unlikely that this overall suite of special features in vision and olfaction, several of which are unique among mammals, could have developed entirely independently in tarsiers and in simians. Further, it is difficult to see why a set of adaptations convergently similar to those of simians should have emerged during the evolution of tarsiers but not during the evolution of other prosimian primates.

Several lines of evidence in fact indicate that tarsiers, in contrast to other nocturnal prosimians (most lemurs and all lorises), have become secondarily nocturnal. Reduction of the olfactory system is generally correlated with the emergence of diurnal habits among primates generally, so the reduced olfactory apparatus of tarsiers (and that of the only nocturnal simian, *Aotus*) is best explained by postulating a diurnal ancestry at some stage in the evolution of the haplorhine primates. Similarly, the presence of a yellow spot and fovea in the retina of tarsiers is

best explained as a retention from some diurnal ancestor that was well adapted for colour vision. The implications of this can be examined by superimposing the distribution of nocturnal and diurnal habits among living primates on an outline phylogenetic tree to indicate the approximate pattern of evolutionary changes (Fig. 12.3; see also Martin, 1979). Figure 12.3 clearly reflects the fact that most strepsirhine primates have nocturnal habits, whereas the haplorhine primates are predominantly diurnal, the only exceptions being *Tarsius* and *Aotus*.

As suggested earlier in this chapter, it is most parsimonious to assume that nocturnal habits were retained by the ancestral primates, which were probably also small-bodied, and that living nocturnal strepsirhines have remained primitive in both respects. Those Malagasy lemurs that are now diurnal have undergone a relatively recent adaptive shift, with independent evolution of increased body size and diurnality in at least two separate lineages. (Indeed, if it is accepted that the large body sizes and relatively small orbits of most subfossil lemurs indicate diurnal habits, there might have been four or more independent origins of diurnality among the Malagasy lemurs.) By contrast, among the haplorhines diurnal habits, commonly associated with relatively large body sizes, predominate; only the relatively small tarsiers and owl monkeys are now nocturnal. In fact, owl monkeys have been reported to be diurnal in part of their range, so their adaptation for nocturnal habits may not be as complete as for most nocturnal strepsirhines and for tarsiers.

Although tarsiers and owl monkeys may have retained nocturnal habits from the ancestral primates, hence remaining primitive in this respect, it seems far more likely that there has been a secondary return to nocturnality. As noted above, this is indicated by the fact that tarsiers and owl monkeys show several features of the visual and olfactory systems that are best explained through original adaptation to diurnal habits. In addition, however, these nocturnal haplorhines lack a special feature that apparently emerged in the common ancestor of the strepsirhines – the reflecting tapetum behind the

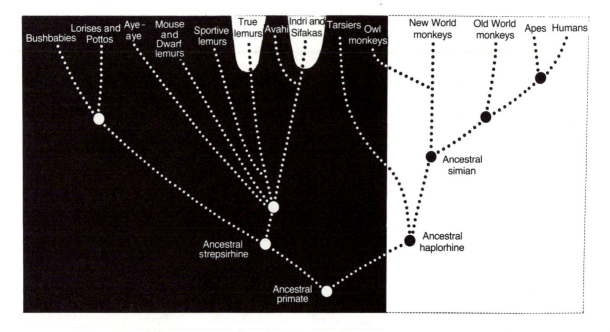

Figure 12.3. Outline phylogenetic tree of the primates, indicating the fundamental distinction between nocturnal species (black zone) and diurnal species (white zone). Note that the separation indicated applies only to living species; fossil primates have not been taken into account. It is probable, however, that the ancestral primates and the ancestral strepsirhines were nocturnal, and that there was a shift from nocturnal to diurnal habits in the emergence of the ancestral haplorhines. (Reprinted from Martin, 1979, with kind permission of Academic Press, New York.)

retina. Given that the tapetum apparently functions as an aid to vision in dim light conditions and that different kinds of tapetum have been developed in various mammalian groups adapted for nocturnal activity, it is initially surprising that tarsiers and owl monkeys lack a reflecting tapetum. However, if it is accepted that both are derived from a diurnally adapted ancestral haplorhine, the absence of such a tapetum is explicable. It is conceivable that prior adaptation of the retina for diurnal conditions in ancestral haplorhines in some way ruled out the subsequent development of a tapetum. Certainly, in comparison to the eyes of strepsirhine primates, those of tarsiers seem to have become adapted in a difference way for nocturnal vision – namely, through a radical increase in the size of the retina (and hence in the number of photoreceptors) well beyond the level in nocturnal strepsirhines. Tarsiers have by far

the largest eyes, relative to body size, of all living primates (see Figs 7.6 and 7.7). The relative size of the eyes in owl monkeys has not increased to such an extreme, but this primate certainly has significantly larger eyes than other simians and they fall at the upper end of the distribution for nocturnal strepsirhines (Fig. 7.7), perhaps indicating a more recent transition to diurnal habits than in the case of tarsiers. It should be noted that the nocturnal habits of owl monkeys can apparently be traced back at least to the Miocene (Setoguchi and Rosenberger, 1987; Martin, 1987). This sets a lower limit on the timespan over which the owl monkey lineage has been nocturnal and suggests that the development of nocturnality in the tarsier lineage, which probably began at an even earlier stage than in owl monkeys, may well date back to the early Tertiary.

The hypothetical pattern of evolution of

nocturnal and diurnal habits among primates shown in Fig. 12.3 also has implications for the body size of ancestral haplorhines. As discussed in Chapter 4, it is an empirical fact that diurnal habits are relatively rare in small mammals and are common only among mammals weighing more than 5 kg, which happens to be the modal body weight for haplorhine primates (Fig. 4.15). The modal body weight of nocturnal primates is around 500 g. Assuming that the distribution of body weights among living nocturnal primates is broadly comparable to that present among the earliest primates, there has been a tenfold increase in modal body weight during haplorhine evolution. If a shift from nocturnal to diurnal habits accompanied the evolution of the ancestral haplorhines from the ancestral primates, it is therefore likely that there was also a significant increase in body size, probably to at least 1 kg and possibly to 5 kg or more. Incidentally, this also suggests that a secondary return to nocturnal habits in the evolution of tarsiers was accompanied by some reduction in body size (dwarfing).

With respect to locomotor adaptations, there is little to indicate a specific link between tarsiers and simians. In many respects, the locomotor system of tarsiers has remained relatively primitive. On the other hand, there are clearly some specializations, particularly in relation to the vertical-clinging-and-leaping habit of tarsiers. It was at one time suggested that the vertical-clinging-and-leaping habits of tarsiers represented a primitive feature for primates of modern aspect (Napier, 1967a; Napier and Walker, 1967), as this pattern is also found in some lemurs and in some bushbabies. However, it now seems more likely that ancestral primates were merely adapted for a hindlimb-dominated pattern of arboreal locomotion involving at least some leaping (see Chapter 10). Although this hindlimb-dominated pattern provided a basis from which vertical-clinging-and-leaping could subsequently develop, ancestral primates lacked the extreme morphological specializations found in most modern vertical-clingers-and-leapers. The most likely explanation is that vertical-clinging-and-leaping evolved in parallel in lemurs

(probably more than once), in bushbabies and in tarsiers.

Superficially (e.g. in the extreme development of the tarsus and in simple proportions) the hindlimbs of tarsiers seem to be closely similar to those of bushbabies in many respects and it might seem possible, therefore, that bushbabies and tarsiers are linked phylogenetically. However, closer examination shows that the hindlimb morphology of *Tarsius* is radically different from that of bushbabies (Jouffroy, Berge and Niemitz, 1984; Gebo, 1987a). Whereas bushbabies have a morphological pattern that allows rotation of the tarsus about a longitudinal axis, and hence inversion and eversion of the foot, such rotation is ruled out in tarsiers. This means that tarsiers are not well adapted for grasping a vertical support with their hindlimbs and explains why a secondary supporting function of the tail has developed. The superficial resemblances between tarsiers and bushbabies in hindlimb proportions and other features are probably attributable to convergent evolution and there is therefore no basis for suggesting a phylogenetic link between these two groups of primates.

There is, however, a further implication of the special hindlimb morphology of tarsiers. The elimination of the capacity for inversion and eversion in the hindfoot that clearly took place at some stage in the evolution of tarsiers suggests adaptation for terrestrial locomotion, or at least adaptation for locomotion over relatively flat substrates in which the importance of grasping of variable supports was reduced. A unique feature of the hindlimb of tarsiers, compared to that of other living primates, is the partial fusion of the fibula with the tibia. Such fusion is related to restriction of the ankle joint to a hinge action in various terrestrially adapted mammals (e.g. rabbits and elephant-shrews), and fusion of the fibula and tibia in tarsiers can also be explained on this basis. The possibility therefore arises that the relatively large-bodied haplorhine ancestor that eventually gave rise to tarsiers was not only diurnal but also adapted for some degree of terrestrial adaptation. This would explain the unusual features of the hindlimb of tarsiers and it could also account for the fact that medium-sized simians (monkeys) are uniformly characterized by quadrupedal locomotion. It should also be noted that lower limb bones from the Oligocene fossil site at Fayum in Egypt attributed to the early simian *Apidium* also show (presumably convergently) fusion of the fibula and tibia, thus confirming that some reduction of the capacity for inversion and eversion of the foot may have characterized early haplorhines.

There is also some limited evidence that lemurs and lorises share certain derived features of postcranial morphology. For instance, the digital formula of 4:3:5:2:1 found in both hands and feet of strepsirhine primates is probably derived relative to an ancestral mammalian formula of 3:4:2:5:1, which is still generally present among simian primates. (Tarsiers have this presumed primitive formula in the hand, but the digital formula in the foot is the same as in lemurs and lorises, presumably as a convergent development). Further, it is possible that lemurs and lorises share certain derived character states in the ankle region (Beard, Dagosto *et al.*, 1988; Dagosto, 1988).

Evidence from chromosomes and biomolecular data indicates that tarsiers are monophyletically related to simians. The chromosomal evidence is still preliminary, as the banding patterns of tarsier chromosomes are very unusual, but tarsiers have a large number of chromosome arms (i.e. a high fundamental number), in common with simian primates and in contrast to most strepsirhine primates (see Fig. 11.10). Although some strepsirhine primates have high fundamental numbers, various features indicate that these have been achieved independently. In fact, tarsiers (in contrast to strepsirhines with high fundamental numbers) also resemble simians in having a relatively high content of nuclear DNA (see Fig. 11.13), indicating some special factor in the evolution of haplorhine chromosomes compared to those of strepsirhines. The biomolecular evidence is generally more convincing, although conflicting conclusions have been drawn from immunological studies. Immunodiffusion comparisons of albumins indicate that tarsiers are more closely related to simians than to lemurs and lorises (see

Fig. 11.21; Dene, Goodman and Prychodko, 1976a) and confirm that strepsirhines and haplorhines constitute separate monophyletic groups. However, a phylogenetic tree based on microcomplement fixation tests applied to albumins and transferrins, while indicating that strepsirhines and simians constitute two monophyletic groups, fails to indicate a specific relationship between tarsiers and simians (see Fig. 11.24; Sarich and Cronin, 1976). Amino acid sequence data are still somewhat limited for tarsiers, but analysis of tandem alignment of available data for primates from several proteins indicates that *Tarsius* is closer to simians than to strepsirhines (Goodman, Romero-Herrera *et al.*, 1982). This is confirmed by analysis of amino acid sequences in lens α-crystallins (de Jong and Goodman, 1988). Finally, preliminary DNA hybridization data (Bonner, Heinemann and Todaro, 1980) indicate that living strepsirhines and haplorhines constitute separate monophyletic groups that diverged from the common ancestral primate stock. Taken in conjunction with evidence from dental morphology, from reproductive biology, from postcranial features, from the visual system and from the olfactory system, these data contribute to an overwhelming body of evidence indicating that living strepsirhines and haplorhines are members of two separate monophyletic assemblages within the phylogenetic tree of primates.

A further indication of a specific link between tarsiers and simians has been provided by Pollock and Mullin (1987). It has been known for some time that various simian primates are unable to synthesize vitamin C (ascorbic acid) and that they must therefore obtain this essential substance in their natural diets. An inability to synthesize ascorbic acid seems to be a relatively uncommon feature among placental mammals generally, having otherwise been reported only for bats (without exception to date) and for just one rodent (the guinea-pig). In contrast to simian primates, tree-shrews and various species of the loris group can synthesize ascorbic acid. Pollock and Mullin (1987) showed that *Tarsius* lacks the specific enzymatic activity that would indicate an ability to synthesize ascorbic acid, whereas a variety of lemurs clearly possess such activity. It would therefore seem that the inability to synthesize ascorbic acid is a shared derived feature of haplorhine primates (tarsiers and simians), whereas the ability of strepsirhine primates and tree-shrews to synthesize this vitamin is a primitive feature retained from the common ancestral stock of placental mammals. It has been suggested that the loss of the ability to synthesize ascorbic acid might be associated with particular dietary habits leading to an abundant intake of the vitamin, such that a mutation leading to loss of a vital enzyme could spread through a population unhindered (see later). However, the absence of the ability to synthesize ascorbic acid in all bats investigated to date, regardless of dietary habits, indicates that there is no simple correlation between a particular diet and the loss of the ability to synthesize the vitamin. Thus, although the evidence regarding the inability to synthesize ascorbic acid does provide further evidence of a phylogenetic link between *Tarsius* and simians, it is at present impossible to draw any conclusions about the likely dietary habits of ancestral haplorhines.

When it comes to fossil evidence, the picture is far less clear, mainly because many of the characters indicating a fundamental dichotomy between living strepsirhines and living haplorhines are not preserved in the fossil record. In the first place, there is little to link the early Tertiary 'lemuroids' (Adapidae) to living strepsirhine primates. There is no tooth-comb comparable to that possessed by living strepsirhines in any early Tertiary fossil 'lemuroid'. Gingerich (1975) suggested that modification of the lower anterior dentition of *Adapis parisiensis*, such that the canines form a functional unit with the incisors, could indicate a link to the origin of lemurs and lorises. However, derivation of ancestral strepsirhines from *Adapis parisiensis* is unlikely, because *Adapis* was characterized by fusion of the symphysis in the lower jaw (a derived feature lacking from living strepsirhines) and because the postcranial adaptations of *Adapis* (including *Leptadapis*) are in many respects uniquely specialized in comparison to those of other Eocene 'lemuroids'

and to modern strepsirhines (e.g. Dagosto, 1983). As the cranial and dental features of modern strepsirhines are otherwise generally primitive in comparison to those of modern simians, it is difficult to test the hypothesis that early Tertiary 'lemuroids' are phylogenetically related to modern strepsirhine primates. Beard, Dagosto *et al.* (1988) and Dagosto (1988) have noted certain features of the tarsus that may link adapids specifically to lemurs and lorises, but the evidence is limited and the possibility of convergence cannot be ruled out. For the present, it remains possible that early Tertiary 'lemuroids' may have diverged separately from the ancestral stock of primates of modern aspect. By contrast, the Miocene members of the loris group known from Africa and Asia clearly had a tooth-comb and represent an integral part of the adaptive radiation of modern strepsirhines.

Comparable difficulties are encountered in assessing the relationships of early Tertiary 'tarsioids' (Omomyidae). Although they have certain features that may indicate a specific relationship to modern tarsiers, notably in the shape of the upper dental arcade and in the modification of the anterior dentition, they lack clear defining features that would permit recognition as haplorhines. Although it was originally reported (Martin, 1973, 1979) that early Tertiary 'tarsioids' lacked a gap between the upper incisors, indicating that loss of the rhinarium might have taken place, it has since been shown (e.g. Schmid, 1982; Aiello, 1986) that there was in fact a fairly large gap. Accordingly, it is possible (though not definite) that a rhinarium was still present in at least some omomyids and that these primates cannot be regarded as haplorhines in the literal sense of the term.

Given the fact that living tarsiers and early Tertiary 'tarsioids' have relatively primitive cheek teeth, it is difficult to establish a specific phylogenetic link between the omomyids and living haplorhines. Some authors (e.g. Cartmill, 1980; Aiello, 1986) have accordingly suggested that the omomyids may have branched off before the divergence between tarsiers and simians took place (Fig. 12.4, tree a). Schmid (1982) has gone

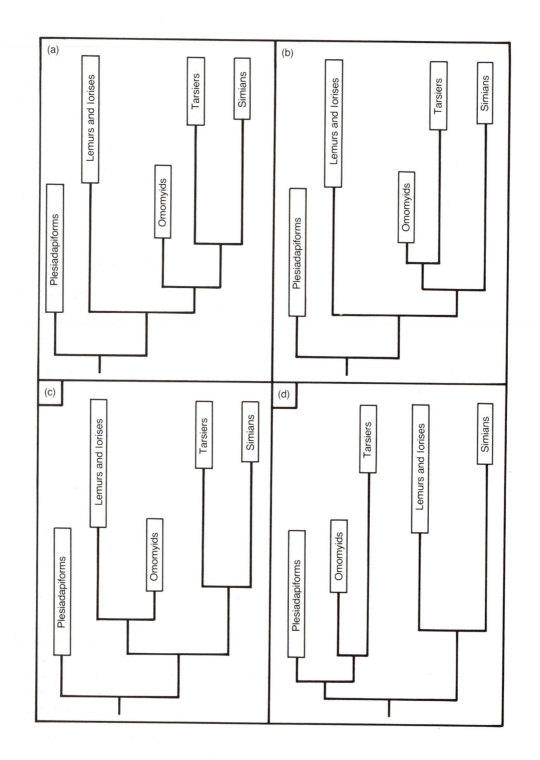

even further in reviving Hürzeler's suggestion (1948) that Eocene 'tarsioids' such as *Necrolemur* may be related to strepsirhines, rather than to haplorhines (Fig.12.4, tree c). This suggestion was apparently reinforced by evidence that *Necrolemur* used its lower incisors to groom its fur (Schmid, 1983). However, similarity to the typical strepsirhine tooth-comb is superficial and it has since emerged that at least one tarsier species, *Tarsius pumilus*, probably uses its lower anterior teeth to groom its fur (Musser and Dagosto, 1987). An even more radical rearrangement of the primate tree was proposed by Gingerich (1976b), who linked tarsiers and omomyids to the 'archaic primates' (plesiadapiforms). For a variety of reasons, he had previously concluded (e.g. see Gingerich, 1973, 1975b) that lemurs and lorises, along with the Eocene 'lemuroids', are more closely related to simians than are tarsiers (Fig. 12.4, tree d). Gingerich went on to propose an alternative classification of primates on this basis, placing 'archaic primates', omomyids and tarsiers in the suborder Plesitarsiiformes and uniting lemurs, lorises and simians in the suborder Simiolemuriformes. Thus, Gingerich did not question the association between omomyids and tarsiers, but rejected instead any specific link between tarsiers and simians. This particular hypothesis met with numerous criticisms and the proposal of a specific link between tarsiers, omomyids and 'archaic primates' was subsequently replaced by a grade concept of primate evolution (Gingerich, 1981b; MacPhee, Cartmill and Gingerich 1983; Gingerich, 1986a). However, Gingerich (1986a) has continued to

maintain that Eocene 'tarsioids' (Omomyidae) may be more closely related to plesidapiforms and that Eocene 'lemuroids' (Adapidae) may be closely related to simian primates.

Szalay (1975a) reviewed evidence from the basicranial region of the primate skull and concluded that there were specific characters indicating that the omomyids, tarsiers and simians form a monophyletic group (Fig. 12.4, tree b). He inferred that a medial point of entry of the internal carotid into the auditory bulla is a shared derived feature of haplorhine primates. This character is linked to the emphasis of the promontorial artery at the expense of the stapedial artery that appears to be a shared derived feature of living haplorhines (Chapter 7; see Fig. 7.22). Szalay also cited 'an increasingly extrabullar position of the ectotympanic' as a derived feature of haplorhines. However, the ectotympanic was clearly located within the bulla in at least some omomyids (e.g. *Necrolemur*), as was recognized by Szalay, whereas an extrabullar ectotympanic is characteristic of the loris group among strepsirhine primates. The position of the ectotympanic therefore provides no clear evidence linking the omomyids to tarsiers and simians. Schmid (1981) subsequently noted that a medial entry of the internal carotid into the bulla is also present in the loris group. Because the medial entry also occurs in several other groups of placental mammals, Schmid concluded that this is, in fact, the primitive condition and that it is the lemurs, the Eocene 'lemuroids' and some 'archaic primates' that have undergone secondary modification such that the point of entry of the internal carotid is now

Figure 12.4. Hypothetical trees indicating possible relationships between the living tarsiers (Tarsiidae), the Eocene 'tarsioids' (Omomyidae), the 'archaic primates' and other primates.

(a) The omomyids are related to the haplorhine group of primates (tarsiers and simians), but branched off before the division between tarsiers and simians (cf. Cartmill, 1980).

(b) The omomyids are specifically related to tarsiers and therefore belong to a monophyletic group containing tarsiers and simians (cf. Simons, 1972; Szalay and Delson, 1979).

(c) The omomyids are related to strepsirhine primates (lemurs and lorises), rather than to haplorhines (cf. Hürzeler, 1948; Schmid, 1982, 1983).

(d) The omomyids and tarsiers are specifically linked to 'archaic primates' (plesiadapiforms), and the strepsirhine primates are related to simians (Gingerich, 1976b, 1986a).

posterolateral. Although it is, indeed, true that the promontorial artery is emphasized at the expense of the stapedial artery in living haplorhine primates, Schmid (1981) observed that *Necrolemur* has a relatively narrow promontory artery and a stapedial artery of comparable diameter, as in the condition presumed to be primitive for primates. Finally, MacPhee and Cartmill (1986) concluded from a thorough review of the basicranial region of the skull that omomyids are not specifically related to tarsiers, although they may be linked to the origins of a monophyletic group containing simians and tarsiers. There is even the possibility that the omomyids themselves might not form a monophyletic group.

It therefore remains uncertain whether omomyids are linked to tarsiers in some way, but there is abundant evidence that the latter are linked to simians. This being the case, the term 'haplorhine' should certainly be restricted to living primates for the time being and it is necessary to keep an open mind about the affinities of the Omomyidae. By the same token, the term 'tarsioids' should be used only in a broad descriptive sense for the Omomyidae. A similar state of affairs exists with use of the term 'lemuroids' for the Adapidae. There is little evidence to link adapids directly with modern strepsirhine primates and it is possible that they diverged separately from the ancestral primate stock. Thus, at present it is only possible to conclude with confidence that definite primates of modern aspect are identifiable in the Eocene and that these may be divided fairly sharply into two groups (Adapidae and Omomyidae). The exact relationships of these two early Tertiary groups to modern primates have yet to be determined.

ARE THE MALAGASY LEMURS MONOPHYLETIC?

It is now widely accepted that the living strepsirhine primates (lemurs and lorises) constitute a monophyletic group. However, there has been considerable discussion about whether, within the strepsirhine assemblage, lemurs and the loris group comprise two distinct monophyletic groups (Fig. 12.5, tree a), or whether some more complex pattern of relationships is involved (Fig. 12.5, tree b). For some time, it was virtually taken for granted that lemurs and the loris group are two distinct 'natural groups', each traceable to a specific ancestral stock. Gregory (1915) noted several characters apparently separating lemurs from the loris group. A brief list of features said to distinguish the two groups was also provided by Weber (1928) in his discussion of primate systematics (Table 12.2). A more extensive list of 39 distinguishing features was later provided by Hill (1936) and several of the most important features from all these sources were cited by Le Gros Clark (1959). But such distinctions were largely based on comparison of representative species from each group, with the lemurs commonly represented by the single genus *Lemur* and the loris group by a single genus such as *Perodicticus*. It was pointed out by Charles-Dominique and Martin (1970) that the distinctions do not clearly separate mouse lemurs and dwarf lemurs (family Cheirogaleidae) from the loris group. In particular, an os planum formed by the ethmoid is present in the medial wall of the orbit in *Microcebus* and *Cheirogaleus*, and in all members of the family Cheirogaleidae the course of the internal carotid follows the same pattern as in the loris group (see Chapter 7). Charles-Dominique and Martin went on to note that field studies of mouse lemurs (*Microcebus*) and bushbabies (*Galago*) had revealed marked similarities in behaviour and ecology between them. Four possible interpretations of these findings were suggested:

1. Mouse lemurs and bushbabies have retained a great number of ancestral primate features that would have been present in a common ancestor of the strepsirhines before the separation of Madagascar from the African mainland.
2. Adaptation to very similar ecological niches led to convergent evolution between mouse lemurs and bushbabies.

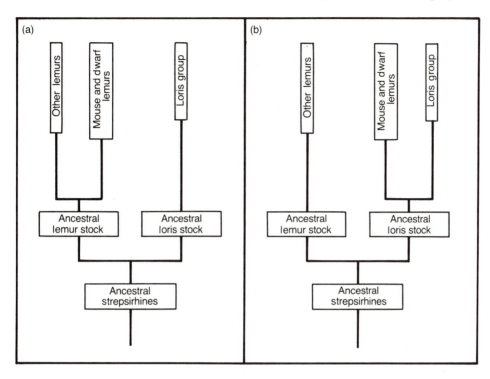

Figure 12.5. Two alternative trees indicating possible phylogenetic relationships among strepsirhine primates (lemurs and lorises).

(a) The lemurs and the loris group constitute two separate monophyletic groups, each traceable to a specific ancestral stock.

(b) The lemurs are not a monophyletic group, as mouse lemurs and dwarf lemurs (Cheirogaleidae) are more closely related to the loris group than to other lemurs.

3. Mouse lemurs and bushbabies in fact shared a specific common ancestry subsequent to the initial adaptive radiation of the strepsirhines.
4. Some combination of the above (e.g. 1 and 2).

It was concluded that the most likely hypothesis is the first, perhaps combined with the second. That is to say, it was inferred that *Microcebus* and *Galago* resemble one another largely because of the retention of primitive features from the ancestral strepsirhine stock, although some convergent similarities might also have accumulated because of their continued occupation of similar ecological niches (see also Martin, 1972b).

There followed a series of papers all concluding that the third interpretation (Fig. 12.5, tree b) is, in fact, correct (e.g. Szalay and Katz, 1973; Groves, 1974; Hoffstetter, 1974; Tattersall and Schwartz, 1974, 1975). Having inferred that mouse and dwarf lemurs (Cheirogaleidae) are phylogenetically related to the loris group, several of these authors then proceeded to propose revised classifications of the strepsirhine primates – for example, with the Cheirogaleidae moved to the superfamily Lorisoidea from the infraorder Lemuroidea, where they had been placed in Simpson's widely used classification (1945). Inclusion of the Cheirogaleidae in the Lorisoidea was, for instance, a feature of the classification of Primates presented by Szalay and Delson (1979). This overall development therefore raises two issues: (1) Is it probable that the Cheirogaleidae

Table 12.2 Characters that supposedly distinguish lemurs from the loris group (after Weber, 1928; see also Charles-Dominique and Martin, 1970)

Lemurs	Loris group
1. Ring-shaped ectotympanic bone within the bulla	1. Ectotympanic bone fused to margin of the bulla
2. Jugal in contact with the lacrimal bone	2. Jugal separated from the lacrimal by the maxilla, to varying degrees
3. Ethmoid does not form an os planum in the medial wall of the orbit	3. Ethmoid forms a prominent os planum in the medial wall of the orbit
4. Ethmoturbinal 1 small; does not cover the maxilloturbinal	4. Ethmotubinal 1 large; covers the maxilloturbinal
5. Internal carotid enters the bulla through the posterior carotid foramen, continues through the bulla and enters the braincase via the basisphenoid	5. Internal carotid artery does not enter the bulla, but runs medially and enters the braincase anteriorly, via the anterior lacerate foramen

did indeed share a specific ancestral stock with the loris group (Fig. 12.5, tree b)? (2) If so, is it advisable to change the classification of primates to reflect this inferred common ancestry?

Among the various morphological characters that have been cited as linking cheirogaleids to the loris group, the most important are present in the orbit and in the ear region. Exposure of the ethmoid in the medial orbital wall is, however, of questionable significance. Whereas all members of the loris group have an os planum, this character is variable within the family Cheirogaleidae, being identifiable in *Allocebus* and *Microcebus* but not in *Cheirogaleus* or *Phaner* (Cartmill, 1978). Further, it is on other grounds unjustifiable to interpret the os planum as a shared derived feature linking cheirogaleids to the loris group. In the first place, it is clear that exposure of the ethmoid in the medial wall of the orbit is determined by a complex relationship between the size of the orbits, their degree of forward rotation and overall skull size (see Chapter 7). Accordingly, an os planum may have been developed in separate lineages. Secondly, there is a possibility that an os planum was already present in the common ancestral stock of the primates and therefore represents a primitive homology in the loris group and some cheirogaleids. Application of the sister-group principle would support this interpretation. Among strepsirhine primates, there are two possible conditions: presence or absence of an os planum. According to the sister-group principle, the primitive condition should be that found in the next most closely related group of mammals, the haplorhine primates. As all living haplorhines have an os planum, this is apparently the primitive condition for primates. Of course, all of the guiding principles used in phylogenetic reconstruction are subject to exceptions (see Chapter 3), so no binding conclusion may be drawn. Nevertheless, it may be stated that the possession of an os planum by members of the loris group and some cheirogaleids is far more likely to be either a primitive retention or a convergent development than a shared derived feature.

The resemblance between the loris group and cheirogaleids in the ear region are far more striking (Cartmill, 1975). It should be noted, however, that the relationship between the ectotympanic ring and bulla does separate lemurs from all members of the loris group. In all lemurs, the ectotympanic ring is located within the bulla, whereas in the loris group the ectotympanic is always fused to the external margin of the bulla (see Fig. 7.18). Although Wood Jones (1929; cited by Charles-Dominique and Martin, 1970) published an illustration of a bulla of *Galago* with an enclosed ectotympanic ring, it seems likely that this was a straightforward error. It is therefore justifiable to cite an enclosed ectotympanic ring as a universally shared feature of lemurs and an externally fused ectotympanic as a universally shared feature of the loris group. Although cheirogaleids may be regarded as somewhat intermediate, in that the ectotympanic ring lies close to the bulla wall (Cartmill, 1975), it is important to remember that body size has a major influence. Because the size of the ear drum decreases only very slowly with decreasing body size, the ectotympanic ring is large in relation to the overall size of the bulla in all small-bodied primates. It may be for this reason that the relatively small cheirogaleids are more similar to members of the loris group than to other lemurs.

Having accepted that the location of the ectotympanic ring represents a good character separating all lemurs from all members of the loris group, it is now necessary to identify the primitive condition. For instance, it is possible that the location of the ectotympanic as a 'free' ring within the bulla is a primitive feature for strepsirhine primates. If this were the case, the possession of an intrabullar ectotympanic by cheirogaleids would not rule out a phylogenetic relationship with the loris group. If, on the other hand, the intrabullar ectotympanic is a shared derived feature of lemurs, a relationship between cheirogaleids and the loris group would be ruled out. Once again, the sister-group principle can be applied. Among strepsirhine primates, there are two different conditions; 'free' intrabullar ectotympanic or fused extrabullar ectotympanic. Among haplorhine primates, which constitute the appropriate sister group, the ectotympanic is in all cases fused to the external margin of the bulla. It is therefore likely that the ancestral

primates had an extrabullar ectotympanic, fused to the margin of the bulla, and that the intrabullar ectotympanic emerged as a derived feature in the common ancestor of lemurs. At first sight, however, this interpretation appears to conflict with the fossil evidence. As far as is known, the ectotympanic ring was located within the bulla in all Eocene 'lemuroids' and 'tarsioids'. This would suggest that the intrabullar ectotympanic was primitive for primates generally and that the fused extrabullar ectotympanic arose independently at least twice (i.e. in the ancestral stock of the loris group and in the ancestral stock of modern haplorhines). Not for the first time, there is a contradiction between inferences from two different guidelines for phylogenetic reconstruction (sister-group principle versus fossil precedence) and no clear resolution is possible.

As usual, it is valuable to consider developmental evidence, for this reveals that the difference between an intrabullar ectotympanic and an extrabullar ectotympanic is relatively limited. All strepsirhines pass through essentially the same initial stages of development of the ear region until the developing petrosal bulla contacts the margin of the ectotympanic ring (which develops earlier). At this stage, in lemurs the petrosal passes over the ring and encloses it, whereas in the loris group the migrating margin of the petrosal remains in contact with the ectotympanic ring. This means that a shift from an intrabullar to an extrabullar ectotympanic (or vice versa) can result from a relatively simple alteration in the developmental sequence. It is a moot point, therefore, whether enclosure of the ring emerged as a convergent feature in Eocene primates of modern aspect and in lemurs, or whether non-enclosure of the ring emerged as a convergent feature of the loris group and living haplorhines. Both possibilities remain open and new evidence on the functional implications of the different locations of the ectotympanic is needed before a conclusion can be reached. One can, however, conclude that the relationship between the ectotympanic and the bulla provides no evidence for a specific relationship between cheirogaleids and the loris group.

The most persuasive evidence for a phylogenetic link between cheirogaleids and the loris group (Fig. 12.5, tree b) comes from the relationship between the internal carotid system and the bulla (Cartmill, 1975; see Fig. 7.22). The suppression of the intrabullar vessels (stapedial artery; promontorial artery) and the special development of the extrabullar ascending pharyngeal artery is an unusual characteristic shared by these two groups of primates, commonly interpreted as a shared derived feature. It should be noted, however, that in the hypothetical primitive condition for primates (Fig. 7.22), all three branches of the internal carotid system were present – stapedial artery, promontorial artery and ascending pharyngeal artery. The general evolutionary trend among primates has been to reduce two of these vessels and to emphasize the third. There is accordingly a possibility that cheirogaleids and the loris group may have convergently developed a pattern in which the ascending pharyngeal artery has been developed at the expense of the other two vessels. Unfortunately, it is not known why there has been a general trend among primates to emphasize one vessel at the expense of the other two. Nor is it known what functional significance attaches to the specific enlargement of the ascending pharyngeal artery, as opposed to enlargement of either of the other two vessels. It is therefore difficult to assess the likelihood of convergence between cheirogaleids and the loris group. Nevertheless, it should be noted that body size may also influence the pattern of internal carotid vessels (see Chapter 7). It is possible that cheirogaleids and members of the loris group, which are all relatively small-bodied primates, may have developed similar patterns independently.

The evidence supposedly linking cheirogaleids to the loris group is therefore questionable. On the other hand, apart from the presence of a 'free' ectotympanic ring within the bulla (which may or may not be a derived feature in primates), there is little morphological evidence indicating that the Malagasy lemurs constitute a monophyletic group. Some evidence is available, however, from the structure and use of the hand. In all lemurs, the number of palmar pads has been reduced to five, through fusion of the first interdigital pad with the thenar pad, whereas members of the loris group have retained the primitive set of six separate pads (see Chapter 10). In addition, lemurs have a distinctive prehensive pattern, whereas the loris group and tarsiers share what is presumably the primitive pattern (Bishop, 1962, 1964).

Given that the morphological evidence is equivocal, particular weight must be attached to supplementary evidence from chromosomes, proteins and genes in attempting to distinguish between the two alternative trees presented in Fig. 12.5. This is particularly necessary because of the marked deficiences of the fossil record for the strepsirhine side of the phylogenetic tree of primates. With the exception of the subfossil lemurs, which should really be counted as part of the modern fauna, there is no fossil record for lemurs on Madagascar. For the loris group, there is some limited fossil evidence from Miocene deposits of Africa and Asia, but their earlier phylogenetic history remains undocumented. In the face of these limitations, it is regrettable that virtually all of the authors who have advocated a specific link between the Cheirogaleidae and the loris group on morphological grounds have paid little attention to evidence from other sources, in some cases dismissing it out of hand. In fact, such evidence provides no support whatsoever for a link between cheirogaleids and the loris group. Taking the chromosomal evidence first of all, the overall reconstruction for primates produced by Dutrillaux, Couturier *et al.* (1986a) clearly indicates that lemurs and the loris group comprise two separate monophyletic groups (see Figs 11.18 and 11.19). This is in full agreement with the results of comprehensive immunological comparisons conducted by Dene, Goodman and Prychodko (1976a), as shown in Fig. 11.21. Although the results of microcomplement fixation studies (Sarich and Cronin, 1976) do not indicate two clear-cut monophyletic groups for lemurs and the loris group (see Fig. 11.24), they also fail to indicate a specific relationship between cheirogaleids and the loris group. Unfortunately, no protein sequence data are available for cheirogaleids, so it is not at present possible to test the alternative trees shown in Fig. 12.5 at a finer level of protein structure. There is, however, clear evidence from DNA hybridization (Bonner, Heinemann and Todaro 1980) that lemurs and the loris group constitute separate monophyletic groups. The evidence as a whole permits confident rejection of the hypothesis that there is a specific phylogenetic relationship between cheirogaleids and members of the loris group.

The lemurs and the loris group were originally treated as distinct groups primarily on geographical grounds. The existence of the Mozambique channel as a major barrier to migration between Africa and Madagascar is obviously an additional factor that must be taken into account in assessing the evolutionary relationships of lemurs. As was argued in Chapter 3, this additional factor provides more support for the interpretation that the lemurs constitute a monophyletic group.

If it is true that cheirogaleids are not specifically related to the loris group, it naturally follows that any new classification based on the assumption of such a relationship has no foundation. Given that the balance of evidence indicates that lemurs and the loris group do

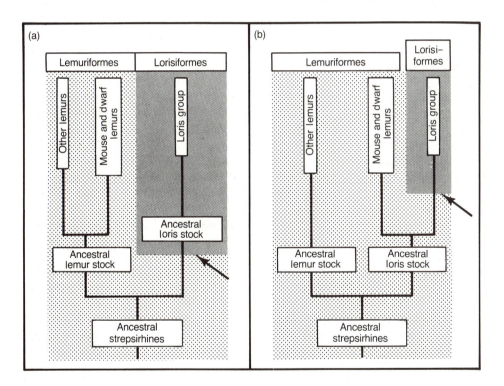

Figure 12.6. The relationship between the two alternative phylogenetic trees shown in Fig. 12.5 and the classical approach to classification. If lemurs and the loris group comprise two separate monophyletic groups (tree a) the division into Lemuriformes and Lorisiformes is justifiable with both classical and cladistic classifications. Nevertheless, in order to classify early fossil forms, it is convenient with a classical classification to designate the Lemuriformes as the basic grade and to define Lorisiformes on the basis of the acquisition of specific defining features (arrow). If, on the other hand, mouse and dwarf lemurs (Cheirogaleidae) are phylogenetically related to the loris group (tree b), a cladistic approach would require a different classification. The classical classification into Lemuriformes and Lorisiformes can be retained, however, because the former still constitute the basic grade. There has merely been a shift in the point at which definitive features of the loris group can be identified (arrow).

comprise two separate monophyletic groups, the classification originally proposed by Simpson (1945) – containing the two separate infraorders Lemuriformes and Lorisiformes – fully meets the requirement of compatibility with inferred phylogenetic relationships. In fact, however, it was never really necessary for a new classification to be proposed. Although the cladistic philosophy of classification requires that any classification should reflect inferred branching points in the phylogenetic tree, this is not a requirement of the classical school of classification. Even if it were true that cheirogaleids were more closely related to the loris group than to other lemurs, a grade classification would still permit a primary division into 'lemurs' and 'lorises', with the boundary between the two groups determined by apparent shared derived features of the loris group (Fig. 12.6). Recent discussions of the subdivisions of the strepsirhine primates therefore provide an object lesson in the dangers attached to the cladistic prescription for classification. While it is perfectly reasonable for comparative morphologists to favour the working hypothesis that cheirogaleids and the loris group shared a specific common ancestor, it was not reasonable to translate this provisional hypothesis with unseemly haste into new classifications. In the event, the hypothesis has proved to be highly improbable and the new classifications are accordingly redundant.

ARE THE SIMIANS MONOPHYLETIC?

It is now widely accepted that the living simian primates (New World monkeys, Old World monkeys, apes and humans) constitute a clearly definable group of primates. The simians are commonly referred to as 'higher primates', partly because humans are included among the simians but also because it is possible to recognize a suite of distinctive simian characteristics. For classificatory purposes, at least, it seems justifiable to refer to a 'simian grade' in contrast to a generally more primitive 'prosimian grade' (see Simpson, 1961). Although the simians are now predominantly regarded as forming a

monophyletic group, because they exhibit many similarities that may be interpreted as shared derived features, there have often been suggestions that the New World and Old World representatives attained the simian grade independently. In other words, it has been proposed that the two groups of simians were derived separately from prosimian ancestors and that numerous morphological features of simians were developed convergently in the New and Old Worlds (e.g. see Simpson, 1945). Indeed, this has often been cited as an example of 'parallel evolution'. A prime reason for such an interpretation has been that the earliest known (Palaeocene/Eocene) primates have been documented almost exclusively from the Northern Hemisphere, whereas fossil primates from the Southern Hemisphere are reliably documented only from the Oligocene onwards. A narrow interpretation of the known fossil record would indicate that primates arose in the Northern Hemisphere and subsequently spread southwards. Naturally, this interpretation seemed to be quite reasonable at a time when it was believed that the continents had always occupied their present-day positions. Given that the known Palaeocene/Eocene forms from the Northern Hemisphere include no representatives of simian grade, it would follow that the New World and Old World forms acquired all or most of their 'simian' features independently.

The main distinguishing features shared by modern simians belong to a number of different functional complexes. In the first place, there is a set of characteristics of the jaws and teeth that sets the simians apart from modern prosimians. The two halves of the lower jaw (dentaries) are always fused firmly together early in life and this is probably linked to the fact that the jaw hinge is consistently located above the occlusal plane of the cheek teeth, rather than virtually coinciding with it as in typical modern prosimians. As has been explained in Chapter 6, fusion of the dentaries to form a U-shaped structure in simian primates provides certain mechanical advantages. This, in turn, may be linked to the fact that in simian primates the palate and the cheek-tooth row are both shorter (relative to overall skull length) than in prosimian primates (see Figs 6.11 and 6.12). The incisor teeth typically have a spatulate form in simians, whereas in living prosimians they are either peg-like or bilaterally flattened. In most cases, the cheek teeth are also relatively advanced, with 'squaring' of the molars (involving loss of the paraconid in lower teeth) and general levelling of the occlusal surface. The only exception to this rule is provided by the marmosets and tamarins (family Callitrichidae), which have probably undergone secondary simplification of their molars in association with marked reduction of body size (see next section). The jaws are also typically relatively deep in simians.

Another feature of simian primates – the early fusion of the midline suture between the two frontal bones of the skull – may well be related to an overall remodelling of the jaw apparatus. All living simian primates also have a virtually complete postorbital plate separating the orbit from the temporal jaw musculature, although this feature was possibly already present in the common ancestral stock of modern haplorhines (tarsiers and simians). The simian primates are further characterized by a generally larger relative brain size than found in prosimians. Although there is a limited degree of overlap between prosimians and simians in this respect, there is a definite grade shift overall (see Figs 8.5 and 8.9), amounting to an average two-fold augmentation of simian brain size in comparison to the typical prosimian condition. This has provided the primary justification in the literature for referrring to the simians as 'higher primates'. The simians are also distinguished from prosimians (including tarsiers) by the possession of a single-chambered, simplex uterus (see Fig. 9.2). As has been noted in Chapter 9, this is a very rare condition among placental mammals and may be regarded with confidence as a shared derived feature of simian primates. It seems highly likely that development of a simplex uterus in the common ancestral stock of simians was associated with a restriction to typical development of a single fetus at a time, particularly as the possession of only one pair of teats is a universal feature of simians. In the sphere of reproduction, formation of the amnion by cavitation is also a shared derived feature of simians.

The morphological evidence indicating that simians are a monophyletic group has been abundantly supported by comparative studies of chromosomes, proteins and genes. The chromosomal tree for primates produced by Dutrillaux, Couturier *et al.* (1986a) indicates a common ancestral stock for simians defined by three chromosomal rearrangements following separation from the inferred ancestral primate stock (see Fig. 11.18). Despite disagreements with respect to certain other aspects of primate relationships, the two phylogenetic trees based on immunological comparisons – the immunological distance tree in Fig. 11.21 (Dene, Goodman and Prychodko, 1976a) and the microcomplement fixation tree in Fig. 11.24 (Sarich and Cronin, 1976a) – agree in clearly indicating a specific ancestral stock for simian primates. The same applies to all of the phylogenetic trees based on analyses of amino acid sequences of proteins, such as that for α- and β-haemoglobins combined (Fig. 11.27), that for myoglobin (Fig. 11.30) and that for a tandem alignment of up to seven proteins (Fig. 11.31). Finally, the evidence from DNA hybridization also indicates a common ancestral stock for simians (Figs 11.35 and 11.36).

The uniform evidence from genes and proteins has exerted a particularly strong influence in convincing primatologists that the modern simian primates comprise a monophyletic group. Several authors have, in fact, proceeded to draw the conclusion that the molecular evidence demonstrates that the common ancestor had already reached a simian 'grade' of organization. Such a conclusion is, however, completely unjustified. The molecular evidence can only indicate the existence of a specific common ancestor for modern simians; it cannot tell us what that ancestor looked like nor where it lived. Thus, it is possible to accept the existence of a common ancestor for simians without necessarily accepting that this ancestor already possessed any of the defining features of modern simian primates listed above. Accordingly, it is erroneous to argue that the molecular evidence necessarily rules out separate derivation of New World and Old World simians from some stock of prosimian grade. Only an analysis of morphological evidence, including due reference to the fossil record, can provide direct information about the likely characteristics of the common ancestor of simians. In fact, as far as is recorded, all reliably identified fossil simians from the Oligocene upwards (beginning with *Branisella* in South America and with the Fayum simians in Africa) had all of the fossilizable defining features of the skull and dentition mentioned above: enlarged braincase (compared with approximately contemporary prosimians); fusion of the frontals; fusion and deepening of the dentaries; raising of the jaw hinge; spatulate incisors; and 'squaring' of the molars. As there is no indication from the fossil record that these features were developed independently in the New and Old Worlds, it is parsimonious to conclude that they were already present in the common ancestor of all simians. In other words, the common ancestor of modern simians had already attained a simian grade in various key features.

The concept of an independent origin of New World simians from prosimian ancestors has often been linked to the suggestion that there might be a phylogenetic relationship between Eocene 'lemuroids' (Adapidae) of North America and New World monkeys (Gregory, 1920a). This suggestion was revived in somewhat modified form by Gingerich (e.g. 1975b, 1980b), who proposed that the Adapidae gave rise both to modern strepsirhines and to modern simians (see also Rasmussen, 1986). Gingerich noted that adapids resemble simians in such features as overall body size, fusion of the dentaries, possession of spatulate incisors, presence of large, interlocking canines and 'squared' molars. Closer examination reveals, however, that the similarities concerned probably amount to a mixture of convergent and primitive features. Body size itself carries little conviction as a shared feature and the fact that adapids were relatively large-bodied in comparison to most

modern prosimians naturally introduces the possibility that they developed a number of simian-like features independently, to meet the demands of increased body size. As explained in Chapter 6, fusion of the dentaries in the lower jaw developed during the evolution of the Adapidae, perhaps separately in several different lineages, and was lacking in the earliest-known representatives (*Pelycodus*, including *Cantius*). Thus, fusion of the lower jaw can only be regarded as a shared derived feature of simians and adapids if the former evolved from a particular lineage of adapids in which this character had already emerged. Similar arguments apply to the development of quadratic molars and of spatulate incisors. The earliest-known adapids (*Pelycodus*) had simple tritubercular upper molars and the lower molars were still relatively primitive in form. The incisors are typically peg-like in adapids and definite spatulate incisors are apparently confined to *Adapis parisiensis*. On the other hand, large, interlocking canines may well be a primitive feature of primates, and perhaps of placental mammals generally, so this would not link adapids specifically to simians. In any case, *Adapis parisiensis*, which has the most simian-like incisors, has reduced canines and Gingerich (1975b) argued that the strepsirhine tooth-comb could be derived from the condition found in this species.

Hence, the simian features cited by Gingerich appeared in mosaic fashion during the evolution of the Adapidae and it is unlikely that simians can be derived from any known specific lineage of adapids. This is particularly true of *Adapis* (including *Leptadapis*), given the unusual features of the postcranial bones attributed to this genus described by Dagosto (1983) and discussed in Chapter 10. Incidentally, it should also be noted that a number of simian-like features emerged independently in certain large-bodied subfossil lemurs. *Archaeolemur*, *Hadropithecus*, *Archaeoindris* and *Palaeopropithecus* all had fused, deep dentaries, spatulate incisors and 'squared' molars. The dentaries were also deep and fused in

Megaladapis. On the other hand, such significant features of simian primates as reduction of the length of the palate and of the cheek-tooth row, fusion of the frontal bones, postorbital closure and expansion of the brain, are lacking from known adapids (as is also true of subfossil lemurs). The origins of the simians are therefore simply undocumented in the known fossil record.

If it is accepted that modern simians can be traced to a common ancestral stock that was already recognizably simian, the question arises as to the geographical location of that stock. It has already been noted above that there are no reliably identifiable simian fossils from early Tertiary deposits of the Northern Hemisphere. One must therefore conclude either that such fossils have for some reason escaped discovery in the relatively well-known deposits of the Northern Hemisphere, or that the origin of the simians was located somewhere in the (as yet) poorly documented Southern Hemisphere. There are, therefore, two main alternative hypotheses: (1) simian primates originated somewhere in the Northern Hemisphere and subsequently spread via separate migration routes to South America and Africa; (2) simian primates originated somewhere in the Southern Hemisphere (probably in Africa) and subsequently migrated directly between the southern continents (i.e. probably from Africa to South America). Although the latter hypothesis was widely discounted when it was believed that the continents had always occupied stable positions, the confirmation of continental drift opened up the possibility that the second hypothesis might be a serious contender (Hoffstetter, 1972). The proposal that there might be some direct link between simians of the Old and New Worlds has received additional support because there is a parallel case among rodents. Several lines of evidence indicate that hystricomorph rodents (porcupines and their relatives) of Africa + Asia and South America form a monophyletic group and it is possible that, as with simian primates, direct migration from Africa to South America provides the best explanation for the current geographical

distribution of the hystricomorphs (Lavocat, 1969).

The two alternative hypotheses for the origins of New World simian primates are discussed in considerable detail in Ciochon and Chiarelli (1980). Although opinions were divided, some general conclusions can be drawn from that discussion. First, there is very strong evidence that the simian primates are a monophyletic group. Second, when seen against a background of continental drift, invasion of South America by early simians from Africa is certainly no less likely than invasion from North America (Tarling, 1980). Although proponents of an invasion from North America have repeatedly emphasized the large sea barrier that would have existed between South America and Africa throughout most of the Tertiary, they have failed to observe that an at least equally large barrier probably separated North America from South America at the relevant time. This, of course, raises the question of the timing of any migration of early simians into South America.

As was noted in Chapter 2, there has been a tendency among palaeontologists to under-estimate times of origin because of the wide-spread practice of dating origins from the first known fossil forms attributable to each group (see later). Gingerich (1980b), for instance, refers to the 'approximate distribution of continental land masses during the late Eocene, when simiiform primates evolved from their adapid ancestors'. If it is assumed that simian primates first emerged in the late Eocene, it is undoubtedly true that migration from Africa to South America would have been relatively unlikely. On the other hand, if simian primates were present in Africa during the early Eocene or during the Palaeocene, a crossing from Africa would have been correspondingly more likely, while any migration of terrestrial mammals from North America to South America would have been faced with a formidable sea barrier in the Caribbean region (Tarling, 1980). At present, there is no good reason to rule out an early origin of simians (or of hystricomorph rodents) in Africa and a migration to South America early in the Tertiary would seem to be the most reasonable explanation.

ARE THE MARMOSETS AND TAMARINS PRIMITIVE OR SPECIALIZED?

With the sole exception of Goeldi's monkey (*Callimico goeldii*), the modern New World monkeys can be sharply divided into two groups. The marmosets and tamarins (*sensu stricto*) are small-bodied, clawed forms that show no overlap in average body weight per species with the larger-bodied 'true' New World monkeys (Cebidae), which have nails on all digits. Marmosets and tamarins also share several other significant features that distinguish them from the cebid monkeys. In the dentition, the third molars are lacking from both upper and lower jaws and the upper molars are simple tritubercular teeth, normally lacking a hypocone. The reproductive biology of marmosets and tamarins is also distinctive in that dizygotic twins are characteristic and the infants are typically carried primarily by group members other than the mother. Goeldi's monkey is intermediate between these two groups of New World monkeys. This species resembles marmosets and tamarins in having claws on all digits except the hallux and a relatively small body weight, but it resembles cebid monkeys in having third molars (albeit markedly reduced in size and lacking a distinct hypocone) and in producing only one infant at a time. *Callimico goeldii* therefore represents an interesting intermediate case bridging between the two main groups of New World monkeys. In fact, recent chromosomal evidence indicates that *Callimico* is more closely related to marmosets and tamarins then to cebids (Dutrillaux, Lombard *et al.*, 1988).

In any reconstruction of a hypothetical common ancestor for the New World monkeys, the question obviously arises as to whether the distinguishing features of marmosets and tamarins are primitive or derived relative to the features of cebid monkeys. Depending on which answer emerges, *Callimico goeldii* can be interpreted either as representing an intermediate stage in the transition from the marmoset/tamarin condition to the cebid condition or as representing a transition in the reverse direction. The interpretation that marmosets and tamarins are relatively primitive, and hence closer to the ancestral condition for New World monkeys generally, has been championed particularly by Hershkovitz (1977). The opposite interpretation, that the cebids are closer to the ancestral condition for New World monkeys, has been supported particularly by Leutenegger (1973), Rosenberger (1977) and Ford (1980). Ford provided a valuable synthetic review of the evidence, indicating quite convincingly that the condition in marmosets and tamarins is most likely to be derived and that the common ancestor of the New World monkeys would have been both larger in size and generally more like typical monkeys.

The small body size characteristic of marmosets and tamarins might, at first sight, appear to be a primitive character, as the general trend is for body size to increase during mammalian evolution ('Cope's Law'). There are, however, some demonstrable exceptions to this general rule in which secondary dwarfing has taken place, notably in the case of recently extinct dwarf elephants. Ford (1980) argues that small body size is likely to be a derived feature for marmosets and tamarins as well, given that all other living simians in both the New and Old Worlds are relatively large-bodied forms and that all known fossil simians are also relatively large. The latter argument is not very convincing, however, because there is an obvious bias against the preservation of small-bodied primates in the fossil record (see Chapter 2). In addition, Setoguchi and Rosenberger (1985) reported, subsequent to Ford's review, fragmentary dental evidence of a small-bodied Miocene New World monkey (*Micodon*) that was claimed – essentially on size grounds – to be a relative of marmosets and tamarins. However, application of the sister-group principle (see Chapter 3) indicates that small body size is likely to be derived for New World monkeys. Given two alternative character states among New World monkeys (small body size versus large body size), the condition found in Old World simians, the relevant sister group, is

Are the marmosets and tamarins primitive or specialized? 681

likely to be primitive. Living Old World simians are uniformly larger than marmosets and tamarins and the same seems to apply to all known fossil forms (although there would, of course, be the usual bias against the discovery of small-bodied Old World simian fossils). It therefore seems likely that dwarfing has indeed occurred in the evolution of marmosets and tamarins and that the common ancestor of New World monkeys was a relatively large-bodied form.

The absence of third molars is a particularly striking feature of marmosets and tamarins. First, it is undoubtedly a derived feature, as the possession of three molars represents the probable ancestral condition both for placental mammals generally and for primates specifically. Second, the possession of only two molars is in fact almost a unique feature among primates, both living and fossil. Assuming that dwarfing has taken place during the evolution of marmosets and tamarins, loss of the third molars can easily be explained. With any morphological parameter, such as size of the teeth, that scales in a negatively allometric fashion with overall body size, secondary reduction in body size is likely to lead initially to 'overscaling' of that parameter because of the phylogenetic inertia of growth programmes. Such 'overscaling' can then be gradually corrected, perhaps in a novel manner, during subsequent evolution. In the case of marmosets and tamarins, correction of overscaled cheek teeth could theoretically have been achieved either by gradual reduction of the dimensions of all molars or by suppression of the third molar (the last to erupt). The fact that the latter has taken place confirms the hypothesis of secondary dwarfing, as marmosets and tamarins now have a cheek-tooth row that, relative to body size, has approximately the length expected relative to other simian primates (see Fig. 6.12). That this should be so in spite of the reduction in number of cheek teeth in marmosets and tamarins indicates that some or all of the remaining cheek teeth are still longer than expected for simians of that body size. Overscaling as a result of secondary dwarfing is also identifiable in other morphological features

of marmosets and tamarins. It was shown in Chapter 7 (Fig. 7.8) that, in comparison to other primates, marmosets and tamarins seem to have unusually large eyes in relation to their orbits.

Small body size, whether primary or secondary, may also be linked to the absence of a hypocone on the upper molars of marmosets and tamarins. As a general rule, small-bodied primates tend to have simpler molars than large-bodied primates. This has yet to be subjected to systematic study, so it is not clear whether larger species acquire more complex molars as a direct consequence of scaling effects in the dentition or whether there is an indirect influence of the general trend for diets to include more-resistant plant parts (e.g. leaves) with increasing body size. In either event, it is only to be expected that the relatively small body size of marmosets and tamarins should be associated with relatively simple molar morphology. If it is true that the common ancestor of marmosets and tamarins had undergone secondary dwarfing, then absence of the hypocone in the upper molars could well be the result of secondary simplification. Once again, it is possible to apply the sister-group principle. All known Old World simians, both living and fossil, have a hypocone, so possession of a hypocone by cebid monkeys is likely to represent a primitive feature for New World monkeys generally (see also reconstructions by Gregory, 1920a and by Rosenberger, 1977). Fusion of the dentaries in the lower jaw in marmosets and tamarins also indicates dwarfing, as this feature is otherwise confined to primates of medium or large body size.

Incidentally, it should be noted that the large-bodied Caribbean subfossil simian *Xenothrix* also lacks third molars. It seems likely that this represents a convergent feature, as there is no other evidence to link *Xenothrix* to marmosets and tamarins. In this case, secondary dwarfism cannot be invoked to account for loss of the third molars and some other explanation must be sought for this unusual condition. Unfortunately, only one specimen, an incomplete lower jaw, has been reported to date and it is not known whether *Xenothrix* had a hypocone on its upper molars.

From what has been said above, however, it can be inferred that a hypocone should have been present in the upper molars of this relatively large-bodied New World simian.

The presence of claws on all digits except the hallux in marmosets and tamarins (and in Goeldi's monkey, *Callimico*) has commonly been interpreted as a primitive feature. At first sight, it would seem to be most parsimonious to assume that their claws were retained in an unbroken sequence from an early reptilian stock, through the ancestral stock of placental mammals and on through the ancestral stock of primates. However, as shown in Chapter 10 (see Fig. 10.12), it is more parsimonious to postulate a common ancestor with nails for the evolutionary radiation of primates of modern aspect. Although this requires an initial transition from claws to nails in the evolution of the common ancestor of primates of modern aspect, only two secondary reversals from nails to claws are then required (in the aye-aye among lemurs and in the common ancestor of marmosets, tamarins and *Callimico*). Further, application of the sister-group principle to simian primates again indicates that possession of nails is primitive for New World monkeys, as all Old World simians have nails rather than claws. As discussed in Chapter 10, a secondary development of claws in a small-bodied common ancestor of marmosets, tamarins and *Callimico* can be explained as a special adaptation for locomotion on relatively broad trunks and branches, perhaps in association with gum-feeding (see Chapter 6).

There are also several indications from the sphere of reproductive biology that marmosets and tamarins have undergone secondary dwarfing. As noted in Chapter 9 (see Fig. 9.2) and in the preceding section, a single-chambered (simplex) uterus is a universal feature of simian primates, probably developed in their common ancestor. This is a highly unusual condition in placental mammals and, as pointed out by Ford (1980), is otherwise found only in certain bats (Phyllostomatidae) and in edentates (anteaters, sloths and armadillos). As was noted by Hamlett and Wilsocki (1934), a simplex uterus would presumably represent an adaptation for

accommodation of a single fetus and it is therefore an unexpected condition in marmosets and tamarins, which typically have twins. Similarly, there is broad matching between the litter size and the number of pairs of teats both in primates and among mammals generally (see Chapter 9), but marmosets and tamarins have only one pair of teats in spite of having twins. Schultz (1948) specifically commented on this as an unexpected feature of marmosets and tamarins. It therefore seems likely that marmosets and tamarins are descended from an ancestral simian stock in which the number of teats and the form of the uterus had become adapted for single births. Thus, although it is true that the occurrence of multiple litters is generally primitive for placental mammals (Chapter 9), twinning can be interpreted as a secondary reversal in marmosets and tamarins. In fact, the fetal development of marmosets and tamarins is clearly highly specialized. During development the twins share a placenta, such that there is a common circulatory system. The twins are dizygotic, so in about half of cases one twin is female and the other is male. In principle, this could generate problems with respect to hormonal control of sex determination and special mechanisms have probably developed to protect the female fetus against masculinization. Twinning in marmosets and tamarins has also introduced other problems requiring special adaptation. The burden of carrying two relatively large infants (see below) is very heavy for the mother and all social groups of marmosets or tamarins hence share the carriage of infants among adults and subadults.

Overall, therefore, there is overwhelming evidence that secondary dwarfing occurred during the evolution of marmosets and tamarins. One outcome has been a secondary return to multiple litters (typically twins). It is, however, necessary to explain why twinning should have emerged as a derived feature of marmosets and tamarins, as this would not seem to be a necessary outcome of dwarfing. A specific hypothesis was proposed by Leutenegger (1973), on the basis of relationships between neonatal body weight and maternal body weight established for primates.

As has been explained in Chapter 9 (see Fig. 9.9), there is a sharp distinction between haplorhines (tarsiers and simians) and strepsirhines (lemurs and lorises) in this respect. At any given maternal body weight, haplorhine mothers produce considerably larger neonates than strepsirhine mothers. As a general approximation, there is a threefold difference in neonatal body weight between haplorhines and strepsirhines at any given maternal body weight. Therefore, in common with other haplorhines, marmosets and tamarins – other things being equal – would be expected to give birth to relatively large neonates. The second part of Leutenegger's argument related to the fact that the scaling of neonatal size to maternal body weight in primates, and in mammals generally, is negatively allometric. In other words, the ratio of neonatal size to maternal size increases progessively with decreasing body size of the mother. The combination of negative allometric scaling with the characteristic of specially large neonatal size in haplorhines means that small-bodied haplorhines produce neonates that are very large in proportion to maternal body size. Potentially, at least, this could give rise to problems during birth. Leutenegger cites evidence that marmosets and tamarins do indeed have obstetric problems in captivity and concludes that twinning represents a special adaptation to limit neonatal body size relative to maternal body size. This hypothesis was cited with approval by Ford (1980).

Although Leutenegger's hypothesis has been widely cited, there are numerous problems associated with it. Most importantly, although the evidence presented may suggest good reasons for restriction of neonatal size in marmosets and tamarins, it does not explain the production of twins. The simplest solution to the potential obstetric problem associated with a large neonate is to give birth to a small neonate at a slightly earlier stage in its development. The production of two infants instead of one is in no way explained by Leutenegger's hypothesis. In any case, although marmosets and tamarins do produce twins, the individual neonate is not consistently smaller than would be predicted

from the best-fit line for haplorhine primates generally (see Fig. 9.9). Although some marmosets and tamarins have relatively small neonates compared to most other haplorhines in general and to most other New World monkeys in particular, the pygmy marmoset (*Cebuella*) in fact shows one of the largest recorded individual neonatal body weights relative to maternal body size. Thus, twinning is not consistently associated with a demonstrable reduction of individual neonatal size in marmosets and tamarins. Indeed, according to Leutenegger's hypothesis *Cebuella*, being the smallest New World monkey, should encounter the most severe obstetric problems and yet the individual neonates are larger than expected in comparison to the best-fit line for simians. On the other hand, the single neonate produced by *Callimico* falls below the best-fit line, showing that a relatively small neonate size can be exhibited by a small-bodied New World monkey in the absence of twinning. Further, *Tarsius* lies within the body size range of marmosets and tamarins and yet produces a single neonate that is almost exactly of the size expected from the best-fit line for haplorhine primates (see Fig. 9.9). Although it is possible that relatively small-bodied haplorhine primate species experience some difficulty at birth because of the relatively large size of the neonate(s), any significant adverse effect on breeding success under natural conditions would surely lead quite rapidly to a reduction in neonate size. The obstetric problems reported for captive colonies of marmosets and tamarins probably represent pathological side-effects of artificial housing conditions, associated with other abnormalities such as a widely reported tendency to produce triplets instead of twins.

It must, in any case, be remembered that a doubling of litter size in the evolution of marmosets and tamarins must have had a marked effect on reproductive output. As population growth is geometric until constrained by available resources, simple doubling of the litter size (without any other change in reproductive parameters) must be associated with a pronounced upward shift in reproductive potential. It is in this context that a convincing explanation for the evolution of twinning in marmosets and tamarins is likely to be found. As explained in Chapter 9, there is an inverse relationship between reproductive potential (as indicated, for example, by the measure r_{max}) and body size (see also Fenchel, 1974; Ross, 1988). Thus, small-bodied species may be expected to have a higher reproductive potential than large-bodied species. Accordingly, dwarfing is likely to be accompanied by adaptations to increase reproductive potential. Such adaptations can affect age at first breeding (typically lower in smaller species), interbirth intervals (typically shorter for small-bodied species), and litter size (typically larger for small-bodied species). The evolution of twinning in marmosets and tamarins can therefore be explained as part of a general shift in reproductive parameters to generate the required higher reproductive capacity. In fact, most marmosets and tamarins seem to lie somewhat above the best-fit line for the scaling of the intrinsic population growth parameter r_{max} to body size in simian primates (C. Ross, personal communication). This indicates that their reproductive potential is probably somewhat higher than would be accounted for by small body size alone. Although this could be explained as a side-effect of twinning that has occurred for some other reason, there is now increasing evidence that marmosets and tamarins, in contrast to cebid monkeys, are typically adapted for occupation of zones of secondary growth rather than primary forest. Such zones would probably require a relatively high reproductive capacity, as reflected in r_{max} (Ross, 1988).

If it is accepted that small body size, and certain directly associated features, are the result of secondary dwarfing in marmosets and tamarins, it follows that Goeldi's monkey (*Callimico*) represents an intermediate stage in the process of dwarfing. The body size of *Callimico* lies at the upper end of the range for marmosets and tamarins and claws are also present. These would seem to represent shared derived features, although convergence cannot be excluded. In other respects, however, Goeldi's monkey resembles cebid monkeys. These resemblances to cebid monkeys, according

to the dwarfing hypothesis, must represent primitive retentions and indicate no special relationship between *Callimico* and cebids. It can therefore be concluded that *Callimico* shared a specific common ancestry with marmosets and tamarins but branched off before the development of certain features such as twinning and loss of the third molars. Some dental reduction has taken place in Goeldi's monkey – the third molars are very small and hypocones are virtually lacking from the upper molars – so this also suggests a specific link with marmosets and tamarins, although convergent evolution remains a possibility. As yet, the only biochemical evidence available to throw light on this problem stems from the immunological comparisons reported by Cronin and Sarich (1975, 1978a). Their findings do indeed indicate that *Callimico* is specifically related to marmosets and tamarins, but the tree generated from the data links *Callimico* more closely to marmosets (*Callithrix* and *Cebuella*) than to tamarins (*Leontopithecus* and *Saguinus*). Such a relationship conflicts with all available morphological evidence and with chromosomal evidence, so it is highly likely that the inferred branching pattern in the immunological tree is incorrect. This discrepancy thus illustrates how errors may result from reconstructing branching patterns from immunological evidence, probably as a result of variation in rates of evolution (see later).

'APES' AND 'MONKEYS' AMONG THE OLD WORLD SIMIANS

It has long been accepted that modern Old World simians (monkeys, apes and humans) belong to a monophyletic group. On morphological grounds, they can be distinguished from New World monkeys ('platyrrhines') by such features as the typical shape of the nose ('catarrhine' condition), the reduction of the dental formula to include only two premolars in the upper and lower jaws, and the possession of an ectotympanic tube in the ear region (see Fig. 7.18). Chromosomal and biochemical evidence has also uniformly indicated that modern Old World simians are monophyletic with respect to all other living primates. Fossil Old World simians from the beginning of the Miocene onwards generally fit the morphological pattern expected for catarrhine primates and differ from that expected for platyrrhine primates. Given that Old World simians are undoubtedly derived in comparison to New World monkeys with respect to the reduction in number of premolars and the structure of the bulla, any fossils with such features can be clearly linked to modern catarrhines. Problems arise, however, with the Oligocene simian fossils from the Fayum deposits in Egypt. In the first place, some forms (*Apidium* and *Parapithecus*) have three premolars in the upper and lower jaws, as in New World simians, and are therefore more primitive than later Old World simians in this respect. Further, although other Fayum simians retain only two premolars in their upper and lower jaws (*Oligopithecus*, *Aegyptopithecus* and *Propliopithecus*), it is known that *Aegyptopithecus* (at least) lacked an ectotympanic tube in the ear region, as do modern New World monkeys. Finally, study of fragmentary postcranial elements (e.g. Conroy, 1976a,b) has revealed that the Fayum simians resemble modern New World monkeys more than modern Old World simians.

Because the various similarities shared by the Fayum simians and New World monkeys are primitive for simians generally, they provide no evidence of a specific phylogenetic link between the two groups. It is, indeed, hardly surprising that the earliest known simians from the Old World should share certain primitive features with New World simians. On the other hand, the presence of primitive simian features in the Fayum forms indicates that they may not be an integral part of the adaptive radiation of modern Old World simians. One must either suggest that the Fayum simians diverged before the split between modern Old World monkeys and apes occurred or propose that certain shared derived features of the modern Old World simians developed independently in monkeys and apes. If, for instance, it is accepted that *Aegyptopithecus* is specifically related to modern apes (e.g. Simons, 1967, 1972; Szalay and

Delson, 1979), it must also be accepted that the tubular ectotympanic was developed independently in Old World monkeys and apes. This latter conclusion may seem to be supported by the fact that the ectotympanic tube was reportedly incomplete in *Pliopithecus*. But the question then arises as to whether (if this observation is accurate) *Pliopithecus* is really related to modern gibbons, as is often stated, or whether this Miocene form also branched away before the modern monkeys and apes diverged. Because of the common practice of attempting to link all known fossil forms directly with living forms, there has been a tendency to associate the Fayum simians too directly with the evolution of modern Old World simians. They are best treated as a separate category, at least for the time being.

It has long been the practice to distinguish between 'monkeys' (Cercopithecoidea) and 'apes' (Hominoidea) in referring to living Old World simians. Originally, the words 'monkey' and 'ape' were virtually synonymous, but the latter term has gradually been restricted to refer to our immediate relatives among living simians. To the extent that it is possible to recognize shared derived features that link modern apes and humans, this practice is defensible. As was observed in Chapter 2, however, the *Scala naturae* has long exerted an influence on interpretations in this respect. It has often been assumed that, because living apes are our closest relatives, they must necessarily be more advanced than monkeys. In other words, wherever a difference in character states was found between monkeys and apes, it was automatically assumed that the character states of apes were necessarily derived. Such a blanket assumption is quite unjustified and each character must, of course, be examined in turn to distinguish primitive from derived conditions (e.g. see Harrison, 1982, 1987).

As far as living Old World simians are concerned, it is relatively easy to separate monkeys and hominoids (apes and humans) on the basis of apparent shared derived features. The living Old World monkeys are particularly characterized by the possession of bilophodont molar teeth (see Fig. 6.21), whereas the living hominoids are distinguished by loss of the tail, expansion of the sacrum to include five or six vertebrae and specialization of the wrist to permit greater mobility (see Chapter 10). All of these adaptations are, in principle, recognizable in fossil forms and it is therefore potentially possible to trace the separation between monkeys and apes back through the fossil record. Unfortunately, however, there is no clear dental criterion permitting unequivocal recognition of early apes in the fossil record. Although modern great apes and fossil 'great apes' of the *Dryopithecus* group (including *Proconsul*, *Gigantopithecus* and *Sivapithecus*) share an apparently derived feature in exhibiting a Y5 pattern in their lower molars (Figs 6.22 and 6.23), this pattern is not found in modern gibbons or in fossil 'lesser apes' such as *Limnopithecus* and *Pliopithecus*. All of these latter forms have five-cusped lower molar teeth, but in the absence of the Y5 pattern (which involved an outward shift of the hypoconulid cusp) this is merely a primitive feature for Old World simians generally. By contrast, during the development of bilophodonty, the hypoconulid cusp was lost from the lower molars of Old World monkeys as a derived feature of that group. Given that most fossil simians of the Old World are known exclusively from teeth and jaws, it is accordingly very difficult to determine whether

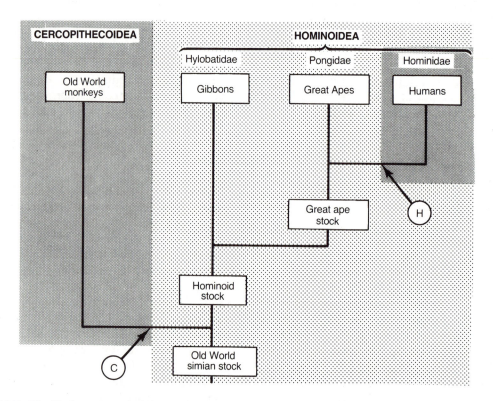

Figure 12.7. The likely pattern of phylogenetic relationships among Old World simians, showing the implications for a classical classification using the 'grade' concept. The Old World monkeys (Cercopithecoidea) can be defined on dental grounds by the emergence of bilophodonty (arrow C), while the human lineage (Hominidae) can also be defined on dental grounds (arrow H). Early apes, on the other hand, had no obvious defining features of their dentitions and it is therefore appropriate to treat the apes as the basic grade for purposes of classification. One implication of this is that early Old World simians of uncertain affinities are probably best allocated to the Hominoidea, rather than to the Cercopithecoidea.

an early fossil form with five-cusped lower molars is specifically allied to the modern apes or is merely dentally primitive.

Because of the difficulties involved in identifying the early relatives of modern apes on dental grounds, reconstructions of phylogenetic relationships among Old World simians (and hence of any cladistic classifications attached to them) are currently in turmoil. It would therefore be premature to suggest any particular pattern of relationships between known fossil forms and modern apes. Indeed, as many fossil forms will probably never be documented by anything other than dental evidence, uncertainty may reign indefinitely. This being the case, with a classical classification it is arguably best to treat the 'apes' (Hominoidea) as the basic category among the Old World simians, with the 'monkeys' (Cercopithecoidea) recognized as a separate grade definable by the presence of bilophodonty. Among the Hominoidea, of course, the Hominidae must also be recognized as a distinct grade, definable on the basis of a set of derived features (Fig. 12.7). Relatively primitive members of the Old World simian group are accordingly 'apes' rather than 'monkeys'.

A TIMESCALE FOR PRIMATE EVOLUTION

One of the most important issues in the reconstruction of primate phylogeny concerns the provision of a timescale to accompany any inferred tree. As has been emphasized in Chapter 2, fossil evidence, at least at the present time, provides the only source of direct information permitting the inference of a timescale for any phylogenetic tree (see also Fig. 3.12). In principle, it is possible to infer the date of a particular branching point in a phylogenetic tree through reference to one or (preferably) more specific fossil species that are believed to lie close to that branching point. It must be emphasized, however, that great caution must be exercised in the inference of the times of branching points from individual fossil species. In the first place, the positions of those fossils in the tree must themselves be inferred and are subject to the same kind of uncertainty as that applying to

inferred relationships among living forms. In fact, most fossil species are known only from fragmentary evidence (see Chaper 2) and it follows, therefore, that the few characters identifiable in a fragmentary fossil species may seem to fit the prescription for an inferred ancestral stock far more closely than would be the case with the many characters identifiable in a substantial fossil species. Thus, an error will often be made in concluding that a given fragmentary fossil species is close to a given branching point in a phylogenetic tree.

More importantly, however, it is simply not justifiable to assume that a particular fossil species is likely to coincide exactly with a particular branching point in the tree. Given the fact that the effective sampling level for the fossil record of primate species, and probably for the mammalian fossil record in general, is probably no better than 1% on average (see Fig. 2.1), the likelihood of finding actual ancestral species is very small. In other words, in most cases any relevant fossil species is bound to be somewhat distant from the branching point to be dated (Fig. 12.8). Because a fossil will be recognized as a relative of a group of living species only if it has a feature diagnostic of that group, it is almost inevitable that the fossil will be more recent than the common ancestor shared with the living species. A particularly extreme example of this is provided by the strepsirhine primates (lemurs and lorises). Apart from the relatively recent subfossil forms, there is no fossil record for the lemurs in Madagascar. The first fossil primates that can be clearly identified as strepsirhines because they have a tooth-comb are Miocene lorisids from Africa and from the Indian subcontinent. As the fossil record for primates in Africa before the Miocene is limited to the single, peripheral Oligocene site of Fayum in North Africa, the times of origin and initial diversification of the strepsirhine primates are simply unknown. If the time of divergence between the lemurs and lorises is dated from the first known fossil forms (the Miocene lorisids), it must be concluded that the divergence took place only 20 million years (Ma) ago. It is almost impossible to reconcile such a recent date with

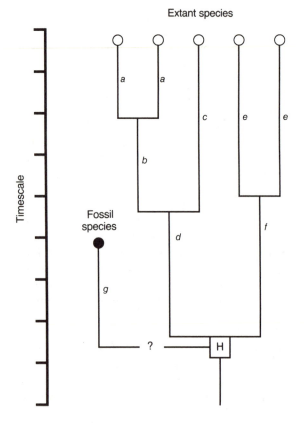

Extant species

Timescale

Fossil
species

a *a*

b

c *e* *e*

d *f*

g

? H

Figure 12.8. Schematic illustration of the uncertainty involved in inferring a date for a hypothetical ancestral stock (H) from the earliest known, apparently related fossil species. Although a reliable absolute date may be available for the fossil species, the time interval (*g*) separating the fossil from the ancestral stock is unknown. Determination of lengths of branches linking the extant species (*a, b, c, d, e, f*) depends both on the accuracy of the date inferred for H and on rates of change of the characters compared.

biogeographical evidence and evidence from the average rate of evolution of myoglobin, for instance, indicates that this date is less than half that expected in comparison to dates accepted for other primates (Romero-Herrera, Lehmann *et al.*, 1973).

Unfortunately, there is no objective means of determining the interval of time by which a particular branching point predates the closest available fossil species used for dating. Hence, it is really only permissible to state that dates for

branching points inferred directly from known fossil species represent minimum dates. In practice, it is likely that actual dates exceed these minimum dates by a considerable margin, and, in some cases (as with strepsirhine primates), it may be that minimum dates should be at least doubled to give actual dates. It is exceedingly difficult to set reliable maximum dates for divergence if the fossil record is very patchy, as is usually the case.

Because the margin of underestimation of the actual date of divergence is uncertain in any particular case, palaeontologists have, as a rule, reasonably preferred not to suggest actual dates. Romero-Herrera, Lehmann *et al.* (1978), for instance, recognized that dates of the first known representatives of each group taken directly from the fossil record can only be minimum dates, but preferred not to estimate actual dates because such a procedure is 'open ended'. As an indirect result of such caution, however, there has been a widespread tendency to equate inferred minimum dates of divergence with actual dates of divergence. This creeping tendency has been exacerbated by the general inclination to treat the available fossil record as more complete than it is in reality and this, in turn, has led to widely accepted notions that have no proper foundation. For instance, it is common practice to infer that the first known fossil species that can be allocated to the order Primates indicates both the time and the geographic site of origin of primates. Thus it is that the single lower molar tooth known from the late Cretaceous 'archaic primate' *Purgatorius ceratops* is often cited as demonstrating that the primates originated in North America 65 Ma ago. Further, the first known primates of modern aspect date only from the early Eocene, less than 55 Ma ago, and the origin of primates of modern aspect would accordingly be traced back only to this time. In fact, however, it is only justifiable to conclude that primates of modern aspect originated at least 55 Ma ago. The possibility must be considered that primates of modern aspect originated during the latter part of the Cretaceous and that, for a combination of reasons, no relevant fossil evidence has yet been uncovered. It must also be remembered that the relatively large sample of fossil primates recovered from the Eocene (see Fig. 2.1) is attributable to global climatic warming at that time. This permitted the subtropical/tropical primates to extend their geographical range into the Northern Hemisphere. During the cooler periods before and after this time, however, we must seek fossil evidence in the Southern Hemisphere, where the fossil record of mammals is generally poorly documented.

Early evolution of placental mammals

There are several very good reasons for inferring that the origins of placental mammals were probably considerably earlier than is generally assumed (see Fig. 4.6). First and foremost, if the mammalian fossil record is as incomplete as suggested in Chapter 2, direct dating of any divergence point by reference to the closest known fossil form must inevitably lead to marked underestimation in most cases. It must be remembered that there is a relatively 'abrupt' appearance of various types of placental mammals in the known fossil record during the Palaeocene and Eocene (Gregory, 1920c). Intermediate forms are scarce and in many cases missing altogether. Indeed, this phenomenon is prevalent in the fossil record for animals generally. As was noted by Schaeffer (1948): 'Most orders, classes and phyla of invertebrates and vertebrates appear abruptly in the geological record with one or more character complexes well differentiated but without any indication of transitional stages.' The customary explanation for this is that the initial phase of the adaptive radiation of various animal groups, such as the placental mammals, took place relatively rapidly and that intermediate forms were accordingly short-lived. An alternative explanation in the case of the placental mammals, however, is that the initial phase of their evolution was considerably longer than generally believed and is particularly poorly documented because of a combination of at least two factors. First, given that early placental mammals were probably quite small and relatively rare, it is only to be expected that a poorly sampled fossil record would reveal very little of this stage of

mammalian evolution. To this must be added the fact that the early evolutionary history of placental mammals has been reconstructed very largely from fossils found in the Northern Hemisphere. Our knowledge of early mammalian evolution in the southern continents – especially in Africa – for the period from the mid-Jurassic to the mid-Cretaceous is exceptionally poor. If the early evolution of the placental mammals took place largely in the Southern Hemisphere, the origins of the modern orders might date back considerably farther than is generally believed.

Two striking cases of the 'abrupt' appearance of major groups of placental mammals in the known fossil record are provided by the even-toed (artiodactyl) group of hoofed mammals and by the bats. All artiodactyls have a unique 'double-pulley' system in the ankle region of their hindlimbs; bats have wings (Simpson, 1955). The peculiar structure of the artiodactyl hindlimb is fully identified in early Eocene representatives (Schaeffer, 1948), but no intermediate stages have yet been discovered. Fully formed bats are also found in the Eocene, and once again no intermediate fossil forms have yet been found to document the evolution of their wings. Comparably, if less spectacularly, the first primates of modern aspect from the Eocene already have a suite of characters linking them to modern primates, but no convincing earlier relatives of the modern primates have yet been discovered.

The middle Eocene fossil site of Messel in West Germany is of particular importance with respect to documentation of early representatives of various mammalian orders, particularly because traces of soft body parts and even remnants of stomach contents are identifiable for some specimens (Heil, von Koenigswald et al., 1987). At least 32 mammalian species, including two marsupial species belonging to a single order alongside representatives of 12 orders of placental mammals, have already been discovered. The rich finds include more than 50 individual microchiropteran bat specimens and detailed study has shown that these bats consumed a spectrum of arthropod prey roughly comparable to that consumed by their modern relatives. In addition to this, they already possessed well-developed adaptations for echolocation. Hence, all of the major adaptations that characterize typical modern microchiropteran bats were already present in their Eocene relatives. Striking finds of early placental mammals include a virtually complete skeleton of a pangolin, *Eomanis waldi*, that is remarkably similar to its modern relatives (Storch, 1978). Even more striking is the discovery of an essentially complete skeleton identified as that of a tamandua, *Eurotamandua joresi* (Storch, 1981). Not only is this skeleton closely similar to that of modern tamanduas, reflecting extensive specializations for feeding on ants and/or termites, but its discovery in the Eocene deposits of Messel was totally un-expected. Modern tamanduas are entirely restricted to South and Central America and their evolutionary history, along with that of other edentates (i.e. armadillos and sloths), was previously assumed to be confined to this area of the New World. The isolated discovery of a tamandua skeleton in the geographically far distant Messel deposits on the one hand illustrates the great gaps that still exist in our knowledge of the early evolutionary history of the placental mammals and on the other hand poses great problems with respect to biogeographical origins. (It is therefore understandable that some authors have denied that *Eurotamandua* is a tamandua – e.g. Novacek, 1982b.) Given the relative positions of the continents during the Cretaceous and early Tertiary (see Chapter 4), it is difficult to explain how fully developed tamanduas could exist in Europe in the Eocene and yet nowadays be confined to South and Central America. It is clear from these and other finds at Messel both that the southern continents, especially Africa, must have played a larger part in the early evolution of placental mammals than is generally recognized (see Storch, 1984, 1986). It is also clear that the adaptive radiation of placental mammals began much earlier than commonly stated. Storch (1986) has specifically suggested that the earliest origins of primates and of hoofed

mammals should be traced to as yet unknown progenitors in Africa. Given the continuing uncertainties attaching to the extremely patchy fossil record of mammals, inference of times of origin is a risky business indeed.

The concept of the molecular clock

The practical problems inherent in inferring from palaeontological evidence an accurate timescale for the phylogenetic tree of primates have been somewhat eclipsed by potentially very exciting developments in the field of molecular biology, indicating the possibility of devising a 'molecular clock' (Lewin, 1988). The basic concept, first suggested by Zuckerkandl and Pauling (1962), is that mutational changes in the base sequence of DNA, and hence in the amino acid sequences of proteins prescribed by individual genes, may accumulate (i.e. become fixed) at an approximately constant rate. Accordingly, given an inferred phylogenetic tree based on appropriate molecular evidence (sequence data for proteins and/or DNA), it is theoretically possible to interpret the numbers of sequence differences accumulated along individual branches in terms of relative times of origin. Unfortunately, there is no known means of inferring absolute dates directly from molecular data, so any tree based on such data must be calibrated by dating at least one of the branching points on the basis of the fossil record. Following the premise that rates of change in genes and proteins are equal along all lineages, it is then possible to infer dates for other branching points.

The concept of the molecular clock has achieved a certain notoriety in the field of primate evolution because of a long-simmering dispute between certain molecular biologists and several primate palaeontologists over the timing of the divergence between humans and the African great apes (chimpanzees and gorillas). When the first attempts were made to apply the concept of the molecular clock to the phylogenetic tree for primates, many primate palaeontologists accepted that *Ramapithecus* was the earliest known member of the hominid lineage (e.g. see Simons and Pilbeam, 1965). At that time, it was commonly inferred that the divergence between humans and the African great apes must have occurred before the date for the earliest known specimens of *Ramapithecus* – at least 14 Ma ago (e.g. Simons, 1964, 1967; Pilbeam, 1966, 1972). It was also suggested that the ancestry of the African great apes could be traced back to *Dryopithecus* species, particularly in the case of the gorilla, which was linked to *Dryopithecus major* (Simons, 1967; Pilbeam and Uzzell, 1971). This extended the division between African apes and humans back to at least 20 Ma. Indeed, it was even suggested that the divergence between great apes and humans might be traced back to Oligocene times (Simons, 1965). Palaeontological interpretations indicating a divergence between great apes and humans in the Miocene persisted for a considerable period thereafter. Walker (1976), for instance, indicated that divergence between hominids and African great apes took place at about 17 Ma ago.

A major problem arose, however, when Sarich and Wilson (1967a) claimed that such early dates were incompatible with dates inferred for a phylogenetic tree based on immunological data derived from serum albumins. Their tree had been calibrated with a single date of 30 Ma ago for the divergence between Old World monkeys and hominoids (apes and humans). This, according to their calculations, yielded a date of only 5 Ma ago for the divergence between the African great apes and humans (Fig. 12.9). In a subsequent paper (Wilson and Sarich, 1969), it was argued that immunological comparisons of transferrins, analysis of amino acid sequences of haemoglobins and preliminary information from DNA hybridization studies all yielded similar results. Taking a divergence date of 30 Ma ago for the separation between Old World monkeys and apes, the date of the divergence between African apes and humans was inferred to be 4–5 Ma ago.

The conclusion that the African great apes and humans are far more closely related than had previously been assumed was further reinforced by other studies. For instance, Doolittle and Mross (1970) showed that the fibrinopeptides A

and B are identical in chimpanzees and humans, even though these proteins seem to have evolved rapidly among mammals generally. This and other findings led to the paper of King and Wilson (1975) in which it was estimated that there is about 99% genetic similarity between humans and chimpanzees overall. King and Wilson also noted that the degree of genetic dissimilarity between chimpanzees and humans is less than that found between species of the same genus in certain other groups of animals (e.g. frogs). As a result, there has been increasing support for the notion that humans and African apes diverged far more recently than previously believed and that primate palaeontologists had misinterpreted the fossil evidence.

As has been explained in Chapter 2, it has now emerged that *Ramapithecus* is probably not related specifically to hominids after all and may, indeed, be related to the orang-utan. *Ramapithecus* cannot, therefore, be taken as an indicator of the minimum time of divergence for the hominid lineage. It might therefore seem justifiable to conclude that a far more recent date of around 5 Ma ago for the divergence between humans and African great apes, originally suggested by Sarich and Wilson (1976a), is correct. This conclusion has, indeed, been drawn by numerous authors, including some recent converts among primate palaeontologists who previously believed *Ramapithecus* to be a hominid. In certain quarters, this outcome has been hailed as a triumph for the molecular clock hypothesis and as a major setback for primate palaeontology, indicating – at least in some cases – that palaeontological inference is simply unreliable and must now take a back-seat position. Such an interpretation does not stand up to close examination, however, and there are several important lessons to be learned from the apparent conflict between certain applications of the molecular clock concept and the inferences made by several primate palaeontologists.

Calibration of the clock

Before embarking on a review of the basic principles of the molecular clock, it is necessary

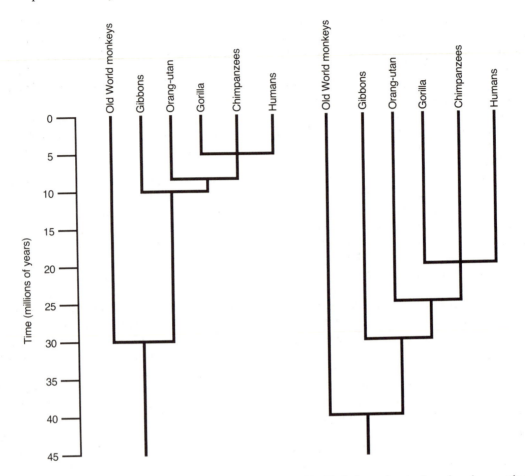

Figure 12.9. Comparison of inferred divergence times for Old World simian primates based on immunological evidence (left) and on palaeontological evidence (right). The immunological tree (after Sarich and Wilson, 1967a) is based on the concept that the immunological 'index of dissimilarity' (IND) is related to time (T) by the formula: $\log \text{IND} = kT$. The tree has been 'calibrated' with a single date of 30 Ma ago for the time of separation between Old World monkeys and hominoids (apes and humans). The palaeontological tree (after Pilbeam, 1972) was based on an assessment of a large body of relevant fossil evidence.

to emphasize two points of elementary logic. First, rejection of a phylogenetic tree incorporating *Ramapithecus* as an early hominid by no means demonstrates that an alternative timescale such as that advocated by Sarich and Wilson (1967b) is correct. Second, as has been noted above, fossil evidence currently provides the only direct source of evidence for attaching absolute dates to phylogenetic trees. Any tree based on molecular evidence must be calibrated with at least one date (and preferably with several

dates) from the fossil record. It is therefore straightforwardly illogical to accept one date based on palaeontological evidence for the calibration of a molecular tree and then to proceed to use that tree to question any other date based on exactly the same kind of evidence. If palaeontological evidence is inherently unreliable, then the calibration date must be just as questionable as any secondarily inferred date. In fact, Sarich and Wilson (1967a) were obliged to use palaeontological dating in another context as well. Lengths of branches within their hominoid tree were calculated using the following formula: $\log \mathrm{IND} = kT$ (where IND is their empirically measured index of immunological dissimilarity and T is time). This formula had been previously inferred through comparison between indices of immunological dissimilarity and inferred times of divergence for a wide range of vertebrate species. There is clearly an element of circularity involved if inferred palaeontological dates are used to determine a formula relating immunological dissimilarity to time, and values determined from the formula are subsequently used to question palaeontological dates.

In the case of the molecular tree presented by Sarich and Wilson (1967a), the calibration date used is, in palaeontological terms, one of the most questionable for the entire primate tree. (These authors themselves admitted that a date of 30 Ma for the divergence between Old World monkeys and hominoids would be regarded as too recent by many authors.) The time of divergence of the monkeys and apes among Old World simians is simply not documented in any direct fashion in the fossil record and can be inferred only very approximately from known fossil species. Molar teeth with the bilophodont pattern characteristic of Old World monkeys have been reported from early Miocene deposits of Africa (*Prohylobates* and *Victoriapithecus*), so the division between monkeys and apes must have taken place at least by that time. The first known Old World simian fossils are documented from the early Oligocene Fayum site of Egypt, but this is so far the only source of substantial primate fossil material for the entire continent of Africa before Miocene times. According to some interpretations, the division between monkeys and apes may already be documented among the Fayum fossils (Simons, 1972; Szalay and Delson, 1979), in which case it would be reasonable to conclude that the Old World monkeys and apes probably diverged at least during the earliest Oligocene. In fact, however, too much reliance has been placed on interpretation of fossil primates from this single site and we simply do not know what primates were present elsewhere in Africa during the Oligocene or in the entire continent before that time. It cannot be ruled out, therefore, that Old World monkeys and apes diverged during the Eocene or even earlier. The fossil record of the later hominoid primates (Miocene apes) is far better documented than that of the earliest simians. Accordingly, it would surely be more appropriate to calibrate a molecular clock using palaeontological dates inferred for hominoid divergence points and thence to infer the time of divergence of the earliest simians (see also Uzzell and Pilbeam, 1971).

In any event, as a molecular tree must be calibrated with at least one date from the fossil record, the only way to achieve reliably dated trees is through a genuinely cooperative venture involving both palaeontologists and molecular biologists (e.g. see Romero-Herrera, Lehmann *et al.*, 1973). (This is particularly necessary in cases where a relationship between a measure of immunological distance and elapsed time must also be established.) Both fields must overcome problems associated with rates of evolution and if there is apparent conflict in the conclusions reached it is necessary to examine all of the methods involved. The uncertainties surrounding inference from morphological evidence have been abundantly illustrated in previous chapters. It is now time to consider the assumptions and concepts associated with the molecular clock itself.

Are rates of change constant?

An absolutely fundamental requirement for a molecular clock is that rates of evolution at the molecular level must remain approximately constant. In other words, the 'clock' must keep good time, with no appreciable changes in tempo either during particular phases of the Earth's history or along individual lineages. Two main sources of evidence have been invoked to support the notion that rates of molecular evolution are, indeed, approximately constant. The first source of evidence is theoretical and is embodied in the notion that evolutionary change in DNA takes place predominantly through accumulation of

neutral mutations. (Note: As observed by Fitch and Langley (1976), the concept of the molecular clock does not necessarily depend on the neutral mutation theory, although most authors have linked the two. Even under the action of natural selection, average rates of change in DNA and proteins might conceivably be approximately constant when averaged over long periods of geological time.) The second source of evidence is empirical and relies on repeated claims that numbers of inferred changes along the lineages joining any pair of species to their common ancestor tend to be approximately constant in phylogenetic trees constructed on the basis of molecular evidence.

The idea that most changes occurring at the molecular level during the evolutionary diversification of organisms are neutral, rather than filtered by the action of natural selection, owes its origin to developments in population genetics (Kimura and Crow, 1964; Kimura, 1968a; Kimura and Ohta, 1971; Lewontin, 1974). It has already been mentioned in Chapter 11 that various electrophoretic studies of proteins have revealed a far higher level of polymorphism than was originally expected. The existence of high levels of genetic variation within populations is theoretically of great importance in relation to the 'cost' of natural selection. This is particularly true of **balanced polymorphism** in which there is maintenance of approximately stable frequencies of alternative forms of a gene (alleles) at a given locus (Kimura and Crow, 1964; Lewontin and Hubby, 1966). In a simple case of natural selection, lower survival of individuals with one allele requires a certain level of 'selective deaths' of such individuals in each generation. With balanced polymorphism maintained (as is commonly believed to be the case) by selection favouring heterozygotes, there must be 'selective deaths' of both homozygotes. The total level of selective deaths required to account for active maintenance of polymorphism and evolutionary replacement of alleles represents the '**genetic load**' (Haldane, 1957) or 'substitutional load' (Kimura, 1968a,b) of a population. It is argued that this load must remain beneath a certain limit because the overall fecundity of the population as a whole will otherwise be too low. For simple natural selection, Haldane calculated that only about one allele replacement per 300 generations in a population would be permissible, as a greater rate of replacement requires too many selective deaths and hence an unbearable 'cost' to the population. But empirical data indicating rates of change in amino acid sequences of proteins and the results of electrophoretic surveys of protein polymorphism for a variety of animal species both suggest far more genetic variability and change than is compatible with Haldane's concept of genetic load.

One way out of the dilemma posed by apparently excessive genetic load is to propose that alternative alleles at most loci are selectively neutral or almost neutral. Hence, changes over time will occur predominantly through chance fixation of single alleles at individual loci, rather than through elimination of less-favoured alleles by natural selection. Clearly, if most change at the level of the individual gene occurs through random fixation of alleles, rates of evolution are more likely to be approximately constant than if environmentally dependent natural selection is the major agency of change. Indeed, calculations indicate that the rate of fixation of neutral mutations should be determined uniquely by the natural rate of mutation and is independent of population size (Kimura, 1968a,b; King and Jukes, 1969). Hence, neutral mutation theory can provide direct underpinning for the concept of a molecular clock (Kimura, 1969, 1983; King and Jukes, 1969).

The concept of genetic load is well established in the field of quantitative genetics, but it is not without its problems. In the first place, the calculations involved are based on the assumption that the effects of selection on individual loci are multiplicative. In other words, it is assumed that the selective deaths attributable to individual gene loci can be mathematically combined into an overall measure of genetic load. But natural selection cannot act on individual gene loci in isolation; it must act on whole organisms (Maynard Smith, 1968; Lewontin, 1974). Thus, except in cases where selection affects different genes at separate times in an individual's life, differential death due to natural selection is likely to depend on complex interactions of many genes, rather than on multiplicative combination of levels of selection against individual genes. Accordingly, neither relatively rapid evolution through natural selection nor the high levels of genetic poly-morphism recorded for many animal species need necessarily imply such high genetic loads as has often been inferred. Maynard Smith (1968) has proposed an alternative 'threshold model', in which natural selection commonly acts against individuals with a combination of relatively deleterious alleles (see also Sved, 1968). Hence, one must also question the theoretical proposition that most evolutionary change at the molecular level occurs through fixation of alleles that are selectively neutral. It is necessary to examine the empirical evidence with a critical eye to determine whether rates of change at the molecular level are in practice relatively constant in accordance with the prediction from neutral mutation theory.

At the very beginning of the debate about the molecular clock hypothesis, considerable influence was exerted by the seminal paper of King and Jukes (1969), dealing with so-called 'non-Darwinian evolution'. The basic tenet of this article was that the patterns of evolutionary change recorded for whole organisms do not necessarily apply at the level of the individual gene, as many mutations of DNA may be selectively neutral. For instance, it is often the case that a single mutation within a triplet of nucleotide bases produces no change in the corresponding amino acid, because of the 'redundancy' of the nuclear code. This is particularly true of changes in the third nucleotide base of a codon triplet, as can be seen from Fig. 11.4. To take just one example of such a change, the base sequences CUA, CUC, CUG and CUU all code for the amino acid leucine (Leu). Although some amino acids correspond to only one base triplet (e.g. methionine corresponds uniquely to AUG), as many as six different triplets may code for a single amino acid (e.g. in addition to the four codons listed above, UUA and UUG also code for leucine). Overall,

about a quarter of single base mutations are 'synonymous' in the sense that changes in the DNA sequence are not reflected in changes in the coded amino acid sequence. It would therefore seem at first sight that synonymous mutations are highly likely to be selectively neutral. King and Jukes (1969) cited several preliminary studies indicating far more rapid change in DNA than would be expected from amino acid sequences alone. This suggested that there is, indeed, much change at the level of DNA that is not reflected in protein structure and is therefore likely to be 'neutral'. King and Jukes also went on to conduct an interesting simulation, using the base sequences corresponding to the amino acid compositions established for 53 different proteins from various vertebrates (mainly mammals). The bases were reshuffled in a random order and the amino acids corresponding to the base triplets were then determined. It was found that there was a very good correlation ($r = 0.89$) between the frequencies of the different amino acids present in the original sample and the frequencies in the sample generated from the randomly reshuffled bases. This might be taken to indicate that there is a large random element in the organization of nucleotide bases in DNA. However, it must be remembered that the possible permutations of the four nucleotide bases (A, C, G and U) are severely limited, and their implications for levels of occurrence of amino acids are also limited, particularly given the degeneracy of the code. Accordingly, the result obtained by King and Jukes is not very convincing without some more penetrating statistical analysis.

In addition to the known degeneracy of the genetic code, there are other reasons for expecting numerous changes in DNA to remain undetectable at the level of protein structure. The main reason is that there are several lines of evidence indicating that a very large pro-portion (97–99%) of the DNA in the mammalian genome does not code for proteins. Various classes of non-coding repetitive DNA have already been mentioned in Chapter 11. King and Jukes (1969) commented on the existence of such repetitive DNA and emphasized that the heterochromatic regions of chromosomes appear to be virtually devoid of specific genetic information. More recently, it has emerged that non-coding, slightly modified copies of coding genes ('pseudogenes') are relatively common and that the internal structure of individual genes is also more complex than originally believed. It is now known that each coding gene typically incorporates non-coding regions (**introns**) intercalated between coding regions (**exons**). In principle, it is to be expected that changes in repetitive DNA, in non-coding copies of genes and in the non-coding introns might all be selectively neutral, because they are not translated into amino acid sequences of proteins.

In fact, changes in DNA that are not reflected in protein structure are irrelevant for many studies relating to the molecular clock, because such investigations depend on comparison of proteins and not on direct comparison of DNA sequences. (Although direct studies of DNA are increasingly being used to investigate the molecular clock, broad surveys of primates have yet to be conducted). In practice, therefore, it is usually necessary to ask whether changes in amino acid sequences of proteins are neutral, rather than considering the possibility of neutral changes in DNA sequences themselves. King and Jukes, in addition to making a strong case for effectively neutral changes in DNA, also reviewed a number of studies indicating that some changes in protein structure may be selectively neutral. For instance, they mentioned *in vitro* studies indicating that differences in the amino acid sequences of certain proteins (e.g. cytochrome *c* and haemoglobin) do not seem to affect observed function (but see Perutz, 1983). The loss of the ability to synthesize vitamin C in haplorhine primates and in certain rodents, mentioned earlier in this chapter, was also interpreted as a neutral, 'nonadaptive' change, arising in an environment where sufficient amounts of vitamin C were available to render superfluous the ability to synthesize the vitamin. King and Jukes also specifically referred to the studies of Sarich and Wilson (1967b) on immunological cross-reactions of primate albumins and emphasized the 'remarkable

constancy' in inferred rates of evolutionary change. This constancy was taken as providing empirical support for the hypothesis that much evolutionary change in proteins occurs through the fixation of neutral mutations. Subsequently, there have been repeated claims that the rate of change of amino acid sequences of proteins is approximately constant over time and this provides one of the main pillars of support for current advocates of the molecular clock.

Discussion of rates of change in relation to the molecular clock has commonly been conducted on the basis that the phylogenetic tree inferred from a given data set is necessarily correct and that the numbers of changes attributed to individual branches are inherently reliable. There are, however, good reasons for questioning these assumptions. First, it has already been noted (see Chapter 3) that variation in rates of evolution gives rise to major problems in accurate reconstruction of phylogenetic relationships because of potential confusion between primitive and derived homologous features. Second, trees derived from molecular data are generated with the explicit use of the principle of parsimony. Yet there is no justification for believing that evolution has necessarily taken the most direct course and, in some cases at least, it is possible that application of the parsimony principle may lead to reconstruction of a more parsimonious but erroneous tree in which inferred rates of evolution exhibit spurious constancy (e.g. see Fig. 3.14). Thus, it is essential to consider the accuracy of phylogenetic reconstructions themselves as well as the implications of implied rates of change. It is not correct to assume that phylogenetic trees can be determined 'unambiguously' from molecular data (Sarich and Wilson, 1973) and that such trees can then be used as a reliable basis for discussion of rates of change.

One of the most satisfactory ways of examining the accuracy of reconstructions based on protein sequence data is to conduct simulations in which phylogenetic trees are generated for a hypothetical protein according to certain basic principles of molecular evolution. Using the artificial sequence data for the descendants at the end points of a tree generated in this way, an attempt can then be made to reconstruct the tree. A simulation of this kind was devised by Tateno and Tajima (1986) using trees containing 8 or 16 descendants generated from an ancestral 'gene' with 100 codons. In fact, their model explicitly incorporated a constant probability of nucleotide substitution, so average rates of change were held constant and problems of phylogenetic reconstruction should have been minimized. (It remains to be seen how the incorporation of realistic levels of variation in the rate of nucleotide substitution under the influence of natural selection would affect the reconstruction of simulated phylogenetic trees.) Various tree-construction methods were then applied to infer the tree corresponding to each set of artificially generated descendant proteins, in order to assess their reliability in reconstructing the actual branching pattern and in estimating branch lengths. Incidentally, it should be noted that the numbers of nucleotide substitutions inferred for any given lineage must be corrected to allow for the probability of multiple substitutions (including reverse substitutions) at any given site, which should conform to a Poisson distribution. Using an appropriate formula, adjustment can be made to a matrix of distances between species (Jukes and Cantor, 1969; Holmquist, 1972). Tateno and Tajima conducted such an adjustment to the matrix of nucleotide 'distances' between the descendants in any comparison.

The results obtained by Tateno and Tajima showed that, even with a model involving a constant probability of nucleotide substitution, the accuracy of phylogenetic reconstruction was surprisingly low. With only eight descendants, the probability of correct reconstruction of the phylogenetic tree was between 10 and 50%, according to the tree-building method used, and with a larger set of 16 descendants, the probability of correct reconstruction ranged between 0 and 5%! Thus, even if rates of nucleotide substitution are set at a uniform level, it is very difficult to reconstruct the correct phylogenetic tree from a given set of molecular

data for a relatively small number of species. Problems arise because of stochastic variation in the occurrence of nucleotide substitutions, in spite of a constant probability of substitution overall. As a result, artificially generated trees do show some variation in rate of change along individual branches and reconstruction from descendant sequences is rendered unreliable.

Tateno and Tajima (1986), in fact, went on to show that the accuracy of reconstruction was increased, and the degree of variation in rates of change along individual branches was reduced, when several 'genes' (2, 5 or 10) were used instead of just one. This result, which is to be expected from the fact that the effects of stochastic variation are mitigated by including more information, further underpins the general rule in phylogenetic reconstruction that accuracy is increased by maximizing the number of characters considered (see Chapter 3). Further, Tateno and Tajima's (1986) simulation study showed that, even if a constant probability of nucleotide substitution is postulated, there are three different sources of error affecting reconstruction of phylogenetic trees: (1) experimental error (i.e. error in the data set), (2) stochastic error of nucleotide substitution and (3) the methodological error of tree-construction. They concluded that 'no tree construction method is perfect, and all methods often give a wrong tree even under such a simple evolutionary process as one driven by neutral mutation'.

Many reconstructions of phylogenetic trees using molecular data have (implicitly or explicitly) assumed roughly constant rates of evolution. It is therefore circular to argue that the results of such reconstructions provide empirical support for constant rates of evolution and hence for evolution through neutral mutation (see also Uzzell and Pilbeam, 1971). If, on the other hand, natural selection exerts a major influence on the fixation of nucleotide substitutions, marked fluctuations in rate of change are to be expected, at least over the short term, and the problems involved in tree-construction must inevitably increase. Quantitative tree-construction methods currently in use are, in most cases at least, adversely affected by variation in rates of change along individual branches, as is to be expected from the principles discussed in Chapter 3. Thus, if rates of change are shown to vary markedly in practice, it is not only the 'molecular clock' that is called into question; the branching patterns of phylogenetic trees based on molecular data themselves become suspect. It is therefore of paramount importance to examine the empirical data to see whether there is evidence for major disparities in rates of fixation of nucleotide substitutions between lineages and hence for the action of selection.

In fact, there is abundant empirical evidence of inconstancy of rates of evolution from evolutionary trees based on all kinds of molecular evidence (comparative immunology, protein sequences and DNA base sequences). This is true even though there is a strong possibility that the tree-construction methods themselves have commonly tended to minimize apparent differences in rates. Although rates of molecular evolution may be approximately constant when averaged over very long periods of geological time, numerous cases are known in which individual species or groups of species are distinguished by markedly slower rates of evolution.

As has already been noted in Chapter 11, several different studies have revealed that the inferred rate of evolution of serum albumin for the tree-shrew is considerably lower than that inferred for primates generally. Similarly, among simian primates the owl monkey (*Aotus*) stands out as a form in which serum albumin has apparently evolved very slowly, while slow lorises (*Nycticebus*) and lemurs (*Lemur*) show relatively slow rates in comparison to *Homo* (see Goodman, 1962b; Sarich and Wilson 1966; Sarich, 1970b). Overall, differences in rates by a factor of at least two seem to be relatively common. Hence, as noted by Radinsky (1978), even those investigators who have argued most strongly for overall constancy of rates of evolution in proteins have acknowledged individual exceptions. However, it may be argued that when such exceptions occur they are

easily recognized. Sarich and Wilson (1967) and Sarich (1970b) have, indeed, used the slower rates demonstrated for *Aotus*, *Nycticebus* and *Lemur* to illustrate the principle of 'internal consistency' in trees based on immunological data (Fig. 12.10; see also Fig. 11.22, case 2). The argument is that the slower rate of evolution of albumin in *Aotus* is immediately evident because the owl monkey shows smaller-than-expected immunological distances from all other simian primates with which it is compared. (This provides the basis for the 'relative rate test' subsequently emphasized by Sarich and Cronin (1976) and by Wilson, Carlson and White (1977).) Accordingly, the argument continues, any other rate differences in individual primate lineages would also be readily apparent because

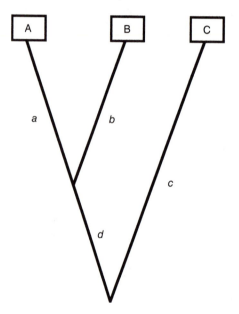

Figure 12.10. Illustration of the principle of 'internal consistency' with respect to rates of change. If comparative data suggest that species A has evolved more slowly than species B since their divergence from a common ancestor (i.e. branch length *a* is shorter than branch length *b*), this can be tested with reference to an outgroup (C). If A has, indeed, evolved more slowly than B, the distance ($a + d + c$) should be shorter than the distance ($b + d + c$). The test is, of course, based on the questionable assumption that the basic tree shown has been correctly reconstructed in spite of the slower rate of change along the lineage leading to A.

of such systematic disparities. However, it may be that *Aotus* is simply an extreme case and that more subtle variations in rates between lineages are largely obscured by the method of tree-construction used (see Uzzell and Pilbeam, 1971). Further, until we have an adequate explanation for the slower rates of change inferred for the albumins of tree-shrews and *Aotus*, the theoretical basis for interpretation of rates generally must remain dangerously weak. Can one really accept the proposition that unknown selective forces have slowed the rate of evolution of albumins specifically in tree-shrews and owl monkeys, while albumins in other species have changed at a uniform rate because they have been modified only through the steady accumulation of neutral mutations?

There is also abundant evidence for marked variation in rates of evolution when amino acid sequences of proteins – and nucleotide base sequences inferred from them – are considered. A general analysis by Langley and Fitch (1974) and by Fitch and Langley (1976) showed quite clearly that inferred rates of evolution for a variety of proteins fluctuate more than would be expected from neutral mutation theory and specific studies of individual proteins have confirmed this. For instance, Romero-Herrera, Lehmann *et al.* (1978) clearly showed for myoglobin sequences of mammals and some other vertebrates that rates of evolution must vary at least to some extent between lineages. This finding is particularly convincing because these authors applied a great variety of approaches to tree-building and demonstrated considerable variation in rates in all cases. With one parsimonious tree, several palaeontological dates for inferred mammalian divergence points were used to examine the constancy of rates across lineages and several extreme cases were identified. In particular, a very slow rate of substitution of nucleotide bases was inferred for the tree-shrew in comparison to all other placental mammals in the sample. To take just one example, the lineage leading to the ox (an artiodactyl) indicated an inferred rate of substitution three times higher than that for the lineage leading to the tree-shrew. Further, on

the basis of the calibration dates used, it was concluded that the average rate of evolution for myoglobin seems to have been markedly greater for both mammals and birds over the past 80 Ma in comparison with the previous 200 Ma.

Using a somewhat different approach, Goodman, Romero-Herrera *et al.* (1982) used a single calibration date of 90 Ma ago for the common ancestor of modern placental mammals to infer dates for particular ancestral nodes in trees constructed from sequence information for several different proteins (α-haemoglobin, β-haemoglobin, α-crystallin, carbonic anhydrases, cytochrome *c*, fibrinopeptides A and B, and myoglobin). For the ancestral node of Old World simians (monkeys, apes and humans), the inferred dates ranged from 3.1 Ma ago for carbonic anhydrase II to 26.2 Ma ago for α-haemoglobin, representing an eight-fold difference in timing. The mean inferred date was 14.1 ± 6.8 Ma, yielding a coefficient of variation of $\pm 48.2\%$. Similar variability was found with dates for other ancestral nodes inferred from individual proteins. An extreme case was found with the order Rodentia, for which α-crystallin and cytochrome *c* sequences indicated a common ancestor at 0 and 5.3 Ma ago, respectively, whereas α-haemoglobin and β-haemoglobin indicated dates of 63 and 67 Ma ago, respectively. As predicted by Fitch and Langley (1976), Goodman, Romero-Herrera *et al.* (1982) found that the clock model worked better when all proteins were combined into a single calibrated tree, but striking inconsistencies were still found. For instance, a divergence date of only 1.3 Ma ago for *Pan* and *Homo* is indicated! On a different tack, Goodman *et al.* used several divergence dates indicated by the palaeontological record to calibrate nodes in a phylogenetic tree based on all proteins combined, to check apparent constancy of rates in the evolution of vertebrates. This indicated that vertebrate proteins underwent two periods of relatively rapid change, one 500–300 Ma ago (rates 5–7 times higher than average) and the other 90–40 Ma ago (rates 10–20 times higher than average). These two periods were interpreted as times of large-scale adaptive radiation, with the implication that rates of protein evolution may be slower once lineages generated in such a radiation have become established.

Overall, therefore, studies based on protein sequence data show that there are quite marked differences in inferred rates of evolution along individual lineages, regardless of which particular tree is selected as optimal. Such variation in rates commonly extends to a two-fold or three-fold difference when averaged over long periods of evolutionary time, and in some cases even greater variation is found. As has been noted by Romero-Herrera, Lehmann *et al.* (1978) and Andrews (1986), among others, variation at this level will have a particular influence on inferred dates for relatively short branches.

One of the most striking and convincing demonstrations of differences in rate of fixation of nucleotide substitutions has emerged from the DNA hybridization studies of Bonner, Heinemann and Todaro (1980). In comparison to all other primates examined, the single-copy DNA of the lemurs tested (*Cheirogaleus, Lemur, Microcebus* and *Propithecus*) was consistently found to be much less divergent than expected. This suggests that, for some as yet unexplained reason, the overall rate of nucleotide base replacement in the entire complement of single-copy DNA among lemurs has been about half that in other primates (see Fig. 11.36). This finding is incompatible with any simple model of evolutionary change in DNA taking place predominantly through the accumulation of neutral mutations. As the effect is averaged across the entire fraction of DNA believed to code for proteins and applies to all the lemurs examined, it is most unlikely that mere stochastic variation is responsible.

Correcting for generation times

As lack of uniformity in rates of molecular evolution between lineages is an established fact, it is necessary to find some explanation for this, in so far as it can be shown to exceed the level expected because of stochastic fluctuation. At

first sight, such variation might seem to be incompatible with neutral mutation theory; but it is conceivable that other factors might influence rates of change even without the operation of natural selection. A prime candidate for such a factor is **generation time**. In principle, it might be expected that species with a slow breeding history might exhibit lower rates of molecular evolution than rapidly breeding species, as the rate of spread of genes through a population should be influenced by the rate at which individuals achieve reproductive maturity. Goodman (1962b), in an early discussion of quantitative immunological distances between primate species, noted that rates of change might differ between lineages and stated: 'An obvious parameter is the average rate of reproduction or the number of generations separating the heterologous species from the common ancestor.' Kimura (1968a), in fact, repeatedly mentioned an expected effect of generation times in his seminal paper on neutral mutation theory and seems to have taken it for granted that this factor should be taken into account in evolutionary models based on neutral mutation theory (see also Kimura and Ohta, 1971). King and Jukes (1969) also referred to the influence of generation times as an integral feature of their discussion of 'non-Darwinian evolution'. Consequently, it has been claimed that, according to neutral mutation theory, generation time should affect rates of molecular evolution (Read, 1975). This being the case, one might try to account for the relatively small degree of molecular difference between humans and African apes, in comparison to other primates and to other mammals in general, in terms of their long generation times (e.g. see Lovejoy, Burstein and Heiple, 1972).

Some authors have specifically attempted to make corrections for differential generation times in calibrating trees based on molecular data. For instance, Kohne (1975) corrected for estimated generation times in his analysis of molecular evolution in primates based on DNA hybridization data (see also: Kohne, 1970; Kohne, Chiscon and Hoyer, 1972). Without any correction for generation times, the raw distances between species assessed from DNA hybridization (see Fig. 11.35) would indicate a divergence date of only 3.7 Ma ago for *Pan* and *Homo*, taking a conservative calibration date of 65 Ma for the common ancestor of primates of modern aspect. However, if the branch lengths in Fig. 11.35 are adjusted to allow for expected differences in generation time, the same calibration date would indicate a divergence date of 12–13 Ma ago for *Pan* and *Homo*, much closer to reasonable estimates based on palaeontological evidence. Kohne (1970, 1975) justified such correction using inferred generation times on the basis of an apparent disparity in rates of evolution of DNA between rodents and primates. He accepted a date of 10 Ma ago for the divergence between mouse and rat, and on this basis suggested that the single-copy DNA of these rodents appeared to have changed 10–15 times more rapidly than in primates (albeit taking a date of 80 Ma ago for the common ancestor of primates of modern aspect, some 25% earlier than generally proposed by primate palaeontologists). Correction for generation times removed this apparent disparity in rates of evolution between rodents and primates.

A few other researchers have also made explicit attempts to correct for expected effects of generation times in inferring times of divergence from comparative molecular data. Lovejoy, Burstein and Heiple (1972), for instance, made adjustments to the immunological distance data of Sarich and Wilson (1967a,b) and to the DNA hybridization data of Kohne (1970) for Old World simians. Changes in generation length over time were inferred by assuming a consistent relationship between relative brain size and generation time. When an appropriate correction was made, Lovejoy *et al.* obtained a divergence date of approximately 14 Ma ago for African apes and humans, relative to a calibration date of 30 Ma ago for the divergence between Old World monkeys and apes. Benveniste and Todaro (1976), in their study of DNA hybridization in simian primates, also made an adjustment for generation times. In their phylogenetic tree for Old World simians, they allowed for an effect of

longer generation times in great apes and humans in inferring dates of divergence. Using the same calibration date of 30 Ma ago for the separation between Old World monkeys and apes, they inferred a date of 12 Ma ago for the divergence between African apes and humans. (Incidentally, even without a correction for inferred differences in generation time, Benveniste and Todaro determined a date of 8 Ma ago for the divergence between African apes and humans, which is notably earlier than the date suggested by Sarich and Wilson (1967a,b) from their immunological comparisons of albumins.)

Most authorities who have used comparative molecular data to reconstruct phylogenetic relationships between primates have made no attempt to correct for the effects of differential generation times. In many cases, the issue has simply been ignored; but several authors have actively rejected the need to make any correction. Prominent among these have been Sarich, Wilson and their colleagues (e.g. Sarich, 1972; Sarich and Wilson, 1973; Sarich and Cronin, 1977; Wilson, Carlson and White, 1977), who have repeatedly argued that disparities in rates due to differential generation times would be easily identifiable in reconstructed trees. The principle of internal consistency of trees based on quantitative molecular data requires that any lineage marked by a slow rate of evolution – for whatever reason – should show a relatively small number of changes in comparison with all other lineages (see Fig. 12.10). It has been noted above that the serum albumins of tree-shrews and of owl monkeys (*Aotus*) seem to have undergone relatively slow evolutionary change and that this is clearly reflected by smaller-than-expected immunological distances from all other primates with which they have been compared. Accordingly, if (for example) great apes and humans have undergone relatively little molecular change because of their comparatively long generation times, this should be indentifiable in terms of smaller-than-expected distances from all other primates. However, a tree based on immunological comparison of serum albumins (Sarich and Wilson, 1967a,b), provided no evidence that less change has occurred along the lineages leading to great apes and humans than in other primate lineages. Indeed, tree-shrews have shorter generation times than any primate, so the inferred slower evolutionary change for their albumin runs directly counter to expectation. Further, it is difficult to explain the apparent deceleration in evolution of serum albumin in *Aotus*, which has a shorter generation time than, for example, great apes and humans.

Sarich (1972) also specifically rejected the argument advanced by Kohne (1970) that rates of change in DNA have been much faster in rodents than in primates. He tested this proposition by comparing the serum albumins of primates, rodents and carnivores. If it were, indeed, true that molecular change has occurred more rapidly in rodents than in primates because of the shorter generation times, it follows that the immunological distance between rodents and carnivores should be greater than that between primates and carnivores. In fact, however, the immunological distance between primates and carnivores was found to be about the same as that between rodents and carnivores. Hence, there is no evidence from immunological comparisons of albumins for a deceleration in rate of evolution in primates in comparison to rodents and hence no evidence for an effect of longer generation times in primates. Sarich accordingly dismissed Kohne's date of 10 Ma ago for the divergence between rat and mouse as far too recent and inferred a date of 35–40 Ma ago instead, assuming a constant rate of albumin evolution in mammals generally. In support of this early divergence date, Romer (1966) was cited as indicating the time of divergence between the families Cricetidae and Muridae (the latter including mice and rats) as about 30 Ma ago. But Romer's phylogenetic tree actually gives no indication of the (necessarily more recent) time of separation between rats and mice within the family Muridae and the date of 35–40 Ma ago suggested by Sarich is in no way supported by palaeontological evidence. Such an early date for the divergence between mice and rats would appear to be highly unlikely and this leads one to

suspect that in this case, at least, there is a serious anomaly in the 'molecular clock' based on immunological comparisons of albumins.

It should also be noted that the arguments advanced by Sarich and Wilson to rule out any influence of generation time on the rate of change in proteins over time are entirely empirical in nature. As pointed out above, a direct influence of generation time is to be expected from the original formulation of neutral mutation theory (Kimura, 1968a,b) and there is, therefore, no sound theoretical basis for expecting constant rates of evolutionary change regardless of differences in generation time. Read (1975) stated that the issue regarding the potential influence of generation times had been 'resolved', but it is difficult to accept this conclusion in the absence of a sound theoretical framework. Until a mathematical model is proposed in which evolution through fixation of neutral mutations can be predicted to occur at uniform rates in all lineages in spite of variation in generation times, no satisfactory link can be claimed between neutral mutation theory and the concept of the molecular clock. In this context, it is unfortunate that Kimura (1983) made virtually no mention of generation times in his review of neutral mutation theory.

Sarich and Wilson (1973) also applied their 'relative rate test' (see Fig. 12.10) to prosimians and simians, by using carnivores as an outgroup. Overall, it is to be expected that generation times have been shorter in prosimians than in simians, and the prediction is that rates of change should have been slower in the latter. However, comparison with carnivores indicates that, if anything, the immunological distance separating them from simian primates is somewhat greater than that separating them from prosimians. In the same paper, Sarich and Wilson specifically examined the correction for differential generation times proposed by Lovejoy, Burstein and Heiple (1972). They rejected the proposed correction procedure because an excessively high rate of change (for which there is no empirical evidence) would be required for the lineage leading to tree-shrews. However, it should be noted that for less-extreme cases it emerged that it is very difficult to rule out an effect of generation times.

In a further application of the 'relative rate test', Sarich and Cronin (1977) subsequently pointed out that there was no empirical evidence from the DNA hybridization data of Benveniste and Todaro (1976) to support the notion that the rate of change had been slower in great apes and humans than in gibbons and Old World monkeys. Contrary to the expectation based on a postulated effect of generation time, the inferred genetic distance between Old World monkeys and great apes or humans appears to be no less than the inferred genetic distance between Old World monkeys and gibbons. However, in a reply appended to the comments by Sarich and Cronin (1977), Benveniste and Todaro point out that it would be very difficult to demonstrate convincingly the presence or absence of an effect of generation times given the available data. Indeed, if the African apes and humans diverged only 5 Ma ago, as suggested by Sarich and Wilson (1967a,b), exceptionally accurate data would be required to test an effect of generation time.

There is, unfortunately, a major problem inherent in all practical discussions of generation time in that direct data are available only for the rate of reproduction of living forms. Although modern great apes and humans have longer generation times than all other living primates, it is not clear how far this observation may be extrapolated back into the past. As has been noted in Chapter 9, there is a general scaling relationship between reproductive parameters and body size, such that (other things being equal) generation time should increase in a fairly predictable fashion with body size across species. Hence, it may be accepted as a general rule that average generation times have probably been shorter for prosimian primates throughout their evolutionary history than for simian primates, as average body size has been typically greater in the latter. Further, there are also grade differences among mammalian groups. For instance, primates seem to be specifically adapted for low reproductive turnover, relative to body size, in comparison to most other placental mammals (see Chapter 9). Hence, it may also be accepted as a general rule that average generation times have probably been longer for primates throughout their evolutionary history than for many other placental mammals. As a more specific level, however, it is difficult to specify generation times. This is, for instance, true for a comparison between the lineages leading to Old World monkeys and great apes. Although it is probably reasonable to assume that longer generation times characterized the latter stages of evolution of great apes in comparison to monkeys, it is likely that early members of both lineages were similar in both body sizes and generation times. In other words, it is difficult to specify when great apes began to have significantly longer generation times than monkeys. Thus, accurate correction for the effect of differential generation times is problematic. As a rule, however, the difference in generation times is likely to be smaller when averaged over lineages than would be expected from a direct comparison of living species. Accordingly, most published discussions (which have failed to take this into account) have exaggerated the expected effect of differential generation times. The possibility should also be borne in mind that there may have been some progressive change in generation times over all lineages – for example, a general tendency for lower reproductive turnover in mammals generally. Such a general trend over time would not be identifiable with the 'relative rate test' of Sarich and Wilson, because this simply examines the total amount of change along lineages and cannot assess how the change is apportioned within a lineage (see also Gingerich, 1985a).

Of course, if natural selection exerts a major influence on rates of fixation of mutations, differences between lineages are to be expected. It is well known that there are differences between proteins in overall rates of change per amino acid site and also that within a given protein some sites are more prone to change than others (e.g. see Romero-Herrera, Lehmann *et al.*, 1978). This alone indicates that natural selection influences overall rates of change according to the degree of conservation required of individual proteins or of sites within proteins for appropriate function. It is hence to be expected that a given protein may evolve at different rates at different times in its evolutionary history, depending on changes in its functional context. In this respect, an interesting hypothesis was advanced by Goodman (1961, 1962b), initially to account for the apparent slower evolution of serum albumins in primates. He linked the rate of evolution of proteins to the degree of invasiveness of the placenta, arguing that with increasing invasiveness there is an increasing risk of formation of maternal antibodies to fetal proteins differing from those of the mother in their amino acid sequence. It is therefore to be expected that in mammals with an invasive (haemochorial placenta) there should be greater selection against change in protein sequence than in mammals with a non-invasive (epitheliochorial) placenta. This would apply only to those proteins present in the fetal bloodstream and not to proteins developed only later in life. However, although Goodman claimed to have evidence of the expected slow-down in protein evolution in simian primates (monkeys, apes and humans), which all have haemochorial placentation (see Chapter 9), this was not confirmed by the 'relative rate test' conducted by Sarich and Wilson (1973) on strepsirhines and simians. As strepsirhine primates (lemurs and lorises) have non-invasive epitheliochorial placentation (Chapter 9), the prediction from Goodman's hypothesis is that protein evolution should have taken place more rapidly in strepsirhines than in simians. But Sarich and Wilson found no evidence for more rapid evolution of proteins in strepsirhines. Further, differences in placentation cannot account for the pattern of evolution of single-copy DNA in primates identified by Bonner, Heinemann and Todaro (1980), as a slow rate of change was inferred not for haplorhine primates (tarsiers and simians), which have invasive haemochorial placentation, but for lemurs, which have non-invasive epitheliochorial placentation. Hence, there is little empirical

support for the hypothesis that the type of placentation influences the rate of evolution of proteins in general. Nevertheless, it is possible that the hypothesis might apply in specific cases, notably to proteins that easily pass between the maternal and fetal circulations during pregnancy, and it is important that such possible selective factors should be considered in discussions of rates of evolution.

Relationship between molecular distance and time

An entirely different factor that might affect inferred rates of change along lineages is the nature of the relationship between molecular distance measures and time. A common assumption has been that various measures of molecular distance, such as the 'immunological distance' of Sarich and Wilson (1967a,b), defined as the logarithm of the index of dissimilarity, or distances based on sequence comparisons of proteins and DNA, are linearly related to time. This assumption is associated with neutral mutation theory in that a constant linear relationship between molecular distance and time is likely to exist if changes in nucleotide base sequences are predominantly neutral. By contrast, if most changes are subject to natural selection there are reasons to expect that systematic departures from linearity might occur. In particular, it might be expected that individual molecules would change more rapidly during the initial phase of adaptive radiation of a group of animals and then remain more stable following a period of functional consolidation. In other words, the rate of change of individual molecules subjected to natural selection would be expected to decrease over time, giving a curvilinear relationship. Hence, although there may be a regular relationship between molecular distance and time, simple linear calibration of trees using a single date from the fossil record may be unjustifiable.

The point has been made repeatedly that empirical tests of the relationship between molecular distance and time, as indicated by palaeontological dates, fail to support the assumption that simple linearity exists (Read and

Lestrel, 1970; Read, 1975; Gingerich, 1985b; Andrews 1986). In the first place, it should be noted that even a simple model of molecular evolution with constant rates of change along lineages does not predict a simple linear relationship between observed differences in sequences of nucleotide bases or amino acids. Because of the problem of multiple (including reverse) changes at a given site, it is necessary to adjust the raw data for differences between pairs of species to allow for 'hidden' changes (Jukes and Cantor, 1969). It is only the adjusted values that should theoretically exhibit a linear relationship with time (Read, 1975; Gingerich, 1985b). Ideally, therefore, the values in a distance matrix based on differences in sequences of amino acids or nucleotide bases should be adjusted before any further analyses. This is important when it comes to the assessment of immunological data, as in this case there is no direct information on the numbers of differences in amino acid sequences between species and no adjustment is usually made to the data prior to analysis. It is therefore unrealistic to expect a linear relationship between raw data on immunological distances between species and time of divergence. In fact, because of the problem of 'hidden' nucleotide replacements, approaches that are merely based on numbers of differences in amino acids between species will increasingly underestimate the actual number of changes for earlier dates of origin. Thus, the data on immunological dissimilarity originally provided by Sarich and Wilson (1967a,b) actually underestimated the numbers of changes in the earlier parts of the tree. Taking their method of inferring times of origin, an appropriate adjustment would therefore indicate an even more recent divergence between African apes and humans than the date of 5 Ma ago indicated by their calculations (Uzzell and Pilbeam, 1971). It is therefore particularly important to examine the relationship between measures of molecular distance between species and elapsed time.

Read and Lestrel (1970) specifically examined the proposed relationship between inferred times of origin (T) and the index of immunological dissimilarity (IND) defined by Sarich and Wilson

(1967a,b). To do this, they took the same set of T and IND values for various vertebrate species (bullfrog, turtle, caiman, duck, pigeon, ostrich, turkey and chicken) originally used to calibrate the equation used by Sarich and Wilson. As the value for the bullfrog seemed to be aberrant, Read and Lestrel dropped the value for this species from the comparison. They then fitted two alternative equations – a simple linear equation (IND $= kT + 1$) and a power function (IND $- 1 = kT^{\alpha}$) – to the values for the remaining species. Both matched the data better than the equation used by Sarich and Wilson (i.e. IND $= e^{kT}$, or log IND $= kT$). They concluded that the power function seemed to provide the best model overall, although they noted that the available data are simply inadequate to permit a reliable choice between alternative models. Taking a calibration date of 37 Ma ago for the divergence between Old World monkeys and apes, the data on IND given by Sarich (1968) yielded a divergence time between African apes and humans of 21 Ma ago. Read (1975) subsequently further analysed the data and concluded that the most satisfactory results were obtained with a power function that indicated a divergence between African great apes and humans about 10 Ma ago. This conclusion was backed up with the supplementary observation that backward extrapolation of a regression of cranial capacity against time for hominids would indicate a similar date, but this is not a compelling argument (see Chapter 8).

Uzzell and Pilbeam (1971) questioned one of the reasons given by Read and Lestrel (1970) for eliminating the data for the bullfrog and stated that the formula advocated by Sarich and Wilson (1967a,b) did, in fact, give the best fit to the complete data set. However, as was later noted by Read and Lestrel (1972), the immunological data for the bullfrog are clearly anomalous relative to any line fitted to the data. More importantly, they also pointed out that no really convincing statistical discrimination between alternative formulae had been provided. It is, in fact, a relatively simple matter to check whether there is a simple linear relationship between immunological dissimilarity (IND) and inferred dates of divergence, or whether some more complex relationship (e.g. involving a power formula) is involved. One can apply the test for isometry commonly used in allometric analysis (see appendix to Chapter 4). The allometric equation, which expresses the curvilinear relationship between any given parameter (Y) and a measure of body size (X), is a power function of the form: $Y = kX^{\alpha}$. Isometry is a special case in which there is a simple linear relationship between Y and X of the form: $Y = kX$, and in which the value of the exponent (α) is accordingly 1. A test of linearity can therefore be conducted using a bivariate plot of log Y against log X. If the relationship between Y and X is genuinely linear, the slope of the line (representing the exponent value α in the original equation) will be indistinguishable from 1. If, on the other hand, the slope of the line is significantly different from 1, it can be concluded that the relationship is non-linear. It is possible to test the linearity of the relationship between any given measure of molecular distance (e.g. immunological distance, ID) and time in a comparable fashion (see also Corruccini, Cronin and Ciochon, 1979; Corruccini, Baba *et al.*, 1980).

Gingerich (1985b) took three palaeontological divergence dates that he regarded as reliably established: (1) between Old World monkeys and hominoids (apes and humans) at 25 Ma ago; (2) between Old World simians and New World simians at 40 Ma ago; and (3) between strepsirhine primates (lemurs and lorises) and simians at 55 Ma ago. He then used these three inferred divergence dates to test the relationship between molecular distance and time, using the data on immunological distance ($= \log$ IND) provided by Sarich (1968, 1970b) for serum albumins. It emerged that a power function gave a better fit to the data, taking the form: $ID = 0.27 \times T^{1.49}$. In other words, the rate of molecular change would seem to have decreased progressively over time. Hence, calibration of a primate tree using an equation requiring a simple linear relationship between immunological distance and time apparently leads to overestimation of the times of early divergence

points and underestimation of the times of more recent divergence points. Using the empirically determined power formula given above, instead of the simple linear formula, Gingerich (1985a) inferred a date of about 9 Ma ago for divergence between humans and African great apes, rather than the figure of 4–5 Ma ago given by Sarich and Wilson. Gingerich applied this approach to several other data sets for molecular distances, including the DNA hybridization data of Benveniste and Todaro (1976) and in every case found a better fit using a power function with an exponent value greater than 1. Accordingly, dates inferred for the divergence between great apes and humans were also earlier.

The isometry test can also be applied to the immunological distance data for carnivores and ungulates presented by Carlson, Wilson and Maxson (1978). When the data provided in their graph are replotted on logarithmic coordinates, it is found that the exponent value is greater than 1, namely 1.13 (Fig. 12.11). Hence, a power function rather than a linear function better describes the relationship between immunological distance and time indicated by their data. Incidentally, it should be noted that if the isometry test indicates an exponent value greater than 1 with data that have not been adjusted for the effects of superimposed ('hidden') mutations, this runs directly counter to expectation. Given a linear relationship between actual numbers of amino acid differences between species and time, a plot of observed numbers of differences (which increasingly underestimates changes with increasing time since separation) should give an exponent value of less than 1.

There is, of course, a problem inherent in any analysis of the nature of the relationship between molecular distance and time in that inference of divergence dates from the fossil record (Fig. 12.8) is itself subject to considerable uncertainty. As a general rule, it is likely that palaeontologists have tended to underestimate times of divergence, because these have traditionally been linked very closely to the earliest known representatives for each group. It is also likely that such underestimation of divergence times

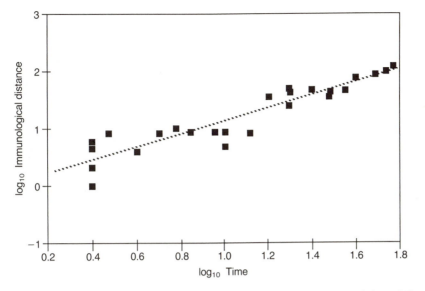

Figure 12.11. Double logarithmic plot of immunological distance (*ID*) against inferred time of divergence (*T*) for a sample of 25 pairs of carnivores and ungulates (data extracted from Carlson, Wilson and Maxson, 1978). This plot shows that the relationship between immunological distance and time is probably curvilinear, as an exponent value of 1.13 is indicated.

$$\log_{10} ID = 1.13 \log_{10} T + 0.004$$

$$ID = 1.01\, T^{1.13} \qquad (r = 0.93)$$

Determination of a timescale for the phylogenetic tree for carnivores and ungulates based on the assumption of a linear relationship between immunological distance and time would lead to progressive underestimation of times of divergence for more recent branching points.

tends to become more pronounced with increasing antiquity of ancestral nodes, because the early fossil record of any group is often particularly poorly documented. In other words, the relationship between dates of divergence inferred by palaeontologists and actual dates may itself be curvilinear. The three calibration dates for primates taken by Gingerich (1985b) are, for instance, probably too recent. The divergence between Old World monkeys and hominoids (apes and humans) may well have taken place at least 35 Ma ago, rather than at 25 Ma ago. Divergence between Old World simians and New World simians probably took place at least 55 Ma ago, rather than at 40 Ma ago. Finally, the separation between strepsirhine primates (lemurs and lorises) might date back as far as 90 Ma ago rather than to only 55 Ma ago (see Fig. 12.1). If the relationship between the

dates suggested here and those proposed by Gingerich (1985b) is examined with the isometric test, it is found to be non-linear, with an exponent value of 1.18. Hence, exponent values for the relationship between molecular distance and elapsed time would be lower if the divergence times indicated in Fig. 12.1 are roughly correct. Such an adjustment, however, does not entirely eliminate the non-linearity between molecular distance and time identified by Gingerich. For that to be the case, it would be necessary to take even earlier divergence dates for New and Old World simians and for strepsirhines, assuming that the date of 35 Ma ago accepted for divergence between Old World monkeys and hominoids is approximately correct.

With the data on carnivores and ungulates provided by Carlson, Wilson and Maxson (1978), the exponent value indicated by the isometry test is lower than that determined by Gingerich (1985b) for the primate data. Accordingly, it would be possible to obtain a linear relationship between immunological distance and time by progressively increasing inferred divergence times for the more-ancient branching points. In this case, therefore, increasing inferred divergence dates to obtain a linear relationship between immunological distance and time would bring the early evolution of the carnivores and ungulates into the range suggested in Chapter 4 (Fig. 4.6), with the initial divergence of the placental mammals beginning around 140 Ma ago.

Uncertainties involved in the inference of divergence times from the fossil record may also be responsible for certain anomalies in molecular distances noted by various authors. For instance, Romero-Herrera, Lehmann *et al.* (1978) pointed out that, according to their analysis, the average rate of evolution of myoglobin seems to have been far greater for both mammals and birds over the past 80 Ma than during the previous 220 Ma. Goodman, Romero-Herrera *et al.* (1982), on the other hand, noted from their analysis of several proteins aligned in tandem that their inferred date for the initial divergence between Mammalia and Aves (taking a calibration date of 90 Ma ago for the common ancestor of the placental mammals) seems to be be too recent: around 180 Ma instead of 300 Ma ago. It is possible, however, that both observations can be explained on the grounds that there happens to be a relatively good fossil record for the early radiation of reptiles (many of which were large-bodied forms), during which the original separation between the lineages ultimately leading to birds and mammals occurred. The earliest mammals and perhaps the earliest birds were relatively small forms and for the mammals, at least, the first two-thirds of their evolutionary history is very poorly documented (Chapter 4). It is therefore likely that the initial divergence between the lineages ultimately leading to birds and mammals has been more-or-less correctly dated at about 300 Ma ago, but that the dates of the ancestral nodes for modern birds and modern mammals have been underestimated by a

considerable margin. Recalibration of the data of Goodman, Romero-Herrera *et al.* (1982) with a date of 135 Ma ago for the origin of the placental mammals (Fig. 4.6) in fact yields a date of 270 Ma ago for the initial divergence between the lineages leading to birds and mammals. Such a recalibration also removes the apparent anomaly noted by Romero-Herrera, Lehmann *et al.* (1978) in rates of evolution during the earlier part of mammalian history (300–80 Ma ago) and the later part (80 Ma ago to the present).

One of the major problems involved in testing the linearity of the relationship between molecular distance and time of separation is that the times of divergence inferred from the fossil record are themselves subject to doubt. If, as has been suggested above, there is any systematic distortion in inferred times of divergence, this will inevitably lead to distortion of the inferred relationship between molecular distance and time. However, a neat technique for checking the linearity of molecular change with time without recourse to inferred date of divergence was introduced by Corruccini, Baba *et al.* (1980). They pointed out that, if all proteins change in a linear fashion with time, there must also be linear matching between proteins. Multivariate analysis of data for several proteins (e.g. albumin, α-haemoglobin, β-haemoglobin, myoglobin and transferrin) for a sample of primate species showed that there were considerable departures from linearity in the relationships between molecules for primates generally, although for simian primates there was a closer approximation to linearity. It is therefore obvious that for primates generally there can be no simple relationship between molecular distance and time, although it is possible that this is largely because of a major change during the origin of simian primates.

Poor timekeeping by the clock

Taking all of the evidence relating to the concept of the 'molecular clock', it is possible to draw certain conclusions. First and foremost, it is perfectly clear that the so-called 'clock' does not keep very good time. Only in the very broadest sense can it be claimed that rates of evolution of a particular protein, or of the sequence of DNA coding for that protein, remain approximately constant over long periods of geological time. There are a sufficient number of well-documented exceptions to show that, at least in some cases, there are marked departures from uniform rates. To some extent, such departures may be due to purely stochastic effects, but is seems likely that natural selection and other factors have also played a part. There is in any case some empirical evidence that the rate of change of individual proteins may decrease systematically over time, such that there is no linear relationship between numbers of fixed mutations and time. Further, given that there is no consistent colinear relationship in degrees of change across proteins among primates generally (Corruccini, Baba *et al.*, 1980), it is apparent that different proteins evolve not only at different rates but with different patterns over time. Accordingly, it is unrealistic to expect a simple linear relationship between degree of change in a protein and geological time.

There is also the tricky problem of mutations that are 'hidden' because of multiple, superimposed changes and reversed changes at particular sites. Such 'hidden' mutations are not included in simple molecular distance measures between proteins. If no correction is made for their occurrence, raw data on molecular distance (e.g. measures of degrees of immunological cross-reaction) must necessarily exhibit a curvilinear relationship with time even if the actual numbers of fixed mutations are linearly related to time elapsed. Thus, any linear 'clock' that is based on raw molecular distances with no correction for superimposed mutations has a serious theoretical flaw. There is also a problem in that many authors have linked the concept of the 'molecular clock' directly to neutral mutation theory. In principle, this theory predicts that rates of change in proteins should be approximately constant (after allowing for stochastic variation), but arguably only in relation to numbers of generations and not to time itself. Unless a framework can be developed that predicts a linear relationship between

corrected numbers of fixed mutations and geological time, rather than number of generations, the concept of the molecular clock as used by many authors has no theoretical foundation.

IMPLICATIONS FOR PRIMATE CLASSIFICATION

One major conclusion reached in Chapter 3 is that, in order to be of general utility, any classificatory scheme for the order Primates should be compatible with commonly inferred patterns of phylogenetic relationships while offering a maximum degree of stability over time. It would be naive to expect this conclusion to meet with universal acceptance. Many authors have expended a great deal of time and effort – not to say passion – developing new 'cladistic' classifications, in the attempt to provide direct structural counterparts of particular hypotheses regarding the evolutionary relationships among primates. As a general rule, such phylogenetic hypotheses have been based on a remarkably small selection of characters and are therefore only weakly supported. Partly for this reason, a colourful variety of alternative classifications has been proposed, particularly in recent years. Because of this, it is an unfortunate fact that such novel classifications have contributed little (if anything) either to our understanding of primate phylogenetic relationships or to practical applications of primate taxonomy. Indeed, especially for uninitiated readers struggling to find their way in an increasingly complex field, it is probably true to say that the plethora of competing classifications of primates now available has been more of a hindrance than an aid to communication. The way out of this dilemma is to retain as far as possible a well-established classical classification and to make only those changes that are absolutely necessary. Accordingly, the outline classification presented in Table 3.1 – extracted from the general mammalian classification originally proposed by Simpson (1945) and incorporating a few modifications, such as those suggested by Simons

(1972) – will be used as a basis for the following discussion. Possible implications for key features of the pattern of phylogenetic relationships reviewed earlier in this chapter will be examined in turn.

Tree-shrews

There is now abundant evidence that tree-shrews are simply unrelated to primates. Although in some quarters the view persists that tree-shrews may be the closest relatives of primates among placental mammals, there is no convincing evidence that this is the case. It therefore seems appropriate to exclude tree-shrews from the order Primates (e.g. see Simons, 1972; Szalay and Delson, 1979; Napier and Napier, 1985) and to allocate them to their own order Scandentia (see Butler, 1972, 1980; Corbet and Hill, 1980), to reflect their wide degree of separation from primates. Even if reliable evidence of an early specific relationship between tree-shrews and primates were to emerge, there would still be no convincing case for including tree-shrews in the order Primates. The order Primates can be most satisfactorily defined to the exclusion of tree-shrews, regardless of the relationships of the latter, and the separation between tree-shrews and primates must in any case be very ancient.

A similar conclusion applies to the fruit-bats (Megachiroptera). It is theoretically possible (although on present evidence unlikely) that the two groups of bats (Microchiroptera and Megachiroptera) originated quite separately during the early adaptive radiation of placental mammals. If this proves to be the case and if it proves that the Megachiroptera (fruit-bats) are specifically related to primates, there is still no pressing reason to include them in the order Primates. A more appropriate solution, which would require a minimal degree of change to existing classifications such as that of Simpson (1945), would be to treat the Microchiroptera and the Megachiroptera as separate orders, rather than as suborders of the single order Chiroptera as at present.

'Archaic primates'

The question of the 'archaic primates' (infra-order Plesiadapiformes) is more problematic. Although much evidence has been assembled in previous chapters to challenge the interpretation that plesiadapiforms have a specific connection with primates, there are at least some characters of the cheek teeth and of the ear region that possibly indicate a very early link. The continued retention of Plesiadapiformes in the order Primates might therefore be justifiable, at least as a provisional measure. Some authors have recognised the distinctive nature of the Plesiadapiformes by referring to other primates as 'primates of modern aspect' (Simons, 1972) or as 'euprimates' (Szalay and Delson, 1979). It seems highly likely that the 'primates of modern aspect' represent a monophyletic group, regardless of the relationships of the Plesiadapiformes (Fig. 12.1), so some clear taxonomic separation of the 'archaic primates' would seem to be advisable. Such an approach has the merit that it emphasizes both the distinctiveness of the 'archaic primates' and the tenuous nature of their relationship to primates of modern aspect. However, the distinction between 'archaic primates' and 'primates of modern aspect' has seldom been indicated in formal classificatory terms. Further, inclusion of the Plesiadapiformes in the order Primates almost automatically leads investigators (especially palaeontologists) to regard 'archaic primates' as a suitable starting point for tracing the ancestry of primates of modern aspect. This approach is firmly rejected here. Very little of any lasting benefit has been gained by including these so-called 'archaic primates' in discussions of the origins of primates of modern aspect. To the contrary, such a procedure has led to certain questionable notions, such as the widespread but probably erroneous belief that the order Primates arose in North America at the close of the Cretaceous. Given the large degree of uncertainty about the phylogenetic affinities of the Plesiadapiformes, a reasonable solution is to retain them as a separate infraorder within the order Primates, but bearing the warning label *incertae sedis*.

Primates of modern aspect

There is now very little doubt about the integrity of the primates of modern aspect as a monophyletic group. They can be traced to a common ancestral stock that already possessed a number of distinctive features acquired during a relatively long period of separation from the ancestral stock of the placental mammals. Unfortunately, the earliest origins of the primates of modern aspect are not documented by the existing fossil record, but this gap in our knowledge at least brings with it the advantage that all true primates, from the early Eocene onwards, can be defined as a cohesive group. Of course, if fossil remains of direct relatives of currently known primates are eventually recovered from earlier deposits (Palaeocene and late Cretaceous), the relatively clear separation between primates of modern aspect and other placental mammals will disappear as the price of increased knowledge. The characters that distinguish Eocene primates of modern aspect and their more recent relatives from other placental mammals must, of course, have arisen gradually over millions of years. As is so often the case in evolutionary history, recognition of the primates of modern aspect as a clear-cut group is a (hopefully temporary) convenient illusion attributable to a regrettable gap in the fossil record. Be that as it may, present problems relating to the classification of primates of modern aspect are essentially confined to internal subdivisions within the group.

There is now overwhelming evidence that living primates of modern aspect fall into two distinct groups: strepsirhines (lemurs and lorises) and haplorhines (tarsiers and simians). It might, therefore, seem appropriate to reflect this well-established division in the formal classification of the order Primates, as (for example) in the classification advocated by Szalay and Delson (1979). There are, however, some thorny problems associated with the earliest known

fossil primates of modern aspect – the primarily Eocene 'lemuroids' (Adapidae) and 'tarsioids' (Omomyidae).

The affinities of the Adapidae are still uncertain. Although they have numerous features indicating, beyond reasonable doubt, that they are primates of modern aspect, these 'lemuroids' cannot be linked with certainty to any other group of primates, living or fossil. There has been a general tendency to assume that adapids are linked to modern strepsirhine primates, and this assumption has been an implicit or explicit feature of several classificatory schemes proposed for the order Primates. For example, in the classification advocated by Simons (1972), the Adapidae are included in the infraorder Lemuriformes (see Table 3.1). In fact, however, virtually all of the characters shared by adapids and by modern strepsirhines would seem to be primitive retentions from the common ancestral stock of primates of modern aspect. Beard, Dagosto *et al.* (1988) have suggested that adapids and strepsirhines share a few derived features in the ankle region, but this would seem to be the only evidence of a possible direct phylogenetic link. Gingerich, by contrast, has repeatedly suggested (e.g. Gingerich, 1975a, 1976b, 1981b, 1986a) that adapids are related not only to strepsirhine primates but also to simian primates (monkeys, apes and humans). As has been explained above, this suggestion was originally linked to the proposal that tarsiers and omomyids are specifically related to the plesiadapiforms. It also led to yet another classification of the order Primates, with a primary subdivision into the suborders Plesitarsiiformes and Simiolemuriformes. Although Gingerich subsequently retracted his suggestion of a specific phylogenetic link between tarsiers, omomyids and plesiadapiforms, which is, of course, quite incompatible with the concept that the primates of modern aspect constitute a monophyletic assemblage, he has continued to favour the interpretation that adapids are linked to simian primates as well as to strepsirhines. This interpretation is rejected here, on the grounds that any similarities between adapids and simians are attributable to a combination of convergent adaptation and retention of primitive features from the ancestral stock of primates of modern aspect. At present, the safest conclusion is that adapids are primates of modern aspect, but that their affinities to other primates are uncertain.

A similar problem exists with respect to relationships between the modern haplorhine primates (tarsiers and simians) and the early Tertiary 'tarsioids' (Omomyidae). For some time, it has been widely accepted that the omomyids are directly related to modern tarsiers. Indeed, this interpretation is reflected by the general practice of including the Omomyidae in the infraorder Tarsiiformes (Simpson, 1945; Simons, 1972; Szalay and Delson, 1979; see also Table 3.1). If omomyids are the closest relatives of modern tarsiers, it follows that they must be part of any monophyletic group of primates including both tarsiers and simians. But, as has been noted above, increasing doubts are now being expressed about the phylogenetic affinity between omomyids and tarsiers. At the very least, it seems that the resemblance between the fossil 'tarsioids' and modern tarsiers is relatively limited. In particular, there was a median gap between the upper incisors in at least some omomyids, which suggests that a rhinarium was still present. Accordingly, the term 'haplorhine' is not applicable in the descriptive sense to omomyids. It is possible that the omomyids are nevertheless directly related to modern tarsiers and that the loss of the rhinarium occurred independently in the evolution of the tarsiers and of the common ancestral stock of the simians (monkeys, apes and humans). It seems more likely, however, that omomyids branched off at some stage before the divergence between tarsiers and simians, which would therefore still constitute a monophyletic group characterized, among other things, by the loss of the rhinarium. The problem is that there is now no clear consensus about the relationships of the Omomyidae. They may be directly related to modern tarsiers (Simons, 1972; Szalay, 1975a, 1976; Szalay and Delson, 1979); they may be early offshoots from the lineage leading to

the common ancestor of tarsiers and simians (Cartmill, 1980; Aiello, 1986); or they may even be related to strepsirhine primates (Hürzeler, 1948; Schmid, 1981, 1982, 1983). Although it would seem, on balance, that the second possibility is the strongest, it would be unwise to petrify any particular hypothesis in the form of a rigid classificatory decision at this stage.

The essential difficulty with any division of primates into 'strepsirhines' and 'haplorhines' is that it is really applicable only to living forms (see also: Simons, 1974c; Cartmill and Kay, 1978). The great majority of the inferred shared derived characters involved in this division, notably those linking tarsiers to simians, are non-fossilizable. A major problem therefore arises in attempting to apply the distinction between strepsirhines and haplorhines to the fossil record, especially in the case of fragmentary fossils for which only dental evidence is available. Indeed, it is primarily for this reason that the affinities of the Adapidae and Omomyidae with living primates remain subject to so much uncertainty. There is therefore little point in establishing a classification that requires formal application of labels such as 'strepsirhine' and 'haplorhine' to fossil primates. Thus, although there is abundant evidence for living forms that lemurs are specifically related to lorises and that tarsiers are related to simians, the traditional classificatory division between Prosimii (prosimians) and Anthropoidea (simians) is to be preferred (see also Aiello, 1986). With this classification, all early fossil primates can be classified in the suborder Prosimii and it is not necessary to decide whether a specific relationship with modern strepsirhine or haplorhine primates exists. It might, in fact, be advisable to remove the Adapidae from the Lemuriformes, as this tends to imply that adapids are more closely related to lemurs than the latter are to the loris group. A similar treatment might be accorded to the Omomyidae, given the fact that it now seems likely that tarsiers are more closely related to simians than are omomyids. In practice, however, there is no pressing need to make any changes to the relevant parts of the outline classification given in Table 3.1.

The major limitation of a primary subdivision

of the order Primates into the two suborders Prosimii and Anthropoidea is that is does not indicate to the casual reader that lemurs are probably specifically related to lorises and that tarsiers are probably specifically related to simians. On the other hand, this subdivision is justifiable in terms of a 'grade' model, because tarsiers share numerous primitive features with lemurs and lorises, and simians (monkeys, apes and humans) are clearly definable as a distinct grade on the basis of a set of shared derived features not found in tarsiers. In addition, rather than continually modifying the classification to keep pace with changing interpretations of primate evolution, it is surely more economical to educate users of such a classification to recognize that it is not intended to reflect inferred phylogenetic relationships directly. The grade-based subdivision of the Primates into Prosimii and Anthropoidea, as set out in the classification of Simpson (1945), has stood the test of time precisely because it is compatible with several different phylogenetic trees for the primates. For the same reason, continued use of this classification will ensure relative stability of nomenclature in the future. There remains the apparent problem that it is often convenient to refer to 'strepsirhines' and 'haplorhines' in discussion of living primates, because this provides a useful means of referring to the collective properties of the two groups. However, it is perfectly feasible to refer to living lemurs and lorises as 'strepsirhines' and to living tarsiers and simians as 'haplorhines' without incorporating these terms into a formal classification. In any case, it is also convenient to be able to use the terms 'prosimian' and 'simian' in many discussions of primates, so no classification can be universally convenient.

The classification of the strepsirhine primates has been subject to particular upheaval recently because of the interpretation, widely held among primate morphologists, that the mouse and dwarf lemurs (Cheirogaleidae) are more closely related to the loris group than to other lemurs. In the classification used by Simons (1972), as in Simpson's original classification (1945), all of the Malagasy lemurs were placed in the infra-

order Lemuriformes, whereas the loris group were allocated to the separate infraorder Lorisiformes. Subsequently, several new classifications were proposed in which the Cheirogaleidae were directly associated with the loris group, either in the infraorder Lorisiformes or in the superfamily Lorisoidea (e.g. Schwartz, Tattersall and Eldredge, 1978; Szalay and Delson 1979; Tattersall, 1982). As noted earlier in this chapter, a review of the evidence underlying this classificatory innovation indicates that there were inadequate grounds for proposing any change. The important point, however, is that in a grade-based classification of the primates the innovation was in any case unnecessary (Fig. 12.6). If the infraorder Lemuriformes is treated as a primitive grade category, with the infraorder Lorisiformes separated on the grounds of the acquisition of a set of specific characters, the Cheirogaleidae can be retained within the Lemuriformes even if it is believed that they are more closely related to the loris group. In fact, the case for linking the Cheirogaleidae directly to the loris group was never very strong and it has received no support whatsoever from comparative studies of chromosomes or molecules. Under the circumstances, retention of the original classification would clearly have been the wisest course and continued retention of a division into Lemuriformes and Lorisiformes (Table 3.1) will permit stability in the future even if details of the phylogenetic relationships among lemurs and lorises are found to differ from those accepted here.

The phylogenetic unity of the simian primates (suborder Anthropoidea) is subject to little doubt and a convincing case can be made for a grade distinction from prosimians. Similarly, the division of simians into two basic groups – New World monkeys versus Old World monkeys, apes and humans – has received general acceptance. In Simpson's classification (1945), however, there was a simple tripartite division of the Anthropoidea into the suborders Ceboidea (New World monkeys), Cercopithecoidea (Old World monkeys) and Hominoidea (apes and humans). Such a division fails to reflect the almost universally accepted interpretation that the Old

World simians, both living and fossil, form a cohesive phylogenetic unit. For this reason, it is to be welcomed that Simons (1972) introduced the two terms Platyrrhini and Catarrhini to distinguish the New World simians from the Old World simians as separate infraorders. Grouping of the Cercopithecoidea and the Hominoidea in the infraorder Catarrhini neatly expresses the unity of the Old World simians. (The Oligocene simians from the Egyptian Fayum site, especially the parapithecines, lack certain features present in all later Old World simians, so they are included in this group largely on geographical grounds).

The internal classification of the New World monkeys (infraorder Platyrrhini, superfamily Ceboidea) has, however, presented some problems. Traditionally, both in general discussions and in formal classifications, a distinction has been drawn between the small-bodied, clawed marmosets and tamarins on the one hand and the larger-bodied 'true' monkeys on the other. Allocation of the 'true' monkeys to the family Cebidae and of the marmosets and tamarins to the family Callitrichidae, as set out in Simpson's classification (1945), was continued by Simons (1972). The relationships between the various groups of New World monkeys have been relatively little studied, so questions remain about the degree to which a classification into the two families Callitrichidae and Cebidae reflects likely phylogenetic relationships. In one of the few direct discussions of this topic, Rosenberger (1977) presented a possible phylogenetic tree for the New World monkeys and (largely on dental grounds) raised the possibility that the subfamily Cebinae, containing the squirrel monkeys (*Saimiri*) and the capuchin monkeys (*Cebus*), may actually be more closely related to marmosets and tamarins than to the other monkeys. This hypothesis was subsequently accepted by Szalay and Delson (1979), who adopted a new classification on this basis. They divided the New World monkeys into two families: (1) the Cebidae, containing the subfamilies Cebinae and Callitrichinae (marmosets and tamarins); and (2) the Atelidae, containing the remaining subfamilies of 'true'

New World monkeys. Rosenberger subsequently (1981b) claimed that the Cebinae are, indeed, more closely related to marmosets and tamarins. He also proposed use of the new classificatory scheme, while noting: 'One of the inescapable problems with this revision is that it requires that some familiar terms . . . be used in different ways.' This is an understatement. Whereas the family name 'Cebidae' in previous classifications (e.g. Simpson, 1945; Simons, 1972) meant 'all New World monkeys except marmosets and tamarins', in the alternative classification (Szalay and Delson, 1979; Rosenberger, 1981b) the name 'Cebidae' means almost the exact opposite (marmosets, tamarins and Cebinae). Thus, adoption of the alternative classification requires a major shift in the meaning of the term 'cebid'. Even in the unlikely event that all other primatologists were to adopt the new classification at once, such a shift in meaning must inevitably lead to great confusion because of the vast pre-existing literature in which the term 'cebid' has been used to mean 'New World monkeys other than marmosets and tamarins'.

In fact, the hypothesis that there is a specific relationship between *Cebus*, *Saimiri*, marmosets and tamarins is based exclusively on a particular interpretation of morphological (mainly dental) evidence. As was explicitly recognized by Szalay and Delson (1979), it receives no support from immunological comparisons of proteins (see Figs 11.21, 11.24). In fact, it has since emerged that it is also unsupported by chromosomal studies (Fig. 11.18) or by analysis of a tandem alignment of protein sequences (Fig. 11.31). In other words, the phylogenetic hypothesis on which the alternative classificatory scheme was based is most probably in error. Once again, however, it was not really necessary to propose a new classification even if the hypothesis had been correct. It was argued in Chapters 2, 6, 7 and 9 that the marmosets and tamarins represent specialized dwarf forms (see also Szalay and Delson, 1979; Rosenberger, 1981b). They can therefore be defined as a separate grade (= family Callitrichidae) on the basis of their shared specializations, leaving the family Cebidae as a basic grade category including all other New World monkeys. The relationships between the various subfamilies of the 'true' monkeys of the New World remain to be established satisfactorily. Their classification within the single family Cebidae – as originally defined by Simpson (1945) – accordingly leaves open options for a variety of different phylogenetic schemes, any one of which might turn out to be correct in the event. The possibility suggested by Rosenberger (1977, 1981b) and by Szalay and Delson (1979) is just one of many that are all compatible with a grade classification of the New World monkeys into the two traditional families Cebidae and Callitrichidae.

A special problem is, of course, posed by Goeldi's monkey (*Callimico goeldii*), which is intermediate between typical cebids and typical callitrichids. *Callimico* could, in principle, be placed either in the Cebidae (because it lacks some of the defining features of marmosets and tamarins, such as reduction in number of molar teeth and twinning) or in the Callitrichidae (because it has certain defining features, such as small body size and claws). A third possibility is to place Goeldi's monkey in its own family, Callimiconidae (e.g. see Hershkovitz, 1977). In fact, as *Callimico* was allocated to the Cebidae both by Simpson (1945) and by Simons (1972), retention of this solution is probably preferable in the interests of stability. Nevertheless, there is increasing evidence (e.g. from chromosomes) that there is a definite phylogenetic relationship between Goeldi's monkey and marmosets and tamarins. As there would be no significant practical problems in the allocation of *Callimico* to the Callitrichidae, this minor modification of the classificatory scheme presented in Fig. 3.1 might in due course prove to be justifiable.

Classification of the catarrhine primates (Old World monkeys, apes and humans) is relatively straightforward, although there are some minor problems. The basic division into the two subfamilies Cercopithecoidea (Old World monkeys) and Hominoidea (apes and humans) is widely accepted. It should be noted, however, that – contrary to the expectation aroused by the *Scala naturae* – it is the monkeys rather than the

apes that are clearly defined by a shared derived dental feature – namely, the possession of bilophodont molars. For this reason, the allocation of fossil forms to the Cercopithecoidea is relatively unproblematic. For living forms, at least, there is also a fairly clear division within the Cercopithecoidea into the subfamilies Cercopithecinae (guenon group) and Colobinae (leaf-monkeys). Problems can, however, arise with the inclusion of fragmentary fossil forms, so a separate subfamily might be justifiable for early cercopithecoids such as *Victoriapithecus*.

The apes are not clearly defined by shared derived dental features, but they do share some postcranial features, including loss of the tail. The allocation of fragmentary fossil forms is therefore difficult. There are two solutions that might be adopted here. One solution would be to treat the superfamily Hominoidea as a basic grade grouping, containing all those Old World simian forms that are not clearly recognizable as monkeys. Although this goes against the grain of the pattern of thinking encouraged by the concept of the *Scala naturae*, there is no objective reason why the term 'hominoid' should not indicate the basic grade among catarrhines. Early fossil forms could be allocated to the Hominoidea without any need to decide whether they are 'really monkeys' or 'really apes'. In line with this suggestion, the Oligocene subfamily Parapithecinae would have to be removed from the superfamily Cercopithecoidea and included in the Hominoidea, where it would join the other group of Oligocene simian primates from the Fayum, the Pliopithecinae. This would obviate the need to make the largely artificial distinction between 'monkeys' and 'apes' in the Oligocene and would leave the way open for inclusion of early fragmentary fossil forms that can only be identified as Old World simians (catarrhines). The alternative solution would be to reserve the superfamily Hominoidea for species with the shared derived locomotor features of modern apes and humans and to create a new superfamily for fossil catarrhines that cannot be reliably identified either as apes or as monkeys. Given the fact that the postcranial skeleton is unknown for most fossil catarrhines, however, the superfamily

Hominoidea would then be largely restricted to *Oreopithecus*, modern apes, fossil hominids and modern humans. The first solution is therefore preferred here.

As was mentioned in Chapter 3, a special problem has arisen with respect to classification of the great apes and humans (Martin, 1981a). It is now widely accepted that the African great apes are more closely related to humans than is the orang-utan. With a grade-based classification, this presents no problem, because hominids can be defined on the basis of a set of shared derived features distinguishing them from all great apes. The traditional classification of orang-utans, gorillas and chimpanzees in the family Pongidae and of humans and their fossil relatives in the family Hominidae is therefore perfectly acceptable (Simpson, 1945; Simons, 1972).

Unfortunately, those authors who have felt the need to translate phylogenetic trees directly into classifications have advocated a variety of new schemes that have led to widespread confusion. In many cases, the African great apes have been included in the family Hominidae, thus utterly modifying the meaning of the term 'hominid'. Future students are now faced with the awful prospect of divining what any particular author means by 'hominid'. Worse, they will undoubtedly have to cope with such statements as: 'Bipedal locomotion is a fundamental feature of hominids, excluding chimpanzees and gorillas'. At the end of the day, we must ask ourselves whether this continual upheaval in primate classification is really necessary, when we can continue to use a traditional classification that has proved its worth over a period of more than 40 years. It is perhaps appropriate, then, to end this chapter with a plea to fellow primatologists: Please restrict changes to primate classification to the absolute minimum so that we can concentrate our efforts without distraction on the central, challenging task of reconstructing the course of primate evolution.

References

Abbott, D.H. (1984) Behavioural and physiological suppression of fertility in subordinate marmoset monkeys. *Am. J. Primatol.* **6**, 169–86.

Abel, O. (1929) *Paläobiologie und Stammesgeschichte*. Gustav Fischer Verlag: Jena.

Abel, O. (1935) *Vorzeitliche Lebensspuren*. Gustav Fischer Verlag: Jena.

Adams, C.G. (1981) An outline of Tertiary palaeogeography. In *The Evolving Earth* (ed. Cocks, L.R.M.). Cambridge University Press: Cambridge, pp. 221–35.

Aiello, L.C. (1981a) The allometry of primate body proportions. *Symp. zool. Soc. Lond.* **48**, 331–58.

Aiello, L.C. (1981b) Locomotion in the Miocene Hominoidea. *Symp. Soc. Stud. hum. Biol.* **21**, 63–97.

Aiello, L.C. (1986) The relationship of the Tarsiiformes: a review of the case for the Haplorhini. In *Major Topics in Primate and Human Evolution* (eds Wood, B.A., Martin, L.B. and Andrews, P.J.). Cambridge University Press: Cambridge, pp. 47–65.

Aiello, L.C. and Day, M.H. (1982) The evolution of locomotion in early Hominidae. In *Progress in Anatomy*, vol. 2 (eds Harrison, R.J. and Navaratnam, V.). Cambridge University Press: Cambridge, pp. 81–97.

Albrecht, H. and Dunnett, S.C. (1971) *Chimpanzees in Western Africa*. Piper Verlag: Munich.

Alexander, R.McN. and Maloiy, G.M.O. (1984) Stride lengths and stride frequencies of primates. *J. Zool., Lond.* **202**, 577–82.

Alexander, R.McN., Jayes, A.S., Maloiy, G.M.O. and Wahuta, E.M. (1979). Allometry of limb bones of mammals from shrews (*Sorex*) to elephant (*Loxodonta*). *J. Zool., Lond.* **183**, 291–300.

Alfieri, R., Pariente, G.F. and Solé, P. (1976) Dynamic electroretinography in monochromatic lights and fluorescence electroretinography in lemurs. *Doc. ophthal. Proc. Ser.* **10**, 169–78.

Allin, E.F. (1975) Evolution of the mammalian middle ear. *J. Morph.* **147**, 403–38.

Allman, J.M. (1977) Evolution of the visual system in the early primates. *Progress in Psychobiology, Physiology and Psychology*, vol. 7 (eds Sprague, J.M. and Epstein, A.N.). Academic Press: New York, pp. 1–53.

Allman, J.M. (1982) Reconstructing the evolution of the brain in primates through the use of comparative neurophysiological and neuroanatomical data. In *Primate Brain Evolution* (eds Armstrong, E. and Falk, D). Plenum Press: New York, pp. 13–28.

Altner, G. (1971) Histologische und vergleichend-anatomische Untersuchungen zur Ontogenie und Phylogenie des Handskeletts von *Tupaia glis* (Diard 1820) und *Microcebus murinus* (J.F. Miller 1777). *Folia primatol.* **14** (suppl.), 1–106.

Ameghino, F. (1891) Nuevos restos de mamíferos

fósiles descubietos por Cárlos Ameghino en el Eoceno inferior de la Patagonia austral. *Rev. Argent. Hist. Nat.* **1**, 289–328.

Amoroso, E.C. (1952) Placentation. In *Marshall's Physiology of Reproduction*, vol. 2 (ed. Parkes, A.S.). Longmans, Green: London, pp. 127–311.

Anderson, J.F., Rahn, H. and Prange, H.D. (1979) Scaling of supportive tissue mass. *Q. Rev. Biol.* **54**, 139–48.

Andrews, C.W. (1916) Note on a new baboon (*Simopithecus oswaldi* gen. et sp. n.) from the (?) Pliocene of British East Africa. *Ann. Mag. nat. Hist.* **18**, 410–19.

Andrews, P.J. (1973) Miocene primates (Pongidae, Hylobatidae) of East Africa. PhD thesis, University of Cambridge.

Andrews, P.J. (1974) New species of *Dryopithecus* from Kenya. *Nature, Lond.* **249**, 188–90.

Andrews, P.J. (1978a) A revision of the Miocene Hominoidea of East Africa. *Bull. Br. Mus. nat. Hist., Geol.* **30**, 85–224.

Andrews, P.J. (1978b) Taxonomy and relationships of fossil apes. In *Recent Advances in Primatology*, vol. 3: *Evolution* (eds Chivers, D.J. and Joysey, K.A.). Academic Press: London, pp. 43–56.

Andrews, P.J. (1981) Species diversity and diet in monkeys and apes during the Miocene. *Symp. Soc. Stud. hum. Biol.* **21**, 25–61.

Andrews, P.J. (1983) The natural history of *Sivapithecus*. In *New Interpretations of Ape and Human Ancestry* (eds Ciochon, R.L. and Corruccini, R.S.). Plenum Press: New York, pp. 441–63.

Andrews, P.J. (1985) Family group systematics and evolution among catarrhine primates. In *Ancestors: The Hard Evidence* (ed. Delson, E.). Alan R. Liss: New York, pp. 14–22.

Andrews, P.J. (1986) Molecular evidence for catarrhine evolution. In *Major Topics in Primate and Human Evolution* (eds Wood, B.A., Martin, L.B. and Andrews, P.J.). Cambridge University Press: Cambridge, pp. 107–29.

Andrews, P.J. and Aiello, L.C. (1984) An evolutionary model for feeding and positional behaviour. In *Food Acquisition and Processing in Primates* (eds Chivers, D.J., Wood, B.A. and Bilsborough, A.). Plenum Press: New York, pp. 429–66.

Andrews, P.J. and Cronin, J.E. (1982) The relationships of *Sivapithecus* and *Ramapithecus* and the evolution of the orang-utan. *Nature, Lond.* **297**, 541–6.

Andrews, P.J. and Nesbit Evans, E.M. (1983) Small mammal bone accumulations produced by mammalian carnivores. *Paleobiology* **9**, 289–307.

Andrews, P.J. and Simons, E.L. (1977) A new African Miocene gibbon-like genus, *Dendropithecus* (Hominoidea, Primates) with distinctive postcranial adaptations: its significance to origin of Hylobatidae. *Folia primatol.* **28**, 161–9.

Andrews, P.J. and Tekkaya, I. (1980) A revision of the Turkish Miocene hominoid *Sivapithecus meteai*. *Palaeontology* **23**, 85–95.

Andrews, P.J. and Tobien, H. (1977) New Miocene locality in Turkey with evidence on the origin of *Ramapithecus* and *Sivapithecus*. *Nature, Lond.* **268**, 699–701.

Ankel, F. (1965) Der Canalis sacralis als Indikator für die Länge der Caudal-Region der Primaten. *Folia primatol.* **3**, 263–76.

Antinucci, F. and Visalberghi, E. (1986) Tool use in *Cebus apella*: a case study. *Int. J. Primatol.* **7**, 351–63.

Archer, M. (1978) The nature of the molar–premolar boundary in marsupials and a reinterpretation of the homology of marsupial cheek teeth. *Queensl. Mus. Mem.* **18**, 157–64.

Archer, M., Flannery, T.F., Ritchie, A. and Molnar, R.E. (1985) First Mesozoic mammal from Australia – an early Cretaceous monotreme. *Nature, Lond.* **318**, 363–6.

Ardran, G.M. and Kemp, F.H.A. (1960) Biting and mastication: a cineradiographic study. *Dent. Practnr, Bristol* **11**, 23–34.

Ardran, G.M., Kemp, F.H.A. and Ride, W.D.L. (1958) A radiographic analysis of mastication and swallowing in the domestic rabbit *Oryctolagus cuniculus*. *Proc. zool. Soc. Lond.* **130**, 257–74.

Armstrong, E. (1982a) Mosaic evolution in the primate brain: differences and similarities in the hominoid thalamus. In *Primate Brain Evolution* (eds Armstrong, E. and Falk, D.). Plenum Press: New York, pp. 131–61.

Armstrong, E. (1982b) A look at relative brain size in mammals. *Neurosci. Lett.* **34**, 101–4.

Arnason, U. (1972) The role of chromosomal rearrangement in mammalian speciation with special reference to Cetacea and Pinnipedia. *Hereditas* **70**, 113–18.

Arnason, U. (1974) Comparative chromosome studies in Pinnipedia. *Hereditas* **76**, 179–226.

Arnheim, N. (1983) Concerted evolution of multigene families. In *Evolution of Genes and Proteins* (eds Nei, M. and Koehn, R.K.). Sinauer Associates: Sunderland, Mass., pp. 38–61.

Arrhigi, F.E. and Hsu, T.C. (1971) Localization of heterochromatin in human chromosomes. *Cytogenetics* **10**, 81–6.

Asdell, S.A. (1964) *Patterns of Mammalian Reproduction*. Cornell University Press: New York.

Ashley-Montagu, M.F. (1933) The anthropological significance of the pterion in primates. *Am. J. phys. Anthrop.* **18**, 159–336.

Ashton, E.H. and Oxnard, C.E. (1964) Locomotor patterns in primates. *Proc. zool. Soc. Lond.* **142**, 1–28.

Atassi, M.Z. (1975) Antigenic structure of myoglobin: the complete immunochemical anatomy of a protein and conclusions relating to antigenic structures of proteins. *Immunochemistry* **12**, 423–38.

Atkin, N.B., Mattison, G., Beçak, W. and Ohno, S. (1965) The comparative DNA content of 19 species of placental mammals, reptiles and birds. *Chromosoma (Berl.)* **17**, 1–10.

Avis, V. (1961) The significance of the angle of the mandible: an experimental and comparative study. *Am. J. phys. Anthrop.* **19**, 55–61.

Avis, V. (1962) Brachiation: the crucial issue for man's ancestry. *Southw. J. Anthrop.* **18**, 119–48.

Ayala, F.J., Tracy, M.L., Barr, L.F., McDonald, J.F. and Pérez-Salas, S. (1974) Genetic variation in natural populations of five *Drosophila* species and the hypothesis of selective neutrality of protein polymorphisms. *Genetics* **77**, 343–84.

Baba, M.L., Darga, L.L. and Goodman, M. (1979) Immunodiffusion systematics of the Primates. Part V: The Platyrrhini. *Folia primatol.* **32**, 207–38.

Baba, M.L., Goodman, M., Dene, H. and Moore, G.W. (1975) Origins of the Ceboidea viewed from an immunological perspective. *J. hum. Evol.* **4**, 89–102.

Bachmann, K. (1972) Genome size in mammals. *Chromosoma (Berl.)* **37**, 85–93.

Baldwin, J.D. and Baldwin, J.I. (1981) The squirrel monkeys, genus *Saimiri*. In *Ecology and Behavior of Neotropical Primates* (eds Coimbra-Filho, A.F. and Mittermeier, R.A.). Academia Brasileira de Ciências: Rio de Janeiro, pp. 277–330.

Barlow, D.J., Edwards, M.S. and Thornton, J.M. (1986) Continuous and discontinuous protein antigenic determinants. *Nature, Lond.* **322**, 747–8.

Barnabas, J., Goodman, M. and Moore, G.W. (1971) Evolution of haemoglobin in primates and other therian mammals. *Comp. Biochem. Physiol.* **39**, 455–82.

Barnabas, J., Goodman, M. and Moore, G.W. (1972) Descent of mammalian alphaglobin chain sequences investigated by the maximum parsimony method. *J. molec. Biol.* **69**, 249–78.

Barnett, C.H. (1970) Talocalcaneal movements in mammals. *J. Zool. Lond.,* **160**, 1–7.

Barnicot, N.A. and Wade, P.T. (1970) Protein structure and the systematics of Old World monkeys. In *Old World Monkeys* (eds Napier, J.R. and Napier, P.H.). Academic Press: New York, pp. 227–60.

Baron, G., Frahm, H.D., Bhatnager, K.P. and Stephan, H. (1983) Comparison of brain structure volumes in Insectivora and Primates. III. Main olfactory bulb. *J. Hirnforsch.* **24**, 551–68.

Barrie, P.A., Jeffreys, A.J. and Scott, A.F. (1981) Evolution of the β-globin gene cluster in man and the primates. *J. molec. Biol.* **149**, 319–36.

Barron, E.J., Harrison, C.G.A., Sloan, J.L. and Hay, W.W. (1981) Paleogeography, 180 million years ago to the present. *Eclogae geol. Helv.* **74**, 443–70.

Bauchop, T. (1978) Digestion of leaves in vertebrate arboreal folivores. In *The Ecology of Arboreal Folivores* (ed. Montgomery, G.G.). Smithsonian Institution Press: Washington, pp. 193–204.

Bauchop, T. and Martucci, R.W. (1968) Ruminant-like digestion of the langur monkey. *Science* **161**, 698–700.

Bauchot, R. (1978) Encephalization in vertebrates: a new mode of calculation for allometry coefficients and isoponderal indices. *Brain Behav. Evol.* **15**, 1–18.

Bauchot, R. and Stephan, H. (1966) Données nouvelles sur l'encéphalisation des insectivores et des prosimiens. *Mammalia* **30**, 160–96.

Bauchot, R. and Stephan, H. (1967) Encéphales et moulages endocraniens de quelques insectivores et primates actuels. *Coll. int. C.N.R.S.* **163**, 575–86.

Bauchot, R. and Stephan, H. (1969) Encéphalisation et niveau évolutif chez les simiens. *Mammalia* **33**, 225–75.

Bauer, K. (1974) Comparative analysis of protein determinants in primatological research. In *Prosimian Biology* (eds Martin, R.D., Doyle, G.A. and Walker, A.C.). Duckworth: London, pp. 915–36.

Beadle, G.W. and Tatum, E.L. (1941) Genetic control of biochemical reactions in *Neurospora*. *Proc. natn. Acad. Sci., U.S.A.* **27**, 499–506.

Beard, J.M. and Goodman, M. (1976) The haemoglobins of *Tarsius bancanus*. In *Molecular Anthropology* (eds Goodman, M. and Tashian, R.E.). Plenum Press: New York, pp. 239–55.

Beard, J.M., Barnicot, N.A. and Hewett-Emmett, D. (1976) α and β chains of the major haemoglobin and a note on the minor component of *Tarsius*. *Nature, Lond.* **259**, 338–40.

Beard, K.C. (1987) Functional anatomy of the wrist in *Smilodectes gracilis*, with comments on adapid hand evolution. *Am. J. phys. Anthrop.* **72**, 177.

Beard, K.C., Dagosto, M., Gebo, D.L. and Godinot, M. (1988) Interrelationships among primate higher taxa. *Nature, Lond.* **331**, 712–14.

Bearder, S.K. and Doyle, G.A. (1974) Ecology of bushbabies, *Galago senegalensis* and *Galago crassicaudatus*, with some notes on their behaviour in the field. In *Prosimian Biology* (eds Martin, R.D., Doyle, G.A. and Walker, A.C.). Duckworth: London, pp. 109–30.

Bearder, S.K. and Martin, R.D. (1980a) *Acacia* gum and its use by bushbabies, *Galago senegalensis* (Primates, Lorisidae). *Int. J. Primatol.* 1, 103–28.

Bearder, S.K. and Martin, R.D. (1980b) The social organization of a nocturnal primate revealed by radio tracking. In *A Handbook on Biotelemetry and Radio Tracking* (eds Amlaner, C.J. and MacDonald, D.W.). Pergamon Press: Oxford, pp. 633–48.

Bedford, J.M. (1978) Anatomical evidence for the epididymis as the prime mover in the evolution of the scrotum. *Am. J. Anat.* 152, 483–508.

Bedford, J.M. (1979) Evolution of the sperm maturation and sperm storage functions of the epididymis. In *The Spermatozoon* (eds Fawcett, D.W. and Bedford, J.M.). Urban and Schwarzenberg: Baltimore, pp. 7–21.

Beecher, R.M. (1977) Function and fusion at the mandibular symphysis. *Am. J. phys. Anthrop.* 47, 325–36.

Beecher, R.M. (1979a) Functional significance of the mandibular symphysis. *J. Morph.* 159, 117–30.

Beecher, R.M. (1979b) Evolution of the mandibular symphysis in the Notharctinae: implications for anthropoid origins. *Am. J. phys. Anthrop.* 50, 418.

Beecher, R.M. (1983) Evolution of the mandibular symphysis in Notharctinae (Adapidae, Primates). *Int. J. Primatol.* 4, 99–112.

Behrensmeyer, A.K. and Hill, A.P. (eds) (1980) *Fossils in the Making.* University of Chicago Press: Chicago.

Ben Shaul, D.M. (1962) The composition of the milk of wild animals. *Int. Zoo Yb.* 4, 333–42.

Bender, M.A. and Chu, E.H.Y. (1963) The chromosomes of primates. In *Evolutionary and Genetic Biology of Primates* (ed. Buettner-Janusch, J.). Academic Press: New York, pp. 261–310.

Bender, M.A. and Metler, L.E. (1958) Chromosome studies in primates. *Science* 128, 186–90.

Bengtsson, B.O. (1975) Mammalian chromosomes similar in length are also similar in shape. *Hereditas* 79, 287–92.

Benirschke, K. and Layton, W. (1969) An early twin blastocyst of the golden lion marmoset, *Leontopithecus rosalia. Folia primatol.* 10, 131–8.

Benirschke, K. and Miller, C.J. (1982) Anatomical and functional differences in the placenta of primates. *Biol. Reprod.* 26, 29–53.

Bennejeant, C. (1953) Les dentures temporaires des primates. *Bull. Mém. Soc. Anthrop., Paris, sér. 10*, 4, 11–44.

Bensley, B.A. (1901a) On the question of an arboreal ancestry of the Marsupialia, and the interrelationships of the mammalian subclasses. *Am. Nat.* 35, 117–38.

Bensley, B.A. (1901b) A theory of the origin and evolution of the Australian Marsupialia. *Am. Nat.* 35, 245–69.

Benton, M.J. (1985) First marsupial fossil from Asia. *Nature, Lond.* 318, 313.

Benveniste, R.E. and Todaro, G.J. (1976) Evolution of type C viral genes: evidence for an Asian origin of man. *Nature, Lond.* 261, 101–8.

Berge, C. and Jouffroy, F.K. (1986) Morphofunctional study of *Tarsius*'s foot as compared to the galagines: what does an 'elongate calcaneus' mean? In *Current Perspectives in Primate Biology* (eds Taub, D.M. and King, F.A.). Van Nostrand Reinhold: New York, pp. 146–56.

Bick, Y.A.E. and Jackson, W.D. (1967) DNA content of monotremes. *Nature, Lond.* 215, 192–3.

Bickham, J.W. (1981) Two-hundred-million-year-old-chromosomes: deceleration of the rate of karyotypic evolution in turtles. *Science* 212, 1291–3.

Bickham, J.W. and Baker, R.J. (1979). Canalization model of chromosomal evolution. *Bull. Carnegie Mus. nat. Hist.* 13, 70–84.

Biegert, J. (1961) Volarhaut der Hände und Füsse. In *Primatologia*, vol. 2 (eds Hofer, H., Schultz, A.H. and Starck, D.). Karger: Basel, pp. 1–326.

Bigelow, R.S. (1958) Classification and phylogeny. *Syst. Zool.* 7, 49–59.

Bishop, A. (1962) Control of the hand in lower primates. *Ann. N.Y. Acad. Sci.* 102, 316–37.

Bishop, A. (1964) Use of the hand in lower primates. In *Evolutionary and Genetic Biology of Primates*, vol. 2 (ed. Buettner-Janusch, J.). Academic Press: New York, pp. 133–225.

Black, D. (1927) On a lower molar hominid tooth from Chou-kou-tien deposit. *Palaeont. sin. D* 7, 1–28.

Blackett, P.M.S. (1961) Comparison of ancient climates with the ancient latitudes deduced from rock magnetic measurements. *Proc. R. Soc. Lond. A* 263, 1–30.

Blakeslee, B. and Jacobs, G.H. (1985) Color vision in the ring-tailed lemur (*Lemur catta*). *Brain Behav. Evol.* 26, 154–66.

Blumenbach, J.F. (1791) *Handbuch der Naturgeschichte* (4th edition). Johann Christian Dietrich: Göttingen.

Bluntschli, H. (1931) *Homunculus patagonicus* und die ihm zugereihten Fossilfunde aus den Santa-Cruz-

Schichten Patagoniens. Eine morphologische Revision an Hand der Originalstücke in der Sammlung Ameghino zu la Plata. *Morph. Jb.* **67**, 811–92.

Boaz, N.T. and Cronin, J.E. (1985) A new classification of the Catarrhini. *Am. J. phys. Anthrop.* **66**, 146.

Bock, W.J. (1973) Philosophical foundations of classical evolutionary classification. *Syst. Zool.* **22**, 375–92.

Bodmer, W.F. (1975) Analysis of linkage by somatic cell hybridization and its conservation by evolution. *Symp. Soc. Stud. hum. Biol.* **14**, 53–61,

Bodmer, W.F. (1981) Gene clusters, genome organization, and complex phenotypes. When the sequence is known, what will it mean? *Am. J. hum. Genet.* **33**, 664–82.

Bohlin, B. (1951) Some mammalian remains from Shih-ehr-ma-ch'eng, Hui-hui-p'u area, Western Kansu. *Sino-Swed. Exp. Publ., Stockh.* **10**, 1–47.

Böker, H. (1927) Die Entstehung der Wirbeltiertypen und der Ursprung der Extremitäten. *Z. Morph. Anthrop.* **26**, 1–58.

Böker, H. and Pfaff, R. (1931) Die biologische Anatomie der Fortbewegung auf dem Boden und ihre phylogenetische Abhängigkeit vom primären Baumklettern bei den Säugetieren. *Morph. Jb.* **68**, 496–540.

Bonner, T.I., Heinemann, R. and Todaro, G.J. (1980) Evolution of DNA sequences has been retarded in Malagasy primates. *Nature, Lond.* **286**, 420–3.

Boskoff, K.J. (1978) The estrous cycle of the brown lemur, *Lemur fulvus. J. Reprod. Fertil.* **54**, 313–18.

Bown, T.M. (1976) Affinities of *Teilhardina* (Primates, Omomyidae) with description of a new species from North America. *Folia primatol.* **25**, 62–72.

Bown, T.M. and Gingerich, P.D. (1973) The Paleocene primate *Plesiolestes* and the origin of Microsyopidae. *Folia primatol.* **19**, 1–8.

Bown, T.M. and Kraus, M.J. (1979) Origin of the tribosphenic molar and metatherian and eutherian dental formulae. In *Mesozoic Mammals* (eds Lillegraven, J.A., Kielan-Jaworowska, Z. and Clemens, W.A.). University of California Press: Berkeley, pp. 172–91.

Bown, T.M. and Rose, K.D. (1976) New early Tertiary primates and a reappraisal of some Plesiadapiformes. *Folia primatol.* **26**, 109–38.

Bown, T.M. and Simons, E.L. (1984a) First record of marsupials (Metatheria: Polyprotodonta) from the Oligocene of Africa. *Nature, Lond.* **308**, 447–9.

Bown, T.M. and Simons, E.L. (1984b) Reply to Jaeger and Martin: African marsupials – vicariance or dispersion? *Nature, Lond.* **312**, 379–80.

Boycott, B.B. and Dowling, J.E. (1969) Organization of the primate retina: light microscopy. *Phil. Trans. R. Soc. Lond. B* **255**, 109–84.

Boyd, J.D. (1932) The classification of the upper lip in mammals. *J. Anat., Lond.* **67**, 409–16.

Boyde, A. (1971) Comparative histology of mammalian teeth. In *Dental Morphology and Evolution* (ed. Dahlberg, A.A.). University of Chicago Press: Chicago, pp. 81–94.

Boyden, A.A. (1953) Fifty years of systematic serology. *Syst. Zool.* **2**, 19–30.

Boyden, A.A. (1958) Comparative serology: aims, methods, and results. In *Serological and Biochemical Comparisons* (ed. Cole, W.H.). Rutgers University Press: New Brunswick, pp. 1–24.

Boyden, A.A. (1959) Serology as an aid to systematics. *Proc. int. Congr. Zool. (Lond.)* **15**, 120–2.

Boyer, S.H., Crosby, E.F., Noyes, A.N., Fuller, G.F., Leslie, S.E., Donaldson, L.J., Vrablik, G.R., Schaefer, E.W. and Thurman, T.F. (1971) Primate haemoglobins: some sequences and some proposals concerning evolution and mutation. *Biochem. Genet.* **5**, 405–48.

Brand, H.M. (1980) Influence of season on birth distribution in marmosets and tamarins. *Lab. Anim.* **14**, 301–2.

Brand, H.M. (1981) Urinary oestrogen excretion in the female cotton-topped tamarin (*Saguinus oedipus oedipus*). *J. Reprod. Fert.* **62**, 467–73.

Brandt, A. (1867) Sur le rapport du poids du cerveau à celui du corps chez différents animaux. *Bull. Soc. Imp. nat., Moscow* **40**, 525–43.

Bridges, C.B. (1919) Duplication. *Anat. Rec.* **15**, 357–8.

Bridges, C.B. (1935) Salivary chromosome maps, with a key to the banding of the chromosomes of *Drosophila melanogaster. J. Hered.* **26**, 60–4.

Brink, A.S. (1957) Speculations on some advanced mammalian characters in the higher mammal-like reptiles. *Palaeont. afr.* **4**, 77–96.

Brink, A.S. (1958) Note on a new skeleton of *Thrinaxodon liorhinus. Palaeont. afr.* **6**, 15–22.

Britten, R.J. and Davidson, E.H. (1971) Repetitive and non-repetitive DNA sequences and a speculation on the origins of evolutionary novelty. *Q. Rev. Biol.* **46**, 111–33.

Britten, R.J. and Kohne, D.E. (1968) Repeated sequences in DNA. *Science* **161**, 529–40.

Broca, P. (1861) Sur le volume et la forme du cerveau, suivant les individus et suivant les races. *Bull Mém. Soc. Anthrop., Paris, sér. 1* **2**, 139–204 and 301–21.

Brodman, K. (1909) *Vergleichende Lokalisationslehre der Grosshirnrinde in ihren Prinzipien dargestellt auf Grund des Zellenbaus*. Barth: Leipzig.

Brody, S. (1945) *Bioenergetics and Growth*. Rheinhold: New York.

Brody, S., Procter, R.C. and Ashworth, U.S. (1934) Basal metabolism, endogenous nitrogen, creatinine and neutral sulphur excretions as functions of body weight. *Missouri Agr. Exp. Sta. Res. Bull.* **220**, 1–40.

Broom, R. (1898) A contribution to the comparative anatomy of the mammalian organ of Jacobson. *Trans. R. Soc. Edinb.* **39**, 231–55.

Broom, R. (1902) On the organ of Jacobson in the elephant shrew (*Macroscelides proboscideus*). *Proc. zool. Soc. Lond.* **1902**, 224–8.

Broom, R. (1913) On evidence of mammal-like dental succession in cynodont reptiles. *Bull. Am. Mus. nat. Hist.* **32**, 465–8.

Broom, R. (1915a) On the organ of Jacobson and its relations in the 'Insectivora'. Part I. *Tupaia* and *Gymnura. Proc. zool. Soc. Lond.* **1915**, 157–62.

Broom, R. (1915b) On the organ of Jacobson and its relations in the 'Insectivora'. Part II. *Talpa, Centetes* and *Chrysochloris. Proc. zool. Soc. Lond.* **1915**, 347–54.

Broom, R. (1937) On some new Pleistocene mammals from limestone caves of the Transvaal. *S. Afr. J. Sci.* **33**, 750–68.

Broom, R. (1938) The Pleistocene anthropoid apes of South Africa. *Nature, Lond.* **142**, 377–9.

Broom, R. and Robinson, J.T. (1949) A new type of fossil baboon, *Gorgopithecus major. Proc. zool. Soc. Lond.* **119**, 374–83.

Brown, F., Harris, J., Leakey, R.E.F. and Walker, A.C. (1985) Early *Homo erectus* skeleton from west Lake Turkana, Kenya. *Nature, Lond.* **316**, 788–92.

Brown, H., Sanger, F. and Kitai, R. (1955) Pig and sheep insulins. *Biochem. J.* **60**, 556–65.

Bruce, E.J. and Ayala, F.J. (1979) Phylogenetic relationships between man and the apes: electrophoretic evidence. *Evolution* **33**, 1040–56.

Brumbach, R.A., Staton, R.D., Benjamin, S.A. and Lang, C.M. (1971) The chromosomes of *Aotus trivirgatus* Humboldt 1812. *Folia primatol.* **15**, 264–73.

Buchanan, D.B., Mittermeier, R.A. and van Roosmalen, M.G.M. (1981) The saki monkeys, genus *Pithecia*. In *Ecology and Behavior of Neotropical Primates* (eds Coimbra-Filho, A.F. and Mittermeier, R.A.). Academia Brasileira de Ciências: Rio de Janeiro, pp. 391–417.

Buchardt, B. (1978) Oxygen isotope palaeo-temperatures from the Tertiary period in the North Sea area. *Nature, Lond.* **275**, 121–3.

Budnitz, N. and Dainis, K. (1975) *Lemur catta*: ecology and behavior. In *Lemur Biology* (eds Tattersall, I. and Sussman, R.W.). Plenum Press: New York, pp. 219–35.

Buettner-Janusch, J. (1964) The breeding of galagos in captivity and some notes on their behaviour. *Folia primatol.* **2**, 93–110.

Buttner-Janusch, J. (1966) *Origins of Man: Physical Anthropology*. John Wiley: New York.

Buettner-Janusch, J. (1973) *Physical Anthropology: A Perspective*. John Wiley: New York.

Buettner-Janusch, J. and Andrews, R.J. (1962) The use of the incisors by primates in grooming. *Am. J. phys. Anthrop.* **20**, 127–9.

Bugge, J. (1972) The cephalic arterial system in the insectivores and the primates with special reference to the Macroscelidoidea and Tupaioidea and the insectivore–primate boundary. *Z. Anat. Entw.-Gesch.* **135**, 279–300.

Bugge, J. (1974) The cephalic arterial system in insectivores, primates, rodents and lagomorphs, with special reference to the systematic classification. *Acta anat.* **87**, (suppl. **62**), 1–160.

Bullard, E.C., Everett, J.E. and Smith, A.G. (1965) The fit of continents around the Atlantic. *Phil. Trans. R. Soc. Lond. A.* **258**, 41–51.

Bush, G.L. (1975) Modes of animal speciation. *Ann. Rev. Ecol. Syst.* **6**, 339–64.

Bush, G.L., Case, S.M., Wilson, A.C. and Patton, J.L. (1977) Rapid speciation and chromosomal evolution in mammals. *Proc. natn. Acad. Sci. U.S.A.* **74**, 3942–6.

Butler, H. (1964) The reproductive biology of a strepsirhine (*Galago senegalensis senegalensis*). *Int. Rev. gen. exp. Zool.* **1**, 241–96.

Butler, H. (1974) Evolutionary trends in primate sex cycles. *Contr. Primatol.* **3**, 2–35.

Butler, P.M. (1937) Studies of the mammalian dentition. I: The teeth of *Centetes ecaudatus* and its allies. *Proc. zool. Soc. Lond.* **107**, 103–32.

Butler, P.M. (1939) Studies of the mammalian dentition: differentiation of the postcanine dentition. *Proc. zool. Soc. Lond.* **109**, 1–36.

Butler, P.M. (1952a) The milk molars of the Perissodactyla, with remarks on molar occlusion. *Proc. zool. Soc. Lond.* **121**, 777–817.

Butler, P.M. (1952b) Molarization of the premolars in the Perissodactyla. *Proc. zool. Soc. Lond.* **121**, 819–43.

Butler, P.M. (1956) The ontogeny of molar pattern. *Biol. Rev.* **31**, 30–70.

Butler, P.M. (1961) Relationships between upper and lower molar patterns. *K. VI. Acad. Wet. Lett. Sch. Kunst. Belg.* **1**, 117–26.

Butler, P.M. (1963) Tooth morphology and primate evolution. In *Dental Anthropology* (ed. Brothwell, D.R.). Pergamon Press: Oxford, pp. 1–13.

Butler, P.M. (1972) The problem of insectivore classification. In *Studies in Vertebrate Evolution* (eds Joysey, K.A. and Kemp, T.S.). Oliver and Boyd: Edinburgh, pp. 253–65.

Butler, P.M. (1973) Molar wear facets of Tertiary North American primates. *Symp. 4th int. Congr. Primatol.* **3**, 1–27.

Butler, P.M. (1978) Molar cusp nomenclature and homology. In *Development, Function and Evolution of Teeth* (eds Butler, P.M. and Joysey, K.A.). Academic Press: London, pp. 439–53.

Butler, P.M. (1980) The tupaiid dentition. In *Comparative Biology and Evolutionary Relationships of Tree Shrews* (ed. Luckett, W.P.). Plenum Press: New York, pp. 171–204.

Butler, P.M. (1982) Some problems of the ontogeny of tooth patterns. In *Teeth: Form, Function and Evolution* (ed. Kurtén, B.). Columbia University Press: New York, pp. 44–51.

Butler, P.M. (1986) Problems of dental evolution in the higher primates. In *Major Topics in Primate and Human Evolution* (eds Wood, B.A., Martin, L.B. and Andrews, P.J.). Cambridge University Press: Cambridge, pp. 89–106.

Butler, P.M. and Kielan-Jaworowska, Z. (1973) Is *Deltatheridium* a marsupial? *Nature, Lond.* **245**, 105–6.

Butler, P.M. and Mills, J.R.E. (1959) A contribution to the odontology of *Oreopithecus*. *Bull. Br. Mus. nat. Hist., Geol.* **4**, 3–26.

Bygott, J.D. (1972) Cannibalism among wild chimpanzees. *Nature, Lond.* **238**, 410–11.

Byrd, K.E. (1981) Sequences of dental ontogeny and callitrichid taxonomy. *Primates* **22**, 103–18.

Cachel, S.M. (1979) A functional analysis of the primate masticatory system and the origin of the anthropoid post-orbital septum. *Am. J. phys. Anthrop.* **50**, 1–18.

Cai, J., Kuang, P., Sun, G., Tian, Y., Chang, W., Li, X. and Luo, Z. (1984) Visual discrimination under different illumination in the tree shrews (*Tupaia belangeri chinensis*), slow lorises (*Nycticebus coucang*) and rhesus monkeys (*Macaca mulatta*). *Int. J. Primatol.* **5**, 326.

Cain, A.J. (1954) *Animal Species and their Evolution*. Hutchinson University Library: London.

Cain, A.J. (1959) The post-Linnaean development of taxonomy. *Proc. Linn. Soc. Lond.* **170**, 234–43.

Cain, A.J. and Harrison, G.A. (1960) Phyletic weighting. *Proc. zool. Soc. Lond.* **135**, 1–31.

Calder, W.A. (1984) *Size, Function, and Life History*. Harvard University Press: Cambridge, Mass.

Campbell, C.B.G. (1966a) Taxonomic status of tree shrews. *Science* **153**, 436.

Campbell, C.B.G. (1966b) The relationships of the tree shrews: the evidence of the nervous system. *Evolution* **20**, 276–81.

Campbell, C.B.G. (1974) On the phyletic relationships of the tree shrews. *Mammal Rev.* **4**, 125–43.

Campbell, C.B.G. (1975) The central nervous system: its uses and limitations in assessing phylogenetic relationships. In *Phylogeny of the Primates* (eds Luckett, W.P. and Szalay, F.S.). Plenum Press: New York, pp. 183–97.

Cappetta, H., Jaeger, J.-J., Sabatier, M., Sudre, J. and Vianey-Liaud, M. (1978) Découverte dans le Paléocène du Maroc des plus anciens mammifères euthériens d'Afrique. *Géobios* **11**, 257–63.

Carlson, S.S., Wilson, A.C. and Maxson, R.D. (1978) Reply to Radinsky: Do albumin clocks run on time? *Science* **200**, 1183–5.

Carlsson, A. (1922) Über die Tupaiidae und ihre Beziehungen zu den Insectivora und den Prosimiae. *Acta zool, Stockh.* **3**, 227–70.

Carpenter, C.R. (1934) A field study of the behavior and social relations of the howling monkeys (*Alouatta palliata*). *Comp. Psychol. Monogr.* **10**(2), 1–168.

Carpenter, C.R. (1941) The menstrual cycle and body temperature in two gibbons. *Anat. Rec.* **79**, 291–6.

Carrick, F.N. and Setchell, B.P. (1977) The evolution of the scrotum. In *Reproduction and Evolution* (eds Calaby, J.H. and Tyndale-Biscoe, C.H.). Australian Academy of Sciences: Canberra, pp. 165–70.

Carrier, D.R. (1987) The evolution of locomotor stamina in tetrapods: circumventing a mechanical constraint. *Paleobiology* **13**, 326–41.

Cartmill, M. (1970) The orbits of arboreal mammals: a reassessment of the arboreal theory of primate evolution. PhD thesis, University of Chicago.

Cartmill, M. (1971) Ethmoid component in the orbit of primates. *Nature, Lond.* **232**, 566–7.

Cartmill, M. (1972) Arboreal adaptations and the origin of the order Primates. In *The Functional and Evolutionary Biology of Primates* (ed. Tuttle, R.H.). Aldine-Atherton: Chicago, pp. 97–122.

Cartmill, M. (1974a) *Daubentonia, Dactylopsila* and klinorhynchy. In *Prosimian Biology* (eds Martin, R.D., Doyle, G.A. and Walker, A.C.). Duckworth: London, pp. 655–70.

Cartmill, M. (1974b) Rethinking primate origins. *Science* **184**, 436–43.

Cartmill, M. (1974c) Pads and claws in arboreal locomotion. In *Primate Locomotion* (ed. Jenkins, F.A.). Academic Press: New York, pp. 45–83.

Cartmill, M. (1975) Strepsirhine basicranial structures and the affinities of the Cheirogaleidae. In *Phylogeny of the Primates* (eds Luckett, W.P. and Szalay, F.S.). Plenum Press: New York, pp. 313–54.

Cartmill, M. (1978) The orbital mosaic in prosimians and the use of variable traits in systematics. *Folia primatol.* **30**, 89–114.

Cartmill, M. (1979) The volar skin of primates: its frictional characteristics and their functional significance. *Am. J. phys. Anthrop.* **50**, 497–510.

Cartmill, M. (1980) Morphology, function, and evolution of the anthropoid postorbital septum. In *Evolutionary Biology of the New World Monkeys and Continental Drift* (eds Ciochon, R.L. and Chiarelli, A.B.). Plenum Press: New York, pp. 243–74.

Cartmill, M. (1981) Hypothesis testing and phylogenetic reconstruction. *Z. zool. Syst. Evol.-Forsch.* **19**, 73–96.

Cartmill, M. (1982) Basic primatology and prosimian evolution. In *A History of American Physical Anthropology, 1930–1980* (ed. Spencer, F.). Academic Press: New York, pp. 147–86.

Cartmill, M. and Gingerich, P.D. (1978) An ethmoid exposure (os planum) in the orbit of *Indri indri* (Primates, Lemuriformes). *Am. J. phys. Anthrop.* **48**, 535–8.

Cartmill, M. and Kay, R.F. (1978) Cranio-dental morphology, tarsier affinities, and primate suborders. In *Recent Advances in Primatology*, vol. 3: *Evolution* (eds Chivers, D.J. and Joysey, K.A.). Academic Press: London, pp. 205–14.

Cartmill, M. and MacPhee, R.D.E. (1980) Tupaiid affinities: the evidence of the carotid arteries and cranial skeleton. In *Comparative Biology and Evolutionary Relationships of Tree Shrews* (ed. Luckett, W.P.). Plenum Press: New York, pp. 95–132.

Cartmill, M. and Milton, K. (1977) The lorisiform wrist joint and the evolution of 'brachiating' adaptations in the Hominoidea. *Am. J. phys. Anthrop.* **47**, 249–72.

Cartmill, M., MacPhee, R.D.E. and Simons, E.L. (1981) Anatomy of the temporal bone in early anthropoids, with remarks on the problem of anthropoid origins. *Am. J. phys. Anthrop.* **56**, 3–21.

Castenholz, A. (1965) Über die Struktur der Netzhautmitte bei Primaten. *Z. Zellforsch.* **65**, 646–61.

Catchpole, H.R. and Fulton, J.F. (1943) The oestrous cycle of *Tarsius*. *J. Mammal.* **24**, 90–3.

Cauna, N. (1954) Nature and function of the papillary ridges of the digital skin. *Anat. Rec.* **119**, 449–68.

Cauna, N. (1956) Nerve supply and nerve endings in Meissner's corpuscles. *Am. J. Anat.* **99**, 315–50.

Cavalli-Sforza, L.L. and Edwards, A.W.F. (1967) Phylogenetic analysis, models and estimation procedures. *Am. J. hum. Genet.* **19**, 233–57.

Cave, A.J.E. (1967) Observations on the platyrrhine nasal fossa. *Am. J. phys. Anthrop.* **26**, 277–88.

Cave, A.J.E. (1973) The primate nasal fossa. *Biol. J. Linn. Soc.* **5**, 377–87.

Chang, Y.-y., Wu, M.-l. and Liu, C.-j. (1973) New discovery of *Gigantopithecus* teeth from Wuming, Kwangsi. *Kexue Tongebao (Sci. Bull.)* **18**, 130–3.

Changeux, J.-P. (1983) *L'Homme Neuronal*. Fayard: Paris.

Charles-Dominique, P. (1972) Ecologie et vie sociale de *Galago demidovii* (Fischer 1808, Prosimii). *Z. Tierpsychol., Beiheft* **9**, 7–41.

Charles-Dominique, P. (1975) Nocturnality and diurnality: an ecological interpretation of these two modes of life by an analysis of the higher vertebrate fauna in tropical forest ecosystems. In *Phylogeny of the Primates* (eds Luckett, W.P. and Szalay, F.S.). Plenum Press: New York, pp. 69–88.

Charles-Dominique, P. (1977a) *Ecology and Behaviour of Nocturnal Primates* (transl. by Martin, R.D.). Duckworth: London.

Charles-Dominique, P. (1977b) Urine marking and territoriality in *Galago alleni* (Waterhouse 1837 – Lorisoidea, Primates). A field study by radiotelemetry. *Z. Tierpsychol.* **43**, 113–38.

Charles-Dominique, P. and Hladik, C.M. (1971) Le lépilémur du sud de Madagascar: ecologie, alimentation et vie sociale. *Terre et Vie* **25**, 3–66.

Charles-Dominique, P. and Martin, R.D. (1970) Evolution of lorises and lemurs. *Nature, Lond.* **227**, 257–60.

Charles-Dominique, P., Cooper, H.M., Hladik, A., Hladik, C.M., Pages, E., Pariente, G.F., Petter-Rousseaux, A., Petter, J.-J. and Schilling, A. (eds) (1980) *Nocturnal Malagasy Primates*. Academic Press: New York.

Chiarelli, B.A. (1971) Comparative cytogenetics in primates and its relevance for human cytogenetics. In *Comparative Genetics in Monkeys, Apes and Man* (ed. Chiarelli, B.A.). Academic Press: New York, pp. 276–304.

Chiarelli, B.A. (1974) The chromosomes of the prosimians. In *Prosimian Biology* (eds Martin, R.D., Doyle, G.A. and Walker, A.C.). Duckworth: London, pp. 871–80.

Chivers, D.J. (1974) The siamang in Malaya. *Contr. Primatol.* **4**, 1–335.

Chivers, D.J. and Hladik, C.M. (1980) Morphology of the gastrointestinal tract in primates. Comparisons with other mammals in relation to diet. *J. Morph.* **166**, 337–86.

Chopra, S.R.K. and Vasishat, R.N. (1979) Sivalik fossil tree shrews from Haritalyangar, India. *Nature, Lond.* **281**, 214–5.

Chopra, S.R.K., Kaul, S. and Vasishat, R.N. (1979) Miocene fossil tree shrews from the Indian Sivaliks. *Nature, Lond.* **281**, 213–14.

Chow, M.-c. (1961) A new tarsioid primate from the Lushi Eocene, Honan. *Vert. palasiat.* **5**, 1–5.

Christie-Linde, A. (1914) On the cartilago palatina and the organ of Jacobson in some mammals. *Morph. Jb.* **48**, 343–64.

Chu, E.H.Y. and Bender, M.A. (1961) Chromosome cytology and evolution in primates. *Science* **133**, 1399–405.

Ciochon, R.L. (1983) Hominoid cladistics and the ancestry of modern apes and humans: a summary statement. In *New Interpretations of Ape and Human Ancestry* (eds Ciochon, R.L. and Corruccini, R.S.). Plenum Press: New York, pp. 783–843.

Ciochon, R.L. and Chiarelli, A.B. (eds) (1980) *Evolutionary Biology of the New World Monkeys and Continental Drift.* Plenum Press: New York.

Ciochon, R.L. and Corruccini, R.S. (1975) Morphometric analysis of platyrrhine femora with taxonomic implications and notes on two fossil forms. *J. hum. Evol.* **4**, 193–217.

Ciochon, R.L. and Corruccini, R.S. (1977) The phenetic position of *Pliopithecus* and its phylogenetic relationship to the Hominoidea. *Syst. Zool.* **26**, 290–9.

Ciochon, R.L. and Corruccini, R.S. (eds) (1983) *New Interpretations of Ape and Human Ancestry.* Plenum Press: New York.

Clemens, W.A. (1971) Mammalian evolution in the Cretaceous. In *Early Mammals* (eds Kermack, D.M. and Kermack, K.A.). Academic Press: London, pp. 165–80.

Clemens, W.A. (1974) *Purgatorius*, an early paromomyid primate (Mammalia). *Science* **184**, 903–6.

Clemens, W.A. (1979) Marsupialia. In *Mesozoic Mammals* (eds Lillegraven, J.A., Kielan-Jaworowska, Z. and Clemens, W.A.). University of California Press: Berkeley, pp. 192–220.

Clemens, W.A. and Kielan-Jaworowska, Z. (1979) Multituberculata. In *Mesozoic Mammals* (eds Lillegraven, J.A., Kielan-Jaworowska, Z. and Clemens, W.A.). University of California Press: Berkeley, pp. 99–149.

Clutton-Brock, T.H. (1974) Primate social organization and ecology. *Nature, Lond.* **250**, 539–42.

Clutton-Brock, T.H. (1975) Ranging behaviour of red colobus (*Colobus badius tephrosceles*) in the Gombe National Park. *Anim. Behav.* **23**, 706–22.

Clutton-Brock, T.H. and Harvey, P.H. (1977) Species differences in feeding and ranging behaviour in primates. In *Primate Ecology* (ed. Clutton-Brock, T.H.). Academic Press: London, pp. 557–79.

Cochet, C., Creau-Goldberg, N., Turleau, C. and de Grouchy, J. (1982) Gene mapping of *Microcebus murinus* (Lemuridae). A comparison with man and *Cebus capucinus* (Cebidae). *Cytogenet. Cell Genet.* **33**, 213–21.

Cocks, L.R.M. (ed.) (1981) *The Evolving Earth.* Cambridge University Press: Cambridge.

Coimbra-Filho, A.F. and Maia, A. de A. (1979) A sazonalidade do processo reprodutivo em *Leontopithecus rosalia* (Linnaeus, 1766) (Callitrichidae, Primates). *Rev. Brasil. Biol.* **39**, 643–51.

Coimbra-Filho, A.F. and Mittermeier, R.A. (1976) Exudate-eating and tree-gouging in marmosets. *Nature, Lond.* **262**, 630.

Coimbra-Filho, A.F. and Mittermeier, R.A. (1977) Tree-gouging, exudate-eating, and the short-tusked condition in *Callithrix* and *Cebuella*. In *The Biology and Conservation of the Callitrichidae* (ed. Kleiman, D.G.). Smithsonian Institution Press: Washington, pp. 105–15.

Colbert, E.H. (1969) *Evolution of the Vertebrates* (2nd edition). John Wiley: New York.

Colillas, O. and Coppo, J. (1978) Breeding *Alouatta caraya* in Centro Argentino de Primates. In *Recent Advances in Primatology*, vol. 2: *Conservation* (eds Chivers, D.J. and Lane-Petter, W.). Academic Press: London, pp. 201–14.

Comings, D.E. (1972) Evidence for ancient tetraploidy and conservation of linkage groups in mammalian chromosomes. *Nature, Lond.* **238**, 455–7.

Comings, D.E. (1978) Mechanisms of chromosome banding and implications for chromosome structure. *Ann. Rev. Genet.* **12**, 25–46.

Conaway, C.H. (1971) Ecological adaptation and mammalian reproduction. *Biol. Reprod.* **4**, 239–47.

Conaway, C.H. and Sorenson, M.W. (1966) Reproduction in tree shrews. *Symp. zool. Soc. Lond.* **15**, 471–92.

Connolly, C.J. (1950) *External Morphology of the Primate Brain.* C.C. Thomas: Springfield, Illinois.

Conroy, G.C. (1976a) Primate postcranial remains from the Oligocene of Egypt. *Contr. Primatol.* **8**, 1–134.

Conroy, G.C. (1976b) Hallucial tarsometatarsal joint in an Oligocene anthropoid, *Aegyptopithecus zeuxis*. *Nature, Lond.* **262**, 684–6.

Conroy, G.C. (1980) Ontogeny, auditory structures, and primate evolution. *Am. J. phys. Anthrop.* **52**, 443–51.

Conroy, G.C. (1987) Problems of body-weight estimation in fossil primates. *Int. J. Primatol.* **8**, 115–37.

Conroy, G.C. and Fleagle, J.G. (1972) Locomotor behaviour in living and fossil pongids. *Nature, Lond.* **237**, 103–4.

Conroy, G.C. and Rose, M.D. (1983) The evolution of the primate foot from the earliest primates to the Miocene. *Foot and Ankle* **3**, 342–64.

Conroy, G.C. and Wible, J.R. (1978) Middle ear morphology of *Lemur variegatus*. Some implications for primate paleontology. *Folia primatol.* **29**, 81–5.

Conroy, G.C., Schwartz, J.H. and Simons, E.L. (1975) Dental eruption patterns in Parapithecidae (Primates, Anthropoidea). *Folia primatol.* **24**, 275–81.

Cook, C. and Hewett-Emmett, D. (1974) The uses of protein sequence data in systematics. In *Prosimian Biology* (eds Martin, R.D., Doyle, G.A. and Walker, A.C.). Duckworth: London, pp. 937–58.

Coppens, Y., Clark Howell, F., Isaac, G.L. and Leakey, R.E.F. (eds) (1976) *Earliest Man and Environments in the Lake Rudolf Basin.* University of Chicago Press: Chicago.

Corbet, G.E. and Hill, J.E. (1980) *A World List of Mammalian Species.* British Museum (Natural History): London.

Corruccini, R.S. (1975a) Multivariate analysis of *Gigantopithecus* mandibles. *Am. J. phys. Anthrop.* **42**, 167–70.

Corruccini, R.S. (1975b) *Gigantopithecus* and hominids. *Anthrop. Anz.* **35**, 55–7.

Corruccini, R.S., Ciochon, R.L. and McHenry, H.M. (1976) The postcranium of Miocene hominoids: were dryopithecinces merely 'dental apes'? *Primates* **17**, 205–23.

Corruccini, R.S., Cronin, J.E. and Ciochon, R.L. (1979) Scaling analysis and congruence among anthropoid primate macromolecules. *Hum. Biol.* **51**, 167–85.

Corruccini, R.S., Baba, M., Goodman, M., Ciochon, R.L. and Cronin, J.E. (1980) Non-linear macromolecular evolution and the molecular clock. *Evolution* **34**, 1216–19.

Couturier, J., Dutrillaux, B., Turleau, C. and De Grouchy, J. (1982) Comparaisons chromosomiques chez quatre espèces de gibbons. *Ann. Génét.* **25**, 5–10.

Covert, H.H. (1985) Adaptations and evolutionary relationships of the Eocene primate family Notharctidae. PhD thesis, Duke University.

Covert, H.H. (1987) Tarsals of *Cantius mckennai*. *Am. J. phys. Anthrop.* **72**, 189–90.

Covert, H.H. and Kay, R.F. (1980) Dental microwear and diet – implications for early hominoid feeding behavior. *Am. J. phys. Anthrop.* **52**, 216.

Cowles, R.B. (1958) The evolutionary significance of the scrotum. *Evolution* **12**, 417–18.

Cox, C.B. (1970) Migrating marsupials and drifting continents. *Nature, Lond.* **226**, 767–70.

Cox, C.B. (1973) Triassic tetrapods. In *Atlas of Palaeobiogeography* (ed. Hallam, A.). Elsevier Scientific: London, pp. 213–23.

Cox, C.B. (1974) Vertebrate palaeodistributional patterns and continental drift. *J. Biogeogr.* **1**, 75–94.

Crompton, A.W. (1971) The origin of the tribosphenic molar. In *Early Mammals* (eds Kermack, D.M. and Kermack, K.A.). Linnean Society: London, pp. 65–87.

Crompton, A.W. and Hiiemae, K.M. (1970) Molar occlusion and mandibular movements during occlusion in the American opossum *Didelphis marsupialis*. *Zool. J. Linn. Soc., Lond.* **49**, 21–47.

Crompton, A.W. and Jenkins, F.A. (1973) Mammals from reptiles: a review of mammalian origins. *A. Rev. earth planet. Sci.* **1**, 131–54.

Crompton, A.W. and Jenkins, F.A. (1978) Mesozoic mammals. In *Evolution of African Mammals* (eds Cooke, H.B.S. and Maglio, V.). Harvard University Press: Cambridge, Mass., pp. 46–55.

Crompton, A.W. and Jenkins, F.A. (1979) Origin of mammals. In *Mesozoic Mammals* (eds Lillegraven, J.A., Kielan-Jaworowska, Z. and Clemens, W.A.). University of California Press: Berkeley, pp. 59–73.

Crompton, A.W. and Kielan-Jaworowska, Z. (1978) Molar structure and occlusion in Cretaceous therian mammals. In *Studies in the Development, Structure and Function of Teeth* (eds Butler, P.M. and Joysey, K.A.). Academic Press: London, pp. 249–87.

Crompton, A.W., Thexton, A.J., Parker, P. and Hiiemae, K.M. (1977) The activity of the jaw and hyoid musculature in the Virginian opossum (*Didelphis virginiana*). In *The Biology of Marsupials* (eds Stonehouse, B. and Gilmore, D.). Macmillan: London, pp. 287–305.

Crompton, R.H. (1984) Foraging, habitat structure, and locomotion in two species of *Galago*. In *Adaptations for Foraging in Nonhuman Primates* (eds Rodman, P.S. and Cant, J.G.H.). Columbia University Press: New York, pp. 73–111.

Crompton, R.H. and Andau, P.M. (1986) Locomotion and habit utilization in free-ranging *Tarsius bancanus*: a preliminary report. *Primates* **27**, 337–55.

Crompton, R.H. and Andau, P.M. (1987) Ranging, activity rhythms, and sociality in free-ranging

Tarsius bancanus: a preliminary report. *Int. J. Primatol.* **8**, 43–71.

Crompton, R.H., Lieberman, S.S. and Oxnard, C.E. (1987) Morphometrics and niche metrics in prosimian locomotion: an approach to measuring locomotion, habitat and diet. *Am. J. phys. Anthrop.* **73**, 149–77.

Cronin, J.E. and Sarich, V.M. (1975) Molecular systematics of the New World monkeys. *J. hum. Evol.* **4**, 357–75.

Cronin, J.E. and Sarich, V.M. (1978a) Marmoset evolution: the molecular evidence. *Primates Med.* **10**, 12–19.

Cronin, J.E. and Sarich, V.M. (1978b) Primate higher taxa – the molecular view. In *Recent Advances in Primatology,* vol. 3 (eds Chivers, D.J. and Joysey, K.A.). Academic Press: London, pp. 287–9.

Cronin, J.E. and Sarich, V.M. (1980) Tupaiid and Archonta phylogeny: the macromolecular evidence. In *Comparative Biology and Evolutionary Relationships of the Tree Shrews* (ed. Luckett, W.P.). Plenum Press: New York, pp. 293–312.

Cronin, J.E., Boaz, N.T., Stringer, C.B. and Rak, Y. (1981) Tempo and mode in hominid evolution. *Nature, Lond.* **292**, 113–22.

Crusafont-Pairo, M. and Hürzeler, J. (1961) Les pongidés fossiles d'Espagne. *C.R. Acad. Sci., Paris* **254**, 582–4.

Dagosto, M. (1983) Postcranium of *Adapis parisiensis* and *Leptadapis magnus* (Adapiformes, Primates). *Folia primatol.* **41**, 49–101.

Dagosto, M. (1988) Implications of postcranial evidence for the origin of euprimates. *J. hum Evol.* **17**, 35–56.

Darlington, C.D. and Haque, A. (1955) Chromosomes of monkeys and men. *Nature, Lond.* **175**, 32.

Darney, K.J. and Franklin, L.E. (1982) Analysis of the estrous cycle of the laboratory-housed Senegal galago (*Galago senegalensis senegalensis*). Natural and induced cycles. *Folia primatol.* **37**, 106–26.

Dart, R. (1925) *Australopithecus africanus*; the man-ape of South Africa. *Nature, Lond.* **115**, 195–9.

Dartnall, H.J.A., Arden, G.B., Ikeda, H., Luck, C.P., Rosenberg, C.M., Pedler, H. and Tansley, K. (1965) Anatomical, electrophysiological and pigmentary aspects of vision in the bush baby: an interpretive study. *Vision Res.* **5**, 399–424.

Darwin, C. (1859) *On the Origin of Species by Means of Natural Selection in the Struggle for Life.* Murray: London.

Dashzeveg, D.T. and McKenna, M.C. (1977) Tarsioid primate from the early Tertiary of the Mongolian People's Republic. *Acta palaeont. pol.* **22**, 119–37.

Datta, P.M. (1981) The first Jurassic mammals from India. *Zool. J. Linn. Soc. Lond.* **73**, 307–12.

Datta, P.M., Yadagira, P. and Rao, B.R.J. (1978) Discovery of early Jurassic micromammals from upper Gondwana sequence of Pranhita Godivari Valley. *J. Indian geol. Soc.* **19**, 64–8.

Davis, D.D. (1938) Notes on the anatomy of the treeshrew *Dendrogale*. *Field Mus. nat. Hist., zool. Ser.* **20**, 383–404.

Davis, D.D. (1962) Mammals of the lowland rain-forest of North Borneo. *Bull. natn. Mus. Singapore* **31**, 1–129.

Davis, P.R. and Napier, J.R. (1963) A reconstruction of the skull of *Proconsul africanus*. *Folia primatol.* **1**, 20–8.

Dawson, G.A. (1977) Composition and stability of social groups of the tamarin, *Saguinus oedipus geoffroyi* in Panama: ecological and behavioural implications. In *The Biology and Conservation of the Callitrichidae* (ed. Kleiman, D.G.). Smithsonian Institution Press: Washington, pp. 23–37.

Dawson, T.J. and Hulbert, A.J. (1970) Standard metabolism, body temperature and surface areas of Australian marsupials. *Am. J. Physiol.* **218**, 1233–8.

Day, M.H. (1985) Hominid locomotion – from Taung to the Laetoli footprints. In *Hominid Evolution: Past, Present and Future* (ed. Tobias, P.V.). Alan R. Liss: New York, pp. 115–27.

Day, M.H. and Napier, J.R. (1964) Fossil foot bones. *Nature, Lond.* **201**, 969–70.

Day, M.H. and Wickens, E.H. (1980) Laetoli Pliocene hominid footprints and bipedalism. *Nature, Lond.* **286**, 385–7.

Day, M.H. and Wood, B.A. (1968) Functional affinities of the Olduvai hominid 8 talus. *Man* **3**, 440–55.

Dayhoff, M.O. (1969) *Atlas of Protein Sequence and Structure*. National Biomedical Research Foundation: Silver Spring, Maryland.

de Beer, G.R. (1937) *The Development of the Vertebrate Skull*. Oxford University Press: Oxford.

de Beer, G.R. (1958) *Embryos and Ancestors*. Oxford University Press: Oxford.

de Boer, L.E.M. (1973) Cytotaxonomy of the Lorisoidea (Primates: Prosimii). I: Chromosome studies and karyological relationships in the Galagidae. *Genetica* **44**, 155–93.

de Boer, L.E.M. (1974) Cytotaxonomy of the Platyrrhini (Primates). *Genen Phaenen* **17**, 1–115.

de Bonis, L. (1983) Phyletic relationships of Miocene hominoids and higher primate classification. In *New Interpretations of Ape and Human Ancestry* (eds Ciochon, R.L. and Corruccini, R.S.). Plenum Press: New York, pp. 625–49.

de Bonis, L. and Melentis, J. (1976) Les dryopithécinés de Macédoine (Grèce). Leur place dans l'évolution des primates hominoïdes du Miocène. In *Les Plus Anciens Hominidés* (eds Tobias, P.V. and Coppens, Y.). C.N.R.S.: Paris, pp. 26–38.

de Bonis, L. and Melentis, J. (1977a) Un nouveau genre de primate hominoïde dans le Vallésien (Miocène supérieur) de Macédoine. *C.R. Acad. Sci., Paris, D* **284**, 1393–6.

de Bonis, L. and Melentis, J. (1977b) Les primates hominoïdes du Vallésien de Macédoine (Grèce). Etude de la mâchoire inférieure. *Géobios* **10**, 849–85.

de Bonis, L. and Melentis, J. (1978) Les primates hominoïdes du Miocène supérieur de Macédoine: étude de la mâchoire supérieure. *Ann. paléont. (Vert.)* **64**, 185–202.

de Bonis, L. and Melentis, J. (1980) Nouvelles remarques sur l'anatomie d'un primate hominoïde du Miocène: *Ouranopithecus macedoniensis*. Implications sur la phylogénie des hominidés. *C. R. Acad. Sci., Paris D* **290**, 755–8.

de Bonis, L., Bouvrain, G., Geraads, D. and Melentis, J. (1974) Première decouverte d'un primate hominoïde dans le Miocène supérieur de Macédoine (Grèce). *C.R. Acad. Sci., Paris, D* **278**, 3063–6.

de Grouchy, J., Turleau, C. and Finaz, C. (1978) Chromosomal phylogeny of the primates. *A. Rev. Genet.* **12**, 289–328.

de Jong, W.W. (1982) Eye lens proteins and vertebrate phylogeny. In *Macromolecular Sequences in Systematics and Evolutionary Biology* (ed. Goodman, M.). Plenum Press: New York, pp. 75–114.

de Jong, W.W. and Goodman, M. (1988) Anthropoid affinities of *Tarsius* supported by α-crystallin sequences. *J. hum. Evol.* **17**, 575–82.

Decker, R.L. and Szalay, F.S. (1974) Origins and function of the pes in the Eocene Adapidae (Lemuriformes, Primates). In *Primate Locomotion* (ed. Jenkins, F.A.). Academic Press: New York, pp. 261–91.

Deininger, P.L. and Schmid, C.W. (1976) Thermal stability of human DNA and chimpanzee DNA heteroduplexes. *Science* **194**, 846–8.

Delson, E. (1975a) Evolutionary history of the Cercopithecidae. *Contr. Primatol.* **5**, 167–217.

Delson, E. (1975b) Paleoecology and zoogeography of the Old World monkeys. In *Primate Functional Morphology and Evolution* (ed. Tuttle, R.) Mouton: The Hague, pp. 37–64.

Delson, E. (1979) *Prohylobates* (Primates) from the early Miocene of Libya: a new species and its implications for cercopithecoid origins. *Géobios* **12**, 725–33.

Delson, E. (ed.) (1985) *Ancestors: The Hard Evidence*. Alan Liss: New York.

Delson, E. (1986) An anthropological enigma: historical introduction to the study of *Oreopithecus bambolii*. *J. hum. Evol.* **15**, 523–31.

Delson, E. and Andrews, P.J. (1975) Evolution and interrelationships of the catarrhine primates. In *Phylogeny of the Primates* (eds Luckett, W.P. and Szalay, F.S.). Plenum Press: New York, pp. 405–46.

Delson, E. and Plopsor, D. (1975) *Paradolichopithecus*, a large terrestrial monkey (Cercopithecidae, Primates) from the Plio-Pleistocene of Southern Europe and its importance for mammalian biochronology. *Proc. sess. Reg. Comm. Mediterr. Neogene Stratig., Bratislava* **6**, 91–6.

Dene, H.T., Goodman, M. and Prychodko, W. (1976a) Immunodiffusion evidence on the phylogeny of the primates. In *Molecular Anthropology* (eds Goodman, M., Tashian, R.E. and Tashian, J.H.). Plenum Press: New York, pp. 171–95.

Dene, H.T., Goodman, M. and Prychodko, W. (1976b) Immunodiffusion systematics of the primates. III: The Strepsirhini. *Folia primatol.* **25**, 35–61.

Dene, H.T., Goodman, M. and Prychodko, W. (1978) An immunological examination of the systematics of Tupaioidea. *J. Mammal.* **59**, 697–706.

Dene, H.T., Goodman, M. and Prychodko, W. (1980) Immunodiffusion systematics of the primates. IV: Lemuriformes. *Mammalia* **44**, 211–23.

Dene, H.T., Goodman, M., Prychodko, W. and Matsuda, G. (1980) Molecular evidence for the affinities of the Tupaiidae. In *Comparative Biology and Evolutionary Relationships of the Tree Shrews* (ed. Luckett, W.P.). Plenum Press: New York, pp. 269–91.

Depéret, C. (1889) Sur le *Dolichopithecus ruscinensis*, nouveau singe fossile du Pliocène du Rousillon. *C.R. Acad. Sci., Paris* **109**, 982–3.

Detwiler, S.R. (1939) Comparative studies upon the eyes of nocturnal lemuroids, monkeys and man. *Anat. Rec.* **74**, 129–45.

Detwiler, S.R. (1940) The eye of *Nyticebus tardigrada*. *Anat. Rec.* **76**, 295–301.

Dixson, A.F. (1982) The owl monkey (*Aotus trivirgatus*). In *Reproduction in New World Primates* (ed. Hearn, J.P.). MTP: Lancaster, pp. 71–113.

Dixson, A.F. (1986) Plasma testosterone concentrations during postnatal development in the male common marmoset. *Folia primatol.* **47**, 166–70.

Dixson, A.F. and Fleming, D. (1981) Parental behaviour and infant development in owl monkeys (*Aotus trivirgatus griseimembra*). *J. Zool., Lond.* **194**, 25–39.

Dobzhansky, Th. (1950) Mendelian populations and their evolution. *Am. Nat.* **84**, 401–18.

Dobzhansky, Th., Ayala, F.J., Stebbins, G.L. and Valentine, J.W. (1977) *Evolution.* W.H. Freeman: San Fransisco.

Dollo, L. (1893) Les lois de l'évolution. *Bull. Soc. belge Géol. Palaeont. Hydrol.* **7**, 164–66.

Dollo, L. (1899) Les ancêtres des marsupiaux etaient-ils arboricoles? *Trav. Stat. zool. géol. Wimereux* **7**, 188–203.

Doolittle, R.F. (1974) Prosimian biology and protein evolution. In *Prosimian Biology* (eds Martin, R.D., Doyle, G.A. and Walker, A.C.). Duckworth: London, pp. 959–70.

Doolittle, R.F. and Mross, G.A. (1970) Identity of chimpanzee with human fibrinopeptides. *Nature, Lond.* **225**, 643–4.

Doolittle, R.F. and Sapienza, C. (1980) Selfish genes, the phenotype paradigm and genome evolution. *Nature, Lond.* **284**, 601–3.

Doran, A.H.G. (1879) Morphology of the mammalian ossicula auditus. *Trans. Linn. Soc. Lond.* **1**, 371–497.

Dostal, A. and Zapfe, D.H. (1986) Zahnschmelz-prismen von *Mesopithecus pentelicus* Wagner, 1839, im Vergleich mit rezenten Cercopitheciden (Primates, Cercopithecidae). *Folia primatol.* **46**, 235–51.

Doyle, G.A. (1979) Development of behavior in prosimians with special reference to the lesser bushbaby, *Galago senegalensis moholi*. In *The Study of Prosimian Behavior* (eds Doyle, G.A. and Martin, R.D.). Academic Press: New York, pp. 157–206.

Drucker, G.R. (1984) The feeding ecology of the Barbary macaque and cedar forest conservation in the Moroccan Moyen Atlas. In *The Barbary Macaque* (ed. Fa, J.E.). Plenum Press: New York, pp. 135–64.

D'Souza, F. (1974) A preliminary field report on the lesser tree-shrew *Tupaia minor*. In *Prosimian Biology* (eds Martin, R.D., Doyle, G.A. and Walker, A.C.). Duckworth: London, pp. 167–82.

D'Souza, F. and Martin, R.D. (1974) Maternal behaviour and the effect of stress in tree-shrews. *Nature, Lond.* **251**, 309–11.

du Toit, A.L. (1937) *Our Wandering Continents: An Hypothesis of Continental Drifting.* Oliver and Boyd: Edinburgh and London.

Dubois, E. (1894) *Pithecanthropus erectus. Eine menschenähnliche Übergangsform aus Java.* Landsdruckerei: Batavia.

Dunbar, R.I.M. (1977) Feeding ecology of gelada baboons: a preliminary report. In *Primate Ecology* (ed. Clutton-Brock, T.H.). Academic Press: London, pp. 251–73.

Dunbar, R.I.M. (1983) Theropithecines and hominids: contrasting solutions to the same ecological problem. *J. hum. Evol.* **12**, 647–58.

Dunbar, R.I.M. (1984) *Reproductive Decisions: An Economic Analysis of Gelada Baboon Social Strategies.* Princeton University Press: New Jersey.

Duncan, T. and Stuessy, T.F. (eds) (1984) *Cladistics: Perspectives on the Reconstruction of Evolutionary History.* Columbia University Press: New York.

Dutrillaux, B. (1975) *Sur la Nature et l'Origine des Chromosomes Humains.* Expansion: Paris.

Dutrillaux, B. (1979) Chromosomal evolution in primates: tentative phylogeny from *Microcebus murinus* (prosimian) to man. *Hum. Genet.* **48**, 251–314.

Dutrillaux, B. (1986) Le role des chromosomes dans l'évolution: une nouvelle interprétation. *Ann. Génét.* **29**, 69–75.

Dutrillaux, B. and Couturier, J. (1981) The ancestral karyotype of platyrrhine monkeys. *Cytogenet. Cell Genet.* **30**, 232–42.

Dutrillaux, B. and Couturier, J. (1983) The ancestral karyotype of Carnivora: a comparison with that of platyrrhine monkeys. *Cytogenet. Cell Genet.* **35**, 200–8.

Dutrillaux, B. and Rumpler, Y. (1980) Chromosome banding analogies between a prosimian (*Microcebus murinus*), a platyrrhine (*Cebus capucinus*) and man. *Am. J. phys. Anthrop.* **52**, 133–7.

Dutrillaux, B. and Rumpler, Y. (1988) Absence of chromosomal similarities between tarsiers (*Tarsius syrichta*) and other primates. *Folia primatol.* **50**, 130–3.

Dutrillaux, B., Couturier, J. and Fosse, A.-M. (1980). The use of high resolution banding in comparative cytogenetics: comparison between man and *Lagothrix lagotricha* (Cebidae). *Cytogenet. Cell Genet.* **27**, 45–51.

Dutrillaux, B., Couturier, J. and Viégas-Pequignot, E. (1981) Chromosomal evolution in primates. In *Chromosomes Today*, vol. 7 (eds Bennet, M.D., Bobrow, M. and Hewitt, G.M.). George Allen and Unwin: London, pp. 176–91.

Dutrillaux, B., Rethoré, M.O. and Lejeune, J. (1975) Comparaison du caryotype de l'orang-outan

(*Pongo pygmaeus*) à celui de l'homme, du chimpanzé et du gorille. *Ann. Génét.* **18**, 153–61.

Dutrillaux, B., Viégas-Pequignot, E. and Couturier, J. (1980) Très grande analogie de marquage chromosomique entre le lapin (*Oryctolagus cuniculus*) et les primates dont l'homme. *Ann. Génét.* **23**, 22–5.

Dutrillaux, B., Couturier, J., Viégas-Pequignot, E. and Muleris, M. (1982) Cytogenetic aspects of primate evolution. In *Human Genetics.* Part A: *The Unfolding Genome* (eds Bonne-Tamir, B. and Cohen, T.). Allan R. Liss: New York, pp. 183–94.

Dutrillaux, B., Couturier, J., Muleris, M., Rumpler, Y. and Viégas-Pequignot, E. (1986a) Relations chromosomiques entre sous-ordres et infra-ordres et schéma évolutif général des primates. *Mammalia* **50**, 108–21.

Dutrillaux, B., Couturier, J., Sabatier, L., Muleris, M. and Prieur, M. (1986b) Inversions in evolution of man and closely related species. *Ann. Génét.* **29**, 195–202.

Dutrillaux, B., Lombard, M., Carroll, J.B. and Martin, R.D. (1988) Chromosomal affinities of *Callimico goeldii* (Platyrrhini) and characterization of a Y-autosome translocation in the male. *Folia primatol.* **50**, 230–6.

Dutrillaux, B., Viégas-Pequignot, E., Couturier, J. and Chauvier, G. (1978) Identity of euchromatic bands from man to Cercopithecidae: *Cercopithecus aethiops, Cercopithecus sabaeus, Erythrocebus patas* and *Miopithecus talapoin.* *Hum. Genet.* **45**, 283–96.

Dutta, A.K. (1975) Micromammals from Siwaliks. *Indian Min.* **29**, 76–7.

Eager, R.M.C. (1976) *The Geological Column* (5th revised edition). The Manchester Museum: Manchester.

Eaglen, R.H. (1984) Incisor size and diet revisited: the view from a platyrrhine perspective. *Am. J. phys. Anthrop.* **64**, 263–75.

Eaglen, R.H. (1985) Behavioral correlates of tooth eruption in Malagasy lemurs. *Am. J. phys. Anthrop.* **66**, 307–15.

Eaton, G.G., Slob, A. and Resko, J.A. (1973) Cycles of mating behaviour, oestrogen and progesterone in the thick-tailed bushbaby (*Galago crassicaudatus crassicaudatus*) under laboratory conditions. *Anim. Behav.* **21**, 309–15.

Eck, C.G. (1977) Diversity and frequency of Omo group Cercopithecoidea. *J. hum. Evol.* **6**, 55–63.

Eck, R.V. and Dayhoff, M.O. (1966) *Atlas of Protein Sequence and Structure.* National Biomedical Research Foundation: Silver Spring, Maryland.

Eckhardt, R.B. (1973) *Gigantopithecus* as a hominid ancestor. *Anthrop. Anz.* **34**, 1–8.

Eckhardt, R.B. (1975) *Gigantopithecus* as a hominid. In *Paleoanthropology, Morphology and Paleoecology* (ed. Tuttle, R.). Mouton: The Hague, pp. 105–27.

Eckstein, P. and Zuckerman, S. (1956) Morphology of the reproductive tract. In *Marshall's Physiology of Reproduction*, vol. 1, part 1 (ed. Parkes, A.S.). Longman's Green: London, pp. 43–155.

Edinger, T. (1929) Die fossilen Gehirne. *Ergebn. Anat. Entwickl.-Gesch.* **28**, 1–249.

Egozcue, J. (1969) Primates. In *Comparative Mammalian Cytogenetics* (ed. Benirschke, K.). Springer Verlag: New York, pp. 357–89.

Egozcue, J. (1970) The chromosomes of the lesser bushbaby (*Galago senegalensis*) and the greater bushbaby (*Galago crassicaudatus*). *Folia primatol.* **12**, 236–40.

Egozcue, J. (1971) A possible case of centric fission in a primate. *Experientia* **27**, 969–71.

Egozcue, J. (1974) Chromosomal evolution in prosimians. In *Prosimian Biology* (eds Martin, R.D., Doyle, G.A. and Walker, A.C.). Duckworth: London, pp. 857–63.

Egozcue, J. (1975) *Animal Cytogenetics. IV: Chordata 4, Mammalia II, Placentalia 5, Primates.* Gebrüder Borntraeger: Berlin.

Ehara, A. (1969) Zur Phylogenese und Funktion des Orbitaseitenrandes der Primaten. *Z. Morph. Anthrop.* **60**, 263–71.

Eicher, D.L. (1968) *Geologic Time.* Prentice-Hall: New Jersey.

Eisenberg, J.F. (1981) *The Mammalian Radiations: An Analysis of Trends in Evolution, Adaptation and Behaviour.* Athlone Press: London.

Eisenhart, W.L. (1975) A review of the fossil cercopithecoids from Makapansgat and Sterkfontein, South Africa. *Am. J. phys. Anthrop.* **42**, 299.

Eldredge, N. and Cracraft, J. (1980) *Phylogenetic Patterns and the Evolutionary Process: Method and Theory in Comparative Biology.* Columbia University Press: New York.

Eldredge, N. and Gould, S.J. (1972) Punctuated equilibria: an alternative to phyletic gradualism. In *Models in Paleobiology* (ed. Schopf, T.J.M.). Freeman, Cooper: San Francisco, pp. 82–115.

Eldredge, N. and Tattersall, I. (1975) Evolutionary models, phylogenetic reconstruction, and another look at hominid phylogeny. *Contr. Primatol.* **5**, 218–42.

Elgar, M.A. and Harvey, P.H. (1987) Basal metabolic rates in mammals: allometry, phylogeny and ecology. *Functional Ecol.* **1**, 25–36.

Elias, H. and Schwartz, D. (1971) Cerebro-cortical surface areas, volumes, lengths of gyri and their interdependence in mammals, including man. *Z. Säugetierk.* **36**, 147–63.

Ellefson, J.O. (1968) Territorial behaviour in the common white-handed gibbon, *Hylobates lar* Linn.. In *Primates: Studies of Adaptation and Variability* (ed. Jay, P.). Holt, Rinehart and Winston: New York, pp. 180–99.

Elliot, D.G. (1913) *A Review of the Primates* (3 vols). American Museum of Natural History: New York.

Elliot, D.H., Colbert, E.H., Breed, W.J., Jensen, J.A. and Powell, J.S. (1970) Triassic tetrapods from Antarctica: evidence for continental drift. *Science* **169**, 1197–201.

Elliot Smith, G. (1901) The natural subdivision of the cerebral hemisphere. *J. Anat. Physiol.* **35**, 431–54.

Elliot Smith, G. (1902) On the morphology of the brain in the Mammalia, with special reference to that of the lemurs, recent and extinct. *Trans. Linn. Soc., Lond. (Zool).* **8**, 319–432.

Elliot Smith, G. (1910) On the impossibility of instituting exact homologies between the sulci called 'calcarine' in various primates. *Anat. Anz.* **36**, 486–7.

Elliot Smith, G. (1927) *Essays on the Evolution of Man* (2nd edition). Oxford University Press: Oxford.

Erikson, G.E. (1963) Brachiation in New World monkeys and in anthropoid apes. *Symp. zool. Soc. Lond.* **10**, 135–63.

Estabrook, G.F. (1979) Some concepts for the estimation of evolutionary relationships in systematic botany. *Syst. Bot.* **3**, 146–58.

Estes, R.D. (1972) The role of the vomeronasal organ in mammalian reproduction. *Mammalia* **36**, 315–41.

Eudey, A.A. (1978) Earth-eating by macaques in Western Thailand: a preliminary analysis. In *Recent Advances in Primatology*, vol. 1: *Behaviour* (eds Chivers, D.J., and Herbert, J.). Academic Press: London, pp. 351–3.

Evans, C.S. and Goy, R.W. (1968) Social behaviour and reproductive cycles in captive ring-tailed lemurs (*Lemur catta*). *J. Zool., Lond.* **156**, 181–97.

Evans, F.G. (1942) The osteology and relationships of the elephant shrews (Macroscelididae). *Bull. Am. Mus. nat. Hist.* **80**, 85–125.

Every, R.G. (1970) Sharpness of teeth in man and other primates. *Postilla* no. 143, 1–30.

Every, R.G. (1974) Thegosis in prosimians. In *Prosimian Biology* (eds Martin, R.D., Doyle, G.A. and Walker, A.C.). Duckworth: London, pp. 579–619.

Every, R.G. and Kühne, W.F. (1970) Funktion und Form der Säugerzähne: I. Thegosis, Usur und Druckusur. Z. Säugetierk. **35**, 247–52.

Every, R.G. and Kühne, W.G. (1971) Bimodal wear of mammalian teeth. In *Early Mammals* (eds Kermack, D.M. and Kermack, K.A.). Linnean Society: London, pp. 23–8.

Ewer, R.F. (1963) Reptilian tooth replacement. *News Bull. zool. Soc. S. Afr.* **4**, 4–9.

Falk, D. (1978) External neuroanatomy of the Old World monkeys (Cercopithecoidea). *Contr. Primatol.* **15**, 1–95.

Farris, J.S. (1972) Estimating phylogenetic trees from distance matrices. *Am. Nat.* **106**, 645–68.

Felsenstein, J. (1978) The number of evolutionary trees. *Syst. Zool.* **27**, 27–33.

Felsenstein, J. (1983) Parsimony in systematics: biological and statistical issues. *A. Rev. Ecol. Syst.* **14**, 313–33.

Fenchel, T. (1974) Intrinsic rate of natural increase: the relationship with body size. *Oecologia* **14**, 317–26.

Ferembach, D. (1958) Les limnopithèques du Kenya. *Ann. Paléont.* **44**, 149–249.

Ferraz de Oliveira, L. and Ripps, H. (1968) The 'area centralis' of the owl monkey (*Aotes trivirgatus*). *Vision Res.* **8**, 223–8.

Fiedler, W. (1956) Übersicht über das System der Primates. In *Primatologia*, vol. 1 (eds Hofer, H., Schultz, A.H. and Starck, D.). Karger: Basel, pp. 1–266.

Fitch, W.M. (1970) Distinguishing homologous from analogous proteins. *Syst. Zool.* **19**, 99–113.

Fitch, W.M. (1984) Cladistic and other methods: problems, pitfalls, and potentials. In *Cladistics* (eds Duncan, T. and Stuessy, T.F.). Columbia University Press: New York, pp. 1–266.

Fitch, W.M. and Langley, C.H. (1976) Evolutionary rates in proteins: neutral mutations and the molecular clock. In *Molecular Anthropology* (eds Goodman, M. and Tashian, R.E.). Plenum Press: New York, pp. 197–219.

Fitch, W.M. and Margoliash, E. (1967) The construction of phylogenetic trees – a generally applicable method utilizing estimates of the mutation distance obtained from cytochrome-c-sequences. *Science* **155**, 279–84.

Fitch, W.M., Margoliash, E. and Gould, K.S. (1969) The construction of phylogenetic trees. II. How well do they reflect past history? *Brookhaven Symp. Biol.* **21**, 217–42.

Fleagle, J.G. (1974) Dynamics of a brachiating siamang (*Hylobates (Symphalangus) syndactylus*). *Nature, Lond.* **248**, 259–60.

Fleagle, J.G. (1975) A small gibbon-like Hominoid from the Miocene of Uganda. *Folia primatol.* **24**, 1–15.

Fleagle, J.G. (1978) Size distributions of living and fossil primate faunas. *Paleobiology.* **4**, 67–76.

Fleagle, J.G. (1984) Are there any fossil gibbons? In *The Lesser Apes: Evolutionary and Behavioural Biology* (eds Preuschoft, H., Chivers, D.J.

Brockelmann, W.Y. and Creel, N.). Edinburgh University Press: Edinburgh, pp. 431–47.

Fleagle, J.G. (1985) New primate fossils from Colhuehuapian deposits at Gaiman and Sacanana, Chubut Province, Argentina. *Ameghiniana* **21**, 266–74.

Fleagle, J.G. (1986) The fossil record of early catarrhine evolution. In *Major Topics in Primate and Human Evolution* (eds Wood, B.A., Martin, L.B. and Andrews, P.J.). Cambridge University Press: Cambridge, pp. 130–49.

Fleagle, J.G. and Bown, T.M. (1983) New primate fossils from late Oligocene (Colhuehapian) localities of Chubut province, Argentina. *Folia primatol.* **41**, 240–66.

Fleagle, J.G. and Simons, E.L. (1982a) Skeletal remains of *Propliopithecus chirobates* from the Egyptian Oligocene. *Folia primatol.* **39**, 161–77.

Fleagle, J.G. and Simons, E.L. (1982b) The humerus of *Aegyptopithecus zeuxis*, a primitive anthropoid. *Am. J. phys. Anthrop.* **59**, 175–94.

Fleagle, J.G. and Kay, R.F. (1983) New interpretations of the phyletic position of Oligocene hominoids. In *New Interpretations of Ape and Human Ancestry* (eds Ciochon, R.L. and Corruccini, R.S.). Plenum Press: New York, pp. 181–210.

Fleagle, J.G. and Kay, R.F. (1985) The paleobiology of catarrhines. In *Ancestors: The Hard Evidence* (ed. Delson, E.). Alan R. Liss: New York, pp. 23–36.

Fleagle, J.G., Kay, R.F. and Simons, E.L. (1980) Sexual dimorphism in early anthropoids. *Nature, Lond.* **287**, 328–9.

Fleagle, J.G., Simons, E.L. and Conroy, G.C. (1975) Ape limb bones from the Oligocene of Egypt. *Science* **189**, 135–7.

Fleagle, J.G., Bown, T.M., Obradovitch, J.D. and Simons, E.L. (1986) Age of the earliest African anthropoids. *Science* **234**, 1247–9.

Fleagle, J.G., Powers, D.W., Conroy, G.C. and Watters, J.P. (1987) New fossil platyrrhines from Santa Cruz Province, Argentina. *Folia primatol.* **48**, 65–77.

Fleagle, J.G., Stern, J.T., Jungers, W.L., Susman, R.L., Vangor, A.K. and Wells, J.P. (1981) Climbing: a biomechanical link with brachiation and with bipedalism. *Symp. zool. Soc. Lond.* **48**, 359–75.

Foerg, R. (1982) Reproduction in *Cheirogaleus medius. Folia primatol.* **39**, 49–62.

Fogden, M.P.L. (1974) A preliminary field study of the western tarsier, *Tarsius bancanus* Horsefield. In *Prosimian Biology* (eds Martin, R.D., Doyle, G.A. and Walker, A.C.). Duckworth: London, pp. 151–65.

Fontaine, R. (1981) The uakaris, genus *Chiropotes*. In *Ecology and Behavior of Neotropical Primates* (eds Coimbra-Filho, A.F. and Mittermeier, R.A.). Academia Brasileira de Ciências: Rio de Janeiro, pp. 443–93.

Ford, C.E. and Hamerton, J.L. (1956) The chromosomes of man. *Nature, Lond.* **178**, 1020.

Ford, S.M. (1980) Callitrichids as phyletic dwarfs, and the place of Callitrichidae in Platyrrhini. *Primates* **21**, 31–43.

Ford, S.M. and Morgan, G.S. (1986) A new ceboid femur from the late Pleistocene of Jamaica. *J. vert. Paleont.* **6**, 281–9.

Forrest, M.G., Caithard, A.M. and Bertrand, J.A. (1973) Evidence of testicular activity in early infancy. *J. clin. Endocr. Metab.* **37**, 148–51.

Forster, A. (1916) Die Mm contrahentes und interossei manus in der Säugetierreihe und beim Menschen. *Arch. Anat. Physiol.* **1916**, 101–378.

Forsyth Major, C.I. (1880) Beiträge zur Geschichte der fossilen Pferde, insbesondere Italiens. *Abh. schweiz. paläont. Ges.* **7**, 1–154.

Forsyth Major, C.I. (1893) On *Megaladapis madagascariensis*, an extinct gigantic lemuroid from Ambolisatra. *Phil. Trans. R. Soc. Lond. B* **185**, 15–38.

Forsyth Major, C.I. (1896) Preliminary notice on fossil monkeys from Madagascar. *Geol. Mag. n.s.* **3**, 433–6.

Forsyth Major, C.I. (1899) Exhibition of, and remarks upon, some skulls of Malagasy lemurs. *Proc. zool. Soc. Lond.* **1899**, 987–8.

Forsyth Major, C.I. (1900) A summary of our present knowledge of extinct primates from Madagascar. *Geol. Mag. n.s.* **7**, 492–9.

Forsyth Major, C.I. (1901) On some characters of the skull in lemurs and monkeys. *Proc. Zool. Soc. Lond.* **1901**, 129–53.

Fortau, R. (1918) *Contribution à l'Étude des Vertébrés Miocènes de l'Egypte.* Survey Department: Cairo.

Fossey, D. and Harcourt, A.H. (1977) Feeding ecology of free-ranging mountain gorilla (*Gorilla gorilla beringei*). In *Primate Ecology* (ed. Clutton-Brock, T.H.). Academic Press: London, pp. 415–47.

Francke, U. and Oliver, N. (1978) Quantitative analysis of high-resolution trypsin-Giemsa bands on human prometaphase chromosomes. *Hum. Genet.* **45**, 137–65.

Franzen, J.L. (1983) Senckenberg-Grabungen 1982 in der Grube Messel. *Natur u. Museum* **113**, 148–51.

Franzen, J.L. (1987) Ein neuer Primate aus dem Mitteleozän der Grube Messel (Deutschland, S-Hessen). *Cour. Forsch.-Inst. Senckenberg* **91**, 151–87.

Frayer, D.W. (1973) *Gigantopithecus* and its relationship to *Australopithecus*. *Am. J. phys. Anthrop.* **39**, 413–26.

Freedman, L. (1976) South African fossil Cercopithecoidea: a reassessment including a description of new material from Makapansgat, Sterkfontein and Taung. *J. hum. Evol.* **5**, 297–315.

Freedman, L. and Stenhouse, N.S. (1972) The *Parapapio* species of Sterkfontein, Transvaal, South Africa. *Palaeont. afr.* **14**, 93–111.

Freeland, W.J. and Janzen, D.H. (1974) Strategies in herbivory by mammals: the role of plant secondary compounds. *Am. Nat.* **108**, 269–89.

Freese, C.H. and Oppenheimer, J.R. (1981) The capuchin monkeys, genus *Cebus*. In *Ecology and Behavior of Neotropical Primates* (eds Coimbra-Filho, A.F. and Mittermeier, R.A.). Academia Brasileira de Ciências: Rio de Janeiro, pp. 331–90.

Freitas, L. and Seuánez, H. (1982) Chromosome heteromorphisms in *Cebus apella*. *J. hum. Evol.* **10**, 173–80.

Friant, M. (1947) L'état de la dentition d'un lémurien nouveau-né (*Lepilemur leucopus* F. Major). *Rév. Stomat.* **48**, 597–9.

Friday, A.E. (1980) The status of immunological distance data in the construction of phylogenetic classifications: a critique. In *Chemosystematics: Principles and Practice* (eds Bisby, F.A., Vaughan, J.G. and Wright, C.A.). Academic Press: New York, pp. 289–304.

Galileo, G. (1638) *Dialogues Concerning Two New Sciences* (transl. by Crew, H. and De Salvio, A., 1914). Macmillan: New York.

Ganzhorn, J.U., Abraham, J.P. and Razanahoera-Rakotomalala, M. (1985) Some aspects of the natural history of food selection of *Avahi laniger*. *Primates* **26**, 452–63.

Garber, P.A., Moya, L. and Malaga, C. (1984) A preliminary field study of the moustached tamarin monkey (*Saguinus mystax*) in northeastern Peru: questions concerned with the evolution of a communal breeding system. *Folia primatol.* **42**, 17–33.

Gardner, R.C. and La Duke, J.C. (1979) An estimate of phylogenetic relationships within the genus *Crusea* (Rubiaceae) using character compatability analysis. *Syst. Bot.* **3**, 179–96.

Gartlan, J.S. (1970) Preliminary notes on the ecology and behavior of the drill, *Mandrillus leucophaeus* Ritgen, 1824. In *Old World Monkeys* (eds Napier, J.R. and Napier, P.H.). Academic Press: New York, pp. 445–80.

Gaudry, A. (1862) *Animaux Fossiles et Géologie de l'Attique*. F. Savy: Paris.

Gaudry, A. (1878) *Les Enchainements du Monde Animal dans les Temps Géologiques*, vol. 1: *Mammifères Tertiaires*. F. Savy: Paris.

Gaudry, A. (1890) Le dryopithèque. *Mém. Soc. géol. Fr. pal. Mém.* **1**, 1–11.

Gaupp, E. (1913) Die Reichertsche Theorie (Hammer-, Amboss- und Kieferfrage). *Arch. Anat. Physiol., Anat. Abt., Suppl.-Band* **13**, 1–416.

Gautier-Hion, A. (1978) Food niches and coexistence in sympatric primates in Gabon. In *Recent Advances in Primatology*, vol. 1: *Behaviour* (eds Chivers, D.J. and Herbert, J.). Academic Press: London, pp. 269–86.

Gautier-Hion, A. and Gautier, J.-P. (1978) Le singe de Brazza: une stratégie originale. *Z. Tierpsychol.* **46**, 84–104.

Gazin, C.L. (1958) A review of the Middle and Upper Eocene Primates of North America. *Smithson. misc. Colln* **136**, 1–112.

Gazin, C.L. (1965) An endocranial cast of the Bridger middle Eocene primate *Smilodectes gracilis*. *Smithson. misc. Colln* **149**, 1–13.

Gebo, D.L. (1986a) Miocene lorisids – the foot evidence. *Folia primatol.* **47**, 217–25.

Gebo, D.L. (1986b) Anthropoid origins – the foot evidence. *J. hum. Evol.* **15**, 421–30.

Gebo, D.L. (1987a) Functional anatomy of the tarsier foot. *Am. J. phys. Anthrop.* **73**, 9–31.

Gebo, D.L. (1987b) Implications of pedal adaptations for anthropoid origins. *Am. J. phys. Anthrop.* **72**, 202.

Gebo, D.L. (1987c) Locomotor diversity in prosimian primates. *Am. J. Primatol.* **13**, 271–81.

George, R.M. (1977) The limb musculature of the Tupaiidae. *Primates* **18**, 1–34.

Gérard, P. (1932) Études sur l'ovogenèse et l'ontogenèse chez les lémuriens du genre *Galago*. *Arch. Biol.* **43**, 93–151.

Gervais, P. (1872) Sur un singe fossile, d'espèce non encore décrite, qui a été découvert au Monte Bamboli (Italie). *C.R. Acad. Sci., Paris* **74**, 1217–23.

Gidley, J.W. (1919) Significance of divergence of the first digit in the primitive mammalian foot. *J. Wash. Acad. Sci.* **9**, 273–80.

Gidley, J.W. (1923) Paleocene primates of the Fort Union, with discussion of relationships of Eocene primates. *Proc. U.S. Nat. Mus.* **63**, 1–38.

Gilbert, A.N. (1986) Mammary number and litter-size in Rodentia: the 'one-half rule'. *Proc. natn. Acad. Sci. U.S.A.* **83**, 4828–30.

Gillman, J. and Gilbert, C. (1946) The reproductive cycle of the chacma baboon (*Papio ursinus*) with special reference to the problem of menstrual irregularities as assessed by the behavior of the sex skin. *S. Afr. J. med. Sci.* **11**, 1–54.

Gingerich, P.D. (1971) Cranium of *Plesiadapis*. *Nature, Lond.* **232**, 566.

Gingerich, P.D. (1973) Anatomy of the temporal bone in the Oligocene anthropoid *Apidium* and the origin of the Anthropoidea. *Folia primatol.* **19**, 329–37.

Gingerich, P.D. (1974) Dental function in the Palaeocene primate *Plesiadapis*. In *Prosimian Biology* (eds Martin, R.D., Doyle, G.A. and Walker, A.C.). Duckworth: London, pp. 531–41.

Gingerich, P.D. (1975a) Systematic position of *Plesiadapis*. *Nature, Lond.* **253**, 111–13.

Gingerich, P.D. (1975b) Dentition of *Adapis parisiensis* and the evolution of lemuriform primates. In *Lemur Biology* (eds Tattersall, I. and Sussman, R.W.). Plenum Press: New York, pp. 65–80.

Gingerich, P.D. (1976a) Paleontology and phylogeny: patterns of evolution at the species level in Early Tertiary mammals. *Am. J. Sci.* **276**, 1–28.

Gingerich, P.D. (1976b) Cranial anatomy and evolution of early Tertiary Plesiadapidae (Mammalia, Primates). *Mus. Pal. Univ. Mich., Pap. Paleont.* **15**, 1–140.

Gingerich, P.D. (1977a) Patterns of evolution in the mammalian fossil record. In *Patterns of Evolution as Illustrated in the Fossil Record* (ed. Hallam, A.). Elsevier: Amsterdam, pp. 469–500.

Gingerich, P.D. (1977b) New species of Eocene primates and the phylogeny of European Adapidae. *Folia primatol.* **28**, 60–80.

Gingerich, P.D. (1977c) Radiation of Eocene Adapidae in Europe. *Géobios Mém. spéc.* **1**, 165–82.

Gingerich, P.D. (1977d) Homologies of the anterior teeth in Indriidae and a functional basis for dental reduction in primates. *Am. J. phys. Anthrop.* **47**, 387–93.

Gingerich, P.D. (1979) The stratophenetic approach to phylogeny reconstruction in vertebrate paleontology. In *Phylogenetic Analysis and Paleontology* (eds Cracraft, J. and Eldredge, N.). Columbia University Press: New York, pp. 41–77.

Gingerich, P.D. (1980a) Dental and cranial adaptations in Eocene Adapidae. *Z. Morph. Anthrop.* **71**, 135–42.

Gingerich, P.D. (1980b) Eocene Adapidae, paleobiogeography, and the origin of South American Platyrrhini. In *Evolutionary Biology of the New World Monkeys and Continental Drift* (eds Ciochon, R.L. and Chiarelli, A.B.). Plenum Press: New York, pp. 123–38.

Gingerich, P.D. (1981a) Cranial morphology and adaptations in Eocene Adapidae. I. Sexual dimorphism in *Adapis magnus* and *Adapis parisiensis*. *Am. J. phys. Anthrop.* **56**, 217–34.

Gingerich, P.D. (1981b) Early Cenozoic Omomyidae and the evolutionary history of tarsiiform primates. *J. hum. Evol.* **10**, 345–74.

Gingerich, P.D. (1984a) Punctuated equilibria – where is the evidence? *Syst. Zool.* **33**, 335–8.

Gingerich, P.D. (1984b) Paleobiology of tarsiiform primates. In *Biology of Tarsiers* (ed. Niemitz, C.). Gustav Fischer Verlag: Stuttgart, pp. 33–44.

Gingerich, P.D. (1985a) Molecular evolutionary clocks. *Science* **229**, 8.

Gingerich, P.D. (1985b) Nonlinear molecular clocks and ape–human divergence times. In *Hominid Evolution: Past, Present and Future* (ed. Tobias, P.V.). Alan Liss: New York, pp. 411–16.

Gingerich, P.D. (1986a) *Plesiadapis* and the delineation of the order Primates. In *Major Topics in Primate and Human Evolution* (eds Wood, B.A., Martin, L.B. and Andrews, P.J.). Cambridge University Press: Cambridge, pp. 32–46.

Gingerich, P.D. (1986b) Early Eocene *Cantius torresi* – oldest primate of modern aspect from North America. *Nature, Lond.* **319**, 319–21.

Gingerich, P.D. and Haskin, R.A. (1981) Dentition of early Eocene *Pelycodus jarrovii* (Mammalia, Primates) and the generic attribution of species formerly referred to *Pelycodus*. *Contr. Mus. Paleont. Univ. Mich.* **25**, 327–37.

Gingerich, P.D. and Martin, R.D. (1981) Cranial morphology and adaptations in Eocene Adapidae. II. The Cambridge skull of *Adapis parisiensis*. *Am. J. phys. Anthrop.* **56**, 235–57.

Gingerich, P.D. and Rose, K.D. (1979) Anterior dentition of the Eocene condylarth *Thryptacodon*: convergence with the tooth comb of lemurs. *J. Mammal.* **60**, 16–22.

Gingerich, P.D. and Sahni, A. (1979) *Indraloris* and *Sivaladapis*: Miocene adapid primates from the Siwaliks of India and Pakistan. *Nature, Lond.* **279**, 415–16.

Gingerich, P.D. and Sahni, A. (1984) Dentition of *Sivaladapis nagrii* (Adapidae) from the Late Miocene of India. *Int. J. Primatol.* **5**, 63–79.

Gingerich, P.D. and Schoeninger, M.J. (1977) The fossil record and primate phylogeny. *J. hum. Evol.* **6**, 483–505.

Gingerich, P.D. and Simons, E.L. (1977) Systematics, phylogeny, and evolution of Early Eocene Adapidae (Mammalia, Primates) in North America. *Contr. Mus. Paleont. Univ. Mich.* **24**, 245–79.

Gingerich, P.D. and Smith, B.H. (1985) Allometric scaling in the dentition of primates and insectivores. In *Size and Scaling in Primate Biology* (eds Jungers, W.L.). Plenum Press: New York, pp. 257–72.

Gingerich, P.D., Smith, B.H. and Rosenberg, K. (1982) Allometric scaling in the dentition of primates and prediction of body weight from tooth size in fossils. *Am. J. phys. Anthrop.* **58**, 81–100.

Ginsburg, L. (1975) Le pliopithèque des Faluns helvétiens de la Touraine et de l'Anjou. *Coll. int. C.N.R.S.* **218**, 877–86.

Giolli, R.A. and Tigges, J. (1970) The primary optic pathways and nuclei of primates. In *The Primate Brain* (eds Noback, C.R. and Montagna, W.). Appleton-Century-Crofts: New York, pp. 29–54.

Giusto, J.P. and Margulis, L. (1981) Karyotypic fission theory and the evolution of Old World monkeys and apes. *Biosystems* **13**, 267–302.

Glander, K.E. (1982) The impact of plant secondary compounds on primate feeding behavior. *Yb. phys. Anthrop.* **25**, 1–18.

Glover, T.D. and Sale, J.B. (1968) The reproductive system of the male rock hyrax (*Procavia* and *Heterohyrax*). *J. Zool., Lond.* **156**, 351.

Godinot, M. (1984) Un nouveau genre temoignant de la diversité des adapinés (Primates, Adapidae) à l'Éocène terminal. *C.R. Acad. Sci. Paris, sér. II.* **299**, 1291–6.

Godinot, M. and Jouffroy, F.K. (1984) La main d'*Adapis* (primate, adapidé). In *Actes du Symposium Paléontologique Georges Cuvier* (eds Buffetaut, E., Mazin, J.M. and Salmon, E.). Ville de Montbeliard: Montbeliard, pp. 221–42.

Goffart, M., Missotten, L., Faidherbe, J. and Watillon, M. (1976) A duplex retina and the electroretinogram in the nocturnal *Perodicticus potto*. *Arch. int. Physiol. Biochem.* **84**, 493–516.

Goldman, D., Giri, P.R. and O'Brien, S.J. (1987) A molecular phylogeny of the hominoid primates as indicated by two-dimensional protein electrophoresis. *Proc. natn. Acad. Sci. U.S.A.* **84**, 3307–11.

Goldizen, A.W. and Terborgh, J. (1986) Cooperative polyandry and helping behavior in saddle-backed tamarins (*Saguinus fuscicollis*). In *Primate Ecology and Conservation* (eds Else, J.G. and Lee, P.C.). Cambridge University Press: Cambridge, pp. 191–8.

Goldschmidt, R. (1940) *The Material Basis of Evolution*. Yale University Press: New Haven.

Goodall, A.G. (1977) Feeding and ranging behaviour of a mountain gorilla group (*Gorilla gorilla beringei*) in the Tshibinda-Kahuzi Region, Zaire. In *Primate Ecology* (ed. Clutton-Brock, T.H.). Academic Press: London, pp. 449–79

Goodall, J. (1968) The behaviour of free-living chimpanzees in Gombe Stream Reserve. *Anim. Behav. Monogr.* **1**, 165–311.

Goodall, J., Bandora, A., Bergmann, E., Busse, C. Matama, H., Mpongo, E., Pierce, A. and Riss, D.

(1979) Intercommunity interactions in the chimpanzee population of the Gombe National Park. In *The Great Apes* (eds Hamburg, D.A. and McCown, E.R.). Benjamin/Cummings: Menlo Park, California, pp. 13–54.

Goodman, L. and Wislocki, G.B. (1935) Cyclical uterine bleeding in a New World monkey (*Ateles geoffroyi*). *Anat. Rec.* **61**, 379–87.

Goodman, M. (1961) The role of immunological differences in the phyletic development of human behavior. *Hum. Biol.* **33**, 131–62.

Goodman, M. (1962a) Immunochemistry of the primates and primate evolution. *Ann. N.Y. Acad. Sci.* **102**, 219–34.

Goodman, M. (1962b) Evolution of immunologic species specificity of human serum proteins. *Hum. Biol.* **34**, 104–50.

Goodman, M. (1963a) Man's place in the phylogeny of the primates as reflected in serum proteins. In *Classification and Human Evolution* (ed. Washburn, S.L.). Wenner-Gren Foundation: New York, pp. 204–34.

Goodman, M. (1963b) Serological analysis of the systematics of recent hominoids. *Hum. Biol.* **35**, 377–436.

Goodman, M. (1966) Phyletic position of tree shrews. *Science* **153**, 1550.

Goodman, M. (1967) Deciphering primate phylogeny from macromolecular specificities. *Am. J. phys. Anthrop.* **26**, 255–76.

Goodman, M. (1968) Evolution of catarrhine primates at the macromolecular level. *Primates Med.* **1**, 10–26.

Goodman, M. (1973) The chronicle of primate phylogeny contained in proteins. *Symp. zool. Soc. Lond.* **33**, 339–75.

Goodman, M. (1975) Protein sequence and immunological specificity: their role in phylogenetic studies of primates. In *Phylogeny of the Primates* (eds Luckett, W.P. and Szalay, F.S.). Plenum Press: New York, pp. 219–48.

Goodman, M. (1976) Toward a genealogical description of the primates. In *Molecular Anthropology* (eds Goodman, M. and Tashian, R.E.). Plenum Press: New York, pp. 321–53.

Goodman, M. and Moore, G.W. (1971) Immunodiffusion systematics of the primates. I. The Catarrhini. *Syst. Zool.* **20**, 19–62.

Goodman, M. and Tashian, R.E. (eds) (1976) *Molecular Anthropology: Genes and Proteins in the Evolutionary Ascent of the Primates*. Plenum Press: New York.

Goodman, M., Czelusniak, J. and Beeber, J.E. (1985) Phylogeny of primates and other eutherian orders: a cladistic analysis using amino acid and nucleotide sequence data. *Cladistics* **1**, 171–85.

Goodman, M., Moore, G.W. and Matsuda, G. (1975) Darwinian evolution in the genealogy of haemoglobin. *Nature, Lond.* **253**, 603–8.

Goodman, M., Barnabas, J., Matsuda, G. and Moore, G. (1971) Molecular evolution in the descent of man. *Nature, Lond.* **233**, 604–13.

Goodman, M., Czelusniak, J., Moore, G., Romero-Herrera, A.E. and Matsuda, G. (1979) Fitting the gene lineage into its species lineage. A parsimony strategy illustrated by cladograms constructed from globin sequences. *Syst. Zool.* **28**, 132–63.

Goodman, M., Farris, W., Moore, G.W., Prychodko, W., Poulik, E. and Sorenson, M. (1974) Immuno-diffusion systematics of the primates II: findings on *Tarsius*, Lorisidae and Tupaiidae. In *Prosimian Biology* (eds Martin, R.D., Doyle, G.A. and Walker, A.C.). Duckworth: London, pp. 881–90.

Goodman, M., Moore, G.W., Barnabas, J. and Matsuda, G. (1974) The phylogeny of human globin genes investigated by the maximum parsimony method. *J. molec. Evol.* **3**, 1–48.

Goodman, M., Romero-Herrera, A., Dene, H., Czelusniak, J. and Tashian, R.E. (1982) Amino acid sequence evidence on the phylogeny of primates and other eutherians. In *Macromolecular Sequences in Systematics and Evolutionary Biology* (ed. Goodman, M.). Plenum Press: New York, pp. 115–91.

Gould, E. (1978) The behavior of the moonrat, *Echinosorex gymnurus* (Erinaceidae) and the pentail tree shrew, *Ptilocercus lowii* (Tupaiidae) with comments on the behavior of other Insectivora. *Z. Tierpsychol.* **48**, 1–27.

Gould, S.J. (1966) Allometry and size in ontogeny and phylogeny. *Biol. Rev.* **41**, 587–640.

Gould, S.J. (1975) Allometry in primates, with emphasis on scaling and the evolution of the brain. *Contr. Primatol.* **5**, 244–92.

Gould, S.J. (1977) *Ontogeny and Phylogeny*. Belknap Press (Harvard University Press): Cambridge, Mass.

Gould, S.J. and Eldredge, N. (1977) Punctuated equilibria: the tempo and mode of evolution reconsidered. *Paleobiology* **3**, 115–51.

Graham, A.L. (1981) Plate tectonics. In *The Evolving Earth* (ed. Cocks, R.L.M.). Cambridge University Press: Cambridge, pp. 165–77.

Graham, C.E. (1981) Menstrual cycle of the great apes. In *Reproductive Biology of the Great Apes* (ed. Graham, C.E.). Academic Press: New York, pp. 1–43.

Grand, T.I. (1972) A mechanical interpretation of terminal branch feeding. *J. Mammal.* **53**, 198–201.

Grandidier, G. (1904) Un lémurien fossile de France, le *Pronycticebus gaudryi*. *Bull. Mus. Hist. nat., Paris* **10**, 9–13.

Gray, J. (1944) Studies in the mechanics of the tetrapod skeleton. *J. exp. Biol.* **20**, 88–116.

Gray, J.E. (1825) An outline of an attempt at the disposition of Mammalia into tribes and families, with a list of the genera apparently appertaining to each tribe. *Ann. Philos.* **10**, 337–44.

Greaves, W.S. (1985) The mammalian postorbital bar as a torsion-resisting helical strut. *J. Zool., Lond.* A **207**, 125–36.

Greenfield, L.O. (1979) On the adaptive pattern of *Ramapithecus*. *Am. J. phys. Anthrop.* **50**, 527–48.

Gregory, W.K. (1910) The orders of mammals. *Bull. Am. Mus. nat. Hist.* **27**, 1–524.

Gregory, W.K. (1913) Relationship of the Tupaiidae and of Eocene lemurs, especially *Notharctus*. *Bull. geol. Soc. Am.* **24**, 247–52.

Gregory, W.K. (1915) I. On the relationship of the Eocene lemur *Notharctus* to the Adapidae and to other primates. II. On the classification and phylogeny of the Lemuroidea. *Bull. geol. Soc. Am.* **26**, 419–46.

Gregory, W.K. (1916) Studies on the evolution of the Primates. I. The Cope–Osborn 'Theory of Trituberculy' and the ancestral molar pattern of the Primates. *Bull. Am. Mus. nat. Hist.* **35**, 239–57.

Gregory, W.K. (1920a) On the structure and relations of *Notharctus*, an American Eocene primate. *Mem. Am. Mus. nat. Hist. n.s.* **3**, 49–243.

Gregory, W.K. (1920b) The origin and evolution of the human dentition: a palaeontological review. Part I. Stages of ascent from the Silurian fishes to the mammals of the age of reptiles. *J. dent. Res.* **2**, 89–183.

Gregory, W.K. (1920c) The origin and evolution of the human dentition: a palaeontological review. Part II. Stages of ascent from the Paleocene placental mammals to the lower primates. *J. dent. Res.* **2**, 215–83.

Gregory, W.K. (1920d) The origin and evolution of the human dentition: a palaeontological review. Part III. Nature's earlier experiments in evolving large-eyed and short-jawed primates. *J. dent. Res.* **2**, 357–433.

Gregory, W.K. (1920e) The origin and evolution of the human dentition: a palaeontological review. Part IV. The dentition of the higher primates and their relationships with man. *J. dent. Res.* **2**, 607–717.

Gregory, W.K. (1921) The origin and evolution of the human dentition: a palaeontological review. Part V. Later stages in the evolution of the human dentition; with a final summary and a bibliography. *J. dent. Res.* **3**, 87–228.

Gregory, W.K. (1922) *The Origin and Evolution of the Human Dentition: A Palaeontological Review.* Williams and Wilkins: Baltimore.

Gregory, W.K. (1928) The upright posture of man: a review of its origin and evolution. *Proc. Am. philos. Soc.* **67**, 339–74.

Gregory, W.K. (1951) *Evolution Emerging: A Survey of Changing Patterns from Primeval Life to Man* (2nd edition). Macmillan: New York.

Gregory, W.K. and Hellman, M. (1927) The dentition of *Dryopithecus* and the origin of man. *Anthrop. Pap. Am. Mus. nat. Hist.* **28**, 1–123.

Gregory, W.K. and Simpson, G.G. (1926) Cretaceous mammal skulls from Mongolia. *Am. Mus. Novit.* no. **225**, 1–20.

Grine, F.E. and Vrba, E.S. (1980) Prismatic enamel: a preadaptation for mammalian diphyodonty. *S. Afr. J. Sci.* **76**, 134–41.

Grine, F.E., Krause, D.W. and Martin, L.B. (1985) The ultrastructure of *Oreopithecus bambolii* tooth enamel: systematic implications. *Am. J. phys. Anthrop.* **66**, 177–8.

Grine, F.E., Vrba, E.S. and Cruickshank, A.R.I. (1979) Enamel prisms and diphyodonty: linked apomorphies of Mammalia. *S. Afr. J. Sci.* **75**, 114–20.

Grosser, O. (1909) *Vergleichende Anatomie und Entwicklungsgeschichte der Eihäute und der Placenta.* Wilhelm Braumüller: Vienna.

Grosser, O. (1927) *Frühentwicklung, Eihautbildung und Placentation des Menschen und der Säugetiere.* J.F. Bergmann: Munich.

Groves, C.P. (1974) Taxonomy and phylogeny of prosimians. In *Prosimian Biology* (eds Martin, R.D., Doyle, G.A. and Walker, A.C.). Duckworth: London, pp. 449–73.

Gunnell, G.F. and Gingerich, P.D. (1987) Skull and partial skeleton of *Plesiadapis cookei* from the Clark's Fork Basin. *Am. J. phys. Anthrop.* **72**, 206.

Gurche, J.A. (1982) Early primate brain evolution. In *Primate Brain Evolution* (eds Armstrong, E. and Falk, D.). Plenum Press: New York, pp. 227–46.

Haeckel, E. (1866) *Generelle Morphologie der Organismen: Allgemeine Grundzüge der organischen Formen-Wissenschaft, mechanisch begründet durch die von Charles Darwin reformierte Descendenz-Theorie* (2 vols). Georg Reimer: Berlin.

Hafleigh, A.S. and Williams, C.A. (1966) Antigenic correspondence of serum proteins among the primates. *Science* **151**, 1530–5.

Haines, R.W. (1955) The anatomy of the hand of certain insectivores. *Proc. zool. Soc. Lond.* **125**, 761–77.

Haines, R.W. (1958) Arboreal or terrestrial ancestry of placental mammals. *Q. Rev. Biol.* **33**, 1–23.

Haldane, J.B.S. (1957) The cost of natural selection. *J. Genet.* **55**, 511–24.

Hall, K.R.L. (1965) Behaviour and ecology of the wild patas monkey, *Erythrocebus patas*, in Uganda. *J. Zool., Lond.* **148**, 15–87.

Hall, R.D., Beattie, R.J. and Wyckoff, G.H. (1977) Weight gains and sequence of dental eruptions in infant owl monkeys (*Aotus trivirgatus*). In *Nursery Care of Nonhuman Primates* (ed. Ruppenthal, G.C.). Plenum Press: New York, pp. 321–8.

Hallam, A. (ed.) (1973) *Atlas of Palaeobiogeography.* Elsevier: New York.

Hallam, A. (1975) *Jurassic Environments.* Cambridge University Press: Cambridge.

Hall-Craggs, E.C.B. (1965) An osteometric study of the hind limb of the Galagidae. *J. Anat., Lond.* **99**, 119–26.

Hamasaki, D.I. (1967) An anatomical and electrophysiological study of the retina of the owl monkey, *Aotes trivigatus*. *J. comp. Neurol.* **130**, 163–74.

Hambrey, M.J. and Harland, W.B. (1981) The evolution of climates. In *The Evolving Earth* (ed. Cocks, R.L.M.). Cambridge University Press: Cambridge, pp. 137–52.

Hamerton, J.L. (1963) Primate chromosomes. *Symp. zool. Soc. Lond.* **10**, 211–19.

Hamerton, J.L. (1968) Robertsonian translocations in man: evidence for pre-zygotic selection. *Cytogenetics* **7**, 260–76.

Hamerton, J.L., Canning, N., Ray, M. and Smith, S. (1975) A cytogenetic survey of 14,069 newborn infants. I. Incidence of chromosome abnormalities. *Clin. Genet.* **8**, 223–43.

Hamilton, E.L. (1965) *Applied Geochronology.* Academic Press: London.

Hamlett, G.W.D. and Wislocki, G.B. (1934) A proposed classification for types of twins in mammals. *Anat. Rec.* **61**, 81–96.

Hanson, E.D. (1977) *The Origin and Early Evolution of Animals.* Pitman: London.

Haq, B.U. and van Eysinga, F.W.B. (1987). *Geological Time Table.* Elsevier: New York.

Harcourt, A.H. (1979) The social relations and group structure of wild mountain gorillas. In *The Great Apes* (eds Hamburg, D.A. and McCown, E.R.). Benjamin/Cummings: Menlo Park, California, pp. 187–92.

Harding, R.S.O. (1981) An order of omnivores: nonhuman primate diets in the wild. In *Omnivorous Primates: Gathering and Hunting in Human Evolution* (eds Harding, R.S.O. and Teleki, G.). Columbia University Press: New York, pp. 191–214.

Harlé, E. (1899) Nouvelles pièces de dryopithèque et quelques coquilles, de Saint-Gaudens (Haute Garonne). *Bull. Soc. Géol. Fr. sér.* 3. **27**, 304–10.

Harms, J.W. (1956) Fortpflanzungsbiologie. In *Primatologia,* vol. 1 (eds Hofer, H., Schultz, A.H. and Starck, D.). Karger: Basel, pp. 561–660.

Harnden, D.G., Klinger, H.P., Jensen, J.T. and Kaelbling, M. (1985) *An International System for Human Cytogenetic Nomenclature (1985). Report of the Standing Committee on Human Cytogenetic Nomenclature.* Karger: Basel.

Harrison, G.A. and Weiner, J.S. (1964) Human Evolution. In *Human Biology* (eds Harrison, G.A., Weiner, J.S., Tanner, J.M. and Barnicot, N.A.). Oxford University Press: Oxford, pp. 3–98.

Harrison, J.L. (1951) Squirrels for bird-watchers. *Malay. Nat. J.* **5**, 134–54.

Harrison, J.L. (1955) Data on the reproduction of some Malayan mammals. *Proc. zool. Soc. Lond.* **125**, 445–60.

Harrison, T. (1982) Small-bodied apes from the Miocene of East Africa. PhD thesis, University of London.

Harrison, T. (1986) A reassessment of the phylogenetic relationships of *Oreopithecus bambolii* Gervais. *J. hum. Evol.* **15**, 541–83.

Harrison, T. (1987) The phylogenetic relationships of the early catarrhine primates: a review of the current evidence. *J. hum. Evol.* **16**, 41–80.

Hartman, C.G. (1932) Studies in the reproduction of the monkey *Macacus (Pithecus) rhesus,* with special reference to menstruation and pregnancy. *Contr. Embryol. Carnegie Inst. Wash.* **23**, 1–161.

Harvey, P.H. and Mace, G.M. (1982) Comparison between taxa and adaptive trends: problems of methodology. In *Current Problems in Sociobiology* (eds King's College Sociobiology Group). Cambridge University Press: Cambridge, pp. 343–61.

Harvey, P.H., Martin, R.D. and Clutton-Brock, T.H. (1987) Life histories in comparative perspective. In *Primate Societies* (eds Smuts, B.B., Cheney, D.L., Seyfarth, R.M., Wrangham, R.W. and Struhsaker, T.T.). University of Chicago Press: Chicago, pp. 181–96.

Hassler, R. (1966) Comparative anatomy of the central visual systems in day- and night-active primates. In *Evolution of the Forebrain* (eds Hassler, R. and Stephan, H.). Thieme Verlag: Stuttgart, pp. 419–34.

Hearn, J.P. (1977) The endocrinology of reproduction in the common marmoset, *Callithrix jacchus.* In *The Biology and Conservation of the Callitrichidae* (ed. Kleiman, D.G.). Smithsonian Institution Press: Washington, pp. 163–71.

Hearn, J.P. (1983) The common marmoset (*Callithrix jacchus*). In *Reproduction in New World Primates* (ed. Hearn, J.P.). MT Press: Lancaster, pp. 183–215.

Hecht, M.K. (1976) Phylogenetic inference and methodology as applied to the vertebrate record. *Evol. Biol.* **9**, 335–63.

Hecht, M.K. and Edwards, J.L. (1977) The methodology of phylogenetic inference above the species level. In *Major Patterns in Vertebrate Evolution* (eds Hecht, M.K., Goody, P.C. and Hecht, B.M.). Plenum Press: New York, pp. 3–51.

Heil, R., von Koenigswald, W., Lippmann, H.G., Graner, D. and Heunisch, C. (1987) *Fossilien der Messel-Formation.* Hessisches Landesmuseum: Darmstadt.

Heitz, E. (1928) Das Heterochromatin der Moose. *Jb. wiss. Bot.* **69**, 762–818.

Hellman, M. (1928) Racial characters in human dentition. *Proc. Am. philos. Soc.* **67**, 157–74.

Heltne, P.G., Wojcik, J.F. and Pook, A.G. (1981) Goeldi's monkey, genus *Callimico.* In *Ecology and Behavior of Neotropical Primates* (eds Coimbra-Filho, A.F. and Mittermeier, R.A.). Academia Brasileira de Ciências: Rio de Janeiro, pp. 169–209.

Hemmingsen, A.M. (1950) The relation of standard (basal) energy metabolism to total fresh weight of living organisms. *Rep. Steno Mem. Hosp.* **4**, 7–58.

Hemmingsen, A.M. (1960) Energy metabolism as related to body size and respiratory surfaces, and its evolution. *Rep. Steno Mem. Hosp.* **9**, 1–110.

Henckel, K.O. (1928) Das Primordialcranium von *Tupaia* und der Ursprung der Primaten. *Z. Anat. Entw.-Gesch.* **86**, 204–27.

Henderson, A.S., Warburton, D., Megraw-Ripley, S. and Atwood, K.C. (1977) The chromosomal location of rDNA in selected lower primates. *Cytogenet. Cell Genet.* **19**, 281–302.

Hendrickx, A.G. and Kraemer, D.C. (1969) Observations on the menstrual cycle, optimal mating time and pre-implantation embryos of the baboon, *Papio anubis* and *Papio cynocephalus.* *J. Reprod. Fert. (Suppl.)* **6**, 119–28.

Henkel, S. and Krebs, B. (1977) Der erste Fund eines Säugetier-Skeletts aus der Jura-Zeit. *Umschau Wiss. Technol.* **77**, 217–18.

Henkel, S. and Krusat, G. (1980) Die Fossillagerstätte in der Kohlengrube Guimarota (Portugal) und der erste Fund eines Docodontiden-Skeletts. *Berl. geowiss. Abh. A* **20**, 209–16.

Hennig, W. (1950) *Grundzüge einer Theorie der Phylogenetischen Systematik.* Deutscher Zentral-Verlag: Berlin.

Hennig, W. (1965) Phylogenetic systematics. *Ann. Rev. Entomol.* **10**, 97–116.

Hennig, W. (1979) *Phylogenetic Systematics* (reprint of 1966 edition with a foreword by Rosen, D.E., Nelson, G. and Patterson, C.). University of Illinois Press: Urbana.

Hershkovitz, P. (1970) Notes on Tertiary platyrrhine monkeys and description of a new genus from the late Miocene of Colombia. *Folia primatol.* **12**, 1–37.

Hershkovitz, P. (1971) Basic crown patterns and cusp homologies of mammalian teeth. In *Dental Morphology and Evolution* (ed. Dahlberg, A.A.). University of Chicago Press: Chicago, pp. 95–150.

Hershkovitz, P. (1974a) A new genus of Late Oligocene monkey (Cebidae, Platyrrhini) with notes on postorbital closure and platyrrhine evolution. *Folia primatol.* **21**, 1–35.

Hershkovitz, P. (1974b) The entotympanic bone and the origin of higher primates. *Folia primatol.* **22**, 237–42.

Hershkovitz, P. (1977) *Living New World Primates (Platyrrhini), with an Introduction to Primates*, vol. 1. University of Chicago Press: Chicago.

Hewett-Emmett, D., Venta, P.J. and Tashian, R.E. (1982) Features of gene structure, organization, and expression that are providing unique insights into molecular evolution and systematics. In *Macromolecular Sequences in Systematics and Evolutionary Biology* (ed. Goodman, M.). Plenum Press: New York, pp. 357–405.

Hiiemae, K.M. (1976) Masticatory movements in primitive mammals. In *Mastication* (eds Anderson, D.J. and Joysey, K.A.). Academic Press: New York, pp. 105–18.

Hiiemae, K.M. (1978) Mammalian mastication, a review of the activity of the jaw muscles and the movements they produce in chewing. In *Development, Function and Evolution of Teeth* (eds Butler, P.M. and Joysey, K.A.). Academic Press: New York, pp. 359–98.

Hiiemae, K.M. and Ardran, G.M. (1968) A cinefluorographic study of mandibular movement during feeding in the rat (*Rattus norvegicus*). *J. Zool., Lond.* **154**, 139–54.

Hiiemae, K.M. and Kay, R.F. (1973) Evolutionary trends in the dynamics of primate mastication. *Symp. 4th Int. Congr. Primatol.* **3**, 28–64.

Hildebrand, M. (1967) Symmetrical gaits of primates. *Am. J. phys. Anthrop.* **26**, 119–30.

Hildwein, G. (1972) Métabolisme énergétique de quelques mammifères et oiseaux de la forêt équatoriale. *Arch. Sci. physiol.* **26**, 379–400.

Hildwein, G. and Goffart, M. (1975) Standard metabolism and thermoregulation in a prosimian, *Perodicticus potto*. *Comp. Biochem. Physiol.* **50A**, 201–12.

Hill, J.P. (1932) The developmental history of the primates. *Phil. Trans. R. Soc. Lond. B* **221**, 45–178.

Hill, J.P. (1965) On the placentation of *Tupaia*. *J. Zool., Lond.* **146**, 278–304.

Hill, W.C.O. (1936) The affinities of the lorisoids. *Ceylon J. Sci. B* **19**, 287–314.

Hill, W.C.O. (1953) *Primates: Comparative Anatomy and Taxonomy,* vol. 1: *Strepsirhini.* Edinburgh University Press: Edinburgh.

Hill, W.C.O. (1962) *Primates: Comparative Anatomy and Taxonomy*, vol. 5: *Cebidae, Part B*. Edinburgh University Press: Edinburgh.

Hill, W.C.O. (1972) *Evolutionary Biology of the Primates*. Academic Press: London.

Hill, W.C.O. and Rewell, R.E. (1954) The caecum of monotremes and marsupials. *Trans. zool. Soc. Lond.* **28**, 185–240.

Hladik, C.M. (1973) Alimentation et activité d'un groupe de chimpanzés réintroduits en forêt gabonaise. *Terre et Vie* **27**, 343–413.

Hladik, C.M. (1981) Diet and the evolution of feeding strategies among forest primates. In *Omnivorous Primates: Gathering and Hunting in Human Evolution* (eds Harding, R.S.O. and Teleki, G.). Columbia University Press: New York, pp. 215–54.

Hladik, C.M. and Guegen, L. (1974) Géophagie at nutrition minérale chez les primates sauvages. *C. R. Acad. Sci., Paris, D* **279**, 1393–5.

Hladik, C.M., Charles-Dominique, P. and Petter, J.-J. (1980) Feeding strategies of five nocturnal prosimians in the dry forest of the west coast of Madagascar. In *Nocturnal Malagasy Primates: Ecology, Physiology and Behaviour* (eds Charles-Dominique, P. *et al.*). Academic Press: New York, pp. 41–73.

Hodos, W. and Campbell, C.B.G. (1969) *Scala naturae*: why there is no theory in comparative psychology. *Psychol. Rev.* **76**, 337–50.

Hofer, H.O. (1980) The external anatomy of the oro-nasal region of primates. *Z. Morph. Anthrop.* **71**, 233–49.

Hofer, H.O. and Wilson, J.A. (1967) An endocranial cast of an early Oligocene primate. *Folia primatol.* **5**, 148–52.

Hoffstetter, R. (1969) Un primate de l'Oligocène inférieur sud-americain *Branisella boliviana* gen. et sp. nov.. *C.R. Acad. Sci., Paris, D* **269**, 434–7.

Hoffstetter, R. (1972) Relationships, origins, and history of the ceboid monkeys and caviomorph rodents: a modern reinterpretation. In *Evolutionary Biology*, vol. 6 (eds Dobzhansky, Th., Hecht, M.K. and Steere, W.C.). Appleton-Century-Crofts: New York, pp. 323–47.

Hoffstetter, R. (1974) Phylogeny and geographical deployment of the primates. *J. hum. Evol.* **3**, 327–50.

Hofman, M.A. (1982) Encephalization in mammals in relation to the size of the cerebral cortex. *Brain Behav. Evol.* **20**, 84–96.

Holloway, R.L. and Post, D.G. (1982) The relativity of relative brain measures and hominid mosaic evolution. In *Primate Brain Evolution* (eds Armstrong, E. and Falk, D.). Plenum Press: New York, pp. 57–76.

Holmquist, W. (1972) Empirical support for a stochastic model of evolution. *J. molec. Evol.* **1**, 211–22.

Holmquist, W. (1976) Random and nonrandom processes in the molecular evolution of higher organisms. In *Molecular Anthropology* (eds Goodman, M., Tashian, R.E. and Tashian, J.H.). Plenum Press: New York, pp. 89–116.

Holt, A.B., Renfree, M.B. and Cheek, D.B. (1981) Comparative aspects of brain growth: a critical evaluation of mammalian species used in brain growth research with emphasis on the Tammar wallaby. In *Fetal Brain Disorders* (eds Hetzel, B.S. and Smith, R.M.). Elsevier/North-Holland Biomedical Press, pp. 17–43.

Holt, A.B., Cheek, D.B., Mellits, E.D. and Hill, D.E. (1975) Brain size and the relation of the primate to the nonprimate. In *Fetal and Postnatal Cellular Growth: Hormones and Nutrition* (ed. Cheek, D.B.). John Wiley: New York, pp. 23–44.

Hopson, J.A. (1971) Postcanine development in the gomphodont cynodont Diademodon. In *Evolving Mammals* (eds Kermack, D.M. and Kermack, K.A.). Academic Press: London, pp. 1–21.

Hopson, J.A. (1973) Endothermy, small size, and the origin of mammalian reproduction. *Am. Nat.* **107**, 446–52.

Hopwood, A.T. (1933) Miocene primates from Kenya. *J. Linn, Soc. Lond. (Zool.)* **38**, 437–64.

Horton, J.C. (1984) Cytochrome oxidase patches: a new cytoarchitectonic feature of monkey visual cortex. *Phil. Trans. R. Soc. Lond. B* **304**, 199–253.

Howarth, M.K. (1981) Palaeogeography of the Mesozoic. In *The Evolving Earth* (ed. Cocks, R.L.M.). Cambridge University Press: Cambridge, pp. 197–220.

Howell, A.B. (1944) *Speed in Animals, Their Specializations for Running and Leaping.* Hafner: New York.

Howell, F.C. (1967) Recent advances in human evolutionary studies. *Q. Rev. Biol.* **42**, 471–513.

Howell, F.C. (1969) Remains of Hominidae from Pliocene/Pleistocene formations in the lower Omo basin. *Nature, Lond.* **223**, 1234–9.

Howell, F.C. (1978) Hominidae. In *Evolution of African Mammals* (eds Maglio, V.J. and Cooke, H.B.S.). Harvard University Press: Cambridge, Mass., pp. 154–248.

Howells, W.W. (1973) *Evolution of the Genus Homo.* Addison-Wesley: Reading, Mass.

Howells, W.W. (1974) Neanderthals: names, hypotheses and scientific method. *Am. Anthrop.* **76**, 24–38.

Hsu, C.-h., Han, K.-x and Wang, L.-h. (1974) Discovery of Gigantopithecus teeth and associated fauna in western Hopei. *Vert. palasiat.* **12**, 293–309.

Hubrecht, A.A.W. (1898) Über die Entwicklung der Placenta von Tarsius und Tupaia, nebst Bemerkungen über deren Bedeutung als haemopoeitische Organe. *Proc. int. Congr. Zool.* **4**, 345–411.

Hubrecht, A.A.W. (1908) Early ontogenetic phenomena in mammals and their bearing on our interpretation of the phylogeny of the vertebrates. *Q. J. microsc. Sci.* **53**, 1–181.

Hubel, D.H. and Wiesel, T.N. (1967). Receptive fields and functional architecture of monkey striate cortex. *J. Physiol., Lond.* **195**, 215–43.

Hull, D.L. (1967) Certainty and circularity in evolutionary taxonomy. *Evolution* **2**, 174–89.

Hunt, R.M. (1974) The auditory bulla in Carnivora: an anatomical basis for reappraisal of carnivore evolution. *J. Morph.* **143**, 21–76.

Hürzeler, J. (1946) Zur Charakteristik, systematischen Stellung, Phylogenese und Verbreitung der Necrolemuriden aus dem europäischen Eocaen. *Eclogae geol. Helv., Basel* **39**, 352–4.

Hürzeler, J. (1948) Zur Stammesgeschichte der Necrolemuriden. *Schweiz. palaont. Abh.* **66** (1), 1–46.

Hürzeler, J. (1949) Neubeschreibung von Oreopithecus bambolii Gervais. *Schweiz. paläont. Abh.* **66** (5), 3–20.

Hürzeler, J. (1954) Contribution á l'odontologie et à la phylogénèse du genre Pliopithecus. *Ann. Paléont. (Vert.)* **40**, 5–63.

Hürzeler, J. (1958) Oreopithecus bambolii Gervais: a preliminary report. *Verh. naturf. Ges. Basel.* **69**, 1–48.

Hürzeler, J. (1968) Questions et réflexions sur l'histoire des anthropomorphes. *Ann. Paléont. (Vert.)* **54**, 195–233.

Huxley, J.S. (1932) *Problems of Relative Growth.* Methuen: London.

Huxley, J.S. (1957) The three types of evolutionary process. *Nature, Lond.* **180**, 454.

Huxley, J.S. (1958) Evolutionary processes and taxonomy with special reference to grades. *Uppsala Univ. Arssks.* **6**, 21–39.

Huxley, T.H. (1880) On the application of the laws of evolution to the arrangement of the Vertebrata, and more particularly of the Mammalia. *Proc. zool. Soc. Lond.* **1880**, 649–61.

Hylander, W. (1975) Incisor size and diet in anthropoids with special reference to Cerco-pithecidae. *Science* **189**, 1095–8.

Hylander, W. (1979a) Mandibular function in *Galago crassicaudatus* and *Macaca fascicularis*: an *in vivo* approach to stress analysis of the mandible. *J. Morph.* **159**, 253–96.

Hylander, W. (1979b) Functional significance of primate mandibular form. *J. Morph.* **160**, 223–39.

Hylander, W. and Johnson, K.R. (1985) Temporalis and masseter muscle function during incision in macaques and humans. *Int. J. Primatol.* **6**, 289–322.

Imai, H.T. and Crozier, R.H. (1980) Quantitative analysis of directionality in mammalian karyotype evolution. *Am. Nat.* **116**, 537–69.

Ioannou, J.M. (1966) The oestrous cycle of the potto. *J. Reprod. Fert.* **11**, 455–7.

Izard, M.K. (1987) Lactation length in three species of *Galago. Am. J. Primatol.* **13**, 73–6.

Izard, M.K. and Rasmussen, D.T. (1985) Reproduction in the slender loris (*Loris tardigradus malabaricus*). *Am. J. Primatol.* **8**, 153–65.

Izard, M.K., Wright, P.C. and Simons, E.L. (1985) Gestation length in *Tarsius bancanus. Am. J. Primatol.* **9**, 327–31.

Jacobs, G.H. (1977) Visual capacities of the owl monkey (*Aotus trivirgatus*) – I. Spectral sensitivity and color vision. *Vision Res.* **17**, 811–20.

Jacobs, G.H. (1981) *Comparative Color Vision*. Academic Press: New York.

Jacobs, L.L. (1980) Siwalik fossil tree shrews. In *Comparative Biology and Evolutionary Relation-ships of Tree Shrews* (ed. Luckett, W.P.). Plenum Press: New York, pp. 205–16.

Jacobs, L.L. (1981) Miocene lorisid primates from the Pakistan Siwaliks. *Nature, Lond.* **289**, 585–7.

Jaeger, J.J. and Martin, M. (1984) African marsupials – vicariance or dispersion? *Nature, Lond.* **312**, 379.

James, W.W. (1960) *The Jaws and Teeth of Primates*. Pitman Medical: London.

Jane, J.A., Campbell, C.B.G. and Yashon, D. (1965) Pyramidal tract: a comparison of two prosimian primates. *Science* **147**, 153–5.

Jay, P. (1965) The common langur of north India. In *Primate Behavior: Field Studies of Monkeys and Apes* (ed. DeVore, I.). Holt, Rinehart & Winston: New York, pp. 197–249.

Jenkins, F.A. (ed.) (1974a) *Primate Locomotion*. Academic Press: New York.

Jenkins, F.A. (1974b) Tree shrew locomotion and the origins of primate arborealism. In *Primate Locomotion* (ed. Jenkins, F.A.). Academic Press: New York, pp. 85–115.

Jenkins, F.A. (1981) Wrist rotation in primates: a critical adaptation for brachiators. *Symp. zool. Soc. Lond.* **48**, 429–51.

Jenkins, F.A. and Crompton, A.W. (1979) Triconodonta. In *Mesozoic Mammals* (eds Lillegraven, J.A., Kielan-Jaworowska, Z. and Clemens, W.A.). University of California Press: Berkeley, pp. 74–90.

Jenkins, F.A. and Krause, D.W. (1983) Adaptations for climbing in North American multituberculates (Mammalia). *Science* **220**, 712–15.

Jenkins, F.A. and McClearn, D. (1984) Mechanisms of hind foot reversal in climbing mammals. *J. Morph.* **182**, 197–219.

Jenkins, F.A. and Parrington, F.R. (1976) The postcranial skeletons of the Triassic mam-mals *Eozostrodon*, *Megazostrodon* and *Erythrotherium. Phil. Trans. R. Soc. Lond. B* **273**, 387–431.

Jerison, H.J. (1970) Gross brain indices and the analysis of fossil endocasts. In *The Primate Brain* (eds Noback, C.R. and Montagna, W.). Appleton-Century-Crofts: New York, pp. 225–44.

Jerison, H.J. (1973) *Evolution of the Brain and Intelligence*. Academic Press: New York.

Jerison, H.J. (1977) The theory of encephalization. *Ann. N.Y. Acad. Sci.* **299**, 146–60.

Jerison, H.J. (1979) Brain, body and encephalization in early primates. *J. hum. Evol.* **8**, 615–35.

Jerison, H.J. (1982) Allometry, brain size, cortical surface and convolutedness. In *Primate Brain Evolution* (eds Armstrong, E. and Falk, D.). Plenum Press: New York, pp. 77–84.

Jewett, D.A. and Dukelow, W.R. (1972) Cyclicity and gestation length of *Macaca fascicularis. Primates* **13**, 327–30.

Johannsen, W. (1909) *Elemente der Exakten Erblichkeitslehre*. Gustav Fischer Verlag: Jena.

Johanson, D.C. and Taieb, M. (1976) Plio-Pleistocene hominid discoveries in Hadar, Ethiopia. *Nature, Lond.* **260**, 293–7.

Johanson, D.C. and White, T.D. (1979) A systematic assessment of early African hominids. *Science* **203**, 321–30.

Johanson, D.C., White, T.D. and Coppens, Y. (1978) A new species of the genus *Australopithecus* (Primates, Hominidae) from the Pliocene of eastern Africa. *Kirtlandia* **28**, 1–14.

Johanson, D.C., Masao, F.T., Eck, G.C., White, T.D., Walter, R.C., Kimbel, W.H., Asfaw, B., Manega, P., Ndessokia, P. and Suwa, G. (1987) New partial skeleton of *Homo habilis* from Olduvai Gorge, Tanzania. *Nature, Lond.* **327**, 205–9.

John, B. (1981) Chromosome change and evolutionary change: a critique. In *Evolution and Speciation: Essays in Honor of M.J.D. White* (eds Atchley, W.R. and Woodruff, D.S.). Cambridge University Press: Cambridge, pp. 23–51.

Johnson, G.L. (1901) Contributions to the comparative anatomy of the mammalian eye, chiefly based on ophthalmoscopic examination. *Phil. Trans. R. Soc. Lond. B* **194**, 1–82.

Johnson, M. and Everitt, B. (1980) *Essential Reproduction*. Blackwell: Oxford.

Johnson, R.H. (1899) Pads on the palm and sole of the human fetus. *Am. Nat.* **33**, 729–34.

Johnston, G.W., Dreizen, S. and Levy, B.M. (1970) Dental development in the cotton ear marmoset (*Callithrix jacchus*). *Am. J. phys. Anthrop.* **33**, 41–8.

Jolly, A. (1966) *Lemur Behavior: A Madagascan Field Study*. University of Chicago Press: Chicago.

Jolly, A. (1972a) Troop continuity and troop spacing in *Propithecus verreauxi* and *Lemur catta* at Berenty (Madagascar). *Folia primatol.* **17**, 335–62.

Jolly, A. (1972b) Hour of birth in primates and man. *Folia primatol.* **18**, 108–21.

Jolly, A. (1973) Primate birth hour. *Int. Zoo Yb.* **13**, 391–7.

Jolly, C.J. (1967) The evolution of the baboons. In *The Baboon in Medical Research* (ed. Vagtborg, H.). University of Texas Press: Austin, pp. 427–57.

Jolly, C.J. (1970a) *Hadropithecus*, a lemuroid small-object feeder. *Man, n.s.* **5**, 525–9.

Jolly, C.J. (1970b) The seedeaters: a new model of hominid differentiation based on a baboon analogy. *Man, n.s.* **5**, 5–26.

Jolly, C.J. (1972) The classification and natural history of *Theropithecus (Simopithecus)* (Andrews, 1916) baboons of the African Plio-Pleistocene. *Bull. Br. Mus. nat. Hist., Geol.* **22**, 1–122.

Jolly, C.J. (ed.) (1978) *Early Hominids of Africa*. Duckworth: London.

Jones, A.E. (1965). The retinal structure of (*Aotes trivirgatus*) the owl monkey. *J. comp. Neurol.* **125**, 19–27.

Jones, J.S. (1981) An uncensored page of fossil history. *Nature, Lond.* **293**, 427–8.

Jones, J.S. (1986) The origin of *Homo sapiens*: the genetic evidence. In *Major Topics in Primate and Human Evolution* (eds Wood, B.A., Martin, L.B. and Andrews, P.J.). Cambridge University Press: Cambridge, pp. 317–30.

Jones, T.R. (1937) A new fossil primate from Sterkfontein, Krugersdorp, Transvaal. *S. Afr. J. Sci.* **33**, 709–28.

Jouffroy, F.-K. (1960) Caractères adaptatifs dans les proportions des membres chez les lémurs fossiles. *C.R. Acad. Sci., Paris* **251**, 2756–7.

Jouffroy, F.-K. (1962) La musculature des membres chez les lémuriens de Madagascar: étude descriptive et comparative. *Mammalia* **26**, 1–326.

Jouffroy, F.-K. (1975) Osteology and myology of the lemuriform postcranial skeleton. In *Lemur Biology* (eds Tattersall, I. and Sussman, R.W.). Plenum Press: New York, pp. 149–92.

Jouffroy, F.-K., Berge, C. and Niemitz, C. (1984) Comparative study of the lower extremity in the genus *Tarsius*. In *Biology of Tarsiers* (ed. Niemitz, C.). Gustav Fischer: Stuttgart, pp. 167–90.

Joysey, K.A. (1978) An appraisal of molecular sequence data as a phylogenetic tool, based on the evidence of myoglobin. In *Recent Advances in Primatology*, vol. 3: *Evolution* (eds Chivers, D.J. and Joysey, K.A.). Academic Press, London, pp. 57–67.

Jukes, T.H. and Cantor, C.R. (1969) Evolution and protein molecules. In *Mammalian Protein Metabolism* (ed. Munro, H.N.). Academic Press: New York, pp. 21–132.

Jungers, W.L. (1977) Hindlimb and pelvic adaptations to vertical climbing and clinging in *Megaladapis*, a giant subfossil prosimian from Madagascar. *Yb. phys. Anthrop.* **20**, 508–24.

Jungers, W.L. (1978) The functional significance of skeletal allometry in *Megaladapis* in comparison to living prosimians. *Am. J. phys. Anthrop.* **49**, 303–14.

Jungers, W.L. (ed.) (1985) *Size and Scaling in Primate Biology*. Plenum Press: New York.

Jungers, W.L. (1987) Body size and morphometric affinities of the appendicular skeleton in *Oreopithecus bambolii* (IGF 11778). *J. hum. Evol.* **16**, 445–56.

Kaas, J.H., Guillery, R.W. and Allman, J.M. (1972) Some principles of organization in the dorsal lateral geniculate nucleus. *Brain Behav. Evol.* **6**, 253–99.

Kaas, J.H., Huerta, M.F., Weber, J.T. and Harting, J.K. (1978) Patterns of retinal terminations and laminar organization of the lateral geniculate nucleus of primates. *J. comp. Neurol.* **182**, 517–14.

Kälin, J. (1961a) Sur les primates de l'Oligocène inférieur d'Egypte. *Ann. Paléont. (Vert.)* **47**, 1–48.

Kälin, J. (1961b) Sur les primates de l'Oligocène inférieur d'Egypte. *Coll. Int. C.N.R.S.* **104**, 433–9.

Kälin, J. (1962) Über *Moeripithecus markgrafi* Schlosser und die phyletischen Vorstufen der Bilophodontie der Cercopithecoidea. *Bibl. primatol.* **1**, 32–42.

Kaudern, W. (1910) Über einige Ähnlichkeiten zwischen *Tupaja* und den Halbaffen. *Anat. Anz.* **37**, 561–73.

Kaudern, W. (1911) Studien über die männlichen

Geschlechtsorgane von Insectivoren und Lemuriden. *Zool. Jb.* **31**, 1–106.

Kay, R.F. (1977) The evolution of molar occlusion in the Cercopithecidae and early catarrhines. *Am. J. phys. Anthrop.* **46**, 327–52.

Kay, R.F. (1981) The nut-crackers: a new theory of the adaptations of the Ramapithecinae. *Am. J. phys. Anthrop.* **55**, 141–51.

Kay, R.F. (1982) *Sivapithecus simonsi*, a new species of Miocene hominoid with comments on the phylogenetic status of the Ramapithecinae. *Int. J. Primatol.* **3**, 113–74.

Kay, R.F. (1984) On the use of anatomical features to infer foraging behavior in extinct primates. In *Adaptations for Foraging in Nonhuman Primates* (eds Rodman, P.S. and Cant, J.G.H.). Columbia University Press: New York, pp. 21–53.

Kay, R.F. and Cartmill, M. (1974) Skull of *Palaechthon nacimienti*. *Nature, Lond.* **252**, 37–8.

Kay, R.F. and Cartmill, M. (1977) Cranial morphology and adaptation of *Palaechthon nacimienti* and other Paromomyidae (Plesiadapoidea, Primates), with a description of a new genus and species. *J. hum. Evol.* **6**, 19–53.

Kay, R.F. and Covert, H.H. (1984) Anatomy and behaviour of extinct primates. In *Food Acquisition and Processing in Primates* (eds Chivers, D.J., Wood, B.A. and Bilsborough, A.). Plenum Press: New York, pp. 467–508.

Kay, R.F. and Hiiemae, K.M. (1974) Mastication in *Galago crassicaudatus*: a cinefluorographic and occlusal study. In *Prosimian Biology* (eds Martin, R.D., Doyle, G.A. and Walker, A.C.). Duckworth: London, pp. 501–30.

Kay, R.F. and Simons, E.L. (1980) Comments on the adaptive strategy of the first African anthropoids. *Z. Morph. Anthrop.* **71**, 143–8.

Kay, R.F. and Simons, E.L. (1983a) A reassessment of the relationship between later Miocene and subsequent Hominoidea. In *New Interpretations of Ape and Human Ancestry* (eds Ciochon, R.L. and Corruccini, R.S.). Plenum Press: New York, pp. 577–624.

Kay, R.F. and Simons, E.L. (1983b) Dental formulae and dental eruption patterns in Parapithecidae (Primates, Anthropoidea). *Am. J. phys. Anthrop.* **62**, 363–75.

Kay, R.F., Fleagle, J.G. and Simons, E.L. (1981) A revision of the Oligocene apes from the Fayum Province, Egypt. *Am. J. phys. Anthrop.* **55**, 293–322.

Kay, R.F., Madden, R.H., Plavcan, J.M., Cifelli, R.L. and Diaz, J.G. (1987) *Stirtonia victoriae*, a new species of Miocene Colombian primate. *J. hum. Evol.* **16**, 173–96.

Keith, A. (1899) On the chimpanzees and their relationship to the gorilla. *Proc. zool. Soc. Lond.* **1899**, 296–312.

Keith, A. (1923) Man's posture: its evolution and disorders. *Br. Med. J.* **1**, 451–4, 499–502, 545–8, 587–90, 624–6, 669–72.

Kemp, T.S. (1982) *Mammal-like Reptiles and the Origin of Mammals*. Academic Press: London.

Kemp, T.S. (1983) The relationships of mammals. *Zool. J. Linn. Soc.* **77**, 353–84.

Kermack, K.A. and Haldane, J.B.S. (1950) Organic correlation and allometry. *Biometrika* **37**, 30–41.

Kermack, D.M. and Kermack, K.A. (1984) *The Evolution of Mammalian Characters*. Croom Helm: London.

Kermack, K.A., Mussett, F. and Rigney, H.W. (1981) The skull of *Morganucodon*. *Zool. J. Linn. Soc., Lond.* **71**, 1–158.

Kern, J.A. (1964) Observations on the habits of the proboscis monkey, *Nasalis larvatus* (Wurmb), made in the Brunei Bay area, Borneo. *Zoologica, N.Y.* **49**, 183–92.

Keverne, E.B. (1981) Do Old World primates have oestrus? *Malay. appl. Biol.* **10**, 119–26.

Kielan-Jaworowska, Z. (1977) Results of the Polish–Mongolian palaeontological expeditions. Part VIII: Evolution of the therian mammals in the Late Cretaceous of Asia. Part II. Postcranial skeleton in *Kennalestes* and *Asioryctes*. *Palaeont. pol.* **37**, 65–83.

Kielan-Jaworowska, Z. (1978) Results of the Polish–Mongolian palaeontological expeditions. Part VIII: Evolution of the therian mammals in the Late Cretaceous of Asia. Part III. Postcranial skeleton in the Zalambdalestidae. *Palaeont. pol.* **38**, 5–41.

Kielan-Jaworowska, Z. (1981) Results of the Polish–Mongolian palaeontological expeditions. Part IX: Evolution of the therian mammals in the Late Cretaceous of Asia. Part IV. Skull structure in *Kennalestes* and *Asioryctes*. *Palaeont. pol.* **42**, 25–78.

Kielan-Jaworowska, Z., Bown, T.M. and Lillegraven, J.A. (1979) Eutheria. In *Mesozoic Mammals* (eds Lillegraven, J.A., Kielan-Jaworowska, Z. and Clemens, W.A.). University of California Press: Berkeley, pp. 221–58.

Kielan-Jaworowska, Z., Crompton, A.W. and Jenkins, F.A. (1987) The origin of egg-laying mammals. *Nature, Lond.* **326**, 871–3.

Kihlström, J.E. (1972) Period of gestation and body weight in some placental mammals. *Comp. Biochem. Physiol.* **43**, 673–9.

Kimura, M. (1968a) Evolutionary rate at the molecular level. *Nature, Lond.* **217**, 624–6.

Kimura, M. (1968b) Genetic variability maintained in a finite population due to mutational production of neutral and nearly neutral isoalleles. *Genet. Res.* **11**, 247–69.

Kimura, M. (1969) The rate of molecular evolution considered from the standpoint of population genetics. *Proc. natn. Acad. Sci. USA.* **63**, 1181–8.

Kimura, M. (1983) *The Neutral Theory of Molecular Evolution.* Cambridge University Press: Cambridge.

Kimura, M. and Crow, J.F. (1964) The number of alleles that can be maintained in a finite population. *Genetics* **49**, 725–38.

Kimura, M. and Ohta, T. (1971) Protein poly-morphism as a phase of molecular evolution. *Nature, Lond.* **229**, 467–9.

Kimura, T., Okada, M. and Ishida, H. (1979) Kinesiological characteristics of primate walking: its significance in human walking. In *Environment, Behavior and Morphology: Dynamic Interactions in Primates* (eds Morbeck, M.E., Preuschoft, H. and Gomberg, N.). Gustav Fischer: New York, pp. 297–311.

King, B.F. (1984) The fine structure of the placenta and chorionic vesicles of the bush baby, *Galago crassicaudatus. Am. J. Anat.* **169**, 101–16.

King, B.F. (1986) Morphology of the placenta and fetal membranes. In *Comparative Primate Biology*, vol. 3: *Reproduction and Development* (eds Dukelow, W.R. and Erwin, J.). Alan R. Liss: New York, pp. 311–31.

King, J.L. (1926) Menstrual records and vaginal smears in a selected group of normal women. *Contr. Embryol. Carnegie Inst. Wash.* **95**, 79–94.

King, J.L. and Jukes, T.H. (1969) Non-Darwinian evolution. *Nature, Lond.* **164**, 788–98.

King, M.-C. and Wilson, A.C. (1975) Evolution at two levels in humans and chimpanzees. *Science* **188**, 107–16.

Kinzey, W.G. (1974) Ceboid models for the evolution of the hominoid dentition. *J. hum. Evol.* **3**, 193–203.

Kinzey, W.G., Rosenberger, A.L. and Ramirez, M. (1975) Vertical clinging and leaping in a neotropical anthropoid. *Nature, Lond.* **255**, 327–8.

Kirsch, J.A.W. (1969) Serological data and phylogenetic inference: the problem of rates of change. *Syst. Zool.* **18**, 296–311.

Kirsch, J.A.W. (1977a) The classification of marsupials with special reference to karyotypes and serum proteins. In *The Biology of Marsupials* (ed. Hunsaker, D.). Academic Press: New York, pp. 1–50.

Kirsch, J.A.W. (1977b) The six-percent solution: second thoughts on the adaptedness of the Marsupialia. *Am. Sci.* **65**, 276–88.

Klatsky, H. (1940) A cinefluorographic study of the human masticatory apparatus in function. *Am. J. Orthodont.* **25**, 205–10.

Kleiber, M. (1947) Body size and metabolic rate. *Physiol. Rev.* **27**, 511–41.

Kleiber, M. (1961) *The Fire of Life: An Introduction to Animal Energetics.* John Wiley: New York.

Kleiman, D.G. (1977a) Monogamy in mammals. *Q. Rev. Biol.* **52**, 39–69.

Kleiman, D.G. (ed.) (1977b) *The Biology and Conservation of the Callitrichidae.* Smithsonian Institution Press: Washington.

Klein, L.L. and Klein, D.B. (1977) Feeding behaviour of the Colombian spider monkey. In *Primate Ecology* (ed. Clutton-Brock, T.H.). Academic Press: London, pp. 153–81.

Kluge, A.G. (1984) The relevance of parsimony to phylogenetic inference. In *Cladistics* (eds Duncan, T. and Stuessy, T.F.). Columbia University Press: New York, pp. 24–38.

Kluge, A.G. and Farris, J.S. (1969) Quantitative phyletics and the evolution of anurans. *Syst. Zool.* **18**, 1–32.

Knight, D. (1981) *Ordering the World: A History of Classifying Man.* Burnett Books/André Deutsch: London.

Kohne, D.E. (1970) Evolution of higher-organism DNA. *Q. Rev. Biophys.* **3**, 327–75.

Kohne, D.E. (1975) DNA evolution data and its relevance to mammalian phylogeny. In *Phylogeny of the Primates* (eds Luckett, W.P. and Szalay, F.S.). Plenum Press: New York, pp. 249–61.

Kohne, D.E., Chiscon, J.A. and Hoyer, B.H. (1972) Evolution of primate DNA sequences. *J. hum. Evol.* **1**, 627–44.

Kollman, M. (1920) L'os planum des lémuriens. *Bull. Soc. Linn. Normandie, sér.* 7 **2**, 216–19.

Kollman, M. (1925) Études sur les lémuriens. La fosse orbito-temporale et l'os planum. *Mém. Soc. Linn. Normandie (Caen), sect. Zool.* (n.s.) **1**, 1–20.

Kolmer, W. (1930) Zur Kenntnis des Auges der Primaten. *Z. Anat. Entw.-Gesch.* **93**, 679–722.

Kowalski, K. and Zapfe, H. (1974) *Pliopithecus antiquus* (Blainville, 1839) (Primates, Mammalia) from the Miocene of Przeworno in Silesia (Poland). *Acta zool. cracov.* **19**, 19–30.

Kraglievich, J.L. (1951) Contribuciones al conocimiento de los primates fosiles de la Patagonia. I. Diagnosis prévia de un nuevo primate fosil del Oligoceno superior (Colhuehuapiano) de Gaimán, Chubut. *Comm. Inst. Nac. Invest. Cienc. Nat. (Buenos Aires), Cienc. Zool.* **11**, 55–82.

Kraus, M.J. (1979) Eupantotheria. In *Mesozoic Mammals* (eds Lillegraven, J.A., Kielan-

Jaworowska, Z. and Clemens, W.A.). University of California Press: Berkeley, pp. 162–71.

Kretzoi, M. (1975) New ramapithecines and *Pliopithecus* from the lower Pliocene of Rudabanya in north-eastern Hungary. *Nature, Lond.* **257**, 578–81.

Kuhn, A. (1957) *Grundriss der Allgemeinen Zoologie* (12th revised edition). Georg Thieme Verlag: Stuttgart.

Kuhn, H.-J. (1964) Zur Kenntnis von Bau und Funktion des Magens der Schlankaffen (Colobinae). *Folia primatol.* **2**, 193–221.

Kummer, H. (1968) *Social Organization of Hamadryas Baboons.* University of Chicago Press: Chicago.

Kurtén, B. (1973) Early Tertiary land mammals. In *Atlas of Palaeobiogeography* (ed. Hallam, A.). Elsevier Scientific: London, pp. 437–42.

Kuypers, H.G.J.M. (1968) Corticospinal connections: postnatal development in the rhesus monkey. *Science* **138**, 678–80.

Lamberton, C. (1934) Contribution à la connaissance de la faune subfossile de Madagascar (lémuriens et ratites). *Mém. Acad. malgache (n.s.)* **17**, 1–168.

Lamberton, C. (1936) Nouveaux lémuriens fossiles du groupe des propithèques et l'intérêt de leur découverte. *Bull. Mus. Hist. nat., Paris* **2**, 370–3.

Lamberton, C. (1938) Contribution à la connaissance de la faune subfossile de Madagascar. Note III: Les hadropithèques. *Bull. Acad. malgache* **20**, 1–44.

Lamberton, C. (1939) Contribution à la connaissance de la faune subfossile de Madagascar. Notes IV à VIII: Lémuriens et cryptoproctes. *Mém. Acad. malgache* **27**, 5–203.

Lamberton, C. (1941) Contribution à la connaissance de la faune subfossile de Madagascar. Note IX: Oreille osseuse des lémuriens. *Mém. Acad. malgache* **35**, 1–132.

Lancaster, J.B. and Lee, R.B. (1965) The annual reproductive cycle in monkeys and apes. In *Primate Behavior: Field Studies of Monkeys and Apes* (ed. DeVore, I.). Holt, Rinehart & Winston: New York, pp. 486–513.

Lande, R. (1979a) Quantitative genetic analysis of multivariate evolution applied to brain:body size allometry. *Evolution* **33**, 402–16.

Lande, R. (1979b) Effective deme sizes during long-term evolution estimated from rates of chromosomal rearrangement. *Evolution* **33**, 234–51.

Lang, C.M. (1967) The estrous cycle of nonhuman primates: a review of the literature. *Lab. Anim. Care* **17**, 172–9.

Langham, N.P.E. (1982) The ecology of the common tree shrew *Tupaia glis* in peninsular Malaysia. *J. Zool., Lond.* **197**, 323–44.

Langley, C.H. and Fitch, W.M. (1974) An examination of the constancy of the rate of molecular evolution. *J. molec. Evol.* **3**, 161–77.

Lartet, E. (1856) Note sur un grand singe fossile qui se rattache au groupe des singes supérieurs. *C.R. Acad. Sci., Paris* **43**, 219–23.

Lasiewski, R.C. and Dawson, W.R. (1967) A re-examination of the relation between standard metabolic rate and body weight in birds. *Condor* **69**, 13–23.

Latter, B.D.H. (1980) Genetic differences within and between populations of the major human subgroups. *Am. Nat.* **116**, 220–37.

Lavocat, R. (1969) Le systématique des rongeurs hystricomorphes et la dérive des continents. *C.R. Acad. Sci. Paris, D* **269**, 1496–7.

Le Gros Clark, W.E. (1924a) Notes on the living tarsier (*Tarsius spectrum*). *Proc. zool. Soc. Lond.* **1924**, 216–23.

Le Gros Clark, W.E. (1924b) The myology of the tree-shrew (*Tupaia minor*). *Proc. zool. Soc. Lond.* **1924**, 461–97.

Le Gros Clark, W.E. (1924c) On the brain of the tree-shrew (*Tupaia minor*). *Proc. zool. Soc. Lond.* **1924**, 1053–74.

Le Gros Clark, W.E. (1925) On the skull of *Tupaia*. *Proc. zool. Soc. Lond.* **1925**, 559–67.

Le Gros Clark, W.E. (1926) On the anatomy of the pen-tailed tree-shrew (*Ptilocercus lowii*). *Proc. zool. Soc. Lond.* **1926**, 1179–309.

Le Gros Clark, W.E. (1931) The brain of *Microcebus murinus*. *Proc. zool. Soc. Lond.* **1931**, 463–86.

Le Gros Clark, W.E. (1934a) On the skull structure of *Pronycticebus gaudryi*. *Proc. zool. Soc. Lond.* **1934**, 19–27.

Le Gros Clark, W.E. (1934b) *Early Forerunners of Man.* Baillière, Tindall and Cox: London.

Le Gros Clark, W.E. (1936) The problem of the claw in primates. *Proc. zool. Soc. Lond.* **1936**, 1–24.

Le Gros Clark, W.E. (1945a) Deformation patterns in the cerebral cortex. In *Essays on Growth and Form Presented to D'Arcy Wentworth Thompson* (eds Le Gros Clark, W.E. and Medawar, P.B.). Clarendon Press: Oxford, pp. 1–22.

Le Gros Clark, W.E. (1945b) Note of the palaeontology of the human brain. *J. Anat., Lond.* **79**, 123–6.

Le Gros Clark, W.E. (1956) A Miocene lemuroid skull from East Africa. *Fossil Mammals Afr.* **9**, 1–6.

Le Gros Clark, W.E. (1959) *The Antecedents of Man.* Edinburgh University Press: Edinburgh. (NB. This book was reissued in 1962 and 1971, but with only minor changes from the definitive 1959 edition.)

Le Gros Clark, W.E. (1967) *Man-Apes or Ape-Men?*

The Story of Discoveries in Africa. Holt, Rinehart & Winston: New York.

Le Gros Clark, W.E. and Leakey, L.S.B. (1950) Diagnoses of East African Miocene Hominoidea. *Q. J. geol. Soc. Lond.* **105**, 260–2.

Le Gros Clark, W.E. and Leakey, L.S.B. (1951) The Miocene Hominoidea of East Africa. *Fossil Mammals Afr.* **1**, 1–117.

Le Gros Clark, W.E. and Thomas, D.P. (1951) Associated jaws and limb bones of *Limnopithecus macinnesi. Fossil Mammals Afr.* **3**, 1–27.

Le Gros Clark, W.E. and Thomas, D.P. (1952) The Miocene lemuroids of East Africa. *Fossil Mammals Afr.* **5**, 1–20.

Leakey, L.S.B. (1959) A new fossil site from Olduvai. *Nature, Lond.* **184**, 491–3.

Leakey, L.S.B., Tobias, P.V. and Napier, J.R. (1964) A new species of genus *Homo* from Olduvai Gorge. *Nature, Lond.* **202**, 7–9.

Leakey, M.D. and Harris, J.M. (eds) (1987) *Laetoli: A Pliocene Site in Northern Tanzania.* Clarendon Press: Oxford.

Leakey, M.D. and Hay, R.L. (1979) Pliocene footprints in the Laetolil Beds at Laetoli, northern Tanzania. *Nature, Lond.* **278**, 317–23.

Leakey, M.G. (1985) Early Miocene cercopithecids from Buluk, northern Kenya. *Folia primatol.* **44**, 1–14.

Leakey, M.G. and Leakey, R.E.F. (1973a) Further evidence of *Simopithecus* (Mammalia, Primates) from Olduvai and Olorgesailie. *Fossil Vertebrates Afr.* **3**, 101–20.

Leakey, M.G. and Leakey, R.E.F. (1973b) New large Pleistocene Colobinae (Mammalia, Primates) from East Africa. *Fossil Vertebrates Afr.* **3**, 121–38.

Leakey, R.E.F. (1969) New Cercopithecidae from the Chemeron beds of Lake Baringo, Kenya. *Fossil Vertebrates Afr.* **1**, 53–69.

Leakey, R.E.F. and Leakey, M.G. (1986a) A new Miocene hominoid from Kenya. *Nature, Lond.* **324**, 143–6.

Leakey, R.E.F. and Leakey, M.G. (1986b) A second new Miocene hominoid from Kenya. *Nature, Lond.* **324**, 146–8.

Leakey, R.E.F. and Leakey, M.G. (1987) A new Miocene small-bodied ape from Kenya. *J. hum. Evol.* **16**, 369–87.

Leche, W. (1892) Studien über die Entwicklung des Zahnsystems bei den Säugethieren. *Morph. Jb.* **19**, 502–47.

Leche, W. (1895) Zur Entwicklungsgeschichte des Zahnsystems der Säugethiere, zugleich ein Beitrag zur Stammesgeschichte dieser Thiergruppe. I: Ontogenie. *Bibl. zool., Stuttgart* **6** (17), 1–160.

Leche, W. (1897) Untersuchungen des Zahnsystems lebender und fossiler Halbaffen. *Festschr. Gegenbaur, Leipzig* **3**, 125–66.

Leche, W. (1907) Zur Entwicklungsgeschichte des Zahnsystems der Säugethiere, zugleich ein Beitrag zur Stammesgeschichte dieser Thiergruppe. 2 Teil: Phylogenie. 2. Heft: Die Familien der Centetidae, Solenodontidae und Chrysochloridae. *Bibl. zool., Stuttgart* **20**, 1–157.

Leche, W. (1915) Zur Frage nach der stammes-geschichtlichen Bedeutung des Milchgebisses bei den Säugetieren. *Zool. Jb., Abt. 1* **38**, 275–370.

Lehmann, H., Romero-Herrera, A.E., Joysey, K.A. and Friday, A.E. (1974) Comparative structure of myoglobin: primates and tree-shrew. *Ann. N.Y. Acad. Sci.* **241**, 380–91.

Leutenegger, W. (1973) Maternal–fetal weight relationships in primates. *Folia primatol.* **20**, 280–93.

Levan, A., Fredga, K. and Sandberg, A.A. (1964) Nomenclature for centromeric position on chromosomes. *Hereditas* **52**, 201–20.

Lewin, R. (1988) Molecular clocks turn a quarter of a century. *Science* **239**, 561–3.

Lewis, E.B. (1950) The phenomenon of position effect. *Adv. Genet.* **3**, 73–116.

Lewis, G.E. (1934) Preliminary notice of new man-like apes from India. *Am. J. Sci.* **27**, 161–79.

Lewis, O.J. (1964a) The evolution of the long flexor muscles in the leg and foot. *Rev. gen. exp. Zool.* **1**, 165–85.

Lewis, O.J. (1964b) The homologies of the mammalian tarsal bones. *J. Anat., Lond.* **98**, 195–208.

Lewis, O.J. (1969) The hominoid wrist joint. *Am. J. phys. Anthrop.* **30**, 251–68.

Lewis, O.J. (1971a) Brachiation and the early evolution of the Hominoidea. *Nature, Lond.* **230**, 577–9.

Lewis, O.J. (1971b) The contrasting morphology found in the wrist joints of semibrachiating monkeys and brachiating apes. *Folia primatol.* **16**, 248–56.

Lewis, O.J. (1972a) Osteological features characterizing the wrists of monkeys and apes, with a reconsideration of the region in *Dryopithecus (Proconsul) africanus. Am. J. phys. Anthrop.* **36**, 45–58.

Lewis, O.J. (1972b) Evolution of the hominoid wrist. In *Functional and Evolutionary Biology of Primates* (ed. Tuttle R.). Aldine: Chicago, pp. 207–22.

Lewis, O.J. (1974) The wrist articulations of the Anthropoidea. In *Primate Locomotion* (ed. Jenkins, F.A.). Academic Press: New York, pp. 143–69.

Lewis, O.J. (1980a) The joints of the evolving foot. Part I. The ankle joint. *J. Anat., Lond.* **130**, 527–43.

Lewis, O.J. (1980b) The joints of the evolving foot. Part II. The intrinsic joints. *J. Anat., Lond.* **130**, 833–57.

Lewis, O.J. (1980c) The joints of the evolving foot. Part III. The fossil evidence. *J. Anat., Lond.* **131**, 275–98.

Lewis, O.J. (1981) Functional morphology of the joints of the evolving foot. *Symp. zool. Soc. Lond.* **46**, 169–88.

Lewis, O.J. (1983) The evolutionary emergence and refinement of the mammalian pattern of foot architecture. *J. Anat., Lond.* **137**, 21–45.

Lewis, O.J. (1985a) Derived morphology of the wrist articulations and theories of hominoid evolution. Part I. The lorisine joints. *J. Anat., Lond.* **140**, 447–60.

Lewis, O.J. (1985b) Derived morphology of the wrist articulations and theories of hominoid evolution. Part II. The midcarpal joints of higher primates. *J. Anat., Lond.* **142**, 151–72.

Lewontin, R.C. (1974) *The Genetic Basis of Evolutionary Change.* Columbia University Press: New York.

Lewontin, R.C. and Hubby, J.L. (1966) A molecular approach to the study of genetic heterozygosity in natural populations. II. Amount of variation and degree of heterozygosity in natural populations of *Drosophila pseudobscura. Genetics* **54**, 595–609.

Libby, W.F. (1965) *Radiocarbon Dating* (2nd edition). University of Chicago Press: Chicago.

Lillegraven, J.A. (1972) Ordinal and familial diversity of Cenozoic mammals. *Taxon* **21**, 261–74.

Lillegraven, J.A. (1979) Reproduction in Mesozoic mammals. In *Mesozoic Mammals* (eds Lillegraven, J.A., Kielan-Jaworowska, Z. and Clemens, W.A.). University of California Press: Berkeley, pp. 259–76.

Lillegraven, J.A., Kielan-Jaworowska, Z. and Clemens, W.A. (eds) (1979) *Mesozoic Mammals: The First Two-Thirds of Mammalian History.* University of California Press: Berkeley.

Lim, B.L. (1967) Note on the food habits of *Ptilocercus lowii* Gray (pentail tree-shrew) and *Echinosorex gymnurus* Raffles (moonrat) in Malaya with remarks on 'ecological labelling' by parasite patterns. *J. Zool., Lond.* **152**, 375–9.

Liming, S., Yingying, Y. and Xingsheng, D. (1980) Comparative cytogenetic studies on the red muntjac, Chinese muntjac and their F_1 hybrids. *Cytogenet. Cell Genet.* **26**, 22–7.

Lindburg, D.G. (1987) Seasonality of reproduction in primates. In *Comparative Primate Biology*, vol. 3:

Reproduction and Development (eds Dukelow, W.R. and Erwin, J.). Alan R. Liss: New York, pp. 167–218.

Linnaeus, C. (1758) *Systema Naturae per Regna Tria Naturae, Secundam Classes, Ordines, Genera, Species cum Characteribus, Synonymis, Locis.* Laurentii Sylvii: Stockholm. (This is the zoologically definitive 10th edition.)

Loesch, D.Z. and Martin, N.G. (1984) Finger ridge patterns and tactile sensitivity. *Ann. hum. Biol.* **11**, 113–24.

Long, C.A. (1969) The origin and evolution of mammary glands. *Bioscience* **19**, 519–23.

Long, C.A. (1972) Two hypotheses on the origin of lactation. *Am. Nat.* **106**, 141–4.

Lorenz, G.F. (1927) Über Ontogenese und Phylogenese der Tupajahand. *Morph. Jb.* **58**, 431–9.

Lorenz von Liburnau, L.R. (1899) Einen fossilen Anthropoiden von Madagaskar. *Anz. K. Akad. Wiss., Wien* **19**, 255–7.

Lorenz von Liburnau, L.R. (1900) Über einige Reste ausgestorbener Primaten von Madagaskar. *Denkschr. K. Akad. Wiss., Wien, math.-naturwiss. Kl.* **70**, 243–54.

Lovejoy, A.O. (1960) *The Great Chain of Being.* Harper and Row: New York. (Reprint of 1936 edition.)

Lovejoy, C.O., Burstein, A.H. and Heiple, K.G. (1972) Primate phylogeny and immunological distance. *Science* **176**, 803–5.

Luchterhand, K., Kay, R.F. and Madden, R.H. (1986) *Mohanamico hershkovitzi*, gen. et sp. nov., un primate du Miocène moyen d'Amérique du sud. *C.R. Acad. Sci., Paris, sér. 2* **303**, 1753–8.

Luckett, W.P. (1969) Evidence for the phylogenetic relationships of the tree shrews (family Tupaiidae) based on the placenta and foetal membranes. *J. Reprod. Fert., Suppl.* **6**, 419–33.

Luckett, W.P. (1974a) The comparative development and evolution of the placenta in primates. *Contr. Primatol.* **3**, 142–234.

Luckett, W.P. (1974b) The phylogenetic relationships of the prosimian primates: evidence from the morphogenesis of the placenta and foetal membranes. In *Prosimian Biology* (eds Martin, R.D., Doyle, G.A. and Walker, A.C.). Duckworth: London, pp. 475–88.

Luckett, W.P. (1975) Ontogeny of the fetal membranes and placenta: their bearing on primate phylogeny. In *Phylogeny of the Primates* (eds Luckett, W.P. and Szalay, F.S.). Plenum Press: New York, pp. 157–82.

Luckett, W.P. (1980a) The suggested evolutionary relationships and classification of tree shrews.

In *Comparative Biology and Evolutionary Relationships of Tree Shrews* (ed. Luckett, W.P.). Plenum Press: New York, pp. 3–31.

Luckett, W.P. (ed) (1980b) *Comparative Biology and Evolutionary Relationships of Tree Shrews.* Plenum Press: New York.

Luckett, W.P. (1980c) The use of reproductive and developmental features in assessing tupaiid affinities. In *Comparative Biology and Evolutionary Relationships of Tree Shrews* (ed. Luckett, W.P.). Plenum Press: New York, pp. 245–66.

Luckett, W.P. and Maier, W. (1982) Development of deciduous and permanent dentition in *Tarsius* and its phylogenetic significance. *Folia primatol.* **37**, 1–36.

Lumsden, A.G.S. (1979) Pattern formation in the molar dentition of the mouse. *J. Biol. buccale* **7**, 77–103.

Lumsden, A.G.S. and Osborn, J.W. (1977) The evolution of chewing: a dentist's view of palaeontology. *J. Dentist.* **5**, 269–87.

Lyon, M.W. (1913) Treeshrews: an account of the mammalian family Tupaiidae. *Proc. U.S. natn. Mus., U.S.A.* **45**, 1–188.

Ma, N.S.F. (1981) Chromosome evolution in the owl monkey. *Am. J. phys. Anthrop.* **54**, 294–304.

Ma, N.S.F., Rossan, R.N., Kelley, S.T., Harper, J.S., Bedard, M.T. and Jones, T.C. (1978) Banding patterns of the chromosomes of two new karyotypes of the owl monkey, *Aotus*, captured in Panama. *J. med. Primatol.* **7**, 146–55.

MacArthur, R.H. and Wilson, E.O. (1967) *The Theory of Island Biogeography.* Princeton University Press: Princeton, New Jersey.

MacKinnon, J.R. (1974) The behaviour and ecology of wild orang-utans (*Pongo pygmaeus*). *Anim. Behav.* **22**, 3–74.

MacKinnon, J.R. and MacKinnon, K.S. (1980) The behaviour of wild spectral tarsiers. *Int. J. Primatol.* **1**, 361–79.

MacLarnon, A.M. (1987) Size relationships of the spinal cord and associated skeleton in primates. PhD thesis, University of London.

MacLarnon, A.M., Chivers, D.J. and Martin, R.D. (1986) Gastro-intestinal allometry in primates including new species. In *Primate Ecology and Conservation* (eds Else, J.G. and Lee, P.C.). Cambridge University Press: Cambridge, pp. 75–85.

MacLarnon, A.M., Martin, R.D., Chivers, D.J. and Hladik, C.M. (1986) Some aspects of gastro-intestinal allometry in primates and other mammals. In *Définition et Origines de l'Homme*

(ed. Sakka, M.). Éditions du C.N.R.S: Paris, pp. 293–302.

MacPhee, R.D.E. (1977) Ontogeny of the ectotympanic–petrosal plate relationships in strepsirhine prosimians. *Folia primatol.* **27**, 245–83.

MacPhee, R.D.E. (1981) Auditory regions of primates and eutherian insectivores: morphology, ontogeny and character analysis. *Contr. Primatol.* **18**, 1–282.

MacPhee, R.D.E. and Cartmill, M. (1986) Basicranial structures and primate systematics. In *Comparative Primate Biology*, vol. 1: *Systematics, Evolution and Anatomy* (eds Swindler, D.R. and Erwin, J.). Alan R. Liss: New York, pp. 219–75.

MacPhee, R.D.E. and Jacobs, L.L. (1986) *Nycticeboides simpsoni* and the morphology, adaptations, and relationships of Miocene Siwalik Lorisidae. In *Vertebrates, Phylogeny, and Philosophy* (eds Flanagan, K.M. and Lillegraven, K.A.). Contributions to Geology Special Papers 3, University of Wyoming: Laramie, Wyoming, pp. 131–61.

MacPhee, R.D.E., Cartmill, M. and Gingerich, P.D. (1983) New Palaeogene primate basicrania and definition of the order Primates. *Nature, Lond.* **301**, 509–11.

MacRoberts, M.H. and MacRoberts, B.R. (1966) The annual reproductive cycle of the Barbary ape (*Macaca sylvana*) in Gibraltar. *Am. J. phys. Anthrop.* **25**, 299–304.

Mahboubi, M., Ameur, R., Crochet, J.Y. and Jaeger, J.J. (1983) Première découverte d'un marsupial en Afrique. *C. R. Acad. Sci. Paris, D* **297**, 415–18.

Mahé, J. (1972) Craniométrie des lémuriens. Analyse multivariables – phylogénie. Thése d'état, Université de Paris.

Mahé, J. (1976) Craniométrie des lémuriens. Analyse multivariables – phylogénie. *Mém. Mus. natn. Hist. nat., Paris, C* **32**, 1–342.

Mai, L. (1983) A model of chromosome evolution and its bearing on cladogenesis in the Hominoidea. In *New Interpretations of Ape and Human Ancestry* (eds Ciochon, R.L. and Corruccini, R.S.). Plenum Press: New York, pp. 87–114.

Maier, W. (1970) Neue Ergebnisse der Systematik und der Stammesgeschichte der Cercopithecoidea. *Z. Säugetierk.* **35**, 193–214.

Maier, W. (1971a) New fossil Cercopithecoidea from the Lower Pleistocene cave deposits of the Makapansgat Limeworks, South Africa. *Palaeont. Afr.* **13**, 69–108.

Maier, W. (1971b) Two new skulls of *Parapapio antiquus* from Taung and a suggested phylogenetic arrangement of the genus *Parapapio*. *Ann. S. Afr. Mus.* **59**, 1–16.

Maier, W. (1972) The first complete skull of *Simopithecus darti* from Makapansgat, South Africa, and its systematic position. *J. hum. Evol.* **1**, 395–405.

Maier, W. (1977) Die bilophodonten Molaren der Indriidae (Primates) – ein evolutionsmorphologischer Modellfall. *Z. Morph. Anthrop.* **68**, 307–44.

Mandel, P., Métais, P. and Cuny, S. (1950) Les quantités d'acide désoxypentose-nucléique par leucocyte chez diverses espèces de mammifères. *C. R. Acad. Sci., Paris* **231**, 1172–4.

Manfredi-Romanini, M.G. (1972) The nuclear DNA content and area of primate lymphocytes as a cytotaxonomic tool. *J. hum. Evol.* **1**, 23–40.

Manfredi-Romanini, M.G., de Boer, L.E.M., Chiarelli, B. and Tinozzi-Massari, S. (1972) DNA nuclear content in the cytotaxonomy of *Galago senegalensis* and *Galago crassicaudatus*. *J. hum. Evol.* **1**, 473–6.

Manley, G.H. (1966) Reproduction in lorisoid primates. *Symp. zool. Soc. Lond.* **15**, 493–509.

Manley, G.H. (1974) Functions of the external genital glands of *Perodicticus* and *Arctocebus*. In *Prosimian Biology* (eds Martin, R.D., Doyle, G.A. and Walker, A.C.). Duckworth: London, pp. 313–29.

Margoliash, E. and Fitch, W.M. (1968) Evolutionary variability of cytochrome *c* primary structures. *Proc. N. Y. Acad. Sci.* **151**, 359–81.

Margoliash, E., Fitch, W.M. and Dickerson, R.S. (1968) Molecular expression of evolutionary phenomena in the primary and tertiary structure of cytochrome *c*. *Brookhaven Symp. Biol.* **21**, 259–305.

Marks, J. (1983) Hominoid cytogenetics and evolution. *Yb. phys. Anthrop.* **26**, 131–59.

Martin, L.B. (1985) Significance of enamel thickness in hominoid evolution. *Nature, Lond.* **314**, 260–3.

Martin, L.B. and Andrews, P.J. (1984) The phyletic position of *Graecopithecus freybergi* Koenigswald. *Cour. Forsch.-Inst. Senckenberg* **69**, 25–40.

Martin, P.G. and Hayman, D.L. (1967) Quantitative comparisons between the karyotypes of Australian marsupials from three different superfamilies. *Chromosoma (Berl.)* **20**, 290–310.

Martin, R.D. (1967) Behaviour and taxonomy of tree-shrews (Tupaiidae). DPhil thesis, University of Oxford.

Martin, R.D. (1968a) Towards a new definition of primates. *Man* **3**, 377–401.

Martin, R.D. (1968b) Reproduction and ontogeny in tree-shrews (*Tupaia belangeri*) with reference to their general behaviour and taxonomic relationships. *Z. Tierpsychol.* **25**, 409–532.

Martin, R.D. (1969) The evolution of reproductive mechanisms in primates. *J. Reprod. Fert. (Suppl.)* **6**, 49–66.

Martin, R.D. (1972a) A preliminary field-study of the lesser mouse lemur (*Microcebus murinus* J.F. Miller 1777). *Z. Tierpsychol., Beiheft* **9**, 43–89.

Martin, R.D. (1972b) Adaptive radiation and behaviour of the Malagasy lemurs. *Phil. Trans. R. Soc. Lond., B* **264**, 295–352.

Martin, R.D. (1972c) A laboratory breeding colony of the lesser mouse lemur. In *Breeding Primates* (ed. Beveridge, W.I.B.). Karger: Basel, pp. 161–71.

Martin, R.D. (1973) Comparative anatomy and primate systematics. *Symp. zool. Soc. Lond.* **33**, 301–37.

Martin, R.D. (1975) The bearing of reproductive behavior and ontogeny on strepsirhine phylogeny. In *Phylogeny of the Primates* (eds Luckett, W.P. and Szalay, F.S.). Plenum Press: New York, pp. 265–97.

Martin, R.D. (1976) A zoologist's view of research on reproduction. *Symp. zool. Soc. Lond.* **40**, 283–319.

Martin, R.D. (1978) Major features of prosimian evolution: a discussion in the light of chromosomal evidence. In *Recent Advances in Primatology*, vol. 3: *Evolution* (eds Chivers, D.J. and Joysey, K.A.). Academic Press: London, pp. 3–26.

Martin, R.D. (1979) Phylogenetic aspects of prosimian behavior. In *The Study of Prosimian Behavior* (eds Doyle, G.A. and Martin, R.D.). Academic Press: New York, pp. 45–77.

Martin, R.D. (1980) Adaptation and body size in primates. *Z. Morph. Anthrop.* **71**, 115–24.

Martin, R.D. (1981a) Phylogenetic reconstruction *versus* classification: the case for clear demarcation. *Biologist* **28**, 127–32.

Martin, R.D. (1981b) Well-groomed predecessors. *Nature, Lond.* **289**, 536.

Martin, R.D. (1981c) Relative brain size and metabolic rate in terrestrial vertebrates. *Nature, Lond.* **293**, 57–60.

Martin, R.D. (1981d) Field studies of primate behaviour. *Symp. zool. Soc. Lond.* **46**, 287–336.

Martin, R.D. (1982) Allometric approaches to the evolution of the primate nervous system. In *Primate Brain Evolution* (eds Armstrong, E. and Falk, D.). Plenum Press: New York, pp. 39–56.

Martin, R.D. (1983) *Human Brain Evolution in an Ecological Context* (52nd James Arthur Lecture on the Evolution of the Human Brain). American Museum of Natural History: New York.

Martin, R.D. (1984) Scaling effects and adaptive strategies in mammalian lactation. *Symp. zool. Soc. Lond.* **51**, 87–117.

Martin, R.D. (1986a) Primates: a definition. In *Major Topics in Primate and Human Evolution* (eds Wood, B.A., Martin, L.B. and Andrews, P.J.). Cambridge University Press: Cambridge, pp. 1–31.

Martin, R.D. (1986b) Are fruit-bats primates? *Nature, Lond.* **320**, 482–3.

Martin, R.D. (1987) Long night for owl monkeys. *Nature, Lond.* **326**, 639–40.

Martin, R.D. and Harvey, P.H. (1985) Brain size allometry: ontogeny and phylogeny. In *Size and Scaling in Primate Biology* (ed. Jungers, W.L.). Plenum Press: New York, pp. 147–73.

Martin, R.D. and MacLarnon, A.M. (1985) Gestation period, neonatal size and maternal investment in placental mammals. *Nature, Lond.* **313**, 220–3.

Martin, R.D. and MacLarnon, A.M. (1988) Comparative quantitative studies of growth and reproduction. *Symp. zool. Soc. Lond.* **60**, 39–80.

Martin, R.D., Rivers, J.P.W. and Cowgill, U.M. (1976) Culturing mealworms as food for animals in captivity. *Int. Zoo Yb.* **16**, 63–70.

Martin, R.D., Chivers, D.J., MacLarnon, A.M. and Hladik, C.M. (1985) Gastrointestinal allometry in primates and other mammals. In *Size and Scaling in Primate Biology* (ed. Jungers, W.L.). Plenum Press: New York, pp. 61–89.

Martin, S.L., Zimmer, E.A., Kan, Y.W. and Wilson, A.C. (1980) Silent delta-globin gene in Old World monkeys. *Proc. natn. Acad. Sci. U.S.A.* **77**, 3563–6.

Maslin, T.P. (1952) Morphological criteria of phyletic relationships. *Syst. Zool.* **1**, 49–70.

Mason, W.A. (1968) Use of space by *Callicebus* groups. In *Primates: Studies in Adaptation and Variability* (ed. Jay, P.C.). Holt: New York, pp. 200–16.

Matthew, W.D. (1904) The arboreal ancestry of the Mammalia. *Am. Nat.* **38**, 811–18.

Matthew, W.D. (1909) The Carnivora and the Insectivora of the Bridger Basin, middle Eocene. *Mem. Am. Mus. nat. Hist.* **9**, 291–559.

Matthew, W.D. (1937) Paleocene faunas of San Juan Basin, New Mexico. *Trans. Am. Philos. Soc.* **30**, 1–510.

Matthew, W.D. and Granger, W. (1915) A revision of the lower Eocene Wasatch and Wind River faunas. Part IV, Entelonychia, Primates, Insectivora (part). *Bull. Am. Mus. nat. Hist.* **34**, 429–83.

Matthew, W.D. and Granger, W. (1924) New insectivores and ruminants from the Tertiary of Mongolia, with remarks on the correlation. *Am. Mus. Novit. no.* 105, 1–7.

Matthey, R. (1945) L'évolution de la formule chromosomale chez les vertébrés. *Experientia* **1**, 50–6 and 78–86.

Matthey, R. (1949) *Les Chromosomes des Vertébrés*. F. Rouge: Lausanne.

Matthey, R. (1973a) Les nombres diploides des euthériens. *Mammalia* **37**, 394–421.

Matthey, R. (1973b) The chromosome formulae of eutherian mammals. In *Cytotaxonomy and Vertebrate Evolution* (eds Chiarelli, A.B. and Capanna, E.). Academic Press: London, pp. 531–616.

Maynard Smith, J. (1968) Haldane's dilemma and the rate of evolution. *Nature, Lond.* **219**, 1114.

Maynard Smith, J. (1981) Macroevolution. *Nature, Lond.* **289**, 13–14.

Mayr, E. (1942) *Systematics and the Origin of Species*. Columbia University Press: New York.

Mayr, E. (1963) *Animal Species and Evolution*. Belknap Press, Harvard University Press: Cambridge, Mass.

Mayr, E. (1969) *Principles of Systematic Zoology*. McGraw-Hill: New York.

Mayr, E. (1974) Cladistic analysis or cladistic classification? *Z. zool. Syst. Evol.-Forsch.* **12**, 94–128.

McArdle, J.E. (1981) Functional morphology of the hip and thigh of the Lorisiformes. *Contr. Primatol.* **17**, 1–132.

McDowell, S.B. (1958) The Greater Antillean insectivores. *Bull. Am. Mus. nat. Hist.* **115**, 113–214.

McElhinny, M.W. (1973) *Palaeomagnetism and Plate Tectonics*. Cambridge University Press: Cambridge.

McGhee, R.B. and Frank, A.A. (1968) On the stability properties of quadrupedal creeping gaits. *Math. Biosci.* **3**, 331–51.

McGrew, W.C. (1974) Tool use by wild chimpanzees in feeding upon driver ants. *J. hum. Evol.* **3**, 501–8.

McGuinness, E. and Allman, J.M. (1985) Organization of the visual system in tarsiers. *Am. J. phys. Anthrop.* **66**, 200.

McHenry, H.M. (1975) Fossils and the mosaic nature of human evolution. *Science* **190**, 425–31.

McKenna, M.C. (1963) New evidence against tupaioid affinities of the mammalian family Anagalidae. *Am. Mus. Novit.* no. 2158, 1–16.

McKenna, M.C. (1966) Paleontology and the origin of primates. *Folia primatol.* **4**, 1–25.

McKenna, M.C. (1975) Toward a phylogenetic classification of the Mammalia. In *Phylogeny of the Primates* (eds Luckett, W.P. and Szalay, R.S.). Plenum Press: New York, pp. 21–46.

McKusick, V.A. (1984) The human gene map. In *Genetic Maps*, vol. 3 (ed. O'Brien, S.J.). Cold Spring Harbor Press: New York, pp. 417–41.

McMahon, T.A. and Bonner, J.T. (1983) *On Size and Life*. Scientific American Books: New York.

McNab, B.K. (1978a) The evolution of endothermy in the phylogeny of mammals. *Am. Nat.* **112**, 1–27.

McNab, B.K. (1978b) Energetics of arboreal folivores: physiological problems and ecological consequences of feeding on an ubiquitous food supply. In *The Ecology of Arboreal Folivores* (ed. Montgomery, G.G.). Smithsonian Institution: Washington, pp. 153–72.

McNab, B.K. (1980) Food habits, energetics, and the population biology of mammals. *Am. Nat.* **116**, 106–24.

McNab, B.K. (1986) The influence of food habits on the energetics of eutherian mammals. *Ecol. Monogr.* **56**, 1–19.

Medway, Lord (1978) *The Wild Mammals of Malaya and Singapore* (2nd edition). Oxford University Press: London.

Meier, B., Albignac, R., Peyriéras, A., Rumpler, Y. and Wright, P. (1987) A new species of *Hapalemur* (Primates) from South East Madagascar. *Folia primatol.* **48**, 211–15.

Meister, W. and Davis, D.D. (1956) Placentation of the pigmy tree-shrew *Tupaia minor*. *Fieldiana, Zool.* **35**, 73–84.

Meister, W. and Davis, D.D. (1958) Placentation of the terrestrial tree-shrew (*Tupaia tana*). *Anat. Rec.* **132**, 541–53.

Melnick, D.J. and Kidd, K.K. (1985) Genetic and evolutionary relationships among Asian macaques. *Int. J. Primatol.* **6**, 123–60.

Mendel, F.C. (1979) The wrist joint of two-toed sloths and its relevance to brachiating adaptations in the Hominoidea. *J. Morph.* **162**, 413–24.

Mendel, J. (G.) (1866) Versuche über Pflanzen-Hybriden. *Verh. Natur-Vereins Brünn* **4**, 3–57.

Miao, D. and Lillegraven, J.A. (1986) Discovery of three ear ossicles in a multituberculate mammal. *Nat. Geog. Res.* **2**, 500–7.

Midlo, C. (1934) Form of the hand and foot in primates. *Am. J. phys. Anthrop.* **19**, 337–89.

Miller, D.A. (1977) Evolution of primate chromosomes: man's closest relative may be the gorilla, not the chimpanzee. *Science* **198**, 1116–24.

Miller, R.A. (1932) Evolution of the pectoral girdle and forelimb in primates. *Am. J. phys. Anthrop.* **17**, 1–56.

Miller, R.A. (1945) The ischial callosities of primates. *Am. J. Anat.* **76**, 67–91.

Mills, J.R.E. (1955) Ideal dental occlusion in the Primates. *Dent. Practnr.* **6**, 47–61.

Mills, J.R.E. (1963) Occlusion and malocclusion of the teeth of primates. In *Dental Anthropology* (ed. Brothwell, D.R.). Pergamon Press: Oxford, pp. 29–51.

Mills, J.R.E. (1978) The relationship between tooth patterns and jaw movements in the Hominoidea.

In *Development, Function and Evolution of Teeth* (eds Butler, P.M. and Joysey, K.A.). Academic Press: London, pp. 341–53.

Milton, K. (1980) *The Foraging Strategy of Howler Monkeys: A Study in Primate Economics.* Columbia University Press: New York.

Milton, K. (1984) Habitat, diet and activity patterns of free-ranging woolly spider monkeys (*Brachyteles arachnoides* E. Geoffroy 1806). *Int. J. Primatol.* **5**, 491–514.

Mitchell, A.R., Gosden, J.R. and Ryder, O.A. (1981) Satellite DNA relationships in man and the primates. *Nucleic Acids Res.* **9**, 3235–49.

Mittermeier, R.A. and Fleagle, J.G. (1976) The locomotor and postural repertoires of *Ateles geoffroyi* and *Colobus guereza* and a re-evaluation of the locomotor category semi-brachiation. *Am. J. phys. Anthrop.* **45**, 235–56.

Mivart, St. G.J. (1867) On the appendicular skeleton of the primates. *Phil. Trans. R. Soc. Lond.* **157**, 299–430.

Mivart, St. G.J. (1873) On *Lepilemur* and *Cheirogaleus* and on the zoological rank of the Lemuroidea. *Proc. zool. Soc. Lond.* **1873**, 484–510.

Miyamoto, M.M. and Goodman, M. (1986) Biomolecular systematics of eutherian mammals: phylogenetic patterns and classification. *Syst. Zool.* **35**, 230–40.

Mollet, O.D. van der Spuy (1947) Fossil mammals from the Makapan Valley, Potgietersrust. I. Primates. *S. Afr. J. Sci.* **43**, 295–303.

Mollison, T. (1910) Die Körperproportionen der Primaten. *Morph. Jb.* **42**, 79–304.

Molnar, S. and Gantt, D.G. (1977) Functional implications of primate enamel thickness. *Am. J. phys. Anthrop.* **46**, 447–54.

Moore, G.W., Barnabas, J. and Goodman, M. (1973) A method for constructing maximum parsimony ancestral amino acid sequences on a given network. *J. theor. Biol.* **38**, 459–85.

Moore, W.J. (1981) *The Mammalian Skull.* Cambridge University Press: Cambridge.

Morbeck, M.E. (1975) *Dryopithecus africanus* forelimb. *J. hum. Evol.* **4**, 39–46.

Morgan, T.H. (1910) Chromosomes and heredity. *Am. Nat.* **44**, 449–96.

Morgan, T.H. (1911) An attempt to analyse the constitution of the chromosomes on the basis of sex-linked inheritance in *Drosophila. J. exp. Zool.* **11**, 365–413.

Morgan, T.H. (1926) *The Theory of the Gene.* Yale University Press: New Haven.

Morton, D.J. (1924) Evolution of the human foot: II. *Am. J. phys. Anthrop.* **7**, 1–52.

Moss-Salentijn, L. (1978) Vestigial teeth in the rabbit,

rat and mouse: their relationship to the problem of lacteal dentitions. In *Development, Function and Evolution of Teeth* (eds Butler, P.M. and Joysey, K.A.). Academic Press: London, pp. 13–29.

Mossman, H.W. (1937) Comparative morphogenesis of the fetal membranes and accessory uterine structures. *Contr. Embryol. Carnegie Inst. Wash.* **26**, 129–246.

Mossman, H.W. (1953) The genital system and the fetal membranes as criteria for mammalian phylogeny and taxonomy. *J. Mammal.* **24**, 289–98.

Mountcastle, V.B. (1978) An organizing principle for cerebral function: the unit module and the distributed system. In *The Mindful Brain* (eds Edelman, G.M. and Mountcastle, V.B.). MIT Press: Cambridge, Mass., pp. 7–50.

Muleris, M., Dutrillaux, B. and Chauvier, G. (1983) Mise en évidence d'une fission centromérique héterozygote chez un mâle *Theropithecus gelada* et comparaison chomosomique avec les autres Papioninae. *Génét. Sél. Evol.* **15**, 177–84.

Müller, A.H. (1951) Grundlagen der Biostratonomie. *Abh. deutsch. Akad. Wiss., Kl. Math. algem. Naturwiss.* **3**, 1–147.

Müller, E.F. (1979) Energy metabolism, thermoregulation and water budget in the slow loris (*Nycticebus coucang*, Boddaert, 1785). *Comp. Biochem. Physiol., A* **64**, 109–19.

Müller, E.F. (1983) Wärme- und Energiehaushalt bei Halbaffen (Prosimiae). *Bonn. zool. Beitr.* **34**, 29–71.

Müller, E.F. (1985) Basic metabolic rates in primates – the possible role of phylogenetic and ecological factors. *Comp. Biochem. Physiol., A* **81**, 707–11.

Müller, E.F., Nieschalk, U. and Meier, B. (1985) Thermoregulation in the slender loris (*Loris tardigradus*). *Folia primatol.* **44**, 216–26.

Muller, J. (1935) The orbito temporal region of the skull of the Mammalia. *Arch. néerl. Zool.* **1**, 118–259.

Müller-Schwarze, D. (1983) Scent glands in mammals and their functions. In *Advances in the Study of Mammalian Behavior* (eds Eisenberg, J.F. and Kleiman, D.G.). American Society of Mammalogists: Shippensburg, Pennsylvania, pp. 150–97.

Munro, H.N. (1969) Evolution of protein metabolism in mammals. In *Mammalian Protein Metabolism*, vol. 3 (ed. Munro, H.N.). Academic Press: New York, pp. 133–82.

Murray, P. (1975) The role of cheek pouches in cercopithecine monkey adaptive strategy. In *Primate Functional Morphology and Evolution* (ed. Tuttle, R.H.). Mouton: The Hague, pp. 151–94.

Musich, P.R., Brown, F.L. and Maio, J.J. (1980) Highly repetitive component alpha and related alphoid DNAs in man and monkeys. *Chromosoma (Berl.)* **80**, 331–48.

Musser, G.G. and Dagosto, M. (1987) The identity of *Tarsius pumilus*, a pygmy species endemic to the montane mossy forest of Central Sulawesi. *Am. Mus. Novit.* no. 2867, 1–53.

Muybridge, E. (1899) *Animals in Motion.* Chapman and Hall: London.

Myers, R.H. and Shafer, D.A. (1978) Hybrid offspring of a mating of gibbon and siamang. *Science* **205**, 308–10.

Nadler, R.D. (1975) Cyclicity in tumescence of the perineal labia of female lowland gorillas. *Anat. Rec.* **181**, 791–8.

Nagle, C.A. and Denari, J.H. (1983) The cebus monkey (*Cebus apella*). In *Reproduction in New World Primates* (ed. Hearn, J.P.). MTP: Lancaster, pp. 39–67.

Napier, J.R. (1960) Studies of the hands of living primates. *Proc. zool. Soc. Lond.* **134**, 647–57.

Napier, J.R. (1961) Prehensility and opposability in the hands of primates. *Symp. zool. Soc. Lond.* **5**, 115–32.

Napier, J.R. (1963) Brachiation and brachiators. *Symp. zool. Soc. Lond.* **10**, 183–94.

Napier, J.R. (1967a) Evolutionary aspects of primate locomotion. *Am. J. phys. Anthrop.* **27**, 333–42.

Napier, J.R. (1967b) The antiquity of human walking. *Sci. Am.* **216**(4), 56–66.

Napier, J.R. (1971) *The Roots of Mankind.* George Allen and Unwin: London.

Napier, J.R. and Davis, P.R. (1959) The forelimb skeleton and associated remains of *Proconsul africanus*. *Fossil Mammals Afr.* **16**, 1–69.

Napier, J.R. and Napier, P.H. (1967) *A Handbook of Living Primates.* Academic Press: London.

Napier, J.R. and Napier, P.H. (1985) *The Natural History of the Primates.* British Museum (Natural History). London.

Napier, J.R. and Walker, A.C. (1967) Vertical clinging and leaping – a newly recognised category of locomotor behaviour of primates. *Folia primatol.* **6**, 204–19.

Nash, W.G. and O'Brien, S.J. (1982) Conserved subregions of homologous G-banded chromosomes between orders in mammalian evolution: carnivores and primates. *Proc. natn. Acad. Sci. U.S.A.* **79**, 6631–5.

Nei, M. (1972) Genetic distance between populations. *Am. Nat.* **106**, 283–92.

Nei, M. (1982) Evolution of human races at the gene level. In *Human Genetics. Part A: The Unfolding Genome* (eds Bonne-Tamir, B., Cohen, T. and

Goodman, R.M.). Alan R. Liss: New York, pp. 167–81.

Nei, M. (1987) *Molecular Evolutionary Genetics*. Columbia University Press: New York.

Nelson, G.J. (1974) Classification as an expression of phylogenetic relationships. *Syst. Zool.* **22**, 344–59.

Newman, I.M. and Hendrickx, A.G. (1984) Fetal development in the normal thick-tailed bushbaby (*Galago crassicaudatus panganiensis*). *Am. J. Primatol.* **6**, 337–55.

Niemitz, C. (1979) Outline of the behaviour of *Tarsius bancanus*. In *The Study of Prosimian Behavior* (eds Doyle, G.A. and Martin, R.D.). Academic Press: New York, pp. 631–60.

Niemitz, C. (ed.) (1984a) *Biology of Tarsiers*. Gustav Fischer Verlag: Stuttgart.

Neimitz, C. (1984b) An investigation and review of the territorial behaviour and social organization of the genus *Tarisus*. In *Biology of Tarsiers* (ed. Niemitz, C.). Gustav Fischer Verlag: Stuttgart, pp. 117–27.

Nishida, T. (1979) The social structure of chimpanzees in the Mahale Mountains. In *The Great Apes* (eds Hamburg, D.A. and McCown, E.R.). Benjamin/ Cummings: Menlo Park, California, pp. 73–122.

Noback, C.R. (1975) The visual system of primates in phylogenetic studies. In *Phylogeny of the Primates* (eds Luckett, W.P. and Szalay, F.S.). Plenum Press: New York, pp. 199–218.

Noback, C.R. and Laemle, L.K. (1970) Structural and functional aspects of the visual pathways of primates. In *The Primate Brain* (eds Noback, C.R. and Montagna, W.). Appleton-Century-Crofts: New York, pp. 55–81.

Noback, C.R. and Moskowitz, N. (1963) The primate nervous system: functional and structural aspects in phylogeny. In *Evolutionary and Genetic Biology of Primates* (ed. Buettner-Janusch, J.). Academic Press: New York, pp. 131–77.

Noback, C.R. and Shriver, J.E. (1966) Phylogenetic and ontogenetic aspects of the lemniscal systems and the pyramidal system. In *Evolution of the Forebrain* (eds Hassler, R. and Stephan, H.). Georg Thieme Verlag: Stuttgart, pp. 316–25.

Novacek, M.J. (1977a) Aspects of the problem of variation, origin, and evolution of the eutherian auditory bulla. *PalaeoBios* **24**, 1–42.

Novacek, M.J. (1977b) A review of Paleocene and Eocene Leptictidae (Eutheria: Mammalia) from North America. *Mammal Rev.* **7**, 131–49.

Novacek, M.J. (1980) Cranioskeletal features in tupaiids and selected Eutheria as phylogenetic evidence. In *Comparative Biology and Evolutionary Relationships of Tree Shrews* (ed. Luckett, W.P.). Plenum Press: New York, pp. 35–93.

Novacek, M.J. (1982a) The brain of *Leptictis dakotensis*, an Oligocene leptictid from North America. *J. Paleont.* **56**, 1177–86.

Novacek, M.J. (1982b) Information for molecular studies from anatomical and fossil evidence on higher eutherian phylogeny. In *Macromolecular Sequences in Systematic and Evolutionary Biology* (ed. Goodman, M.). Plenum Press: New York, pp. 3–41.

Novacek, M.J. (1986a) The skull of leptictid insectivorans and the higher-level classification of eutherian mammals. *Bull. Am. Mus. nat. Hist.* **183**, 1–112.

Novacek, M.J. (1986b) The primitive eutherian dental formula. *J. vert. Paleont.* **6**, 191–6.

Novacek, M.J. and Wyss, A. (1986) Origin and transformation of the mammalian stapes. In *Vertebrates, Phylogeny, and Philosophy* (eds Flanagan, K.M. and Lillegraven, J.A.). Contributions to Geology, Special Paper 3, University of Wyoming: Laramie, Wyoming, pp. 35–53.

Nuttall, G.H.F. (1904) *Blood Immunity and Blood Relationship*. Cambridge University Press: Cambridge.

O'Brien, S.J. and Nash, W.G. (1982) Genetic mapping in mammals: chromosome maps of domestic cat. *Science* **216**, 257–65.

O'Connor, B.L. (1976) *Dryopithecus (Proconsul) africanus*: quadruped or non-quadruped? *J. hum. Evol.* **5**, 279–83.

Oftedal, O.T. (1984) Milk composition, milk yield and energy output at peak lactation: a comparative review. *Symp. zool. Soc. Lond.* **51**, 33–85.

Ogden, T.E. (1974) The morphology of retinal neurons of the owl monkey *Aotes*. *J. comp. Neurol.* **153**, 399–428.

Odgen, T.E. (1975) The receptor mosaic of *Aotus trivirgatus*: distribution of rods and cones. *J. comp. Neurol.* **163**, 193–202.

Ohno, S. (1970) *Evolution by Gene Duplication*. Springer Verlag: Berlin.

Olson, E.C. (1944) Origin of mammals based on the cranial morphology of therapsid suborders. *Geol. Soc. Am. Spec. Pap.* **55**, 1–136.

Olson, E.C. (1959) The evolution of mammal characters. *Evolution* **13**, 344–53.

Ordy, J. and Keefe, J. (1965) Visual acuity and retinal specialization in primates: a comparison of the transitional primate tree shrew with the diurnal rhesus monkey. *Anat. Rec.* **151**, 394.

Ordy, J.M. and Samorajski, T. (1968) Visual acuity and ERG-CFF in relation to the morphologic organisation of the retina among diurnal and nocturnal primates. *Vision Res.* **8**, 1205–25.

Orgel, L.E. and Crick, F.H.C. (1980) Selfish DNA: the ultimate parasite. *Nature, Lond.* **284**, 604–7.

Osborn, H.F. (1888) The nomenclature of the mammalian dental cusps. *Am. Nat.* **22**, 926–8.

Osborn, H.F. (1907) *Evolution of Mammalian Molar Teeth to and from the Triangular Type* (ed. by Gregory, W.K.). Macmillan: New York.

Osborn, J.W. (1970) New approach to Zahnreihen. *Nature, Lond.* **225**, 343–6.

Osborn, J.W. (1973) The evolution of dentitions. *Am. Sci.* **61**, 548–59.

Osborn, J.W. (1978) Morphogenetic gradients: fields versus clones. In *Development, Function and Evolution of Teeth* (eds Butler, P.M. and Joysey, K.A.). Academic Press: London, pp. 171–201.

Osborn, J.W. and Lumsden, A.G.S. (1978) An alternative to 'thegosis' and a re-examination of the ways in which mammalian molars work. *Neues Jb. Geol. Paläontol. Abh.* **156**, 371–92.

Ouchterlony, O. (1953) Antigen–antibody reactions in gels. IV. Types of reactions in coordinated systems of diffusion. *Acta path. microbiol. scand.* **32**, 231–40.

Owen, R. (1848) Report on the archetype and homologies of the vertebrate skeleton. *Rep. Meeting Br. Ass. Advmt Sci.* **16**, 169–340.

Oxnard, C.E. (1973) Some locomotor adaptations among lower primates: implications for primate evolution. *Symp. zool. Soc. Lond.* **33**, 255–299.

Packer, D.J. (1987) The influence of carotid arterial sounds on hearing sensitivity in mammals. *J. Zool., Lond.* **211**, 547–60.

Pagel, M.D. and Harvey, P.H. (1988) The taxon-level problem in the evolution of mammalian brain size: facts and artifacts. *Am. Nat.* **132**, 344–59.

Palay, S.L. (1967) Principles of cellular organization in the nervous system. In *The Neurosciences: A Study Program* (eds Quarton, G.C., Melnechuk, T. and Schmitt, F.O.). Rockfeller University Press: New York, pp. 24–31.

Paluska, E. and Korinek, J. (1960). Studium der antigenen Eiweissverwandtschaft zwischen Menschen und einigen Primaten mit Hilfe neuerer immunbiologischer Methoden. *Z. Immun.-Forsch. exp. Ther.* **119**, 244–57.

Pan, Y. and Wu, R. (1986) A new species of *Sinoadapis* from the hominoid site, Lufeng. *Acta anthrop. sin.* **5**, 31–40.

Panchen, A.L. (1982) The use of parsimony in testing phylogenetic hypotheses. *Zool. J. Linn. Soc., Lond.* **74**, 305–28.

Panzer, W. (1932) Beiträge zur biologischen Anatomie des Baumkletterns der Säugetiere. I. Das Nagel-Kralle-Problem. *Z. Anat. Entw.-Gesch.* **98**, 147–98.

Pariente, G.F. (1979) The role of vision in prosimian behavior. In *The Study of Prosimian Behavior* (eds Doyle, G.A. and Martin, R.D.). Academic Press: New York, pp. 411–59.

Parker, W.K. (1885) On the structure and development of the skull in Mammalia. Part III: Insectivora. *Phil. Trans. R. Soc. Lond.* **176**, 121–275.

Parrington, F.R. (1971) On the Upper Triassic mammals. *Phil Trans. R. Soc. Lond.* B **261**, 231–72.

Passingham, R.E. (1975) The brain and intelligence. *Brain Behav. Evol.* **11**, 1–15.

Passingham, R.E. (1978) Brain size and intelligence in primates. In *Recent Advances in Primatology*, vol. 3: *Evolution* (eds Chivers, D.J. and Joysey, K.A.). Academic Press: London, pp. 85–6.

Passingham, R.E. (1981) Primate specialization in brain and intelligence. *Symp. zool. Soc. Lond.* **46**, 361–88.

Paterson, H.E.H. (1973) Animal and plant speciation studies in Western Australia. *J. R. Soc. West. Aust.* **56**, 31–6.

Paterson, H.E.H. (1982) Darwin and the origin of species. *S. Afr. J. Sci.* **78**, 272–5.

Paterson, H.E.H. (1985) The recognition concept of species. In *Species and Speciation* (ed. Vrba, E.S.). Transvaal Museum Monographs: Pretoria, pp. 21–9.

Patterson, B. (1956) Early Cretaceous mammals and the evolution of mammalian molar teeth. *Fieldiana, Geol.* **13** (1), 1–105.

Patterson, B. (1968) The extinct baboon, *Parapapio jonesi*, in the early Pleistocene of northwestern Kenya. *Breviora* **282**, 1–4.

Paulli, S. (1900) Über die Pneumaticität des Schädels bei den Säugetieren. *Morph. Jb.* **28**, 147–78, 179–251, 483–564.

Payne, P.R. and Wheeler, E.F. (1968) Comparative nutrition in pregnancy and lactation. *Proc. Nutr. Soc.* **27**, 129–38.

Pedler, C. (1963) The fine structure of the tapetum cellulosum. *Exp. Eye Res.* **2**, 185–95.

Pei, W.C. (1957) Discovery of *Gigantopithecus* mandibles and other material in Liu-Cheng district of Central Kwangsi in South China. *Vert. Palasiat.* **1**, 65–71.

Pellicciari, C., Formenti, D., Redi, C.A. and Manfredi-Romanini, M.G. (1982) DNA content variability in primates. *J. hum. Evol.* **11**, 131–41.

Perry, J.S. (1971) *The Ovarian Cycle of Mammals*. Oliver & Boyd: Edinburgh.

Perutz, M.F. (1983) Species adaptation in a protein molecule. *Molec. Biol. Evol.* **1**, 1–28.

Petter, J.-J. (1962) Recherches sur l'écologie et

l'éthologie des lémuriens malgaches. *Mém. Mus. natn. Hist. nat., A (Zool.)* **27**, 1–146.

Petter, J.-J., and Peyrieras, A. (1970) Nouvelle contribution à l'étude d'un lémurien malgache, le aye-aye (*Daubentonia madagascariensis* E. Geoffroy). *Mammalia* **34**, 167–93.

Petter, J.-J., Albignac, R. and Rumpler, Y. (1977) Mammifères lémuriens (Primates prosimiens). *Faune de Madagascar* **44**, 1–513.

Petter, J.-J., Schilling, A. and Pariente, G. (1971) Observations éthologiques sur deux lémuriens malgaches nocturnes, *Phaner furcifer* et *Microcebus coquereli. Terre et Vie* **25**, 285–327.

Petter-Rousseaux, A. (1964) Reproductive physiology and behavior of the Lemuroidea. In *Evolutionary and Genetic Biology of the Primates* (ed. Buettner-Janusch, J.). Academic Press: New York, pp. 91–132.

Pettigrew, J.D. (1986) Flying primates? Megabats have the advanced pathway from eye to midbrain. *Science* **231**, 1304–6.

Phillips, C.G. (1971) Evolution of the corticospinal tract in primates with special reference to the hand. *Proc. 3rd int. Congr. Primatol., Zürich* **2**, 2–23.

Phillips, C.G. and Porter, R. (1977) *Corticospinal Neurons: Their Role in Movement.* Academic Press: London.

Pianka, E.R. (1970) On r- and K-selection. *Am. Nat.* **104**, 592–7.

Pickford, M. (1986) Sexual dimorphism in *Proconsul. Hum. Evol.* **1**, 111–48.

Pickford, M. (1987) The chronometry of the Cercopithecoidea of East Africa. *Hum. Evol.* **2**, 1–17.

Pilbeam, D.R. (1966) Notes on *Ramapithecus*, the earliest known hominid, and *Dryopithecus. Am. J. phys. Anthrop.* **25**, 1–5.

Pilbeam, D.R. (1967) Man's earliest ancestors. *Science J.* **3**, 47–53.

Pilbeam, D.R. (1969) Tertiary Pongidae of East Africa: evolutionary relationships and taxonomy. *Peabody Mus. J.* **31**, 1–185.

Pilbeam, D.R. (1970) *Gigantopithecus* and the origins of Hominidae. *Nature, Lond.* **225**, 516–19.

Pilbeam, D.R. (1972) *The Ascent of Man: An Introduction to Human Evolution.* Macmillan: New York.

Pilbeam, D.R. (1982) New hominoid skull material from the Miocene of Pakistan. *Nature, Lond.* **295**, 232–4.

Pilbeam, D.R. and Gould, S.J. (1974) Size and scaling in human evolution. *Science* **186**, 892–901.

Pilbeam, D.R. and Walker, A.C. (1968) Fossil monkeys from the Miocene of Napak, northeast Uganda. *Nature, Lond.* **220**, 657–60.

Pilbeam, D.R., Meyer, G.E., Badgley, C., Rose, M.D., Pickford, M.H.L., Behrensmeyer, A.K. and Ibrahim Shah, S.M. (1977) New hominoid primates from the Siwaliks of Pakistan and their bearing on hominoid evolution. *Nature, Lond.* **270**, 689–95.

Pilgrim, G.E. (1927) A *Sivapithecus* palate and other primate fossils from India. *Mem. Geol. Surv. India (Paleont. ind.), n.s.* **14**, 1–26.

Pirie, A. (1959) Riboflavin in the tapetum of the bushbaby. *Nature, Lond.* **183**, 985.

Pirie, A. (1966) The chemistry and structure of the tapetum lucidum in animals. In *Aspects of Comparative Ophthalmology* (ed. Graham-Jones, O.). Pergamon Press, Oxford, pp. 57–68.

Piveteau, J. (ed.) (1957) *Traité de Paléontologie.* Vol. VII: *Primates; Paléontologie Humaine.* Masson: Paris.

Platnick, N.I. and Nelson, G.J. (1978) A model of analysis for historical biogeography. *Syst. Zool.* **27**, 1–16.

Pocock, R.I. (1918) On the external characters of the lemurs and of *Tarsius. Proc. zool. Soc. Lond.* **1918**, 19–53.

Poirier, F.E. (1969) Behavioural flexibility and intertroop variation among Nilgiri langurs (*Presbytis johnii*) of South India. *Folia primatol.* **11**, 119–33.

Pollock, J.I. (1975) Field observations on *Indri indri*: a preliminary report. In *Lemur Biology* (eds Tattersall, I. and Sussmann, R.W.). Plenum Press: New York, pp. 287–311.

Pollock, J.I. (1977) The ecology and sociology of feeding in *Indri indri*. In *Primate Ecology* (ed. Clutton-Brock, T.H.). Academic Press: London, pp. 37–69.

Pollock, J.I. (1979) Spatial distribution and ranging behavior in lemurs. In *The Study of Prosimian Behavior* (eds Doyle, G.A. and Martin, R.D.). Academic Press: New York, pp. 359–409.

Pollock, J.I. and Mullin, R.J. (1987) Vitamin C biosynthesis in prosimians: evidence for the anthropoid affinities of *Tarsius. Am. J. phys. Anthrop.* **73**, 65–70.

Polson, M.C. (1968) Spectral sensitivity and color vision in *Tupaia glis.* PhD thesis, Indiana University, Bloomington.

Polyak. S. (1957) *The Vertebrate Visual System.* University of Chicago Press: Chicago.

Pond, C.M. (1977) The significance of lactation in the evolution of mammals. *Evolution* **31**, 177–99.

Poorman, P.A., Cartmill, M. and MacPhee, R.D.E. (1985) The G-banded karyotype of *Tarsius bancanus* and its implications for primate phylogeny. *Am. J. phys. Anthrop.* **66**, 215.

Portmann, A. (1939) Nesthocker und Nestflüchter als Entwicklungszustände von verschiedener Wertigkeit bei Vögeln und Säugern. *Rev. suisse Zool.* **46**, 385–90.

Portmann, A. (1941) Die Tragzeiten der Primaten und die Dauer der Schwangerschaft beim Menschen: Ein Problem der vergleichenden Biologie. *Rev. suisse Zool.* **48**, 511–18.

Portmann, A. (1952) Besonderheiten und Bedeutung der menschlichen Brutpflege. *Ciba-Zeitschr.* **11**, 4758–63.

Portmann, A. (1962) Cerebralisation und Ontogenese. *Medizin. Grundlagenforsch.* **4**, 1–62.

Portmann, A. (1965) Über die Evolution der Tragzeit bei Säugetieren. *Rev. suisse Zool.* **72**, 658–66.

Potter, B. (1986) The allometry of primate skeletal weight. *Int. J. Primatol.* **7**, 457–66.

Potts, M. (1965) Implantation: an electron microscope study with special reference to the mouse. PhD thesis, University of Cambridge.

Prange, H.D., Anderson, J.F. and Rahn, H. (1979) Scaling of skeletal mass to body mass in birds and mammals. *Am. Nat.* **113**, 103–22.

Presley, R. (1978) Ontogeny of some elements of the auditory bulla in mammals. *J. Anat., Lond.* **126**, 428.

Presley, R. (1979) The primitive course of the internal carotid artery in mammals. *Acta anat.* **103**, 238–44.

Preuschoft, H. (1973) Body posture and locomotion in some East African Miocene Dryopithecinae. In *Human Evolution* (ed. Day, M.H.). Taylor and Francis: London, pp. 13–46.

Preuschoft, H. and Weinmann, W. (1973) Biomechanical investigations of *Limnopithecus* with special reference to the influence exerted by body weight on bone thickness. *Am. J. phys. Anthrop.* **38**, 241–50.

Preuschoft, H., Chivers, D.J., Brockelman, W.Y. and Creel, N. (eds) (1984) *The Lesser Apes: Evolutionary and Behavioural Biology.* Edinburgh University Press: Edinburgh.

Preuss, T.M. (1982) The face of *Sivapithecus indicus*: description of a new, relatively complete, specimen from the Siwaliks of Pakistan. *Folia primatol.* **38**, 141–57.

Prince, J.H. (1953) Comparative anatomy of the orbit. *Br. J. physiol. Optics* **10**, 144–54.

Prost, J.H. (1965) A definitional system for the classification of primate locomotion. *Am. Anthrop.* **67**, 1198–214.

Prost, J.H. (1980) Origin of bipedalism. *Am. J. phys. Anthrop.* **52**, 175–90.

Prothero, J.W. and Sundsten, J.W. (1984) Folding of the cerebral cortex in mammals: a scaling model. *Brain Behav. Evol.* **24**, 152–67.

Quinet, G.E. (1966) *Teilhardina belgica*, ancêtre des Anthropoidea de l'ancien monde. *Bull. Inst. R. Sci. nat. Belg.* **42**, 1–14.

Quiroga, J.C. (1980) The brain of the mammal-like reptile *Probainognathus jenseni* (Therapsida, Cynodontia). A correlative paleoneurological approach to the cortex at the reptile–mammal transition. *J. Hirnforsch.* **21**, 299–336.

Qumsiyeh, M.B., Hamilton, M.J. and Schlitter, D.A. (1987) Problems in using Robertsonian rearrangements in determining monophyly: examples from the genera *Tatera* and *Gerbillus*. *Cytogenet. Cell Genet.* **44**, 198–208.

Radinsky, L.B. (1967) Relative brain size: a new measure. *Science* **155**, 836–8.

Radinsky, L.B. (1968) A new approach to mammalian cranial analysis, illustrated with examples of prosimian primates. *J. Morph.* **124**, 167–80.

Radinsky, L.B. (1970) The fossil evidence of prosimian brain evolution. In *The Primate Brain* (eds Noback, C.R. and Montagna, W.). Appleton-Century-Crofts: New York, pp. 209–24.

Radinsky, L.B. (1972) Endocasts and studies of primate brain evolution. In *The Functional and Evolutionary Biology of Primates* (ed. Tuttle, R.). Aldine-Atherton: Chicago, pp. 175–84.

Radinsky, L.B. (1974a) The fossil evidence of anthropoid brain evolution. *Am. J. phys. Anthrop.* **41**, 15–27.

Radinsky, L.B. (1974b) Prosimian brain morphology: functional and phylogenetic implications. In *Prosimian Biology* (eds Martin, R.D., Doyle, G.A. and Walker, A.C.). Duckworth: London, pp. 781–98.

Radinsky, L.B. (1977) Early primate brains: facts and fiction. *J. hum. Evol.* **6**, 79–86.

Radinsky, L.B. (1978) Do albumin clocks run on time? *Science* **200**, 1182–3.

Ralls, K. (1971) Mammalian scent marking. *Science* **171**, 443–9.

Rasmussen, D.T. (1986) Anthropoid origins: a possible solution to the Adapidae–Omomyidae paradox. *J. hum. Evol.* **15**, 1–12.

Rathbun, G.B. and Gache, M. (1980) Ecological survey of the night monkey, *Aotus trivirgatus*, in Formosa Province, Argentina. *Primates* **21**, 211–19.

Raup, D.M. and Stanley, S.M. (1978) *Principles of Paleontology.* W. H. Freeman: San Francisco.

Read, D.W. (1975) Primate phylogeny, neutral mutations and molecular clocks. *Syst. Zool.* **24**, 209–21.

Read, D.W. and Lestrel, P.E. (1970) Hominid phylogeny and immunology: a critical appraisal. *Science* **168**, 578–80.

Read, D.W. and Lestrel, P.E. (1972) Phyletic divergence dates of hominoid primates: a note. *Evolution* **26**, 669–70.

Reichert, C. (1837) Über die Visceralbogen der Wirbeltiere im allgemeinen und deren Metamorphosen bei den Vögeln und Säugetieren. *Arch. Anat. Physiol. Wiss. Med.* **1837**, 120–222.

Remane, A. (1921) Zur Beurteilung der fossilen Anthropoiden. *Zentbl. Miner. Geol. Paläont.* **11**, 335–9.

Remane, A. (1956a) Paläontologie und Evolution der Primaten. In *Primatologia*, vol. 1 (eds Hofer, H., Schultz, A.H. and Starck, D.). Karger: Basel, pp. 267–378.

Remane, A. (1956b) *Die Grundlagen des natürlichen Systems, der vergleichenden Anatomie und der Phylogenetik.* Geest und Portig: Leipzig.

Remane, A. (1960) Zähne und Gebiss. In *Primatologia*, vol. 3, part 2 (eds Hofer, H., Schultz, A.H. and Starck, D.). Karger: Basel, pp. 637–846.

Reng, R. (1977) Die Placenta von *Microcebus murinus* Miller. *Z. Säugetierk.* **42**, 201–4.

Reynolds, V. and Reynolds, F. (1965) Chimpanzees of the Budongo Forest. In *Primate Behavior: Field Studies of Monkeys and Apes* (ed. DeVore, I.). Holt: New York, pp. 368–424.

Richard, A.F. (1974) Patterns of mating in *Propithecus verreauxi*. In *Prosimian Biology* (eds Martin, R.D., Doyle, G.A. and Walker, A.C.). Duckworth: London, pp. 49–75.

Richard, A.F. (1977) The feeding behaviour of *Propithecus verreauxi*. In *Primate Ecology* (ed. Clutton-Brock, T.H.). Academic Press: London, pp. 72–96.

Richard, A.F. (1985) *Primates in Nature.* W.H. Freeman: New York.

Ride, W.D.L. (1970) *A Guide to the Native Mammals of Australia.* Oxford University Press: Melbourne.

Ride, W.D.L., Sabrosky, C.W., Bernardi, G. and Melville, R.V. (eds) (1985) *International Code of Zoological Nomenclature* (3rd edition). British Museum (Natural History): London.

Riek, E.F. (1970) Lower Cretaceous fleas. *Nature, Lond.* **227**, 746–7.

Riesenfeld, A. (1975) Volumetric determinations of metatarsal robusticity in a few living primates and in the foot of *Oreopithecus. Primates* **16**, 9–15.

Rightmire, G.P. (1986) Stasis in *Homo erectus* defended. *Paleobiology* **12**, 324–5.

Rijksen, H.D. (1978) *A Field Study on Sumatran Orang Utans* (Pongo pygmaeus abelii *Lesson 1827*). *Ecology, Behaviour and Conservation.* H. Veenman and Zonen: Wageningen.

Rimoli, R. (1977) Una nueva especie de monos (Cebidae: Saimirinae? *Saimiri*) de la Hispaniola. *Cuad. del Cendia (Univ. Auton. Santo Domingo)* **242** (1), 1–16.

Ripley, S. (1967) The leaping of langurs: a problem in the study of locomotor adaptation. *Am. J. phys. Anthrop.* **26**, 149–70.

Robbins, L.W., Moulton, M.P. and Baker, R.J. (1983) Extent of geographic range and magnitude of chromosomal evolution. *J. Biogeogr.* **10**, 533–41.

Roberts, D.F. (1953) Body weight, race and climate. *Am. J. phys. Anthrop.* **11**, 533–58.

Robertson, W.R.B. (1916) Taxonomic relationships shown in the chromosomes of Tettigidae and Acrididae: V-shaped chromosomes and their significance in Acrididae, Locustidae and Gryllidae: chromosomes and variation. *J. Morph.* **27**, 129–331.

Robinson, J.A. and Bridson, W.E. (1978) Neonatal hormone patterns in the macaque. I. Steroids. *Biol. Reprod.* **19**, 773–8.

Robinson, J.A. and Goy, R.W. (1986) Steroid hormones and the ovarian cycle. In *Comparative Primate Biology*, vol 3: *Reproduction and Development* (eds Dukelow, W.R. and Erwin, J.). Alan, R. Liss: New York, pp. 63–91.

Robinson, J.T. (1972) *Early Hominid Posture and Locomotion.* University of Chicago Press: Chicago.

Robinson, J.T. and Allin, E.F. (1966) On the Y of the *Dryopithecus* pattern of mandibular molar teeth. *Am. J. phys. Anthrop.* **25**, 323–4.

Robinson, J.T. and Steudel, K. (1973) Multivariate discriminant analyses of dental data bearing on early hominid affinities. *J. hum. Evol.* **2**, 509–28.

Robinson, P.L. (1971) A problem of faunal replacement on Permo-Triassic continents. *Palaeontology* **14**, 131–53.

Roderick, T.H., Lalley, P.A., Davisson, M.T., O'Brien, S.J., Womack, J.E., Creau-Goldberg, N., Echard, G. and Moore, K.L. (1984) Comparative genetic mapping in mammals: report of the International Committee. *Cytogenet. Cell Genet.* **37**, 312–39.

Rodieck, R.W. (1973) *The Vertebrate Retina: Principles of Structure and Function.* W. H. Freeman: San Francisco.

Rodieck, R.W. (1988) The primate retina. In *Comparative Primate Biology*, vol. 4: *Neurosciences* (eds Steklis, H.D. and Erwin, J.). Alan R. Liss: New York, pp. 203–78.

Rodman, P.S. (1973) Population composition and adaptive organization among orang-utans of the Kutai Reserve. In *Comparative Ecology and Behaviour of Primates* (eds Crook, J.H. and Michael, R.P.). Academic Press: New York, pp. 171–209.

Rohen, J.W. (1962) Sehorgan. In *Primatologia*, vol 2, part 6 (eds Hofer, H., Schultz, A.H. and Starck, D.). S. Karger: Basel, pp. 1–210.

Rohen, J.W. and Castenholz, A. (1967) Über die Zentralization der Retina bei Primaten. *Folia primatol.* **5**, 92–147.

Rollinson, J.M.M. (1975) Interspecific comparisons of locomotor behaviour and prehension in eight species of African forest monkey. PhD thesis, University of London.

Rollinson, J.M.M. and Martin, R.D. (1981) Comparative aspects of primate locomotion, with special reference to arboreal cercopithecines. *Symp. zool. Soc. Lond.* **48**, 377–427.

Romer, A.S. (1945) *Vertebrate Paleontology* (2nd edition). Chicago University Press: Chicago.

Romer, A.S. (1955) *The Vertebrate Body* (2nd edition). W.B. Saunders: Philadelphia.

Romer, A.S. (1956) *Osteology of the Reptiles*. Chicago University Press: Chicago.

Romer, A.S. (1966) *Vertebrate Paleontology* (3rd edition). Chicago University Press: Chicago.

Romer, A.S. (1968) *Notes and Comments on Vertebrate Paleontology*. Chicago University Press: Chicago.

Romer, A.S. (1969) Cynodont reptile with incipient mammalian jaw articulation. *Science* **166**, 881–2.

Romer, A.S. and Price, L.I. (1940) Review of the Pelycosauria. *Geol. Soc. Am. Spec. Pap.* **28**, 1–538.

Romero-Herrera, A.E., Lehmann, H., Joysey, K.A. and Friday, A.E. (1973) Molecular evolution of myoglobin and the fossil record: a phylogenetic synthesis. *Nature, Lond.* **246**, 389–95.

Romero-Herrera, A.E., Lehmann, H., Castillo, O., Joysey, K.A. and Friday, A.E. (1976a) Myoglobin of orangutan as a phyletic enigma. *Nature, Lond.* **261**, 162–4.

Romero-Herrera, A.E., Lehmann, H., Joysey, K.A. and Friday, A.E. (1976b) Evolution of myoglobin amino acid sequences in primates and other vertebrates. In *Molecular Anthropology* (eds Goodman, M., Tashian, R.E. and Tashian, J.H.). Academic Press: New York, pp. 289–300.

Romero-Herrera, A.E., Lehmann, H., Joysey, K.A. and Friday, A.E. (1978) On the evolution of myoglobin. *Phil Trans. R. Soc. Lond. B* **283**, 61–163.

Rose, K.D. (1975) The Carpolestidae: early Tertiary primates from North America. *Bull. Mus. comp. Zool., Harv.* **147**, 1–74.

Rose, K.D. (1987) Climbing adaptations in the early Eocene mammal *Chriacus* and the origin of the Artiodactyla. *Science* **236**, 314–16.

Rose, K.D. and Bown, T.M. (1984) Gradual phyletic evolution at the generic level in early Eocene omomyid primates. *Nature, Lond.* **309**, 250–2.

Rose, K.D. and Fleagle, J.G. (1981) The fossil history of nonhuman primates in the Americas. In *Ecology and Behavior of Neotropical Primates* (eds Coimbra-Filho, A.F. and Mittermeier, R.A.). Academia Brasileira de Ciências: Rio de Janeiro, pp. 111–67.

Rose, K.D. and Krause, D.W. (1984) Affinities of the primate *Altanius* from the early Tertiary of Mongolia. *J. Mammal.* **65**, 721–6.

Rose, K.D. and Walker, A.C. (1985) The skeleton of early Eocene *Cantius*, oldest lemuriform primate. *Am. J. phys. Anthrop.* **66**, 73–89.

Rose, K.D., Walker, A.C. and Jacobs, L.L. (1981) Function of the mandibular tooth comb in living and extinct mammals. *Nature, Lond.* **289**, 583–5.

Rose, M.D. (1973) Quadrupedalism in primates. *Primates* **14**, 337–58.

Rosen, D.E. (1978) Vicariant patterns and historical explanation in biogeography. *Syst. Zool.* **27**, 159–88.

Rosen, D.E. (1979) Fishes from the uplands and intermontane basins of Guatemala: revisionary studies and comparative biogeography. *Bull. Am. Mus. nat. Hist.* **162**, 267–376.

Rosenberger, A.L. (1977) *Xenothrix* and ceboid phylogeny. *J. hum. Evol.* **6**, 461–81.

Rosenberger, A.L. (1979) Cranial anatomy and implications of *Dolichocebus*: a late Oligocene ceboid primate. *Nature, Lond.* **279**, 416–18.

Rosenberger, A.L. (1980) Gradistic views and adaptive radiation of platyrrhine primates. *Z. Morph. Anthrop.* **71**, 157–63.

Rosenberger, A.L. (1981a) A mandible of *Branisella boliviana* (Platyrrhini, Primates) from the Oligocene of South America. *Int. J. Primatol.* **2**, 1–7.

Rosenberger, A.L. (1981b) Systematics: the higher taxa. In *Ecology and Behavior of Neotropical Primates* (eds Coimbra-Filho, A.F. and Mittermeier, R.A.). Academia Brasileira de Ciências: Rio de Janeiro, pp. 9–27.

Rosenberger, A.L. (1985) In favor of the *Necrolemur*-tarsier hypothesis. *Folia primatol.* **45**, 179–94.

Rosenberger, A.L. (1986) Platyrrhines, catarrhines and the anthropoid transition. In *Major Topics in Primate and Human Evolution* (eds Wood, B.A., Martin, L.B. and Andrews, P.J.). Cambridge University Press: Cambridge, pp. 66–88.

Rosenberger, A.L. and Delson, E. (1985) The dentition of *Oreopithecus bambolii*: systematic and paleobiological implications. *Am. J. phys. Anthrop.* **66**, 222.

Rosenberger, A.L. and Strasser, E. (1985) Toothcomb origins: support for the grooming hypothesis. *Primates* **26**, 73–84.

Rosenberger, A.L. and Szalay, F.S. (1980) On the tarsiiform origins of the Anthropoidea. In *Evolutionary Biology of the New World Monkeys and Continental Drift* (eds Cochon, R.C. and Chiarelli, A.B.). Plenum Press: New York, pp. 139–57.

Rosenberger, A.L., Strasser, E. and Delson, E. (1985) Anterior dentition of *Notharctus* and the adapid–anthropoid hypothesis. *Folia primatol.* **44**, 15–39.

Ross, C. (1988) The intrinsic rate of natural increase and reproductive effort in primates. *J. Zool., Lond.* **214**, 199–219.

Roux, G.H. (1947) The cranial development of certain Ethiopian insectivores and its bearing on the mutual affinities of the group. *Acta zool., Stock.* **28**, 165–397.

Rowell, T.E. (1970) Reproductive cycles of two *Cercopithecus* monkeys. *J. Reprod. Fert.* **22**, 321–38.

Rowell, T.E. and Chalmers, N.R. (1970) Reproductive cycles of the mangabey *Cercocebus albigena*. *Folia primatol.* **12**, 264–72.

Ruben, J.A., Bennett, A.F. and Hisaw, F.L. (1987) Selective factors in the origin of the mammalian diaphragm. *Paleobiology* **13**, 54–9.

Rudder, B.C.C. (1979) The allometry of primate reproductive patterns. PhD thesis, University of London.

Rudran, R. (1973) Adult male replacement in one-male troops of purple-faced langurs (*Presbytis senex senex*) and its effect on population structure. *Folia primatol.* **19**, 166–92.

Rumpler, Y., Couturier, J., Warter, S. and Dutrillaux, B. (1983) The karyotype of *Galago crassicaudatus* is ancestral for lorisiforms. *Folia primatol.* **40**, 227–31.

Runcorn, S.K. (ed.) (1962) *Continental Drift.* Academic Press: London.

Rusconi, C. (1933) Nuevos restos de monos fosiles del terciario antiquo de la Patagonia. *Ann. Soc. Cient. Argen.* **116**, 286–9.

Ruse, M. (1979) Falsifiability, consilience, and systematics. *Syst. Zool.* **28**, 530–6.

Russell, D.E. (1964) Les mammifères paléocènes d'Europe. *Mém. Mus. natn. Hist. Nat., Paris (n.s.), C* **13**, 1–324.

Russell, D.E., Louis, P. and Savage, D.E. (1967) Primates of the French early Eocene. *Univ. Calif. Publs geol. Sci.* **73**, 1–46.

Russell, R.J. (1977) The behavior, ecology, and environmental physiology of a nocturnal primate, *Lepilemur mustelinus* (Strepsirhini, Lemuri-formes, Lepilemuridae). PhD thesis, Duke University, North Carolina.

Ruvolo, M. (1982) Genetic evolution in the African guenon monkeys (Primates, Cercopithecinae). PhD thesis, Harvard University.

Ruvolo, M. (1988) Genetic evolution in the African guenons. In *A Primate Radiation: Evolutionary Biology of the African Guenons* (eds Gautier-Hion, A., Bourlière, F., Gautier, J.-P. and Kingdon, J.). Cambridge University Press: Cambridge, pp. 127–39.

Rylands, A.B. (1981) Preliminary field observations on the marmoset *Callithrix humeralifer intermedius* (Hershkovitz, 1977) at Dardanelos, Rio Aripuano, Mato Grosso. *Primates* **22**, 46–59.

Saban, R. (1953) Presence de l'ethmoïde (os planum) dans le paroi orbitaire des Erinacéidae. *Bull. Mus. natn. Hist. nat., Paris* **25**, 127–9.

Saban, R. (1956/57) Les affinités du genre *Tupaia* Raffles 1821, d'après les caractères de la tête osseuse. *Ann. Paléont.* **42**, 169–224; **43**, 1–44.

Saban, R. (1963) Contribution à l'étude de l'os temporal des primates. *Mém. Mus, natn. Hist. nat., Paris (n.s.), A (Zool.)* **29**, 1–377.

Saban, R. (1975) Structure of the ear region in living and subfossil lemurs. In *Lemur Biology* (eds Tattersall, I. and Sussman, R.W.). Plenum Press: New York, pp. 83–109.

Sacher, G.A. (1982) The role of brain maturation in the evolution of the primates. In *Primate Brain Evolution* (eds Armstrong, E. and Falk, D.). Plenum Press: New York, pp. 97–112.

Sacher, G.A. and Staffeldt, E. (1974) Relation of gestation time to brain weight for placental mammals. *Am. Nat.* **108**, 593–615.

Sailer, L.D., Gaulin, S.J.C., Boster, J.S. and Kurland, J.A. (1985) Measuring the relationship between dietary quality and body size in primates. *Primates* **26**, 14–27.

Samorajski, T., Ordy, J.M. and Keefe, J.R. (1966) Structural organization of the retina in the tree shrew (*Tupaia glis*). *J. Cell Biol.* **28**, 489–504.

Sarich, V.M. (1968) Hominid origins: an immunological view. In *Perspectives on Human Evolution* (eds Washburn, S.L. and Jay, P.C.). Holt, Rinehart & Winston: New York, pp. 94–121.

Sarich, V.M. (1970a) Molecular data in systematics. In *Old World Monkeys* (eds Napier, J.R. and Napier, P.H.). Academic Press: New York, pp. 17–24.

Sarich, V.M. (1970b) Primate systematics with special reference to Old World monkeys. In *Old World Monkeys* (eds Napier, J.R. and Napier, P.H.). Academic Press: New York, pp. 175–226.

Sarich, V.M. (1972) Generation time and albumin evolution. *Biochem. Genet.* **7**, 205–12.

Sarich, V.M. and Cronin, J.E. (1976) Molecular systematics of the primates. In *Molecular Anthropology* (eds Goodman, M. and Tashian, R.E.). Plenum Press: New York, pp. 141–70.

Sarich, V.M. and Cronin, J.E. (1977) Generation length and rates of hominoid molecular evolution. *Nature, Lond.* **269**, 354.

Sarich, V.M. and Wilson, A.C. (1966) Quantitative immunochemistry and the evolution of primate albumins: micro-complement fixation. *Science* **154**, 1563–6.

Sarich, V.M. and Wilson, A.C. (1967a) Immunological time scale for hominid evolution. *Science* **158**, 1200–3.

Sarich, V.M. and Wilson, A.C. (1967b) Rates of albumin evolution in primates. *Proc. natn. Acad. Sci., U.S.A.* **58**, 142–8.

Sarich, V.M. and Wilson, A.C. (1973) Generation time and genomic evolution in primates. *Science* **179**, 1144–7.

Savage, D.E. (1975) Cenozoic – the primate episode. *Contr. Primatol.* **5**, 2–27.

Savage, D.E. and Russell, D.E. (1983) *Mammalian Paleofaunas of the World.* Addison Wesley: Reading, Mass.

Savage, D.E. and Waters, B.T. (1978) A new omomyid primate from the Watch formation of southern Wyoming. *Folia primatol.* **30**, 1–29.

Scanlon, C.E., Chalmers, N.R. and Monteiro da Cruz, M.A.O. (1988) Changes in the size, composition, and reproductive condition of wild marmoset groups (*Callithrix jacchus jacchus*) in North East Brazil. *Primates* **29**, 295–305.

Schaeffer, B. (1941) The morphological and functional evolution of the tarsus in amphibians and reptiles. *Bull. Am. Mus. nat. Hist.* **78**, 395–472.

Schaeffer, B. (1947) Notes on the origin and function of the artiodactyl tarsus. *Am. Mus. Novit.* no. 1356, 1–24.

Schaeffer, B. (1948) The origin of a mammalian ordinal character. *Evolution* **2**, 164–75.

Schaeffer, B., Hecht, M.K. and Eldredge, N. (1972) Paleontology and phylogeny. *Evol. Biol.* **6**, 31–46.

Schaffer, J. (1940) *Die Hautdrüsenorgane der Säugetiere.* Urban und Schwartzenberg: Berlin.

Schaffer, W.M. (1974) Optimal reproductive effort in fluctuating environments. *Am. Nat.* **108**, 783–98.

Schaller, G.B. (1963) *The Mountain Gorilla: Ecology and Behavior.* University of Chicago Press: Chicago.

Schankler, D.M. (1981) Local extinction and ecological re-entry of early Eocene mammals. *Nature, Lond.* **293**, 135–8.

Schilling, A. (1970) L'organe de Jacobson du lémurien malgache *Microcebus murinus* (Miller, 1907).

Mém. Mus. natn. Hist. nat. Paris (n.s), A (Zool.) **61**, 203–80.

Schlaginhaufen, O. (1905) Das Hautleistensystem der Primatenplanta unter Mitberücksichtigung der Palma. *Morph. Jb.* **33**, 577–671; **34**, 1–125.

Schlosser, M. (1887) Die Affen, Lemuren, Chiropteren, Insectivoren, Marsupialier, Creodonten, und Carnivoren des europäischen Tertiärs und deren Beziehungen zu ihren lebenden und fossilen aussereuropäischen Verwandten. *Beitr. Paläont. Öst.-Ungarns Orients* **6**, 1–227.

Schlosser, M. (1907) Beitrag zur Osteologie und systematischen Stellung der Gattung *Necrolemur*, sowie zur Stammesgeschichte der Primaten überhaupt. *Neues. Jb. Miner. Geol. Paläont. Stuttgart (Festband)* **1907**, 197–226.

Schlosser, M. (1910) Über einige fossile Säugetiere aus dem Oligocän von Ägypten. *Zool. Anz.* **35**, 500–8.

Schlosser, M. (1911) Beiträge zur Kenntnis der oligozänen Landsäugetiere aus dem Fayum, Ägypten. *Beitr. Paläont. Geol. Ost.-Ungarns Orients* **24**, 51–167.

Schlosser, M. (1924) Fossil primates from China. *Palaeont. Sin.* **1** (2), 1–16.

Schmid, C.W. and Jelenek, W.R. (1982) The Alu family of dispersed repetitive sequences. *Science* **216**, 1065–70.

Schmid, P. (1978) Die archaischen Primaten. *Z. Morph. Anthrop.* **69**, 1–6.

Schmid, P. (1979) Evidence of microchoerine evolution from Dielsdorf (Zürich Region, Switzerland) – a preliminary report. *Folia primatol.* **31**, 301–11.

Schmid, P. (1981) Comparison of Eocene nonadapids and *Tarsius*. In *Primate Evolutionary Biology* (eds Chiarelli, A.B. and Corruccini, R.S.). Springer-Verlag: Berlin, pp. 6–13.

Schmid, P. (1982) Die systematische Revision der europäischen Microchoeridae Lydekker, 1887 (Omomyiformes, Primates). PhD thesis, University of Zürich.

Schmid, P. (1983) Front dentition of the Omomyiformes (Primates). *Folia primatol.* **40**, 1–10.

Schmidt-Nielsen, K. (1972) *How Animals Work.* Cambridge University Press: Cambridge.

Schmidt-Nielsen, K. (1984) *Scaling: Why is Animal Size so Important?* Cambridge University Press: Cambridge.

Schneider, R. (1958) Zunge und weiche Gaumen. In *Primatologia*, vol. 3, part 1 (eds Hofer, H., Schultz, A.H and Starck, D.). Karger: Basel, pp. 61–126.

Schultz, A.H. (1933) Die Körperproportionen der erwachsenen catarrhinen Primaten, mit spezieller

Berücksichtigung der Menschenaffen. *Anthrop. Anz.* **10**, 154–85.

Schultz, A.H. (1935) Eruption and decay of the permanent teeth in primates. *Am. J. phys. Anthrop.* **19**, 489–581.

Schultz, A.H. (1937) Proportions, variability and asymmetries in the long bones of the limbs and clavicles in man and apes. *Hum. Biol.* **9**, 281–328.

Schultz, A.H. (1940) The size of the orbit and of the eye in primates. *Am. J. phys. Anthrop.* **26**, 389–408.

Schultz, A.H. (1948) The number of young at a birth and the number of nipples in primates. *Am. J. phys. Anthrop., n.s.* **6**, 1–23.

Schultz, A.H. (1956) Postembryonic age changes. In *Primatologia,* vol. 1 (eds Hofer, H., Schultz, A.H and Starck, D.). Karger: Basel, pp. 887–964.

Schultz, A.H. (1960) Einige Beobachtungen und Masse am Skelett von *Oreopithecus* im Vergleich mit anderen catarrhinen Primaten. *Z. Morph. Anthrop.* **50**, 136–49.

Schultz, A.H. (1962) The relative weights of the skeletal parts in adult primates. *Am. J. phys. Anthrop., n.s.* **20**, 1–10.

Schultz, A.H. (1963a) The relative weights of the foot skeleton and its main parts in primates. *Symp. zool. Soc. Lond.* **10**, 199–206.

Schultz, A.H. (1963b) Relationships between the lengths of the main parts of the foot skeleton in primates. *Folia primatol.* **1**, 150–71.

Schultz, A.H. (1969) *The Life of Primates.* Weidenfeld and Nicolson: London.

Schultz, A.H. (1970) The comparative uniformity of the Cercopithecoidea. In *Old World Monkeys* (eds Napier, J.R. and Napier, P.H.). Academic Press: New York, pp. 39–51.

Schwalbe, G. (1915) Über den fossilen Affen *Oreopithecus bambolii,* zugleich ein Beitrag zur Morphologie der Zähne der Primaten. *Z. Morph. Anthrop.* **19**, 149–254.

Schwartz, J.H. (1974a) Observations on the dentition of the Indriidae. *Am. J. phys. Anthrop.* **41**, 107–14.

Schwartz, J.H. (1974b) Premolar loss in the primates: a re-investigation. In *Prosimian Biology* (eds Martin, R.D., Doyle, G.A. and Walker, A.C.). Duckworth: London, pp. 621–40.

Schwartz, J.H. (1974c) Dental development and eruption in the prosimians and its bearing on their evolution. PhD, thesis, Columbia University.

Schwartz, J.H. (1975a) Development and eruption of the premolar region of prosimians and its bearing on their evolution. In *Lemur Biology* (eds Tattersall, I. and Sussman, R.W.). Plenum Press: New York, pp. 41–63.

Schwartz, J.H. (1975b) Re-evaluation of the morphocline of molar appearance in the primates. *Folia primatol.* **23**, 290–307.

Schwartz, J.H. (1978) Dental development, homologies, and primate phylogeny. *Evol. Theory* **4**, 1–32.

Schwartz, J.H. (1980) A discussion of dental homology with reference to primates. *Am. J. phys. Anthrop.* **52**, 463–80.

Schwartz, J.H. and Tattersall, I. (1985) Evolutionary relationships of living lemurs and lorises (Mammalia, Primates) and their potential affinities with European Eocene Adapidae. *Anthrop. Pap. Am. Mus. nat. Hist.* **60**, 1–100.

Schwartz, J.H., Tattersall, I. and Eldredge, N. (1978) Phylogeny and classification of the primates revisited. *Yb. phys. Anthrop.* **21**, 95–133.

Setoguchi, T. (1985) *Kondous laventicus,* a new ceboid primate from the Miocene of La Venta, Colombia, South America. *Folia primatol.* **44**, 96–101.

Setoguchi, T. (1986) Relations between morphology and function of the dentition in the *Stirtonia–Alouatta* lineage. In *Current Perspectives in Primate Biology* (eds Taub, D.M. and King, F.A.). Van Nostrand Reinhold: New York, pp. 201–13.

Setoguchi, T. and Rosenberger, A.L. (1985) Miocene marmosets: first fossil evidence. *Int. J. Primatol.* **6**, 615–25.

Setoguchi, T. and Rosenberger, A.L. (1987) A fossil owl monkey from La Venta, Colombia. *Nature, Lond.* **326**, 692–4.

Setoguchi, T., Shigehara, N., Rosenberger, A.L. and Alberto, C.G. (1986) Primate fauna from the Miocene La Venta, in the Tatacoa Desert, Department of Hurla, Colombia. *Caldasia* **15**, 761–73.

Seuánez, H.N. (1979) *The Phylogeny of Human Chromosomes.* Springer-Verlag: Berlin.

Seuánez, H.N., Evans, H.J., Martin, D.E. and Fletcher, J. (1979) Inversion of chromosome-2 that distinguishes between Bornean and Sumatran orang-utans. *Cytogenet. Cell Genet.* **23**, 137–40.

Seuánez, H.N., Fletcher, J., Evans, H.J. and Martin, D.E. (1976) Chromosome rearrangement in an orang-utan studied with Q-banding, C-banding and G-banding techniques. *Cytogenet. Cell Genet.* **17**, 26–34.

Shafer, D.A., Myers, R.H. and Saltzman, D. (1984) Biogenetics of the siabon (gibbon–siamang hybrids). In *The Lesser Apes: Evolutionary and Behavioural Biology* (eds Preuschoft, H., Chivers, D.J., Brockelman, W.Y. and Creel, N.). Edinburgh University Press: Edinburgh, pp. 486–97.

Shea, B.T. (1985) Ontogenetic allometry and scaling: a discussion based on the growth and form of the skull in African apes. In *Size and Scaling in Primate Biology* (ed. Jungers, W.L.). Plenum Press: New York, pp. 175–205.

Shea, B.T. (1987) Reproductive strategies, body size, and encephalization in primate evolution. *Int. J. Primatol.* **8**, 139–56.

Shelford, R.W.C. (1916) *A Naturalist in Borneo.* Fisher Unwin: London.

Shigehara, N. (1980) Epiphyseal union, tooth eruption and sexual maturation in the common tree shrew, with reference to its systematic problem. *Primates* **21**, 1–19.

Shipman, P. (1981) *Life History of a Fossil: An Introduction to Taphonomy and Paleoecology.* Harvard University Press: Cambridge, Mass.

Shoshani, J. (1986) Mammalian phylogeny: comparison of morphological and molecular results. *Molec. Biol. Evol.* **3**, 222–42.

Shriver, J.W. and Noback, C.R. (1967) Color vision in the tree shrew (*Tupaia glis*). *Folia primatol.* **6**, 161–9.

Sibley, C.G. and Ahlquist, J.E. (1984) The phylogeny of the hominoid primates, as indicated by DNA–DNA hybridization. *J. molec. Evol.* **20**, 2–15.

Simons, E.L. (1960) *Apidium* and *Oreopithecus. Nature, Lond.* **186**, 824–6.

Simons, E.L. (1961a) Notes on Eocene tarsioids and a revision of some Necrolemurinae. *Bull. Br. Mus. nat. Hist., Geol.* **5**, 45–69.

Simons, E.L. (1961b) The phyletic position of *Ramapithecus. Postilla* no. 57, 1–9.

Simons, E.L. (1962a) A new Eocene primate genus, *Cantius,* and a revision of some allied European lemuroids. *Bull. Br. Mus. nat. Hist., Geol.* **7**, 1–36.

Simons, E.L. (1962b) Two new primate species from the African Oligocene. *Postilla* no. 64, 1–12.

Simons, E.L. (1962c) Fossil evidence relating to the early evolution of primate behavior. *Ann. N. Y. Acad. Sci.* **102**, 282–95.

Simons, E.L. (1963) A critical reappraisal of Tertiary primates. In *Evolutionary and Genetic Biology of Primates* (ed. Buettner-Janusch, J.) Academic Press: New York, pp. 65–129.

Simons, E.L. (1964) The early relatives of man. *Sci. Am.* **211**(1), 50–62.

Simons, E.L. (1965) New fossil apes from Egypt and the initial differentiation of Hominoidea. *Nature, Lond.* **205**, 135–9.

Simons, E.L. (1966) In search of the missing link. *Discovery, Yale Peabody Mus.* no. 1, 24–30.

Simons. E.L. (1967) The earliest apes. *Sci. Am.* **217**(6), 28–35.

Simons, E.L. (1969) Miocene monkey (*Prohylobates*) from northern Egypt. *Nature, Lond.* **223**, 687–9.

Simons, E.L. (1972) *Primate Evolution: An Introduction to Man's Place in Nature.* Macmillan: New York.

Simons, E.L. (1974a) The relationships of *Aegyptopithecus* to other primates. *Ann. geol. Surv. Egypt* **4**, 149–56.

Simons, E.L. (1974b) *Parapithecus grangeri* (Parapithecidae, Old World higher primates). New species from the Oligocene of Egypt and the initial differentiation of Cercopithecoidea. *Postilla* no. 166, 1–12.

Simons, E.L. (1974c) Notes on early Tertiary prosimians. In *Prosimian Biology* (eds Martin, R.D., Doyle, G.A. and Walker, A.C.). Duckworth: London, pp. 415–33.

Simons, E.L. (1976a) Relationships between *Dryopithecus, Sivapithecus* and *Ramapithecus* and their bearing on hominid origins. In *Les Plus Anciens Hominidés* (eds Tobias, P.V. and Coppens, Y.). C.N.R.S.: Paris, pp. 60–9.

Simons, E.L. (1976b) The nature of the transition in the dental mechanism from pongids to hominids. *J. hum. Evol.* **5**, 500–28.

Simons, E.L. (1985) Origins and characteristics of the first hominoids. In *Ancestors: The Hard Evidence* (ed. Delson, E.). Alan R. Liss: New York, pp. 37–41.

Simons, E.L. (1986) *Parapithecus grangeri* of the African Oligocene: an archaic catarrhine without lower incisors. *J. hum. Evol.* **15**, 205–13.

Simons, E.L. (1987) New faces of *Aegyptopithecus* from the Oligocene of Egypt. *J. hum. Evol.* **16**, 273–89.

Simons, E.L. and Bown, T.M. (1985) *Afrotarsius chatrathi,* first tarsiiform primate (?Tarsiidae) from Africa. *Nature, Lond.* **313**, 475–7.

Simons, E.L. and Chopra, S.R.K. (1969) *Gigantopithecus* (Pongidae, Hominoidea) – a new species from north India. *Postilla* no. 138, 1–18.

Simons, E.L. and Delson, E. (1978) Cercopithecidae and Parapithecidae. In *Evolution of African Mammals* (eds Maglio, V.J. and Cooke, H.B.S.). Harvard University Press: Cambridge, Mass., pp. 100–19.

Simons, E.L. and Ettel, P.C. (1970) *Gigantopithecus, Sci. Am.* **222**(1), 76–85.

Simons, E.L. and Fleagle, J. (1973) The history of extinct gibbon-like primates. *Gibbon and Siamang* **2**, 121–48.

Simons, E.L. and Kay, R.F. (1983) *Qatrania,* a new basal anthropoid from the Fayum, Oligocene of Egypt. *Nature, Lond.* **304**, 624–6.

Simons, E.L. and Pilbeam, D.R. (1965) Preliminary revision of the Dryopithecinae (Pongidae, Anthropoidea). *Folia primatol.* **3**, 81–152.

Simons, E.L. and Pilbeam, D.R. (1971) A gorilla-sized ape from the Miocene of India. *Science* **173**, 23–7.

Simons, E.L. and Pilbeam, D.R. (1972) Hominid paleoprimatology. In *The Functional and Evolutionary Biology of Primates* (ed. Tuttle, R.H.). Aldine-Atherton: Chicago, pp. 36–62.

Simons, E.L. and Russell, D.E. (1960) Notes on the cranial anatomy of *Necrolemur*. *Breviora* **127**, 1–14.

Simons, E.L., Andrews, P. and Pilbeam, D.R. (1978) Cenozoic apes. In *Evolution of African Mammals* (eds Maglio, V.J. and Cooke, H.B.S.). Harvard University Press: Cambridge, Mass., pp. 120–46.

Simons, E.L., Bown, T.M. and Rasmussen, D.T. (1986) Discovery of two additional prosimian primate families (Omomyidae, Lorisidae) in the African Oligocene. *J. hum. Evol.* **15**, 431–7.

Simons, E.L., Kay, R.F. and Fleagle, J.G. (1980) Recently discovered Oligocene apes from Egypt. *Am. J. phys. Anthrop.* **52**, 279.

Simpson, G.G. (1927) Mesozoic Mammalia. IX: The brain of Jurassic mammals. *Am. J. Sci.* **214**, 259–68.

Simpson, G.G. (1931a) A new classification of mammals. *Bull. Am. Mus. nat. Hist.* **59**, 259–93.

Simpson, G.G. (1931b) A new insectivore from the Oligocene, Uan Gochu horizon, of Mongolia. *Am. Mus. Novit.* no. 505, 1–22.

Simpson, G.G. (1935a) The Tiffany Fauna, Upper Palaeocene. II – Structure and relationships of *Plesiadapis*. *Am. Mus. Novit.* no. 816, 1–30.

Simpson, G.G. (1935b) The Tiffany Fauna, Upper Palaeocene. III – Primates, Carnivora, Condylarthra and Amblypoda. *Am. Mus. Novit.* no. 817, 1–28.

Simpson, G.G. (1936) Studies on the earliest mammalian dentitions. *Dental Cosmos* **78**, 791–800 and 940–953.

Simpson, G.G. (1937) The Fort Union of the Crazy Mountain field, Montana, and its mammalian faunas. *Bull. U.S. nat. Mus.* **169**, 1-287.

Simpson, G.G. (1940) Studies on the earliest primates. *Bull. Am. Mus. nat. Hist.* **77**, 185–212.

Simpson, G.G. (1943) Mammals and the nature of continents. *Am. J. Sci.* **241**, 1–31.

Simpson, G.G. (1945) The principles of classification and a classification of mammals. *Bull. Am. Mus. nat. Hist.* **85**, 1–350.

Simpson, G.G. (1950) *The Meaning of Evolution*. Yale University Press: New Haven.

Simpson, G.G. (1955) The Phenacolemuridae, new family of early Primates. *Bull. Am. Mus. nat. Hist.* **105**, 411–42.

Simpson, G.G. (1961) *Principles of Animal Taxonomy*. Columbia University Press: New York.

Simpson, G.G. (1963) The meaning of taxonomic statements. In *Classification and Human Evolution* (ed. Washburn, S.L.). Aldine: Chicago, pp. 1–31.

Simpson, G.G. (1965) Long-abandoned views. *Science* **147**, 1397.

Simpson, G.G (1967) The Tertiary lorisiform primates of Africa. *Bull. Mus. comp. Zool. Harv.* **136**, 39–61.

Slaughter, B. (1981) The Trinity therians (Albian, Mid-Cretaceous) as marsupials and placentals. *J. Paleont.* **55**, 682–3.

Sly, D.L., Harbaugh, S.W., London, W.T. and Rice, J.M. (1983) Reproductive performance of a laboratory breeding colony of patas monkeys (*Erythrocebus patas*). *Am. J. Primatol.* **4**, 23–32.

Smith, A.G. and Briden, J.C. (1977) *Mesozoic and Cenozoic Paleocontinental World Maps*. Cambridge University Press: Cambridge.

Smith, A.G., Hurley, A.M. and Briden, J.C. (1981) *Phanerozoic Paleocontinental World Maps*. Cambridge University Press: Cambridge.

Smith, F.H. and Spencer, F. (ed) (1984) *The Origins of Modern Humans: A World Survey of the Fossil Evidence*. Alan R. Liss: New York.

Smith, R.J. (1981) Interspecific scaling of maxillary canine size and shape in female primates: relationships to social structure and diet. *J. hum. Evol.* **10**, 165–73.

Sneath, P.H.A. and Sokal, R.R. (1973) *Numerical Taxonomy: The Principles and Practice of Numerical Classification*. W.H. Freeman: San Francisco.

Sokal, R.R. and Michener, C.D. (1958) A statistical method for evaluating systematic relationships. *Univ. Kansas Sci. Bull.* **38**, 1409–38.

Sonntag, C.F. (1925) Comparative anatomy of the tongues of the Mammalia: summary, classification and phylogeny. *Proc. zool. Soc. Lond.* **1925**, 701–62.

Soulié, J. and de Grouchy, J. (1981) A cytogenetic survey of 110 baboons (*Papio cynocephalus*). *Am. J. phys. Anthrop.* **56**, 107–13.

Soulié, J. and de Grouchy, J. (1982) Of rabbit and man: comparative gene mapping. *Hum. Genet.* **60**, 172–5.

Spatz, W.B. (1966) Zur Ontogenese der Bulla tympanica von *Tupaia glis* Diard 1820 (Prosimiae, Tupaiiformes). *Folia primatol.* **4**, 26–50.

Spatz, W.B. (1968) Die Bedeutung der Augen für die sagittale Gestaltung des Schädels von *Tarsius* (Prosimiae, Tarsiiformes). *Folia primatol.* **9**, 22–40.

Sprankel, H. (1965) Untersuchungen an *Tarsius*. I. Morphologie des Schwanzes nebst ethologischen Bermerkungen. *Folia primatol.* **3**, 153–88.

Sprankel, H. (1969) Comparative microscopic studies of nail plate and surrounding soft tissue of some Hominoidea. In *Recent Advances in Primatology* (ed. Hofer, H.). Karger: Basel, pp. 82–6.

Stains, H. (1959) Use of the calcaneum in studies of taxonomy and food habits. *J. Mammal.* **40**, 392–401.

Standing, H.F. (1908) On recently discovered subfossil primates from Madagascar. *Trans. zool. Soc. Lond.* **18**, 59–162.

Stanley, S.M. (1976) Stability of species in geologic time. *Science* **192**, 267–9.

Stanley, S.M. (1978) Chronospecies' longevities, the origin of genera, and the punctuational model of evolution. *Paleobiology* **4**, 26–40.

Stanley, S.M. (1979) *Macroevolution: Pattern and Process.* W.H. Freeman: Oxford.

Stanley, S.M. (1981) *The New Evolutionary Timetable.* Basic Books: New York.

Starck, D. (1956) Primitiventwicklung und Plazentation der Primaten. In *Primatologia*, vol. 1 (eds Hofer, H., Schultz, A.H. and Starck, D.). Karger: Basel, pp. 723–886.

Starck, D. (1962) Die Evolution des Säugetiergehirns. *Sber. J.W. Goethe-Univ. Frankfurt am Main* **1**, 7–60.

Starck, D. (1975) The development of the chondrocranium in primates. In *Phylogeny of the Primates* (eds Luckett, W.P. and Szalay, F.S.). Plenum Press: New York, pp. 127–55.

Staton, R.D. (1967) Karyotypes and the phylogeny of New World primates. *Mammal. Chrom. Newsl.* **8**, 203–19.

Stearns, S.C. (1976) Life-history tactics: a review of the ideas. *Q. Rev. Biol.* **51**, 3–47.

Stearns, S.C. (1977) The evolution of life history traits. *A. Rev. Ecol. Syst.* **8**, 145–71.

Steele, D.G. (1973) Dental variability in the tree shrews (Tupaiidae). *Symp. 4th Int. Congr. Primatol.* **3**, 154–79.

Stehlin, H.G. (1912) Die Säugetiere des schweizerischen Eocäns. Kritischer Katalog der Materialien. Siebenter Theil, erste Hälfte: *Adapis. Abh. schweiz. paläont. Ges.* **38**, 1165–298.

Stehlin, H.G. (1916) Die Säugetiere des schweizerischen Eocäns. Kritischer Katalog der Materialien. Siebenter Theil, zweite Hälfte. *Abh. schweiz. paläont. Ges.* **41**, 1299–552.

Steiner, H. (1942) Der Aufbau des Säugetier-Carpus und -Tarsus nach neueren embyologischen Untersuchungen. *Rev. suisse Zool.* **49**, 217–23.

Stephan, H. (1972) Evolution of primate brains: a comparative anatomical investigation. In *The Functional and Evolutionary Biology of Primates.* (ed. Tuttle, R.H.). Aldine-Atherton: Chicago, pp. 155–74.

Stephan, H. and Andy, O.J. (1969) Quantitative comparative neuroanatomy of primates: an attempt at a phylogenetic interpretation. *Ann. N.Y. Acad. Sci.* **167**, 370–87.

Stephan, H., Bauchot, R. and Andy, O.J. (1970) Data on size of the brain and of various brain parts in insectivores and primates. In *The Primate Brain* (eds Noback, C.R. and Montagna, W.). Appleton-Century-Crofts: New York, pp. 289–97.

Stephan, H., Frahm, H. and Baron, G. (1981) New and revised data on volume of brain structures in insectivores and primates. *Folia primatol.* **35**, 1–29.

Stephan, H., Frahm, H. and Bauchot, R. (1977) Vergleichende Untersuchungen an den Gehirnen madagassischer Halbaffen. Encephalisation und Macromorphologie. *J. Hirnforsch.* **18**, 115–47.

Stern, J.T. (1971) Functional myology of the hip and thigh of cebid monkeys and its implications for the evolution of erect posture. *Bibl. primatol.* **14**, 1–319.

Stern, J.T. and Jungers, W.L. (1985) Body size and proportions of the locomotor skeleton in *Oreopithecus bambolii. Am. J. phys. Anthrop.* **66**, 233.

Stern, J.T. and Oxnard, C.E. (1973) Primate locomotion: some links with evolution and morphology. In *Primatologia,* vol. 4 (eds Hofer, H., Schultz, A.H. and Starck, D.). Karger: Basel, pp. 723–886.

Stern, J.T. and Susman, R.L. (1981) Electromyography of the gluteal muscles in *Hylobates, Pongo* and *Pan:* implications for the evolution of hominid bipedality. *Am. J. phys. Anthrop.* **55**, 153–66.

Stern, J.T. and Susman, R.L. (1983) The locomotor anatomy of *Australopithecus afarensis. Am. J. phys. Anthrop.* **60**, 279–317.

Steven, D.H. (ed.) (1975a) *Comparative Placentation: Essays in Structure and Function.* Academic Press: London.

Steven, D.H. (1975b) Anatomy of the placental barrier. In *Comparative Placentation: Essays in Structure and Function* (ed. Steven, D.H.). Academic Press: London, pp. 25–57.

Steven, D.H. and Morriss, G. (1975) Development

of the foetal membranes. In *Comparative Placentation: Essays in Structure and Function* (ed. Steven, D.H.). Academic Press: London, pp. 58–86.

Stevenson, M.F.(1978) The behaviour and ecology of the common marmoset (*Callithrix jacchus*) in its natural environment. In *Biology and Behaviour of Marmosets* (eds Rothe, H., Wolters, H.J. and Hearn, J.P.). Eigenverlag Hartmut Rothe: Göttingen, p. 298.

Stirton, R.G. (1951) Ceboid monkeys from the Miocene of Colombia. *Univ. Calif. Publs, Bull. Dep. geol. Sci.* **28**, 315–56.

Stirton, R.G. and Savage, D.E. (1951) A new monkey from the La Venta Miocene of Colombia. *Comp. Estud. Geol. Ofic. Colombia, Serv. Geol. Nac. Bogota* **7**, 345–56.

Stockholm Conference (1978) An international system for human cytogenetic nomenclature. *Cytogenet. Cell Genet.* **21**, 309–409.

Storch, G. (1978) *Eomanis waldi*, ein Schuppentier aus dem Mittel-Eozän der 'Grube Messel' bei Darmstadt (Mammalia, Pholidota). *Senckenb. lethaea* **59**, 503–29.

Storch, G. (1981) *Eurotamandua joresi*, ein Myrmecophagide aus dem Eozän der 'Grube Messel' bei Darmstadt (Mammalia, Xenarthra). *Senckenb. lethaea* **61**, 247–89.

Storch, G. (1984) Die alttertiäre Säugetierfauna von Messel – ein paläobiographisches Puzzle. *Naturwissenschaften* **71**, 227–33.

Storch, G. (1986) Die Säuger von Messel: Würzeln auf vielen Kontinenten. *Spektrum Wiss.* no. 6, 48–65.

Straus, W.L. (1961) Primate taxonomy and *Oreopithecus*. *Science* **133**, 760–61.

Straus, W.L. (1963) The classification of *Oreopithecus*. In *Classification and Human Evolution* (ed. Washburn, S.L.). Aldine: Chicago, pp. 146–77.

Strauss, F. (1978a) The ovoimplantation of *Microcebus murinus* Miller (Primates, Lemuroidea, Strepsirhini). *Am. J. Anat.* **152**, 99–110.

Strauss, F. (1978b) Eine Neuuntersuchung der Implantation und Placentation bei *Microcebus murinus*. *Mitt. Naturforsch. Ges. Bern* **35**, 107–19.

Streier, K.B. (1987) Socioecology of woolly spider monkeys, or muruquis (*Brachyteles arachnoides*). *Am. J. phys. Anthrop.* **72**, 259.

Stringer, C.B. (1974) Population relationships of later Pleistocene hominids: a multivariate study of available crania. *J. arch. Sci.* **1**, 317–42.

Stringer, C.B. and Andrews, P.J. (1988) Genetic and fossil evidence for the origin of modern humans. *Science* **239**, 1263–8.

Stromer, E. (1913) Mitteilungen über die Wirbeltierreste aus dem Mittlepliozän des Natrontales (Ägypten). I. Affen. *Z. deutsch. Geol. Ges. Abh.* **65**, 349–61.

Struhsaker, T.T. (1969) Correlates of ecology and social organization among African cercopithecines. *Folia primatol.* **11**, 80–118.

Struhsaker, T.T. (1975) *The Red Colobus Monkey*. University of Chicago Press: Chicago.

Struhsaker, T.T. and Oates, J.F. (1975) Comparison of the behavior and ecology of red colobus and black-and-white colobus monkeys in Uganda: a summary. In *Primate Functional Morphology and Evolution* (ed. Tuttle, R.H.). Mouton: The Hague, pp. 103–23.

Suckling, J.A., Suckling, E.E. and Walker, A.C. (1969) Suggested function of the vascular bundles in the limbs of *Perodicticus potto*. *Nature, Lond.* **221**, 379–80.

Sudre, M. (1975) Un prosimien du Paléogène ancien du Sahara nord-occidental: *Azibius trerki* n. g., n. sp.. *C. R. Acad. Sci., Paris, D* **280**, 1539–42.

Susman, R.L., Stern, J.T. and Jungers, W.L. (1984) Arboreality and bipedality in the Hadar hominids. *Folia primatol.* **43**, 113–56.

Sussman, R.W. (1974) Ecological distinctions in sympatric species of *Lemur*. In *Prosimian Biology* (eds Martin, R.D., Doyle, G.A. and Walker, A.C.). Duckworth: London, pp. 75–108.

Sussman, R.W. (1977) Feeding behaviour of *Lemur catta* and *Lemur fulvus*. In *Primate Ecology* (ed. Clutton-Brock, T.H.). Academic Press: London, pp. 1–36.

Sussman, R.W. and Tattersall, I. (1976) Cycles of activity, group composition and diet of *Lemur mongoz mongoz* Linnaeus, 1766 in Madagascar. *Folia primatol.* **26**, 270–83.

Suzuki, A. (1965) An ecological study of wild Japanese monkeys in snowy areas, focused on their food habits. *Primates* **6**, 31–72.

Suzuki, K., Nagai, H., Hayama, S. and Tanate, H. (1985) Anatomical and histological observations on the stomach of François' leaf monkeys (*Presbytis francoisi*). *Primates* **26**, 99–103.

Sved, J.A. (1968) Possible rates of gene substitution in evolution. *Am. Nat.* **102**, 283–93.

Swindler, D.R. (1976) *Dentition of Living Primates*. Academic Press: New York.

Szalay, F.S. (1966) The tarsus of the Paleocene leptictid *Prodiacodon* (Insectivora, Mammalia). *Am. Mus. Novit.* no. 2267, 1–13.

Szalay, F.S. (1968a) The Picrodontidae, a family of early primates. *Am. Mus. Novit.* no. 2329, 1–55.

Szalay, F.S. (1968b) The beginnings of primates. *Evolution* **22**, 19–36.

Szalay, F.S. (1969) The Mixodectidae, Microsyopidae and the insectivore–primate transition. *Bull. Am. Mus. nat. Hist.* **140**, 195–330.

Szalay, F.S. (1970) Late Eocene *Amphipithecus* and the origins of catarrhine primates. *Nature, Lond.* **227**, 355–7.

Szalay, F.S. (1971) The Eocene adapid primates *Agerina* and *Pronycticebus*. *Am. Mus. Novit.* no. 2466, 1–19.

Szalay, F.S. (1972a) Cranial morphology of the Early Tertiary *Phenocolemur* and its bearing on primate phylogeny. *Am. J. phys. Anthrop., n.s.* **36**, 56–76.

Szalay, F.S. (1972b) Hunting-scavenging proto-hominids: a model for hominid origins. *Man* **10**, 420–9.

Szalay, F.S. (1975a) Phylogeny of primate higher taxa: the basicranial evidence. In *Phylogeny of the Primates* (eds Luckett, W.P. and Szalay, F.S.). Plenum Press: New York, pp. 91–125.

Szalay, F.S. (1975b) Early primates as a source for the taxon Dermoptera. *Am. J. phys. Anthrop., n.s.* **42**, 332–3.

Szalay, F.S. (1975c) Where to draw the nonprimate–primate taxonomic boundary. *Folia primatol.* **23**, 158–63.

Szalay, F.S. (1976) Systematics of the Omomyidae (Tarsiiformes, Primates). Taxonomy, phylogeny and adaptations. *Bull. Am. Mus. nat. Hist.* **156**, 157–450.

Szalay, F.S. (1977) Phylogenetic relationships and a classification of the eutherian Mammalia. In *Major Patterns in Vertebrate Evolution* (eds Hecht, M.K., Goody, P.C. and Hecht, B.M.). Plenum Press: New York, pp. 315–74.

Szalay, F.S. and Berzi, A. (1973) Cranial anatomy of *Oreopithecus*. *Science* **180**, 183–5.

Szalay, F.S. and Decker, R.L. (1974) Origins, evolution, and function of the tarsus in Late Cretaceous Eutheria and Paleocene primates. In *Primate Locomotion* (ed. Jenkins, F.A.). Academic Press: New York, pp. 223–59.

Szalay, F.S and Delson, E. (1979) *Evolutionary History of the Primates*. Academic Press: New York.

Szalay, F.S. and Drawhorn, G. (1980) Evolution and diversification of the Archonta in an arboreal milieu. In *Comparative Biology and Evolutionary Relationships of Tree Shrews* (ed. Luckett, W.P.). Plenum Press: New York, pp. 133–69.

Szalay, F.S. and Katz, C.C. (1973) Phylogeny of lemurs, galagos and lorises. *Folia primatol.* **19**, 88–103.

Szalay, F.S. and Langdon, J.H. (1986) The foot of *Oreopithecus*: an evolutionary assessment. *J. hum. Evol.* **7**, 585–621.

Szalay, F.S. and Seligsohn, D. (1977) Why did the strepsirhine tooth comb evolve? *Folia primatol.* **27**, 75–82.

Szalay, F.S. and Wilson, J.A. (1976) Basicranial morphology of the early Tertiary tarsiiform *Rooneyia* from Texas. *Folia primatol.* **25**, 288–93.

Szalay, F.S., Tattersall, I. and Decker, R.L. (1975) Phylogenetic relationships of *Plesiadapis* – post-cranial evidence. *Contr. Primatol.* **5**, 136–66.

Szarski, H. (1970) Changes in the amount of DNA in cell nuclei during vertebrate evolution. *Nature, Lond.* **226**, 651–2.

Szarski, H. (1976) Cell size and nuclear DNA content in vertebrates. *Int. Rev. Cytol.* **44**, 93–111.

Szarski, H. (1980) A functional and evolutionary interpretation of brain size in vertebrates. *Evol. Biol.* **13**, 149–74.

Tanner, J.M. (1963) Regulation of growth in size in mammals. *Nature, Lond.* **199**, 845–50.

Tansley, K. (1965) *Vision in Vertebrates*. Chapman and Hall: London.

Tarling, D.H. (1980) The geological evolution of South America with special reference to the last 200 million years. In *Evolutionary Biology of the New World Monkeys and Continental Drift* (eds Ciochon, R.L. and Chiarelli, A.B.). Plenum Press: New York, pp. 1–41.

Tarling, D.H. and Tarling, M.P. (1971) *Continental Drift: A Study of the Earth's Moving Surface*. G. Bell: London.

Tarlo, L.B.H. (1964) Tooth-replacement in the mammal-like reptiles. *Nature, Lond.* **201**, 1081–2.

Tateno, Y. and Tajima, F. (1986) Statistical properties of molecular tree construction methods under the neutral mutation model. *J. molec. Evol.* **23**, 354–61.

Tattersall, I. (1970) *Man's Ancestors: An Introduction to Primate and Human Evolution*. John Murray: London.

Tattersall, I. (1973) Subfossil lemuroids and the 'adaptive radiation' of the Malagasy lemurs. *Trans. N.Y. Acad. Sci.* **35**, 314–24.

Tattersall, I. (1975) *The Evolutionary Significance of Ramapithecus*. Burgess: Minneapolis.

Tattersall, I. (1982) *The Primates of Madagascar*. Columbia University Press: New York.

Tattersall, I. (1984) The tree-shrew, *Tupaia*: a 'living model' of the ancestral primate? In *Living Fossils* (eds Eldredge, N. and Stanley, S.M.). Springer Verlag: New York, pp. 32–7.

Tattersall, I. and Schwartz, J.H. (1974) Craniodental morphology and the systematics of the Malagasy lemurs (Primates, Prosimii). *Anthrop. Pap. Am. Mus. nat. Hist.* **52**, 139–92.

Tattersall, I. and Schwartz, J.H. (1975) Relationships

among the Malagasy lemurs: the craniodental evidence. In *Phylogeny of the Primates* (eds Luckett, W.P. and Szalay, F.S.). Plenum Press: New York, pp. 299–312.

Teaford, M.F. and Walker, A.C. (1984) Quantitative differences in dental microwear between primate species with different diets and a comment on the presumed diet of *Sivapithecus*. *Am. J. phys. Anthrop.* **64**, 191–200.

Teilhard de Chardin, P. (1921/22) Les mammifères de l'Eocène inférieur français et leurs gisements. *Ann. Paléont.* **10**, 171–6; **11**, 1–108.

Teilhard de Chardin, P. (1927) Les mammifères de l'Eocène inférieur de la Belgique. *Mém. Mus. R. Hist. nat. Belg.* **36**, 1–33.

Teilhard de Chardin, P. (1938) The fossils from Locality 12 of Choukoutien. *Palaeont. sin., n.s.* **114**, 2–46.

Teissier, G. (1948) La relation d'allométrie, sa signification statistique et biologique. *Biometrics* **4**, 14–53.

Tekkaya, I. (1974) A new species of Tortonian anthropoid (Primates, Mammalia) from Anatolia. *Bull. Miner. Res. Explor. Inst. Turkey* **83**, 148–65.

Teleki, G. (1973) *The Predatory Behavior of Wild Chimpanzees*. Bucknell University Press: Lewisburg, Pa.

Termier, H. and Termier, G. (1960) *Atlas de Paléogéographie*. Masson: Paris.

Terborgh, J. (1983) *Five New World Primates: A Study in Comparative Ecology*. Princeton University Press: New Jersey.

Tettenborn, U. and Gropp, A. (1970) Meiotic nondisjunction in mice and mouse hybrids. *Cytogenetics* **9**, 272–83.

Thenius, E. and Hofer, H. (1960) *Stammesgeschichte der Säugetiere*. Springer-Verlag: Berlin.

Thorington, R.W. (1968a) Observations of squirrel monkeys in a Colombian forest. In *The Squirrel Monkey* (ed. Rosenblum, L.A. and Cooper, R.W.). Academic Press: New York, pp. 69–85.

Thorington, R.W. (1968b) Observations of the tamarin *Saguinus midas*. *Folia primatol.* **9**, 95–8.

Thorndike, E.E. (1968) A microscopic study of the marmoset claw and nail. *Am. J. phys. Anthrop., n. s.* **28**, 247–62.

Tigges, J. (1963) Untersuchungen über den Farbsinn von *Tupaia glis* (Diard 1820). *Z. Morph. Anthrop.* **53**, 109–23.

Tigges, J. and Tigges, M. (1987) Termination of retinofugal fibers and lamination pattern in the lateral geniculate nucleus of the gibbon. *Folia primatol.* **48**, 186–94.

Tijo, H. and Levan, A. (1956) The normal chromosome number of man, *Hereditas* **42**, 1–6.

Tilson, R.L. (1977) Social organization of Simakobu monkeys (*Nasalis concolor*) in Siberut Island, Indonesia. *J. Mammal.* **58**, 202–12.

Tilson, R.L. and Tenaza, R.R. (1976) Monogamy and duetting in an Old World monkey. *Nature, Lond.* **263**, 320–1.

Tobias, P.V. (1967) *Olduvai Gorge,* vol. 2: *The Cranium of* Australopithecus (Zinjanthropus) boisei.Cambridge University Press: Cambridge.

Tobias, P.V. (1973) New developments in hominid palaeontology in South and East Africa. *A. Rev. Anthrop.* **2**, 311–34.

Tobias, P.V. (ed.) (1985) *Hominid Evolution: Past, Present and Future*. Alan R. Liss: New York.

Todd, N.B. (1967) A theory of karyotypic fissioning, genetic potentiation and eutherian evolution. *Mammal. Chrom. Newsl.* **8**, 268–79.

Towe, A.L. (1973) Relative numbers of pyramidal tract neurons in mammals of different sizes. *Brain Behav. Evol.* **7**, 1–17.

Towe, A.L. (1975) Notes on the hypothesis of columnar organization in somatosensory cerebral cortex. *Brain Behav. Evol.* **11**, 16-47.

Trevor, J.C. (1963) The history of the word 'brachiator' and a problem of authorship in primate nomenclature. *Symp. zool. Soc. Lond.* **10**, 197–8.

Trinkaus, E. (1981) Neanderthal limb proportions and cold adaptation. In *Aspects of Human Evolution* (ed. Stringer, C.B.). Taylor and Francis: London, pp. 187–224.

Tripp, H.R.H. (1970) Reproduction in the Macroscelididae, with special reference to ovulation. PhD thesis, University of London.

Turleau, C., de Grouchy, J. and Klein, M. (1972) Phylogénie chromosomique de l'homme et des primates hominiens (*Pan troglodytes, Gorilla gorilla* et *Pongo pygmaeus*). Essai de reconstruction du caryotype de l'ancêtre commun. *Ann. Génét. (Paris)* **15**, 225–40.

Turleau, C., Creau-Goldberg, N., Cochet, C. and de Grouchy, J. (1983) Gene mapping of the gibbon: its position in primate evolution. *Hum. Genet.* **64**, 65–72.

Tutin, C.E.G. and Fernandez, M. (1985) Foods consumed by sympatric populations of *Gorilla g. gorilla* and *Pan t. troglodytes* in Gabon: some preliminary data. *Int. J. Primatol.* **6**, 27–43.

Tuttle, R.H. (1967) Knuckle-walking and the evolution of hominoid hands. *Am. J. phys. Anthrop., n. s.* **26**, 171–206.

Tuttle, R.H. (1969a) Knuckle-walking and the problem of human origins. *Science* **166**, 953–61.

Tuttle, R.H. (1969b) Quantitative and functional

studies on the hands of the Anthropoidea. I. The
Hominoidea. *J. Morph.* **128**, 309–64.

Tuttle, R.H. (1975) Parallelism, brachiation and
hominoid phylogeny. In *Phylogeny of the Primates*
(eds Luckett, W.P. and Szalay, F.S.). Plenum
Press: New York, pp. 447–80.

Tyndale-Biscoe, H. (1973) *Life of Marsupials.* Edward
Arnold: London.

Uzzell, T. and Pilbeam, D.R. (1971) Phyletic
divergence dates of hominoid primates: a
comparison of fossil and molecular data. *Evolution*
25, 615–35.

van Andel, T.H. (1985) *New Views on an Old Planet:
Continental Drift and the History of the Earth.*
Cambridge University Press: Cambridge.

van der Horst, C.J. (1949) The placentation of *Tupaia
javanica. Proc. K. ned. Akad. Wet.* **52**, 1205–13.

van der Klaauw, C.J. (1929) On the development
of the tympanic region of the skull in the
Macroscelididae. *Proc. zool. Soc. Lond.* **1929**,
491–560.

van der Klaauw, C.J. (1931) The auditory bulla in
some fossil mammals, with a general introduction
to this region of the skull. *Bull. Am. Mus. nat. Hist.*
62, 1–352.

Van Horn, R.N. (1980) Seasonal reproductive
patterns in primates. *Progr. Reprod. Biol.* **5**,
181–221.

Van Horn, R.N. and Eaton, G.G. (1979)
Reproductive physiology and behavior in
prosimians. In *The Study of Prosimian Behavior*
(eds Doyle, G.A. and Martin, R.D.). Academic
Press: New York, pp. 79–122.

van Kampen, P.N. (1905) Die Tympanalgegend des
Säugetierschädels. *Morph. Jb.* **34**, 321–722.

van Roosmalen, M.G.M., Mittermeier, R.A. and
Milton, K. (1981) The bearded sakis, genus
Chiropotes. In *Ecology and Behavior of
Neotropical Primates* (eds Coimbra-Filho, A.F.
and Mittermeier, R.A.). Academia Brasileira de
Ciências: Rio de Janeiro, pp. 419–41.

Van Valen, L. (1964) A possible origin for rabbits.
Evolution **18**, 484–91.

Van Valen, L. (1965) Tree shrews, primates, and
fossils. *Evolution* **19**, 137–51.

Van Valen, L. (1971) Adaptive zones and orders of
mammals. *Evolution* **25**, 420–8.

Van Valen, L. and Sloan, R.E. (1965) The earliest
primates. *Science* **150**, 743–5.

Vandebroek, G. (1961) The comparative anatomy of
the teeth of lower and non-specialized mammals.
K. VI. Acad. Wet. Lett. Sch. Kunst. Belg. **1**,
215–313 and **2**, 1–181.

Verhaart, W.J.C. (1970) The pyramidal tract in
primates. In *The Primate Brain* (eds Noback, C.R.

and Montagna, W.). Appleton-Century-Crofts:
New York, pp. 83–108.

Verheyen, W.N. (1962) Contribution à la craniologie
comparée des primates, les genres *Colobus* Illiger
1811 et *Cercopithecus* Linné 1758. *Ann. Mus. R.
Afr. centr. Tervuren, sér. 8, Sci. Zool.* **105**, 1–256.

Vilensky, J.A. (1978) The function of ischial
callosities. *Primates* **119**, 363–9.

Villalta, J.F. and Crusafont-Pairo, M. (1944) Dos
nuevos antropomorfos del Mioceno español y su
situación dentro de la moderna sistemática de los
simidos. *Notas Commun. Inst. Geol. Miner.
España* **13**, 91–139.

Vogel, C. (1968) The phylogenetic evaluation of some
characters and some morphological trends in the
evolution of the skull in catarrhine primates. In
Taxonomy and Phylogeny of Old World Primates
(ed. Chiarelli, B.). Rosenberg and Sellier: Turin,
pp. 21–55.

Vogel, F., Kopun, M. and Rathenberg, R. (1976)
Mutation and molecular evolution. In *Molecular
Anthropology* (eds Goodman, M. and Tashian,
R.E.). Plenum Press: New York, pp.13–33.

von Baer, K.E. (1828) *Entwicklungsgeschichte der
Thiere: Beobachtung und Reflexion.* Bornträger:
Königsberg.

von Beyrich, H. (1861) Über *Semnopithecus
pentelicus. Phys. Abh. K. Akad. Wiss., Berl.* **1860**,
1–26.

von Holst, D. (1969) Sozialer Stress bei Tupajas
(*Tupaia belangeri*). *Z. vergl. Physiol.* **63**, 1–58.

von Holst, D. (1974) Social stress in the tree-shrew:
its causes and physiological and ethological
consequences. In *Prosimian Biology* (eds
Martin, R.D., Doyle, G.A. and Walker, A.C.).
Duckworth: London, pp. 389–411.

von Koenigswald, G.H.R. (1935) Eine fossile
Säugetierfauna mit *Simia* aus Südchina. *Proc. K.
ned. Akad. Wet.* **38**, 872–9.

von Koenigswald, G.H.R. (1952) *Gigantopithecus
blacki* von Koenigswald, a giant fossil hominoid
from the Pleistocene of southern China. *Anthrop.
Pap. Am. Mus. nat. Hist.* **43**, 291–326.

von Koenigswald, G.H.R. (1969) Miocene
Cercopithecoidea and Oreopithecoidea from the
Miocene of East Africa. *Fossil Vertebrates Afr.*
1, 39–51.

von Koenigswald, W. (1979) Ein Lemurenrest aus
dem eozänen Ölschiefer der Grube Messel bei
Darmstadt. *Paläont. Z.* **53**, 63–76.

von Koenigswald, W. (1985) Der dritte Lemurenrest
aus dem mitteleozänen Ölschiefer der Grube
Messel bei Darmstadt. *Carolinea* **42**, 145–8.

von Koenigswald, W. and Schierning, H.-P. (1987)
The ecological niche of an extinct group of

mammals, the early Tertiary apatemyids. *Nature, Lond.* **326**, 595–7.

Wagner, A. (1839) Fossile Überreste von einem Affenschädel und anderen Säugetieren aus Griechenland. *Gelehrt. Anz. Bayr. Akad. Wiss. München* **8**, 305–12.

Walker, A.C. (1967a) Locomotor adaptations in recent and fossil Madagascan lemurs. PhD thesis, University of London.

Walker, A.C. (1967b) Patterns of extinction among the subfossil Madagascan lemuroids. In *Pleistocene Extinctions* (eds Martin, P.S. and Wright, H.E.). Yale University Press: New Haven, pp. 425–32.

Walker, A.C. (1969a) The locomotion of the lorises, with special reference to the potto. *E. Afr. Wildl. J.* **7**, 1–5.

Walker, A.C. (1969b) New evidence from Uganda regarding the dentition of Miocene Lorisidae. *Uganda J.* **32**, 90–1.

Walker, A.C. (1970) Post-cranial remains of the Miocene Lorisidae of East Africa. *Am. J. phys. Anthrop.* **33**, 249–61.

Walker, A.C. (1974a) Locomotor adaptations in past and present prosimians. In *Primate Locomotion* (ed. Jenkins, F.A.). Academic Press: New York, pp. 349–81.

Walker, A.C. (1974b) A review of the Miocene Lorisidae of East Africa. In *Prosimian Biology* (eds Martin, R.D., Doyle, G.A. and Walker, A.C.). Duckworth: London, pp. 435–47.

Walker, A.C. (1976) Splitting times among hominoids deduced from the fossil record. In *Molecular Anthropology* (eds Goodman, M. and Tashian, R.E.). Plenum Press: New York, pp. 63–77.

Walker, A.C. (1978) Prosimian primates. In *Evolution of African Mammals* (eds Maglio, V.J. and Cooke, H.B.S.). Harvard University Press: Cambridge, Mass., pp. 90–9.

Walker, A.C. (1979) Prosimian locomotor behavior. In *The Study of Prosimian Behavior* (eds Doyle, G.A. and Martin, R.D.). Academic Press: London, pp. 543–65.

Walker, A.C. and Andrews, P.J. (1973) Reconstruction of the dental arcades of *Ramapithecus wickeri. Nature, Lond.* **244**, 313–14.

Walker, A.C. and Pickford, M. (1983) New postcranial fossils of *Proconsul africanus* and *Proconsul nyanzae.* In *New Interpretations of Ape and Human Ancestry* (eds Ciochon, R.L. and Corrucchini, R.S.). Plenum Press: New York, pp. 325–51.

Walker, A.C. and Rose, M.D. (1968) Fossil hominoid vertebra from the Miocene of Uganda. *Nature, Lond.* **217**, 980–1.

Walker, A.C., Hoeck, H.N. and Perez, L. (1978) Microwear of mammalian teeth as an indicator of diet. *Science* **201**, 908–10.

Walker, A.C., Falk, D., Smith, R. and Pickford, M. (1983) The skull of *Proconsul africanus:* reconstruction and cranial capacity. *Nature, Lond.* **305**, 525–7.

Walker, A.C., Leakey, R.E.F., Harris, J.M. and Brown, F.H. (1986) 2.5-Myr *Australopithecus boisei* from west of Lake Turkana, Kenya. *Nature, Lond.* **322**, 517–22.

Walker, E.P. (ed.) (1968) *Mammals of the World* (2nd edition). John Hopkins Press: Baltimore.

Warburton, F.E. (1967) The purposes of classification. *Syst. Zool.* **16**, 241–5.

Ward, S.C. and Pilbeam, D.R. (1983) Maxillofacial morphology of Miocene hominoids from Africa and Indo-Pakistan. In *New Interpretations of Ape and Human Ancestry* (eds Ciochon, R.L. and Corrucchini, R.S.). Plenum Press: New York, pp. 211–38.

Washburn, S.L. (1950) The analysis of primate evolution with particular reference to the origin of man. *Cold Spring Harbor Symp. quant. Biol.* **15**, 67–78.

Washburn, S.L. (1957) Ischial callosities as sleeping adaptations. *Am. J. phys. Anthrop.* **15**, 269–80.

Waterman, P.G. (1984) Food acquisition and processing as a function of plant chemistry. In *Food Acquisition and Processing in Primates* (eds Chivers, D.J., Wood, B.A. and Bilsborough, A.). Plenum Press: New York, pp. 171–211.

Watson, D.M.S. (1913) Further notes on the skull, brain and organs of special sense of *Diademodon. Ann. Mag. nat. Hist.* **12**, 217–28.

Watson, D.M.S. (1931) On the skull of a bauriamorph reptile. *Proc. zool. Soc. Lond.* **1931**, 1163–205.

Watson, J.D. and Crick, F.H.C. (1953) Molecular structure of nucleic acids: a structure for deoxyribose nucleic acid. *Nature, Lond.* **171**, 737–8.

Watson, J.D., Tooze, J. and Kurtz, D.T. (1983) *Recombinant DNA: A Short Course.* Scientific American Books: New York.

Weale, R.A. (1966) Why does the human retina possess a fovea? *Nature, Lond.* **212**, 255–6.

Weber, M. (1927) *Die Säugetiere: Einführung in die Anatomie und Systematik der recenten und fossilen Mammalia,* vol. 1: *Anatomischer Teil* (2nd edition). Gustav Fischer Verlag: Jena.

Weber, M. (1928) *Die Säugetiere: Einführung in die Anatomie und Systematik der recenten und fossilen Mammalia,* vol. 2: *Systematischer Teil* (2nd edition). Gustav Fischer Verlag: Jena.

Weber, R. (1950) Transitorische Verschlüsse von

Fernsinnesorganen in der Embryonalperiode bei Amnioten. *Rev. suisse Zool.* **57**, 19–108.

Webster, K.E. (1973) Thalamus and basal ganglia in reptiles and birds. *Symp. zool. Soc. Lond.* **33**, 169–203.

Wegener, A. (1915) *Die Entstehung der Kontinente und Ozeane.* F. Viehweg: Braunschweig.

Weidenreich, F. (1921) Der Menschenfuss. *Z. Morph. Anthrop.* **22**, 51–282.

Weidenreich, F. (1943) The skull of *Sinanthropus pekinensis*; a comparative study on a primitive hominid skull. *Palaeont. Sin. D* **10**, 1–484.

Weidenreich, F. (1945) Giant early man from Java and South China. *Anthrop. Pap. Am. Mus. nat. Hist.* **40**, 1–134.

Weigelt, I. (1933) Neue Primaten aus der mitteleozänen (oberlutetischen) Braunkohle des Geiseltales. *Nova Acta Leopold., Halle (n.s.)* **1**, 97–153; 321–3.

Weir, B.J. and Rowlands I.W. (1973) Reproductive strategies in mammals. *A. Rev. Ecol. Syst.* **4**, 139–63.

Weiss, M.L., Goodman, M., Prychodko, W., Moore, G.W. and Tanaka, T. (1973) An analysis of macaque systematics using gene frequency data. *J. hum. Evol.* **1**, 41–8.

Westergaard, B. (1980) Evolution of the mammalian dentition. *Mém. Soc. géol. Fr., n.s.* **139**, 191–200.

Westergaard, B. (1983) A new detailed model for mammalian dentitional evolution. *Z. zool. Syst. Evolut. -Forsch.* **21**, 68–78.

Western, D. (1979) Size, life history and ecology in mammals. *Afr. J. Ecol.* **17**, 185–204.

Whipple, I.L. (1904) The ventral surface of the mammalian cheiridium, with special reference to the conditions found in man. *Z. Morph. Anthrop.* **7**, 261–368.

White, M.J.D. (1968) Models of speciation. *Science* **159**, 1065–70.

White M.J.D. (1973) *Animal Cytology and Evolution* (3rd edition). Cambridge University Press: Cambridge.

White, M.J.D. (1978) *Modes of Speciation.* W.H. Freeman: San Francisco.

Wible, J.R. (1983) The internal carotid artery in early eutherians. *Acta palaent. pol.* **28**, 281–93.

Wiley, E.O. (1975) Karl Popper, systematics and classification. *Syst. Zool.* **24**, 233–42.

Wiley, E.O. (1981) *Phylogenetics: The Theory and Practice of Phylogenetic Systematics.* John Wiley: New York.

Williams, E.E. and Koopman, K. (1952) West Indian fossil monkeys. *Am. Mus. Novit.* no. 1546, 1–16.

Williamson, P.G. (1981) Palaeontological documentation of speciation in Cenozoic molluscs from Turkana basin. *Nature, Lond.* **193**, 437–43.

Wilson, A.C. and Sarich, V.M. (1969) A molecular time scale for human evolution. *Proc. natn. Acad. Sci., U.S.A.* **63**, 1088–93.

Wilson, A.C., Carlson, S.S. and White, T.J. (1977) Biochemical evolution. *A. Rev. Biochem.* **46**, 573–639.

Wilson, A.C., Bush, G.L., Case, S.M. and King, M.-C. (1975) Social structuring of mammalian populations and rate of chromosomal evolution. *Proc. natn. Acad. Sci. U.S.A.* **72**, 5061–5.

Wilson, J.A. (1966) A new primate from the earliest Oligocene, West Texas, preliminary report. *Folia primatol.* **4**, 227–40.

Wilson, J.A. and Stevens, M.S. (1986) Fossil vertebrates from the latest Eocene, Skyline channels, Trans-Pecos Texas. In *Vertebrates, Phylogeny, and Philosophy* (eds Flanagan, K.M. and Lillegraven J.A.). Contributions to Geology, Special Paper 3, University of Wyoming: Laramie, Wyoming, pp. 221–35.

Wilson, J.A. and Szalay, F.S. (1976) New adapid primate of European affinities from Texas. *Folia primatol.* **25**, 294–312.

Winkelmann, R.K. (1963) Nerve endings in the skin of primates. In *Evolutionary and Genetic Biology of Primates*, vol. 1 (ed. Buettner-Janusch, J.). Academic Press: New York, pp. 229–59.

Winkelmann, R.K. (1964) Nerve endings of the North American opossum (*Didelphis virginiana*). A comparison with nerve endings of primates. *Am. J. phys. Anthrop., n.s.* **22**, 253–8.

Winkelmann, R.K. (1965) Innervation of the skin: notes on a comparison of primate and marsupial nerve endings. In *Biology of the Skin and Hair Growth* (eds Lyne, A.G. and Short, B.F.). Elsevier: New York, pp. 171–82.

Winter, J.S.D., Faiman, C., Hobson, W.C. and Reyes, F.I. (1980) The endocrine basis of sexual development in the chimpanzee. *J. Reprod. Fertil., Suppl.* **28**, 131–8.

Wislocki, G.B. (1929) On the placentation of primates, with a consideration of the phylogeny of the placenta. *Contr. Embryol. Carnegie Inst. Wash.* **20**, 51–80.

Wislocki, G.B. (1939) Observations on twinning in marmosets. *Am. J. Anat.* **64**, 445–84.

Wolf, R., O'Connor, R. and Robinson, J. (1977) Cyclic changes in plasma progestins and oestrogens in squirrel monkeys. *Biol. Reprod.* **17**, 228–31.

Wolfe, J.A (1978) A paleobotanical interpretation of Tertiary climates in the northern hemisphere. *Am. Sci.* **66**, 694–703.

Wolff, R.G. (1984) New specimens of the primate *Branisella boliviana* from the early Oligocene of Salla, Bolivia. *J. vert. Paleontol.* **4**, 570–74.

Wolfson, A. (1954) Sperm storage at lower than body temperature outside the body cavity in some passerine birds. *Science* **120**, 68–71.

Wolin, L.R. and Massopust, L.C. (1970) Morphology of the primate retina. In *The Primate Brain* (eds Noback, C.R. and Montagna, W.). Appleton-Century-Crofts: New York, pp. 1–27.

Wolpoff, M.H. (1984) Evolution in *Homo erectus*: the question of stasis. *Paleobiology* **10**, 389–406.

Wolpoff, M.H. (1986) Stasis in the interpretation of evolution in *Homo erectus*: a reply to Rightmire. *Paleobiology* **12**, 325–8.

Wood, B.A. (1978) *Human Evolution*. Chapman and Hall: London.

Wood Jones, F. (1916) *Arboreal Man*. Edward Arnold: London.

Wood Jones, F. (1917) The structure of the orbito-temporal region of the skull of *Lemur*. *Proc. zool. Soc. Lond.* **1917**, 323–9.

Wood Jones, F. (1929) *Man's Place among the Mammals*. Edward Arnold: London.

Wood Jones, F. (1941) *The Principles of Anatomy as Seen in the Hand*. Baillière, Tindall and Cox: London.

Wood Jones, F. (1949) *Structure and Function as Seen in the Foot*. Baillière, Tindall and Cox: London.

Wood Jones, F. and Lambert, V.F. (1939) The occurrence of the lemurine form of ectotympanic in a primitive marsupial. *J. Anat., Lond.* **74**, 72–5.

Woodburne, M.O. and Zinsmeister, W.J. (1982) Fossil land mammals from Antarctica. *Science* **218**, 284–6.

Wooding, G.L. and Doolittle, R.F. (1972) Primate fibrinopeptides: evolutionary significance. *J. hum. Evol.* **1**, 553–63.

Wortman, J.L. (1903/04) Studies of Eocene Mammalia in the Marsh collections, Peabody Museum. Part 2: Primates. *Am. J. Sci., ser. 4* **15**, 163–76, 399–414, 419–36; **16**, 345–68; **17**, 23–33, 133–40, 203–14.

Wrangham, R.W. (1977) Feeding behaviour of chimpanzees in Gombe National Park, Tanzania. In *Primate Ecology* (ed. Clutton-Brock, T.H.). Academic Press: London, pp. 504–38.

Wrangham, R.W. (1979) Sex differences in chimpanzee dispersion. In *The Great Apes* (eds Hamburg, D.A. and McCown, E.R.). Benjamin/Cummings: Menlo Park, California, pp. 481–90.

Wright, P.C. (1981) The night monkeys, genus *Aotus*. In *Ecology and Behavior of Neotropical Primates* (eds Coimbra-Filho, A.F. and Mittermeier, R.A.). Academia Brasileira de Ciências: Rio de Janeiro, pp. 211–40.

Wu, R. (1957) *Dryopithecus* teeth from Keiyuan, Yunnan Province. *Vert. palasiat.* **1**, 25–32.

Wu, R. (1958) New materials of *Dryopithecus* from Keiyuan, Yunnan. *Vert. palasiat.* **2**, 31–43.

Wu, R. (1962) The mandibles and dentition of *Gigantopithecus*. *Palaeont. sin.* **11**, 1–94.

Wu, R. (1984) The crania of *Ramapithecus* and *Sivapithecus* from Lufeng, China. *Cour. Forsch.-Inst. Senckenberg* **69**, 41–8.

Wu, R. (1987) A revision of the classification of the Lufeng great apes. *Acta anthrop. sin.* **6**, 265–71.

Wu, R. and Pan, Y. (1984) A late Miocene gibbon-like primate from Lufeng, Yunnan Province. *Acta anthrop. sin.* **3**, 185–94.

Wu, R. and Pan, Y. (1985a) A new adapid primate from the Lufeng Miocene, Yunnan. *Acta anthrop. sin.* **4**, 1–7.

Wu, R. and Pan, Y. (1985b) Preliminary observation on the cranium of *Laccopithecus robustus* from Lufeng, Yunnan with reference to its phylogenetic relationship. *Acta anthrop. sin.* **4**, 7–12.

Wurster, D.H. and Benirschke, K (1970) Indian muntjak, *Muntiacus muntjak*: a deer with a low diploid chromosome number. *Science* **168**, 1364–6.

Xu, Q. and Lu, Q. (1979) The mandibles of *Ramapithecus* and *Sivapithecus* from Lufeng, Yunnan. *Vert. palasiat.* **17**, 1–13.

Xu, Q. and Lu, Q. (1980) The Lufeng ape skull and its significance. *China Reconstructs* **29**, 56–7.

Young, J.Z. (1957) *The Life of Mammals*. Clarendon Press: Oxford.

Young, W.C. and Yerkes, R.M. (1943) Factors influencing the reproductive cycle in the chimpanzee, the period of adolescent sterility and related problems. *Endocrinology.* **33**, 121–54.

Yunis, J.J. and Prakash, O. (1982) The origin of man: a chromosomal pictorial legacy. *Science* **215**, 1525–30.

Zapfe, H. (1958) The skeleton of *Pliopithecus (Epipliopithecus) vindobonensis* Zapfe and Hürzeler. *Am. J. phys. Anthrop.* **16**, 441–58.

Zapfe, H. (1960) Die Primatenfunde aus der miozänen Spaltenfüllung von Neudorf an der March (Devinská Nová Ves), Tschechoslowakei. Mit Anhang: Der Primatenfund aus dem Miozän von Klein Hadersdorf in Niederösterreich. *Schweiz. paläont. Abh.* **78**, 1–293.

Zapfe, H. (1963) Lebensbild von *Megaladapis edwardsi* (Grandidier). *Folia primatol.* **1**, 178–87.

Zeller, U.A. (1986a) The systematic relations of tree shrews: evidence from skull morphogenesis. In *Primate Evolution* (eds Else, J.G. and Lee, P.C.). Cambridge University Press: Cambridge, pp. 273–80.

Zeller, U.A. (1986b) Ontogeny and cranial

morphology of the tympanic region of the Tupaiidae, with special reference to *Ptilocercus*. *Folia primatol.* **47**, 61–80.

Zeuner, F.E. (1958) *Dating the Past* (4th revised edition). Methuen: London.

Zieg, J., Silverman, M., Hilmen, M. and Simon, M. (1977) Recombinational switch for gene expression. *Science* **196**, 170–2.

Ziegler, A.C. (1969) A theoretical determination of tooth succession in the therapsid *Diademodon*. *J. Paleont.* **43**, 771–8.

Zingeser, M.R. (1969) Cercopithecoid canine tooth honing mechanisms. *Am. J. phys. Anthrop.* **31**, 205–13.

Zuckerkandl, E. (1963) Perspectives in molecular anthropology. In *Classification and Human Evolution* (ed. Washburn, S.L.). Wenner-Gren Foundation: New York, pp.243-72.

Zuckerkandl, E. (1965) The evolution of hemoglobin. *Sci. Am.* **212** (5), 110–18.

Zuckerkandl, E. and Pauling, L. (1962) Molecular disease, evolution, and genetic heterogeneity. In *Horizons in Biochemistry* (eds Kasha, M. and Pullman, N.). Academic Press: New York, pp. 189–225.

Zuckerkandl, E. and Pauling, L. (1965) Evolutionary divergence and convergence in proteins. In *Evolving Genes and Proteins* (eds Bryson, V. and Vogel, H.J.). Academic Press: New York, pp. 97–166.

Zuckerman, S. (1932) *The Social Life of Monkeys and Apes*. Kegan Paul, Trench and Trubner: London.

Author index

Page numbers in italics refer to the Reference list. Page numbers in bold refer to illustrations.

Taxonomic index

Page numbers in italics refer to illustrations, figures and tables

Subject index

Page numbers in italics refer to illustrations, figures and tables.